# BASIC ELECTRICITY AND ELECTRONICS

## Other books

By Charles A. Schuler

ELECTRONICS: Principles and Applications

Activities Manual for ELECTRONICS: Principles and Applications

INDUSTRIAL ELECTRONICS AND ROBOTICS (with William L. McNamee)

By Richard J. Fowler

ELECTRICITY: Principles and Applications

Activities Manual for ELECTRICITY: Principles and Applications

# BASIC ELECTRICITY AND ELECTRONICS

Charles A. Schuler

*California University of Pennsylvania*

Richard J. Fowler

*Western Washington University*

Illustrations by

Jay D. Helsel

*California University of Pennsylvania*

McGRAW-HILL BOOK COMPANY

New York • Atlanta • Dallas • St. Louis • San Francisco
Auckland • Bogotá • Guatemala • Hamburg • Lisbon
London • Madrid • Mexico • Milan • Montreal • New Delhi
Panama • Paris • San Juan • São Paulo • Singapore
Sydney • Tokyo • Toronto

Sponsoring Editor: Gordon Rockmaker
Editing Supervisor: Arthur Pomponio
Design and Art Supervisor: Nancy Axelrod
Production Supervisor: Albert Rihner

Text Designer: Edward Butler
Cover Photo: Wolfson Photography Inc.

**Library of Congress Cataloging-in-Publication Data**

Schuler, Charles A.
Basic electricity and electronics.

Includes index.
1. Electric engineering. 2. Electronics.
I. Fowler, Richard J. II. Title.
TK146.S32 1988 621.3 87-2922
ISBN 0-07-055627-X



2 3 4 5 6 7 8 9 0 VNHVNH 8 9 4 3 2 1 0 9

ISBN 0-07-055627-X

# Contents

# Preface

This text is designed for introductory courses in electricity and electronics covering dc and ac circuits and both linear and digital electronics. Basic algebra and trigonometry are the only prerequisites.

We live in a highly technological society. There is an expanding need for technical personnel who are well-versed in the theory and applications of electricity and electronics. It is extremely important that beginning students who are preparing for careers in these fields are exposed to the best possible learning materials.

Electricity is a somewhat abstract field. It has its own vocabulary, units of measure, laws, rules, and a myriad of applications. It must be presented in a logical format with smooth transitions from topic to topic. Laws and concepts must be intermingled with applications or they will seem vague to many students and will be quickly forgotten. The first circuits course is often the most difficult for students. The ideas come too fast, seem somewhat disconnected, and are difficult to relate to what is already known. Our job as authors (and teachers) was to make the material as readable and accurate as possible and to organize it in a logical sequence of topics. The topics were carefully developed, with each preceding topic serving as a building block for subsequent discussions.

Every chapter in this text is divided into major sections, and each of these concludes with a self-test to allow the student to check his or her progress. The test also serves to reinforce the material just studied. Illustrative examples with complete step-by-step solutions are included wherever appropriate. Each chapter ends with a summary of concepts and equations, review questions, and review problems. A comprehensive glossary appears at the end of the text. There are ample opportunities for students to become involved in applying the essentials and for measuring their own progress and understanding.

This text is practical. It focuses on the knowledge required by technical personnel. It discusses fault analysis and ties it to circuit theory to properly begin the development of the analytical skills needed by problem solvers and troubleshooters.

Many concepts are much easier to understand if good illustrations are available. The figures in this text have been carefully designed and prepared and relate directly to, and support, the text. The use of a second color helps emphasize key features and makes the material interesting and inviting.

An accompanying experiments manual was developed concurrently with this text and is completely correlated to it. It provides activities, experiments, and a broad assortment of BASIC computer programs designed to involve the student in practical hands-on experiences.

We have taken considerable care to ensure that this text is as error-free as possible, technologically up to date, interesting, and pedagogically sound. We welcome comments, suggestions, and corrections from both students and instructors.

The experiences of many years of classroom teaching are reflected in this work. We have derived much inspiration from the students and colleagues with whom we have been associated over the years. This text is gratefully dedicated to those people. Finally, to our families, we owe special thanks and appreciation for relinquishing the enormous amount of time that a project of this magnitude demands of an author.

*Charles A. Schuler*
*Richard J. Fowler*

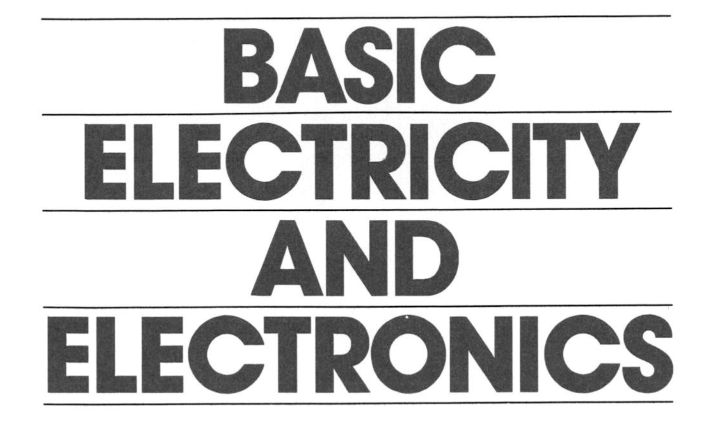
BASIC
ELECTRICITY
AND
ELECTRONICS

# Chapter 1 Introductory Concepts

Electricity is a technical subject. As such, it is intimately involved with numbers, equations, calculations, and systems of measurement. This chapter introduces what electricity is and how numbers and measuring units are used to describe its characteristics and behavior.

## 1-1 SI UNITS

Most of the countries of the world use the metric system of measurement. The metric system offers tremendous advantages because it is almost universal and because the various sized portions of each unit are multiples of 10 when compared to each other and to the base unit. Thus, it is a decimal system and forever eliminates clumsy relationships such as 12 inches (in.) = 1 foot (ft), 3 ft = 1 yard (yd), and 5280 ft = 1 mile (mi).

The international system of units, usually abbreviated SI, was developed from the metric system of measurement in 1960. It is mainly based on the older metric meter-kilogram-second-ampere (mksa) system which uses base units of meters, kilograms, seconds, and amperes. The SI system has become the accepted standard and was officially adopted by the Institute of Electrical and Electronics Engineers (IEEE) in 1965. Therefore, all electrical and electronic measurement should use the SI system. However, there are carryovers from the old English system of measurements. For example, many physical standards used in electronics are derived from the inch. The contacts on an elec-

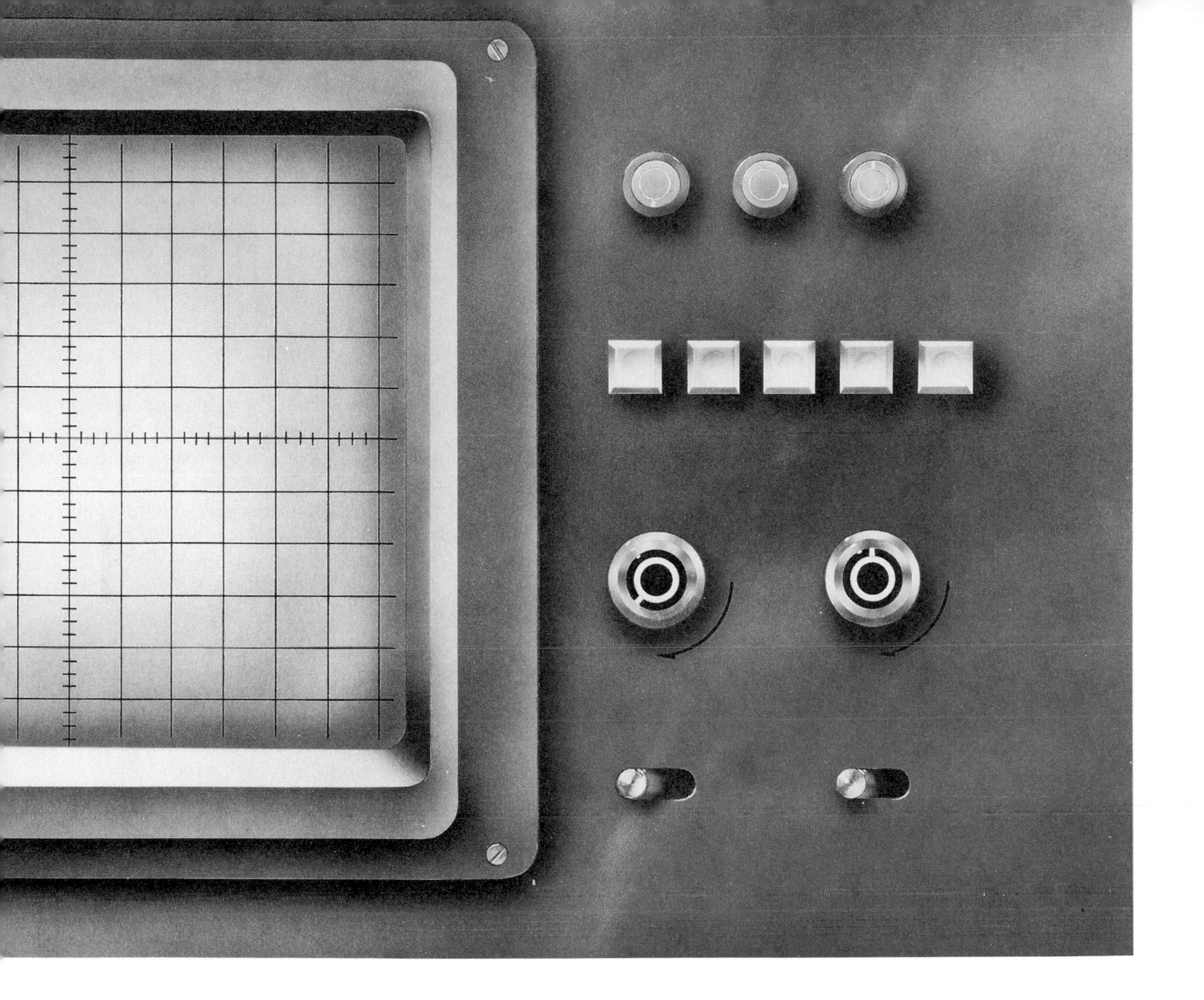

tronic component or a circuit board connector are often spaced at inch intervals. Figure 1-1 shows an integrated circuit socket. The contact spacing is based on tenths of inches. Figure 1-2 shows an edge connector for a printed circuit board. The standard contact spacings are specified in inches. It is unreasonable to expect a 100 percent conversion to SI standards in every aspect of electrical and electronic measurements. However, electrical quantities dealing with phenomena such as the flow of electricity, its pressure, and its ability to produce heat are totally based on the SI system.

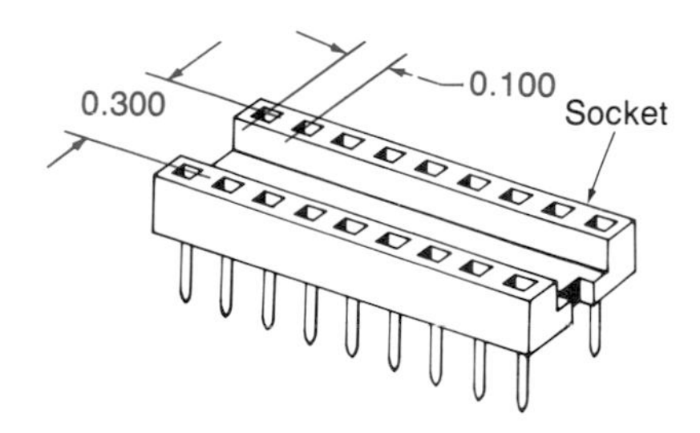


**FIGURE 1-1 An integrated circuit socket.**

The SI system is particularly convenient in technical areas where energy calculations are common. This is because only one unit, the joule (J), is used to quantify all types of energy including electric, atomic, chemical, heat, and mechanical. The joule is covered in detail later. Now, thanks to the SI system, electrical specialists can communicate with scientists, mechanical engineers, and chemists without the burden of unique systems of measurements.

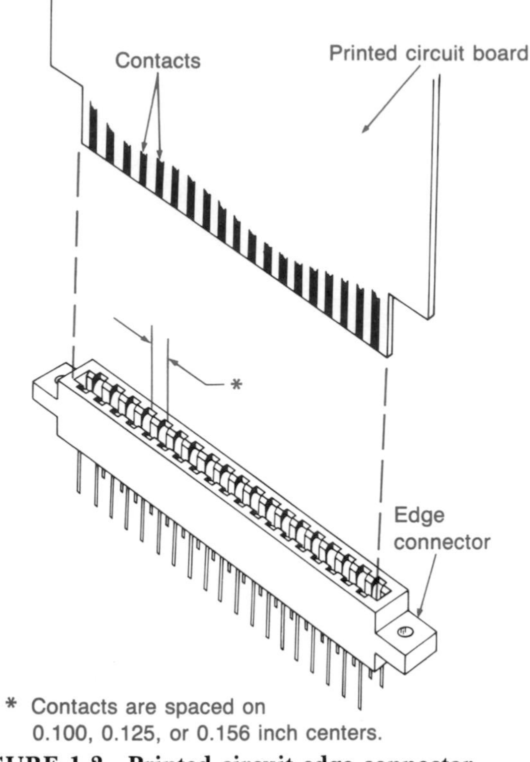


**FIGURE 1-2 Printed circuit edge connector.**

Table 1-1 is a comparison of the base units among three systems of measurement. The SI system is essentially the same as the metric mksa system, as shown by that column heading. The major difference is that the original mksa system used the Celsius temperature scale. When the SI system was extracted from the mksa system, the Kelvin temperature scale was adopted in favor of the Celsius scale. This deviation from the mksa system is carried into some areas of electricity and electronics. Most manufacturers continue to specify the temperature characteristics of their assemblies and component parts in degrees Celsius even though the IEEE has officially adopted the SI system. On the other hand, the equivalent noise temperature of some electronic devices is specified in degrees Kelvin. The English system is sometimes referred to as the U. S. Customary system. This is because England has officially adopted the SI system as mandatory, while the United States has only recognized the SI system and not officially mandated it. The centimeter-gram-second (cgs) system is also metric. It is convenient for measurements in small-scale systems since it is based on smaller base units of length and mass.

Base units, such as those found in Table 1-1, are the foundation for all measurements. In mechanics, all additional units can be derived from three base units which are length, mass, and time. By adding a base unit for electric current flow, the additional units needed for measurements in electricity can be derived. When the base units of temperature, luminous intensity, and amount of substance are added, any and all scientific and technical units for any type of measurement may be derived.

Table 1-2 shows examples of derived SI units. Area and volume are derived from the base unit of length. Velocity and acceleration are derived from

**Table 1-1 A Comparison of Base Units**

| Quantity | SI, mksa | English (U.S. Customary) | cgs |
|---|---|---|---|
| Length | Meter, m = 100 cm | Foot, ft =0.3048 m | Centimeter, cm 2.54 cm = 1 in |
| Mass | Kilogram, kg =1000 g | Slug =14.6 kg | Gram, g |
| Time | Second, s | Second | Second |
| Electric current | Ampere, A | Ampere | Ampere |
| Temperature | Kelvin, K =273.15 + °C | Fahrenheit, F =($\frac{9}{5}$ °C) + 32 | Celsius, C =$\frac{5}{9}$ (°F − 32) |
| Luminuous intensity | Candela, cd | | |
| Amount of substance | Mole, mol | | |

**Table 1-2 Examples of Derived SI Units**

| Quantity | Derivation | Units |
|---|---|---|
| Area | Length × width | Square meters, $m^2$ |
| Volume | Length × width × height | Cubic meters, $m^3$ |
| Velocity | Distance/time | Meters per second, m/s |
| Acceleration | Distance/time/time | Meters per second per second, $m/s^2$ |
| Force | Mass × acceleration | Newtons (kilograms × $m/s^2$), kg × $m/s^2$ |
| Energy and work | Force × distance | Joules (newtons × meters), N × m |
| Power | Energy/time | Watts (joules per second), J/s |
| Pressure | Force/area | Pascals (newtons per $m^2$), $N/m^2$ |
| Electric charge | Current × time | Coulombs (amperes × seconds), A × s |
| Electric potential | Energy/charge | Volts (joules per coulomb), J/C |
| Electric resistance | Potential/current | Ohms (volts per ampere), V/A |
| Capacitance | Charge/potential | Farads (coulombs per volt), C/V |
| Inductance | Potential/current/time | Henrys (volts per ampere per second), V/A/s |

the base units of distance (length) and time. Force is derived from mass (a base unit) and from acceleration (a derived unit). Energy (or work) is derived from force and distance. Obviously, derived units can be used to derive additional units. Several of the derived units in Table 1-2 deal with electrical quantities. These units are explained and applied later.

The base units are arbitrary. They have evolved over years of scientific work and investigations into natural phenomena. The meter was originally defined as one ten-millionth of the distance, at sea level, from either pole to the equator. Later, the meter was defined as 1,650,763.73 wavelengths, in a vacuum, of the krypton 86 atom. The most recent definition of the meter is the distance light travels in a vacuum during 1/299,792,458 of a second. Scientific and technical work must be repeatable, even at independent locations. A unified measurement system is one of the key elements for achieving repeatability.

Since the U. S. Customary system of units will likely remain in existence for the foreseeable future, people who work in technical areas will continue to be faced with making conversions from one system to another.

**EXAMPLE 1-1**

The front panel of an electric control device is 3 in. high and 7.5 in. wide. What are its dimensions in centimeters? What is its area in square centimeters?

**Solution**

Table 1-1 shows that 1 in. is equal to 2.54 cm. Therefore, the inch dimensions can be converted to centimeter dimensions by multiplying them by 2.54.

$$3 \text{ in.} \times 2.54 \text{ cm/in.} = 7.62 \text{ cm}$$
$$7.5 \text{ in.} \times 2.54 \text{ cm/in.} = 19.05 \text{ cm}$$

Note that the inch units cancel, leaving the answers in centimeter units. The area is now found by

$$7.62 \text{ cm} \times 19.05 \text{ cm} = 145.16 \text{ cm}^2$$

**EXAMPLE 1-2**

The upper temperature limit of an electronic computer is specified as being 323.15 K. Convert this to degrees Celsius and also to degrees Fahrenheit.

**Solution**

Table 1-1 shows that the Kelvin temperature is found by adding 273.15 to the Celsius temperature. Therefore, the Celsius temperature may be derived by subtracting 273.15 from the Kelvin value.

$$323.15 - 273.15 = 50° \text{ C}$$

The Fahrenheit temperature can now be found by

$$\left(\frac{9}{5} \times 50\right) + 32 = 122° \text{ F}$$

## Self-Test

**1.** Which system of measurement has been officially adopted by the IEEE?

**2.** There are carryovers from older systems of measurement. For example, the spacings of component and circuit board connectors are often measured in ______.

**3.** Identify the two units of temperature measurement most often used in electricity and electronics.

**4.** Which metric system favors measurements for small-scale systems?

**5.** Identify the SI base units that are used to derive the additional units needed in electricity.

**6.** The performance of electronic components is often specified at 25° C. Convert this temperature to degrees Fahrenheit.

**7.** A circuit board is 12 cm × 8 cm. What are the board's measurements in inches?

## 1-2 SCIENTIFIC NOTATION

The numbers used in electricity range from very small to very large. It is quite cumbersome to write, speak, and manipulate very small and very large numbers. A *power-of-ten notation system* is widely used to represent such numbers and to greatly lessen the burden of working with them. As the following list shows, 10 is the *base:*

$10^{-6} = 0.000001$ = one-millionth
$10^{-5} = 0.00001$ = one hundred-thousandth
$10^{-4} = 0.0001$ = one ten-thousandth
$10^{-3} = 0.001$ = one-thousandth
$10^{-2} = 0.01$ = one-hundredth
$10^{-1} = 0.1$ = one-tenth
$10^{0} = 1$ = one
$10^{1} = 10$ = ten
$10^{2} = 100$ = hundred
$10^{3} = 1000$ = thousand
$10^{4} = 10{,}000$ = ten thousand
$10^{5} = 100{,}000$ = hundred thousand
$10^{6} = 1{,}000{,}000$ = million

Expressing numbers in powers of 10 with one nonzero digit to the left of the decimal point is called *scientific notation.* For example, to convert 0.00000483 to scientific notation, first change the number so that it has one nonzero digit to the left of the decimal point. This involves moving the decimal point six places to the right as shown below (an asterisk is used to mark the original position of the decimal point):

0*000004.83

Now, the number must be multiplied by some power of 10 to restore the number to its original value. When the decimal point is moved to the left, the required power of 10 is positive. When the decimal point is moved to the right, the required power of 10 is negative. The example moved the decimal point six places to the right, so the power of 10 (also called the exponent) is negative 6, producing the scientific notation:

$4.83 \times 10^{-6}$ which is equal to 0.00000483

Suppose we wish to express 483,000 using scientific notation. First, we convert the number so it has one digit to the left of the decimal. This is shown below and again an asterisk is used to show the original location of the decimal point:

4.83000 *

The decimal point had to be moved five places to the left. Therefore, the power of 10 is +5, and the scientific notation for 483,000 is

$4.83 \times 10^{+5}$ or, more commonly, $4.83 \times 10^{5}$

Exponents without a minus sign are understood to be positive.

### EXAMPLE 1-3

Change the following numbers to scientific notation:

12,720,000
0.000023
628
0.5
13,900
1.83

### Solution

$12{,}720{,}000 = 1.272 \times 10^{7}$
$0.000023 = 2.3 \times 10^{-5}$
$628 = 6.28 \times 10^{2}$
$0.5 = 5 \times 10^{-1}$
$13{,}900 = 1.39 \times 10^{4}$
$1.83 = 1.83 \times 10^{0}$ ($10^{0}$ is not normally used)
(1.83 is the scientific notation)

**EXAMPLE 1-4**
Convert the following numbers to standard form:

$1 \times 10^9$

$4.56 \times 10^{-3}$

$9.871 \times 10^5$

$3.9 \times 10^{-7}$

**Solution**
To convert from scientific notation, the power of 10 must be eliminated. Move the decimal point to the right for positive exponents and to the left for negative exponents.

$$1 \times 10^9 = 1{,}000{,}000{,}000$$
$$4.56 \times 10^{-3} = 0.00456$$
$$9.871 \times 10^5 = 987{,}100$$
$$3.9 \times 10^{-7} = 0.00000039$$

The exponents must be handled properly when arithmetic operations are performed with scientific notation. The following rules for exponents apply:

**1.** To perform addition or subtraction, all numbers must have the same exponent (magnitude and sign). The columns must be aligned according to the decimal points. The sum or the difference will have the same exponent as the numbers being added.

**2.** To perform multiplication, the exponents are added.

**3.** To perform division, the exponent of the divisor is subtracted from the exponent of the dividend.

**4.** To extract a root, the exponent is divided by the root.

**5.** To raise to a power, the exponent is multiplied by the power.

**EXAMPLE 1-5**
Add $2.1 \times 10^3$, $3.68 \times 10^2$, and $4 \times 10^{-1}$.

**Solution**
Make the necessary conversions so that all of the numbers are multiplied by the same power of 10. In this case, it is convenient to use $10^3$ since it is the largest power and the answer will be in scientific notation:

$$\begin{array}{r} 2.1 \times 10^3 \\ +0.368 \times 10^3 \\ +0.0004 \times 10^3 \\ \hline 2.4684 \times 10^3 \end{array}$$

Note that the columns are aligned according to the decimal points and that the answer is multiplied by the same power of 10.

**EXAMPLE 1-6**
Subtract $4.5 \times 10^4$ from $1 \times 10^5$.

**Solution**
Once again, we choose the largest power of 10 as the common exponent:

$$\begin{array}{r} 1.00 \times 10^5 \\ -0.45 \times 10^5 \\ \hline 0.55 \times 10^5 \end{array}$$

Notice that the answer is not in scientific notation. It would have been better to use $10^4$ as the common exponent. All is not lost, however. It is a simple matter to move the decimal point of the answer one place to the right and make the exponent one digit smaller:

$$0.55 \times 10^5 = 5.5 \times 10^4$$

**EXAMPLE 1-7**
Perform the following multiplications and express the answers in scientific notation:

$(3.5 \times 10^2) \times (2.1 \times 10^4)$

$(5.1 \times 10^3) \times (4.8 \times 10^{-5})$

$(1.9 \times 10^{-1}) \times (3 \times 10^{-3})$

**Solution**
For the first multiplication problem, $3.5 \times 2.1 = 7.35$ and the sum of the exponents is 6. Therefore, the answer is $7.35 \times 10^6$. For the second problem, $5.1 \times 4.8 = 24.48$ and the sum of the exponents is $10^{-2}$. This gives us $24.48 \times 10^{-2}$, which is not scientific notation. Shifting the decimal one place to the left and making the exponent one digit larger yields $2.448 \times 10^{-1}$. In the last multiplication problem, $1.9 \times 3 = 5.7$ and the sum of the exponents is $-4$, which produces an answer of $5.7 \times 10^{-4}$.

$$3.5 \times 10^2 \times 2.1 \times 10^4 = 7.35 \times 10^6$$
$$5.1 \times 10^3 \times 4.8 \times 10^{-5} = 2.448 \times 10^{-1}$$
$$1.9 \times 10^{-1} \times 3 \times 10^{-3} = 5.7 \times 10^{-4}$$

**EXAMPLE 1-8**

Perform the following divisions and express the answers using scientific notation:

$$\frac{4.3 \times 10^5}{8.6 \times 10^3}$$

$$\frac{9.9 \times 10^2}{3 \times 10^{-2}}$$

$$\frac{-5 \times 10^{-2}}{2 \times 10^{-3}}$$

**Solution**

In the first problem, dividing 4.3 by 8.6 gives 0.5, and subtracting the exponent of the divisor from the exponent of the dividend gives 2 for a result of $0.5 \times 10^2$, or $5 \times 10^1$ in scientific notation. In the second problem, dividing 9.9 by 3 gives 3.3, and subtracting exponents produces a positive 4 since $+2 - (-2) = +4$. The answer is therefore $3.3 \times 10^4$. For the last division problem, dividing $-5$ by 2 yields $-2.5$, and subtracting the exponents gives $+1$ for an answer of $-2.5 \times 10^1$.

$$\frac{4.3 \times 10^5}{8.6 \times 10^3} = 5 \times 10^1$$

$$\frac{9.9 \times 10^2}{3 \times 10^{-2}} = 3.3 \times 10^4$$

$$\frac{-5 \times 10^{-2}}{2 \times 10^{-3}} = -2.5 \times 10^1$$

**EXAMPLE 1-9**

Find the square root of $3.6 \times 10^5$.

**Solution**

To find a square root, it is necessary to divide the exponent by 2. An exponent of 5 will not produce an integer (whole number) result. Therefore, it is best to change the number to an even power of 10. For example, $36 \times 10^4$ is equivalent and the exponent is now evenly divisible by 2. The square root of 36 is 6 and $\frac{4}{2} = 2$, which gives an answer of $6 \times 10^2$.

**EXAMPLE 1-10**

Raise $4 \times 10^2$ to the third power.

**Solution**

The example can be written as $4^3 \times (10^2)^3$. $4 \times 4 \times 4 = 64$ and $2 \times 3 = 6$ for a result of $64 \times 10^6$. Converting to scientific notation gives $6.4 \times 10^7$.

## Self-Test

**8.** Change these numbers to scientific notation:

| | |
|---|---|
| 186,000 | 0.154 |
| 0.0003 | 1,200,000,000 |
| 357.2 | 0.0000000005 |

**9.** Change these numbers to standard form:

| | |
|---|---|
| $4.9 \times 10^2$ | $2 \times 10^{-1}$ |
| $1.1 \times 10^{-3}$ | $5.5 \times 10^4$ |
| $5.678 \times 10^5$ | $1.9 \times 10^{-4}$ |

**10.** Perform the following additions and express the answers using scientific notation:

$(5.5 \times 10^1) + (1.7 \times 10^2)$

$(3.83 \times 10^1) + (2 \times 10^{-1})$

$(2 \times 10^4) + (3 \times 10^3) + (4 \times 10^2)$

**11.** Perform the following subtractions and express the answers using scientific notation:

$(4.4 \times 10^{-2}) - (4)$

$(7.65 \times 10^3) - (-1.2 \times 10^2)$

$(-4 \times 10^{-5}) - (2.2 \times 10^{-6})$

**12.** Perform the following multiplications and express the answers using scientific notation:

$(5 \times 10^2) \times (2.6 \times 10^{-3})$

$(3.32 \times 10^{-2}) \times (1 \times 10^3)$

$(4.4 \times 10^6) \times (8 \times 10^5)$

**13.** Perform the following divisions and express the answers using scientific notation:

$\frac{6.6 \times 10^9}{3 \times 10^6}$ $\quad$ $\frac{1.71 \times 10^5}{4.5 \times 10^{-2}}$

$\frac{9 \times 10^{-6}}{2 \times 10^4}$

**14.** Take the square root of $8.1 \times 10^3$.

**15.** Raise $2 \times 10^2$ to the fourth power.

## 1-3 PREFIXES

Electrical quantities vary considerably in magnitude. Prefixes are used to modify measuring units by making them multiple or submultiple units. A multiple unit is larger than a nonmodified unit and a submultiple unit is smaller. In the SI system, the prefixes are arranged in increments of three powers of 10 as shown in Table 1-3. The SI system does

**Table 1-3 Prefixes**

| Prefix | Multiple-Submultiple | Power-of-10 Form | SI Symbol |
|---|---|---|---|
| Giga | 1,000,000,000 | $10^9$ | G |
| Mega | 1,000,000 | $10^6$ | M |
| Kilo | 1000 | $10^3$ | k |
| Milli | 0.001 | $10^{-3}$ | m |
| Micro | 0.000001 | $10^{-6}$ | $\mu$ |
| Nano | 0.000000001 | $10^{-9}$ | n |
| Pico | 0.000000000001 | $10^{-12}$ | p |

not recognize the older metric prefixes of deca ($10^1$) and centi ($10^{-2}$), and these prefixes are not used with electrical units.

If a prefix is added to a unit, a new unit is effectively formed. When a multiple or submultiple unit is raised to a power, the power applies to the new unit and not to the original unit. As an example, a millimeter (mm) is a unit that equals one-thousandth of a meter. The area of a small surface may be measured in square millimeters. Note that $10\ mm^2 = 10\ (mm)^2$ and does not equal $10\ m(m^2)$.

Prefixes can be used to avoid using numbers smaller than 0.1 or greater than 1000. As an example, if a current flow is 0.004 ampere, it is often expressed in terms of *milliamperes* to avoid using a number smaller than 0.1. As another example, a value such as 1500 watts can be expressed in *kilowatts* to avoid the use of a number greater than 1000.

### EXAMPLE 1-11

Convert 0.004 ampere to milliamperes and determine how it would be written in abbreviated form. How would it be written using scientific notation?

**Solution**

Table 1-3 shows that the prefix milli is a submultiple and represents 0.001 or $10^{-3}$. When a measurement unit is converted to a submultiple or smaller unit by the use of a prefix, the number must become larger. The relationship in this case is three decimal places. Therefore, amperes are converted to milliamperes by moving the decimal point three places to the right:

$$0.004 \text{ ampere} = 4 \text{ milliamperes}$$

Table 1-3 shows that the prefixes, with the exception of giga and mega, are symbolized with lowercase letters. A current of 4 milliamperes is abbreviated as 4 mA. Using scientific notation, the current would be expressed as $4 \times 10^{-3}$ ampere or $4 \times 10^{-3}$ A. Note that abbreviations are not pluralized by adding an "s" even though they are often pronounced or written as plurals. Lowercase "s" is the SI symbol for the second and would cause confusion if used.

### EXAMPLE 1-12

Convert 0.000004 ampere to microamperes and to scientific notation.

**Solution**

Once again, the conversion is from a unit to a submultiple (a smaller unit). In this case, the decimal point must be moved six places to the right:

$$0.000004 \text{ ampere} = 4 \text{ microamperes } (4\ \mu A)$$

Using scientific notation, the current would be expressed as $4 \times 10^{-6}$ A.

### EXAMPLE 1-13

Convert 3400 amperes to kiloamperes and to scientific notation.

**Solution**

This involves changing from a unit to a multiple unit (to a larger unit). The decimal point must be moved three places to the left.

$$3400 \text{ amperes} = 3.4 \text{ kiloamperes } (3.4\ kA)$$

Using scientific notation, the current would be expressed as $3.4 \times 10^3$ A.

Thus far, the examples have produced agreement between scientific notation and the multiple and submultiple units. This is not always the case. For example, a current of 25 $\mu$A could be written as $25 \times 10^{-6}$ A, which is not in scientific notation. It could be converted to $2.5 \times 10^{-5}$ A, but there is no submultiple unit to represent $10^{-5}$. Both representations are correct. However, the method preferred by electrical engineers and technicians is to use the units shown in Table 1-3 when possible. In this system, 25 $\mu$A is preferred over $2.5 \times 10^{-5}$ A. The system of recording values in terms of the multiple units and submultiple units given in Table 1-3 is known as *engineering notation*. As can be seen from this table, engineering notation uses powers of 10 with exponents that are multiples of 3, such as 3, 6, 9, 12, −3, −6, −9, and −12. Thus the number itself will not always contain just one digit to the left of the decimal point.

**EXAMPLE 1-14**
Express 0.015 A using both engineering notation and scientific notation.

**Solution**
Recall that multiples and submultiples can be used to place numbers in the range from 0.1 to 1000. The milliampere unit will produce a number in the desired range. Using engineering notation gives

$$0.015 \text{ A} = 15 \times 10^{-3} \text{ A} = 15 \text{ mA}$$

Converting to scientific notation gives

$$0.015 \text{ A} = 1.5 \times 10^{-2} \text{ A}$$

It is important to represent measurements in terms of their base units when performing calculations. Most formulas and equations assume base units. If a prefix is used, eliminate it by using the appropriate power of 10.

**EXAMPLE 1-15**
The quantity 150 nanoamperes (150 nA) is to be used in a calculation. Convert the quantity to an appropriate form.

**Solution**
The prefix nano represents $10^{-9}$. Therefore, the measurement can be converted to

$$150 \text{ nA} = 150 \times 10^{-9} \text{ A}$$

It may also be converted to $1.5 \times 10^{-7}$ A, which may be more convenient, depending on how the calculation is to be performed. In any case, remember that prefixes must normally be eliminated before quantities are used in calculations.

Compound prefixes should not be used. Examples of compound prefixes are millimicro and micromicro. Use nano instead of millimicro and pico rather than micromicro.

Scientific notation and engineering notation both provide convenient ways to deal with very large and very small quantities. They eliminate long strings of zeros in numbers such as 55,000,000 and 0.00000038. It is very easy to miss a decimal place with such numbers and they are therefore error-prone. It helps to separate such numbers into groups of three. A comma or a space can be used to separate the groups. For example, 55,000,000 may also be written as 55 000 000 and 0.00000038 may be written as 0.000 000 38. Grouping is fine, but the use of powers of 10 is preferred.

## Self-Test

**16.** The SI system of prefixes is arranged in increments of ______ powers of 10.

**17.** How many meters are there in a kilometer?

**18.** How many millimeters are there in a kilometer?

**19.** Convert 1460 watts to kilowatts.

**20.** Convert 321 millivolts to scientific notation.

**21.** Convert $2 \times 10^{-5}$ ampere to engineering notation.

**22.** Convert 100 milliwatts to watts.

**23.** Convert 0.22 megawatt to watts.

**24.** What must normally be done to units with prefixes before they can be used in calculations?

**25.** Convert 5 ms into a form appropriate for use in calculations.

## 1-4 RESOLUTION AND ROUNDING

Electrical measurements are subject to errors. Calculations based on measurements should not introduce significant additional error. Also, calculations should not be based on false accuracy. False accuracy results when data are interpreted to have a degree of resolution beyond what is actually present.

The best way to minimize error and false accuracy is to recognize and use *significant digits*. A digit is significant if it can be considered to be reliable. For example, look at Fig. 1-3. It shows a typical meter used to measure electrical quantities. Note that the meter has a calibrated scale. The scale is made up of major and minor calibrations called

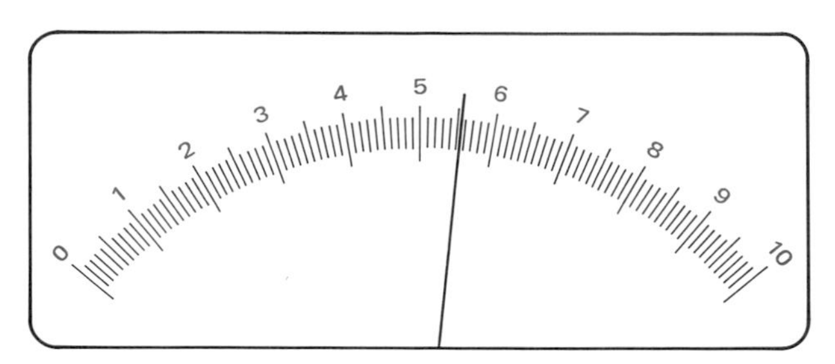


**FIGURE 1-3 Reading a meter.**

*divisions.* Each major division in the figure represents one unit. Therefore, if the meter measures volts, each major division equals 1 V. The minor divisions represent a tenth of one unit. Looking very closely at the meter reading, one could estimate a reading of 5.56 units. Or, a second look might produce an estimate of 5.57 units. One might then be tempted to average the two estimates and record a reading of 5.565 units. Generally, this is a wasted effort and produces false accuracy. The number 5.565 proposes to have four significant digits, but they are not all reliable.

How should a meter such as the one shown in Fig. 1-3 be read? Generally, the minor divisions on this type of an instrument represent the smallest reliable quantity that should be read or recorded. In other words, if the instrument in the figure is a voltmeter, it should be read to the nearest tenth of a volt. The proper reading in this case would be 5.6 V, which has two significant digits because both of them can be considered to be reliable. It would be a waste of effort and misleading to record or use a value of 5.565 V. The number 5.565 proposes to have four significant digits, so the accuracy would be false in this case. Another factor which must be considered is the accuracy of the meter. It will usually be rated as some percentage of the full-scale value of the meter. The meter in Fig. 1-3 has a full-scale value of 10. If it is rated at + or − 2 percent, the error is + or − 0.02 × 10 = 0.2, which is equal to plus or minus two minor divisions. Some meters may be better, with accuracies of 1 or 0.5 percent. Then again, some may be worse, with an accuracy of only 5 percent. In any case, the futility of extracting and using a reading such as 5.565 should now be clear.

Suppose the meter pointer in Fig. 1-3 is exactly over the major division marker below the 5. How should the reading be recorded? Should it be recorded as 5, 5.0, or 5.00? Are all of these readings the same? The reading should be recorded with the appropriate number of significant digits (as 5.0 in this case). However, 5, 5.0, and 5.00 do not have the same number of significant digits. A value of 5 has one significant digit, 5.0 has two significant digits, and 5.00 has three significant digits. When you report a quantity such as 5 V, you are actually stating that the voltage was obtained in such a manner that it is closer to 5 than it is to 4 or to 6. When you report it as 5.0 V, your statement is different because you are saying it is closer to 5.0 than it is to 4.9 or to 5.1. It is important to report and use readings properly.

False accuracy can also occur when data are manipulated without regard to significant digits. Suppose it is necessary to add some electrical quantities as shown below:

$$\begin{array}{r} 22.3 \\ +14.042 \\ +\ 9. \\ \hline 45.342 \end{array}$$

It would not be proper to report 45.342 as a result. The value 9 has only one significant digit. Its reliability ends with the units column. When numbers are added, the sum should not be reported with any digits to the right of the column where any last reliable digit appears. Therefore, the sum must be reported as 45, which has only two significant digits, as opposed to 45.342, which proposes to have five significant digits but does not (false accuracy). Suppose the value 9 in the above example was recorded as 9.00. How should the sum be recorded now? The limiting number is now 22.3, where the 3 is its last reliable number and appears in the tenths column. The sum must be reported as 45.3 in this case.

The number 5184 has four significant digits. How many significant digits are there in the number 5000? One may be tempted to answer four since there are four digits to the left of the decimal point. However, the decimal point has absolutely nothing to do with how many significant digits there are in any number. All we know for sure about the number 5000 is that it is closer to 5000 than it is to 4000 or to 6000. Scientific notation can be used to clearly indicate which digits are significant in a number such as 5000:

$5 \times 10^3$ (one significant digit)

$5.0 \times 10^3$ (two significant digits)

$5.00 \times 10^3$ (three significant digits)

$5.000 \times 10^3$ (four significant digits)

How many significant digits are there in the number 0.000045? Once again, the decimal point has nothing to do with significant digits. Therefore, the 0s between the decimal point and the 4 are placeholders and are not significant. This number has only two significant digits. If the number is written as 0.00004500, it has four significant digits. The trailing zeros are significant because they serve no other purpose. They are not needed for placeholding (marking the position of the decimal point). Numbers less than 1 may also be written in scientific notation with the number of significant digits clearly specified:

$4.5 \times 10^{-5}$ (two significant digits)

$4.50 \times 10^{-5}$ (three significant digits)

$4.500 \times 10^{-5}$ (four significant digits)

If you use a calculator (or a computer) to divide 7.89 by 4.56, depending on the resolution of the calculator, you might obtain an answer of 1.730263158. This number shows 10 significant digits but will represent false accuracy if reported. Generally, an answer cannot have more significant digits than the least precise number used in the calculations. Therefore, the answer should be rounded off to three significant digits and reported as 1.73. The rules of rounding are as follow:

**1.** If the first number dropped is less than 5, the retained digits are unchanged. When 1.730263158 is rounded to three significant digits, the first number dropped is a 0. Therefore, 1.73 is retained with no change.

**2.** If the first number dropped is 5 or more, add 1 to the retained digits. If 1.730263158 is rounded to five significant digits, the first number dropped is a 6. Therefore, add 1 and report 1.7303 as the answer.

**EXAMPLE 1-16**

The odometer in your car shows 286.4 miles (mi) since the last fill-up. The fuel pump registers 9.8 gallons (gal) when you refill the tank. Calculate the miles per gallon and report the answer using the proper number of significant digits.

**Solution**

Dividing 286.4 by 9.8 produces 29.2244898 on a typical calculator. The answer should be reported with only two significant digits because the least precise number used in the calculation is 9.8. Therefore, the correct answer is 29 mi/gal.

Most technical work in electricity and electronics utilizes three significant digits. You should maintain four significant digits when making chain calculations. This is of no concern if you are using a calculator or a computer to store all of the intermediate results since these instruments usually hold many more than four significant digits. If you must record an intermediate value that will be reentered into the calculations, hold one more significant digit than you intend to report. If you are using the language BASIC on a computer, it may be easy to format the output of numbers in scientific notation with the desired number of significant digits. For example,

```
PRINT USING "##.##↑↑↑↑";N
```

may be used with some versions of BASIC to print the value of the variable *N* in scientific notation with three significant digits.

## Self-Test

**26.** Examine Fig. 1-3. Suppose the pointer is closest to the major division under the 8. How should the meter value be reported?

**27.** Examine Fig. 1-3. Suppose the pointer is midway between the major division 6 and the next minor division to its right. How should the reading be reported?

**28.** Add the values 8.01, 19.5, 3.224, and 14.2. Report your answer with an appropriate number of significant digits.

**29.** How many significant digits are there in each of the following numbers?

| | |
|---|---|
| 1189.0 | $2 \times 10^{-6}$ |
| 0.0153 | 0.0004 |
| 1000.0 | $7.17 \times 10^{3}$ |
| 1000 | |

**30.** Suppose the number 800 has three significant digits. How should it be reported?

**31.** Round the following numbers to three significant digits:

| | |
|---|---|
| 1184.32 | 100.8 |
| 0.01236 | 100.2 |
| 3.8442 | 1.333333 |

**32.** Divide 4732 by 1.8 and record the answer with the appropriate number of significant digits.

**33.** Multiply 1.46 by 3.42 and report the answer using the appropriate number of significant digits.

## 1-5 THE STRUCTURE OF MATTER

The electrical behavior of materials varies according to their physical makeup. All matter is formed from tiny building blocks called atoms. The atoms, in turn, are built from subatomic particles. An element is a fundamental type of matter in which all of the atoms in the material are the same. An atom represents the smallest unit of an element that retains the properties of that element. If an atom is broken down, the subatomic particles that result will not have the properties of the original element. Copper, gold, silicon, carbon, and oxygen are all examples of elements. Compounds are formed from two or more different kinds of atoms. The smallest unit of a compound that retains its properties is known as a molecule. Water is a compound and is composed of hydrogen and oxygen atoms.

The size of an atom is an abstract concept since the atomic scale is so tiny compared with the things we normally deal with. The total number of hydrogen and oxygen atoms in a thimble of water is approximately $1 \times 10^{22}$. This is an extremely large number. To put it in perspective, consider that the age of the universe is roughly $1 \times 10^{17}$ s.

Figure 1-4 shows the atomic structure of helium, which is one of the most basic elements. The central part of the atom is known as its nucleus. All atomic nuclei, with the exception of hydrogen, are made up of neutral particles called *neutrons* and positively charged particles called *protons*. The nucleus of a hydrogen atom consists of a single proton. The outer portion of the atom is made up of orbiting particles called *electrons*. The electrons have a negative charge. Opposite charges attract, and the negative electrons are attracted to the positive protons in the nucleus. This electric force of attraction holds the electrons in their orbits. This is similar to the way that the gravitational attraction of the sun holds the planets in orbit. The mass of a

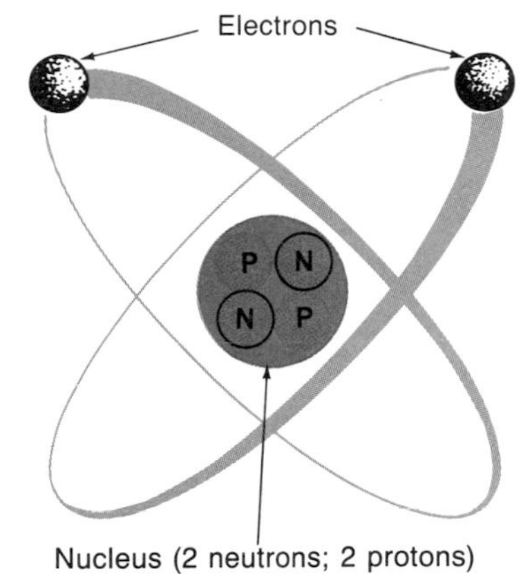


**FIGURE 1-4 Helium atom.**

proton is 1837 times the mass of an electron. Neutron mass and proton mass are about equal. This means that the vast majority of the mass of any atom is contained in its nucleus.

Neutrons, protons, and electrons are subatomic particles. These particles are not unique according to their parent atom. There is no difference between electrons from one atom to another even though the atoms may be from different elements.

The elements that make up all matter can be arranged in order of increasing mass. Hydrogen is the lowest in mass and has an atomic number of 1. This number is equal to the number of protons in the nucleus. Helium is next with an atomic number of 2, and Fig. 1-4 shows two protons in its nucleus. Uranium is the highest mass atom that occurs in nature and has an atomic number of 92. Even more complicated atoms, with masses greater than uranium, can be created in laboratories by bombarding materials with subatomic particles. These materials are sometimes referred to as manufactured elements.

The number of orbiting electrons normally equals the number of protons in the nucleus. This makes the net electric charge on any atom zero. This is demonstrated by the helium atom shown in Fig. 1-4 and by the copper atom shown in Fig. 1-5. The copper atom has four major electron orbits. The sum of the electrons in all of the orbits is equal to 29, which is also the number of protons in the nucleus. Therefore, the net charge on a copper atom is zero.

It is possible to remove an outer electron from an atom. When this happens, the atom becomes a positive ion because it has one more positive

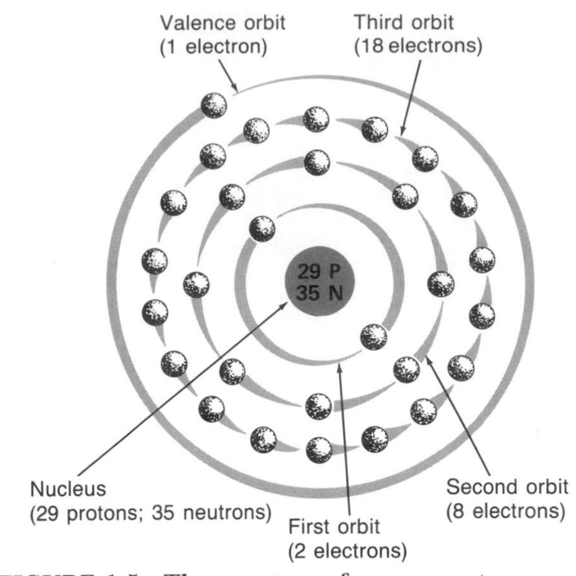


**FIGURE 1-5 The structure of a copper atom.**

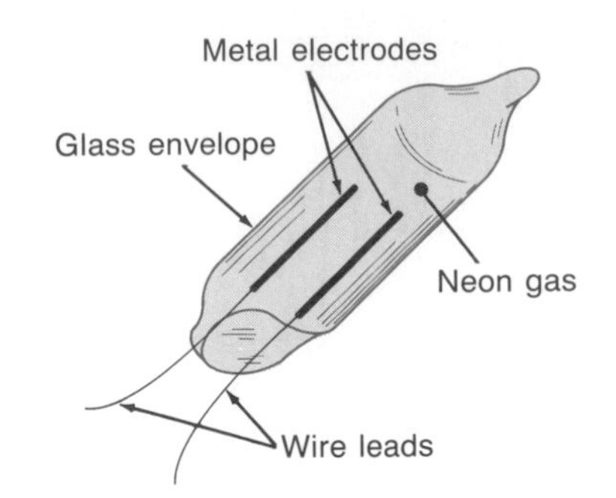


**FIGURE 1-6 Neon lamp.**

charge in its nucleus than it has negative electrons. It is also possible to force an extra electron into the outer orbit of an atom. This makes the atom a negative ion because it will have one more negative electron than it has protons. Ionization may be used to make materials electrically active that are normally inactive. The neon glow lamp shown in Fig. 1-6 is an example. If sufficient electric force is applied across the wire leads, the neon gas in the envelope ionizes and becomes electrically active. The lamp glows when it is active.

A considerable electric force is required to ionize neon gas. Without this force, the neon atoms are electrically neutral and are not active. Copper is a more active electrical material and can be used to readily support an electric current which is made up of a flow of subatomic electrons. Refer to Fig. 1-5. The outer orbit is called the *valence orbit.* It contains only one electron. On a relative scale, the spacing between the nucleus and the valence electron is vast. If the copper atom could be magnified until the electrons were as large as coins, the valence electron would be several kilometers away from the nucleus. This relatively large distance dictates that the valence electron is only weakly attracted to the positive charges in the nucleus. Very little energy is needed to move the valence electron away from the atom. It is also easy to push an extra electron into the valence orbit.

An electron that escapes the valence orbit is a free electron. Figure 1-7 shows how free electrons can be used to form a current flow. For simplicity, only the valence orbits are shown because only the outermost electrons can take part in the flow of current. The conductor is made up of a material, such as copper, that has weakly bound valence electrons. The valence electrons are easy to free and are therefore available to support the flow of current. Good electric conductors contain many, many free electrons. For example, 1 $cm^3$ of copper

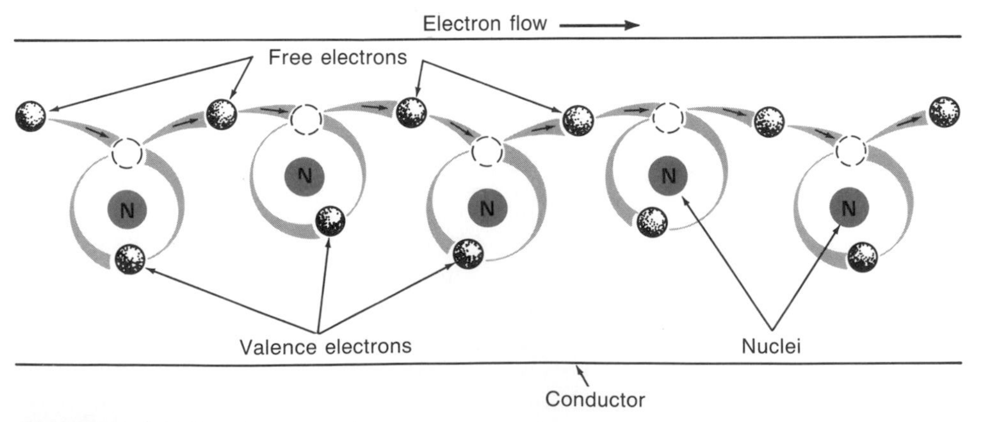


**FIGURE 1-7 Electric current.**

has about $8.5 \times 10^{22}$ free electrons at room temperature. Copper is a good electric conductor since it readily supports the flow of current.

Silver has an atomic structure that is similar to copper. It has the most free electrons of any element and is the very best electric conductor. However, silver is far too expensive for most applications. Some critical electronic circuits use silver plating to enhance conduction. Gold is another material with many free electrons. Copper is a better electric conductor than gold, but gold is more stable and is sometimes used to plate electric contacts in those applications where corrosion must be minimized. Gold is very expensive and it is used sparingly. Aluminum is also an electric conductor. Figure 1-8 shows the structure of an aluminum atom. One might guess that it would be a better conductor than copper since it has three valence electrons. However, the valence electrons in aluminum are more strongly attracted to the nuclei and not as many free electrons are available. Even though it is not as good a conductor, aluminum is considerably less expensive and is used in place of copper in some electrical applications. The best electric conductors have one, two, or three valence electrons and exhibit large numbers of free electrons at room temperature.

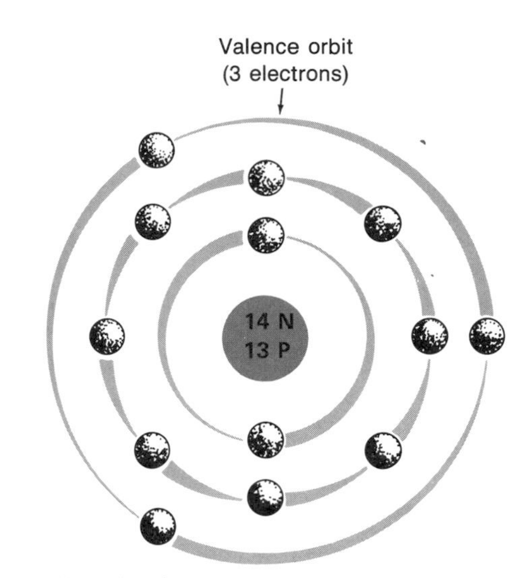


**FIGURE 1-8 Aluminum atom.**

Conductors are used in electrical applications to support current flow just as pipes are used to conduct fluids or gases. A pipe has an open center area which allows the material to flow. The flow is confined to the center area by the wall of the pipe. A wire is a similar structure, but the flow is supported by a solid conductor such as copper. The flow is confined to the conductor by another solid layer called the insulation. Figure 1-9 shows an insulated conductor. The current flow is confined to the conductor by the insulation around it. The insulation is a material with very tightly bound valence electrons. There are very few free electrons in the insulation and the flow is confined to the

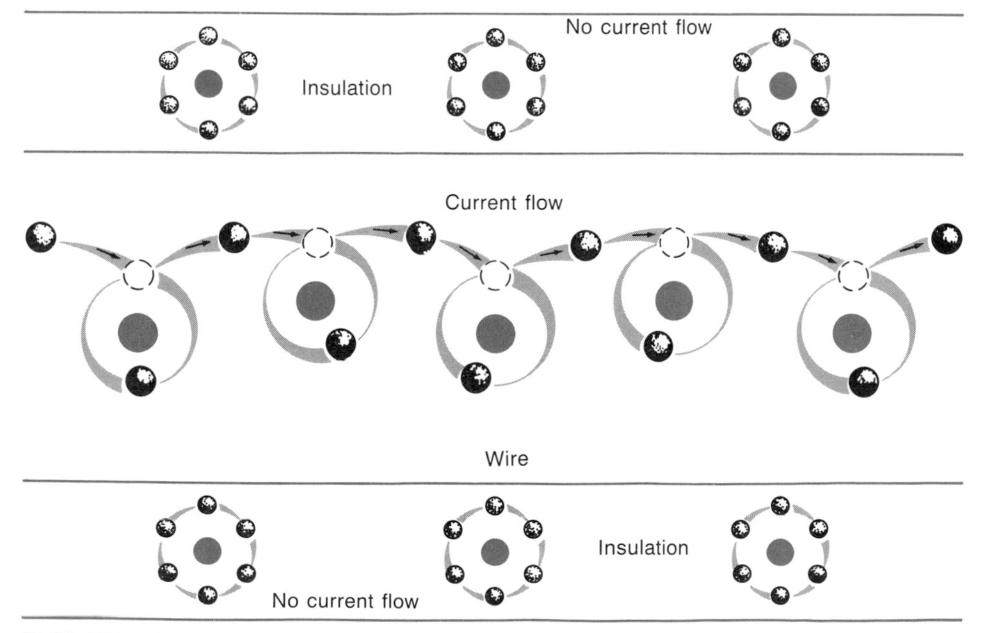


**FIGURE 1-9 An insulated conductor.**

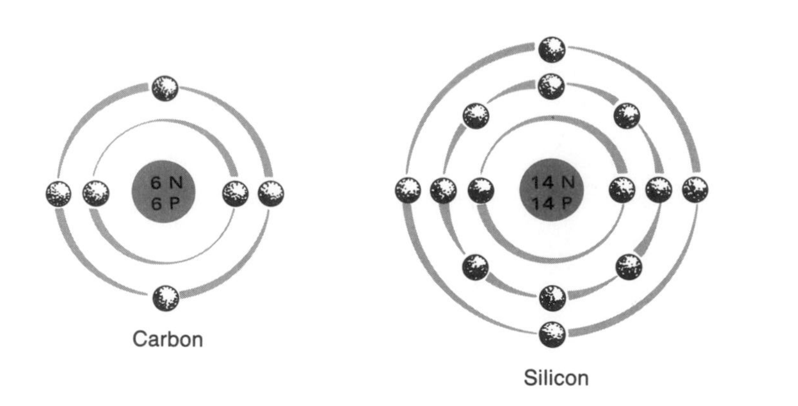


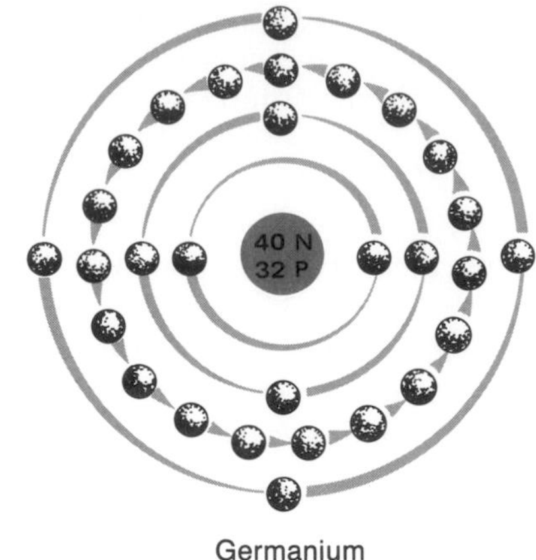


**FIGURE 1-10 Semiconductors.**

wire on the inside. Compounds such as plastic or rubber are often used to insulate the conductors used in electrical work.

In addition to conductors and insulators, semiconductors are materials with intermediate characteristics. These materials have four valence electrons. Figure 1-10 shows the atomic structure of carbon, silicon, and germanium. Carbon is used to make resistors and some sliding contacts such as motor brushes. A resistor is a component that supports the flow of current better than an insulator but not as well as a conductor. Resistors are treated in detail in later chapters. Silicon is used to make most solid-state electronic devices such as diodes, transistors, and integrated circuits. These are also covered in later chapters. Germanium is used to make a small percentage of modern diodes and transistors. Other semiconducting compounds, such as gallium arsenide, are also employed in some solid-state devices.

## Self-Test

**34.** When all the atoms that make up a material are identical, that material is known as a(n) ______.

**35.** The smallest unit of any element that retains the properties of that element is called a(n) ______.

**36.** The smallest unit of a compound such as water that retains the characteristics of the compound is called a(n) ______.

**37.** The vast majority of the mass of an atom is located in its ______.

**38.** The orbital electrons are attracted to the ______ located in the nucleus.

**39.** Electrons have a ______ charge.

**40.** The atomic number of silicon is 14. How many protons are there in the nucleus of a silicon atom? How many orbital electrons are there?

**41.** The net electric charge on an atom is ______.

**42.** When an electron is removed from the valence orbit of an atom, a ______ ______ results.

**43.** When an electron is added to the valence orbit of an atom, a ______ ______ is formed.

**44.** The net electric charge on an ion is ______ or ______.

**45.** The only electrons that can take part in current flow are the ones from the ______ orbit.

**46.** Materials, such as copper, with large numbers of free electrons are classified as ______.

**47.** Which electric conductor is sometimes used in place of copper to reduce cost?

**48.** Insulators are materials with ______ free electrons.

**49.** Materials with conduction properties that fall in between insulators and conductors are classified as ______.

## 1-6 THE NATURE OF ELECTRICITY

We have seen that electric forces control the structure of atoms. A lightning storm is an impressive display of electricity at work in nature. As you read this, your nervous system and brain communicate via tiny electric impulses. Electricity is a vital and

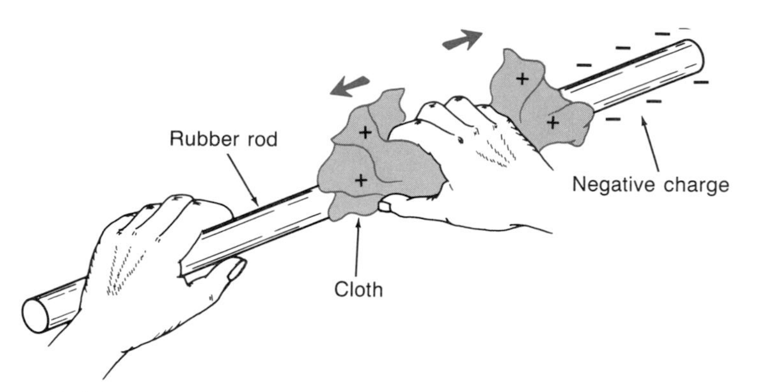


FIGURE 1-11 Charging by friction.

integral part of our bodies and, for that matter, of the entire physical universe.

Long ago, the Greeks noted that elektron (amber) would attract bits of straw after being rubbed. The Greeks had no way to explain this force of attraction. Our understanding of atomic structure provides the answers. Figure 1-11 shows how a rubber rod can be "electrified" by friction. The rubbing action removes valence electrons from the cloth and deposits them on the rod. The extra electrons give the rod a net negative charge. If a glass rod is rubbed with silk, it is also charged but with opposite polarity. Some of the valence electrons in the glass are transferred to the silk. This provides the glass rod with a positive charge. If the two charged rods are hung in stirrups and placed near each other, it will be obvious that a force of attraction exists between them. As in the atom, opposite charges attract.

Gravitational forces also exist between bodies. However, there are two major differences between gravitational forces and electric forces. First, gravitational forces become significant only when at least one of the bodies has an enormous mass, such as an entire planet. A second major difference is that electric forces can be repulsive as well as attractive. The electric force of attraction between the two charged rods is far more significant than the gravitational forces between them. If two rubber rods are charged with cloth and placed near each other, they repel one another. The law of charges states that "like charges repel and opposite charges attract."

Figure 1-12 shows an instrument, called an electroscope, that exploits the law of charges. It has two very thin metal leaves suspended from a metal stem. If the leaves are charged, they will have the same polarity since they are connected together at the top. The leaves will therefore repel as shown in Fig. 1-13 where the electroscope has been charged by contacting the metal sphere with a charged rod. Charging by contact shares the charge between the rod and the metal structure of the electroscope. If the rod is negatively charged, some of the extra electrons leave the rod and move to the metal sphere, the metal stem, and the metal leaves. Now the entire metal structure also has extra electrons and has received a negative charge from the rod. If the rod is withdrawn, the electroscope remains charged, as shown in Fig. 1-14. Similar results are obtained if the electroscope is contacted by a positively charged body. The leaves will repel and remain apart when the positive body is withdrawn. However, the polarity of the charge will be positive in this case. This is because the positive body removes some of the valence electrons from the electroscope when contact is made.

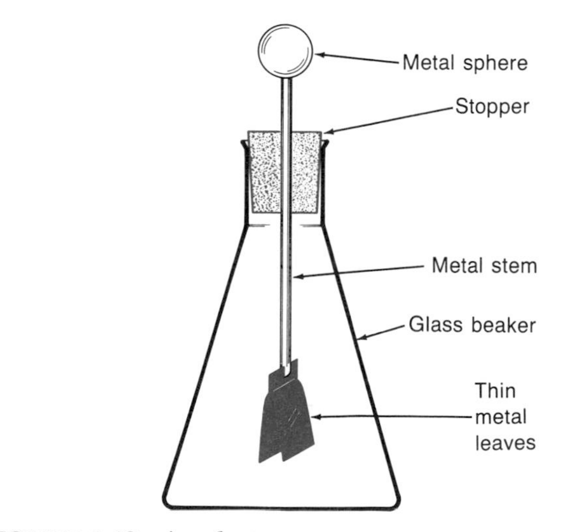

FIGURE 1-12 An electroscope.

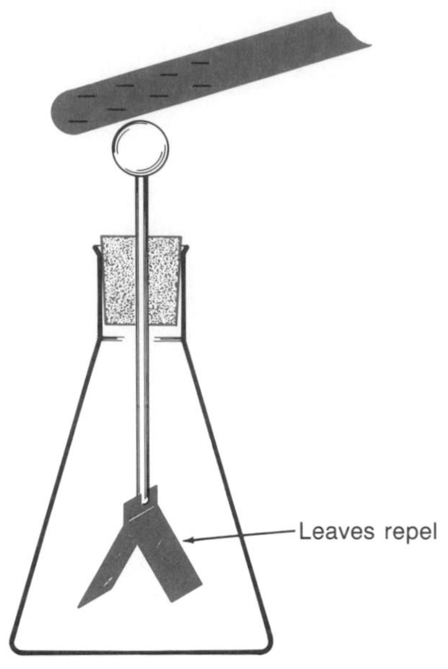


FIGURE 1-13 Charging by contact.

Once the electroscope is charged, some other experiments are possible to demonstrate some of the principles of electricity. Suppose a metal sphere mounted on an insulating handle is brought into contact with the charged electroscope of Fig. 1-14. The leaves will partially collapse at the moment of contact. The charge has been shared between the bodies, and some of the charge has been taken away from the electroscope. If a larger sphere is used, the collapse of the leaves is more complete because the larger object takes yet more of the charge away. If a very large object is used, the charge is distributed in such a way as to provide total collapse of the leaves in the electroscope. For example, if a conductor is connected from the earth to the sphere of the electroscope, the discharge is complete for all practical purposes. The process of sharing charges with the earth is known as *grounding*.

If a charged electroscope is not contacted, it retains its charge for some time. This is because the air around the metal parts is a good insulator. However, we learned that gases can become electrically active by the process of ionization. This is easily demonstrated by bringing a flame near the metal sphere of a charged electroscope. A flame contains many free ions, and they will drain the charge from the electroscope. Similar experiments can be conducted with radioactive sources which ionize the air around them.

Figure 1-15 shows that it is also possible to charge by electrostatic induction. This time, the charged rod is positioned near the sphere of the electroscope. The negative rod has an electrostatic field around it that can influence other charges at a distance. In this case, the field is negative and repels the electrons in the sphere. The electrons move down the metal stem and onto the leaves. Note that the charge induced in the sphere is opposite in polarity to the charge source which is the negative rod. Charging by electrostatic induction is different in this respect since an opposite charge has been produced in that part of the object near-

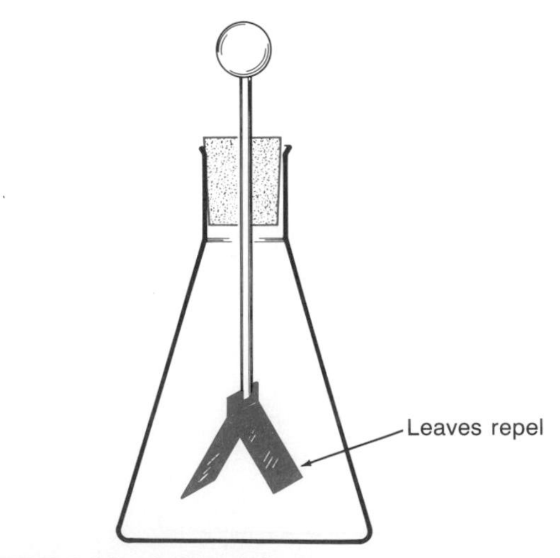


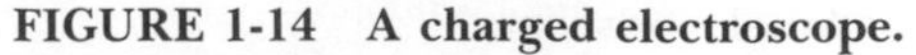
FIGURE 1-14 A charged electroscope.

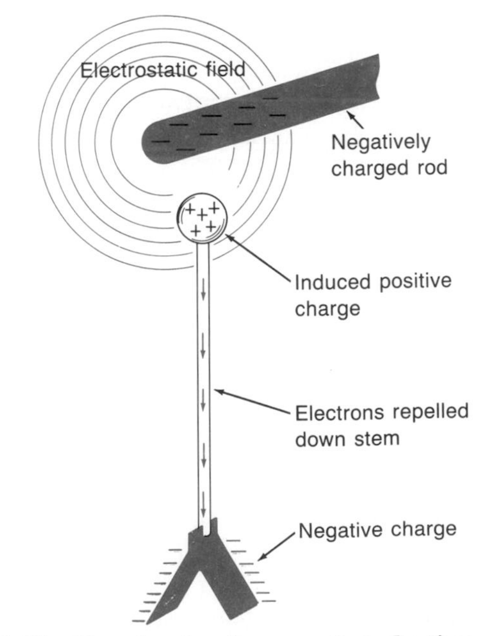


FIGURE 1-15 Charging by electrostatic induction.

est the charge source. The second difference is that the leaves of the electroscope collapse immediately when the negative rod is withdrawn. The induced charge equalizes when the electrostatic field is withdrawn. The electroscope does not remain charged as it did when it was charged by contact.

The electrostatic induction concept leads to another experiment, shown in Fig. 1-16. Two metal rods are placed end to end and in contact with each other. A positively charged body is brought near the right end of the rods. A charge is induced as shown in Fig. 1-16*a*. Figure 1-16*b* shows that the metal rods are then separated. Now, when the charged body is removed, the charges cannot equalize and each half remains charged as shown in Fig. 1-16*c*. Each half can now be tested with an electroscope to verify that it is charged. Finally, the halves are joined again as shown in Fig. 1-16*d*, and the charges equalize. An electroscope can be used to verify that no charge is present on the assembly or on either half. This experiment aids our understanding of the concept of *polarity*. The words *positive* and *negative* are used because the forces exhibited by oppositely charged objects tend to cancel.

An electroscope can be used to identify polarity. If the electroscope is initially charged with a known polarity, it reacts differently to charged objects brought near it, depending on the polarity of their charge. If the electroscope is initially charged negatively, a negative object causes additional separation of the leaves. This is because the object forces additional electrons down the stem and adds to the negative charge that is already there. On the other hand, if the object is positively charged, the leaves collapse somewhat as the object nears. This happens as the positive charge attracts electrons up the stem and removes some of the original charge from the leaves.

Figure 1-17 shows a pith ball apparatus which is also useful to investigate electric forces. Figure 1-18 shows what happens when a negative object is brought near the ball. In Fig. 1-18*a*, the negative object induces a positive charge on the side of the ball nearest the object. Opposites attract, so the ball is attracted to the rod. In Fig. 1-18*b*, the ball contacts the charge source. Now, extra electrons are transferred to the ball. Note that the ball is now negatively charged. Finally, Fig. 1-18*c* shows the ball repelled by the charge source. This is because like charges repel. This experiment is interesting because it shows that charging by electrostatic induction produces an opposite charge which attracts and that charging by contact produces a like charge which repels. You should now understand why a charged comb attracts bits of paper which then fly away after touching the comb.

A storm can produce extremely large charges. A cloud is a mass of water vapor. A thundercloud is a

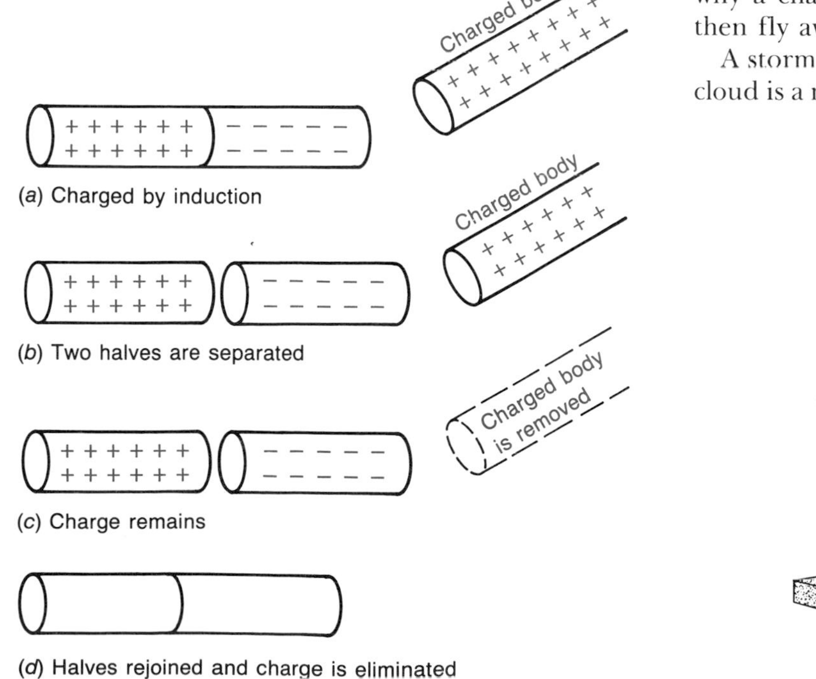


FIGURE 1-16 Electrostatic induction experiment.

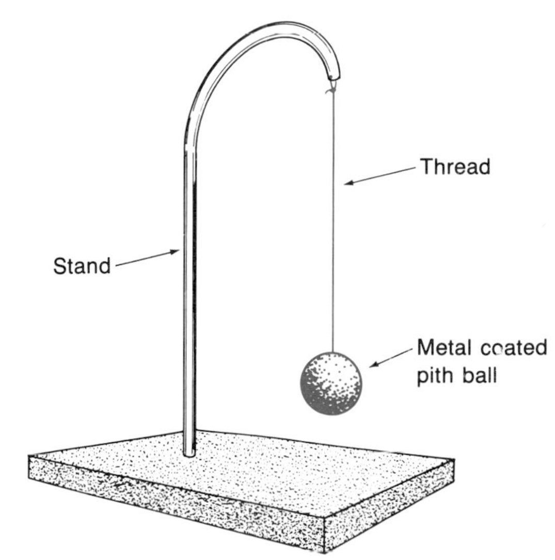

FIGURE 1-17 Pith ball apparatus.

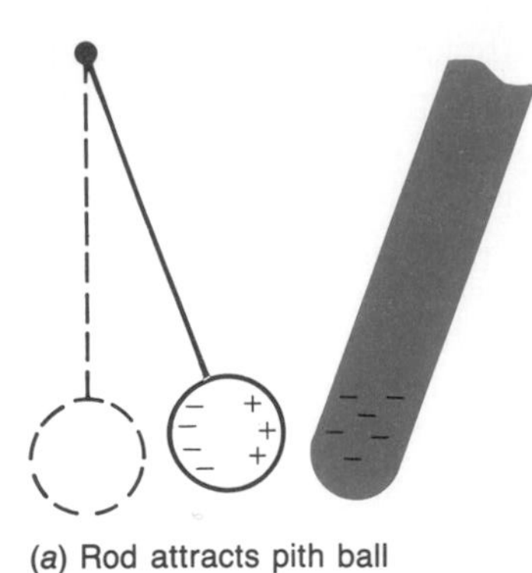

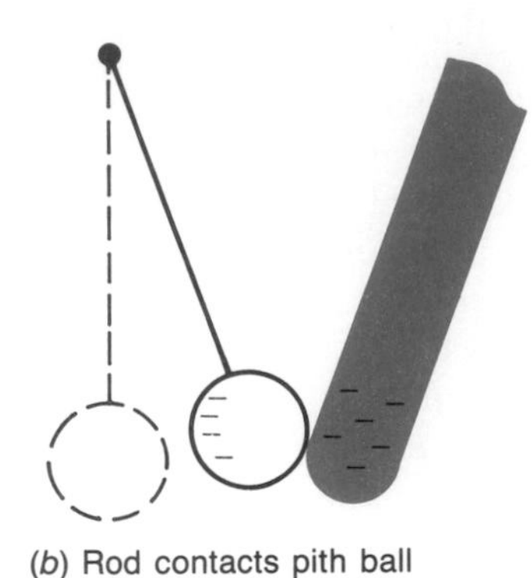

(*a*) Rod attracts pith ball (*b*) Rod contacts pith ball (*c*) Rod repels pith ball

**FIGURE 1-18 Pith ball experiment.**

cloud that has become charged by rising air masses. Figure 1-19 shows a positively charged cloud over a home which has induced a negative electrostatic charge on the top of the structure. If the difference between the two charges is high enough, ionization of the atmosphere occurs and a discharge results. Electrons rush from the structure to the cloud in an attempt to equalize the charges. This discharge is called *lightning*. Lightning can be very destructive. Structural damage and fire often result. Figure 1-20 shows a lightning protection system. The ground rods and metal conductors electrically connect the earth with the rods at the top of the structure. This connection makes it more difficult for a thundercloud to induce an opposite charge onto the roof of the building. The lightning rods tend to dissipate the charge into the nearby atmosphere. By reducing the induced charge, it is possible to decrease the probability of a strike. If a strike does occur, the electron current flows in the metal grounding system because its parts are good conductors. This usually protects the building from damage.

**FIGURE 1-19 Electrostatically induced charge on a home.**

Many buildings do not have lightning rods and grounding systems. If they are equipped with an antenna for radio or television reception, a static discharge device such as the one shown in Fig. 1-21 is recommended. The ground clamp is mounted to a metal structure which is connected via a metal conductor to the earth. The lead-in wire passes through the device on the way to the receiver. The device acts to limit the buildup of an electrostatic charge at the antenna. It does this by allowing electrons to move between the antenna and the connection to the earth when the charge reaches a certain level.

The study of charged bodies and the resulting forces is often called electrostatics or the study of

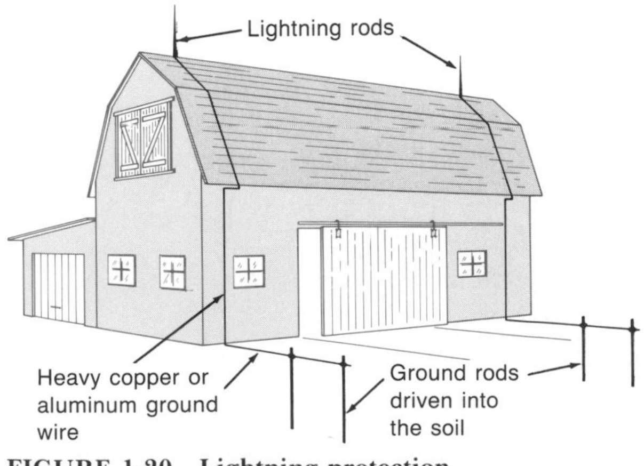


**FIGURE 1-20 Lightning protection.**

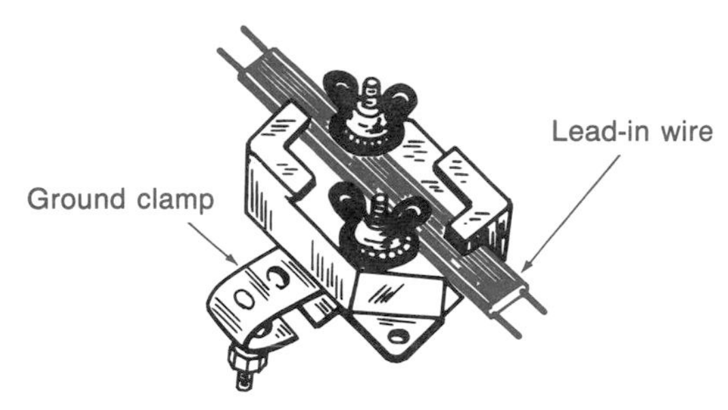


FIGURE 1-21 Static discharge device.

static electricity. There are useful applications for static electricity beyond the experiments and phenomena discussed thus far. Spray painting is a technique that provides a superior finish for many products. Overspray, as shown in Fig. 1-22, is wasteful and dangerous. The overspray can be significantly reduced by using a negative charge on the spray gun and a positive charge on the object to be painted. Figure 1-23 shows how static charges can be used to remove dirt and dust from air. The particles to be removed pick up a positive charge from one plate and are then attracted to the negatively charged collector. Another application of electrostatics is in the manufacture of abrasive paper as shown in Fig. 1-24. The opposite charges attract the abrasive particles to the adhesive coated paper to provide a dense, uniform coating.

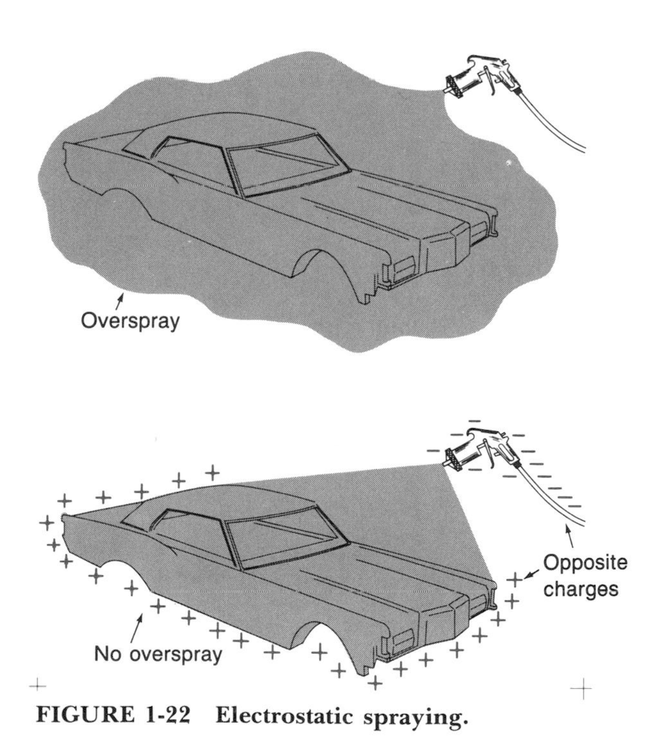


FIGURE 1-22 Electrostatic spraying.

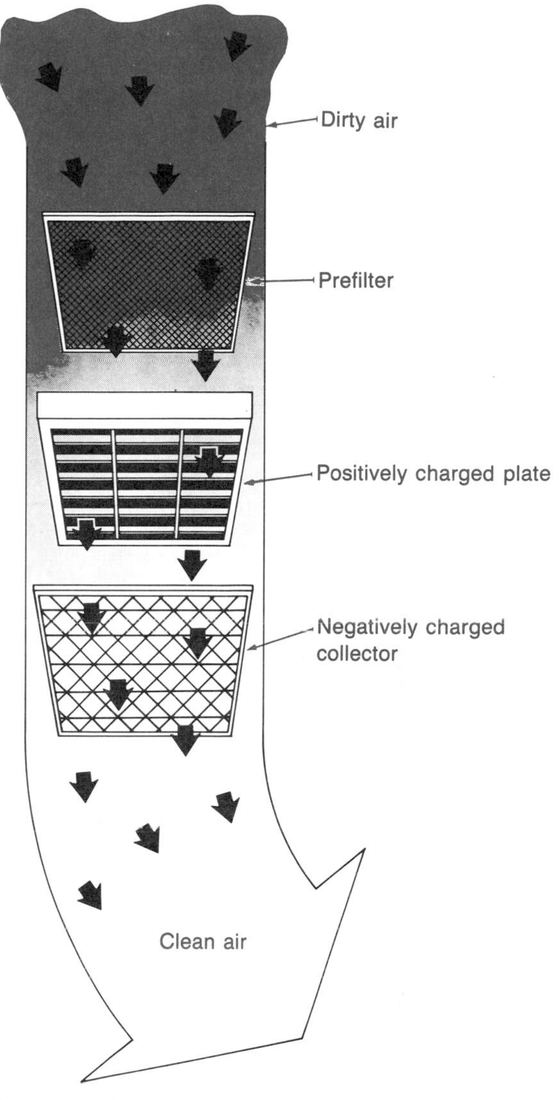


FIGURE 1-23 Electrostatic air filter.

Static forces are significant. Figure 1-25 shows the action resulting from the force of repulsion between two negatively charged pith balls. The angle of displacement suggests that the electrostatic force that is repelling the balls is comparable in magnitude to the gravitational force of the earth that is pulling down on the balls. To gain some insight into how electrostatic forces compare to gravitational forces, it is useful to review Newton's law of universal gravitation:

$$F_g = G\frac{m_1 m_2}{r^2}$$

where $F_g$ = force of attraction due to gravity

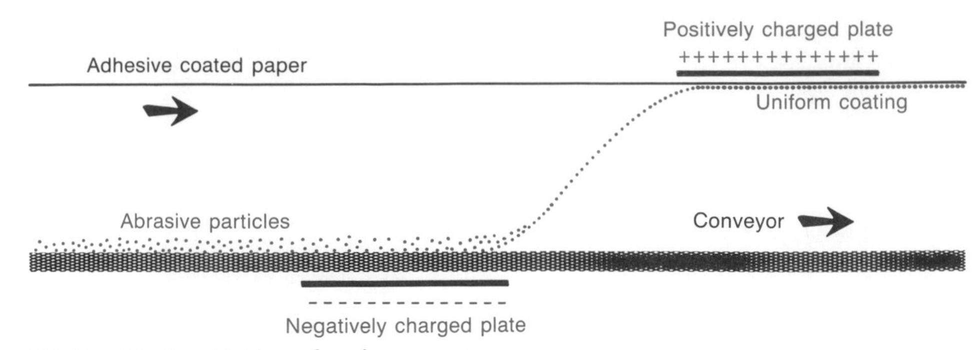


**FIGURE 1-24 Making abrasive paper.**

$G$ = gravitational constant
$m_1, m_2$ = masses of two bodies involved
$r$ = distance between two bodies

The mass of the earth is on the order of $10^{25}$ kg, and the mass of a pith ball is less than 1 g. With this in mind, look again at Fig. 1-25. The downward pull on the pith balls is due to the gravitational force of the entire earth acting on them. We may therefore conclude that the repulsive force between the balls in enormously greater than the gravitational attraction between them. The mass of the balls is very small and they are relatively close together. Therefore, the electric force that acts to repel the balls is on the order of $10^{28}$ times greater than the gravitational force that tends to attract them to each other. Another comparison is the electric attraction between a proton and an electron, which is $10^{39}$ times greater than the gravitational attraction between them.

Coulomb's law of electric interaction gives us a method of quantifying the force of attraction or repulsion between two charged bodies:

$$F_e = K_e \frac{q_1 q_2}{r^2}$$

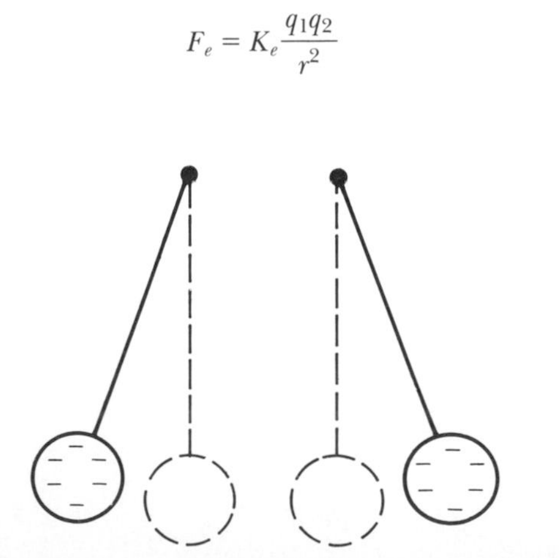

**FIGURE 1-25 Two negatively charged pith balls.**

where $F_e$ = force in newtons*
$K_e = 9 \times 10^9$ (a proportionality constant)
$q_1, q_2$ = charges in coulombs
$r$ = distance in meters

The coulomb (C) is the mksa unit of charge. It is a very large unit. If you rub a balloon on your sleeve, the charge will only approach $1 \times 10^{-6}$ C. If 1-C charges were attainable, the forces could be enormous. Two 1-C charges separated by a distance of 1 m would produce a force of $9 \times 10^9$ N (approximately 1 million tons).

---

**EXAMPLE 1-17**

Calculate the force between two charged bodies separated by a distance of 0.5 m if the charge on one body is 0.5 μC and a like charge of 1.25 μC exists on the other body. Also determine if the force is one of attraction or repulsion.

**Solution**

Using Coulomb's law and applying the data given,

$$F_e = 9 \times 10^9 \frac{(0.5 \times 10^{-6})(1.25 \times 10^{-6})}{0.5^2}$$
$$= 2.25 \times 10^{-2} \text{ N}$$

Since the charges are alike, the force is one of repulsion.

---

The charge on a single electron is known as an elementary charge and it is equal to $1.6 \times 10^{-19}$ C.

*One newton (1 N) is the force required to accelerate a one-kilogram (1 kg) mass one meter per second per second (1 m/s²).

The number of electrons or elementary charges in a coulomb can be found by

$$\frac{1}{1.6 \times 10^{-19}} = 6.25 \times 10^{18} \text{ electrons}$$

We will refer again to the coulomb unit of charge in the next chapter when discussing and developing the definitions for some of the base and derived units used in electricity and electronics.

## Self-Test

**50.** Opposite charges produce a force of ______.

**51.** Like charges produce a force of ______.

**52.** The metal structure of an electroscope is charged by touching it with a positive rod. The polarity of charge produced in the electroscope will be ______.

**53.** There are two neutral metal spheres mounted on insulating handles. One is larger than the other. Which would produce the most discharge when touched to a charged electroscope?

**54.** The process of sharing charges with the earth is known as ______.

**55.** A charge is electrostatically induced in an electroscope by a negative rod. What is the polarity of the induced charge at the point nearest the rod? At a point near the other end of the electroscope's metal structure?

**56.** What will the leaves of the electroscope in Question 55 do when the negative rod is withdrawn? What if the charge had been by contact rather than by induction?

**57.** Suppose a positively charged metal rod is brought into contact with a negatively charged metal rod. If the original charges were of equal magnitude, the net charge after contact would be ______.

**58.** What will the leaves of a positively charged electroscope show as a positively charged body draws near?

**59.** What will the leaves of a negatively charged electroscope show as a positively charged body draws near?

**60.** Compared to electrostatic forces, gravitational forces are usually much ______.

**61.** Ionized air tends to ______ a charged electroscope.

**62.** Calculate the force between two charged bodies if they are separated by 0.1 m and if each has a charge of 1 $\mu$C. How does the force vary according to the distance between the two bodies?

## SUMMARY

1. The SI system of measurements has been officially adopted by the Institute of Electrical and Electronic Engineers (IEEE).
2. Some physical measurements, such as those dealing with lead or contact spacings, are still based on the inch.
3. Anyone who works with electricity and electronics will be confronted with conversions from one measuring system to another.
4. Scientific notation is a power-of-10 system that is very convenient when working with very large or very small numbers.
5. Scientific notation places one nonzero digit to the left of the decimal point.
6. Engineering notation is a power-of-10 system based on standard prefixes.
7. Engineering notation is preferred for technical work in electricity and electronics. For example, a current would be expressed as $6.5 \times 10^{-5}$ A using scientific notation and as $65 \times 10^{-6}$ A (which is 65 $\mu$A) using engineering notation.
8. Reporting answers with more significant digits than the original data is known as false accuracy.

9. A digit is considered significant if it is known to be reliable.

10. Most technical calculations in electricity and electronics use three significant digits.

11. The electrical behavior of materials is based on subatomic particles called electrons. The electrons involved in any electrical activity are called valence electrons.

12. Electrons have a negative charge and protons have a positive charge.

13. The number of electrons in a material is normally equal to the number of protons which produces a net charge of zero.

14. When an object gains or loses some electrons, it becomes electrically unbalanced and is said to be charged.

15. Like charges repel and unlike charges attract.

16. Good conductors have an abundance of free valence electrons and good insulators have tightly bound valence electrons. Semiconductors have intermediate characteristics.

17. Coulomb's law is used to calculate the force between charged bodies.

18. The SI unit of charge is the coulomb. It is a very large unit and represents $6.25 \times 10^{18}$ elementary charges (electrons).

## CHAPTER REVIEW QUESTIONS

**1-1.** What is the abbreviation for the official measurement system used in electricity and electronics?

**1-2.** What is the base unit of length in the cgs system?

**1-3.** The atomic number of an atom predicts the number of _______ in its nucleus.

**1-4.** An atom has an atomic number of 32. How many electrons does it have?

**1-5.** Suppose an atom gains an extra electron. What is the atom called?

**1-6.** Gases must be _______ in order to conduct electric current.

**1-7.** Conductors, such as copper, are characterized by an abundance of _______ electrons.

**1-8.** Current flow is confined to a conductor by surrounding the conductor with an _______.

**1-9.** Carbon, silicon, and germanium are examples of _______ materials.

**1-10.** Like charges _______ and unlike charges _______.

**1-11.** An object can be charged by contact or by electrostatic _______.

**1-12.** An electroscope with an initial negative charge is approached by a positively charged body. What will the leaves of the instrument indicate?

**1-13.** What is the mksa unit of charge?

## CHAPTER REVIEW PROBLEMS

**1-1.** An electric connector is specified as having contacts spaced on 0.156-in. centers. What is the spacing in millimeters?

**1-2.** The upper temperature limit of many solid-state electronic devices is 100°C. Convert this temperature to degrees Fahrenheit.

**1-3.** Evaluate the following expression:

$$\frac{(3 \times 10^3)(4 \times 10^{-2})}{6 \times 10^{-1}} + \frac{8 \times 10^{-1}}{2 \times 10^{-2}}$$

**1-4.** Convert 0.00123 to scientific notation.

**1-5.** Convert 1,140,000 to scientific notation.

**1-6.** Evaluate the following expression:

$$(4 \times 10^2)^2$$

**1-7.** Evaluate the following expression:

$$(6.4 \times 10^3)^{1/2}$$

(*Note:* Raising a number to the $\frac{1}{2}$ power is the same as taking the square root of that number.)

**1-8.** Convert $1.09 \times 10^{-2}$ to standard form.

**1-9.** Change 3.88 $\mu$A to scientific notation.

**1-10.** Express 3.88 $\mu$A in standard form.

**1-11.** Change 0.15 kV to volts.

**1-12.** Add 50 mA to 1.20 A and express the answer in standard form.

**1-13.** Convert $1.35 \times 10^4$ V to engineering notation.

**1-14.** Convert 75 $\mu$V to scientific notation.

**1-15.** How many significant digits are there in the number 1,000,004?

**1-16.** How many significant digits are there in the number 10.0032?

**1-17.** How many significant digits are there in the number 1.000?

**1-18.** Suppose the result of a calculation is 1000 and all of the digits are significant. How should this result be reported?

**1-19.** Round off 11,135 ohms ($\Omega$) to three significant digits and express the result using engineering notation.

**1-20.** Round off 0.03455 A to three significant digits and express the result with engineering notation.

## ANSWERS TO SELF-TESTS

**1.** SI (international system of units)

**2.** inches

**3.** Celsius and Kelvin

**4.** cgs

**5.** meter, kilogram, second, and ampere

**6.** 77°

**7.** 4.72 by 3.15 in.

**8.** $1.86 \times 10^5$, $3 \times 10^{-4}$, $3.572 \times 10^2$, $1.54 \times 10^{-1}$, $1.2 \times 10^9$, $5 \times 10^{-10}$

**9.** 490, 0.0011, 567,800, 0.2, 55,000, 0.00019

**10.** $2.25 \times 10^2$, $3.85 \times 10^1$, $2.34 \times 10^4$

**11.** $-3.956$, $7.77 \times 10^3$, $-4.22 \times 10^{-5}$

**12.** 1.3, $3.32 \times 10^1$, $3.52 \times 10^{12}$

**13.** $2.2 \times 10^3$, $4.5 \times 10^{-10}$, $3.8 \times 10^6$

**14.** $9 \times 10^1$

**15.** $1.6 \times 10^9$

**16.** 3

**17.** $1 \times 10^3$

**18.** $1 \times 10^6$

**19.** 1.46 kW

**20.** $3.21 \times 10^{-1}$ V

**21.** 20 $\mu$A

**22.** 0.1 W

**23.** $2.2 \times 10^5$ W

**24.** Remove the prefixes by moving the decimal point or by converting to a power of 10.

**25.** $5 \times 10^{-3}$ s

**26.** 8.0

**27.** 6.1

**28.** 44.9

**29.** 1189.0 (5), 0.0153 (3), 1000.0 (5), 1000 (1, 2, 3, or 4), $2 \times 10^{-6}$ (1), 0.0004 (1), $7.17 \times 10^3$ (3)

**30.** $8.00 \times 10^2$

**31.** $1.18 \times 10^3$, 0.0124 or $1.24 \times 10^{-2}$, 3.84, 101, $1.00 \times 10^2$, 1.33

**32.** $2.6 \times 10^3$

**33.** 4.99

**34.** element

**35.** atom

**36.** molecule

**37.** nucleus

**38.** protons

**39.** negative

**40.** 14; 14

**41.** zero

**42.** positive ion

**43.** negative ion

**44.** positive, negative

**45.** valence
**46.** conductors
**47.** aluminum
**48.** few
**49.** semiconductors
**50.** attraction
**51.** repulsion
**52.** positive
**53.** the large sphere
**54.** grounding
**55.** positive, negative
**56.** they will collapse; they would remain apart
**57.** zero
**58.** additional repulsion (leaves diverge more)
**59.** less repulsion (leaves diverge less)
**60.** smaller
**61.** discharge
**62.** 0.9 N; as the inverse square

# Chapter 2 Electric Circuits and Quantities

Circuits are paths for the transfer of electric energy. Electric energy may be converted into other useful energy forms such as heat, light, or mechanical energy. The term *circuit* may also be used to refer to the collection of electric and electronic parts used to make up a device or a system. This chapter discusses circuits and how they are illustrated, their component parts, and some basic circuit measurements.

## 2-1 CIRCUITS AND DIAGRAMS

Figure 2-1 shows an elementary electric circuit. Two oppositely charged electroscopes are discharged by a length of wire. If the charge magni-

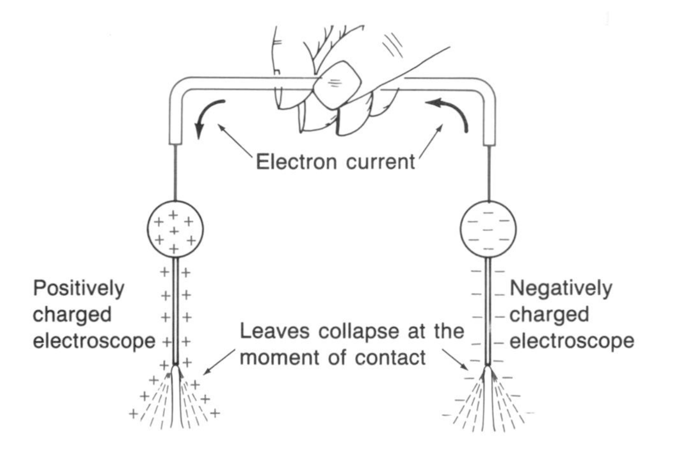


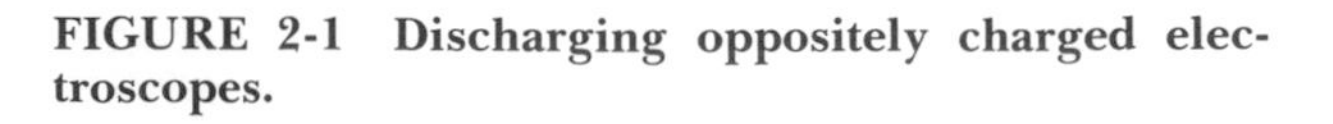
**FIGURE 2-1 Discharging oppositely charged electroscopes.**

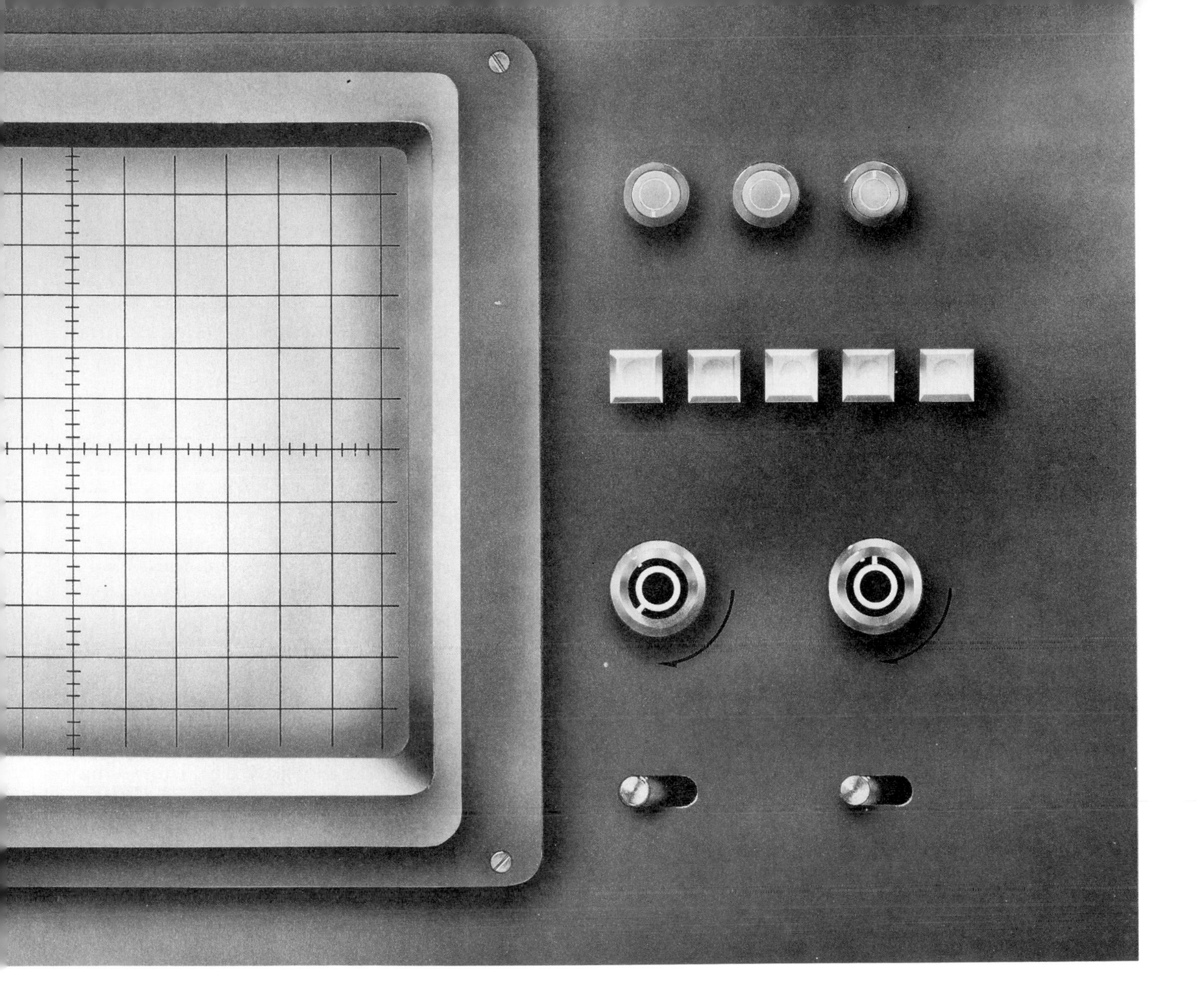

tude is equal before contact, both sets of leaves will indicate total discharge after contact is made. The leaves collapse due to a flow of electron current from the negative structure to the positive structure. It only takes a brief period of time for all of the excess electrons in the negative electroscope to move through the wire and cancel the electron deficiency in the positive electroscope. During that period of time, an *electron current* is flowing through an elementary *circuit,* which consists of a charge source and a length of wire. Please note that the direction of current is from the negative structure to the positive structure.

Figure 2-2 shows a more practical electric circuit. A generator provides the source of electric energy. For now, think of it as a replacement for two oppositely charged electroscopes. Since both charge

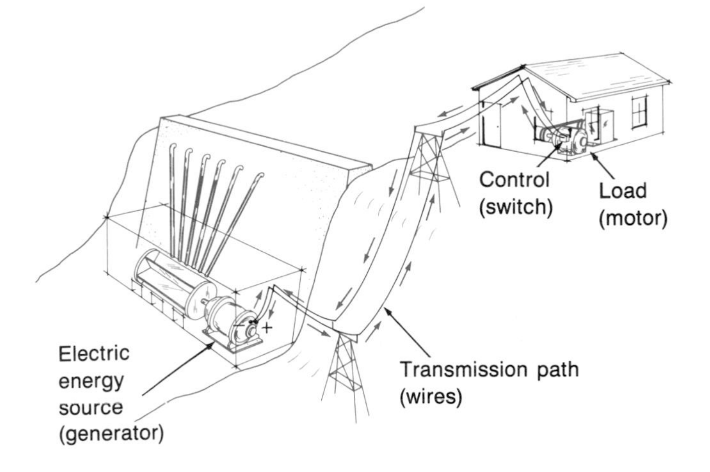


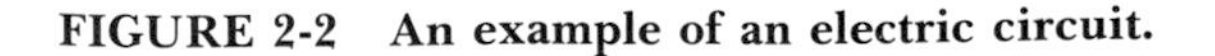

**FIGURE 2-2 An example of an electric circuit.**

sources are at one location (inside the generator), it takes two wires to transmit the electric energy to another location. One wire supports the electrons leaving the negative charge source, and the other supports the electrons returning to the positive charge source. We find a load at the other location. A load is a device to convert the electric energy to some useful output. In this case, it is a motor, but it could just as well be a light, a heater, or some other load device. We also find a means of controlling the circuit at the load location. The control device is a switch in this case. Opening the switch stops the flow of current and also stops the transfer of electric energy from the generator to the motor. Of course, the motor stops turning when the switch is opened. Circuits can be broken down into four major parts: (1) an energy source, (2) a path for the current to flow, (3) some form of control, and (4) a load. The control device can be eliminated in those cases where the circuit can be on all of the time.

Most practical electric and electronic circuits use a number of components to achieve the desired operation. For example, the part of the circuit that fulfills the control function may be based on several components. Diagrams are necessary to show how components are interconnected. Figure 2-3 shows an example of a *schematic* diagram. It uses *symbols* to represent the various components that make up the circuit. Compare it with Fig. 2-4,

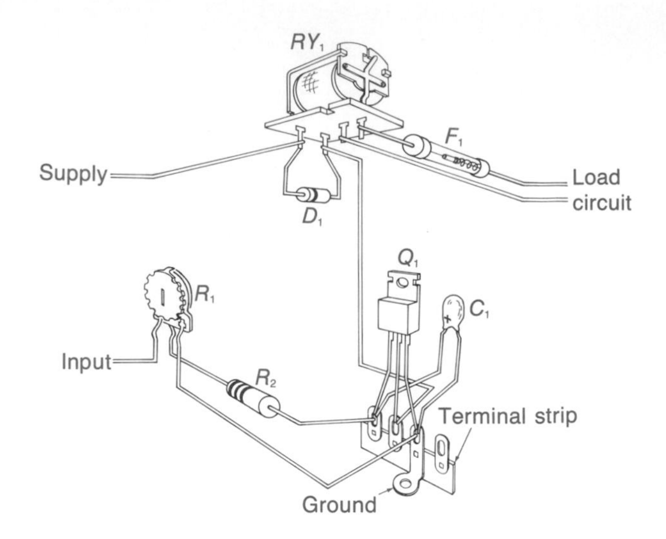


**FIGURE 2-4 Pictorial diagram.**

which shows a *pictorial* diagram of the same circuit. Schematic diagrams are favored for most applications because they are more convenient to draw and easier to understand by experienced people. For those who are not experienced, pictorial diagrams are easier to understand. For this reason, manufacturers may prepare pictorial diagrams to assist nontechnical personnel to install and maintain electric equipment. Figure 2-5 shows an exam-

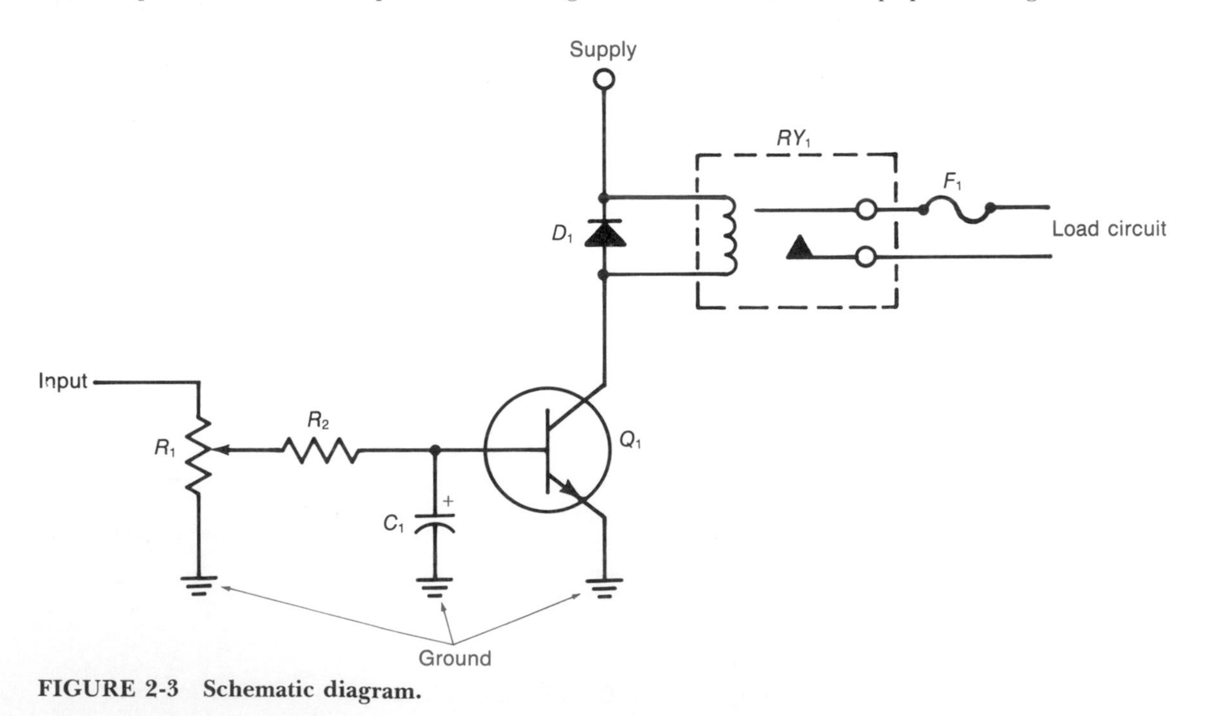


**FIGURE 2-3 Schematic diagram.**

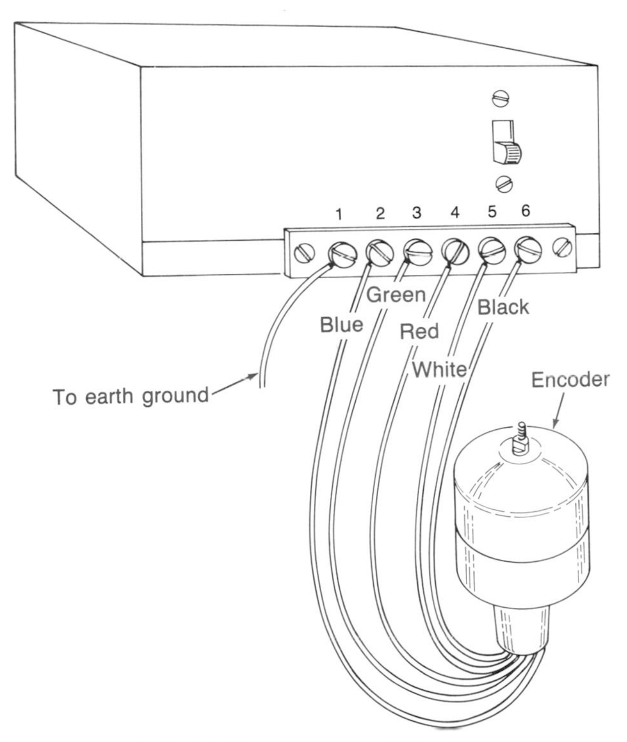


**FIGURE 2-5 Pictorial connection diagram.**

ple of a pictorial connection diagram intended to assist in the proper interconnection of two electric devices. Combination diagrams are also used in some cases. Note that the diagram shown in Fig. 2-6 uses some schematic symbols and some pictorial drawings to represent the various parts of the circuit.

Pictorial diagrams, such as the one shown in Fig. 2-4, may reasonably represent the way a circuit looks when assembled using *chassis wiring techniques*. Many circuits are assembled using another method called the *printed wiring technique*. Figure 2-7 shows an example of a printed circuit. The wire leads from each component pass through holes in the board and are soldered to the metal foil. The metal foil replaces the wires used to interconnect the parts when the chassis wiring technique is used. The board itself is a good insulator such as fiberglass. Thus, any current flowing in a foil trace is confined to that trace and the components soldered to it. Some printed circuits have metal traces on both sides of the board, and some even have one or more metal patterns sandwiched in between the top and bottom surfaces. These are called *multilayered boards* and are used in computer-

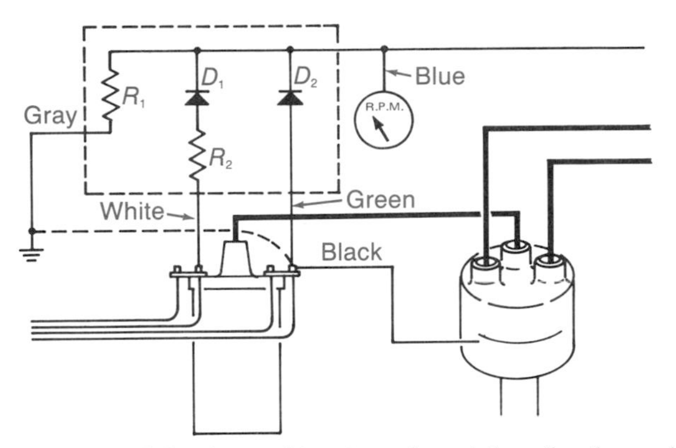


**FIGURE 2-6 A combination pictorial and schematic diagram.**

type circuits where many connections are required from component to component.

Not all schematic symbols represent an actual component part. For example, *ground* symbols are shown in Fig. 2-3. A ground symbol may represent a connection to the earth. If so, it is called an *earth* ground. Or, it may represent a connection to some common part of an electric circuit, in which case it is called a *common* ground. If the circuit depicted in Fig. 2-4 is built on a metal chassis, the chassis can serve as the common ground. In this case, mounting the terminal strip on the chassis provides a common ground connection point. Other terminal strips mounted at other locations may provide ad-

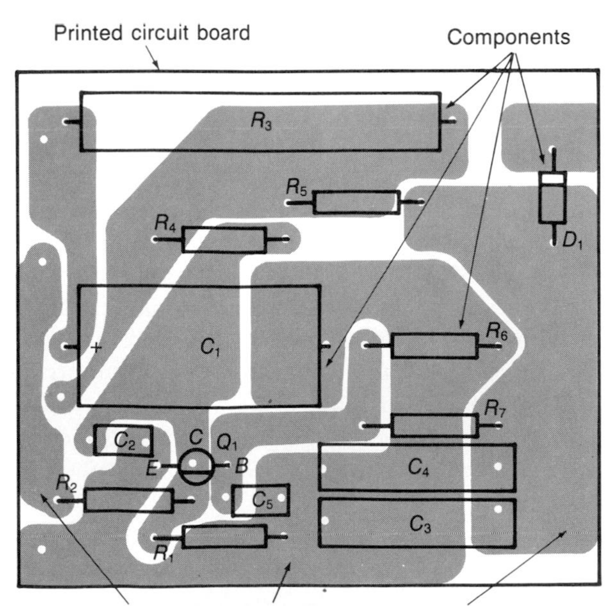


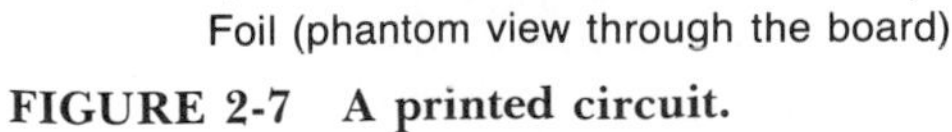


**FIGURE 2-7 A printed circuit.**

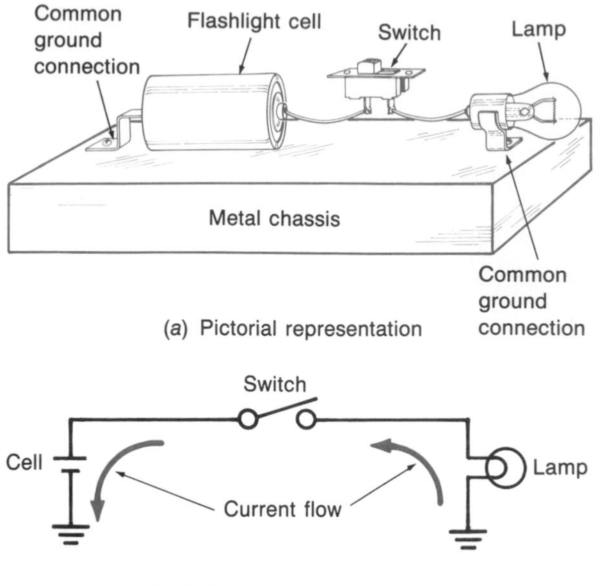


(a) Pictorial representation

(b) Schematic representation

**FIGURE 2-8 An example of a common ground.**

ditional common ground connection points. These grounds are all electrically connected together because the metal chassis is a good conductor. Current might flow into one ground connection and out of another. For example, Fig. 2-8 shows how a common ground could be used to eliminate one wire in a simple circuit. The common ground supports the flow of current from the negative side of the cell to the lamp. Earth grounds and common grounds are sometimes connected together. For example, a metal rod could be driven into the soil and connected through a wire to the chassis shown in Fig. 2-8.

Because of their convenience, schematic diagrams are used most of the time in electricity and electronics. Through experience, you will learn to relate these diagrams to the physical circuits that they represent. Figure 2-9 will assist you. It shows the symbols, and in some cases the alternate symbols, for some of the most popular components. It also shows some examples of the typical appearance of these components. How many of the components have you seen before, and how many of the symbols can you identify?

## Self-Test

**1.** If a conductor connects a negative charge source to a positive charge source, the electron current will flow from the ______ source to the ______ source.

**2.** List the four major parts of an electric circuit.

**3.** Which type of diagram uses symbols to represent the component parts of the circuit?

**4.** Which type of circuit diagram is preferred for nontechnical personnel because it is less abstract?

**5.** Name the type of circuit assembly that interconnects components with foil traces instead of with wires.

**6.** Refer to Fig. 2-8. The cell serves as the ______, the wire and the chassis serve as the ______, the switch provides ______, and the lamp acts as the ______.

## 2-2 ELECTRIC CURRENT FLOW

Figure 2-10 shows a very basic electric circuit. It consists of a battery, some wires, and a lamp. The battery provides a positive and a negative charge source in one convenient package. An electrochemical reaction inside the battery produces an excess of electrons (negative charge) at the bottom terminal and a deficiency of electrons (positive charge) at the top terminal. If the lamp and wires are intact, the battery supplies a continuous flow of current until it is chemically exhausted. Obviously, the battery is a much more practical source of current than two charged electroscopes. Note that the battery symbol is composed of alternate line lengths. The outermost long line always represents the positive terminal, and the outermost short line always represents the negative terminal. The + and − polarity markings shown in Fig. 2-10 are therefore not necessary.

The electron current, shown in Fig. 2-10, leaves the negative terminal, travels through the circuit, and enters the positive terminal. Long ago, scientists who investigated electrical phenomena guessed that "something" was moving in the electric circuits. Unfortunately, they did not know about atomic structure and electrons. Benjamin Franklin proposed that the current flows from positive to negative. Later, when atomic structure was revealed, electron flow from negative to positive was proposed. Today, Franklin current flow is known as *conventional current.* Conventional current is still used by practically all engineers and physicists. Many technicians use *electron current.* This book uses electron flow.

| NAME | SYMBOL | ALTERNATE SYMBOL | PHYSICAL EXAMPLES |
|---|---|---|---|
| Resistor |  |  | 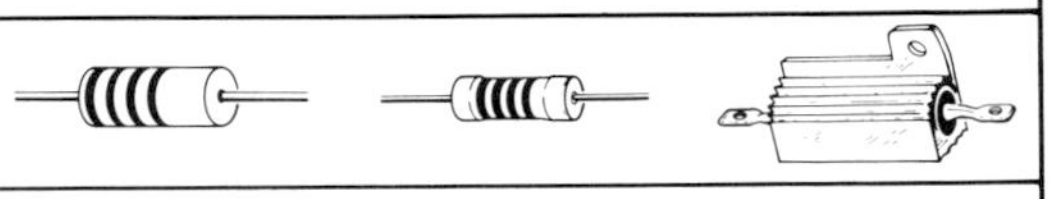 |
| Variable resistor (potentiometer) |  | 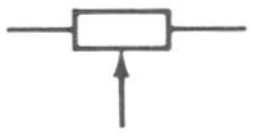 |  |
| Variable resistor (rheostat) |  | 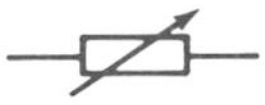 |  |
| Light dependent resistor | 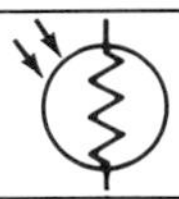 |  λ |  |
| Voltage dependent resistor (varistor) |  |  |  |
| Capacitor | 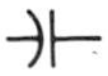 |  |  |
| Variable capacitor |  |  | 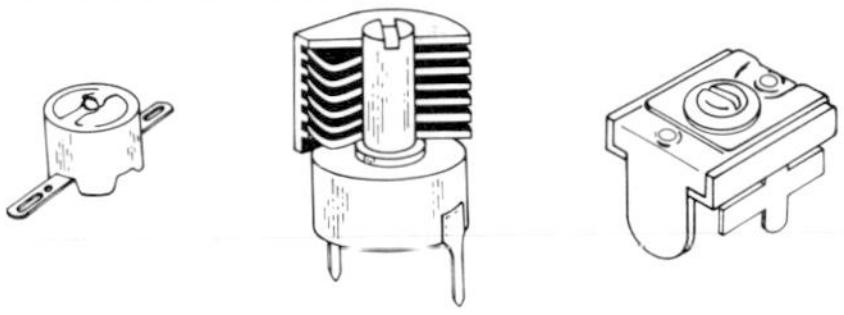 |
| Electrolytic capacitor | 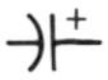 + | |  + |
| Inductor |  | | |
| Inductor with iron core |  | | 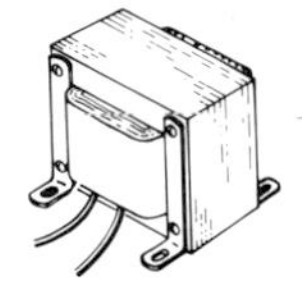 |
| Variable inductor |  | | 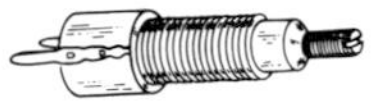 |
| Transformer with iron core |  | | 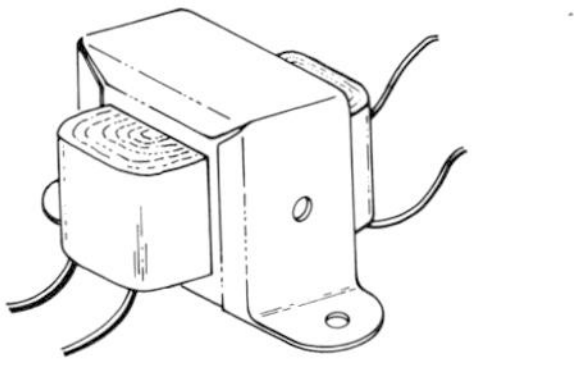 |

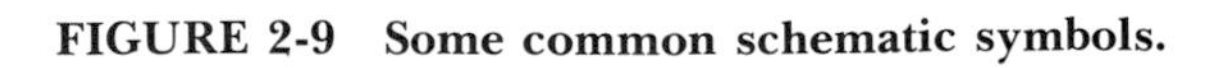

**FIGURE 2-9 Some common schematic symbols.**

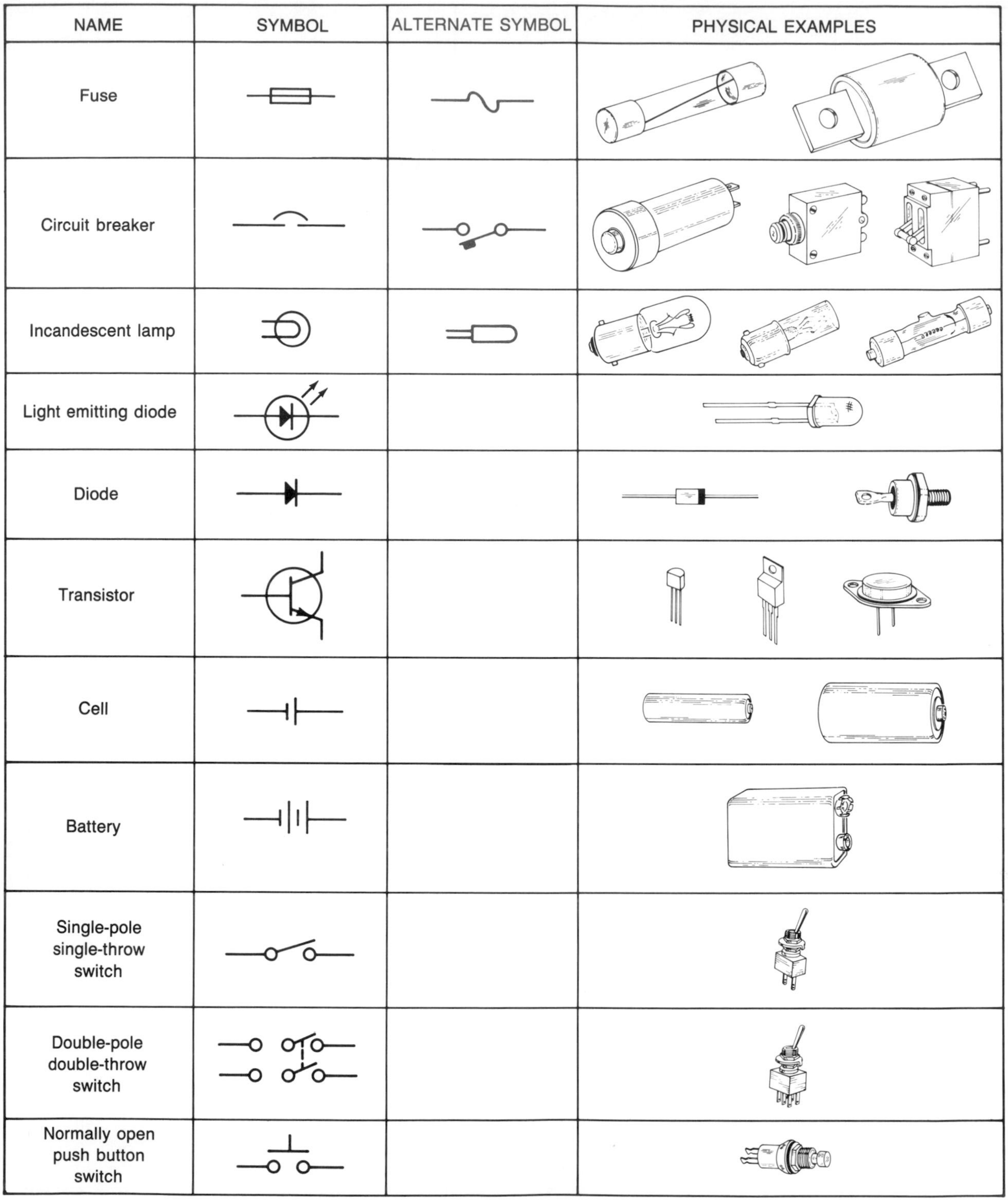

| NAME | SYMBOL | ALTERNATE SYMBOL | PHYSICAL EXAMPLES |
|---|---|---|---|
| Fuse | | | |
| Circuit breaker | | | |
| Incandescent lamp | | | |
| Light emitting diode | | | |
| Diode | | | |
| Transistor | | | |
| Cell | | | |
| Battery | | | |
| Single-pole single-throw switch | | | |
| Double-pole double-throw switch | | | |
| Normally open push button switch | | | |

**FIGURE 2-9** ***(Continued)***

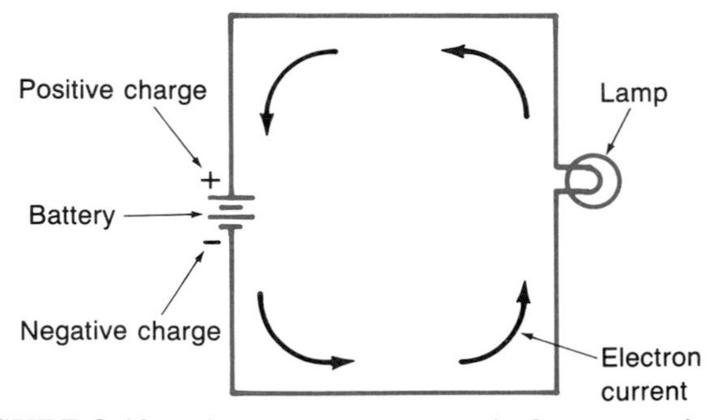


**FIGURE 2-10 The electron current is from negative to positive.**

You are urged to be flexible in regard to the direction of current flow. Some enjoy arguing the various merits of the two systems. Both flow systems are equally useful for understanding, analyzing, troubleshooting, and designing circuits. In Chap. 9, you will learn that you cannot always immediately determine the direction of flow. In order to begin an analysis of some circuits, you will be forced to assume the direction of current flow. If your assumption is wrong, the algebraic sign of your answers will indicate which assumed directions are incorrect. The most that will be required is to change some arrowheads on your circuit diagram.

Engineers use conventional current flow, and engineers are usually the experts who establish schematic symbols. For this reason, those schematic symbols that contain flow arrows indicate *conventional current flow,* not electron current flow. Figure 2-11 shows two examples.

The *magnitude* of the current flow is related to the number of current carriers passing a reference point in a given period of time. Current is due to the motion of charges. These charges are usually electrons, but they can also be ions as in the case of current flow in a gas or liquid. The charges can also be holes as in the case of certain semiconductor materials. Hole current is explained in Chap. 23. For now, we focus our attention on electron current.

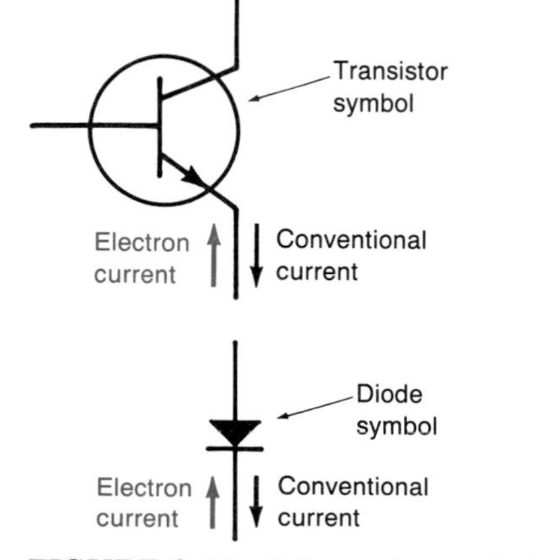


**FIGURE 2-11 Schematic symbol arrows point in the direction of conventional current flow.**

The SI units for charge and time are the coulomb (C) and the second (s), respectively. Current flow is a rate and is defined in terms of coulombs per second. Do not confuse current with velocity, which is a different kind of rate. The drift velocity of electrons in a circuit is surprisingly slow. Drift velocity varies with the magnitude of the current, the size of the conductor, and the number of free electrons in the conductor. It is typically around 0.1 mm/s. Thus, it takes a given electron about 10,000 s (2.78 h) to move 1 m in a circuit. In spite of this, electric circuits are very fast acting. When the switch is closed in a physically long circuit, the current begins moving almost instantaneously at points located many meters from the switch.

Figure 2-12 shows a basic circuit. Point A has been chosen arbitrarily. The amount of current is 1 ampere if 1 C of charge passes point A in 1 s. The relationship between current, charge, and time is expressed as

$$I = \frac{Q}{t}$$

where $I$ = letter symbol for current
$Q$ = letter symbol for charge
$t$ = letter symbol for time

The relationship may also be expressed as

$$1\ \mathrm{A} = \frac{1\ \mathrm{C}}{1\ \mathrm{s}}$$

where A = the abbreviation for amperes
C = the abbreviation for coulombs
s = the abbreviation for seconds

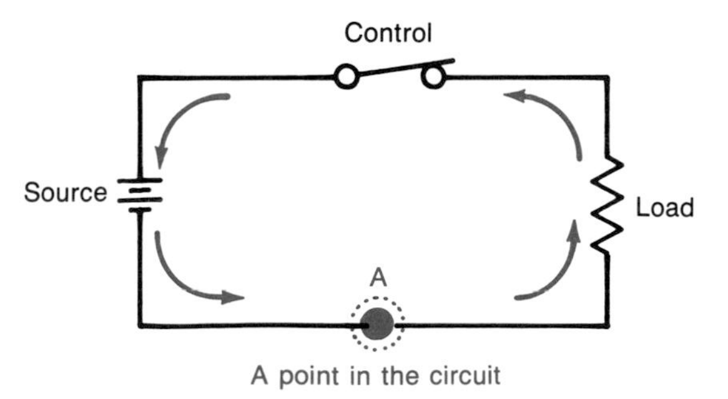


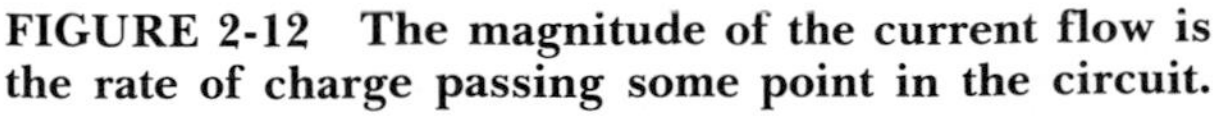

**FIGURE 2-12 The magnitude of the current flow is the rate of charge passing some point in the circuit.**

**EXAMPLE 2-1**

How much current is flowing in a circuit where $6.25 \times 10^{18}$ electrons move past a given point in 1 s?

**Solution**

We learned in Chap. 1 that $6.25 \times 10^{18}$ elementary charges are contained in 1 C. Therefore, $Q$ is equal to 1 C and the current flow is 1 C/s or 1 A.

**EXAMPLE 2-2**

How much current is flowing in a circuit where $1.27 \times 10^{15}$ electrons move past a given point in 100 ms?

**Solution**

To calculate the current, the charge must be expressed in coulombs. This is accomplished by

$$Q = \frac{1.27 \times 10^{15}}{6.25 \times 10^{18}}$$
$$= 2.03 \times 10^{-4} \text{ C}$$

Now the current can be found:

$$I = \frac{2.03 \times 10^{-4}}{100 \times 10^{-3}}$$
$$= 2.03 \times 10^{-3} \text{ A (or 2.03 mA)}$$

**EXAMPLE 2-3**

How long does it take 50 $\mu$C of charge to pass a point in a circuit if the current flow is 15 mA?

**Solution**

Transposing the equation gives

$$t = \frac{Q}{I}$$
$$= \frac{50 \times 10^{-6}}{15 \times 10^{-3}}$$
$$t = 3.33 \times 10^{-3} \text{ s (or 3.33 ms)}$$

**EXAMPLE 2-4**

Calculate the quantity of charge that will be transferred by a current flow of 10 A over a 1-h period.

**Solution**

$$Q = I \times t$$
$$= 10 \times 60 \times 60$$
$$Q = 3.6 \times 10^{4} \text{ C (or 36 kC)}$$

## Self-Test

**7.** How can the positive terminal of a battery be identified on a schematic diagram?

**8.** What is the direction of electron current in a circuit?

**9.** What is the direction of conventional current in a circuit?

**10.** If Fig. 2-12 used conventional current flow, would the direction be clockwise or counterclockwise?

**11.** The arrows used in schematic symbols often indicate the direction of ______ current flow.

**12.** The ampere is a unit for measuring ______.

**13.** Define the ampere unit.

**14.** What is the letter symbol for current?

**15.** What is the abbreviation for ampere?

**16.** Calculate the current flow in a circuit where $1.55 \times 10^{11}$ electrons pass a point in 1 $\mu$s.

**17.** Calculate the time required for 1 $\mu$C of charge to pass a point if the current flow is 20 $\mu$A.

**18.** How much charge is transferred by a current flow of 10 mA in 1 min?

## 2-3 MEASURING CURRENT

The ability to make safe and accurate circuit measurements is an integral part of the knowledge that you must acquire. Your first experiences with measurements will most likely be with low-energy circuits. These circuits are not likely to cause electric shock, burns, or severe equipment damage if mistakes are made. However, you will eventually progress to circuits that can present a hazard if the correct procedures are not followed.

The human body is controlled by minute electric impulses. If a finger touches a hot surface, the nerves in the finger send an electric signal to the brain. The brain processes the information, makes a decision, and sends impulses to the appropriate muscles to withdraw the finger from the heat. An external source of current, if allowed to flow through the body tissue, may override the normal body signals and cause unpredictable and dangerous reactions. Two of the most serious are ventricular fibrillation and paralysis of the breathing mechanism (asphyxiation). Ventricular fibrillation disrupts the normal rhythmic pumping action of the heart and is usually fatal. Other consequences of electric shock include severe muscle contractions and burns. The involuntary contraction of muscles can cause a fall or other serious accidents.

It is obvious that external currents must not be allowed to take a path through the body. Technicians must exercise caution to remain insulated from the circuits they are working on. It is also important to remain insulated from ground since current can pass through the body and flow to ground in many cases. The following procedures are recommended to avoid electric shock and related accidents:

**1.** When possible, turn the circuit off and verify that it is off before working on it. Make sure that all components which are capable of storing electric energy have been discharged.

**2.** Keep tools and test equipment in good condition. Repair or replace frayed leads, damaged probes and handles, and worn or cracked insulation.

**3.** Do not work around electric equipment if your clothing is damp or wet. Water makes clothing and human skin a better electric conductor. Damp floors should also be avoided.

**4.** Never place both hands in a circuit. A hand-to-hand shock is particularly dangerous because the current can pass through the chest and heart.

**5.** Never place one hand on a metal cabinet, a chassis, or any grounded object while working in a circuit with the other hand. This could also lead to a hand-to-hand shock.

**6.** Do not work with rings, bracelets, or other metal items on your hands or wrists.

**7.** Make sure there is adequate illumination around the work area.

**8.** Do not work if you are taking any type of medicine that could make you drowsy or impair your vision or concentration.

**9.** Do not remove equipment grounds or use adapters that defeat a ground connection.

**10.** Do not modify circuits or equipment unless you know exactly what the effects of the modification will be.

**11.** Always follow proscribed procedures. Use all relevant literature. When in doubt, research and study the equipment before you proceed.

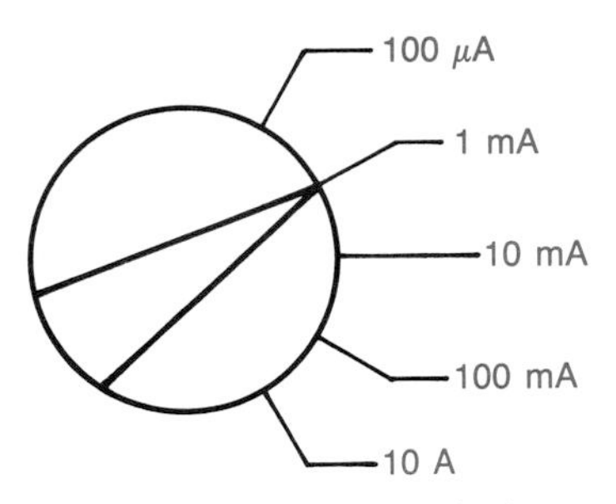


**FIGURE 2-13 Ammeter range switch.**

A device used to measure current in a circuit is called an *ammeter*. A current function might be available as one of several measurement options in other instruments such as a multimeter. There are often several ranges that can be selected for measuring current. Figure 2-13 shows the current ranges that might be found on an ammeter or multimeter. The range must be used in conjunction with the meter reading when measuring current.

### EXAMPLE 2-5

How much current is indicated when the range switch is set as shown in Fig. 2-13 and the meter appears as shown in Fig. 2-14?

#### Solution

The setting of the range switch indicates the *full-scale* value of the meter reading. On a 1-mA range, the meter would indicate 1 mA if the pointer reached exactly full scale. Since the meter face in Fig. 2-14 is calibrated from 0 to 10, it is necessary to divide the reading by 10 to obtain the current:

$$I = \frac{7.5\text{ mA}}{10} = 0.75\text{ mA}$$

The indicated current is 0.75 mA.

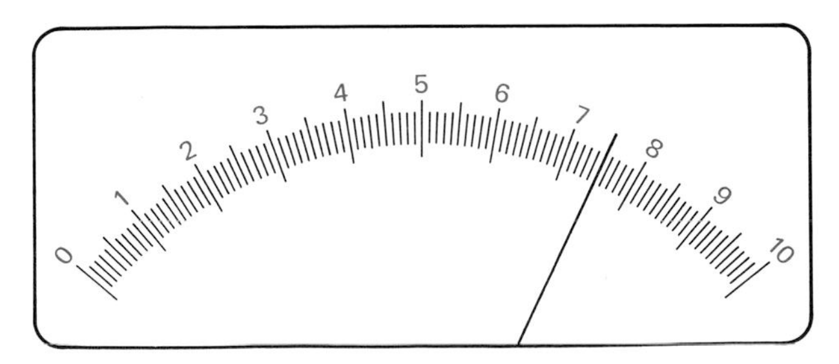


**FIGURE 2-14 Reading an analog meter.**

**EXAMPLE 2-6**

What is the current flow if the meter appears as shown in Fig. 2-14 and the switch is set to the 100-mA range?

**Solution**

The full-scale value of the meter is now 100 mA. Therefore, it is necessary to multiply the reading by 10:

$$7.5 \text{ mA} \times 10 = 75 \text{ mA}$$

The current flow is 75 mA.

Many meter faces have more than one scale. When working with this type of meter, be sure to use the proper scale. Always remember that the range switch indicates the full-scale value of the meter. If you have selected a 50-$\mu$A range, examine the meter face to find a scale that ranges from 0 to 50. Or, you may have to use a 0 to 5 scale and multiply the reading by 10. Also, examine the meter face to determine if any of the scales are restricted to certain ranges and functions of the meter. Using the wrong scale can cause a significant amount of error.

Meters of the type shown in Fig. 2-14 are *analog* instruments. Figure 2-15 shows an example of the display on a *digital* instrument. Digital meters are usually easier to use than analog meters. It is not necessary to interpret the position of a pointer on a scale. It is also not necessary to divide or multiply readings by 10. The reading shown in Fig. 2-15 might be obtained on a 2-mA range. In this case it would be interpreted as 1.342 mA. If the next higher range is 20 mA, the digital display would show 1.34 if that range was selected. Note that we would lose a significant figure but the reading would still be correct because the decimal point would be automatically shifted to the right. You might be wondering why the display would not show 01.34 when the range is switched. This is because leading zeros are automatically blanked in meters of this type.

In the last example, useful readings were obtained on both the 2- and the 20-mA ranges. However, the 2-mA range produced one more significant digit. It is considered the range for best resolution for that particular example. Best resolution is obtained by selecting the smallest range that does not produce overflow. Overflow occurs on the display shown in Fig. 2-15 when the measured value exceeds 1.999. Different meters indicate overflow in various ways, such as by an overflow light, a blanked display, or a flashing display. An excellent way to find the range for best resolution is to start on the highest range and begin moving to lower ranges until an overflow is indicated. Then, go back to the next higher range. The same thing is true for an analog instrument. However, analog instruments overflow by moving the pointer past full scale and possibly banging the pointer against a stop. This can damage the meter, so this procedure is not recommended for analog instruments. The range for best resolution in an analog meter can be found by starting on the highest range and then switching to lower ranges until the current can be read with reasonable accuracy. Then, switch to the next lower range only if it is high enough to include the measured current.

Measuring current flow is similar to measuring liquid flow. Figure 2-16 shows that fluid flow is measured by breaking the tubing at some point and then inserting the flowmeter. Current measurement is shown in Fig. 2-17. Note that the circuit must be broken, and the two ammeter leads are then connected across the break. The current now

**FIGURE 2-15 Example of a digital display.**

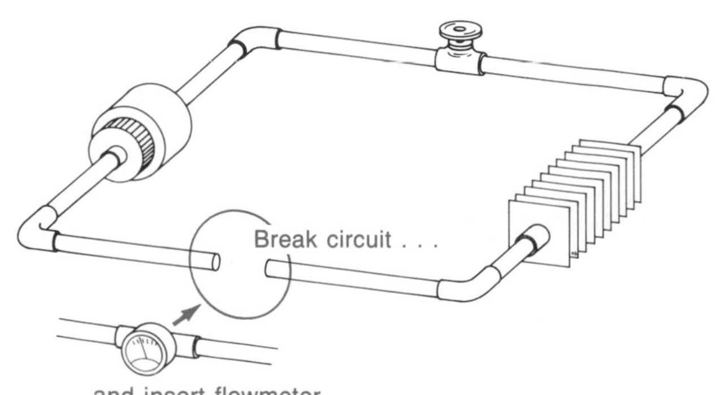


**FIGURE 2-16 Measuring flow rate in a fluid circuit.**

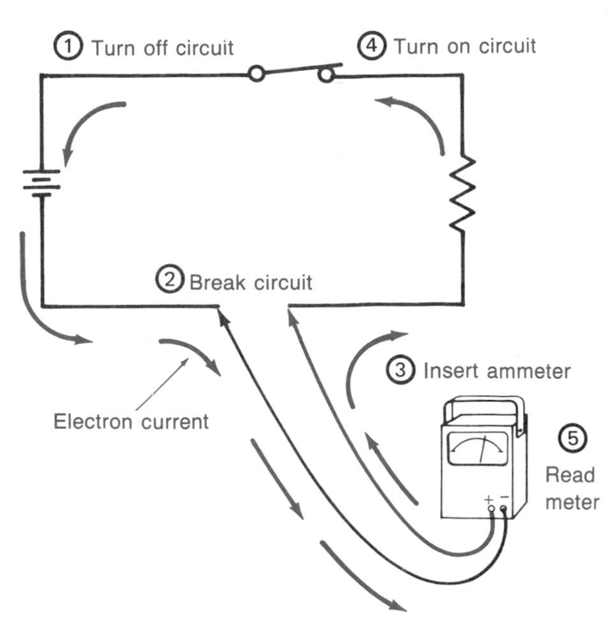


**FIGURE 2-17 Measuring current in a circuit.**

flows through the meter as well as through the circuit. The *polarity* markings of the meter terminals are important. Figure 2-17 shows that the negative lead of the meter is connected to the negative side of the break and that the positive lead of the meter is connected to the positive side of the break. By simply looking at the battery, one can determine which side of the break is positive and which is negative. Remember that the short line represents the negative terminal and the long line represents the positive terminal.

A reverse connection causes an analog meter to deflect backward. This may cause the pointer to bang up against the left-hand stop with enough force to damage the meter. Most digital meters have autopolarity. They produce a correct reading when connected backward, but a minus sign appears in front of the numbers to indicate reverse polarity. Meter leads are color-coded to help you connect them properly. The positive lead is red and the negative lead is black. Make sure that they are plugged into the meter properly. Also, check to see if the meter has a polarity reversal switch. It is used to reverse the internal connection of the positive and negative leads.

To avoid a mess, you would probably remember to turn the fluid circuit off in Fig. 2-16 before cutting into the tubing. This same practice is recommended when measuring current flow. Connecting an ammeter to a live high-energy circuit is very dangerous. Use safe work practices with all circuits, and you will never develop dangerous habits.

Figure 2-18 shows that it is possible to measure the current flow in a part of a circuit. Once again, it is necessary to break the circuit. The meter has been connected to measure the current through $R_1$.

The most important point to remember is, break the circuit and connect the meter leads with the correct polarity to the break points. Many, many ammeters and circuits have been damaged because this was not done. For example, if the circuit is not broken as shown in Fig. 2-18 and the ammeter is connected across $R_1$, an abnormally high current will result and the meter could be severely damaged.

## Self-Test

**19.** Suppose that the pointer in Fig. 2-14 is on the 2 calibration. How much current would this indicate for the following settings of the range switch?

| | |
|---|---|
| 100 μA | 100 mA |
| 1 mA | 1000 mA |
| 10 mA | 10 A |

**20.** An examination of a meter face shows multiple scales of 0–2.5, 0–5, and 0–10. Which of these scales must be used in conjunction with the 250-mA range? What must be done to the meter reading to obtain the correct current?

**21.** A digital meter is set on a 200-mA range, and the display blanks when the circuit is turned on. Why?

**22.** The actual current in a circuit is 75 mA. Which of the following ranges will provide the best resolution?

**a.** 0 to 1 mA
**b.** 0 to 10 mA
**c.** 0 to 100 mA
**d.** 0 to 1 A

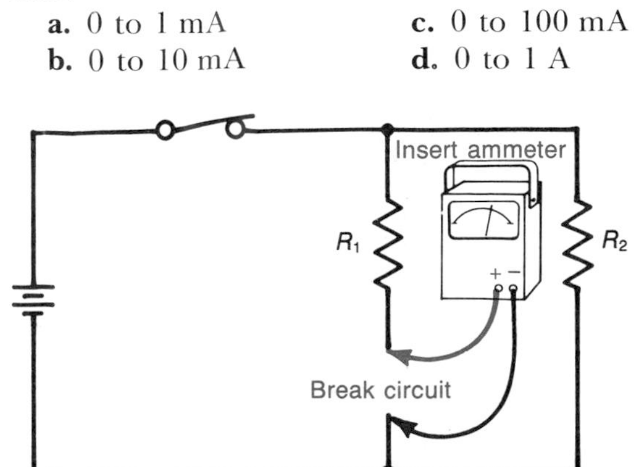


**FIGURE 2-18 Measuring current in part of a circuit.**

**23.** To measure current, the circuit should be turned off and then ________ before connecting the ammeter.

**24.** An analog ammeter deflects backward. What is wrong?

**25.** A digital ammeter is connected to a circuit backward. Assuming the meter has autopolarity, what will the display indicate?

**26.** What would happen if the meter in Fig. 2-18 was connected across $R_2$?

## 2-4 ELECTRIC POTENTIAL

*Potential* is associated with the ability to do work. Work involves force and distance, and force is what is required to accelerate or decelerate a mass:

$$F = M \times A$$

where $F$ = force, measured in newtons, N
$M$ = mass, measured in kilograms, kg
$A$ = acceleration, measured in meters per second per second, $m/s^2$

A force of 1 N is required to accelerate a 1-kg mass 1 $m/s^2$.

Weight and mass are often confused in the SI system. Weight is the force created by gravity acting on a mass. If the gravitational field strength is known, the weight of an object can be found by

$$W = M \times g$$

where $W$ = weight, N
$M$ = mass, kg
$g$ = gravitational field strength, $m/s^2$

The gravitational field strength of our planet, at the standard location, is equal to 9.81 $m/s^2$. Therefore, a 1-kg mass will weigh 9.81 N at the standard location.

When a force is applied over some distance, work has been done:

$$W = F \times D$$

where $W$ = work done, J
$F$ = force exerted, N
$D$ = distance, m

Energy is the ability to do work. Energy and work share the joule unit. An energy source with a potential of 1 J can accomplish one newton-meter (1 N · m) of work.

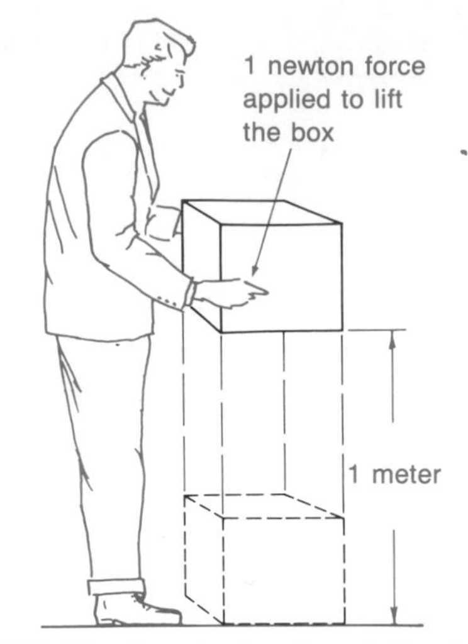


**FIGURE 2-19 One joule of work has been done.**

Figure 2-19 shows a man lifting a box. He exerts a force of 1 N over a distance of 1 m. He has done 1 J of work. The box has the potential to do work when raised to its new position. If the box is dropped, it will fall and release its potential energy.

Electric energy sources also have the potential to do work. We have learned that two oppositely charged bodies have the potential to cause current to flow. A *potential difference* exists across the charged bodies or across the terminals of a battery. This potential difference is proportional to the work the charged bodies or the battery can accomplish when a current flows. The unit of potential difference is the *volt:*

$$V = \frac{W}{Q}$$

where $V$ = potential difference, V
$W$ = work, J
$Q$ = charge, C

An electric energy source has a potential difference of 1 V if it provides 1 J of energy with every coulomb of charge that it delivers.

**EXAMPLE 2-7**

Calculate the potential difference of an energy source that provides 50 mJ of energy for every microcoulomb of charge that flows.

**Solution**

Entering the given data into the equation yields

$$V = \frac{50 \times 10^{-3}}{1 \times 10^{-6}}$$
$$= 50 \text{ kV}$$

**EXAMPLE 2-8**

What quantity of charge must be delivered by a battery with a potential difference of 100 V to do 500 J of work?

**Solution**

Transposing the equation gives

$$Q = \frac{W}{V}$$
$$= \frac{500}{100}$$
$$Q = 5 \text{ C}$$

**EXAMPLE 2-9**

How much work will be done by an electric energy source with a potential difference of 3 kV that delivers a current of 1 A for 1 min?

**Solution**

Recall that 1 A of current represents a charge transfer rate of 1 C/s. Therefore, the total charge $Q$ will be 60 C for a 1-min period. Now, transposing and solving gives

$$W = Q \times V$$
$$= 60 \times 3 \times 10^3$$
$$W = 180 \text{ kJ}$$

Note that the work accomplished is directly proportional to the voltage of the energy source. This is why utility companies transfer electric energy at as high a voltage as is feasible. Voltage is the work potential assigned per unit of charge. The higher the voltage, the more work that is done by every coulomb of charge that flows. When large amounts of work must be done, higher voltages are required. A practical example is a flashlight. A small pocket flashlight might use a single 1.5-V cell. Figure 2-20 shows that this cell provides each coulomb of charge with 1.5 J of energy. This energy is changed to heat and light energy by the resistance of the lamp. A large flashlight might have six 1.5-V cells as shown if Fig. 2-21. Each cell adds 1.5 J of energy to every coulomb of charge that flows. This means that the resistance of the lamp converts 6 × 1.5 or 9 J of energy into heat and light for every coulomb of charge that passes through the lamp. The lamp in this flashlight therefore reaches a higher temperature and produces much more light for a given charge transfer rate.

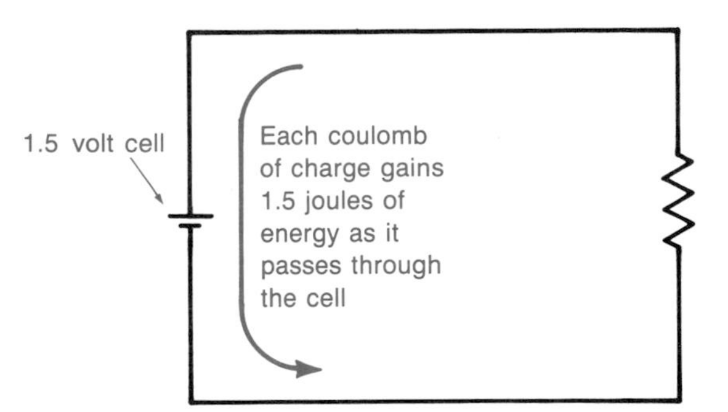


**FIGURE 2-20 A circuit with a 1.5-V energy source.**

The popular way to refer to a potential difference is as a *voltage*. This is an instance where the name of the unit has replaced the measured quantity. An analogy would be to refer to weight as "poundage." There is nothing wrong with this practice as long as you do not forget that a voltage is really a potential difference. Always remember that *a voltage is really a potential difference and must be referenced across two points.*

The letter symbol for potential difference is $V$. In the past the symbol $E$ was used to refer to *electromotive force* or emf. The use of $E$ is still widely used when referring to a source of electric energy; however, $V$ is now the accepted standard when referring to an energy source. Figure 2-22 shows a circuit with a cell and a resistor. The cell provides a potential difference of 1.5 V. Since it is an energy

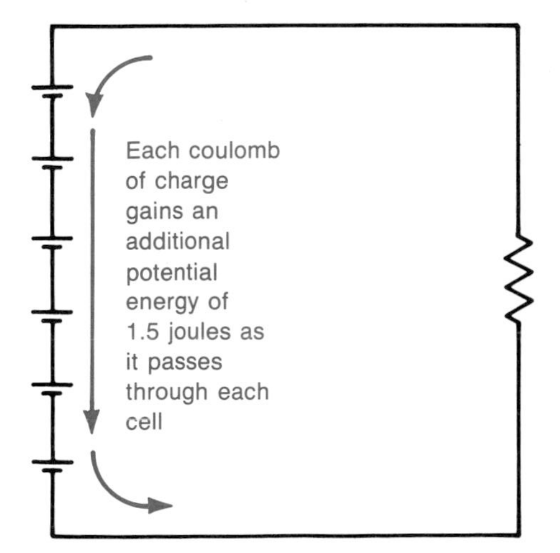


**FIGURE 2-21 Six 1.5-V cells in series.**

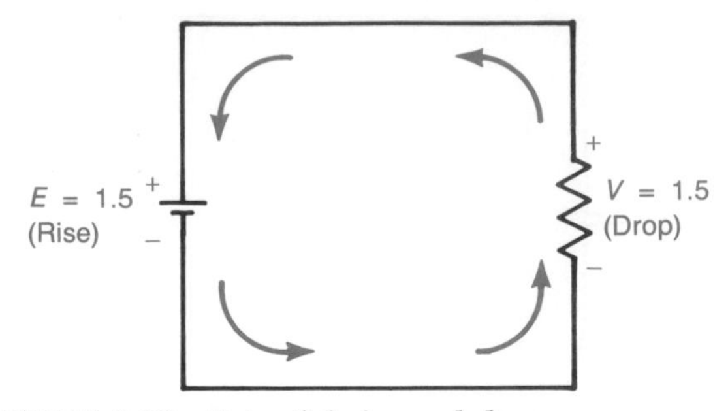


**FIGURE 2-22 Potential rise and drop.**

source, there is a *rise* in potential associated with the cell. The cell's potential difference represents an emf so the symbol $E$ could be used. The resistor in Fig. 2-22 is also associated with a potential difference. Since it is a consumer (converter) of energy, there is a *drop* in potential across the resistor. The symbol $E$ would not be appropriate when referring to drops in potential. Hereafter, this book will use the accepted standard of the symbol $V$ to refer to rise or drop.

Now, we can combine the idea of potential rise and drop with the popular term "voltage." It is customary to refer to the potential difference across the cell in Fig. 2-22 as a *voltage rise* and to the potential difference across the resistor as a *voltage drop.*

Figure 2-23 shows that the potential difference can be zero, even though charges are involved. In Fig. 2-23*a*, the two spheres are identical and have the same positive charge. There is no potential difference across the spheres. If a wire is connected from one sphere to the other, no current flows. Figure 2-23*b* shows a circuit with two cells. The ground symbol is used to establish a reference point. Point A is at +1.5 V with respect to the ground reference and point B is also at +1.5 V with respect to ground. The potential difference from A to B is (+1.5) − (+1.5) = 0. Thus, there is no potential difference across the resistor and no current flow. If you have ever inserted one cell backward in a two-cell flashlight, you have already experienced this concept. Does the flashlight work?

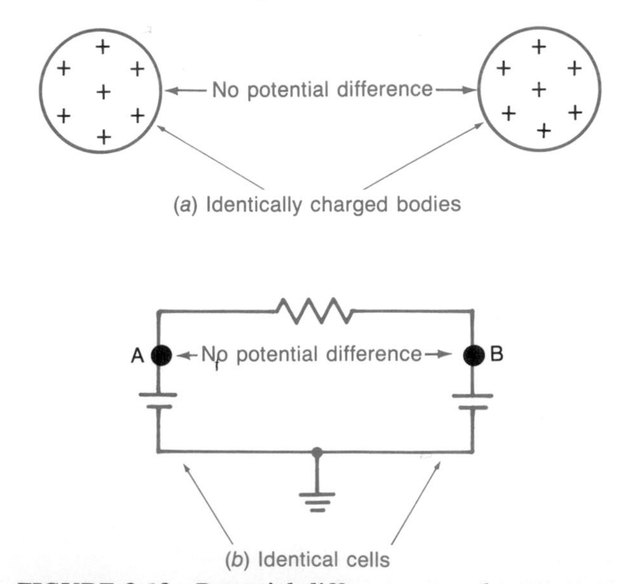


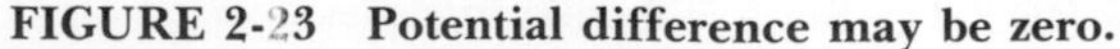
**FIGURE 2-23 Potential difference may be zero.**

Look at Fig. 2-24. It shows a circuit with a cell and three resistors. Each of the resistors will drop some of the potential produced by the cell. Since a voltage drop is really a potential difference, it is always necessary to clearly understand the two points across which the drop is being specified. Resistor $R_1$ will drop some voltage. This drop can be specified as $V_{AB}$ since point A is one end of the resistor and point B is the other. Or, it can be specified as $V_{R_1}$. The drop across $R_2$ can be specified as $V_{BC}$ or $V_{R_2}$. Sometimes, a voltage is specified at one point. There is no such thing as a voltage at one point. Always remember that the term voltage refers to a potential difference across two points. In cases where a single point is specified, some reference must be used as the other point. Figure 2-24 shows a ground connection at the bottom of the circuit. Unless noted otherwise, the ground or common point in any circuit is the reference when specifying a voltage at some other point. For example, $V_B$ is interpreted as the potential difference from point B to ground.

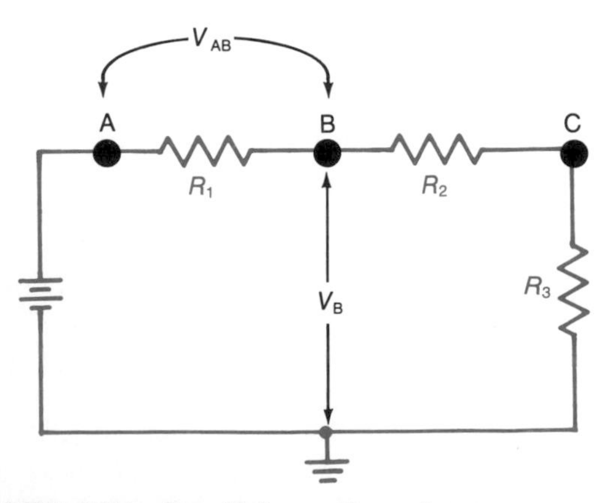

**FIGURE 2-24 Specifying voltage drops.**

## Self-Test

**27.** How much energy is expended when a force of 10 N is exerted over a distance of 15 m?

**28.** What is the unit of potential difference in an electric circuit?

**29.** Calculate the potential difference across a resistor if it converts 15 J of energy for every coulomb of charge that passes through it.

**30.** The potential difference across a battery is 9 V. How much charge must it deliver to do 50 J of work?

**31.** A 300-V energy source delivers 500 mA for 1 h. How much energy does this represent?

**32.** Why is a high-voltage source required when a considerable amount of work must be done?

**33.** What does the term *voltage* refer to?

**34.** What are the two letter symbols for potential difference?

**35.** Which letter symbol is used to represent the potential difference across a load?

**36.** Refer to Fig. 2-24. Would the potential difference of the battery be referred to as a voltage rise or as a voltage drop?

**37.** Refer to Fig. 2-24. Would the potential difference across $R_3$ be referred to as a voltage rise or as a voltage drop?

**38.** Refer to Fig. 2-24. Would $V_C$ be equal to $V_{R_3}$?

## 2-5 RESISTANCE AND OHM'S LAW

A simple analogy for voltage (potential difference) is to consider it as a pressure. Flow is proportional to pressure: the more pressure, the more flow. In an electric circuit, the flow is called *current.* Years ago, Georg Simon Ohm noticed that he could measure a consistent current in a circuit every time he connected it. He also noticed that the current was proportional to the applied voltage. Doubling the voltage doubled the current, and tripling the voltage tripled the current. This led him to the conclusion that, for a given circuit, the voltage-to-current ratio is a constant:

$$k = \frac{V}{I}$$

where $k$ = Ohm's proportionality constant
$V$ = applied potential in volts
$I$ = current in amperes

The proportionality constant $k$ could be given the unit volts per ampere, but instead it has been given the name ohm in honor of Georg Simon Ohm. The letter symbol for the ohm unit is the Greek capital letter omega ($\Omega$). The ohm is the SI unit for *resistance,* and the following equation is called *Ohm's law:*

$$R = \frac{V}{I}$$

where $R$ = circuit resistance, $\Omega$
$V$ = potential difference, V
$I$ = current, A

In older texts, Ohm's law is sometimes written with the letter symbol $E$ in place of the $V$ in the numerator. It is used when the potential difference refers to an electromotive force rather than the voltage drop across a resistance. However, the symbol $V$ has been established as the standard for representing all voltages. A circuit with a potential difference of 1 V and a resistance of 1 $\Omega$ has a current flow of 1 A.

**EXAMPLE 2-10**

Calculate the resistance of a circuit with a potential difference of 15 V and a current flow of 10 mA.

**Solution**

Entering the given information into the Ohm's law equation:

$$R = \frac{15}{10 \times 10^{-3}} = 1.5\ \text{k}\Omega$$

**EXAMPLE 2-11**

Calculate the voltage drop across a 560-$\Omega$ resistor that is conducting a current of 100 mA.

**Solution**

Rearranging Ohm's law gives

$$V = I \times R$$
$$= 100 \times 10^{-3} \times 560$$
$$V = 56\ \text{V}$$

**EXAMPLE 2-12**

Calculate the current flow through a 390-Ω resistor with a potential difference of 18 V.

**Solution**

Once again, rearranging Ohm's law:

$$I = \frac{V}{R}$$

$$= \frac{18}{390} = 0.0462$$

$$I = 46.2 \text{ mA}$$

Resistance is an opposition to the flow of current. If the resistance of a circuit is doubled, the current is reduced to one-half. If the resistance is tripled, the current is reduced to one-third, and so on. Resistance can be very useful when the flow of current must be controlled. For example, look at Fig. 2-25. A component called a *rheostat* has been added to a motor circuit. A rheostat is an adjustable resistor. The current flow must overcome the resistance in the circuit. As the rheostat is adjusted for more resistance, less current flows and the motor slows down. As the rheostat is adjusted for less resistance, the current flow increases and the motor speeds up. Resistance can also be used to dim lights, control loudness, and perform many other useful circuit functions.

In some circumstances, resistance is undesirable. Figure 2-26 shows that the conductors used to carry the motor current have resistance. If the motor is used in an application where it must develop maximum output at all times, the conductor resistance is undesirable. Conductor resistance prevents the motor from developing its full output, and some of the electric energy is wasted since heat is produced by the current flowing in the wires. In severe cases, the wires could be damaged

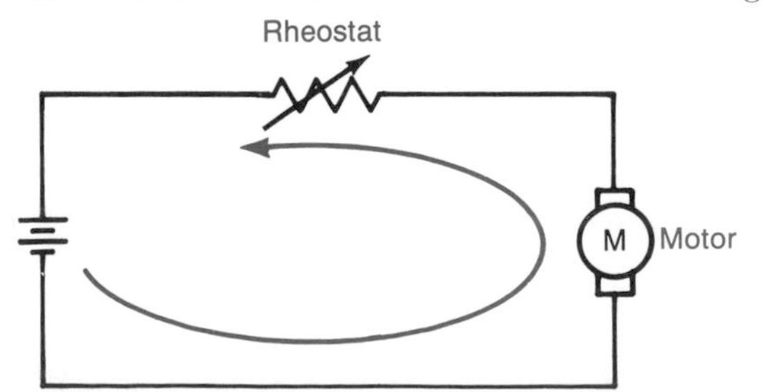


FIGURE 2-25 Using resistance to control a circuit.

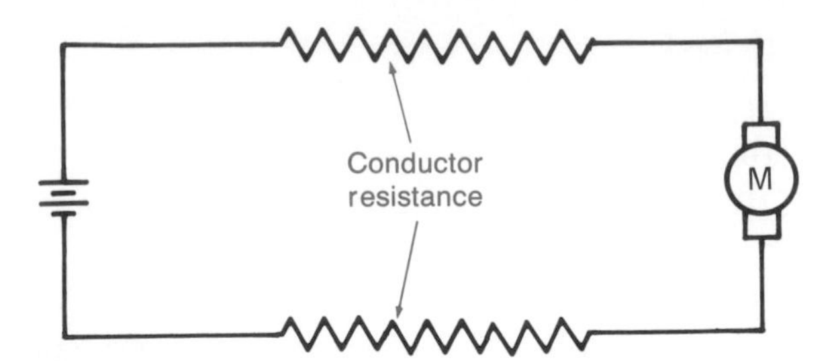


FIGURE 2-26 An example of undesirable resistance.

by the heat or a fire could result. Conductor resistance becomes a significant problem when the wires are very long and high currents must flow.

What causes resistance? We learned in Chap. 1 that good conductors, such as copper, have an enormous number of valence electrons. These valence electrons are only weakly attracted to the nuclei of their atoms. At temperatures above absolute zero (0 K), they gain enough thermal energy to break free. When a potential difference is applied, the free electrons begin to drift from negative to positive. Collisions between the moving electrons and the atomic structure of the metal result. These collisions convert some of the electric energy to heat energy. The collisions also create an opposition to current flow. All conductors, even the best ones such as copper and silver, have some resistance to electric current flow. For example, the copper wire that is typically used in buildings for energizing electric outlets has a resistance of 5.31 mΩ/m.

Even though copper has some resistance, it does not have enough for those applications where significant amounts of opposition are needed. Remember, sometimes resistance is desirable. Resistors are electric components that can offer significant opposition to the flow of current. Some resistors offer millions of ohms of opposition. Resistors are made from carbon or some other material that is a semiconductor or a poor conductor of electricity.

## Self-Test

**39.** In an electric circuit, the current flow is directly proportional to the ______.

**40.** In an electric circuit, the current flow is inversely proportional to the ______.

**41.** A simple analogy for potential difference (voltage) is to consider it as a ______.

**42.** The circuit characteristic that opposes the flow of current is called ______.

**43.** The equation $V = I \times R$ is known as ______.

**44.** Calculate the resistance for the circuit shown in Fig. 2-27.

**45.** How much current would flow in the circuit of Fig. 2-27 if the resistance value were cut in half?

**46.** Calculate the value for $V$ in Fig. 2-28.

**47.** How much current would flow in the circuit of Fig. 2-28 if the battery voltage were doubled?

**48.** Calculate the current flow in the circuit shown in Fig. 2-29.

**49.** Why is the current flow in Fig. 2-29 clockwise rather than counterclockwise as it is in Fig. 2-28?

## 2-6 MEASURING VOLTAGE AND RESISTANCE

The instrument used to measure potential difference is called a *voltmeter*. Voltmeters have much in common with ammeters, but they are connected differently. Figure 2-30 shows the use of a voltmeter. Please note the polarity of the meter connections: + is connected to + and − is connected to −. In Fig. 2-30*a*, $V_B$, the electromotive force of the battery, is measured. Note that the leads of the voltmeter are connected directly across the battery terminals. The circuit is *not* broken as it is when using an ammeter. In Fig. 2-30*b*, $V_L$, the voltage drop across the resistor, is measured. The resistor is considered the *load* in this circuit so the drop is called $V_L$. Once again, the circuit is not broken, and the meter leads are connected directly across the resistor leads. In many cases, there is no practical difference between the two voltmeter connections shown in Fig. 2-30. The voltmeter reads the same for both connections. Figure 2-31 shows an example where there is a difference. There is significant conductor resistance and the motor draws a substantial current. In a case such as this, $V_L$ will be less than $V_B$.

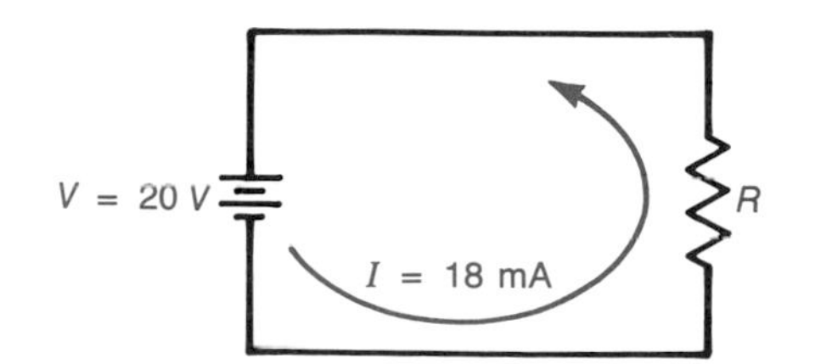


**FIGURE 2-27 Circuit for Self-Test.**

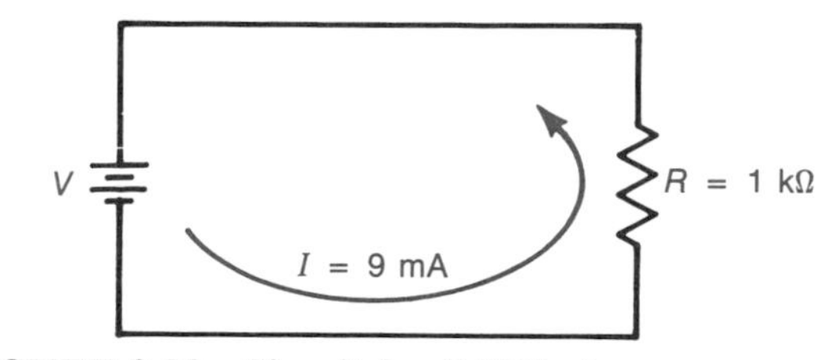

**FIGURE 2-28 Circuit for Self-Test.**

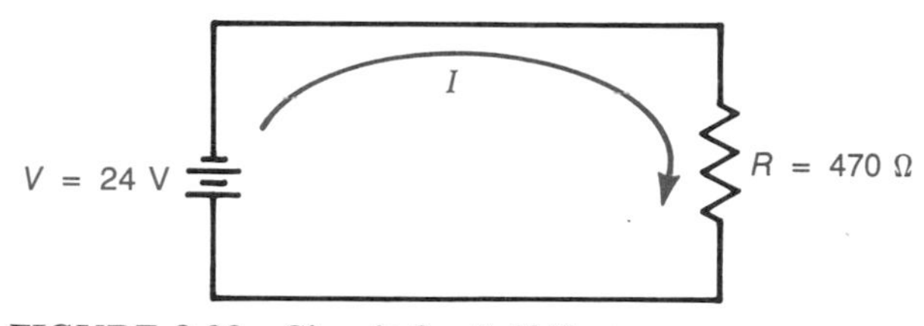


**FIGURE 2-29 Circuit for Self-Test.**

**EXAMPLE 2-13**

Calculate the voltage loss in the conductors shown in Fig. 2-31 if the motor current is 15 A and the total conductor resistance is 860 mΩ.

**Solution**

Ohm's law provides the answer:

$$V = I \times R$$
$$= 15 \times 860 \times 10^{-3}$$
$$V = 13 \text{ V}$$

The loss in overcoming the resistance of the conductors is 13 V. This means that the potential dif-

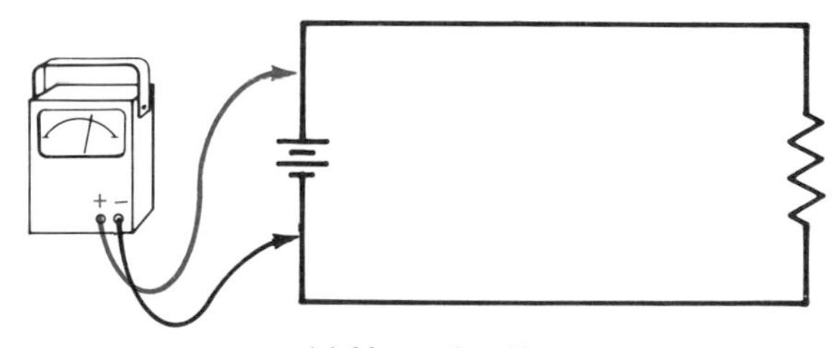


(*a*) Measuring $V_B$

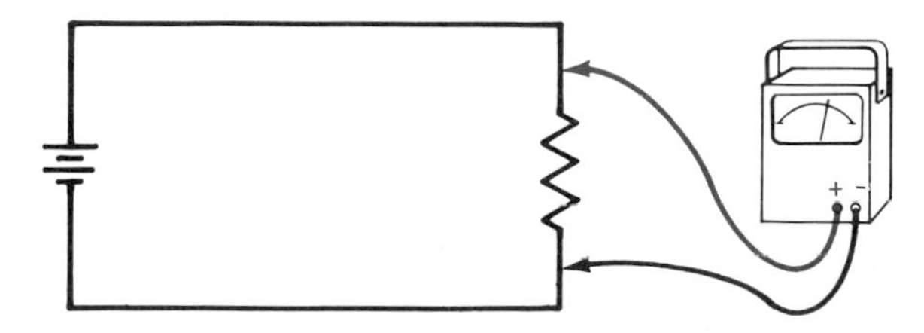


(*b*) Measuring $V_L$

**FIGURE 2-30 Measuring potential difference.**

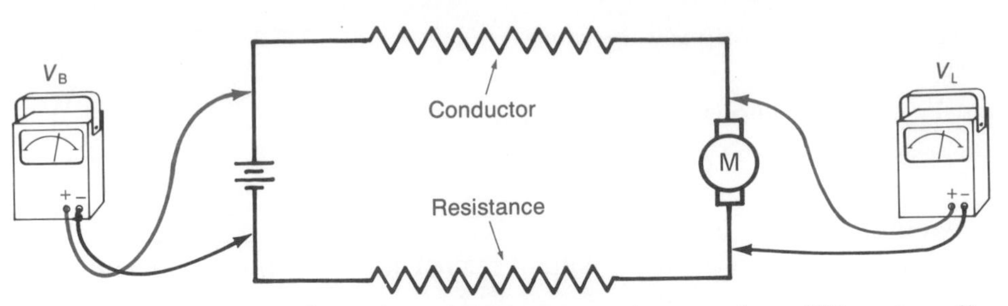


**FIGURE 2-31 A case where the voltmeter connections produce different readings.**

ference measured at the motor terminals will be 13 V less than the potential difference measured at the energy source.

Refer to Fig. 2-30 again. When circuits use low currents (in the milliampere range or less) and when the conductors are short (a meter or less), the source reading $V_B$ and the load reading $V_L$ will be the same. This is because the voltage drop in the conductors is insignificant. Therefore, you may use either connection with the same results. When currents are significant (in the ampere range) and the circuits are long, the readings may be different due to voltage drop across the conductors.

Figure 2-32 shows the measurement of individual voltage drops. Once again, notice the polarity of each connection. The voltage drop across $R_1$ is measured with the meter leads connected across $R_1$. The voltage drop across $R_2$ is measured across $R_2$ and so on. Potential difference is much easier to measure than current flow because the circuit does not have to be broken for meter insertion. All that is required is to decide which two points the reading is to be taken across and then connect the meter leads at those points. This brings us to an alternative to current measurements. If the value of a resistance is known, Ohm's law and a voltmeter can be used to find the current flowing through that resistance.

**EXAMPLE 2-14**

If the voltmeter in Fig. 2-32*a* is indicating 4.9 V and if $R_1$ is a 680-Ω resistor, how much current is flowing in $R_1$?

**Solution**

Once again, we use Ohm's law:

$$I = \frac{V}{R}$$

$$= \frac{4.9}{680} = 0.0072$$

$$I = 7.2 \text{ mA}$$

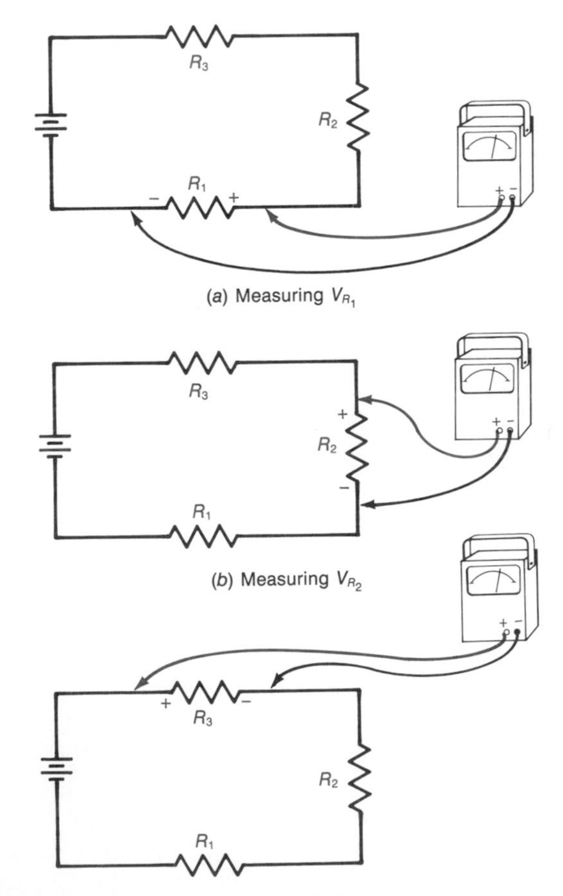

**FIGURE 2-32 Measuring individual voltage drops.**

Using Ohm's law and a voltmeter to find current flow is not always practical. There may be paths in a circuit where there is no known resistance across which a voltage drop can be measured. Also, the resistance may not be what it is supposed to be. If the resistance value is wrong, the calculation will be based on incorrect data and the derived current will be in error. In spite of these limitations, many technicians use this technique often in lieu of an actual current measurement.

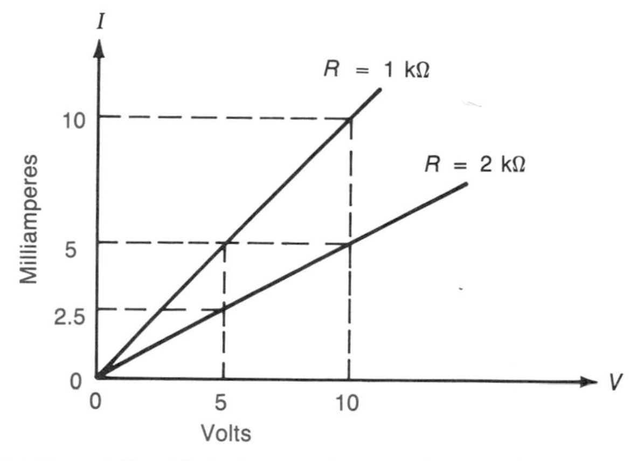


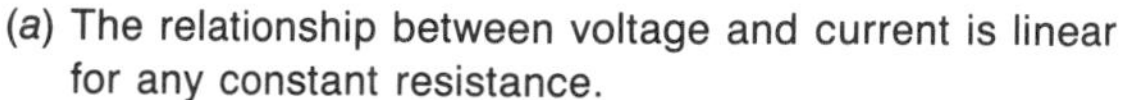
(*a*) The relationship between voltage and current is linear for any constant resistance.

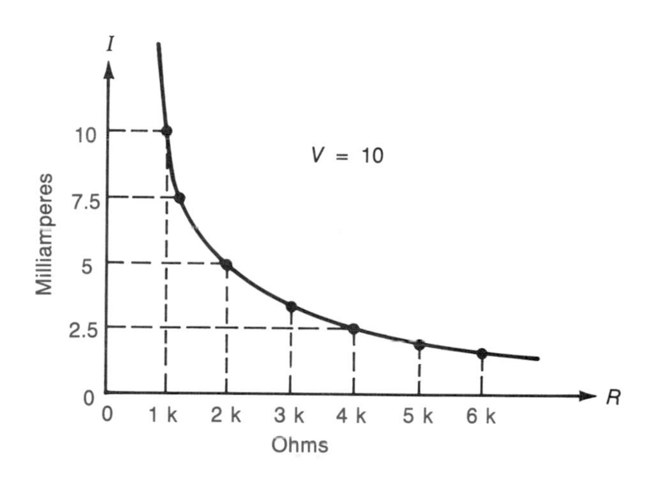


(*b*) The relationship between resistance and current is non-linear for any fixed voltage.

**FIGURE 2-34 Ohm's law graphs.**

Voltmeter ranges, analog scales, digital readouts, and the range for best resolution all follow the same guidelines presented earlier in this chapter for current measurements. The safety precautions are also the same. Voltmeter polarity is simply a matter of + to + and − to −. The polarity can be predicted from the current flow. For example, in Fig. 2-32*b*, the electron current is flowing up through $R_2$. Electron current flows from − to +, so the polarity markings shown can be derived by looking at the direction of current flow. They can also be derived by looking at the source (the battery). The bottom end of $R_2$ eventually connects with the negative side of the battery, and the top end eventually connects with the positive side.

The instrument for measuring resistance is called an *ohmmeter*. Figure 2-33 shows the scale for a typical analog ohmmeter. The scale is *nonlinear*. The pointer travel from 0 to 1 is greater than the pointer travel from 500Ω to 1 kΩ. This is because an ohmmeter responds to the current that flows through the resistance being measured. Figure 2-34*a* shows a volt-ampere graph for two fixed resistors. A 1-kΩ resistor conducts 5 mA at 5 V and 10 mA at 10 V. A 2-kΩ resistor conducts 2.5 mA at 5 V and 5 mA at 10 V. The relationship between voltage and current is *linear* (straight-line) for any constant resistance. Figure 2-34*b* shows an ohm-ampere graph for a fixed potential difference of 10 V. A 1-kΩ resistor conducts 10 mA, a 2-kΩ resistor conducts 5 mA, and a 4-kΩ resistor conducts 2.5 mA. The relationship between resistance and current is nonlinear for any fixed voltage. Compare the curve in Fig. 2-34*b* with the ohmmeter scale in Fig. 2-33. The graph shows that maximum current flows when the resistance approaches 0. This is why most ohmmeters indicate 0 Ω at the right-hand end of the scale. The graph also shows that maximum current change occurs when the resistance is low. For example, the current changes 5 mA for a resistance change from 1 to 2 kΩ and only 0.333 mA for a resistance change from 5 to 6 kΩ. This is why the right-hand portion of the ohmmeter scale is expanded when compared to the left-hand portion.

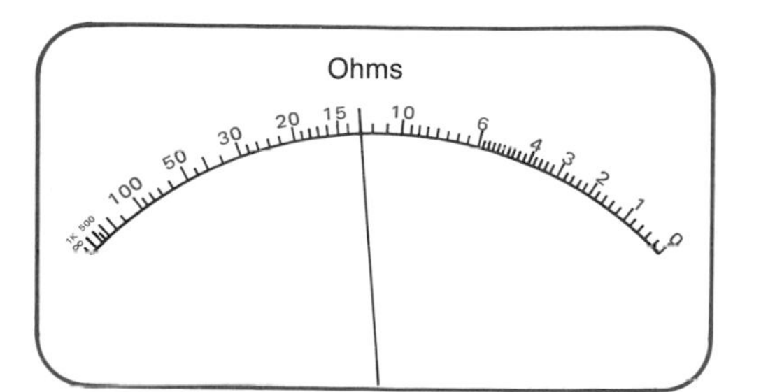


**FIGURE 2-33 A typical analog ohmmeter.**

Ohmmeters are usually multirange instruments. A typical set of ranges includes $R \times 1$, $R \times 10$,

$R \times 100$, $R \times 1$ kΩ, and $R \times 10$ kΩ. The letter symbol for resistance is $R$. The meter indication shown in Fig. 2-33 would be interpreted as 13 Ω on the $R \times 1$ range, 130 Ω on the $R \times 10$ range, 1.3 kΩ on the $R \times 100$ range, and so on. Suppose you wish to measure a 1.5-kΩ resistor. Which range is best? Look at Fig. 2-33. The 1-kΩ end of the scale is so crowded, due to the nonlinearity, that the $R \times 1$ range is nearly useless. Even if the 1.5-kΩ resistor is several thousand ohms high in value, the meter indication will be very little different than it would be if the resistor value were correct. The $R \times 1$ kΩ range would produce better results. Deviations as small as 200 Ω will produce a pointer deflection of one minor calibration. The $R \times 100$ range provides the best results. Deviations as small as 100 Ω produce a pointer deflection of one minor calibration. The range for best resolution on an analog ohmmeter is the one that places the pointer closest to midscale.

The accuracy of an analog ohmmeter is degraded if the meter is not properly adjusted. With the ohmmeter leads not touching each other or anything else, the pointer should be at the infinity mark at the far left end of the scale. If it is not, the meter should be adjusted. This is usually a mechanical adjustment on the meter movement itself. Refer to the operator's manual. Next, select the range for best resolution and firmly touch or clip the ohmmeter leads together. The pointer should travel to the 0 calibration at the right end of the scale. If it does not, adjust the 0-Ω control. If the pointer will not reach zero, the meter probably needs a new cell or battery. No adjustments are required with most digital ohmmeters. However, it is still good practice to touch the leads together to verify a zero indication. The reading may not be zero on the lowest range. It will probably be something near 0.3 Ω. This is due to the residual resistance of the leads, connections, and a protection device inside the meter. You should subtract the residual reading from the actual reading when testing low-resistance components. For example, a 1-Ω reading represents a component resistance of 0.7 Ω if the residual value is 0.3 Ω ($1 - 0.3 = 0.7$).

Resistance must never be measured in a circuit that is energized. Ohmmeters use an internal energy source to produce a current flow in the component under test and to drive the pointer up the scale. If an ohmmeter is connected to a battery or to a circuit that is on, it may be severely damaged.

Components should be isolated for ohmmeter tests. If they are not, the ohmmeter may provide a current to more than one device. This will cause an erroneously low reading. Refer to Fig. 2-35. The resistance of $R_3$ is to be measured. At least one of its leads should be opened to separate it from $R_2$. This prevents the ohmmeter from sending a current through both devices.

The ohmmeter must also be isolated from your fingers when measuring large values of resistance. Human skin resistance is on the order of several hundred kilohms. This can produce a significant error if the fingers become part of the circuit when making measurements on a high range.

All of the measurement functions mentioned in this chapter can be found in one instrument called a VOM. These letters stand for volt-ohm-milliammeter, which are the most prevalent functions on these multipurpose instruments. The digital counterpart may be called a DMM, for digital multimeter. This chapter has introduced the general concepts of circuit measurements. Consult the operator's manual for any special features that your particular instrument may have.

## Self-Test

**50.** Voltmeters are easier to use than ammeters because the circuit does not have to be ________ to insert the meter.

**51.** Refer to Fig. 2-31. How much less will $V_L$ measure than $V_B$ if the current is 20 A and the conductor resistance is 500 mΩ?

**52.** Refer to Fig. 2-32*c*. How much current is flowing in $R_3$ if the meter reads 4.8 V and the resistance of $R_3$ is 1.8 kΩ?

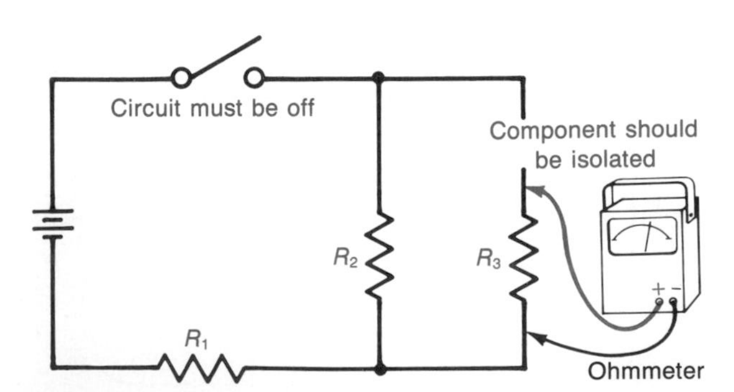


**FIGURE 2-35 Measuring resistance.**

**53.** Suppose that an analog voltmeter is connected across a resistor and the pointer deflects backward. What is wrong?

**54.** Describe the shape of a volt-ampere plot for a fixed value of resistance.

**55.** Suppose that the pointer of an ohmmeter is on the 5.2 calibration mark. How much resistance does this indicate if the meter is set to an $R \times 10\ \text{k}\Omega$ range?

**56.** Refer to Fig. 2-33. Which range would provide the best resolution when measuring a 120-Ω resistor?

**57.** What would the resolution be for the correct range in Question 56?

**58.** What must be done before an analog ohmmeter can be zeroed? (Assume that the infinity adjustment has already been checked.)

**59.** A digital ohmmeter measures 0.2 Ω when its leads are connected together. This reading is called the ______ resistance.

**60.** Ohmmeters must not be used in circuits that are ______.

## SUMMARY

1. The four major parts of a circuit are the energy source, a path for current to flow, control, and a load.
2. Schematic diagrams use symbols to represent the component parts of circuits.
3. Pictorial diagrams use picture views of components and circuit wiring.
4. Printed circuits consist of metal foil traces used to interconnect components on an insulated board.
5. A connection to the soil is called an earth ground.
6. The metal chassis or framework of a device is often used as a common ground.
7. The schematic symbols for cells and batteries represent the positive terminal with a long line and the negative terminal with a short line.
8. Electron current flows from negative to positive and conventional current flows from positive to negative.
9. The flow arrows that are a part of some schematic symbols point in the direction of conventional current flow.
10. The SI unit of current flow is the ampere and the letter symbol is A. One ampere is equal to one coulomb per second.
11. Electric impulses travel at nearly the speed of light in a circuit, while the drift velocity of a given electron is much, much less.
12. The instrument used to measure current is called an ammeter. It may also be called a milliammeter or a microammeter.
13. Always follow safe practices when working on electric circuits.
14. Choose the range for best resolution when using multirange instruments. When measuring current or voltage, this will be the most sensitive range that does not result in overload or overflow.
15. Digital instruments indicate overflow by blanking the display, blinking the display, or turning on an indicator.

**16.** The circuit must be broken and the ammeter leads connected at the break points when measuring current.

**17.** Reverse polarity causes an analog meter to deflect backward. Most digital meters react to reverse polarity by showing – to the left of the number display.

**18.** Work is force times distance. The SI unit of work or energy is the joule, which is equal to one newton-meter.

**19.** The unit of potential difference is the volt. Voltage is represented by the letter symbol $V$ though in older literature the letter $E$ is still used. One volt is equal to one joule per coulomb.

**20.** Large potential differences are advantageous when large amounts of work must be done by electric circuits.

**21.** Voltage is a term often used in place of potential difference. It only has meaning when it is referenced or measured across two points in a circuit. If only one point is referenced, then ground or common must be used as the other point.

**22.** The words *voltage rise* may be used to refer to the potential difference of an energy source. The words *voltage drop* refer to the potential difference across a load or resistance.

**23.** Pressure is a simple analogy for voltage.

**24.** The current flow in a circuit is directly proportional to voltage.

**25.** Resistance is the opposition to the flow of current.

**26.** The SI unit of resistance is the ohm, symbolized by the Greek capital letter omega. One ohm is equal to one volt per ampere.

**27.** The relationship $V = I \times R$ is known as Ohm's law. In older texts it is sometimes written as $E = I \times R$ when referring to an electromotive force.

**28.** Resistance may be used to control the amount of current in a circuit.

**29.** Resistance is generally undesirable in conductors. An ideal conductor would have zero resistance.

**30.** Voltmeters are used to measure potential difference.

**31.** The circuit is not broken when using a voltmeter.

**32.** A voltmeter may be used to determine current flow by measuring the drop across a known resistance and then applying Ohm's law.

**33.** An ohmmeter is used to measure resistance.

**34.** Ohmmeter scales are typically nonlinear.

**35.** The range for best resolution on an analog ohmmeter is one that places the pointer nearest to midscale.

**36.** Ohmmeters are checked for zero by connecting their leads together.

**37.** Ohmmeters must never be used in circuits that are energized.

**38.** Components should be isolated from the rest of the circuit for ohmmeter testing.

## CHAPTER REVIEW QUESTIONS

**2-1.** Refer to Fig. 2-36. Identify each schematic symbol.

a. ______ d. ______ g. ______ j. ______
b. ______ e. ______ h. ______ k. ______
c. ______ f. ______ i. ______ l. ______

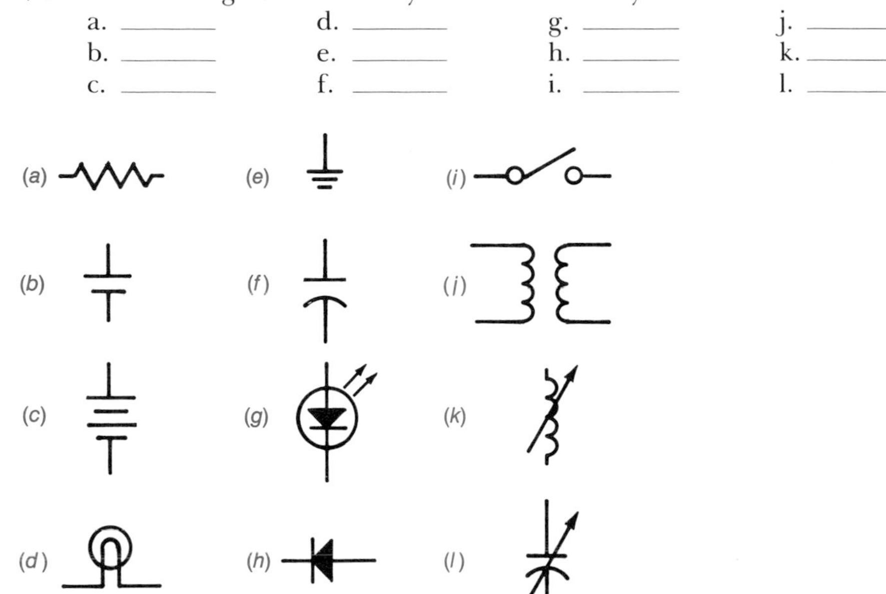


**FIGURE 2-36 Illustration for Chapter Review Question 2-1.**

**2-2.** A connection to the soil is known as a(n) ______ ground.

**2-3.** A connection to a metal chassis or frame is known as a(n) ______ ground.

**2-4.** The symbols shown in Fig. 2-36 are used in ______ diagrams.

**2-5.** Where is the positive terminal located on the cell and battery symbols shown in Fig. 2-36?

**2-6.** When the flow is assumed to be from positive to negative, it is known as ______ current.

**2-7.** When the flow is assumed to be from negative to positive, it is known as ______ current.

**2-8.** Current flow is equal to ______ per second.

**2-9.** The unit of current flow is the ______.

**2-10.** The instrument used to measure current flow is a(n) ______.

**2-11.** Suppose an analog meter is set at its 0 to 500-mA range. What is the current flow if the pointer is on the 3 calibration mark on a 0 to 5 scale?

**2-12.** Suppose the meter in Question 2-11 is switched to a 0 to 5-A range. What scale position should the pointer move to, assuming the measured current does not change?

**2-13.** What is wrong with using the range indicated in Question 2-12?

**2-14.** A circuit must be ______ to measure current flow.

**2-15.** If an analog meter deflects backward, the ______ is not correct.

**2-16.** The SI unit of energy is the ______.

**2-17.** The SI unit of work is the ______.

**2-18.** The SI unit of potential difference is the ______.

**2-19.** The potential difference across a source of energy can be referred to as a voltage ______.

**2-20.** The potential difference across a load is often referred to as a voltage ______.

**2-21.** Assuming no difference in current flow, will a high-voltage source deliver more energy in a given time period when compared to a low-voltage source?

**2-22.** A voltage always refers to ______ points in a circuit.

**2-23.** Refer to Fig. 2-23. How would $V_A$ be measured?

**2-24.** The SI unit of resistance is the ______ and its symbol is the Greek letter ______.

**2-25.** Why is a voltmeter easier to use than an ammeter?

**2-26.** Refer to Fig. 2-24. Where should a meter be connected to measure $V_{R_2}$?

**2-27.** Is the scale shown in Fig. 2-33 linear?

**2-28.** Refer to Fig. 2-33. Which ohmmeter range would provide the best resolution for measuring a resistance of 890 Ω?

**2-29.** What would the resolution be for the correct range in Question 2-28?

**2-30.** How is an ohmmeter checked for zero?

**2-31.** Ohmmeters must not be connected to circuits that are ______.

**2-32.** What is a VOM?

## CHAPTER REVIEW PROBLEMS

**2-1.** Calculate the current flow in a circuit where $6.25 \times 10^{18}$ electrons pass a point in 100 ms.

**2-2.** How long will it take a battery to deliver a 0.5-C charge if the current flow is 10 mA?

**2-3.** How much charge will be delivered per minute by a current flow of 375 mA?

**2-4.** How much work has been done when a 5.5-N force is exerted over a distance of 100 m?

**2-5.** Calculate the potential difference of an energy source that provides 6.8 J for every millicoulomb of charge that it delivers.

**2-6.** How much charge must be delivered by a 12-V battery to do 350 J of work?

**2-7.** Calculate the resistance of a lamp that conducts 0.1 A of current with a potential difference of 120 V.

**2-8.** A motor has a resistance of 32 Ω. Calculate its voltage drop when conducting 4.7 A.

**2-9.** An electric heater has a resistance of 11 Ω. How much current will it draw from a 120-V source?

**2-10.** A compressor motor draws 20 A when energized through an extension cord connected to a 115-V source. Calculate the potential difference across the motor terminals if the extension cord resistance is 750 mΩ.

**2-11.** Refer to Fig. 2-25. A voltmeter is connected across the rheostat and indicates a drop of 25 V. If the resistance of the rheostat is 5 Ω, how much current is flowing?

## ANSWERS TO SELF-TESTS

**1.** negative, positive
**2.** source, path, control, load
**3.** schematic
**4.** pictorial
**5.** printed circuit
**6.** source, path, control, load
**7.** The positive terminal is the long outside line.
**8.** from negative to positive
**9.** from positive to negative
**10.** clockwise
**11.** conventional
**12.** current
**13.** 1 C of charge passing a point in 1 s
**14.** $I$
**15.** A
**16.** 24.8 mA
**17.** 0.05 s
**18.** 0.6 C
**19.**

| | |
|---|---|
| 20 μA | 20 mA |
| 0.2 mA | 200 mA |
| 2 mA | 2 A |

**20.** 0–2.5; multiply by 100
**21.** The circuit current is greater than 200 mA.
**22.** c
**23.** broken
**24.** polarity (it is connected backward)
**25.** a minus sign
**26.** An excessive current would flow through the meter and possibly damage it.
**27.** 150 J
**28.** volt
**29.** 15 V
**30.** 5.56 C
**31.** 540 kJ
**32.** A high-voltage source assigns more joules to every coulomb of charge that flows.
**33.** a potential difference
**34.** $E$ and $V$
**35.** $V$
**36.** voltage rise
**37.** voltage drop
**38.** yes
**39.** potential difference (voltage)
**40.** resistance
**41.** pressure
**42.** resistance
**43.** Ohm's law
**44.** 1.11 kΩ
**45.** 36 mA
**46.** 9 V
**47.** 18 mA
**48.** 51.1 mA
**49.** The polarity of the battery is reversed.
**50.** broken (opened)
**51.** 10 V
**52.** 2.7 mA
**53.** The polarity is reversed.
**54.** It will be linear (a straight line).
**55.** 52 kΩ
**56.** $R \times 10$
**57.** 10 Ω
**58.** The leads must be connected together.
**59.** residual
**60.** energized

# Chapter 3 Power and Energy

Electric energy can be obtained by converting other energy forms. The vast amounts of electric energy consumed in the world come indirectly from the combustion of fuels, nuclear fission, and the natural movement of water and air, and directly from electrochemical reactions and solar radiation. Electric energy is nearly ideal. It is relatively easy to transmit and distribute from one location to another. In some forms it is extremely portable. It is easily converted to other energy forms such as mechanical, heat, and light. It is essentially nonpolluting, although its production is often associated with environmental problems. This chapter continues with the development of the basic measuring units used in electricity and introduces the concepts of power, energy, and efficiency.

## 3-1 POWER

Work is accomplished when energy is converted from one form to another. *Power* is the rate at which energy is converted. Power may also be defined as *the rate of doing work.* The word power is often used interchangeably with the word energy. This is unfortunate because it is not correct. For example, electric energy sources are usually referred to as "power supplies." Another example is that electric utility companies are often called "power companies" and people even talk about their "power bills." Strictly speaking, these terms are not correct. A power supply does not supply power; it supplies energy (or work). The electric utilities do not sell power. They sell energy, and their invoices are energy bills, not power bills.

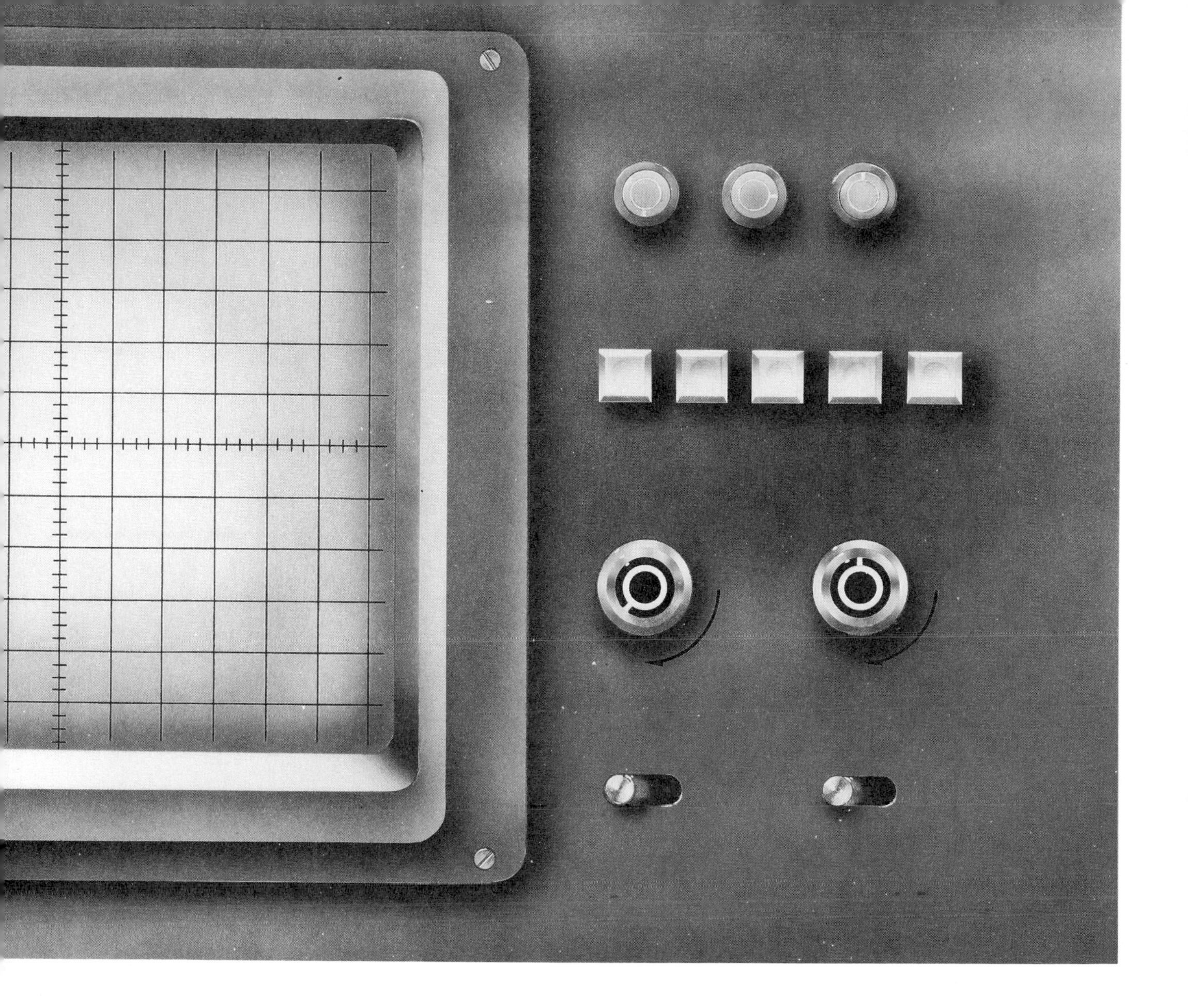

The SI unit of power is the watt and its letter symbol is (W):

$$P = \frac{W}{t}$$

where $P$ = power in watts, W
$W$ = energy converted in joules, J
$t$ = time in seconds, s

One watt of power is equal to one joule per second (1 J/s).

**EXAMPLE 3-1**
Calculate the power involved in lifting a 640-kg mass a distance of 100 m in 1 min. Assume a gravitational acceleration of 9.81 m/s$^2$.

**Solution**
The first step is to calculate the force required to lift the mass. Force ($F$) is equal to mass ($M$) times acceleration ($A$):

$$\begin{aligned} F &= M \times A \\ &= 640 \text{ kg} \times 9.81 \text{ m/s}^2 \\ F &= 6.28 \text{ kN} \end{aligned}$$

Next, the amount of work is calculated. Work ($W$) is equal to force ($F$) times distance ($D$):

$$\begin{aligned} W &= F \times D \\ &= 6.28 \times 10^3 \times 100 \\ W &= 628 \text{ kJ} \end{aligned}$$

Finally, the power is found:

$$P = \frac{W}{t}$$

$$= \frac{628 \times 10^3}{60}$$
$$P = 10.5 \text{ kW}$$

Since power is rated in joules per second, it is possible to express electric power in terms of voltage and current. You will recall that voltage (potential difference) is equal to joules per coulomb and that current is equal to coulombs per second. Power can therefore be expressed as the product of voltage and current:

$$P = V \times I$$
$$= \frac{\text{joules}}{\cancel{\text{coulomb}}} \times \frac{\cancel{\text{coulombs}}}{\text{second}}$$
$$P = \frac{\text{joules}}{\text{second}}$$

A 1-V energy source that delivers a current of 1 A provides 1 W of power.

**EXAMPLE 3-2**

Calculate the power for a 120-V energy source that delivers 15 A of current.

**Solution**

Substituting the data into the equation,

$$P = V \times I$$
$$= 120 \times 15$$
$$P = 1.8 \text{ kW}$$

**EXAMPLE 3-3**

How much current will be required from a 120-V energy source to produce 10.5 kW of power?

**Solution**

Rearranging the power equation,

$$I = \frac{P}{V}$$
$$= \frac{10.5 \times 10^3}{120}$$
$$I = 87.5 \text{ A}$$

A current of 87.5 A requires very low conductor resistance to avoid substantial voltage drop in the conductors. This reinforces the concept of using higher voltages when considerable work must be done. However, now the concept can be refined to include accomplishing the work in a reasonable amount of time. An elevator requires a power of 10.5 kW to lift a 640-kg mass 100 m in 1 min. It would require only one-sixtieth of that amount to do the same work in 1 h. Not many people would tolerate 1-h elevator rides. The only practical alternatives are to use very large conductors or a higher voltage energy source. For example, with a 480-V source, the required current is 21.9 A to develop 10.5 kW. Conductors rated to handle 21.9 A are much smaller and more reasonable in cost than conductors rated to handle 87.5 A.

**EXAMPLE 3-4**

Calculate the voltage required to develop 10.5 kW with 5 A of current.

**Solution**

Rearranging the power equation,

$$V = \frac{P}{I}$$
$$= \frac{10.5 \times 10^3}{5}$$
$$= 2.1 \times 10^3$$
$$V = 2.1 \text{ kV}$$

Power is often used in conjunction with the word *dissipation*. Power dissipation usually refers to the rate at which electric energy is converted to heat energy. A resistor in an active circuit may feel warm to the touch. This is due to the conversion of electric energy. The resistor dissipates the heat energy by transferring it to surrounding structures and to the atmosphere. A resistor that is dissipating quite a bit of power may become hot enough to burn a finger. It takes time to transfer (dissipate) the heat energy. High power means a rapid conversion of electric energy to heat energy, and the resistor will reach a high temperature. If the power dissipation is extreme, the resistor reaches an excessive temperature and is damaged. Resistors and many other components have maximum power dissipation ratings. You must be able to calculate power dissipation to determine if components are adequately rated for a given set of operating conditions.

**EXAMPLE 3-5**

A 0.1-Ω resistor is rated at 5 W. Is this resistor safe when conducting a current of 10 A?

**Solution**

We have been using the product of voltage and current to calculate power. In this case, the current is known but not the voltage. Ohm's law can be used first to find the voltage drop across the resistor:

$$V = I \times R$$
$$= 10 \times 0.1$$
$$V = 1 \text{ V}$$

Now, applying the power equation,

$$P = V \times I$$
$$= 1 \times 10$$
$$P = 10 \text{ W}$$

The resistor is not safe since the calculated power exceeds the dissipation rating.

It is possible to solve the previous example by preparing another equation. This equation allows some problems to be solved in one step rather than in two. Since $V = I \times R$, we can make a substitution in the power equation:

$$P = (I \times R) \times I$$
$$= I^2 \times R$$

Using this equation to check the previous example, we find

$$P = 10^2 \times 0.1$$
$$= 10 \text{ W}$$

Another possibility is to substitute $V/R$ for $I$ in the power equation:

$$P = V \times \frac{V}{R}$$
$$= \frac{V^2}{R}$$

**EXAMPLE 3-6**

Calculate the power dissipation in a 2.2-kΩ resistor that has a voltage drop of 18 V.

**Solution**

We could solve for the current using Ohm's law and then multiply the current by the voltage drop. Instead, we will use the derived power formula:

$$P = \frac{V^2}{R}$$
$$= \frac{18^2}{2.2 \times 10^3}$$
$$P = 0.15 \text{ W}$$

The power equations based on Ohm's law substitutions reduce the number of steps required to solve some problems. They also provide answers for some problems where the two-step technique will not work.

**EXAMPLE 3-7**

What is the maximum safe current flow in a 47-Ω, 2-W resistor?

**Solution**

Ohm's law will not work because we do not know the voltage or the current. We need the power equation that does not contain a voltage term:

$$P = I^2 \times R$$

Solving for $I$:

$$I = \sqrt{\frac{P}{R}}$$
$$= \sqrt{\frac{2}{47}}$$
$$I = 0.21 \text{ A}$$

**EXAMPLE 3-8**

What is the maximum voltage that can be applied across a 100-Ω, 10-W resistor in order to keep within the resistor's power rating?

**Solution**

Once again, it is necessary to use a power formula based on an Ohm's law substitution:

$$P = \frac{V^2}{R}$$

Solving for $V$:

$$V = \sqrt{P \times R}$$
$$= \sqrt{10 \times 100}$$
$$V = 31.6 \text{ V}$$

Although the SI system of measurements has been officially adopted, some units from the U.S. customary system remain popular. The horsepower (hp) is an example of a unit that has persisted:

$$1 \text{ hp} = 550 \text{ ft} \cdot \text{lb/s}$$

A conversion from the horsepower unit to the watt unit is often required in electrical work:

$$1 \text{ pound-force (lbf)} = 4.45 \text{ N}$$
$$1 \text{ ft} = 0.305 \text{ m}$$
$$1 \text{ ft} \cdot \text{lb} = 4.45 \text{ N} \times 0.305 \text{ m} = 1.36 \text{ N} \cdot \text{m}$$
$$1 \text{ hp} = (550 \times 1.36 \text{ N} \cdot \text{m})/\text{s} = 746 \text{ N} \cdot \text{m/s} = 746 \text{ W}$$

One horsepower is equal to 746 W.

**EXAMPLE 3-9**

An 8-hp device is energized from a 240-V source. Calculate its current demand.

**Solution**

The first step is to convert 8 hp to watts:

$$8 \times 746 = 5.97 \text{ kW}$$

Now, the current can be found from the power formula:

$$I = \frac{P}{V} = \frac{5.97 \times 10^3}{240}$$
$$I = 24.9 \text{ A}$$

## Self-Test

1. Power is the ______ at which work is done.
2. The watt is the ______ unit of power.
3. One watt is equal to ______.
4. What does a "power supply" actually provide?
5. How much energy is involved in exerting a 15-N force over a distance of 50 m? Calculate the power required to complete this action in 10 s.
6. How much power would be required in Question 5 to accomplish the same work in 5 s? Would the amount of energy required be any different?
7. Calculate the power developed by a 12-V battery that delivers 50 A of current.
8. How much current will be required from a 50-V source to deliver 100 W of power?
9. A 6.8-kΩ, $\frac{1}{4}$-W resistor shows a potential difference of 40 V. Is it safe?
10. What is the resistance of a 120-V, 100-W lamp?
11. What is the maximum safe current through a 1-MΩ, $\frac{1}{2}$-W resistor?
12. Calculate the current required by a $\frac{1}{2}$-hp device energized by a 120-V source.

## 3-2 ENERGY

Work requires energy. Electric energy is sometimes rated in units of watt-seconds (watts times seconds) or kilowatthours (kilowatts times hours). Since power is the rate at which energy is converted, when power is multiplied by time, the amount of energy can be determined:

$$P \times t = \frac{W}{\cancel{t}} \times \frac{\cancel{t}}{1} = W$$

The watt-second is a unit of work or energy. It is equal to one watt times one second, or one joule.

**EXAMPLE 3-10**

Calculate the energy converted in 24 h by a 100-W lamp.

**Solution**

Enter the data into the equation:

$$W = P \times t$$
$$= 100 \times 24 \times 60 \times 60$$
$$W = 8.64 \text{ MW} \cdot \text{s, or } 8.64 \text{ MJ}$$

The watt-second (joule) is a rather small unit of energy for large systems such as a building or a home. The electric utilities use a much larger energy unit for their measurement and billing procedures. This unit is the *kilowatthour* (kWh). The relationship of the kilowatthour to the watt-second can be found by

$$1 \text{ kWh} = 1 \times 10^3 \times 60 \times 60 = 3.6 \text{ MW} \cdot \text{s}$$

Thus, a kilowatthour is equal to 3.6 MJ. The cost of electric energy varies from location to location. Six cents per kilowatthour is an average figure.

### EXAMPLE 3-11

What will the energy cost to operate a 100-W lamp continuously for 1 yr assuming a cost of 6 cents per kilowatthour?

**Solution**

Multiply the power consumption in kilowatts by the hours of operation and then by \$0.06 (6 cents):

$$\text{Cost} = \text{kilowatts} \times \text{hours} \times \text{rate}$$
$$= 0.1 \times 365 \times 24 \times 0.06$$
$$\text{Cost} = \$52.56$$

This shows that a device with moderate power ratings can cost a significant amount when allowed to run continuously. Money can be saved by turning devices off when they are not required.

Table 3-1 shows some typical power ratings for household devices. With some devices, quite a bit of variation is found. For example, different model air conditioners may vary from 1000 to 5000 W. Most devices have a plate or a label that states the power rating as well as an energy efficiency rating. Consult this information when calculating the costs of operation.

**Table 3-1 Typical Power Ratings of Household Devices**

| Household Devices | Typical Power Rating, W |
|---|---|
| Clock | 3 |
| Radio | 5 |
| Videocassette recorder | 50 |
| Personal computer | 75 |
| AM-FM stereo receiver and amplifier | 100 |
| Color television receiver | 100 |
| Fan | 200 |
| Refrigerator | 400 |
| Clothes washer | 400 |
| Iron | 1,000 |
| Toaster | 1,000 |
| Space heater | 1,500 |
| Air conditioner | 2,000 |
| Water heater (electric) | 5,000 |
| Clothes dryer (electric) | 6,000 |
| Range (electric) | 7,000 |
| Furnace (electric) | 20,000 |

### EXAMPLE 3-12

How much will it cost per year to operate an electric clock at 6¢ per kwhr?

**Solution**

Consulting Table 3-1 shows a typical power rating of 3 W:

$$\text{Cost} = 0.003 \times 365 \times 24 \times 0.06$$
$$= \$1.58$$

### EXAMPLE 3-13

An electric furnace is on 50 percent of the time in the month of January. Calculate the heating cost for this month.

**Solution**

Consulting Table 3-1 shows a typical power rating of 20,000 W:

$$\text{Cost} = 20 \times 31 \times 12 \times 0.06$$
$$= \$446.40$$

### EXAMPLE 3-14

How much does it cost to dry a load of clothes in an electric dryer if it runs for 30 min?

**Solution**

Table 3-1 shows a typical power rating of 6000 W:

$$\text{Cost} = 6 \times 0.5 \times 0.06$$
$$= \$0.18$$

It should be clear now that the electric utilities bill their customers for energy and not for power. When power is multiplied by time, the time unit cancels, leaving energy. Electric energy is measured in joules (or watt-seconds, which is the same unit) or in kilowatthours. There is nothing wrong with the terms power supply or power company as long as you clearly understand that the word power refers to the rate of doing work and is not to be confused with work or the energy required to do it.

## Self-Test

**13.** What is the result when power is multiplied by time?

**14.** A 47-Ω resistor has a current flow of 100 mA for 24 min. How much energy is converted?

**15.** A 4000-W air conditioner runs 50 percent of the time in the month of August. What is the cost of operation for the month of August if the energy costs 6 cents per kilowatthour?

## 3-3 EFFICIENCY

The efficiency of a device or of a circuit is a comparison of the useful output to the input. The law of conservation of energy states that "energy cannot be created or destroyed but can be converted from one form to another." Some of the energy in electric circuits may be converted into a form that is not useful. Figure 3-1 shows an electric motor. Its purpose is to convert electric energy into mechanical energy. It does this, but it also converts some portion of the input energy into heat. The heat output of a motor is not useful. A perfect motor would convert all of the electric energy into useful mechanical output.

The letter symbol for efficiency is the Greek lowercase letter eta ($\eta$), and it is the ratio of useful output to the input:

$$\eta = \frac{\text{useful output}}{\text{input}}$$

Efficiency is often based on the ratio of work output (useful) ($o$) to work input ($i$):

$$\eta = \frac{W_o}{W_i}$$

### EXAMPLE 3-15

Calculate the efficiency for the motor in Fig. 3-1 if the input is 2280 J and the useful output at the motor shaft is 1750 J.

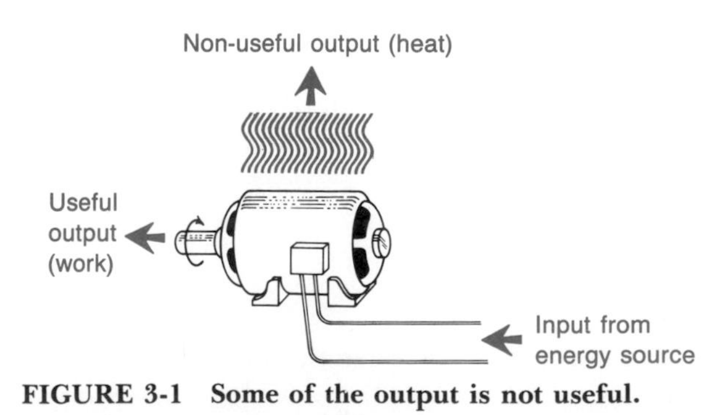


**FIGURE 3-1 Some of the output is not useful.**

**Solution**

The ratio is

$$\eta = \frac{1750\ \text{J}}{2280\ \text{J}} = 0.768$$

The answer 0.768 is dimensionless. That is, the joule units cancel, leaving a pure number. Efficiency is often expressed as a percentage. This is easily accomplished by multiplying the ratio by 100 percent:

$$\eta = \frac{W_o}{W_i} \times 100\%$$

Thus $\eta = 0.768 \times 100 = 76.8\%$

### EXAMPLE 3-16

The useful output of a machine is 6800 J for an input of 8500 J. What is the efficiency of the machine in percent?

**Solution**

Multiply the ratio of useful output to input by 100 percent:

$$\eta = \frac{6800}{8500} \times 100\% = 80\%$$

Work can be expressed as power times time. This leads to another method of calculating efficiency:

$$\eta = \frac{P_o \times \cancel{t}}{P_i \times \cancel{t}} = \frac{P_o}{P_i}$$

where $P_o$ = useful output power
$P_i$ = input power

Again, the efficiency may be expressed in percentage form by multiplying by 100 percent.

### EXAMPLE 3-17

Calculate the efficiency of a motor that produces a mechanical output of 1 hp while taking 1 kW from the energy source. Express the answer in percentage form.

**Solution**

We cannot divide watts into horsepower. First, the output power must be converted to watts or the input power must be converted to horsepower. One horsepower = 746 W:

$$\eta = \frac{746}{1000} \times 100\%$$
$$= 74.6\%$$

Some electric devices are nearly 100 percent efficient. A space heater is an example. Practically all of the input energy is converted to heat energy. In this case, the heat is useful output. An ideal motor would convert all of its input to mechanical output. Unfortunately, this is not possible. The motor converts some of the energy into heat. Lights have the same limitation. An ideal light would convert all of the input energy to output in the visible spectrum. If you have ever touched an active 100-W light bulb, you know that they also produce quite a bit of heat.

Efficiency is important for several reasons. First, it is never desirable to waste energy on nonuseful output. Second, poor efficiency usually means a significant temperature rise. High temperature is one of the major limiting factors in producing reliable electric and electronic devices. Circuits and devices that run hot are more likely to fail. Third, the heat has to be dissipated. It has to be transferred to the atmosphere or to some other mass. Heat removal can become quite involved in high-power circuits and adds to the cost and size of the equipment.

## Self-Test

**16.** The efficiency of a device is determined by the ratio of its useful work output to its work ______.

**17.** A motor develops 22,500 J of mechanical output for an input of 28,700 J. Calculate its percent efficiency.

**18.** For Question 17, how much work does the motor fail to convert to useful output? What happens to this work?

**19.** A motor is 62 percent efficient. Calculate its mechanical output in horsepower if it takes 1200 W from an energy source.

## 3-4 ENERGY SOURCES

Cells and batteries are convenient energy sources for portable and low-power devices. Some examples of cells are shown in Fig. 3-2*a*. The D cell is the largest of those shown, and the AAA cell is the smallest. Many other sizes are available. The cells used in watches are often smaller than a button. The schematic symbol is the same for any size cell. Most cells develop a potential difference of approximately 1.5 V. However, depending on the exact materials used to make the cell, the potential difference can range from 1.2 to 3.8 V.

Cells of the same chemical type develop the same voltage regardless of their size. A D cell is substantially larger than an AAA cell but develops the same voltage. There is a difference in current capacity, however. Since the D cell is physically larger, it will supply a given amount of current for a longer period of time. The current is produced by an electrochemical process. The materials inside the cell are chemically changed by the current, and eventually no active materials are left. A "dead cell" is one that is chemically exhausted or ineffective due to some fault.

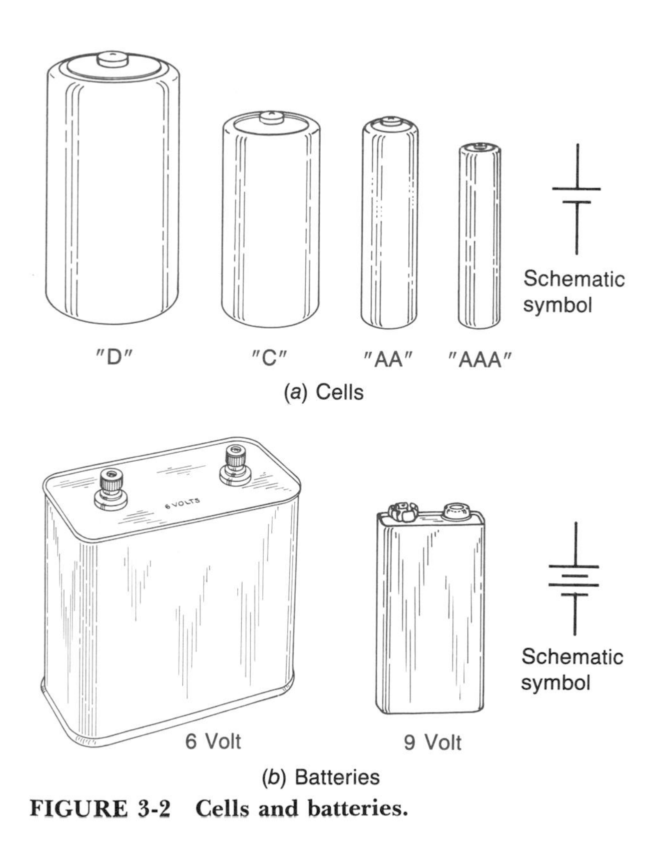


**FIGURE 3-2 Cells and batteries.**

Figure 3-2*b* shows two examples of batteries along with the schematic symbol. As the symbol implies, batteries are made up of more than one cell. Each cell in the battery adds to the overall potential difference that is developed. A 6-V battery has four internal cells (4 × 1.5 V = 6 V) and a 9-V battery has six cells (6 × 1.5 V = 9 V). The symbol is the same regardless of the number of cells used in the battery. The symbol also shows that the positive side of one cell must connect to the negative side of the next cell. These connections are inside the battery case. Additional information about cells and batteries is presented in a later chapter.

Wall outlets are another obvious source of electric energy. Large rotating machines called *generators* produce the energy that is present at the outlet. Wall outlets are connected to the generators through the power grid, which is a complex array of transmission lines and distribution circuits. The energy that is available at the outlet is different from the energy provided by cells and batteries. It is usually specified as 120 V, 60 hertz (Hz) ac. The abbreviation ac stands for *alternating current.* The current alternates because the polarity periodically reverses. Look at Fig. 3-3. The generator produces one polarity and one direction of flow for some period of time. Then, the polarity and the direction of flow both reverse and remain that way for another period of time. Both the polarity and the direction of flow *alternate* on a periodic basis. The *frequency* determines how long it takes one cycle consisting of two alternations to be completed. The standard frequency is 60 Hz (60 cycles per second) in the United States. Alternating current is covered in more detail in subsequent chapters.

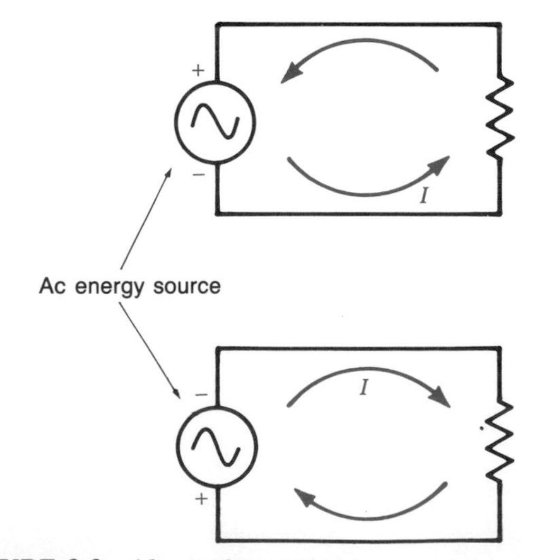


**FIGURE 3-3 Alternating current.**

Many electric and electronic circuits require direct current (dc). The direction of flow does not periodically reverse in dc circuits. Cells and batteries provide direct current but are not practical for many applications. It is possible to change alternating current to direct current. Figure 3-4 shows a typical dc power supply designed for bench use. It is designed to be plugged into an ordinary wall outlet and provide direct current for testing and analysis. It is much more convenient and economical (over a period of time) than cells or batteries. The voltage output from most bench supplies is adjustable. Note that the schematic symbol shows an arrow added to the battery symbol. Common ranges are 0 to 5, 15, 25, or 50 V. Many bench power supplies have a fixed output in addition to one or two adjustable outputs. The output terminals are usually clearly labeled to indicate the voltage, the range, and the polarity.

Most bench supplies are relatively easy to use. The one shown in Fig. 3-4 has four controls:

1. A meter control to select one of two current ranges or a voltage range
2. A control to set the maximum output current
3. A control to adjust the output voltage
4. An on-off switch

The only control that requires much explanation is the current adjust. This feature protects the supply and the circuit under test from possible damage due to excessive current flow. The current is adjusted for the desired maximum with the leads coming from the supply output connected together. This puts a very low resistance load on the

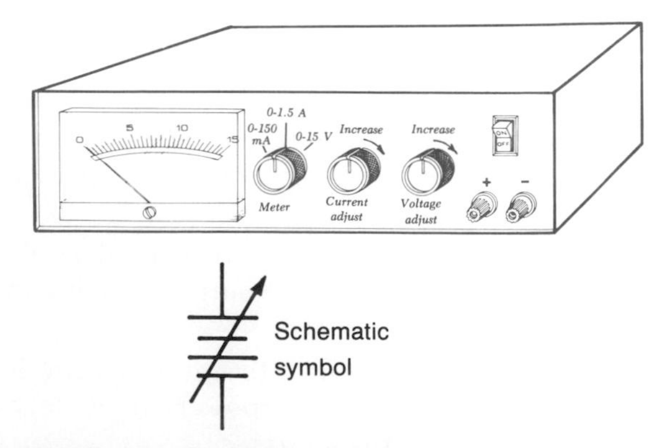


**FIGURE 3-4 Typical bench dc power supply.**

supply because the leads have very little resistance. By Ohm's law, as the resistance approaches zero, the current approaches a very high value:

$$I = \frac{V}{R}$$

(As $R$ approaches 0, $I$ approaches infinity.)

A circuit with very low load resistance is often called a *short circuit.* Short circuits can cause extremely high current flow and may damage power supplies, wiring, and components. The supply shown in Fig. 3-4 is *current-limited* to prevent damage due to short circuits. Never connect power supply leads together unless the supply is current-limited. If the supply is current-limited, the leads may be "shorted" together and the desired maximum current set. After the current is set, the supply is connected to the circuit under test. If the circuit is incorrect or has a fault, the supply automatically limits the current to the preset limit. Some supplies have an overload lamp or a current limit lamp that comes on when this happens. If the supply has a meter, it shows a drop in output voltage when the current limit is reached. Refer to the power supply manual for additional operating details.

## Self-Test

**20.** The physical size of a cell determines its ______.

**21.** A battery has individual cells that are internally connected negative to ______.

**22.** An automotive battery uses six cells, each of which develops 2.1 V. What is the battery voltage?

**23.** A battery provides direct current, while a wall outlet provides ______ current.

**24.** Suppose a bench supply is set for 15 V and is adjusted to current-limit at 1 A. How much current will flow when the following resistors are connected one at a time to its output terminals?

**a.** 470 Ω  **d.** 4.7 Ω
**b.** 100 Ω  **e.** 2.2 Ω
**c.** 10 Ω

**25.** A short circuit is one with abnormally low load ______.

**26.** It is safe to short power supply leads together only if the supply is ______.

## 3-5 PROTECTIVE DEVICES

A short circuit can cause severe equipment damage and may even start a fire. The abnormally high currents that often accompany a short circuit can produce high power dissipation and high temperature. The wires in a building can get hot enough to ignite the structural materials. Remember, wires do have some resistance. The power dissipated in a resistance varies as the square of the current:

$$P = I^2R$$

If the current doubles, the power increases four times. If the current quadruples, the power increases 16 times. Many circuits require *overcurrent* protection.

*Fuses* are intentionally weakened circuit components that open when the current reaches a dangerous level. Figure 3-5 shows how fuses work.

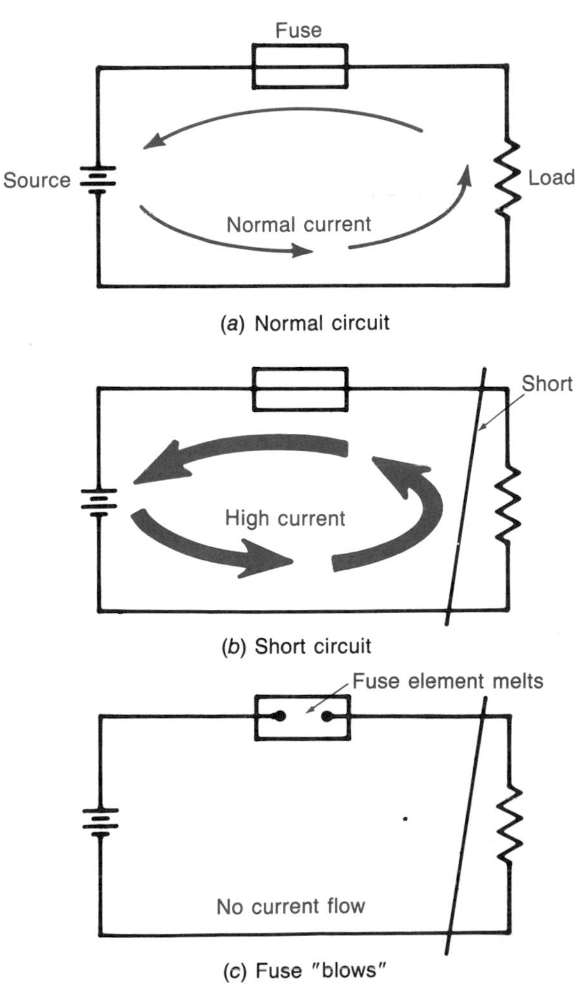


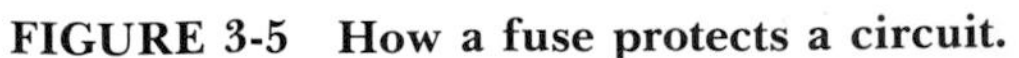

**FIGURE 3-5 How a fuse protects a circuit.**

There is an element inside the fuse that has resistance. When the current flow is normal, the power dissipation is not high enough to melt the element. Figure 3-5*b* shows a short circuit which causes an abnormally high flow of current. The power dissipation in the fuse element increases and it reaches a high temperature. Eventually, the element melts and the circuit is opened (Fig. 3-5*c*). When a fuse melts, it is said to have "blown."

Figure 3-6 shows some examples of fuses used to protect electric and electronic equipment. The AG designation is a matter of history. It originally meant an "automobile glass" fuse. Not all fuses of this type have glass tubes. Some use Bakelite, fiber, or ceramic materials. These are designated AB. Today, fuses of this type are used in many applications in addition to automotive ones. The number that comes before the AG or AB designation specifies the physical size of the fuse. Table 3-2 lists the numbers and their sizes. Note that the dimensions are in inches. A 3AG fuse has a glass tube and is $\frac{1}{4}$ in. in diameter and $1\frac{1}{4}$ in. long. A 4AB fuse has a nonglass tube (usually ceramic) and is $\frac{9}{32}$ in. in diameter and $1\frac{1}{4}$ in. long.

**Table 3-2 Fuse Sizes**

| Size | Diameter, in. | Length, in. |
|---|---|---|
| 1 | $\frac{1}{4}$ | $\frac{5}{8}$ |
| 3 | $\frac{1}{4}$ | $1\frac{1}{4}$ |
| 4 | $\frac{9}{32}$ | $1\frac{1}{4}$ |
| 5 | $\frac{13}{32}$ | $1\frac{1}{2}$ |
| 7 | $\frac{1}{4}$ | $\frac{7}{8}$ |
| 8 | $\frac{1}{4}$ | 1 |
| 9 | $\frac{1}{4}$ | $1\frac{7}{16}$ |

In addition to physical size, fuses are rated according to *current* and *voltage* and by their *fusing characteristic.* The current rating of a fuse should exceed the maximum normal circuit current by at least 25 percent. The fuse may have to be rated even higher for operation at temperatures much higher than 25° C. The type 3AG *normal-blo* fuse shown in Fig. 3-6 can be obtained with current ratings from 0.01 to 10 A. The current rating is stamped on one of the metal ends or printed on the fuse body.

When the power source is capable of very high current, the voltage rating of a fuse must be at least equal to the circuit voltage. This rating is very conservative for most applications because a fuse will open a circuit at its voltage rating for fault currents up to 10,000 A. Very few circuits can develop that much current even when shorted. Therefore, it is a common and safe practice to use fuses in circuits with voltages far in excess of the fuse ratings. Most 3AG and 3AB fuses are rated 250 V. The voltage rating is marked on the fuse in a manner similar to the current rating.

The fusing characteristic is determined by blowing time versus current flow. This fuse specification must be within the safe time-temperature characteristic of the circuit or device that is protected. For example, suppose a motor is protected by a fuse. A motor has a significant thermal mass. It takes a fair amount of time for the motor to heat up and reach a dangerous temperature. On the other hand, an electronic component may have an extremely small thermal mass. It rapidly reaches a damaging temperature when excessive current flows through it. Therefore, a fuse that protects a motor can safely blow much more slowly than a fuse that must protect an electronic component.

Figure 3-6 includes a *slo-blo* fuse. This designates a fuse with a long blowing time. Such fuses are often used to protect motors which have a large inrush current. Some motors draw five times their running current when starting. A fast-blowing fuse cannot be used to adequately protect such a load. It would blow every time the motor started. Of course, a fuse with a much higher current rating

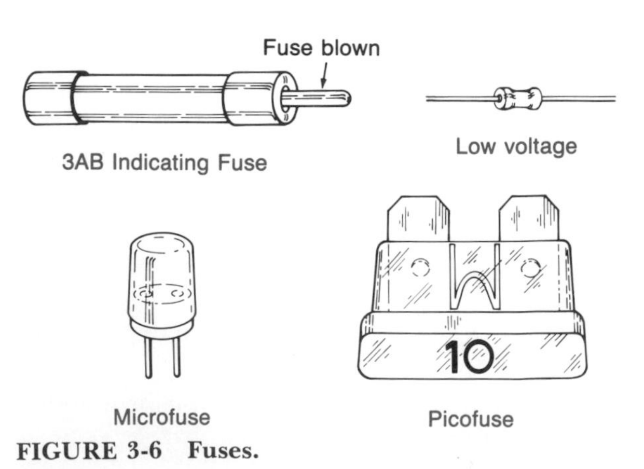


**FIGURE 3-6 Fuses.**

could be used to allow the motor to start. However, then a constant overload on the motor may not blow the fuse and the motor may eventually burn up. The advantage of a slo-blo fuse is that it allows momentary surges but will open eventually if the overload persists. Some slow-blowing fuses use a solder joint that is under spring tension. The solder joint supports a momentary overcurrent. If the overcurrent persists, the solder melts and the spring pulls the joint apart and opens the circuit.

Figure 3-7 shows the relationship of blowing time to current. With a 500 percent overcurrent, a slo-blo fuse takes about 2 s to open the circuit. A normal action fuse, with the same overcurrent, opens the circuit in 0.01 s, and a rectifier fuse opens the circuit in 1 ms. Rectifier fuses are fast-acting and designed to protect solid-state electronic devices which have a small thermal mass.

All fuses must be downrated when used in high-temperature environments. This is most important with slo-blo types. For example, a 1-A slo-blo fuse must be downrated to 0.6 A for operation at 100° C. Normal and fast-blowing 1-A fuses are downrated to 0.85 A at 100° C.

Figure 3-6 also shows some other fuse types. The indicating fuse has a pin which extends from the end of the fuse when it is blown. The pin may be used to activate an alarm circuit. The low-voltage fuse shown was originally developed for the automotive industry and is now used in a variety of applications. The current rating is marked with a large numeral which is easy to read. These fuses are rated at 32 V and are available in current ratings from 3 to 30 A. Microfuses and picofuses are designed for applications where space is a factor. They are ideal for installation on printed circuit boards. They are usually rated at 32 or 125 V and have current ratings from 0.05 to 15 A. Figure 3-6 is far from complete. There are many other styles, shapes, and sizes. High-voltage fuses are also available with ratings up to 10,000 V.

Fuse selection is not a trivial process. Many factors must be considered for proper protection. A fuse is a safety valve. If a fuse is replaced with another having different characteristics or ratings, the safety of the product or the circuit may be impaired. Consult the manufacturer's literature when in doubt concerning a proper replacement. Some fuses have wire leads and are soldered into the circuit. Others have metal terminals and are placed in fuse holders such as those shown in Fig. 3-8. Make absolutely certain that the power is turned off when replacing a fuse.

*Circuit breakers* are also used to protect against overcurrent. Figure 3-9 shows some examples. Circuit breakers have the advantage of being able to open a circuit by "tripping." A tripped breaker can be reset by pushing a button or by throwing a toggle. Also available are automatic circuit breakers that reset themselves after they cool down.

Figure 3-10 shows the principle of operation of a thermal circuit breaker. The circuit current flows through a special bimetallic strip. Bimetallic strips are made up of two different metal alloys, each having a different coefficient of expansion. When a piece of metal gets hot, it tends to increase its physical dimensions. The metal that makes up the bottom of the bimetallic strip in Fig. 3-10 has a larger coefficient of expansion. It increases its length more than the piece that makes up the top

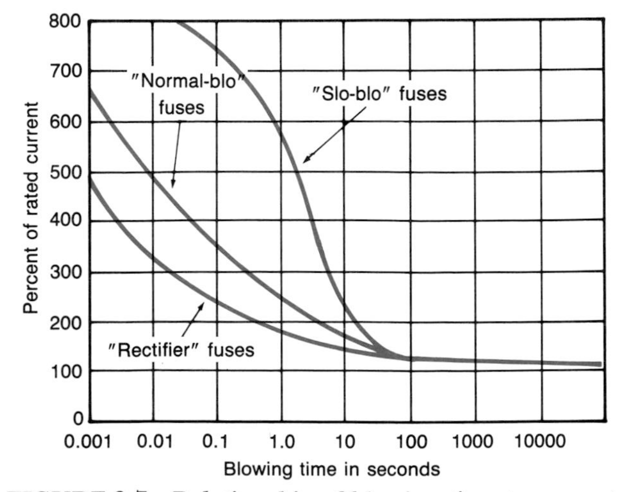


**FIGURE 3-7 Relationship of blowing time to percent of rated current.**

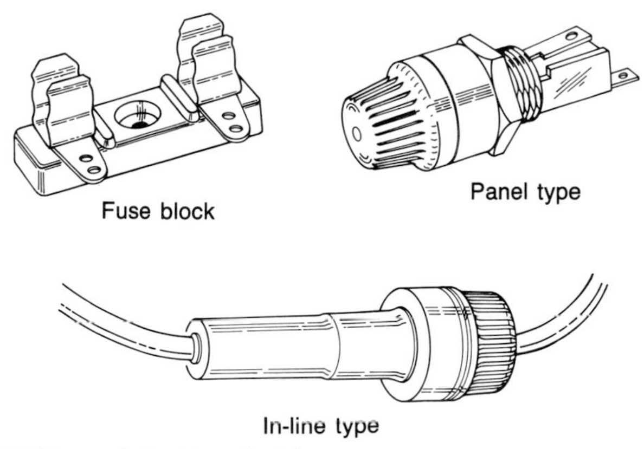

**FIGURE 3-8 Fuse holders.**

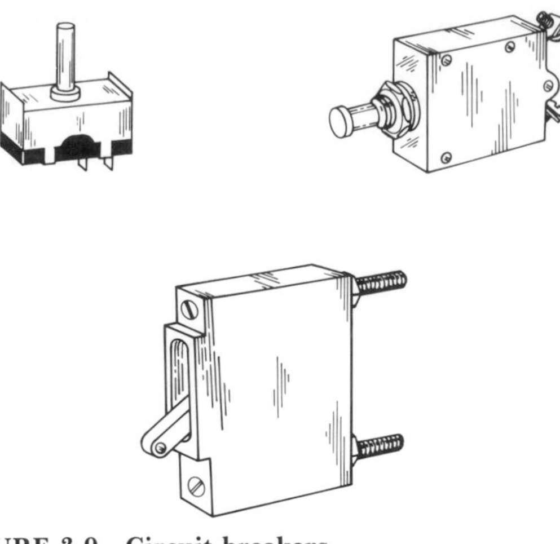

FIGURE 3-9 Circuit breakers.

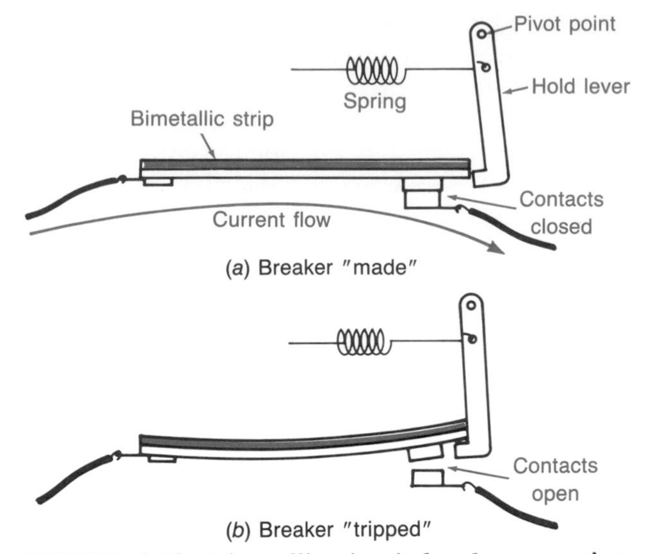


FIGURE 3-10 Bimetallic circuit breaker operation.

of the strip. When the strip gets hot, it bends upward due to the increased expansion of its bottom member. Excessive current flow heats the strip, causing it to bend up and open the contacts. The spring-loaded hold lever then moves into a position to catch the strip and prevent it from remaking contact when it cools down. A reset button must be pushed to move the hold lever aside and allow the contacts to close again. If the hold lever is eliminated, the breaker resets automatically when the bimetallic strip cools. Other circuit breakers work on a magnetic principle, and some use both thermal and magnetic principles for improved circuit protection. Magnetism is covered in a later chapter.

## Self-Test

**27.** Fuses and circuit breakers provide ______ protection for circuits.

**28.** A fuse element has resistance. It blows when the current through it heats it to a high temperature. Which equation describes the blowing mechanism?

**29.** Refer to Table 3-2. How would a ceramic tube fuse that is $\frac{1}{4}$ in. in diameter and $1\frac{1}{4}$ in. long be designated?

**30.** In addition to physical size, name three fuse ratings.

**31.** Refer to Fig. 3-7. How long will it take a slo-blo fuse to open if it is rated at 1 A and it is conducting 2 A? How long will it take a rectifier fuse to open under the same conditions?

**32.** Is it ever safe to use a 250-V fuse in a 350-V circuit?

**33.** Which fusing characteristic is most appropriate for protecting a device such as a motor that demands a surge of current when first turned on?

**34.** Fuses must be ______ when operated in high-temperature environments.

**35.** Thermal circuit breakers use a ______ strip to open the circuit.

**36.** Circuit breakers may be manual reset or ______ reset.

**37.** A circuit breaker that is open is said to be ______.

## SUMMARY

1. Power is the rate of doing work or the rate of converting energy.
2. The SI unit of power is the watt and is equal to one joule per second.
3. Electric power is equal to the product of voltage and current:

$$P = V \times I$$

4. The word *dissipation* usually refers to the conversion of electric energy to heat energy in a resistance.
5. The greater the power dissipation, the greater the temperature rise in a conductor or in a component.
6. Ohm's law can be used to derive other forms of the power equation:

$$P = I^2R$$

$$P = \frac{V^2}{R}$$

7. The horsepower is an old unit that is still used today:

$$1 \text{ hp} = 550 \text{ lb} \cdot \text{ft/s}$$

8. 1 hp = 746 W.
9. Energy = power × time.
10. Electric energy can be measured in watt-seconds (joules) or kilowatt-hours (kWh).
11. The utility companies bill their customers for energy, not for power.
12. Efficiency is the useful output divided by the input:

$$\eta = \frac{W_o}{W_i}$$

$$\eta = \frac{P_o}{P_i}$$

13. Efficiency may be stated in percent form.
14. A battery is made up of two or more cells.
15. The schematic symbol for a battery shows two cells regardless of the actual number of cells used.
16. A cell or a battery supplies direct current (dc) while a wall outlet supplies alternating current (ac).
17. The frequency of alternating current at wall outlets in the United States is 60 Hz.
18. A *short circuit* is one with a very low load resistance.
19. Short circuits may cause extremely high fault currents.
20. Fuses are used to protect circuits and devices from overcurrent.
21. A *blown* fuse is one that has opened.
22. Fuses have four important characteristics: (a) physical size, (b) current rating, (c) voltage rating, and (d) fusing characteristic.
23. Slo-blo fuses have a longer blowing time than normal-blo fuses.
24. All fuses, especially the slo-blo types, must be downrated for operation at high temperatures.

## CHAPTER REVIEW QUESTIONS

**3-1.** The standard frequency for power distribution in the United States is ______ Hz.

**3-2.** Which fusing characteristic is desirable for a load that draws a surge of current when first energized?

**3-3.** Circuit breakers ______ when excess current flows through them.

**3-4.** Thermal circuit breakers use a ______ strip to sense current flow.

**3-5.** Circuit breakers are available in ______ and automatic reset models.

## CHAPTER REVIEW PROBLEMS

**3-1.** An electric load converts 15 J of energy in 100 ms. Calculate the power.

**3-2.** An energy source develops 168 V and supplies 475 mA of current. Calculate its power output.

**3-3.** How much current must be provided by a 12.6-V battery to deliver 395 W of power?

**3-4.** How much voltage is required to develop 746 W of power at a current flow of 14.8 A.

**3-5.** Calculate the maximum safe current through a 0.1-Ω, 5-W resistor.

**3-6.** Calculate the maximum safe voltage across a 4.7-kΩ, 0.25-W resistor.

**3-7.** A motor is rated at 2 hp output. What is its output in kilowatts?

**3-8.** Calculate the energy used by a 50-W lamp in 1 h. State your answer in joules.

**3-9.** Calculate the daily cost of operating a 1500-W heater that runs continuously if the energy is priced at 5 cents per kilowatthour.

**3-10.** An electric lamp develops 45 W of useful output for 85 W of input. Calculate its percent efficiency.

**3-11.** Calculate the percent efficiency of a motor that produces $\frac{3}{4}$ hp of mechanical output for 850 W of input.

**3-12.** What is the output voltage of a battery that uses four 1.5-V cells?

**3-13.** A 12-V power supply is set to current-limit at 225 mA. What is the *smallest* value of load resistance that can be connected to the supply before it goes into current limiting?

## ANSWER TO SELF-TESTS

**1.** rate
**2.** SI
**3.** one joule per second
**4.** energy (or work)
**5.** 750 J, 75 W
**6.** 150 W; no
**7.** 600 W
**8.** 2 A
**9.** yes
**10.** 144 Ω
**11.** 0.71 mA
**12.** 3.1 A
**13.** energy or work
**14.** 677 W · s, or 677 J
**15.** $89.28

**16.** input

**17.** 78.4 percent

**18.** 6200 J; most of it is converted to heat

**19.** 1 hp

**20.** current capacity

**21.** positive

**22.** 12.6 V

**23.** alternating

**24.** **a.** 31.9 mA
**b.** 150 mA
**c.** 1 A
**d.** 1 A
**e.** 1 A

**25.** resistance

**26.** current-limited

**27.** overcurrent

**28.** $P = I^2R$

**29.** 3AB

**30.** current, voltage, fusing characteristic

**31.** 20 s; 0.6 s

**32.** yes

**33.** slo-blo

**34.** downrated

**35.** bimetallic

**36.** automatic

**37.** tripped

# Chapter 4 Electrical Materials

The materials used in electricity and electronics can be broadly divided into three major types: (1) conductors, (2) semiconductors, and (3) insulators. Each material has characteristics that make it useful for certain functions in constructing circuits and devices. It has already been mentioned that conductors are used to support the flow of current and insulators are used to confine the current to the desired path. This chapter presents a more detailed look at common electrical materials.

## 4-1 CHARACTERISTICS

The atoms that make up solid materials are held together by *bonds*. The bonds can be ionic, covalent, or metallic. With ionic bonding, electrostatically charged atoms (ions) are held together by the attraction of opposite charges. Ionic bonds lead to relatively simple structures such as sodium chloride (table salt). Covalent bonds are based on electron sharing among neighboring atoms. A valence orbit with eight electrons is considered stable. Nature always seeks stability, and atoms with fewer than eight valence electrons will attempt to arrange themselves in such a way that the valence electrons can be shared to achieve a stable count of eight. Metallic bonds are based on a structure of positive metal ions surrounded by a cloud of electrons.

Covalent bonds and metallic bonds account for the bulk of the materials used in electric circuits and devices. Insulators are predominantly covalent

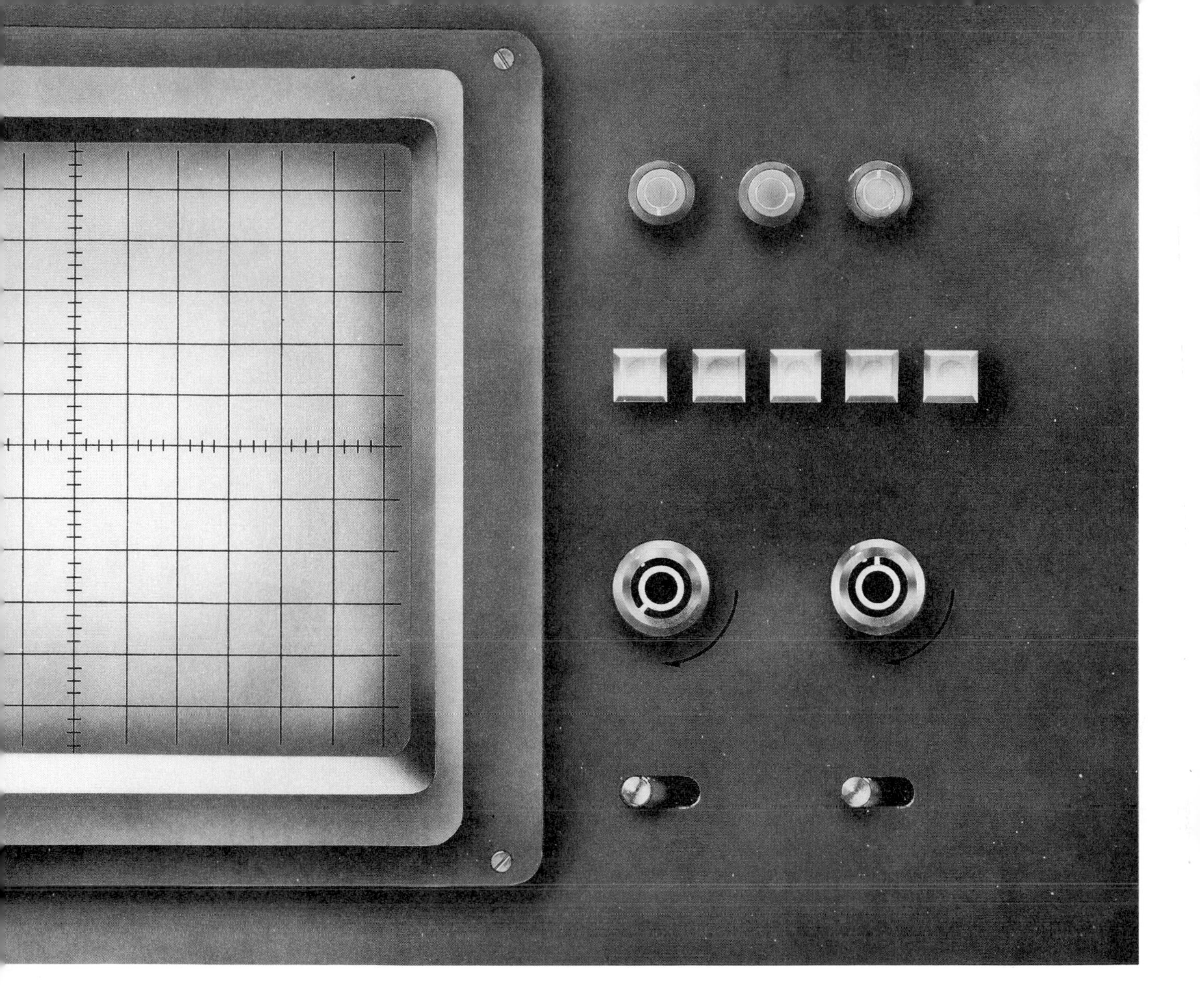

compounds. The valence electrons are tightly locked into bonds with neighboring atoms and are not available to support the flow of current. Semiconductors are also covalent materials and also tend toward insulation rather than conduction. They can be altered chemically to make some limited number of current carriers available. Semiconductors are covered in detail in a subsequent chapter. Metallic bonds form conductors. There are many free electrons available to provide good thermal and electrical conductivity.

Figure 4-1 shows energy band diagrams for conductors, semiconductors, and insulators. The lowest energy level for an electron is next to the nucleus of the atom. The electrons in intermediate orbits are at a higher energy level and the electrons in the valence orbit are at the highest energy level. Figure 4-1*a* shows that the valence band of a conductor overlaps with the conduction band. This means that the valence electrons are at a sufficient energy level to support the flow of electric current. Figure 4-1*b* shows an energy gap between the valence band and the conduction band for semiconductors. The valence electrons are not at an energy level high enough to support conduction. Figure 4-1*c* shows an even wider energy gap for insulators. The valence electrons fall quite short of being able to support conduction.

Conductors conduct and insulators insulate because of the way the atoms are bonded together. There are no absolutes, however. We already know that the very best conductors have some resistance.

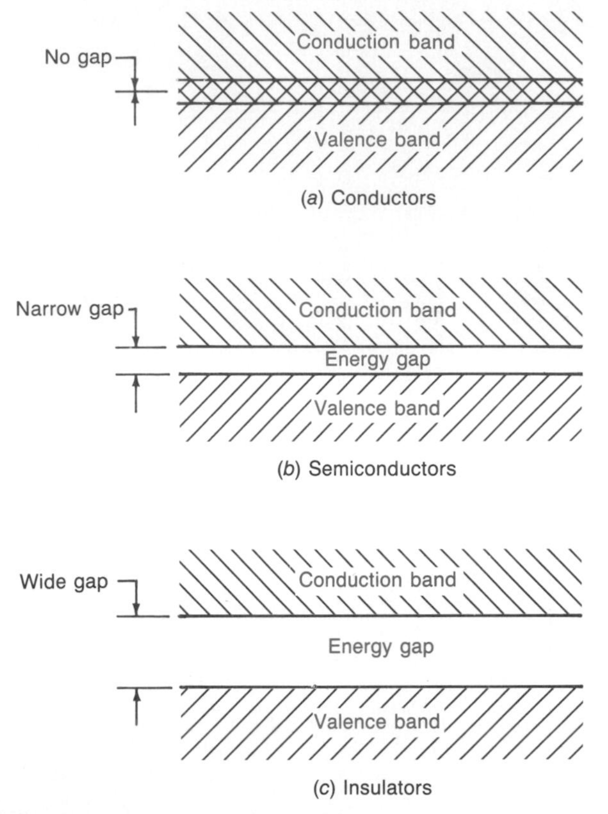


**FIGURE 4-1 Energy band diagrams.**

Insulators are not perfect either, and even the best exhibit some conduction. The valence electrons in an insulator can gain enough energy (thermal or electric) to jump the gap and enter the conduction band as shown in Fig. 4-2. At room temperature, there is enough thermal energy for a few electrons to be available to support current flow. This flow is called *leakage current* and is illustrated in Fig. 4-3. The leakage current is such a small percentage of the load current that it is often insignificant. Leakage does increase as insulation deteriorates with age and as it becomes contaminated with moisture and other materials. Increased leakage can interfere with the proper and safe operation of equipment. Leakage testers may be used to determine the integrity of insulation materials. They are sometimes called "hipot" testers because they use high potentials. Ordinary ohmmeters are not adequate for all leakage tests because they use a potential difference that is too low to detect some types of insulation defects.

Insulators do fail if subjected to excessive potential difference. With high voltage applied, some of the valence electrons gain enough energy to jump the energy gap and move into the conduction band. They rapidly accelerate toward the positive side of the energy source. Collisions with other atoms occur, and additional electrons gain enough energy to move into the conduction band. An avalanche of electrons is produced, which breaks down the insulator and destroys it. Table 4-1 lists the breakdown voltages for some popular insulating materials. The ratings are in kilovolts per centimeter. The thicker the insulating material, the greater the breakdown rating.

**Table 4-1 Typical Breakdown Voltages at 25°C**

| Material | kV/cm* |
|---|---|
| Mica | 2000 |
| Glass | 850 |
| Teflon | 500 |
| Quartz | 400 |
| Plexiglas | 350 |
| Polyvinylchloride | 300 |
| Polyethylene | 275 |
| Rubber | 270 |
| Polystyrene | 250 |
| Neoprene | 200 |
| Nylon | 160 |
| Transformer oil | 150 |
| Bakelite | 140 |
| Porcelain | 75 |
| Dry air (at atmospheric pressure) | 30 |

*Varies with exact chemical composition of material.

**EXAMPLE 4-1**

Calculate the insulation breakdown rating of an electric wire that is coated with 0.15 mm of rubber.

**Solution**

Table 4-1 shows that rubber is rated at 270 kV/cm. Millimeters can be converted to centimeters by multiplying by 0.1:

$$\text{Breakdown} = 0.1 \times 0.15\ \cancel{\text{cm}} \times \frac{270\ \text{kV}}{\cancel{\text{cm}}}$$
$$= 4.05\ \text{kV}$$

The wire would not be considered safe at 4 kV. Insulation materials are ordinarily derated by a factor of 10 or more. This particular wire would most likely be rated at less than 400 V.

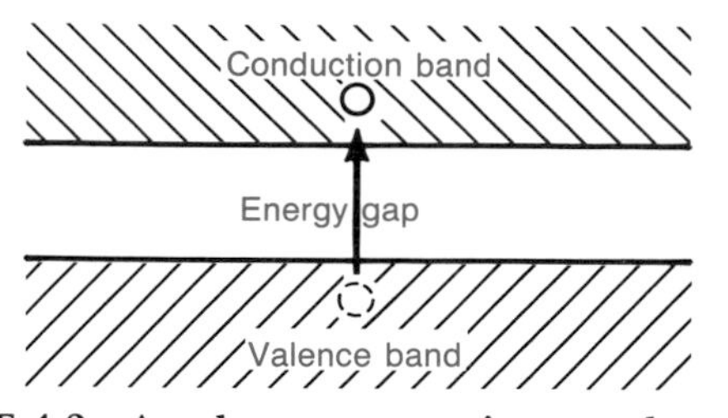


**FIGURE 4-2 An electron can gain enough energy to jump the energy gap and enter the conduction band.**

## EXAMPLE 4-2

Calculate the voltage required to arc across a 1-in. air gap.

### Solution

An *arc* results when air breaks down and supports the flow of electric current. The current flow heats the air molecules and light is produced. Table 4-1 lists a breakdown rating of 30 kV/cm for air. One inch is equal to 2.54 cm:

$$\text{Breakdown} = 2.54 \times 30 \text{ kV} = 76 \text{ kV}$$

Covalent materials have a *negative temperature coefficient.* This includes both insulators and semiconductors. The temperature coefficient is negative because there is an inverse relationship between resistance and temperature. The resistance drops as temperature goes up. Figure 4-4 shows temperature coefficients in graphical form. Leakage current is larger at higher temperatures. Increased thermal energy allows more electrons to jump the energy gap and become available to support the flow of current. Materials at a high temperature also tend to change chemically. For example, if a wire overheats, its insulation can become charred due to increased oxidation. The insulating capabilities are drastically and permanently reduced. A pungent odor often provides a clue when this occurs.

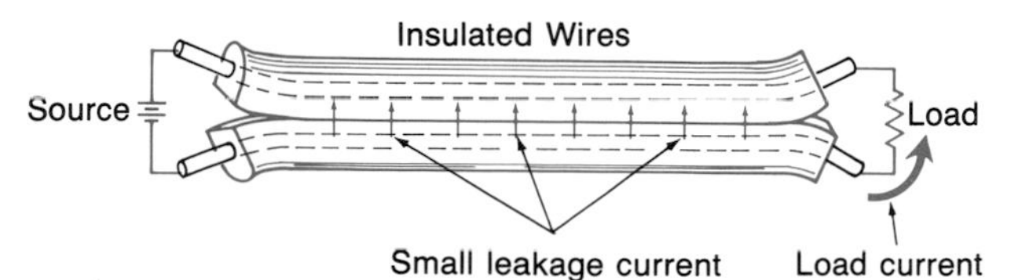


**FIGURE 4-3 A small leakage current flows in the insulation.**

Figure 4-5 shows some examples of insulators. Standoffs are used to support conductors or devices and provide insulation to the mounting plane. Feedthroughs are used to make a connection through sheet material, such as metal, and provide isolation from the sheet. Grommets allow a wire to pass through sheet material, provide insulation, and protect the wire from sharp edges. Strain reliefs are similar to grommets but clamp the wire and hold it securely in place. Caps, boots, and tubing are used to insulate splices, clips, and wires. Wire nuts are used to make solderless splices and eliminate the need for insulating tape. Solid-state devices, such as transistors, must often be mounted to a heat sink or a metal chassis to transfer thermal energy. Insulating washers and bushings may be required to electrically isolate the solid-state device from the mounting plane. The washers used to insulate solid-state devices are often made of mica or Teflon.

Insulation is an important part of electric circuits. It provides proper and safe operation. Look for signs of burning, cracking, deteriorating, and missing insulation. Take corrective action when necessary.

## Self-Test

**1.** Which two types of chemical bonding account for most of the materials used in electricity and electronics?

**2.** Which type of bonding makes the most electrons available to support current flow?

**3.** In an insulator or semiconductor, the valence electrons are separated from the conduction band by an _______.

**4.** Leakage current tends to _______ as temperature increases.

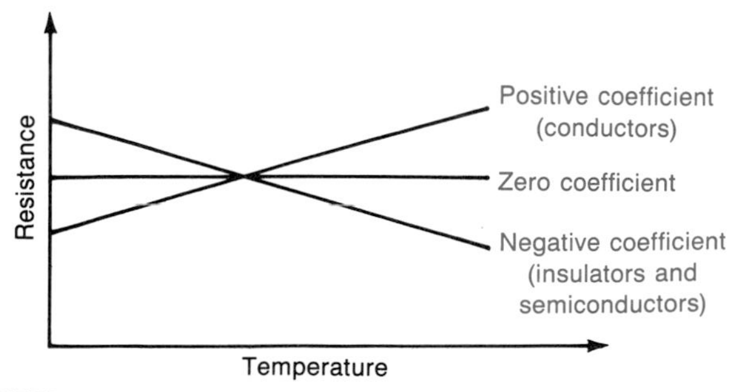


**FIGURE 4-4 Temperature coefficients.**

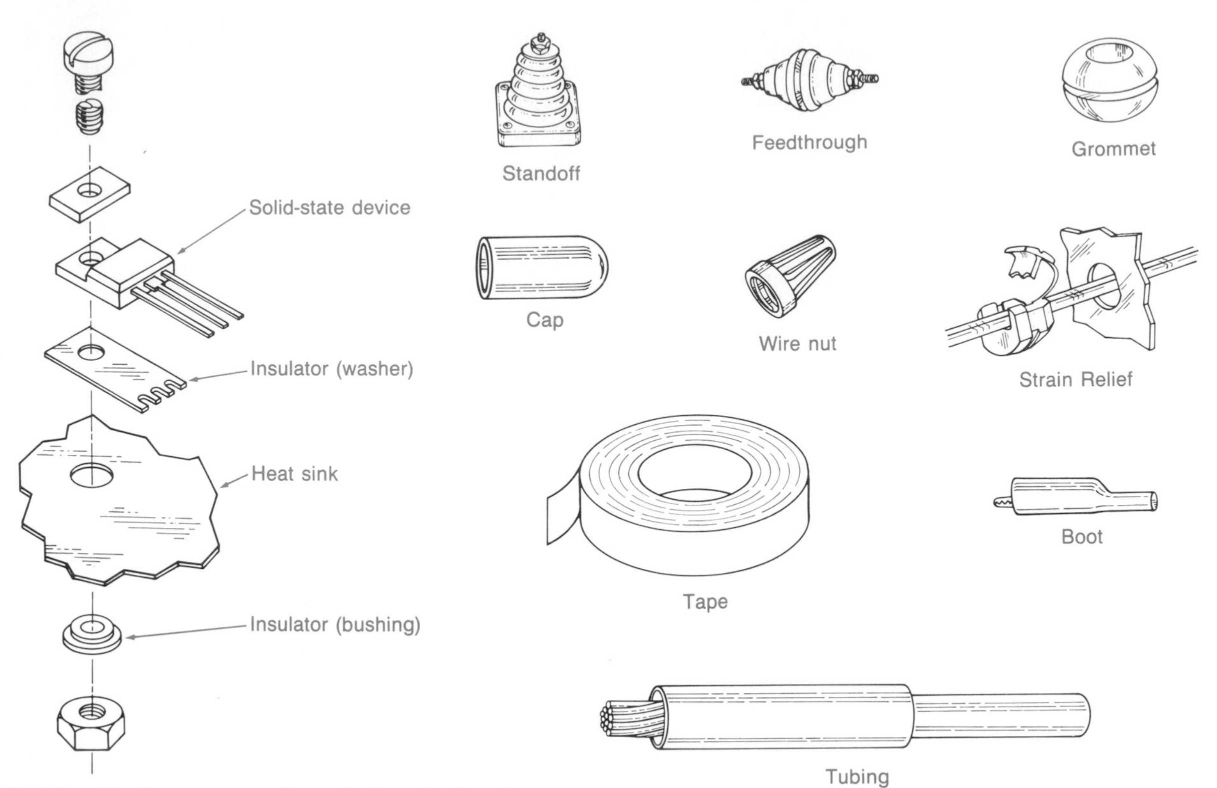


**FIGURE 4-5 Some examples of electric insulators.**

**5.** Insulators and semiconductors have a ______ temperature coefficient.

**6.** Breakdown results when an insulator is subjected to excess ______.

**7.** Refer to Table 4-1. Calculate the breakdown point for a mica washer that is 0.003 in. thick.

## 4-2 RESISTIVITY

The *resistivity* of a material is based on the resistance of a defined volume of that material. The letter symbol for resistivity is $\rho$ (the Greek letter rho). Resistivity may also be called specific resistance. Figure 4-6 shows that resistivity can be based on a cubic centimeter of the material. The SI unit of resistivity is based on the cubic meter. This is a rather large volume for evaluating common electrical materials, so the cubic centimeter is more commonly used. Table 4-2 lists the resistivity in ohm-centimeters ($\Omega \cdot$ cm) for some of the materials used in electricity and electronics. It also lists values in units of circular mil–ohms per foot (cmil $\cdot\ \Omega$/ft) for the conductor group. These values are utilized later in this section.

**Table 4-2 Resistivity of Selected Electrical Materials**

| Material | $\rho$ (resistivity) at 20°C | |
|---|---|---|
| | $\Omega \cdot$ cm | cmil $\cdot\ \Omega$/ft |
| **Conductors** | | |
| Silver | $1.65 \times 10^{-6}$ | 9.9 |
| Copper | $1.72 \times 10^{-6}$ | 10.37 |
| Gold | $2.44 \times 10^{-6}$ | 14.7 |
| Aluminum | $2.83 \times 10^{-6}$ | 17.0 |
| Tungsten | $5.49 \times 10^{-6}$ | 33.0 |
| Nickel | $7.81 \times 10^{-6}$ | 47.0 |
| Iron | $1.23 \times 10^{-5}$ | 74.0 |
| Constantan | $4.90 \times 10^{-5}$ | 295 |
| Nichrome | $9.97 \times 10^{-5}$ | 600 |
| **Semiconductors** | | |
| Carbon | $3.49 \times 10^{-3}$ | |
| Germanium | $4.70 \times 10^{1}$ | |
| Silicon | $6.40 \times 10^{4}$ | |
| **Insulators** | | |
| Polyvinylchloride | $>10^{10}$ | |
| Mica | $>10^{12}$ | |
| Teflon | $>10^{15}$ | |
| Quartz | $>10^{17}$ | |

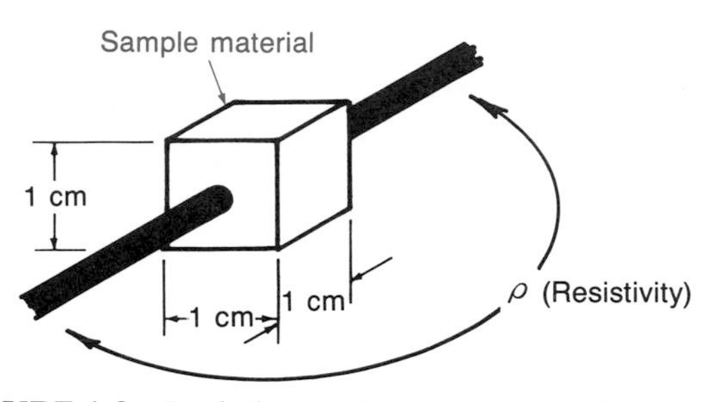


**FIGURE 4-6 Resistivity is based on the resistance of a volume of the sample material.**

The inverse of resistivity is called *conductivity*. The letter symbol for conductivity is $\sigma$ (the Greek letter sigma):

$$\sigma = \frac{1}{\rho}$$

Conductivity is expressed in siemens per meter (S/m) in the SI system.

Table 4-2 shows that the range of resistivities is very large. For example, 1 $cm^3$ of silver has a resistance of $1.65 \times 10^{-6}\ \Omega$, while 1 $cm^3$ of quartz has a resistance greater than $10^{17}\ \Omega$. This represents a range of $10^{23}$. There is even a considerable range within the conductor group. Nichrome (an alloy of nickel, chrome, and iron) has a resistivity that is approximately 58 times greater than that of copper. For this reason, nichrome wire is called resistance wire and is used to make resistors and heating elements. Table 4-2 also shows that semiconductors are intermediate. Their resistivities are greater than those of the conducting group and less than those in the insulating group.

It is possible to calculate the resistance of a structure with the following equation:

$$R = \frac{\rho l}{A}$$

where $R$ = resistance, $\Omega$
$\rho$ = resistivity, $\Omega \cdot$ cm
$l$ = length, cm
$A$ = cross-sectional area, $cm^2$

---

**EXAMPLE 4-3**

Calculate the resistance of a copper trace on a printed circuit board if the trace is 10 cm long, 0.003 cm thick, and 0.2 cm wide.

**Solution**

Table 4-2 lists the resistivity of copper as $1.72 \times 10^{-6}\ \Omega \cdot$ cm. Entering the data into the equation gives

$$R = \frac{1.72 \times 10^{-6} \times 10}{0.003 \times 0.2} = 28.7\ \text{m}\Omega$$

---

Note that if the trace were 20 cm long, the resistance would be twice as much. The resistance of a material is directly proportional to its length. The resistance can be decreased by making the trace wider or thicker, which would increase the area. This is an important point. When conductors are long, it is often necessary to increase their cross-sectional area to maintain a low resistance and a low voltage drop.

---

**EXAMPLE 4-4**

Calculate the power dissipated by a nichrome heating element that is 600 cm long, has a cross-sectional area of 0.01 $cm^2$, and is connected to a 120-V energy source.

**Solution**

The first step is to find the resistance of the heating element. Table 4-2 lists the resistivity of nichrome as $9.97 \times 10^{-5}\ \Omega \cdot$ cm. Entering the data into the equation gives

$$R = \frac{9.97 \times 10^{-5} \times 600}{0.01} = 5.98\ \Omega$$

Now the power can be calculated with

$$P = \frac{V^2}{R} = \frac{120^2}{5.98}$$
$$P = 2.4\ \text{kW}$$

---

The resistivities listed in Table 4-2 are for a temperature of 20°C. A heating element that dissipates several kilowatts is obviously going to reach a much higher temperature than 20°C. Therefore, our calculations are not realistic. Conductors, including nichrome, have a positive temperature coefficient. Their resistance goes up as temperature goes up. The heating element will reach a resistance greater than 5.98 Ω when it is operating. Temperature coefficients are covered in Sec. 4-3 of this chapter.

**EXAMPLE 4-5**
Calculate the resistance of a copper cube that is 1 m per side.

**Solution**
There are 100 cm in a meter. Entering the data into the equation gives

$$R = \frac{1.72 \times 10^{-6} \times 100}{100 \times 100} = 1.72 \times 10^{-8}\,\Omega$$

This is the SI resistivity for copper. To convert the ohm-centimeter resistivities in Table 4-2 to ohm-meters, multiply by $1 \times 10^{-2}$.

*AWG* is the abbreviation for *American wire gage.* It is also known as the *Brown and Sharpe (B & S) gage.* It is a nonmetric system and is based on the *mil,* which is one-thousandth of an inch. Wire is usually round, and the cross-sectional area is measured in units of *circular mils* (cmils). Figure 4-7*a* shows that a wire with a diameter of 1 mil has a cross-sectional area equal to 1 cmil. The circular mil area of round wire is equal to the square of the diameter in mils:

$$\text{CMA} = D^2$$

where CMA = circular mil area
$D$ = diameter in mils

**EXAMPLE 4-6**
What is the area in circular mils of a wire that is 0.04 in. in diameter?

**Solution**
Convert 0.04 in. to mils (thousandths of an inch) and square the results:

$$\begin{aligned}\text{CMA} &= (0.04 \times 1000)^2 \\ &= 1600\end{aligned}$$

Figure 4-7*b* shows that the area of a wire with a square cross section is greater than the area of a round wire with the same diameter. Square mils and circular mils are not equal. A conversion factor can be found with the standard equation for the area of a circle:

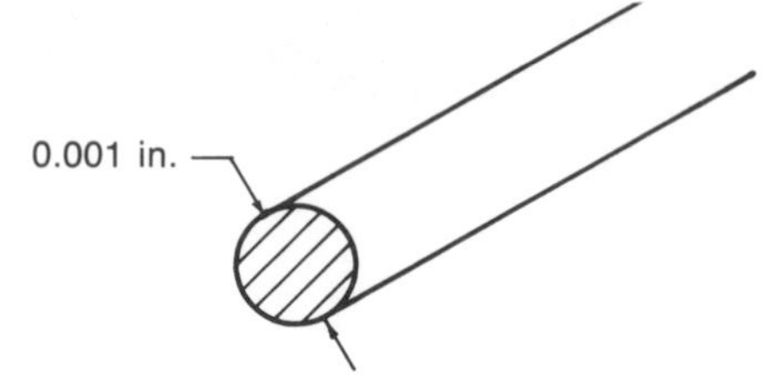


(*a*) A wire with a diameter of one-thousandth of an inch has a circular mil area of one.

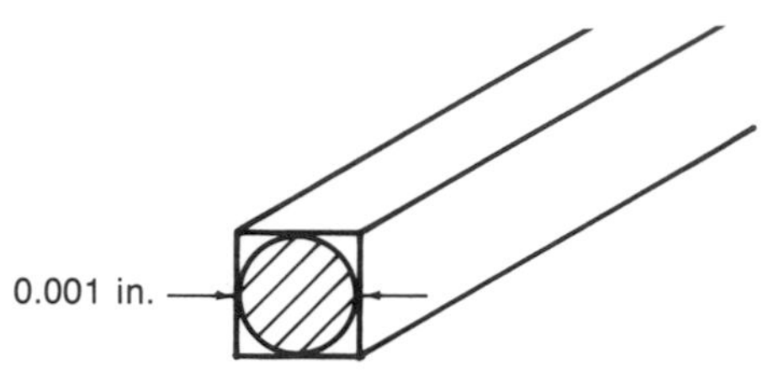


(*b*) A wire having a square cross section with a diameter of one-thousandth of an inch has a circular mil area greater than one.

**FIGURE 4-7 Circular mil area.**

$$\begin{aligned}A &= \pi\left(\frac{D}{2}\right)^2 \\ &= \frac{\pi D^2}{4}\end{aligned}$$

Dividing by $D^2$ (the area of a square) gives

$$\text{Square mil area} = \text{CMA} \times \frac{\pi}{4}$$

And, the inverse relationship

$$\text{CMA} = \text{square mil area} \times \frac{4}{\pi}$$

**EXAMPLE 4-7**
Find the CMA of a rectangular wire that measures $0.15 \times 0.035$ in. Also determine what diameter of round wire would be required to provide the same current-carrying capacity.

**Solution**
Determine the area of the wire in rectangular mils and then apply the conversion factor:

$$\begin{aligned}\text{Square mil area} &= (0.15 \times 1000) \times (0.035 \times 1000) \\ &= 5250\end{aligned}$$

$$\text{CMA} = 5250 \times \frac{4}{\pi} = 6685$$

Current-carrying capacity is based on the cross-sectional area of the wire. A round wire with an area of 6685 cmils will have the same current-carrying capacity. Rearranging the CMA equation:

$$D = \sqrt{\text{CMA}} = \sqrt{6685} = 82 \text{ mils (or 0.082 in.)}$$

The standard sizes of copper wire are shown in Table 4-3. Note that the wire becomes smaller as the gage number becomes larger. The wire table supplies useful data for each standard gage. For example, the resistance per unit length is important to predict voltage drop in long runs. Wire tables are also used to select a gage that will safely support the amount of current flow.

**EXAMPLE 4-8**

An electric motor must be operated at a distance of 200 ft from the nearest outlet. If the motor draws 15 A, select a gage of wire for an extension cord that will limit the line drop to 5 V.

**Solution**

Ohm's law is used to determine the maximum resistance of the extension cord:

$$R = \frac{V}{I} = \frac{5}{15}$$
$$R = 0.33\ \Omega$$

Next, the resistance per 1000 ft must be calculated so that Table 4-3 can be utilized. An extension cord has two current-carrying conductors. Therefore, the total wire length is 2 × 200 or 400 ft:

$$\frac{1000}{400} \times 0.33 = 0.83\ \Omega \text{ per 1000 ft}$$

Referring to the wire table shows that a no. 9 wire is required. It can be difficult to find wire in the odd gage sizes. A no. 8 wire would work but would be rather heavy and expensive. This illustrates the problem associated with operating substantial electric loads with long extension cords.

**EXAMPLE 4-9**

The coils used in certain electric devices, such as transformers, are often designed to have 700 cmils/A. What gage wire will be required for a transformer coil rated at 1.5 A.

**Solution**

Using the design criterion of 700 cmils/A:

$$700 \times 1.5 = 1050 \text{ cmils}$$

Table 4-3 shows that no. 20 wire comes close to the calculated value. Since odd gage wire may be difficult to find, no. 20 wire would probably be used to construct the transformer.

The design rule of 700 cmils/A is used in devices where the wire is coiled and adjacent turns are near each other. In fact, they are usually touching each other. This leads to considerable heat buildup. If the area is much less than 700 cmils/A, the insulation may be damaged by excess temperature. The wire in a building or in an appliance will probably run cooler and can safely conduct more current. General-purpose circuits in buildings are rated at 20 A and use no. 12 wire, which equates to approximately 330 cmils/A.

Wire tables for materials other than copper are rare. Refer back to Table 4-2 on page 74. The resistivities for common conductors are given in units of circular mil–ohms per foot. This provides a means of generating some of the same information found in the copper wire table for materials other than copper.

**EXAMPLE 4-10**

What is the resistance of a no. 20 nichrome wire that is 50 ft long?

**Solution**

We can use Table 4-3 to find the CMA of a no. 20 wire and Table 4-2 to find the resistivity of nichrome in circular mil–ohms per foot. Entering this information into the resistivity equation gives

**Table 4-3 Copper Wire Table**

| AWG (B&S) Gage | Standard Metric Size, mm | Diameter, mils | Cross-Sectional Area cmils | in.$^2$ | Ω per 1000 ft at 20°C (68°F) | Lb per 1000 ft | Ft/lb |
|---|---|---|---|---|---|---|---|
| 0000 | 11.8 | 460.0 | 211,600 | 0.1662 | 0.04901 | 640.5 | 1.561 |
| 000 | 11.0 | 409.6 | 167,800 | 0.1318 | 0.06180 | 507.9 | 1.968 |
| 00 | 9.0 | 364.8 | 133,100 | 0.1045 | 0.07793 | 402.8 | 2.482 |
| 0 | 8.0 | 324.9 | 105,500 | 0.08289 | 0.09827 | 319.5 | 3.130 |
| 1 | 7.1 | 289.3 | 83,690 | 0.06573 | 0.1239 | 253.3 | 3.947 |
| 2 | 6.3 | 257.6 | 66,370 | 0.05213 | 0.1563 | 200.9 | 4.977 |
| 3 | 5.6 | 229.4 | 52,640 | 0.04134 | 0.1970 | 159.3 | 6.276 |
| 4 | 5.0 | 204.3 | 41,740 | 0.03278 | 0.2485 | 126.4 | 7.914 |
| 5 | 4.5 | 181.9 | 33,100 | 0.02600 | 0.3133 | 100.2 | 9.980 |
| 6 | 4.0 | 162.0 | 26,250 | 0.02062 | 0.3951 | 79.46 | 12.58 |
| 7 | 3.55 | 144.3 | 20,820 | 0.01635 | 0.4982 | 63.02 | 15.87 |
| 8 | 3.15 | 128.5 | 16,510 | 0.01297 | 0.6282 | 49.98 | 20.01 |
| 9 | 2.80 | 114.4 | 13,090 | 0.01028 | 0.7921 | 39.63 | 25.23 |
| 10 | 2.50 | 101.9 | 10,380 | 0.008155 | 0.9989 | 31.43 | 31.82 |
| 11 | 2.24 | 90.74 | 8,234 | 0.006467 | 1.260 | 24.92 | 40.12 |
| 12 | 2.00 | 80.81 | 6,530 | 0.005129 | 1.588 | 19.77 | 50.59 |
| 13 | 1.80 | 71.96 | 5,178 | 0.004067 | 2.003 | 15.68 | 63.80 |
| 14 | 1.60 | 64.08 | 4,107 | 0.003225 | 2.525 | 12.43 | 80.44 |
| 15 | 1.40 | 57.07 | 3,257 | 0.002558 | 3.184 | 9.858 | 101.4 |
| 16 | 1.25 | 50.82 | 2,583 | 0.002028 | 4.016 | 7.818 | 127.9 |
| 17 | 1.12 | 45.26 | 2,048 | 0.001609 | 5.064 | 6.200 | 161.3 |
| 18 | 1.00 | 40.30 | 1,624 | 0.001276 | 6.385 | 4.917 | 203.4 |
| 19 | 0.90 | 35.89 | 1,288 | 0.001012 | 8.051 | 3.899 | 256.5 |
| 20 | 0.80 | 31.96 | 1,022 | 0.0008023 | 10.15 | 3.092 | 323.4 |
| 21 | 0.71 | 28.46 | 810.1 | 0.0006363 | 12.80 | 2.452 | 407.8 |
| 22 | 0.63 | 25.35 | 642.4 | 0.0005046 | 16.14 | 1.945 | 514.2 |
| 23 | 0.56 | 22.57 | 509.5 | 0.0004002 | 20.36 | 1.542 | 648.4 |
| 24 | 0.50 | 20.10 | 404.0 | 0.0003173 | 25.67 | 1.223 | 817.7 |
| 25 | 0.45 | 17.90 | 320.4 | 0.0002517 | 32.37 | 0.9699 | 1,031.0 |
| 26 | 0.40 | 15.94 | 254.1 | 0.0001996 | 40.81 | 0.7692 | 1,300 |
| 27 | 0.355 | 14.20 | 201.5 | 0.0001583 | 51.47 | 0.6100 | 1,639 |
| 28 | 0.315 | 12.64 | 159.8 | 0.0001255 | 64.90 | 0.4837 | 2,067 |
| 29 | 0.280 | 11.26 | 126.7 | 0.00009953 | 81.83 | 0.3836 | 2,607 |
| 30 | 0.250 | 10.03 | 100.5 | 0.00007894 | 103.2 | 0.3042 | 3,287 |
| 31 | 0.224 | 8.928 | 79.70 | 0.00006260 | 130.1 | 0.2413 | 4,145 |
| 32 | 0.200 | 7.950 | 63.21 | 0.00004964 | 164.1 | 0.1913 | 5,227 |
| 33 | 0.180 | 7.080 | 50.13 | 0.00003937 | 206.9 | 0.1517 | 6,591 |
| 34 | 0.160 | 6.305 | 39.75 | 0.00003122 | 260.9 | 0.1203 | 8,310 |
| 35 | 0.140 | 5.615 | 31.52 | 0.00002476 | 329.0 | 0.09542 | 10,480 |
| 36 | 0.125 | 5.000 | 25.00 | 0.00001964 | 414.8 | 0.07568 | 13,210 |
| 37 | 0.112 | 4.453 | 19.83 | 0.00001557 | 523.1 | 0.06001 | 16,660 |
| 38 | 0.100 | 3.965 | 15.72 | 0.00001235 | 659.6 | 0.04759 | 21,010 |
| 39 | 0.090 | 3.531 | 12.47 | 0.000009793 | 831.8 | 0.03774 | 26,500 |
| 40 | 0.080 | 3.145 | 9.888 | 0.000007766 | 1049.0 | 0.02993 | 33,410 |

$$R = \frac{\rho l}{A}$$
$$= \frac{600 \times 50}{1022}$$
$$R = 29.4\ \Omega$$

## Self-Test

**8.** The SI unit of resistivity is the ________.

**9.** The reciprocal of resistivity is called ________.

**10.** Calculate the resistance of a copper bar 1.5 cm × 0.5 cm × 8.5 m long.

**11.** The resistance of a material ________ as its length decreases.

**12.** The resistance of a material ________ as its cross-sectional area increases.

**13.** Convert the resistivity of tungsten from Table 4-2 into SI units.

**14.** A round wire is 0.162 in. in diameter. What is its CMA?

**15.** Refer to Table 4-3. What is the AWG number for the wire described in Question 14?

**16.** A wire with a square cross section has a diameter of 0.080 in. What is its area in square mils? In circular mils?

**17.** What AWG size would come closest to having the same current-carrying capacity as the wire described in Question 16?

**18.** A 10-A load is operated from a no. 14 copper extension cord that is 100 ft in length. What is the load potential difference if the source develops 120 V?

**19.** What is the resistance of a no. 10 aluminum wire that is 150 ft long? How does this compare to a copper wire of the same size?

**20.** Why is the current-carrying capacity of wire reduced when it is used to build devices such as transformer coils?

## 4-3 TEMPERATURE COEFFICIENT

Almost all conductors have a positive temperature coefficient. Their resistance increases as temperature increases. The alloy constantan is an exception with a temperature coefficient near zero. Table 4-4 lists the temperature coefficients of some common metals at 0°C. The following equation is useful for predicting resistance at one temperature when the resistance at another temperature is known:

**Table 4-4 Temperature Coefficients of Selected Metals at 0°C**

| Material | Temperature Coefficient, Ω/°C per Ohms Resistance 0°C |
|---|---|
| Aluminum | 0.00420 |
| Brass (annealed) | 0.00208 |
| Copper (annealed) | 0.00426 |
| Gold | 0.00365 |
| Iron | 0.00618 |
| Lead | 0.00466 |
| Mercury | 0.00068 |
| Nichrome | 0.00044 |
| Nickel | 0.006 |
| Platinum | 0.0037 |
| Silver | 0.00411 |
| Steel (soft) | 0.00458 |
| Tin | 0.00458 |
| Tungsten | 0.0049 |
| Zinc | 0.0040 |

$$R_1 = R_2 \times \frac{1 + \alpha T_1}{1 + \alpha T_2}$$

where $R_1$ = resistance at $T_1$
$R_2$ = resistance at $T_2$
$\alpha$ = temperature coefficient at 0°C
$T_1, T_2$ = temperatures, °C

### EXAMPLE 4-11

A copper wire has a resistance of 0.135 Ω at 20°C. What will its resistance be at 100°C?

**Solution**

Table 4-4 lists the temperature coefficient of copper as 0.00426. Entering the data into the equation gives

$$R_1 = 0.135 \times \frac{1 + (0.00426 \times 100)}{1 + (0.00426 \times 20)}$$
$$= 0.177\ \Omega$$

The resistance in a conductor is due to collisions between the current-carrying electrons and the positive ions that make up the metallic bonds. The ions vibrate due to thermal energy. Increased vibration produces more collisions. Absolute zero is the total absence of thermal energy. There will be very few collisions at temperatures near absolute zero, and the resistance should also approach zero. If the numerator in the temperature coefficient equation equates to zero, then $R_1$ will equal zero. This condition is known as *superconductivity*. It is a simple matter to determine the temperature at which this will occur:

$$1 + \alpha T_1 = 0$$

Solving for $T_1$ and using the temperature coefficient of copper gives

$$T_1 = \frac{-1}{0.00426} = -235°\text{C}$$

Absolute zero is −273°C. The value derived in the foregoing procedure is the inferred value of absolute zero. Refer to Fig. 4-8. The temperature coefficient is equal to the slope of a straight-line approximation of the actual resistance versus temperature response of copper. The curve does reach zero resistance at −273°C, and the straight line reaches zero at −235°C. The actual curve is close enough to linear so that reasonable accuracy is obtained near 0°C. However, the error increases at temperatures far removed from 0°C.

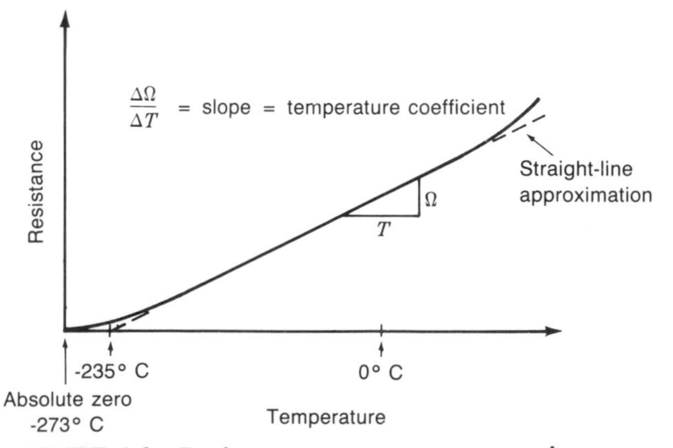


**FIGURE 4-8 Resistance versus temperature in a copper conductor.**

---

**EXAMPLE 4-12**

A 120-V, 100-W lamp has a tungsten filament with 9.7 Ω of resistance at 25°C. Calculate the operating temperature of the filament.

**Solution**

The operating resistance of a lamp is much greater than its cold resistance. The operating resistance is found by

$$R = \frac{V^2}{P} = \frac{120^2}{100}$$
$$R = 144\ \Omega$$

Table 4-4 lists the temperature coefficient of tungsten as 0.0049. The temperature coefficient equation can be rearranged to solve for $T_1$:

$$T_1 = \frac{R_1(1 + \alpha T_2)}{\alpha R_2} - \frac{1}{\alpha} = \frac{144(1 + 0.0049 \times 25)}{0.0049 \times 9.7} - \frac{1}{0.0049}$$
$$T_1 = 3200°\text{C}$$

---

This answer is based on a straight-line approximation similar to the situation for copper shown in Fig. 4-8. The approximation is accurate for temperatures near 0°C. The melting point of tungsten is 3382°C. It would not be feasible to run a filament near its melting point. The actual operating temperature of the lamp is less than 3200°C.

Although the calculation for the lamp temperature is perhaps a poor approximation, the other facts are correct. The resistance does increase from roughly 10 to 144 Ω as the filament comes up to operating temperature. This represents a resistance increase greater than 14 times. Of course, this means that the lamp will conduct more than 14 times its operating current at the moment of turn-on. This is known as a *surge current*. You have probably noticed that electric lamps often burn out at the moment of turn-on. This is due to the surge current.

A standard reference temperature is 20°C. If you consult other sources, you are likely to find temperature coefficients listed for a temperature of 20°C. They will be slightly different from those shown in Table 4-4. For example, the temperature coefficient of copper at 20°C is 0.0039. The following equation applies for a 20° reference temperature:

$$R_X = R(1 + \alpha \Delta T)$$

where $R_X$ = resistance at the operating temperature
$R$ = resistance at 20°C
$\alpha$ = temperature coefficient at 20°C
$\Delta T$ = difference in operating temperature from 20°C

**EXAMPLE 4-13**

Rework Example 4-11 using the 20°C temperature coefficient and equation.

**Solution**

Entering the temperature coefficient 0.0039 and the data from Example 4-11 into the equation produces

$$R_X = 0.135(1 + 0.0039 \times 80) = 0.177\ \Omega$$

This agrees with the results obtained previously. The advantage of Table 4-4 and the first equation presented is that the known resistance may be at any temperature.

**EXAMPLE 4-14**

A carbon electrode has a resistance of 0.125 Ω at 20°C. The temperature coefficient of carbon is −0.0005 at 20°C. What will the resistance of the electrode be at 85°C?

**Solution**

Entering the data into the equation gives

$$R_X = 0.125(1 - 0.0005 \times 65) = 0.121\ \Omega$$

Note that the resistance decreases due to the negative temperature coefficient of carbon which is a semiconductor.

## Self-Test

**21.** Refer to Table 4-4. A platinum wire has a resistance of 6.83 Ω at −20°C. What is its resistance at 150°C?

**22.** Superconductivity occurs at a temperature of ________.

**23.** The flow in an electric lamp at the moment of turn-on is called a ________ current.

**24.** The error in temperature coefficient calculations ________ as the actual temperature departs farther from the reference temperature.

**25.** Semiconductors have a ________ temperature coefficient.

## 4-4 WIRE AND CABLE

There is a wide array of wire and cable types available for various applications. Some applications are regulated by the National Electric Code® (NEC), which is published by the National Fire Protection Association (NFPA) and has been adopted by the American National Standards Institute (ANSI) and by many governmental organizations. The wiring in and to buildings is usually required to meet the specifications of the NEC. Figure 4-9 shows some examples of building wire and cable. Commercial and industrial wiring is usually required to be protected by metal tubing called *conduit.* Residential wiring does not require conduit, unless local regulations require it, and typically uses the nonmetallic-type cable shown in Fig. 4-9. The protective ground wire in this type of cable is not provided with a separate insulation but is protected by the thermoplastic jacket. Service entrance cable is used in overhead connections to supply buildings. It often uses aluminum conductors to reduce cost and weight. It uses a neutral conductor that is wrapped around the separately insulated conductors. Cable similar in appearance to the nonmetallic and service entrance types is also available for direct burial in the ground.

Most of the wires used in electronic applications are copper and are coated with another metal such as tin, tin alloy, lead, or silver. Electronic wiring is not normally regulated by the NEC but may be required to meet various military specifications or specifications set forth by other agencies. Many of

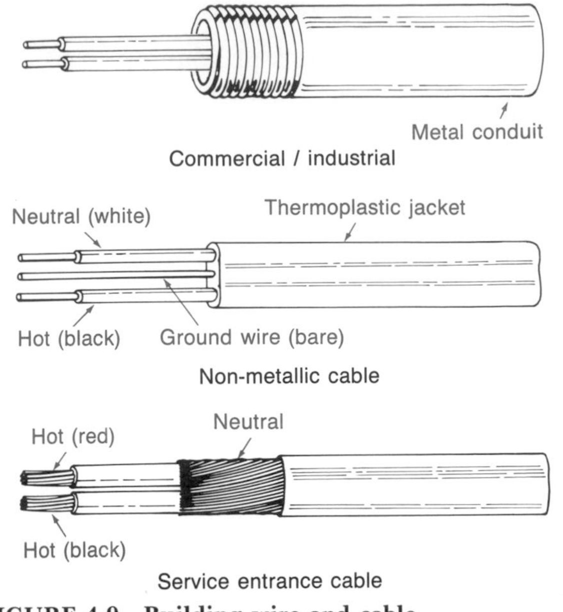


**FIGURE 4-9 Building wire and cable.**

the conductors used in electronics are made up of several strands of wire. Stranded wire is more flexible and vibration resistant than solid wire. Figure 4-10 shows a typical stranded wire. Note that seven strands are present. As additional layers are added, the number of strands per layer increases by six. Figure 4-11 shows the total number of strands and the number of strands per layer.

---

**EXAMPLE 4-15**

If a wire is made up of seven strands, each with a diameter of 0.0192 in., what is the total CMA of the wire and its equivalent gage? What is its current-carrying capacity?

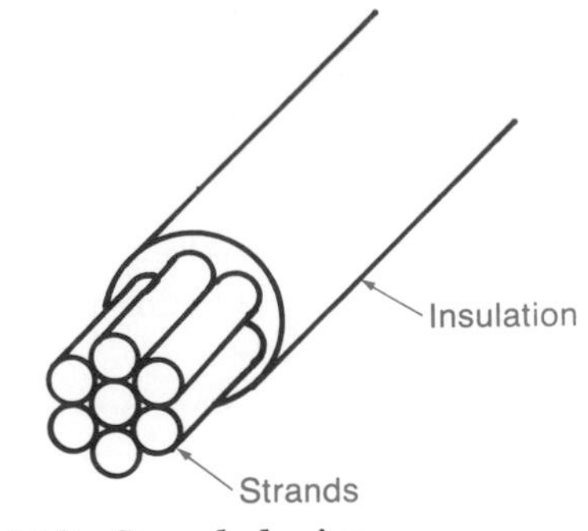


**FIGURE 4-10 Stranded wire.**

**Solution**

Find the CMA of one strand by converting the diameter to mils and squaring:

$$\text{CMA} = (0.0192 \times 1000)^2$$
$$= 369$$

Next, multiply by the number of strands to find the total CMA:

$$\text{Total CMA} = 7 \times 367$$
$$= 2583$$

Referring back to Table 4-3, it can be determined that this wire has approximately the same CMA as a no. 16 solid wire. It would therefore have the same current-carrying capacity as a no. 16 wire if all other factors are the same.

---

The current capacity of any given wire is dependent on several conditions:

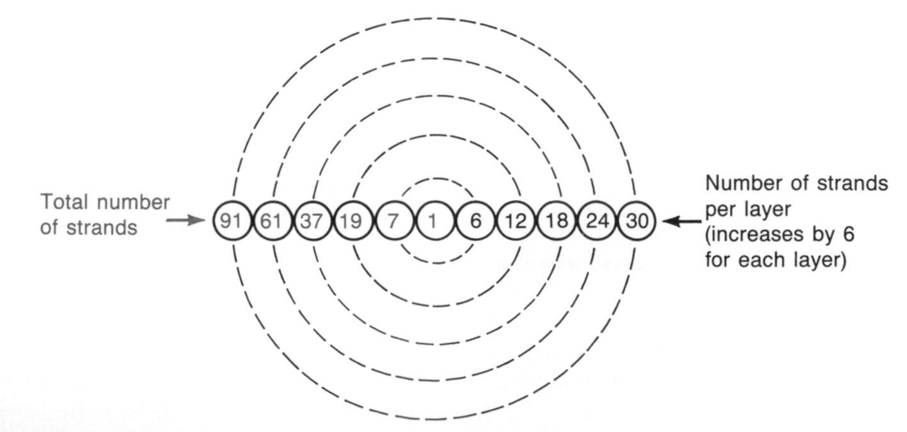


**FIGURE 4-11 Number of strands.**

**Table 4-5 Characteristics of Wire Insulating Materials**

| Material | Maximum Temperature, °C | Abrasion Resistance | Flexibility | Acid Resistance | Hydraulic Fluid Resistance | Flame Resistance |
|---|---|---|---|---|---|---|
| Teflon | 260 | Excellent | Good | Excellent | Excellent | Nonflammable |
| Polyethylene | 75 | Good | Good | Excellent | Fair | Slow-burning |
| Asbestos | 500 | Excellent | Fair | Poor | Good | Nonflammable |
| PVC | 105 | Good | Good | Excellent | Good | Self-extinguishing |
| Nylon | 105 | Excellent | Good | Poor | Good | Self-extinguishing |
| Silicone rubber | 200 | Poor | Excellent | Poor | Fair | Slow-burning |
| Butyl rubber | 90 | Poor | Excellent | Good | Fair | Slow-burning |

**1.** The wire gage. The smaller the gage number, the greater the CMA and the current capacity.

**2.** The insulation. The type of insulating material determines the maximum operating temperature of the wire and how much current it can safely support. Table 4-5 lists some of the important characteristics of insulation materials.

**3.** The proximity of other conductors. Bundled conductors add to the heat buildup and decrease the current capacity.

**4.** Ambient temperature. A high ambient temperature detracts from the ability of a wire to throw off thermal energy and therefore decreases its capacity.

**5.** Environmental conditions. A wire that must pass through combustible materials will have a lower safe operating temperature and reduced capacity. This is the case in residential wiring. Also, a wire in free air will have an increased current capacity.

**6.** Allowable voltage drop. Although a wire may operate safely at a given current, it may drop excessive voltage.

**Table 4-6 Current Rating of Wires in Amperes**

| AWG | Bundled | Free Air | Confined Space |
|---|---|---|---|
| 24 | 3 | 3.52 | 2.11 |
| 22 | 5 | 5 | 3 |
| 20 | 7.5 | 8.33 | 5 |
| 18 | 10 | 15.4 | 9.24 |
| 16 | 13 | 19.4 | 11.34 |
| 14 | 17 | 31.2 | 18.72 |
| 12 | 23 | 40 | 24 |
| 10 | 33 | 55 | 33 |
| 8 | 46 | 75 | 45 |
| 6 | 60 | 100 | 60 |

Even though there are many variables, Table 4-6 is offered as a rough guide to the current rating of several wire gages. It is not to be applied universally. For example, the NEC specifies a maximum current of 20 A for a no. 12 wire. This is more conservative than the values shown in Table 4-6 due to the potential of fires in the wood frame type of construction used in many residences.

Figure 4-12 shows examples of wires and cables for applications other than building wiring. Some of these applications include

**1.** Lamp cord. Used for lamps and appliances such as radio receivers, television receivers, clocks, etc.

**2.** Twisted pair. Used for low-voltage control, signaling, sensors, data communications, etc.

**3.** Jacketed twisted pair. Similar applications to those for the twisted pair.

**4.** Multiconductor. Used for remote control, data communications, instrumentation, sensors, etc.

**5.** Multipaired shielded. Similar applications to those for multiconductor cable. Offers the advantage of being less susceptible to electric noise.

**6.** Shielded pair. Used for audio signals, data communications, instrumentation, sensors, etc. The shield reduces noise.

**7.** Coaxial. Used for audio signals, video signals, radio frequencies, high-speed data communications, sensors, etc.

**8.** Twin-axial. Used for high-speed data communications, video, etc.

**9.** Double-shielded. Used where low-level signals must be protected from extraneous noise and where high-frequency signals must be confined to the cable as much as possible.

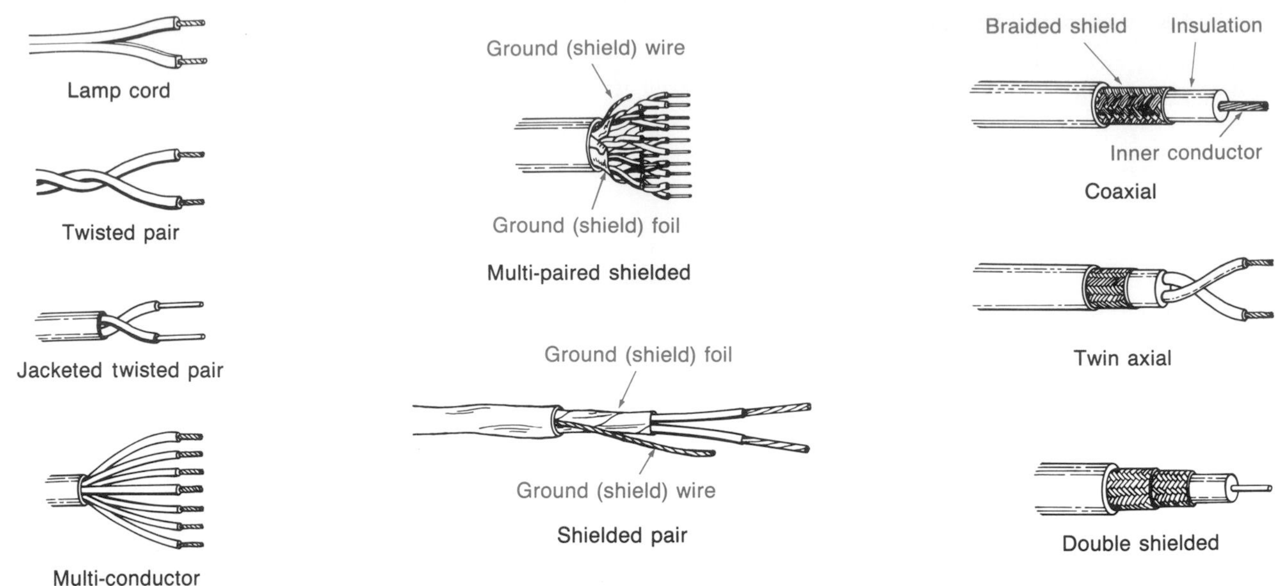


**FIGURE 4-12 Common wires and cables.**

The increasing application of digital devices, such as computers and computer peripherals, has produced a need for neat and economical ways to interconnect circuits and devices. The ribbon cable shown in Fig. 4-13 is one popular solution. This cable consists of stranded wires that are embedded in and separated by an insulating material. Special press-on connectors are available that save considerable installation time.

When a group of wires and/or cables are routed from one point to another, they are often secured together in an assembly called a harness. This makes for a much neater arrangement and provides protection for the conductors. Wire harnesses used to be prepared by lacing the conductors together with a wax-impregnated cord. Today, lacing is being replaced with other materials such as heat-shrink tubing, plastic spiral wrap, plastic molding compounds, and cable ties. Figure 4-14 shows some of the devices used to harness and secure wires and cables.

Another type of conductor that must be mentioned is magnet wire. It is used to make coils for devices such as motors, transformers, and relays. It is insulated by a dip-coating process. After coating, the wire is baked in an oven. This process produces a thin but tough insulating coating. The coating must be thin to pack as much wire as possible into a given area, which is important in coil construction. Some of the coatings are nearly transparent, which gives some magnet wire an appearance similar to bare copper wire.

There are many factors to be considered in selecting wire and cable. Some applications may require special attributes such as the ability to resist fungus. Cost, size, and weight are usually important. Ease of soldering and applying connectors must be considered. This is covered in the next section. Tensile strength may also be a factor. Copper alloy wires are available that have a much higher tensile strength and retain about 85 percent of the conductivity of annealed copper.

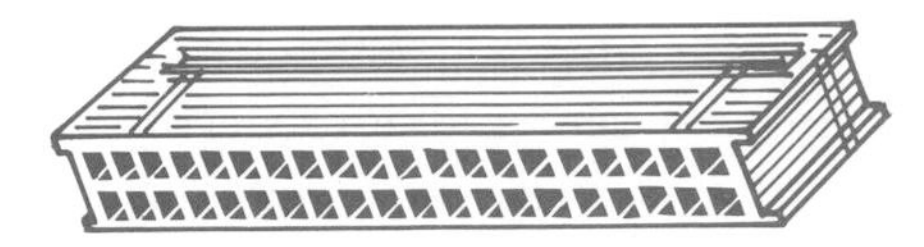


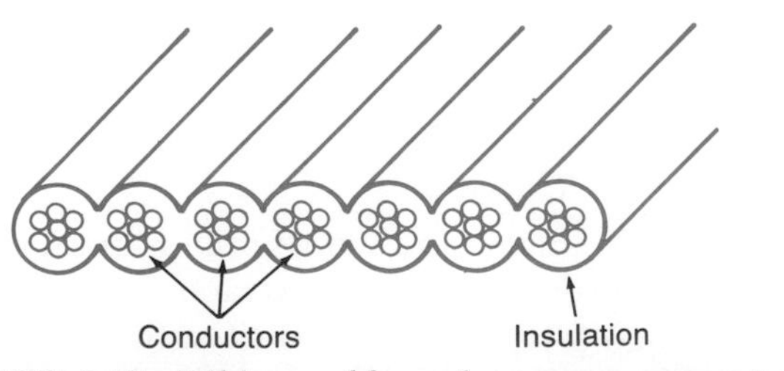


**FIGURE 4-13 Ribbon cable and press-on connector.**

## Self-Test

**26.** What is the abbreviation for the wiring code published by the NFPA?

**27.** What is the name of the metal tubing used to protect wiring used in commercial and industrial buildings?

**28.** Most electronic conductors are made from the metal ______.

**29.** Stranded wire has the advantage of better ______ and resistance to vibration damage when compared to solid wires.

**30.** Refer to Table 4-3. A wire is made up of 61 strands, each having a diameter of 0.0082 in. What is the nearest equivalent AWG size?

**31.** What is the name given to electronic cable with a center conductor surrounded by insulation and then by a braided outer conductor?

**32.** Various conductors tied together are called a ______.

**33.** What type of wire is used to make coils?

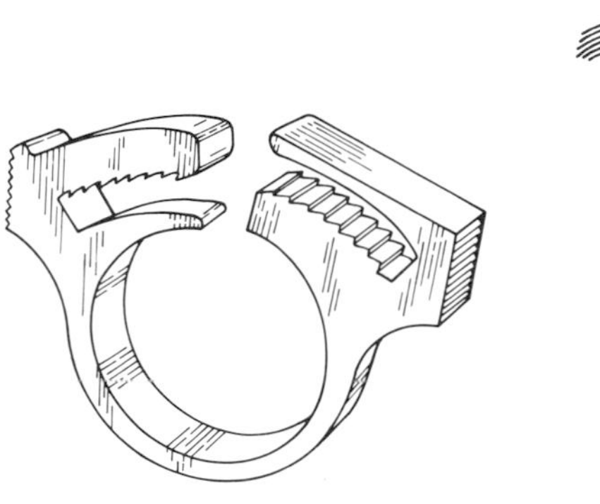

**FIGURE 4-14 Devices used to secure wire and cable.**

## 4-5 CONNECTIONS AND CONNECTORS

*Soldering* is a widely applied method of making electric connections. The solder used in most electrical and electronic work is a fusible alloy of tin and lead. The solder and the materials to be soldered are heated to approximately 230°C (446° F). The solder alloy melts and dissolves a small amount of the surface material of the objects being joined. A new alloy is created at the juncture of the solder and the material.

Electronic solder is usually a 60/40 alloy. The first number refers to the percentage of tin and the second number to the percentage of lead. A 60/40 alloy is desirable because it is very near the *eutectic* point. Refer to Fig. 4-15. Except for the eutectic, all alloys of tin and lead pass through a plastic range when they are heated. A eutectic solder requires the least heat to achieve the liquid state and provides good fusion at the interface of the objects being joined. Tin costs more than lead, and "economy" solders are available with alloys such as 40/60 and 50/50. These solders should not be used in electronic work. The additional heat that they require can damage circuit boards and components.

An essential requirement for solder used in electrical and electronic work is that the *flux* be noncorrosive. Flux is a material used to break down the oxide film that tends to form on metal surfaces. Without flux, it is difficult or impossible for the solder to dissolve the surface of the objects being joined. Resin is a common noncorrosive flux. It attacks the oxide film but does not corrode metal.

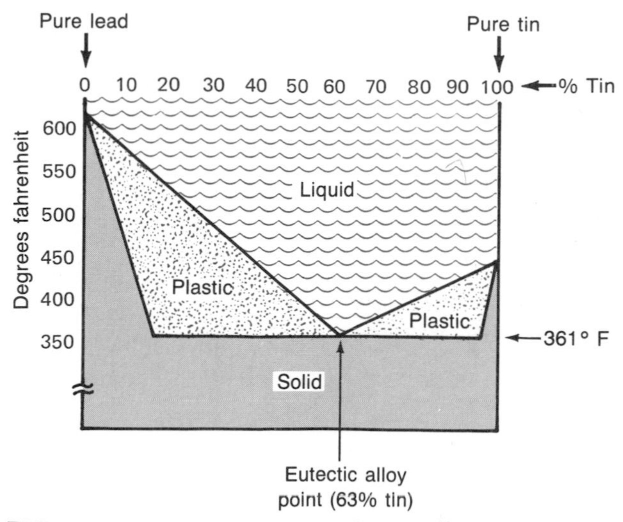


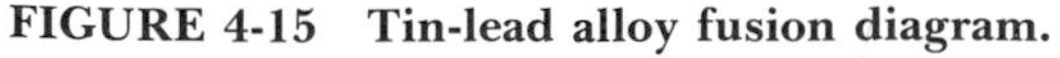


**FIGURE 4-15 Tin-lead alloy fusion diagram.**

Most solder looks like round wire but has a hollow core with flux in it. Flux is also sold as a separate item and can be applied to the joint before soldering. This is necessary if solder without flux is used.

Soldering is a skill based on knowledge and practice. Poor solder joints have caused many circuit malfunctions. The first requisite to a good solder joint is clean metal. Bare copper solders very well, but only if it is bright and shiny. Unfortunately, copper oxidizes rapidly. The oxide film resists the soldering process. Flux helps, but a well-oxidized wire is still difficult to solder properly. This is why almost all electronic conductors made from copper are coated with another metal such as tin. Noncoated copper surfaces should be brightened with a light abrasive material immediately before soldering. An ordinary pencil eraser works well for flat surfaces. Fine steel wool is also useful for some applications, but the debris left behind must be removed or it may cause short circuits.

The second requisite to good soldering is the application of the correct amount of heat for the correct period of time. Refer to Fig. 4-16. The soldering iron is applied to the joint and then the solder is applied next. The iron and the solder must be positioned so that all of the materials involved come up to soldering temperature. When the solder starts to flow, the solder wire is advanced into the joint to provide the correct amount. When the joint has the correct amount of solder, the solder is withdrawn. The iron is removed last when the joint has the correct appearance. Nothing should be moved until the joint cools down. Forced cooling, such as blowing on the joint or using a cooling liquid, is not recommended.

The correct amount of solder has been used when the joint appears shiny and all surfaces are coated. The solder should not be flowing down the lug or appear as a large blob at the joint. As shown in Fig. 4-16, a printed circuit solder joint should show that the solder has formed a fillet between the component lead and the foil.

The correct amount of heat also produces a shiny joint. If the soldering iron is too small or is not applied long enough, a dull joint results. These are called "cold solder joints" and are a prominent cause of malfunctions. An iron that is too hot or that is applied too long can also produce a joint that is dull in appearance. Component damage and damage to insulating materials are other consequences of excess heat. Most workers prefer a small (approximately 20-W) pencil-type iron for printed circuit work. Irons rated at 50 W or more are desirable for joints involving larger wire sizes and heavy lugs. Soldering guns are acceptable for repair work involving wires and lugs. They are designed for intermittent duty only. They should never be used on printed circuit boards or on other delicate soldering jobs.

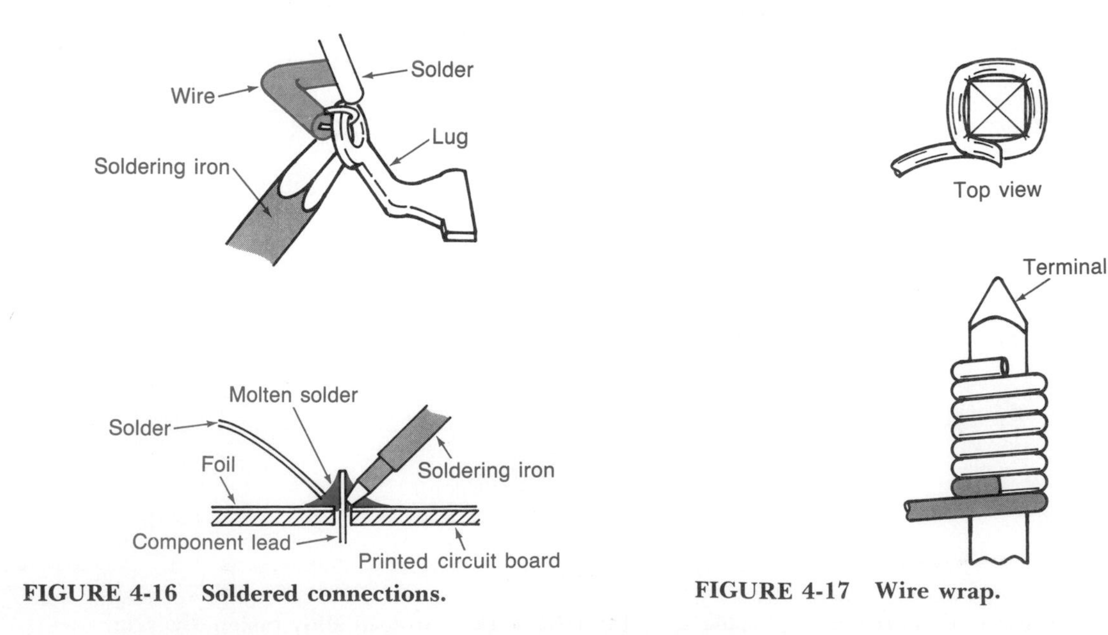


**FIGURE 4-16 Soldered connections.**

**FIGURE 4-17 Wire wrap.**

Wire wrap is a method of making connections that do not require solder. It is used for prototyping and limited production runs. Figure 4-17 shows two views of a wire-wrap connection. The posts are square, and metal adhesion occurs at the corners when the wire is wrapped tightly. Guns and pencils are used to wrap at least six turns of wire around the post. The solid copper wire ranges from no. 30 gage to no. 24 gage and is silver- or tin-coated.

Figure 4-18 shows some examples of heavy-duty electric connectors. The conductors are securely clamped by the pressure of a screw or nut. Some metals, such as copper and aluminum, produce a chemical action when in contact with each other. Special electric connectors may be used to join dissimilar conductors, and pastes that retard the chemical action are also available. Other connectors, such as all but one of those shown in Fig. 4-19, may be crimped to the conductor. Solder lugs are usually used to make ground connections in electric and electronic equipment.

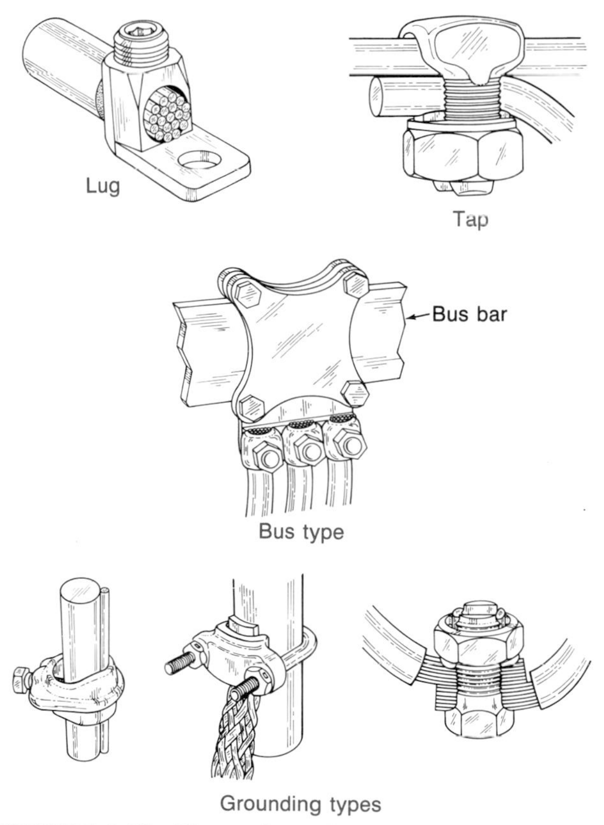


**FIGURE 4-18 Heavy-duty electric connectors.**

There are quite a variety of electronic connectors. Figure 4-20 provides a limited sample. The male part is often called a *plug* and the female part a *jack*. The UHF and BNC connectors are used in conjunction with coaxial cable. Banana and tip connectors are often used to connect test leads to instruments. The DB-25 connector is very popular for connecting digital equipment such as computer terminals.

## Self-Test

**34.** A 40/60 alloy solder contains ______ percent tin.

**35.** The alloy preferred for electronic soldering is ______.

**36.** A eutectic alloy has the ______ possible melting temperature.

**37.** The flux used in electronic soldering must be of the ______ type.

**38.** A good solder joint requires that the objects being joined are ______ and at the proper temperature.

**39.** The desired power rating for a soldering iron or pencil for printed circuit work is around ______ W.

**40.** What joining technique is solderless and uses several turns of solid wire around a square post?

**41.** The male part of a connector may be called a plug, while the female part is the ______.

**42.** Joining copper and aluminum is not recommended unless a paste or special ______ is used.

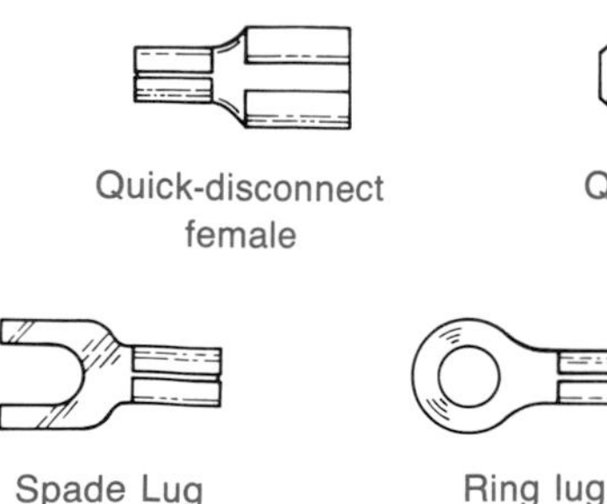


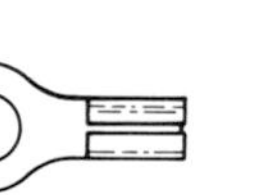

**FIGURE 4-19 Examples of lugs and quick-disconnects.**

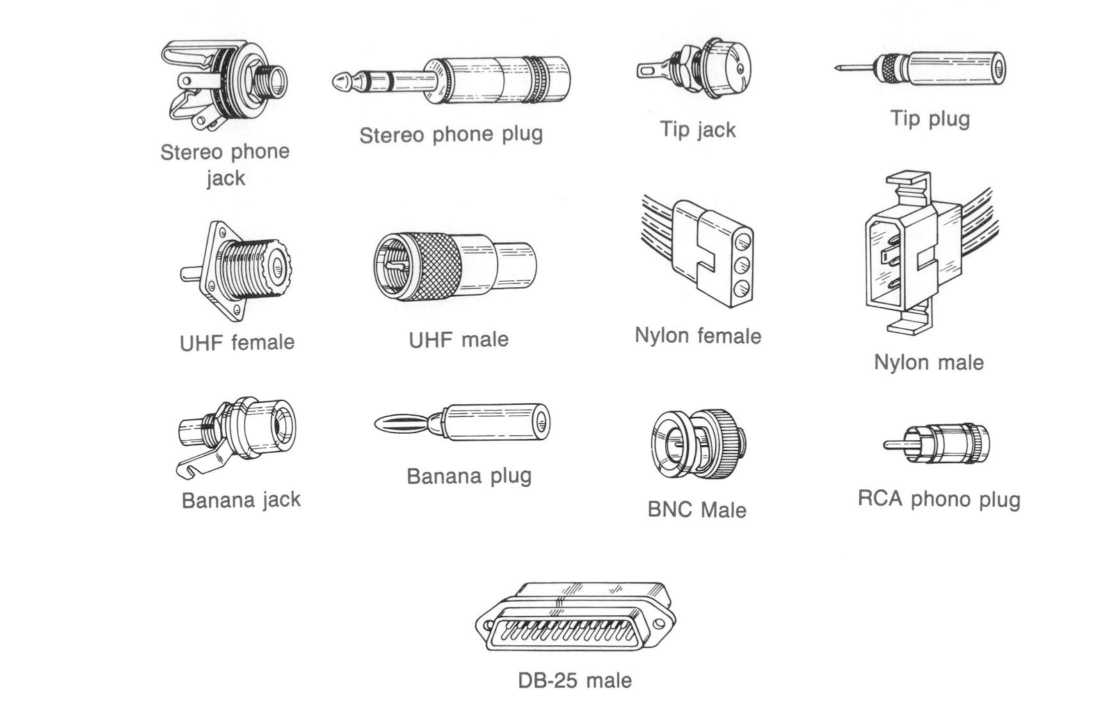


**FIGURE 4-20 Some common connectors used in electronics.**

## SUMMARY

1. Conductors have metallic bonding, while insulators and semiconductors have covalent bonding.
2. The conduction band overlaps the valence band in a conductor.
3. There is an energy gap between the conduction band and the valence band in semiconductors and insulators.
4. An electron may gain enough thermal or electric energy to jump the energy gap and enter the conduction band in an insulator or a semiconductor.
5. The leakage current in an insulator is normally small enough to be ignored.
6. Breakdown occurs when too much voltage is applied across an insulator.
7. When the resistance of a material drops with increasing temperature, that material has a negative temperature coefficient.
8. Resistivity (or specific resistance) is based on the resistance in ohms of a specified volume of some material.
9. The SI unit of resistivity is the ohm-meter.
10. The reciprocal of resistivity is conductivity.

11. The SI unit of conductivity is siemens per meter.

12. The resistance of a structure increases as its length increases.

13. The resistance of a structure decreases as its cross-sectional area increases.

14. AWG is the abbreviation for American wire gage. It is also sometimes called the Brown and Sharpe gage.

15. The AWG is based on the mil, which is equal to one-thousandth of an inch.

16. The CMA (circular mil area) of a round wire is equal to the square of the wire diameter in mils.

17. As the AWG number gets smaller, the wire diameter gets larger.

18. Larger current flow requires a smaller gage number.

19. When the resistance and the temperature coefficient of a material are known, the resistance of the material can be predicted for various temperatures.

20. The resistance of a conductor approaches zero as its temperature approaches absolute zero.

21. Temperature coefficient calculations are based on straight-line approximations. The error increases at the temperature extremes.

22. Electric lamps draw a surge current many times their operating current.

23. The National Electric Code sets the standards for much of the wiring in buildings in the United States.

24. Commercial and industrial wiring may be placed in metal tubes called conduit.

25. Stranded wire is more flexible than solid wire.

26. The wire used in electronics is usually copper that has been coated with another metal to reduce oxidation and improve solderability.

27. The current-carrying capacity of a wire is dependent on (1) its gage, (2) its insulation, (3) its proximity to other conductors, (4) the ambient temperature, (5) the environment, and (6) the allowable voltage drop.

28. A wire harness may be used to bundle various conductors together.

29. Magnet wire is used to wind coils for devices such as motors and transformers.

30. Many of the connections in electronics are soldered.

31. Electronic solder is usually composed of 60% tin and 40% lead.

32. The flux used in electric and electronic soldering must be noncorrosive.

33. Good soldering requires clean metal, flux, and the correct temperature.

34. Wire-wrap joints are based on metal adhesion and do not require solder.

35. Male connectors may be called plugs, while female connectors may be called jacks.

## CHAPTER REVIEW QUESTIONS

4-1. Flow in an insulator is called ______ current.

4-2. The breakdown rating of an insulating material ______ as its thickness increases.

4-3. Semiconductors and insulators show ______ resistance as their temperature is decreased.

4-4. Conductors show ______ resistance as their temperature is decreased.

4-5. What does the abbreviation NEC stand for?

4-6. 60/40 solder has ______ % lead and ______ % tin.

4-7. A eutectic alloy is one that has the ______ melting point.

4-8. The flux used in electronic soldering must never be ______.

4-9. Pure copper can be very difficult to solder due to an ______ film on its surface.

4-10. Name two solderless methods of connecting wires.

## CHAPTER REVIEW PROBLEMS

4-1. Dry air is rated at a breakdown of 30 kV/cm. Calculate the voltage required to jump across a 1-mm air gap.

4-2. The resistance of 100 m of wire is 0.560 Ω. What would the resistance be for 300 m of the same wire?

4-3. What would the resistance be for 100 m of wire made from the same metal as in Question 4-2, but with twice the cross-sectional area?

4-4. Refer to Table 4-2. Calculate the resistance of an iron wire at 20°C that is 0.5 cm wide × 0.1 cm thick and is 10 m long.

4-5. What is the CMA of a round wire with a diameter of 0.057 in.?

4-6. What is the CMA of a rectangular conductor that is 0.100 in. × 0.009 in. in cross section?

4-7. Use Table 4-3 to determine the resistance of a no. 40 copper wire that is 10 ft long.

4-8. Use Tables 4-2 and 4-3 to determine the resistance of a no. 40 nichrome wire that is 10 ft long.

4-9. Refer to Table 4-4. A nickel wire has a resistance of 0.491 Ω at 10°C. What is its resistance at −50°C?

4-10. A wire is made up of 19 round strands, each of which has a diameter of 0.015 in. Based on 300 cmils/A, what is the current capacity of this wire?

## ANSWERS TO SELF-TESTS

**1.** covalent and metallic
**2.** metallic
**3.** energy gap
**4.** increase
**5.** negative
**6.** voltage (potential difference)
**7.** 15 kV
**8.** ohm-meter
**9.** conductivity
**10.** 1.95 mΩ
**11.** decreases
**12.** decreases
**13.** $5.49 \times 10^{-8}\ \Omega \cdot \text{m}$
**14.** 26,244
**15.** no. 6
**16.** 6400 mils$^2$, 8149 cmils
**17.** no. 11
**18.** 115 V
**19.** 0.25 Ω; copper would be less (0.15 Ω)
**20.** because of heat buildup
**21.** 11.5 Ω
**22.** absolute zero (−273°C)
**23.** surge
**24.** increases
**25.** negative
**26.** NEC
**27.** conduit
**28.** copper
**29.** flexibility
**30.** no. 14
**31.** coaxial
**32.** harness
**33.** magnet wire
**34.** 40
**35.** 60/40
**36.** lowest
**37.** noncorrosive
**38.** clean
**39.** 20
**40.** wire wrap
**41.** jack
**42.** connector

# Chapter 5 Resistive Components

Resistance has been discussed as a desirable characteristic in insulators, where it should be as high as possible, and as an undesirable characteristic in conductors, where it should be as low as possible. *Resistors* are components with intermediate values of resistance. They serve many useful functions such as (1) providing a voltage drop, (2) providing a current limit, and (3) dissipating (converting) electric energy. Some components have resistance that varies with a condition such as temperature or light. They form the basis for many of the sensors used in various applications. This chapter examines various resistive components.

## 5-1 FIXED RESISTORS

*Fixed resistors* are usually two-terminal devices. They are available in a wide variety of types, sizes, and resistance values. Figure 5-1 shows the construction of a carbon composition resistor. Carbon is a semiconductor and serves as a resistance element for many applications. The carbon may also be deposited on an insulating tube to form the resistance element. Carbon film resistors use this construction technique.

Figure 5-2 shows that carbon resistors are available in several standard power ratings ranging from $\frac{1}{8}$ to 2 W. Note that the power rating is a func-

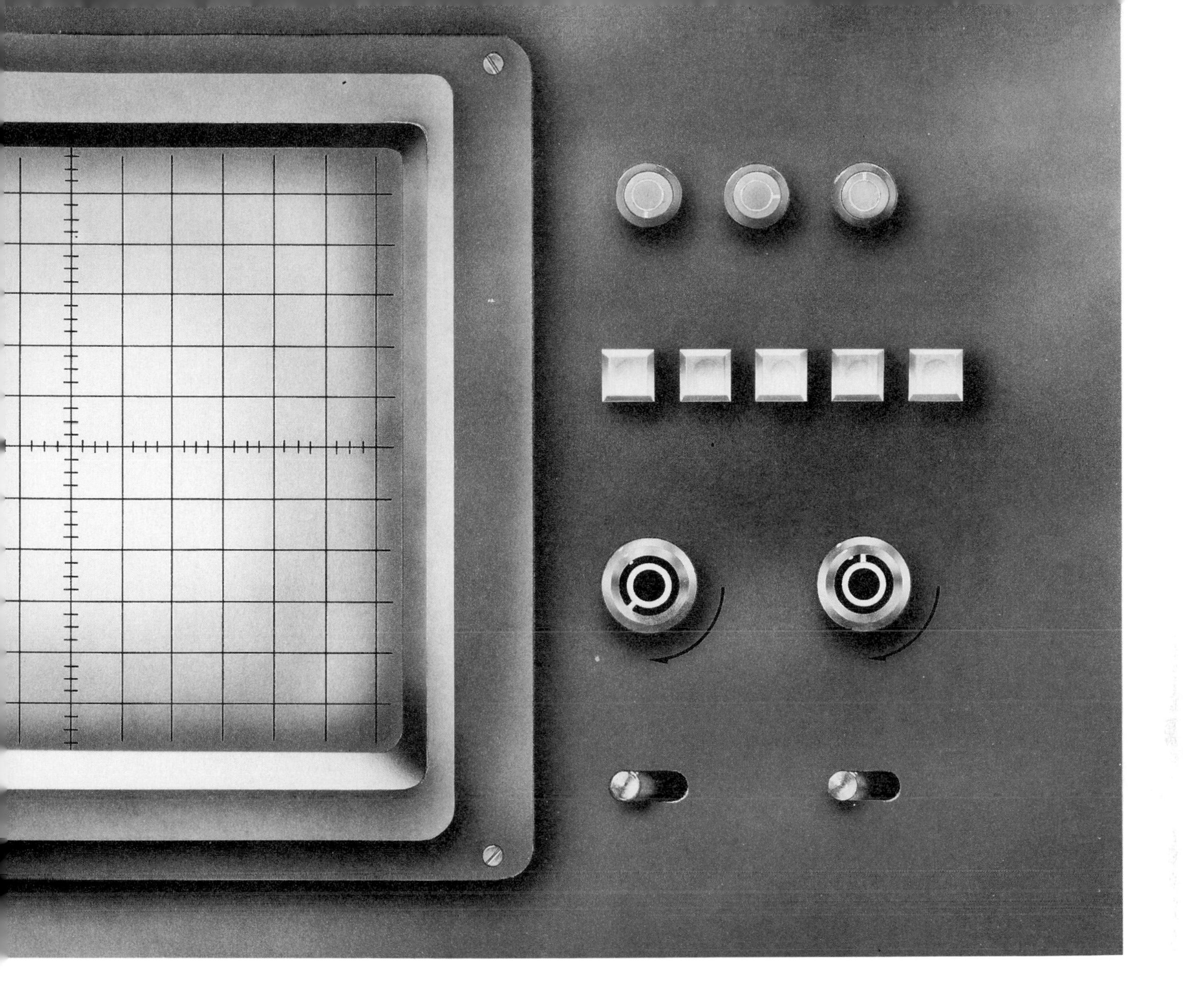

tion of the physical size of the resistor. A larger resistor is able to throw off (dissipate) more heat

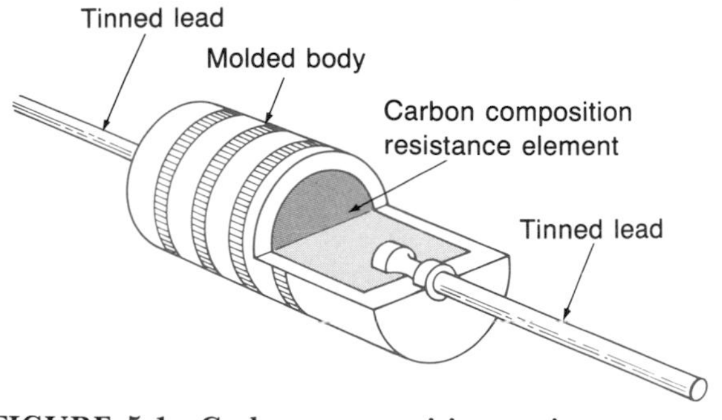


**FIGURE 5-1 Carbon composition resistor construction.**

than a smaller one. The *ambient* temperature also determines how much heat a resistor can give off. This is the temperature of the air surrounding the resistor. Figure 5-3 is a power derating curve. It shows that carbon composition resistors must be derated for ambient temperatures above 70°C.

#### EXAMPLE 5-1

A 120-Ω, 2-W carbon composition resistor drops 13 V and is located in an environment that reaches a temperature of 100°C. Is it safe?

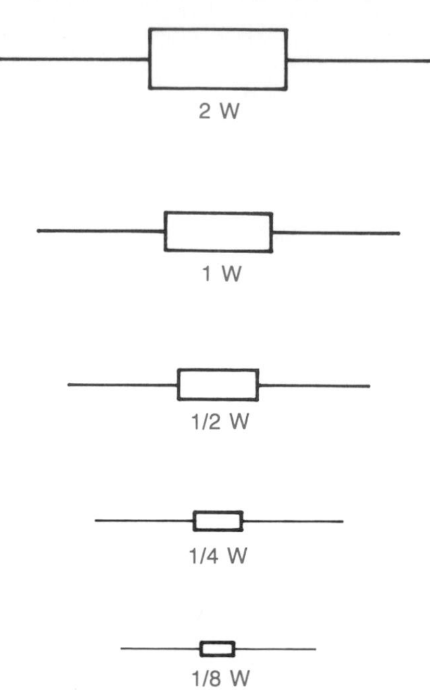

**FIGURE 5-2 Full-size outlines of carbon resistors.**

### Solution

The power dissipation can be found by

$$P = \frac{V^2}{R}$$
$$= \frac{13^2}{120}$$
$$P = 1.41 \text{ W}$$

Figure 5-3 shows that the power rating drops to 60 percent at 100°C. Sixty percent of 2 W is

$$2 \text{ W} \times 0.60 = 1.2 \text{ W}$$

The resistor is not safe since 1.41 W is greater than 1.2 W.

---

Wire-wound resistors are able to safely operate at higher temperatures than carbon types. Figure 5-4*a* shows that a 2-W wire-wound resistor is the same physical size as a 1-W carbon type. A wide band is often used on the package to indicate the type of construction. It is important to take notice of such details when replacing components. Figure 5-4*b* shows a style of wire-wound resistor that is available with power ratings that range from 12 to 225 W. Intermediate ratings of 25, 50, and 100 W are also available. The 12-W units are $1\frac{3}{4}$ in. long × $\frac{5}{6}$ in. in diameter, and the 225-W resistors are $10\frac{1}{2}$ in. long × $1\frac{1}{8}$ in. in diameter. As before, the power rating is directly related to the physical size. Figure 5-4*c* shows that adjustable wire-wound resistors are also manufactured.

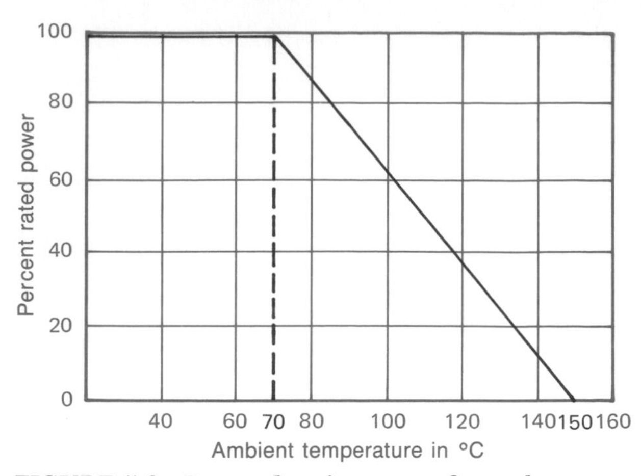


**FIGURE 5-3 Power derating curve for carbon composition resistors.**

The trend is to make many electronic circuits as compact as possible. Resistors have not escaped that trend. Figure 5-5 shows a single in-line pack-

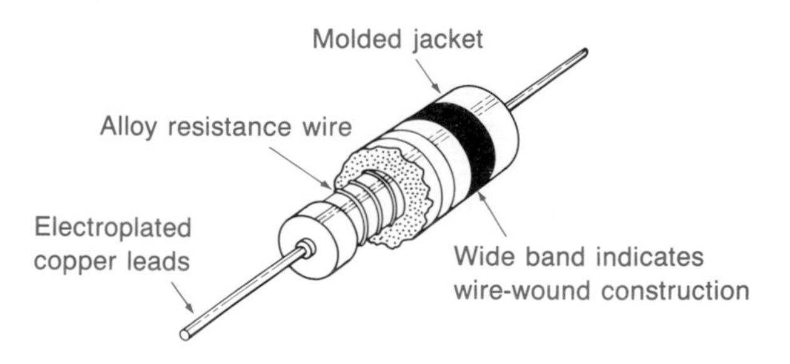


(*a*) 2-W type in a 1-W carbon size package

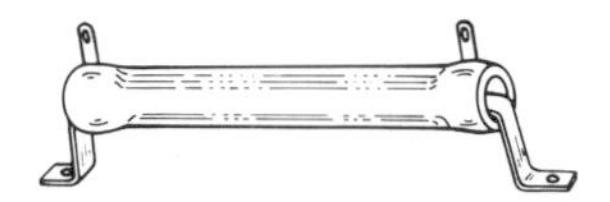

(*b*) 12-W to 225-W fixed type

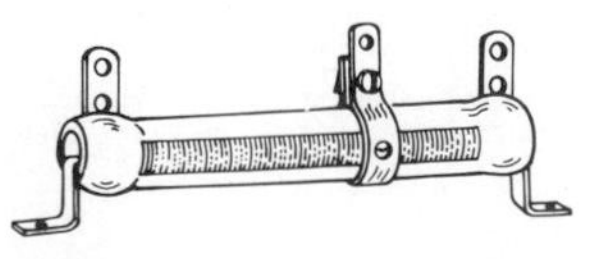

(*c*) 12-W to 100-W adjustable type

**FIGURE 5-4 Wire-wound resistors.**

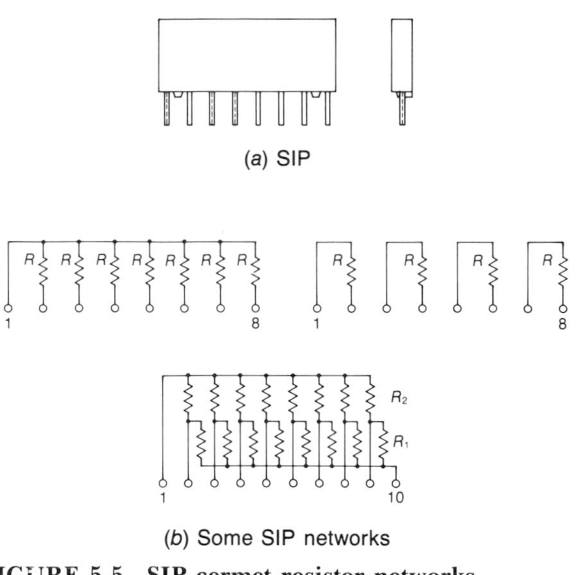


**FIGURE 5-5 SIP cermet resistor networks.**

age (SIP) that may house four or more resistors. Figure 5-5*a* shows that the pins on the package are arranged in a single line. SIPs typically have 6, 8, or 10 pins. Figure 5-5*b* shows some of the many ways that the resistors are arranged in SIPs. Dual in-line package (DIP) resistor networks are also available. Figure 5-6*a* shows that the dual in-line package has its pins arranged in two lines, and Fig. 5-6*b* shows some DIP networks. Both SIPs and DIPs lend themselves to printed circuit construction and are also used to house other types of electronic components such as solid-state devices.

The resistor elements in SIPs and DIPs are not carbon, nor are they wound from resistance wire. They are "cermet," which is a contraction for ceramic and metal. These elements are formed by depositing a thin film of metal such as nichrome or chromium cobalt on a ceramic substrate. Table 5-1 lists the typical characteristics for the various types of fixed resistors. There is considerable variation in some of the characteristics. For example, the temperature coefficient for carbon composition resistors can be as large as −8000 ppm (parts per million) per degree Celsius. Metal film resistors are much more stable with a temperature coefficient no worse than ±175 ppm/°C.

### EXAMPLE 5-2

A carbon composition resistor has a resistance of 10,000 Ω at 20°C. Calculate its resistance at 100°C if its temperature coefficient is −4000 ppm/°C.

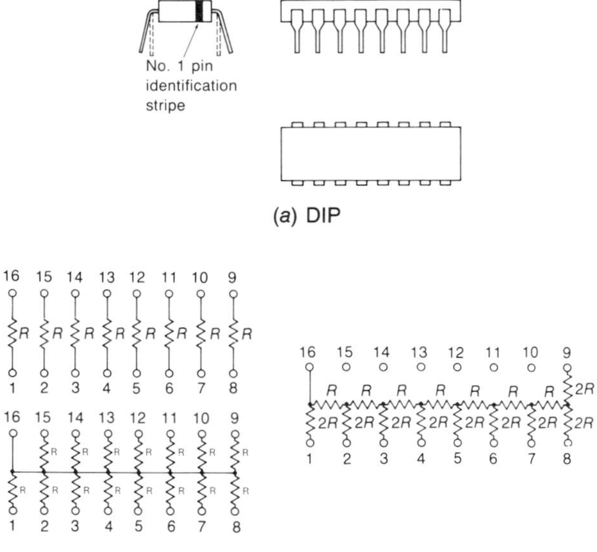


**FIGURE 5-6 DIP cermet resistor networks.**

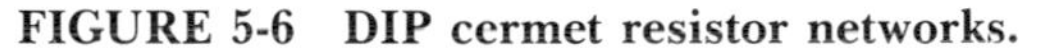

#### Solution

The temperature change is 100 − 20 = 80°. The change in resistance is found by multiplying the resistance at 20° times the temperature change times the temperature coefficient:

$$\Delta R = \frac{(10{,}000)\ (80)\ (-4000)}{1{,}000{,}000} \qquad \text{(where } \Delta \text{ is the symbol for "change in")}$$

$$= -3200\ \Omega$$

The resistance at 100°C will be 10,000 − 3200 = 6800 Ω. It should be obvious that carbon resistors are not the best choice for circuits where temperature stability is important.

### EXAMPLE 5-3

A metal film resistor has a resistance of 10,000 Ω at 20°C. Calculate its resistance at 100°C if its temperature coefficient is +175 ppm/°C.

#### Solution

The change in temperature is 80°C, and the change in resistance is

$$\Delta R = \frac{(10{,}000)\ (80)\ (+175)}{1{,}000{,}000}$$

$$= +140\ \Omega$$

The resistance at 100°C is 10,140 Ω.

**Table 5-1 Typical Characteristics of Fixed Resistors**

| Type of Construction | Resistance Range | Maximum Power Rating, W | Tolerance, % | Temperature Coefficient, ppm/° C | Comments |
|---|---|---|---|---|---|
| Carbon composition | 1 Ω to 22 MΩ | 2 | 5–20 | −200 to −8000 | General-purpose applications; economical; resistance tends to increase with age; not recommended where flaming is possible. |
| Carbon film | 10 Ω to 22 MΩ | 10 | 2–10 | −200 to −1000 | General purpose; more economical than carbon composition types. |
| Metal film | 10 Ω to 3 MΩ | 10 | 0.1–2 | ±25 to ±175 | Fair precision and high stability; high-voltage types with very high resistance values are available; produce less noise than carbon types; recommended for ac applications and have good high-frequency characteristics; recommended where flaming in carbon types must be avoided. |
| Power wire-wound | 0.1 Ω to 180 kΩ | 225 | 5–10 | Less than ±260 | Used where high dissipation is required and where high-frequency operation is unimportant; power dissipation depends on heat sink or air flow around the device. |
| Precision wire-wound | 0.1 Ω to 800 kΩ | 15 | 0.01–1 | Varies with resistance value | Not suitable above 50 kHz; used in precision circuits such as instrumentation. |

## Self-Test

**1.** The power rating of a resistor is directly related to its ________.

**2.** Refer to Fig. 5-3. A 1-W, 1000-Ω resistor conducts 30 mA. What is its maximum safe environmental temperature?

**3.** A wide band on a resistor package indicates that it is a ________ type.

**4.** A popular package with two rows of pins that is well suited to printed circuit applications is called a ________.

**5.** A carbon composition resistor has a resistance

of 150 Ω at 20°C. Calculate its resistance at 80°C if its temperature coefficient is −1000 ppm/°C.

**6.** What type of resistor has the highest power ratings?

**7.** What type of resistor has the poorest temperature stability?

**8.** What type of resistor has the best precision?

## 5-2 RESISTOR COLOR CODE

The power rating of a resistor can often be determined by its physical size, while its resistance value may be printed on the body of the resistor. Printing is not practical on small resistors, and color bands are often used instead. Figure 5-7 shows the location of the color bands used to indicate the value and the tolerance of the resistor, and Table 5-2 lists the values assigned to each color. The resistor color code is one of many standards approved by the Electronic Industries Association (EIA) and also by the various branches of the military (MIL).

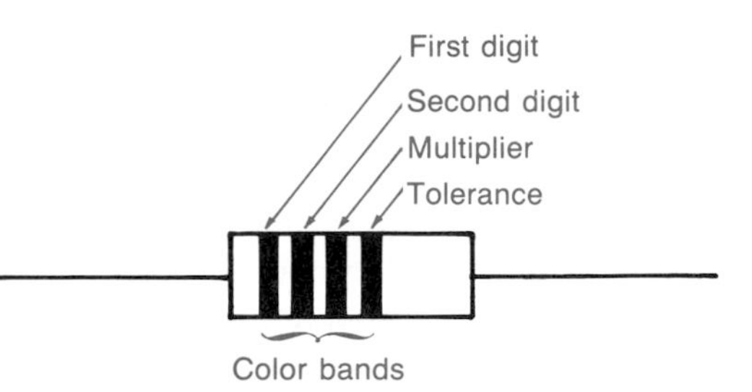


**FIGURE 5-7 Color bands are used to indicate the resistance in ohms and the tolerance in percent.**

**Table 5-2 EIA-MIL Resistor Color Code**

| Color | Digit Value | Multiplier Value | Tolerance Value |
|---|---|---|---|
| Black | 0 | $10^0$ | — |
| Brown | 1 | $10^1$ | — |
| Red | 2 | $10^2$ | — |
| Orange | 3 | $10^3$ | — |
| Yellow | 4 | $10^4$ | — |
| Green | 5 | $10^5$ | — |
| Blue | 6 | $10^6$ | — |
| Violet | 7 | $10^7$ | — |
| Gray | 8 | $10^8$ | — |
| White | 9 | $10^9$ | — |
| Gold | — | $10^{-1}$ | ±5% |
| Silver | — | $10^{-2}$ | ±10% |

NOTE: EIA-Electronic Industries Association; MIL-military.

The resistor color code uses the first two color bands to designate the two significant figures of the resistance value. The third band is a multiplier. The fourth band indicates the tolerance of the resistor. The resistance indicated by the color code is known as the *nominal* value. The actual resistance will fall in a range of values. The smaller the tolerance value, the smaller the range of actual resistance. A resistor with a nominal value of 100 Ω and a tolerance of ±10 percent may have an actual resistance anywhere from 90 to 110 Ω, assuming that it is in tolerance. If its actual value falls below 90 Ω or above 110 Ω; the resistor is out of tolerance.

### EXAMPLE 5-4

A resistor has the following order of color bands: red, red, red, gold. Determine its nominal resistance and its range of resistance values.

**Solution**

The first three color bands determine the nominal resistance value:

| red | red | | red | |
|---|---|---|---|---|
| 2 | 2 | × | $10^2$ | = 2200 Ω or 2.2 kΩ |

The tolerance band is gold, which indicates ±5 percent:

$$2200\ \Omega \times \pm 0.05\ \Omega = \pm 110\ \Omega$$

$$2200\ \Omega + 110\ \Omega = 2310\ \Omega \quad \text{(the upper limit)}$$

$$2200\ \Omega - 110\ \Omega = 2090\ \Omega \quad \text{(the lower limit)}$$

### EXAMPLE 5-5

What color bands will be found on a resistor with a nominal value of 390 Ω and a tolerance of ±10 percent?

**Solution**

Convert the significant figures and the multiplier to colors:

| 3 | 9 | $10^1$ |
|---|---|---|
| orange | white | brown |

Silver is used to indicate a tolerance of ±10 percent. The color bands will be orange, white, brown, and silver.

The colors gold and silver are used to denote resistor tolerance. If a resistor has no fourth band, it has a tolerance of ±20 percent. Such resistors used to be fairly commonplace but are rare today. Table 5-2 shows that gold and silver may also be used as multipliers. These multipliers are necessary to code resistors with a nominal value less than 10 Ω.

**EXAMPLE 5-6**

What is the nominal value of a resistor with bands of yellow, violet, gold, and gold?

**Solution**

In this case, notice that the third band is gold. This denotes a multiplier of $10^{-1}$:

$$\left|\begin{matrix}\text{yellow}\\4\end{matrix}\right| \left|\begin{matrix}\text{violet}\\7\end{matrix}\right| \times \left|\begin{matrix}\text{gold}\\10^{-1}\end{matrix}\right| = 4.7\ \Omega$$

The fourth band is also gold, which indicates a tolerance of ±5 percent.

Resistors are also rated for reliability. Industrial-grade resistors are intended for general-purpose applications. Military-grade resistors must meet more demanding specifications. For example, some resistors are specified as having a maximum failure rate of 0.001 percent per 1000 h of operation. This reliability level is indicated by a yellow fifth band on the body of the resistor.

Table 5-3 lists the standard nominal resistance values for carbon composition resistors. The values range from 1 Ω to 22 MΩ. Five percent tolerance resistors are available in all the values shown in the table, while 10 percent tolerance resistors are available only in those values shown in bold type. Standard values are spaced almost equally on a *logarithmic* scale. A logarithm is a power of some base number. For example, $10^2 = 100$ and "2" is the base 10 logarithm of the number 100. The logarithmic spacing of component values is a systematic way of providing reasonable coverage of all necessary values. Consider the standard value 2700. It is bracketed by the 10 percent values of 2200 and 3300 as shown in Table 5-3. Figure 5-8 shows that for a nominal value of 2700, the actual range is from 2430 to 2970 considering the tolerance of ±10 percent. Figure 5-8 also shows that a nominal value of 3300 ranges down to 2970 and that a nominal value of 2200 ranges up to 2420. Reasonably continuous coverage is provided by the logarithmic spacing of standard values.

When circuits are designed, the calculated values are seldom standard. Selecting the nearest standard value provides adequate performance in most cases. Special values may be ordered, or standard values can be combined to produce special values. However, these two procedures entail extra cost. Many of the values used in this book are nonstandard. This is because real calculations also include nonstandard values and because nonstandard values often produce easier arithmetic.

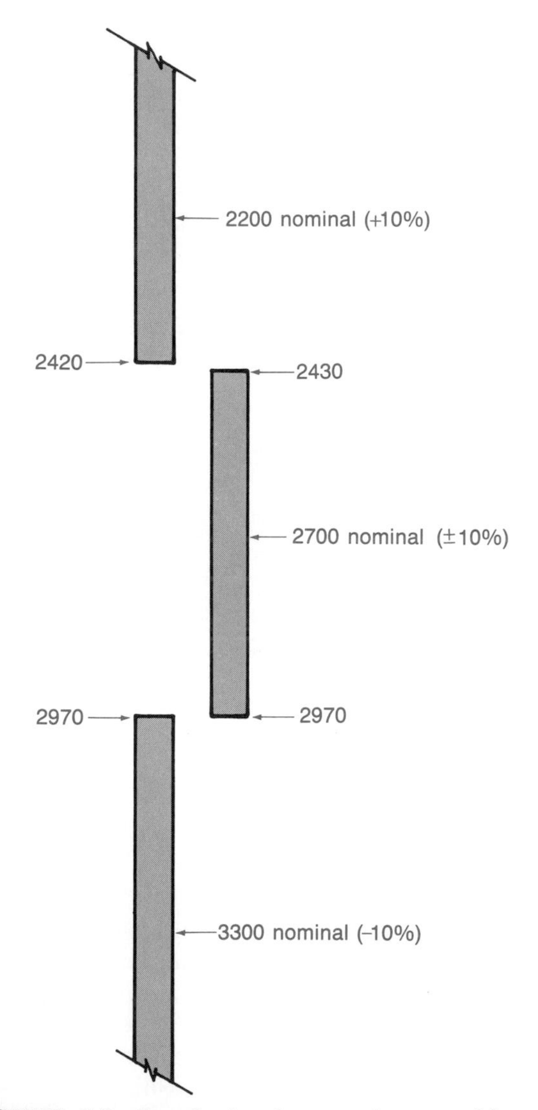


**FIGURE 5-8 Standard value spacing provides reasonably continuous coverage.**

**Table 5-3 Standard Nominal Resistance Values for Carbon Composition Resistors**

| Ohms | | | | | | | |
|---|---|---|---|---|---|---|---|
| **1.0** | 5.1 | **27** | 130 | **680** | 3,600 | **18,000** | 91,000 |
| 1.1 | **5.6** | 30 | **150** | 750 | **3,900** | 20,000 | **100,000** |
| **1.2** | 6.2 | **33** | 160 | **820** | 4,300 | **22,000** | 110,000 |
| 1.3 | **6.8** | 36 | **180** | 910 | **4,700** | 24,000 | **120,000** |
| **1.5** | 7.5 | **39** | 200 | **1,000** | 5,100 | **27,000** | 130,000 |
| 1.6 | **8.2** | 43 | **220** | 1,100 | **5,600** | 30,000 | **150,000** |
| **1.8** | 9.1 | **47** | 240 | **1,200** | 6,200 | **33,000** | 160,000 |
| 2.0 | **10** | 51 | **270** | 1,300 | **6,800** | 36,000 | **180,000** |
| **2.2** | 11 | **56** | 300 | **1,500** | 7,500 | **39,000** | 200,000 |
| 2.4 | **12** | 62 | **330** | 1,600 | **8,200** | 43,000 | **220,000** |
| **2.7** | 13 | **68** | 360 | **1,800** | 9,100 | **47,000** | |
| 3.0 | **15** | 75 | **390** | 2,000 | **10,000** | 51,000 | |
| **3.3** | 16 | **82** | 430 | **2,200** | 11,000 | **56,000** | |
| 3.6 | **18** | 91 | **470** | 2,400 | **12,000** | 62,000 | |
| **3.9** | 20 | **100** | 510 | **2,700** | 13,000 | **68,000** | |
| 4.3 | **22** | 110 | **560** | 3,000 | **15,000** | 75,000 | |
| **4.7** | 24 | **120** | 620 | **3,300** | 16,000 | **82,000** | |
| **Megohms** | | | | | | | |
| 0.24 | 0.43 | 0.75 | 1.3 | 2.4 | 4.3 | 7.5 | 13.0 |
| **0.27** | **0.47** | **0.82** | **1.5** | **2.7** | **4.7** | **8.2** | **15.0** |
| 0.30 | 0.51 | 0.91 | 1.6 | 3.0 | 5.1 | 9.1 | 16.0 |
| **0.33** | **0.56** | **1.0** | **1.8** | **3.3** | **5.6** | **10.0** | **18.0** |
| 0.36 | 0.62 | 1.1 | 2.0 | 3.6 | 6.2 | 11.0 | 20.0 |
| **0.38** | **0.68** | **1.2** | **2.2** | **3.9** | **6.8** | **12.0** | **22.0** |

NOTE: ±10% resistors are available in bold values only; ±5% resistors are available in all values listed.

## Self-Test

**9.** Determine the nominal value and the range for the following resistors:

**a.** red-violet-yellow-gold
**b.** brown-black-black-silver
**c.** yellow-violet-brown-gold
**d.** brown-red-red-silver
**e.** green-blue-gold-gold

**10.** Determine the color bands for the following resistors:

**a.** 1 MΩ, ±5 percent
**b.** 27 kΩ, ±10 percent
**c.** 1.2 Ω, ±5 percent
**d.** 680 kΩ, ±5 percent
**e.** 1.5 kΩ, ±10 percent

**11.** A yellow fifth band on a resistor denotes its ______.

**12.** Five percent resistors are available in about ______ as many standard values as are 10 percent resistors.

**13.** Standard values are almost evenly spaced on a logarithmic scale to provide ______ coverage.

## 5-3 VARIABLE RESISTORS

*Variable resistors* are used where adjustment of circuit performance is required. They may be used to vary volume, intensity, speed, and voltage, and to calibrate various instruments. A variable resistor consists of a resistance element and a contact that moves along the element. The resistance element may be conductive plastic, carbon, wire-wound, or cermet. The moving contact is called a *wiper* and is usually made of metal. The wiper may be driven directly by a shaft or indirectly by a threaded rod.

Figure 5-9 shows the physical appearance of some variable resistors. Most of the units shown are called *potentiometers,* or *pots.* This is due to their ability to provide a varying potential difference from the wiping contact (usually the center lead) to

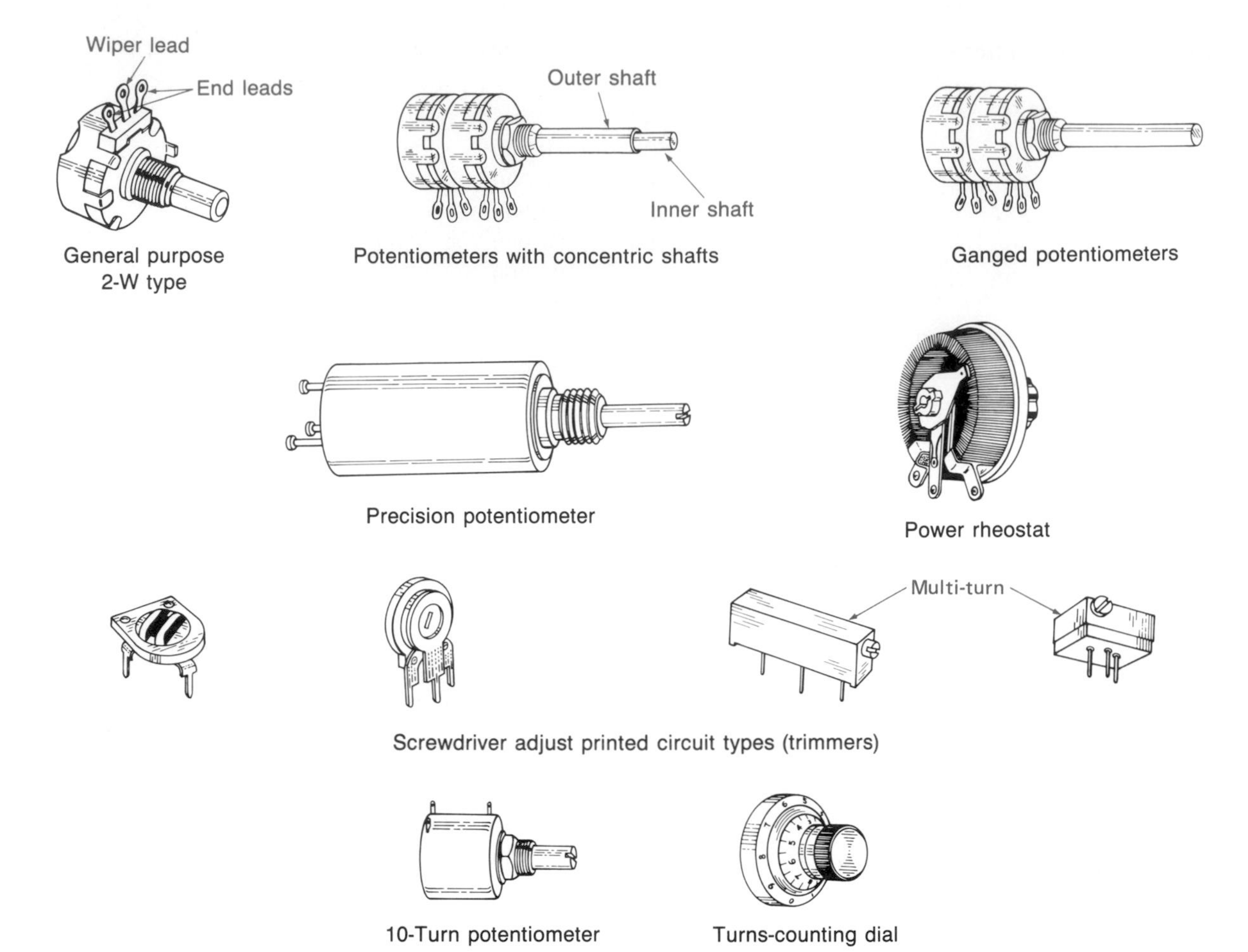


**FIGURE 5-9 Variable resistors.**

either end lead. Potentiometers may be *ganged* in groups of two or more. Figure 5-10 shows the schematic symbols for a single potentiometer and for two ganged potentiometers. In all cases, an arrow is used to indicate the wiping contact. The ganged pots have a broken line that connects the wipers. Both wipers are moved by a single control shaft. Figure 5-9 shows the appearance of a typical pair of ganged potentiometers. A similar physical appearance is presented by potentiometers with concentric control shafts. These are used to conserve panel space on equipment and are not the same as ganged controls. Concentric controls are fitted with a dual knob arrangement to allow separate adjustment for each wiping contact, while ganged controls are simultaneously driven by a single control knob.

The screwdriver adjust pots (trimmers) shown in Fig. 5-9 are usually mounted on printed circuit boards. They are used in applications where only occasional adjustments are required. The multiturn types use a threaded mechanism to move the wiper. Several turns of the screw are required for complete wiper travel. They are easier to adjust when good resolution is required.

The 10-turn potentiometer shown at the bottom of Fig. 5-9 is designed for front-panel mounting. It also provides good resolution and may be fitted with a turns-counting dial to indicate the relative position of the wiping contact. Multiturn pots and counting dials are expensive and are used in demanding applications such as precision instruments.

The precision potentiometer shown in Fig. 5-9 is also a costly item. These are available with linearity tolerances as small as ±0.1 percent. They are used when it is necessary to precisely convert shaft angle to resistance.

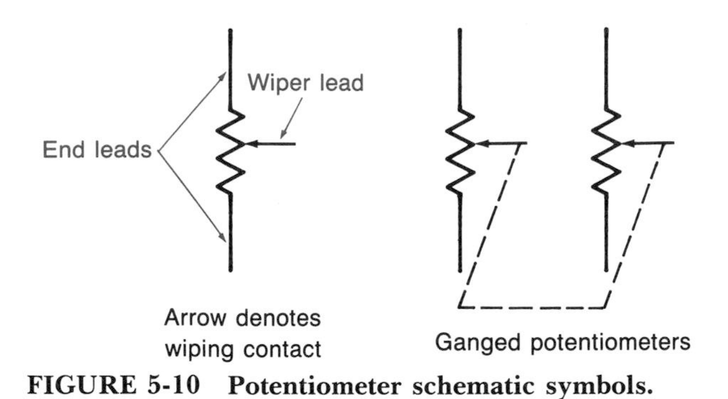


**FIGURE 5-10 Potentiometer schematic symbols.**

A *linear taper* potentiometer is one that has a straight-line relationship between shaft rotation and the resistance from one end lead to the wiper lead. Figure 5-11 illustrates the potentiometer taper. A linear taper shows a 1:1 correlation between the percentage of rotation and the percentage of resistance. The resistance percentage is 50 percent at 50 percent rotation, 70 percent at 70 percent rotation, and so on. Potentiometers are also manufactured with logarithmic tapers. A pot with a log taper will show 10 percent resistance at 50 percent rotation. A true log taper is not practical in all applications since the first 30 percent of rotation produces almost no resistance change. The modified log taper shows 20 percent resistance at 50 percent rotation and eliminates the "dead range." Human hearing is approximately logarithmic for loudness, and modified log pots are used for volume controls.

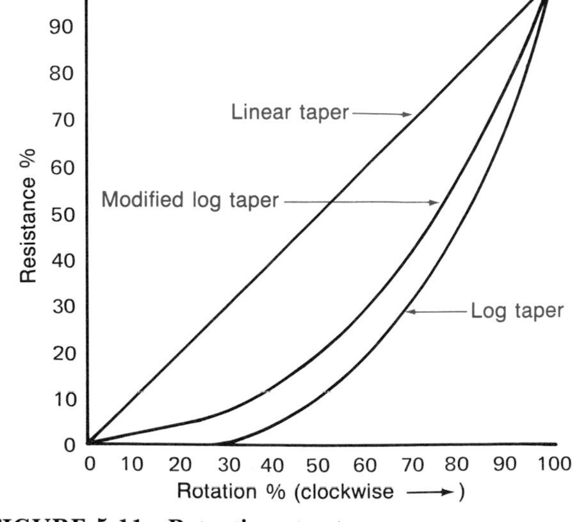


**FIGURE 5-11 Potentiometer taper.**

Figure 5-9 also shows a power rheostat. Note that it has three terminals. Rheostats are two-terminal adjustable resistors. Figure 5-12 shows the rheostat schematic symbol. It also shows that a three-terminal rheostat can be converted to a two-terminal device by connecting the wiper lead to either end lead. Having three terminals allows two different control actions. When terminal 1 is connected to the wiper (terminal 2), clockwise rotation provides increasing resistance. When terminal 3 is connected to the wiper, clockwise rotation gives decreasing resistance.

People expect clockwise rotation of a control to increase the controlled quantity (such as speed or intensity). Likewise, they expect counterclockwise rotation to provide a decrease. Depending on how the circuit works, an increase in the controlled quantity may require an increase in resistance or a decrease in resistance. This is why it is convenient to have three-terminal rheostats. They can always be configured so that clockwise rotation increases the controlled quantity.

Potentiometers having a linear taper can also be configured to allow either increasing or decreasing resistance with clockwise rotation. However, log taper pots cannot because the resistance-versus-rotation performance would be inverted. For this reason, potentiometers are manufactured with the following taper types:

1. Linear
2. Clockwise modified logarithmic
3. Counterclockwise modified logarithmic
4. Clockwise exact logarithmic
5. Counterclockwise exact logarithmic

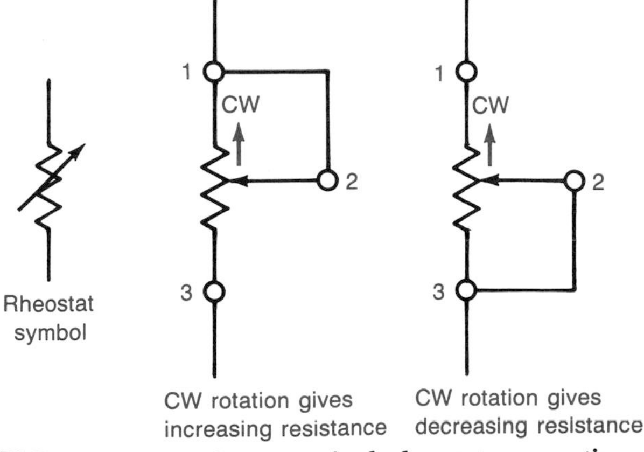


**FIGURE 5-12 Three-terminal rheostat connections.**

## Self-Test

**14.** The moving contact in a variable resistor is called a ______.

**15.** When multiple potentiometers are driven by a single shaft, they are said to be ______.

**16.** Dual knobs are used with ______ control shafts.

**17.** A potentiometer shows 50 percent resistance at 50 percent rotation. Identify its taper.

**18.** A potentiometer shows 10 percent resistance at 50 percent rotation. Identify its taper.

**19.** A potentiometer shows 20 percent resistance at 50 percent rotation. Identify its taper.

**20.** True or false? A three-terminal rheostat can be connected to provide either increasing or decreasing resistance with counterclockwise rotation.

**21.** True or false? A clockwise modified log pot can be connected to provide the same performance as a counterclockwise modified log pot.

## 5-4 OTHER RESISTIVE DEVICES

The resistors that have been discussed to this point are *linear* devices. Figure 5-13 shows that the volt-ampere characteristic curve for a 1-kΩ resistor is a straight line. Ohm's law can be used to verify some data points:

$$I = \frac{5\text{ V}}{1\text{ k}\Omega} = 5\text{ mA}$$

$$I = \frac{2\text{ V}}{1\text{ k}\Omega} = 2\text{ mA}$$

$$I = \frac{-4\text{ V}}{1\text{ k}\Omega} = -4\text{ mA}$$

The last calculation shows a negative voltage and a negative current. The current is negative because it flows in the opposite direction due to the polarity being reversed across the resistor. The polarity reversal is designated by $-V$. The volt-ampere characteristic curve for a linear resistor is a straight line that passes through the origin. The origin is the point on the graph in Fig. 5-13 where $V = 0$ and $I = 0$. The slope of the line is set by the resistance value. The curve for a 2-kΩ resistor shows half as much slope as the curve for a 1-kΩ resistor.

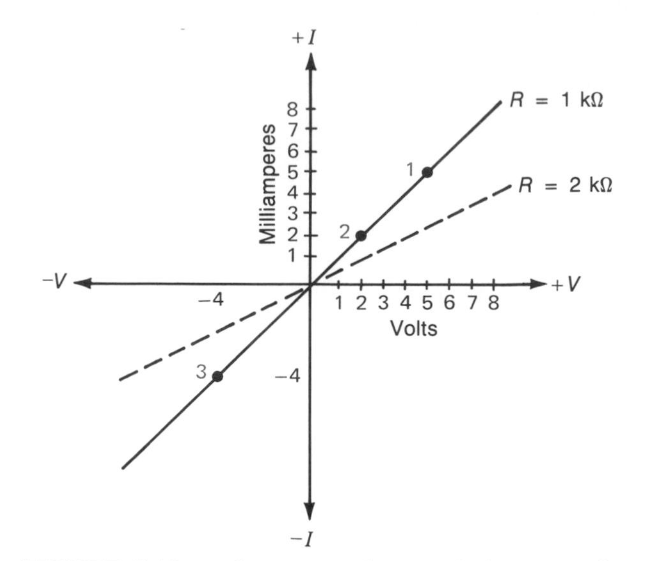


**FIGURE 5-13 Volt-ampere characteristic curves for linear resistors.**

Not all resistive devices are linear. Figure 5-14 shows the characteristic curve for a *varistor*. Varistors are voltage-dependent resistors. They show a decreasing resistance with increasing voltage. They are semiconductor devices and may be made from silicon carbide or zinc oxide. The curve in Fig. 5-14 is for a metallic-oxide varistor (MOV) device. The curve shows a very small slope and thus a very high resistance from the origin to either knee. Once the

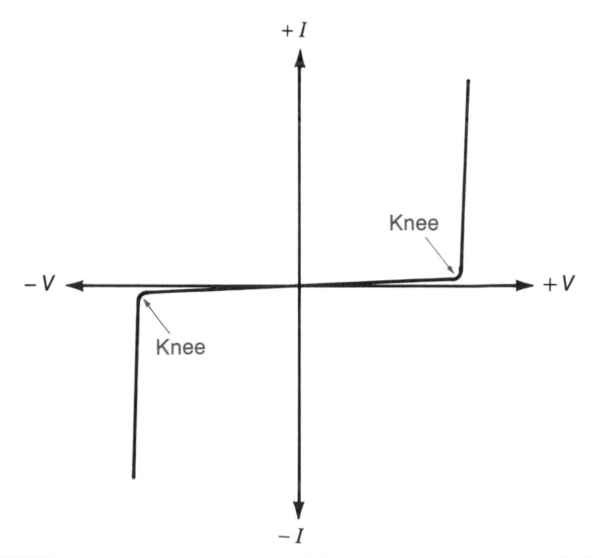


**FIGURE 5-14 Metallic-oxide varistor characteristic curve.**

knee voltage is exceeded, the curve shows a dramatic increase in slope and a drop in resistance.

MOVs can be used to protect equipment from voltage surges. As long as the voltage applied is less than the knee voltage, the varistor remains in its high-resistance state and draws very little current. A surge that exceeds the knee voltage drives the varistor into its low-resistance state. The varistor then conducts a relatively large amount of current and dissipates much of the surge as heat. Without a varistor in the circuit, the surge dissipates in other devices and may damage them. Varistors are often built into sensitive circuits, such as computers, or they can be added to the line that supplies the equipment. Outlet strips are available with built-in surge protection. Figure 5-15 shows the physical appearance of some typical MOVs along with the schematic symbol. There is also an alternate schematic symbol which shows a resistance element with a nonlinear curve drawn through it.

*Thermistors* are another type of nonlinear resistance device. They are thermally sensitive resistors. They have a pronounced negative temperature coefficient and are made from oxides of manganese, nickel, cobalt, or strontium. Barium titanate thermistors with a positive temperature coefficient are also available but are not as widely applied. Figure 5-16 shows a characteristic curve for a typical negative temperature coefficient thermistor. The device resistance drops as its temperature increases due to increased power dissipation. Ohm's law can be used to find the resistance at two operating points:

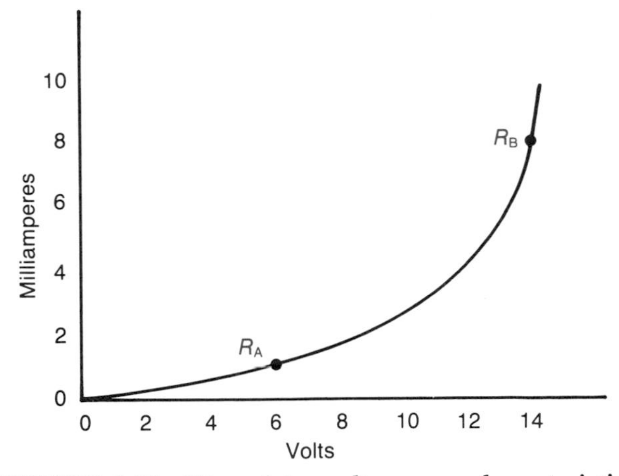


**FIGURE 5-16 Thermistor volt-ampere characteristic curve.**

$$R_A = \frac{V_A}{I_A} = \frac{6\ \text{V}}{1\ \text{mA}} = 6\ \text{k}\Omega$$

$$R_B = \frac{V_B}{I_B} = \frac{14\ \text{V}}{8\ \text{mA}} = 1.75\ \text{k}\Omega$$

The power dissipation at the two operating points is found by

$$P = V \times I = 6\ \text{V} \times 1\ \text{mA} = 6\ \text{mW}$$
$$= 14\ \text{V} \times 8\ \text{mA} = 112\ \text{mW}$$

Increasing power dissipation raises the temperature of the thermistor, which causes its resistance to decrease. This is known as the *self-heating mode*.

Thermistors can be used in the self-heating mode to limit surge current. We have discussed how certain devices, such as incandescent lamps, draw large currents when first turned on. Thermistors allow a gradual buildup of current. They present a relatively high resistance when first turned on. This high resistance limits current flow. As they heat up, their resistance drops to some low value, allowing more current to flow in the circuit.

Thermistors are also used in the external heating mode. This mode allows them to sense the temperature of another device or their environment.

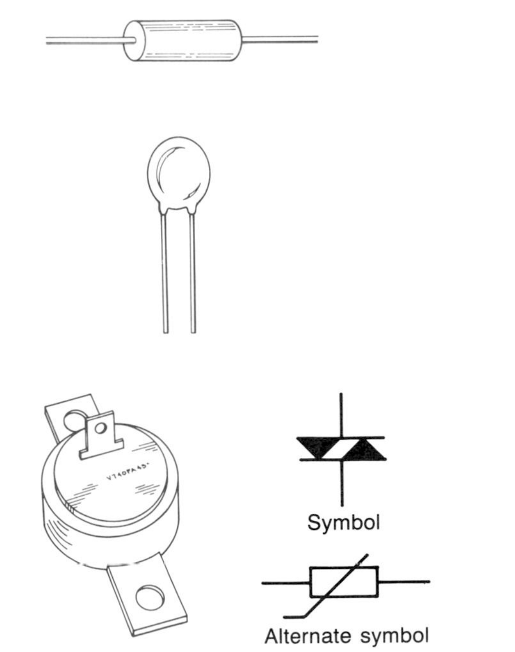


**FIGURE 5-15 Typical MOVs and schematic symbol.**

Figure 5-17 shows how the resistance of a thermistor drops with temperature. Figure 5-18 shows an application for the external heating mode. It is a circuit that issues an alert for excessive temperature. When the thermistor is cool, its resistance is high and only a small current flows through it and the relay coil. When the thermistor gets hot, its resistance drops and enough current flows in the relay coil to close the contacts. This completes the circuit for the signal lamp or buzzer. Note the schematic symbol for the thermistor in Fig. 5-18. Figure 5-19 shows the physical appearance of some thermistors.

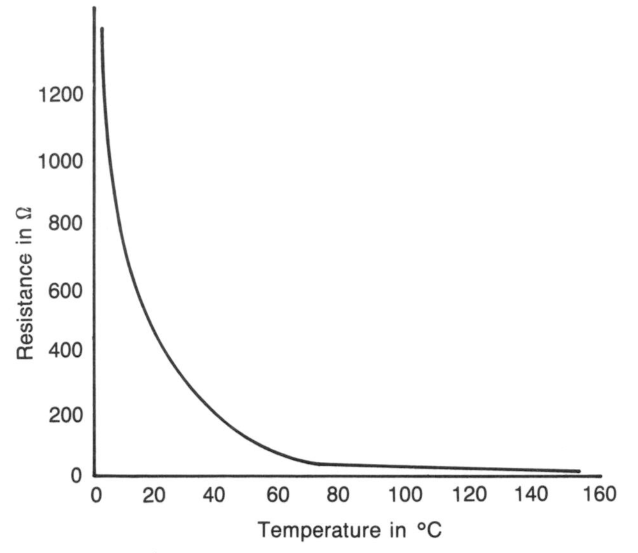


**FIGURE 5-17 Thermistor resistance versus temperature.**

Figure 5-20 shows the schematic symbol and the typical appearance of a light-dependent resistor (LDR). LDRs are made from materials such as cadmium selenide or cadmium sulfide. LDRs are also known as photoconductive cells. They show a pronounced decrease in resistance with increasing illumination as illustrated by the graph of Fig. 5-21. The horizontal axis of the graph is calibrated in lux (lx), the SI unit for surface illumination. Figure 5-22 shows an application for an LDR. The relay contacts are *normally closed* in this circuit. When little or no current flows in the relay coil, the contacts are closed and the auxiliary lighting is on. This oc-

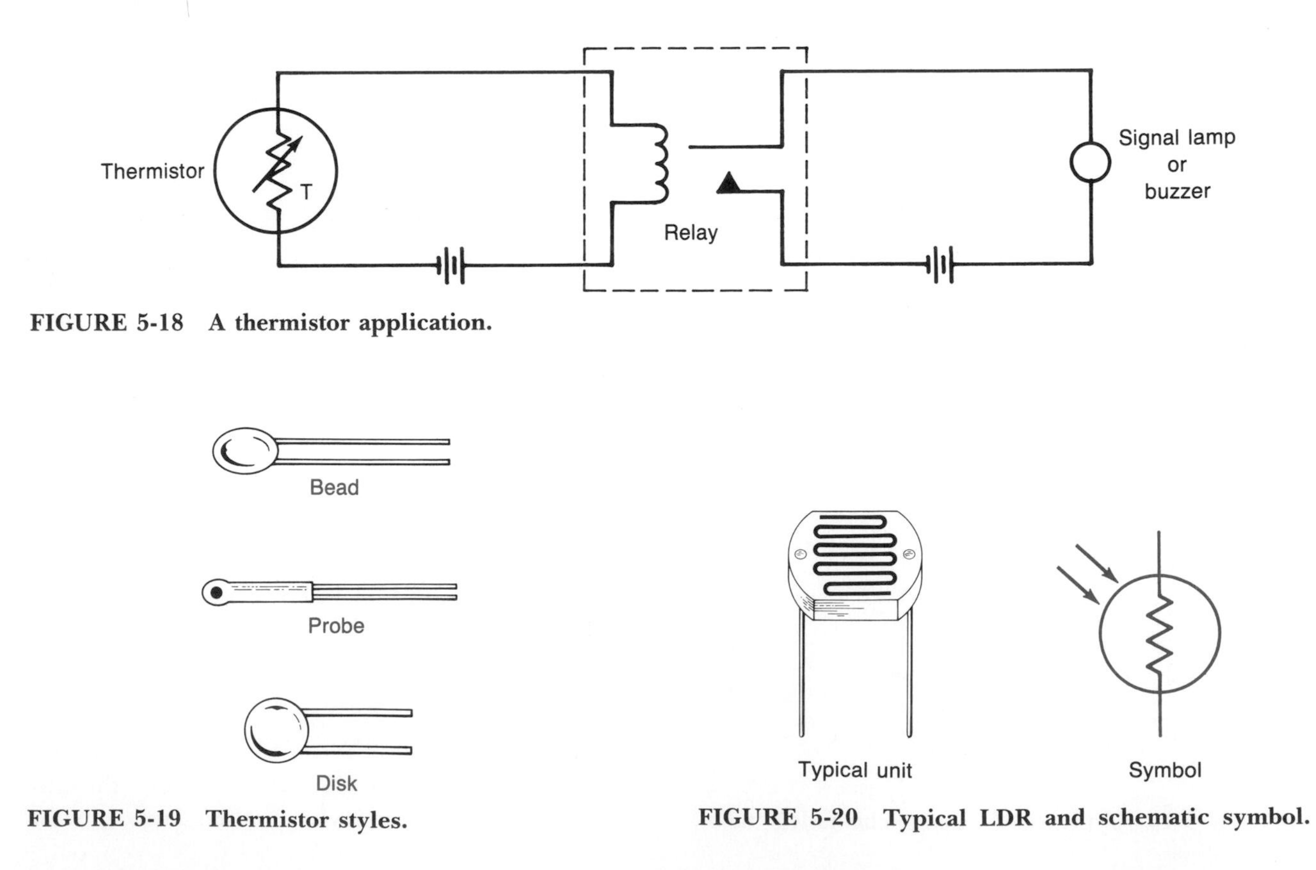


**FIGURE 5-18 A thermistor application.**

**FIGURE 5-19 Thermistor styles.**

**FIGURE 5-20 Typical LDR and schematic symbol.**

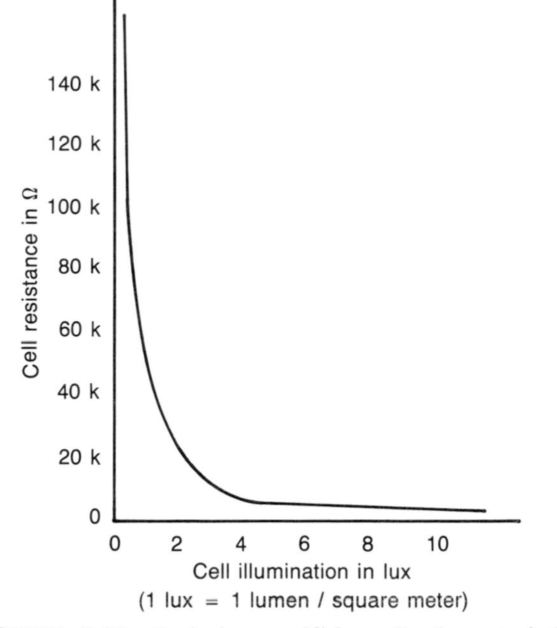


**FIGURE 5-21 Cadmium sulfide cell characteristic curve.**

curs when the cell is in darkness and its resistance is very high. A cadmium sulfide cell shows several megohms of resistance in darkness, and the relay coil current will be practically zero. When the cell is illuminated, its resistance drops and more current flows in the coil. When enough current flows, the contacts open and the auxiliary lighting is turned off. The cell in Fig. 5-22 must not "see" the output of the auxiliary lights. If it does, the circuit will cycle on and off during periods of ambient darkness.

## Self-Test

**22.** The volt-ampere characteristic curve for a linear resistor is a ______ line that passes through the origin of the graph.

**23.** The slope of the volt-ampere characteristic curve of a device is inversely related to the device's ______.

**24.** When potential difference across an MOV is less than the knee voltage, the current through the MOV should be very ______.

**25.** MOVs are often used to protect circuits and equipment from ______ surges.

**26.** Refer to Fig. 5-16. What is the thermistor's resistance when it is conducting 4 mA?

**27.** The resistance of the thermistor in Fig. 5-16 is ______ when it is operating near the origin.

**28.** What happens to the current flow in the relay coil of Fig. 5-18 as the thermistor cools?

**29.** The relay contacts in Fig. 5-18 are normally ______.

**30.** An LDR shows decreasing resistance with ______ illumination.

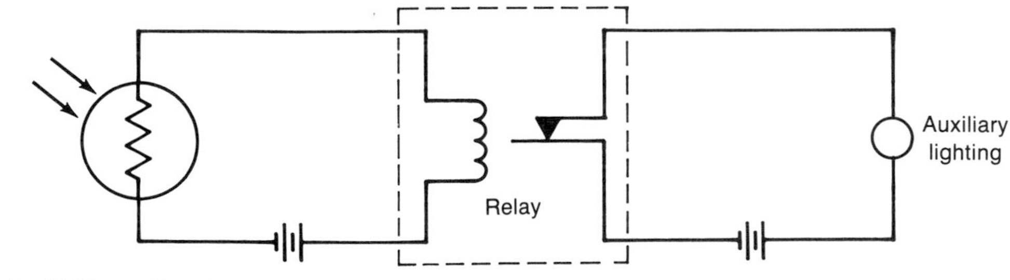


**FIGURE 5-22 LDR application.**

## SUMMARY


1. The power rating of a resistor is directly related to its physical size.
2. Resistors that operate in high environmental temperatures may have to be derated.
3. Wire-wound resistors safely operate at higher temperatures than carbon types.
4. Carbon resistors are available with power ratings to 2 W, while wire-wound resistors are available up to 225 W.
5. A wide band on a resistor body indicates that it is a wire-wound type.
6. Resistors intended for printed circuits may be mounted in SIPs or DIPs.
7. SIP and DIP resistors use cermet elements.
8. Metal film resistors have much better temperature stability than carbon types.
9. Carbon resistors have a temperature coefficient as large as −8000 ppm/°C.
10. The resistance of a resistor is often marked on its body with color bands.
11. The indicated value of a resistor is known as its nominal value.
12. The actual value of a resistor can vary above or below its nominal value by an amount equal to its tolerance.
13. Some resistors use a fifth color band to indicate their reliability.
14. Resistors that are ±5 percent are manufactured in approximately twice as many standard values as ±10 percent resistors.
15. Standard resistor values are spaced almost equally on a logarithmic scale to provide nearly continuous coverage of all resistance values.
16. The moving contact in a variable resistor is called the wiper.
17. Ganged variable resistors are driven by a single shaft.
18. Two variable resistors may be mounted together and have concentric control shafts.
19. Screwdriver adjust potentiometers are also called trimmers.
20. A linear taper potentiometer has a straight-line relationship between rotation angle and resistance.
21. Pots are available with log and modified log tapers.

**22.** A three-terminal variable resistor can be connected as a two-terminal rheostat.

**23.** Log and modified log pots are made in both clockwise and counter-clockwise models.

**24.** The volt-ampere plot for a linear resistor is a straight line.

**25.** Varistors are nonlinear resistors that show decreasing resistance with increasing voltage.

**26.** Thermistors are nonlinear resistors that show decreasing resistance with increasing temperature.

**27.** LDRs show decreasing resistance with increasing illumination.

## CHAPTER REVIEW QUESTIONS

**5-1.** The power rating of a resistor is a function of its ______.

**5-2.** A resistor with a wide band on its body is a ______ type.

**5-3.** Which resistor type is the poorest choice for circuits that must have good temperature stability?

**5-4.** What does a yellow fifth band on a resistor indicate?

**5-5.** Standard resistance values are spaced almost equally on a ______ scale.

**5-6.** The moving contact in a variable resistor is identified by an ______ on the schematic symbol.

**5-7.** True or false? Ganged potentiometers have concentric control shafts.

**5-8.** Trimmer pots are adjusted with a ______.

**5-9.** A linear 1-kΩ pot is rotated to its midposition. What resistance value should be measured across its end contacts? From either end to the center contact?

**5-10.** Linear, log, and modified log are terms that describe the ______ of a pot.

**5-11.** Why do rheostats often have three terminals when their schematic symbols show only two?

**5-12.** True or false? A clockwise log pot can be substituted for a counter-clockwise log pot by reversing the end connections.

**5-13.** True or false? A linear pot can provide proper clockwise or counter-clockwise performance by reversing the end connections.

**5-14.** What will the volt-ampere plot for a short circuit look like?

**5-15.** What will the volt-ampere plot for an open circuit look like?

**5-16.** Which nonlinear device is used to protect circuits from excess voltage?

**5-17.** What does MOV abbreviate?

**5-18.** Which nonlinear device can be used to sense temperature?

**5-19.** Which nonlinear device can be used to prevent turn-on current surge?

**5-20.** A relay coil is conducting no current and its contacts are closed. What type of contacts are they?

**5-21.** True or false? The resistance of a photoconductive cell in total darkness is typically several hundred ohms.

## CHAPTER REVIEW PROBLEMS

**5-1.** Refer to Fig. 5-3. What is the maximum safe current in a 890-Ω, $\frac{1}{2}$-W carbon resistor operating in an 85°C environment?

**5-2.** A carbon composition resistor has a temperature coefficient of −3000 ppm/°C. If its resistance is 100 kΩ at 30°C, what is it at 70°C?

**5-3.** A metal film resistor has a temperature coefficient of −100 ppm/°C. If its resistance is 100 kΩ at 30°C, what is it at 70°?

**5-4.** Find the nominal value and the resistance range for each of the following color codes:

**a.** orange-black-gold-gold
**b.** yellow-violet-black-silver
**c.** brown-red-brown-gold
**d.** brown-gray-orange-gold
**e.** orange-orange-green-silver

**5-5.** Determine the color code for each of the following resistors:

**a.** 1.5 Ω, ±5 percent
**b.** 11 Ω, ±5 percent
**c.** 220 Ω, ±10 percent
**d.** 180 kΩ, ±10 percent
**e.** 10 MΩ, ±5 percent

## ANSWERS TO SELF-TESTS

1. size
2. 80°C
3. wire-wound
4. DIP
5. 141 Ω
6. wire-wound
7. carbon composition
8. precision wire-wound
9. **a.** 270,000; 256,500 to 283,500
   **b.** 10; 9 to 11
   **c.** 470; 446.5 to 493.5
   **d.** 1200; 1080 to 1320
   **e.** 5.6; 5.32 to 5.88
10. **a.** brown-black-green-gold
    **b.** red-violet-orange-silver
    **c.** brown-red-gold-gold
    **d.** blue-gray-yellow-gold
    **e.** brown-green-red-silver
11. reliability
12. twice
13. continuous
14. wiper
15. ganged
16. concentric
17. linear
18. log
19. modified log
20. true
21. false
22. straight
23. resistance
24. small
25. voltage
26. 3 kΩ
27. greatest
28. it decreases
29. open
30. increasing

# Chapter 6 Series Circuits

There are various ways that electric components are interconnected to form circuits. Circuit laws are used to determine quantities such as current flow, voltage drop, and power dissipation for each component in a circuit. You must be able to recognize the types of circuits and apply the appropriate laws when solving circuits. This chapter introduces the series circuit and discusses the laws that govern it.

## 6-1 CURRENT IN A SERIES CIRCUIT

Examination of a chain shows that the links are in series. One link connects to the next link, which connects to the next link, and so on. Series circuits are similar. One component connects to the next and so on. Examine Fig. 6-1*a*. This is a series circuit. All of the components are connected end to end just as the links in a chain. Notice how this differs from the *parallel* circuit shown in Fig. 6-1*b*. Parallel circuits are covered in Chap. 7. Figure 6-1*c* shows a *series-parallel* circuit. Resistor $R_1$ is in series with the parallel combination of $R_2$ and $R_3$. Series-parallel circuits are covered in Chap. 8. It is essential that we be able to identify how components are connected.

Figure 6-2 illustrates a very important law for series circuits: *The current is the same in any part of a series circuit.* The current $I_1$ runs through resistor $R_1$, $I_2$ is the current through $R_2$, and $I_3$ is the current through $R_3$:

$$I_1 = I_2 = I_3$$

The above equation is specific to Fig. 6-2. The gen-

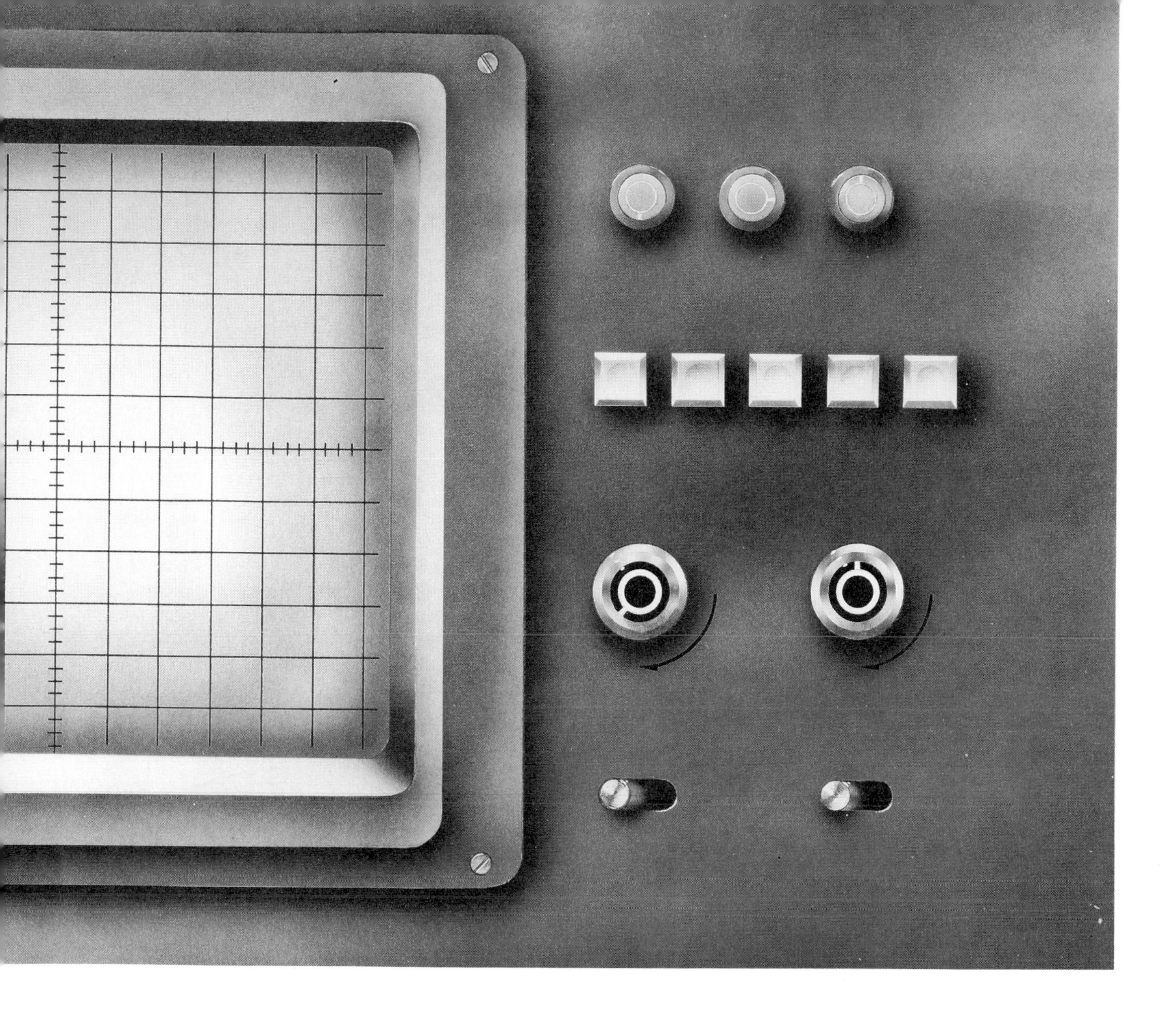

eral equation for current in a series circuit is

$$I = I_1 = I_2 = I_3 = \cdots = I_N$$

The general equation shows that the current is the same for all parts of the circuit up to and including the $N$th or last part of the circuit.

Ohm's law can be used to find the current flow in a series circuit:

$$I = \frac{V}{R_T}$$

where $I$ = current, A
$V$ = applied potential, V
$R_T$ = total resistance of circuit, Ω

The general equation for the total resistance of a series circuit is

$$R_T = R_1 + R_2 + R_3 + \cdots + R_N$$

**EXAMPLE 6-1**
Find the current flow for the circuit shown in Fig. 6-3.

**Solution**
The total resistance is found by adding

$$R_T = 100\ \Omega + 200\ \Omega + 300\ \Omega = 600\ \Omega$$

The current is found by Ohm's law:

$$I = \frac{6\ \text{V}}{600\ \Omega} = 10\ \text{mA}$$

**EXAMPLE 6-2**
Find the current for the circuit shown in Fig. 6-4.

**Solution**
Again, the total resistance must be found. Note

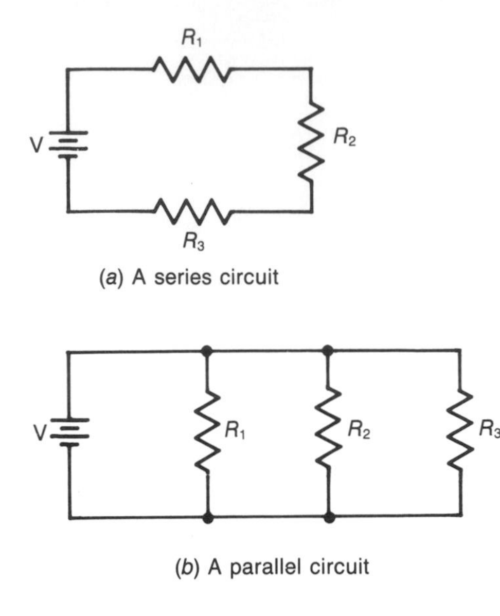


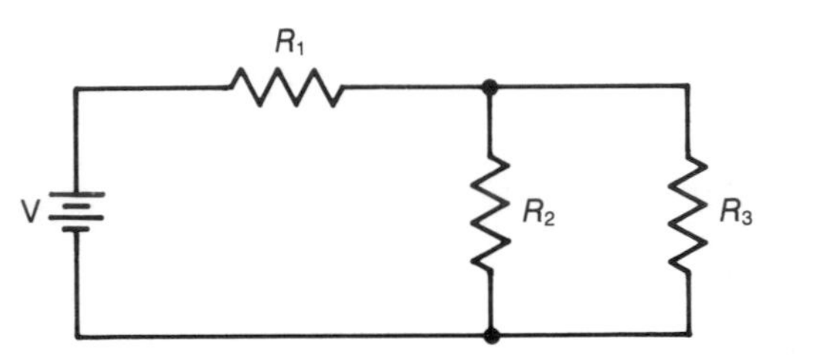


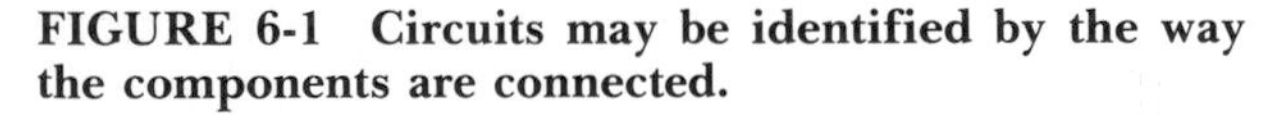

(c) A series-parallel circuit

**FIGURE 6-1 Circuits may be identified by the way the components are connected.**

that some of the resistor values are in ohms and some are in kilohms. All of the resistor values may be expressed in kilohms and then added:

$$R_T = 0.35\ \text{k}\Omega + 1\ \text{k}\Omega + 0.15\ \text{k}\Omega + 2.2\ \text{k}\Omega + 0.3\ \text{k}\Omega$$
$$= 4\ \text{k}\Omega$$

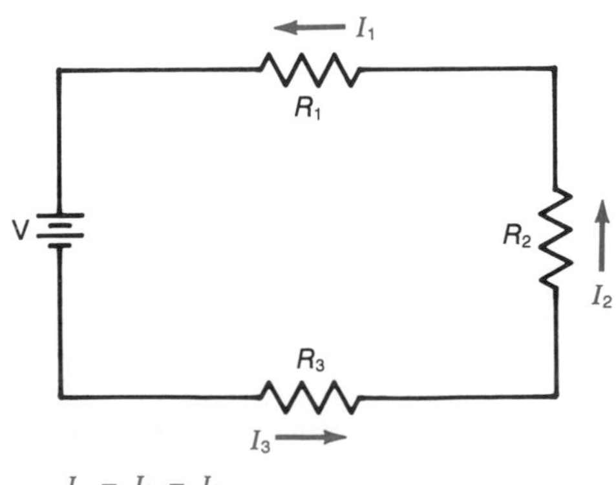


$I_1 = I_2 = I_3$

**FIGURE 6-2 Current is the same in any part of a series circuit.**

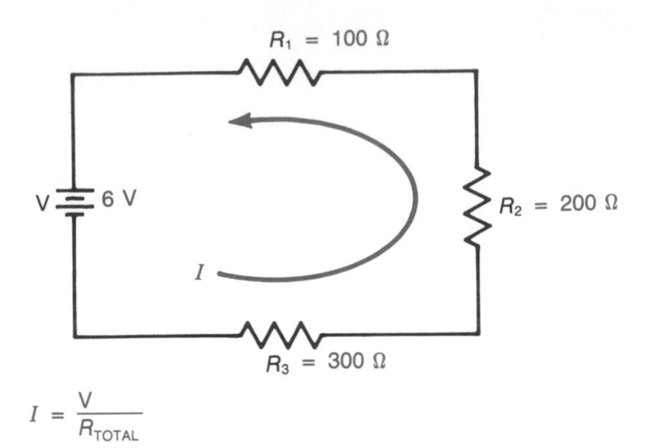


$$I = \frac{V}{R_{TOTAL}}$$

**FIGURE 6-3 The current in a series circuit is determined by the applied voltage and the total resistance.**

Applying Ohm's law gives

$$I = \frac{60\ \text{V}}{4 \times 10^3\ \Omega} = 15\ \text{mA}$$

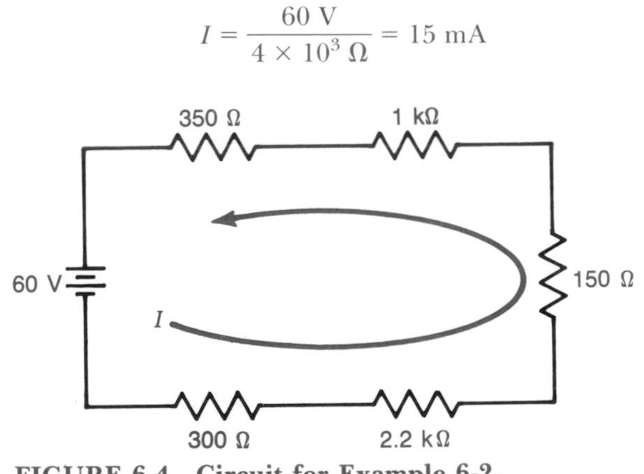


**FIGURE 6-4 Circuit for Example 6-2.**

## Self-Test

1. The current flow is the ________ in all parts of a series circuit.
2. The total resistance of series resistors can be found by ________ the individual resistances.
3. Refer to Fig. 6-5*a*. Will the direction of electron current be clockwise or counterclockwise?
4. Refer to Fig. 6-5*b*. Will the direction of electron current be clockwise or counterclockwise?
5. Calculate the current flow for Fig. 6-5*a*.
6. Calculate the current flow for Fig. 6-5*b*.
7. Calculate the current flow for Fig. 6-5*c*.

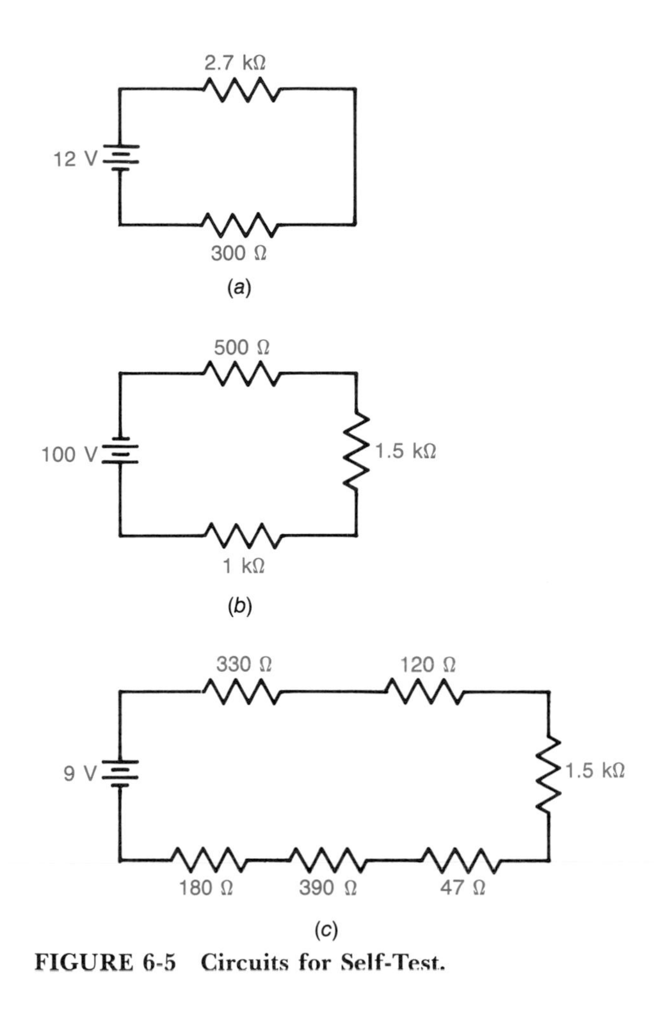


**FIGURE 6-5 Circuits for Self-Test.**

## 6-2 KIRCHHOFF'S VOLTAGE LAW

One of the most significant aspects of series circuits is described by Kirchhoff's voltage law, which states the following: *The algebraic sum of the voltage rises and drops in any closed loop is equal to zero.*

This is a formal statement of the law. A more "friendly" translation is: *The sum of the voltage drops is equal to the applied voltage in a series circuit.*

### EXAMPLE 6-3

Find the individual voltage drops and their sum for the circuit shown in Fig. 6-6.

**Solution**

Begin by finding the total resistance of the circuit:

$$R_T = 50\ \Omega + 100\ \Omega + 150\ \Omega = 300\ \Omega$$

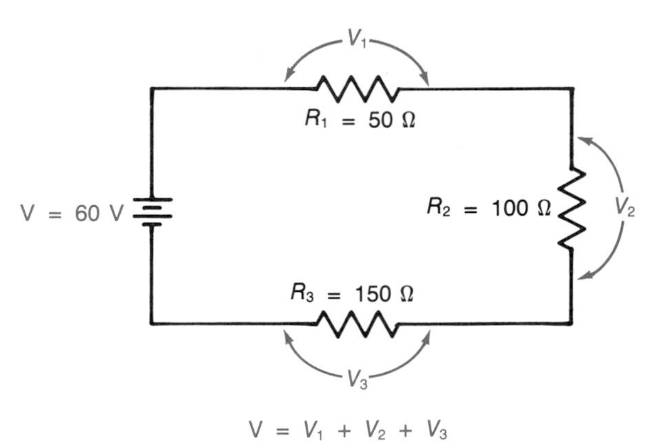


$V = V_1 + V_2 + V_3$

**FIGURE 6-6 The sum of the voltage drops in a series circuit is equal to the applied voltage.**

Next, use Ohm's law to find the current:

$$I = \frac{60\ \text{V}}{300\ \Omega} = 200\ \text{mA}$$

Now we can find the drop across each resistor. Current is constant in a series circuit, so the same 200 mA flows through each resistor. Ohm's law gives

$$V_1 = I \times R_1 = 200 \times 10^{-3} \times 50 = 10\ \text{V}$$
$$V_2 = I \times R_2 = 200 \times 10^{-3} \times 100 = 20\ \text{V}$$
$$V_3 = I \times R_3 = 200 \times 10^{-3} \times 150 = 30\ \text{V}$$

The sum of the voltage drops is

$$V_T = V_1 + V_2 + V_3 = 10 + 20 + 30 = 60\ \text{V}$$

Note that the sum of the drops $V_T$ is equal to the applied voltage. This is a circuit law and must always be true. You should also notice that the voltage drop is proportional to the resistance. The 150-Ω resistor produces three times the drop of the 50-Ω resistor, and the 100-Ω resistor produces twice as much drop as the 50-Ω resistor.

A series circuit may have more than one voltage source. Figure 6-7 shows an example. Voltage sources in series may *aid* or *oppose* one another. The sources in Fig. 6-7 are aiding. Note that the electron current *I* flows out of the negative terminal and into the positive terminal of both voltage sources. Both sources produce a rise in potential for the direction of electron current flow. The voltages are added in cases such as this.

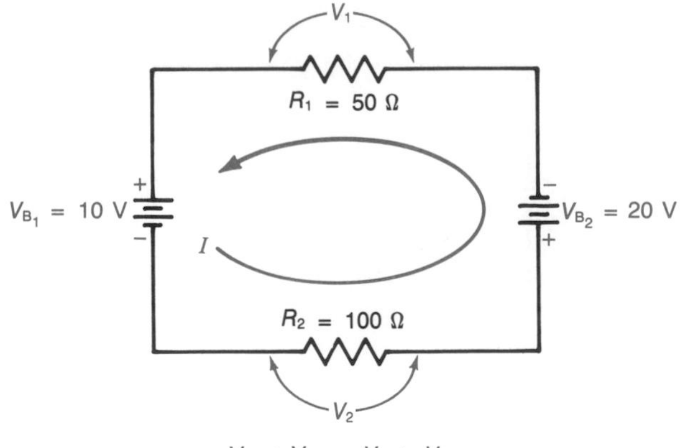


**FIGURE 6-7** **A series circuit with aiding voltage sources.**

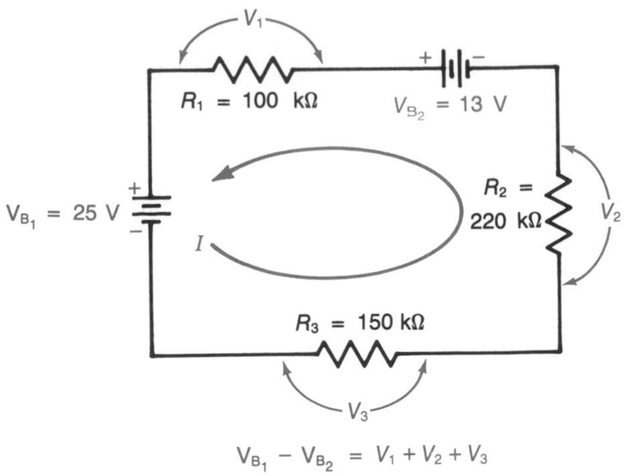

$V_{B_1} - V_{B_2} = V_1 + V_2 + V_3$

**FIGURE 6-8** **A series circuit with opposing voltage sources.**

---

**EXAMPLE 6-4**

Find the current flow, the voltage drops, and the sum of the drops for the circuit shown in Fig. 6-7.

**Solution**

Find the total resistance:

$$R_T = 50\ \Omega + 100\ \Omega = 150\ \Omega$$

Find the current:

$$I = \frac{10\text{ V} + 20\text{ V}}{150\ \Omega} = 200\text{ mA}$$

Find the voltage drops:

$$V_1 = 200 \times 10^{-3} \times 50 = 10\text{ V}$$
$$V_2 = 200 \times 10^{-3} \times 100 = 20\text{ V}$$

The sum of the drops:

$$10 + 20 = 30\text{ V}$$

---

**EXAMPLE 6-5**

Find the current, the voltage drops, and the sum of the voltage drops for the circuit shown in Fig. 6-8.

**Solution**

Find the total resistance:

$$R_T = 100\text{ k}\Omega + 220\text{ k}\Omega + 150\text{ k}\Omega = 470\text{ k}\Omega$$

Find the current:

$$I = \frac{25\text{ V} - 13\text{ V}}{470 \times 10^3\ \Omega} = 25.5\ \mu\text{A}$$

Find the drops:

$$V_1 = 25.5 \times 10^{-6} \times 100 \times 10^3 = 2.55\text{ V}$$
$$V_2 = 25.5 \times 10^{-6} \times 220 \times 10^3 = 5.62\text{ V}$$
$$V_3 = 25.5 \times 10^{-6} \times 150 \times 10^3 = 3.83\text{ V}$$

Find the sum of the drops:

$$2.55 + 5.62 + 3.83 = 12\text{ V}$$

---

The sum of the drops is equal to the sum of the voltage sources. This satisfies Kirchhoff's voltage law. The sum of the source voltages is sometimes called the *effective* voltage.

Figure 6-8 shows a series circuit with *opposing* voltage sources. The direction of current is set by the *dominant* polarity in cases such as these. Source $V_{B_1}$ is greater than source $V_{B_2}$. Its polarity is dominant, and it determines the direction of current. The effective voltage is the difference between the sources when they oppose.

The sum of the drops is equal to the effective applied voltage (25 V − 13 V).

A circuit may have three or more sources with one or more opposing polarities. In these cases, sum all the sources that tend to produce clockwise flow. Then sum all the sources that tend to produce counterclockwise flow. The larger of the two values is the dominant polarity and dictates the direction of current. The effective voltage is equal to the difference between the two values.

Some circuits may be more readily analyzed by reducing them to a simplified but *equivalent* circuit. This technique is shown in Fig. 6-9. The original circuit is at the left. It contains two sources and three resistors. The equivalent circuit is shown at the right. It contains a single source and a single resistor. The current flow in the equivalent circuit is the same as the flow in the original. The equivalent source voltage is determined by subtracting $V_{B_2}$ from $V_{B_1}$ because they are opposing in this example. The polarity is established by the dominant source $V_{B_1}$. The equivalent resistance is found by adding the individual resistance values. Solving the equivalent circuit for current flow is a matter of applying Ohm's law. The current value can then be transferred back to the original circuit to calculate the individual drops. The value of equivalent circuits will be obvious later in Chap. 9 when more complex networks are presented.

Figure 6-10 shows how to assign polarity to the voltage drops in a series circuit. The source establishes the direction of electron current. When current overcomes resistance, a drop in potential occurs. The drop in potential is from negative to positive when electron current is used. Note, in Fig. 6-10, that the electron flow is from negative to positive in the case of each resistor. This is always the case with resistors because they must produce a drop in potential. The flow in the source is from positive to negative. This represents a rise in potential. Algebra requires a rise to have a positive value and a drop to have a negative value. This leads us to an equation for a series loop that is more closely aligned with the formal statement of Kirchhoff's voltage law presented at the beginning of this section.

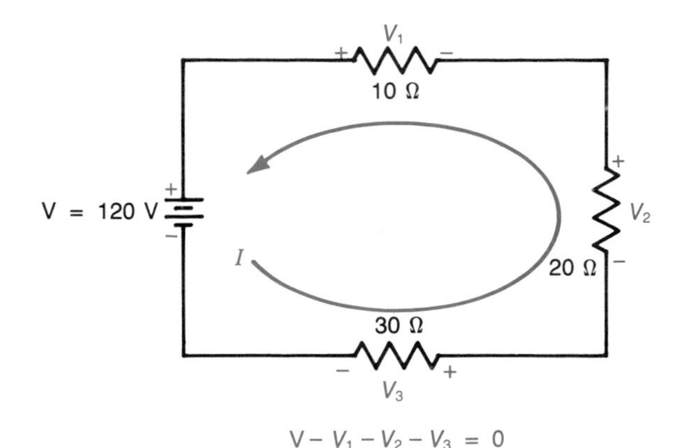


**FIGURE 6-10 Assigning polarity to the voltage drops.**

## EXAMPLE 6-6

Prepare a formal loop equation using the values for the circuit shown in Fig. 6-10.

### Solution

The total resistance is the sum of the resistor values or 60 Ω. The current flow is equal to 120 V divided by 60 Ω or 2 A. The voltage drops will be 20, 40, and 60. The equation, based on these values, is

$$+120 - 20 - 40 - 60 = 0$$

The algebraic sum of the voltage rises and drops in any closed loop is equal to zero. The only rise is due to the 120-V battery. This value is positive. The resistors each produce a voltage drop. These values are negative. The algebraic sum around the series loop is zero.

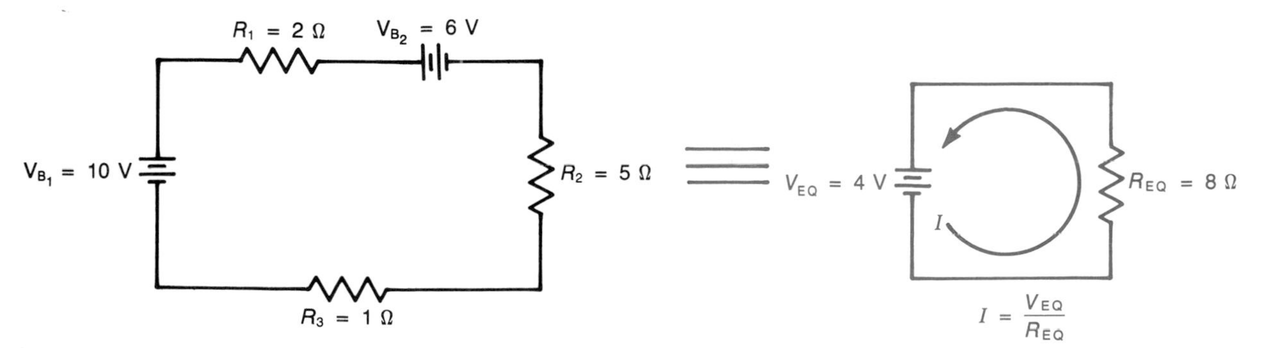


**FIGURE 6-9 An equivalent circuit may be used to simplify analysis.**

Refer back to Fig. 6-8. Voltage $V_{B_2}$ is a drop in this circuit since the current flows from negative to positive. Voltage $V_{B_2}$ is therefore assigned a negative sign. If battery $V_{B_2}$ is a rechargeable type, it will charge in this circuit. It will convert some electric energy to chemical energy and the rest to heat. If it is not a rechargeable type or is already fully charged, it will change all of the electric energy to heat. In either case, it converts electric energy and produces a voltage drop.

Ohm's law tells us that $V = IR$ and therefore $IR$ can be substituted for $V$ in a loop equation. Figure 6-11 shows a series loop where the direction of electron current flow has been assigned incorrectly. It will be shown that this does not produce any serious problems when writing or solving loop equations.

**EXAMPLE 6-7**

Write a loop equation for the circuit shown in Fig. 6-11. Use the indicated direction of flow and substitute $IR$ for all instances of $V$ in the equation. Explain the consequences of the algebraic sign when the equation is solved for $I$.

**Solution**

The assigned direction of flow shows the current through the battery to be from negative to positive. This represents a drop and the algebraic sign must be negative:

$$-30 - IR_1 - IR_2 - IR_3 = 0$$
$$-30 - 10I - 20I - 30I = 0$$
$$-30 - 60I = 0$$
$$-60I = 30$$
$$I = -0.5 \text{ A}$$

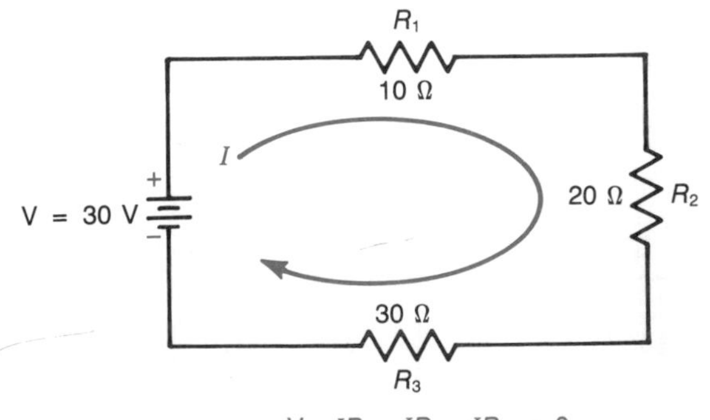


**FIGURE 6-11 A negative current results when the assumed direction of current is incorrect.**

The current flow is correct as far as its absolute value is concerned. The minus sign tells us the assumed direction of flow is incorrect. The current flow in Fig. 6-11 is 0.5 A and flows counterclockwise around the loop. The polarity of the drops must be assigned according to the correct direction of flow and not according to the assumed direction which was incorrect in this case. This does not require rewriting the loop equation. Simply assign the drops based on counterclockwise flow. For example, the drop across $R_3$ in Fig. 6-11 is positive at its right end and negative at its left end.

## Self-Test

**8.** The source voltage in a series circuit is equal to the ________ of the voltage drops.

**9.** A series circuit has one source and five resistors, all of different values. Where will the largest voltage drop be produced?

**10.** For the same situation described in Question 9, where will the smallest drop be produced?

**11.** A series circuit has three resistors and one 5-V source. If drop 1 is 1 V and drop 2 is 2 V, what is the value of drop 3?

All of the following questions refer to Fig. 6-12.

**12.** For circuit a, find all of the drops.

**13.** What is the direction of electron current in circuit a?

**14.** Is the top of $R_2$ in circuit a positive or negative with respect to the bottom of $R_2$?

**15.** For circuit b, find all of the drops.

**16.** What is the direction of electron current in circuit b?

**17.** For circuit c, find all of the drops.

**18.** Is the top of $R_2$ in circuit c positive or negative with respect to the bottom of $R_2$?

**19.** Write a formal loop equation for circuit c. Assume a clockwise flow and substitute the resistance value times the current for each voltage drop.

**20.** For circuit d, find all of the drops.

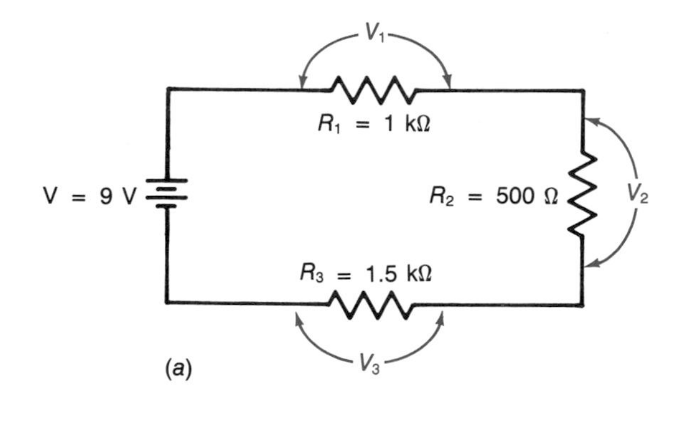


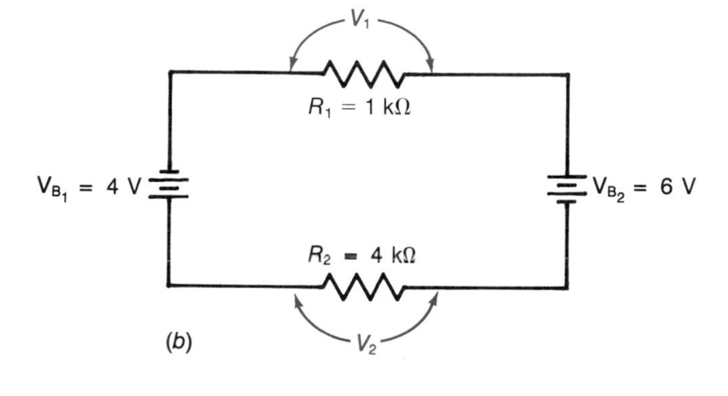


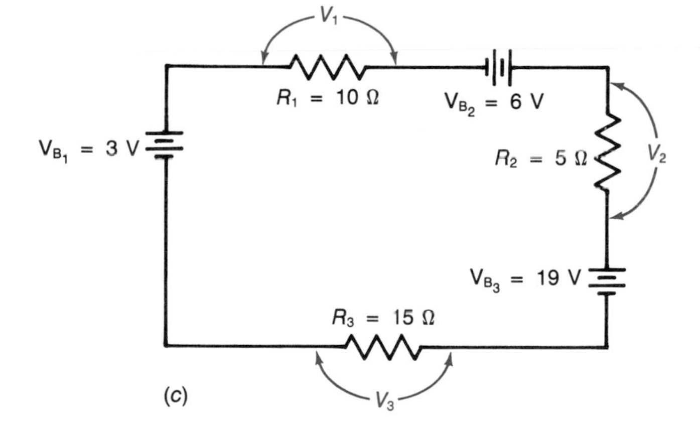


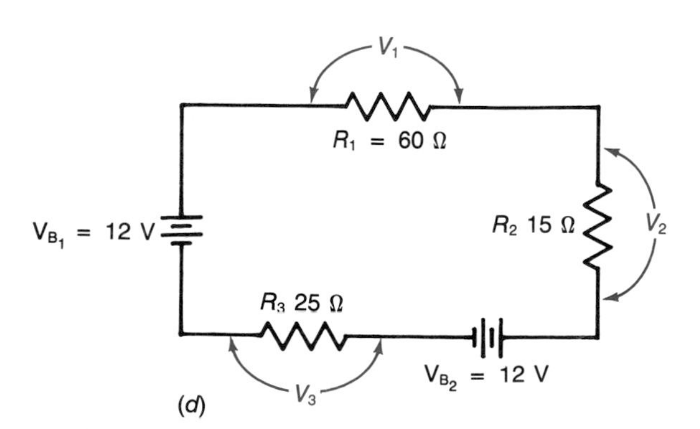


**FIGURE 6-12 Circuits for Self-Test.**

## 6-3 VOLTAGE DIVIDERS

Series resistors are often used to divide a voltage into smaller parts. The energy source for a circuit may develop only a single voltage. When a circuit requires several voltages, they can often be obtained from a single source by using a divider. The voltage divider rule is a useful shortcut to determine how series resistors will divide voltage. Figure 6-13 shows that the voltage drop across any resistor in a series string can be found by dividing the value of the resistor by the total string resistance and multiplying by the source voltage. This technique eliminates the need to solve for the current flow. The derivation for the drop across $R_1$ in Fig. 6-13 is straightforward:

$$V_{R_1} = IR_1$$

$$I = \frac{V}{R_T}$$

Substituting the second equation into the first gives

$$V_{R_1} = \frac{V}{R_T} \times R_1 = \frac{V \times R_1}{R_T}$$

**EXAMPLE 6-8**

Find the drops in Fig. 6-13 using the voltage divider rule.

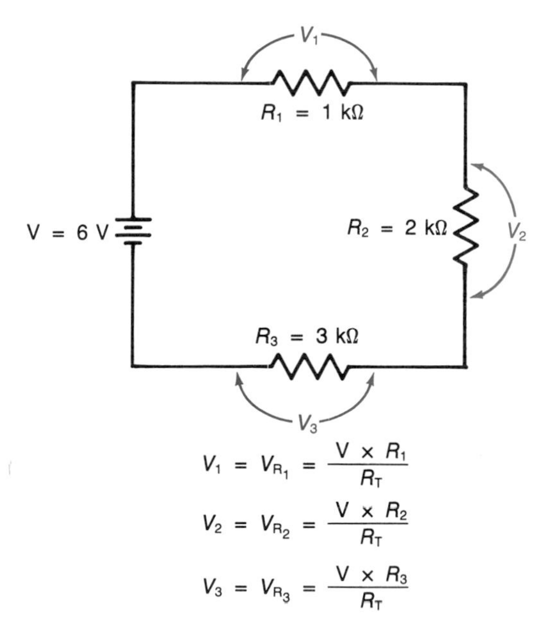

**FIGURE 6-13 The voltage divider rule.**

**Solution**

The total resistance is equal to 6 kΩ. This is divided into the value of each individual resistor, which is also in kilohms. In this case, the kilohm units cancel and need not be entered into the calculations:

$$V_1 = \frac{6 \times 1}{6} = 1\ \text{V}$$
$$V_2 = \frac{6 \times 2}{6} = 2\ \text{V}$$
$$V_3 = \frac{6 \times 3}{6} = 3\ \text{V}$$

---

The sum of the drops is equal to the applied voltage, which satisfies Kirchhoff's voltage law.

Potentiometers (pots) are adjustable voltage dividers. Refer to Fig. 6-14. When the wiper arm is in the center position of a pot with a linear taper, the voltage drops are equal:

$$V_{AB} = V_{BC} = 5\ \text{V}$$

Note that two equal-value resistances divide the applied voltage $V$ into two equal parts. You can verify this with the voltage divider rule:

$$V_{Drop} = \frac{V \times R}{2R} = \frac{V}{2}$$

When the wiper arm in Fig. 6-14 is moved up (toward A), $V_{AB}$ decreases and $V_{BC}$ increases. The sum of the two drops is equal to 10 V and either drop is adjustable from 0 to 10 V.

Voltages $V_{AB}$ and $V_{BC}$ are examples of double-subscript notation. This notation clearly specifies two points for a voltage. The second subscript is considered the reference point. Voltage $V_{AB}$ would be positive in Fig. 6-14 because point A is positive with respect to point B. Voltage $V_{BC}$ would also be positive since point B is positive with respect to point C. Do not confuse the use of "positive" in this context with the algebraic rules set forth earlier for writing loop equations. Reporting $V_{AB}$ as +5 simply means that point A is 5 V positive with respect to point B. The algebraic sign of this drop would still be negative when writing a loop equation.

*Grounds* are often used as reference points for specifying voltages. Figure 6-15 shows three ground symbols. There are different symbols for earth grounds and common grounds. However, the earth ground symbol is often used in place of the common ground symbol on schematics, even when there is no connection to the soil. Of course, a common ground may be connected to an earth ground as shown in the partial schematic of Fig. 6-16. Figure 6-17 shows a more typical approach to the use of ground symbols when schematics are drawn. It is understood that all common grounds are electrically "tied" together.

A *common ground* is a point in a circuit that is common to many of the components used in that circuit. Physically, it could be a wire, a metal strap, or an area of foil on a printed circuit. The important idea is that it is used as a reference point. Figure 6-18 shows a voltage divider with a common ground. This provides another way of specifying voltages. For example, what does $V_A$ mean? There is no such thing as voltage at a single point. Two points are necessary to reference or measure any voltage. When a single point is referenced, the common automatically becomes the other point. Voltage $V_A$ in Fig. 6-18 is the same as $V_{AC}$, and $V_B$ is the same as $V_{BC}$. They are both positive with respect to common:

$$V_A = +6\ \text{V}$$
$$V_B = +3\ \text{V}$$

---

**EXAMPLE 6-9**

Specify the voltages for all of the labeled points in Fig. 6-19.

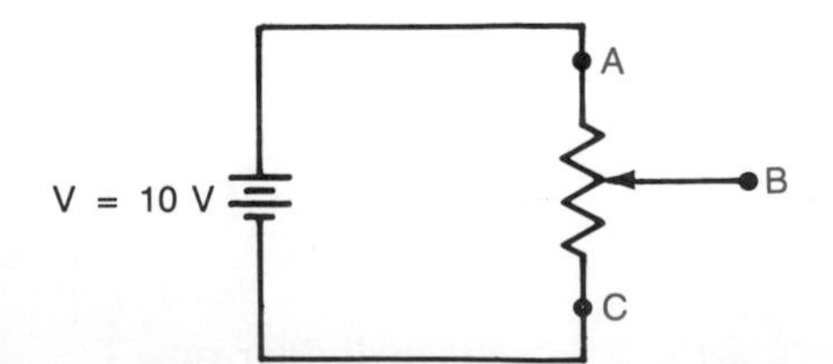


**FIGURE 6-14 Potentiometer voltage divider.**

**FIGURE 6-15 Ground symbols.**

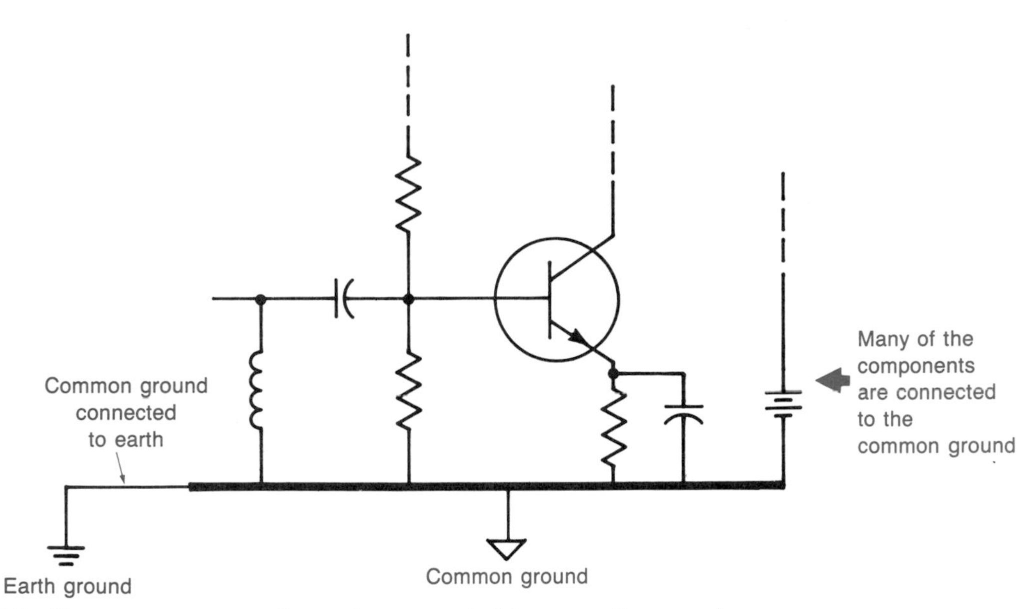


**FIGURE 6-16 The common ground may be connected to an earth ground.**

**Solution**

The voltage divider rule could be used to determine the drops. Note that the resistors are equal in value. When all the resistors in a string are equal in value, each must drop an equal part of the applied voltage. Three equal-value resistors will each drop one-third of the applied voltage or 10 V in this case. The common is point C. Points A and B are positive with respect to common, and point D is *negative* with respect to common. Therefore,

$$V_A = +20 \text{ V}$$
$$V_B = +10 \text{ V}$$
$$V_C = 0 \text{ V}$$
$$V_D = -10 \text{ V}$$

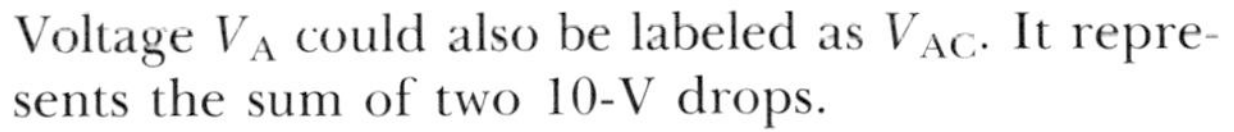

Voltage $V_A$ could also be labeled as $V_{AC}$. It represents the sum of two 10-V drops.

Each resistor in a series circuit or in a voltage divider changes electric energy to heat energy.

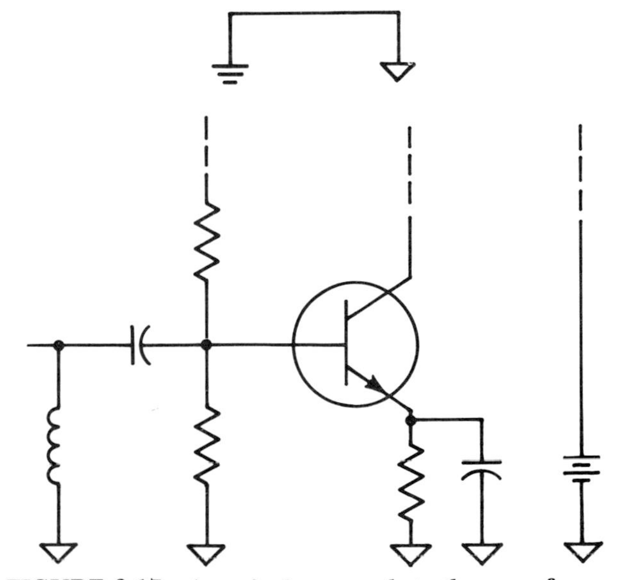

**FIGURE 6-17 A typical approach to the use of common ground symbols.**

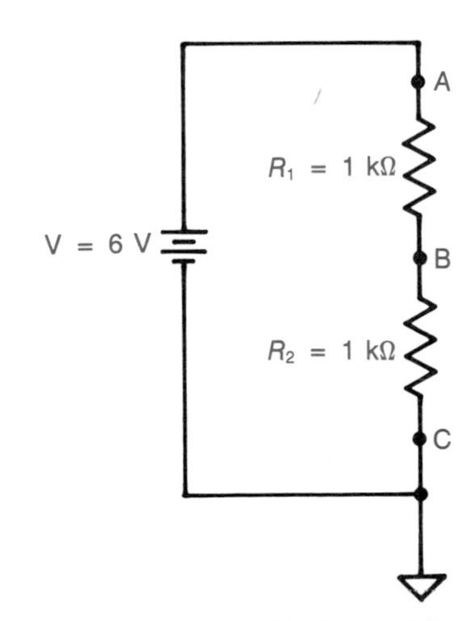


**FIGURE 6-18 A voltage divider with a common ground.**

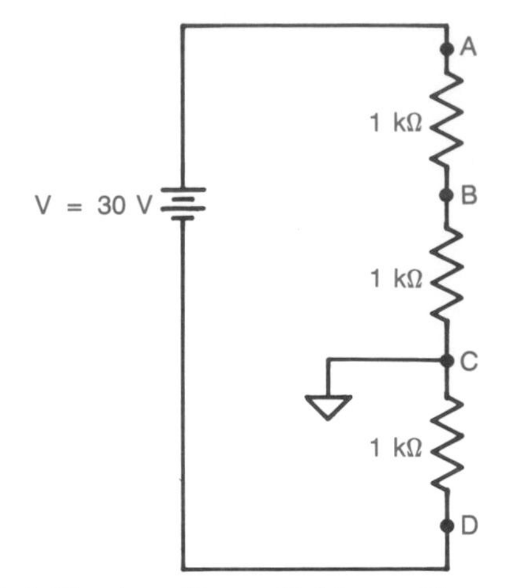


FIGURE 6-19 A voltage divider with a common ground between resistors.

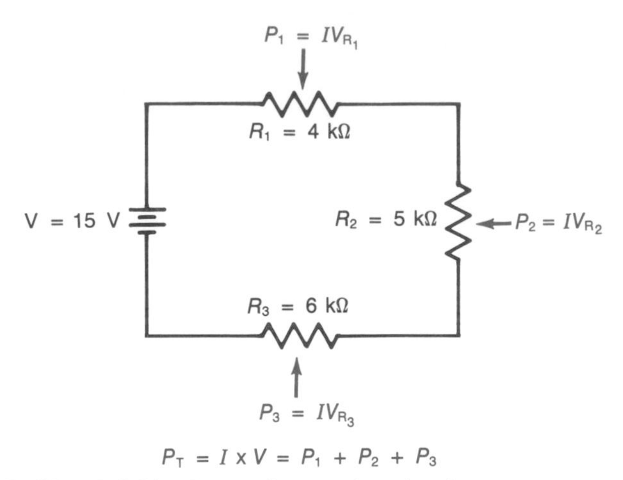


FIGURE 6-20 Power in a series circuit.

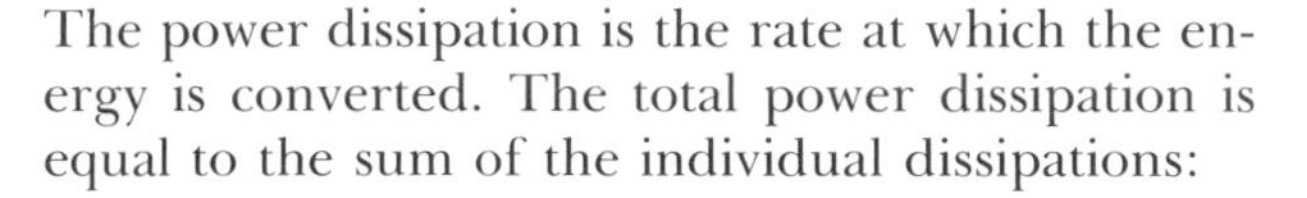

The power dissipation is the rate at which the energy is converted. The total power dissipation is equal to the sum of the individual dissipations:

$$P_T = P_1 + P_2 + P_3 + \cdots + P_N$$

---

**EXAMPLE 6-10**

Calculate the individual dissipations for Fig. 6-20 and verify that their sum is equal to the total power dissipation.

**Solution**

The voltage divider rule could be used to find the individual drops, and then the individual dissipations could be found with $V^2/R$. In this case, we elect to find the current first:

$$I = \frac{V}{R_T} = \frac{15}{15\ \text{k}\Omega} = 1\ \text{mA}$$

Then $I^2R$ could be used next to find the individual dissipations. In this case, we elect to find the voltage drops next:

$$V_{R_1} = 1\ \text{mA} \times 4\ \text{k}\Omega = 4\ \text{V}$$
$$V_{R_2} = 1\ \text{mA} \times 5\ \text{k}\Omega = 5\ \text{V}$$
$$V_{R_3} = 1\ \text{mA} \times 6\ \text{k}\Omega = 6\ \text{V}$$

Now find the individual dissipations:

$$P_1 = 1\ \text{mA} \times 4\ \text{V} = 4\ \text{mW}$$
$$P_2 = 1\ \text{mA} \times 5\ \text{V} = 5\ \text{mW}$$
$$P_3 = 1\ \text{mA} \times 6\ \text{V} = 6\ \text{mW}$$
$$\text{Total dissipation} = 15\ \text{mW}$$

The total dissipation can be verified by multiplying the current times the applied voltage:

$$P_T = 1\ \text{mA} \times 15\ \text{V} = 15\ \text{mW}$$

The total dissipation is equal to the sum of the individual dissipations.

---

## Self-Test

**21.** The voltage divider rule is a technique that eliminates the need to solve for ______ when calculating drops in a series circuit.

**22.** Refer to Fig. 6-14. What happens to $V_{BC}$ as the wiper arm is moved toward A?

**23.** Four 1-kΩ resistors are in series and connected to a source that develops $X$ volts. What is the drop across one of the resistors?

**24.** Refer to Fig. 6-18. What is the value of $V_{BA}$? $V_B$?

**25.** The total power supplied by the source in a series circuit is equal to the ______ of the individual dissipations.

**26.** Refer to Fig. 6-21 and find
$V_{AB}$ $V_{DE}$
$V_A$ $V_E$

**27.** What is the total power dissipation in Fig. 6-21?

**28.** Which resistor in Fig. 6-21 dissipates the most power?

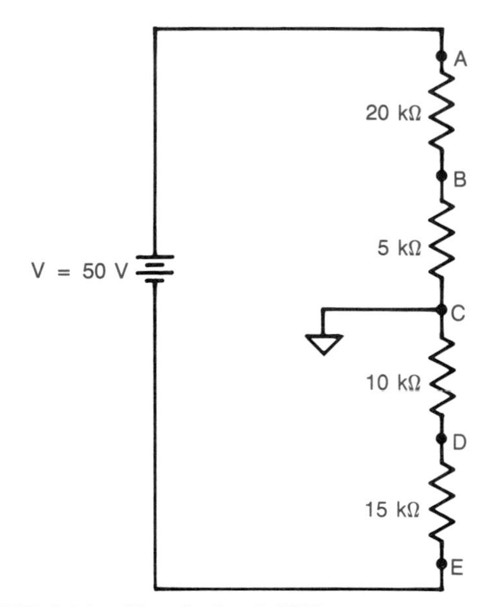


FIGURE 6-21 Circuit for Self-Test.

## 6-4 APPLICATIONS OF SERIES CIRCUIT CONCEPTS

There are various applications for series components. For example, Fig. 6-22 shows a light-emitting diode (LED) circuit. A series resistor is used to limit the current flow in the diode to some desired value. Without the series resistor, the current would be excessive and the diode would be destroyed. Current limiting is a common application for series resistors.

### EXAMPLE 6-11

The specifications for the LED in Fig. 6-22 show that the desired brightness occurs at a diode current of 15 mA with a drop across the diode of 1.6 V. Calculate the value of $R_L$ and its required power rating.

**Solution**

Current is constant in a series circuit. The current through $R_L$ must be the same as the diode current, 15 mA in this case. Ohm's law is used to calculate the resistance of $R_L$. The drop across $R_L$ is the difference between the source voltage and the diode drop:

$$R = \frac{5 - 1.6}{15 \text{ mA}} = 227 \ \Omega$$

The resistor dissipation can be found by

$$P = (15 \text{ mA})^2 \times 227 \ \Omega = 0.051 \text{ W}$$

The power rating of the resistor actually used in a circuit is usually at least twice the calculated value. This rule of thumb produces reliable circuits. A $\frac{1}{8}$-W resistor will be adequate in this case. The nominal resistance value would be 220 Ω, which is the closest standard value.

The use of a series resistor in applications such as Fig. 6-22 can be viewed in two different ways. One is to consider the resistor as a current limiter. The other is as a means to drop excess voltage. The LED requires 1.6 V, and $R_L$ is used to drop the remainder of the supply voltage. Both viewpoints are the same because a resistor produces a voltage drop that is proportional to the current flowing through it. Figure 6-23 shows another application for a series resistor. A rheostat is used to adjust the speed of a motor. When the rheostat is set for maximum resistance, the circuit current is minimum and the motor speed is also minimum. From the voltage viewpoint, when the rheostat resistance is maximum, the drop across it is maximum and the drop across the motor is minimum and so is the motor speed.

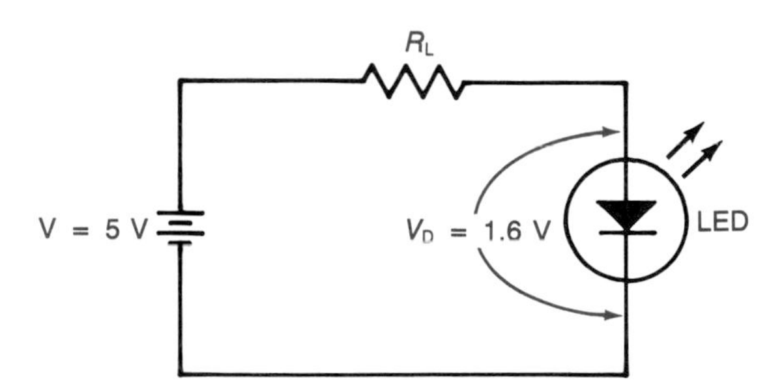


FIGURE 6-22 Using a series resistor to drop excess voltage or limit current flow.

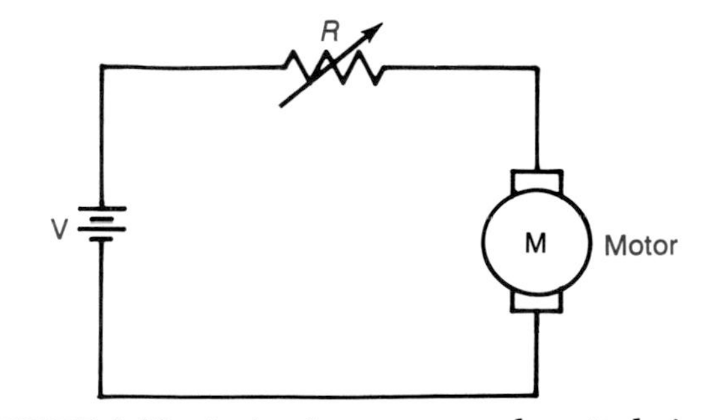


FIGURE 6-23 A simple motor speed control circuit.

Simple control circuits, such as the one shown in Fig. 6-23, are not suited to high-power applications. The efficiency of the circuit is poor because much of the electric energy is converted to heat energy in the rheostat. For example, if the rheostat resistance is equal to the motor resistance, half of the power supplied will be converted to heat in the rheostat. This results in an efficiency of 50 percent, which may not be acceptable when several watts or more are involved.

Series resistance that does not show on schematic diagrams may be a significant factor. Figure 6-24 shows that there is a series resistance associated with a real voltage source. An ideal voltage source is what is normally represented on a schematic such as is shown in Fig. 6-24*a*. Figure 6-24*b* shows that a real voltage source has an internal resistance $R_I$. The internal resistance of a source can make a significant difference in how a circuit performs.

---

**EXAMPLE 6-12**

Calculate the short circuit current for both voltage sources shown in Fig. 6-24.

**Solution**

A short circuit assumes 0 Ω. This produces an impossible situation in the case of an ideal voltage source. Ohm's law will not work:

$$I = \frac{9}{0} = ?$$

Division by zero is an undefined operation. Mathematically, it has no meaning. If the real voltage source is shorted from terminals A to B, Ohm's law does work:

$$I = \frac{9}{2} = 4.5 \text{ A}$$

---

All real voltage sources have an internal resistance that is greater than 0 Ω. When they are short-circuited, the current flow is determined by their voltage and internal resistance.

---

**EXAMPLE 6-13**

Calculate the terminal voltages for both sources in Fig. 6-24 for a 1-kΩ load.

**Solution**

The ideal voltage source has an internal resistance of 0 Ω. Its terminal voltage will be 9 V whether a load is connected or not. The real voltage source has an internal resistance. The internal resistance forms a voltage divider with any load connected to its terminals. The voltage divider rule can be used to find the load voltage which is the same as the terminal voltage:

$$V_{AB} = \frac{9 \times 1000}{1002} = 8.98 \text{ V}$$

---

The terminal voltage of the real voltage source is very close to the terminal voltage of the ideal voltage source for a 1000-Ω load.

---

**EXAMPLE 6-14**

Find the terminal voltage for both sources in Fig. 6-24 with a 10-Ω load.

**Solution**

As before, the ideal voltage source will maintain 9 V at its terminals. For the real voltage source

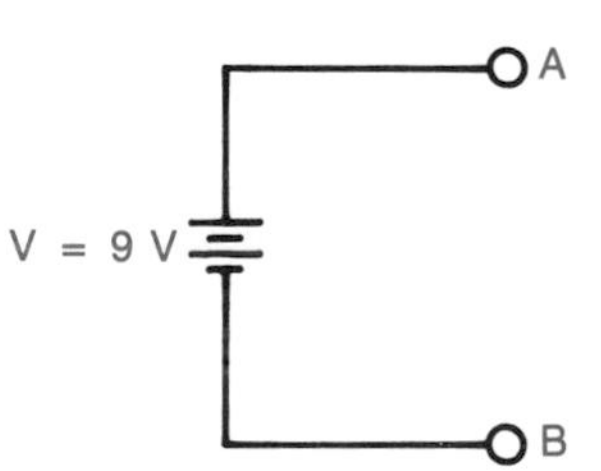


(*a*) Ideal voltage source

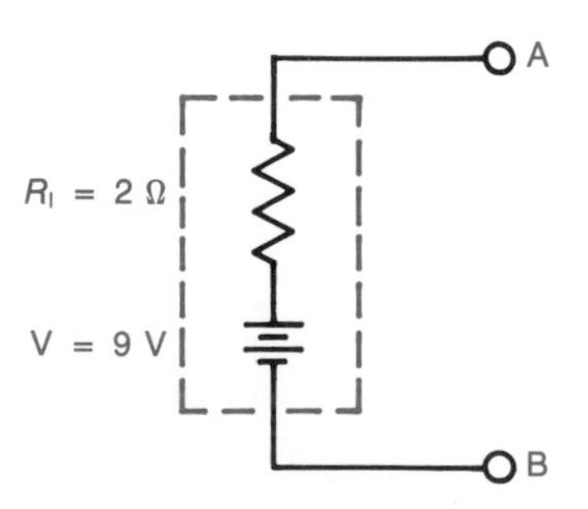


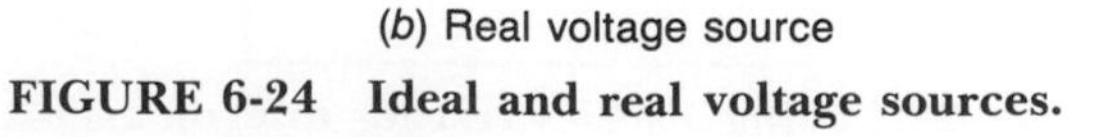

(*b*) Real voltage source

**FIGURE 6-24 Ideal and real voltage sources.**

$$V_{AB} = \frac{9 \times 10}{12} = 7.5 \text{ V}$$

The real voltage source produces a terminal voltage that is significantly less than the ideal voltage source when the load resistance is low. *A real voltage source can be treated as an ideal voltage source only when its internal resistance is significantly less than the load resistance.*

Figure 6-25 shows another application of series circuit concepts. The constant-voltage charging circuit uses an ideal voltage source to charge a battery. The constant-current circuit uses a much higher source voltage and a series resistor. It could be viewed as a real source with an 89-Ω internal resistance. The battery being charged is assumed to develop 11 V in the discharged condition and 12 V in the charged condition and has an internal resistance of 0.01 Ω.

**EXAMPLE 6-15**

Calculate the charging current for the discharged and charged state for both circuits in Fig. 6-25.

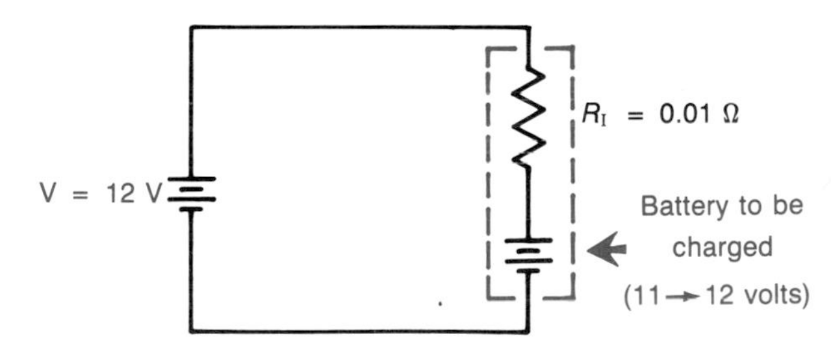


(*a*) Constant voltage charging circuit

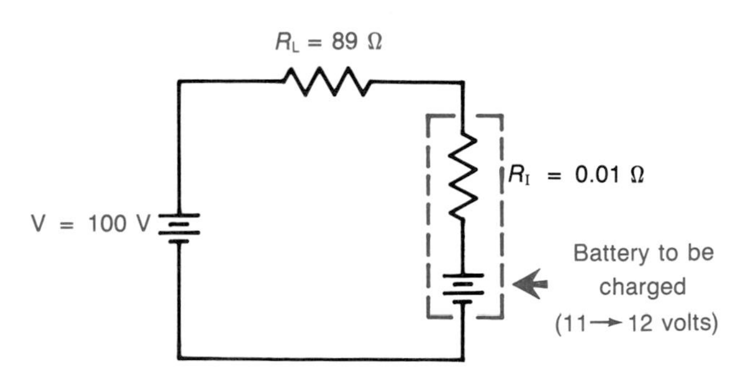


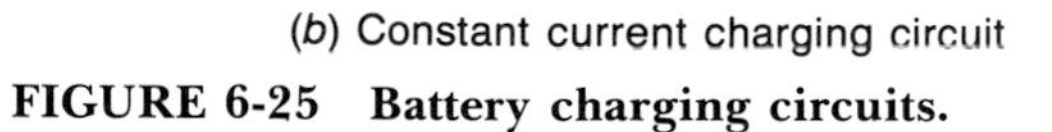

(*b*) Constant current charging circuit

**FIGURE 6-25 Battery charging circuits.**

**Solution**

The constant-voltage circuit in the discharged state is

$$I = \frac{12 - 11}{0.01} = 100 \text{ A}$$

In the charged state

$$I = \frac{12 - 12}{0.01} = 0 \text{ A}$$

The constant-current circuit in the discharged state is

$$I = \frac{100 - 11}{89.01} = 1 \text{ A}$$

In the charged state

$$I = \frac{100 - 12}{89.01} = 0.99 \text{ A}$$

The constant-voltage circuit starts off at a high current which eventually drops to zero as the battery reaches full charge. The constant-current circuit charges at a rate that is only slightly dependent on the battery condition. A voltage source with a series resistance can perform as a *current source.* Current sources are covered in more detail in Chap. 9. Figure 6-26 provides a graphical comparison of the two charging circuits.

Switches are components that are usually connected in series. When a switch is closed, current can flow through the switch and the other parts connected in series with it. When a switch is open, its current drops to zero and so does the current in the components connected in series with it. Switches provide on-off control and other circuit functions. Figure 6-27 shows the physical appearance of some popular switches. Figure 6-28 shows some switch schematic symbols and their verbal descriptions.

Single-pole single-throw (SPST) switches provide simple on-off control. Figure 6-29 shows how two single-pole double-throw (SPDT) switches can be used to give on-off control from two different locations. The load is off with the switches in the positions shown. If either switch is thrown, the load comes on. It can be turned off again by throwing either switch. Figure 6-30 shows how a double-pole double-throw (DPDT) switch can be used to reverse a permanent magnet motor (not all motors can be reversed with this approach). When the switch is thrown to the forward position, the posi-

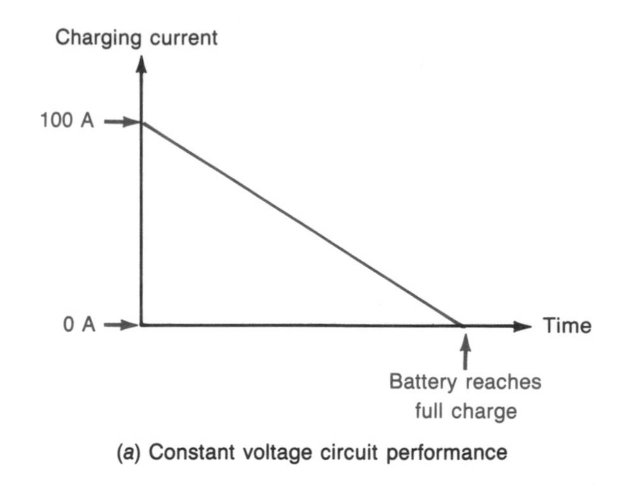

**FIGURE 6-26 Performance of the charging circuits.**

tive terminal of the source is connected to the top of the motor. When the switch is thrown to the reverse position, the positive terminal of the source is routed to the bottom of the motor. The motor polarity determines its direction of rotation. Some switches, such as the rotary and thumbwheel units shown in Fig. 6-27, provide diverse and complex circuit control.

## Self-Test

**29.** Calculate the value for a series resistor used to operate an LED from a 12-V source. Assume a diode drop of 1.6 V and a current of 20 mA.

**30.** What is the dissipation of the resistor in Question 29? What dissipation rating should the resistor have?

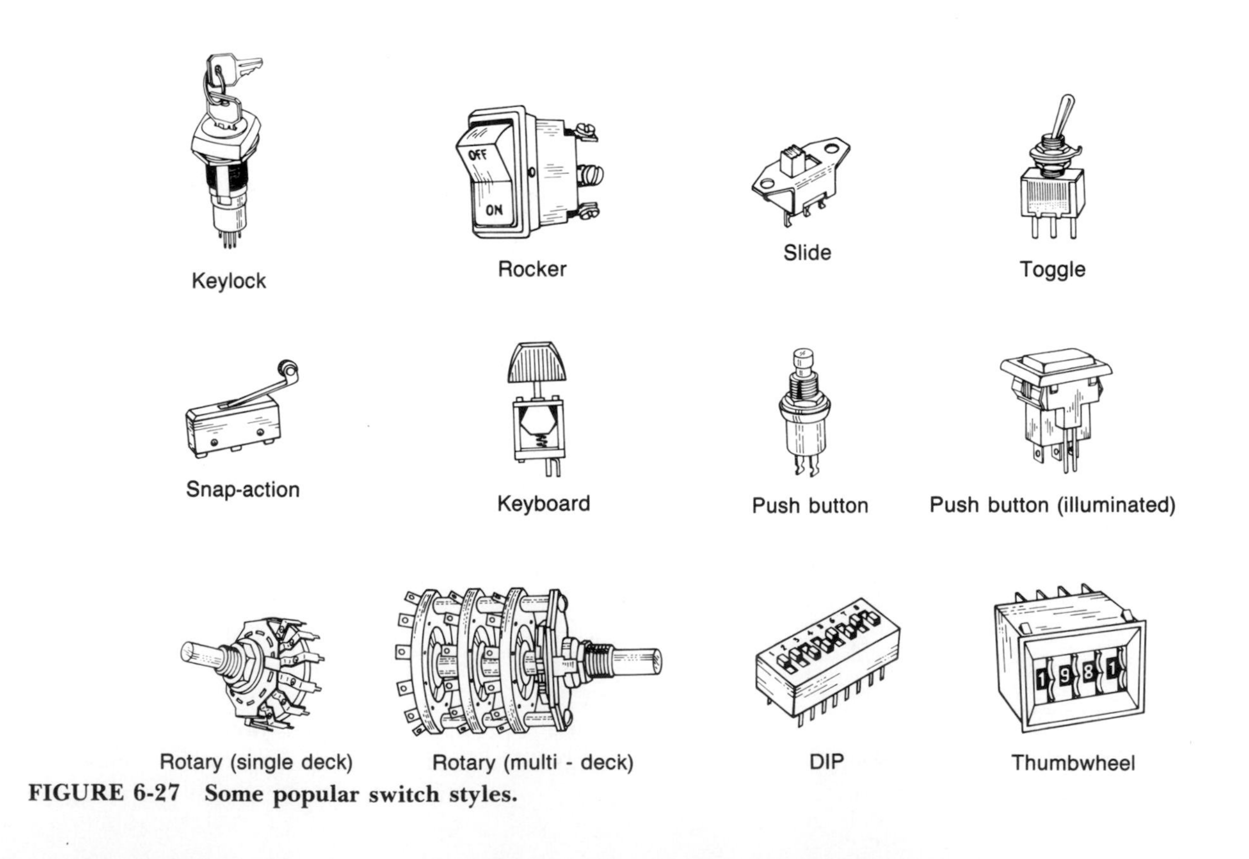


**FIGURE 6-27 Some popular switch styles.**

| Description | Symbol |
|---|---|
| Single pole single throw (SPST) | |
| Single pole double throw (SPDT) | |
| Double pole single throw (DPST) | |
| Double pole double throw (DPDT) | |
| Momentary contact (normally open) | N.O. |
| Push button (normally closed) | N.C. |
| Rotary | |

**FIGURE 6-28 Some common switches and their symbols.**

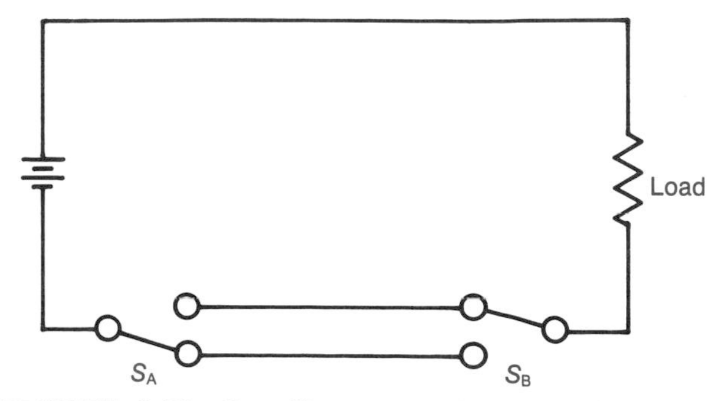


**FIGURE 6-29 On-off control of a load from two points.**

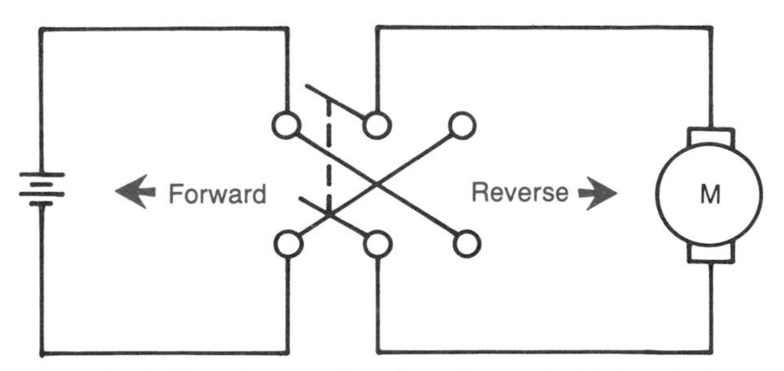


**FIGURE 6-30 An application for a DPDT switch.**

**31.** Refer to Fig. 6-23. Calculate the efficiency of the control circuit if the rheostat is set to 50 Ω and the motor resistance is 25 Ω.

**32.** A 12-V source has an internal resistance of 0.75 Ω. What voltage will it produce across a 5-Ω load?

**33.** What is the short circuit current for the source in Question 32?

**34.** A voltage source can be considered ideal when its internal resistance is much ______ than the load resistance.

**35.** Refer to Fig. 6-25*b*. Calculate the current flow if the battery being charged is shorted.

**36.** Refer to Fig. 6-28. Which switch would be best suited for a charging circuit with four selectable charge rates?

## 6-5 FAULT ANALYSIS IN SERIES CIRCUITS

A component in a series circuit may develop infinite resistance. A fault of this type is called an *open*. An open component is also said to be "burned out." Figure 6-31 shows a series circuit with an open resistor. Because of the open, there can be no current flow in the circuit. With no current flowing, the resistors that are normal will not drop any voltage. This is easily demonstrated with Ohm's law:

$$V_{\mathrm{DROP}} = I \times R = 0 \times R = 0$$

Resistors $R_1$, $R_2$, $R_3$, and $R_5$ in Fig. 6-31 are all normal. They will not drop any voltage. Kirchhoff's voltage law must be satisfied. The loop equation is

$$V - V_{\mathrm{AB}} - V_{\mathrm{BC}} - V_{\mathrm{CD}} - V_{\mathrm{DE}} - V_{\mathrm{EF}} = 0$$
$$120 - 0 - 0 - 0 - 120 - 0 = 0$$

The entire source potential appears across the open component. Fault analysis in a series circuit with an open is based on these facts:

**1.** When a series circuit is open, measuring across the source with a voltmeter produces a normal or high reading. The reading may be high because the open circuit does not load the source as a normal circuit does. The lack of load current eliminates any drop across the internal resistance of the source, and thus the terminal voltage of the source may measure higher than normal.

**2.** When a series circuit is open, measuring the current with an ammeter produces a reading of zero.

**3.** When a series circuit is open, a voltmeter indicates zero drop across all normal loads and other normal components such as fuses and closed switches. When the voltmeter is placed across the open component, the entire source voltage is read.

**4.** If a series circuit has two or more opens, a voltmeter reads zero when connected across any single component other than the source. If both $R_3$ and $R_4$ in Fig. 6-31 are open, every one of the resistors shows a zero drop. A voltmeter can be applied in various ways in such a case to find the open components. For example, the negative lead could be connected to point F. Then the positive lead can be connected to point A, followed by point B, and so on. The full source voltage is indicated at each point until the far side of one of the open components is encountered. This occurs at point D in our example, assuming that $R_3$ is open. The voltmeter indicates zero when point D is probed.

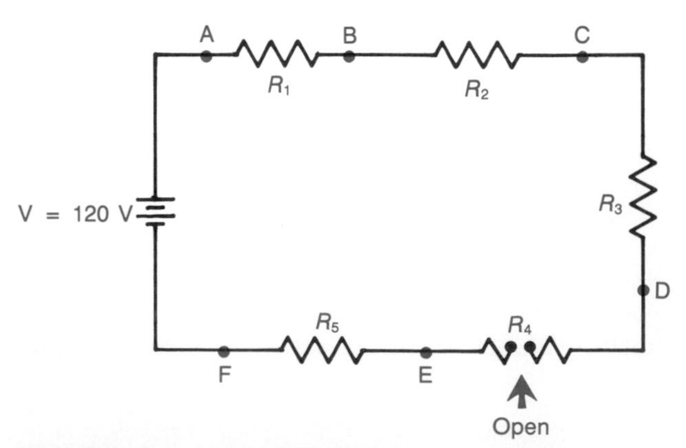


**FIGURE 6-31 A series component may open.**

Series circuits may also develop another type of fault called a *short*. A short is an unwanted path caused by an insulation failure or perhaps by a solder splash on a circuit board. Components may also short, although this type of failure is rare in resistors. Figure 6-32*a* shows a series circuit with a partial short. An unwanted path has connected $R_1$ to $R_3$ and has eliminated $R_2$ from the circuit.

### EXAMPLE 6-16

Calculate both the normal current and the fault current for Fig. 6-32*a*. Also calculate the normal drops and dissipations and the fault drops and dissipations.

#### Solution

The normal current can be found by Ohm's law:

$$I = \frac{9\text{ V}}{5\ \Omega + 10\ \Omega + 15\ \Omega} = 0.3\text{ A}$$

The normal drops can also be found by Ohm's law:

$$V_1 = 0.3 \times 5 = 1.5\text{ V}$$
$$V_2 = 0.3 \times 10 = 3\text{ V}$$
$$V_3 = 0.3 \times 15 = 4.5\text{ V}$$

The sum of the drops is equal to 9 V, which satisfies Kirchhoff's voltage law. The normal power dissipations can be found by multiplying the drops by the current flow:

$$P_1 = 1.5 \times 0.3 = 0.45\text{ W}$$
$$P_2 = 3 \times 0.3 = 0.9\text{ W}$$
$$P_3 = 4.5 \times 0.3 = 1.35\text{ W}$$

The sum of the dissipations is equal to 2.7 W, the total dissipation. This can be checked by

$$P_T = V \times I = 9 \times 0.3 = 2.7\text{ W}$$

The fault current is found by using Ohm's law. Because of the short, only $R_1$ and $R_3$ are in the circuit:

$$I = \frac{9\text{ V}}{5\ \Omega + 15\ \Omega} = 0.45\text{ A}$$

The fault current is *greater* than the normal current. The voltage drops across the 5- and 15-Ω resistors are

$$V_1 = 0.45 \times 5 = 2.25\text{ V}$$
$$V_3 = 0.45 \times 15 = 6.75\text{ V}$$

The sum of these two drops is 9 V, which is equal to the supply voltage. Nothing drops across $R_2$ because of the short. The voltage drops across $R_1$ and

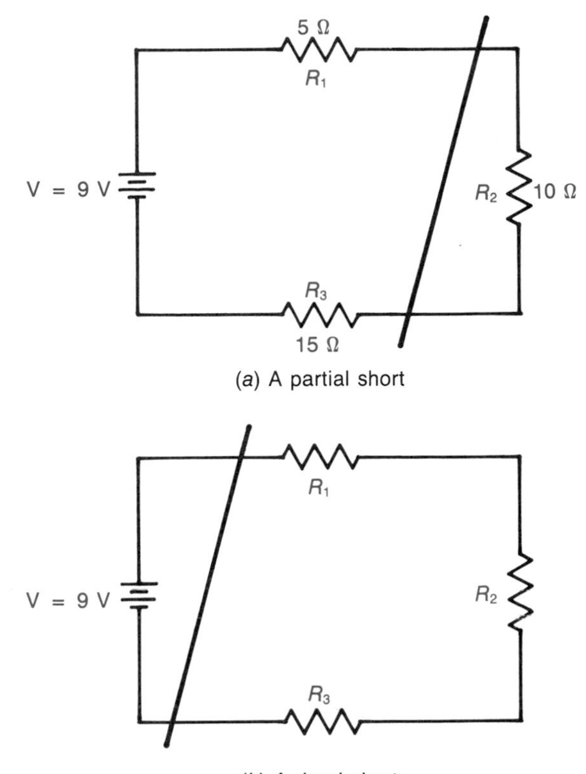


**FIGURE 6-32 Shorts in series circuits.**

$R_3$ are *greater* than in the normal circuit. The fault dissipations in $R_1$ and $R_3$ are found next:

$$P_1 = 2.25 \times 0.45 = 1.01 \text{ W}$$
$$P_3 = 6.75 \times 0.45 = 3.04 \text{ W}$$

The dissipations in $R_1$ and $R_3$ are *greater* than in the normal circuit.

---

Fault analysis in a series circuit with a partial short is based on these facts:

1. A partial short causes the current to be greater than normal.
2. The supply voltage measures normal or less than normal. The increased current flow may cause additional drop across the internal resistance of the supply, making its terminal voltage lower than normal.
3. The voltage drop across components that are not shorted will be higher than normal.
4. The voltage drop across components that are shorted will be zero.
5. The power dissipation in components that are not shorted will be greater than normal.

The last fact is important because it explains why many circuits show more than one fault. A partial short may cause another component to burn out due to abnormally high dissipation. Now the circuit has two faults: a partial short and an open. A technician may find the open component and replace it. The replacement may burn out in a short period of time since the original fault has still not been corrected.

Figure 6-32*b* shows a "dead" short. This is a case where all of the normal loads have been removed by the unwanted path. The only thing to limit the flow of current in these cases is the internal resistance of the source and the resistance of the conductors and of the short itself. The fault current can be quite high. In fact, if there are no overcurrent devices in the circuit, drastic results may occur. Smoke, fires, explosions, and conductors burned in half are some of the consequences of dead shorts in circuits with no protection. This is one of the reasons why it is so important to never tamper with safety devices no matter how tempting a situation may be.

Opens and shorts are examples of extreme circuit failures. Fault analysis also includes dealing with situations that are in between and an understanding of the effects of component tolerance. Consider Fig. 6-33*a*. It shows a voltage divider made up of two 1-kΩ resistors. The *nominal* voltage at point A is 5 V. If both resistors are at the upper end of their tolerance range (1.1 kΩ), the voltage at point A is the same as nominal, or 5 V. Or, if both resistors are at the bottom end of their tolerance range (0.9 kΩ), point A is still at 5 V. However, if one resistor is at one end of its tolerance range and the other resistor is at the opposite end of its tolerance range, a maximum error in $V_A$ occurs. *Worst-case analysis* is a series of calculations to determine the limits of circuit performance.

Figure 6-33*b* shows the worst-case low reading that can be expected for $V_A$. Notice that $R_1$ is at the upper end of its tolerance range and $R_2$ is at the lower end, which makes $V_A$ equal to 4.5 V. Figure 6-33*c* shows that the worst-case high reading occurs when the tolerances slide in the opposite direction, and $V_A$ equals 5.5 V when this occurs. Ignoring source tolerance, the worst-case analysis of a simple two-resistor voltage divider shows that the

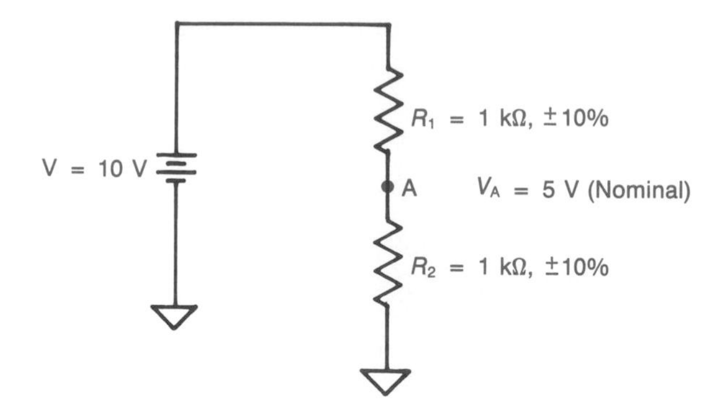


(*a*) Nominal performance

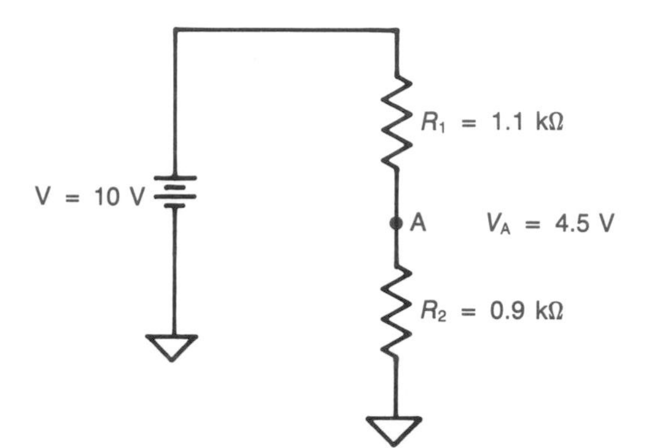


(*b*) Worst case low performance

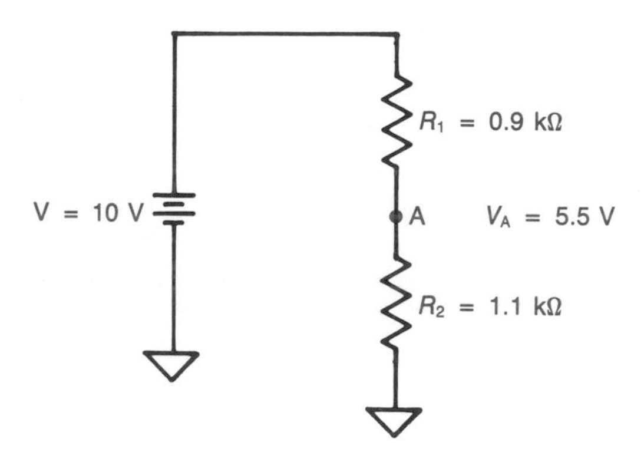


(*c*) Worst case high performance

**FIGURE 6-33 Nominal and worst-case performance in a voltage divider.**

voltage error can be as much as the resistance error. When 10 percent resistors are used, the voltage error can be as much as 10 percent. If the source voltage also varies, then the error can be more than 10 percent.

Errors can be greater than resistor tolerance in some cases even when the source voltage does not contribute to the error. For example, three 1-kΩ resistors will divide a source voltage into three equal parts. Figure 6-34 shows a situation where two of the nominal 1-kΩ resistors are at the low end of their tolerance range and the third is at the high end. Therefore, the *nominal* value for $V_A$ in Fig. 6-34 is 10 V and the *actual* value is 11.38 V. The percentage of error is found by

$$\text{Error percentage} = \frac{\text{error}}{\text{nominal}} \times 100$$
$$= \frac{1.38}{10} \times 100$$
$$\text{Error percentage} = 13.8$$

Worst-case analysis shows that errors can accumulate. A series circuit having more components can have greater errors than a circuit with fewer components.

Luckily, it is unusual for all of the components in a circuit to have tolerance errors that produce as much error as the worst-case analysis predicts. However, it does happen occasionally. Schematics often show the desired readings at various points in a circuit. They also may indicate how much error is acceptable. It is important to make note of this information when analyzing any circuit for a fault. A 20 percent error may be acceptable in one situation and a 5 percent error not acceptable in another situation. Components change with time. Carbon composition resistors have a tendency to increase in resistance value and many eventually go out of tolerance. This is also worth remembering when analyzing a circuit.

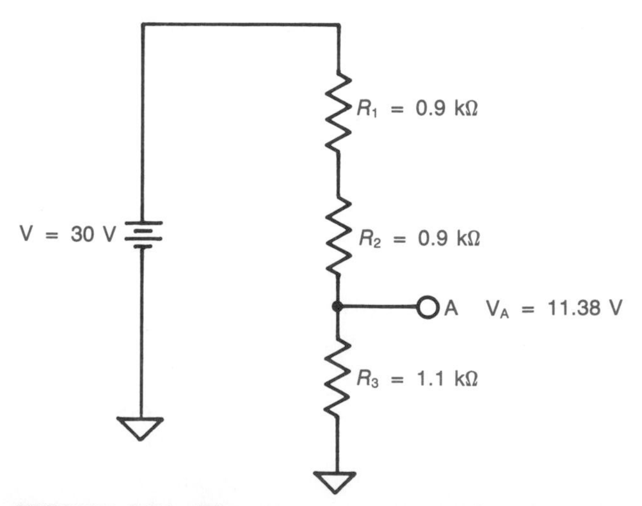


**FIGURE 6-34 The worst case can produce an error greater than the component tolerance.**

## Self-Test

**37.** Refer to Fig. 6-31 and assume that all of the components are normal except $R_2$, which is open. What will the following voltages be?

$V_{AB}$ $V_{DE}$
$V_{BC}$ $V_{EF}$
$V_{CD}$ $V_{AF}$

**38.** Refer to Fig. 6-31 and assume that all components are normal except $R_2$ and $R_4$, which are open. What will the following voltages be?

$V_{AF}$ $V_{AE}$
$V_{BF}$ $V_{AD}$
$V_{CF}$ $V_{DE}$

**39.** A series circuit has an open and it also has a fuse. If the fuse is intact, what voltage drop will appear across it?

**40.** A series circuit has a partial short. What voltage drop can be expected across the shorted component or components?

**41.** A series circuit has a partial short. Will the voltage drop across the unshorted components be normal, high, or low?

**42.** A voltage divider consists of two 390-Ω, ±5 percent resistors connected across a 10-V source. Calculate the nominal and worst-case output voltages from this divider.

**43.** A series circuit has a partial short. Will the power dissipations in the unshorted components be normal, high, or low?

## SUMMARY

1. Current is constant in a series circuit:

$$I = I_1 = I_2 = I_3 = \cdots = I_N$$

2. The total resistance in a series circuit is equal to the sum of the individual resistances:

$$R_T = R_1 + R_2 + R_3 + \cdots + R_N$$

3. The current flow in a series circuit can be found with Ohm's law:

$$I = \frac{V}{R_T}$$

4. The sum of the drops in a series circuit is equal to the applied voltage:

$$V = V_1 + V_2 + V_3 + \cdots + V_N$$

5. Series voltage sources may aid or oppose one another. When they aid, the effective voltage is found by adding. When they oppose, the effective voltage is found by subtracting and the direction of flow is set by the largest voltage.

6. A formal loop equation takes the following form:

$$V - V_1 - V_2 - V_3 - \cdots - V_N = 0$$

and the $IR$ values may be substituted for the drops, giving

$$V - IR_1 - IR_2 - IR_3 - \cdots - IR_N = 0$$

7. If the assigned direction of flow is incorrect when writing a loop equation, the algebraic sign of $I$ will be negative when the equation is solved but the magnitude of $I$ will be correct if no other mistake was made.

8. The voltage divider rule provides a shortcut for finding drops in a series circuit:

$$V_X = \frac{V \times R_X}{R_T}$$

**9.** Potentiometers are adjustable voltage dividers.

**10.** When double-subscript notation is used for a voltage, such as $V_{AB}$, the second subscript is the reference point.

**11.** When a single point in a circuit is called out to specify a voltage, the circuit common or ground automatically becomes the reference point.

**12.** When all of the resistors in a series circuit are equal, they will each drop an equal share of the voltage. When they are not equal, the largest resistance produces the greatest drop and the smallest resistance the smallest drop.

**13.** The total power dissipation in a series circuit is equal to the sum of the individual dissipations:

$$P_T = P_1 + P_2 + P_3 + \cdots + P_N$$

**14.** Series resistors are often used to drop excess voltage or limit current flow.

**15.** An ideal voltage source has zero internal resistance. All real voltage sources have an internal resistance greater than zero.

**16.** A voltage source can be treated as ideal when its internal resistance is significantly less than its load resistance.

**17.** A voltage source with a large series resistance (or a high internal resistance) can act as a current source.

**18.** Switches can provide simple on-off control or more complex circuit functions.

**19.** When a series circuit has an open component, all of the source voltage appears across the open.

**20.** The current is zero in a series circuit with one or more opens.

**21.** The current flow in a series circuit with a partial short will be higher than normal and so will the drops across any unshorted resistors. The power dissipation in the unshorted resistors will also be greater than normal, which may lead to their failure.

**22.** Component tolerance determines circuit performance to some extent. Worst-case analysis is a technique used to find the limits of circuit performance.

## CHAPTER REVIEW QUESTIONS

**6-1.** What is constant in a series circuit?

**6-2.** What is the direction of electron current in Fig. 6-35*a*?

**6-3.** The sum of the drops in Fig. 6-35*a* must equal ______ V.

**6-4.** The polarity of the 1.2-kΩ resistor in Fig. 6-35*a* is such that its left end will be ______ with respect to its right end.

**6-5.** What is the direction of electron current in Fig. 6-35*b*?

**6-6.** Which battery in Fig. 6-35*b* is being charged?

**6-7.** What is the direction of the electron current flow in Fig. 6-35*c*?

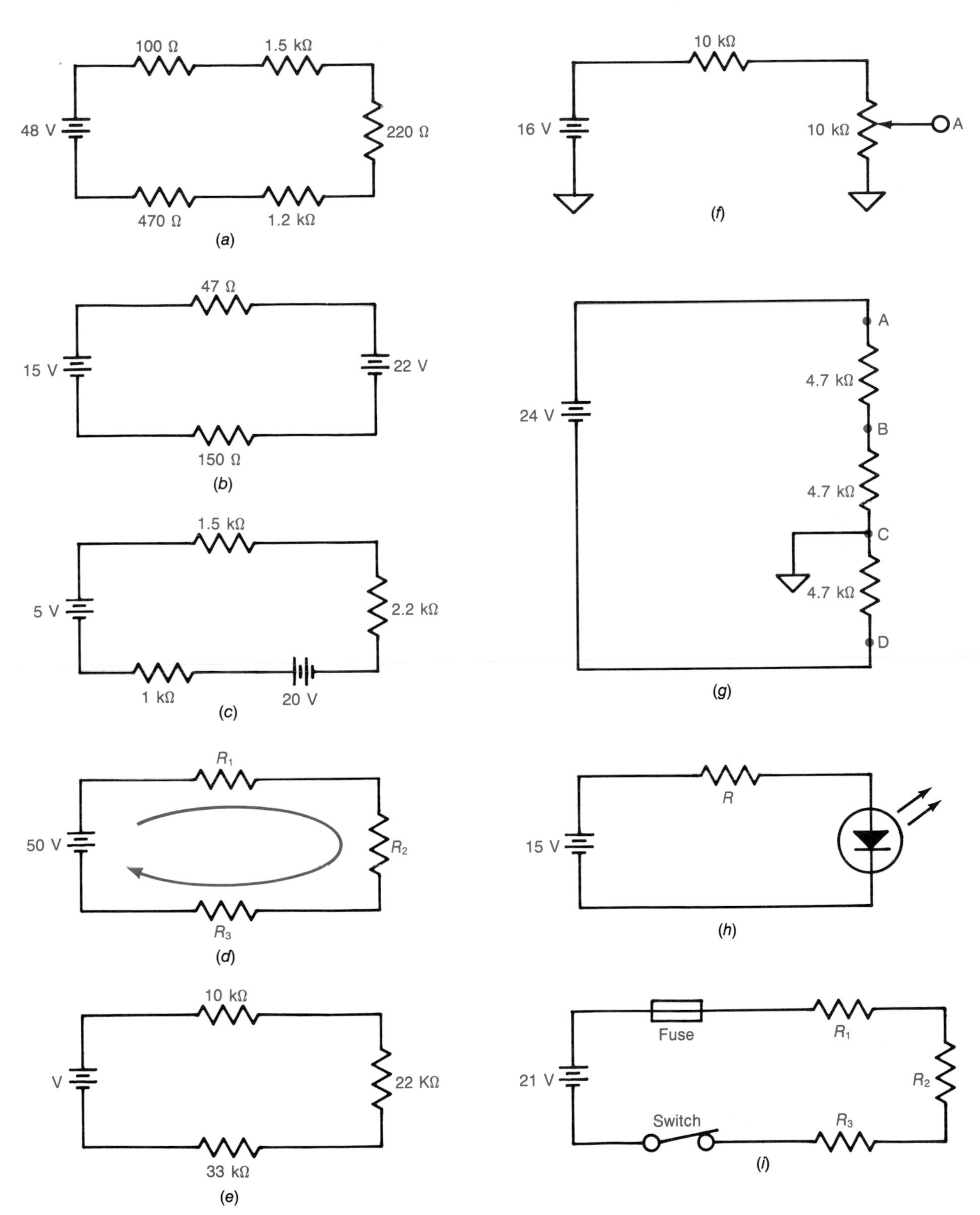


**FIGURE 6-35 Circuits for Chapter Review Questions and Problems.**

**6-8.** The simplified equivalent circuit for Fig. 6-35*c* will contain a single ________-V source and a single ________-Ω resistor.

**6-9.** The actual polarity of the drop across $R_2$ in Fig. 6-35*d* is such that its top lead is ________ with respect to its bottom lead.

**6-10.** Where should the wiper arm of the 10-kΩ pot in Fig. 6-35*f* be set for minimum output?

**6-11.** The total power dissipation in a series circuit is equal to the ________ of the individual dissipations.

**6-12.** What is the abbreviation for the switch shown in Fig. 6-35*i*?

**6-13.** Suppose the fuse in Fig. 6-35*i* is blown. If the resistors are all of the same value, what will each resistor drop?

**6-14.** Suppose that all of the parts in Fig. 6-35*i* are normal and that the switch is open. What will a voltmeter indicate when connected across the switch?

**6-15.** Refer to Fig. 6-35*i*. If a short appears across $R_2$, the circuit current will be ________ than normal until the fuse blows.

**6-16.** What voltage drop will appear across $R_2$ in Question 6-15 before the fuse blows?

**6-17.** If the fuse in Fig. 6-35*i* has been replaced with one having too large a current rating, could a partial short cause damage to one or more components?

**6-18.** If the fuse is blown in Fig. 6-35*i* and the switch is open, what voltage drop will appear across the fuse?

## CHAPTER REVIEW PROBLEMS

**6-1.** What is the total resistance in Fig. 6-35*a*?

**6-2.** How much current flows in Fig. 6-35*a*?

**6-3.** Calculate the drop across the 100-Ω resistor in Fig. 6-35*a*.

**6-4.** Calculate the drop across the 1.5-kΩ resistor in Fig. 6-35*a*.

**6-5.** What is the drop across the 150-Ω resistor in Fig. 6-35*b*?

**6-6.** What is the drop across the 1-kΩ resistor in Fig. 6-35*c*?

**6-7.** Write a formal loop equation for Fig. 6-35*d* using the indicated direction of flow and substitute $IR$ values for the drops.

**6-8.** Solve the equation of Problem 6-7 for $I$ using resistor values of 10, 15, and 25 Ω.

**6-9.** Interpret the sign of your answer for Problem 6-8.

**6-10.** What *percentage* of $V$ will be dropped by the 22-kΩ resistor in Fig. 6-35*e*? *Hint:* Use the voltage divider rule.

**6-11.** What is the *range* of $V_A$ in Fig. 6-35*f*?

**6-12.** The pot in Fig. 6-35*f* has a linear taper and is set at its center of rotation. What is $V_A$?

**6-13.** What is $V_{AB}$ in Fig. 6-35*g*?

**6-14.** What is $V_{BA}$ in Fig. 6-35*g*?

**6-15.** What is $V_D$ in Fig. 6-35*g*?

**6-16.** What is $V_A$ in Fig. 6-35*g*?

**6-17.** Calculate the individual and the total dissipations for Fig. 6-35*g*.

**6-18.** The LED in Fig. 6-35*h* drops 1.6 V. Calculate a value for $R$ that will limit the current flow to 20 mA.

**6-19.** What is the power dissipation in $R$ in Problem 6-18?

**6-20.** A source develops 12 V and has an internal resistance of 10 Ω. Calculate its terminal voltage for a load of 1 kΩ.

**6-21.** What is the terminal voltage for the source in Problem 6-20 for a 10-Ω load?

**6-22.** If all three resistors in Fig. 6-35*i* are 1 kΩ, ±5 percent, calculate the nominal and worst-case limits for the drop across any one of the resistors.

## ANSWERS TO SELF-TESTS

**1.** same

**2.** adding

**3.** counterclockwise

**4.** clockwise

**5.** 4 mA

**6.** 33.3 mA

**7.** 3.51 mA

**8.** sum

**9.** across the largest resistor

**10.** across the smallest resistor

**11.** 2 V

**12.** $V_1 = 3$ V, $V_2 = 1.5$ V, $V_3 = 4.5$ V

**13.** counterclockwise

**14.** positive

**15.** $V_1 = 2$ V, $V_2 = 8$ V

**16.** clockwise

**17.** $V_1 = 3.33$ V, $V_2 = 1.67$ V, $V_3 = 5$ V

**18.** negative

**19.** $19 - 3 - 6 - 15I - 10I - 5I = 0$

**20.** $V_1 = 0$, $V_2 = 0$, $V_3 = 0$

**21.** current

**22.** it increases

**23.** $X/4$

**24.** −3 V, +3 V

**25.** sum

**26.** $V_{AB} = 20$ V  $V_{DE} = 15$ V
$V_A = 25$ V  $V_E = -25$ V

**27.** 50 mW

**28.** the 20-kΩ resistor

**29.** 520 Ω

**30.** 0.208 W, $\frac{1}{2}$ W

**31.** 33.3 percent

**32.** 10.4 V

**33.** 16 A

**34.** less

**35.** 1.12 A

**36.** rotary

**37.** $V_{AB} = 0$  $V_{DE} = 0$
$V_{BC} = 120$ V  $V_{EF} = 0$
$V_{CD} = 0$  $V_{AF} = 120$ V

**38.** $V_{AF} = 120$ V  $V_{AE} = 120$ V
$V_{BF} = 120$ V  $V_{AD} = 0$
$V_{CF} = 0$  $V_{DE} = 0$

**39.** zero

**40.** zero

**41.** high

**42.** 5 V, 5.25 V, 4.75 V

**43.** high

# Chapter 7 Parallel Circuits

The series circuit is one basic configuration and the parallel circuit is the other. This chapter deals with the characteristics of parallel circuits and the laws that govern them. Conductance, current dividers, applications, and fault analysis of parallel circuits are also covered in this chapter.

## 7-1 VOLTAGE AND CURRENT IN PARALLEL CIRCUITS

Components are in parallel when they are connected across one another. Figure 7-1*b* shows a parallel circuit. Compare it with the series circuit shown in Fig. 7-1*a*. Figure 7-1*c* shows both series and parallel connections, which are covered in Chap. 8. Now look at Fig. 7-2. It shows that there may be variations in the way parallel circuits are shown in schematic form. It is important to be able to recognize how components are connected so the proper laws and procedures may be applied. Even though the drawings in Fig. 7-2 may appear different, the same parallel relationship applies in each case.

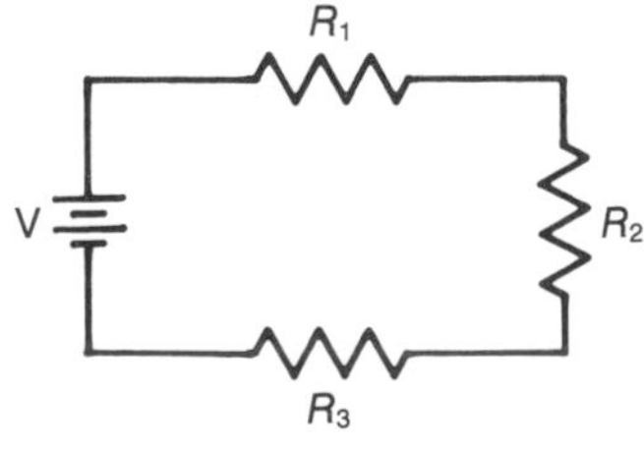


(*a*) A series circuit

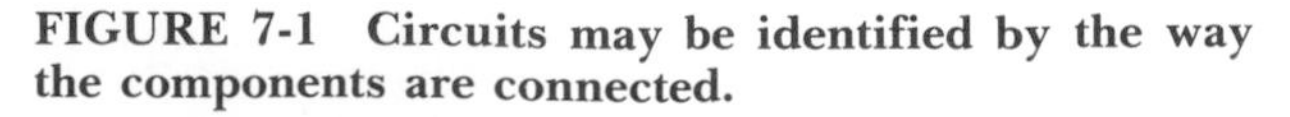

**FIGURE 7-1 Circuits may be identified by the way the components are connected.**

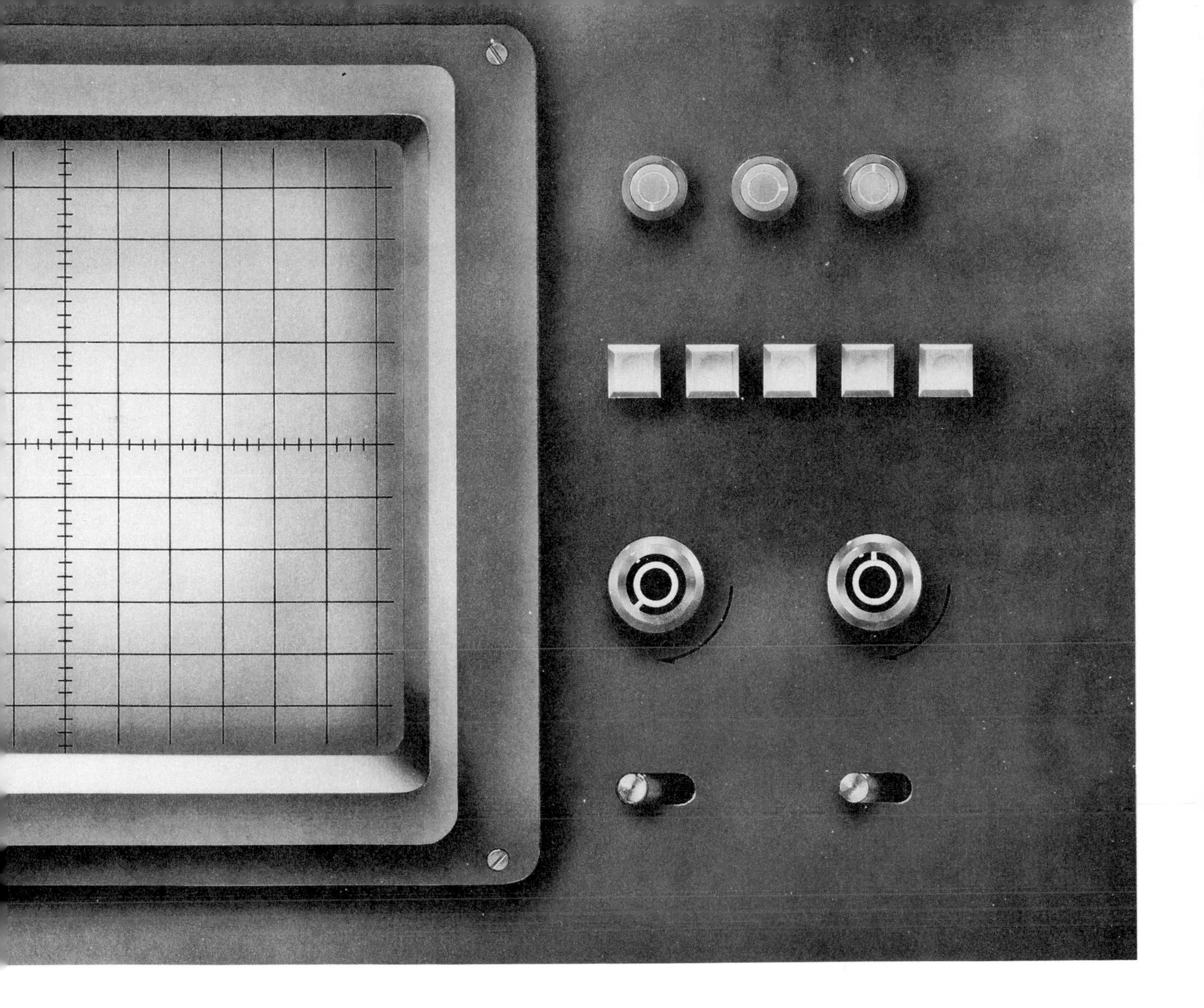

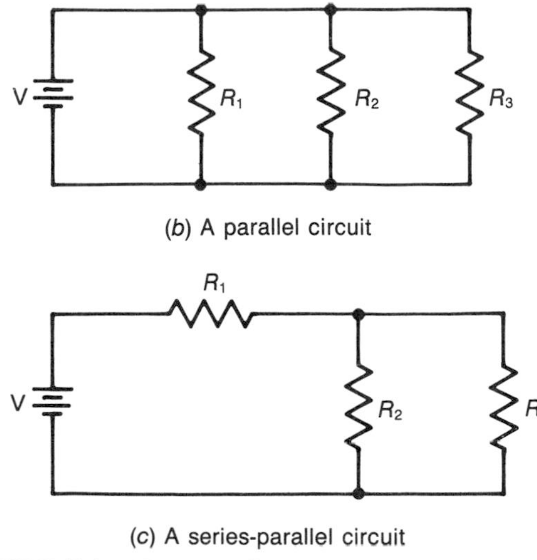


(*b*) A parallel circuit



(*c*) A series-parallel circuit

**FIGURE 7-1** (***Continued***)

Electric circuits have *nodes*. A node is a point where two or more components are connected together. Nodes are also called *junctions*. Figure 7-3 shows a circuit with two nodes. Node A is where the positive terminal of the battery and the top end of the resistors are connected. Node B is the connection point for the negative terminal of the battery and the bottom of the resistors. The circuit at the bottom of the illustration is electrically identical to the top circuit but is drawn differently. Node A can be considered to be any point along the top of the circuit. Node B can be considered to be any point along the bottom of the circuit. This is based on the assumption that the conductors which connect the components are perfect (have no resistance). Being perfect, they will not drop any voltage. Therefore, a constant voltage appears from

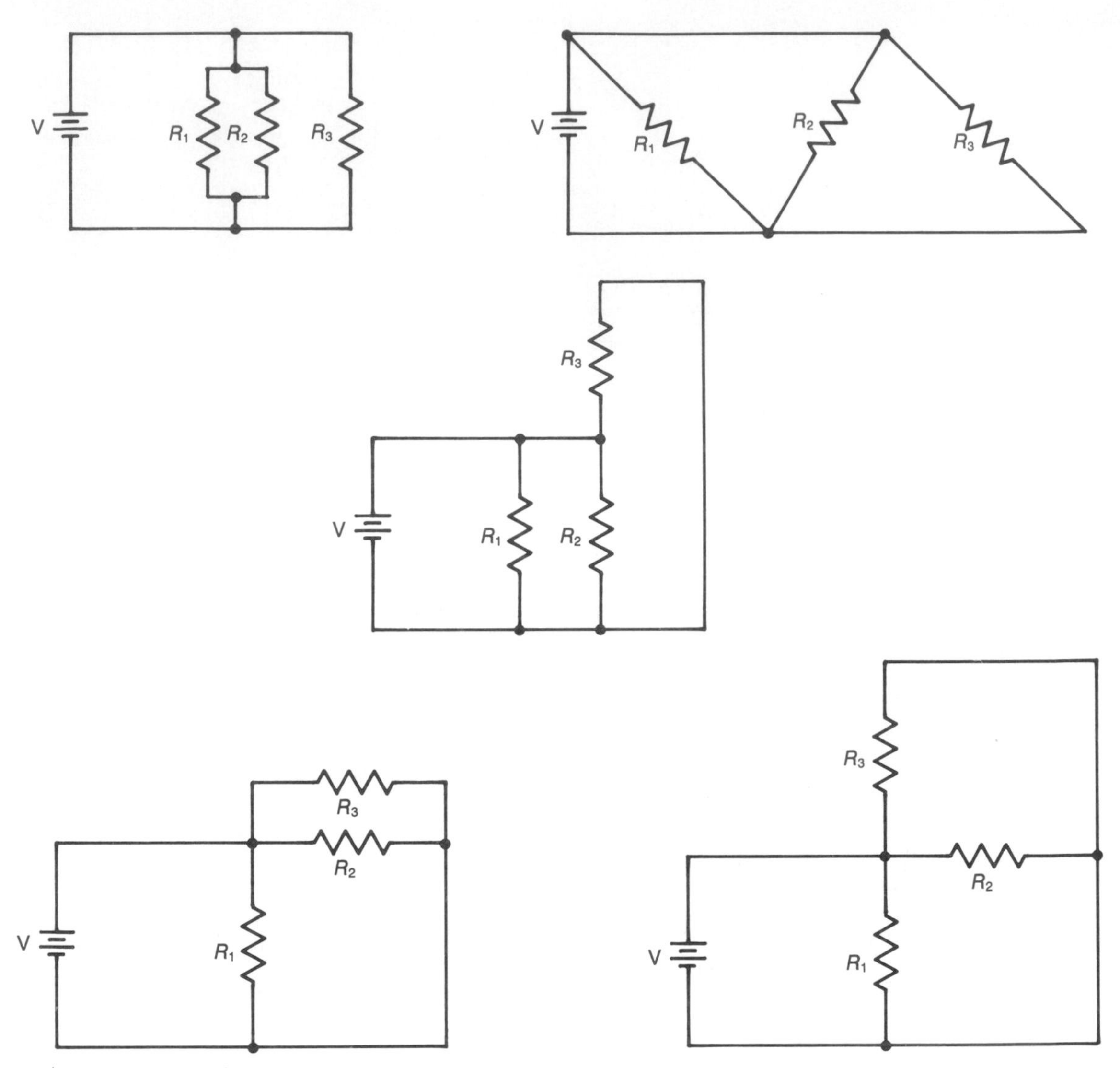


FIGURE 7-2 **$R_1$, $R_2$, and $R_3$ are in parallel in each circuit.**

node A to node B in Fig. 7-3. This leads us to one of the basic laws for parallel circuits: *The voltage is constant across parallel-connected components:*

$$V_1 = V_2 = V_3 = \cdots = V_N$$

A parallel path is often called a *branch.* Figure 7-4 shows a circuit with three resistors in parallel. There are three branch currents in this circuit, and they are labeled $I_1$, $I_2$, and $I_3$.

**EXAMPLE 7-1**

Find the branch currents for Fig. 7-4.

**Solution**

Since voltage is constant across parallel components, the same 40 V appears across each of the three resistors in Fig. 7-4. Ohm's law can be used to find each of the branch currents:

$$I_1 = \frac{40\text{ V}}{5\ \Omega} = 8\text{ A}$$

$$I_2 = \frac{40\text{ V}}{10\ \Omega} = 4\text{ A}$$

$$I_3 = \frac{40\text{ V}}{20\ \Omega} = 2\text{ A}$$

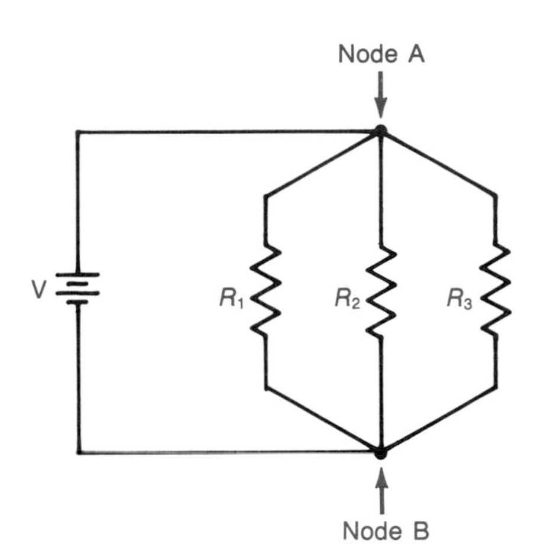


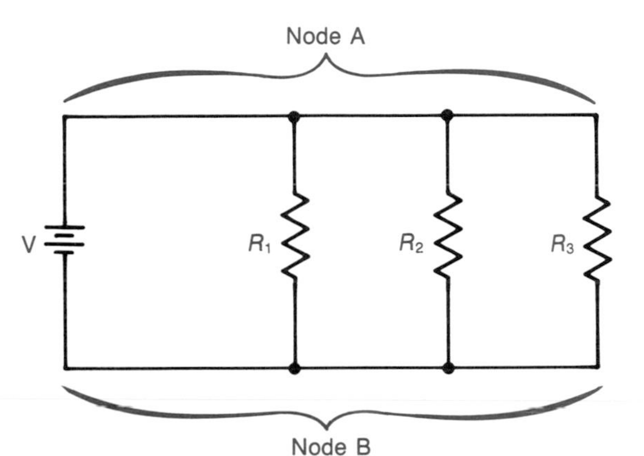


**FIGURE 7-3 A node is a point in a circuit where components are interconnected.**

Note that the branch currents are not equal. This is one of the fundamental differences when parallel circuits are compared to series circuits. The current flow is constant in all parts of a series circuit.

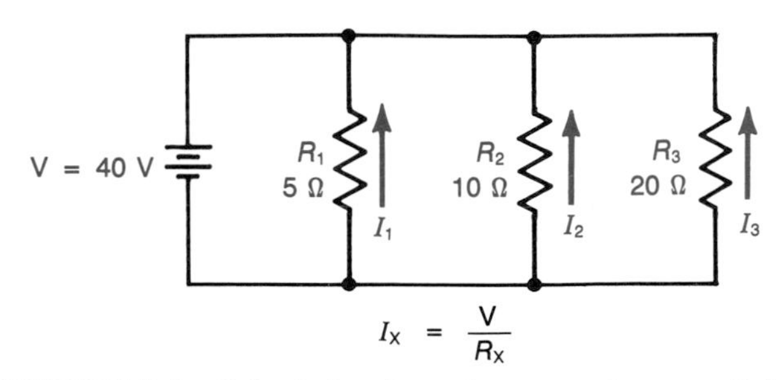


**FIGURE 7-4 Calculating branch current in a parallel circuit.**

The branch with the least resistance has the most current flow in a parallel circuit.

In addition to branch currents, a parallel circuit has a total current flow. Figure 7-5 shows that the total current flow in a parallel circuit can be found by summing the branch currents. This is another one of the fundamental laws that govern parallel circuits:

$$I_T = I_1 + I_2 + I_3 + \cdots + I_N$$

The total electron current in Fig. 7-5 leaves the negative side of the battery. It then divides among the three branches. Some of it becomes $I_1$, some becomes $I_2$, and some becomes $I_3$. The three branch currents merge at the top of the circuit to form the total flow entering the positive terminal of the battery.

### EXAMPLE 7-2

Find the total current for Fig. 7-5 if the battery develops 12 V and each resistor has a value of 24 Ω.

**Solution**

Voltage is constant in a parallel circuit. Twelve volts will appear across each of the 24-Ω resistors. The branch currents will be equal in this case and may be found using Ohm's law:

$$I = \frac{12\text{ V}}{24\ \Omega} = 0.5\text{ A}$$

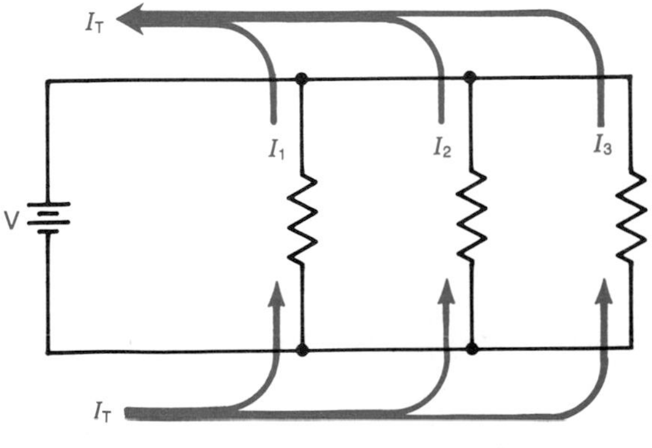


**FIGURE 7-5 Calculating total current in a parallel circuit.**

The total current is equal to the sum of the branch currents:

$$I_T = 0.5 \text{ A} + 0.5 \text{ A} + 0.5 \text{ A} = 1.5 \text{ A}$$

The total current is three times the value of any one of the branch currents. Even though the branch currents are equal in this example, current flow is not constant at all points in the circuit as it is in series circuits.

Kirchhoff's current law is a formal statement concerning current flow in circuits. It tells us that the algebraic sum of the currents at any junction is equal to zero. Figure 7-6 shows an analogy in a fluid circuit. There are three inlets and one outlet. All of the flow rates are in liters per minute. The sum of the inlet flow must be equal to the outlet flow. If inlet flow is assigned a positive sign and outlet flow a negative sign, a flow equation can be written as

$$+1 + 2 + 3 - 6 = 0$$

Figure 7-7 shows a node (junction) in an electric circuit. The currents entering the node are assigned a positive sign, and the currents leaving the node are assigned a negative sign. The algebraic sum of the node currents is equal to zero.

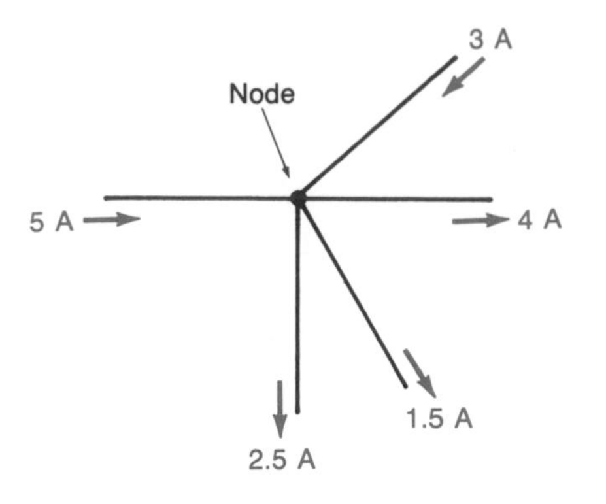


$$+5 \text{ A} + 3 \text{ A} - 4 \text{ A} - 1.5 \text{ A} - 2.5 \text{ A} = 0$$

**FIGURE 7-7 The algebraic sum of the node currents equals zero.**

### EXAMPLE 7-3

Find $I_X$ in Fig. 7-8. Determine if it is a current entering the junction or a current leaving the junction.

#### Solution

Write an equation based on Kirchhoff's current law:

$$+1.75 \text{ A} - 3.25 \text{ A} + 3.5 \text{ A} + I_X = 0$$
$$I_X = -2.0 \text{ A}$$

Current $I_X$ is equal to 2 A. The negative sign indicates that it flows away from the junction.

Kirchhoff's current law is most useful in parallel circuits, just as Kirchhoff's voltage law is most useful in series circuits. However, the laws apply universally. Figure 7-9*a* shows that series circuits also have nodes. The current entering any node is equal to current leaving that node. This satisfies the current law. Figure 7-9*b* shows that parallel circuits have loops. The voltage rise in any loop is equal to the voltage drop. This satisfies the voltage law.

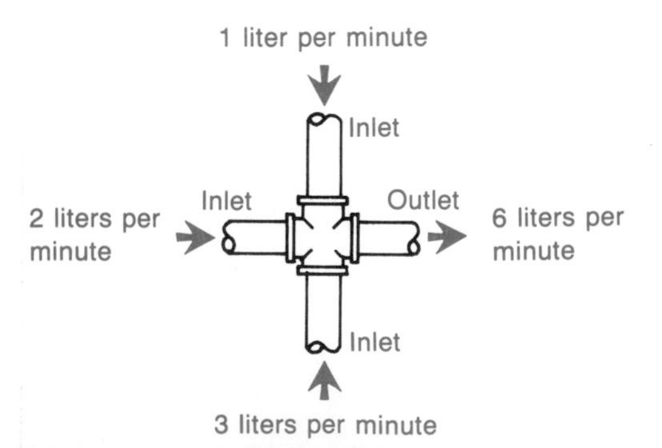


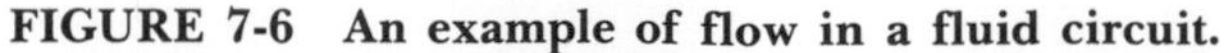

**FIGURE 7-6 An example of flow in a fluid circuit.**

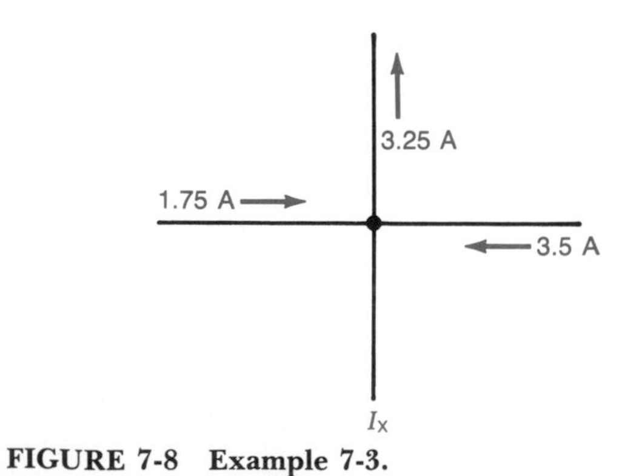


**FIGURE 7-8 Example 7-3.**

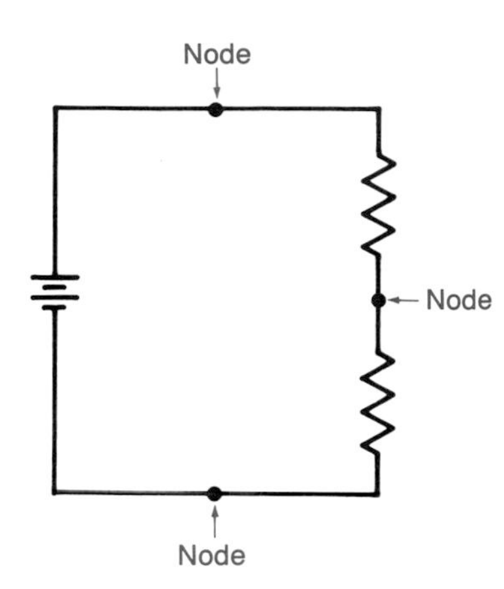


(a) Series circuits have nodes

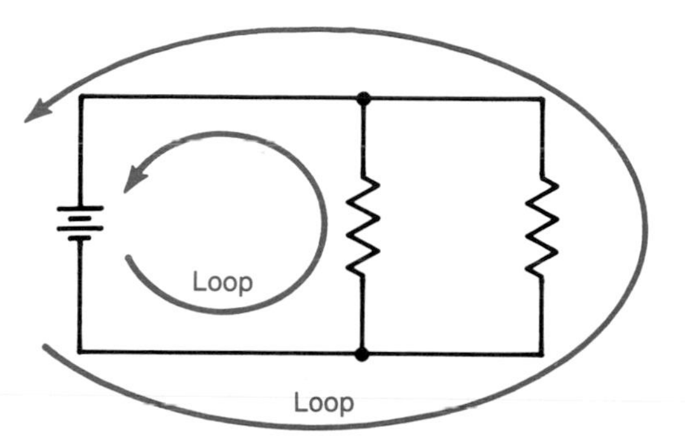


(b) Parallel circuits have loops

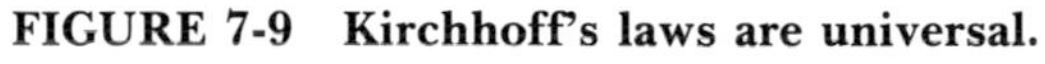

**FIGURE 7-9 Kirchhoff's laws are universal.**

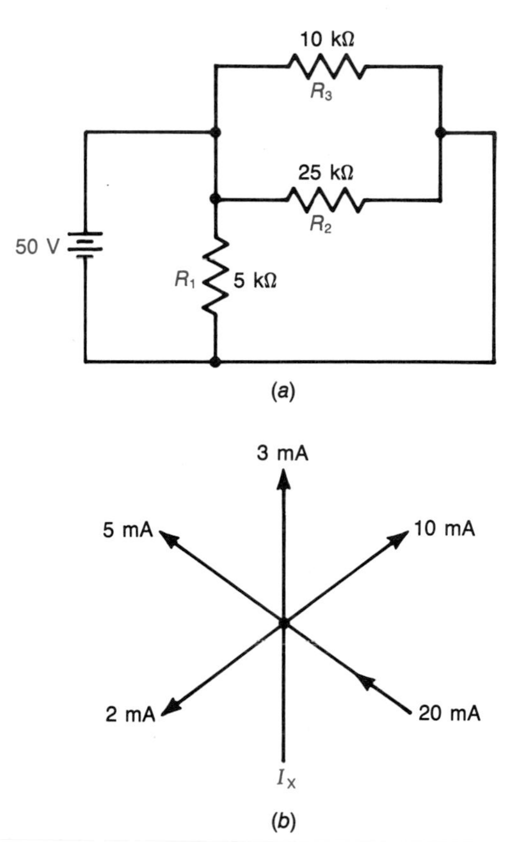

**FIGURE 7-10 Self-Test.**

## Self-Test

1. The current is constant in a ______ circuit.
2. The voltage is constant across ______ components.
3. Three unequal resistances are connected in parallel. Which of the three will have the least current flow?
4. A point in a circuit where components are interconnected is known as a node or a ______.
5. True or false? Kirchhoff's voltage law applies only to series circuits.
6. True or false? Kirchhoff's current law applies only to parallel circuits.
7. Calculate the branch currents for Fig. 7-10*a*.
8. Calculate the total current for Fig. 7-10*a*.
9. Calculate $I_X$ for Fig. 7-10*b* and find its direction.

## 7-2 CONDUCTANCE

*Conductance* is a measure of how well a component or circuit supports the flow of electric current. It is the opposite of resistance. A circuit with high conductance has low resistance, and a circuit with low conductance has high resistance. Conductance and resistance are related as reciprocals:

$$G = \frac{1}{R}$$

where $G$ = conductance, siemens (S)
$R$ = resistance, Ω

### EXAMPLE 7-4

What is the conductance of a length of wire that has a resistance of 0.5 Ω?

**Solution**

Taking the reciprocal:

$$G = \frac{1}{0.5\ \Omega} = 2\ \text{S}$$

Conductance can be used instead of resistance in Ohm's law. Beginning with Ohm's law,

$$R = \frac{V}{I}$$

Then taking the reciprocal of both sides,

$$\frac{1}{R} = \frac{I}{V}$$

Substituting $G$ for $1/R$,

$$G = \frac{I}{V}$$

**EXAMPLE 7-5**

Calculate the current flow for the circuit shown in Fig. 7-11.

**Solution**

Solving the conductance form of Ohm's law for current gives

$$I = V \times G$$

Substituting the values from Fig. 7-11,

$$I = 5\text{ V} \times 2\text{ S} = 10\text{ A}$$

Conductance is convenient in parallel circuits because the total conductance of parallel branches is equal to the sum of the individual branch conductances:

$$G_T = G_1 + G_2 + G_3 + \cdots + G_N$$

**EXAMPLE 7-6**

Find the total conductance, total current, and total resistance for the circuit shown in Fig. 7-12.

**Solution**

The first step is to find the individual branch conductances:

$$G_1 = \frac{1}{10\ \Omega} = 0.1\text{ S}$$

$$G_2 = \frac{1}{5\ \Omega} = 0.2\text{ S}$$

$$G_3 = \frac{1}{2\ \Omega} = 0.5\text{ S}$$

The total conductance is found by summing the branch conductances:

$$G_T = 0.1\text{ S} + 0.2\text{ S} + 0.5\text{ S} = 0.8\text{ S}$$

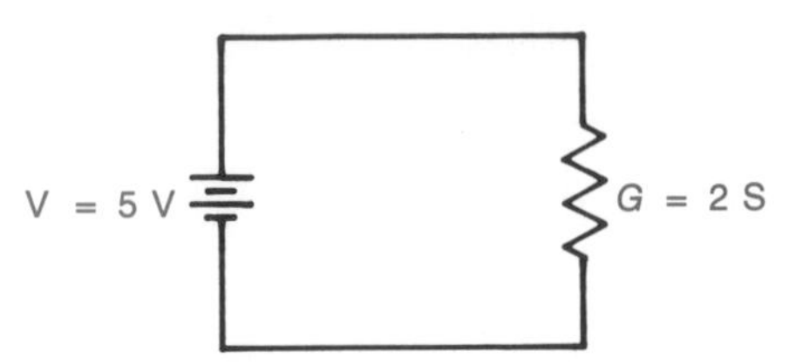


**FIGURE 7-11 Circuit for Example 7-5.**

Ohm's law is used to find the total current:

$$I_T = V \times G_T = 10\text{ V} \times 0.8\text{ S} = 8\text{ A}$$

The total resistance can be found by taking the reciprocal of the total conductance:

$$R_T = \frac{1}{0.8\text{ S}} = 1.25\ \Omega$$

This example shows that it is possible to find the total conductance or the total resistance for parallel components. The total conductance may also be referred to as the equivalent conductance, and the total resistance may be referred to as the equivalent resistance. Figure 7-13 shows equivalent circuits for the original circuit of Fig. 7-12. Figure 7-13*a* shows that the three parallel branches can be replaced with an equivalent single conductance, and Fig. 7-13*b* shows the three parallel branches reduced to an equivalent single resistance.

Summing the branch conductances provides a way to find the total current in parallel circuits without first finding each branch current. However, it will be informative to look at the branch current method again. Assuming a parallel circuit with three branches, the total current can be found with

$$I_T = \frac{V}{R_1} + \frac{V}{R_2} + \frac{V}{R_3}$$

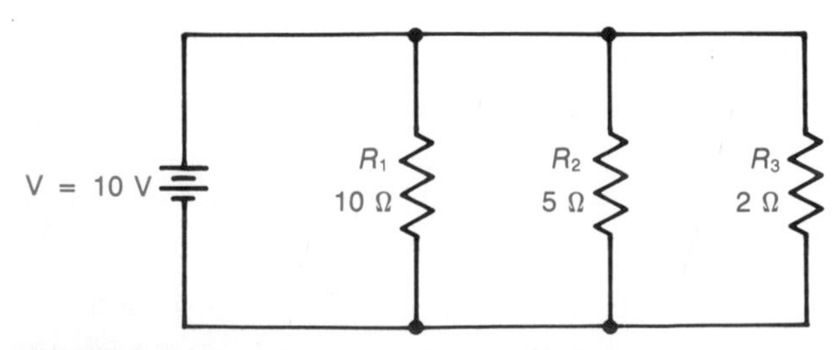


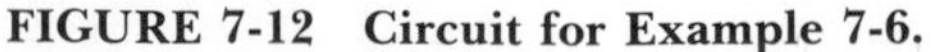

**FIGURE 7-12 Circuit for Example 7-6.**

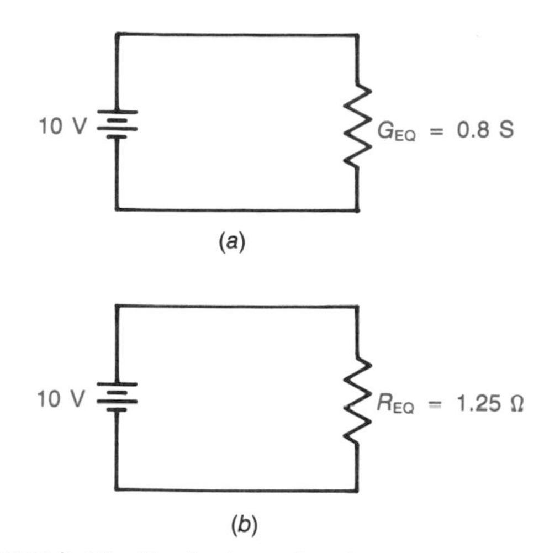


**FIGURE 7-13 Equivalent circuits.**

The total current is also equal to the supply voltage divided by the total resistance. Therefore,

$$\frac{V}{R_T} = \frac{V}{R_1} + \frac{V}{R_2} + \frac{V}{R_3}$$

Dividing all terms by $V$ gives

$$\frac{1}{R_T} = \frac{1}{R_1} + \frac{1}{R_2} + \frac{1}{R_3}$$

Thus, the reciprocals (which are conductances) can be summed to find the reciprocal of the total resistance (which is the total conductance). The above equation can also be solved for $R_T$ by taking the reciprocal of both sides of the equation:

$$R_T = \frac{1}{1/R_1 + 1/R_2 + 1/R_3}$$

This equation is widely applied for reducing two or more parallel resistors to a single equivalent resistance.

## EXAMPLE 7-7

Find the equivalent resistance and the total current for the circuit shown in Fig. 7-14.

### Solution

The total or equivalent resistance is found with the reciprocal equation:

$$R_T = \frac{1}{1/1\ \text{k}\Omega + 1/2\ \text{k}\Omega + 1/2.5\ \text{k}\Omega + 1/10\ \text{k}\Omega} = 500\ \Omega$$

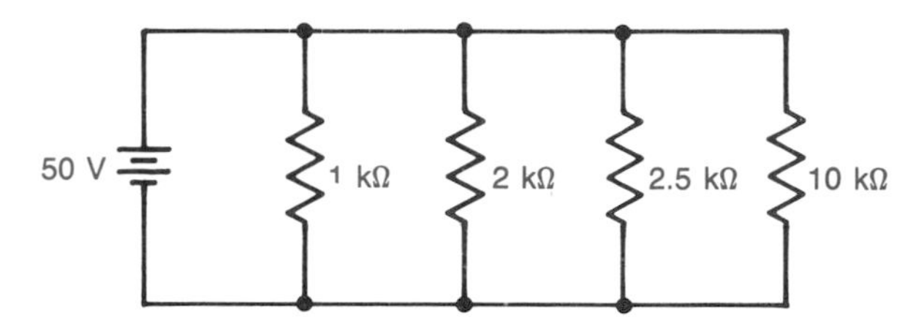


**FIGURE 7-14 Circuit for Example 7-7.**

Note that the equivalent resistance is less than the smallest resistor in the circuit. This is *always* the case when dealing with parallel resistors. The total current flow is found with Ohm's law:

$$I_T = \frac{50\ \text{V}}{500\ \Omega} = 0.1\ \text{A}$$

Calculators make it easy to deal with reciprocals. When a parallel circuit contains only two resistors, another equation is often used to find the total resistance. The other equation is easier to solve when a calculator is not available and can be derived from the reciprocal equation:

$$R_T = \frac{1}{1/R_1 + 1/R_2}$$

Multiplying both the numerator and denominator of both fractions by $R_1R_2$ to find a common denominator gives

$$R_T = \frac{1}{(R_1R_2)/(R_1R_1R_2) + (R_1R_2)/(R_2R_1R_2)}$$

Simplifying and adding,

$$R_T = \frac{1}{(R_2 + R_1)/(R_1R_2)}$$

Inverting the denominator and multiplying,

$$R_T = \frac{R_1R_2}{R_1 + R_2}$$

The result is often called the *product-over-sum equation*. Its use is limited to two resistors in parallel.

## EXAMPLE 7-8

Use the product-over-sum equation to find the equivalent resistance for Fig. 7-15. Also find the total current flow.

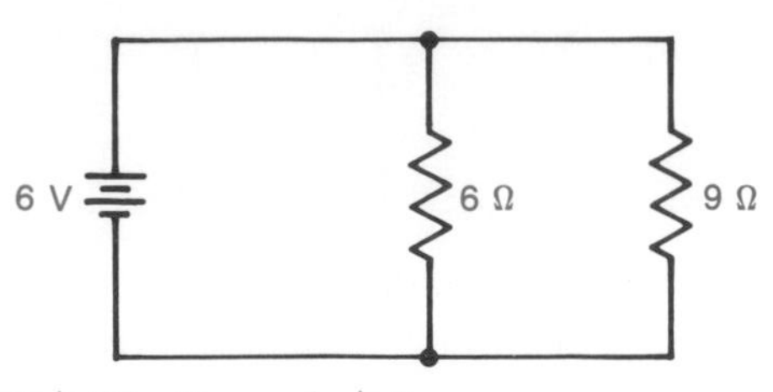


FIGURE 7-15 Example 7-8.

**Solution**

Applying the equation,

$$R_T = \frac{6\ \Omega \times 9\ \Omega}{6\ \Omega + 9\ \Omega} = 3.6\ \Omega$$

The total resistance is less than the smallest resistor, as it should be. The total current is found with Ohm's law:

$$I_T = \frac{6\ \text{V}}{3.6\ \Omega} = 1.67\ \text{A}$$

When resistors of equal value are in parallel, a shortcut is available. The equivalent resistance is equal to the value of one resistor divided by the number of resistors in the circuit. Two 10-Ω resistors in parallel have an equivalent resistance of 5 Ω. Three 90-Ω resistors in parallel have a total resistance of 30 Ω. Four 100-Ω resistors in parallel have a total resistance of 25 Ω and so on.

When the conductances of parallel branches are known, the total conductance can be found by adding. What should be done for conductances in series? The reciprocal of each conductance can be added to find the total resistance. Then the reciprocal of the total resistance can be taken to find the total conductance. The following equation allows series conductances to be reduced to an equivalent (total) conductance:

$$G_T = \frac{1}{1/G_1 + 1/G_2 + 1/G_3 + \cdots + 1/G_N}$$

**EXAMPLE 7-9**

Find the current flow for Fig. 7-16.

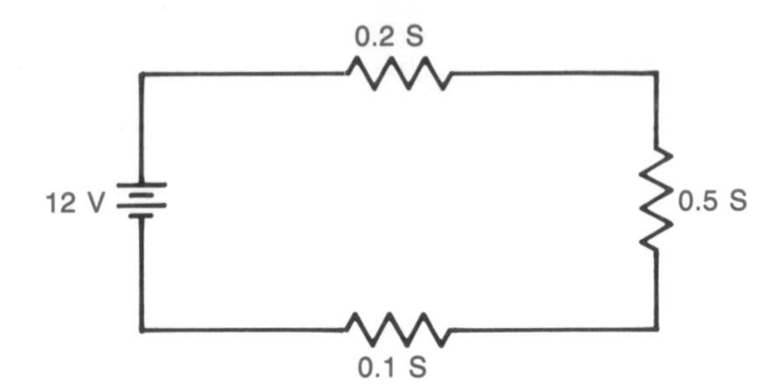


FIGURE 7-16 Example 7-9.

**Solution**

Find the total conductance of the circuit:

$$G_T = \frac{1}{1/0.2\ \text{S} + 1/0.5\ \text{S} + 1/0.1\ \text{S}} = 0.059\ \text{S}$$

The total conductance is less than the smallest conductance in the circuit. This is always the case for series conductances. Applying Ohm's law gives the current flow:

$$I = V \times G = 12\ \text{V} \times 0.059\ \text{S} = 0.708\ \text{A}$$

When conductances of equal value are in series, a shortcut is available. Divide the value of one conductance by the number of conductances in series. Two 4-S conductances in series have a total conductance of 2 S. Three 27-S conductances in series have an equivalent conductance of 9-S and so on.

## Self-Test

**10.** The reciprocal of resistance is called _______.

**11.** Find $G$ in Fig. 7-17*a*.

**12.** Find $G_T$ in Fig. 7-17*b*.

**13.** The equivalent resistance of a parallel circuit is always less than the _______ resistor in the circuit.

**14.** Find $R_{eq}$ for Fig. 7-17*c*.

**15.** Find $I_T$ for Fig. 7-17*d*.

**16.** What is the total resistance of three 24-Ω resistors in parallel?

**17.** What is the total conductance of two 0.1-S conductances in series?

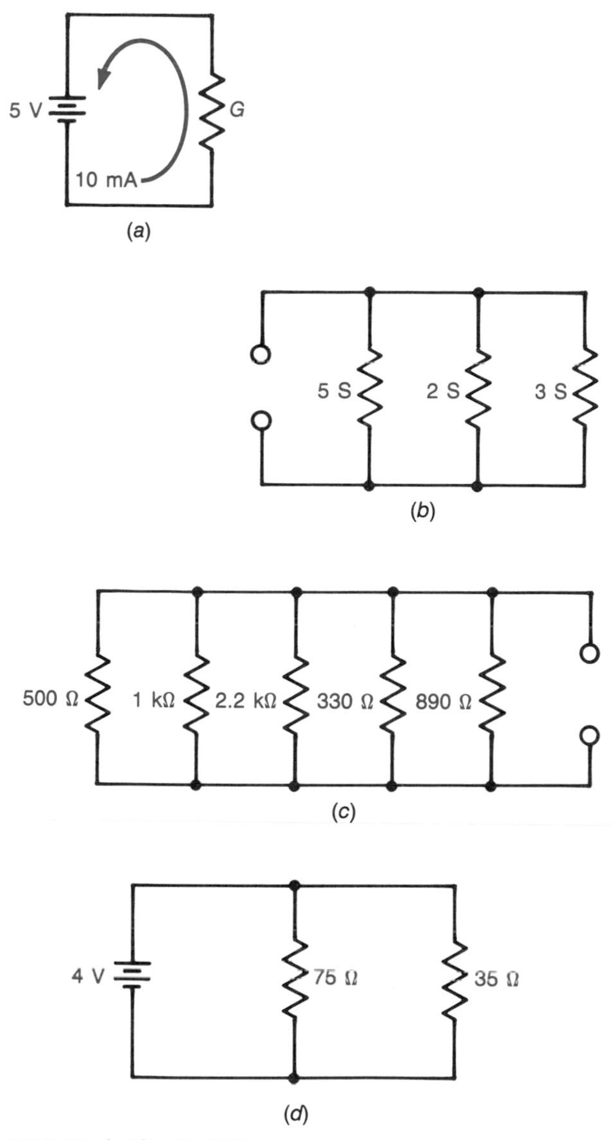

**FIGURE 7-17 Self-Test.**

## 7-3 THE CURRENT DIVIDER RULE

Resistors in parallel divide current flow. Figure 7-18 shows that the total current divides into two branch currents when two resistors are in parallel. The sum of the branch currents must be equal to the total current according to Kirchhoff's current law. The circuit laws that have been presented thus far will allow the branch currents to be calculated when the total current and the branch resistances are known.

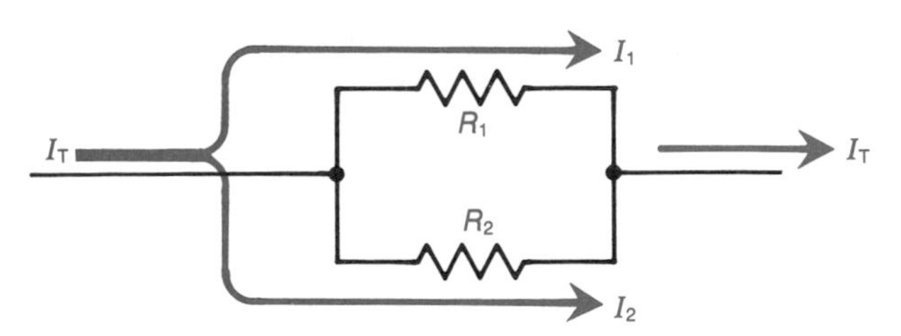


**FIGURE 7-18 Parallel resistors divide current.**

**EXAMPLE 7-10**

Calculate the branch currents for Fig. 7-19.

**Solution**

The first step is to find the total or equivalent resistance. Since $R_1 || R_2$, the product-over-sum equation can be used. The notation $R_1 || R_2$ is a shorthand way of declaring that the resistors are in parallel.

$$R_T = \frac{8\ \Omega \times 2\ \Omega}{8\ \Omega + 2\ \Omega} = 1.6\ \Omega$$

The network in Fig. 7-19 can be reduced to a single resistor of 1.6 Ω. Ohm's law is used to find the drop across the equivalent resistance:

$$V = 3\ \text{A} \times 1.6\ \Omega = 4.8\ \text{V}$$

The same voltage drop must appear across the original resistors. Ohm's law is used to calculate the current through each resistor:

$$I_1 = \frac{4.8\ \text{V}}{8\ \Omega} = 0.6\ \text{A}$$

$$I_2 = \frac{4.8\ \text{V}}{2\ \Omega} = 2.4\ \text{A}$$

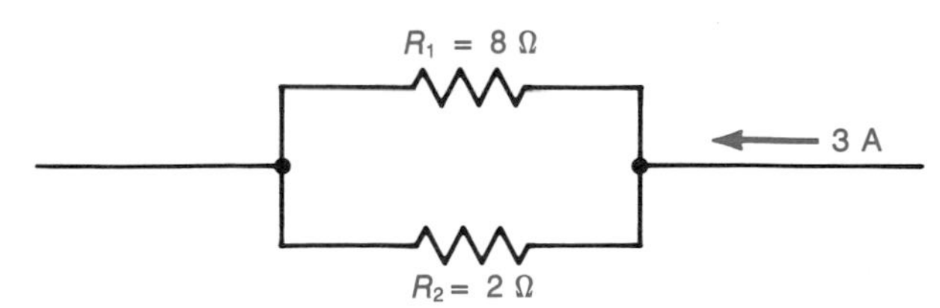


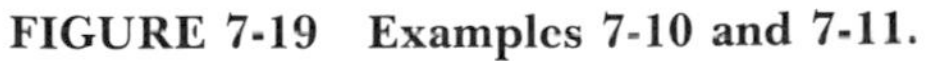
**FIGURE 7-19 Examples 7-10 and 7-11.**

*The majority of the current flows through the path with the least resistance.* The 2-Ω resistor has four times the current flow as the 8-Ω resistor. The sum of the branch currents is equal to the total current, which satisfies Kirchhoff's current law (0.6 A + 2.4 A = 3 A).

This process of finding branch currents works fine, but there is a faster way. Refer again to Fig. 7-18. By Ohm's law,

$$V_1 = I_1 R_1 \quad \text{and} \quad V_2 = I_2 R_2$$

Voltage is constant in a parallel circuit, thus

$$V_1 = V_2$$

By the law of equalities,

$$I_1 R_1 = I_2 R_2$$

From Kirchhoff's current law,

$$I_2 = I_T - I_1$$

Substituting for $I_2$,

$$I_1 R_1 = R_2 (I_T - I_1)$$

Rearranging and solving for $I_1$,

$$I_1 R_1 + I_1 R_2 = R_2 I_T$$
$$I_1 (R_1 + R_2) = R_2 I_T$$
$$I_1 = \frac{R_2 I_T}{R_1 + R_2}$$

This is known as the *current divider equation.* It is a shortcut for finding the branch current when two resistors of known value are in parallel and the total current is given. Using the same technique, it can also be proved that

$$I_2 = \frac{R_1 I_T}{R_1 + R_2}$$

Summarizing the current divider rule: To find a branch current when two resistors are in parallel, multiply the resistance of the *other* branch times the total current and divide by the sum of the resistances.

**EXAMPLE 7-11**
Calculate the branch currents for Fig. 7-19 using the current divider rule.

**Solution**
Applying the rule,

$$I_1 = \frac{2\ \Omega \times 3\ \text{A}}{8\ \Omega + 2\ \Omega} = 0.6\ \text{A}$$
$$I_2 = \frac{8\ \Omega \times 3\ \text{A}}{8\ \Omega + 2\ \Omega} = 2.4\ \text{A}$$

The results are the same as obtained by the method demonstrated in Example 7-10. The current divider rule is a faster and easier way of finding the branch currents.

There is another current divider rule that is based on branch conductances. It can be used to find branch currents for two or more conductances in parallel:

$$I_X = \frac{I_T G_X}{G_T}$$

where $I_X$ = current in branch $X$, A
$I_T$ = total current, A
$G_X$ = conductance of branch $X$, S
$G_T$ = total conductance, S

**EXAMPLE 7-12**
Find the branch currents for Fig. 7-20 using the current divider rule for parallel conductances.

**Solution**
Applying the rule,

$$I_1 = \frac{4\ \text{A} \times 0.5\ \text{S}}{1\ \text{S}} = 2\ \text{A}$$
$$I_2 = \frac{4\ \text{A} \times 0.3\ \text{S}}{1\ \text{S}} = 1.2\ \text{A}$$
$$I_3 = \frac{4\ \text{A} \times 0.2\ \text{S}}{1\ \text{S}} = 0.8\ \text{A}$$

The sum of the branch currents is equal to the total current, satisfying Kirchhoff's current law. The branch currents are proportional to the branch conductances. The most current flows in the branch with the greatest conductance, and the least current flows in the branch with the smallest conductance.

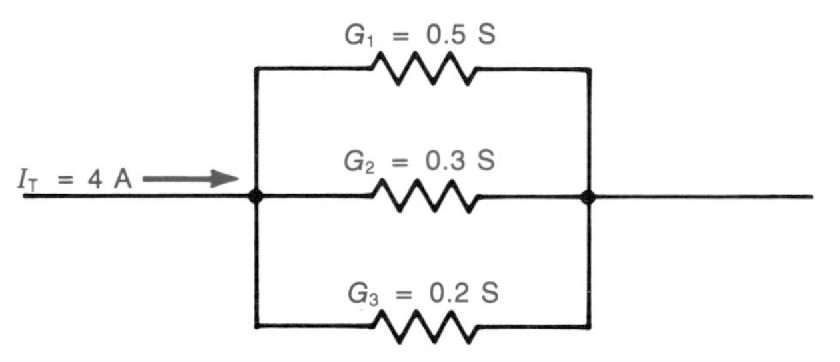


FIGURE 7-20 Example 7-12.

When resistances or conductances of equal value are in parallel, another shortcut can be applied to find the branch currents. Divide the total current by the number of resistances or conductances to find the branch currents. For example, three 100-Ω parallel resistors will divide a total current of 1.5 A into three branch currents of 0.5 A each. As another example, four 100-μS conductances in parallel will divide a total current of 80 mA into four branch currents of 20 mA each.

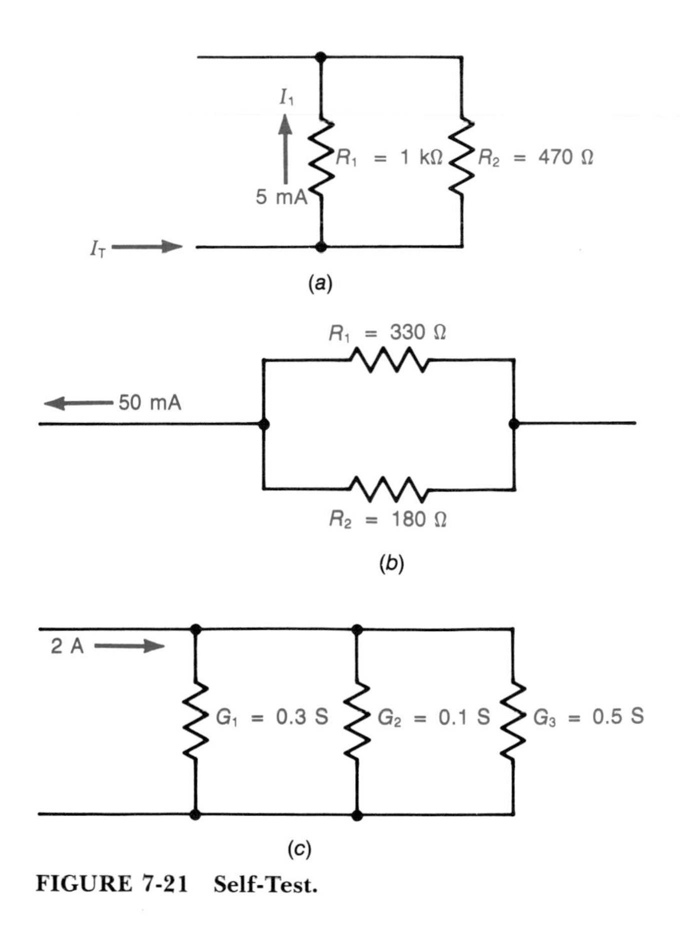


FIGURE 7-21 Self-Test.

## Self-Test

**18.** What does the notation $R_X||R_Y||R_Z$ indicate?

**19.** Refer to Fig. 7-21*a*. Calculate $I_{R_2}$.

**20.** Refer to Fig. 7-21*b*. Calculate both branch currents.

**21.** When a current divides at a junction, the majority of flow will take the path of ______ resistance.

**22.** Refer to Fig. 7-21*c*. Calculate all three branch currents.

**23.** When a current divides at a junction, the majority of flow will take the path of ______ conductance.

**24.** Five resistors of equal value are in parallel. If one of the resistors conducts 15 mA, what is the total current flow entering the parallel network?

## 7-4 APPLICATIONS OF PARALLEL CIRCUIT CONCEPTS

Identical voltage sources may be placed in parallel to provide a greater current capacity. For example, two automobile storage batteries can be used to start an engine. Figure 7-22 shows two 12-V batteries connected in parallel. If the starter motor represents a 400-A load, each battery will supply half the current or 200 A. Not only are higher currents available from parallel sources, but the voltage drop associated with heavy load currents is also decreased. Real voltage sources have some value of internal resistance. The connection shown in Fig. 7-22 places the internal resistances of the batteries in parallel for an effective internal resistance that is one-half the value for a single battery. The parallel batteries will more closely approach an ideal voltage source. Another benefit is that two batteries in parallel will supply a given load current for twice the time when compared to a single battery before discharge is reached. However, do not forget that the source voltages must be identical to allow a parallel connection. If one source develops more voltage than the other, it will deliver energy, possibly at a damaging rate, to the other source.

Conductors may be placed in parallel to increase current capacity. Heavier wire gages are required to safely conduct large amounts of current flow,

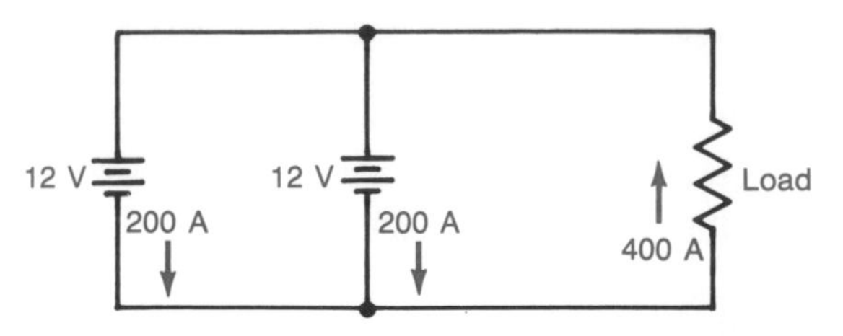


FIGURE 7-22 Voltage sources may be connected in parallel to provide a greater current capacity.

and the voltage drop associated with long conductors and high currents may not be acceptable. Parallel conductors provide an increase in current capacity and a decrease in voltage drop. Components may also be parallel-connected to allow greater current flow. Figure 7-23 shows two transistors connected in parallel. Each supports half of the load current. The parallel connection allows transistors, and other devices, to support currents beyond the ratings of a single device.

Figure 7-24 shows another application for the parallel connection. A resistor, called a shunt, is wired in parallel with an ammeter. The word *shunt* is synonymous with the word parallel. Shunts are devices or paths that are connected in parallel with some other device or circuit. The shunt in Fig. 7-24 allows the measurement of currents beyond the full-scale rating of the meter.

**EXAMPLE 7-13**

Calculate the resistance and power rating of the shunt required in Fig. 7-24. Assume that the meter has an internal resistance of 100 Ω, a full-scale rating of 1 mA, and that currents up to 1 A must be measured.

**Solution**

The meter requires 1 mA for full-scale deflection. The total current entering the junction can be as high as 1 A. Kirchhoff's current law tells us that the difference must flow through the shunt:

$$I_S = \text{the shunt current}$$
$$I_S = 1\text{ A} - 1\text{ mA} = 999\text{ mA}$$

Ohm's law can be used to find the drop across the meter at full-scale deflection. Full-scale deflection occurs when the meter is conducting 1 mA:

$$V_M = 1\text{ mA} \times 100\ \Omega = 0.1\text{ V}$$

Voltage is constant in a parallel circuit. The voltage across the shunt is the same as the voltage across the meter. Ohm's law can now be used to solve for the shunt resistance:

$$R = \frac{0.1\text{ V}}{999\text{ mA}} = 0.1\ \Omega$$

The shunt current and the shunt voltage can be used to calculate its power dissipation:

$$P = 0.1\text{ V} \times 999\text{ mA} = 99.9\text{ mW}$$

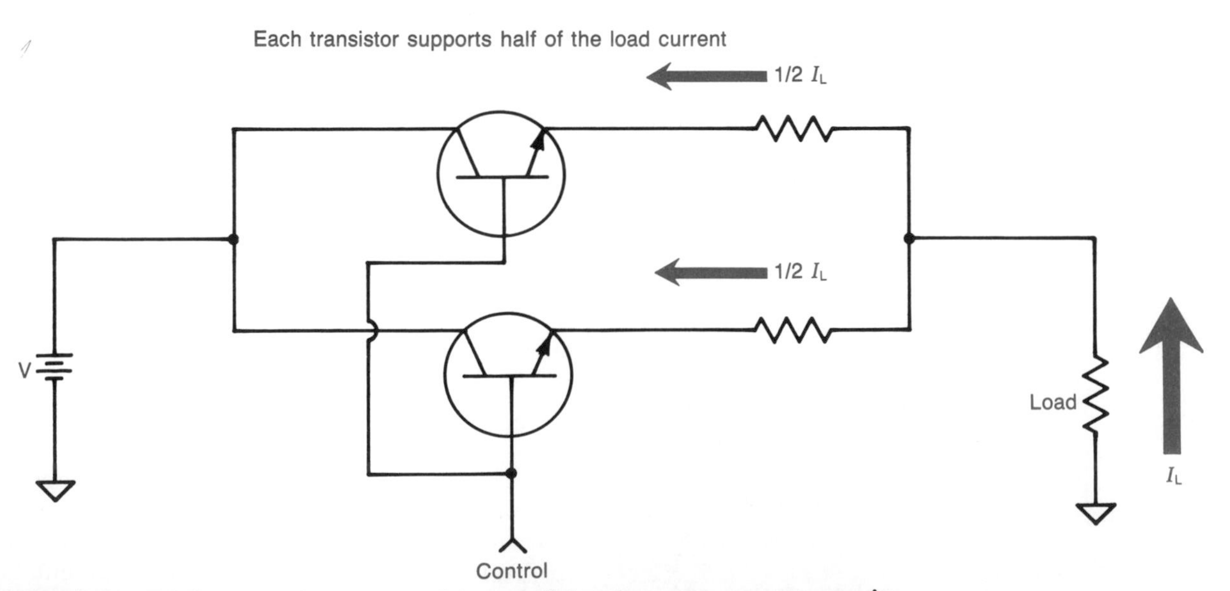


FIGURE 7-23 Devices may be connected in parallel to increase current capacity.

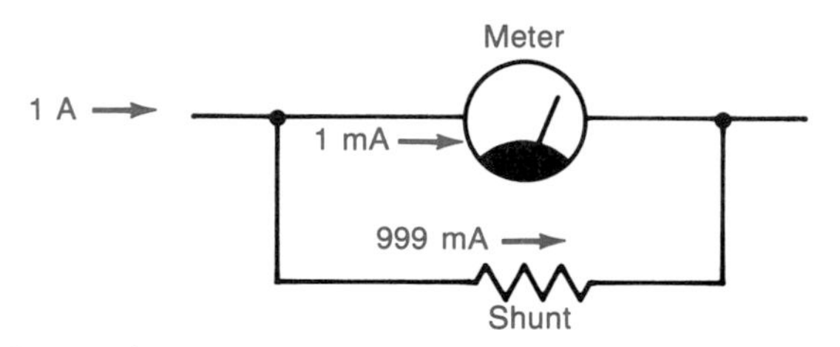


**FIGURE 7-24 Using a shunt to increase ammeter range.**

The parallel connection is widely applied in power distribution. Figure 7-25 shows that electric outlets are wired in parallel. Voltage is constant in parallel circuits. This means that the same voltage, ignoring conductor loss, is available at all of the outlets. Figure 7-26 shows another example of power distribution. The devices are integrated circuits. Each has 16 pins. One of the pins must be grounded, and another must be connected to +5 V to power each device. All of the ground pins are connected to the ground trace on the circuit board. All of the +5 pins are connected to the +5-V trace on the circuit board. The integrated circuits are in parallel as far as power distribution is concerned. The word *bus* is often used in electronic circuits. The power bus distributes power to the various components in the circuit. In computers, a data bus is used to send information back and forth between various devices that have their data pins connected in parallel.

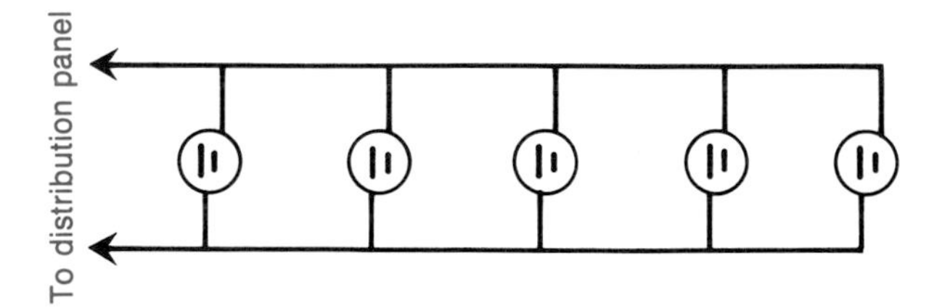


**FIGURE 7-25 Electric outlets are wired in parallel.**

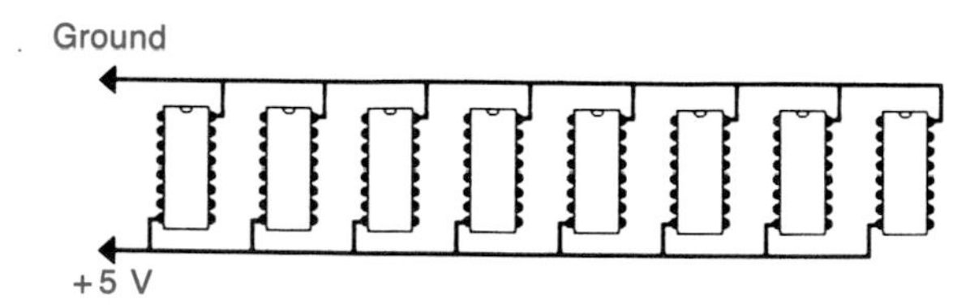


**FIGURE 7-26 Integrated circuits wired in parallel across the power bus.**

The parallel connection may be used to decrease the resistance of one part of a circuit. In a similar fashion, a resistor can be shunted by another resistor to obtain a nonstandard value.

---

**EXAMPLE 7-14**

A resistor in a circuit has a value of 560 Ω. It is necessary to decrease its resistance to 344 Ω. Calculate the required resistance for a shunt that will accomplish this.

**Solution**

The product-over-sum equation can be rearranged, giving

$$R_2 = \frac{R_1 R_T}{R_1 - R_T} = \frac{560 \times 344}{560 - 344} = 892\ \Omega$$

The nearest standard value is 890 Ω.

---

Power in a parallel circuit follows the same rule as for series circuits. The total power is equal to the sum of the individual dissipations.

---

**EXAMPLE 7-15**

Show that the sum of the individual power dissipations in Fig. 7-27 is equal to the total power dissipation.

**Solution**

A constant 15 V appears across each resistor in Fig. 7-27. $V^2/R$ is convenient to determine the individual power dissipations:

$$P_1 = \frac{15^2}{1\ \text{k}\Omega} = 0.225\ \text{W}$$

$$P_2 = \frac{15^2}{2\ \text{k}\Omega} = 0.113\ \text{W}$$

$$P_3 = \frac{15^2}{3\ \text{k}\Omega} = 0.075\ \text{W}$$

$$P_4 = \frac{15^2}{4\ \text{k}\Omega} = 0.056\ \text{W}$$

$$P_T = 0.225\ \text{W} + 0.113\ \text{W} + 0.075\ \text{W} + 0.056\ \text{W} = 0.469\ \text{W}$$

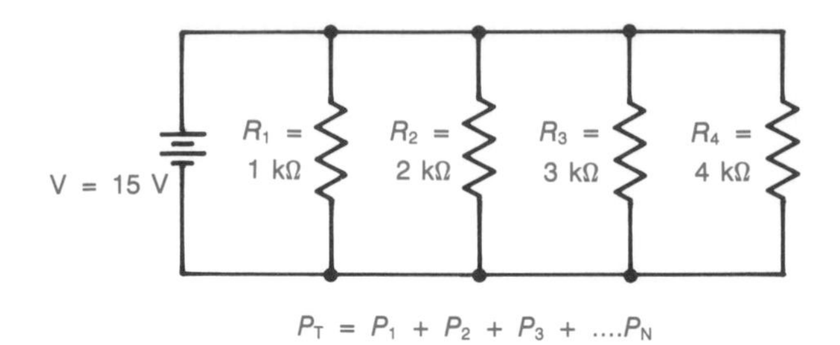


**FIGURE 7-27 The power total dissipation in a parallel circuit is equal to the sum of the individual power dissipations.**

The total resistance of Fig. 7-27 can be found with the reciprocal equation:

$$R_T = \frac{1}{1/1\ \text{k}\Omega + 1/2\ \text{k}\Omega + 1/3\ \text{k}\Omega + 1/4\ \text{k}\Omega} = 480\ \Omega$$

The total power is equal to $V^2/R_T$:

$$P_T = \frac{15^2}{480\ \Omega} = 0.469\ \text{W}$$

This agrees with the results obtained by summing the individual power dissipations.

### EXAMPLE 7-16

A $\frac{1}{2}$-W, 890-Ω resistor is connected in parallel with a $\frac{1}{2}$-W, 560-Ω resistor. Determine the maximum safe dissipation for the parallel pair.

#### Solution

The answer is *not* 1 W. Even though the total power dissipation in parallel circuits is equal to the sum of the individual power dissipations, the power ratings cannot be added when unequal resistances are connected in parallel. Since the voltage is constant, the 560-Ω resistor will conduct more current and run hotter than the 890-Ω resistor. Therefore, the maximum safe voltage across the 560-Ω resistor is found first:

$$V = \sqrt{PR} = \sqrt{0.5 \times 560} = 16.7\ \text{V}$$

This allows us to calculate the power dissipation in the 890-Ω resistor when the 560-Ω resistor is at maximum power dissipation:

$$P = \frac{16.7^2}{890\ \Omega} = 0.315\ \text{W}$$

The maximum safe power dissipation for the parallel pair is the sum of the dissipations:

$$P_T = 0.5\ \text{W} + 0.315\ \text{W} = 0.815\ \text{W}$$

## Self-Test

**25.** True or false? It is acceptable to connect a 6-V source in parallel with a 12-V source.

**26.** Two batteries are connected in parallel. Each has an internal resistance of 0.1 Ω. What is the internal resistance of the pair?

**27.** A shunt is a device that is connected in ______ with some other device or circuit.

**28.** A meter has an internal resistance of 50 Ω and requires 10 mA for full-scale deflection. Calculate the shunt resistance required to allow the meter to measure currents up to 100 mA.

**29.** True or false? The word *bus* implies series-connected devices.

**30.** It is desired to decrease the resistance of a 270-Ω resistor to 149 Ω. Calculate a parallel resistance value that will accomplish this. If both resistors are rated at 1 W, calculate the maximum safe dissipation for the parallel pair.

**31.** Twenty-five 100-W lights are connected in parallel across a 120-V line. Calculate the total current flow.

## 7-5 FAULT ANALYSIS IN PARALLEL CIRCUITS

One or more branches of a parallel circuit may develop an open. Figure 7-28 shows a parallel circuit with an open. Resistor $R_3$ has burned out and now has infinite resistance. The circuit will show these symptoms:

1. Branch current $I_3$ will be equal to zero because $R_3$ is open.
2. The total current $I_T$ will be lower than normal.
3. The voltage will be normal or possibly somewhat higher than normal. This is because the total current is less than normal, so the drop across the internal resistance of the source will be smaller.

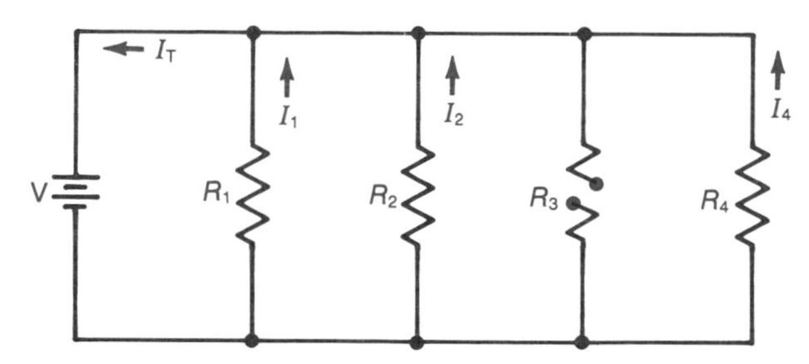


**FIGURE 7-28 An open in a parallel circuit.**

**4.** The operation of branches without opens will be normal. If $R_1$, $R_2$, and $R_4$ are lamps, they will be on. If they represent motors, they will run.

**5.** The open device will not operate. If $R_3$ is a lamp, it will be out. If it is a motor, it will not run.

Figure 7-29 shows a parallel circuit with a short. A "dead" short is one that approaches zero ohms. The circuit symptoms with a dead short are the following:

**1.** The voltage will be zero.

**2.** The total current flow will be much greater than normal. It is labeled as $I_{FAULT}$ in Fig. 7-29. The only limit on current flow is the internal resistance of the energy source and the resistance of the conductors. The source and the conductors may be damaged if no overcurrent protection is available.

**3.** No current will flow in the normal branches. All of the current will flow in the short or in the shorted branch.

**4.** None of the loads will operate. If they are lamps, they will all be out. If they are motors, none will run.

If the fault is not a *dead* short, the symptoms may be somewhat different:

**1.** The voltage may be much less than normal. It will depend on the resistance of the short and the internal resistance of the source. For example, if the short resistance is equal to the source resistance, the voltage will be approximately half of normal. The exact value will be determined by the internal resistance of the source, the short resistance, the resistance of the other branches, and conductor resistance.

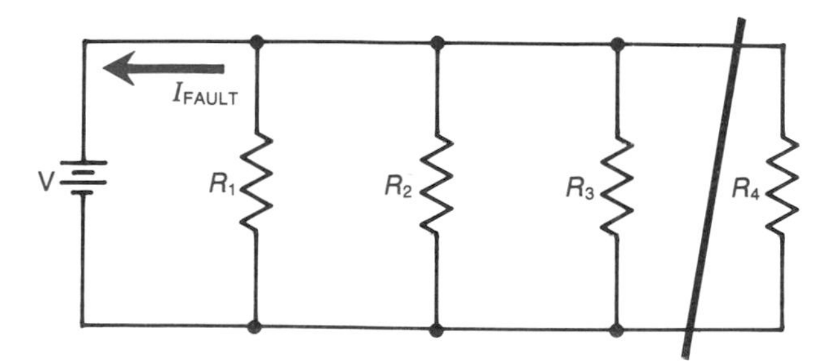


**FIGURE 7-29 A short in a parallel circuit.**

**2.** The total current flow will be greater than normal.

**3.** The current flow in the normal branches will be less than normal.

**4.** The normal branches may operate, but not as they are supposed to. If they are lamps, they may be dim. If they are motors, they may run slowly or stall under load.

Both dead shorts and partial shorts can cause extensive damage. Figure 7-30 shows a parallel circuit with overcurrent protection. Fuses $F_1$, $F_2$, and $F_3$ are branch fuses. They protect against damage that could be caused by a short in one of the three loads ($R_1$, $R_2$, or $R_3$). Fuse $F_M$ is the main fuse. It protects against defects, such as a short from A to B, that would not cause any of the branch fuses to blow. If a branch fuse blows, only that branch will stop operating. The others should continue normal operation. If the main fuse blows, all of the branches are shut down.

The main fuse also protects the conductors in those cases where too many loads are active at one time. Most kitchen circuits in homes and apartments are fuse- or breaker-protected at 20 A. Consider the specifications for the following kitchen appliances:

| | | |
|---|---|---|
| Coffee maker | 525 W | 4.38 A |
| Microwave oven | 1500 W | 12.50 A |
| Toaster | 1000 W | 8.33 A |

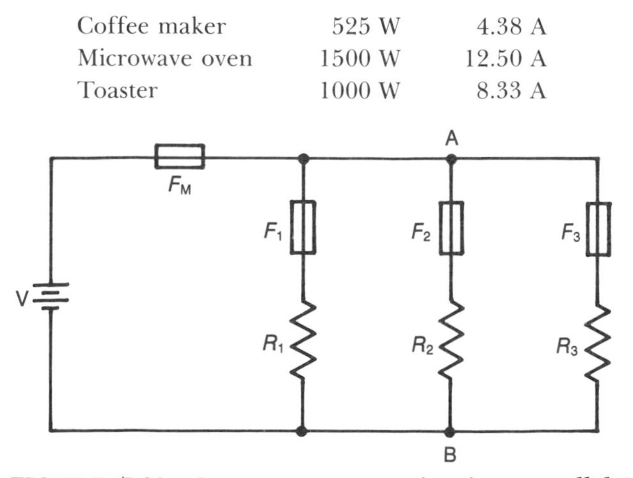


**FIGURE 7-30 Overcurrent protection in a parallel circuit.**

Each appliance produces a branch current when operating. When all three are operating, the sum of the branch currents is 25.2 A, which exceeds the safe capacity of the wire running from the power panel to the kitchen. Sometimes a fuse blows or a breaker trips even though there are no shorts.

Figure 7-31 shows how switches can be used in parallel circuits. The branch switches provide independent on-off control for each load. The main switch $S_M$ provides master control for the circuit.

Fault analysis may require that current flow be measured. Current measurement involves breaking the circuit and inserting an ammeter across the break points. In a series circuit, the current flow is constant so the circuit can be broken any place that is convenient. In a parallel circuit, the current flow is not constant. Look at Fig. 7-32*a*. Total current $I_T$ enters the node and four branch currents leave the node. Depending on how the circuit is broken, different currents and combinations of currents may be measured. Figure 7-32*b* shows the circuit with the node spread out along the bottom conductor. The node is broken at the *X* points and an ammeter is inserted. Because of the point chosen for the break, the meter will indicate the sum of two branch currents ($I_3$ and $I_4$).

Resistance measurements may also be made in parallel circuits. In some cases, equivalent resistance readings may be misleading. Figure 7-33 shows an example. The equivalent resistance can be found with the product-over-sum equation:

$$R_{EQ} = \frac{1\text{ k}\Omega \times 22\text{ k}\Omega}{1\text{ k}\Omega + 22\text{ k}\Omega} = 0.957\text{ k}\Omega$$

The ohmmeter should read 957 Ω if both resistors are nominal. If $R_1$ has a tolerance of +5 percent, its actual value could range from 950 to 1050 Ω. Resistor $R_2$ could be open (have infinite resistance), and the correct ohmmeter reading could still be obtained. Thus, it is not possible to verify the integrity of $R_2$ in Fig. 7-33 with the ohmmeter connected as shown. At least one lead must be

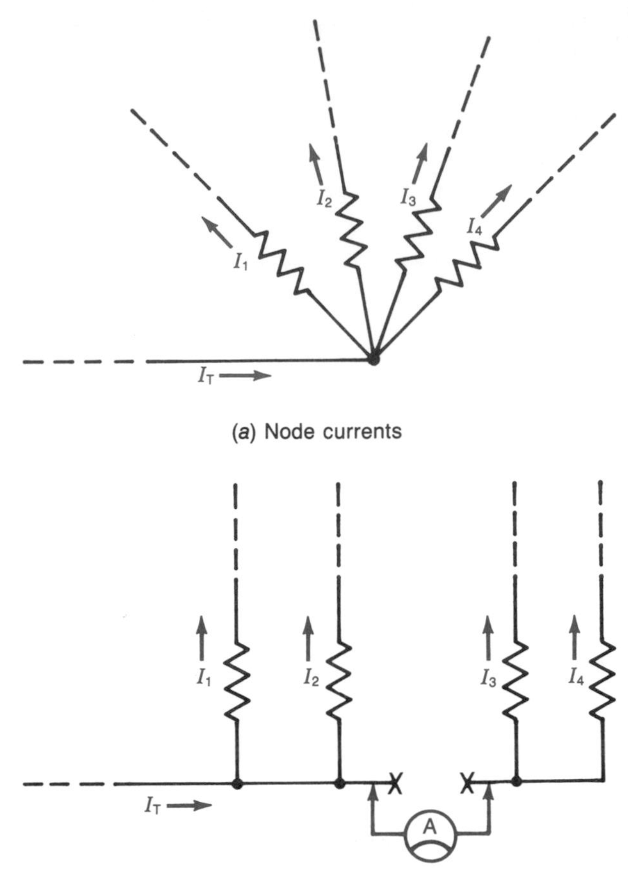


**FIGURE 7-32 Measuring current in parallel circuits.**

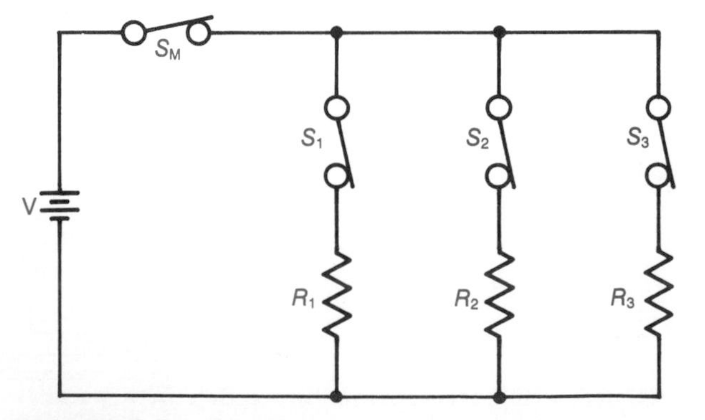


**FIGURE 7-31 Control in a parallel circuit.**

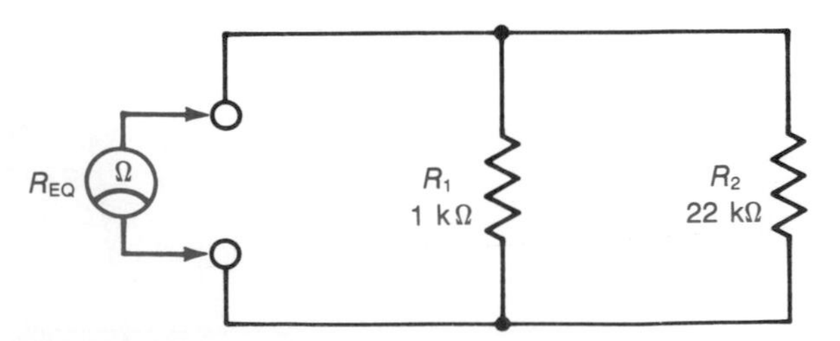


**FIGURE 7-33 Resistor $R_2$ is "swamped" by $R_1$ as far as $R_{EQ}$ is concerned.**

opened so that $R_2$ can be checked as an individual resistance.

Resistor $R_2$ is swamped by $R_1$ in Fig. 7-33 as far as the equivalent resistance is concerned. The resistance of $R_2$ has only about as much effect on the total resistance as the tolerance error of $R_1$. The smallest resistor in a parallel network has the greatest effect on the equivalent resistance of the network. If it is quite a bit smaller in value than the other resistors, it *swamps* or minimizes any effect that they might have.

Swamping also occurs in series circuits. However, the part that does the swamping in a series circuit is the one with the *greatest* resistance. Refer back to Fig. 7-23 for a moment. There is a swamping resistor connected in series with each transistor. The resistors are used to ensure reasonably equal currents in the parallel-connected transistors. The resistors are greater in value than the internal resistances of the transistors, and they swamp any minor differences in the transistors so that the branch currents are nearly equal even when the transistors are not well matched.

## Self-Test

**32.** True or false? If one branch of a parallel circuit develops an open, the other branches will operate normally.

**33.** True or false? If one branch of a parallel circuit develops a dead short, the other branches will not operate.

**34.** True or false? A short across the source terminals in a parallel circuit will cause all of the branch fuses to blow.

**35.** Refer to Fig. 7-31. Switch $S_M$ will provide on-off control for only those loads that have their branch switches set in the ______ position.

**36.** Refer to Fig. 7-28. How must an ammeter be connected to measure $I_1$?

**37.** A 100-Ω resistor is in parallel with a 4.7-kΩ resistor. Which one swamps the other as far as $R_{eq}$ is concerned?

**38.** A 100-Ω resistor is in series with a 4.7-kΩ resistor. Which one swamps the other as far as $R_{eq}$ is concerned?

## SUMMARY

1. Nodes, or junctions, are points in a circuit where two or more components are connected.
2. The voltage is constant in parallel circuits:

$$V_1 = V_2 = V_3 = \cdots = V_N$$

3. Parallel paths are called branches.
4. The total current in a parallel circuit is equal to the sum of the branch currents:

$$I_T = I_1 + I_2 + I_3 + \cdots + I_N$$

5. Kirchhoff's current law states that the algebraic sum of the junction currents is equal to zero. A positive sign is applied to currents entering the junction and a negative sign to currents leaving the junction:

$$+I_1 \pm I_2 \pm I_3 \pm \cdots \pm I_N = 0$$

6. Conductance is the measure of how well a device supports the flow of current. It is the reciprocal of resistance:

$$G = \frac{1}{R}$$

7. The unit of conductance is the siemen (letter symbol S).

8. Ohm's law can be solved for conductance:

$$G = \frac{I}{V}$$

9. The equivalent conductance of a parallel circuit is equal to the sum of the branch conductances:

$$G_{eq} = G_1 + G_2 + G_3 + \cdots + G_N$$

10. The equivalent resistance of parallel resistors can be found with the reciprocal equation:

$$R_{eq} = \frac{1}{1/R_1 + 1/R_2 + 1/R_3 + \cdots + 1/R_N}$$

11. The product-over-sum equation can be used to find the equivalent resistance of two parallel devices:

$$R_{eq} = \frac{R_1 \times R_2}{R_1 + R_2}$$

12. When resistors of equal value are in parallel, $R_{eq}$ can be found by dividing the resistance value by the number of resistors.

13. The value of $R_{eq}$ is *always* less than the smallest resistor in a parallel circuit.

14. The reciprocal equation can be used to find $G_{eq}$ when the conductances are connected in series:

$$G_{eq} = \frac{1}{1/G_1 + 1/G_2 + 1/G_3 + \cdots + 1/G_N}$$

15. Equivalent conductance $G_{eq}$ is always less than the smallest conductance in series circuits.

16. The notation $R_1||R_2$ indicates that $R_1$ and $R_2$ are in parallel.

17. The current divider rule provides a way to find the current in one branch by multiplying the resistance of the *other* branch times the total current and dividing by the sum of the resistances:

$$I_1 = \frac{R_2 I_T}{R_1 + R_2} \quad \text{or} \quad I_2 = \frac{R_1 I_T}{R_1 + R_2}$$

18. The conductance current divider rule can be used to find the currents for two or more branches:

$$I_X = \frac{I_T G_X}{G_{eq}}$$

19. Identical voltage sources may be connected in parallel to provide a greater current capacity.

20. Conductors and components may be connected in parallel for increased current capacity.

21. Shunts are parallel-connected devices.

22. Ammeters may be shunted to increase their full-scale current.

23. Electric outlets are wired in parallel.

**24.** The word *bus* implies parallel-connected devices.

**25.** The total power dissipation in a parallel circuit is equal to the sum of the individual power dissipations:

$$P_T = P_1 + P_2 + P_3 + \cdots + P_N$$

**26.** An open in one branch of a parallel circuit does not affect the operation of the other branches. However, the total current will be less than normal.

**27.** A short in a parallel circuit affects all of the branches.

**28.** A small resistance in parallel with a large resistance has the majority influence for determining $R_{eq}$ and effectively swamps the large resistance.

## CHAPTER REVIEW QUESTIONS

**7-1.** What is constant in series circuits?

**7-2.** What is constant in parallel circuits?

**7-3.** The branch with the least resistance will have the ______ current flow in a parallel circuit.

**7-4.** True or false? Kirchhoff's laws are universal: they apply to both series and parallel circuits.

**7-5.** What does the notation $R_1||R_2||R_3$ indicate?

**7-6.** True or false? A 6-V source and a 9-V source may be connected in parallel for increased current capacity.

**7-7.** Four motors with internal circuit breakers are connected in parallel. What will happen to the performance of the other three motors if one of the breakers opens?

**7-8.** Four motors with no overcurrent protection are wired in parallel. What will happen if one motor develops a dead short?

**7-9.** Refer to Fig. 7-34*h*. Which component swamps the circuit as far as $R_{eq}$ is concerned?

**7-10.** Refer to Fig. 7-34*i*. Which component swamps the circuit as far as $R_{eq}$ is concerned?

## CHAPTER REVIEW PROBLEMS

**7-1.** Refer to Fig. 7-34*a*. Calculate the branch currents.

**7-2.** Refer to Fig. 7-34*a*. Calculate the total current.

**7-3.** Find $I_X$ in Fig. 7-34*b*.

**7-4.** Is $I_X$ in Problem 7-3 entering or leaving the junction?

**7-5.** Calculate the total conductance in Fig. 7-34*c*.

**7-6.** Calculate the total current flow in Fig. 7-34*c*.

**7-7.** A component has a resistance of 25 Ω. Find its conductance.

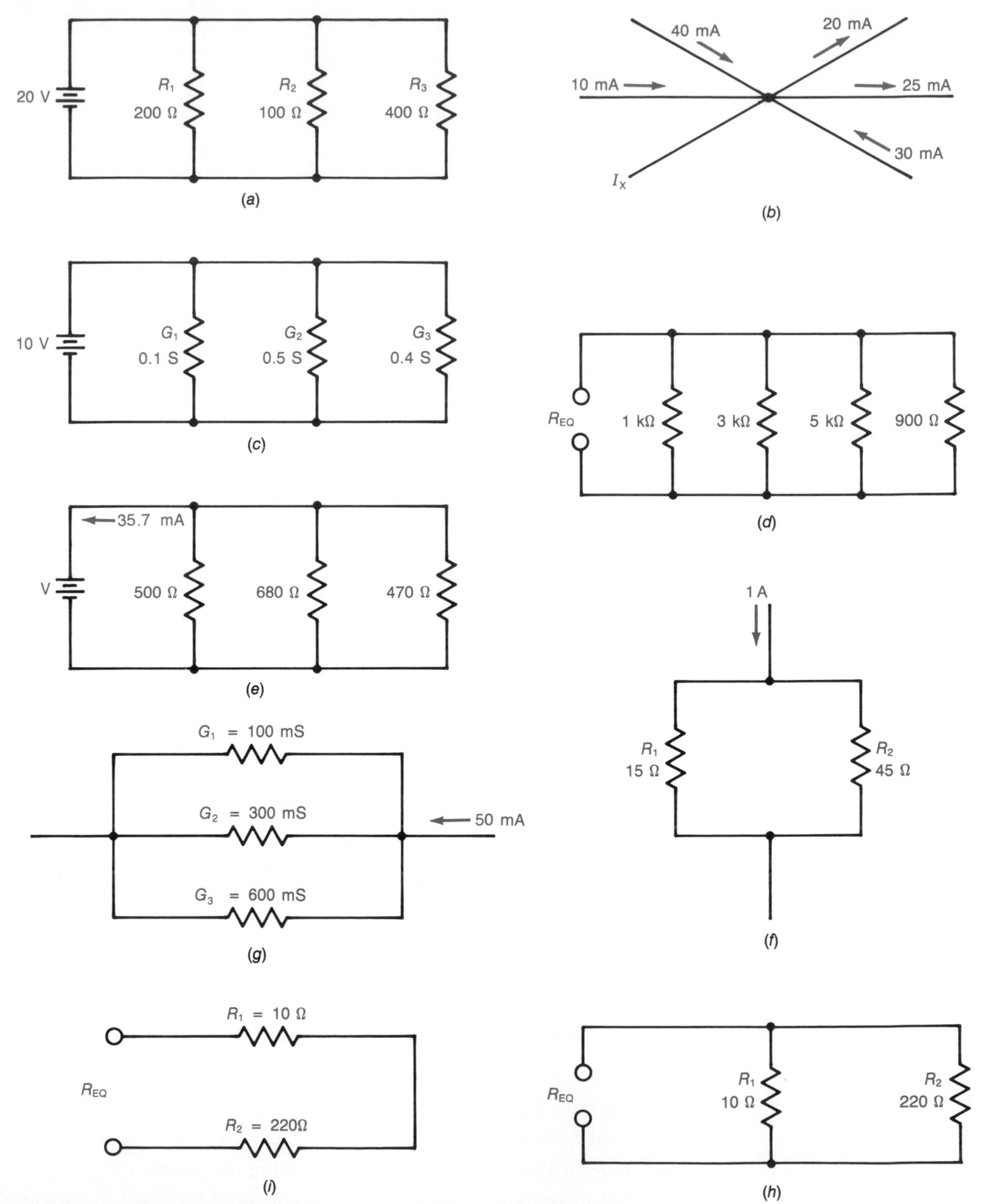


**FIGURE 7-34 Circuits for Chapter Review Questions and Problems.**

**7-8.** Find $R_{eq}$ for Fig. 7-34*d*.

**7-9.** Find $G_{eq}$ for Fig. 7-34*d*.

**7-10.** Find the value of the source voltage for Fig. 7-34*e*.

**7-11.** Find $G_{eq}$ for three 0.5-S conductances connected in series.

**7-12.** Find $R_{eq}$ for six 60-Ω resistances connected in parallel.

**7-13.** Find $I_1$ and $I_2$ for Fig. 7-34*f*.

**7-14.** Find $I_1$, $I_2$, and $I_3$ for Fig. 7-34*g*.

**7-15.** An ammeter requires 100 mA for full-scale deflection and has an internal resistance of 50 Ω. Calculate a shunt that will allow the meter to measure currents up to 1 A.

**7-16.** Calculate the full-scale power dissipation in the shunt for Problem 7-15.

**7-17.** It is desired to reduce the resistance of a 890-Ω resistor to 270 Ω. Calculate the value for a parallel resistor to accomplish this.

**7-18.** If both resistors in Problem 7-17 are rated at 1 W, what is the maximum safe dissipation for the parallel combination?

**7-19.** A 120-V circuit has branch loads of 1 kW, 900 W, 400 W, and 250 W. What is the total current flow?

## ANSWERS TO SELF-TESTS

**1.** series

**2.** parallel

**3.** the one with the greatest resistance

**4.** junction

**5.** false

**6.** false

**7.** $I_1 = 10$ mA, $I_2 = 2$ mA, $I_3 = 5$ mA

**8.** $I_T = 17$ mA

**9.** 0; since it is zero there is no direction

**10.** conductance

**11.** 2 mS

**12.** 10 S

**13.** smallest

**14.** 131 Ω

**15.** 168 mA

**16.** 8 Ω

**17.** 0.05 S

**18.** $R_X$, $R_Y$, and $R_Z$ are in parallel

**19.** 10.6 mA

**20.** $I_1 = 17.6$ mA, $I_2 = 32.4$ mA

**21.** least

**22.** $I_1 = 0.667$ A, $I_2 = 0.222$ A, $I_3 = 1.11$ A

**23.** greatest

**24.** 75 mA

**25.** false

**26.** 0.05 Ω

**27.** parallel

**28.** 5.56 Ω

**29.** false

**30.** 330 Ω, 1.82 W

**31.** 20.8 A

**32.** true

**33.** true

**34.** false

**35.** closed (on)

**36.** break at a point where only $I_1$ flows; i.e., in the vertical lead above or below $R_1$

**37.** the 100-Ω resistor

**38.** the 4.7-kΩ resistor

# Chapter 8 Series-Parallel Circuits

Series-parallel circuits are combinations of the series and parallel connections discussed in Chaps. 6 and 7. They can be analyzed by identifying the series sections and the parallel sections. Series laws are then applied to the series sections and parallel laws to the parallel sections. This chapter treats the methods used to simplify, reduce, and solve series-parallel circuits for current flows, voltage drops, and power dissipations. It also covers applications and fault analysis for series-parallel circuits.

## 8-1 APPLYING LAWS TO SERIES-PARALLEL CIRCUITS

Figure 8-1 shows three types of circuits. The series circuit (*a*) has the resistors connected end to end. The parallel circuit (*b*) has the resistors connected across one another. The series-parallel circuit (*c*) has $R_2$ and $R_3$ connected in parallel, and this combination is in series with $R_1$.

### EXAMPLE 8-1

Solve the circuit shown in Fig. 8-2 for all voltage drops, current flows, and power dissipations.

**Solution**

Series-parallel circuits can be solved by finding total resistance (or conductance) and then applying Ohm's and Kirchhoff's laws. The process involves identifying the series components and the parallel components and then combining them to find the equivalent resistance. For example, $R_3$ in Fig. 8-2 is

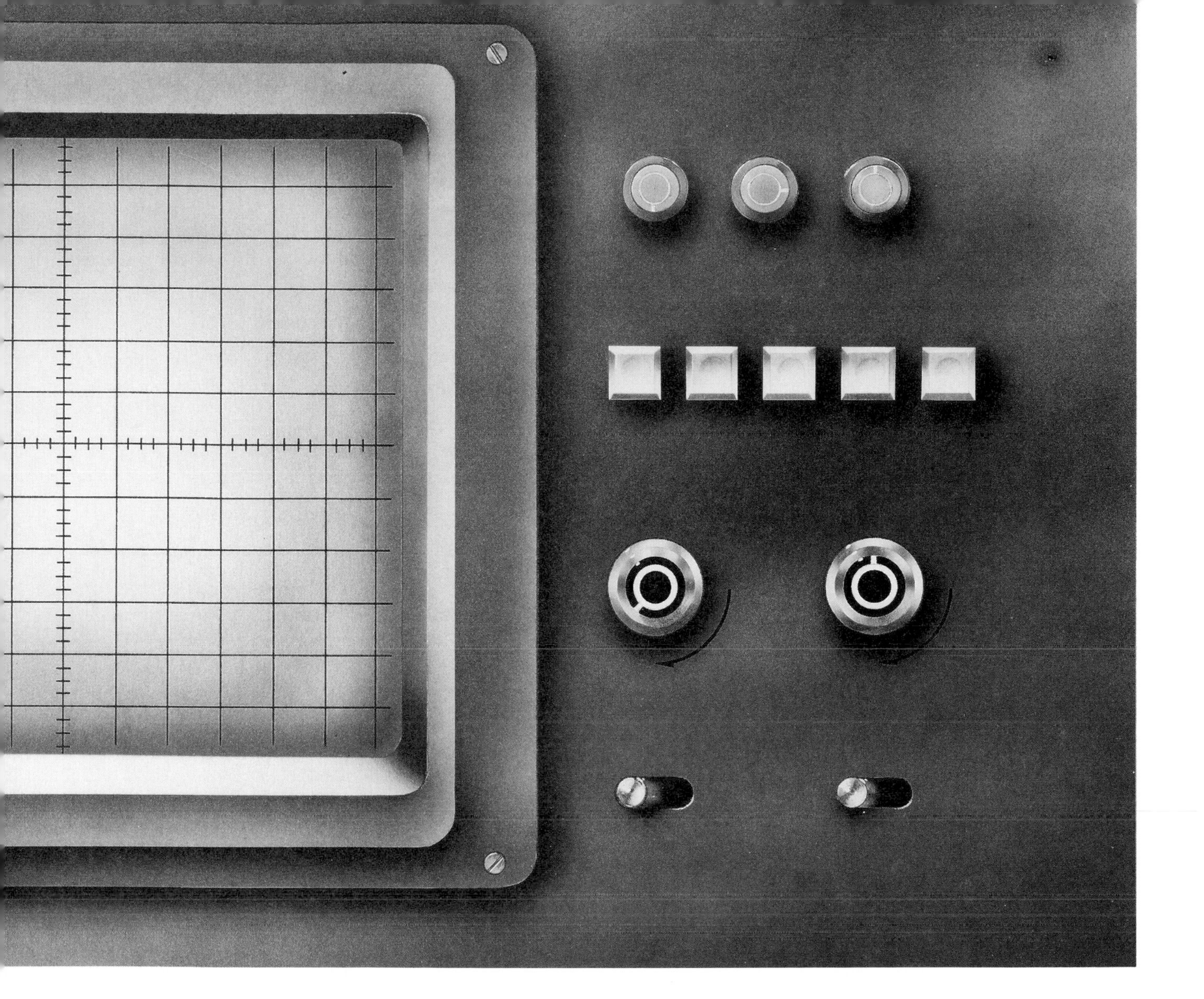

not in series with $R_1$ due to the complicating connection of $R_2$. However, $R_2$ and $R_1$ are in parallel because there are no complicating connections. These two resistors can be combined with the product-over-sum equation:

$$R_T = \frac{60 \times 40}{60 + 40} = 24\ \Omega$$

If $R_1$ and $R_2$ are replaced with a single 24-Ω resistor, it will be in series with $R_3$. Series resistances are additive:

$$R_{eq} = 12 + 24 = 36\ \Omega$$

All three resistors in Fig. 8-2 can be replaced with an equivalent resistance of 36 Ω. Ohm's law can now be used to find the total current flow:

$$I_T = \frac{18\text{ V}}{36\ \Omega} = 0.5\text{ A}$$

In Fig. 8.2, 0.5 A flows in the circuit, but where? This current is not what flows in the 40-Ω resistor, nor in the 60-Ω resistor. These two currents have yet to be found. The fact is that 0.5 A is the total current because it was found by dividing the source voltage by the equivalent resistance for all of the resistances in the circuit. Therefore, 0.5 A flows in the battery and also through $R_3$. Now, Ohm's law can be used to find the drop across $R_3$:

$$V_2 = 0.5\text{ A} \times 12\ \Omega = 6\text{ V}$$

Kirchhoff's voltage law is used to find the other drop $V_1$:

$$V_1 = 18\text{ V} - 6\text{ V} = 12\text{ V}$$

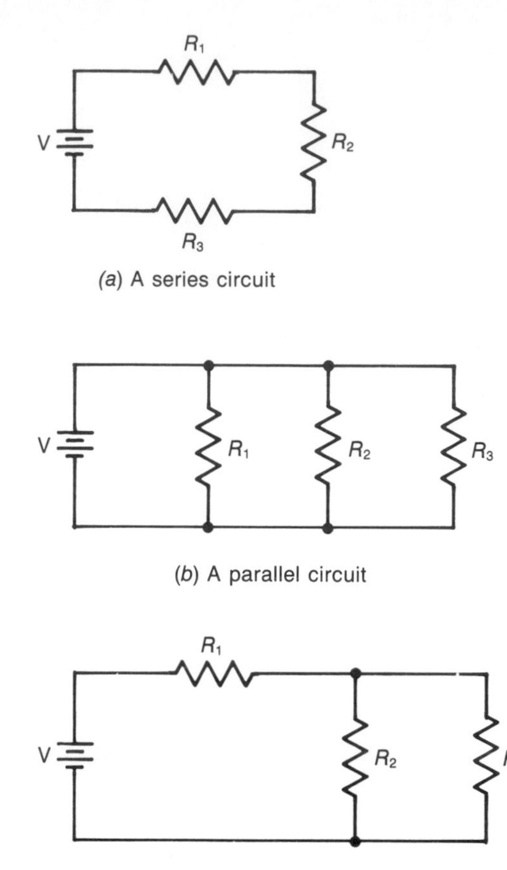


**FIGURE 8-1 Circuits may be identified by the way the components are connected.**

Note that $V_1$ is a single drop in Fig. 8-2. It is constant across the two parallel resistors. Branch currents $I_1$ and $I_2$ are solved next by using Ohm's law (they could also be found with the current divider rule):

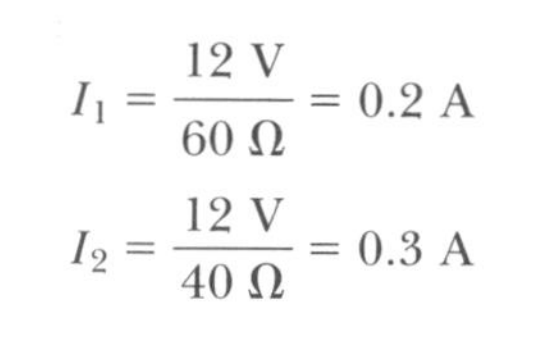

$$I_1 = \frac{12\text{ V}}{60\ \Omega} = 0.2\text{ A}$$

$$I_2 = \frac{12\text{ V}}{40\ \Omega} = 0.3\text{ A}$$

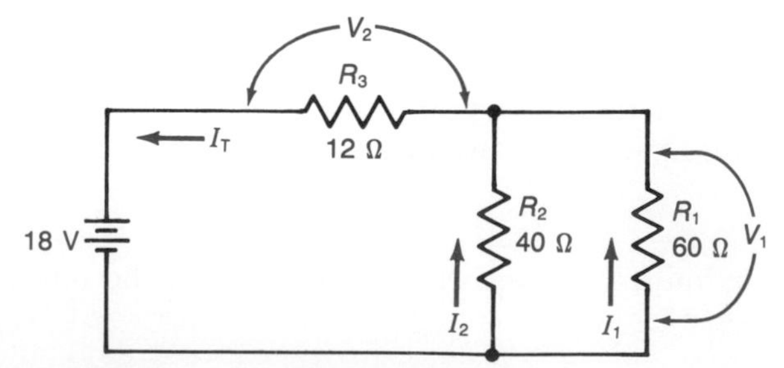


**FIGURE 8-2 Circuit for Example 8-1.**

The sum of $I_1$ and $I_2$ is equal to the total current (0.2 A + 0.3 A = 0.5 A). This satisfies Kirchhoff's current law. The last step is to find all of the power dissipations. We will begin with the total power dissipation which is equal to the applied voltage times the total current:

$$P_T = 18\text{ V} \times 0.5\text{ A} = 9\text{ W}$$

The individual power dissipations are calculated by multiplying the voltage drop across each resistor times the current through it:

$$P_1 = 12\text{ V} \times 0.2\text{ A} = 2.4\text{ W}$$

$$P_2 = 12\text{ V} \times 0.3\text{ A} = 3.6\text{ W}$$

$$P_3 = 6\text{ V} \times 0.5\text{ A} = 3.0\text{ W}$$

---

Universal circuit laws and rules apply to all circuits, regardless of how the components are arranged. The rule which states that the total power dissipation is equal to the sum of the individual power dissipations is universal. It applies to series circuits, parallel circuits, and series-parallel circuits:

$$9\text{ W} = 2.4\text{ W} + 3.6\text{ W} + 3.0\text{ W}$$

---

**EXAMPLE 8-2**

Solve the circuit shown in Fig. 8-3 for all voltage drops, all current flows, and all power dissipations.

**Solution**

Begin by looking for series and parallel combinations. Resistors $R_2$ and $R_3$ are in series. They can be replaced with a single resistor $R_S$ by adding their resistances:

$$R_S = 3\text{ k}\Omega + 7\text{ k}\Omega = 10\text{ k}\Omega$$

The total 10-kΩ resistance is in parallel with $R_1$.

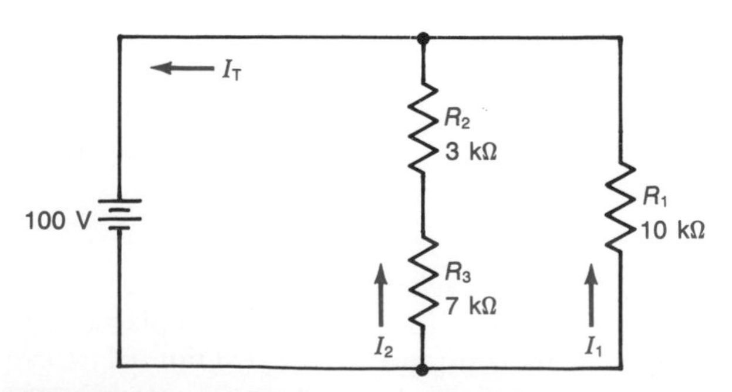


**FIGURE 8-3 Circuit for Example 8-2.**

When equal-valued resistors are in parallel, the shortcut can be used to find $R_{eq}$:

$$R_{eq} = \frac{10\text{ k}\Omega}{2} = 5\text{ k}\Omega$$

All three resistors can be replaced with an equivalent resistance of 5 kΩ. Ohm's law is used next to find the total current flow:

$$I_T = \frac{100\text{ V}}{5\text{ k}\Omega} = 20\text{ mA}$$

There are often several options as to which circuit law or rule to use at a given point in a circuit solution. The voltage divider rule could be used at this point to determine how $R_2$ and $R_3$ divide up the 100-V source. Or, the current divider rule could be used to find $I_1$ and $I_2$. Each branch has 10 kΩ of resistance; therefore, we can expect the 20-mA total current to divide equally in this case. We will use yet another approach. Notice, in Fig. 8-3, that the 100-V source is applied directly across $R_1$. This means that Ohm's law can be used to find $I_1$:

$$I_1 = \frac{100\text{ V}}{10\text{ k}\Omega} = 10\text{ mA}$$

Kirchhoff's current law can be used now to find $I_2$:

$$I_2 = 20\text{ mA} - 10\text{ mA} = 10\text{ mA}$$

The current through $R_2$ and $R_3$ is 10 mA. Ohm's law can be used to find the drops:

$$V_{R_3} = 10\text{ mA} \times 7\text{ k}\Omega = 70\text{ V}$$
$$V_{R_2} = 10\text{ mA} \times 3\text{ k}\Omega = 30\text{ V}$$

Note that the sum of these drops is equal to the applied voltage. This satisfies Kirchhoff's voltage law. The last part of the solution is to find the power dissipations. We begin with the total power dissipation:

$$P_T = V_T \times I_T = 100\text{ V} \times 20\text{ mA} = 2\text{ W}$$
$$P_{R_1} = V_T \times I_1 = 100\text{ V} \times 10\text{ mA} = 1\text{ W}$$
$$P_{R_2} = V_{R_2} \times I_2 = 30\text{ V} \times 10\text{ mA} = 0.3\text{ W}$$
$$P_{R_3} = V_{R_3} \times I_2 = 70\text{ V} \times 10\text{ mA} = 0.7\text{ W}$$

And, as a check,

$$2\text{ W} = 1\text{ W} + 0.3\text{ W} + 0.7\text{ W}$$

Real-life problems often require a different approach to using circuit laws and rules. Never be hesitant to dig in and solve for whatever information that you can. Good problem solvers know that there are situations where a clear path to the required solution cannot be immediately established. They also know that any additional information which can be produced by using the appropriate techniques often paves the way to an ultimate solution.

**EXAMPLE 8-3**

Solve the circuit shown in Fig. 8-4 for the battery voltage and the total power dissipation.

**Solution**

Examination of Fig. 8-4 shows that the drop across $R_2$ is given, along with all of the resistor values. Ohm's law can be used to find the current through $R_2$:

$$I_{R_2} = \frac{10\text{ V}}{5\text{ k}\Omega} = 2\text{ mA}$$

Voltage is constant across parallel sections. The same 10 V appears across $R_1$, allowing us to calculate its current:

$$I_{R_1} = \frac{10\text{ V}}{10\text{ k}\Omega} = 1\text{ mA}$$

Kirchhoff's current law provides the total current for the circuit:

$$I_T = 2\text{ mA} + 1\text{ mA} = 3\text{ mA}$$

Inspection of the circuit reveals that the total current must flow through $R_3$. Ohm's law is used to find the drop across $R_3$:

$$V_{R_3} = 3\text{ mA} \times 10\text{ k}\Omega = 30\text{ V}$$

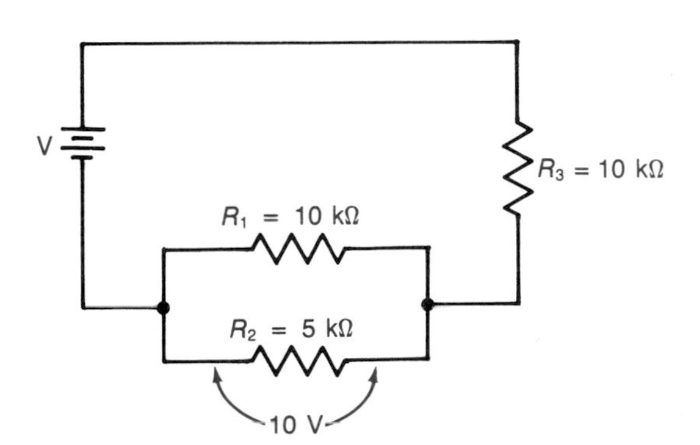


**FIGURE 8-4 Circuit for Example 8-3.**

Kirchhoff's voltage law tells us that the applied (or total) voltage is equal to the sum of the drops:

$$V_T = 10\ \text{V} + 30\ \text{V} = 40\ \text{V}$$

Finally, the power equation gives us the total dissipation:

$$P_T = I_T \times V_T = 3\ \text{mA} \times 40\ \text{V} = 120\ \text{mW}$$

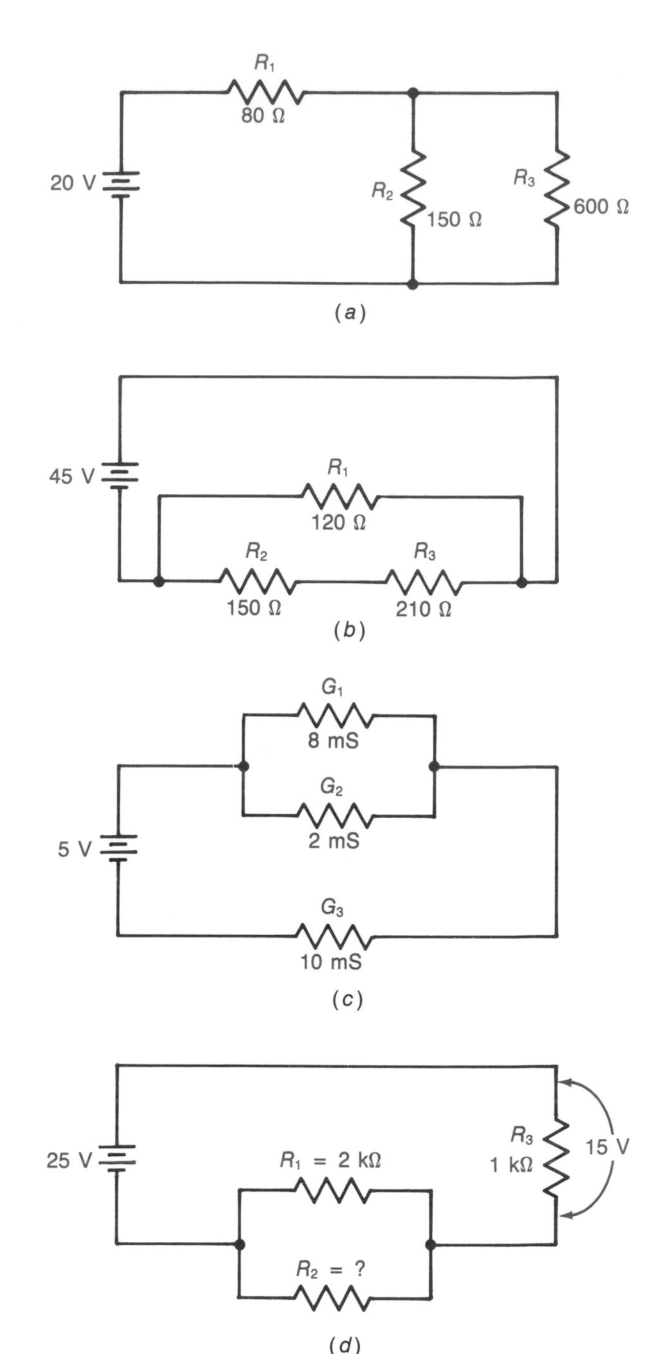

**FIGURE 8-5 Circuits for Self-Test.**

## Self-Test

1. The circuit rule that relates total power dissipation to the individual power dissipations is ______ because it applies to all circuits.
2. Refer to Fig. 8-5*a*. Calculate the voltage drop across $R_1$.
3. Refer to Fig. 8-5*a*. Calculate the power dissipation in $R_3$.
4. Refer to Fig. 8-5*b*. Calculate the current through $R_2$.
5. Refer to Fig. 8-5*b*. Calculate the total power.
6. Refer to Fig. 8-5*c*. Find the voltage drop across $G_3$.
7. Refer to Fig. 8-5*c*. Find the power dissipation in $G_1$.
8. Refer to Fig. 8-5*d*. Find the value of $R_2$.

## 8-2 ANALYSIS OF SERIES-PARALLEL CIRCUITS BY REDUCTION

Many series-parallel circuits can be analyzed by reducing them to an equivalent circuit containing a single source and a single load. The circuit is examined to find any occurrences of series or parallel sections. The series or parallel sections are replaced with a single component by using the appropriate equations. The process is repeated until the equivalent resistance or conductance of the entire circuit is found. Then Ohm's law can be used to find the total current flow. Finally, the circuit is expanded back to its original form to establish the various voltage drops and current flows.

Reduction often works best by beginning with those circuit components that are the farthest away from the source. Figure 8-6 shows an example. Inspection of the original circuit at the top of the illustration reveals that $R_7$, $R_8$, and $R_9$ are in series. These resistances can be combined into one by adding. This leads to the first level of reduction where $R_A$ replaces the original resistors. Looking at the partially reduced circuit, it is obvious that $R_A$ is in parallel with $R_6$. The product-over-sum equation can be used to combine these two resistors into a single resistor called $R_B$, which is shown at the next level of reduction. Resistors $R_2$, $R_3$, and $R_4$ are in parallel. They may be combined with the reciprocal equation to a single value, called $R_C$ in the next level of reduction. $R_C$, $R_5$, $R_B$, and $R_{10}$ are in series and are combined into $R_D$. This leaves only two resistors, which are in parallel. Resistors $R_1$ and $R_D$ can be combined to form $R_{eq}$, and the reduction is complete.

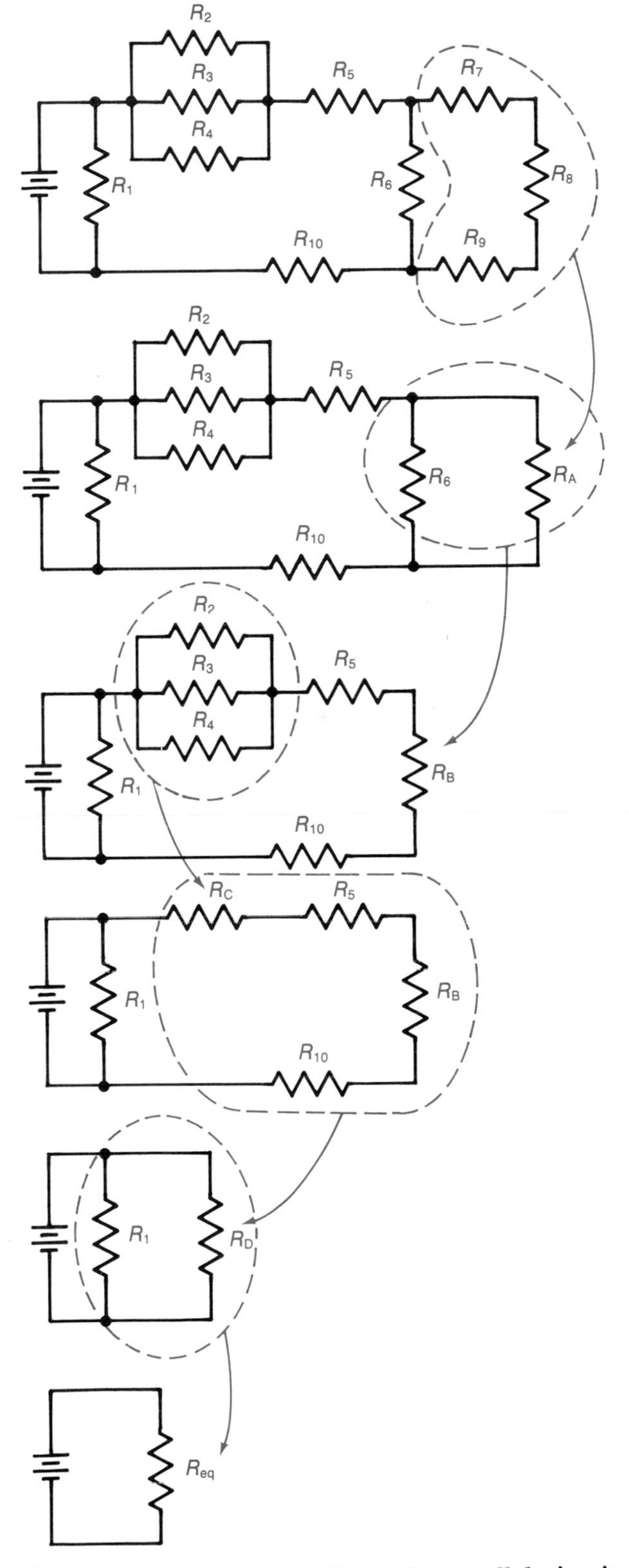


**FIGURE 8-6 Reduction of a series-parallel circuit.**

## EXAMPLE 8-4

Analyze the circuit shown in Fig. 8-7. Find all voltage drops and current flows.

### Solution

The circuit can be reduced to a single equivalent resistance. We will begin with $R_5$ and $R_6$, which are in series:

$$R_S = 15\ \Omega + 15\ \Omega = 30\ \Omega$$

This total of 30 Ω is in parallel with $R_4$. The parallel combination is called $R_P$ which is found next:

$$R_P = \frac{15\ \Omega \times 30\ \Omega}{15\ \Omega + 30\ \Omega} = 10\ \Omega$$

Resistors $R_1$ and $R_2$ are in series:

$$R_S = 25\ \Omega + 5\ \Omega = 30\ \Omega$$

The 30-Ω total is in parallel with $R_3$, which is also 30 Ω. Since equal-valued resistors are involved, the equivalent resistance is half that value, or 15 Ω. There are only two resistances left, and they are in series:

$$R_{eq} = 15\ \Omega + 10\ \Omega = 25\ \Omega$$

Ohm's law is used to find the total current flow:

$$I_T = \frac{50\ \text{V}}{25\ \Omega} = 2\ \text{A}$$

Now it is time to expand the circuit and find the individual drops and current flows. Examination of Fig. 8-7 reveals that $I_T$ flows only in the battery and some of the conductors. None of the resistors supports $I_T$. The current divider rule can be used

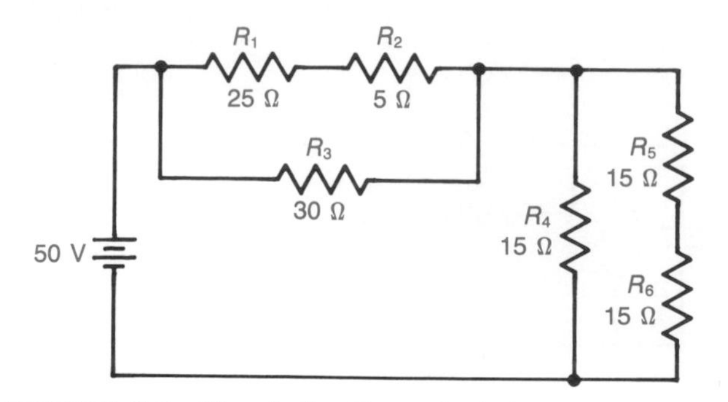


**FIGURE 8-7 Circuit for Example 8-4.**

to determine how the current splits, or the intermediate results obtained during reduction can be used along with Ohm's law. For example, resistors $R_4$, $R_5$, and $R_6$ were originally combined into a single value of 10 Ω. If the three resistors were replaced with an equivalent resistance of 10 Ω, *this equivalent resistance would support* $I_T$. It is important that you verify this by looking at the circuit carefully. This is one of the general methods of expanding back to the original circuit after $I_T$ has been found. The drop is

$$V = 2\text{ A} \times 10\ \Omega = 20\text{ V}$$

This is the drop that would appear across a 10-Ω resistor replacing $R_4$, $R_5$, and $R_6$. Where does it appear in the original circuit? Looking at Fig. 8-7, it can be seen that the entire 20-V drop will appear across $R_4$. It is also apparent that the 20-V drop will be divided into two drops by $R_5$ and $R_6$. Ohm's law is used to find the current in $R_4$:

$$I_{R_4} = \frac{20\text{ V}}{15\ \Omega} = 1.33\text{ A}$$

Kirchhoff's current law gives us the current through $R_5$ and $R_6$:

$$I_{R_{5,6}} = 2\text{ A} - 1.33\text{ A} = 0.67\text{ A}$$

Since $R_5$ and $R_6$ are equal, they divide the 20-V drop equally. This is easily verified with Ohm's law:

$$V = 0.67\text{ A} \times 15\ \Omega = 10\text{ V}$$

To complete the analysis, $R_1$ and $R_2$ can be viewed as a single 30-Ω resistor since they are in series. Resistor $R_3$ is also 30 Ω so the current will divide equally among these two branches. The three drops can be found with Ohm's law:

$$V_{R_1} = 1\text{ A} \times 25\ \Omega = 25\text{ V}$$
$$V_{R_2} = 1\text{ A} \times 5\ \Omega = 5\text{ V}$$
$$V_{R_3} = 1\text{ A} \times 30\ \Omega = 30\text{ V}$$

Note that the drops across $R_1$ and $R_2$ sum to equal the drop across $R_3$.

---

Verification of circuit solutions is a very important part of any analysis. It is a way to detect and eliminate mistakes, and it provides additional experience with circuit laws and rules. Reworking the circuit using the same techniques is not the best approach to verification because some types of mistakes will not be caught this way. There are ways to use independent techniques when verifying circuits. For example, the circuit shown in Fig. 8-7 could be solved for the total power dissipation (using $I_T$ and the battery voltage) and all of the individual dissipations. If the sum of the individual dissipations equals the total dissipation, this would verify the analysis. Figure 8-8 shows another approach. All of the drops from Example 8-4 have been labeled on the circuit. Kirchhoff's voltage law is used to verify the solution. This law states that the algebraic sum of the rises and drops in any closed loop must be equal to zero. Loop 1 includes the battery rise and the drops across $R_1$, $R_2$, $R_5$, and $R_6$. The equation is satisfied. Loop 2 includes the battery rise and the drops across $R_3$ and $R_4$. This loop equation is also satisfied. Other loop equations can be written for this circuit, but the two shown are adequate since they include every drop.

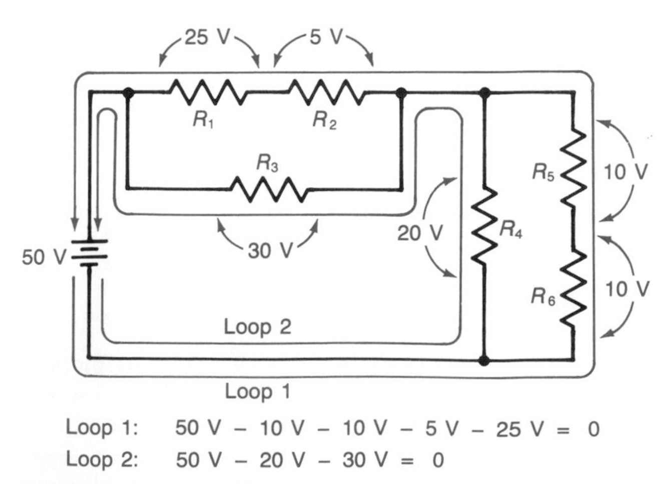


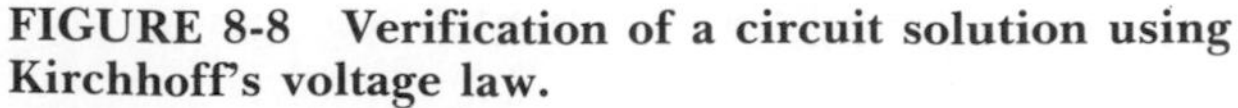


**FIGURE 8-8 Verification of a circuit solution using Kirchhoff's voltage law.**

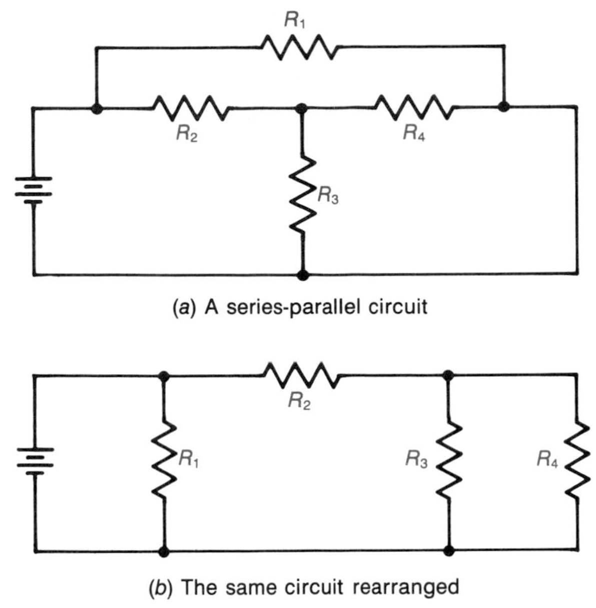


**FIGURE 8-9 Some circuits are "simplified" by rearranging the way they are drawn.**

Series-parallel circuits may be drawn in such a way that it is not readily apparent how they can be reduced. Figure 8-9*a* shows such a circuit. It may be difficult to visualize the simple parallel relationship of two of the resistors because of the way the circuit is drawn. The same circuit is shown in Fig. 8-9*b* with its parts rearranged. Note that it is now more obvious that $R_3$ and $R_4$ are in parallel. The redrawn circuit also makes it easier to determine the steps required to reduce the circuit to a single resistor. The key to Fig. 8-9 is seeing that $R_1$ is connected directly across the source. Figure 8-10 shows another example. The circuit at the top is not drawn in the most friendly manner. The circuit is rearranged at the bottom to make the reduction process more obvious. Never hesitate to redraw a circuit to make analysis easier. It is a valuable technique.

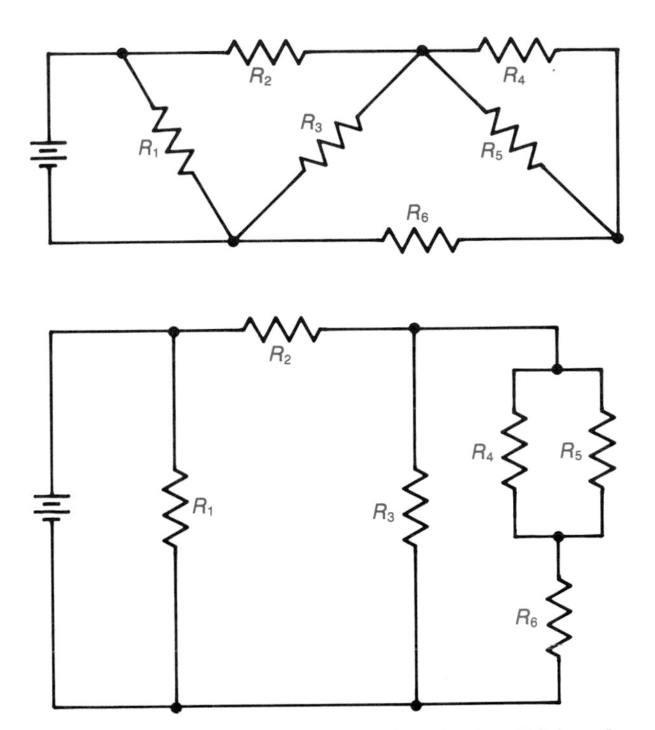


**FIGURE 8-10 Another example of simplifying by rearranging the parts.**

## EXAMPLE 8-5

Solve the circuit shown in Fig. 8-11*a* for all current flows and voltage drops.

### Solution

The circuit is redrawn in Fig. 8-11*b* to aid the analysis. The terminal voltages are replaced with batteries. Note that the ground reference is maintained in the redrawn circuit. Take a moment to satisfy yourself that Fig. 8-11*b* is equivalent to Fig. 8-11*a*. The first step is to determine how the two sources interact as far as $R_1$ is concerned. They are series-aiding. Therefore, the total voltage across $R_1$ is the sum of 12 V and 9 V, or 21 V. Ohm's law can be used to find the current through $R_1$ and the 12-V battery:

$$I_{R_1} = \frac{21\ \text{V}}{42\ \Omega} = 0.5\ \text{A}$$

Resistors $R_3$ and $R_4$ are in series. They can be reduced to a single value by adding:

$$R_S = 22\ \Omega + 14\ \Omega = 36\ \Omega$$

The quantity 36 Ω appears in parallel with $R_2$. The product-over-sum equation can be used to combine the values:

$$R_p = \frac{18\ \Omega \times 36\ \Omega}{18\ \Omega + 36\ \Omega} = 12\ \Omega$$

The 9-V battery is across 12 Ω, and the current flow is

$$I = \frac{9\ \text{V}}{12\ \Omega} = 0.75\ \text{A}$$

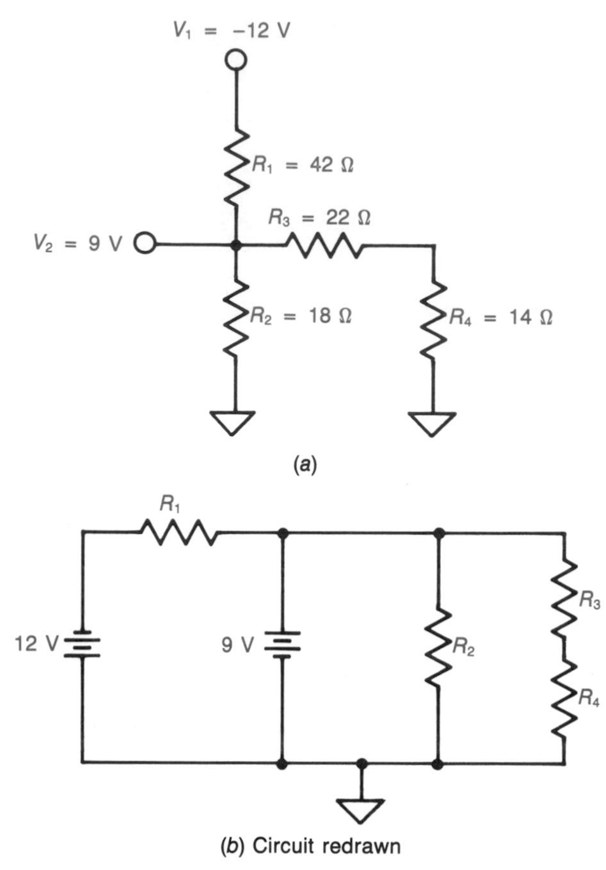


**FIGURE 8-11 Circuit for Example 8-5.**

The current flow in the 9-V battery can be found by Kirchhoff's current law:

$$I_{B_2} = 0.5\text{ A} + 0.75\text{ A} = 1.25\text{ A}$$

The current divider rule is used to determine the current in $R_2$ and the current in $R_3$ and $R_4$. Note that the *combined* value of $R_3$ and $R_4$ must be used as one of the values in the equation:

$$I_{R_2} = \frac{36\ \Omega \times 0.75\text{ A}}{36\ \Omega + 18\ \Omega} = 0.5\text{ A}$$

$$I_{R_{3,4}} = \frac{18\ \Omega \times 0.75\text{ A}}{36\ \Omega + 18\ \Omega} = 0.25\text{ A}$$

Finally, the drops across $R_3$ and $R_4$ are

$$V_{R_3} = 0.25\text{ A} \times 22\ \Omega = 5.5\text{ V}$$

$$V_{R_4} = 0.25\text{ A} \times 14\ \Omega = 3.5\text{ V}$$

The sum of the drops equals 9 V, as it should.

The circuit of Fig. 8-11 is only partly reducible. In its most reduced form, it contains two batteries and two resistors. This did not prevent us from analyzing the circuit for the currents and voltage drops. However, some circuits are not reducible and *cannot* be analyzed with the techniques presented to this point. Figure 8-12 shows two examples that fit this category. These circuits have no reducible sections. No amount of rearranging will change the situation. Such circuits require more advanced analysis techniques. These techniques are presented in Chap. 9.

## Self-Test

**9.** True or false? Any circuit can be reduced to a single source and a single equivalent resistance or conductance.

**10.** Determine $P_T$ for Fig. 8-13*a*.

**11.** Find $V_{R_3}$ in Fig. 8-13*a*.

**12.** Calculate $I_{R_1}$ for Fig. 8-13*b*.

**13.** Find the power dissipation in $R_3$ in Fig. 8-13*b*.

**14.** Find the current supplied by the battery in Fig. 8-13*c*.

**15.** Determine the drop across $R_6$ in Fig. 8-13*c*.

**16.** Calculate the total current flow for Fig. 8-13*d*.

**17.** Find $V_A$ for Fig. 8-13*d*.

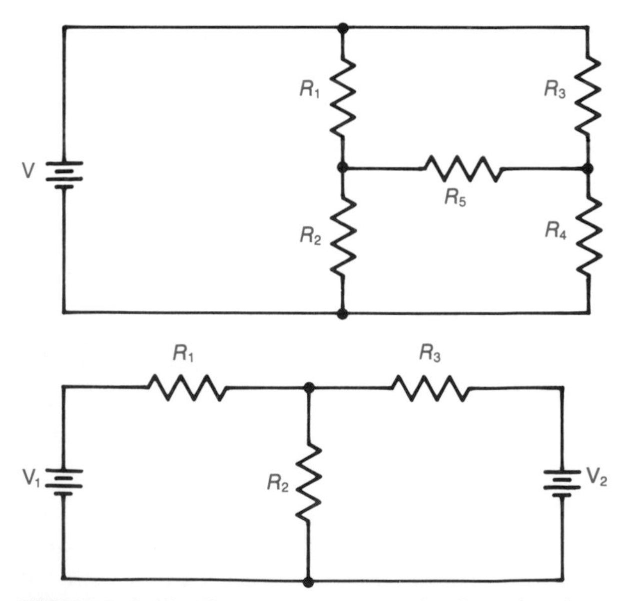


**FIGURE 8-12 Examples of nonreducible circuits requiring more advanced analysis techniques.**

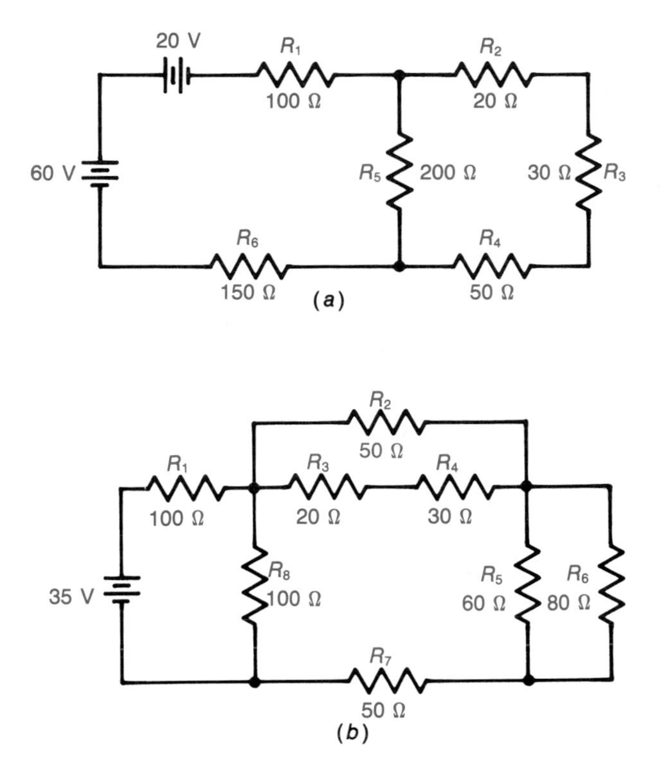


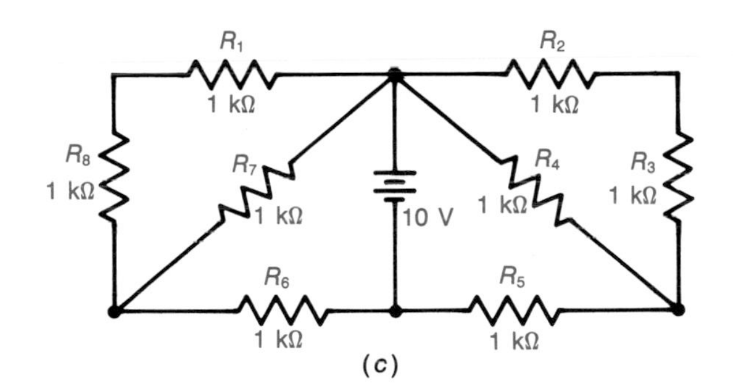


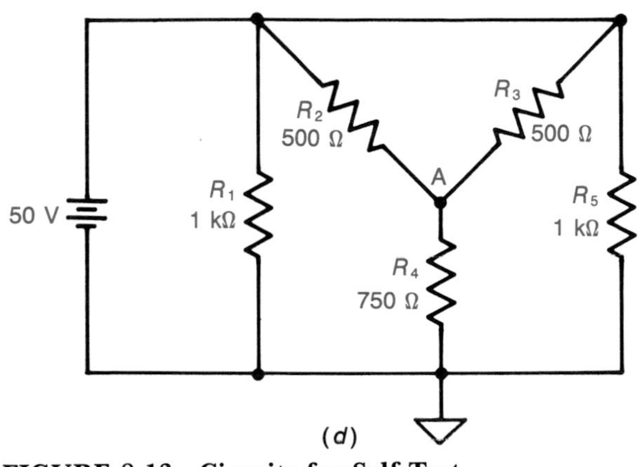


**FIGURE 8-13 Circuits for Self-Test.**

## 8-3 APPLICATIONS OF SERIES-PARALLEL CIRCUITS

Voltage dividers that supply load current are series-parallel circuits. If they are designed without taking the load current into account, the load voltage will be less than desired. Figure 8-14 shows an example. The load is $R_L$. It requires 5 V. Two 1-kΩ resistors have been selected to divide the 10-V source into two equal drops. This circuit will approach 5 V only when $R_L$ has a very high value of resistance compared to the resistors used in the divider. Suppose $R_L$ requires 5 mA. Ohm's law can be used to calculate its resistance:

$$R_L = \frac{5\ \text{V}}{5\ \text{mA}} = 1\ \text{k}\Omega$$

The 1-kΩ load appears in parallel with the bottom 1-kΩ resistor in the divider. This produces an equivalent resistance of 500 Ω. The circuit will not divide the 10-V source into two equal parts. The voltage divider equation shows that the actual load voltage will be

$$V_{RL} = \frac{10\ \text{V} \times 500\ \Omega}{1\ \text{k}\Omega + 500\ \Omega} = 3.33\ \text{V}$$

The performance of the circuit is not acceptable. This is because it was designed without considering the load current.

### EXAMPLE 8-6

Design a voltage divider that will operate the following loads from a 20-V source. Set the bleeder

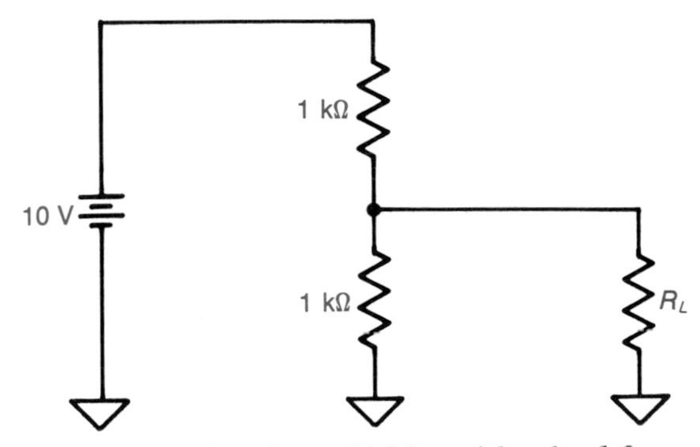


**FIGURE 8-14 A voltage divider with a load forms a series-parallel circuit.**

current equal to 20 percent of the total load current.

5 V at 5 mA
12 V at 10 mA
15 V at 5 mA

**Solution**

A voltage divider that produces a bleeder current requires $N + 1$ resistors, where $N$ is the number of loads. This example has three loads and requires four resistors. The *bleeder current* is typically set in the area of 10 to 20 percent of the load current. Without a bleeder current, the voltage divider outputs go up to the full value of the supply voltage if all of the loads are turned off or disconnected. The ability of a divider circuit to hold voltage constant is known as its *voltage regulation*. Having a bleeder current improves voltage regulation.

The required circuit is shown in Fig. 8-15. The bleeder resistor is $R_1$. It is not required in dividers with no bleeder current. The loads are arranged in ascending order of their voltage requirements, starting at the bottom of the divider network. The bleeder resistor is selected first. Note that it conducts the bleeder current and its drop is equal to the voltage requirement of the first load:

$$I_B = 0.2 \times (5 \text{ mA} + 10 \text{ mA} + 5 \text{ mA}) = 4 \text{ mA}$$

$$R_1 = \frac{5 \text{ V}}{4 \text{ mA}} = 1.25 \text{ k}\Omega$$

$$P_1 = 5 \text{ V} \times 4 \text{ mA} = 20 \text{ mW}$$

The power dissipated in the bleeder resistor is only 20 mW, so even a $\frac{1}{8}$-W resistor is adequate. Resistor $R_2$ is selected next. Looking at Fig. 8-15 shows that it drops 7 V and conducts 9 mA. Seven volts is the difference between load voltage 1 and load voltage 2. Nine milliamperes is the sum of the bleeder current and the current in load 1.

$$R_2 = \frac{7 \text{ V}}{9 \text{ mA}} = 778 \ \Omega$$

$$P_2 = 7 \text{ V} \times 9 \text{ mA} = 63 \text{ mW}$$

Doubling this power dissipation gives just over $\frac{1}{8}$ W. A $\frac{1}{8}$-W resistor would be marginal, and a $\frac{1}{4}$-W resis-

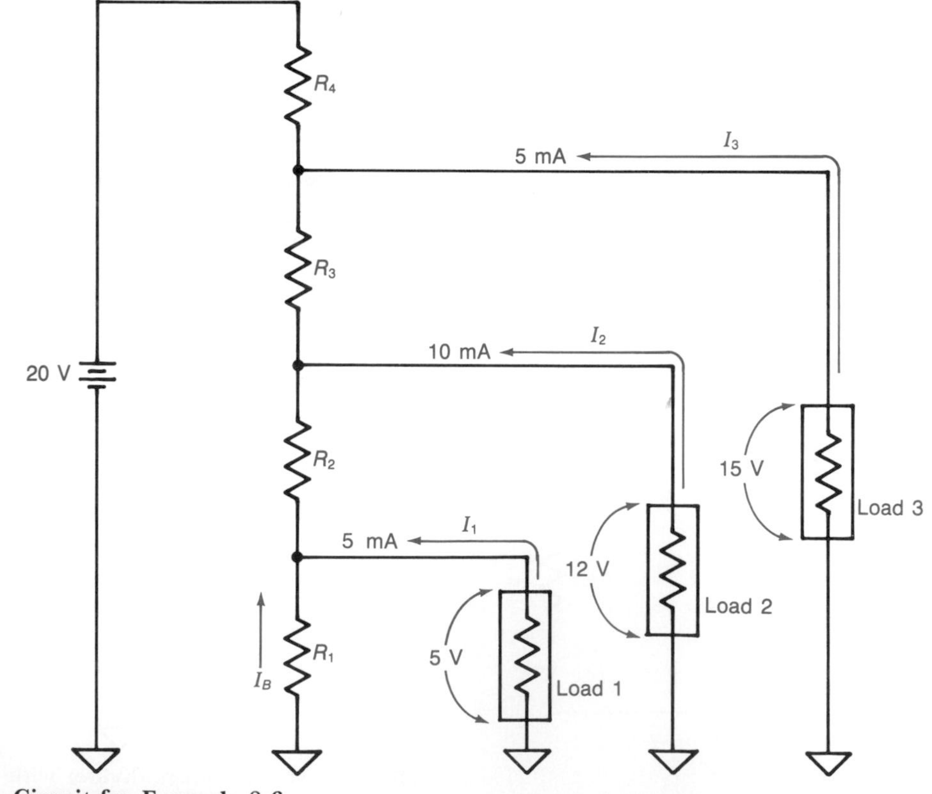


**FIGURE 8-15 Circuit for Example 8-6.**

tor would be better. Resistor $R_3$ is selected next. It drops 3 V and conducts 19 mA. Three volts is the difference between load voltage 2 and load voltage 3. Nineteen milliamperes is the sum of the bleeder current, the current in load 1, and the current in load 2.

$$R_3 = \frac{3\text{ V}}{19\text{ mA}} = 158\ \Omega$$

$$P_3 = 3\text{ V} \times 19\text{ mA} = 57\text{ mW}$$

Doubling this gives 114 mW, and a $\frac{1}{8}$-W resistor is adequate. The last step is to select $R_4$. It drops 5 V and conducts 24 mA. Five volts is the difference between the supply voltage and the third load voltage. Twenty-four milliamperes is the sum of the bleeder current and all of the load currents.

$$R_4 = \frac{5\text{ V}}{24\text{ mA}} = 208\ \Omega$$

$$P_4 = 5\text{ V} \times 24\text{ mA} = 120\text{ mW}$$

Doubling the dissipation gives 240 mW, and a $\frac{1}{4}$-W resistor is required.

*Attenuators* are circuits or devices used to make a circuit voltage smaller. Figure 8-16 shows an attenuator that may be switched in or out and provides a fixed load on the source regardless of the switch setting. Having a fixed load is important for the proper performance of some sources. The switch is shown in the *out* position. The source is applied directly to the load and there is no attenuation. The load voltage and current are maximum, and the source is loaded by 50 Ω. Throwing the switch puts the attenuator resistors in the circuit. The load voltage and current are less in this position, and $R_{eq}$ is approximately 50 Ω.

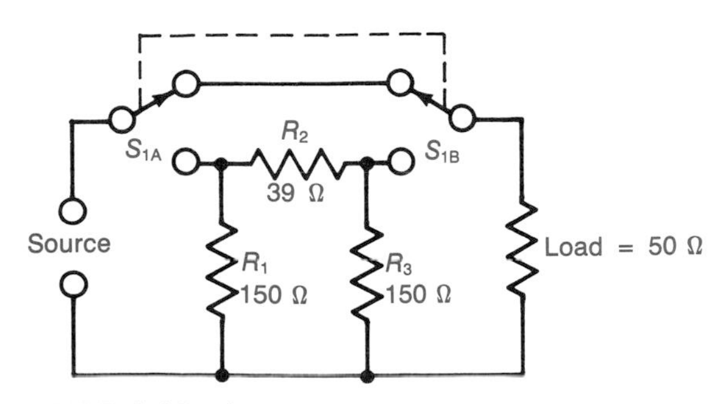


**FIGURE 8-16 An attenuator.**

**EXAMPLE 8-7**

Verify that the attenuator shown in Fig. 8-16 maintains a 50-Ω load on the source, and calculate the load voltage percentage when it is switched in.

**Solution**

With the attenuator switched in, the load appears in parallel with $R_3$:

$$R_P = \frac{50\ \Omega \times 150\ \Omega}{50\ \Omega + 150\ \Omega} = 37.5\ \Omega$$

This resistance appears in series with $R_2$:

$$R_S = 37.5\ \Omega + 39\ \Omega = 76.5\ \Omega$$

Resistance $R_{eq}$ is found by combining with $R_1$:

$$R_{eq} = \frac{150\ \Omega \times 76.5\ \Omega}{150\ \Omega + 76.5\ \Omega} = 50.7\ \Omega$$

This confirms that the load on the source is very close to 50 Ω, whether the attenuator is in or out. Next, we can assume a 100-V source to facilitate calculation of the load voltage percentage. The voltage divider rule can be used to find the load voltage with the attenuator switched in. Note that the combined parallel value of $R_3$ and the load is used in the following equation along with the value of $R_2$:

$$V_L = \frac{100\text{ V} \times 37.5\ \Omega}{39\ \Omega + 37.5\ \Omega} = 49\text{ V}$$

The load voltage is 49 percent of the source voltage with the attenuator switched in.

Figure 8-17 shows a Wheatstone bridge. This circuit is used in many measurement applications. The bridge is considered *balanced* when the voltage at terminal A is equal to the voltage at terminal B. The bridge is also said to be *zeroed or nulled* when this condition is met. The voltage divider rule can be used to describe the terminal voltages:

$$V_A = V_T \times \frac{R_2}{R_1 + R_2}$$

$$V_B = V_T \times \frac{R_4}{R_3 + R_4}$$

If $V_A = V_B$, then

$$V_T \times \frac{R_2}{R_1 + R_2} = V_T \times \frac{R_4}{R_3 + R_4}$$

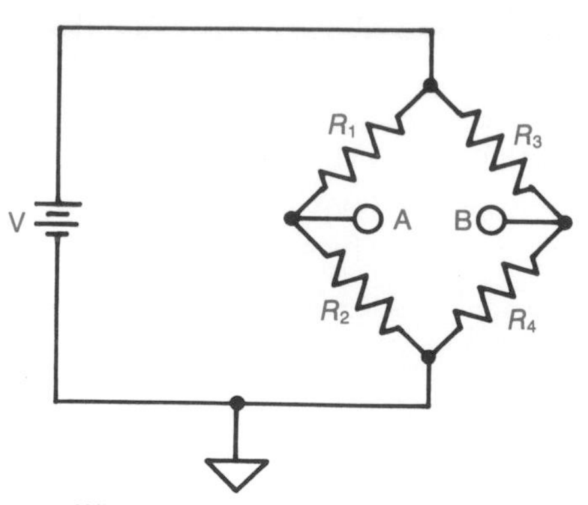


**FIGURE 8-17 The Wheatstone bridge circuit.**

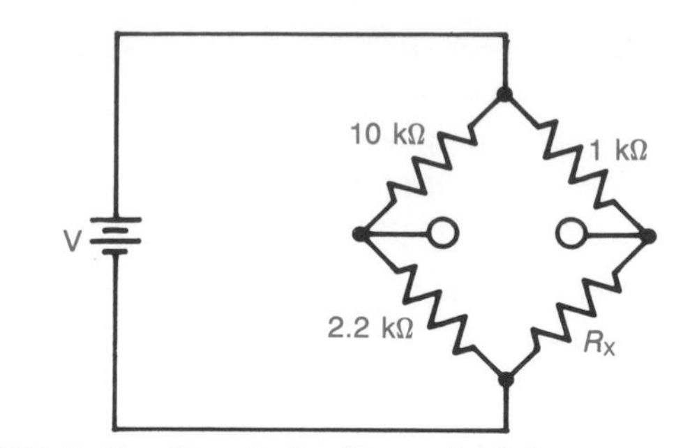


**FIGURE 8-18 Circuit for Example 8-8.**

Dividing both sides by $V_T$ gives

$$\frac{R_2}{R_1 + R_2} = \frac{R_4}{R_3 + R_4}$$

Cross-multiplying provides

$$R_2R_3 + R_2R_4 = R_4R_1 + R_4R_2$$

Subtracting $R_2R_4$ gives

$$R_2R_3 = R_4R_1$$

Which is often rearranged to this form:

$$\frac{R_2}{R_1} = \frac{R_4}{R_3}$$

The Wheatstone bridge is balanced when the ratio of resistances on the left side is equal to the ratio of resistances on the right side.

---

**EXAMPLE 8-8**

Calculate a value for $R_X$ in Fig. 8-18 that will null the bridge.

**Solution**

Applying the Wheatstone bridge equation,

$$\frac{2.2\text{ k}\Omega}{10\text{ k}\Omega} = \frac{R_X}{1\text{ k}\Omega}$$

$$R_X = 220\ \Omega$$

---

Figure 8-19 shows a bridge circuit used in conjunction with a *strain gage*. Strain gages are devices made from wire or metal foil. They are used to measure quantities such as force, weight, and pressure. They show an increase in resistance when stretched. Their resistance increases with stretching because the length of the conductor increases and because the cross-sectional area of the conductor decreases. The following equation was presented in Chap. 4:

$$R = \frac{\rho l}{A}$$

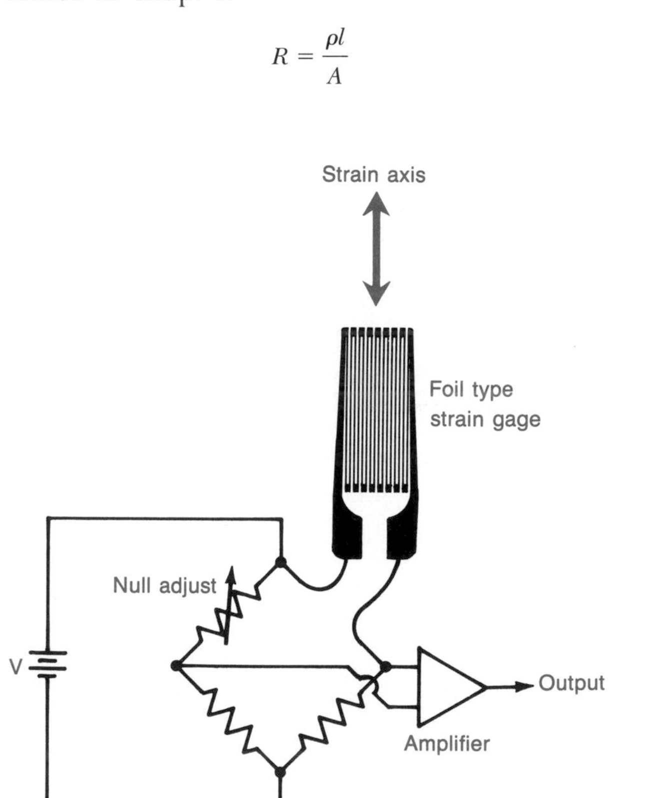


**FIGURE 8-19 A typical application of the bridge circuit.**

where $R$ = resistance, Ω
$\rho$ = resistivity of conductor, Ω · cm
$l$ = length of conductor, cm
$A$ = cross-sectional area of conductor, $cm^2$

The stretching due to strain typically produces only a small increase in resistance. Figure 8-19 includes an *amplifier* to increase the output of the bridge circuit. Amplifiers are covered in Chap. 24. It also includes a null adjust. This variable resistor is normally set for zero output with no strain applied to the gage. Bridge circuits are also used for measuring resistance, temperature, light level, and many other quantities. For example, if the strain gage in Fig. 8-19 is replaced with a thermistor, the bridge output will be proportional to temperature.

Some series-parallel circuits are called *ladder networks* because of the appearance they take on in schematic form. Figure 8-20 shows an $R$-$2R$ ladder network that is often used as the basis for converting *binary data to analog* form. Binary circuits are either on or off, while analog circuits show a large number of in-between situations. An analog current can take on many values. The name $R$-$2R$ is descriptive because the network is made up of resistor values with a 2:1 relationship. Note that the resistors are 5 or 10 kΩ in Fig. 8-20.

The binary number system uses only two symbols: 0 and 1. This number system is explained in more detail in Chap. 26, which treats *digital* circuits. The $R$-$2R$ network converts the switch settings into an analog current marked $I_{OUT}$ in Fig. 8-20. The lowest binary setting is 0000. This is the condition shown with all of the switches thrown to the right. The analog current is 0 at this setting. The following switch settings give evidence that the output current is an analog representation of the binary number set by the switches:

| Binary No. | Equivalent Decimal No. | $I_{OUT}$ |
|---|---|---|
| 0000 | 0 | 0 |
| 0001 | 1 | 0.125 mA |
| 0010 | 2 | 0.250 mA |
| 0011 | 3 | 0.375 mA |
| 0100 | 4 | 0.500 mA |
| 0101 | 5 | 0.625 mA |
| 0110 | 6 | 0.750 mA |
| 0111 | 7 | 0.875 mA |
| 1000 | 8 | 1.000 mA |
| 1001 | 9 | 1.125 mA |
| 1010 | 10 | 1.250 mA |
| 1011 | 11 | 1.375 mA |
| 1100 | 12 | 1.500 mA |
| 1101 | 13 | 1.625 mA |
| 1110 | 14 | 1.750 mA |
| 1111 | 15 | 1.875 mA |

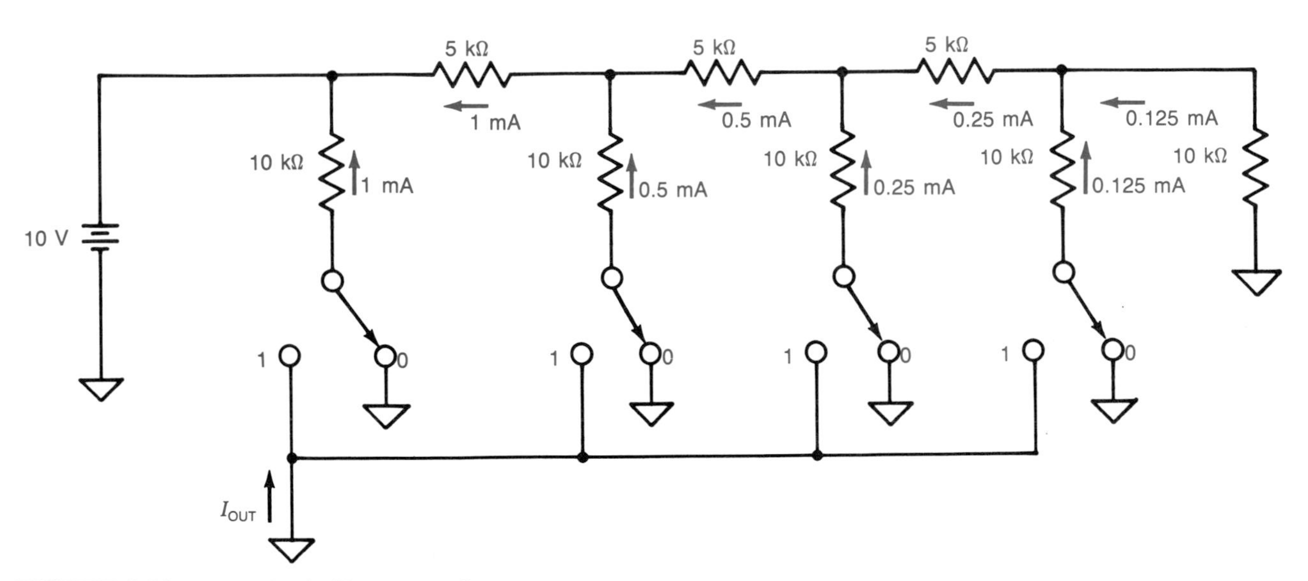


**FIGURE 8-20 An *R*-2*R* ladder network.**

The $R$-$2R$ converter is properly weighted. Each increasing count produces the correct change in output current. For example, the current that corresponds to a decimal count of 6 is 0.750 mA, and the current at a count of 12 is twice that value or 1.500 mA. Digital-to-analog converters make it possible for circuits that are restricted to on-off operation to produce analog outputs such as music or speech. It is fascinating to consider that our favorite music, which represents a rather complex analog signal, can be recorded strictly in the form of 0s and 1s. It is also fascinating that digital computers can create analog outputs such as speech by using circuits based on the principles illustrated by Fig. 8-20.

## Self-Test

**18.** If a voltage divider is designed without consideration for the load current, the actual load voltage will be _______ than desired.

**19.** A bleeder current improves the voltage _______ of divider circuits.

**20.** A voltage divider must provide a bleeder current and develop two load voltages that are less than the source voltage. What is the minimum number of resistors required for this circuit?

**21.** A voltage divider provides 5 V at 100 mA and 12 V at 50 mA from a 20-V source. The desired bleeder current is 15 percent. Calculate the resistance and minimum dissipation rating for the bleeder resistor.

**22.** An attenuator is a circuit or device used to make an electric signal _______.

**23.** True or false? Some attenuators are designed to be switched in and out and always maintain a constant load on the source.

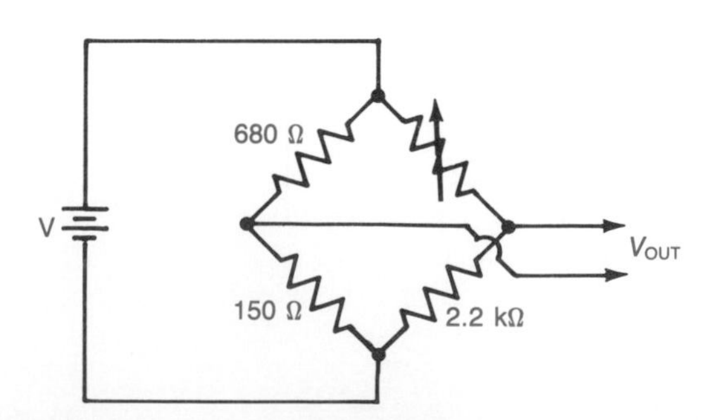


**FIGURE 8-21 Circuit for Self-Test.**

**24.** The circuit shown in Fig. 8-21 is known as a _______ bridge.

**25.** What value will the variable resistor in Fig. 8-21 be set for when the bridge is balanced?

**26.** Assuming the bridge in Fig. 8-21 is nulled and that the source develops 25 V, what is the value of $V_{OUT}$?

**27.** Why are some series-parallel circuits called ladder networks?

**28.** Digital-to-analog converters often use _______ ladder networks.

## 8-4 FAULT ANALYSIS IN SERIES-PARALLEL CIRCUITS

*Fault analysis* is based on a clear understanding of circuit laws and rules. A fault changes a circuit. Some faults produce rather drastic effects. These are often the easiest to diagnose. For example, suppose $R_3$ in Fig. 8-22 burns out. Resistor $R_3$ is in series with the 20-V supply. If it burns out, the supply is effectively disconnected from the other resistors in the divider network and from both loads. The symptoms produced by $R_3$ burning out will be

1. The 20-V supply will measure normal to high.
2. Both load voltages will be zero.
3. No current will flow in any part of the circuit.

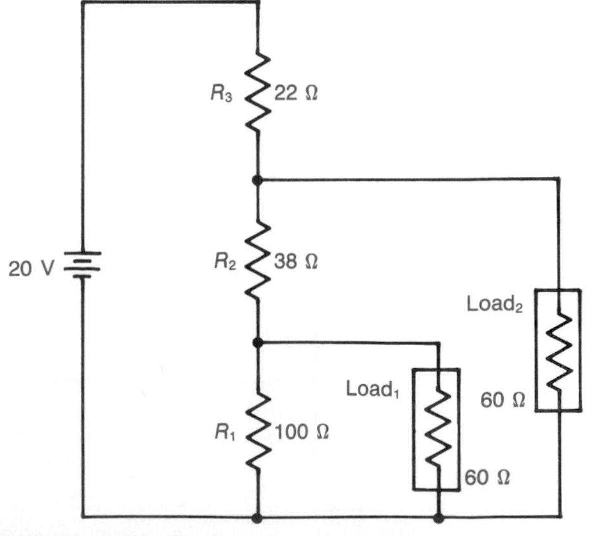


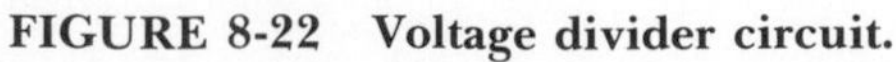

**FIGURE 8-22 Voltage divider circuit.**

Symptom 1 is important. It tells us that the supply is working. Never waste time analyzing circuit components until the supply has been verified. Symptom 2 is also important. It provides evidence that $R_3$ is open. Symptom 3 is important, but measuring current is more troublesome because the circuit must be broken to take the reading. Voltage analysis is used most of the time when troubleshooting because it is faster and easier to take the readings.

$LOAD_2$ in Fig. 8-22 could be shorted. This fault would also cause both load voltages to be zero. Other symptoms are likely to appear in this case. The current in $R_3$ would be much higher than normal. It would run hot and would probably burn up. The extra current in the 20-V supply may cause its output to sag below normal.

**EXAMPLE 8-9**

Determine the normal load voltages and the bleeder current for the circuit shown in Fig. 8-22.

**Solution**

Resistor $R_1$ and $LOAD_1$ are in parallel. They can be combined with the product-over-sum equation:

$$R_P = \frac{100\ \Omega \times 60\ \Omega}{100\ \Omega + 60\ \Omega} = 37.5\ \Omega$$

This resistance is in series with $R_2$:

$$R_S = 37.5\ \Omega + 38\ \Omega = 75.5\ \Omega$$

In parallel with $R_S$ is $LOAD_2$:

$$R_P = \frac{75.5\ \Omega \times 60\ \Omega}{75.5\ \Omega + 60\ \Omega} = 33.4\ \Omega$$

This parallel combination is in series with $R_3$. Now $R_{eq}$ can be found:

$$R_{eq} = 33.4\ \Omega + 22\ \Omega = 55.4\ \Omega$$

Next, the total current flow is found:

$$I_T = \frac{20\ \text{V}}{55.4\ \Omega} = 361\ \text{mA}$$

Resistor $R_3$ supports the total current flow. Its drop is determined next:

$$V_{R_3} = 361\ \text{mA} \times 22 = 7.94\ \text{V}$$

The drop across $LOAD_2$ is determined by

$$V_2 = 20\ \text{V} - 7.94\ \text{V} = 12.1\ \text{V}$$

Finally, the voltage divider equation can be used to find the drop across $LOAD_1$. The parallel value of $LOAD_1$ and $R_1$, along with the value of $R_2$, determines the drop:

$$V_1 = \frac{12.1\ \text{V} \times 37.5\ \Omega}{38\ \Omega + 37.5\ \Omega} = 6\ \text{V}$$

The bleeder current flows through $R_1$, and this resistor also drops 6 V:

$$I_B = \frac{6\ \text{V}}{100\ \Omega} = 60\ \text{mA}$$

The normal load voltages for Fig. 8-22 are 6 and 12.1 V. The bleeder current is 60 mA.

**EXAMPLE 8-10**

Calculate the load voltages for the voltage divider of Fig. 8-22 if $R_1$ burns out.

**Solution**

Figure 8-23 shows the voltage divider redrawn without $R_1$. Now $LOAD_1$ is in series with $R_2$:

$$R_S = 60\ \Omega + 38\ \Omega = 98\ \Omega$$

$LOAD_2$ is in parallel:

$$R_P = 60\ \Omega \parallel 98\ \Omega = 37.2\ \Omega$$

Resistor $R_3$ is in series:

$$R_{eq} = 22\ \Omega + 37.2\ \Omega = 59.2\ \Omega$$

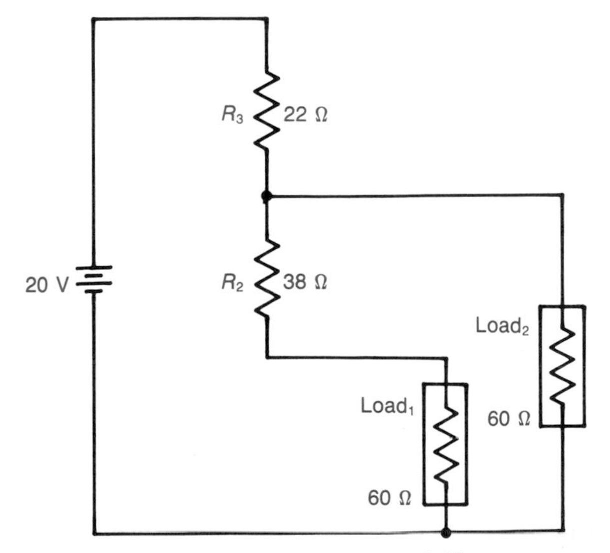


**FIGURE 8-23 Circuit for Example 8-10.**

The total current flow is

$$I_T = \frac{20\text{ V}}{59.2\ \Omega} = 338\text{ mA}$$

The drop across $R_3$ is

$$V_{R_3} = 338\text{ mA} \times 22\ \Omega = 7.43\text{ V}$$

The drop across $\text{LOAD}_2$ is

$$V_2 = 20\text{ V} - 7.43\text{ V} = 12.6\text{ V}$$

The drop across $\text{LOAD}_1$ is

$$V_1 = \frac{12.6\text{ V} \times 60\ \Omega}{60\ \Omega + 38\ \Omega} = 7.71\text{ V}$$

When the bleeder resistor burns out, the bleeder current drops to zero and the load voltages go higher than normal.

**EXAMPLE 8-11**

Calculate the voltage across $\text{LOAD}_2$ in Fig. 8-22 if $\text{LOAD}_1$ is disconnected.

**Solution**

Figure 8-24 shows the circuit with $\text{LOAD}_1$ removed. Resistors $R_1$ and $R_2$ are in series:

$$R_S = 100\ \Omega + 38\ \Omega = 138\ \Omega$$

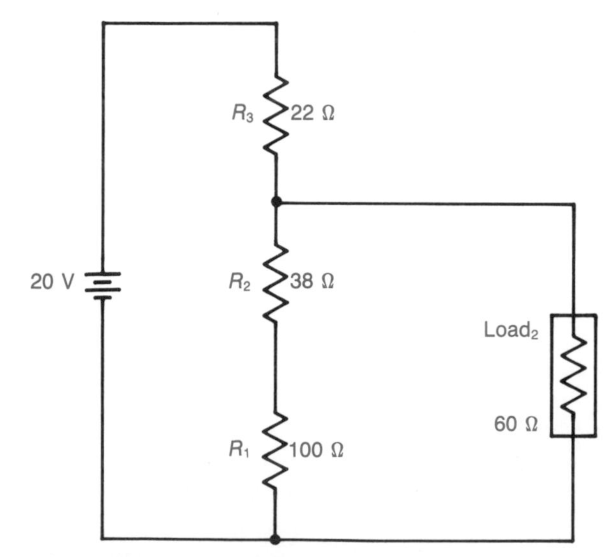


**FIGURE 8-24 Circuit for Example 8-11.**

And $\text{LOAD}_2$ is in parallel:

$$R_P = 60\ \Omega \,||\, 138\ \Omega = 41.8\ \Omega$$

Resistor $R_3$ is in series:

$$R_{eq} = 22\ \Omega + 41.8\ \Omega = 63.8\ \Omega$$

The total current flow is

$$I_T = \frac{20\text{ V}}{63.8\ \Omega} = 313\text{ mA}$$

The drop across $R_3$ is

$$V_{R_3} = 313\text{ mA} \times 22\ \Omega = 6.90\text{ V}$$

The drop across $\text{LOAD}_2$ is

$$V_2 = 20\text{ V} - 6.90\text{ V} = 13.1\text{ V}$$

Disconnecting $\text{LOAD}_1$ from the divider causes the drop across $\text{LOAD}_2$ to increase.

## Self-Test

**29.** Determine the normal load voltages and bleeder current for the circuit shown in Fig. 8-25*a*.

**30.** Assume that $R_2$ in Fig. 8-25*a* is open. Calculate the load voltages.

**31.** Assume that $\text{LOAD}_1$ in Fig. 8-25*a* is shorted. Calculate the load voltages.

**32.** Find the $\text{LOAD}_2$ voltage in Fig. 8-25*a* if $\text{LOAD}_1$ is disconnected.

**33.** What load does the 100-V source in Fig. 8-25*b* have on it with the attenuator switched in and out?

**34.** What is the load voltage in Fig. 8-25*b* with the attenuator switched in?

**35.** What is the load voltage in Fig. 8-25*b* with the attenuator switched in if $R_3$ is open? What is the load on the source with these conditions?

**36.** Calculate the normal output voltage for Fig. 8-25*c*.

**37.** Find $V_{out}$ in Fig. 8-25*c* if $R_3$ is open.

**38.** Find $V_{out}$ in Fig. 8-25*c* if $R_1$ and $R_4$ are both open.

**39.** What is $V_{out}$ in Fig. 8-25*c* if $R_1$ and $R_3$ are both open?

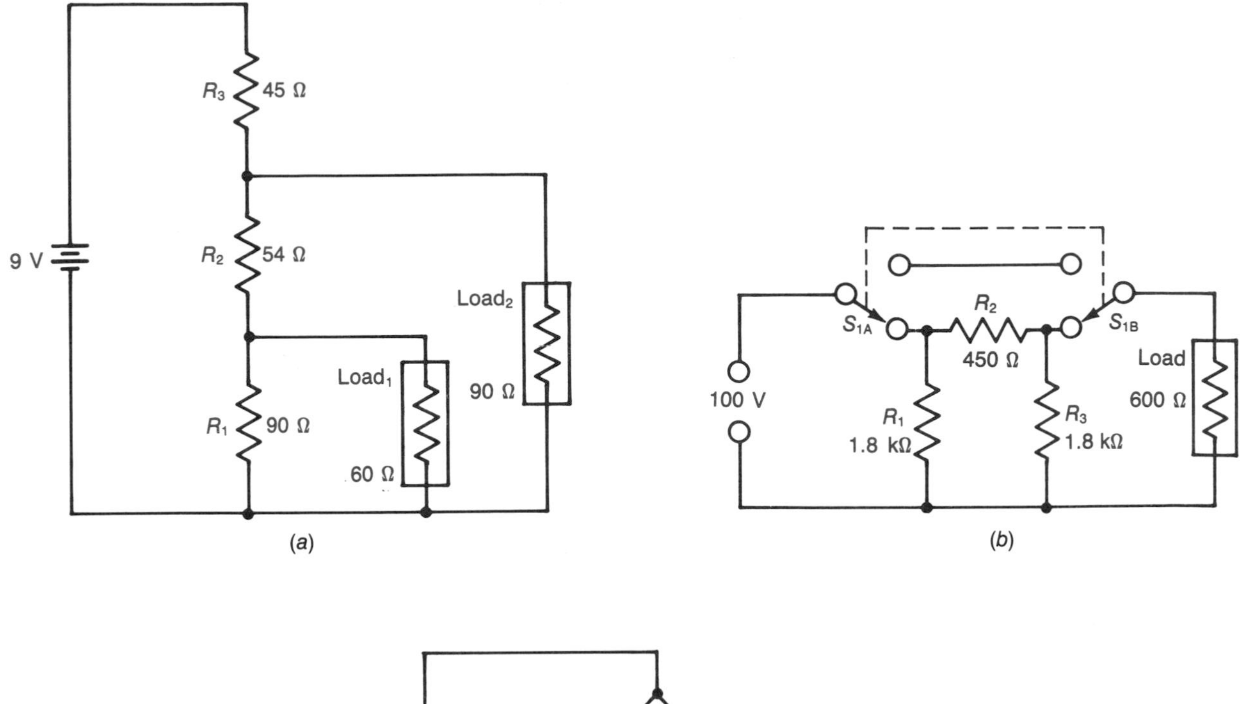


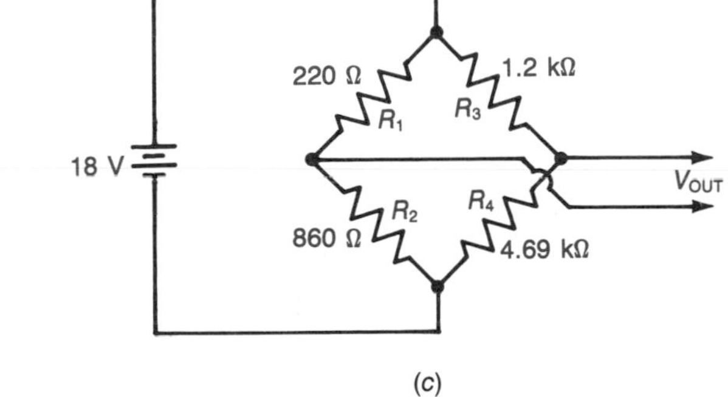


**FIGURE 8-25 Circuits for Self-Test.**

## SUMMARY

**1.** Series-parallel circuits can be analyzed by applying series circuit laws to the series sections and parallel circuit laws to the parallel sections.

**2.** Many series-parallel circuits can be reduced to a single equivalent resistance (or conductance).

**3.** The total current flow in a series-parallel circuit can be calculated by using Ohm's law with the value of $R_{eq}$ and the applied voltage.

**4.** The total power in a series-parallel circuit is equal to the sum of the individual power dissipations:

$$P_T = P_1 + P_2 + P_3 + \cdots + P_N$$

This rule is universal since it applies to all circuits.

**5.** When analysis by reduction is used, the circuit is expanded back to its original form to find the individual currents, voltage drops, and power dissipations.

**6.** Verification of the values for a solved circuit is most effective when independent techniques are used.

**7.** Some series-parallel schematics can be "simplified" by rearranging the component parts.

**8.** There are series-parallel circuits that cannot be reduced and must be analyzed using more advanced techniques.

**9.** Voltage dividers that supply load current are series-parallel circuits.

**10.** A voltage divider may be designed with a bleeder current so that it places a load on the source even when all loads are disconnected.

**11.** The bleeder current is typically from 10 to 20 percent of the load current.

**12.** Voltage dividers with bleeder current require $N + 1$ resistors where $N$ is the number of loads.

**13.** Adding a bleeder resistor improves the voltage regulation of a divider.

**14.** Attenuators are circuits or devices used to make electric signals smaller.

**15.** The Wheatstone bridge circuit is balanced when its output voltage is zero. This condition is met when the ratio of resistances on one side of the bridge is equal to the ratio on the other side:

$$\frac{R_X}{R_Y} = \frac{R_A}{R_B}$$

**16.** The Wheatstone bridge circuit is often used to make measurements.

**17.** A strain gage is a metal foil or wire device that shows an increase in resistance when stretched.

**18.** A ladder network is a series-parallel circuit with a schematic diagram that resembles a ladder.

**19.** The $R$-$2R$ ladder network can be used to convert binary numbers to corresponding analog currents.

## CHAPTER REVIEW QUESTIONS

**8-1.** Voltage dividers may use a bleeder resistor to improve voltage ______.

**8-2.** A divider that supplies five different load voltages and provides a bleeder current will require ______ resistors.

**8-3.** The circuit shown in Fig. 8-26*e* is known as a ______ bridge.

**8-4.** When $V_{XY} = 0$ in Fig. 8-26*e*, the bridge is said to be ______.

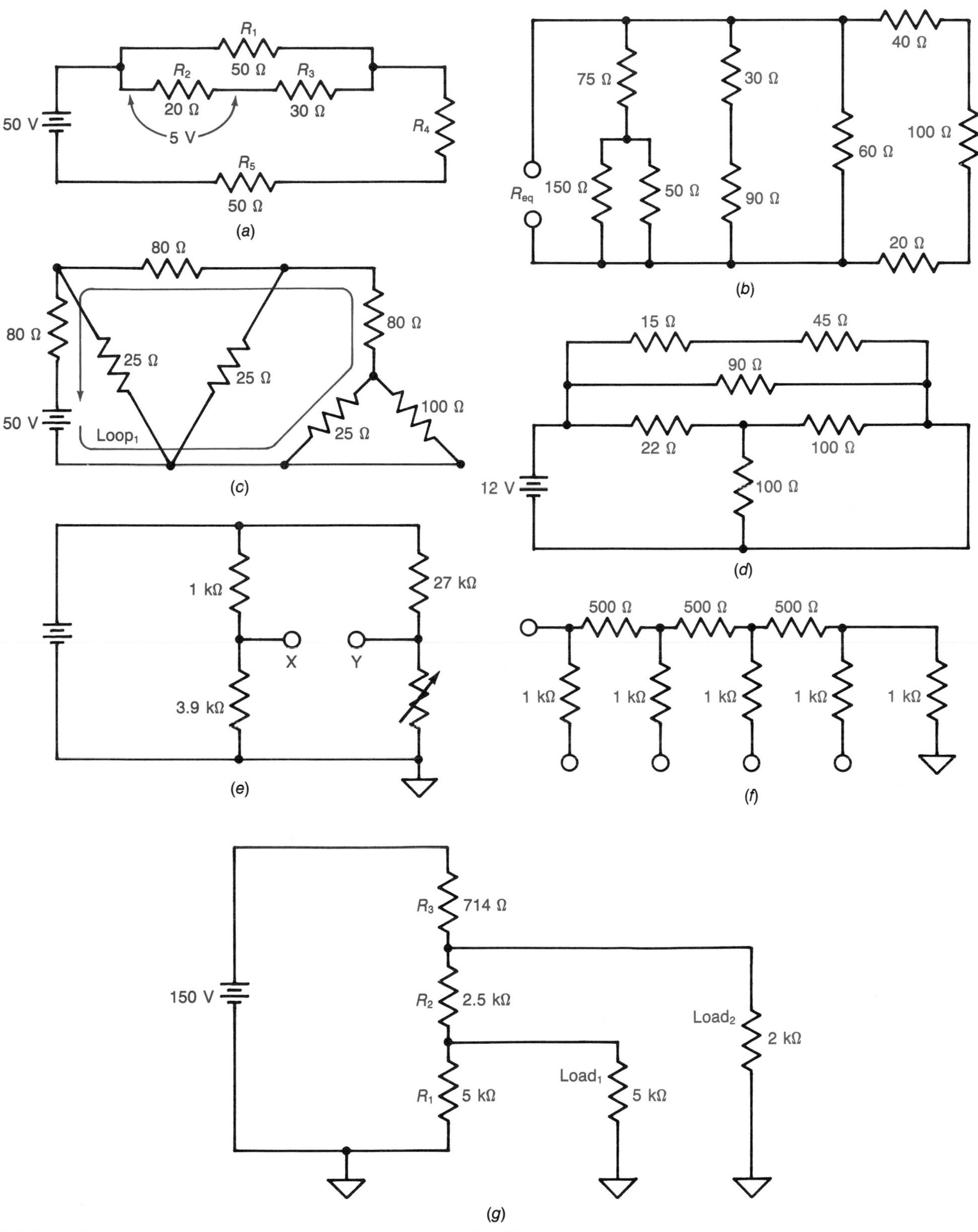


**FIGURE 8-26 Circuits for Review Questions and Problems.**

**8-5.** A strain gage is a component that shows an increase in ______ with elongation.

**8-6.** Amplifiers make electric signals larger while ______ make them smaller.

**8-7.** Identify the circuit shown in Fig. 8-26*f*.

**8-8.** List an application for the circuit shown in Fig. 8-26*f*.

**8-9.** Which component in Fig. 8-26*g* is a bleeder resistor?

## CHAPTER REVIEW PROBLEMS

**8-1.** Find the resistance of $R_4$ in Fig. 8-26*a*.

**8-2.** Find $R_{eq}$ for Fig. 8-26*b*.

**8-3.** Calculate $P_T$ for Fig. 8-26*c*.

**8-4.** Write the equation for loop 1 in Fig. 8-26*c* using the battery voltage and the actual voltage drops.

**8-5.** Find $I_T$ in Fig. 8-26*d*.

**8-6.** What is the drop across the 15-Ω resistor in Fig. 8-26*d*?

**8-7.** Design a voltage divider with a 10 percent bleeder current that will provide 6 V at 300 mA from a 9-V source. Calculate resistor values and their dissipation ratings.

**8-8.** What value must the variable resistor in Fig. 8-26*e* be set to in order to make $V_X = V_Y$?

**8-9.** Calculate both load voltages for Fig. 8-26*g*.

**8-10.** Determine the percentage of bleeder current for Fig. 8-26*g*.

**8-11.** Suppose that $LOAD_2$ in Fig. 8-26*g* shorts. Determine both load voltages.

**8-12.** Suppose that $R_2$ in Fig. 8-26*g* opens. Determine both load voltages.

## ANSWERS TO SELF-TESTS

1. universal
2. 8 V
3. 0.24 W
4. 125 mA
5. 22.5 W
6. 2.5 V
7. 50 mW
8. 1 kΩ
9. false
10. 5.05 W
11. 2.53 V
12. 230 mA
13. 60.4 mW
14. 12 mA
15. 6 V
16. 150 mA
17. +37.5 V
18. less
19. regulation
20. 3
21. 222 Ω, $\frac{1}{4}$ W

**22.** smaller

**23.** true

**24.** Wheatstone

**25.** 9.97 kΩ

**26.** 0 V

**27.** because their schematic diagrams resemble a ladder

**28.** $R$-2$R$

**29.** $\text{LOAD}_1$ = 1.8 V, $\text{LOAD}_2$ = 4.5 V, $I_B$ = 20 mA

**30.** $\text{LOAD}_1$ = 0 V, $\text{LOAD}_2$ = 6 V

**31.** $\text{LOAD}_1$ = 0 V, $\text{LOAD}_2$ = 3.86 V

**32.** 4.97 V

**33.** The load on the source is constant at 600 Ω.

**34.** 50 V

**35.** 57.1 V, 663 Ω

**36.** 0 V

**37.** 14.3 V

**38.** 18 V

**39.** 0 V

# Chapter 9 Network Analysis

This chapter covers the additional concepts and theorems that are required to analyze those circuits that cannot be reduced and solved by the methods already presented. There are always a number of different approaches that may be used to solve a circuit. One approach may be particularly well suited for certain situations, while other approaches are more efficient in other situations. As you become familiar with additional techniques, your tools will expand and so will your understanding of circuit behavior.

## 9-1 CURRENT SOURCES

An ideal current source supplies a constant flow to a load, regardless of the resistance of that load. Figure 9-1 shows the schematic symbol for a current source. This particular source supplies 1 A of current to $R_L$. The arrow in the circle indicates the direction of current flow. In this book, the arrow points in the direction of electron current. In other books, the arrow may point in the direction of con-

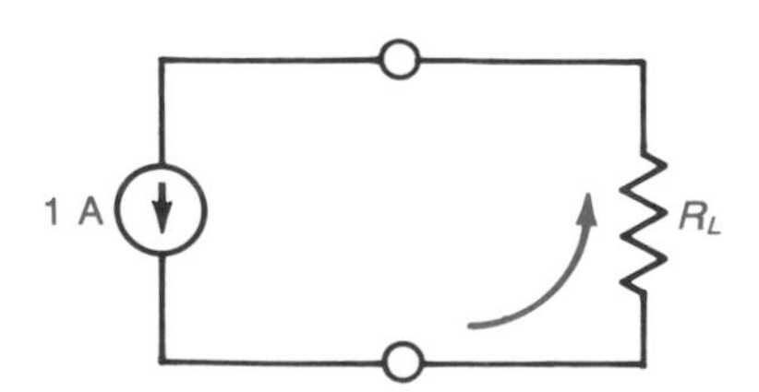


**FIGURE 9-1 The schematic symbol for a current source is a circle with an arrow in it.**

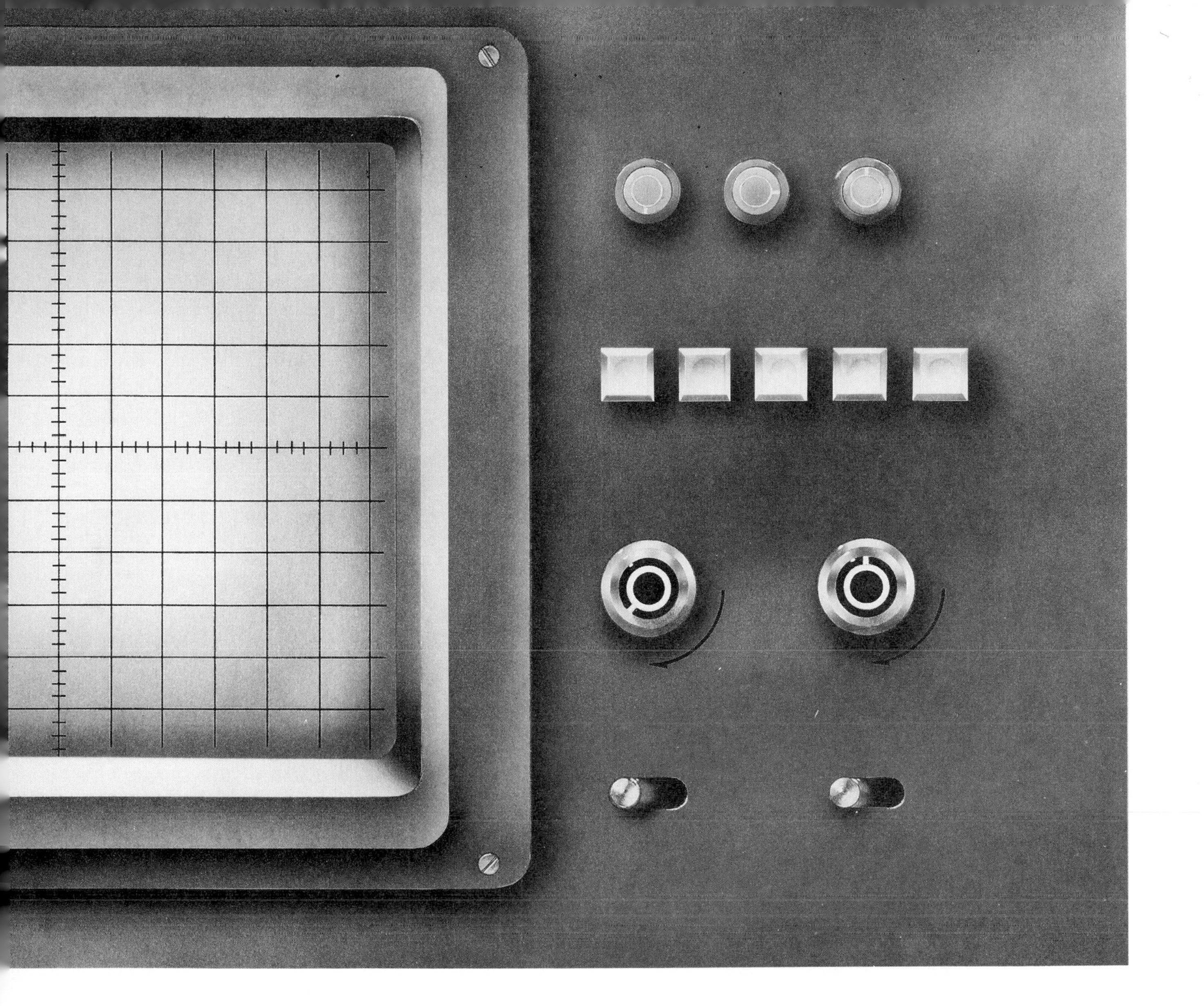

ventional current. It is important to be aware of the convention being used so that the correct polarity of voltage drop can be assigned. Figure 9-2*a* shows the polarity with electron flow, and Fig. 9-2*b* shows the polarity with conventional flow. If you must analyze a circuit containing conventional current sources, mentally reverse the flow arrows and assign the drops from negative to positive as you always do. Your results will be consistent with the conventional current technique.

It has already been shown that there is no such thing as an ideal voltage source. The same thing is true for current sources. Figure 9-3 shows that a real current source has some value of shunt resistance. How closely a current source approaches the ideal is related to this value of shunt resistance. The larger the value of $R_S$, the more closely the current source approaches ideal.

**EXAMPLE 9-1**

Determine the ability of the current source in Fig. 9-3 to maintain a constant load current. You may assume that $R_S$ is 1 kΩ and that $R_L$ varies from 0 Ω to 1 kΩ.

**Solution**

$R_S$ and $R_L$ are in parallel. The current divider rule provides an easy way to find the load current. First, find the load current when $R_L$ is 0 Ω:

$$I_{R_L} = \frac{1\ \text{A} \times 1\ \text{k}\Omega}{1\ \text{k}\Omega + 0\ \Omega} = 1\ \text{A}$$

The load current is 1 A when $R_L = 0\ \Omega$. Next, find the load current when $R_L = 1\ \text{k}\Omega$:

$$I_{R_L} = \frac{1\ \text{A} \times 1\ \text{k}\Omega}{1\ \text{k}\Omega + 1\ \text{k}\Omega} = 0.5\ \text{A}$$

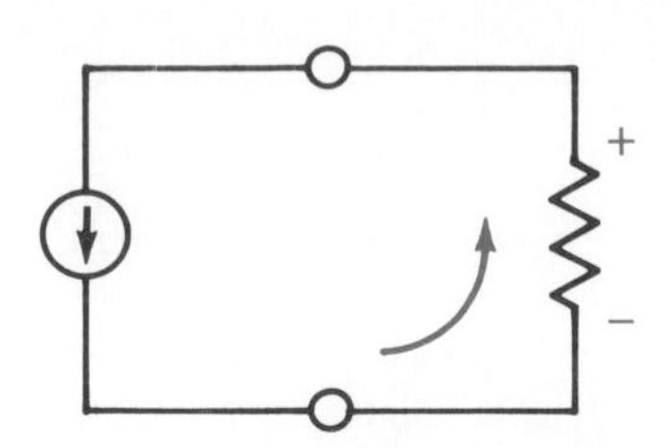


(a) Polarity of drop with electron flow

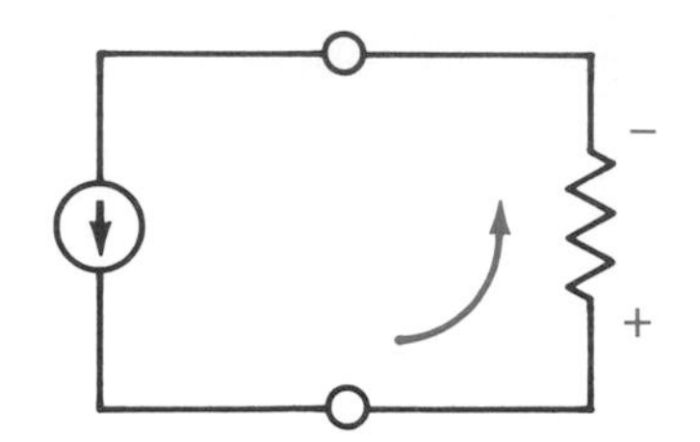


(b) Polarity of drop with conventional flow

**FIGURE 9-2 Comparing current sources by current convention.**

The load current drops to 0.5 A when $R_L$ increases to 1 kΩ. The current source is far from ideal.

## EXAMPLE 9-2

Determine the ability of the current source in Fig. 9-3 to maintain a constant load current if $R_S =$ 100 kΩ and $R_L$ varies from 0 to 1 kΩ.

### Solution

Applying the current divider equation for $R_L =$ 0 Ω:

$$I_{R_L} = \frac{1\text{ A} \times 100\text{ k}\Omega}{100\text{ k}\Omega + 0\ \Omega} = 1\text{ A}$$

And, for $R_L = 1$ kΩ:

$$I_{R_L} = \frac{1\text{ A} \times 100\text{ k}\Omega}{100\text{ k}\Omega + 1\text{ k}\Omega} = 0.99\text{ A}$$

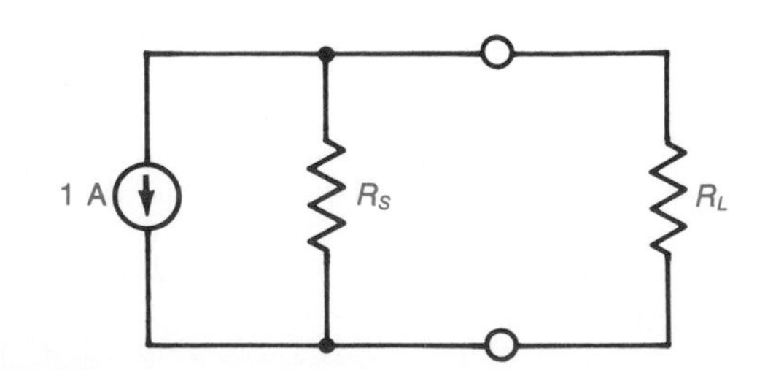


**FIGURE 9-3 A real current source.**

The current source closely approaches being ideal. This is because $R_S$ is now much greater than the maximum value of $R_L$. If $R_L$ were to approach the value of $R_S$, the load current would drop significantly below 1 A.

Figure 9-4 shows a real voltage source. It has an internal resistance $R_S$ that acts in series. As this series resistance becomes smaller, the voltage source more closely approaches the ideal. This is the converse of a real current source where $R_S$ acts in parallel and must become larger for the source to approach the ideal. Even though voltage sources and current sources are different, they have a strong relationship. Figure 9-5 shows two "black boxes." You cannot see what is inside a black box.

## EXAMPLE 9-3

You are given the two black boxes shown in Fig. 9-5 and told that each one contains a source of some type. Can you determine exactly what is in each box?

### Solution

Since there is no way to see inside the boxes, a load is connected to each to determine what it contains. Figure 9-6 shows a 10-Ω load applied to each box. Voltmeters and ammeters are also connected to obtain data. In the case of box A, the 10-Ω resistors act in series for a total resistance of 20 Ω. The ammeter reads

$$I = \frac{10\text{ V}}{20\ \Omega} = 0.5\text{ A}$$

The 10-Ω resistors divide the 10 V equally so the voltmeter reads 5 V. In the case of box B, the 10-Ω resistors divide the current equally so the ammeter reads 0.5 A. The voltmeter reads 5 V due to the current through the external 10-Ω resistor:

$$V = 0.5\text{ A} \times 10\ \Omega = 5\text{ V}$$

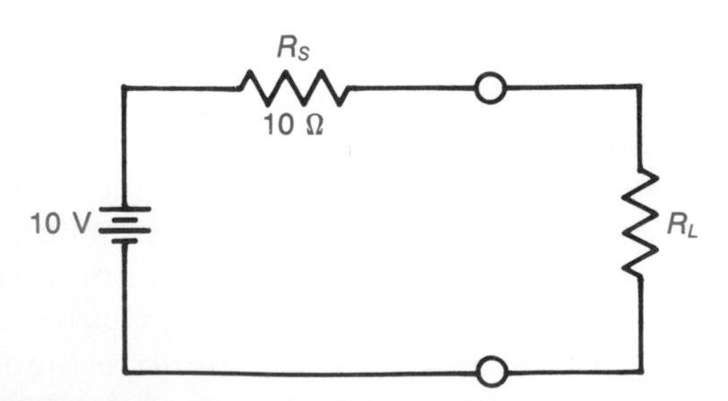


**FIGURE 9-4 A real voltage source.**

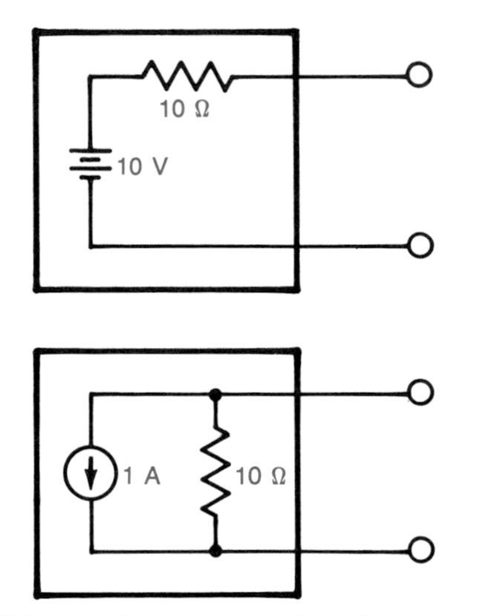


FIGURE 9-5 Real sources in black boxes.

The meter readings are the same for both black boxes so no differences have been found yet. If the external 10-Ω loads are removed, both voltmeters indicate 10 V and, of course, both ammeters read 0. If the external 10-Ω loads are shorted, both ammeters read 1 A and both voltmeters read 0. The boxes *appear to be identical.*

The current source and the voltage source of Fig. 9-5 are considered *duals.* A dual can be found for any real source. The word dual is used because the current source in Fig. 9-5 is not exactly equal to the voltage source. A black box analysis shows them to be so, but there are internal differences. For example, if both boxes are open-circuited, the voltage source delivers no energy, but the current source supplies 1 A to its internal shunt resistance. Power dissipates in the shunt resistance of the current source under open circuit conditions.

To find the dual of a real voltage source, Ohm's law is used to determine the value for the current source. The value of $R_S$ remains the same but appears in shunt. To find the dual of a real current source, Ohm's law is used to determine the value for the voltage source. The value of $R_S$ remains the same but appears in series.

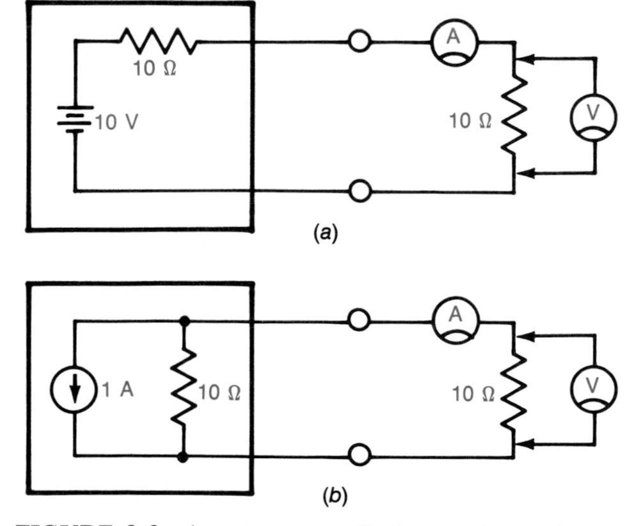


FIGURE 9-6 An attempt to find out what is in the boxes.

### EXAMPLE 9-4

Find duals for the sources shown in Fig. 9-7.

**Solution**

Figure 9-7*a* shows a 500-V source with a series resistance of 1 MΩ. Ohm's law is used to find the value of the dual current source:

$$I = \frac{500 \text{ V}}{1 \text{ M}\Omega} = 0.5 \text{ mA}$$

The dual of Fig. 9-7*a* is a 0.5-mA current source in shunt with a 1-MΩ resistor. Note the high value of

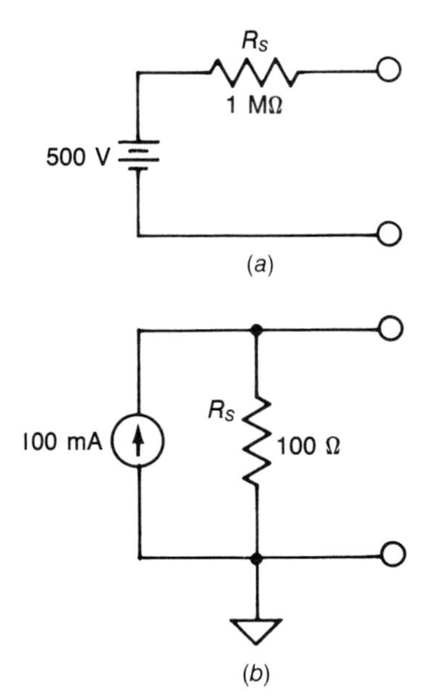


FIGURE 9-7 Sources for Example 9-4.

shunt resistance in this case. This supply closely approaches an ideal current source for loads as high as 100 kΩ. This was pointed out in Chap. 6. Voltage sources with a high series resistance have a constant current characteristic. Figure 9-7*b* shows a 100-mA source in shunt with a 100-Ω resistor. Ohm's law is used to find the value of its dual voltage source:

$$V = 100\text{ mA} \times 100\ \Omega = 10\text{ V}$$

The dual of Fig. 9-7*b* is a 10-V source with a 100-Ω series resistance. The positive terminal of the voltage source must be grounded to match the polarity of the original.

Current sources of different values may *not* be connected in series since current must be constant in a series circuit. Current sources of different values may be parallel-connected. They can be parallel-aiding or parallel-opposing as demonstrated in Fig. 9-8. When they oppose, the load current is equal to the difference of the source currents and its direction is set by the source with the greatest current.

Current sources have several applications. They are used to *model* (represent) certain devices that exhibit a constant-current behavior. Transistors are examples of such devices and are covered in Chap. 24. Current sources may also be used to model equipment such as current-regulated power supplies. These supplies provide a constant current over some limited range of load resistance. The most important application of current sources, as far as this chapter is concerned, is their use in certain network analysis procedures.

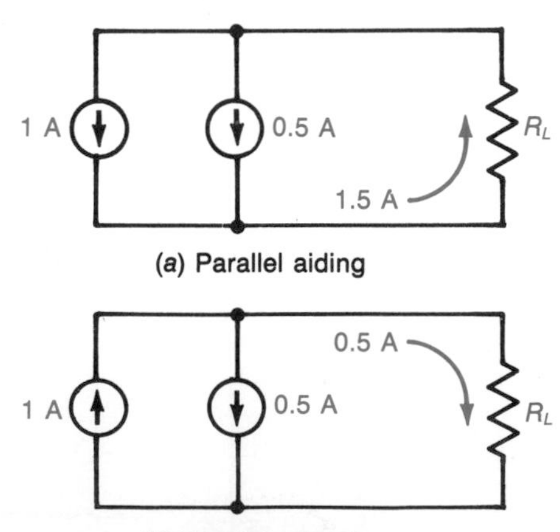


**FIGURE 9-8 Current sources may be connected in parallel.**

## Self-Test

**1.** An ideal ______ source produces constant flow regardless of its load.

**2.** An ideal ______ source produces constant potential regardless of its load.

**3.** Refer to Fig. 9-9*a* and assume the electron current convention. What is the polarity of $V_A$?

**4.** What is the voltage drop across the resistor in Fig. 9-9*a*?

**5.** If the 1-kΩ resistor in Fig. 9-9*a* is replaced with a 1-MΩ resistor, what will happen to $V_A$?

**6.** Will the current source shown in Fig. 9-9*b* closely approach an ideal source for any load? For loads less than 100 Ω?

**7.** What value of voltage source will act as a dual for the source shown in Fig. 9-9*b*? With what value of series resistance?

**8.** Which terminal of the voltage source should be grounded to act as the dual for Fig. 9-9*b*?

**9.** What value will $R_S$ in Fig. 9-9*b* have to be for the source to be considered ideal?

**10.** What value will $R_S$ in Fig. 9-9*c* have to be for the source to be considered ideal?

**11.** What current source would serve as a dual for Fig. 9-9*c* and what would its shunt resistance be?

**12.** Is there a dual for an ideal current source? For an ideal voltage source?

**13.** What is the drop across the resistor in Fig. 9-9*d*? Is the top of the resistor positive or negative with respect to the bottom?

**14.** Can current sources of different values be series-connected?

**15.** Can voltage sources of different values be parallel-connected?

## 9-2 DETERMINANTS

The analysis of complex networks often involves the solution of equations with more than one unknown. Such equations can be solved when the

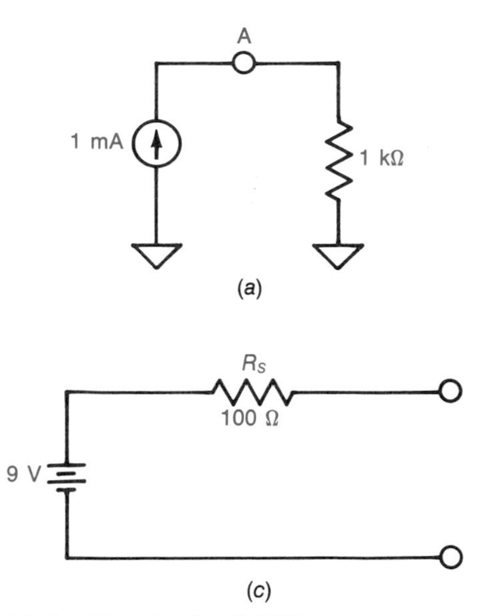


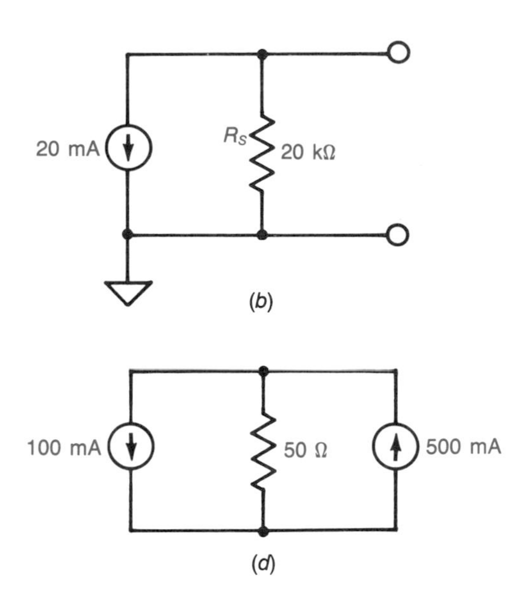


**FIGURE 9-9 Circuits for Self-Test.**

number of independent equations is at least equal to the number of unknowns. They are known as *simultaneous* equations and may be solved by graphing, addition, subtraction, and substitution.

The solution of two or three simultaneous equations can also be achieved by another method that uses *determinants.* A determinant is a numerical value assigned to a square arrangement of numbers called a *matrix.* The determinant method is presented here as a systematic procedure for solving simultaneous equations. The advantage of the determinant method is that it is less cumbersome for three unknowns and is less error prone. The theory behind this method is not presented here but is available in any number of mathematical references. The solution of higher-order determinants (four or more unknowns) is not covered since technical work does not require it.

A 2 × 2 matrix has four numbers arranged in two rows and two columns. The value of such a matrix is called a *second-order determinant* and is equal to the product of the principal diagonal minus the product of the other diagonal. For example,

Other diagonal

$$\begin{vmatrix} a & b \\ c & d \end{vmatrix} \quad \text{value of the matrix} = ad - cb$$

Principal diagonal

---

**EXAMPLE 9-5**

Find the determinant for the 2 × 2 matrix shown below:

$$\begin{vmatrix} 4 & -3 \\ 7 & -1 \end{vmatrix}$$

**Solution**

Multiply the principal diagonal and subtract the product of the other diagonal:

$$4(-1) - 7(-3) = +17$$

---

Second-order determinants can be used to solve simultaneous equations with two unknowns. Consider the following equations:

$$a_1x + b_1y = c_1$$
$$a_2x + b_2y = c_2$$

The unknowns are $x$ and $y$ in these equations. The numbers associated with the unknowns are called the *coefficients.* The coefficients in these equations are $a$ and $b$. The right-hand member of each equation, $c$, is called the *constant.* The coefficients and constants can be arranged as a numerator matrix and as a denominator matrix. The matrix for the numerator is formed by replacing the coefficient of the unknown with the constant. The denominator is called the *characteristic matrix* and is constant in both fractions.

$$x = \frac{\begin{vmatrix} c_1 & b_1 \\ c_2 & b_2 \end{vmatrix}}{\begin{vmatrix} a_1 & b_1 \\ a_2 & b_2 \end{vmatrix}}$$

$$y = \frac{\begin{vmatrix} a_1 & c_1 \\ a_2 & c_2 \end{vmatrix}}{\begin{vmatrix} a_1 & b_1 \\ a_2 & b_2 \end{vmatrix}}$$

---

**EXAMPLE 9-6**

Solve the following simultaneous equations using second-order determinants:

$$3x + 4y = 36$$
$$-3x + y = -6$$

**Solution**

Set up the matrices and find the determinants:

$$x = \frac{\begin{vmatrix} 36 & 4 \\ -6 & 1 \end{vmatrix}}{\begin{vmatrix} 3 & 4 \\ -3 & 1 \end{vmatrix}} = \frac{36 - (-24)}{3 - (-12)} = \frac{60}{15} = 4$$

$$y = \frac{\begin{vmatrix} 3 & 36 \\ -3 & -6 \end{vmatrix}}{\begin{vmatrix} 3 & 4 \\ -3 & 1 \end{vmatrix}} = \frac{-18 - (-108)}{3 - (-12)} = \frac{90}{15} = 6$$

---

Note that the characteristic determinant (denominator) is the same in both cases and needs to be evaluated only once. Also note that the coefficients for $x$ are replaced by the constants when solving for $x$ and that the coefficients for $y$ are replaced by the constants when solving for $y$.

A *third-order determinant* is the value of a $3 \times 3$ matrix. These must also be diagonally multiplied. There are three principal diagonals and three other diagonals. This is because the matrix wraps around like a cylinder. To simplify the multiplication procedure, the first two columns are repeated to the right side of the matrix:

$3 \times 3$ Matrix

$$\begin{vmatrix} 4 & 1 & 2 \\ 3 & 2 & 2 \\ 1 & 2 & 3 \end{vmatrix}$$

With first two columns repeated

$$\left|\begin{array}{ccc|cc} 4 & 1 & 2 & 4 & 1 \\ 3 & 2 & 2 & 3 & 2 \\ 1 & 2 & 3 & 1 & 2 \end{array}\right|$$

The principal diagonals

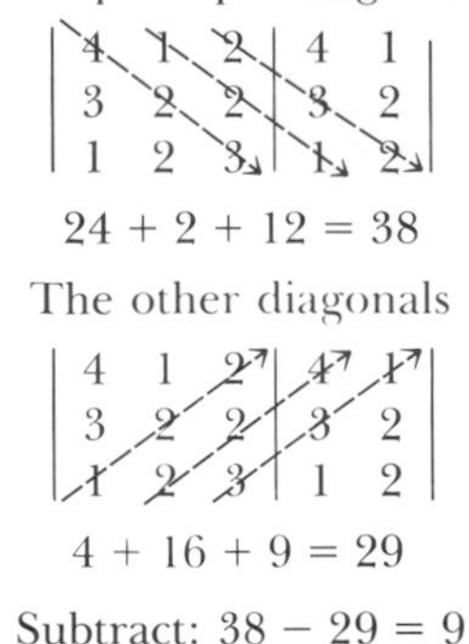

$24 + 2 + 12 = 38$

The other diagonals

$4 + 16 + 9 = 29$

Subtract: $38 - 29 = 9$

The value of the determinant is 9.

Simultaneous equations with three unknowns can be solved with third-order determinants. The coefficients ($a$, $b$, and $c$), the unknowns ($x$, $y$, and $z$), and the constants ($d$) take the following form:

$$a_1x + b_1y + c_1z = d_1$$
$$a_2x + b_2y + c_2z = d_2$$
$$a_3x + b_3y + c_3z = d_3$$

The characteristic matrix forms the denominator and is the same for each fraction. It is formed by the coefficients of the simultaneous equations:

$$\text{Denominator} = \begin{vmatrix} a_1 & b_1 & c_1 \\ a_2 & b_2 & c_2 \\ a_3 & b_3 & c_3 \end{vmatrix}$$

The matrix for each numerator is formed by replacing the coefficient of the unknown with the constant:

$$x = \frac{\begin{vmatrix} d_1 & b_1 & c_1 \\ d_2 & b_2 & c_2 \\ d_3 & b_3 & c_3 \end{vmatrix}}{\text{Denominator}}$$

$$y = \frac{\begin{vmatrix} a_1 & d_1 & c_1 \\ a_2 & d_2 & c_2 \\ a_3 & d_3 & c_3 \end{vmatrix}}{\text{Denominator}}$$

$$z = \frac{\begin{vmatrix} a_1 & b_1 & d_1 \\ a_2 & b_2 & d_2 \\ a_3 & b_3 & d_3 \end{vmatrix}}{\text{Denominator}}$$

**EXAMPLE 9-7**

Solve the following set of simultaneous equations using third-order determinants:

$$3x + 2y - z = 0$$
$$x - y + 2z = -1$$
$$2x - 3y + z = -7$$

**Solution**

Set up the characteristic matrix first and find its determinant:

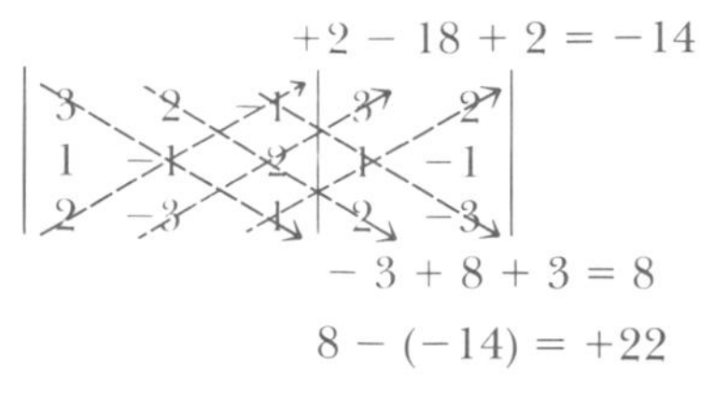

The determinant of the denominator is +22. Next, set up each numerator matrix and solve for the unknowns:

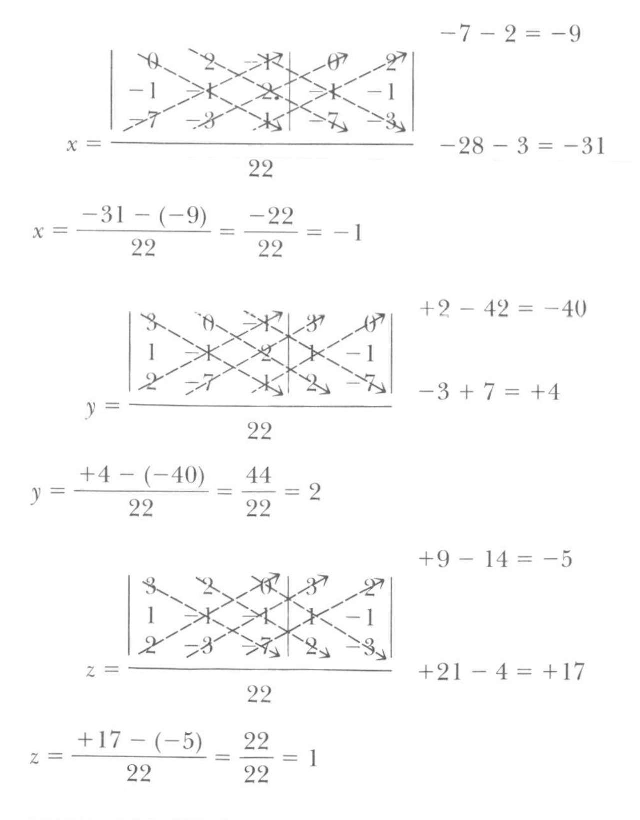

The reader is cautioned that the methods presented here do not work for fourth- and higher-order determinants.

## Self-Test

Solve all of the following equations using determinants:

16. $4x + y = 10$
 $2x - 3y = -9$
17. $x + y = 14$
 $-4x + 3y = 0$
18. $-2x - 2y = -2$
 $x + 6y = 16$
19. $3x + 4y - z = 23$
 $x + 2y + 2z = 12$
 $-2x - 2y - 3z = -17$
20. $4a - b + 2c = 2$
 $-2a + 2b + c = 7$
 $6a + 3b + 3c = 3$
21. $3a - 2b + c = 20$
 $-a + 4b + 2c = -6$
 $2a - b - c = 2$
22. $10 - 15I_1 - 5I_2 + 5I_3 = 0$
 $115 - 5I_1 - 25I_2 - 20I_3 = 0$
 $140 + 5I_1 - 20I_2 - 35I_3 = 0$
 *Hint:* Rearrange the equations so the constants are on the right.

## 9-3 BRANCH ANALYSIS

As used in network analysis, *branch* means a path between two junctions in a circuit. This path may include one or more sources as well as one or more loads. Branch analysis is a technique based on the actual currents in an electric circuit.

The procedure is as follows:

1. Assign a current to each branch using any direction you wish.
2. Determine the polarity of drop across each resistor based on the assigned current direction.
3. Apply Kirchhoff's voltage law around each loop. Stop when all components have been included . . . do not write redundant equations.
4. Apply Kirchhoff's current law to those nodes that will include all branch currents . . . do not write redundant equations.
5. Solve the equations.
6. If the solution of the equations produces any negative currents, those currents have been as-

signed backward. The absolute values of the currents are always correct if no other errors have been made. Reverse any negative currents and the polarity of their associated drops.

**7.** Verify the solution using Ohm's law and Kirchhoff's laws.

---

**EXAMPLE 9-8**

Apply branch analysis to the circuit shown in Fig. 9-10.

**Solution**

The analysis begins by assigning branch currents $I_1$, $I_2$, and $I_3$. The polarity of drop across each resistor is based on the assigned directions as shown in Fig. 9-10. Next, two loop equations are written. One loop involves the 15-V battery, the 10-Ω resistor, and the 100-Ω resistor. The other loop involves the 13-V battery, the 10-Ω resistor, and the 20-Ω resistor. These two loop equations include all components. A third equation could be written for the outside loop but would be redundant. The loop equations are based on Kirchhoff's voltage law:

$$\begin{aligned} &\text{Loop 1:} \quad +15 - 10I_3 - 100I_1 = 0 \\ &\text{Loop 2:} \quad +13 - 10I_3 - \phantom{1}20I_2 = 0 \end{aligned}$$

Look carefully at the loop equations. The battery voltages are both positive. This is because the assigned direction of flow makes them both appear as a *rise* in potential. If the assigned directions had been reversed, both battery voltages would be negative in the loop equations since they would appear as voltage *drops*. The resistor values are multiplied by their associated currents. This is based on Ohm's law, which states that the drop is equal to the current times the resistance. All resistor voltages are negative in the loop equations in this case because of the assigned flow. Start at the 15-V battery and trace the loop. The flow is from negative to positive at each resistor, indicating drops. However, this is not always the case. If $I_3$ had been assigned in the other direction, the polarity across the 10-Ω resistor would be reversed, and the term $10I_3$ would appear as a rise (+ sign) in both loop equations. Node A includes all loop currents. Kirchhoff's current law is used to write the equation:

$$\text{Node A:} \quad I_1 + I_2 - I_3 = 0$$

We now have three equations and three unknowns. Two of the equations are rearranged to allow the orderly application of determinants:

$$\begin{aligned} -100I_1 - 10I_3 &= -15 \\ -20I_2 - 10I_3 &= -13 \\ I_1 + I_2 - I_3 &= 0 \end{aligned}$$

$$I_1 = \frac{\left|\begin{array}{rrr|rr} -15 & 0 & -10 & -15 & 0 \\ -13 & -20 & -10 & -13 & -20 \\ 0 & 1 & -1 & 0 & 1 \end{array}\right|}{\left|\begin{array}{rrr|rr} -100 & 0 & -10 & -100 & 0 \\ 0 & -20 & -10 & 0 & -20 \\ 1 & 1 & -1 & 1 & 1 \end{array}\right|}$$

$$I_1 = \frac{-170 - (+150)}{-2000 - (+1200)} = \frac{-320}{-3200} = 0.1 \text{ A}$$

$$I_2 = \frac{\left|\begin{array}{rrr|rr} -100 & -15 & -10 & -100 & -15 \\ 0 & -13 & -10 & 0 & -13 \\ 1 & 0 & -1 & 1 & 0 \end{array}\right|}{\text{Denominator}}$$

$$I_2 = \frac{-1150 - (+130)}{D} = \frac{-1280}{-3200} = 0.4 \text{ A}$$

$$I_3 = \frac{\left|\begin{array}{rrr|rr} -100 & 0 & -15 & -100 & 0 \\ 0 & -20 & -13 & 0 & -20 \\ 1 & 1 & 0 & 1 & 1 \end{array}\right|}{\text{Denominator}}$$

$$I_3 = \frac{0 - (+1600)}{D} = \frac{-1600}{-3200} = 0.5 \text{ A}$$

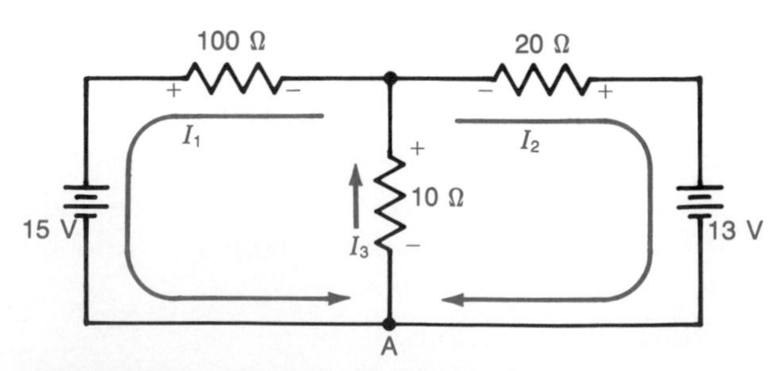


**FIGURE 9-10 Circuits for Example 9-8.**

All of the currents are positive, which means that all of the assigned directions of flow are correct.

---

Figure 9-11 shows the procedure for verifying the answers. The currents are multiplied by the resistor values to obtain the actual drops. The polarities are based on the correct direction of flow. Now, Kirchhoff's voltage law can be applied to determine if each loop is correct:

$$15\text{ V} - 5\text{ V} - 10\text{ V} = 0$$
$$13\text{ V} - 5\text{ V} - 8\text{ V} = 0$$

The outside loop must also be verified. Starting at the 15-V battery and assuming a counterclockwise flow:

$$15\text{ V} - 13\text{ V} + 8\text{ V} - 10\text{ V} = 0$$

Study this equation carefully. It will give you a good sense of how loop equations work. Note that the 13-V battery is written as a drop. This is based on the assumed direction of flow and the battery polarity. Also notice that the voltage across the 20-Ω resistor is entered as a rise. Again, this is a function of the assumed direction of flow and its polarity. Kirchhoff's current law is also satisfied:

$$0.1\text{ A} + 0.4\text{ A} - 0.5\text{ A} = 0$$

A knowledge of source conversion sometimes leads to a simplified analysis procedure. Figure 9-12 shows what the circuit of Example 9-8 looks like when the voltage sources are replaced with current sources. The 100-Ω resistor has been considered as the internal resistance of the 15-V battery and the 20-Ω resistor as the internal resistance of the 13-V battery. The current sources are par-

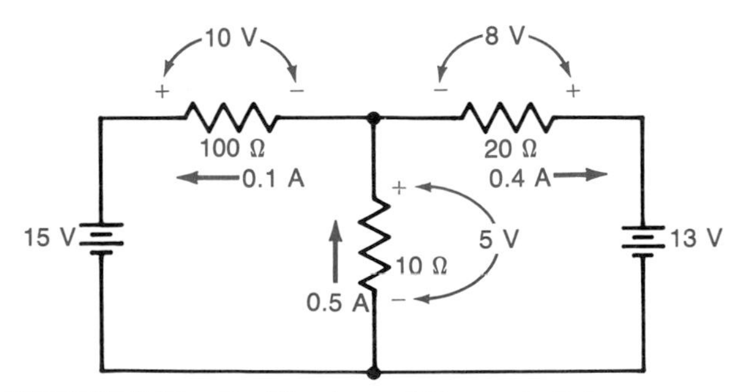


**FIGURE 9-11 Verification of Example 9-8.**

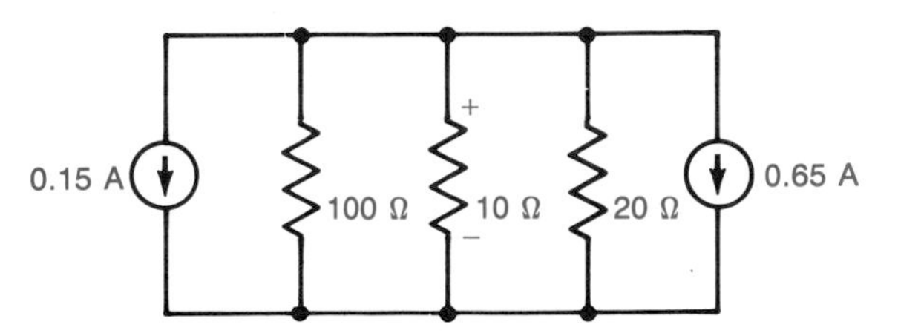


**FIGURE 9-12 Replacing the voltage sources with current sources.**

allel-aiding for a total flow of 0.8 A. The parallel resistors can be combined:

$$100\ \Omega \parallel 10\ \Omega \parallel 20\ \Omega = 6.25\ \Omega$$

The total current flowing through this resistance produces the drop:

$$0.8\text{ A} \times 6.25\ \Omega = 5\text{ V}$$

This 5-V drop can now be "transported" back to the original circuit. It appears across the 10 Ω resistor. Its polarity is negative at the bottom and positive at the top. Knowing this drop is the key to solving the original circuit. Applying Kirchhoff's voltage law produces a 10-V drop across the 100-Ω resistor and an 8-V drop across the 20-Ω resistor. Ohm's law gives the currents, and the solution is complete. Note that the 5-V drop applies only to the 10-Ω resistor in the original circuit.

## Self-Test

**23.** Rework the circuit shown in Fig. 9-10 with $I_1$ and $I_2$ assigned as shown but with $I_3$ reversed. When the equations are solved, what are the magnitudes and algebraic signs of $I_1$, $I_2$, and $I_3$?

**24.** Find all of the branch currents for the circuit shown in Fig. 9-13.

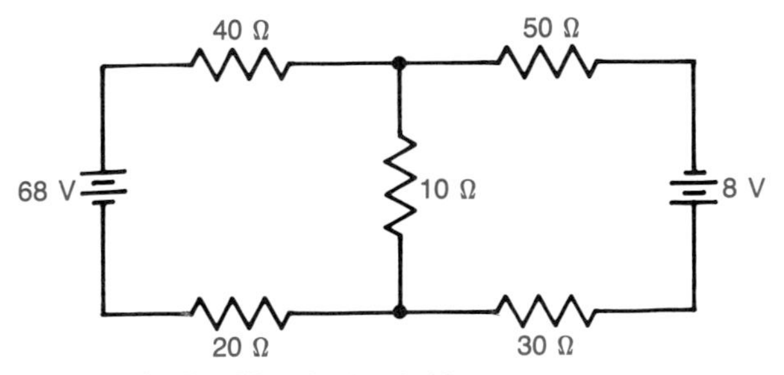


**FIGURE 9-13 Circuit for Self-Test.**

**25.** What is the voltage drop across the 10-Ω resistor in Fig. 9-13?

**26.** What is the polarity across the 10-Ω resistor in Fig. 9-13?

## 9-4 MESH ANALYSIS

Branch analysis is based on real currents. *Mesh analysis* uses abstract currents. Mesh analysis is a better approach, but because it is somewhat abstract, branch analysis has been presented first to develop a better understanding of the general concepts.

Mesh analysis can be handled in several ways. The procedure that follows is a systematic approach that tends to eliminate errors:

**1.** Assign a clockwise direction of flow in each *window*. The windows are the inner loops of the network. The window currents take on the appearance of a mesh fence, hence the name *mesh* analysis.

**2.** Multiply each mesh current by the sum of the resistances in the window.

**3.** Subtract the product of common resistances and adjacent mesh currents.

**4.** Set the terms produced in steps 2 and 3 equal to the total source voltage for each mesh. If a mesh does not have a source, set the terms equal to zero.

**5.** Solve the equations.

**6.** Reverse any currents that solve as negative quantities.

**7.** Find the real branch currents and the voltage drops.

**8.** Verify the solution with Kirchhoff's voltage law.

### EXAMPLE 9-9

Apply mesh analysis to the network shown in Fig. 9-14.

**Solution**

This circuit is repeated from the previous section on branch analysis. It will be interesting to compare the relative ease of mesh analysis with branch analysis. There are two windows and only two equations are required. Another advantage is that resistor polarities do not have to be considered when writing the equations:

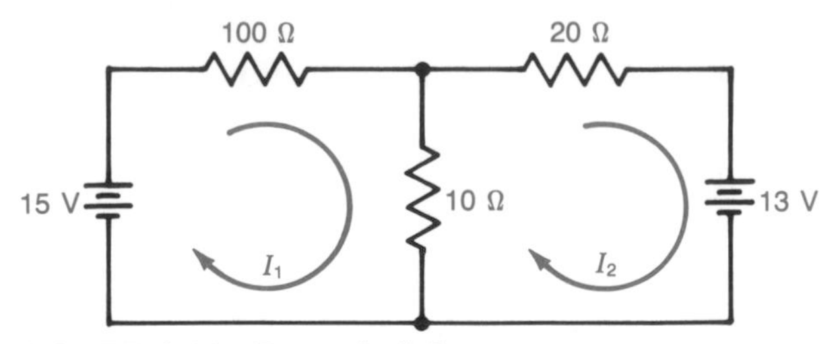

**FIGURE 9-14 Example 9-9.**

$$\text{Mesh 1:} \quad 110I_1 - 10I_2 = -15$$
$$\text{Mesh 2:} \quad -10I_1 + 30I_2 = 13$$

Study the equations. Verify that the sum of the resistances in each mesh is multiplied times the current. Notice that the resistor which is common with the adjacent mesh is multiplied by the adjacent current and that product is subtracted. Finally, note that sources may be entered as a rise or a drop, depending on flow and polarity. Determinants are applied to solve the equations:

$$I_1 = \frac{\begin{vmatrix} -15 & -10 \\ 13 & 30 \end{vmatrix}}{\begin{vmatrix} 110 & -10 \\ -10 & 30 \end{vmatrix}} = \frac{-320}{3200} = -0.1 \text{ A}$$

$$I_2 = \frac{\begin{vmatrix} 110 & -15 \\ -10 & 13 \end{vmatrix}}{\text{Denominator}} = \frac{1280}{3200} = 0.4 \text{ A}$$

Current $I_1$ is negative. Its magnitude is correct, but its direction is backward. When it is reversed, it can be seen that both currents flow up in the 10-Ω branch. The real current in the 10-Ω resistor is found by adding the currents in this case:

$$I = 0.1 + 0.4 = 0.5 \text{ A}$$

When currents oppose in common resistors, the real current is found by subtracting and the direction and polarity are set by the largest current. Ohm's law can now be applied to find all drops, and Kirchhoff's voltage law can be used to verify each mesh. This has already been done for this circuit.

### EXAMPLE 9-10

Solve the circuit shown in Fig. 9-15 using mesh analysis.

**Solution**

The mesh currents are assigned. Three equations are required:

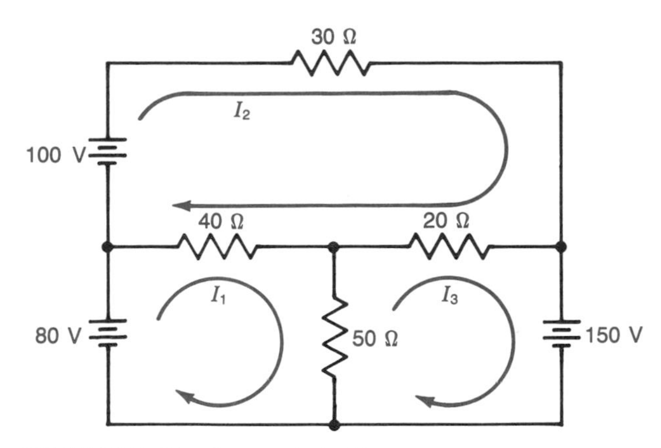


**FIGURE 9-15 Example 9-10.**

$$\begin{aligned}
&\text{Mesh 1:} \quad 90I_1 - 40I_2 - 50I_3 = -80 \\
&\text{Mesh 2:} \quad -40I_1 + 90I_2 - 20I_3 = -100 \\
&\text{Mesh 3:} \quad -50I_1 - 20I_2 + 70I_3 = 150
\end{aligned}$$

Solving the equations produces:

$$\begin{aligned}
I_1 &= -0.5 \text{ A} \\
I_2 &= -1 \text{ A} \\
I_3 &= 1.5 \text{ A}
\end{aligned}$$

Currents $I_1$ and $I_2$ have been assigned backward. Figure 9-16 shows the circuit redrawn with the corrected directions. The *real* currents in the common resistors are found by adding or subtracting the mesh currents. For example, the current in the 50-Ω resistor is the sum of two mesh currents:

$$I_1 + I_3 = 0.5 \text{ A} + 1.5 \text{ A} = 2 \text{ A}$$

Current $I_2$ is greater than $I_1$; therefore, it establishes the direction of current through the 40-Ω resistor:

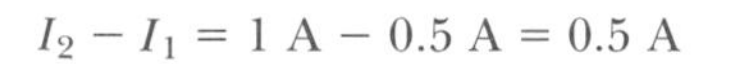

$$I_2 - I_1 = 1 \text{ A} - 0.5 \text{ A} = 0.5 \text{ A}$$

So 0.5 A flows through the 40-Ω resistor from left to right. Finally, $I_2$ and $I_3$ add in the 20-Ω resistor:

$$I_2 + I_3 = 1 \text{ A} + 1.5 \text{ A} = 2.5 \text{ A}$$

The final step is to assign all real polarities and voltage drops. The drops are found with Ohm's law, and the results are shown in Fig. 9-17. Each mesh can now be checked with Kirchhoff's voltage law:

$$\begin{aligned}
&\text{Mesh 1:} \quad 80 \text{ V} - 100 \text{ V} + 20 \text{ V} = 0 \\
&\text{Mesh 2:} \quad 100 \text{ V} - 20 \text{ V} - 50 \text{ V} - 30 \text{ V} = 0 \\
&\text{Mesh 3:} \quad 150 \text{ V} - 100 \text{ V} - 50 \text{ V} = 0
\end{aligned}$$

The law is satisfied and the solution has been verified.

## Self-Test

**27.** Solve the circuit shown in Fig. 9-18*a* for the mesh currents.

**28.** What is the voltage drop across the 1-kΩ resistor in Fig. 9-18*a*?

**29.** Solve the circuit shown in Fig. 9-18*b* for the mesh currents.

**30.** What is the drop across the 200-Ω resistor in Fig. 9-18*b*?

**31.** What is the polarity across the 200-Ω resistor in Fig. 9-18*b*?

**32.** Solve the circuit shown in Fig. 9-18*c* for the mesh currents.

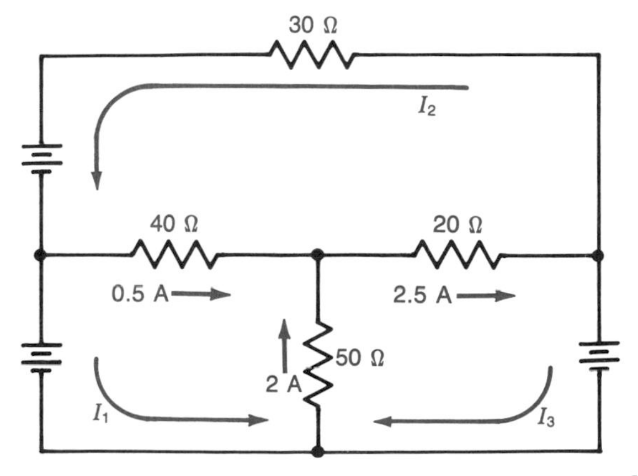


**FIGURE 9-16 Finding the real currents for Example 9-10.**

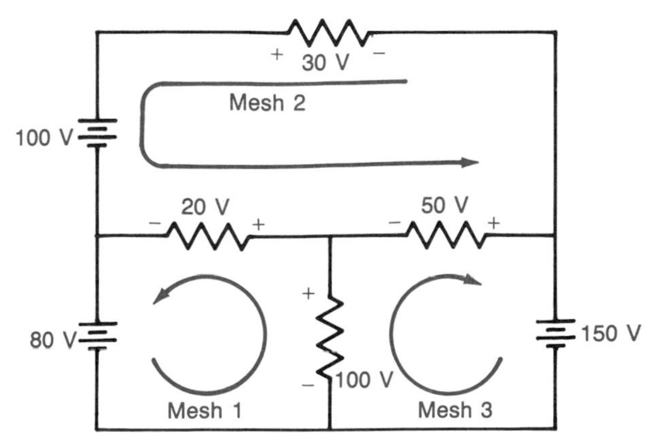


**FIGURE 9-17 Finding the voltage drops for Example 9-10.**

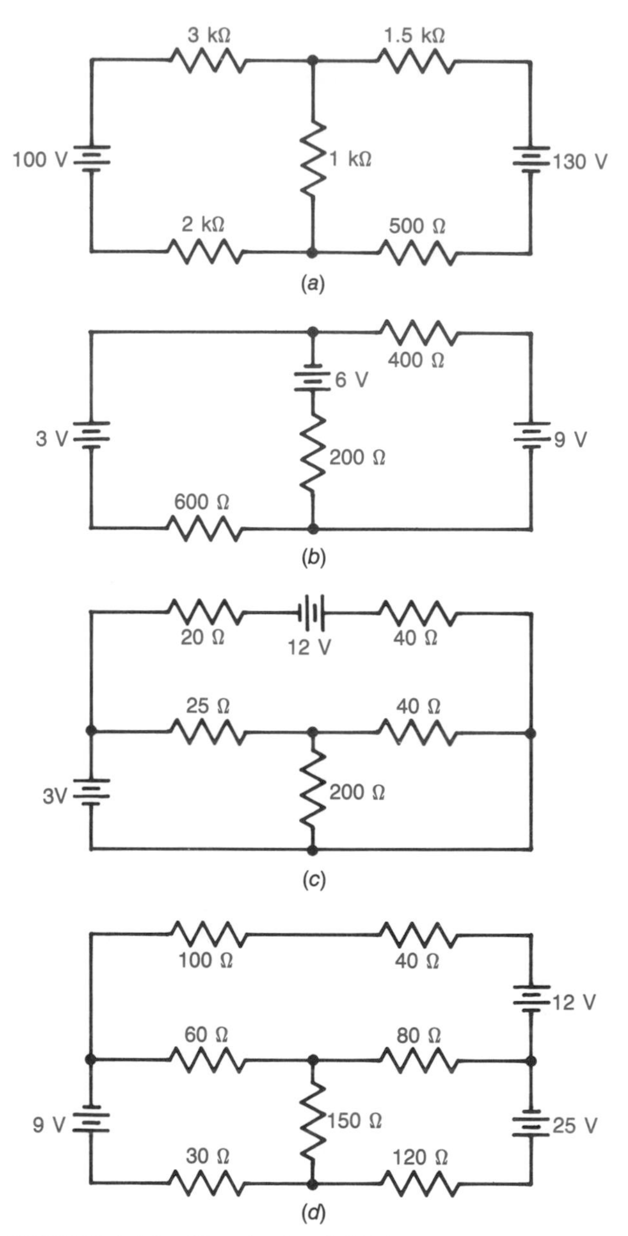


**FIGURE 9-18 Circuits for Self-Test.**

**33.** What is the drop across the 200-Ω resistor in Fig. 9-18*c*?

**34.** Solve the circuit shown in Fig. 9-18*d* for the mesh currents.

**35.** What is the drop across the 150-Ω resistor in Fig. 9-18*d*?

**36.** What is the polarity of the drop across the 150-Ω resistor in Fig. 9-18*d*?

## 9-5 NODAL ANALYSIS

Branch analysis and mesh analysis are based on Kirchhoff's voltage law. *Nodal analysis* is based on Kirchhoff's current law. Nodal analysis is convenient in circuits where various voltages, referenced to a common ground, must be determined. It is also a convenient approach to circuits that have current sources and conductances, rather than voltage sources and resistances. Nodal analysis is commonly used in computer programs that analyze electric and electronic circuits.

As with mesh analysis, there are several ways to apply nodal analysis. A systematic procedure is presented here that helps to eliminate errors:

**1.** Convert any voltage sources to current sources.

**2.** Choose a reference node. This will usually be ground or the node with the greatest number of branches connected to it.

**3.** Assign a label to all nodes other than the reference. These are called the *independent* nodes.

**4.** Write an equation for each independent node. One term will consist of the product of the node voltage and the sum of the conductances connected to that node. *Ignore* any conductances in series with a current source since they cannot affect current.

**5.** Multiply each conductance that connects a node to another independent node times the other node voltage and show these as negative products.

**6.** Set the terms produced in steps 4 and 5 equal to any current source connected to the node. A current leaving a node is shown as a positive value, and a current entering a node is negative. If a node has no current source connected to it, set the terms equal to zero.

**7.** Solve the equations for the node voltages. The polarities are with respect to the reference node.

**8.** Verify the results with Ohm's law and Kirchhoff's current law.

**EXAMPLE 9-11**

Solve the circuit shown in Fig. 9-19 using nodal analysis.

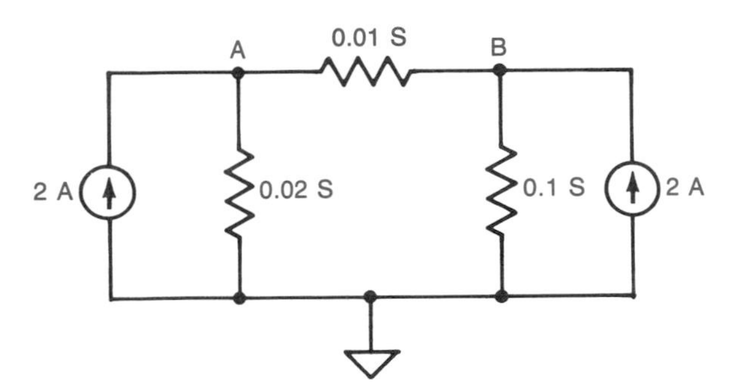


FIGURE 9-19 Circuit for Example 9-11.

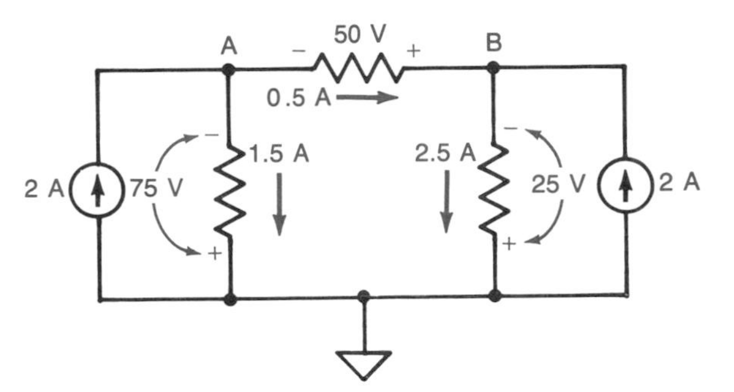


FIGURE 9-20 Verification of Example 9-11.

**Solution**

The sources do not require conversion. Ground is chosen as the reference node, and A and B are the independent nodes. Two equations are required:

$$\text{Node A:}\quad V_A(0.02 + 0.01) - V_B(0.01) = -2$$
$$\text{Node B:}\quad V_B(0.01 + 0.1) - V_A(0.01) = -2$$

Study the equations. They show the product of the node voltage and all conductances connected to that node as a positive quantity. They also show the product of the *connecting* node voltage with the *connecting* conductance as a negative quantity. Both currents enter the nodes in this case, so the right-hand terms of the equations are negative. The equations should be arranged for simultaneous solution:

$$0.03V_A - 0.01V_B = -2$$
$$-0.01V_A + 0.11V_B = -2$$

The solution is

$$V_A = -75 \text{ V}$$
$$V_B = -25 \text{ V}$$

The solution indicates that both voltages are negative with respect to ground. *Nothing needs to be reversed.* All that remains is verification. Figure 9-20 shows the circuit redrawn with the solved voltages. The 75-V drop across the 0.02-S conductance produces a current which can be found with Ohm's law. Recall that current is found by multiplying voltage times conductance:

$$I = VG = 75 \times 0.02 = 1.5 \text{ A}$$

The 25-V drop across the 0.1-S conductance gives a current:

$$I = 25 \times 0.1 = 2.5 \text{ A}$$

The drop across the 0.01-S conductance is the difference between the voltage at node A and node B. Node A is more negative, so the current flows from A to B and the magnitude is

$$I = 50 \times 0.01 = 0.5 \text{ A}$$

Figure 9-20 shows that Kirchhoff's current law is satisfied at both independent nodes:

$$\text{Node A:}\quad 2 \text{ A} - 0.5 \text{ A} - 1.5 \text{ A} = 0$$
$$\text{Node B:}\quad 2 \text{ A} + 0.5 \text{ A} - 2.5 \text{ A} = 0$$

---

**EXAMPLE 9-12**

Write nodal equations for the circuit shown in Fig. 9-21.

**Solution**

Ground is used as the reference node. Three independent nodes are identified and three equations are required:

$$\text{Node A:}\quad V_A(G_1 + G_2) - V_B(G_2) = -I_1$$
$$\text{Node B:}\quad V_B(G_2 + G_3 + G_4) - V_A(G_2) - V_C(G_4) = 0$$
$$\text{Node C:}\quad V_C(G_4 + G_5) - V_B(G_4) = I_2$$

---

Each equation shows the product of the node voltage times the sum of all conductances connected to it. It also shows the negative product of connecting

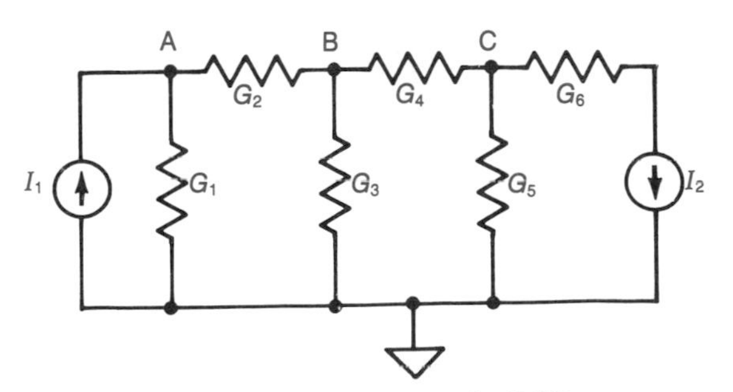


FIGURE 9-21 Circuit for Example 9-12.

conductances and node voltages. The equation for node B is set equal to zero because no current source is connected to it. Please note that $G_6$ is omitted from the equation for node C because *it is in series with a current source*. Also note that the sign of the right-hand term is determined by the direction of the current sources.

### EXAMPLE 9-13

Solve and verify the circuit shown in Fig. 9-22.

**Solution**

Resistances are used rather than conductances. The equations are written as follows:

$$\text{Node A:}\quad V_A\left(\frac{1}{15}+\frac{1}{2.5}\right)-V_B\left(\frac{1}{2.5}\right)=-6$$

$$\text{Node B:}\quad V_B\left(\frac{1}{2.5}+\frac{1}{20}+\frac{1}{6}\right)-V_A\left(\frac{1}{2.5}\right)-V_C\left(\frac{1}{6}\right)=0$$

$$\text{Node C:}\quad V_C\left(\frac{1}{6}+\frac{1}{4}\right)-V_B\left(\frac{1}{6}\right)=2.5$$

Arranging the equations for solution,

$$0.467V_A - 0.4V_B = -6$$
$$-0.4V_A + 0.617V_B - 0.167V_C = 0$$
$$-0.167V_B + 0.417V_C = 2.5$$

The solutions are

$$V_A = -30\text{ V}$$
$$V_B = -20\text{ V}$$
$$V_C = -2\text{ V}$$

Figure 9-23 shows the circuit redrawn with the voltages applied. The currents are calculated with Ohm's law. Kirchhoff's current law is used to verify the results:

Node A: 6 A − 2 A − 4 A = 0
Node B: 4 A − 1 A − 3 A = 0
Node C: 3 A − 0.5 A − 2.5 A = 0

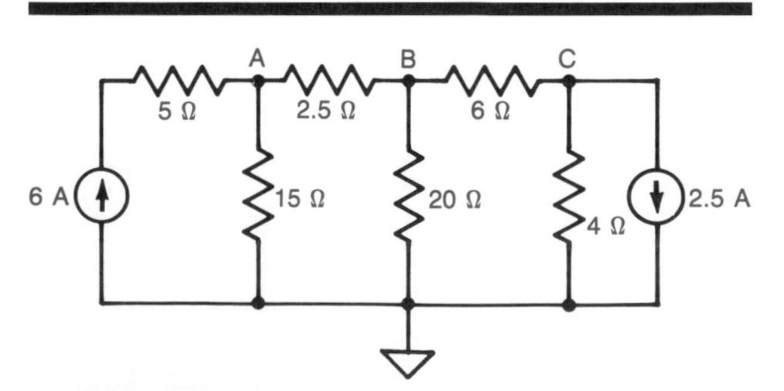


**FIGURE 9-22 Circuit for Example 9-13.**

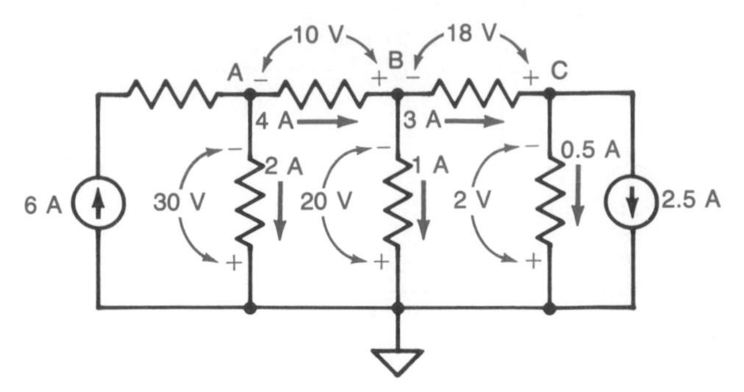


**FIGURE 9-23 Verification of Example 9-13.**

Figure 9-24*a* shows a circuit with an ideal voltage source. This situation may appear to rule out nodal analysis. We know that duals do not exist for ideal sources. However, there is no need to solve for the voltage at node C in Fig. 9-24*a*. It is equal to $-V_1$. Therefore, we can eliminate that node. Figure 9-24*b* shows the circuit with $R_5$ removed. This leaves $R_4$ in series with the voltage source. The voltage source can now be converted to a current source as shown in Fig. 9-24*c*, and equations can be written for nodes A and B. These equations, when solved, give the correct voltages at these nodes. These values can be transported back to the original circuit. If the battery current is required, it can be determined in the original circuit by combining the current in the resistor that was eliminated with the current at node B. It is *not* correct to assume that it will be equal to $I_2$.

## Self-Test

**37.** Refer to Fig. 9-25*a*. Find the node voltages.

**38.** How much current flows through the 0.05-S conductance in Fig. 9-25*a*?

**39.** In which direction does the current in the 0.05-S conductance in Fig. 9-25*a* flow?

**40.** Refer to Fig. 9-25*b*. Find the node voltages.

**41.** What is the current in the 0.3-S conductance in Fig. 9-25*b* and in which direction does it flow?

**42.** What is the current in the 0.4-S conductance in Fig. 9-25*b* and in which direction does it flow?

**43.** Refer to Fig. 9-25*c*. What are the node voltages?

**44.** How much current flows in the battery in Fig. 9-25*c*?

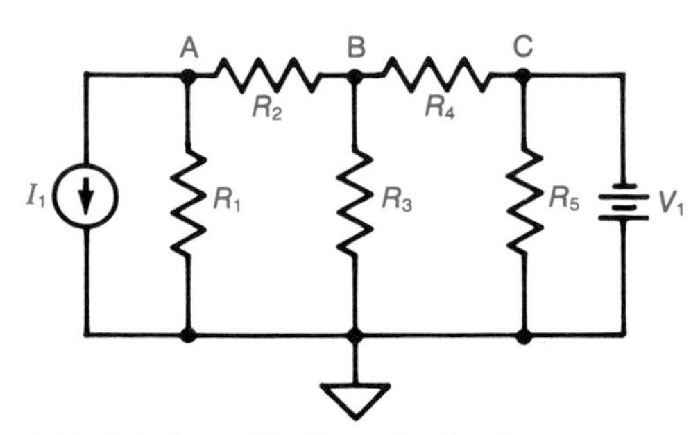


(*a*) Original circuit with an ideal voltage source

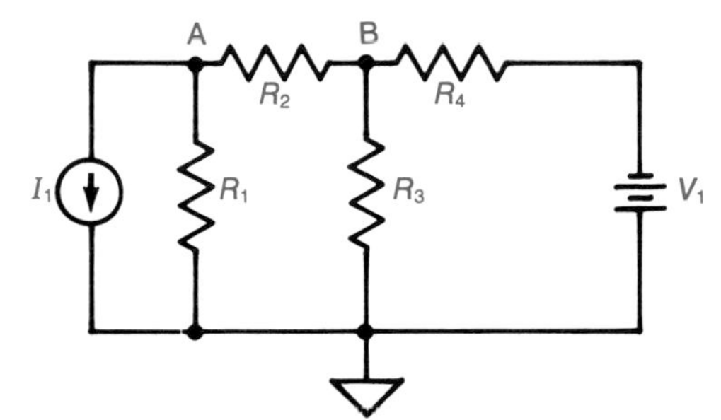


(*b*) Circuit with $R_5$ removed

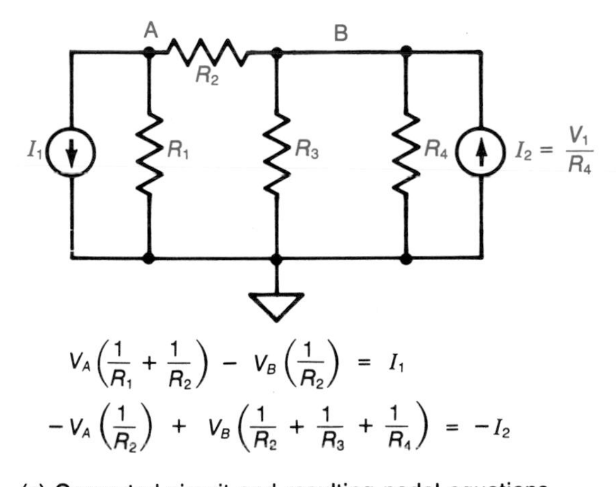


(*c*) Converted circuit and resulting nodal equations

**FIGURE 9-24 Applying nodal analysis to a circuit with an ideal voltage source.**

## 9-6 SUPERPOSITION

The *superposition* theorem provides another method of analyzing multiple source networks and has the advantage of not requiring the solution of simultaneous equations. It achieves this by treating each voltage or current source *independently*. The independent solutions are algebraically *superimposed* to determine the final solution. For a network with $N$ sources, $N$ independent solutions are required before algebraic combination.

A network must be *linear* if it is to be analyzed using superposition. For example, a circuit might contain one or more varistors. These devices have a nonlinear relationship between voltage and current so a varistor circuit may not be analyzed with superposition. Also, the independent solutions of superposition must not be used to calculate power. Power varies as the *square* of voltage or current which is a nonlinear relationship.

> The superposition theorem states: The current through, or the voltage drop across, any component in a linear network is equal to the algebraic sum of the currents or voltages independently produced by each source.

The method of superposition is straightforward:

**1.** Remove the ideal portion of all sources but one. Any voltage source that is removed is replaced

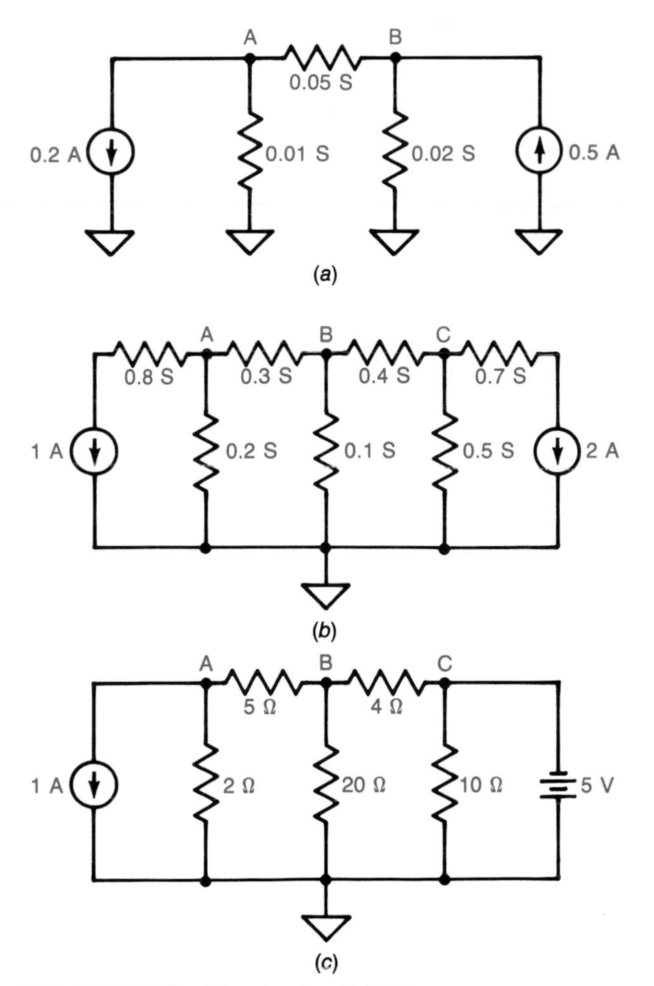

**FIGURE 9-25 Circuits for Self-Test.**

with a short circuit. Any current source that is removed is replaced with an open circuit.

2. Solve the resulting series-parallel circuit for currents, voltage drops, or both. Record the results. Do not solve for any power dissipations.

3. Repeat steps 1 and 2 until the circuit has been solved once for each source.

4. Algebraically superimpose the currents and/or the voltage drops.

5. Calculate any dissipations that are required.

6. Verify the results.

---

**EXAMPLE 9-14**

Apply the superposition theorem to the circuit shown in Fig. 9-26*a*.

**Solution**

The circuit is shown in Fig. 9-26*b* with the 13-V source removed. Note that the source is replaced with a short circuit. The 20-Ω resistor must *not* be removed even if it is the internal resistance of the source. The circuit becomes a series-parallel network, which can be reduced to a single resistor:

$$10\ \Omega \parallel 20\ \Omega = 6.67\ \Omega$$
$$6.67\ \Omega + 100\ \Omega = 107\ \Omega$$

The current is found next:

$$I = \frac{15\ \text{V}}{107\ \Omega} = 141\ \text{mA}$$

Applying the current divider rule shows that the 141 mA splits into 93.8 mA in the 10-Ω branch and 46.9 mA in the 20-Ω branch. These currents are shown in Fig. 9-26*b*. Figure 9-26*c* shows the second independent solution with the 15-V source removed. The currents are determined in the same way as for the first independent solution. Comparing the two solutions shows that both produce a current flowing up in the 10-Ω branch. Since both currents are in the same direction, they are added:

$$I = 93.8\ \text{mA} + 406\ \text{mA} = 500\ \text{mA}$$

The flow in the 15-V battery is found by subtraction, since the independent directions are opposite. The direction is set by the 141-mA current because it is the larger of the two:

$$I = 141\ \text{mA} - 40.6\ \text{mA} = 100\ \text{mA}$$

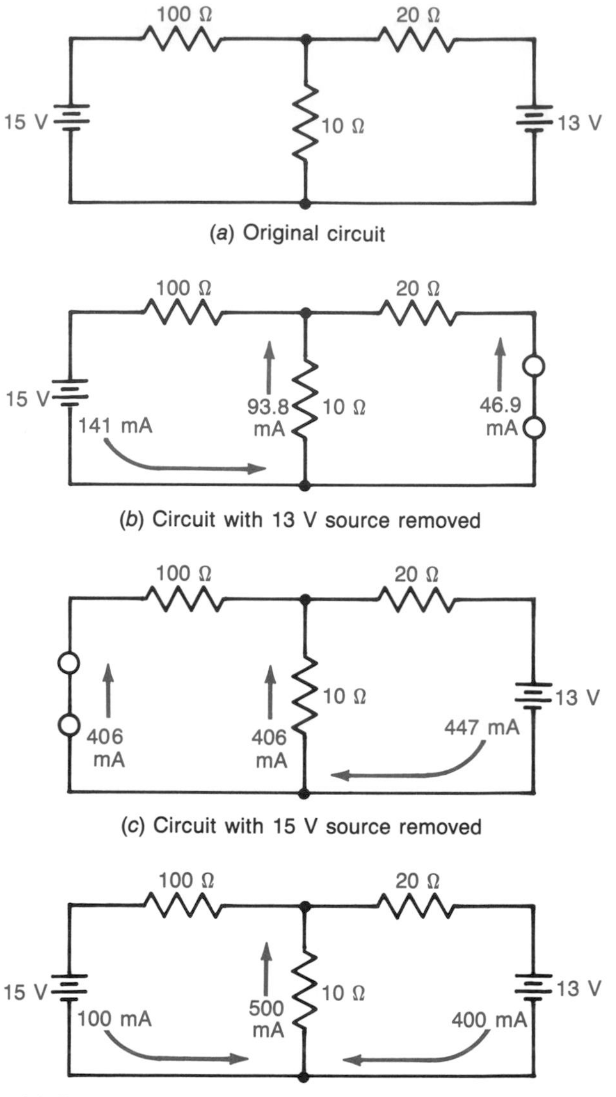


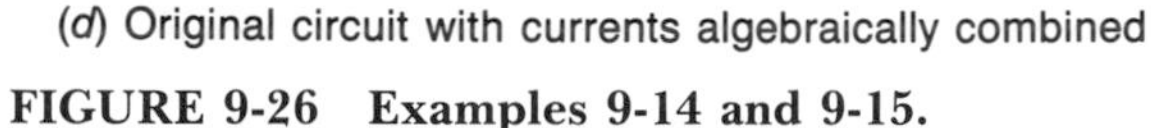
(*d*) Original circuit with currents algebraically combined

**FIGURE 9-26 Examples 9-14 and 9-15.**

The flow in the 13-V battery is also found by subtraction:

$$I = 447\ \text{mA} - 46.9\ \text{mA} = 400\ \text{mA}$$

Figure 9-26*d* shows the original circuit with the superimposed current values. Kirchhoff's current law is satisfied:

$$400\ \text{mA} + 100\ \text{mA} - 500\ \text{mA} = 0$$

Ohm's law can be applied to find the drops, and Kirchhoff's voltage law can be used to verify the loops. This circuit was originally solved in Example 9-8 using branch analysis. It was solved again in Example 9-9 using mesh analysis. You are encouraged to compare the three techniques.

**EXAMPLE 9-15**

Find the power dissipation in the 10-Ω resistor for Fig. 9-26.

**Solution**

The power dissipation may be found *after* the independent solutions have been algebraically combined. The current in the 10-Ω resistor is 500 mA:

$$P = (500 \text{ mA})^2 \times 10 \ \Omega = 2.5 \text{ W} \qquad \text{(Correct)}$$

An attempt to calculate and combine dissipations from the independent solutions will not work:

$$P_1 = (93.8 \text{ mA})^2 \times 10 \ \Omega = 88 \text{ mW}$$
$$P_2 = (406 \text{ mA})^2 \times 10 \ \Omega = 1.65 \text{ W}$$
$$P_T = 88 \text{ mW} + 1.65 \text{ W} = 1.74 \text{ W} \qquad \text{(incorrect)}$$

The attempt fails because power varies as the *square* of the voltage or the current. This relationship is *nonlinear*.

**EXAMPLE 9-16**

Apply the superposition theorem to the circuit shown in Fig. 9-27*a*.

**Solution**

Figure 9-27*b* shows the circuit with the right-hand current source removed. Note that the current source is replaced with an open circuit. The 0.1-S conductance must *not* be removed, even if it is the internal conductance of the current source. The 0.01- and 0.1-S conductances are in series. They can be combined with the reciprocal equation to a single conductance of 9.09 mS. This appears in parallel with the 0.02-S (20-mS) conductance. The current divider rule for conductances gives

$$I = \frac{2 \text{ A} \times 20 \text{ mS}}{20 \text{ mS} + 9.09 \text{ mS}} = 1.375 \text{ A}$$

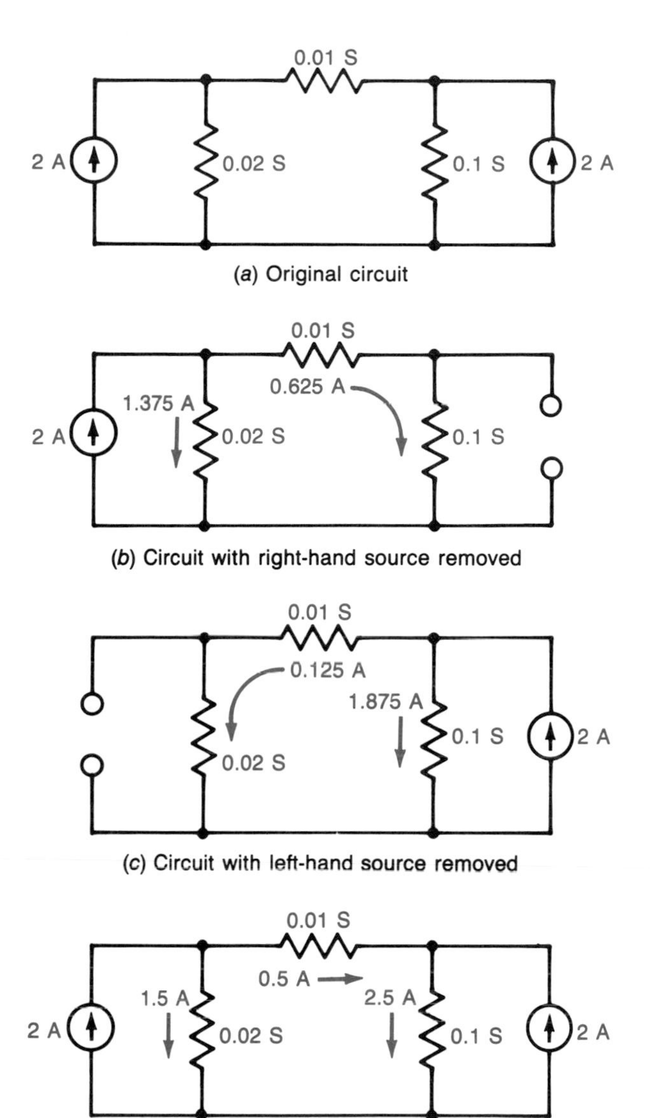


**FIGURE 9-27 Example 9-16.**

The current is 1.375 A in the 0.02-S conductance and 0.625 A in the other branch. Figure 9-27*c* shows the circuit with the left-hand source removed. The 20- and 10-mS conductances are in series for a total conductance of 6.67 mS. The current divider rule gives

$$I = \frac{2 \text{ A} \times 6.67 \text{ mS}}{100 \text{ mS} + 6.67 \text{ mS}} = 0.125 \text{ A}$$

The current in the 100-mS branch is 1.875 A. Figure 9-27*d* shows the original circuit with the currents algebraically combined. The current in the 0.02-S branch is found by adding:

$$I = 1.375 \text{ A} + 0.125 \text{ A} = 1.5 \text{ A}$$

The current in the 0.1-S branch is also found by adding:

$$I = 0.625 \text{ A} + 1.875 \text{ A} = 2.5 \text{ A}$$

The current in the 0.01-S branch is found by subtracting because the independent solutions produced different directions. The direction is set by the larger current:

$$I = 0.625 \text{ A} - 0.125 \text{ A} = 0.5 \text{ A}$$

An inspection of Fig. 9-27*d* reveals that Kirchhoff's current law is satisfied. Ohm's law will find the drops, and Kirchhoff's voltage law must also be satisfied. This circuit was originally solved in Example 9-11 using nodal analysis. You are encouraged to compare the two procedures.

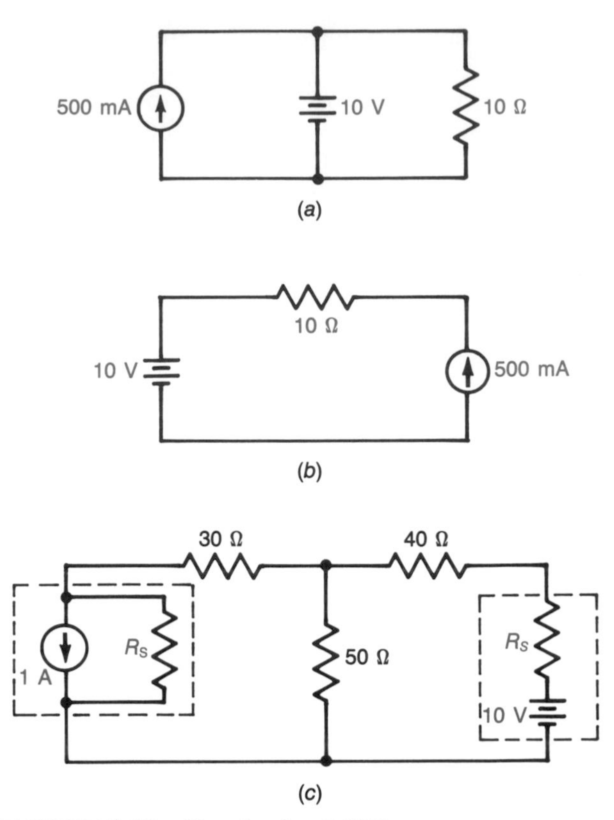


**FIGURE 9-28 Circuits for Self-Test.**

## Self-Test

**45.** Use superposition to find the voltage drop across the 10-Ω resistor in Fig. 9-28*a*.

**46.** What conclusion can you draw concerning your results for Question 45?

**47.** Use superposition to find the current in the 10-Ω resistor for Fig. 9-28*b*.

**48.** What conclusion can you draw concerning your results for Question 47?

**49.** Refer to Fig. 9-28*c*. The internal resistance $R_S$ of the current source is 100 Ω. The internal resistance $R_S$ of the voltage source is 10 Ω. Use superposition to find the power dissipation in the 50-Ω resistor.

## 9-7 THEVENIN'S THEOREM

*Thevenin's theorem* can be used to solve multiple-source and other complex networks. Its best application, however, is to networks where the current, voltage drop, or power dissipation of a single load is of prime concern. It allows a portion of a network to be replaced by a simplified equivalent circuit:

Thevenin's theorem: A two-terminal network can be replaced by an equivalent circuit containing a voltage source and a series resistor.

The procedure for applying this theorem is as follows:

**1.** Remove the load and label the load terminals.

**2.** Find the Thevenin voltage $V_{TH}$ that appears across the load terminals.

**3.** Remove all sources. Ideal voltage sources are replaced with short circuits, and ideal current sources are replaced with open circuits. Any non-ideal source is replaced with a resistor equal to its internal resistance.

**4.** Find the Thevenin resistance $R_{TH}$ looking into the load terminals.

**5.** Draw the Thevenin equivalent circuit with $V_{TH}$, $R_{TH}$, and the load in series.

**6.** Solve the equivalent circuit for the load voltage, the load current, and the load dissipation if required.

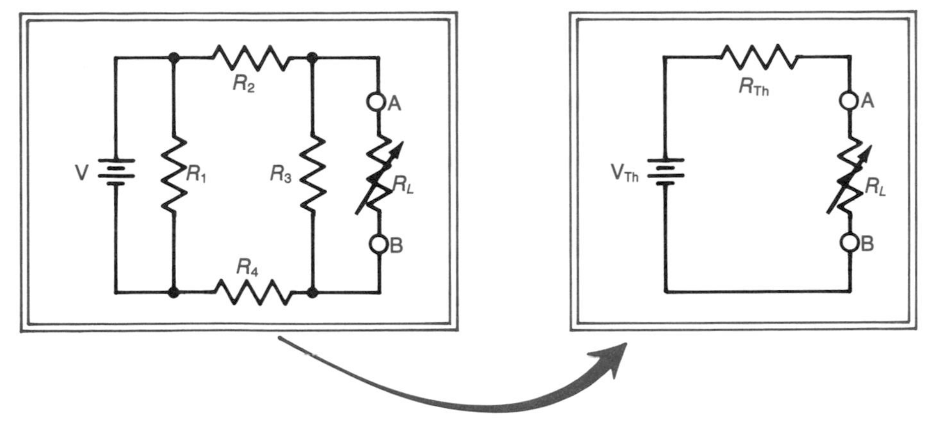


**FIGURE 9-29 A two-terminal network replaced by a Thevenin equivalent circuit.**

Figure 9-29 shows the general method of this theorem. The circuit at the left can be viewed as a two-terminal network. Any circuit can be viewed in this way by considering one component as the load. The terminals of interest are the ones connecting the load to the rest of the circuit. The circuit at the right is the Thevenin equivalent circuit. This circuit is very simple and easy to analyze. Even if the original circuit is readily solvable by other techniques, it may be faster to apply the Thevenin theorem to the circuit. If $R_L$ takes on various values, then the theorem is especially attractive because it is relatively quick and easy to find the load conditions for various values of load resistance using the simplified equivalent circuit.

### EXAMPLE 9-17

Assume that source $V$ in Fig. 9-29 develops 12 V and that all resistors are 10 Ω. Find the Thevenin equivalent circuit and use it to determine the load voltage.

**Solution**

Figure 9-30 shows the application of the theorem in stages. The Thevenin voltage is found in Fig. 9-30*a* by removing the load. The voltage $V_{TH}$ is the one that appears across the load terminals. The 10-Ω resistor in parallel with the battery can be ignored. It will not affect the terminal voltage. The other 10-Ω resistors form a divider, and one-third of the 12-V battery potential appears across the load terminals:

$$V_{TH} = \frac{12\text{ V}}{3} = 4\text{ V}$$

$R_{TH}$ is found next. This procedure is shown in Fig. 9-30*b*. The voltage source is removed and replaced with a short circuit. The 10-Ω resistor in parallel with the short can be ignored. The short connects the top and bottom resistors in series. They appear

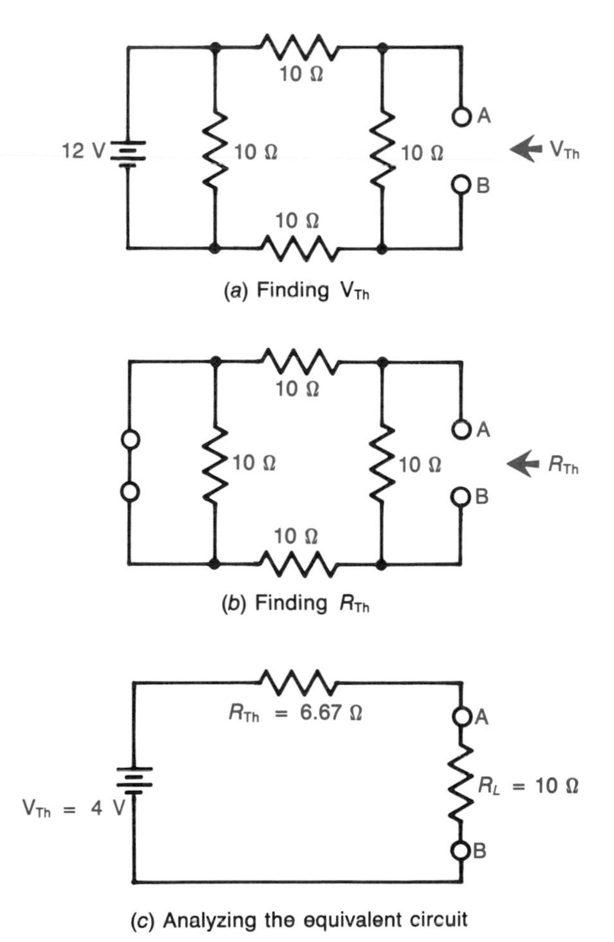


(c) Analyzing the equivalent circuit

**FIGURE 9-30 Example 9-17.**

as a 20-Ω resistance in parallel with the right-hand 10-Ω resistor:

$$R_{TH} = 20\ \Omega \parallel 10\ \Omega = 6.67\ \Omega$$

The equivalent circuit is shown in Fig. 9-30*c*. It is a simple matter to analyze this circuit. The load voltage can be found with the voltage divider rule:

$$V_L = \frac{4\text{ V} \times 10\ \Omega}{10\ \Omega + 6.67\ \Omega} = 2.4\text{ V}$$

The reader is encouraged to solve the original circuit using a different technique to verify the load voltage and to appreciate the relative ease of analyzing the equivalent circuit. Consider the value of this approach in those cases where it is necessary to solve a circuit for many different loads.

**EXAMPLE 9-18**

Find the load current in the bridge circuit of Fig. 9-31 using mesh analysis.

**Solution**

First, we should determine if the bridge is balanced. If it is, the voltage will be zero across the load terminals and the load current will also be zero:

$$\frac{10}{20} \neq \frac{30}{40}$$

The bridge is not balanced, so we will proceed with mesh analysis. Three mesh equations are required,

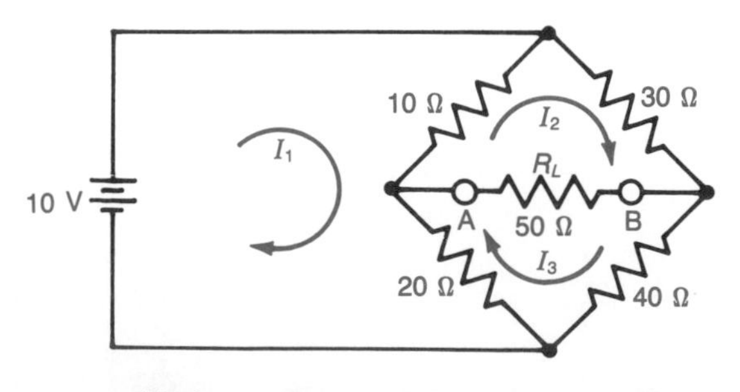


Mesh 1: $30 I_1 - 10 I_2 - 20 I_3 = -10$
Mesh 2: $-10 I_1 + 90 I_2 - 50 I_3 = 0$
Mesh 3: $-20 I_1 - 50 I_2 + 110 I_3 = 0$

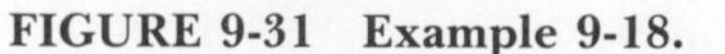
**FIGURE 9-31 Example 9-18.**

one for each window. Figure 9-31 shows the equations and solving them gives

$$I_1 = -0.4774\text{ A}$$
$$I_2 = -0.1355\text{ A}$$
$$I_3 = -0.1484\text{ A}$$

Four significant digits are maintained in this example to better demonstrate the equivalency of the Thevenin approach which follows in the next example. All the currents are negative, so the actual directions are counterclockwise. The direction of $I_2$ is from terminal A to B and the direction of $I_3$ is from terminal B to A. Current $I_3$ is larger, so the load current is from B to A and is equal to the difference between the two mesh currents:

$$I_L = 0.1484\text{ A} - 0.1355\text{ A} = 12.9\text{ mA}$$

The mesh analysis could be continued at this point to find all of the other currents (and voltage drops if required) in the unbalanced bridge circuit. This is not the case when Thevenin's theorem is applied.

**EXAMPLE 9-19**

Use Thevenin's theorem to find the load current in the unbalanced bridge circuit shown in Fig. 9-31.

**Solution**

Figure 9-32 shows the steps in applying the theorem. Voltage $V_{TH}$ is found first by removing the load. The 10- and 20-Ω resistors form a voltage divider that establishes the voltage at terminal A. The 30- and 40-Ω resistors also divide the supply and set the voltage at terminal B. The voltage divider equation gives the drops across the resistors in the bottom of the bridge:

$$V_A = \frac{10\text{ V} \times 20\ \Omega}{10\ \Omega + 20\ \Omega} = 6.667\text{ V}$$

$$V_B = \frac{10\text{ V} \times 40\ \Omega}{30\ \Omega + 40\ \Omega} = 5.714\text{ V}$$

Terminal A is positive with respect to B. Voltage $V_{TH}$ is the difference between the two voltages:

$$V_{TH} = 6.667\text{ V} - 5.714\text{ V} = 0.953\text{ V}$$

The next step is to find $R_{TH}$, and this is shown in Fig. 9-32*b*. The voltage source is removed and replaced with a short circuit. This connects all four

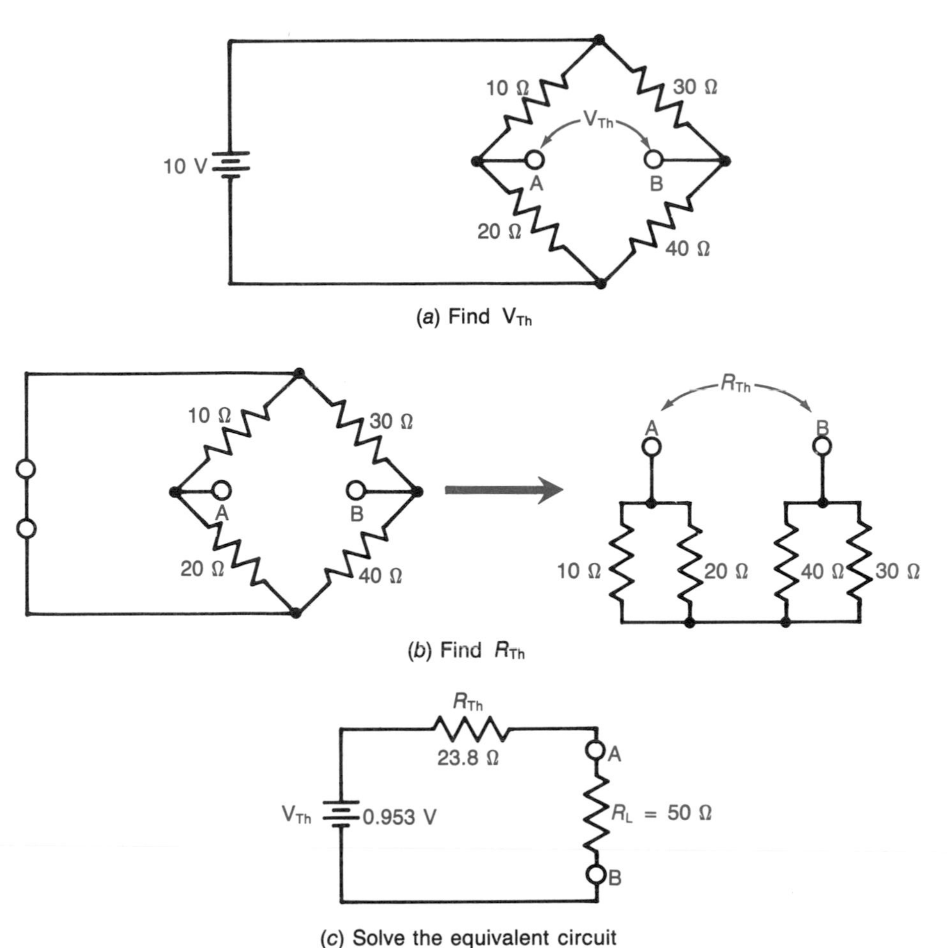


**FIGURE 9-32 Example 9-19.**

resistors to a common point, and the circuit is redrawn to the right to help you visualize the arrangement that results. The 10- and 20-Ω resistors are in parallel. So are the 40- and 30-Ω resistors. These two combinations are formed first:

$$10\ \Omega \parallel 20\ \Omega = 6.667\ \Omega$$
$$40\ \Omega \parallel 30\ \Omega = 17.14\ \Omega$$

Looking into terminals A and B, the combinations are in series:

$$R_{TH} = 6.667\ \Omega + 17.14\ \Omega = 23.81\ \Omega$$

Figure 9-32*c* shows the Thevenin equivalent circuit. It is now a simple matter to find the load current:

$$I_L = \frac{0.953\ \text{V}}{23.81\ \Omega + 50\ \Omega} = 12.9\ \text{mA}$$

This is the same value of current found using mesh analysis. The direction is also the same, from B to A. If $R_L$ changes, it is a simple matter to find the new value of load current using the Thevenin equivalent circuit. With mesh analysis, new equations have to be written and solved.

An unbalanced bridge circuit can be solved for all voltage drops and current flows with two applications of Thevenin's theorem followed by Ohm's and Kirchhoff's laws. Mesh analysis, or some other technique, may be viewed as a more straightforward approach when a complete circuit solution is required. The major advantage of Thevenin's theorem is its ability to reduce a network to a simple equivalent circuit to find the conditions for a single load. In some cases, removal of the load leaves a complex circuit. These cases may be handled by using superposition, nodal analysis, or mesh analysis to find $V_{TH}$ and/or $R_{TH}$.

Thevenin's theorem may also be applied in successive steps. Any two points in a circuit can be chosen, and all of the components to one side of these points can be reduced to an equivalent circuit. Then, the equivalent circuit for the components on the other side may be determined. Complex circuits can be handled in sections with this technique.

## Self-Test

**50.** Find $V_{TH}$ and $R_{TH}$ for the circuit shown in Fig. 9-33*a*.

**51.** Find the load current in Fig. 9-33*a*.

**52.** Find $V_{TH}$ and $R_{TH}$ for Fig. 9-33*b*.

**53.** Calculate the drop across the load resistor in Fig. 9-33*b*.

**54.** Determine $V_{TH}$ and $R_{TH}$ for Fig. 9-33*c*.

**55.** What is the load voltage in Fig. 9-33*c*?

**56.** What is the load polarity in Fig. 9-33*c*?

**57.** Determine the load voltage for Fig. 9-33*c* if $R_L$ is changed to 1 kΩ.

**58.** Determine the load voltage for Fig. 9-33*c* if $R_L$ is changed to 2 kΩ.

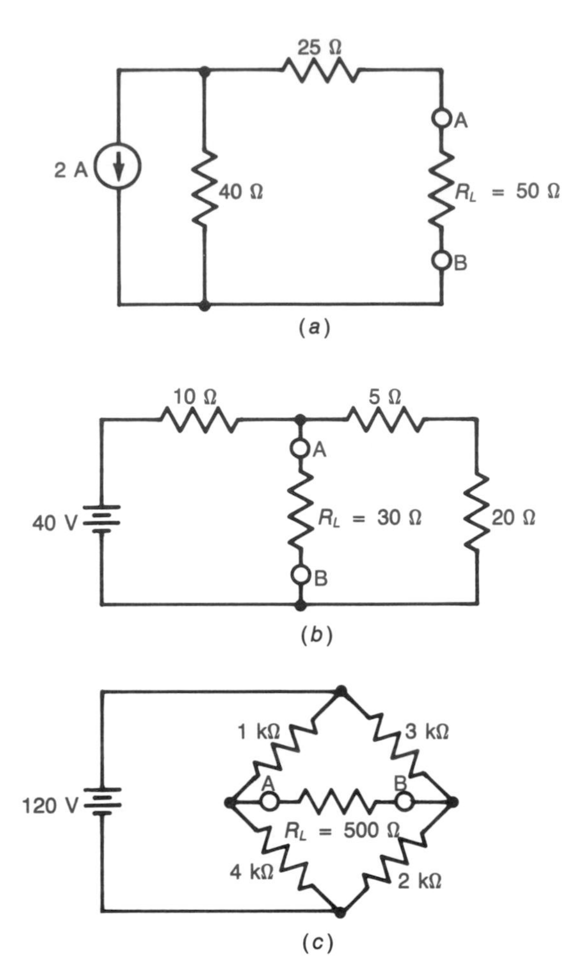


**FIGURE 9-33 Circuits for Self-Test.**

## 9-8 NORTON'S THEOREM

*Norton's theorem* is the dual of Thevenin's theorem. It is used to replace a two-terminal network with a simplified equivalent circuit consisting of a current source and a parallel resistor. Norton's theorem is popular for analyzing transistor circuits. As with the Thevenin equivalent circuit, the Norton equivalent circuit makes it easy to solve for varying load conditions.

> Norton's theorem: A two-terminal network can be replaced by an equivalent circuit containing a current source and a parallel resistor.

The procedure for applying this theorem is as follows:

**1.** Remove the load and label the load terminals.

**2.** Find the Norton current $I_N$ by shorting the load terminals.

**3.** Remove all sources. Ideal voltage sources are replaced with short circuits, and ideal current sources are replaced with open circuits. Any nonideal source is replaced with a resistor equal to its internal resistance.

**4.** Find the Norton resistance $R_N$ looking into the load terminals.

**5.** Draw the Norton equivalent circuit with $I_N$, $R_N$, and the load in parallel.

**6.** Solve the equivalent circuit for the load voltage, the load current, and the load dissipation if required.

### EXAMPLE 9-20

Find the Norton equivalent circuit for Fig. 9-34*a* and use it to solve for the load current and voltage.

**Solution**

Figure 9-34*b* shows the circuit with the load removed and the load terminals shorted. The

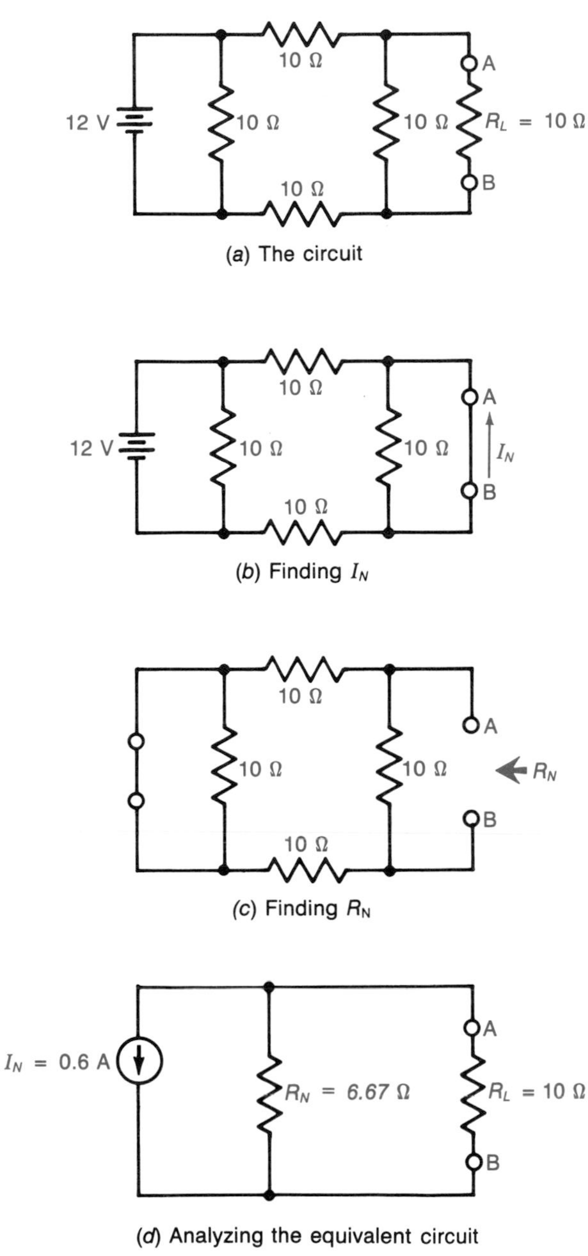


**FIGURE 9-34 Example 9-20.**

Norton current flows in the short. The resistor that is in parallel with the battery has no effect on the Norton current. The resistor that is in parallel with the short also has no effect. The other resistors act in series to limit the current:

$$I_N = \frac{12\text{ V}}{10\ \Omega + 10\ \Omega} = 0.6\text{ A}$$

The Norton resistance is found in Fig. 9-34*c*. The battery is replaced with a short circuit. Two 10-Ω resistors are in series, and this 20-Ω combination is in parallel with 10 Ω:

$$R_N = 20\ \Omega \parallel 10\ \Omega = 6.67\ \Omega$$

Figure 9-34*d* shows the Norton equivalent circuit. The current divider rule can be used to find the load current:

$$I_N = \frac{0.6\text{ A} \times 6.67\ \Omega}{6.67\ \Omega + 10\ \Omega} = 0.24\text{ A}$$

Ohm's law provides the drop across the load resistor:

$$V_{R_L} = 0.24\text{ A} \times 10\ \Omega = 2.4\text{ V}$$

This circuit was solved in Example 9-17 using Thevenin's theorem. The results are the same.

---

**EXAMPLE 9-21**

Verify that the Norton equivalent circuit from Example 9-20 is the dual of the Thevenin equivalent circuit from Example 9-17.

**Solution**

Ohm's law is used to convert a Norton equivalent circuit to a Thevenin equivalent circuit just as a current source is converted to a voltage source:

$$\begin{aligned} V_{TH} &= I_N \times R_N \\ &= 0.6\text{ A} \times 6.67\ \Omega = 4\text{ V} \end{aligned}$$

The dual of the Norton equivalent circuit shown in Fig. 9-34*d* is a Thevenin equivalent circuit consisting of a 4-V battery in series with a 6.67-Ω resistor (see Fig. 9-30*c*).

---

Thevenin's theorem and Norton's theorem are duals. However, one or the other may be easier to use in a given situation. They may also be used in combination with one theorem applied to one part of a network and the other theorem to another part of the network.

---

**EXAMPLE 9-22**

Find the Norton current for the unbalanced bridge circuit shown in Fig. 9-35*a*.

**Solution**

The Norton current is found by shorting the load terminals as shown in Fig. 9-35*b*. This situation is

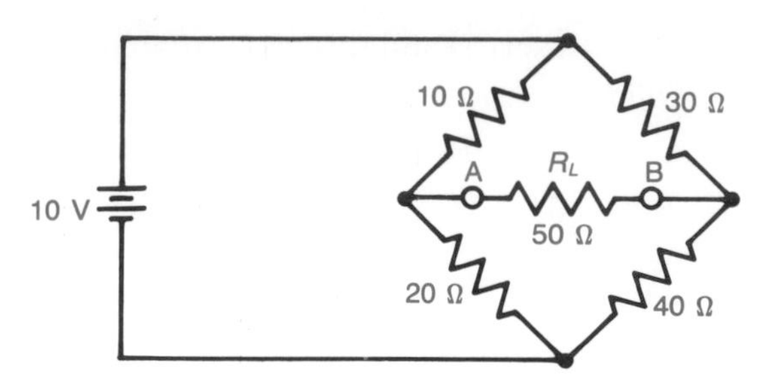


(a) An unbalanced bridge circuit

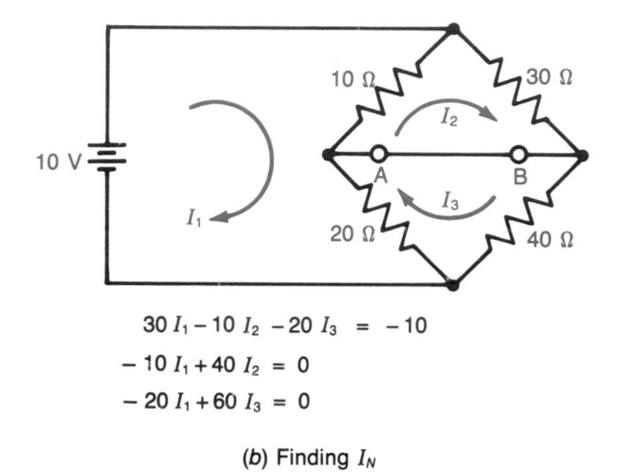


(b) Finding $I_N$

**FIGURE 9-35 Example 9-22.**

more complicated than finding the Thevenin voltage. It would be easier to find the Thevenin voltage and convert it to the Norton current. However, we will proceed to illustrate the point. Mesh analysis is applied to find the current flowing in the short circuit. Solving the three equations gives

$$I_1 = -0.48 \text{ A}$$
$$I_2 = -0.12 \text{ A}$$
$$I_3 = -0.16 \text{ A}$$

The currents have been assigned backward. Current $I_3$ is larger than $I_2$ and flows from B to A. The Norton current also flows from B to A and is equal to the difference between $I_3$ and $I_2$:

$$I_N = 0.16 \text{ A} - 0.12 \text{ A} = 40 \text{ mA}$$

The current in the short circuit of Fig. 9-35*b* can also be found by first determining the total current and then using the current divider equation at the top and bottom nodes of the bridge. Once the currents in the four resistors have been found, Kirchhoff's current law can be used to determine the Norton current. This approach also requires more effort than finding the Thevenin voltage. Referring back to Fig. 9-32*c*, you will find the Thevenin equivalent circuit for the identical bridge circuit. The Thevenin circuit is converted to the Norton circuit with Ohm's law:

$$I_N = \frac{V_{TH}}{R_{TH}} = \frac{0.953 \text{ V}}{23.81\ \Omega} = 40 \text{ mA}$$

When a particular equivalent circuit is needed, it may require less effort to find the dual of that equivalent circuit and convert it.

---

Equivalent circuits may be determined by measurements. Figure 9-36 shows a black box determination of Thevenin and Norton *parameters*. Parameters are the characteristics of a circuit or a device. Current, voltage, and power can be called parameters. The parameters for devices such as transistors can be obtained by measurements. The measurement of $V_{TH}$ is straightforward as shown in Fig. 9-36*a*. An ohmmeter cannot be connected to a circuit containing sources, so $R_{TH}$ and $R_N$ must be

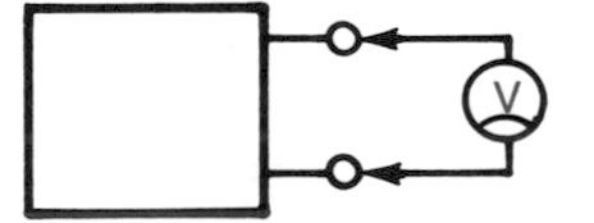


(a) Remove the load and measure $V_{Th}$

(b) Adjust $R_L$ until half the open circuit voltage is indicated.
$R_L = R_{Th} = R_N$

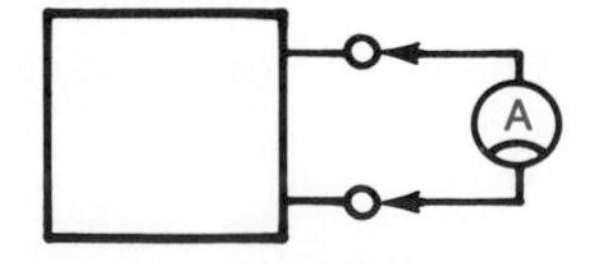


(c) Remove the load and measure $I_N$

**FIGURE 9-36 Black box applications of Thevenin's and Norton's theorems.**

measured indirectly as shown in Fig. 9-36*b*. The variable load is adjusted until the meter indicates half of the open circuit reading. At that point, $R_L = R_{TH} = R_N$ and the load can be removed from the circuit and measured with an ohmmeter. *Current* $I_N$ is measured as shown in Fig. 9-36*c*. The internal resistance of an ammeter is very low and it acts as a short circuit. *Caution:* Depending on the circuit, measurement of $I_N$ may *not* be an acceptable procedure. The circuit, the meter, or both could be damaged.

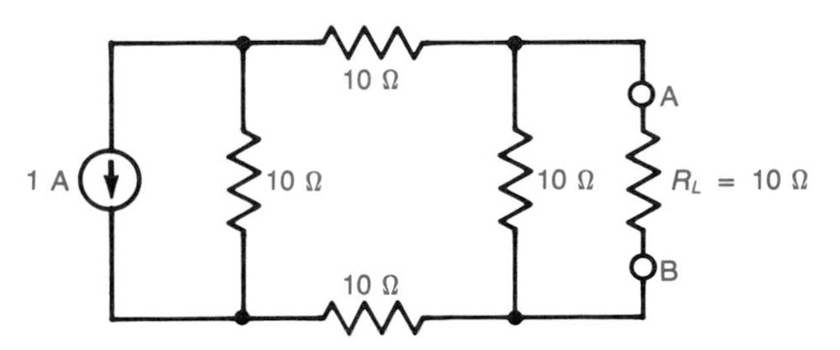


**FIGURE 9-37 Circuit for Self-Test.**

## Self-Test

**59.** Find $I_N$ and $R_N$ for Fig. 9-37.

**60.** Find $I_L$ and $V_L$ for Fig. 9-37.

**61.** Determine $I_L$ for Fig. 9-37 if $R_L = R_N$.

**62.** Is the value of $I_L$ that you found in Question 61 one-half of the Norton current?

**63.** What voltage will appear across terminals A and B in Fig. 9-37 if the load is removed? Is this the Thevenin voltage?

**64.** What voltage will appear across terminals A and B in Fig. 9-37 if the load is equal to the Norton resistance? Is this equal to one-half of the open circuit voltage?

**65.** True or false? When a network is loaded by a resistance equal in value to its Thevenin resistance, the voltage across the load will be equal to one-half of the open circuit voltage.

**66.** True or false? When a network is loaded by a resistance equal in value to its Norton resistance, the current through the load will be equal to one-half of the short circuit current.

## 9-9 MILLMAN'S THEOREM

*Millman's theorem* is a combination of Thevenin's and Norton's theorems. It is used to reduce any number of parallel voltage sources to an equivalent circuit containing only one source. It has the advantage of being easier to apply to some networks than mesh analysis, nodal analysis, or superposition. The procedure is as follows:

1. Identify the load terminals.
2. Convert all voltage sources to current sources.
3. Combine the resistances of the current sources into an equivalent resistance.
4. Combine the current sources into an equivalent current source.
5. Draw an equivalent circuit with the equivalent current source, the equivalent resistance, and the load connected in parallel.
6. Solve the equivalent circuit for the load current, the load voltage, and the load dissipation if required.

### EXAMPLE 9-23

Find the load current for Fig. 9-38*a* using Millman's theorem.

**Solution**

The voltage sources are converted to current sources as shown in Fig. 9-38*b*. The arrow for each current source corresponds to the polarity of each voltage source in the original circuit. The equivalent resistance is found from the parallel combination of the source resistances:

$$R_{eq} = 10\ \Omega \,||\, 5\ \Omega \,||\, 20\ \Omega \,||\, 15\ \Omega = 2.4\ \Omega$$

The equivalent current source is found by algebraically adding the individual sources. A source is considered negative if its direction tends to make terminal A negative with respect to ground:

$$I_{eq} = +1\text{ A} - 4\text{ A} + 0.25\text{ A} + 2\text{ A} = -0.75\text{ A}$$

The negative sign tells us that the equivalent source direction must point away from the ground connection. Figure 9-38*c* shows the equivalent circuit. The current divider equation can be used to find the load current:

$$I_{R_L} = \frac{0.75\text{ A} \times 2.4\ \Omega}{2.4\ \Omega + 5\ \Omega} = 0.243\text{ A}$$

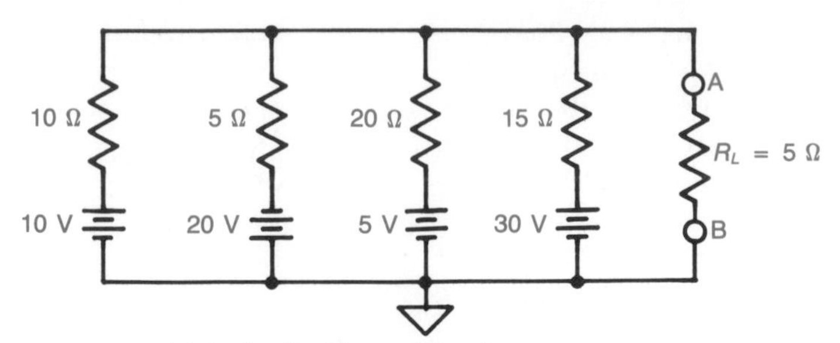


(*a*) A circuit with parallel voltage sources

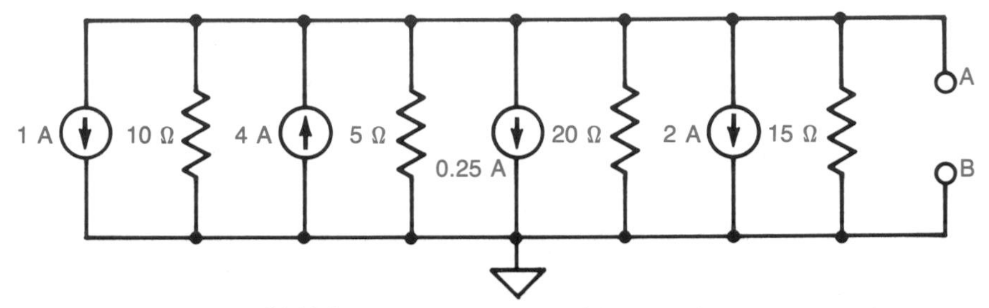


(*b*) Voltage sources converted to current sources

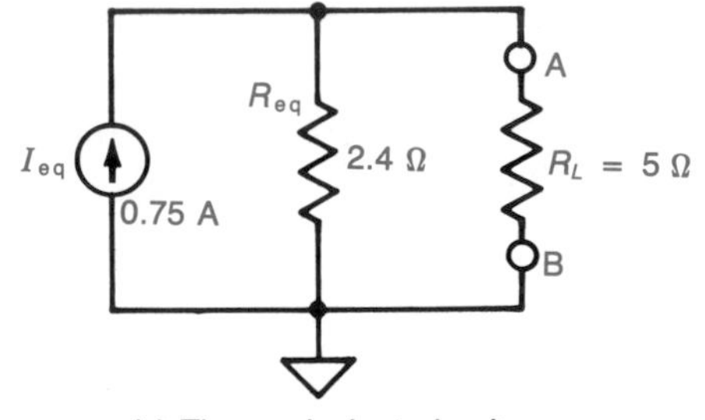


(*c*) The equivalent circuit

**FIGURE 9-38 Example 9-23.**

Millman's theorem can be used to develop a *general* equation for finding an equivalent *voltage* for a network with voltage sources in parallel. First, the equation for the equivalent current is

$$I_{eq} = \pm\frac{V_1}{R_1} \pm \frac{V_2}{R_2} \pm \frac{V_3}{R_3} \pm \cdots \pm \frac{V_N}{R_N}$$

And the equation for the equivalent resistance is

$$R_{eq} = \frac{1}{1/R_1 + 1/R_2 + 1/R_3 + \cdots + 1/R_N}$$

By Ohm's law

$$V_{eq} = I_{eq} \times R_{eq}$$
$$= \frac{\pm V_1/R_1 \pm V_2/R_2 \pm V_3/R_3 \pm \cdots \pm V_N/R_N}{1/R_1 + 1/R_2 + 1/R_3 + \cdots + 1/R_N}$$

## EXAMPLE 9-24

Find an equivalent voltage source for the circuit shown in Fig. 9-38*a* using the general equation.

### Solution

Entering the circuit values into the equation gives

$$V_{eq} = \frac{+10/10 - 20/5 + 5/20 + 30/15}{1/10 + 1/5 + 1/20 + 1/15} = \frac{-0.75}{0.417} = -1.8\text{ V}$$

The equivalent resistance is equal to the parallel combination of the four source resistances. It will be equal to the reciprocal of the denominator of the general equation, or 2.4 Ω in this example. Voltage $V_{eq}$ can be verified by converting the current source shown in Fig. 9-38*c* to a voltage source:

$$V_{eq} = I_{eq} \times R_{eq} = -0.75 \text{ A} \times 2.4 \ \Omega = -1.8 \text{ V}$$

The current is entered as a negative quantity because its direction makes terminal A negative with respect to ground.

---

There is a dual for Millman's theorem, and it is useful for solving circuits with series current sources. The following general equations are useful for finding the current and resistance of the equivalent circuit:

$$I_{eq} = \frac{\pm I_1R_1 \pm I_2R_2 \pm I_3R_3 \pm \cdots \pm I_NR_N}{R_1 + R_2 + R_3 + \cdots + R_N}$$

$$R_{eq} = R_1 + R_2 + R_3 + \cdots + R_N$$

---

**EXAMPLE 9-25**

Find the load current for Fig. 9-39*a* using the dual of Millman's theorem.

**Solution**

Placing the data into the general equation gives

$$I_{eq} = \frac{-0.1(100) + 0.5(150) - 1(50)}{100 + 150 + 50} = \frac{15}{300} = +50 \text{ mA}$$

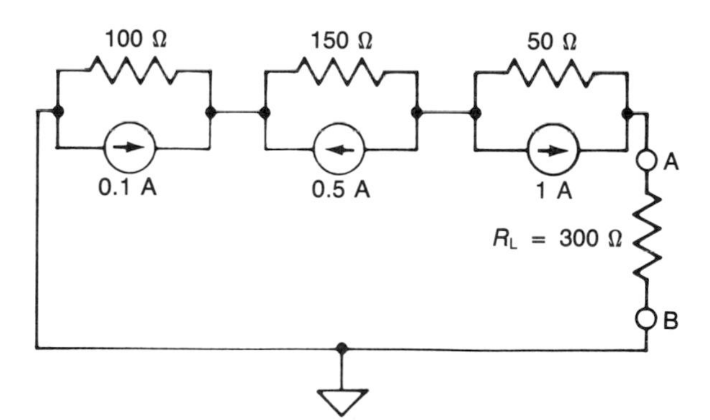


(*a*) A circuit with series current sources

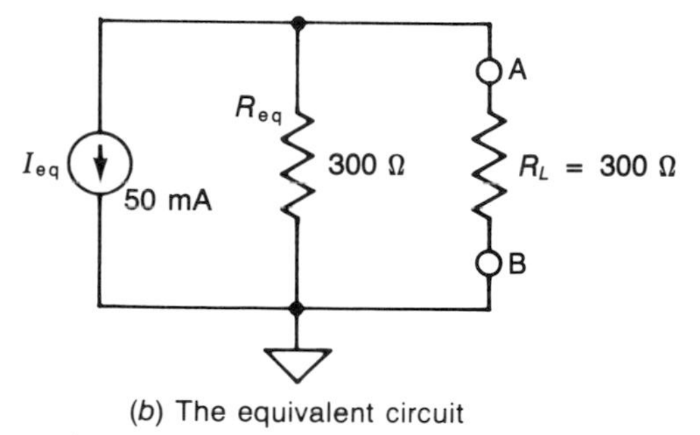


(*b*) The equivalent circuit

**FIGURE 9-39 Example 9-25.**

The equivalent circuit is shown in Fig. 9-39*b*. The current direction makes terminal A positive with respect to ground. The load resistance in this case is equal to the equivalent resistance so the load current is half of the equivalent current:

$$I_L = \frac{50 \text{ mA}}{2} = 25 \text{ mA}$$

---

Ideal voltage sources of different values may not be placed in parallel. Ideal current sources of different values may not be placed in series. Figures 9-38 and 9-39 do not violate these rules. The voltage sources have series resistors and are effectively not ideal. The current sources have parallel resistors and are effectively not ideal.

## Self-Test

**67.** Write, but do not solve, the mesh equations for the circuit shown in Fig. 9-38*a*. Begin at the left window with mesh 1.

**68.** Find $V_{eq}$ and $R_{eq}$ for Fig. 9-40*a*.

**69.** Find $I_{eq}$ for Fig. 9-40*a*.

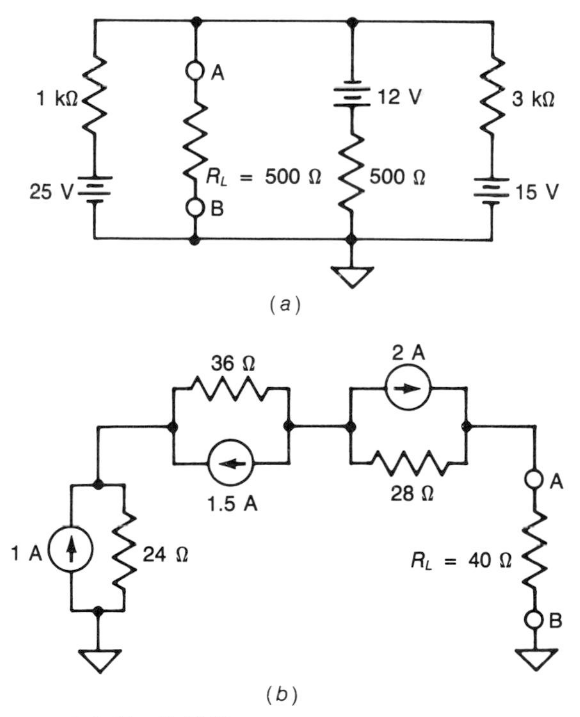

(*b*)

**FIGURE 9-40 Self-Test.**

**70.** What is the current flowing in the load resistor in Fig. 9-40*a*?

**71.** Find $I_{eq}$ and $R_{eq}$ for Fig. 9-40*b*.

**72.** Find $V_{eq}$ for Fig. 9-40*b*.

**73.** What is the drop and polarity of the load in Fig. 9-40*b*?

## 9-10 POWER TRANSFER AND NETWORK TRANSFORMS

Thevenin, Norton, and Millman equivalent circuits can be used to simplify circuit analysis. They can also be used to predict how much power a network can deliver and what value of load resistance will produce maximum power transfer. Maximum power transfer occurs when the load resistance is equal to the equivalent resistance of a network.

**EXAMPLE 9-26**

Determine the maximum power that can be delivered to the load by the circuit shown in Fig. 9-41*a* and the value of load resistance required to achieve this maximum.

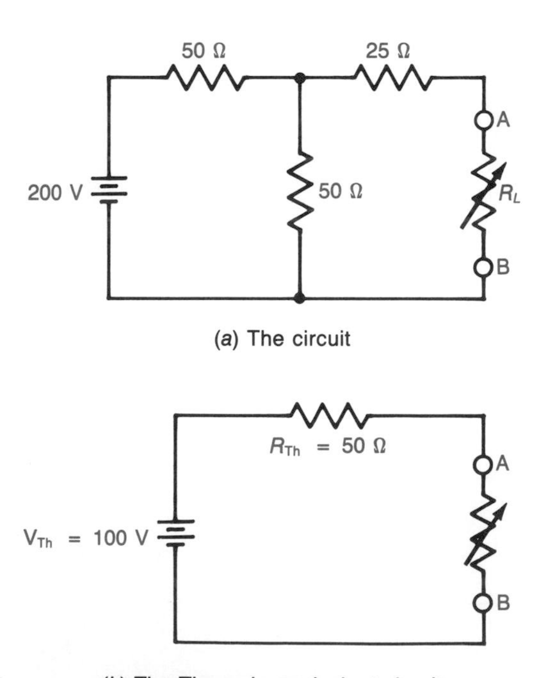


**FIGURE 9-41 Example 9-26.**

**Solution**

Maximum power transfer occurs when $R_L = R_{eq}$. Thevenin's theorem can be used to determine the equivalent network. When $R_L$ is removed from terminals A and B, the two 50-Ω resistors give an open circuit voltage equal to one-half of the battery potential:

$$V_{TH} = \frac{200 \text{ V}}{2} = 100 \text{ V}$$

When the battery is replaced with a short circuit, the 50-Ω resistors are in parallel and this combination appears in series with the 25-Ω resistor:

$$R_{TH} = 50\ \Omega \,||\, 50\ \Omega + 25\ \Omega = 50\ \Omega$$

The Thevenin equivalent circuit is shown in Fig. 9-41*b*. Maximum power transfer occurs when $R_L = 50\ \Omega$. The power delivered to the load can be found by solving the equivalent circuit. With $R_L = R_{TH}$, the voltage drop across the load is equal to half of the Thevenin voltage and the power dissipated in the load is

$$P_{max} = \frac{(V_{TH}/2)^2}{R_{TH}}$$

$$= \frac{(100/2)^2}{50} = 50 \text{ W}$$

The network can deliver a maximum power of 50 W, and this occurs when the load is 50 Ω.

Figure 9-42 shows a graph of load power, load current, and load voltage for the circuit of Fig. 9-41. The horizontal axis ranges from 1 Ω to 10 kΩ. The power curve peaks at 50 W, which occurs when $R_L = 50\ \Omega$. The load current and load voltage are at one-half of their maximum values when the power is at its peak.

The graph in Fig. 9-42 is *semilogarithmic*. The vertical axis is linear and the horizontal axis is logarithmic. Each log cycle represents a 10 times increase in resistance. The first log cycle is from 1 to 10 Ω, the second from 10 to 100 Ω, and so on. The advantage of logarithmic graphing is that a broad range of values can be resolved in a reasonable space. Both semilog and log-log graphs are used in electronics.

Figure 9-43 shows another graph for the circuit of Fig. 9-41. This one plots *efficiency* versus load resistance. Note that the efficiency curve does not

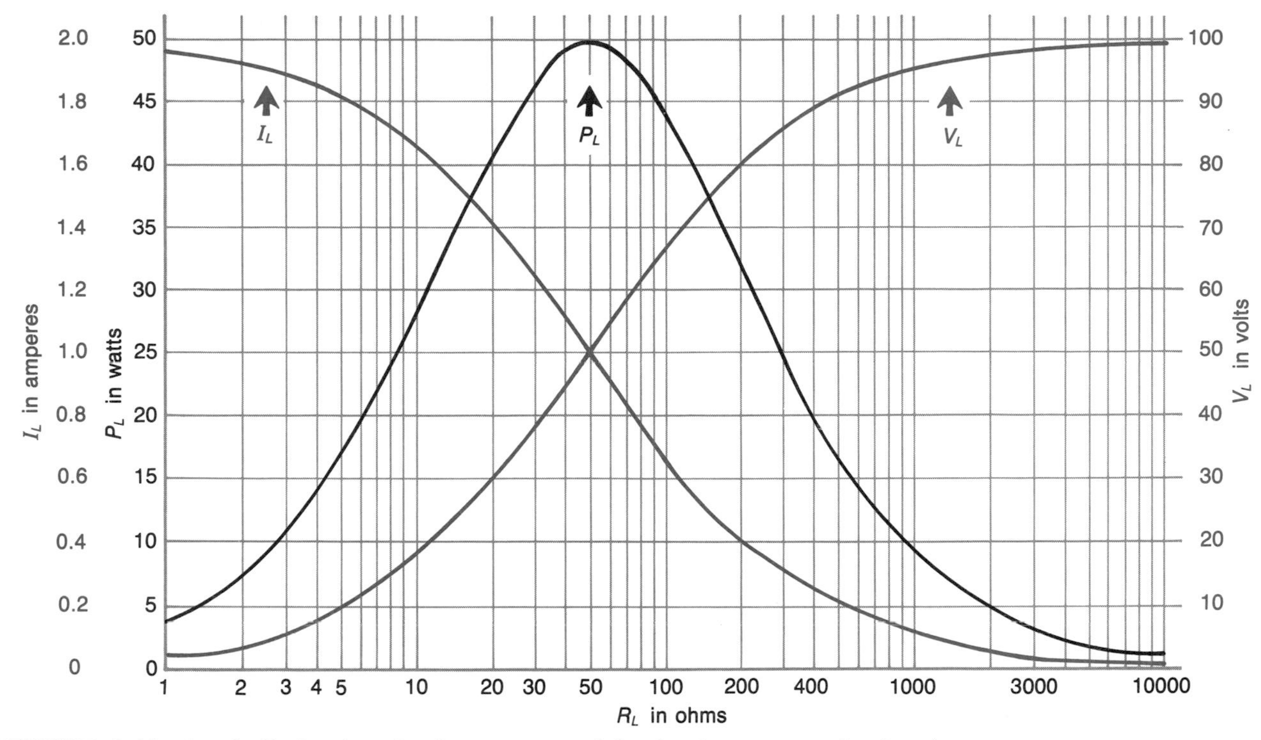

**FIGURE 9-42** **Load dissipation, load current, and load voltage versus load resistance.**

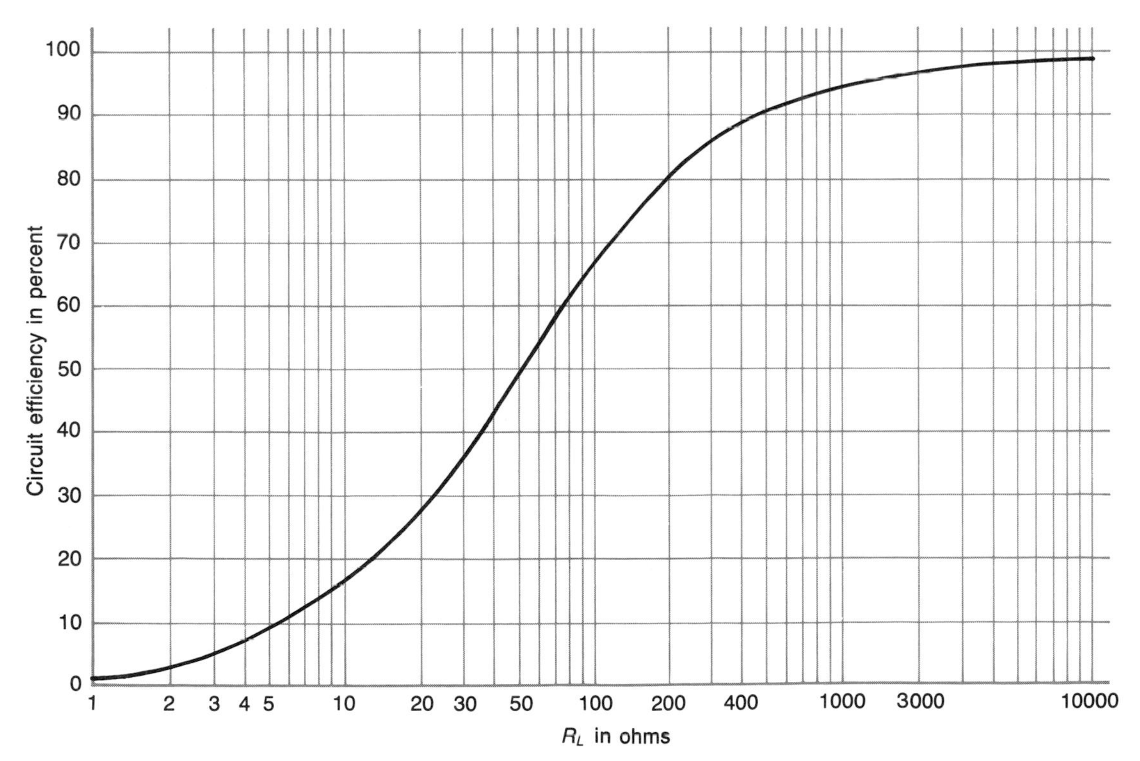

**FIGURE 9-43** **Efficiency versus load resistance.**

peak as the power curve did. The efficiency of a circuit is the ratio of useful (load) power to total power. When the load resistance is equal to the equivalent resistance of the network, one-half of the power supplied by the battery is consumed in the equivalent resistance and the other half is delivered to the load. This gives an efficiency of 50 percent. The graph shows that efficiency improves as $R_L$ increases in value.

### EXAMPLE 9-27

Determine the maximum power that can be delivered by the circuit shown in Fig. 9-44 and the value of load resistance required to achieve this maximum. What is the circuit efficiency when $P_L$ is at maximum?

#### Solution

Figure 9-44*b* shows the Norton equivalent circuit. Maximum power transfer occurs when $R_L = R_N$ and the load current is equal to one-half of the Norton current for a power dissipation of

$$P_{max} = \left(\frac{I_N}{2}\right)^2 \times R_N$$

$$= \left(\frac{0.5}{2}\right)^2 \times 300 = 18.8 \text{ W}$$

(*a*) The circuit

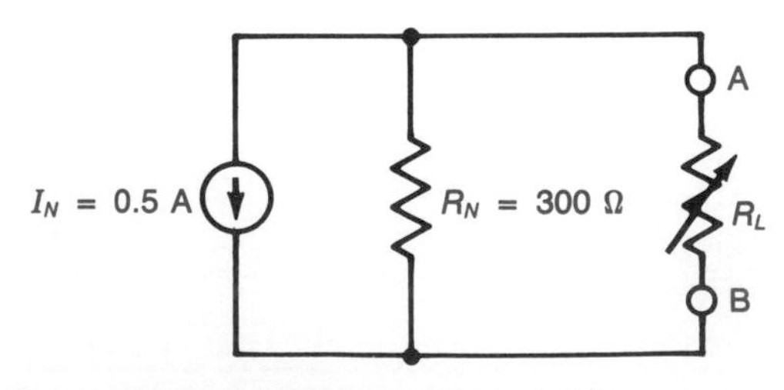


(*b*) The Norton equivalent circuit

**FIGURE 9-44 Example 9-27.**

Half of the power supplied by the current source is delivered to the load for an efficiency of 50 percent.

## Network Transforms

We have seen how theorems are used to reduce networks to a simplified equivalent form. *Transforms* are used to convert a network, or a part of one, from one form to another. Figure 9-45 shows two popular circuit arrangements. The *delta* arrangement looks like the Greek capital letter Δ, and the *wye* arrangement looks like the capital letter Y. Every delta network has an equivalent wye network, and every wye network has an equivalent delta network. Transforming a delta or wye network does not simplify it by reducing it to a single equivalent resistance. However, it is sometimes possible to simplify a circuit by transforming a portion of it.

A delta network can be converted to an equivalent wye network with the following equations. The subscripts are referenced to Fig. 9-45:

$$R_1 = \frac{R_Y R_Z}{R_X + R_Y + R_Z}$$

$$R_2 = \frac{R_X R_Z}{R_X + R_Y + R_Z}$$

$$R_3 = \frac{R_X R_Y}{R_X + R_Y + R_Z}$$

A wye network is converted to an equivalent delta network with the following equations:

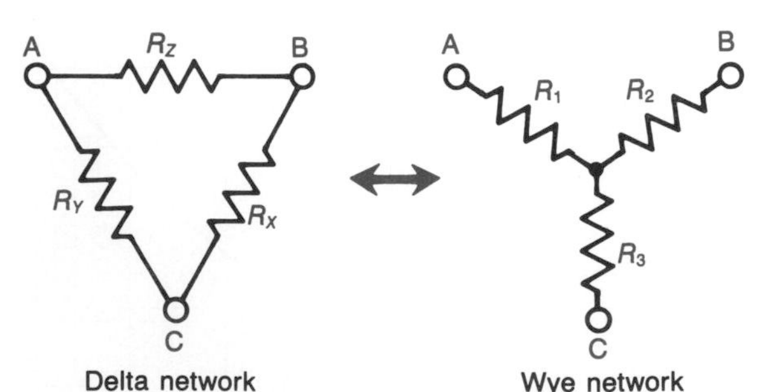


**FIGURE 9-45 A three-terminal network transformed to an equivalent three-terminal network.**

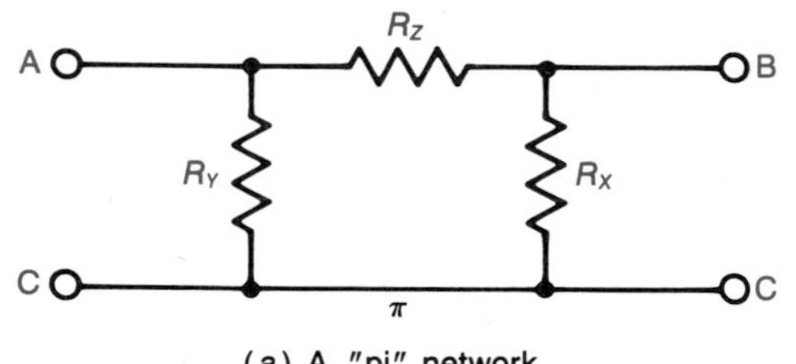


(a) A "pi" network

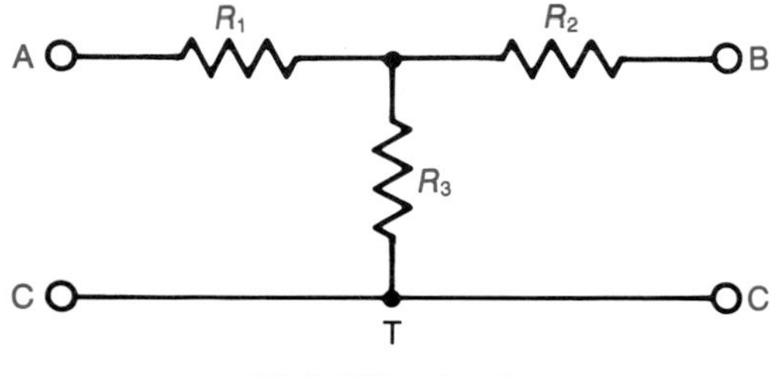


(b) A "T" network

**FIGURE 9-50 Networks are usually given different names in electronic applications.**

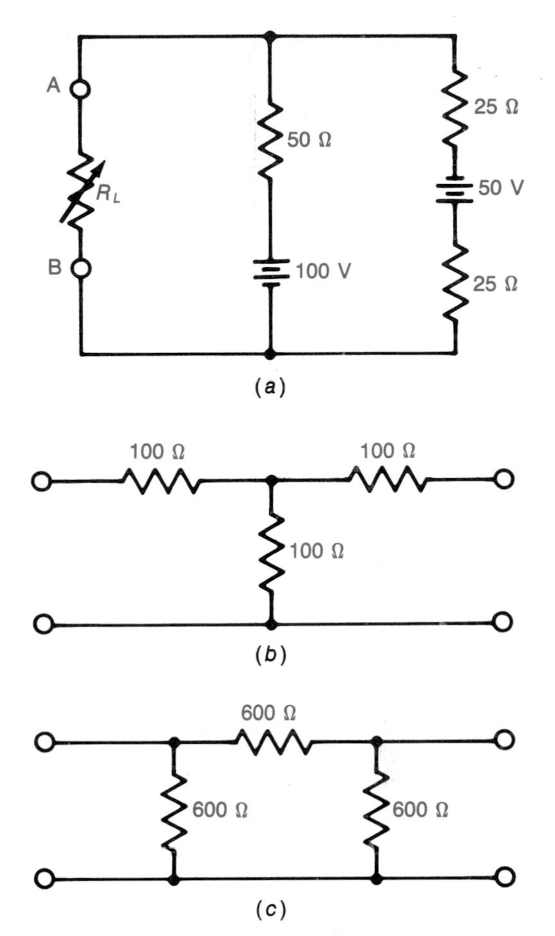


**FIGURE 9-51 Circuits for Self-Test.**

network is usually drawn in the form of a "T" network as shown in Fig. 9-50*b*. The differences are in name only, which can be easily verified by comparing Fig. 9-50 to Fig. 9-45. In electronics, circuits are arranged to show input terminals (usually on the left) and output terminals (usually on the right). There are also terminals that are common to both input and output. The circuits of Fig. 9-50 follow this format. The delta-to-wye transforms can be used to convert pi networks to T networks, and the wye-to-delta transforms can be used to convert T networks to pi networks.

## Self-Test

**74.** A voltage source has an internal series resistance of 75 Ω. What value of load resistor will produce maximum load dissipation?

**75.** A current source has an internal shunt resistance of 1 kΩ. What value of load resistor will produce maximum load dissipation?

**76.** A 120-V source has a series internal resistance of 1 Ω. What is the maximum power that it can deliver to a load?

**77.** A 5-A source has a shunt internal resistance of 100 Ω. What is the maximum power that it can deliver to a load?

**78.** When any source is delivering maximum power to a load, the overall circuit efficiency is ______ percent.

**79.** Refer to Fig. 9-51*a*. What is the maximum power that can be delivered to $R_L$?

**80.** Find the equivalent delta resistances for the circuit shown in Fig. 9-51*b*.

**81.** The circuit shown in Fig. 9-51*b* is called a ______ network.

**82.** Find the equivalent wye resistance for the circuit shown in Fig. 9-51*c*.

**83.** The circuit shown in Fig. 9-51*c* is called a ______ network.

## SUMMARY

1. An ideal current source supplies a constant load current that is independent of the load resistance just as an ideal voltage source supplies a constant load voltage that is independent of the load resistance.
2. A real current source has some value of shunt resistance. The higher this resistance is, the more closely the source approaches being ideal.
3. Every real source has a dual. The conversion between voltage and current sources is based on Ohm's law.
4. Ideal current sources of different values may *not* be placed in series.
5. Ideal current sources may be connected in parallel aiding or parallel opposing.
6. Simultaneous equations may be solved using determinants.
7. A determinant is the numerical value of a square arrangement of numbers called a matrix.
8. Equations with two unknowns can be solved with second-order determinants, and those with three unknowns can be solved with third-order determinants.
9. Branch analysis is based on Kirchhoff's voltage law and the actual circuit currents. Its best application is as a stepping stone to mesh analysis.
10. Mesh analysis is also based on Kirchhoff's voltage law and is considered superior to branch analysis because it requires fewer simultaneous equations and is less error prone.
11. Nodal analysis is based on Kirchhoff's current law. It is a convenient technique where node voltages must be found in reference to a common point such as ground.
12. The application of nodal analysis requires that voltage sources be converted to current sources.
13. The superposition theorem is applicable to multiple source networks. It treats each source independently and combines the independent solutions algebraically for the actual voltages and/or currents.
14. Superposition offers the advantage of not requiring simultaneous equations.
15. In the application of superposition, all sources but one are removed. Voltage sources are replaced with short circuits and current sources with open circuits.
16. Thevenin's theorem is best applied to two-terminal networks where solutions for several values of load resistance are needed.
17. The Thevenin equivalent circuit consists of a voltage source and a series resistance.
18. The Thevenin voltage is found by removing the load from the network.
19. The Thevenin resistance is found by replacing any voltage source with a short circuit and any current source with an open circuit.
20. Norton's theorem is the dual of Thevenin's theorem.
21. The Norton equivalent circuit consists of a current source in parallel with a resistance.

**22.** The Norton current is found by shorting the load terminals.

**23.** Thevenin and Norton parameters can be found by making measurements in actual networks.

**24.** Millman's theorem is a combination of Thevenin's and Norton's theorems.

**25.** Millman's theorem converts parallel voltage sources to current sources and combines them into an equivalent circuit.

**26.** Millman's theorem eliminates the need for simultaneous equations in circuits with parallel voltage sources.

**27.** Millman's theorem has a dual that allows circuits with series current sources to be reduced to an equivalent circuit.

**28.** Real voltage sources of unequal values can be connected in parallel.

**29.** Real current sources of unequal values can be connected in series.

**30.** A load will take maximum power from a network when its resistance is equal to the equivalent resistance of the network.

**31.** Thevenin's, Norton's, and Millman's theorems can be used to advantage in those cases where maximum power transfer is of importance because they provide an equivalent circuit.

**32.** When the equivalent circuit is known, the maximum power that a network can deliver is given by

$$P_{\text{max}} = \frac{(V_{\text{eq}}/2)^2}{R_{\text{eq}}} \qquad \text{or} \qquad P_{\text{max}} = \left(\frac{I_{\text{eq}}}{2}\right)^2 \times R_{\text{eq}}$$

**33.** The efficiency of a resistive network is 50 percent when it delivers maximum power.

**34.** Network efficiency increases as the load resistance increases.

**35.** A delta network can be transformed into an equivalent wye network, and the converse is also true.

**36.** A delta network may be called a pi network. A wye network may be called a T network. The name used depends on how the schematics are arranged and if they are used in electronic applications.

**37.** Transforms may simplify some networks. For example, a nonreducible circuit such as an unbalanced Wheatstone bridge may become reducible after applying the proper transform.

## CHAPTER REVIEW QUESTIONS

**9-1.** As a current source approaches the ideal, its shunt resistance approaches ______.

**9-2.** A current source has a shunt resistance of 100 kΩ. Will it closely approach being ideal for load resistances from 0 to 5 kΩ?

**9-3.** A voltage source with a very high value of series resistance provides a constant ______ characteristic.

**9-4.** Two ideal sources produce currents of 1 and 2 A. They may not be connected in ______.

**9-5.** The solution of equations with three unknowns can be accomplished with ______-order determinants.

**9-6.** True or false? The superposition theorem can be used to solve an unbalanced Wheatstone bridge.

**9-7.** True or false? The superposition theorem is applicable to multiple source circuits.

**9-8.** True or false? The superposition theorem may be used to find independent power dissipations which are then algebraically combined to find the actual power dissipations in a network.

**9-9.** True or false? Norton's theorem is the dual of Thevenin's theorem.

**9-10.** True or false? Thevenin and Norton parameters are abstract and cannot be obtained from real circuits by making measurements.

**9-11.** Assuming that only load conditions must be solved for, why is Millman's theorem a better choice for solving the circuit shown in Fig. 9-52*g*?

**9-12.** What is the efficiency of a resistive network that is delivering maximum power to a load?

**9-13.** True or false? Circuit efficiency is at its best when $R_L = R_{eq}$.

**9-14.** True or false? Power transfer is at its highest when $R_L = R_{eq}$.

## CHAPTER REVIEW PROBLEMS

**9-1.** A source develops 18 V and has an internal resistance of 150 Ω. What is the value and resistance of its dual current source?

**9-2.** Two ideal sources produce currents of 1 and 2 A. If connected in parallel opposition, the effective current will be ______ and the direction will be the same as the ______ source.

**9-3.** Find the determinant for the matrix shown below:

$$\begin{vmatrix} -7 & 3 & 2 \\ 1 & -3 & 4 \\ 0 & 8 & -1 \end{vmatrix}$$

**9-4.** Solve the following equations:

$$\begin{aligned} A + 2B - C &= 6 \\ A + 3B + 3C &= 12 \\ 2A + B + 2C &= 2 \end{aligned}$$

**9-5.** Write the mesh equations for the circuit shown in Fig. 9-52*a*.

**9-6.** Solve the equations from Problem 9-5.

**9-7.** What is the magnitude and polarity of the drop across the 40-Ω resistor in Fig. 9-52*a*?

**9-8.** What is the power dissipation in the 20-Ω resistor of Fig. 9-52*a*?

**9-9.** Write the nodal equations for the circuit shown in Fig. 9-52*b* and arrange them for solution by determinants.

**9-10.** Solve the equations from Problem 9-9.

**9-11.** What is the magnitude and direction of the current in the 0.15-S conductance in Fig. 9-52*b*?

**9-12.** What is the power dissipation in the 0.4-S conductance in Fig. 9-52*b*?

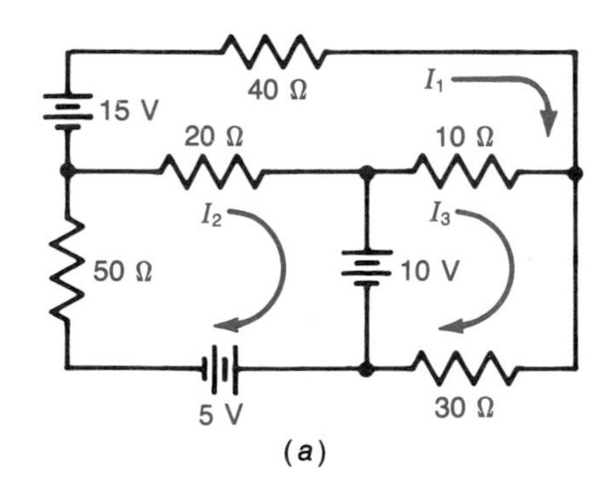


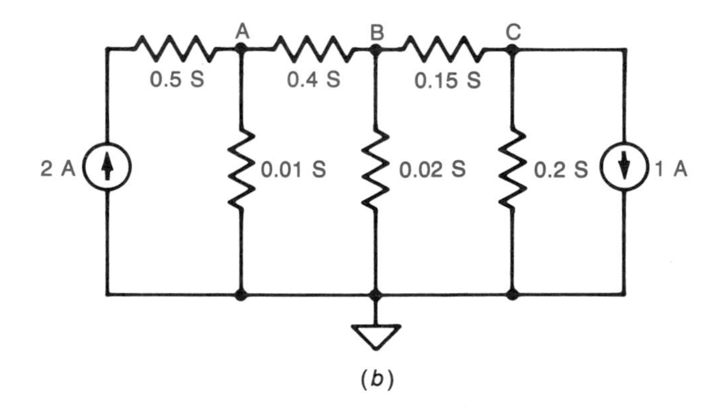


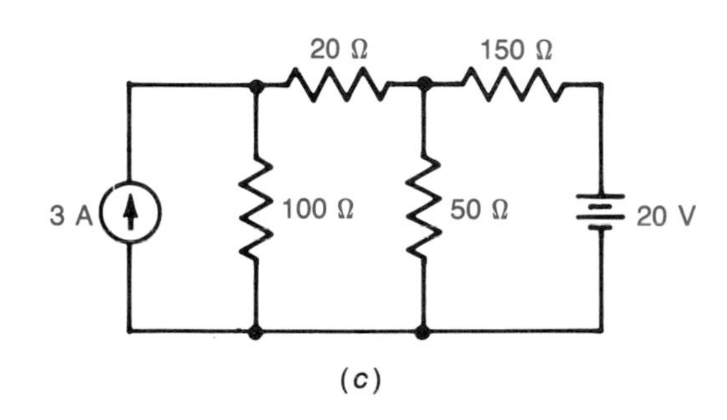


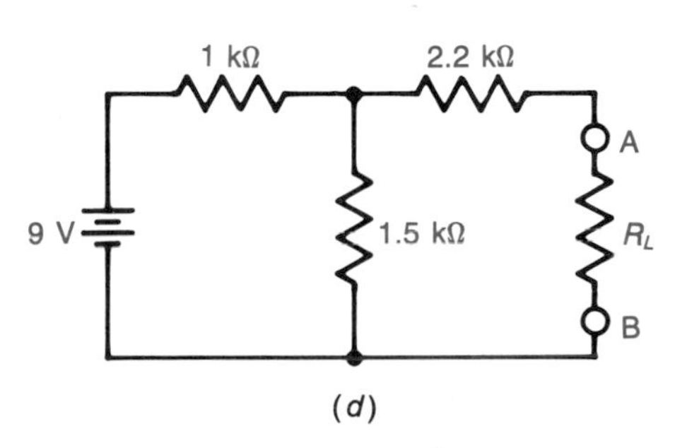


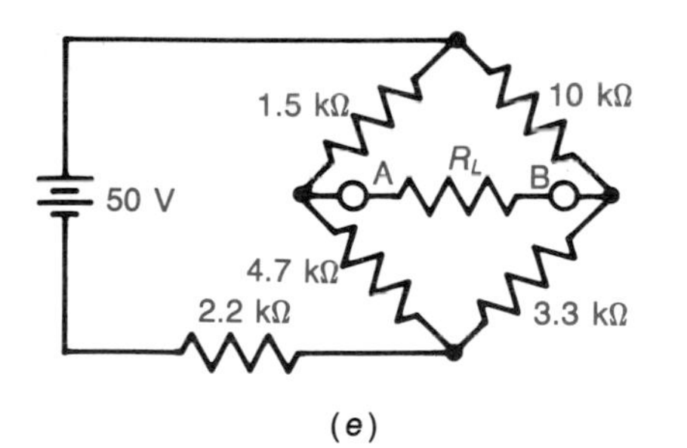


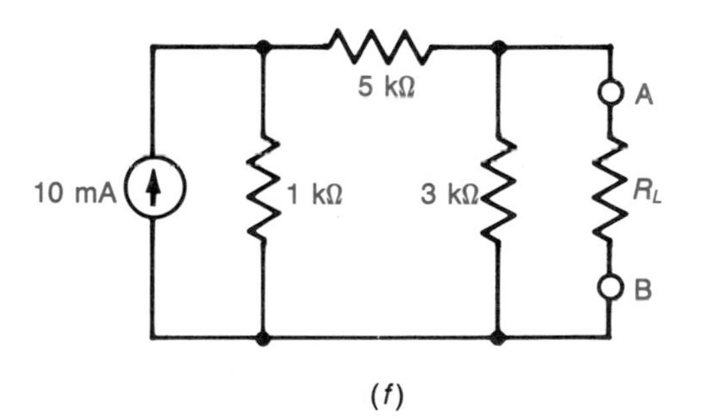


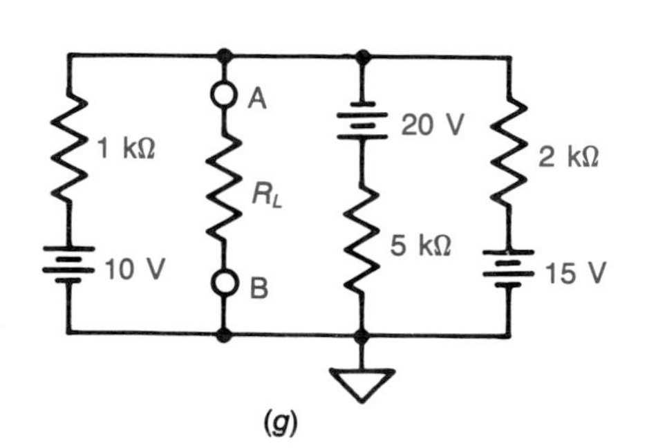


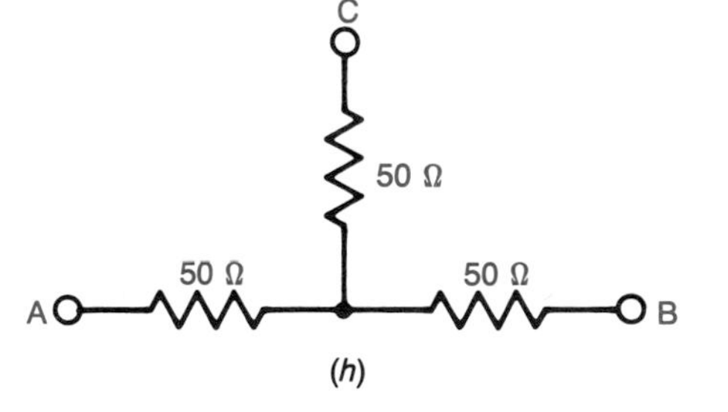


**FIGURE 9-52 Circuits for Chapter Review Questions and Problems.**

**9-13.** Apply the superposition theorem to Fig. 9-52*c*. What are the *independent* currents in the 50-Ω resistor?

**9-14.** What is the magnitude and direction of the actual current in the 50-Ω resistor of Fig. 9-52*c*?

**9-15.** What is the Thevenin voltage for Fig. 9-52*d*?

**9-16.** What is the Thevenin resistance for Fig. 9-52*d*?

**9-17.** What will $V_{AB}$ equal in Fig. 9-52*d* when $R_L$ = 10 kΩ?

**9-18.** Find $V_{TH}$ for Fig. 9-52*e*.

**9-19.** Find $R_{TH}$ for Fig. 9-52*e*.

**9-20.** If $R_L$ = 1 kΩ, find the power it will consume in Fig. 9-52*e*.

**9-21.** Determine $I_N$ for Fig. 9-52*f*.

**9-22.** What is $R_N$ for Fig. 9-52*f*?

**9-23.** If $R_L$ in Fig. 9-52*f* is equal to 2 kΩ, what is $I_L$?

**9-24.** A Norton equivalent circuit shows a current source of 10 μA and a resistance of 1 mΩ. Find $V_{TH}$ and $R_{TH}$ for its dual.

**9-25.** Find $V_{eq}$ and $R_{eq}$ for Fig. 9-52*g*.

**9-26.** If $R_L$ = 1 kΩ in Fig. 9-52*g*, find $V_A$.

**9-27.** What value should $R_L$ be in Fig. 9-52*g* to achieve maximum power transfer to the load?

**9-28.** What is the maximum power that Fig. 9-52*g* can deliver?

**9-29.** A network is reduced to $V_{eq}$ = 10 V and $R_{eq}$ = 50 Ω. What is the maximum power this network can deliver?

**9-30.** A network is reduced to $I_{eq}$ = 500 mA and $R_{eq}$ = 10 Ω. What is the maximum power this network can deliver?

**9-31.** Convert the circuit shown in Fig. 9-52*h* into an equivalent pi network.

## ANSWERS TO SELF-TESTS

**1.** current
**2.** voltage
**3.** negative
**4.** 1 V
**5.** It will increase to 1 kV.
**6.** no, yes
**7.** 400 V, 20 kΩ
**8.** negative
**9.** infinite
**10.** zero
**11.** 90 mA, 100 Ω
**12.** no, no
**13.** 20 V, negative
**14.** no
**15.** no
**16.** $x = 1.5$, $y = 4$
**17.** $x = 6$, $y = 8$
**18.** $x = -2$, $y = 3$
**19.** $x = 4$, $y = 3$, $z = 1$
**20.** $a = -1.5$, $b = 0$, $c = 4$
**21.** $a = 2$, $b = -4$, $c = 6$
**22.** $I_1 = 1$, $I_2 = 2$, $I_3 = 3$
**23.** $I_1 = 0.1$ A, $I_2 = 0.4$ A, $I_3 = -0.5$ A
**24.** $I_1 = 1$ A, $I_2 = 0.2$ A, $I_3 = 0.8$ A
**25.** 8 V
**26.** negative at the bottom
**27.** 10 mA and 40 mA
**28.** 50 V
**29.** 5.45 mA and 23.2 mA
**30.** 5.73 V
**31.** negative at the bottom
**32.** 0.301 A, 0.293 A, 0.250 A
**33.** 1.71 V

**34.** 3.88 mA, 63.96 mA, and 25.42 mA

**35.** 10.2 V

**36.** positive at the bottom

**37.** $V_A = -6.47$, $V_B = -11.8$

**38.** 264 mA

**39.** from right to left

**40.** $V_A = 4.02$, $V_B = 3.37$, $V_C = 3.72$

**41.** The current is 196 mA from right to left.

**42.** The current is 141 mA from left to right.

**43.** $V_A = 0.806$, $V_B = -2.18$, $V_C = -5$

**44.** 1.21 A

**45.** 10 V

**46.** The drop is determined by the voltage source. The current source has no effect.

**47.** 500 mA

**48.** The flow is determined by the current source. The voltage source has no effect.

**49.** 8.26 W

**50.** $V_{TH} = 80$ V, $R_{TH} = 65\ \Omega$

**51.** 0.696 A

**52.** $V_{TH} = 28.6$ V, $R_{TH} = 7.14\ \Omega$

**53.** 23.1 V

**54.** $V_{TH} = 48$ V, $R_{TH} = 2$ kΩ

**55.** 9.6 V

**56.** A is positive with respect to B.

**57.** 16 V

**58.** 24 V

**59.** $I_N = 0.333$ A, $R_N = 7.5\ \Omega$

**60.** $I_L = 0.143$ A, $V_L = 1.43$ V

**61.** $I_L = 0.167$ A

**62.** yes

**63.** 2.5 V, yes

**64.** 1.25 V, yes

**65.** true

**66.** true

**67.** $15I_1 - 5I_2 = -30$
$-5I_1 + 25I_2 - 20I_3 = 25$
$-20I_2 + 35I_3 - 15I_4 = 25$
$-15I_3 + 20I_4 = -30$

**68.** 1.8 V and 300 Ω

**69.** 6 mA

**70.** 2.25 mA

**71.** −0.295 A and 88 Ω

**72.** 26 V

**73.** 8.13 V and terminal A is negative (referenced to ground)

**74.** 75 Ω

**75.** 1 kΩ

**76.** 3.6 kW

**77.** 625 W

**78.** 50

**79.** 6.25 W

**80.** three 300-Ω resistors

**81.** T

**82.** three 200-Ω resistors

**83.** pi

# Chapter 10 Magnetism

There is a strong tie between electricity and magnetism. A study of one invariably gets involved with the other. Most of the electric energy consumed in the world is generated with magnetic rotating machines. A fair percentage of that electric energy is converted back into mechanical energy in motors which are also magnetic machines. In addition to rotating machines, there is a wide range of other useful devices that utilize magnetic properties. These include circuit breakers, relays, loudspeakers, inductors, transformers, information storage devices, certain sensors, and solenoids. This chapter covers the basic theory and some of the practical applications of magnetism.

## 10-1 MAGNETIC FIELDS

Possibly the earliest practical application of magnetism was in navigation. Chinese sailors used natural magnets called lodestones to indicate direction. The earth itself is a magnet. Figure 10-1 shows that a magnetic field emanates from the polar regions and surrounds the planet. This field accounts for the behavior of a compass needle, which is also a magnet. One end of the needle points north and is called the *north-seeking* pole, or simply the north pole. From the behavior of the compass, we can conclude that magnets have *directional* fields.

By convention, the magnetic flow on the outside

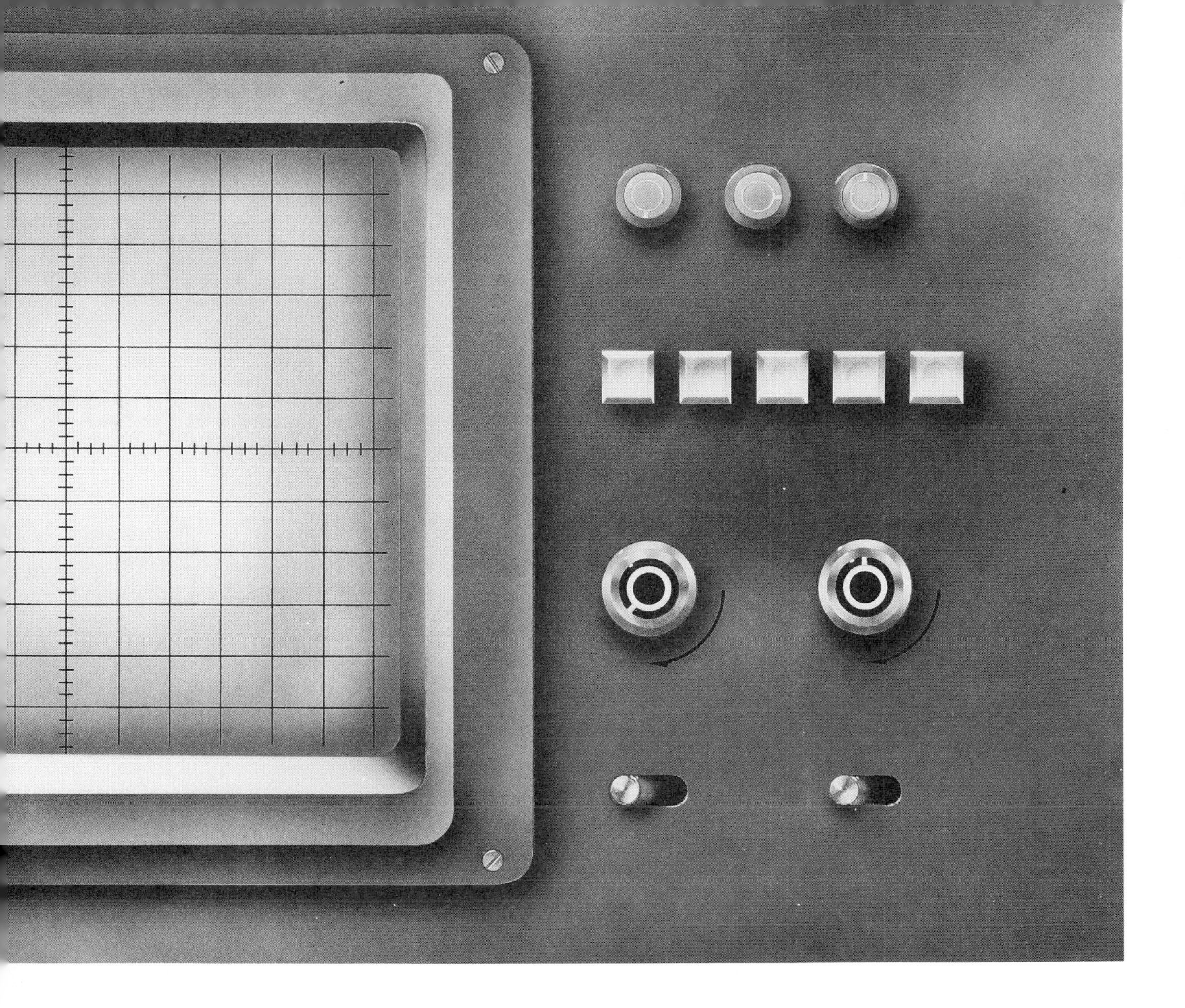

of any magnet is from the north pole to the south pole. Figure 10-1 shows that the north magnetic pole of the earth is located near the south geographic pole and that the south magnetic pole is located near the north geographic pole. The arrows on the lines surrounding the earth show the conventional flow from magnetic north to magnetic south. This flow is conceptual. It is used to predict the polarity behavior of magnetic devices and circuits.

A magnetic field can be viewed as being made up of *lines of force.* This is another conceptual model that aids in dealing with certain phenomena. There is also physical evidence that indicates the existence of these lines. Figure 10-2 shows the iron filings experiment. Fine particles of iron are sprinkled on a sheet of glass which rests on top of a magnet. The particles align themselves to form a "picture" of the force field surrounding the magnet. The field appears to be made up of lines.

Figure 10-3 shows another experiment that maps the field around a magnet. As a compass is placed in different positions around a magnet, the position of the needle takes on a definite pattern. Two important facts can be noted: (1) the needle aligns itself to be parallel with the lines of force, and (2) opposite poles attract. This is also the case with a compass in the earth's magnetic field as shown in Fig. 10-1.

The *poles* of a magnet are those two regions that

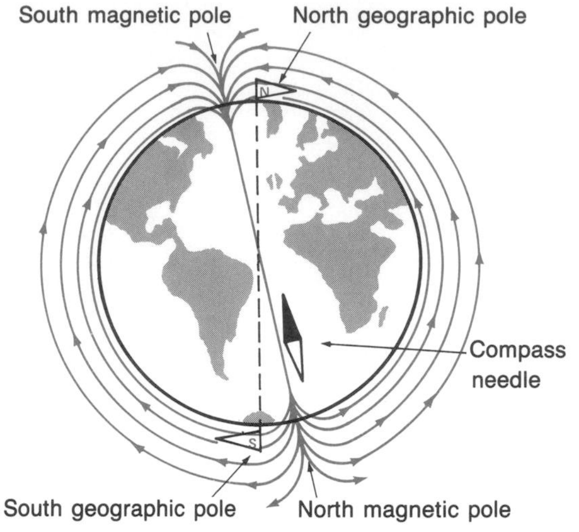


FIGURE 10-1 The earth's magnetic field.

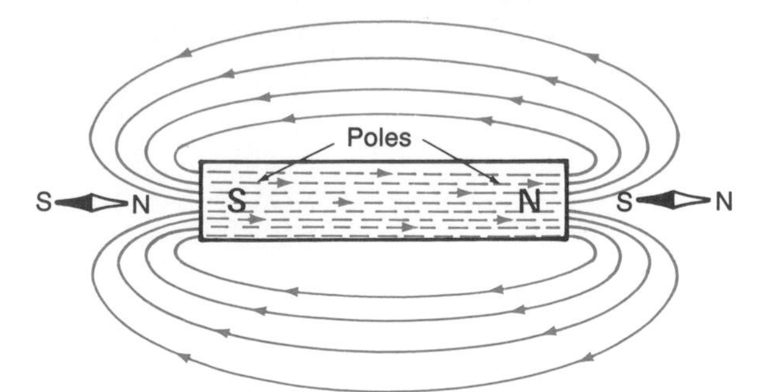


FIGURE 10-3 The field around a bar magnet can be mapped with a compass.

show the greatest concentration of *external* force lines. The word external is important because the lines are continuous. They do not stop at the poles. For every external line of force shown in Fig. 10-3, there is a continuing line inside the bar. Not all magnetic devices have poles. It is advantageous in some devices, such as transformers, to have only an internal field.

Some materials are capable of acting as magnets and others are not. A rigorous explanation of why this is so involves electron spins and other atomic details which will not be presented here. Figure 10-4 offers a simplified explanation. Materials that can be magnetized contain tiny regions called *domains*. The domains may be viewed as individual bar magnets. When a bar of magnetic material is stroked with a magnet, its domains come into alignment. The fields of the individual domains reinforce each other, giving rise to an overall field. When the bar is unmagnetized, its domains are randomly oriented and the fields of individual domains cancel one another.

The bar being magnetized in Fig. 10-4 will show a north pole at the right end and a south pole at the left end. This is due to the direction of motion which leaves the domains oriented as shown. The opposite magnetic polarity can be achieved by reversing the direction of motion. If the domains in the bar retain their alignment when the magnet is removed, the bar is considered a *permanent* magnet. If the domains return to random alignment, the bar reverts to its original condition and is considered a *temporary* magnet. Certain materials are very good at retaining magnetism and are used for making permanent magnets. Other materials are easily magnetized but do not retain a large amount. These are used where temporary magnetism is needed. For example, if a device is needed to pick up steel parts and then release them, temporary magnetism is desirable.

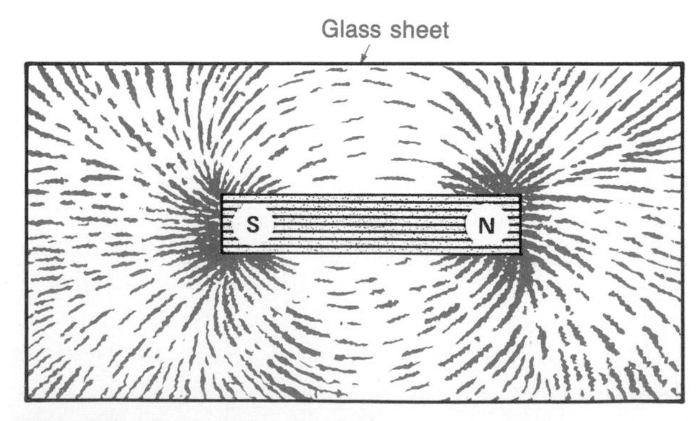


FIGURE 10-2 The iron filings experiment.

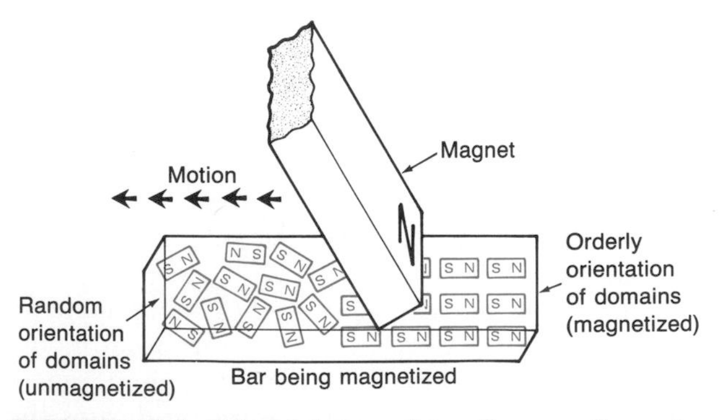


FIGURE 10-4 Magnetizing a bar of magnetic material.

A working understanding of magnetism is assisted by considering the characteristics of magnetic lines:

**1.** They are like stretched rubber bands which tend to be as short as possible.

**2.** They repel each other.

**3.** They will not cross other magnetic lines of force.

The first two characteristics set up an equilibrium of forces. The external field of a magnet tends to be compact, but if more lines are added, the field extends out because the lines repel each other.

Figure 10-5 shows the law of magnetic poles. *Like poles repel,* as shown in Fig. 10-5*a*. The lines are leaving both facing poles. They cannot cross and they repel each other. As the poles are forced closer together, the lines compress even more and the force of repulsion increases. Notice the field distortion between the like poles. *Unlike poles attract,* as shown in Fig. 10-5*b*. The lines coming out of the north pole flow into the south pole, and this occurs at *both* sets of poles. Since the lines are under tension, the magnets are drawn together. Repulsive and attractive forces both follow the inverse square law. That is, if the distance separating the poles is doubled, the force is decreased to one-fourth. If the distance is halved, the force is quadrupled.

Magnetic fields are distorted by magnetic materials in their vicinity. Figure 10-6 shows the influence of a piece of soft iron on the field of a bar magnet. The lines of force tend to pass through the iron rather than through air (or a vacuum). This situation is analogous to an electric current which takes the path of least resistance. Glass is a nonmagnetic material. It does not support magnetic lines of force much differently than air. As shown in Fig. 10-6, the piece of glass does not cause field distortion.

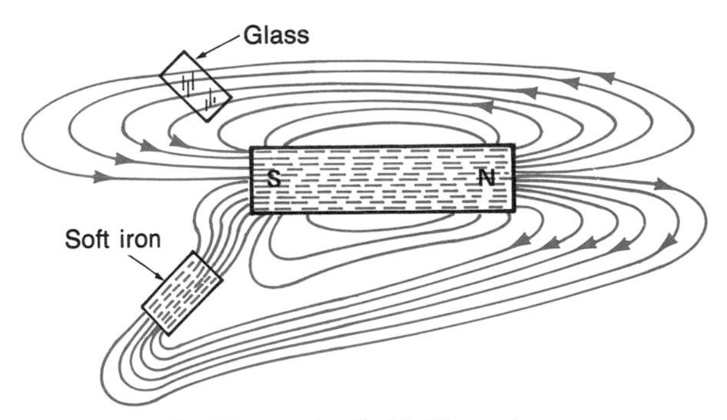

**FIGURE 10-6 Magnetic field distortion.**

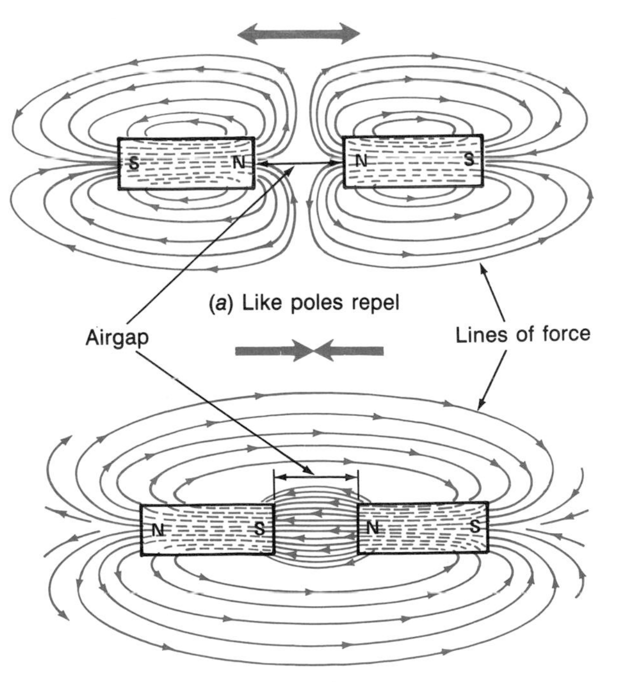


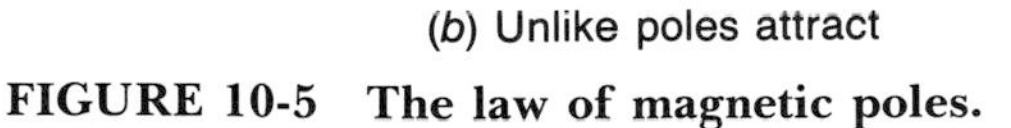
(*b*) Unlike poles attract

**FIGURE 10-5 The law of magnetic poles.**

## Self-Test

**1.** True or false? The lines of force on the outside of a magnet flow from the north pole to the south pole.

**2.** True or false? The south magnetic pole of the earth is near the geographic north pole.

**3.** If a bar magnet is suspended at its center by a string, which one of its poles will point north?

**4.** The regions of a magnet that show the greatest concentration of external lines of force are called the ______.

**5.** True or false? All magnets have poles.

**6.** Refer to Fig. 10-4. If the motion of the magnet is reversed, which end of the bar being magnetized becomes the north pole?

**7.** Which pole of a magnet will be attracted to the south pole of another magnet?

**8.** There is a 100-g force of repulsion between two facing south poles that are separated by a distance of 1 cm. What will the force of repulsion be if the distance is decreased to 0.2 cm?

**9.** True or false? Magnetic lines of force pass through air more readily than through iron.

## 10-2 ELECTROMAGNETISM

Figure 10-7 shows a wire passing through a surface. When the wire is conducting current, a compass will show the presence of a magnetic field around the wire. The field can be mapped by moving the compass to various positions. The field appears to be circular at the surface which is perpendicular to the current flow. When the battery is disconnected, the current flow stops and the field collapses.

The lines shown in Fig. 10-7 travel in a clockwise direction around the conductor. This is indicated by the compass needle as it is moved into various positions around the conductor. If the battery is reversed, the compass needle also reverses. The lines appear to travel counterclockwise around the wire. This proves that there is a relationship between electric polarity and magnetic polarity. The *left-hand rule* can be used to assign the direction of magnetic flow from the direction of electron current. Figure 10-8 shows how it works. Grasp the conductor in your left hand with your thumb pointing in the direction of electron current. Your fingers will point in the direction of the field that circles the wire.

It is sometimes necessary to visualize magnetic effects in circuits where the current is flowing toward or away from an observer. *Dot-and-cross notation* can be used in these cases to indicate the direction of current flow. Figure 10-9*a* shows that current flowing toward an observer is indicated with a dot. The dot is based on the point of an arrow moving toward the observer. Figure 10-9*b* shows that current flowing away from an observer is indicated with a cross. The cross is based on the tail feathers of an arrow moving away from the observer.

Dot-and-cross notation is used in Fig. 10-10 to help explain the forces between two current-carrying conductors. In Fig. 10-10*a*, both wires are marked with a cross. The currents are both flowing

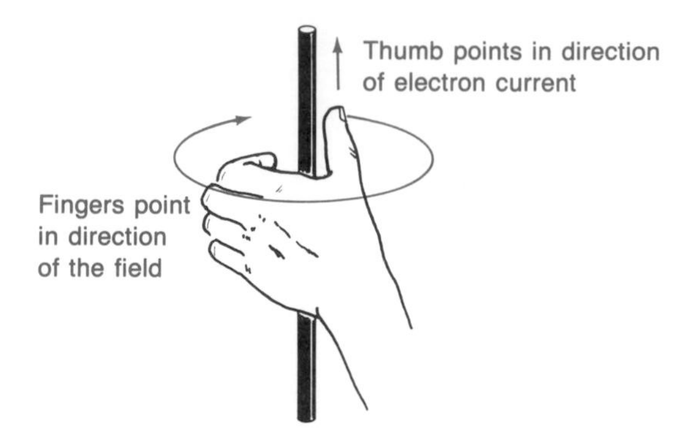


**FIGURE 10-8 The left-hand rule.**

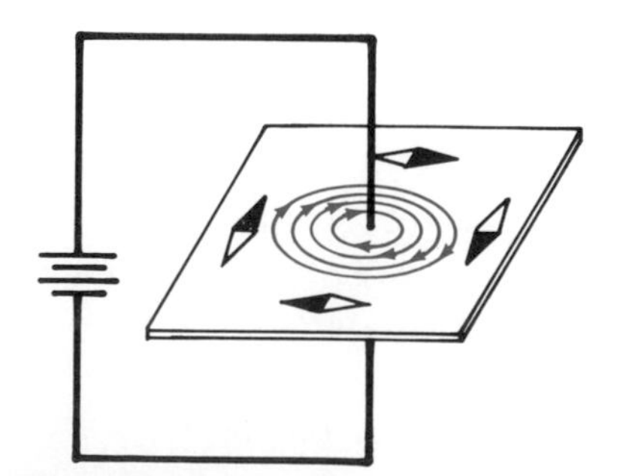

**FIGURE 10-7 There is a magnetic field around a current-carrying conductor.**

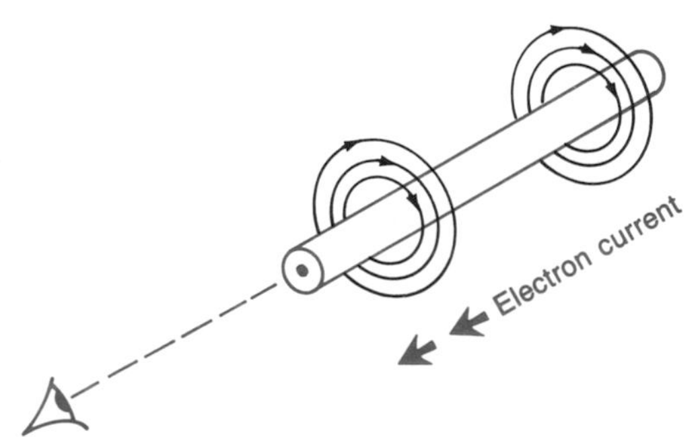


(*a*) Current flowing toward an observer

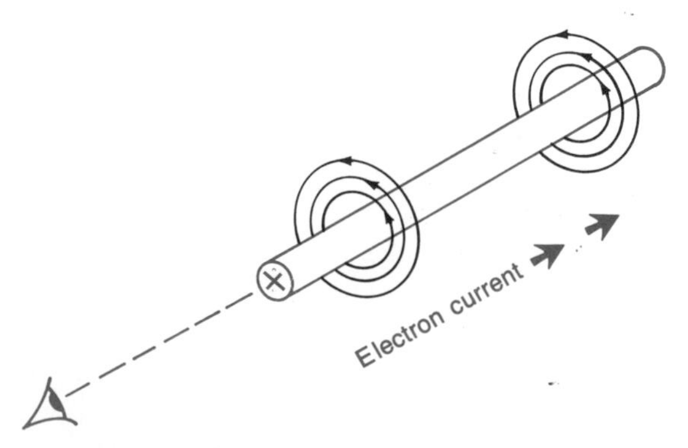


(*b*) Current flowing away from an observer

**FIGURE 10-9 Dot-and-cross notation.**

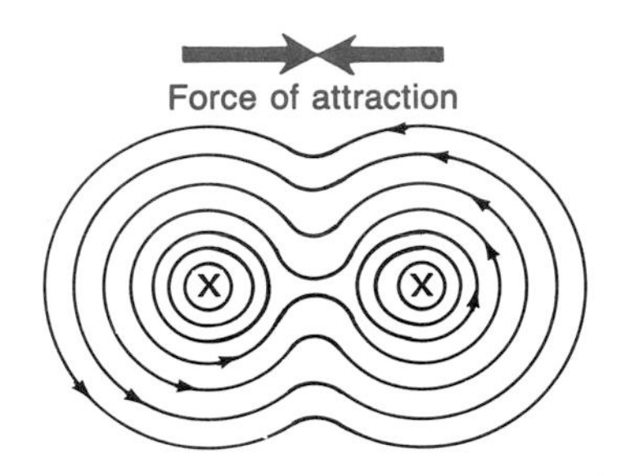


(*a*) Currents flowing in the same direction

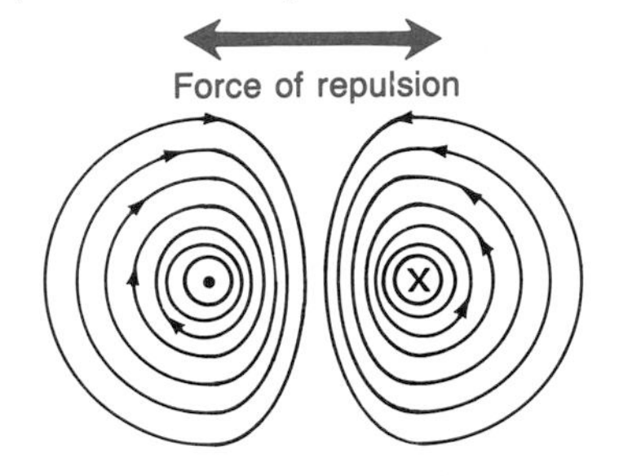


(*b*) Currents flowing in opposite directions

**FIGURE 10-10 Forces between current-carrying conductors.**

away from the observer. Use your left hand to verify that the field around each conductor travels in a counterclockwise direction. The fields join in this case, becoming one field. This results in a force of attraction between the conductors. Figure 10-10*b* shows the currents flowing in opposite directions. The fields cannot join in this case and the lines repel each other. There is a force of repulsion between the current-carrying conductors.

The SI unit of current is the *ampere*. The effect demonstrated in Fig. 10-10 can be used to define the ampere in terms of other SI units. When two straight, parallel conductors of infinite length are carrying 1 A of current and are separated by 1 m, they produce a force of $2 \times 10^{-7}$ N per meter of length in a vacuum. If the current is doubled, the force is doubled. In other words, the strength of an electromagnetic field in a vacuum is directly proportional to the current flow. As we will see in Sec. 10-3, the ampere is also the SI unit for *magnetomotive* force.

The magnetic field around a straight current-carrying conductor has direction, but no poles. The lines are spread out along the length of the conductor so the forces are small unless the current is extremely high. Figure 10-11 shows a single-turn coil. A coil does have poles. The coil

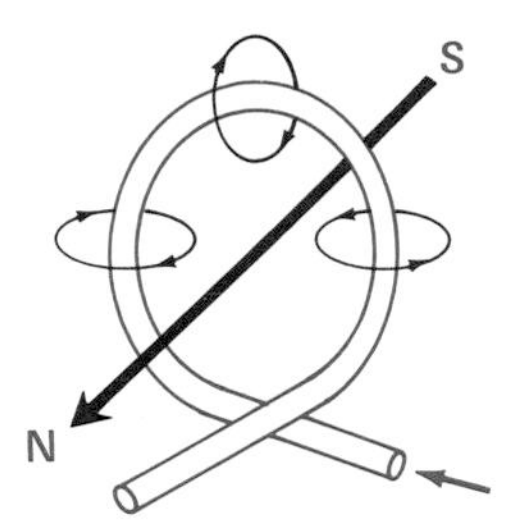


**FIGURE 10-11 A single-turn coil.**

arrangement also concentrates the lines into a smaller space. This increases the magnetic force available. Figure 10-12 shows a coil with more turns. Adding turns to the coil produces a further increase in magnetic force.

The word *flux* is often used to refer to the lines of magnetic force around a magnet or a current-carrying coil. Figure 10-13 shows the field of a long coil (its length is considerably greater than its diameter). The *leakage flux* does not contribute to the intensity of the field at the poles of the coil. Adding turns to a coil is one way to increase the field intensity at the poles. However, as the coil becomes longer, so does the magnetic circuit and the leakage increases. Coils are sometimes wound in layers as shown in Fig. 10-14. This allows many turns of wire in a magnetic circuit of reasonable length. There is less leakage flux and therefore better field intensity at the poles.

Figure 10-15 shows a coil with an iron core. Iron supports magnetic flux much more so than air or a vacuum. Changing the core from air to iron allows many more lines to exist, which greatly increases the field intensity of the coil. The iron core also decreases the flux leakage.

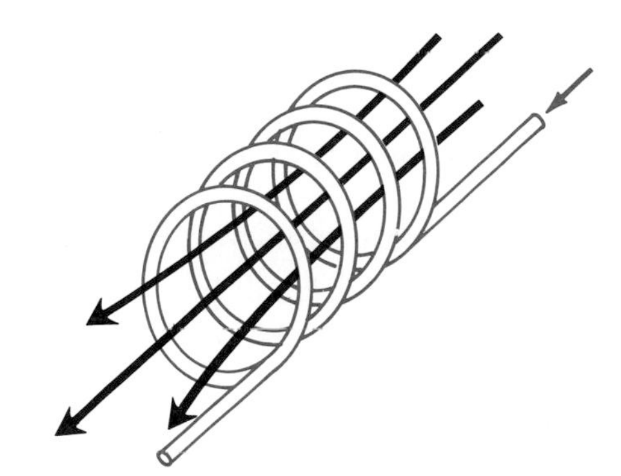

**FIGURE 10-12 A coil with more turns.**

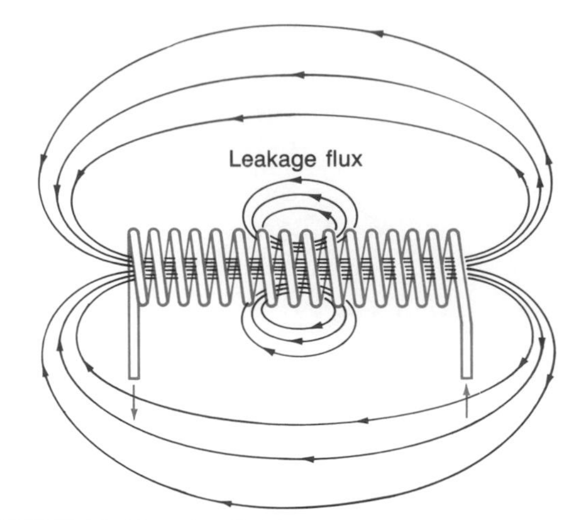


FIGURE 10-13 The field of a long coil.

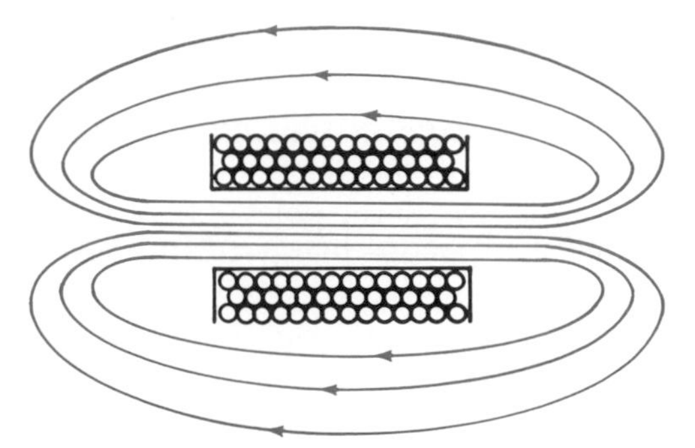

FIGURE 10-14 A multilayer coil.

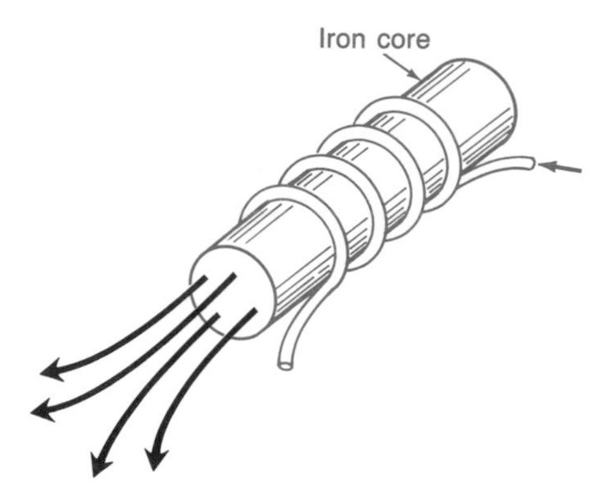


FIGURE 10-15 A coil with an iron core.

In summary, there are four factors which affect the strength of an electromagnet:

1. The amount of current flow
2. The number of turns
3. The size and shape of the magnetic circuit
4. The type of core material

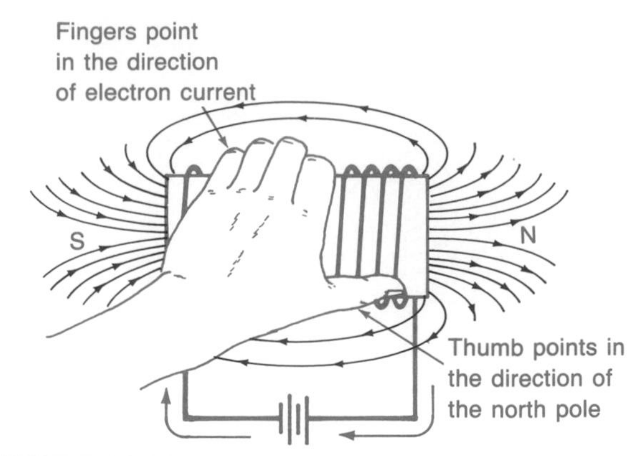


FIGURE 10-16 Applying the left-hand rule to a coil.

There is a left-hand rule for coils. Figure 10-16 shows its application. The coil is grasped in the left hand with the fingers pointing in the direction of electron current. The thumb points in the direction of the magnetic lines and toward the north pole.

## Self-Test

**10.** The field around a conductor increases in strength as the current flow ______.

**11.** Figure 10-17*a* shows the field around a conductor. In which direction is the current flowing?

**12.** Figure 10-17*b* shows a cross section of a coil. Based on dot-and-cross notation, which end of the coil is the north pole?

**13.** Will the two conductors shown in Fig. 10-17*c* attract or repel each other?

**14.** List the four factors that determine the strength of an electromagnet.

**15.** True or false? Leakage flux is desirable when the maximum field strength must be obtained.

**16.** Which end of the coil shown in Fig. 10-17*d* is the north pole?

## 10-3 MAGNETIC UNITS

In an electric circuit, an electromotive force is required to produce current flow. The magnetic counterpart of electromotive force is called *magnetomotive force* (mmf). The letter symbol for mmf is a script $\mathcal{F}$. The SI unit of magnetomotive force is the ampere. However, adding turns of wire to a coil

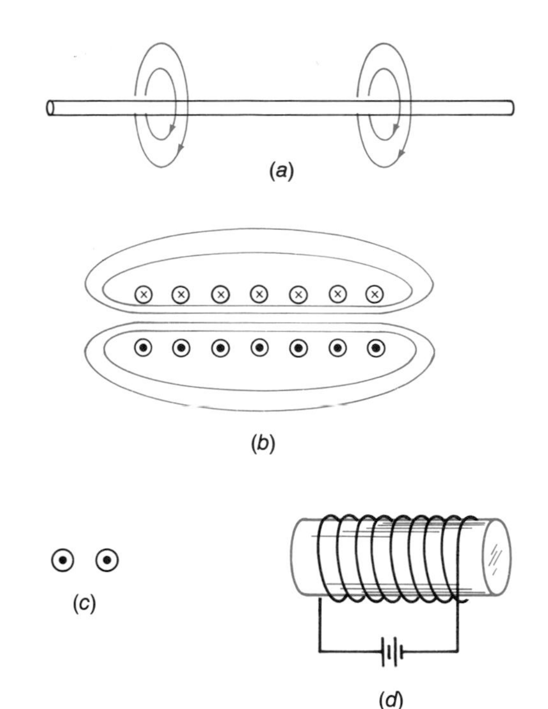


**FIGURE 10-17 Self-Test.**

has the same effect as increasing the current flow. For example, a one-turn coil with 1 A of current flowing in it produces the same mmf as a 10-turn coil with only 0.1 A of current flow. The *effective mmf* of a magnetic circuit is therefore equal to the product of the number of turns and the current in amperes:

$$\mathcal{F} = NI$$

where $\mathcal{F}$ = mmf in ampere-turns (A · t)
$N$ = number of turns
$I$ = current

---

**EXAMPLE 10-1**
What is the mmf developed by a 300-turn coil that is conducting a current of 700 mA?

**Solution**
The mmf is found by multiplying the turns times the current in amperes:

$$\mathcal{F} = NI = 300 \times 0.7 = 210 \text{ ampere-turns, or } 210 \text{ A} \cdot \text{t}$$

---

*Flux* refers to all of the lines in a magnetic circuit. Flux is analogous to the current flow in an electric circuit. The letter symbol for flux is the Greek capital letter phi, Φ. The SI unit of flux is the *weber* (abbreviated Wb), which is defined in terms of flux change. Examine Fig. 10-18. It shows an iron core with two windings (coils). The windings are electrically isolated but share the magnetic circuit. The variable resistor allows the current flow in the right-hand winding to be changed. A change in current produces a change in flux which generates a voltage across the one-turn coil at the left. A flux change of 1 Wb/s generates 1 V across a single-turn coil. The windings and core of Fig. 10-18 form a device called a *transformer*. Transformers are covered in Chap. 20.

*Reluctance* is the opposition to flux in a magnetic circuit. It is analogous to resistance in an electric circuit. It is defined in terms of mmf and flux just as resistance is defined in terms of emf and current in an electric circuit. In fact, the relationship is known as the Ohm's law of magnetic circuits:

$$\mathcal{R} = \frac{\mathcal{F}}{\Phi}$$

where $\mathcal{R}$ = reluctance, (A · t)/Wb
$\mathcal{F}$ = mmf, A · t
$\Phi$ = flux, Wb

---

**EXAMPLE 10-2**
What is the reluctance of a magnetic circuit that has a flux of $4 \times 10^{-5}$ Wb and a mmf of 6 A · t?

**Solution**
Applying Ohm's law for magnetic circuits:

$$\mathcal{R} = \frac{6 \text{ A} \cdot \text{t}}{4 \times 10^{-5} \text{ Wb}} = 1.5 \times 10^{5} \text{ (A} \cdot \text{t)/Wb}$$

---

The physical size of a magnetic circuit is an important factor. Thus, additional units that deal with length and area are required to completely

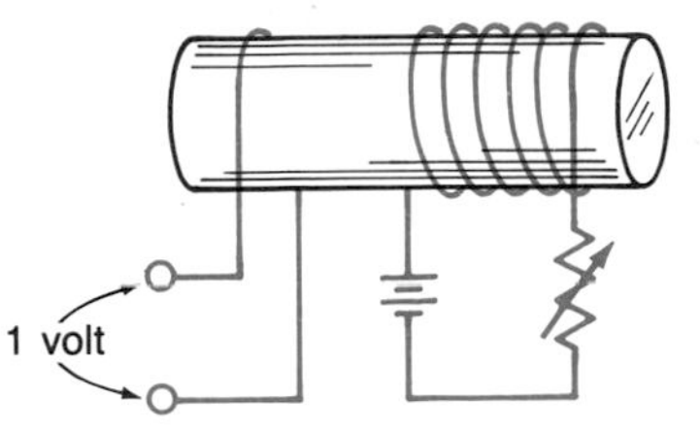


**FIGURE 10-18 A flux change of 1 Wb/s generates 1 V in a single-turn coil.**

describe magnetic circuits. *Magnetizing force* is the amount of mmf available per unit length. Its letter symbol is $H$, which is based on the SI unit of length, the meter:

$$H = \frac{\mathcal{F}}{l}$$

where $H$ = magnetizing force, (A · t)/m
$\mathcal{F}$ = mmf, A · t
$l$ = length of circuit, m

**EXAMPLE 10-3**
What is the magnetizing force in a magnetic circuit that is 6 cm long and has a magnetomotive force of 200 A · t?

**Solution**
The mmf is divided by the length of the circuit in meters:

$$H = \frac{200 \text{ A} \cdot \text{t}}{0.06 \text{ m}} = 3.33 \times 10^3 \text{ (A} \cdot \text{t)/m}$$

*Flux density* is a measure of flux per unit area. The letter $B$ is the symbol for flux density. The SI unit of flux density is the tesla (T), which is equal to one weber per square meter:

$$B = \frac{\Phi}{a}$$

where $B$ = flux density, T
$\Phi$ = flux, Wb
$a$ = area, $\text{m}^2$

**EXAMPLE 10-4**
The pole of a bar magnet measures 1 cm × 2 cm. Assuming a flux of $1 \times 10^{-4}$ Wb, what is the flux density at the pole?

**Solution**
Flux density is found by dividing the flux by the area in square meters:

$$B = \frac{1 \times 10^{-4} \text{ Wb}}{0.01 \text{ m} \times 0.02 \text{ m}} = 0.5 \text{ tesla, or } 0.5 \text{ T}$$

*Permeability* is the ability of a material to support magnetic flux. It is analogous to conductivity, which is the ability of a material to support the flow of current in an electric circuit. The letter symbol for permeability is the Greek lowercase letter, $\mu$. Unfortunately, this is also the symbol for the prefix *micro*. To avoid confusion, the prefix micro is not used when working with magnetic circuits. High-permeability materials are used to build cores for magnetic circuits where large flux densities are required. Permeability is equal to the flux density divided by the magnetizing force:

$$\mu = \frac{B}{H}$$

where $\mu$ = permeability, Wb/(A · t · m)
$B$ = flux density, T
$H$ = magnetizing force, (A · t)/m

**EXAMPLE 10-5**
Calculate the permeability of the core material for the electromagnet shown in Fig. 10-19.

**Solution**
The flux density is given. The magnetizing force must be found:

$$H = \frac{10 \times 20 \text{ A} \cdot \text{t}}{0.05 \text{ m}} = 4 \times 10^3 \text{ (A} \cdot \text{t)/m}$$

The permeability of the core material can now be found:

$$\mu = \frac{5.03 \times 10^{-3} \text{ Wb/m}^2}{4 \times 10^3 \text{ (A} \cdot \text{t)/m}} = 1.26 \times 10^{-6} \text{ Wb/(A} \cdot \text{t} \cdot \text{m)}$$

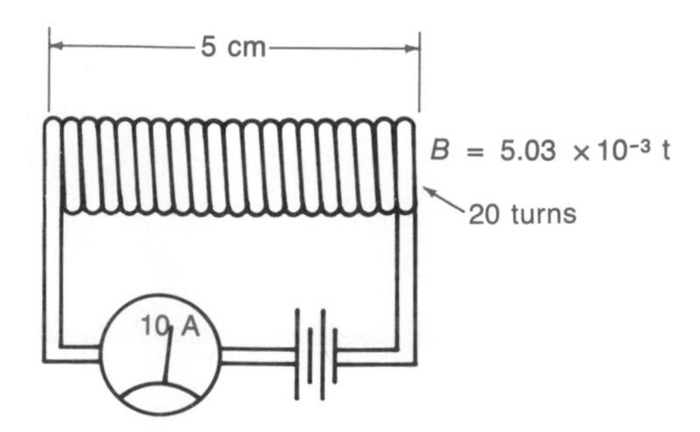


**FIGURE 10-19 Example 10-5.**

This is equal to the permeability of a vacuum. For practical purposes, it is also the permeability of air. The coil in Fig. 10-19 has an air core. The permeability of air, or of a vacuum, is called the *permeability of free space* and has the letter symbol $\mu_o$:

$\mu_o = 4\pi \times 10^{-7}$ Wb/(A · t · m) A fundamental physical constant

Note that this is equivalent to the calculated permeability in the example.

*Relative permeability* is often used to describe magnetic materials. It is represented by the symbol $\mu_r$. It is found by dividing the permeability of a material by the permeability of free space:

$$\mu_r = \frac{\mu}{\mu_o}$$

The units cancel, so relative permeability is dimensionless (a pure number).

Materials can be classified according to relative permeability. *Ferromagnetic* materials have relative permeabilities much greater than 1. Iron alloys, with a $\mu_r$ as large as 7000, are the most widely applied ferromagnetic materials. They are used in the construction of motors, transformers, relays, and other magnetic devices to allow the establishment of a strong flux. Iron ferrite is used in the manufacture of permanent ceramic magnets. Nickel ($\mu_r \approx 50$) and cobalt ($\mu_r \approx 60$) are also ferromagnetic and may be alloyed with iron to make permanent magnets. Materials that have relative permeabilities slightly greater than 1 but much less than 1.1 are considered to be *paramagnetic*. Aluminum and platinum are such materials. Paramagnetic materials are very weakly attracted by a magnetic field, while ferromagnetic materials are strongly attracted.

There are also a few materials with relative permeabilities slightly less than 1. These materials are considered *diamagnetic*. Carbon, copper, and silver are diamagnetic and are very slightly repelled by a magnetic field. This effect is so slight that it can only be detected with an extremely intense field and sensitive instrumentation. Materials with relative permeabilities very close to one are often just classified as *nonmagnetic*. Thus, both paramagnetic and diamagnetic materials can be classified as nonmagnetic. Glass, air, aluminum, wood, and many other materials are nonmagnetic. For practical purposes, they are neither attracted to nor repelled by a magnetic field.

Table 10-1 provides a summary and a useful reference for the magnetic units presented in this section.

## Self-Test

**17.** What is the mmf of a 1000-turn coil with 15 mA of current flow?

**18.** What is the flux in a magnetic circuit with a reluctance of $2 \times 10^5$ (A · t)/Wb and a mmf of 400 A · t?

**19.** A coil is 2 cm long and has 10 turns. What is the magnetizing force of this coil with a current flow of 750 mA?

**20.** A magnetic device has a core with a cross section of 1 in.$^2$. What is the flux density if the flux is 1 mWb? (1 in. = 2.54 cm)

**21.** Assume that the air core in Fig. 10-19 is replaced with a material that has a relative permeability greater than 1. Will the flux density increase?

**22.** The major ferromagnetic material is ______.

**Table 10-1**

| Magnetic Quantity | Symbol | Unit | Electrical Analogy |
|---|---|---|---|
| Magnetomotive force (mmf) | $\mathcal{F}$ | Ampere-turn, A · t | Electromotive force(emf) |
| Flux | $\Phi$ | Weber, Wb | Current |
| Reluctance | $\mathcal{R}$ | Ampere-turns per weber, (A · t)/Wb | Resistance |
| Magnetizing force | $H$ | Ampere-turns per meter, (A · t)/m | Electric field strength |
| Flux density | $B$ | Webers per square meter, 1 tesla = 1 Wb/m$^2$ | Current density |
| Permeability | $\mu$ | Webers per ampere-turn-meter, Wb/(A · t · m) | Conductivity |
| Relational Equations | | | |
| | | Ohm's law for magnetic circuits: $\mathcal{F} = \Phi\mathcal{R}$ | |
| | | Permeability: $\mu = B/H$ | |

## 10-4 MAGNETIZATION CURVES

Most magnetic core materials are made from ferrous materials. The word *ferrous* means *of or pertaining to iron*. The permeability of iron and its various compounds is not a constant. Figure 10-20 illustrates a permeability curve for a typical ferromagnetic material. As the flux density increases from point 1 to point 2 on the curve, the permeability increases. This is due to the behavior of the magnetic domains in the material. As they begin to align, better support is provided for additional flux. From point 2 to point 3 the permeability remains reasonably constant as the domain alignment continues. At point 3, most of the domains have achieved alignment. From point 3 to point 4, the permeability drops because of limited additional alignment. Beyond point 4, the permeability continues decreasing and begins to approach that of a nonmagnetic material. Practically all of the available domains have been aligned, and the term *saturation* is used to describe this situation.

The variation of permeability with flux density creates a design problem. A rearrangement of the equation relating permeability, magnetizing force, and flux density shows

$$B = \mu H$$

Permeability must be known in order to find flux density, yet permeability varies with flux density. This interdependence dictates a graphical approach to magnetic circuit design. Manufacturers of magnetic core materials prepare and supply graphical data for their products.

Figure 10-21 shows *magnetization curves* for three materials. These curves can be used to determine

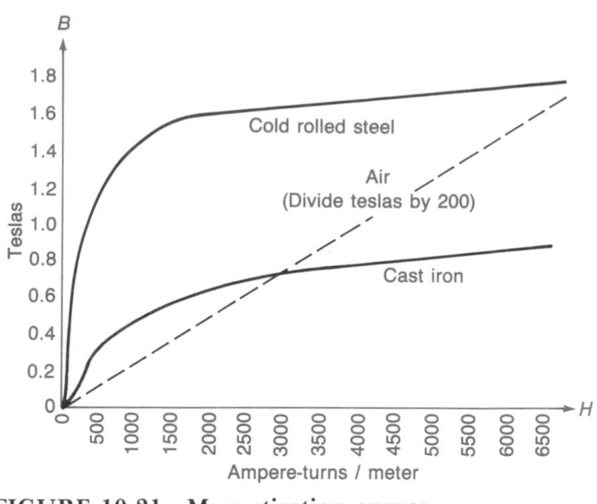


**FIGURE 10-21 Magnetization curves.**

the magnetizing force required to produce a given flux density. They are most often called *BH curves*. Note that the curve for air uses a different vertical scale. Air has a much lower permeability, and its curve would appear almost on the horizontal axis if it used the same vertical scale as the ferrous materials.

The *BH* curves of Fig. 10-21 show that air is a *linear* magnetic material. Its permeability is constant for any value of *H*. The ferrous materials are *nonlinear*. Figure 10-22 shows the regions of a *BH* curve for ferrous materials. The lower knee is caused by the domains starting to align. The linear region represents the continuation of domain alignment with increasing magnetizing force. The upper knee is caused by most of the domains reaching alignment. The *saturation* region is be-

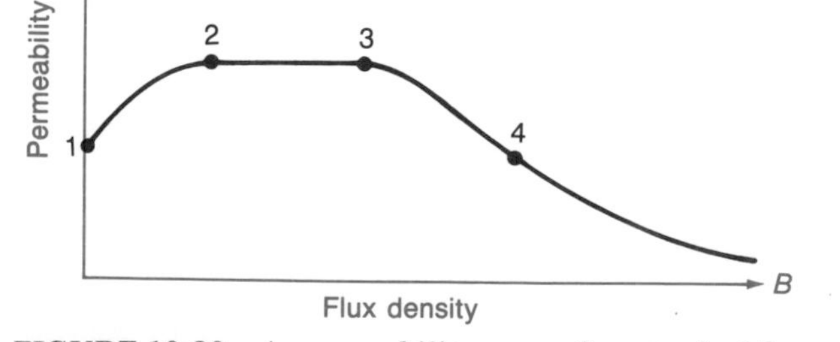

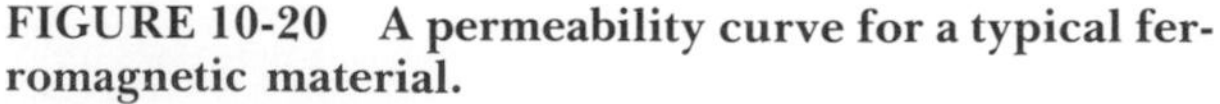

**FIGURE 10-20 A permeability curve for a typical ferromagnetic material.**

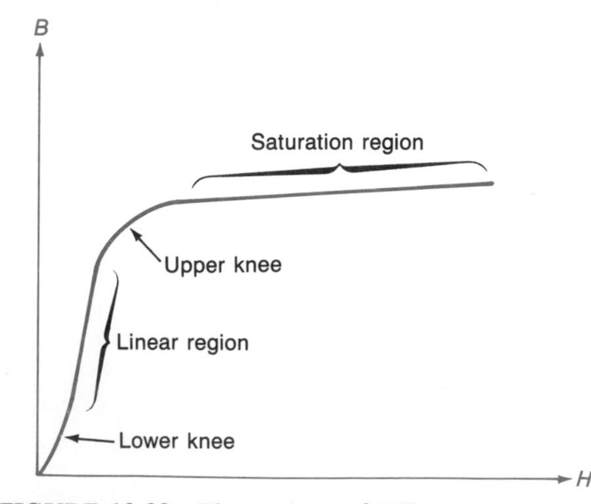


**FIGURE 10-22 The regions of *BH* curve.**

yond the upper knee. When all of the domains of a material reach alignment, no additional contribution to flux can be provided by the material. The slope of the curve in the saturation region approaches the slope of a nonmagnetic material.

Permeability can be "read" from $BH$ curves. Recall that permeability can be found by

$$\mu = \frac{B}{H}$$

Data can be taken from the curves to determine the permeability of a material. Since the curves are nonlinear, there are several methods of extracting the data as shown in Fig. 10-23. *Static* permeability is found by picking a single point on the $BH$ curves. The values of $B$ and $H$ are read from that point, and permeability can be calculated from these two values, as shown in Fig. 10-23*a*. Static permeability is a valid approach for the design of electromagnetic devices that operate at a fixed current, such as some relays and solenoid valves. These devices are covered in Sec. 10-6 of this chapter. *Dynamic* permeability is based on incremental values of $B$ and $H$, as shown in Fig. 10-23*b*. Dynamic permeability is used in the design of devices where the current and flux density swing over a range of values. Some magnetic recording devices fit in this category. *Average* permeability is another approach and is shown in Fig. 10-23*c*. A line is constructed from the origin of the graph to a point tangent with the upper knee. Any point on this straight line can be used to read $B$ and $H$. Average permeability greatly simplifies the design of magnetic devices where the flux density varies over a considerable range. Motors and power transformers can be designed with this approach. It is reasonably accurate, providing that the core is never driven into saturation.

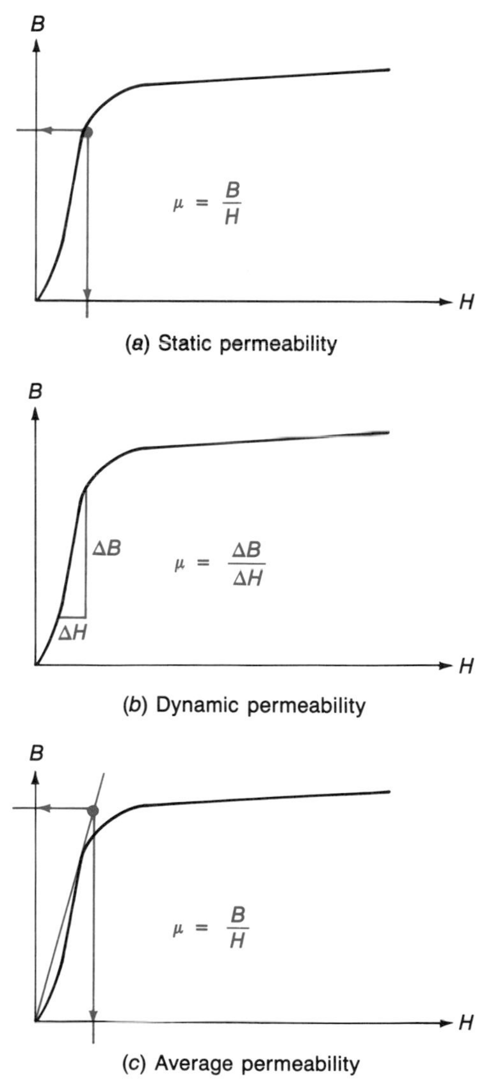


**FIGURE 10-23 Methods of obtaining permeability from $BH$ curves.**

Saturation results in a *significant* decrease in permeability. This can be demonstrated by calculating the dynamic permeability for two regions of a $BH$ curve and comparing them.

## EXAMPLE 10-6

Find the dynamic permeability for the two incremental regions indicated on the $BH$ curve of Fig. 10-24. Also find the relative permeability for each region.

### Solution

First we will find the change in $B$ and $H$ for the linear region of the graph. The values of $\Delta$ are found by subtracting:

$$\Delta B = 1.1\ \text{T} - 0.75\ \text{T} = 0.35\ \text{T}$$
$$\Delta H = 500\ (\text{A}\cdot\text{t})/\text{m} - 250\ (\text{A}\cdot\text{t})/\text{m} = 250\ (\text{A}\cdot\text{t})/\text{m}$$

Permeability is found by dividing the delta values:

$$\mu = \frac{0.35\ \text{T}}{250\ (\text{A}\cdot\text{t})/\text{m}} = 1.4 \times 10^{-3}\ \text{Wb}/(\text{A}\cdot\text{t}\cdot\text{m})$$

Relative permeability is found by dividing by the value of free space:

$$\mu_r = \frac{1.4 \times 10^{-3}}{4\pi \times 10^{-7}} = 1.11 \times 10^3$$

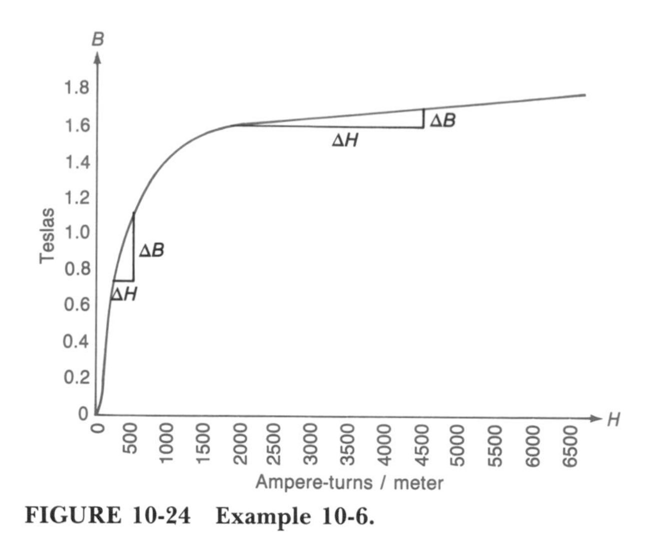


**FIGURE 10-24 Example 10-6.**

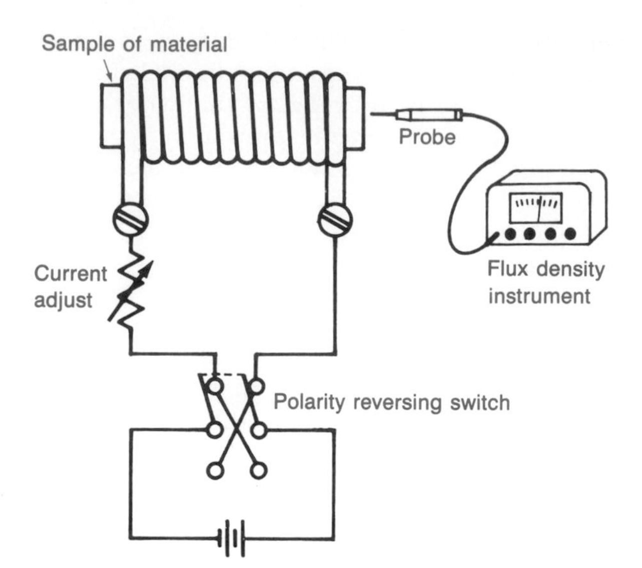


**FIGURE 10-25 Measuring magnetic characteristics.**

Next, find the change in $B$ and $H$ for the saturation region of the graph:

$$\Delta B = 1.7\text{ T} - 1.6\text{ T} = 0.1\text{ T}$$
$$\Delta H = 4500\ (\text{A}\cdot\text{t})/\text{m} - 1800\ (\text{A}\cdot\text{t})/\text{m} = 2700\ (\text{A}\cdot\text{t})/\text{m}$$

Permeability is found by dividing the delta values:

$$\mu = \frac{0.1\text{ T}}{2700\ (\text{A}\cdot\text{t})/\text{m}} = 3.7 \times 10^{-5}\ \text{Wb}/(\text{A}\cdot\text{t}\cdot\text{m})$$

The relative permeability is found by

$$\mu_r = \frac{3.7 \times 10^{-5}}{4\pi \times 10^{-7}} = 29.5$$

---

Before saturation, the material supports magnetic flux over 1000 times better than free space. However, beyond saturation the relative permeability approaches that of a nonmagnetic material.

The current in many electromagnetic devices periodically reverses. This creates the need for a *four-quadrant BH* curve to more completely describe the behavior of core materials in circuits with polarity reversals. Figure 10-25 shows the general method of generating the required data. The double-pole double-throw (DPDT) switch allows the current flowing in the coil to be reversed. Of course, this also reverses the magnetizing force. The variable resistor allows the current and the magnetizing force to be set at various values. An instrument for measuring flux density is also required. Plotting the data obtained from this arrangement results in the *hysteresis loop* shown in Fig. 10-26.

A hysteresis loop is a four-quadrant $BH$ graph. The four quadrants are identified with Roman numerals I through IV in Fig. 10-26. The data are collected in the following manner:

**1.** The measurements begin with the sample completely demagnetized and the variable resistor set to infinity ohms (no current). This results in no flux density ($B = 0$) and no magnetizing force ($H = 0$) and is labeled as point 0 on the graph.

**2.** The resistance is decreased and current starts to flow, which produces a magnetizing force and a

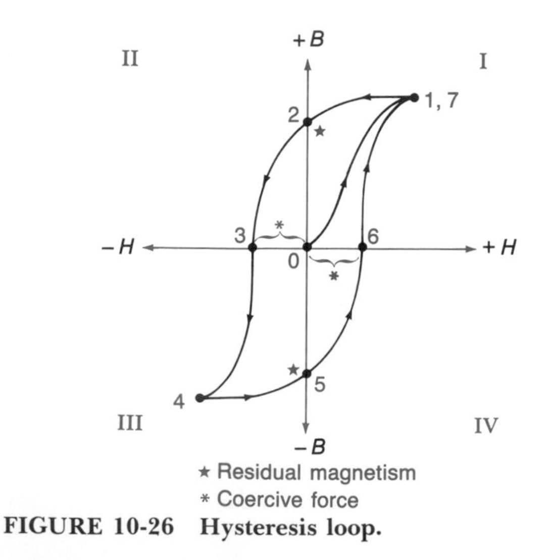


**FIGURE 10-26 Hysteresis loop.**

resulting flux density. Enough readings are taken to plot the portion of the curve from 0 to 1 on the graph. Point 1 is the beginning of saturation.

**3.** The resistance is now increased until the magnetizing force returns to zero again. Enough readings are taken to plot the graph from point 1 to point 2. *Note*: The remaining flux density at point 2 is caused by a number of domains remaining in alignment after the magnetizing force disappears. This is called *residual magnetism.* The ability of a material to retain magnetism is called its *retentivity.*

**4.** The polarity switch is thrown and the resistance is decreased again. This provides the data from point 2 to 3 on the graph. Note that a considerable value of $-H$ is required to return the flux density to zero. This value is known as the *coercive force.*

**5.** The variable resistor is decreased even more, resulting in the data graphed from point 3 to 4. Point 4 is the mirror image of point 1. The magnitudes of $B$ and $H$ are the same, but the polarities are reversed.

**6.** The resistance is increased until the current flow stops again. The data collected result in the curve from point 4 to 5. Point 5 is the residual magnetism and corresponds to point 2.

**7.** The polarity switch is thrown, and the resistance is decreased, providing the data from point 5 to 6. The value of $+H$ at point 6 is the coercive force and corresponds to point 3.

**8.** Finally, the resistance is further decreased, generating the data from point 6 to 7. The hysteresis loop is now complete.

The hysteresis loop shows a "lagging behind" effect. The retentivity of the material provides flux when the magnetizing force has decreased to zero. A coercive force is required to eliminate the residual flux. In circuits where the flux must continuously reverse, this shows up as *heat in the core.* To completely eliminate this loss in the core, an ideal ferromagnetic material would have zero retentivity, as shown in Fig. 10-27. There is no such material, but special silicon-iron alloys have been developed that greatly reduce hysteresis loss in the core. These materials are used extensively in *alternating* (reversing) current motors and transformers. The graph for one of these alloys forms a loop that is more narrow than the one shown in Fig. 10-26.

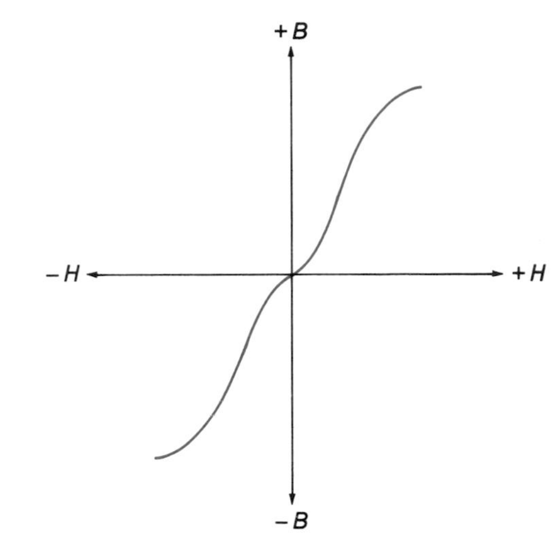


**FIGURE 10-27 Four-quadrant *BH* graph for a theoretical ferromagnetic material with zero retentivity.**

## Self-Test

**23.** When all of the domains in a ferromagnetic core are aligned, the core is in ______.

**24.** Calculate the static permeability for the air curve on the graph shown in Fig. 10-21 where $H = 6000$ (A · t)/m. (Do not forget to compensate for the vertical expansion.)

**25.** What is the relative permeability of air based on your results from Question 24?

**26.** Why will the same answer be obtained in Question 24 if different points on the graph are used?

**27.** What is the static relative permeability for cast iron at the point where $H = 2500$ (A · t)/m in Fig. 10-21?

**28.** Rework Question 27 for the point where $H =$ 500 (A · t)/m.

**29.** Why are the answers different in Questions 27 and 28?

**30.** Refer to Fig. 10-22. Which region of the curve shows the greatest dynamic permeability?

**31.** Refer to Fig. 10-22. Which region shows the smallest dynamic permeability?

**32.** The value of $H$ needed to eliminate the residual magnetism in a core is called the ______ force.

**33.** True or false? A *narrow* hysteresis loop is desirable for transformer iron.

**34.** True or false? A *wide* hysteresis loop is desirable for materials used in the manufacture of permanent magnets.

## 10-5 MAGNETIC CIRCUITS

The reluctance of a magnetic circuit is inversely proportional to the permeability of the material used to support the flux. Table 10-2 compares the opposition offered by magnetic materials to the opposition offered by electrical materials. The equations show that opposition to electric and magnetic flow is directly related to length and inversely related to area. Conductivity is the electrical analog of permeability as noted previously in Table 10-1.

Figure 10-28 shows a long air-core coil. The magnetic flux flows inside the coil and returns outside the coil. When the coil length is appreciably longer than the coil diameter, the reluctance of the circuit is approximately equal to the reluctance of the air core itself. Inspection of the reluctance equation in Table 10-2 shows why. The cross-sectional area of the air core is quite small compared to the area of the return path. The core reluctance is therefore much larger than the reluctance of the return path, even though the return path is longer. The reluctance of the core effectively swamps the magnetic circuit and permits calculations to be based on the core itself.

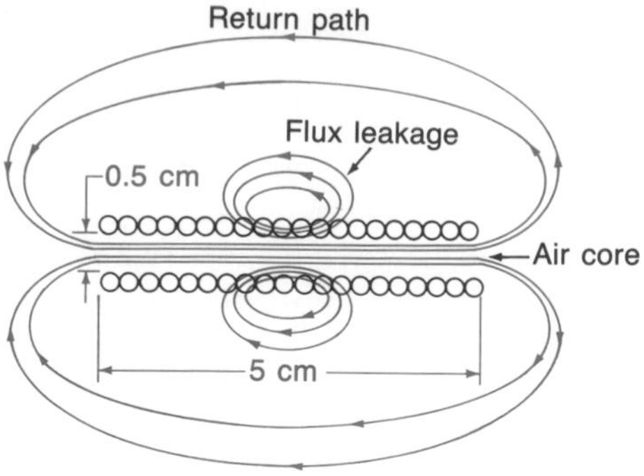

**FIGURE 10-28 The flux of a long air-core coil.**

### EXAMPLE 10-7

The coil in Fig. 10-28 is cylindrical and is appreciably longer than its diameter. Find the reluctance of its air core. Calculate the flux produced by this coil and the flux density, assuming a current flow of 10 A and 20 turns.

#### Solution

The coil cross section is a circle. The radius ($r$) is one-half the diameter. The area of the air core, in square meters, is found by

$$a = \pi r^2 = 3.14 \times (0.25 \times 10^{-2})^2 = 1.96 \times 10^{-5}\ \text{m}^2$$

The reluctance of the air core can be found from its dimensions and the permeability of free space. Using the equation from Table 10-2,

$$\mathcal{R} = \frac{0.05}{4\pi \times 10^{-7} \times 1.96 \times 10^{-5}} = 2.03 \times 10^{9}\ \ (\text{A} \cdot \text{t})/\text{Wb}$$

The mmf is found next:

$$\mathcal{F} = 10\ \text{A} \times 20\ \text{t} = 200\ \text{A} \cdot \text{t}$$

Ohm's law for magnetic circuits establishes the flux:

$$\Phi = \frac{\mathcal{F}}{\mathcal{R}} = \frac{200\ \text{A} \cdot \text{t}}{2.03 \times 10^{9}\ (\text{A} \cdot \text{t})/\text{Wb}} = 9.87 \times 10^{-8}\ \text{Wb}$$

The flux density in the core can be determined by dividing the flux by the area of the core:

$$B = \frac{9.87 \times 10^{-8}\ \text{Wb}}{1.96 \times 10^{-5}\ \text{m}^2} = 5.03 \times 10^{-3}\ \text{T}$$

*Note*: This result is consistent with Example 10-5.

Long air-core coils can be analyzed with this approach. However, there are some important limitations:

**Table 10-2**

| Magnetic Material Opposition | Electrical Material Opposition |
|---|---|
| $\mathcal{R} = \dfrac{l}{\mu a}$ | $R = \dfrac{l}{\sigma a}$ |
| where $\mathcal{R}$ = reluctance, (A · t)/Wb | where $R$ = resistance, Ω |
| $l$ = length, m | $l$ = length, $m$ |
| $\mu$ = permeability, Wb/(A · t · m) | $\sigma$ = conductivity, S/m |
| $a$ = area, $\text{m}^2$ | $a$ = area, $\text{m}^2$ |

**1.** The approach ignores the reluctance of the return path. It is reasonably accurate only when the coil length is at least 10 times the coil diameter.

**2.** There is an appreciable flux leakage in a long air-core coil, as Fig. 10-28 shows. Therefore, the calculated flux density is valid for the center of the coil only.

**3.** The flux density at the poles will be approximately half the calculated value.

**4.** The approach is not valid if the air core is replaced with a high-permeability material such as shown in Fig. 10-29. Although it is not difficult to find the core reluctance, the reluctance of the return path can no longer be ignored. It is difficult to determine the dimensions of the return path, and problems of this type are beyond the scope of this book.

Most practical magnetic circuit problems do not have to deal with long air return paths. The vast majority of the flux is confined to a high-permeability core. Practical circuits can be solved by considering the length and area of the core. A graphical approach, based on *BH* curves for the core material, eliminates the need to calculate reluctance. In most cases, the desired flux is known and the mmf must be found. The general procedure involves finding

1. The cross-sectional area
2. The flux density
3. The magnetizing force (from the *BH* curves)
4. The *average* path length
5. The mmf
6. The required current or number of turns

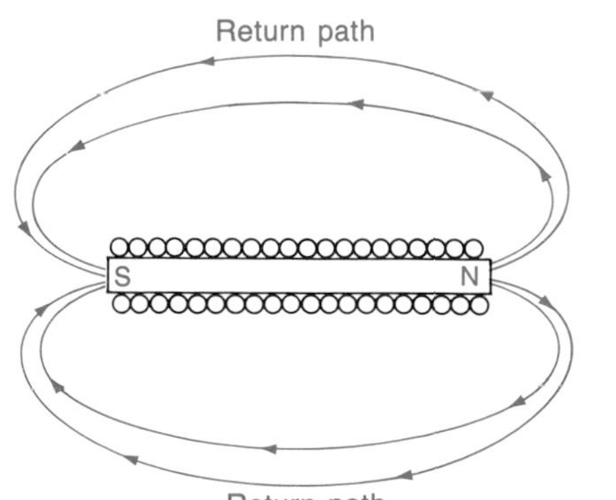


**FIGURE 10-29 A long coil with a ferromagnetic core.**

**EXAMPLE 10-8**

Use the graph shown in Fig. 10-30 to solve the magnetic circuit shown in Fig. 10-31 for the required coil current.

**Solution**

The core is a doughnut-shaped structure known as a *toroid*. Toroid cores do a good job of confining all of the flux to the core itself. The leakage flux is usually so small that it can be ignored. The toroid core shown in Fig. 10-31 has a round cross section. Its area $a$ is found first:

$$a = 3.14 \times (0.5 \times 10^{-2})^2 = 7.85 \times 10^{-5}\ \text{m}^2$$

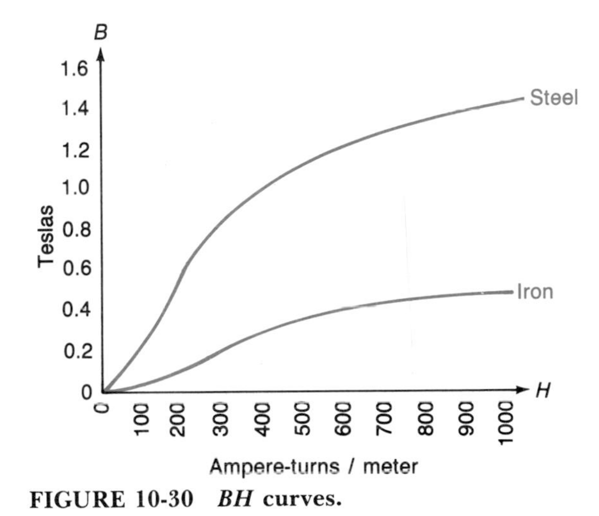


**FIGURE 10-30 *BH* curves.**

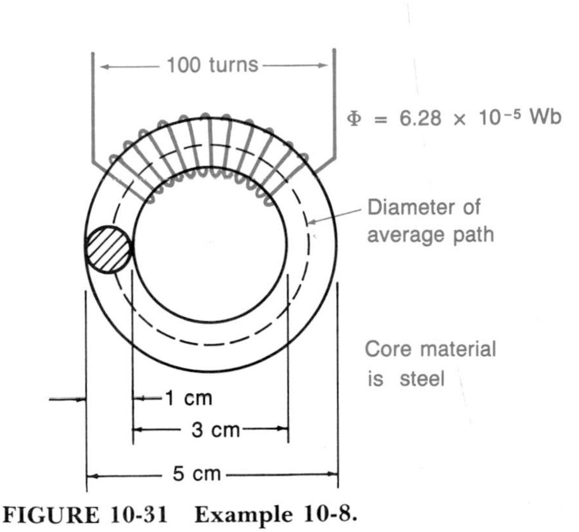


**FIGURE 10-31 Example 10-8.**

The required flux density can now be determined:

$$B = \frac{6.28 \times 10^{-5}\ \text{Wb}}{7.85 \times 10^{-5}\ \text{m}^2} = 0.8\ \text{T}$$

Using the steel curve from Fig. 10-30 shows the required magnetizing force to be

$$H = 275\ (\text{A} \cdot \text{t})/\text{m}$$

The average path length $l$ is found by establishing the middiameter of the toroid core. The opening is 3 cm and the width is 5 cm for a middiameter of 4 cm. The path length in meters is

$$l = \pi D = 3.14 \times 0.04\ \text{m} = 0.126\ \text{m}$$

The mmf is found by multiplying the magnetizing force times the path length:

$$\text{mmf} = 275\ (\text{A} \cdot \text{t})/\text{m} \times 0.126\ \text{m} = 34.6\ \text{A} \cdot \text{t}$$

Finally, the required current is determined by dividing the mmf by the number of turns:

$$I = \frac{34.6\ \text{A} \cdot \text{t}}{100\ \text{t}} = 346\ \text{mA}$$

A current of 346 mA produces a flux of $6.28 \times 10^{-5}$ Wb in the toroid core.

---

Toroids with solid cores are difficult to wind. The wire is placed on a bobbin and passed through the core over and over until the required number of turns is achieved. Many magnetic devices use assembled rectangular cores such as the one shown in Fig. 10-32. The winding is prepared on a separate form, and the core is then assembled into the coil form from separate pieces. This is how most transformers are made. Rectangular cores show some flux leakage. However, it is still small enough to be ignored in most cases.

---

**EXAMPLE 10-9**

Find the required number of turns for the magnetic circuit shown in Fig. 10-32. Use the graph from Fig. 10-30 to determine the magnetizing force.

**Solution**

The procedure is similar to that used for the toroid core example. The cross section of the magnetic circuit is uniform and its area is

$$a = 0.02\ \text{m} \times 0.02\ \text{m} = 4 \times 10^{-4}\ \text{m}^2$$

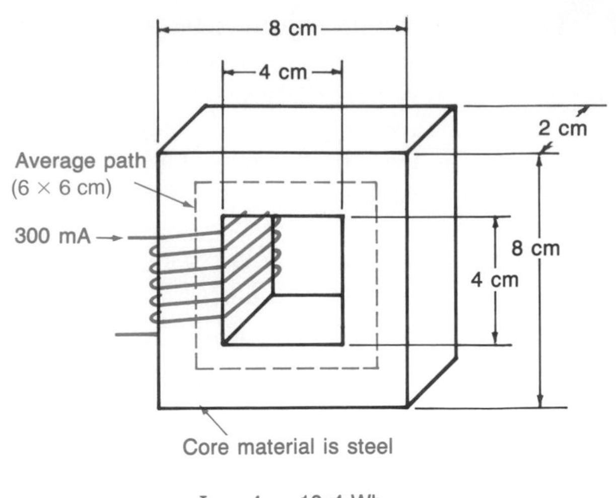

**FIGURE 10-32** **Example 10-9.**

The flux density is found by dividing the flux by the cross-sectional area:

$$B = \frac{4 \times 10^{-4}\ \text{Wb}}{4 \times 10^{-4}\ \text{m}^2} = 1\ \text{T}$$

The magnetizing force required to produce this flux density is determined from the steel curve in Fig. 10-30:

$$H = 400\ (\text{A} \cdot \text{t})/\text{m}$$

The average path length is calculated by adding the rectangular components shown in Fig. 10-32:

$$l = 6\ \text{cm} + 6\ \text{cm} + 6\ \text{cm} + 6\ \text{cm} = 24\ \text{cm} = 0.24\ \text{m}$$

The mmf is based on the magnetizing force and the path length:

$$\text{mmf} = 400\ (\text{A} \cdot \text{t})/\text{m} \times 0.24\ \text{m} = 96\ \text{A} \cdot \text{t}$$

Finally, the number of turns can be found:

$$\frac{96\ \text{A} \cdot \text{t}}{0.3\ \text{A}} = 320\ \text{t}$$

---

Figure 10-33 shows a magnetic circuit based on two different core materials. Circuits of this type are analogous to series electric circuits. For example, ignoring flux leakage, *the flux is constant* in all parts of the circuit. Also, the total reluctance $\mathcal{R}_T$ in

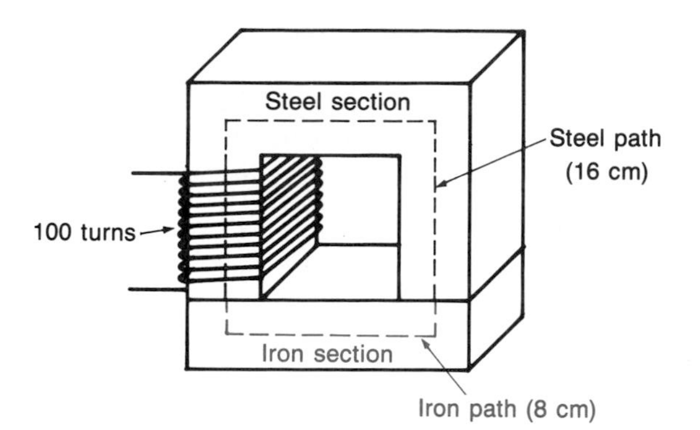


**FIGURE 10-33 A magnetic core composed of two different materials.**

a series magnetic circuit is equal to the sum of the individual reluctances:

$$\mathcal{R}_T = \mathcal{R}_1 + \mathcal{R}_2 + \mathcal{R}_3 + \cdots + \mathcal{R}_n$$

It is possible to analyze series magnetic circuits by finding the individual reluctances. However, Fig. 10-34 shows a better approach. The total mmf can be found by summing all the drops in mmf. A drop in mmf for a section of the circuit is found by multiplying the magnetizing force times the length of that section. This is based on a rearrangement of

$$H = \frac{\mathcal{F}}{l}$$

The equation shown in Fig. 10-34 is the magnetic analog for Kirchhoff's voltage law around a closed electric loop.

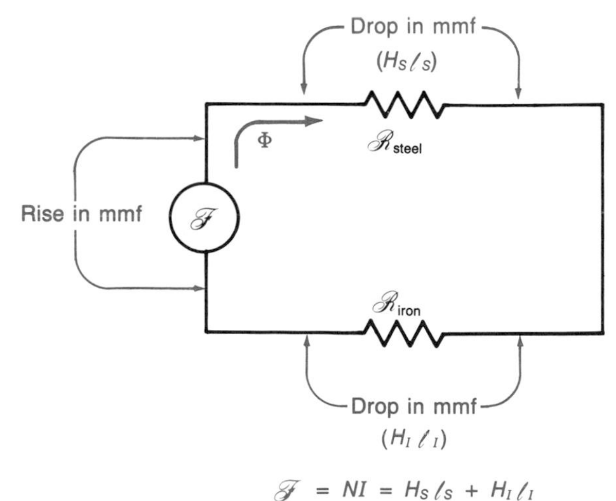


**FIGURE 10-34 The total mmf in a series magnetic circuit is equal to the sum of the drops in mmf.**

---

**EXAMPLE 10-10**

Calculate the current required to establish a flux of $1.2 \times 10^{-4}$ Wb in the magnetic circuit shown in Fig. 10-33. Use the graphical data from Fig. 10-30. The dimensions are the same as the core shown in Fig. 10-32.

**Solution**

The area is the same as in Example 10-9:

$$a = 4 \times 10^{-4}\ \text{m}^2$$

The flux density is

$$B = \frac{1.2 \times 10^{-4}\ \text{Wb}}{4 \times 10^{-4}\ \text{m}^2} = 0.3\ \text{T}$$

The required magnetizing force for each section is read from the graph:

$$H_{\text{steel}} = 125\ (\text{A} \cdot \text{t})/\text{m}$$
$$H_{\text{iron}} = 400\ (\text{A} \cdot \text{t})/\text{m}$$

The average length must be found for each section. Note, in Fig. 10-33, that there are two 1-cm turns in the iron path. The iron path is therefore 6 cm + 2 cm = 8 cm. The total steel path is 5 cm + 6 cm + 5 cm = 16 cm. The path lengths must be in units of meters:

$$\text{Steel} = 0.16\ \text{m}$$
$$\text{Iron} = 0.08\ \text{m}$$

The drop in mmf can now be established for each section of the series circuit:

$$\text{mmf}_{\text{steel}} = 125\ (\text{A} \cdot \text{t})/\text{m} \times 0.16\ \text{m} = 20\ \text{A} \cdot \text{t}$$
$$\text{mmf}_{\text{iron}} = 400\ (\text{A} \cdot \text{t})/\text{m} \times 0.08\ \text{m} = 32\ \text{A} \cdot \text{t}$$

The required rise in mmf is equal to the sum of the drops:

$$\text{mmf} = 20 + 32 = 52\ (\text{A} \cdot \text{t})/\text{m}$$

The necessary current is found by dividing the mmf by the number of turns:

$$I = \frac{52\ \text{A} \cdot \text{t}}{100\ \text{t}} = 520\ \text{mA}$$

---

Electric circuits can have more than one rise in potential, and magnetic circuits can have more than one rise in mmf. Figure 10-35*a* shows a circuit with two reluctances and two sources of mmf. Figure 10-35*b* is an equivalent circuit. The mmf's can

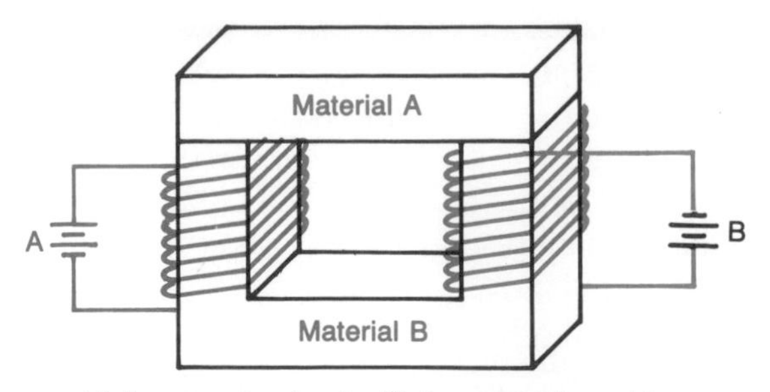


(a) A magnetic circuit with two matrials and two sources of mmf

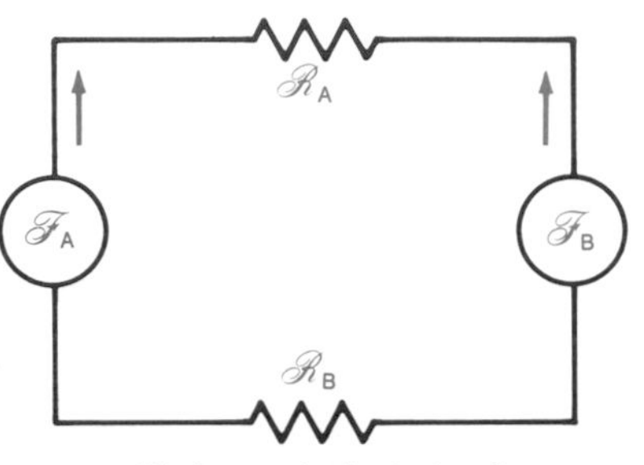

(b) An equivalent circuit

**FIGURE 10-35 A magnetic circuit may have more than one rise in mmf.**

be aiding or opposing just as emf's aid or oppose in series electric circuits. There are many similarities between magnetic and electric circuits. Many of the procedures and theorems for electric circuits can be used in magnetic circuits. However, the superposition theorem cannot be used to properly analyze a circuit that is nonlinear, which includes most magnetic circuits.

Some magnetic circuits have *air gaps.* An air gap is a necessity in a rotating machine such as a motor or a generator. It provides clearance between the fixed and moving parts. Air gaps are also used to prevent saturation in some magnetic devices. Figure 10-36 shows a core with an air gap. The lines of force in an air gap repel each other, causing *fringing* as shown in Fig. 10-37. Fringing can complicate the analysis of magnetic circuits. However, most air gaps are short and the fringing is slight. Fringing can be ignored in these cases, which simplifies the analysis. The flux density in the gap is assumed to be equal to the flux density in the core, which allows the magnetizing force of the gap to be found with the relationship $H = B/\mu_o$:

$$H_{gap} = \frac{B}{4\pi \times 10^{-7}}$$

And the drop in mmf across the gap is

$$\text{mmf}_{gap} = H_{gap} \times \text{gap length}$$

Figure 10-38*a* shows a magnetic circuit with an air gap and a parallel section. Circuits of this type are easiest to deal with by breaking the paths into separate straight-line sections. Figure 10-38*b* shows the equivalent circuit. Note that each straight-line path is represented by its own reluctance. This includes the air gap. Series-parallel magnetic circuits follow the same rules as their electric counterparts. The total flux in Fig. 10-38 is equal to the sum of each branch flux:

$$\Phi_T = \Phi_1 + \Phi_2$$

The drop in mmf across $\mathcal{R}_{AF}$ is equal to the sum of the drops caused by $\Phi_1$ flowing through the reluctances at the right end of the circuit.

Suppose the circuit of Fig. 10-38 must be designed to produce a given flux in the air gap. In other words, $\Phi_1$ is known. The flux density would be found first. The magnetizing force for the air gap would be found next using the permeability of free space. The flux density would be the same for the other reluctances in that part of the circuit:

$$B_{AB} = B_{BC} = B_{gap} = B_{DE} = B_{EF}$$

The *BH* curve for the core material would be used to find the magnetizing force. Then, a loop equation would be written and solved to find the drop across $\mathcal{R}_{AF}$:

$$Hl_{AB} + Hl_{BC} + H_{gap}l_{gap} + Hl_{DE} + Hl_{EF} = Hl_{AF}$$

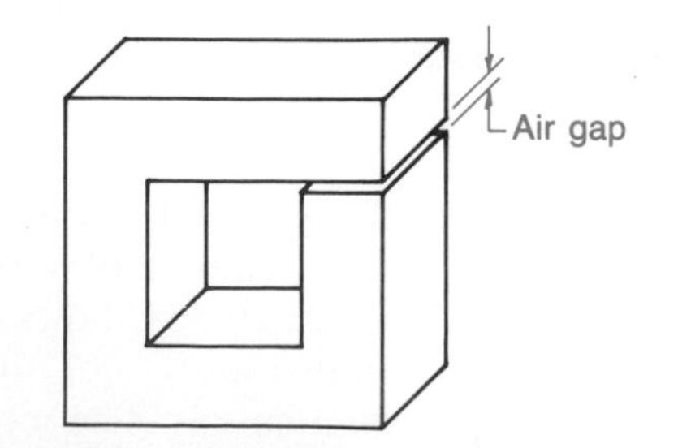


**FIGURE 10-36 An air gap in a magnetic circuit.**

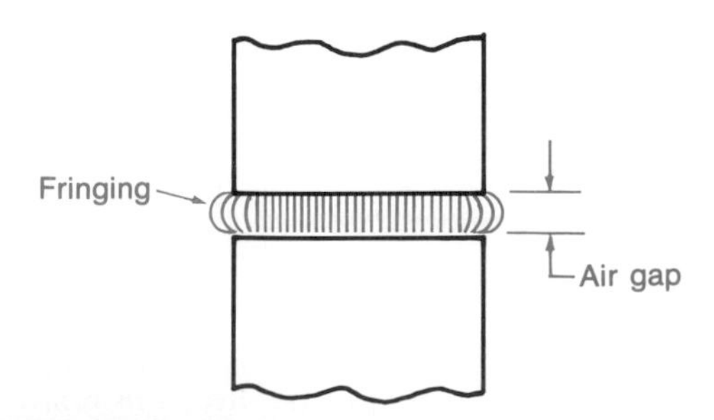


**FIGURE 10-37 Flux fringing at an air gap.**

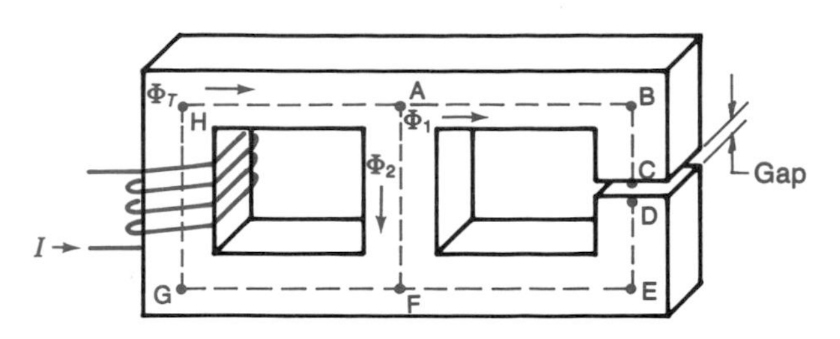

(a) A magnetic circuit

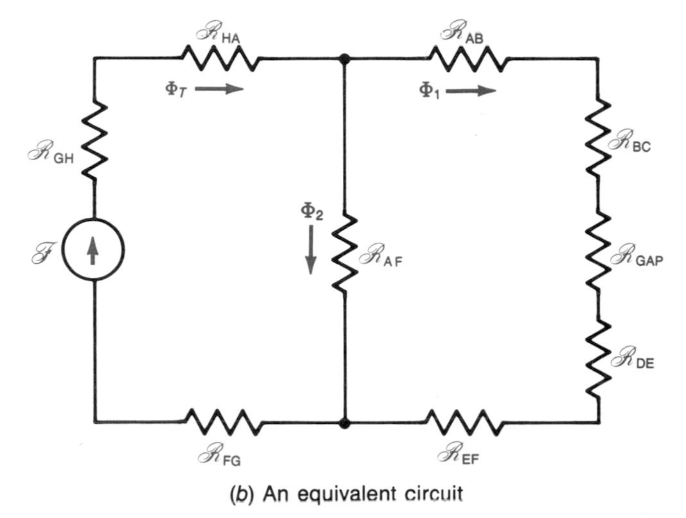

(b) An equivalent circuit

**FIGURE 10-38 A series-parallel approach may be used to analyze some magnetic circuits.**

And dividing this by the length from A to F would give

$$\frac{Hl_{AF}}{l_{AF}} = H_{AF}$$

This would then be applied to the $BH$ curve to find the flux density $B_{AF}$. The flux in path A to F would be found next with

$$\Phi_2 = B_{AF} \times \text{area}$$

And the total flux:

$$\Phi_T = \Phi_1 + \Phi_2$$

Knowing the total flux would allow the flux density to be found for the left-hand portion of the circuit. Once again, the $BH$ curves would be consulted to find the required magnetizing force. Finally, the left-hand loop equation would be generated and solved for the required mmf:

$$\mathcal{F} = NI = H_{GH}l_{GH} + H_{HA}l_{HA} + H_{AF}l_{AF} + H_{FG}l_{FG}$$

## Self-Test

**35.** A cylindrical air-core coil is 10 cm long and has an inside diameter of 1 cm. Calculate the reluctance of its core.

**36.** Assuming 50 turns and a current of 10 A, calculate the flux density at the center of the core for the coil in Question 35.

**37.** What is the approximate flux density at the poles of the coil in Question 36?

**38.** The toroid core shown in Fig. 10-39*a* has a round cross section. Use the graph in Fig. 10-30

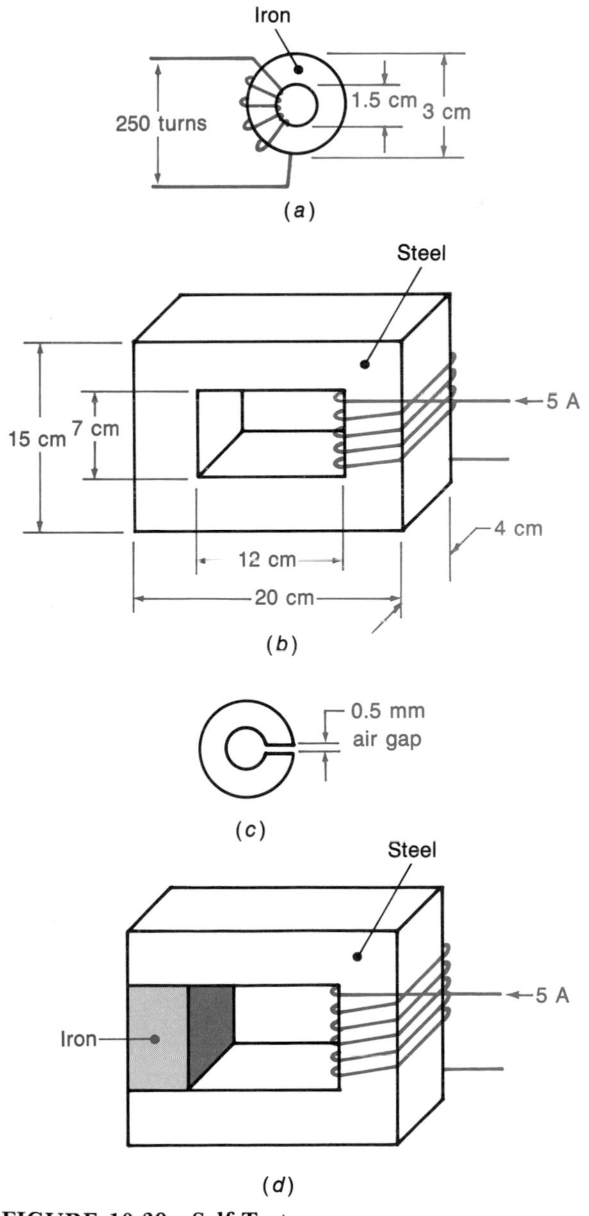

**FIGURE 10-39 Self-Test.**

and calculate the current needed to produce a flux of $10 \times 10^{-6}$ Wb in the core.

**39.** How many turns will be needed to produce a flux of $2.24 \times 10^{-3}$ Wb in the core shown in Fig. 10-39*b*? Use the graph shown in Fig. 10-30.

**40.** Figure 10-39*c* shows a toroid core with an air gap. This core has the same dimensions and the same winding and is made from the same material as the one shown in Fig. 10-39*a*. How much current will be required to produce a flux of $10 \times 10^{-6}$ Wb?

**41.** How many turns will be needed to produce a flux of $6.4 \times 10^{-4}$ Wb in the core of Fig. 10-39*d*? The dimensions are identical to those shown in Fig. 10-39*b*.

## 10-6 APPLICATIONS

Some instruments and devices require magnetic shielding. Figure 10-40 shows a meter movement within a shield. The meter in the illustration uses a permanent magnet and a rotating coil. When a current passes through the meter coil, a second field is generated which interacts with the field produced by the permanent magnet. The interaction produces a twisting force on the coil and drives the pointer up the scale. External fields can add to or subtract from the twisting force and cause errors in instruments of this type. Cathode-ray tubes are also subject to errors from external magnetic fields. These tubes produce a beam of electrons to display information on their screens. An external field can deflect the beam and cause distortion. Oscilloscopes often use shielding around their cathode-ray tubes to prevent this. Magnetic shields are made from materials with high permeability. As Fig. 10-40 illustrates, the high permeability path supports the lines of the external field and prevents them from disrupting the instrument inside the shield.

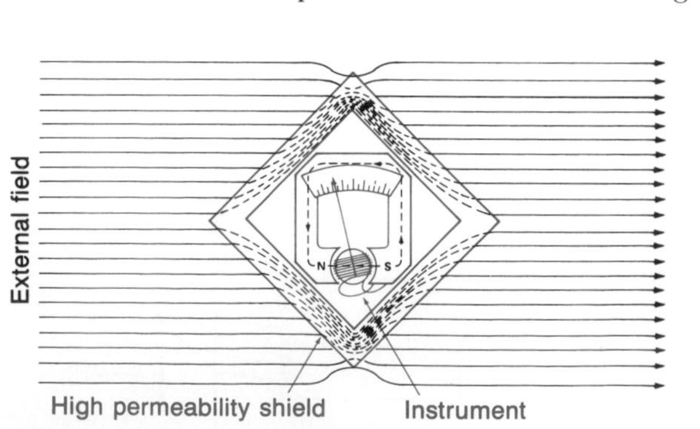


**FIGURE 10-40 A sensitive magnetic instrument may require shielding to reduce the influence of external fields.**

*Transformers* are devices used to transfer electric energy from one circuit to another via a magnetic field. Figure 10-41 shows an example and the schematic symbol. Transformers are operated with the primary winding connected to an electric energy source. A field is generated by the primary coil which links the turns of the secondary coil. The example shown uses an iron core to support the flux, but some transformers use an air core. Transformers may be designed to provide a voltage step-up or step-down function. For example, the ignition coil in an automobile is a transformer that produces as much as 40,000 V when energized from a 12-V battery. Transformers are covered in detail in Chap. 20.

Figure 10-42 shows a *solenoid*. A solenoid is a coil with an iron core and a moveable iron plunger. When the coil is energized, the plunger is attracted by the coil. It "pulls in," and this motion can be used to activate another mechanism. Solenoids are applied in many electrically activated devices such as valves, locks, punches, and marking machines.

*Relays* are widely applied electromagnetic devices. Figure 10-43 shows a typical relay and schematic symbol. When the relay is not energized, the spring keeps the armature away from the coil. This

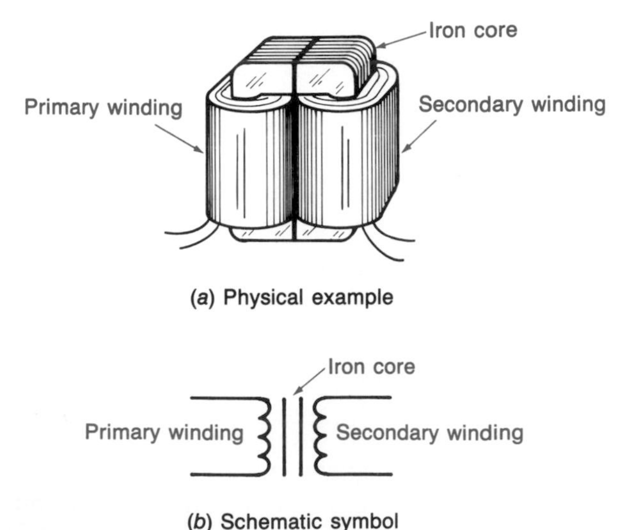


**FIGURE 10-41 Transformer.**

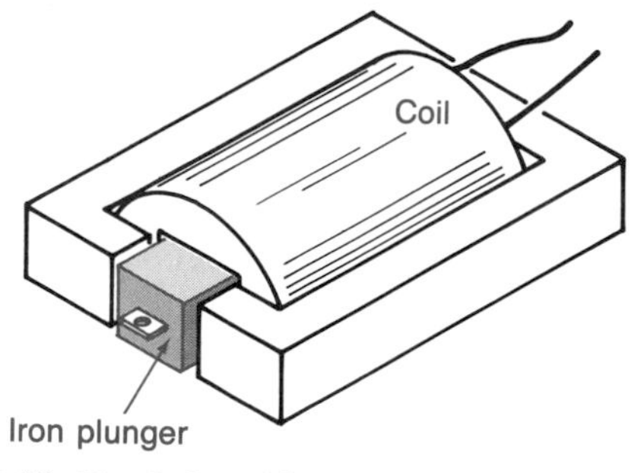


**FIGURE 10-42 Solenoid.**

produces an air gap, and the main contact presses against the normally closed contact. When the relay is energized, the armature is attracted and moves toward the coil. This eliminates the air gap, and the main contact touches the normally open contact and completes that circuit. The circuit with the normally closed contact is opened. The relay acts as a single-pole double-throw switch. Of course, many different contact arrangements are possible.

Relays require a given current for pull-in. Once they pull in, much less current is required to hold them in the closed position. This is because the air gap is eliminated when the armature pulls in. The air gap has quite a bit more reluctance than the iron circuit, and eliminating it means that less mmf is required to overcome the spring tension.

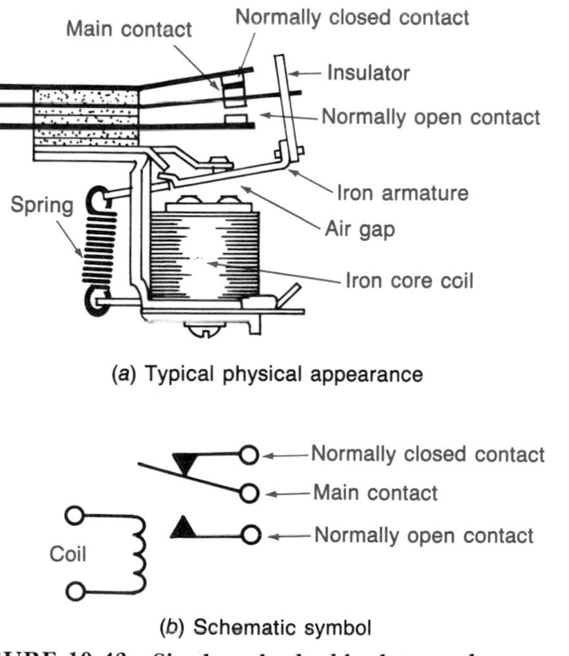


**FIGURE 10-43 Single-pole double-throw relay.**

Magnetic circuit breakers are very similar to relays. The load current flows through the coil. When the current becomes excessive, the spring tension is overcome. The armature moves and opens the circuit contacts. This interrupts the load current and the coil current. A latching mechanism "catches" the armature and prevents it from returning to its original position. The breaker remains tripped until the latch is released. Magnetic circuit breakers react quickly. Their fast tripping action is an advantage for protecting against damage from short circuits. Thermal breakers take more time to trip and offer an advantage for protecting circuits where current surges are normal. Some breakers use both thermal and magnetic principles to provide time delay tripping for moderate overloads and very fast tripping for severe overloads.

Figure 10-44 shows a simplified *series motor*. A magnetic field is set up by the current flowing

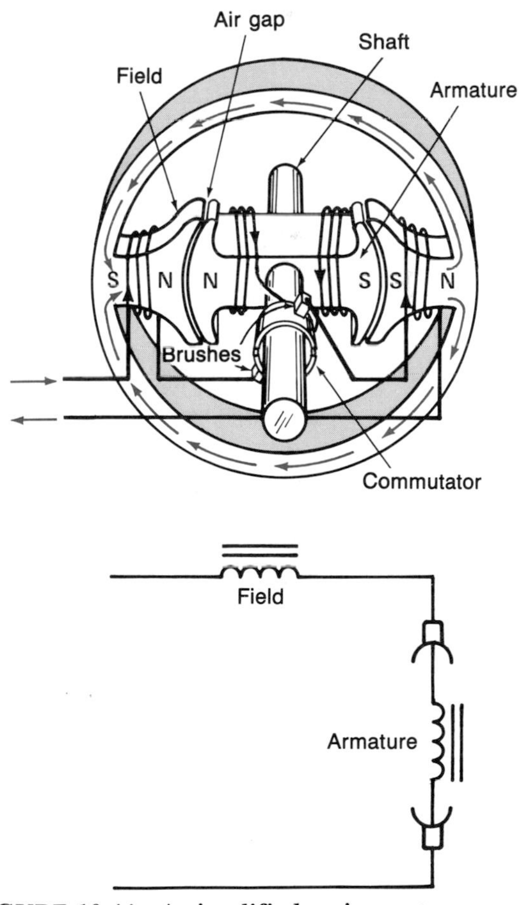


**FIGURE 10-44 A simplified series motor.**

through the field coils. The field circuit is in series with the armature circuit where a second magnetic field is generated. The armature rotates and must be connected by carbon brushes which press against brass commutator segments. The interaction of the two magnetic fields produces a twisting force (torque) at the shaft. As the illustration shows, the fields repel. An armature rotation of 180° still shows repulsion because the commutator segments then contact the opposite brushes and the current flow in the armature is reversed. Thus, the two magnetic fields always react to continue the twisting force at the shaft. There are many types of motors, and Fig. 10-44 is a brief glimpse at one particular design.

The field coils in the motor shown in Fig. 10-44 may be eliminated and replaced with a permanent magnet. Such motors are called *permanent-magnet motors* and are usually employed in low-cost applications.

Loudspeakers are another example of devices which use permanent magnets. As Fig. 10-45 shows, a permanent circular ceramic (or alnico) magnet is used to establish a field through the *voice coil*. The pole pieces are made from high-permeability materials and are arranged to shorten the air gaps as much as possible. The voice coil is connected to the output of an audio amplifier. When the amplifier sends a current through the voice coil, a second field is generated which interacts with the permanent magnetic field. Depending on the direction of current in the voice coil, the coil and cone assembly are driven in or out. The motion of the cone disturbs the air molecules and a sound wave is produced. Headphones have a similar structure, but on a smaller scale.

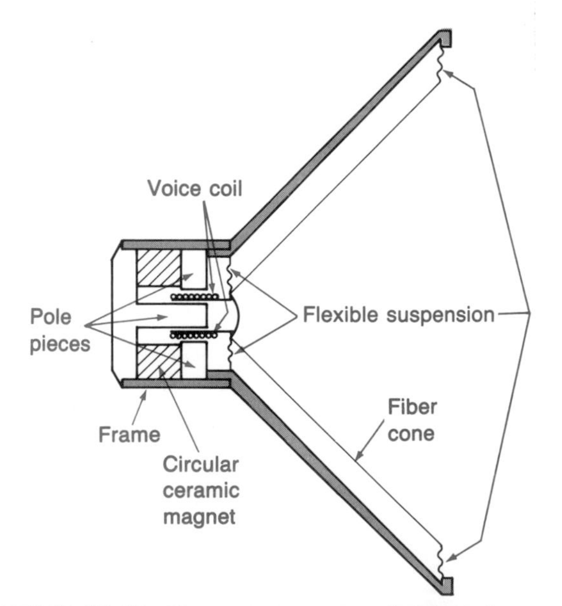


**FIGURE 10-45 Permanent-magnet loudspeaker.**

Some microphones (the dynamic types) are very much like the loudspeaker shown in Fig. 10-45. They are built much smaller, and the coil is connected to the *input* of an amplifier, rather than the output. When sound waves strike the cone structure of the microphone, the coil vibrates in the field of the permanent magnet. A small voltage is generated across the coil and is sent to the amplifier. Some intercoms take advantage of the similarity between speakers and microphones and switch a single device between the two functions.

Magnetic cartridges in record players also use a permanent magnet and a moving coil. The movement comes from the stylus riding in the groove of the record. A small voltage is generated which must be amplified to be useful.

Figure 10-46 shows how magnetic principles can be applied in recording. The tape or disk is coated with iron oxide, a ferromagnetic material. The recording head is connected to the output of an amplifier. The current flow in the head sets up a field and arranges the domains in the coating in a pattern established by the head current. Later, the tape or disk can be played back by using the same head or a separate read head. The coil is connected to the input of an amplifier during playback. As the domain patterns pass under the head, a small voltage is generated which is a replica of the original signal that passed through the recording head.

When a current-carrying conductor or semiconductor is placed in a transverse (perpendicular) magnetic field, the current carriers are deflected. This deflection is known as the *Hall effect* and creates a potential difference across the faces of

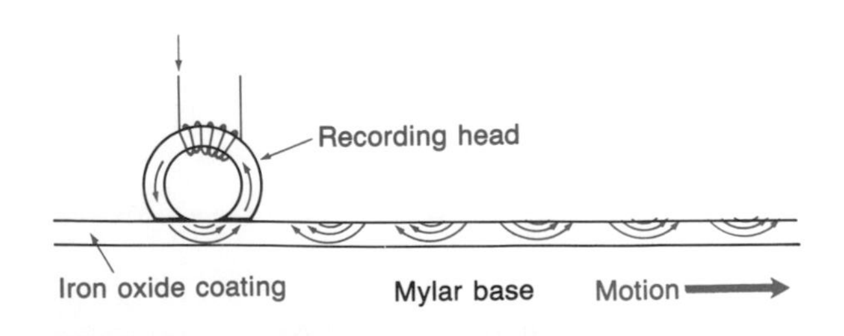

**FIGURE 10-46 Magnetic recording.**

the conductor or semiconductor. This potential difference is called the *Hall voltage,* and it is proportional to the product of the current flow and the flux density of the magnetic field. Figure 10-47 shows a Hall-effect sensor and its output curves. The typical output curve shows an increase in output voltage with flux density. The output voltage drops when the flux direction is reversed. Hall effect devices are used

1. In some switches and keyboards
2. To convert position to voltage
3. To measure angular displacement and velocity
4. To measure current (by sensing the field around a conductor)
5. To measure flux density
6. As thickness gages for nonferrous metals
7. As metal and proximity sensors

The applications of magnetism and electromagnetism are diverse. More information concerning the interaction of electricity and magnetism is presented in following chapters. Some of it will focus on how a voltage is *induced* in a conductor by a magnetic field that shows relative motion to the conductor. Technicians are confronted with many devices and circuits that use magnetic principles. Those who work in the power area are more involved with large devices such as motors and transformers. The U.S. customary system of measurement is still in use in the power area. Table 10-3 is offered as a reference for converting from the SI units presented in this chapter to these units. The unit of mmf is the ampere-turn in both systems. As the table shows, the U.S. customary system uses the inch as the basis for length and area measurements and the *line* as the unit of flux.

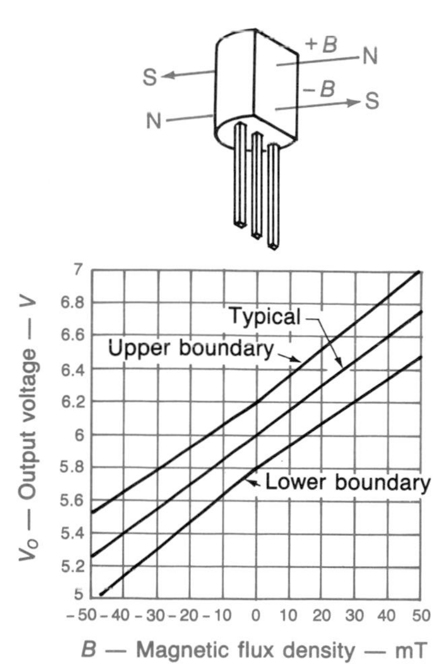


**FIGURE 10-47 Texas Instrument TL173 Hall-effect sensor.**

**Table 10-3 Converting from SI Units to U.S. Customary Units**

| | | |
|---|---|---|
| $l$ | Multiply m × 39.37 | to obtain in. |
| $a$ | Multiply $m^2 \times 1550$ | to obtain $in.^2$ |
| $\Phi$ | Multiply Wb × $10^8$ | to obtain lines |
| $B$ | Multiply $Wb/m^2 \times 6.452 \times 10^4$ | to obtain $lines/in.^2$ |
| $H$ | Multiply $(A \cdot t)/m \times 2.54 \times 10^{-2}$ | to obtain $(A \cdot t)/in.$ |
| $\mu_o$ | $4\pi \times 10^{-7}$ Wb/(A · t · m) = 3.19 lines/(A · t · in.) | |

## Self-Test

42. True or false? Magnetic shields are made from diamagnetic materials.

43. Electric energy in a transformer is transferred from the primary winding to the ______ winding with a magnetic field.

44. True or false? The pull-in current of a relay is significantly more than its holding current.

45. True or false? Thermal circuit breakers react to overloads in less time than magnetic circuit breakers.

46. True or false? Reversing the polarity of the electric supply of the motor shown in Fig. 10-44 will change the direction of shaft rotation.

47. True or false? Suppose the field circuit of the motor shown in Fig. 10-44 is eliminated and replaced with a permanent magnet. The direction of shaft rotation will depend on supply polarity.

48. True or false? It is possible to use a permanent-magnet loudspeaker as a microphone.

49. The Hall voltage increases as flux density ______.

## SUMMARY


1. The north pole of a magnet is the one that seeks geographic north.
2. By convention, the direction of magnetic flow is from the north pole to the south pole in the external field of a magnet.
3. A magnetic field is composed of lines of force.
4. The poles of a magnet are those regions showing the greatest concentration of force lines.
5. Magnetizing a material places its domains in alignment.
6. Permanent magnets retain domain alignment while temporary magnets do not.
7. Magnetic lines exhibit tension, repel other magnetic lines, and will not cross another magnetic line.
8. Like poles repel and unlike poles attract.
9. Magnetic forces follow the inverse square law.
10. Magnetic lines follow the path of least reluctance.
11. A magnetic field exists around current-carrying conductors.
12. The left-hand rule is used to relate electron current direction to flux direction.
13. A dot on the end of a conductor signifies current flowing toward an observer. A cross indicates current flowing away from an observer.
14. The magnetic effect of a current-carrying conductor can be concentrated by winding the conductor into a coil.
15. Some of the lines generated by a long coil do not contribute to the pole flux and are known as leakage flux.
16. The flux generated by a coil can be increased by increasing current, the number of turns, or the permeability of the core material.
17. The SI unit of magnetomotive force (mmf) is the ampere or the ampere-turn. Its letter symbol is $\mathcal{F}$.
18. The SI unit of flux is the weber. A flux change of one weber per second generates one volt in a single-turn coil. The symbol for flux is $\Phi$.
19. Reluctance is the opposition to flux. Its letter symbol is $\mathcal{R}$ and its SI unit is the ampere-turn per weber.
20. The Ohm's law for magnetic circuits is

$$\mathcal{R} = \frac{\mathcal{F}}{\Phi}$$

21. Magnetizing force refers to the mmf per unit of length. Its letter symbol is $H$, and the SI unit is the ampere-turn per meter.
22. Flux density is the flux per unit area. Its letter symbol is $B$, and the SI unit is the tesla, which equals one weber per square meter.
23. Permeability is a measure of how well a material supports flux. Its letter symbol is $\mu$, and the SI unit is the weber per ampere-turn-meter:

$$\mu = \frac{B}{H}$$

**24.** The permeability of free space is a physical constant:

$$\mu_o = 4\pi \times 10^{-7} \text{ Wb/(A} \cdot \text{t} \cdot \text{m)}$$

**25.** Relative permeability is a pure number and is found by

$$\mu_r = \frac{\mu}{\mu_o}$$

**26.** Ferromagnetic materials have relative permeabilities much greater than 1.

**27.** Paramagnetic materials are only slightly better than free space for supporting flux.

**28.** Nonmagnetic materials have a relative permeability close to 1.

**29.** Diamagnetic materials have a relative permeability of less than 1.

**30.** The permeability of ferromagnetic materials varies with flux density.

**31.** When all of the domains in a material are aligned, the material is saturated.

**32.** Magnetization or *BH* curves are used to design magnetic circuits.

**33.** The *BH* curve is linear for air (or a vacuum) and is nonlinear for ferromagnetic materials.

**34.** Since a ferromagnetic material is nonlinear, its permeability can be derived by three approaches: (1) static, (2) dynamic, and (3) average.

**35.** A four-quadrant *BH* curve is called a hysteresis loop.

**36.** Materials with high retentivity have a square (wide) hysteresis loop.

**37.** Materials with low retentivity have a narrow hysteresis loop.

**38.** Hysteresis produces core loss in devices where the current periodically reverses. Special silicon steel alloys are used in these cores to reduce retentivity and hysteresis loss.

**39.** The reluctance of the total magnetic path for air-core coils that are at least 10 times longer than their diameters is approximately equal to the reluctance of the core alone.

**40.** The reluctance of a path can be determined by the permeability of the path material and its dimensions:

$$\mathcal{R} = \frac{l}{\mu a}$$

**41.** The total reluctance in a series circuit is equal to the sum of the individual reluctances:

$$\mathcal{R}_T = \mathcal{R}_1 + \mathcal{R}_2 + \mathcal{R}_2 + \cdots + \mathcal{R}_N$$

**42.** The flux is constant in a series circuit and, if the cross section is uniform, so is the flux density.

**43.** The total rise in mmf around a closed loop is equal to the sum of the drops in mmf.

**44.** Flux fringing can be ignored in short air gaps.

**45.** In a parallel magnetic circuit the total flux is equal to the sum of the branch flux:

$$\Phi_T = \Phi_1 + \Phi_2 + \Phi_3 + \cdots + \Phi_N$$

**46.** A drop in mmf may be found by multiplying the magnetizing force times the path length:

$$\mathcal{F} = Hl$$

**47.** Instruments that are sensitive to magnetic fields may require a high-permeability shield.

**48.** A transformer couples electric energy from one winding to another via a magnetic field.

**49.** Solenoids are commonly used to activate valves and other mechanisms.

**50.** Magnetic circuit breakers trip more quickly than thermal breakers.

**51.** A loudspeaker uses magnetic field interaction to change electric energy to sound energy.

**52.** Magnetic recording devices store information as domain patterns in a ferromagnetic coating.

**53.** A Hall voltage is generated across the faces of a current-carrying conductor in a transverse magnetic field.

## CHAPTER REVIEW QUESTIONS

**10-1.** Two like magnetic poles will show a force of ______ when placed near one another.

**10-2.** True or false? Magnetic flux lines repel each other.

**10-3.** True or false? Magnetic flux passes through a vacuum more readily than through iron.

**10-4.** In which direction is the current flowing in Fig. 10-48*a*?

**10-5.** What is the direction of the flux lines around the conductor shown in Fig. 10-48*b*?

**10-6.** Will the force be one of attraction or repulsion between the two conductors shown in Fig. 10-48*c*?

**10-7.** Which end of the coil shown in Fig. 10-48*d* is the south pole?

**10-8.** Which flux line in Fig. 10-48*d* represents leakage?

**10-9.** What will happen to the flux density at the poles of a coil if its air core is replaced with a ferromagnetic material?

**10-10.** What will happen to the flux density at the poles of a coil if its air core is replaced with a diamagnetic material?

**10-11.** Another name for the *BH* curve shown in Fig. 10-48*e* is a ______ curve.

**10-12.** Which point on the *BH* curve of Fig. 10-48*e* shows the smallest dynamic permeability?

**10-13.** Which point on the *BH* curve of Fig. 10-48*e* is in the saturation region?

**10-14.** Which point on the *BH* curve of Fig. 10-48*e* is in the linear region?

**10-15.** Dividing *B* by *H* for data point 5 in Fig. 10-48*e* produces the ______ permeability for the material.

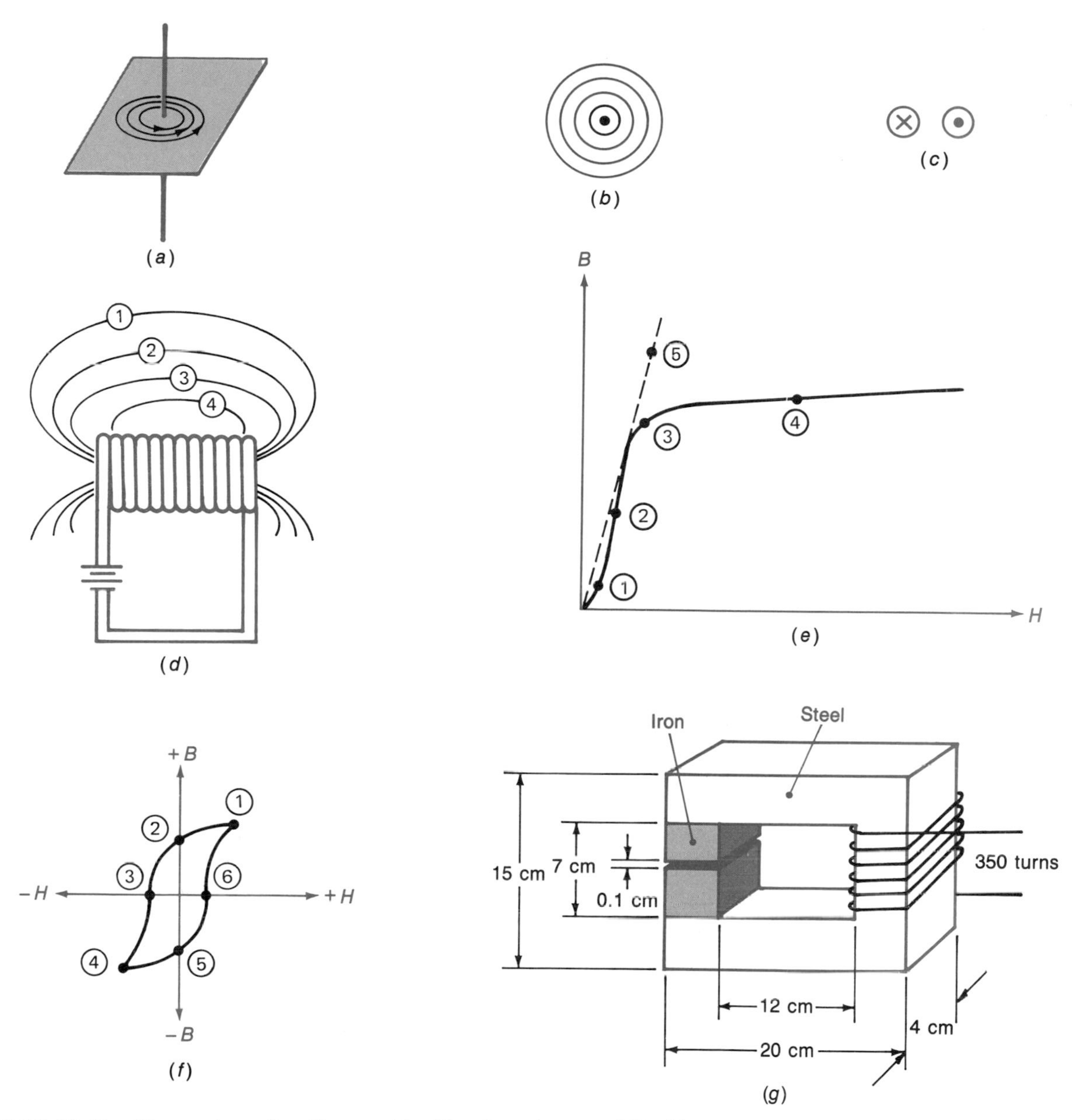


**FIGURE 10-48 Illustrations for Chapter Review Questions and Problems.**

**10-16.** A four-quadrant *BH* curve for a ferromagnetic material is called a ______ loop.

**10-17.** Which point(s) on Fig. 10-48*f* is/are due to retentivity?

**10-18.** Which point(s) on Fig. 10-48*f* correspond(s) to coercive force?

**10-19.** How would the loop in Fig. 10-48*f* appear for a material with much less retentivity?

**10-20.** Silicon steel alloys are used to decrease hysteresis loss in the cores of ______ current devices.

**10-21.** Magnetic shields are made from high-______ materials.

**10-22.** The holding current of a relay is ______ than its pull-in current.

**10-23.** True or false? A thermal breaker will interrupt a short circuit in less time than a magnetic breaker.

**10-24.** True or false? A coil moving in a magnetic field is the basis of operation for magnetic phono cartridges and dynamic microphones.

**10-25.** True or false? The output of a magnetic phono cartridge is called the Hall voltage.

## CHAPTER REVIEW PROBLEMS

**10-1.** Two unlike poles exhibit a force of 20 g when separated by 16 cm. What will the force be if the separation is decreased to 2 cm?

**10-2.** The coil shown in Fig. 10-48*d* is conducting 350 mA and has 15 turns. What is the mmf?

**10-3.** A single-turn coil in a magnetic circuit generates 10 V. What is the flux change in the circuit?

**10-4.** What is the reluctance of a magnetic circuit with a flux of $3.6 \times 10^{-5}$ Wb and a mmf of 4.2 A · t?

**10-5.** How much flux is produced by a magnetic circuit with a reluctance of $1.6 \times 10^{5}$ (A · t)/Wb and a mmf of 85 A · t?

**10-6.** How much mmf is required to produce a flux of 1 mWb in a circuit with a reluctance of $8 \times 10^{4}$ (A · t)/Wb?

**10-7.** What is the magnetizing force in a circuit that is 50 cm long and has a mmf of 500 A · t?

**10-8.** The pole of a magnet measures 5 cm × 4 cm and has a flux of $3 \times 10^{-4}$ Wb. What is the flux density?

**10-9.** A ferromagnetic material has a permeability of $1.26 \times 10^{-3}$ Wb/(A · t · m). What is its relative permeability?

**10-10.** A magnetic circuit has a flux density of 0.8 T with a magnetizing force of 200 (A · t)/m. What is its permeability?

**10-11.** A round air-core coil is 15 cm long and has a 1.5-cm inside diameter. What is the reluctance of its core?

**10-12.** Calculate the flux for the coil in Problem 10-11 if it has 100 turns and a current of 4 A.

**10-13.** What is the flux density at the center of the core for the coil described in Problems 10-11 and 10-12? What is it at the poles?

**10-14.** Use the graph of Fig. 10-30 to find the required current to produce a flux of $1 \times 10^{-4}$ Wb in a steel toroid core with an outside diameter of 6 cm and an inside diameter of 3 cm. The core cross section is round, and the coil has 200 turns.

**10-15.** Use the graph from Fig. 10-30 to determine the necessary current to establish a flux of 0.5 mWb in the air gap of Fig. 10-48(*g*).

## ANSWERS TO SELF-TESTS

1. true
2. true
3. north
4. poles
5. false
6. left
7. north
8. 2.5 kg
9. false
10. increases
11. from right to left
12. right
13. attract
14. current, number of turns, size and shape, the core material
15. false
16. right
17. 15 A · t
18. $2 \times 10^{-3}$ Wb
19. 375 (A · t)/m
20. 1.55 T
21. yes
22. iron
23. saturation
24. $1.26 \times 10^{-6}$ Wb/(A · t · m)
25. 1
26. The graph is linear.
27. 222
28. 476
29. The graph is nonlinear.
30. the linear region
31. the saturation region
32. coercive
33. true
34. true
35. $1.01 \times 10^{9}$ (A · t)/Wb
36. $6.28 \times 10^{-3}$ T
37. $3.14 \times 10^{-3}$ T
38. 92 mA
39. 102 turns
40. 450 mA
41. 23 turns
42. false
43. secondary
44. true
45. false
46. false
47. true
48. true
49. increases

# Chapter 11 Electric Energy Sources

Large rotating machines are the major source of electric energy. They convert mechanical energy into electric energy at megawatt rates. Smaller generators provide energy for vehicles and portable equipment, and serve as emergency sources. Batteries convert chemical energy to electric energy and are convenient for portable devices and as emergency backups in critical applications. Solar cells convert light energy and are applied in remote areas where access to commercial power is not practical or even possible. For example, earth-orbiting communications satellites are energized by arrays of solar cells. In addition to these sources, several other types provide low-level outputs that are useful in control, communication, and instrumentation areas. This chapter covers the theory and characteristics of the major electric energy sources.

## 11-1 ELECTROMAGNETIC INDUCTION AND GENERATORS

When a conductor is moved through a magnetic field, an electromotive force (emf) is generated. This is known as *induction*, and the potential difference that is generated across the conductor is called the *induced emf*. When the conductor is part of a complete circuit, an *induced current* also results. Figure 11-1 shows a conductor moving up between the poles of a horseshoe magnet (named after its shape). A current flows in the wire as long as the

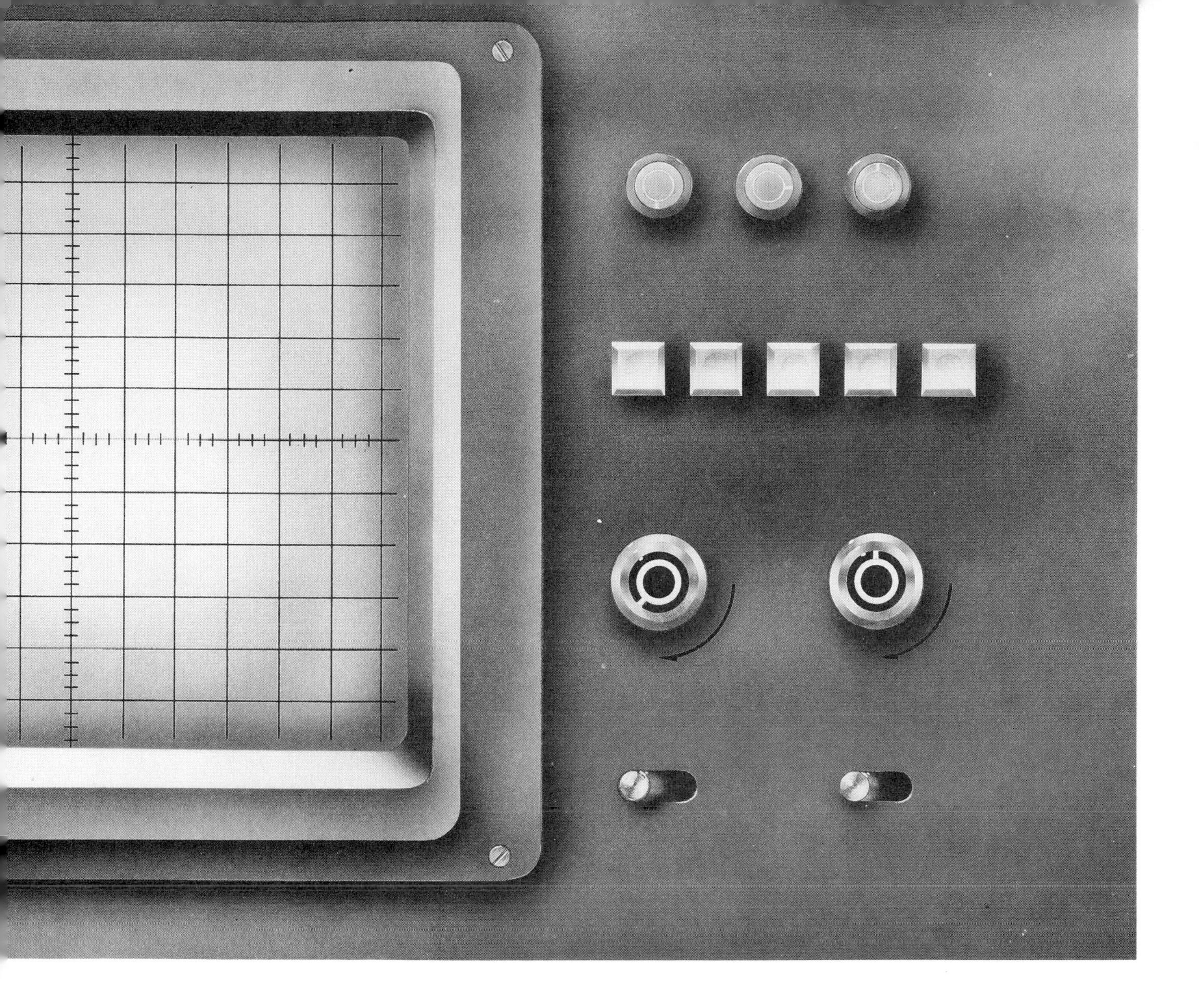

wire is moving and cutting across the flux lines of the magnet.

Figure 11-2 shows an induction experiment using a *zero-center meter.* Meters of this type rest at center scale when the voltage (and/or current) is zero. Applying one polarity to a zero-center meter deflects the pointer in one direction, and applying the opposite polarity causes deflection in the other direction. In Fig. 11-2*a*, the conductor is *stationary* in the magnetic field. There is no induced emf and no current flow. In Fig. 11-2*b*, the conductor is moving up through the field. There is an induced emf, and current flows through the meter as shown. When the conductor is moved down through the field, as shown in Fig. 11-2*c*, the induced emf is of the opposite polarity. The current flows in the other direction and the meter deflects to the right. We can conclude from the experiment that the conductor must be in motion to produce induction and that the induced polarity is related to the direction of motion.

Moving the conductor parallel to the flux lines does not produce electromagnetic induction. For example, if the wire in Fig. 11-2 is moved along a path from the north pole to the south pole, the meter does not deflect. To have induction, the conductor must *cut across* flux lines as it moves.

The induced polarity is a function of conductor motion and the direction of the magnetic flux. If the poles of the magnet in Fig. 11-2 are reversed, so are the results. Moving the conductor up deflects the meter to the right, and moving the con-

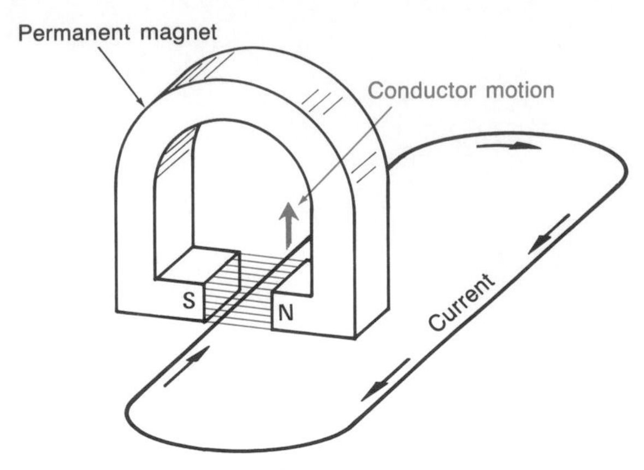


**FIGURE 11-1 Induced current.**

ductor down deflects the meter to the left. The *left-hand generator rule* can be used to predict induction polarity. Figure 11-3 shows how it is applied. The thumb, forefinger, and middle finger are all extended at right angles to one another. The thumb is the motion pointer, that is, it points in the direction that the conductor is moving. This can be *relative motion.* That is, the conductor may be held stationary and the field moved. In this case, the thumb points opposite to the field motion. The index finger is the flux pointer. Remember, flux flows from north to south. The middle finger is the electron current pointer. It also points toward the negative side of the induced voltage. As Fig. 11-3 shows, the current flows from positive to negative because the induced voltage is a *rise* in potential.

Figure 11-4 shows five "snapshots" of a simple machine for generating electric energy. The conductor is formed into a loop which can be rotated in a magnetic field. Starting with the leftmost snapshot, the output is zero. The top and bottom of the loop are not cutting any lines of force at this instant in time. The wires are running parallel to the flux lines, and the induction is zero. Ninety degrees later, the output is at its peak (maximum) value. The conductors are now cutting the maximum number of flux lines. At 180° of rotation, the conductors are running parallel to the flux lines again, and the output drops back to zero. At 270° of rotation, maximum cutting is achieved again, but the current flow in the loop has reversed and so has the induced voltage across the ends of the loop. The illustration shows why. At 90°, the heavy part of the loop is moving up through the field, but at 270°, it is moving down through the field. Therefore, the current periodically reverses in the loop. This is known as *alternating current* (ac). At 360°, the loop is back in its starting position, and the output has dropped back to zero again.

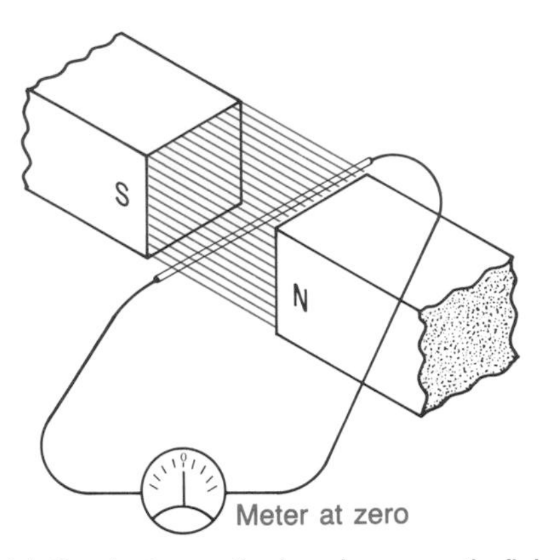


(*a*) Conductor motionless in magnetic field

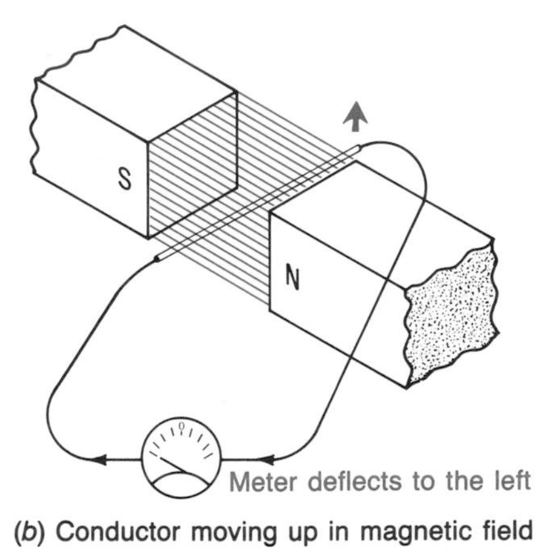


(*b*) Conductor moving up in magnetic field

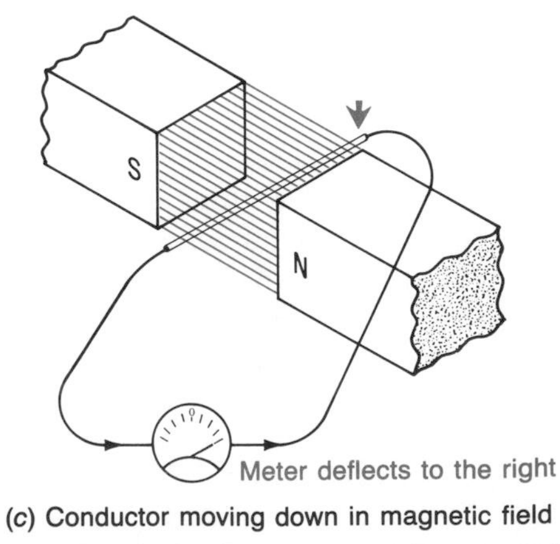


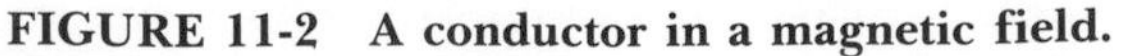

(*c*) Conductor moving down in magnetic field

**FIGURE 11-2 A conductor in a magnetic field.**

The graph in Fig. 11-4 shows how the output current or voltage varies with the angle of rotation of the generator. One complete turn of the shaft produces one *electric cycle*. This cycle begins at zero, peaks at 90°, returns to zero at 180°, peaks with the opposite polarity at 270°, and returns to zero at 360°. The output of a rotating machine follows a *sine* function, and the graph in the illustration is called a *sine wave*. Look at the abbreviated sine table below:

| Angle in Degrees | Sine Function |
|---|---|
| 0 | 0 |
| 90 | 1 |
| 180 | 0 |
| 270 | −1 |
| 360 | 0 |

The table shows the same polarity, the same zeros, and the same peaks as the graph in Fig. 11-4.

The induced voltage is at maximum when the conductor is moving perpendicular to the lines of flux because the maximum number of lines are being cut at that time. Increasing the flux density therefore increases the induced voltage. Moving the conductor (or the field) faster also increases the induced voltage. The induced voltage is also proportional to conductor length. The induced voltage can be predicted by

$$E = Blv$$

where $E$ = induced emf, V
$B$ = flux density, T
$l$ = effective length of conductor, m
$v$ = relative velocity, m/s

The above equation assumes that the conductor is moving perpendicular (90°) to the flux lines. The effective length of the conductor is that portion which takes part in the actual cutting of flux lines.

**EXAMPLE 11-1**

A conductor is effectively 1 m long and is moving at a velocity of 1 m/s in a field with a flux density of 1 T. What is the induced emf in volts?

**Solution**

Applying the data to the equation,

$$E = 1 \times 1 \times 1 = 1 \text{ V}$$

*Note:* This is consistent with the definition of the weber presented in Chap. 10. One tesla is equal to one weber per square meter. The conductor in this problem sees a flux change of 1 Wb/s, and 1 V is generated.

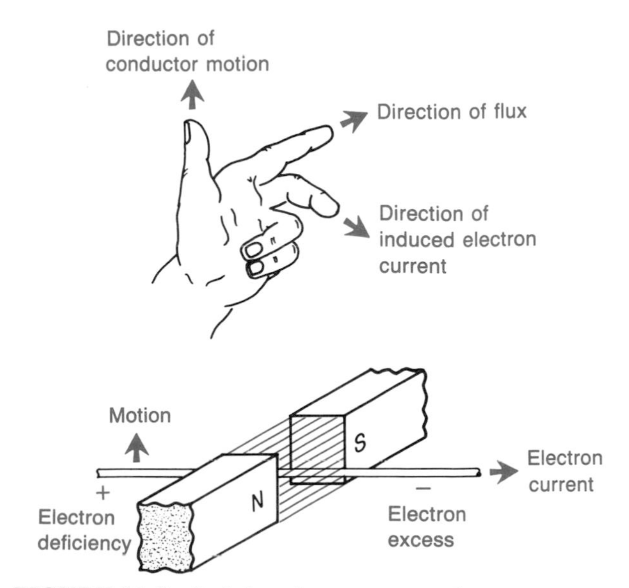


FIGURE 11-3 Left-hand generator rule.

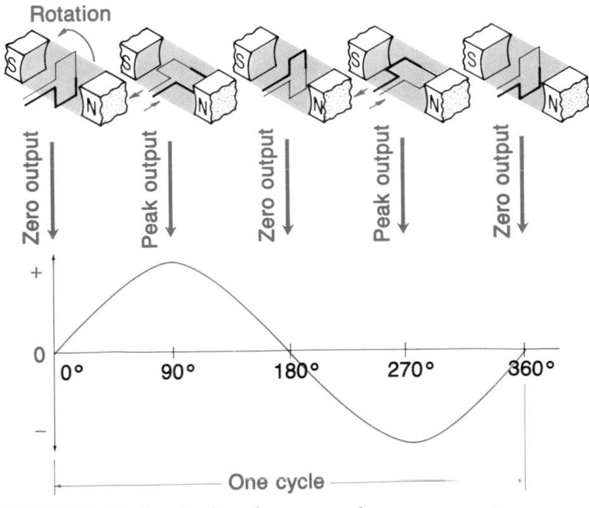


FIGURE 11-4 A simple two-pole ac generator.

In rotating machines, the assumption concerning perpendicular cutting is only valid for two instantaneous angles of rotation. The concept of *instantaneous* voltage can be used to describe the induced voltage:

$$e = Blv \sin \theta$$

where $e$ = instantaneous induced emf, V
$B$ = flux density of field, T
$l$ = effective length of conductor, m
$v$ = relative velocity, m/s
$\theta$ = angle of cutting

**EXAMPLE 11-2**
Calculate the instantaneous voltage at 45° that is induced in a conductor with an effective length of 15 cm moving at a velocity of 10 m/s in a field with a flux density of 0.2 T.

**Solution**
The sine of 45° is 0.707. Applying the data to the equation,

$$e = 0.2 \times 0.15 \times 10 \times 0.707 = 0.212 \text{ V}$$

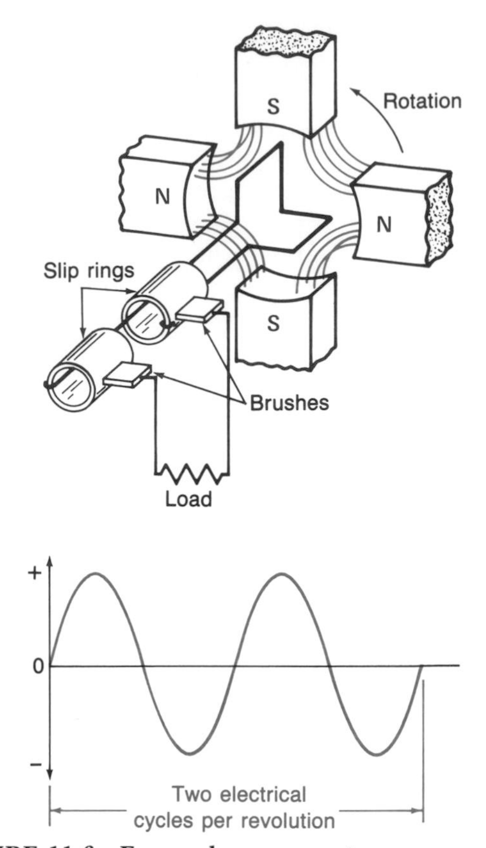


**FIGURE 11-6 Four-pole ac generator.**

Many practical generators must produce hundreds or even thousands of volts. Figure 11-5 shows two ways to increase the output voltage. The permanent magnet is replaced with an electromagnet. Greater flux density can be achieved by an electromagnetic field. The effective length of the conductor is increased by using a coil rather than a single-turn loop. The speed of cutting (and the output voltage) can be increased by turning the shaft faster. Of course, there is a limit to speed, which is especially apparent in larger machines. Figure 11-6 shows another improvement in generator performance. Four magnetic poles are used instead of two. The four-pole machine produces two electric cycles for every shaft revolution. The illustration also shows a practical method for making contact with the rotating loop. Each loop end is soldered to a metal *slip ring*. A carbon brush presses up against each slip ring to make electric contact with the load circuit.

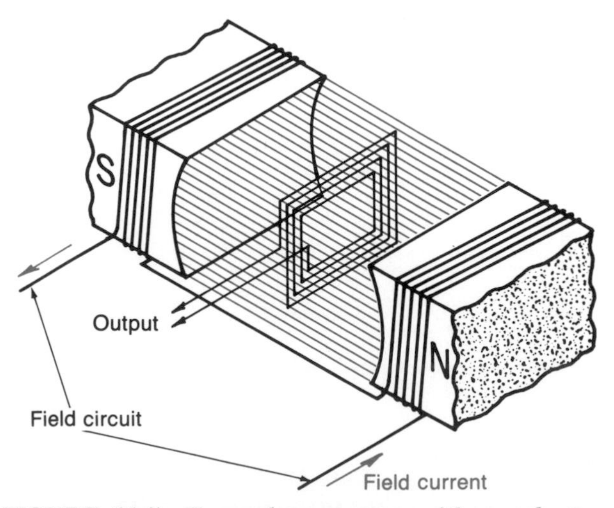


**FIGURE 11-5 Two-pole generator with an electromagnetic field and a multiturn output coil.**

Figure 11-7 shows a different arrangement. The loop ends are soldered to brass *commutator* segments. The segments are insulated from each other. This arrangement produces *direct current* (dc) in the load rather than alternating current. Note that the positive end of the load is always in contact with the part of the loop that is moving up through the field and the negative end of the load with the part of the loop that is moving down.

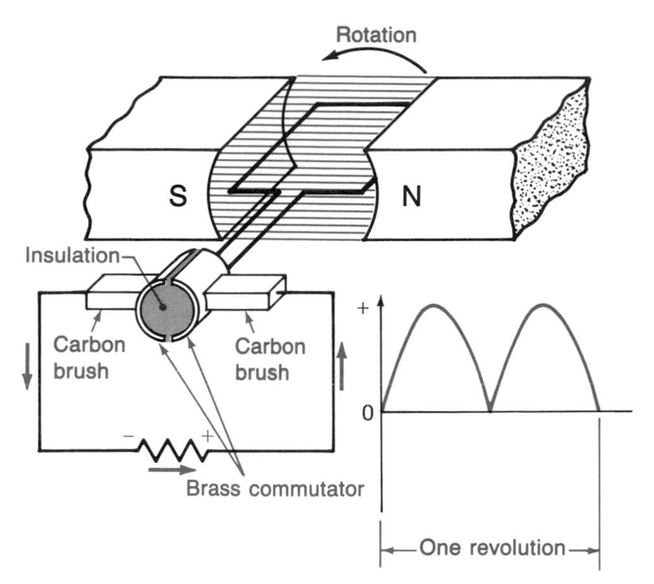


**FIGURE 11-7 Two-pole dc generator.**

Even though the load current in Fig. 11-7 flows in one direction only, it is still not the same as using an energy source such as a battery. The graph shows that the load voltage and current fluctuate between zero and the peak value. This type of current is called *pulsating direct current* and is not acceptable for all loads. For example, most electronic devices will not operate properly on pulsating direct current. Figure 11-8 shows one solution. By using multiple loops and more than two commutator segments, the generator output does not fluctuate as widely. The graph shows how the outputs from the loops superimpose for a reasonably constant voltage and current.

Generators with outputs exceeding several kilowatts use electromagnetic fields. Some smaller units may also use electromagnetic fields. The field circuit can take its energy from an external source, and if it does, the generator is said to be separately or externally *excited*. The other possibility is to use some of the generator output for the field circuit. This method is known as *self-excitation*. The field current is usually controlled to provide the necessary flux density to establish the desired output voltage. Figure 11-9*a* shows a separately excited generator with a rheostat used to control the field current. Figure 11-9*b* shows a self-excited generator which also uses rheostat control. Rheostats work but are not the preferred method of field control. Electronic devices are usually used to control the field circuit. Since these devices and their associated circuits effectively control the generator output voltage, they are often called *voltage regulators*.

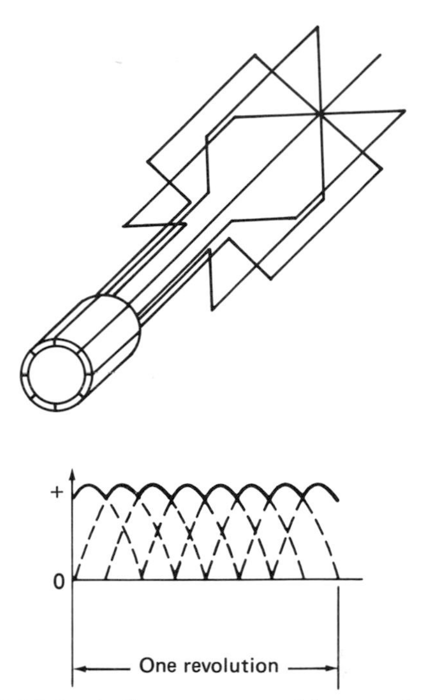


**FIGURE 11-8 A dc generator with more loops and commutator segments.**

Alternating current generators can also use self-excitation, but the field circuit requires direct current. Solid-state devices called *rectifiers* are used to convert the alternating output to direct current for the field circuit. Rectifiers are covered in Chap. 23.

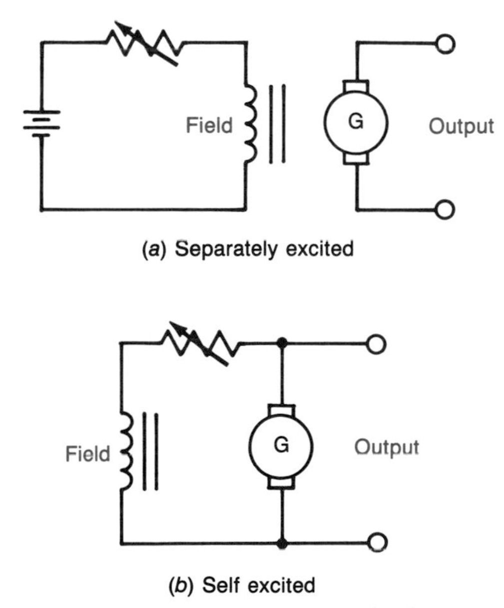


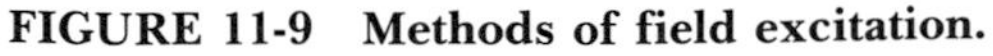
(*b*) Self excited

**FIGURE 11-9 Methods of field excitation.**

There are several important variations used in many generators. One of them is to make the field circuit rotate and the output circuit stationary. Field motion is effectively the same as conductor motion. It offers the advantage of eliminating generator output current through the sliding contacts of slip rings or commutators. When the output circuit does not move, it can be hard-wired to the load. This variation produces a significant improvement in reliability, especially in high current machines. If the field circuit is of the electromagnetic type, it still requires sliding contacts of some type. However, the field current is normally quite a bit less than the output current, so a worthwhile improvement in reliability is still achieved.

## Self-Test

**1.** Refer to Fig. 11-10*a*. What is the polarity of point A with respect to point B?

**2.** Refer to Fig. 11-10*a*. Assuming the conductor is part of a complete circuit, what is the direction of electron current?

**3.** Refer to Fig. 11-10*a*. If the conductor is moved along a path from one pole to the other, the induced emf is zero. Why?

**4.** Refer to Fig. 11-10*b*. The graph is called a _______ wave.

**5.** One revolution of a two-pole ac generator produces one electric _______.

**6.** List the four factors that determine induced emf.

**7.** Identify the graph shown in Fig. 11-10*c*.

**8.** True or false? The output graphed in Fig. 11-10*c* is from a rotating machine using slip rings.

**9.** Generators use circuits called voltage regulators to establish the correct output voltage. These circuits work by controlling _______.

## 11-2 PRIMARY CELLS AND BATTERIES

Figure 11-11 shows a simple electrochemical *primary cell.* It converts chemical energy into electric energy. The chemical process that occurs during discharge is not reversible in a primary cell. When the materials are depleted, they cannot be reformed by charging and the cell must be replaced. *Secondary cells* can be restored by passing a charging current through them; these are covered in Sec. 11-3.

A battery is made up of two or more cells in a single case. The cells are usually connected in series to provide a greater output voltage. The term *battery* is popularly used to refer to cells. For example, many people refer to the cells used in flashlights and watches as batteries. These cells typically develop 1.5 V, while batteries provide a greater output.

All cells have two electrodes. As Fig. 11-11 shows, the *anode* is the negative electrode. It supplies

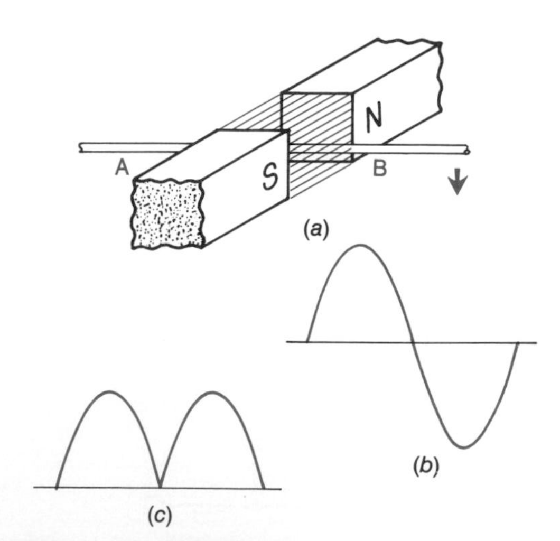


**FIGURE 11-10 Self-Test.**

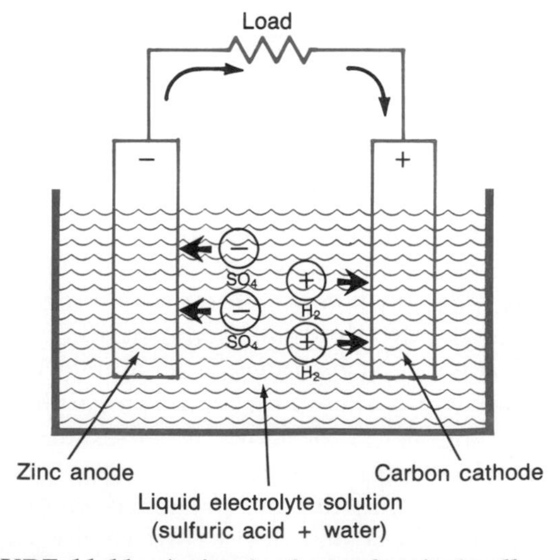


**FIGURE 11-11 A simple electrochemical cell.**

electrons to the external load. The *cathode* is the positive electrode. It provides the return path for the load current. Note that the *internal* flow of a cell is from cathode to anode, and the external flow is from anode to cathode when the cell is discharging.

All cells also have an *electrolyte*. The electrolyte is a chemical that acts on one or both of the electrodes to provide the chemical action necessary for current flow. The electrolyte can be in liquid form (wet cell), in paste form (dry cell), or in gelled form (gel cell). Today, it is possible to seal cells, and the term *dry cell* may be applied to any cell that can be operated in any position without fear of spills. The electrolytes used in cells and batteries are usually acid or alkaline, but salts can also be used.

The cell electrolyte in Fig. 11-11 is a mixture of sulfuric acid ($H_2SO_4$) and water ($H_2O$). The sulfuric acid molecules disassociate (break down) into positive hydrogen ions and negative sulfate ions. The sulfate ions combine with the zinc anode to form zinc sulfate ($ZnSO_4$). This action produces a buildup of electrons on the zinc anode. The hydrogen ions receive electrons from the cathode and become hydrogen gas ($H_2$). Thus the cathode becomes positively charged. During discharge, the zinc sulfate dissolves into the electrolyte and the anode is gradually eaten away. The hydrogen gas forms on the surface of the cathode and eventually escapes into the atmosphere.

If the load circuit is opened in Fig. 11-11, the chemical action ceases. The accumulated negative charge on the anode repels the negative sulfate ions, and the anode stops dissolving into the electrolyte. However, this is true only if the anode is pure zinc. Impurities in the anode can cause small local cells to be set up, and currents will flow between the zinc atoms and the impurity atoms. This is called *local action* and causes the zinc anode to deteriorate even when the cell is not supplying any current to an external load. Local action is one reason why cells and batteries have a *shelf life*. They can only be stored for so long until they discharge themselves and become useless. Local action can be eliminated by using pure anode materials, but this is not practical. It can be reduced by storing cells and batteries at reduced temperatures. Cold storage greatly improves shelf life. It can also be reduced by a process called *amalgamation* whereby mercury is alloyed with the zinc. The mercury effectively separates the impurity atoms from the zinc atoms and local action is retarded. Shelf life can be extended to years by using amalgamated anodes and cool storage temperatures (near 0°C).

Another problem with the cell shown in Fig. 11-11 is that hydrogen gas is an insulator. As this gas collects on the surface of the carbon cathode, the internal resistance of the cell goes up and the ability of the cell to deliver significant current goes down. The collection of hydrogen gas on the cathode is called *polarization*. A depolarizer, such as manganese dioxide ($MnO_2$), may be added to the cell. Depolarizing agents are oxidizers. They provide oxygen atoms to combine with the hydrogen gas and eliminate the bubbles. Depolarizers take time to work. This is why some cells and batteries require rest periods when supplying substantial currents.

*Wet cells*, such as the one shown in Fig. 11-11, are not practical for all applications. They are subject to leakage of highly corrosive electrolyte materials and cannot be operated in all positions. Wet cells can be sealed to achieve dry cell status, but this is more expensive to accomplish. Cells based on paste electrolytes are more widely applied. Figure 11-12 shows the construction of the carbon-zinc cell. This cell is the least expensive and the most popular of the dry types. The anode is zinc and forms the container for the cell. An outer jacket and bottom cover prevent leaks in case the zinc is eaten through. The electrolyte is ammonium chloride, and a carbon rod serves as the positive contact. The ammonium chloride splits into positive ammonium ions and negative chlorine ions. The zinc

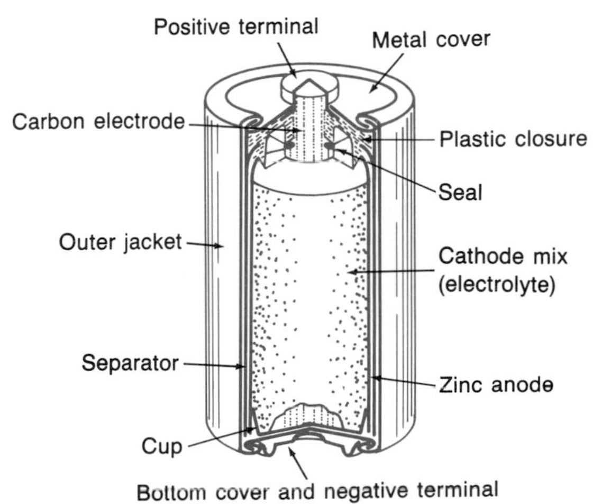


**FIGURE 11-12 The carbon-zinc cell.**

anode dissolves in the electrolyte, giving off positive zinc ions. This leaves an excess of electrons on the anode. The zinc ions combine with the chlorine ions and form zinc chloride which is electrically neutral. The ammonium ions are repelled by the zinc ions going into solution and migrate to the carbon electrode where they pick up electrons and split into ammonia and hydrogen gases. This provides a deficiency of electrons at the positive terminal. Connecting an external load allows electron current to flow from the negative terminal, through the load, and into the positive terminal. The internal chemical action continues as long as the load is connected or until the cell is completely discharged. The practical life of a cell or battery ends before total discharge is reached.

The amount of current that a cell can deliver is a function of its internal resistance. Cell resistance depends on the surface area of the electrodes, the space between them, and the resistance of the electrolyte. Larger cells have lower internal resistance and can deliver more current. Larger cells also have more *capacity* (C). The capacity of cells and batteries is measured in *ampere-hours* (Ah) or *milliampere-hours* (mAh). Capacity is determined by multiplying the load current by the discharge time. Discharge is considered complete when the output voltage drops to 85 percent of maximum. This is known as the *cutoff voltage*. A fresh carbon-zinc cell produces slightly over 1.5 V, and the cutoff voltage is considered to be 1.3 V.

**EXAMPLE 11-3**

Ordinary flashlights use D cells. A typical fresh carbon-zinc D cell can supply 25 mA for 30 h before its output drops to 1.3 V. Calculate the cell capacity.

**Solution**

Multiply the current by the discharge time:

$$\text{Capacity (C)} = 25\text{ mA} \times 30\text{ h} = 750\text{ mAh or } 0.75\text{ Ah}$$

Carbon-zinc cells do not attain their rated capacity at high discharge rates. Polarization raises the internal cell resistance and causes a loss in output voltage and a decrease in output current. Manganese dioxide is added to the cathode mix to remove the buildup of hydrogen gas bubbles but cannot keep up at high discharge rates. Thus, these types of cells and batteries must be rested to achieve their rated capacity when used in high discharge service. In general, all cell types tend to provide less than their rated capacity in high current service.

Carbon-zinc cells should not be stored for extended periods of time. Some local action persists in spite of amalgamation of the anode. Additionally, the paste electrolyte tends to dry out, which increases cell resistance and decreases capacity. The shelf life of carbon-zinc cells is rated in months and is temperature-dependent. They can be stored at a temperature near 0°C, which will extend their shelf life beyond 1 year. When retrieved from cold storage, they should be allowed to come up to room temperature for best performance. Higher temperatures provide even greater output but may also accelerate deterioration.

Some carbon-zinc cells and batteries are rated as *heavy-duty*. These have a construction similar to that shown in Fig. 11-12 but use a zinc chloride electrolyte. This electrolyte has less resistance, which makes the zinc chloride cell a better choice for high current applications. The output voltage does not drop as rapidly and the cell capacity is improved. Heavy-duty cells and batteries cost more than general-purpose carbon-zinc units but may prove more economical for high current applications.

Alkaline-manganese cells and batteries are available in most of the same sizes and voltage ratings as carbon-zinc cells and batteries. They are usually called alkaline cells or alkaline batteries. Compared to the carbon-zinc cell, the alkaline cell is constructed inside out as shown in Fig. 11-13. The center of the cell is made up of a powdered zinc anode and an alkaline electrolyte paste made of potassium hydroxide and zinc oxide. The cathode surrounds the anode and is a manganese dioxide mix that provides rapid depolarization due to the large surface area of the cylinder. This makes alkaline cells better suited to high current applications. They do not require rest periods and can last several times as long as comparable carbon-zinc types in high discharge applications. They are polarity-compatible with carbon-zinc cells, as a comparison of Figs. 11-12 and 11-13 will show. The button top of the alkaline cell contacts the cathode mix and acts as the positive terminal. The bottom cover

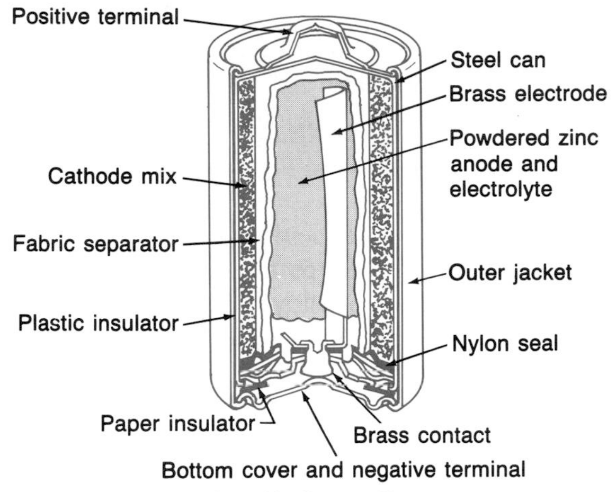


**FIGURE 11-13 The alkaline cell.**

contacts the zinc anode and serves as the negative terminal. The alkaline cell output is 1.5 V which is the same as a carbon-zinc cell.

---

**EXAMPLE 11-4**

A size D carbon-zinc cell will provide 50 mA for 15 h and the same size alkaline cell will provide 50 mA for 56 h before reaching the 1.3-V cutoff. Compare the cell capacities.

**Solution**

Calculate the ampere-hour capacities (C) for both cells:

$$C = 50 \text{ mA} \times 15 \text{ h} = 0.75 \text{ Ah}$$
$$C = 50 \text{ mA} \times 56 \text{ h} = 2.8 \text{ Ah}$$

The alkaline cell provides almost four times the capacity of the carbon zinc cell at a 50-mA discharge rate. In higher current applications, the alkaline cell may provide as much as 10 times the capacity when compared to the same size carbon-zinc cell.

---

Alkaline cells cost about three times as much as carbon-zinc cells but can still be a better bargain for high-drain service. However, for light-drain intermittent applications, they are not as economical as carbon-zinc cells. They also enjoy a longer shelf life (up to 3 years) and are less likely to leak corrosive electrolyte which can damage equipment. Some equipment will not operate properly when alkaline cells or batteries are replaced with carbon-zinc units. The alkaline devices have a lower internal resistance and a flatter discharge curve. A flat discharge curve means that the output voltage remains relatively constant over the life of the cell.

Figure 11-14 shows a mercuric oxide cell that is noted for its extremely flat discharge curve. In fact, these cells are sometimes used as voltage references. A *voltage reference* is a standard for calibrating instruments and may also be used to compare other voltages against. Mercuric oxide cells are commonly called mercury cells and are available in the button style shown in Fig. 11-14 and also in cylindrical and rectangular cases. Mercuric oxide forms the cathode, and amalgamated zinc is used for the anode. The electrolyte may be either potassium hydroxide or sodium hydroxide. During discharge, the zinc anode oxidizes and leaves positive hydrogen ions in the electrolyte. These travel to the cathode and combine with the mercuric oxide to form mercury and water. The anode gains electrons as it oxidizes, and the cathode loses electrons as it gives up oxygen atoms. Cell output is 1.35 or 1.40 V, with the higher voltage types based on a modified cathode made from mercuric oxide and manganese dioxide.

Figure 11-14 shows the anode on top, and consequently the negative terminal is also on top. It is just as feasible to build mercury cells with a reversed internal structure, and both polarity types are available. Mercury cells produce about three times the energy for a given volume when compared to carbon-zinc cells. However, their high

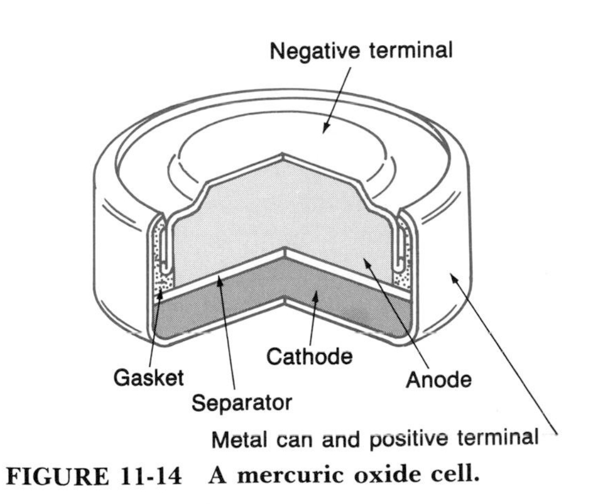


**FIGURE 11-14 A mercuric oxide cell.**

cost limits them to more specialized applications such as medical instruments, smoke alarms, paging receivers, and photographic equipment.

Silver oxide cells are similar to mercury cells but develop an output of 1.5 V under moderate load. The cathode is silver oxide with a small amount of manganese dioxide. They are commonly manufactured in the button style case and are used in light load applications such as watches, cameras, and hearing aids.

Most primary cells use zinc for the anode and potassium hydroxide as the electrolyte. A more recent development in primary cells is to use lithium in place of the zinc. Lithium is the lightest metal and has a specific weight only half that of water. It is also a very active metal, and lithium cells can produce as much as 3.5 V under load. Lithium cells and batteries are quite expensive but can produce up to 10 times the energy per weight and volume compared to alkaline devices. Some types can last over 10 years. General applications for lithium cells and batteries include watches, calculators, memory backup power, security devices, photography, medical devices, emergency systems, data acquisition equipment, telephones, and telecommunication systems.

There are too many lithium cells to cover them in detail, and new types are still being developed. Some use an organic electrolyte, and others use an inorganic electrolyte. Various cathode materials are used, and the list below shows some of the types along with their open circuit voltages:

| | |
|---|---|
| Lithium–chromium dioxide ($Li/CrO_2$) | 3.8 V |
| Lithium–thionyl chloride ($Li/SOCl_2$) | 3.7 V |
| Lithium–iodine (Li/I) | 3.6 V |
| Lithium–manganese dioxide ($Li/MnO_2$) | 3.3 V |
| Lithium–sulfur dioxide ($Li/SO_2$) | 2.9 V |
| Lithium–bismuth trioxide ($Li/Bi_2O_3$) | 2.1 V |

Lithium–chromium dioxide cells boast the highest energy density of all primary cells and approach 1 Wh/cm$^3$. They have a 10-year life expectancy and are used mainly to back up semiconductor memory devices. Lithium–bismuth trioxide cells are only manufactured in button styles and have a loaded output of 1.5 V. This makes them compatible with silver oxide cells, and they are used mainly in watches where they can last as long as 4 years. The other types are general purpose and are used in a multitude of applications. Most are available in button, cylindrical, and rectangular cases. Figure 11-15 shows the construction of a lithium–manganese dioxide cell. Lithium cells are often soldered onto printed circuit boards due to their long life expectancy.

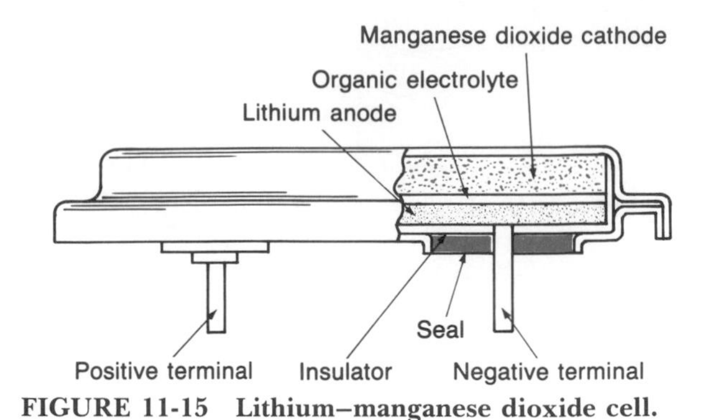

**FIGURE 11-15 Lithium–manganese dioxide cell.**

Figure 11-16 shows the discharge characteristic for a lithium–thionyl chloride cell. Note the unusual voltage response during the first few tenths of an hour of operation. With a 125-mA load, the output drops to 1 V and then recovers to 3.1 V. This is called the *transition period* and may be an unacceptable characteristic for some applications. Some of the other types of lithium cells also exhibit a transition effect. The transition voltage drop is usually moderate or nonexistent for moderate to light load currents. The typical lithium cell is used in applications where the load current is not heavy and the transition effect is not a problem. Figure 11-16 also shows that greater cell capacity is realized when the drain is small. For example, when the load current is 200 $\mu$A, a capacity of 10.8 Ah is realized with a life over 50,000 h (about 6 years).

As mentioned in Sec. 11-2, a battery is an arrangement of several cells. Figure 11-17 shows one method of battery construction. Six 1.5-V carbon-zinc cells are connected in series to produce 9 V. The carbon-to-zinc interfaces form the series connections between the individual cells. A metal strap brings the bottom zinc anode to the top for the negative contact, and the positive terminal contacts the carbon of the top cell.

## Self-Test

**10.** Primary cells and batteries ________ be recharged.

**11.** Which cell electrode serves as the source of electrons for the load circuit?

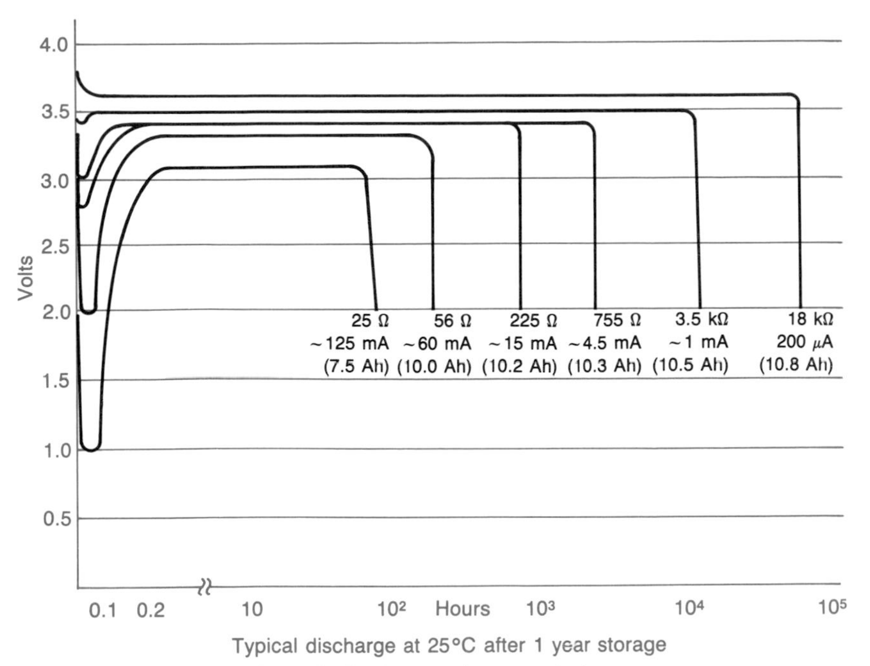


**FIGURE 11-16 Lithium–thionyl chloride cell discharge characteristics.**

**12.** Which cell electrode is deficient in electrons during discharge?

**13.** A cell that can be operated in any position is considered a ______ type.

**14.** True or false? Some cells may require a rest period due to polarization.

**15.** True or false? The anode of a cell may be amalgamated to improve shelf life.

**16.** A radio receiver requires 15 mA when operated at moderate volume. If its battery is rated at 600 mAh, what is the playing time of the receiver?

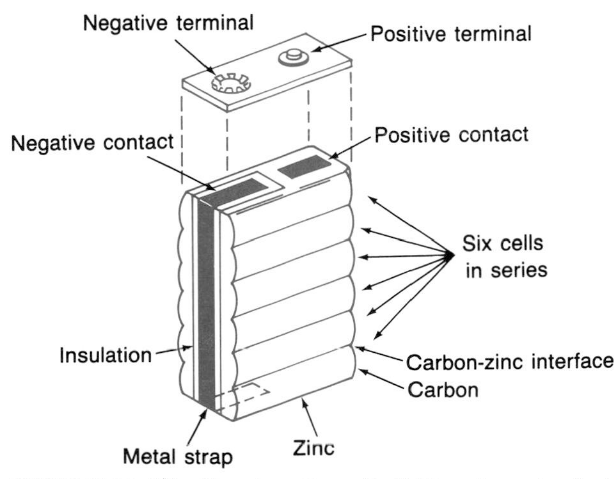


**FIGURE 11-17 Construction of a 9-V carbon-zinc battery.**

**17.** True or false? In general, a cell will deliver greater ampere-hour capacity when operated at high discharge rates.

**18.** True or false? Alkaline cells cost more but are always less expensive in the long run when compared to carbon-zinc cells.

**19.** The highest output voltages are developed by cells with a ______ anode.

**20.** Which type of cell can show a momentary drop in output voltage at the beginning of its discharge cycle?

## 11-3 SECONDARY CELLS AND BATTERIES

Lead-acid batteries are widely used in vehicles. They are *secondary batteries,* and the vehicle generator acts as a charging source when the engine is running. The generator-battery connection is positive-to-positive and negative-to-negative. The generator output voltage exceeds the battery voltage, which results in a current flow through the battery in a direction which is opposite to the flow when the battery is discharging. The charging current reverses the chemical processes that occur during discharge, and the battery is eventually restored to a full-charge condition.

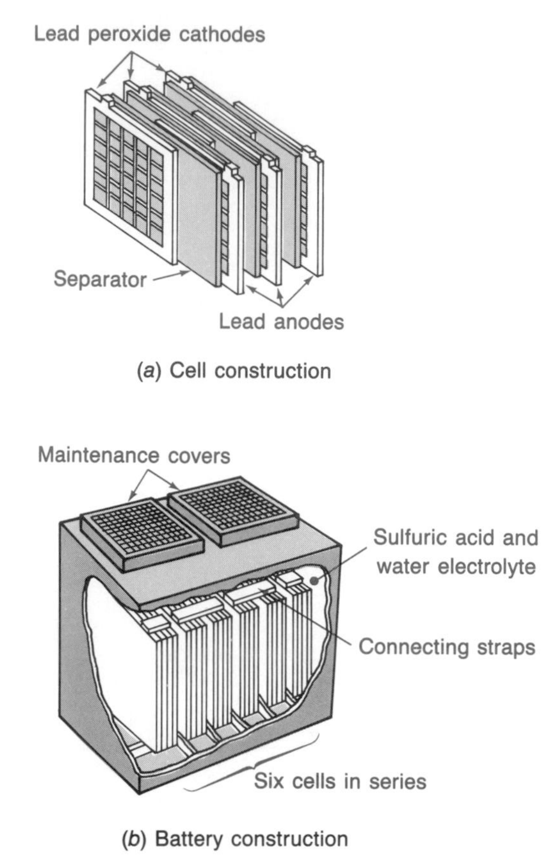


**FIGURE 11-18 Lead-acid automotive battery.**

Figure 11-18 shows the construction of a typical 12-V automotive battery. The battery contains six lead-acid cells connected in series. Each cell develops about 2.1 V under light load for a battery output of 12.6 V. As shown in Fig. 11-18*a*, each cell is constructed of several parallel-connected cathodes and several parallel-connected anodes. This effectively increases the electrode area for low cell resistance and high capacity. Low internal resistance is very important since cranking an automobile engine requires several hundred amperes. In the charged condition, the cathodes are coated with lead peroxide ($PbO_2$) and the anodes are lead (Pb). The electrolyte is a liquid solution of sulfuric acid ($H_2SO_4$) and water ($H_2O$).

Connecting a load to a lead-acid battery starts the chemical reaction. The sulfuric acid molecules separate into negative sulfate ions and positive hydrogen ions. The sulfate ions combine with the lead anode and form lead sulfate ($PbSO_4$). Since the sulfate ions are negative, the anode accumulates a negative charge. The positive hydrogen ions travel to the cathode and combine with the oxygen atoms to form water, and a positive charge accumulates on the cathode. Lead sulfate forms on the surface of the cathode as well as the anode during discharge. The process continues until the load is disconnected or the battery is depleted. Depletion can be caused by all of the sulfuric acid being converted to water or by the surface of both electrodes becoming totally converted to lead sulfate. Most lead-acid batteries are designed so that some acid remains upon total discharge.

Charging a lead-acid battery reverses the chemical process. The lead sulfate at the anode recombines with hydrogen ions to form sulfuric acid which enters the electrolyte, and the anode is gradually restored to lead. The lead sulfate on the cathode combines with water to regenerate the lead peroxide and also adds to the sulfuric acid content of the electrolyte. Charging regenerates the electrodes and restores the acid content of the electrolyte.

The charge is complete when the anode is once again lead, the cathode is lead peroxide, and the electrolyte is back to full strength. A *hydrometer* can be used to test the acid content of the electrolyte to determine the condition of the battery. Refer to Fig. 11-19. A sample of the electrolyte is drawn into the glass tube by squeezing and then releasing the rubber bulb. The float is calibrated and indicates the *specific gravity* of the electrolyte. Specific gravity is the ratio of the weight of a substance to the weight of water. Sulfuric acid is heavier than water, and the specific gravity of the electrolyte is directly related to the charge condition of the cell. In automotive batteries, a specific gravity of 1.26 indicates full charge and a specific gravity of 1.12 indicates total discharge. Most hydrometer floats omit the decimal point, so the actual readings are 1260 and 1120.

It is *not* possible to restore or recharge a lead-acid battery by adding sulfuric acid to the electrolyte. Although this will raise the specific gravity of the electrolyte, it does not restore the electrodes to their charged conditions. The balance of the chemical system is upset by adding acid to the electrolyte, and the battery will no longer perform properly.

A battery can be overcharged. Once full charge is reached, a different chemical reaction takes place if a charging current continues to flow. A

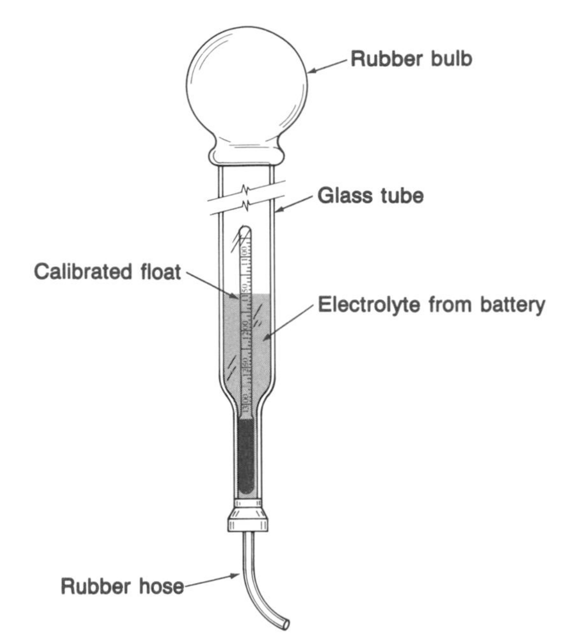


FIGURE 11-19 Hydrometer.

lead-acid battery generates hydrogen and oxygen gas when overcharged. If the battery is vented, the gases escape into the atmosphere. The gases are produced by the overcharge current, which breaks down the water molecules in the electrolyte. Thus, overcharging depletes the water in the electrolyte. Most automotive batteries have caps or maintenance covers (Fig. 11-18*b*) to allow water to be added as needed. Tap water is not recommended since it often contains minerals and other contaminants that can shorten the life of the battery. Distilled water is best.

Lead-acid batteries are potentially hazardous. The hydrogen and oxygen gases that escape during overcharge make an extremely explosive mixture. Such batteries must never be operated in an enclosure that will allow the gases to accumulate. Also, sparks or flames near a lead-acid battery can touch off an explosion. The electrolyte itself is dangerous. It is highly corrosive and can cause skin burns and damage clothing. The electrolyte is particularly damaging to eyes. Any contact with battery electrolyte must be dealt with by flushing the area with large amounts of cool water. If eye contact is involved, seek medical attention immediately after flushing. Wash your hands thoroughly with soap and water after handling lead-acid batteries.

Sealed lead-acid batteries are used in applications other than automotive. Some of these applications include emergency lighting systems, backup power systems, and portable electronic equipment. Figure 11-20 shows a sealed lead-acid cell. A spiral-wound grid assembly is used in conjunction with an absorbent separator which holds most of the electrolyte. This prevents significant leaks even if the seal is broken. A vent at the top of the cell allows pressures up to 60 lb/in.$^2$ to develop inside the cell. This internal pressure greatly aids the recombination of the hydrogen and oxygen gases that are produced in the event of an overcharge. The gas cannot escape, and water is regenerated for the electrolyte. This makes sealed cells and batteries maintenance-free. However, if the cell is charged beyond a C/3 rate (where C represents the cell capacity; the rate C/3 means that the current flow is equal to the ampere-hour capacity divided by 3 h), overcharge can eventually develop enough pressure to open the vent and gases will then escape. Therefore, sealed cells must not be operated in a confined enclosure where the gases could accumulate.

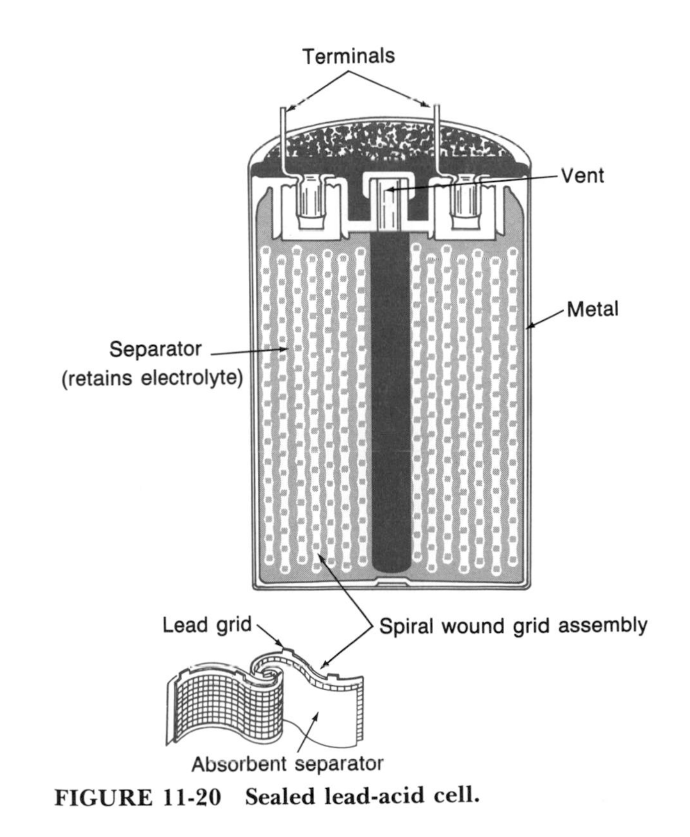


FIGURE 11-20 Sealed lead-acid cell.

**EXAMPLE 11-5**

What amount of current flow ($I$) represents a C/3 rate for a battery rated at 2.1 Ah?

**Solution**

Divide the Ah capacity by 3 h:

$$I = \frac{2.1 \text{ Ah}}{3 \text{ h}} = 0.7 \text{ A}$$

*Note:* A C/3 rate may also be referred to as a 0.33C rate, and a C/10 rate may be specified as a 0.1C rate.

A C/3 rate is common for charging sealed lead-acid cells and batteries. They can be expected to deliver between 200 and 2000 charge-discharge cycles before losing an appreciable portion of their capacity. The actual number varies with the rate of charge and discharge, the depth of discharge, the frequency of charge, and the charging rate. In applications such as emergency lighting, they are discharged only in the event of failure of the main power source. The service expectancy in this type of application is referred to as the *float life*. The typical float charge is a C/500 rate. Some charger circuits are of a constant voltage variety. They tend to supply more charging current when first applied and then less as the cell or battery voltage builds up. This is often called a *taper charge* since the current tapers off with time. Constant voltage chargers are typically rated at 2.5 volts per cell or 2.35 volts per cell when designed for float service.

Sealed lead-acid cells are similar to alkaline cells in terms of their capacity. For example, a D size sealed lead-acid cell is rated at 2.5 Ah as compared to 2.8 Ah for the alkaline type. Due to its higher output voltage, the lead-acid cell has a greater watthour (Wh) rating. The watthour rating is equal to the ampere-hour rating multiplied by the average output voltage.

**EXAMPLE 11-6**

A D alkaline cell produces an average output voltage of 1.35 V, while the same size lead-acid cell averages 2.06 V. Calculate the watthour rating for both cells.

**Solution**

Multiply the ampere-hour rating by the average voltage:

$$\text{Alkaline cell} = 2.8 \text{ Ah} \times 1.35 \text{ V} = 3.78 \text{ Wh}$$
$$\text{Lead-acid cell} = 2.5 \text{ Ah} \times 2.06 \text{ V} = 5.15 \text{ Wh}$$

Lead-acid cells and batteries have very low internal resistances. For example, a D lead-acid cell has an internal resistance of only 10 m$\Omega$ or less, which makes it well suited to high current applications. They should not be shorted or cell and/or circuit damage may result. They also provide a much flatter discharge curve than carbon-zinc or alkaline cells.

Lead-acid cells and batteries should not be discharged below their "end voltage" because their ability to accept a charge may be impaired. The actual end voltage varies with the type of service, but 1.7 volts per cell is an approximate value. Another caution is that they should not be allowed to remain in the discharged state for any length of time. Lead sulfate is water-soluble, and it goes into solution with the electrolyte during deep discharge conditions. Then, upon recharge, the dissolved lead sulfate is converted to sulfuric acid, and a lead precipitate forms which can build up in the separators and short the electrodes.

Sealed lead-acid cells and batteries eliminate some of the disadvantages associated with wet lead-acid units. However, there are applications where they cannot be used. Nickel-cadmium cells and batteries quite often serve in these applications. They are popularly referred to as "ni-cads" and are widely used in electronic and photographic applications where a secondary battery is needed.

Figure 11-21 shows the construction of a cylindrical nickel-cadmium cell. It uses a cadmium and iron oxide anode, a nickel hydroxide and graphite cathode, and a potassium hydroxide electrolyte. A single cell develops an almost constant 1.2 V over 90 percent of its discharge cycle. Figure 11-22 shows the typical discharge characteristics. Cell capacity is a function of the discharge rate and ranges from 90 to 120 percent. A nickel-cadmium D cell is rated at 4 Ah but can deliver 4.8 Ah (120 percent) when it supplies a current of 400 mA (a 0.1C rate). The cutoff point is about 1 volt per cell.

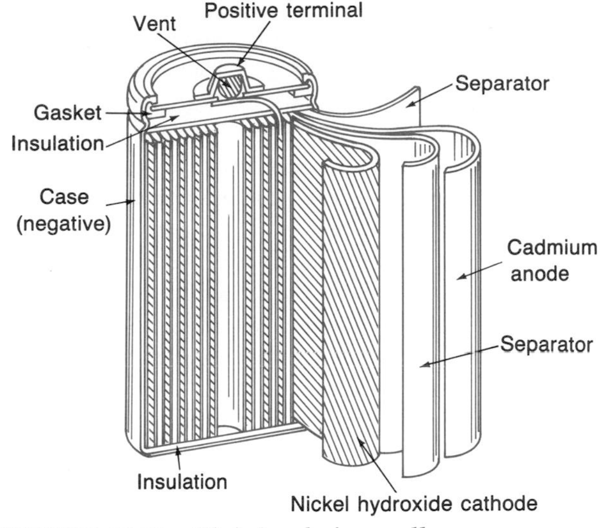


**FIGURE 11-21 Nickel-cadmium cell.**

Nickel-cadmium cells have a low internal resistance and are well suited to high current applications. They must not be shorted because cell or circuit damage can occur. They are typically rated for 500 charge-discharge cycles.

The shelf life of ni-cads varies with storage temperature. The following times are for complete self-discharge:

6 months at 20°C

2 months at 30°C

15 days at 60°C

They may be stored in either a charged or discharged condition. Several cycles of charge and discharge may be required to restore full capacity after an extended storage period. They may also require several deep cycles (drained to 1 volt per cell and then fully charged) after an extended number of shallow cycles. A shallow cycle is when a cell or battery is recharged after having delivered only a small portion of its rated capacity. Too many shallow cycles can cause a *memory effect,* where the cell tends to deliver only that portion of its capacity that it had been delivering.

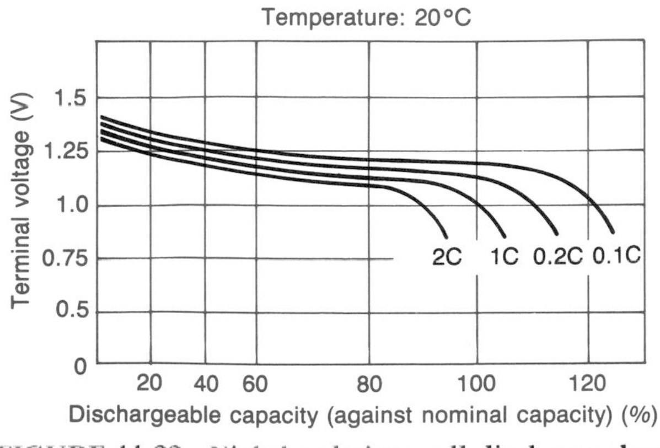


**FIGURE 11-22 Nickel-cadmium cell discharge characteristics.**

Deep discharging of ni-cad batteries containing three or more series cells is a more serious problem than the memory effect. The individual cells in the battery are seldom equal in capacity, and the weakest cell eventually drops to 0 V output. It then becomes a load and reverses in polarity. Polarity reversal often damages the cell. Therefore, it is important to avoid deep discharge of nickel-cadmium batteries.

They can also be damaged by overcharging and by rapid charging. Constant voltage chargers are not recommended. The typical charging circuit supplies a C/10 current. This rate is considered safe because the cell or battery can be left on charge indefinitely with no damage. Nickel-cadmium cells and batteries require about 140 percent of their capacity for complete recharge, and 14 h is the minimum with a C/10 rate. Special quick-charge ni-cad cells and batteries are available. Some of these safely charge at a C/4 rate and require 5.6 h to reach full capacity. However, overcharge is not permissible with these types because the cells begin to heat and are damaged. Quick-charging circuits are designed to provide an automatic setback to a C/10 rate when a certain temperature or cell voltage is reached.

Figure 11-23 compares the discharge curves for the primary and secondary cell types covered in this chapter.

## Self-Test

**21.** In order for a charging source to work, its output voltage must ______ the voltage of the cell or battery to be charged.

**22.** Charging a secondary cell or battery gradually restores the electrodes and the ______ to their charged conditions.

**23.** What is the name of the instrument used to measure the specific gravity of a lead-acid cell?

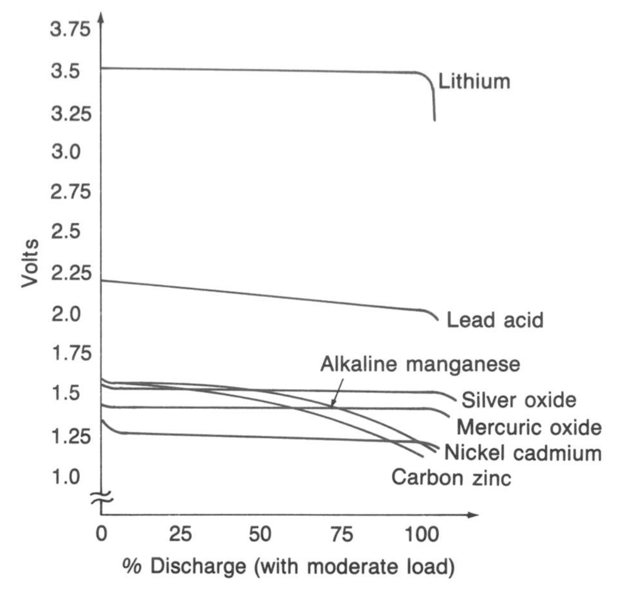


**FIGURE 11-23 Discharge curves for selected cell types.**

**24.** True or false? A discharged lead-acid battery can be restored by adding fresh sulfuric acid to its electrolyte.

**25.** True or false? Overcharging a lead-acid battery generates explosive gases.

**26.** True or false? Sealed lead-acid cells cannot generate external gases.

**27.** What charging current is required for a C/3 rate for a 12.6-V battery rated at 60 Ah?

**28.** What is the watthour rating of a battery that averages 12 V output and has a capacity of 100 Ah?

**29.** How many ni-cad cells are required to build a 12-V battery?

**30.** Refer to Fig. 11-23. Which cell type has the poorest ability to develop a constant output voltage?

**31.** True or false? In general, a cell or battery will produce a greater capacity at low discharge rates than it will at high discharge rates.

## 11-4 OTHER ELECTRIC SOURCES

Fuel cells are similar to the cells and batteries already covered in this chapter. They also produce electric energy by means of chemical reaction but with one important difference: the electrodes are not the reactants. The reactants are fed in on a continuous basis as the electric energy is produced and consumed. Figure 11-24 shows the construction of a hydrooxygen fuel cell. Gas inlets are found on the sides of the cell. The gases pass part way into the porous metal electrodes. The electrolyte solution also passes part way into the electrodes. The gases dissolve in the electrolyte, and the electrolyte side of the anode becomes covered with a thin layer of hydrogen and the electrolyte side of the cathode is covered with a thin layer of oxygen. The layers are not gaseous but are *adsorbed* into the surface of the electrodes. Adsorption is an action whereby a body condenses a gas and holds it on its surface.

The fuel cell electrolyte contains negative hydroxyl ions (OH). When a hydrogen atom moves away from the anode to combine with a hydroxyl ion to form a water molecule ($H_2O$), it leaves an electron behind and the anode accumulates a negative charge. A complementary reaction takes place at the cathode. The oxygen atoms pick up electrons from the cathode and react with the water molecules in the electrolyte to form negative hydroxyl ions. The cathode is left with a positive charge as shown in Fig. 11-24.

One of the problems with hydrooxygen fuel cells is to make the reactions occur at reasonable temperatures. High temperature speeds chemical reactions, but it is not always feasible to operate cells at high temperatures. A platinum *catalyst* may be added to the electrolyte side of the electrodes. A catalyst is a substance that promotes a chemical reaction without itself taking part in the reaction. Platinum is very expensive, and some success has been achieved with nickel oxide catalysts.

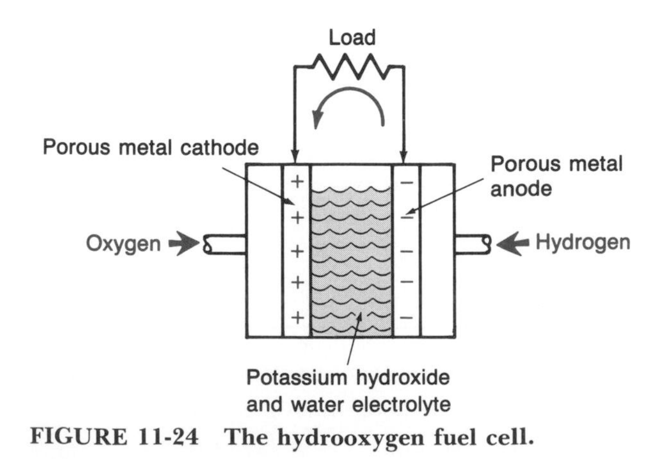


**FIGURE 11-24 The hydrooxygen fuel cell.**

The output of a hydrooxygen fuel cell is 1.23 V. Its theoretical efficiency is 100 percent, and efficiencies as high as 90 percent have been achieved in working cells. Fuel cells have been used in the space program and in underwater applications where they have achieved outputs in the kilowatt range. Acid electrolyte types using hydrocarbon fuels are being developed and are expected to achieve outputs in the megawatt range. Methanol fuel cells and lithium chlorine cells are also being investigated.

## Photovoltaic Sources

The silicon *photovoltaic* cell shown in Fig. 11-25 converts light energy to electric energy. It is called photovoltaic to distinguish it from other photocells which are photoconductive. A *photoconductive* cell's electric resistance changes in response to light input, so it is not usable as an energy source. Photovoltaic cells are energy sources and are also called *solar* cells. Figure 11-26 compares the schematic symbols.

A silicon solar cell is made of P-type and N-type silicon. These materials are covered in detail in Chap. 23. For now, it is adequate to describe them as silicon crystals with inherent electric charges. P-type crystals have positive charges called *holes*, and N-type crystals have negative charges called *electrons*. When light energy enters the P-type silicon in Fig. 11-25, it penetrates to the junction area. The junction is the boundary between the P- and N-type layers. If the light has sufficient energy and the correct wavelength (color), it dislodges valence electrons from the silicon atoms. At the sites where this occurs, holes appear. Holes are positions for electrons and are assigned positive charges. As Fig. 11-25 shows, the dislodged electrons travel to the N side of the junction and are picked up by the metal bottom of the solar cell. The positive holes

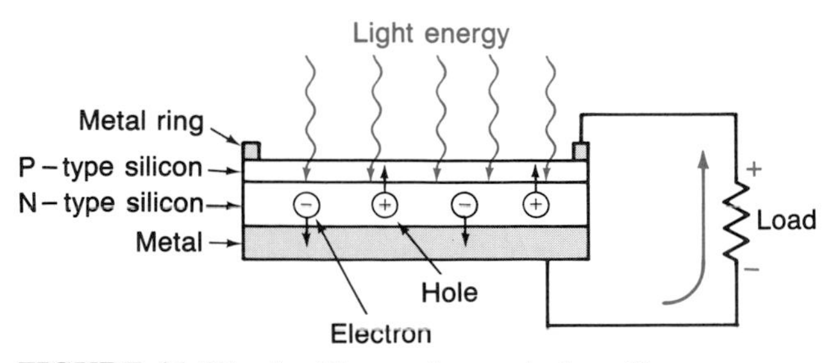

FIGURE 11-25 A silicon photovoltaic cell.

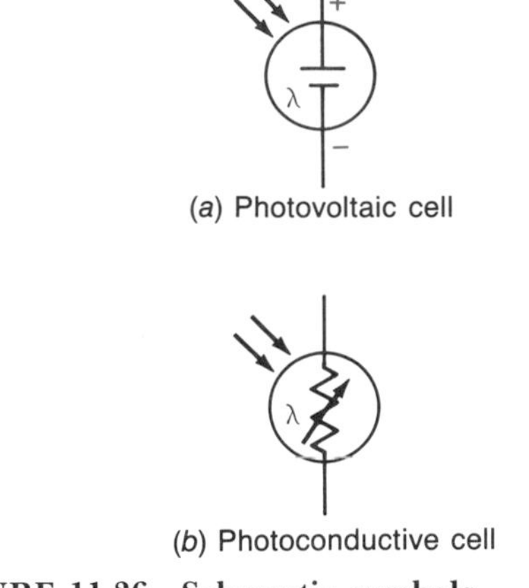


FIGURE 11-26 Schematic symbols.

travel to the P side of the junction and produce a positive charge on the metal ring at the top of the cell. If a load is connected, current flows to equalize the charge imbalance.

Only some of the light energy that enters a solar cell is converted to electric energy. The maximum theoretical efficiency of a silicon solar cell is 25 percent. The cells that are available today range from 10 to 18 percent efficiency. Even though their efficiency is relatively low, silicon solar cells are the only practical source of electric energy for satellites and some types of remote terrestrial equipment. They are manufactured in round, half-round, and various rectangular shapes. The metal back is typically a solder-tinned nickel film. Consequently, these cells are only a few thousandths of an inch thick and are quite fragile. Extreme care must be used when handling them or soldering to them.

A single silicon solar cell develops a no-load output of 0.55 V in full sun conditions. A cell's current and power capabilities are directly related to its size. Solar batteries are series arrangements that provide increased voltage output. Series-parallel arrangements are also used for increased current and voltage output. Solar cells are load-sensitive. Maximum power is transferred when the load resistance is equal to the internal resistance of the cell. Unfortunately, this resistance is a function of light input. Therefore, solar energy systems must be designed for a compromise load resistance. Optimum loading is characterized by a drop in cell output of about 0.1 V when the load is connected. On a clear day, the light energy from the sun is about 1 $kW/m^2$ when the sun is directly overhead.

**EXAMPLE 11-7**

What is the maximum output that can be expected from a round silicon solar cell that is 10 cm in diameter and rated at 15 percent efficiency? What load resistance will result in maximum power transfer for full sun conditions?

**Solution**

The first step is find the area ($a$) of the cell in square meters:

$$a = \pi r^2$$
$$= 3.14 \times (0.05)^2$$
$$a = 7.85 \times 10^{-3}\ \text{m}^2$$

Next, multiply by 1 kW to find the solar power ($P_{\text{sun}}$):

$$P_{\text{sun}} = 1 \times 10^3 \times 7.85 \times 10^{-3} = 7.85\ \text{W}$$

Taking the cell efficiency into account,

$$P_{\text{cell}} = 0.15 \times 7.85\ \text{W} = 1.18\ \text{W}$$

When the cell has an optimum load, its output drops to 0.45 V. The current can be found from the power equation:

$$I = \frac{P}{V} = \frac{1.18\ \text{W}}{0.45\ \text{V}} = 2.62\ \text{A}$$

Finally, Ohm's law provides the optimum load resistance for full sun conditions:

$$R = \frac{0.45\ \text{V}}{2.62\ \text{A}} = 0.172\ \Omega$$

Silicon photovoltaic cells are also used to sense and measure light. For example, they can be used in photographic light meters and in alarm systems. They may also be used in sound movie projectors to sense the light variations produced by the sound track on the film.

## Piezoelectric Sources

The word *piezo* comes from the Greek word for "press." Certain materials produce *piezoelectricity* by the application of pressure. This effect occurs only in certain crystalline insulators. Electric charges appear at the surface of these materials when they are stressed. The piezoelectric effect is reversible; the application of an electric field to a piezoelectric material causes it to distort mechanically. The following materials are piezoelectric:

Sodium potassium tartrate (Rochelle salt)
Ammonium dihydrogen phosphate
Silicon dioxide (quartz)
Barium titanate
Lead zirconate titanate (PZT)

Rochelle salt was one of the first widely applied piezoelectric materials. It was used in microphones and phono cartridges for the purpose of converting mechanical vibrations (sound) into electric signals. Later "crystal" pickups used barium titanate or PZT. Quartz exhibits a rather small output when stressed but forms a very stable crystal. Quartz crystals are widely applied in *oscillator circuits,* where they vibrate hour after hour and month after month with very high accuracy. They are commonly used in electric watches and clocks for timekeeping, in computers where they time events with great precision, and in communications equipment where they control tuning. An ordinary quartz oscillator is stable to one part in 1 million (0.0001 percent), and stabilities of one part in 10 million (0.00001 percent) are attainable by controlling the temperature of the quartz oscillator. Figure 11-27 shows a typical quartz crystal for electronic applications. A metal electrode is fired onto each side of a thin quartz disk. The thickness of the disk determines how rapidly it will vibrate, and rates of hundreds of thousands or millions of times per second are common.

In contrast to quartz, the ceramic-type piezoelectric materials such as barium titanate and PZT can develop extremely high outputs when stressed. The voltage can be in excess of 20 kV, and these materials are used in solid-state ignitors for butane and propane devices, in natural gas stoves and furnaces, and in lighters. A hammer mechanism is used to impact the ceramic crystal when ignition is required. A high-voltage output is produced upon impact, and a spark discharge ignites the gas. The current capability is quite low—in the microampere range.

Piezoelectric *transducers* are also important. A transducer is a device that converts energy from one form to another. Transducers form the basis

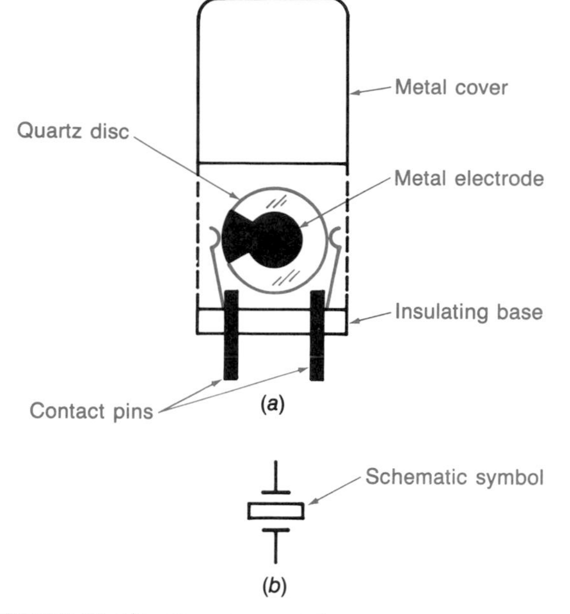


**FIGURE 11-27 Quartz crystal.**

for many measurement systems. Piezoelectric transducers are used both to generate and to detect sound waves. They have applications in sonar, medical sonography, ultrasonic cleaners, high fidelity speakers, and automatic focusing cameras. Piezoelectric transducers are used to sense pressures in industrial and medical applications.

## Thermoelectricity

It is possible to directly convert heat energy to electric energy. A *thermocouple* is shown in Fig. 11-28. This device is based on the *Seebeck effect,* which occurs when two different metals are joined. A voltage appears across the junction of dissimilar metals that is proportional to temperature. All metals are characterized by a cloud of valence electrons. When two different metals interface, the electrons of one material are more energetic than those of the other. The more energetic electrons tend to migrate to the other metal at a rate that is proportional to the temperature of the metals. The Seebeck output is not high, and the process is hardly efficient enough to serve as a practical source of electric energy for many applications. The main application of the Seebeck effect is in thermocouple transducers used to measure temperature in industrial and chemical processes. For example, the output of a chromel-constantan thermocouple is only 75 mV at 1000°C. Amplifiers are usually required to bring the thermocouple output up to a usable level.

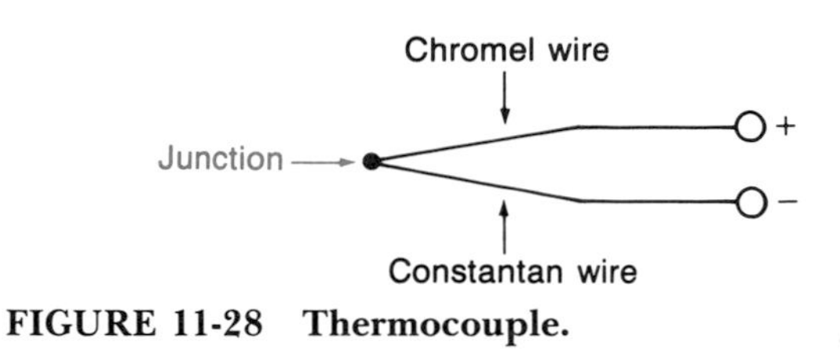


**FIGURE 11-28 Thermocouple.**

Figure 11-29 shows a *thermopile,* which is capable of more significant output. It consists of a series of dissimilar metal junctions. The junctions in the center of the thermopile are heated and run at a much higher temperature than the junctions around the perimeter. The hot junctions develop more voltage than the cold junctions, and the overall output is greater due to series addition. Low-demand electronic circuits have been energized with thermopiles.

The *Peltier effect* is opposite to the Seebeck effect. Forcing a current through a two-junction circuit causes one of the junctions to heat and the other to cool. Sending the current through the circuit in an opposite direction will reverse the hot and cold junctions. Water coolers and other small refrigeration devices have been built from Peltier semicon-

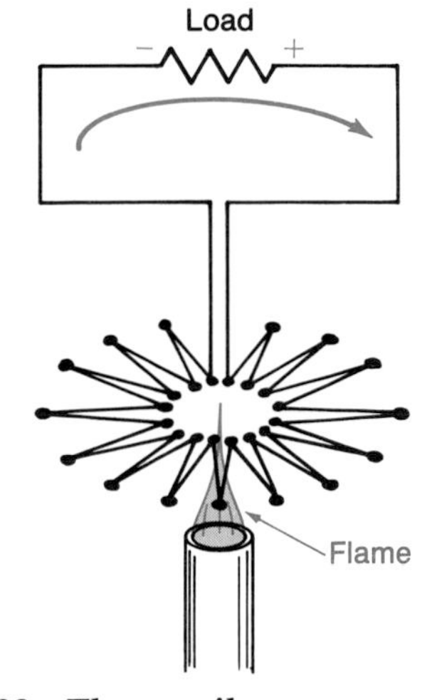

**FIGURE 11-29 Thermopile.**

ductor devices. Their main advantage is simplicity and small size, but they have not been widely applied.

## Self-Test

**32.** A material that promotes a chemical reaction without taking part in the reaction is called a ______.

**33.** True or false? The electrodes in a fuel cell are depleted as the cell supplies current.

**34.** True or false? Both photovoltaic and photoconductive cells convert light energy to electric energy.

**35.** True or false? A solar cell will deliver maximum power to a load when the load resistance is equal to the internal resistance of the cell.

**36.** What is the maximum power output that can be expected from a silicon solar cell that is 12 percent efficient and is 8 cm in diameter?

**37.** Name the effect whereby certain materials produce a voltage when stressed mechanically.

**38.** How does a quartz crystal control timing?

**39.** True or false? A junction of two dissimilar metals will generate a voltage that increases as temperature increases.

**40.** True or false? The phenomenon described in Question 39 is called the Peltier effect.

## SUMMARY

1. Relative motion between a conductor and a magnetic field produces electromagnetic induction in the conductor.

2. The direction of current induced in a conductor can be predicted with the left-hand generator rule.

3. Some rotating generators produce a flow that periodically reverses. This flow is called alternating current (ac).

4. Rotating ac machines produce an output that varies as the sine of the angle of motion. This output is called sinusoidal ac.

5. The instantaneous output from a rotating machine is given by

$$e = B \times l \times v \times \sin \theta$$

The output is proportional to flux density, effective conductor length, conductor (or field) velocity, and the sine of the cutting angle.

6. Rotating generators may use electromagnetic fields for higher flux density.

7. All practical generators use multiturn coils to increase the effective length of the conductor that cuts the magnetic field.

8. The output of a rotating machine can be changed from alternating current to pulsating direct current with the use of a commutator in place of the slip rings.

9. The output of a generator with an electromagnetic field can be controlled by varying the field current.

10. A generator with an electromagnetic field may be self-excited or separately excited.

11. Generators may use voltage regulators to hold their outputs constant.

12. Primary cells and batteries cannot be restored by charging, while secondary cells and batteries can.

13. All cells have two electrodes: a negative anode and a positive cathode.

14. All cells have an electrolyte which is usually acid or alkaline and may be in liquid, paste, or gelled form.

15. A dry cell can be operated in any position without danger of leaks or spills.

16. The maximum storage time for a cell is known as its shelf life.

17. The voltage developed by a cell is determined by the materials it uses for its electrodes and electrolyte.

18. Cell and battery capacity is rated in ampere-hours (Ah) or milliampere-hours (mAh).

19. A cell is considered discharged when its end voltage (or cutoff voltage) is reached.

20. Greater cell capacity is achieved with low discharge rates.

21. Alkaline cells are better suited to high discharge applications than are carbon-zinc cells.

22. Some lithium cells may last in excess of 10 years, and some develop a no-load output as high as 3.8 V.

23. Automotive batteries are of the lead-acid type.

24. The specific gravity of the electrolyte in a lead-acid battery decreases as the battery is discharged and increases as the battery is charged.

25. Lead-acid batteries produce explosive gases when overcharged.

26. The electrolyte in a lead-acid battery is a sulfuric acid solution which is highly corrosive.

27. Sealed lead-acid cells and batteries may be used in some electronic applications.

28. Lead-acid cells have a very low internal resistance and must not be short-circuited.

29. Nickel-cadmium batteries should not be deep-discharged or cell reversal may occur.

30. Cell and battery overcharging can be damaging if the rate exceeds C/10.

31. Nickel-cadmium cells and batteries can develop a memory for repeated shallow cycles.

32. In a fuel cell, the reactants are fed into the cell on a continuous basis.

33. Photovoltaic cells convert light energy to electric energy.

34. The maximum solar radiation at the earth's surface is 1 $kW/m^2$.

35. A typical solar cell is less than 15 percent efficient.

36. Piezoelectric materials convert pressure or stress to electric energy.

37. Quartz crystals are used in many electronic applications to control timing and/or tuning.

38. The direct conversion of heat energy to electric energy at a junction of dissimilar metals is called the Seebeck effect.

**39.** Thermocouples are Seebeck devices and are used in measurement applications.

**40.** When a current is passed through a circuit containing two junctions of dissimilar metals, one of the junctions is cooled. This is known as the Peltier effect.

## CHAPTER REVIEW QUESTIONS

**11-1.** True or false? If the conductor shown in Fig. 11-30*a* is part of a complete circuit, the current direction will be into the page.

**11-2.** True or false? If the conductor in Fig. 11-30*a* moves along a path from one pole to the other (parallel to the flux), then the induced emf will be zero.

**11-3.** Adding more turns to the coil in Fig. 11-30*a* will ______ the induced emf in the conductor.

**11-4.** Slowing the speed of the conductor in Fig. 11-30*a* will ______ the induced emf.

**11-5.** The graph shown in Fig. 11-30*b* is called a ______ wave.

**11-6.** True or false? The graph shown in Fig. 11-30*b* is the expected output from a rotating generator.

**11-7.** True or false? A commutator can be used with rotating generators to change direct current to alternating current.

**11-8.** Refer to Fig. 11-30*c*. How many electric cycles are generated for every 360° turn of the loop?

**11-9.** What type of current will be provided by the machine shown in Fig. 11-30*c*?

**11-10.** If the slip rings are replaced with a commutator, what type of current would be generated by the machine shown in Fig. 11-30*c*?

**11-11.** What type of current is represented by the graph shown in Fig. 11-30*d*?

**11-12.** What type of excitation is used for the generator shown in Fig. 11-30*e*?

**11-13.** What will happen to the output voltage of the generator shown in Fig. 11-30*e* if the resistance of the rheostat is increased?

**11-14.** Electrochemical cells have two electrodes: a positive cathode and a negative ______.

**11-15.** The cell electrodes are immersed in an acid or alkaline ______.

**11-16.** A cell or battery that cannot be recharged is known as a ______ type.

**11-17.** Cells or batteries that can be recharged are the ______ types.

**11-18.** Local action ______ the shelf life of a cell or battery.

**11-19.** A cell may require rest periods during high discharge service due to ______ at its cathode.

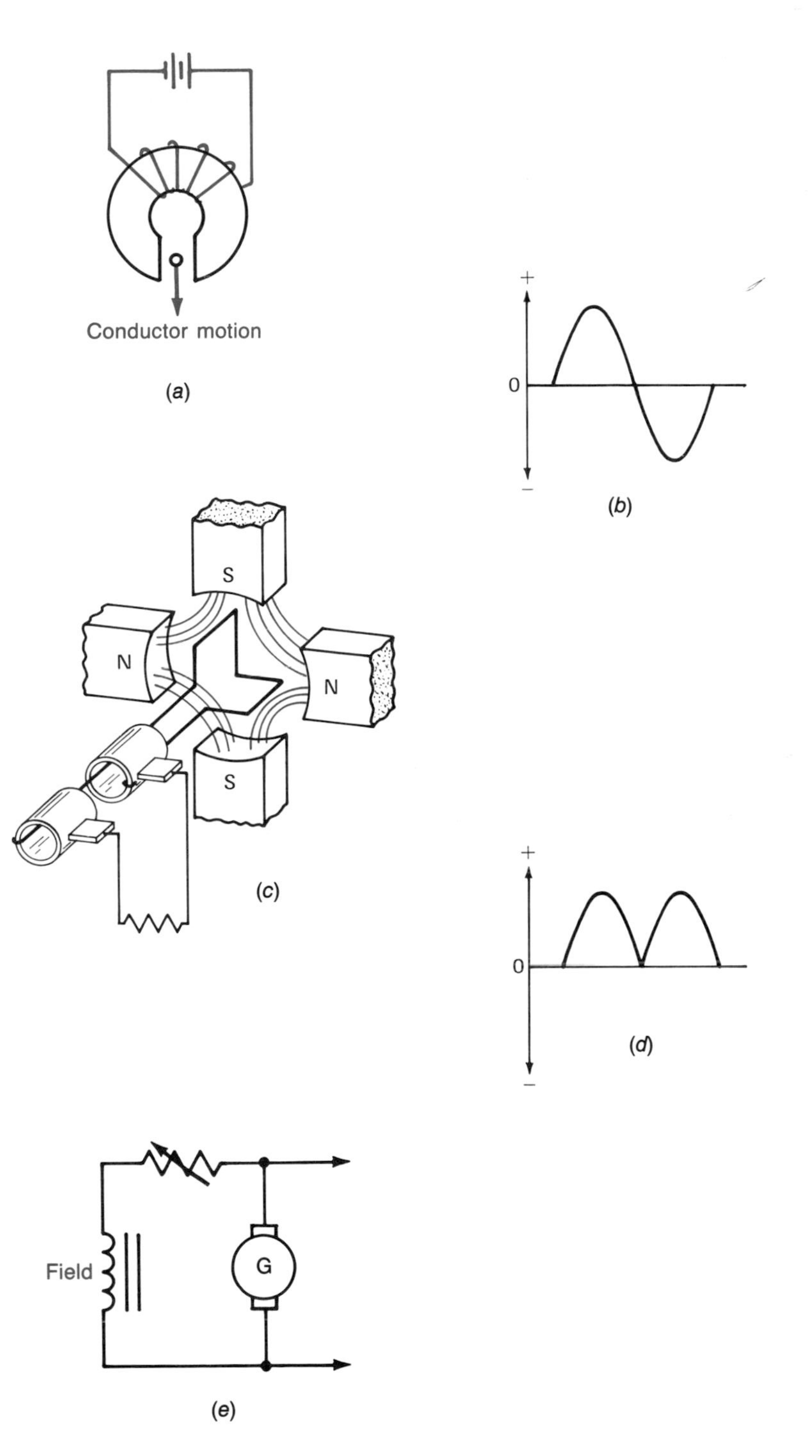


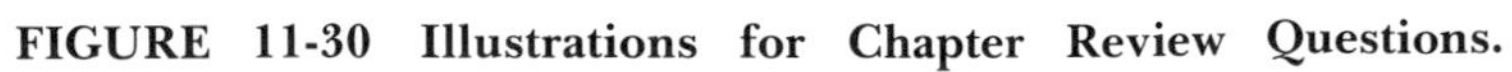

**FIGURE 11-30 Illustrations for Chapter Review Questions.**

**11-20.** A taper charge is provided by constant ______ charging circuits.

**11-21.** A battery charger's output voltage must be ______ than the output voltage of the battery to be charged.

**11-22.** True or false? The electrodes of a fuel cell are slowly decomposed as the cell supplies a load current.

**11-23.** True or false? Silicon solar cells are photovoltaic.

**11-24.** A crystal that converts mechanical pressure or stress to electric energy exhibits the ______ effect.

**11-25.** Transducers convert ______ from one form to another.

**11-26.** Thermocouples exploit the ______ effect.

**11-27.** Solid-state coolers exploit the ______ effect.

**11-28.** An array of thermocouples connected in series is called a ______.

## CHAPTER REVIEW PROBLEMS

**11-1.** Calculate the instantaneous output from a generator with a flux density of 1.5 T, an effective conductor length of 10 m, and a conductor speed of 5 m/s when the angle of cutting is 45°.

**11-2.** A battery can deliver 50 mA for 2 days. What is its capacity?

**11-3.** A battery can deliver 12 V at 150 mA for 4 days. What is its watt-hour rating?

**11-4.** A lead-acid battery is rated at 50 Ah. What charging current is required for a C/3 rate? For a float charge (C/500 rate)?

**11-5.** A completely discharged nickel-cadmium battery is rated at 450 mAh and is to be charged at a constant 45-mA rate. How many hours will be required for a complete recharge?

**11-6.** A 12 percent efficient rectangular solar cell measures 2 cm × 5 cm. What is its maximum output power?

## ANSWER TO SELF-TESTS

**1.** positive
**2.** from left to right
**3.** No flux lines are cut.
**4.** sine
**5.** cycle
**6.** flux density, conductor length, velocity, angle of cutting
**7.** pulsating direct current
**8.** false
**9.** field current
**10.** cannot
**11.** anode
**12.** cathode
**13.** dry
**14.** true
**15.** true
**16.** 40 h
**17.** false
**18.** false
**19.** lithium
**20.** lithium
**21.** exceed
**22.** electrolyte
**23.** hydrometer

**24.** false
**25.** true
**26.** false
**27.** 20 A
**28.** 1.2 kWh
**29.** 10
**30.** carbon-zinc
**31.** true
**32.** catalyst
**33.** false
**34.** false
**35.** true
**36.** 603 mW
**37.** piezoelectric
**38.** by vibrating at a precise rate
**39.** true
**40.** false

# Chapter 12 Alternating Current

The great majority of the electric energy consumed in the world is generated by rotating machines. These machines produce alternating current, which can be sent over long distances with only moderate losses. Alternating current can also be generated by electronic circuits. This chapter covers the terminology, the theory, and the measurements associated with alternating current.

## 12-1 THE SINE WAVE

The abbreviation ac means *alternating current* and ac voltage is interpreted as *alternating voltage*. An ac voltage will cause alternating current to flow as shown in Fig. 12-1. The symbol for a sinusoidal ac generator is a circle with a sine wave in it. The ac generator periodically alternates (reverses its polarity), and when it does the current flowing in the circuit reverses direction.

A *periodic* waveform is one that repeats with time. Figure 12-2 shows three cycles of sinusoidal

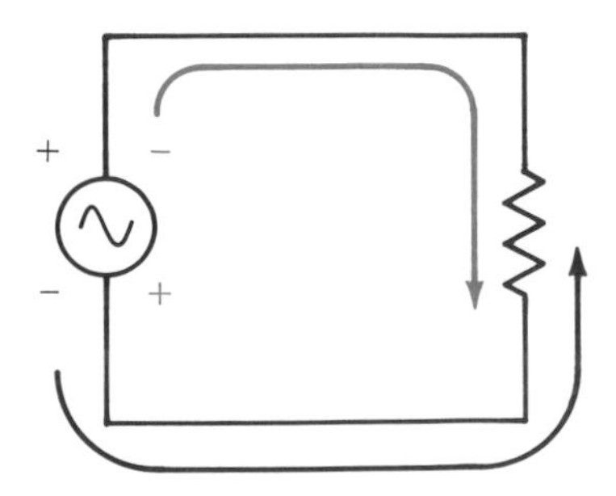

**FIGURE 12-1 The polarity and the flow periodically reverse in ac circuits.**

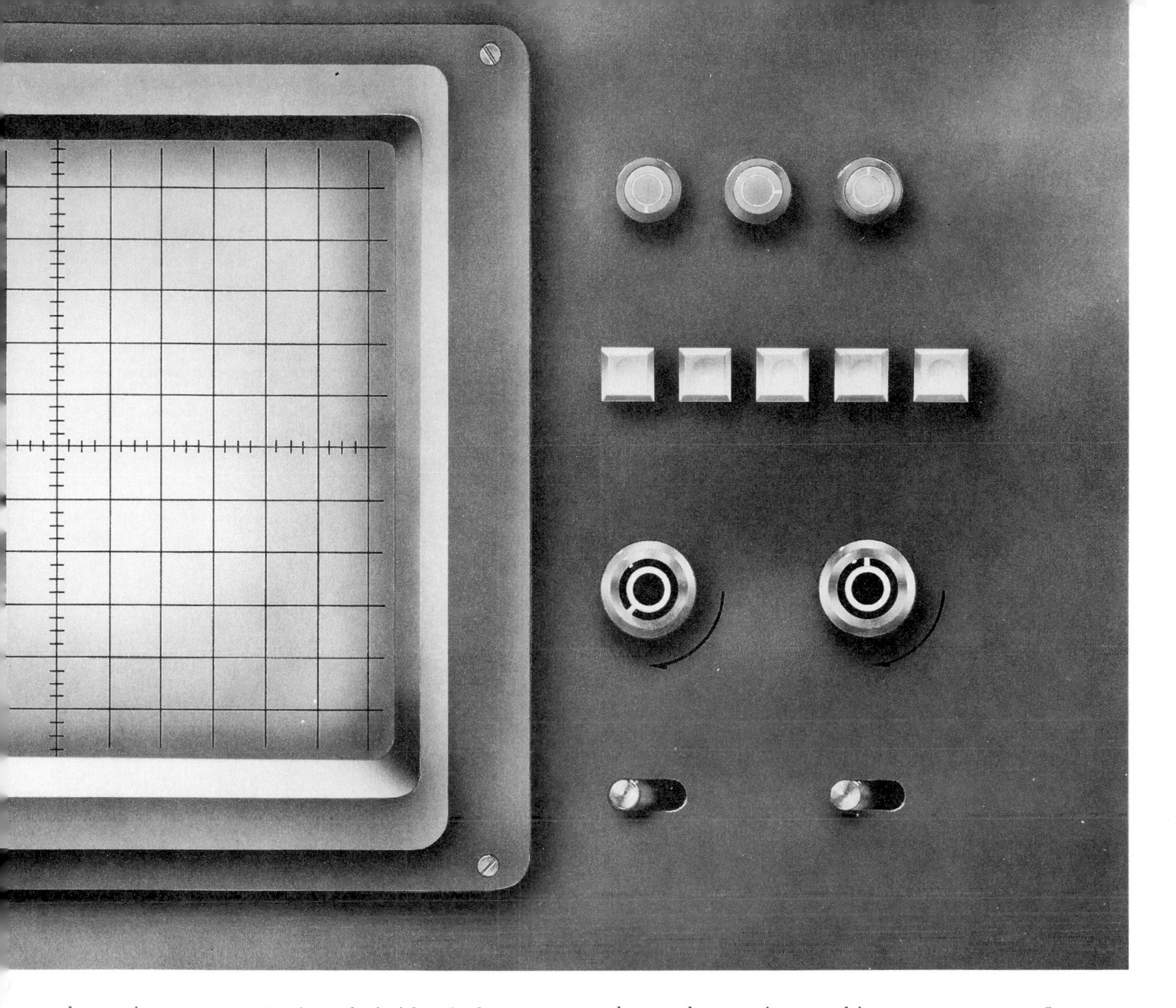

alternating current. Each cycle is identical to any other cycle. Not all periodic waveforms are sinusoidal, and some other types are covered in Sec. 12-5. Sinusoidal alternating current derives its name from the trigonometric sine function. Figure 12-3 shows why rotating machines generate waveforms that follow the sine function. The voltage induced in a conductor varies directly with the speed with which the conductor cuts across magnetic lines of flux. A rotating conductor in a generator can travel at a constant speed, but its output will not be constant. As Fig. 12-3 shows, the linear velocity of a rotating conductor can be split into two component parts: one vertical and the other horizontal. It is the horizontal component of velocity that determines how much voltage will be generated.

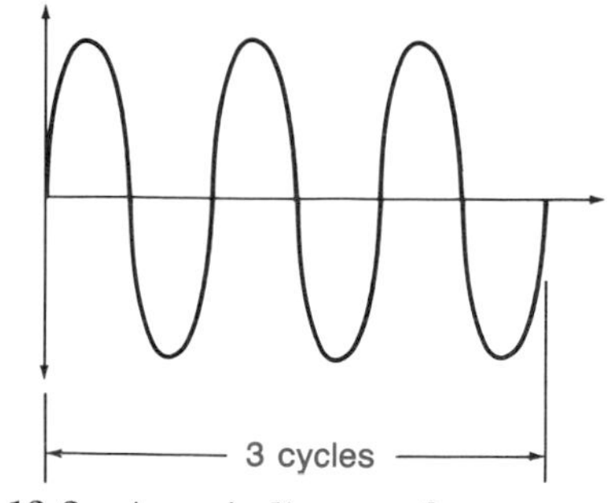


**FIGURE 12-2 A periodic waveform.**

The horizontal component of velocity in Fig. 12-3 varies according to the sine of the angle of conductor rotation. This angle is symbolized by $\theta$ (the Greek letter theta). The horizontal velocity can be determined by drawing a vector representing the linear velocity of the conductor. The length of the vector is proportional to conductor speed.

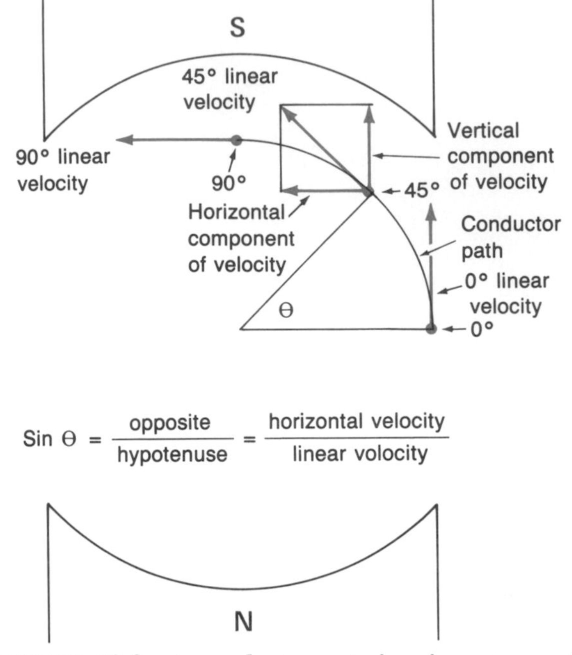


**FIGURE 12-3 A conductor rotating in a magnetic field.**

The vector is drawn *tangent* to the conductor path. At $\theta = 0°$, the linear velocity vector points straight up. At that instant in time, the conductor is moving parallel to the flux lines and no voltage is induced. At $\theta = 90°$, the linear velocity is perpendicular to the flux and maximum voltage is induced. For angles between 0 and 90°, there is both a vertical and a horizontal component of velocity. For example, at 45° they are shown to be equal. Notice that the horizontal component is opposite to angle $\theta$ and that the sine function relates it to the linear velocity vector which forms the hypotenuse of a right triangle. This leads us to the general equation for a sine wave:

$$v = V_{\text{max}} \times \sin \theta$$

where $v$ = instantaneous voltage
$V_{\text{max}}$ = maximum voltage
$\theta$ = angle in electric degrees

A generator with four or more poles will produce more than one electric cycle for every complete rotation of its shaft. Therefore, mechanical degrees of rotation equal the electric degrees only in two-pole machines. This was covered in Chap. 11.

**EXAMPLE 12-1**

What is the instantaneous voltage at 45° for a sine wave with a maximum value of 200 V?

**Solution**

The sine of 45° is 0.707:

$$v = 200 \text{ V} \times 0.707 = 141 \text{ V}$$

**EXAMPLE 12-2**

The maximum current in a sinusoidal ac circuit is 10 A. What is the instantaneous current at 30°?

**Solution**

The current also follows the sine function. The sine of 30° is 0.5:

$$I_i = I_{\text{max}} \times \sin \theta = 10 \text{ A} \times 0.5 = 5 \text{ A}$$

where $I_i$ is the instantaneous current.

Figure 12-4 shows that one sine cycle has 360 electric degrees. The cycle begins at 0° where its magnitude (voltage or current) is 0. At 45°, it has increased to approximately 71 percent of maximum. It achieves its maximum value at 90° and then begins to decrease until it reaches zero again at 180°. At 225°, it is approximately 71 percent of its maximum negative value. It reaches its maximum negative value at 270° and then returns to zero at 360°. One sine cycle has two *alternations:* one positive and the other negative as shown in Fig. 12-5.

As shown in Fig. 12-6, a dc waveform has one steady value. Sinusoidal alternating current has an

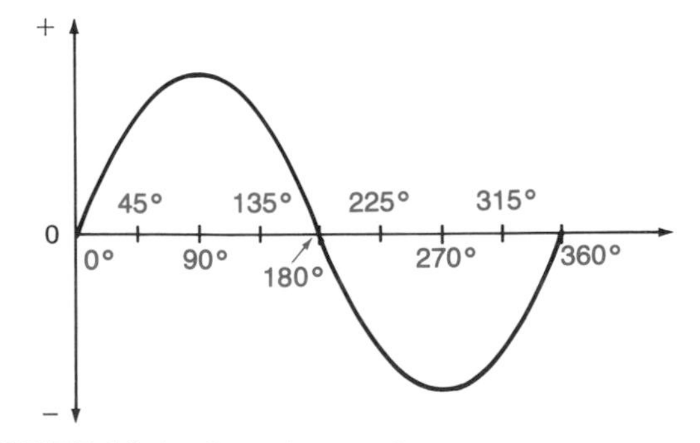


**FIGURE 12-4 One sine cycle.**

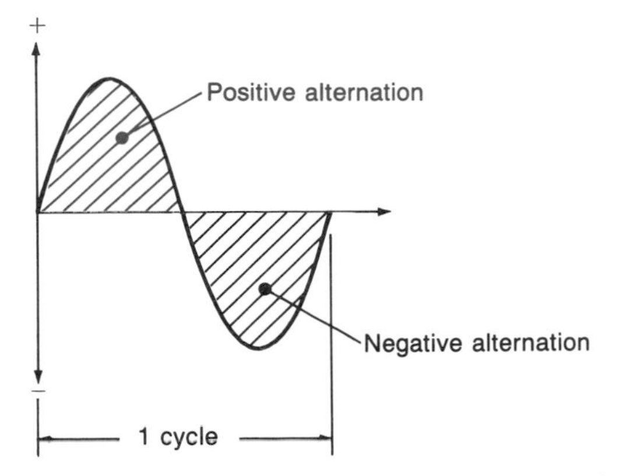


**FIGURE 12-5 A cycle consists of two alternations.**

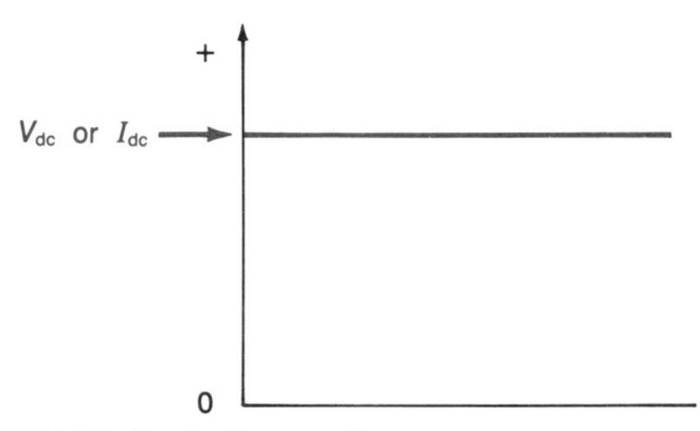


**FIGURE 12-6 A dc waveform.**

infinite number of values. For this reason, there are several ways to measure sine waves. Perhaps the most obvious way is to use its maximum value ($V_{max}$ or $I_{max}$), which is also called its *peak value* ($V_p$ or $I_p$). Another way is to find its average value. The average value can be found by adding a number of instantaneous values, say at 1° intervals, and then dividing by the number of values used. It can also be derived using calculus. If the entire cycle is used, the average value is zero for any sine wave. Of course, this has no purpose, so only one alternation is used when computing the average value. The average value of a sine wave is given by

$$V_{av} = V_p \times 0.637 \qquad \text{or} \qquad I_{av} = I_p \times 0.637$$

---

**EXAMPLE 12-3**

A sinusoidal current has a maximum value of 500 mA. What is its average value? (*Note:* It is understood that the average is being requested for one alternation since it is always zero for the entire cycle.)

**Solution**

Use the constant:

$$I_{av} = 500 \text{ mA} \times 0.637 = 318 \text{ mA}$$

---

The heating value of electric energy is proportional to the square of its voltage or current. This is demonstrated by the familiar power equations:

$$P = \frac{V^2}{R} \qquad \text{and} \qquad P = I^2R$$

The heating value of sinusoidal alternating current can be found by squaring a number of instantaneous values, finding their mean value, and then extracting the square root of the mean. It may also be found by using calculus. The heating value is known as the *root-mean-square* (rms) value or as the *effective* value. The rms value of a sine wave is given by

$$V_{rms} = V_p \times 0.707 \qquad \text{or} \qquad I_{rms} = I_p \times 0.707$$

---

**EXAMPLE 12-4**

What is the effective value of a sine wave with a maximum value of 95 V?

**Solution**

Applying the constant,

$$V_{rms} = 95 \text{ V} \times 0.707 = 67.2 \text{ V}$$

---

**EXAMPLE 12-5**

A resistor eventually reaches a temperature of 65°C when it conducts a direct current of 75 mA. What peak value of sinusoidal alternating current will cause the resistor to reach the same temperature?

**Solution**

The heating effect of rms alternating current is the same as for direct current. Therefore, the rms value of the current must also be 75 mA. This value can be converted to the corresponding peak value by using the reciprocal of 0.707:

$$\frac{1}{0.707} = 1.41$$

The peak value is determined by multiplying the rms current by 1.41:

$$I_p = I_{rms} \times 1.41 = 75 \text{ mA} \times 1.41 = 106 \text{ mA}$$

---

The rms value of alternating current is the standard way of specifying both voltage and current. For example, common household appliances are rated at 120 V ac. This is an rms value. If some other method of measurement is used, it must be specifically stated. Lacking any information to the contrary, always assume that ac values are rms.

The constants 0.637, 0.707, and 1.41 are used quite often in electricity and electronics. They should be memorized. However, if a calculator with a pi key is available, the 0.637 constant can be easily found with

$$0.637 = \frac{2}{\pi}$$

If the calculator has a square root key, the 1.41 constant is given by

$$1.41 = \sqrt{2}$$

And, the 0.707 constant by

$$0.707 = \frac{1}{\sqrt{2}}$$

The peak-to-peak ($V_{p\text{-}p}$ or $I_{p\text{-}p}$) value of an ac waveform is equal to twice its peak value. This method of measurement is used in some electronic applications where the waveform is apt to be viewed on the screen of an oscilloscope. Figure 12-7 summarizes the various ways to measure sinusoidal voltages and the conversion constants. The relationships apply for currents as well as voltages.

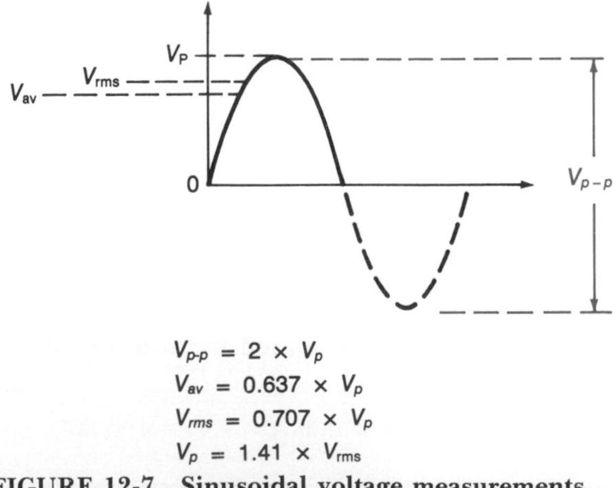


**FIGURE 12-7 Sinusoidal voltage measurements.**

## Self-Test

**1.** When the flow in a circuit is always in the same direction, it is called ______ current.

**2.** When the flow in a circuit periodically reverses, it is called ______ current.

**3.** As a conductor rotates in a vertical magnetic field, the voltage induced in it is proportional to its ______ component of velocity.

**4.** Rotating machines generate ______ alternating current.

**5.** What is the instantaneous output of a generator at 20 electric degrees if its output is 170 V at 90 electric degrees?

**6.** What is the average value of a sine wave with a peak value of 21 V?

**7.** What is the average value of a sine wave over one complete cycle?

**8.** A sinusoidal current has a peak value of 8.5 A. What value of direct current has the same heating effect?

**9.** What is the peak-to-peak value for 120 V ac?

## 12-2 FREQUENCY AND PHASE

If the shaft of an ac generator is turned faster, the generator develops more voltage. It also generates more cycles in a given period of time. The measure of the number of cycles generated per unit of time is called *frequency*. The SI unit of frequency is the hertz (abbreviated Hz), and it is equal to 1 cycle per second:

$$1 \text{ Hz} = 1 \text{ cycle per second}$$

The horizontal axis of a waveform graph can be calibrated in time as well as in electric degrees. Figure 12-8 shows how this type of graph can be used

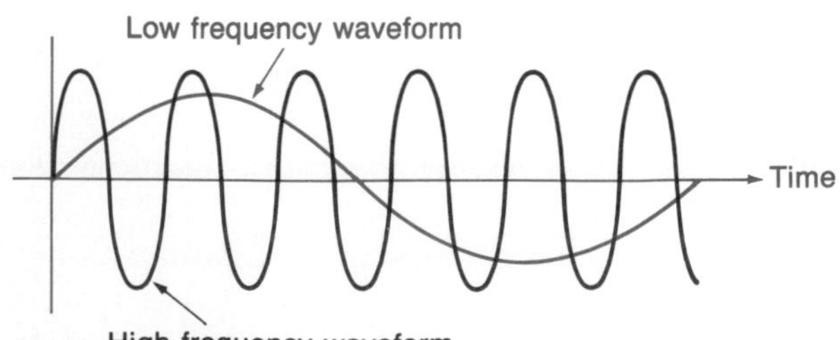


**FIGURE 12-8 Comparing waveforms with different frequencies.**

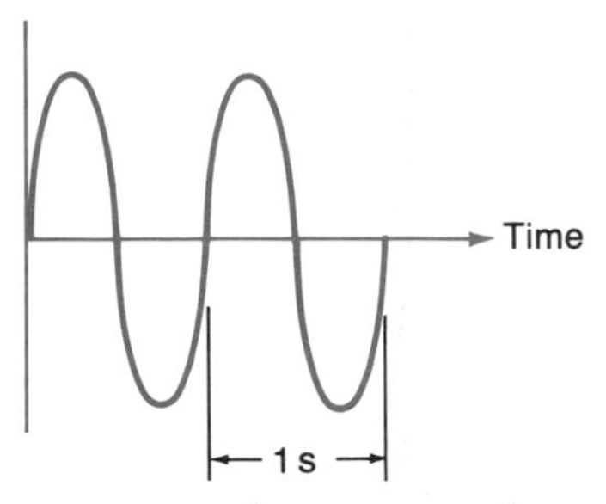


**FIGURE 12-9 One cycle per second.**

to compare two waveforms with different frequencies. The high-frequency waveform shows more cycles for a given time period. The time required to complete one cycle is called the *period* of a waveform. Figure 12-9 shows a waveform with a frequency of 1 Hz. The period of this 1-Hz waveform is 1 s. Frequency and period are related as reciprocals:

$$f = \frac{1}{T} \quad \text{and} \quad T = \frac{1}{f}$$

where $f$ = frequency, Hz
$T$ = period, s

**EXAMPLE 12-6**
Ordinary house current is rated at 120 V, 60 Hz. What is its period?

**Solution**
Taking the reciprocal of the frequency,

$$T = \frac{1}{60} = 16.7 \text{ ms}$$

Figure 12-10 shows the waveform graph for 120 V, 60 Hz ac. Note that the period can be measured from zero-crossing to zero-crossing or from one positive peak to the next. It can also be measured from one negative peak to the next.

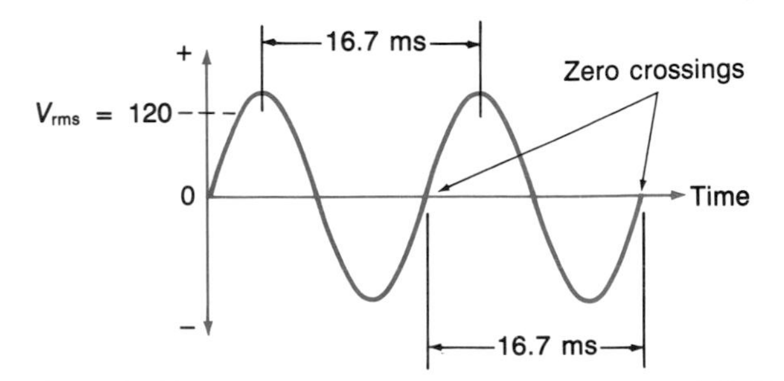


**FIGURE 12-10 A sine wave representing 120 V, 60 Hz ac.**

**EXAMPLE 12-7**
A waveform has a period of 1 $\mu$s. What is its frequency?

**Solution**
Taking the reciprocal,

$$f = \frac{1}{1 \times 10^{-6}} = 1 \times 10^{6} \text{ Hz (1 MHz)}$$

Radio sources send energy waves through space at the speed of light. The distance a wave travels during one period is known as its *wavelength*. Wavelength uses the symbol $\lambda$ (the Greek lowercase letter lambda) and is found by

$$\lambda = vT$$

where $\lambda$ = wavelength, m
$v$ = velocity, m/s
$T$ = period, s

Wavelength may also be determined from velocity and frequency:

$$\lambda = \frac{v}{f}$$

**EXAMPLE 12-8**
Figure 12-11 illustrates the transmission tower for a radio station operating at a frequency of 100 MHz. What is the wavelength of the radio wave?

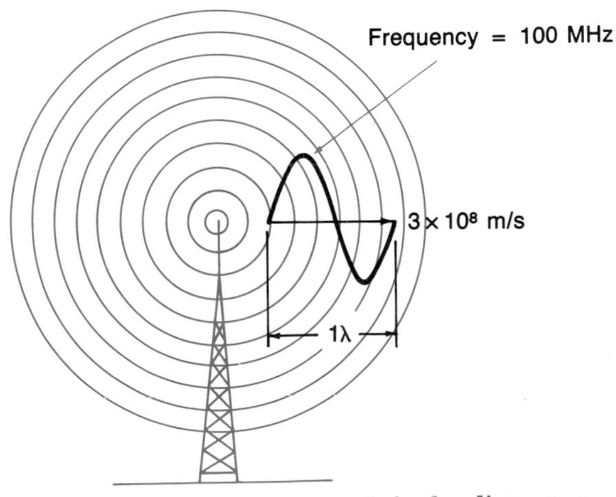


**FIGURE 12-11 One wavelength is the distance traveled during one period.**

**Solution**

As Fig. 12-11 shows, the velocity of a radio wave is $3 \times 10^8$ m/s. The wavelength is found by

$$\lambda = \frac{3 \times 10^8}{100 \times 10^6} = 3 \text{ m}$$

So far, we have seen the horizontal axes of waveform graphs calibrated in degrees and in units of time. They may also be calibrated in *radians*. A radian is an arc equal in length to the radius of the circle it is a part of. It is also the angle (57.3°) subtended by that arc as shown in Fig. 12-12. There are $2\pi$ radians (rad) in 360° (a circle). This is based on the equation for the circumference of a circle:

$$\text{Circumference} = \pi \times \text{diameter} = 2\pi \times \text{radius}$$

Degree measure and radian measure are both used in electricity and electronics. They may be interchanged by

$$\text{Radians} = \text{degrees}\left(\frac{\pi}{180}\right)$$

$$\text{Degrees} = \text{radians}\left(\frac{180}{\pi}\right)$$

Figure 12-13 shows a sine wave graph with the horizontal axis calibrated in radians.

**EXAMPLE 12-9**

What is the instantaneous current for a sine wave with a peak current of 4 A at an angle of 0.68 rad?

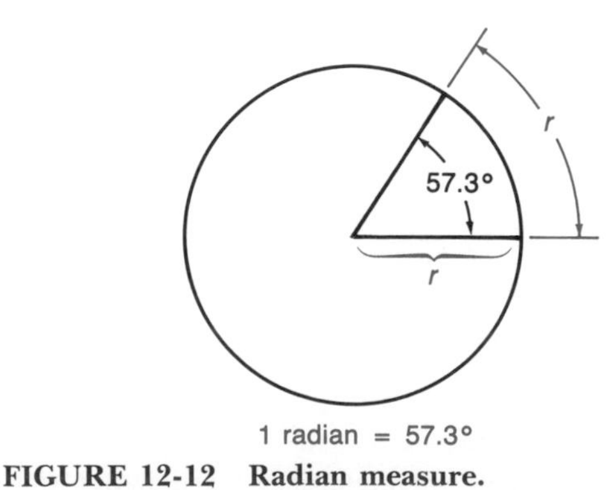


**FIGURE 12-12 Radian measure.**

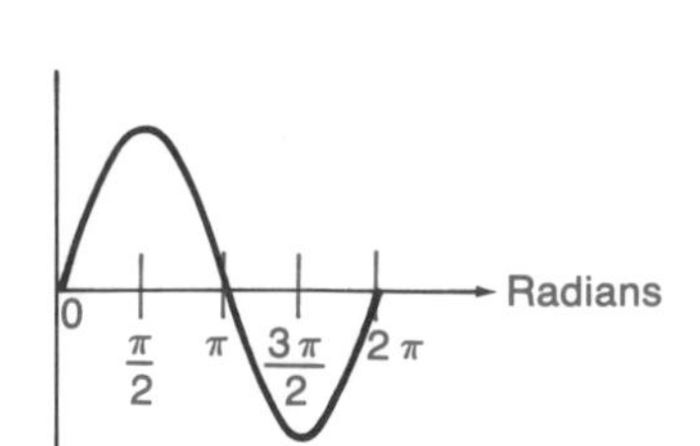


**FIGURE 12-13 There are $2\pi$ rad in 1 cycle.**

**Solution**

Some calculators can operate in a radian mode. The sine of 0.68 rad is 0.629. The angle can be converted from radians to degrees:

$$\text{Degrees} = 0.68 \times \frac{180}{\pi} = 39.0$$

The sine of 39° is 0.629. The instantaneous current is determined by

$$I_i = 4 \text{ A} \times 0.629 = 2.52 \text{ A}$$

As a conductor in a rotating machine generates 1 cycle, it travels $2\pi$ rad. If the shaft turns at a constant rate, the *angular velocity* of the conductor is also constant. Angular velocity is measured in *radians per second* (rad/s). The symbol for angular velocity is $\omega$ (the Greek lowercase letter omega). The angular velocity of a waveform can be found from its frequency:

$$\omega = 2\pi f$$

where $\omega$ = angular velocity, rad/s
$f$ = frequency, Hz

**EXAMPLE 12-10**

Find the angular velocity for 60 Hz ac.

**Solution**

Applying the equation,

$$\omega = 2 \times 3.14 \times 60 = 377 \text{ rad/s}$$

Waveforms of the same frequency may differ in angle. For example, two identical generators can be driven from a common shaft. They would produce the same frequency but may vary in angle.

One of the generators might produce its peak positive output at the same instant in time that the other is at a zero-crossing. An angular reference is used in these situations and the deviation from that reference is called the *phase angle.*

Figure 12-14*a* shows two sine waves that are *in-phase.* If waveform A is chosen as the reference, then waveform B exhibits zero *phase shift.* Both waveforms peak at the same time and cross zero at the same time. Figure 12-14*b* shows a different situation. The waveforms are now 90° out of phase. The peak of one coincides with the zero-crossing of the other. It takes one-fourth of a cycle for a sine wave to vary from peak to zero or from zero to peak. One-fourth of a cycle is 90°. Figure 12-14*c* also shows two waveforms that are 90° out of phase. However, in this case waveform B peaks one-quarter cycle before waveform A. If waveform A is the reference, waveform B shows a 90° *phase lead.* In Fig. 12-14*b*, waveform B shows a 90° *phase lag* with reference to waveform A. Figure 12-14*d* shows two waveforms that are 180° out of phase. Note that the peaks are coincident but are opposite in polarity.

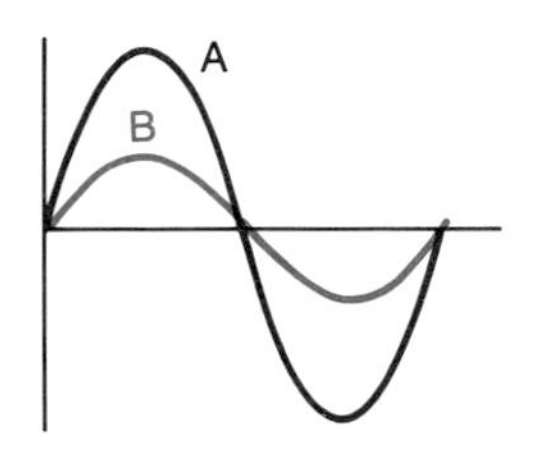


(a) Waveforms in phase

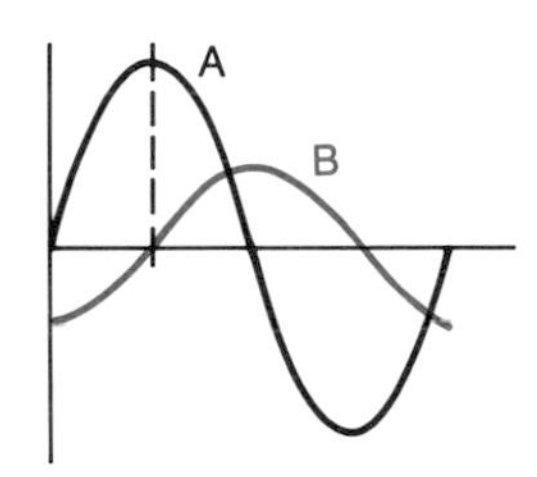


(*b*) Waveforms 90° out of phase (A leads B)

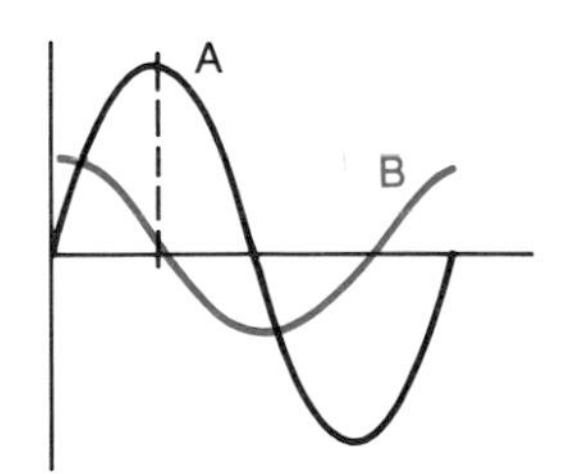


(*c*) Waveforms 90° out of phase (B leads A)

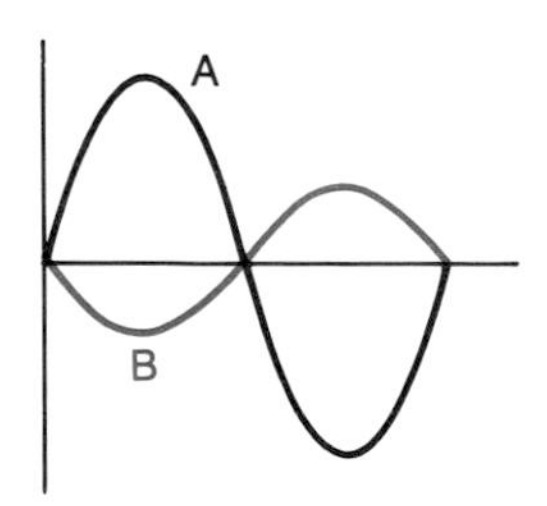


(*d*) Waveforms 180° out of phase

**FIGURE 12-14 Phase relationships.**

## Self-Test

**10.** What is the period of 1 kHz ac?

**11.** An ac waveform has a period of 20 $\mu$s. What is its frequency?

**12.** As frequency increases, period ______.

**13.** What is the wavelength of a radio wave with a frequency of 155 MHz?

**14.** How many radians are there in 1 electric cycle?

**15.** Convert 0.35 rad to degrees.

**16.** Convert 80° to radians.

**17.** Find the angular velocity for 10 kHz ac.

**18.** Refer to Fig. 12-15. What is the phase of waveform B with reference to waveform A?

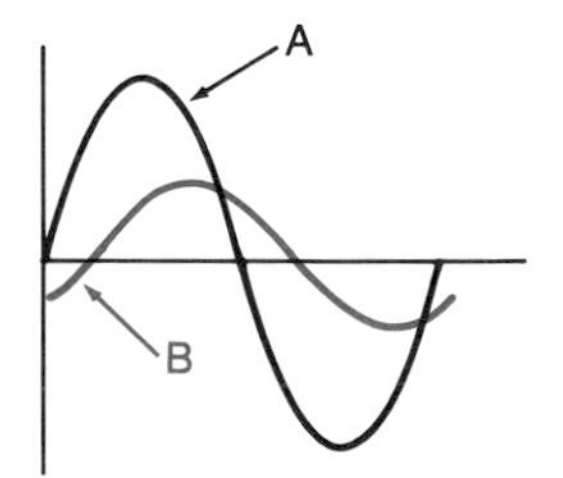


**FIGURE 12-15 Self-Test.**

## 12-3 PHASORS

Figure 12-16 shows a vertical projection of a rotating vector. A *vector* is a line with a length proportional to the magnitude of the quantity it represents and with an angle corresponding to the quantity's direction. A rotating vector can be used to represent alternating current or alternating voltage. The first vector position in Fig. 12-16 is horizontal, pointing to the right. This is known as the *reference position* and represents a phase angle of zero. This vector position has a vertical component of zero. The second vector position is shown for an angle of about 20° and is projected to the right with a positive vertical component. As each vector position is projected to the right, a sine curve is traced by the intersections of the vertical vector components with the corresponding angular positions on the horizontal axis of the waveform graph.

Waveforms can be used to represent the magnitude and phase angle of currents and voltages but are difficult to draw and are not convenient to work with when analyzing ac circuits. *Phasors* are easy to draw and convenient to manipulate using graphical techniques or mathematics. A phasor is an electrical vector. Its length represents the magnitude of voltage or current and its angle represents phase. Refer to Fig. 12-17. By convention, phasors rotate counterclockwise for positive angles and clockwise for negative angles. The phasor shown at 315° is exactly in the same position it would be if it had rotated −45° from the reference position.

Figure 12-18 shows various waveforms along with their corresponding phasor diagrams. In Fig. 12-18*a* the waveforms are 180° out of phase and so are the phasors. Waveform A is in the reference position. In Fig. 12-18*b* the waveforms are 90° out of phase. Once again, waveform A is in the reference position. It peaks $\frac{1}{4}$ cycle before waveform B, which is drawn at −90° in the phasor diagram. Since phasors rotate counterclockwise, the phasor diagram shows that B *phase-lags* A. In Fig. 12-18*c* the waveforms are once again 90° out of phase. However, in this case B is seen to *phase-lead* A by

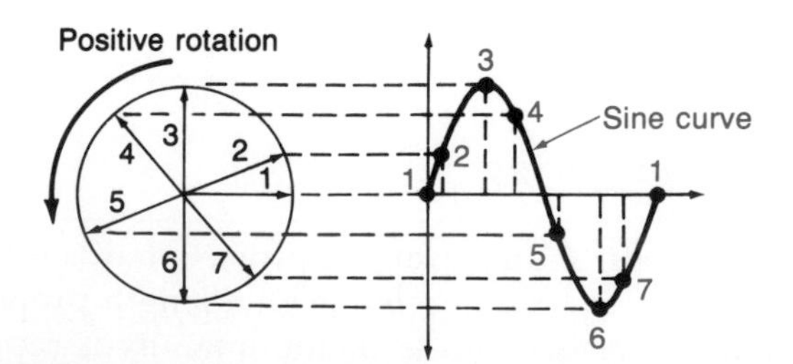


FIGURE 12-16 A vertical projection of a rotating vector.

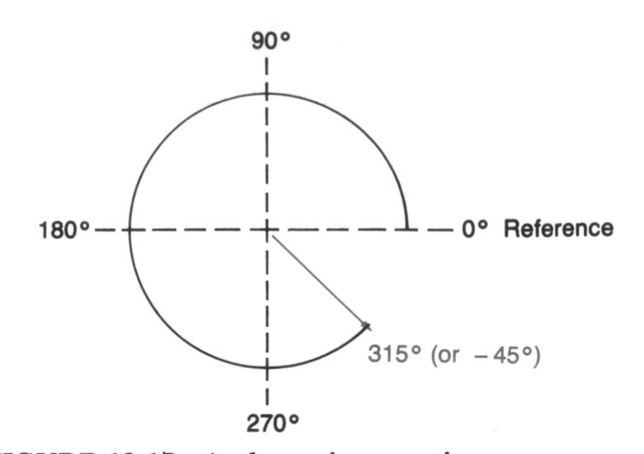

FIGURE 12-17 A phasor is a rotating vector.

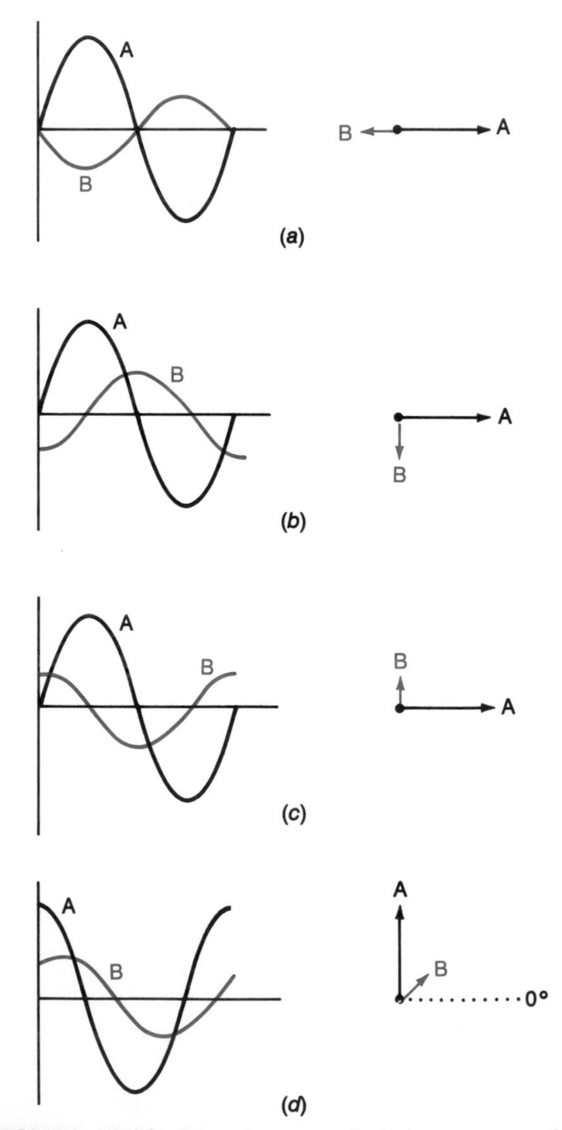


FIGURE 12-18 Waveforms and their corresponding phasor diagrams.

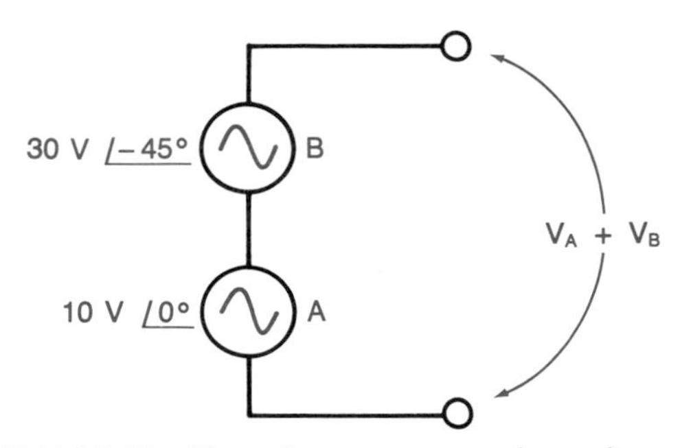


**FIGURE 12-19 Two sine generators in series.**

90°. Figure 12-18*d* shows a situation where neither waveform is in the reference position. Waveform A peaks at the beginning of the waveform graph and is 90° ahead of the reference. Waveform B peaks 45° after waveform A. Phasor B leads the reference by 45° but lags phasor A by 45° degrees.

Figure 12-19 shows two sine generators connected in series. The output of each generator is given in *polar form.* Polar form is a convention where the magnitude of a voltage (or current) is followed by its phase angle. Note the use of the angle symbol (∠) between the numbers. Figure 12-20 shows two methods of summing two ac waveforms. In Fig. 12-20*a*, point-by-point addition is used. Waveform A represents generator A from Fig. 12-19 and waveform B represents generator B. The sum waveform is developed by adding the vertical components of each waveform at several positions along the graph. Notice that the sum waveform is sinusoidal.

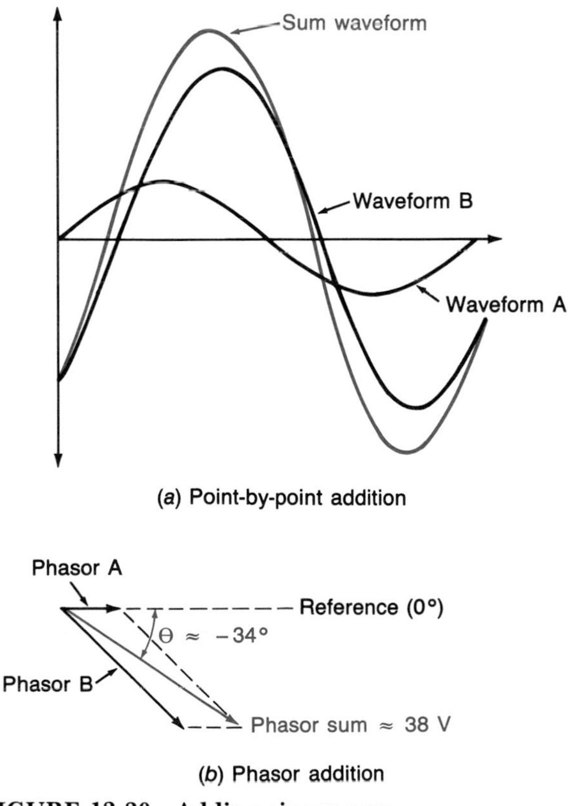


**FIGURE 12-20 Adding sine waves.**

The point-by-point procedure is tedious, and the results are not very accurate unless extreme care is taken. Figure 12-20*b* shows phasor addition. The phasor sum is found by geometric construction of a parallelogram. A *parallelogram* is a four-sided figure whose opposite sides are parallel and equal. The diagonal of the parallelogram is the phasor sum. If the original phasors are accurately drawn, the magnitude of the sum voltage will be proportional to the length of the diagonal. The phase angle of the sum can be measured with a protractor. In polar form, the sum of the phasors in Fig. 12-20*b* is approximately 38 V ∠−34°. An exact mathematical approach is presented in Chap. 19.

Figure 12-21 shows a special case where two generators are 90° out of phase. This special case produces a right-angle phasor diagram which allows the *pythagorean theorem* to be applied in finding the sum, which is usually called the *resultant.* This theorem states that the sum of the squares of the legs of a right triangle is equal to the square of the hypotenuse. Therefore, the hypotenuse of a right triangle is equal to the square root of the sum of the squared sides. As Fig. 12-21 shows, when the parallelogram and its diagonal are constructed, a right triangle is formed. Note the 90° (right) angle and the hypotenuse which is opposite to it. The magnitude of the phasor resultant is found by squaring the sides, adding them, and then taking the square root. The angle of the resultant is found with the trigonometric tangent (tan) function which relates the side opposite to an angle to the side adjacent to it. Since the opposite and adjacent sides are known, the inverse tangent, or *arctan,* function is used to convert the ratio 1.33 to the angle. The arctan function on some calculators may be marked as $\tan^{-1}$.

### EXAMPLE 12-11

Find the resultant of 38 V ∠−90° and 55 V ∠0°. What is the shape of the resultant waveform, assuming that the components are sinusoidal?

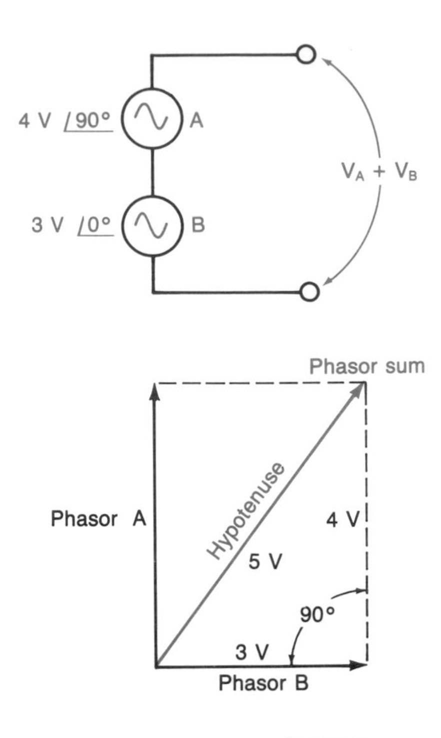


**FIGURE 12-21 Finding the resultant of two right-angle phasors.**

### Solution

Figure 12-22 shows the phasor diagram. The magnitude of the resultant is given by the pythagorean theorem:

$$V_{\text{resultant}} = \sqrt{38^2 + 55^2} = 66.9 \text{ V}$$

The angle of the resultant phasor is

$$\angle = \arctan \frac{38}{55} = 34.6°$$

As Fig. 12-22 shows, the angle should be $-34.6°$ (or $+325.4°$). The 38-V phasor is pointing down and is therefore negative. The algebraic signs have

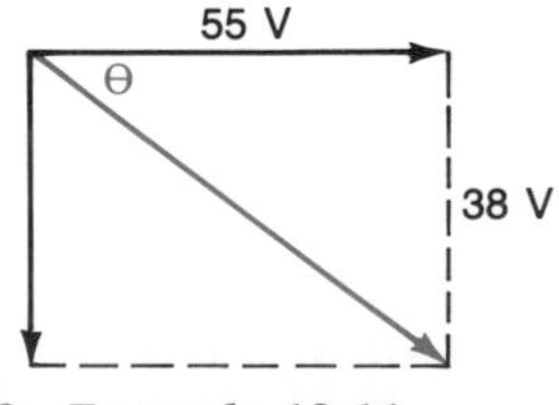


**FIGURE 12-22 Example 12-11.**

no effect on the magnitude calculation since the values are squared. They do, however, affect the angle calculation:

$$\angle = \arctan \frac{-38}{55} = -34.6°$$

In polar form, the phasor sum is 66.9 V $\angle -34.6°$. When the component waveforms are sinusoidal, the resultant waveform is also sinusoidal.

## Self-Test

**19.** If a phasor points straight down, its angle can be given as 270° or ______°.

**20.** Phasors are positioned counterclockwise from the reference axis for ______ angles.

**21.** Refer to Fig. 12-23*a*. Describe the phase relationship between A and B.

**22.** Refer to Fig. 12-23*a*. What is the phase of A with respect to the reference position?

**23.** Refer to Fig. 12-23*a*. What is the phase of B with respect to the reference position?

**24.** Refer to Fig. 12-23*b*. Find $V_{AB}$.

**25.** Find $V_{AB}$ for Fig. 12-23*c*.

**26.** Find $V_{AB}$ for Fig. 12-23*d*.

## 12-4 RESISTIVE CIRCUITS

Alternating current resistive circuits can be analyzed using Ohm's law and the other techniques presented for dc circuit analysis. The instantaneous current $I_i$ in a resistor $R$ is proportional to the instantaneous voltage $v$ across it:

$$I_i = \frac{v}{R}$$

As Fig. 12-24 shows, if the voltage waveform across the resistor is sinusoidal, the current waveform is also sinusoidal. The voltage and current waveforms are in phase.

### EXAMPLE 12-12

A sinusoidal waveform across a 47-Ω resistor has a peak value of 75 V. What is the instantaneous current in the resistor at 60°?

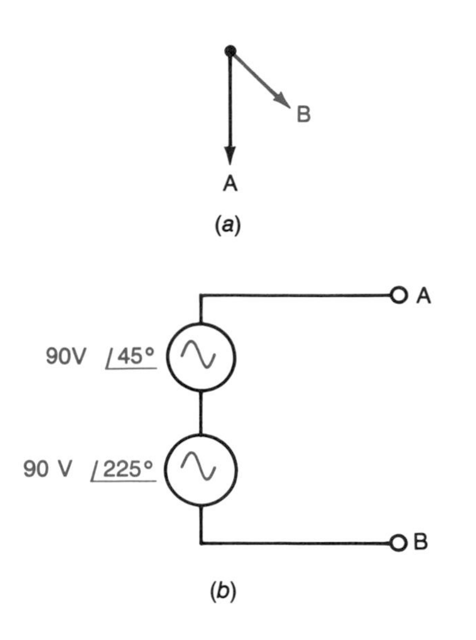


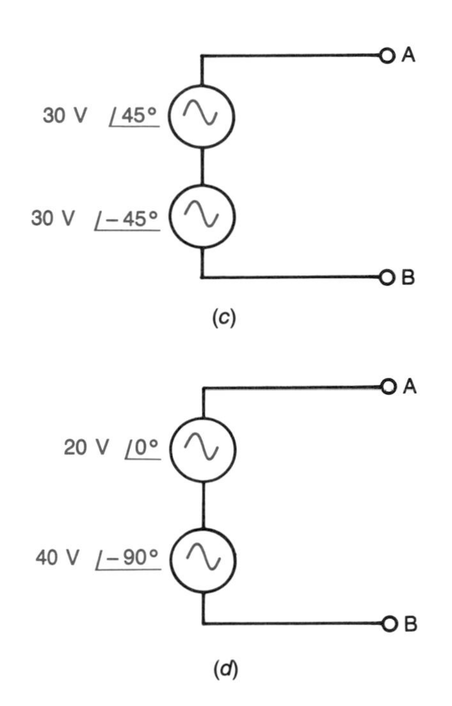


**FIGURE 12-23 Self-Test.**

**Solution**

Find the instantaneous voltage at 60°:

$$v = \sin 60° \times 75\ \text{V} = 65.0\ \text{V}$$

Use Ohm's law:

$$I_i = \frac{65\ \text{V}}{47\ \Omega} = 1.38\ \text{A}$$

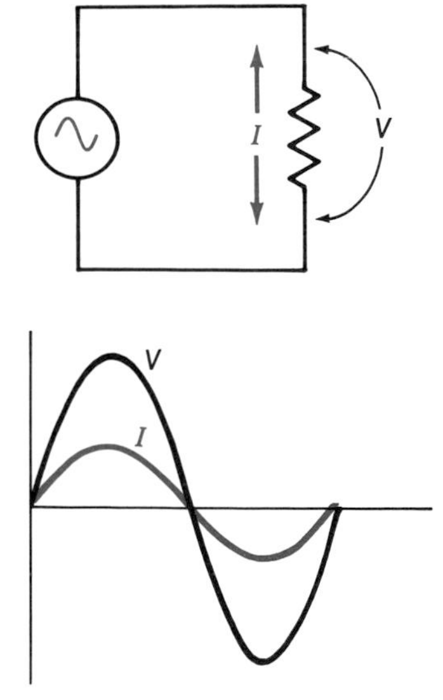


**FIGURE 12-24 The current and voltage are in phase in resistive circuits.**

Any of the amplitude (magnitude) measurement techniques can be used to relate current and voltage in resistive circuits. For example, the rms current can be found with Ohm's law:

$$I_{\text{rms}} = \frac{V_{\text{rms}}}{R}$$

**EXAMPLE 12-13**

The peak-to-peak sinusoidal potential difference across a 150-Ω resistor is 339 V. What is the rms current in the resistor?

**Solution**

Find the rms value of the voltage:

$$V_{\text{rms}} = \frac{V_{p\text{-}p}}{2} \times 0.707 = 120\ \text{V}$$

Use Ohm's law:

$$I = \frac{120\ \text{V}}{150\ \Omega} = 0.8\ \text{A}$$

*Note:* The current is understood to be the effective (rms) value if no other measurement technique is specified.

**EXAMPLE 12-14**

Find the magnitude and the angle of the current flow for $R_2$ in Fig. 12-25.

**Solution**

The superposition theorem provides a convenient technique for this circuit. We begin by replacing the right-hand generator with a short circuit. This places $R_2$ and $R_3$ in parallel, and this combination is in series with $R_1$ for a total resistance of

$$R_T = R_1 + R_2 \,||\, R_3 = 220\ \Omega$$

The current flow is found with Ohm's law:

$$I = \frac{120\ \text{V}}{220\ \Omega} = 0.545\ \text{A} \angle 0^\circ$$

Note that this current is in phase with the 120-V source. The current divider equation can be used to find the current in $R_2$:

$$I = 0.545 \times \frac{200\ \Omega}{200\ \Omega + 300\ \Omega} = 0.218\ \text{A} \angle 0^\circ$$

The 240-V source is now restored, and the 120-V source is replaced with a short circuit. The total resistance is now

$$R_T = R_3 + R_1 \,||R_2 = 275\ \Omega$$

The current flow is

$$I = \frac{240\ \text{V}}{275\ \Omega} = 0.873\ \text{A} \angle 90^\circ$$

Note that this current is in phase with the 240-V source. The flow in $R_2$ is found with the current divider rule:

$$I = 0.873\ \text{A} \times \frac{100\ \Omega}{100\ \Omega + 300\ \Omega} = 0.218\ \text{A} \angle 90^\circ$$

The final step is to superimpose the two currents in $R_2$. Figure 12-25 shows the phasor diagram. The magnitude of the resultant is found with the pythagorean theorem:

$$I_{R_2} = \sqrt{0.218^2 + 0.218^2} = 0.309\ \text{A}$$

The arctan function is used to find the phase angle:

$$\theta = \arctan \frac{0.218}{0.218} = 45^\circ$$

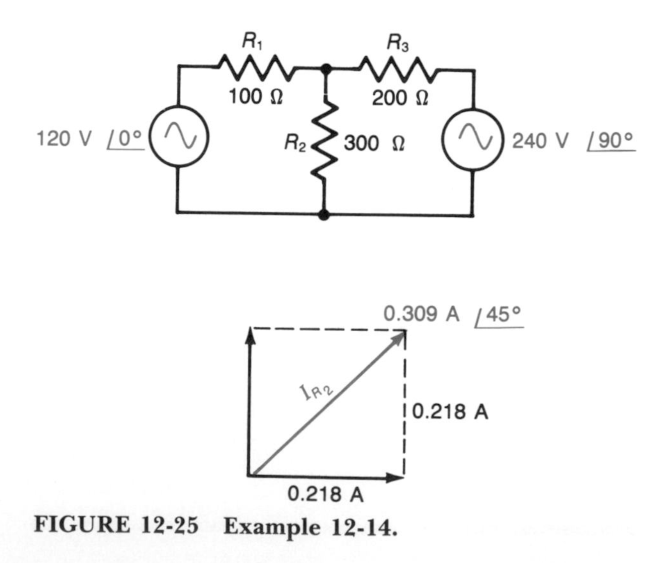


**FIGURE 12-25 Example 12-14.**

Stated in polar form, the current flow in $R_2$ is 0.309 A $\angle 45^\circ$. In some cases, the angle may not be important. An ac current or voltage can be specified as an *absolute value* by using vertical bars around the magnitude:

$$I = |0.309\ \text{A}|$$

The bars are normally eliminated when it is obvious that the absolute value is specified.

Power in ac resistive circuits is equal to the product of voltage and current. Since the instantaneous voltage and current follow the sine function, the instantaneous power follows a sine-squared function. As Fig. 12-26 shows, the power function is always positive when the voltage and current are in phase. The frequency of the power waveform is

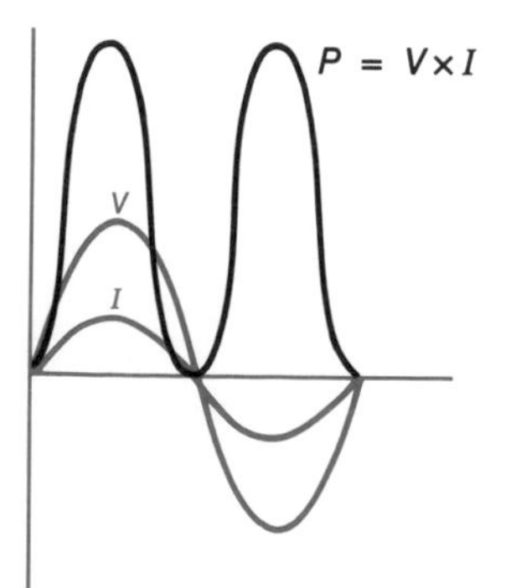


**FIGURE 12-26 Power in a resistive circuit.**

twice the voltage or current frequency. If a lamp is connected to a 60-Hz power source, it receives 120 power pulses per second. Flicker is not a problem because of the thermal lag of the lamp filament (it cannot heat and cool that rapidly) and because of the persistence of human vision.

**EXAMPLE 12-15**

A resistive heating element is designed for 240 V, 60 Hz ac. Calculate its power if its resistance is 28.8 Ω. How will it perform if it is connected to a 240 V dc source?

**Solution**

Use the power equation:

$$P = \frac{V^2}{R} = 2 \text{ kW}$$

Since the element is resistive, it will perform the same as long as the dc voltage is equal to the rms value of the ac voltage.

## Self-Test

**27.** A 500-V sine wave appears across a 10-kΩ resistor. What is the instantaneous current in the resistor at a phase angle of 35°?

**28.** What is the phase relationship between current and voltage in an ac resistive circuit?

**29.** What is the rms current in a 560-Ω resistor when it is connected to a sine generator with a peak output of 20 V?

**30.** Find the magnitude and angle of the current in $R_2$ for Fig. 12-27.

**31.** What is the absolute value of the current for Question 30?

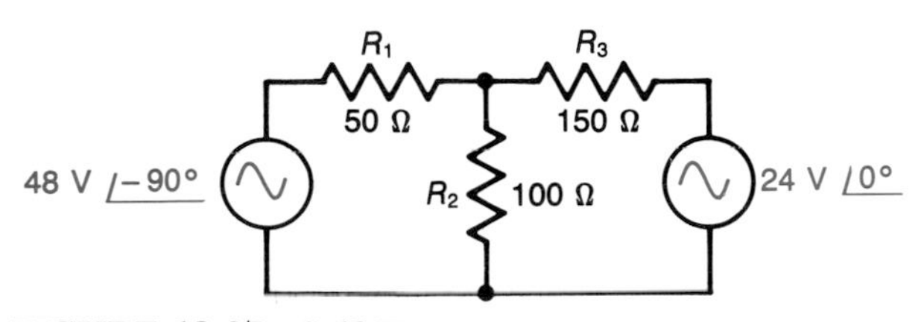


**FIGURE 12-27 Self-Test.**

**32.** A sinusoidal source has a peak-to-peak value of 679 V. Calculate the rms power that it will deliver to a 40-Ω heating element.

## 12-5 ELECTRONIC FUNCTION GENERATORS

Figure 12-28 shows an electronic *function generator.* The name of this instrument is based on its ability to generate several different waveforms. Different periodic waveforms follow different mathematical *functions.* Similar instruments called signal generators are also available, but these usually produce only a sinusoidal output. Not all function generators are the same, but the one shown in Fig. 12-28 provides sine, triangle, square, sawtooth, and pulse functions. Function generators are used for design work, for testing and troubleshooting, and for adjusting some electronic circuits.

There are an infinite number of different waveforms. In theory, it is possible to create any periodic waveform by adding sine waves of the proper frequency, phase, and amplitude. The lowest frequency component of a waveform is called the *fundamental.* The higher frequency components are called *harmonics.* Harmonics can be *odd* or *even.* For example, a waveform may have a fundamental frequency of 1 kHz, a second (even) harmonic at 2 kHz, and a third (odd) harmonic at 3 kHz. Note that harmonics are *integer* multiples of the fundamental frequency. A sine wave has no harmonics; the only frequency present is the fundamental.

Most waveforms other than sinusoidal do have harmonics. Figure 12-29 shows a computer-generated square wave. The computer was programmed

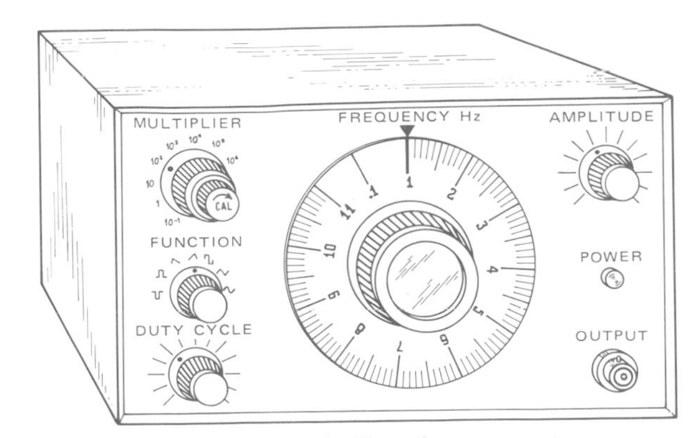


**FIGURE 12-28 Electronic function generator.**

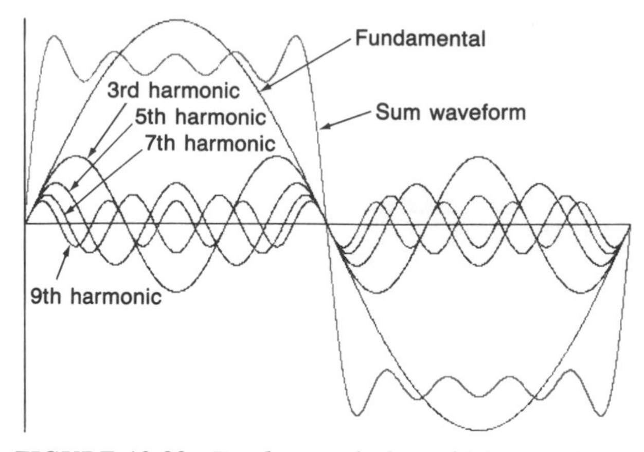


FIGURE 12-29 Fundamental plus odd harmonics.

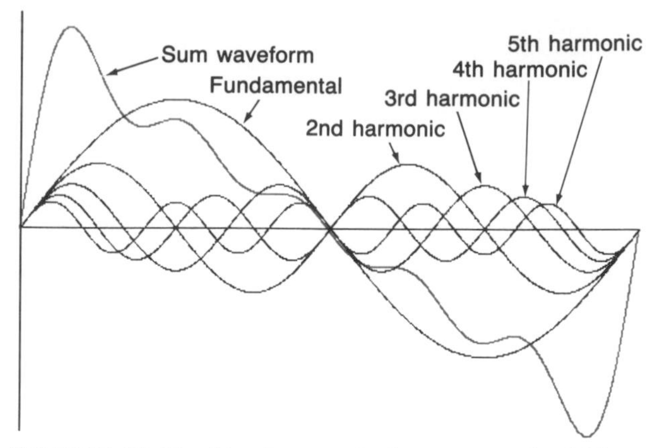


FIGURE 12-31 Fundamental plus even and odd harmonics.

to produce a fundamental sine wave and the first four odd harmonics. It was also programmed to sum all of the sinusoidal components. Note that the sum waveform looks more like a square wave than it does a sine wave even though all of its components are sinusoidal. Figure 12-30 shows an ideal square wave. An ideal square wave contains an infinite number of odd harmonics. The amplitude of each harmonic decreases as the frequency increases, and a reasonably accurate square wave can be produced with a fundamental and the first 10 odd harmonics.

Figure 12-31 shows a computer-generated sawtooth waveform. In this case, the computer was programmed to generate the fundamental plus the first four harmonics. Notice that both the odd and the even harmonics are included. The sum waveform begins to approach a sawtooth shape. Figure 12-32 shows an ideal sawtooth waveform which contains an infinite number of harmonics. A reasonable sawtooth can be generated by using a fundamental and the first 10 harmonics.

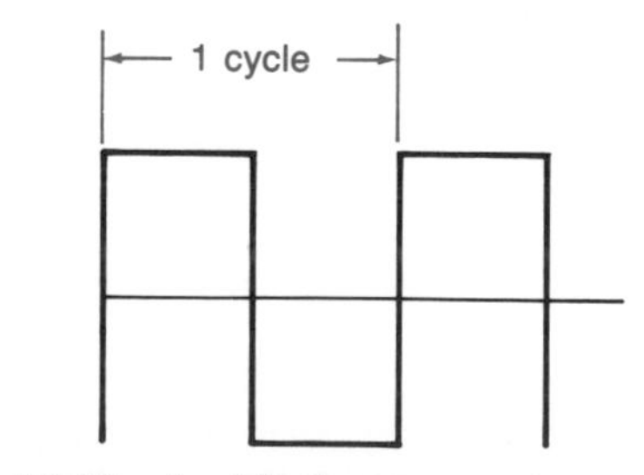


FIGURE 12-30 An ideal square wave.

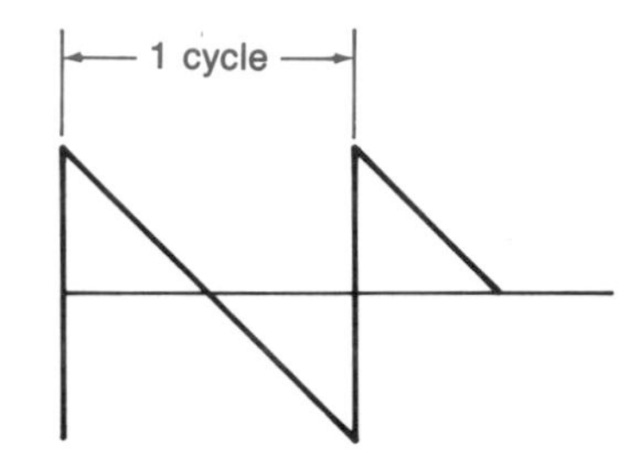


FIGURE 12-32 An ideal sawtooth wave.

An ideal triangle waveform is shown in Fig. 12-33. Like the square wave, it is composed of a fundamental frequency plus the odd harmonics. It differs from the square wave in that the amplitudes of the harmonics follow a different mathematical rule.

A pulse waveform is shown in Fig. 12-34. Pulse waveforms have an important characteristic known as *duty cycle*. The duty cycle is equal to the time that the waveform is positive divided by the time it takes to complete 1 cycle. The duty cycle of the waveform shown in Fig. 12-34 is about 25 percent. It is positive for an interval that is about one-fourth of the waveform's period. A square wave is a special instance of the pulse waveform. Square waves are pulse waveforms with a duty cycle of 50 percent.

**EXAMPLE 12-16**

A 1-kHz pulse waveform is positive 100 $\mu$s each cycle. What is the duty cycle of the waveform?

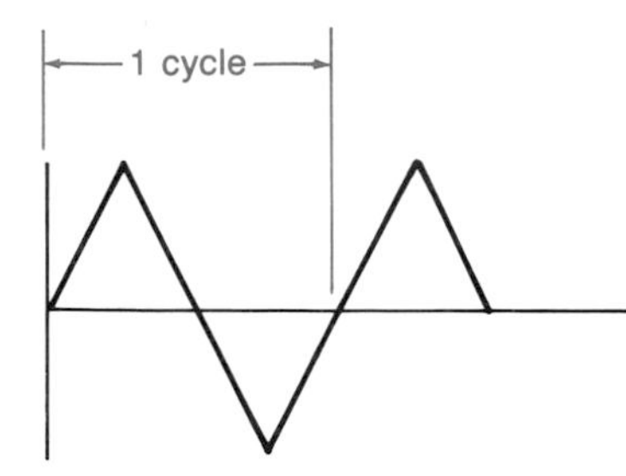


FIGURE 12-33 A triangular wave.

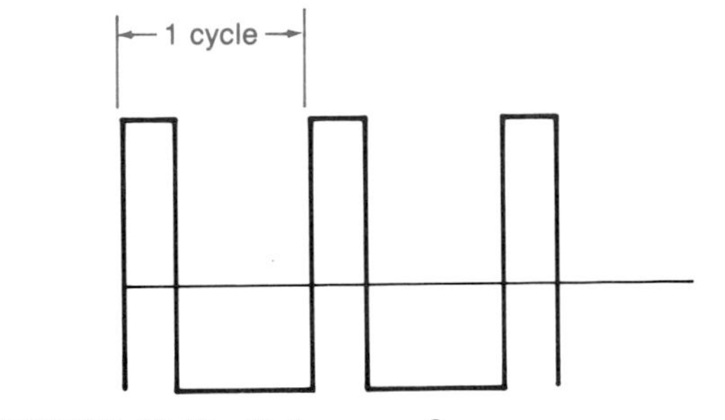


FIGURE 12-34 Pulse waveform.

**Solution**

First, we must find the period of the waveform:

$$T = \frac{1}{f} = \frac{1}{1000} = 1 \text{ ms}$$

The duty cycle is found by

$$\frac{100 \times 10^{-6}}{1 \times 10^{-3}} = 0.10 \text{ or } 10 \text{ percent}$$

---

Most electronic generators are relatively easy to use. As Fig. 12-28 shows, a function control is available to select the desired waveform. A frequency dial, in conjunction with a multiplier control, sets the output frequency. If the dial is set at 1 and the multiplier at $10^3$, the output will be $1 \times 10^3$ or 1 kHz. The generator shown in Fig. 12-28 has a frequency range of 0.01 Hz to 11 MHz. Generators that produce pulse waveforms may also have a duty cycle control.

## Self-Test

**33.** What is the frequency of the fifth harmonic of a 9-kHz waveform?

**34.** Identify a waveform that has no harmonics.

**35.** True or false? Any time a fundamental sine wave is summed with a large number of sine waves at odd harmonic frequencies, a square waveform is produced.

**36.** What is the duty cycle of a 4-kHz pulse waveform that is positive for 125 $\mu$s every cycle?

**37.** The waveform described in Question 36 could also be called a ______.

## SUMMARY

1. A periodic waveform repeats over and over.
2. Rotating ac generators produce a sinusoidal output.
3. The general equations for a sine wave are

$$v = V_p \times \sin \theta$$
$$I_i = I_p \times \sin \theta$$

4. One cycle contains 360 electric degrees.
5. A sine cycle begins with a positive alternation which is followed by a negative alternation.
6. The average value of a sinusoidal voltage or current is zero for 1 cycle and is 63.7 percent of the peak value for one alternation.
7. The heating effect of ac is given by its root-mean-square (rms) value. The rms value is also called the effective value.
8. The rms value of a sine wave is equal to 70.7 percent of its peak value.
9. Alternating voltage and current values are always assumed to be rms unless specifically stated otherwise.

**10.** The peak-to-peak value of a sine wave is equal to twice its peak value.

**11.** Frequency is a measure of the number of cycles generated in 1 s.

**12.** The SI unit of frequency is the hertz (Hz). One hertz equals one cycle per second.

**13.** The time, in seconds, required to complete 1 cycle is called a period.

**14.** Frequency ($f$) and period ($T$) are reciprocals:

$$T = \frac{1}{f}$$

**15.** The distance that a wave travels during one period is called its wavelength ($\lambda$).

**16.** When the velocity ($v$) of a wave is in meters per second, its wavelength in meters is given by

$$\lambda = \frac{v}{f}$$

**17.** The velocity of a radio wave is the same as the velocity of light, which is $3 \times 10^8$ m/s.

**18.** Angles may be measured in radians as well as in degrees. There are $2\pi$ rad in a circle and 1 rad = 57.3°.

**19.** Angular velocity ($\omega$) is measured in radians per second and is found by

$$\omega = 2\pi f$$

**20.** Waveforms that peak with the same polarity at the same instant in time are in phase.

**21.** Waveforms that peak with opposite polarities at the same instant in time are 180° out of phase.

**22.** Phasors are rotating vectors and are commonly used to represent ac waveforms.

**23.** The length of a phasor is proportional to the amplitude of the waveform, and the angle of the phasor is equal to the phase angle of the waveform.

**24.** Phasors, by convention, rotate counterclockwise for positive angles and clockwise for negative angles.

**25.** Phasors can be added by geometric construction (parallelogram method).

**26.** Right-angle phasors can be combined with the pythagorean theorem.

**27.** Alternating current values can be given in polar form:

$$\text{Magnitude} \underline{/\ \text{angle}}$$

**28.** When the angle of an ac value is not required, it may be given in absolute form and vertical bars may be used:

$$|\text{Magnitude}|$$

**29.** The current and voltage waveforms are in phase in resistive ac circuits.

**30.** Ohm's law and the power equations are applied in ac resistive circuits as they are in dc resistive circuits.

**31.** There are two power pulses for every cycle in ac circuits.

**32.** Function generators provide several waveforms. Some possibilities are sine, triangle, square, sawtooth, and pulse.

**33.** Harmonics are integer multiples of a fundamental frequency.

**34.** Most waveforms, other than sinusoidal, contain harmonics.

**35.** Any periodic waveform can be produced by summing sine waves that have the proper frequency, amplitude, and phase.

**36.** The duty cycle of a pulse waveform is found by dividing the pulse time by the period.

**37.** A square wave is a pulse wave with a 50 percent duty cycle.

## CHAPTER REVIEW QUESTIONS

**12-1.** One ac cycle is composed of two ______.

**12-2.** What value of rms voltage will be required to produce a heating effect that is identical to 120 V dc?

**12-3.** If an ac source is specified as 240 V, what type of measurement must be assumed?

**12-4.** Refer to Fig. 12-35*a*. What is the phase relationship of the waveforms?

**12-5.** Refer to Fig. 12-35*b*. What is the phase relationship of the waveforms? Which one leads?

**12-6.** Refer to Fig. 12-35*c*. Which phasor diagram correctly represents Fig. 12-35*b*?

**12-7.** Refer to Fig. 12-35*c*. What is the angle of phasor *B* in diagram number 4?

**12-8.** Refer to Fig. 12-35*d*. Which diagram in Fig. 12-35*c* corresponds to the ac generators?

**12-9.** Identify the waveform shown in Fig. 12-35*g*.

**12-10.** What is the fundamental frequency of a 40-Hz square wave?

**12-11.** Name a waveform that has no harmonics.

## CHAPTER REVIEW PROBLEMS

**12-1.** What is the instantaneous voltage at 50° for a sine wave with a peak value of 10 V?

**12-2.** What is the instantaneous current at 35° for a sine wave with a peak current of 2 A?

**12-3.** A sine wave has a peak value of 60 V. What is its average value taken over 1 cycle? Over one alternation?

**12-4.** What is the effective value of a sine wave with a peak value of 85 V?

**12-5.** What is the peak-to-peak value of a 40-V sine wave?

**12-6.** An ac generator produces 1 cycle in 10 ms. What is its frequency?

**12-7.** What is the period of a 5-MHz waveform?

**12-8.** A radio transmitter operates at 1 Mhz. What is its wavelength?

**12-9.** Convert 25° to radians.

**12-10.** How many degrees are there in 4 rad?

**12-11.** What is the angular velocity of 500 Hz ac?

**12-12.** What is $V_x$ for Fig. 12-35*d*? State your answer in polar form.

**12-13.** What is the absolute value of $V_x$ for Problem 12-12?

**12-14.** Calculate the instantaneous current for Fig. 12-35*e* at a phase angle of 80°.

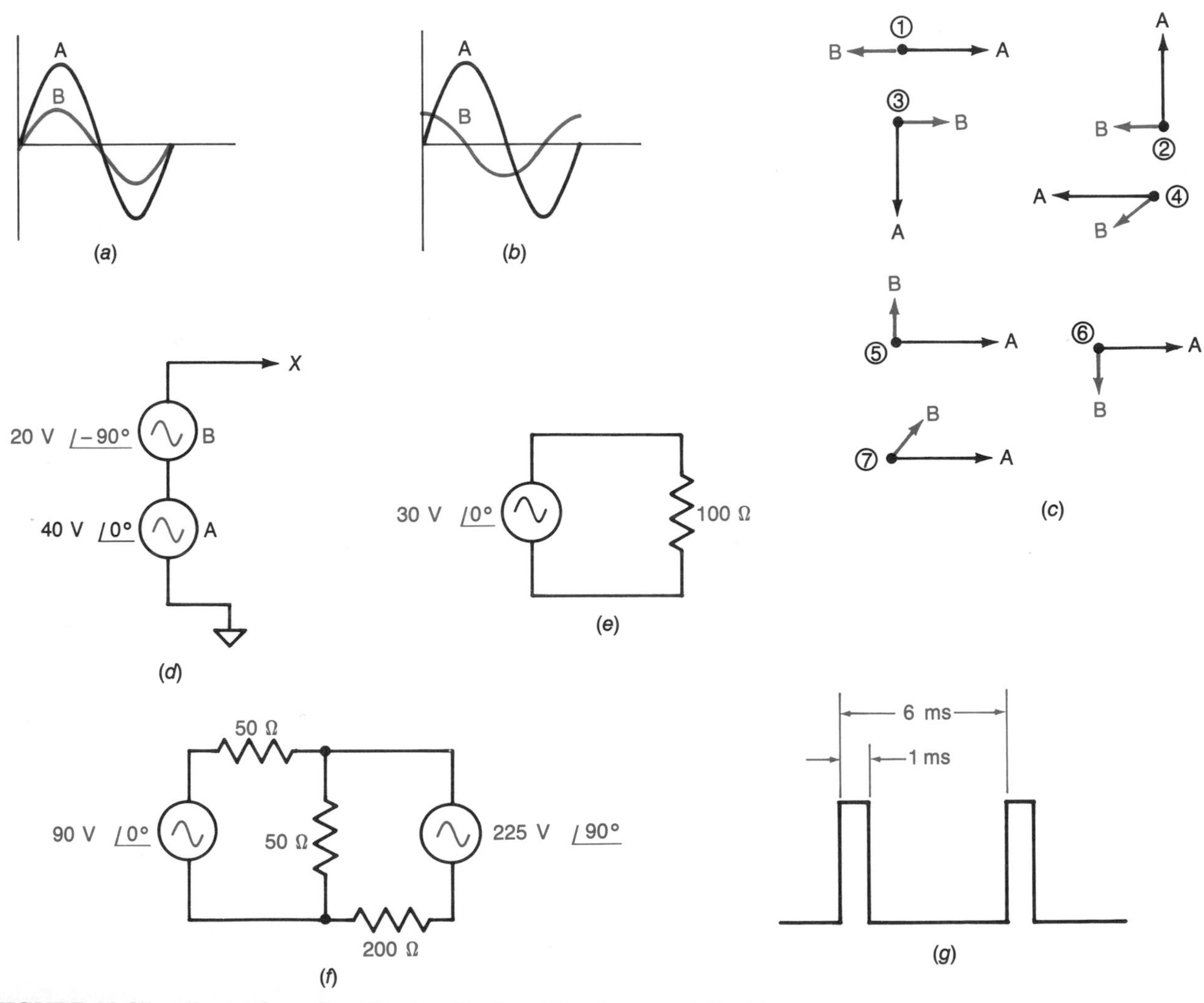


**FIGURE 12-35** **Illustrations for Chapter Review Questions and Problems.**

**12-15.** Calculate the current flow in the 200-Ω resistor shown in Fig. 12-35*f*. State your answer in polar form.

**12-16.** Calculate the power dissipated in the 200-Ω resistor shown in Fig. 12-35*f*.

**12-17.** What is the duty cycle of the waveform shown in Fig. 12-35*g*?

**12-18.** What is the frequency of the waveform shown in Fig. 12-35*g*?

**12-19.** What is the third harmonic of 60 Hz?

## ANSWERS TO SELF-TESTS

**1.** direct
**2.** alternating
**3.** horizontal
**4.** sinusoidal
**5.** 58.1 V
**6.** 13.4 V
**7.** zero
**8.** 6.01 A
**9.** 339 V
**10.** 1 ms
**11.** 50 kHz
**12.** decreases
**13.** 1.94 m
**14.** $2\pi$
**15.** 20.1°
**16.** 1.4 rad
**17.** $6.28 \times 10^4$ rad/s
**18.** It lags by 45°.
**19.** −90
**20.** positive
**21.** B leads A (or A lags B) by 45°
**22.** −90° (or 270°)
**23.** −45° (or 315°)
**24.** zero
**25.** 42.4 V $\angle 0°$
**26.** 44.7 V $\angle -63.4°$
**27.** 40.6 mA
**28.** 0° (in phase)
**29.** 25.3 mA
**30.** 0.265 A $\angle -80.5°$
**31.** 0.265 A
**32.** 1.44 kW
**33.** 45 kHz
**34.** sine
**35.** false
**36.** 50 percent
**37.** square wave

# Chapter 13 Inductance

*Inductance* is that property of an electric circuit that opposes any *change* in current flow. *Inductors* are physical components that exhibit the property of inductance. They are also called coils or chokes. This chapter introduces inductance and inductors.

## 13-1 INDUCTION

Electronic components can be divided into two major groups: *active* and *passive*. Active components include devices such as transistors and integrated circuits. These are covered in Chaps. 24 and 25. Inductors are considered passive elements like resistors. Unlike resistors, inductors have the ability to store electric energy. As we will see, the energy is stored in a magnetic field.

Inductors derive their name from electromagnetic induction, which was discussed in Chaps. 10 and 11. Figure 13-1 shows the methods of achieving electromagnetic induction. One method, shown in Fig. 13-1*a,* is to have relative motion between a conductor and a magnetic field. This is the method employed in rotating generators. Another method, shown in Fig. 13-1*b,* is to vary the current flow in a circuit that is magnetically coupled to another circuit. This is the method employed in transformers. As the current in a transformer changes, so does the flux. The changing flux *induces* a voltage into the other electric circuit. Circuits that are linked by magnetic flux exhibit *mutual inductance.*

Electromagnetic induction is achieved in any circuit when the current changes. This is shown in Fig. 13-1*c*. Any time there is current, there is also

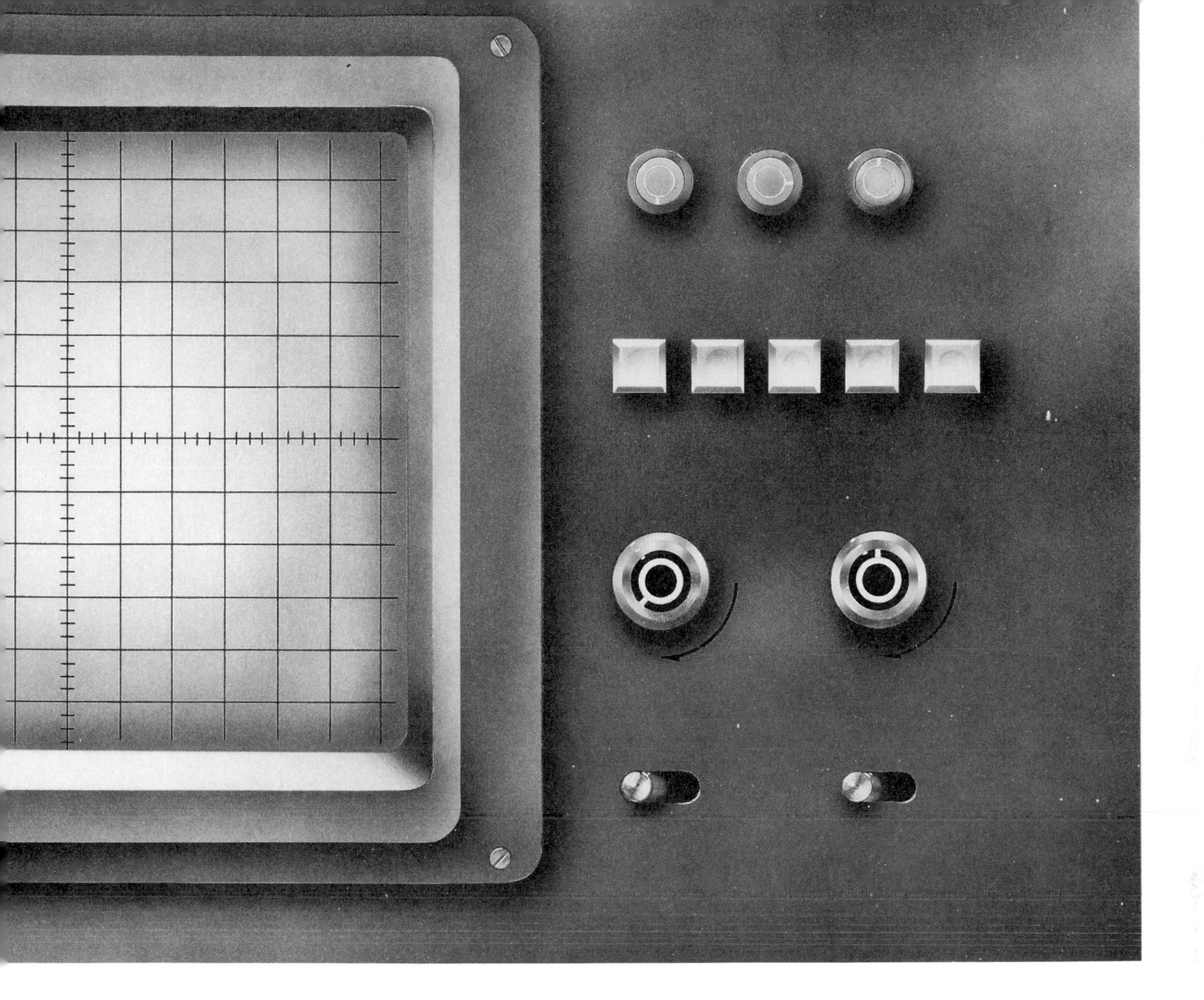

magnetic flux. The flux is proportional to the current flow. When the current changes, so must the flux. The flux change produces electromagnetic induction. This is known as *self-induction;* and conductors exhibit some degree of *self-inductance.*

Suppose that the current flowing in the conductor shown in Fig. 13-1*c* is increasing. The flux lines will be expanding around the conductor. This expansion induces a voltage that opposes the battery voltage as well as the current increase. On the other hand, if the current is decreasing, the flux will be collapsing around the conductor. The induced voltage will be opposite to that shown and will oppose the current decrease. Note that the induced voltage always opposes the change in current. For this reason, the induced voltage is known as a *counter-electromotive force* (cemf) or as a *back-emf.*

*Lenz's law* states that the polarity of an induced emf is such that any current resulting from it will develop a flux that opposes the original change in flux. This law explains the cemf in an inductor and the reaction force in a generator. For example, if the conductor shown in Fig. 13-1*a* is part of a complete circuit, current will flow in it. This current will set up a magnetic field around the conductor which will interact with the flux lines being cut. This interaction creates a mechanical force opposite to the force providing the motion. The net result is that the conductor is more difficult to move through the field when current is being supplied. Likewise, the shaft of a generator is more difficult to turn when the generator supplies a load current.

The magnitude of the voltage is predicted by *Faraday's law,* which states that the emf induced in a conductor is proportional to the rate of change of the flux that links it:

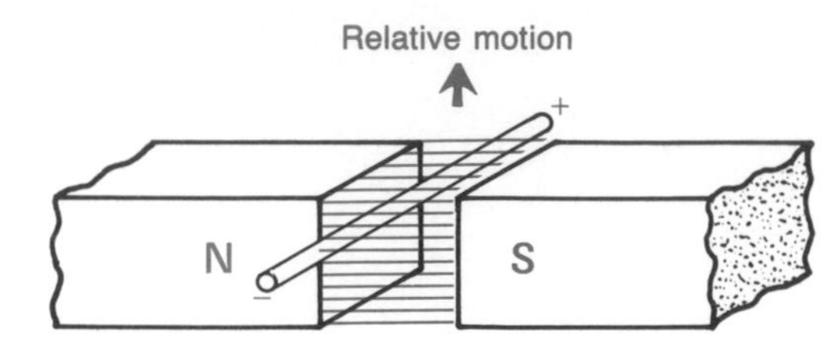


(a) The conductor or the field can be moved.

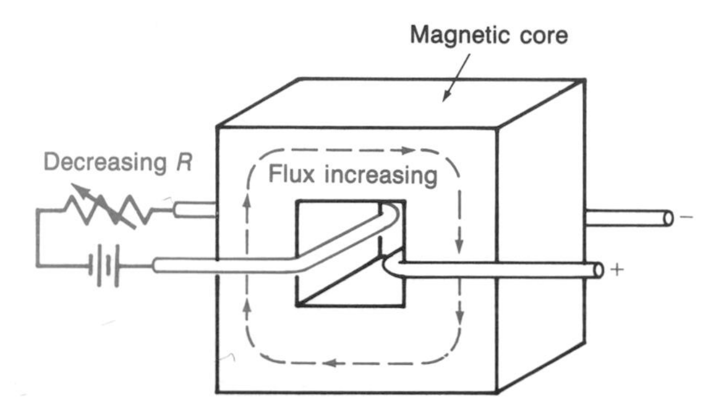


(b) The current in an electrically separate circuit can be varied.

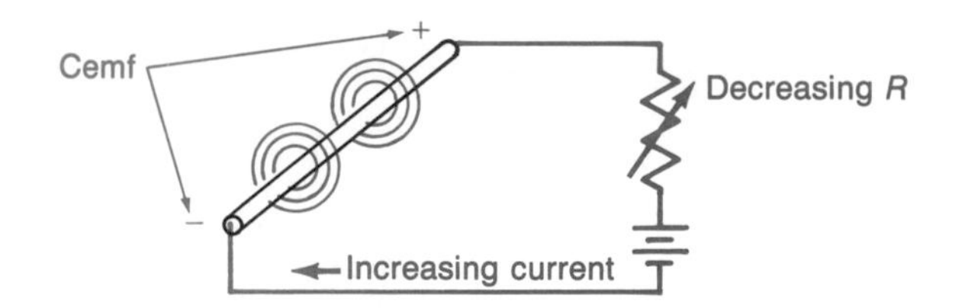


(c) The current in a conductor can be varied.

**FIGURE 13-1 Methods of achieving electromagnetic induction.**

$$V = N \times \frac{\Delta\Phi}{\Delta t}$$

where $V$ = induced emf, V
$N$ = number of turns
$\Delta\Phi$ = flux change, Wb
$\Delta t$ = time change, s

The equation shows that the induced voltage is proportional to the time rate of flux change and the number of turns.

**EXAMPLE 13-1**

How much voltage will be induced in a 500-turn coil when the flux linking it changes from 1.5 to 0.9 Wb in 100 ms?

**Solution**

Applying the equation,

$$V = 500 \times \frac{1.5 - 0.9}{0.1} = 3 \text{ kV}$$

In a case of a transformer, the induced voltage across one winding is directly proportional to the rate of current change in the other winding. Lenz's law predicts the induced polarity. For example, in Fig. 13-2*a* the primary current flow is increasing. The primary flux is also increasing, and a voltage is induced across the load. The load current sets up a flux in the core and is in a direction that opposes the flux increase. In Fig. 13-2*b*, the primary current is decreasing. Note that the load polarity is now reversed, and the secondary current sets up a sustaining flux that opposes the decrease in primary flux.

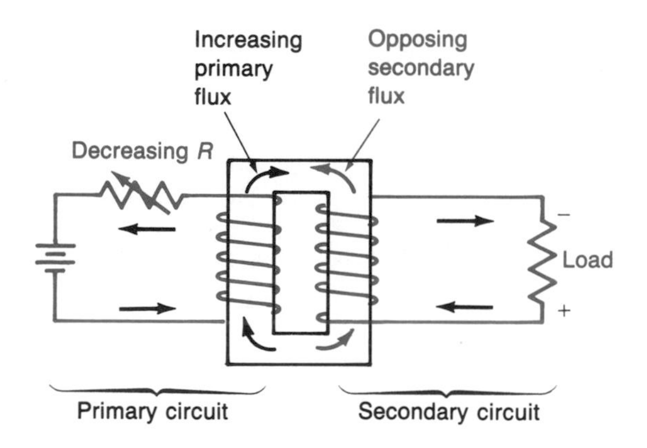


(a) Primary current increasing

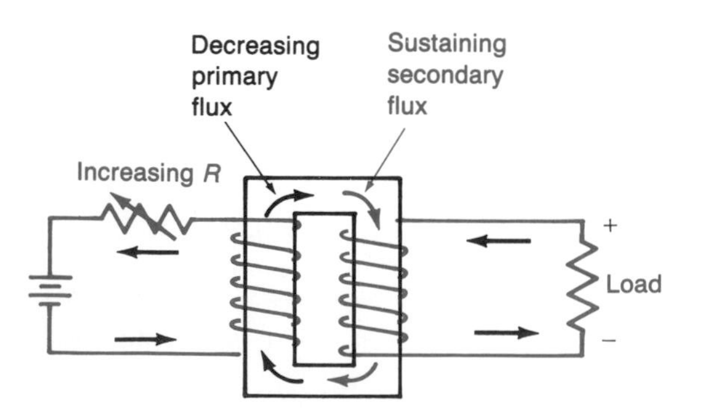


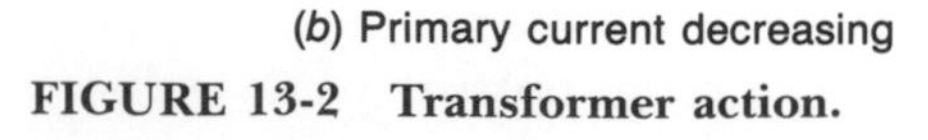

(b) Primary current decreasing

**FIGURE 13-2 Transformer action.**

## Self-Test

**1.** The property of a circuit that opposes any change in current is called ______.

**2.** Transformers exhibit ______ inductance.

**3.** The voltage induced in a conductor by a changing current is known as a ______ electromotive force.

**4.** How much voltage is induced in a 100-turn coil if the flux linking it changes from 0.4 to 0.2 Wb in 1.8 s?

**5.** Refer to Fig. 13-2. What will happen to the magnitude of the load voltage if the variable resistor is changed in a shorter period of time?

## 13-2 SELF-INDUCTION

The voltage induced in a conductor or a coil is proportional to the time rate of current change and to the inductance (letter symbol $L$) of the conductor or coil. The SI unit of inductance is the henry (letter symbol H). A component has an inductance of 1 H if it develops a cemf of 1 V when the current through it changes at a rate of 1 A/s:

$$L = \frac{-V}{\Delta I/\Delta t}$$

where $L$ = inductance, H
$-V$ = induced cemf, V
$\Delta I$ = current change, A
$\Delta t$ = time change, s

### EXAMPLE 13-2

A coil generates a cemf of 65 V when the current through it changes from 0.6 to 0.3 A in 15 ms. What is the inductance of the coil?

**Solution**

A cemf is taken to be negative and $-(-65) = 65$. Applying the equation,

$$L = \frac{65}{(0.6 - 0.3)/15 \times 10^{-3}} = 3.25 \text{ H}$$

The inductance of a straight conductor is very small. If the conductor is formed into a coil, its inductance is increased. Most practical inductors consist of several or many turns of wire. The schematic symbols for inductors recognize this fact, as shown in Fig. 13-3. Iron-core inductors use some ferrous core material to increase the flux and the inductance.

Air-core inductor

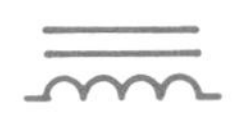

Iron-core inductor

Variable iron-core inductor

**FIGURE 13-3 Inductor symbols.**

The previous equation is often rearranged to solve for the induced voltage:

$$V = -L \times \frac{\Delta I}{\Delta t}$$

The minus sign indicates the opposing nature of the cemf. You may also encounter the following expression:

$$V = -L \times \frac{di}{dt}$$

The term $di/dt$ is a *derivative,* and it expresses the *instantaneous* time rate of current change. The derivative of $\Delta I/\Delta t$ is found by making the time interval vanishingly small. Derivatives can be determined by graphical techniques and by using calculus.

### EXAMPLE 13-3

The current through a 500-mH inductor changes from 400 to 850 mA in 1 ms. What is the magnitude of the induced voltage?

**Solution**

Applying the equation,

$$V = -0.5 \times \frac{0.85 - 0.4}{1 \times 10^{-3}} = -225 \text{ V}$$

Note that the result would have been +225 V if the larger current had been subtracted from the smaller current. The important thing to remember is that the cemf *opposes* the current change. The absolute value of the cemf is normally all that is required.

Figure 13-4 shows an example of current and voltage waveforms in a 10-H inductor. From $t = 0$ to $t = 3$ s, the current is zero, and the induced voltage is also zero. For the time interval from 3 to 6 s, the current is increasing in the inductor at a rate of 1 A/s and the induced voltage is $-10$ V over the period of this interval. From 6 to 9 s, the current is constant at 3 A and the cemf is 0. Finally, from 9 to 12 s, the current decreases at a rate of 1 A/s and the cemf is $+10$ V. Now look at Fig. 13-5. Assuming the same inductance (10 H), the cemf is $-30$ or $+30$ V because the rate of current change is 3 A/s. Finally, look at Fig. 13-6. The current waveform is an "ideal" pulse. An ideal current pulse rises and falls in zero time. Therefore, the value of $dt$ is zero. This situation is not defined by mathematics:

$$V = -L \times \frac{di}{dt} = -10 \times \frac{3}{0} = \text{undefined}$$

There are no ideal pulses in the real world. As a current pulse *approaches* the ideal, the derivative $di/dt$ *approaches* infinity. The cemf waveform for an inductor with a current pulse approaching the ideal will show spikes that reach a very high magnitude.

The practical aspects of the forgoing discussion are as follows:

**1.** All circuits, even those consisting of only straight wires, have some inductance. Therefore, it is impossible to create an instantaneous change in current in any circuit.

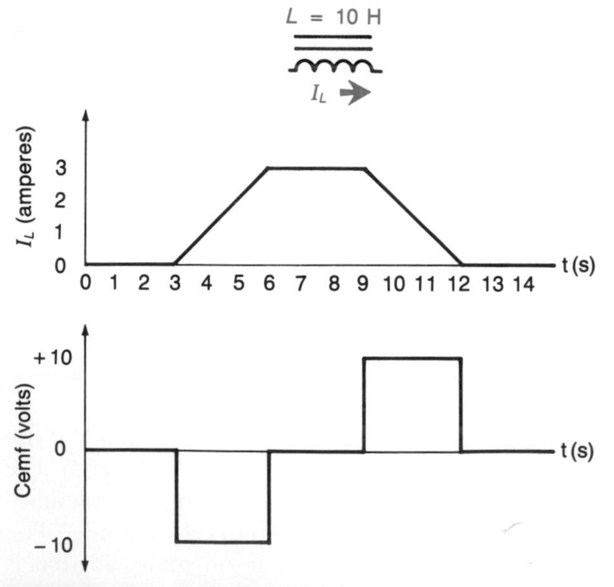


**FIGURE 13-4 Inductor waveforms.**

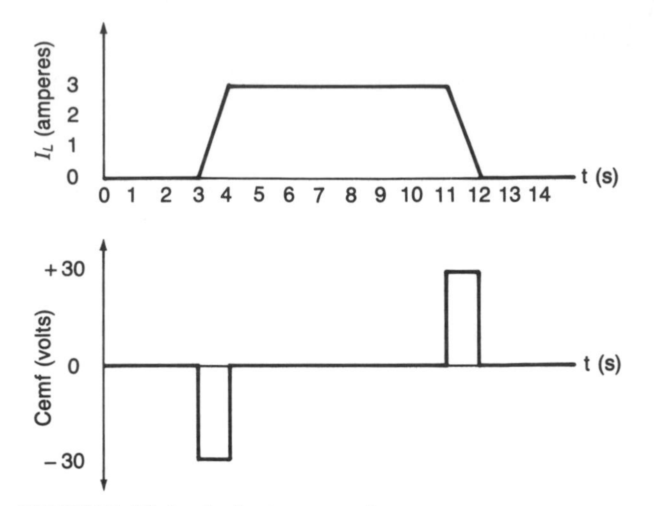


**FIGURE 13-5 Inductor waveforms.**

**2.** Circuits with significant inductance, such as those with coils, offer significant opposition to rapid current changes.

**3.** When the flow in a circuit with significant inductance decreases rapidly, such as when a switch is opened, a large voltage is generated which can cause arcing or component breakdown.

The inductance of a coil depends on the permeability of its core, the number of turns, the cross-sectional area of the core, and the length of the magnetic circuit:

$$L = \mu_r \times 4\pi \times 10^{-7} \times \frac{N^2 a}{l}$$

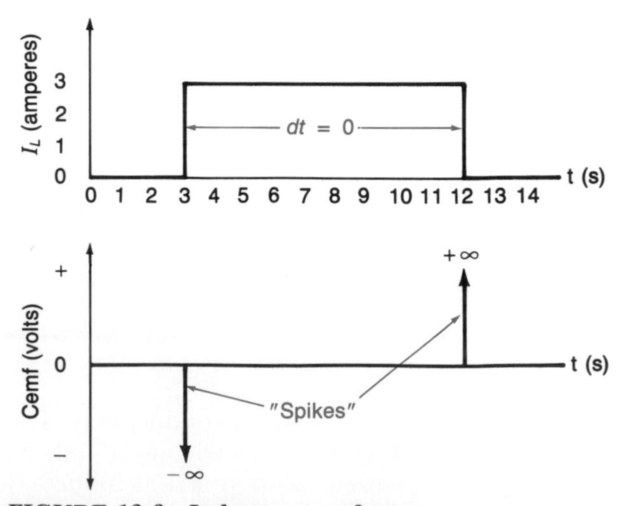


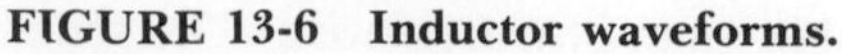

**FIGURE 13-6 Inductor waveforms.**

where $L$ = inductance, H
$\mu_r$ = relative permeability of core
$N$ = number of turns
$a$ = cross-sectional area of core, $m^2$
$l$ = average length of magnetic circuit, m

**EXAMPLE 13-4**
Calculate the inductance for the coil shown in Fig. 13-7.

**Solution**
The average magnetic path length is

$$4 \times 10 \text{ cm} = 40 \text{ cm} = 0.4 \text{ m}$$

The area of the core, in square meters, is

$$(0.02)^2 = 4 \times 10^{-4} \text{ m}^2$$

Applying the equation,

$$L = 850 \times 4\pi \times 10^{-7} \times \frac{500^2 \times 4 \times 10^{-4}}{0.4} = 0.267 \text{ H}$$

The preceding equation provides reasonable accuracy for single-layer air-core coils if their length is at least 10 times the coil diameter. When using this equation for a single-layer air-core coil, $l$ is taken as the length of the coil itself. The length of the magnetic return path is ignored as discussed in Chap. 10. The following approximation is more accurate for predicting the inductance of single-layer air-core coils not having the required length-to-diameter ratio:

$$L \approx \frac{N^2 r^2}{25.4r + 22.9l}$$

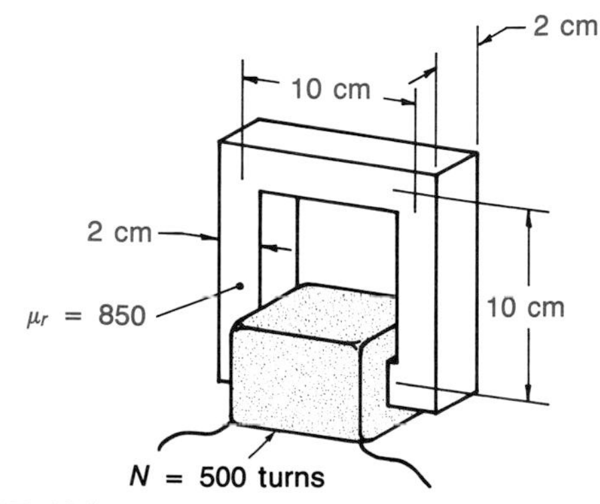

**FIGURE 13-7 Example 13-4.**

where $L$ = inductance, $\mu$H
$N$ = number of turns
$r$ = radius of coil, cm
$l$ = length of coil, cm

**EXAMPLE 13-5**
What is the inductance of a single-layer 10-turn air-core coil that is 1 cm long and 0.5 cm in diameter?

**Solution**
Applying the approximation,

$$L \approx \frac{10^2 \times 0.25^2}{25.4 \times 0.25 + 22.9 \times 1} = 0.214 \ \mu\text{H}$$

As you can see, the inductance of air-core coils is typically quite small. However, this much inductance can be a significant factor in very high frequency circuits where the current changes in a very short period of time.

## Self-Test

**6.** What is the inductance of a coil that develops a cemf of 20 V when the current through it changes 1 A in 50 ms?

**7.** What absolute cemf is generated by a 860-mH inductor when the current through it changes from 200 to 100 mA in 1 $\mu$s?

**8.** It is not possible to create an instantaneous current change in circuit current due to a circuit property called ______.

**9.** Rework Example 13-5 using the inductance equation rather than the approximation. You should recall that the relative permeability of air is 1. What value of inductance is obtained? What is the percentage of error assuming that the approximation provides the nominal inductance?

**10.** Calculate the inductance of a single-layer air-core coil with 50 turns that is 5 cm long and 9.5 cm in diameter. Use both the equation and the approximation. Make the assumption that the approximation approach provides the nominal value and calculate the percentage of error for the equation approach.

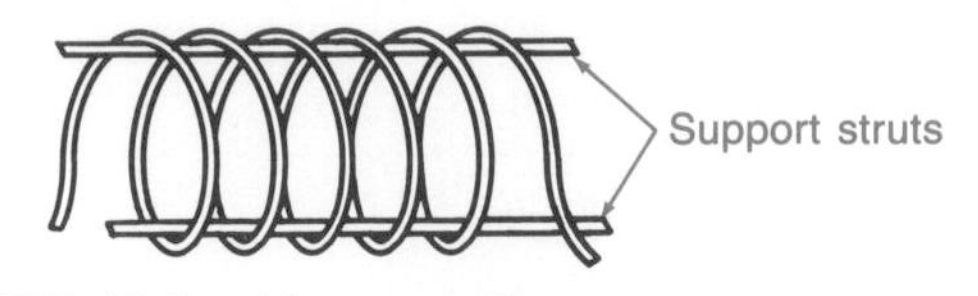


FIGURE 13-8 Air-core inductor.

## 13-3 TYPES OF INDUCTORS

Figure 13-8 shows an *air-core* inductor. Polystyrene struts are used to support the wire. If the wire diameter is sufficiently large, an air-core coil can be self-supporting and the struts eliminated. Air-core inductors are used in applications such as radio transmitters and typically have an inductance value of 1 μH or less. If a coil is wound on a supporting form that has a permeability very close to one, it is still considered to be an air-core inductor. Some inductors of this type use ceramic forms and can be made adjustable with a roller mechanism that contacts one turn on the coil. As the coil is rotated, it acts as a screw and the roller contact point moves, placing more or fewer turns in the circuit. These are called *roller-inductors*.

Figure 13-9 shows the construction and appearance of a typical *iron-core* inductor. The coil is wound on a fiber or plastic form called a *bobbin*. The core is assembled into and around the bobbin and coil assembly using a number of thin E- and I-shaped laminations. Inductors of this style are available with ratings ranging from millihenrys to several henrys. They are used in power supplies and various other electronic applications. Inductor cores can be assembled from other lamination

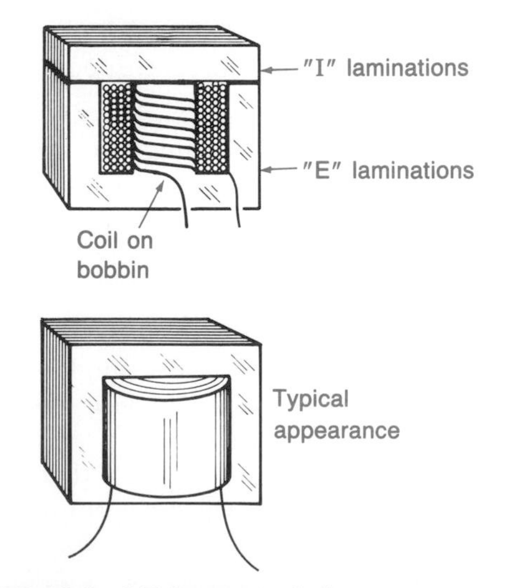


FIGURE 13-9 I-E iron-core inductor.

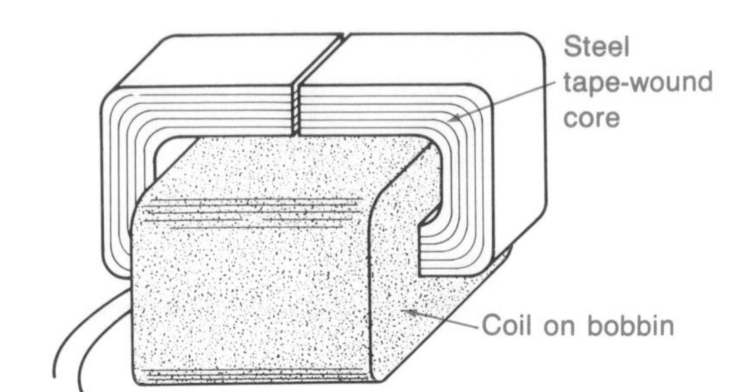


FIGURE 13-10 Tape-wound steel-core inductor.

shapes such as U and I. They may also be constructed from steel or iron tape as shown in Fig. 13-10. After the tape is wound, the core is cut into two halves. The halves are then assembled into the coil, which has been wound on a bobbin. A steel strap around the core holds the halves together.

A *toroidal* inductor is shown in Fig. 13-11. The core can be wound from steel tape or it may be molded from powdered iron or iron ferrite. Tape-wound cores use grain-oriented silicon steel which has a very high permeability. The steel grains achieve a uniform orientation during the rolling process when the steel is manufactured. It is easier to magnetize steel along the axis of grain orientation. Powdered iron cores are made from a mixture of finely divided iron particles and a binding compound. Their relative permeabilities are usually less than 100, which is much less than steel or ferrite core ratings. However, powdered iron cores do not saturate easily, and they are temperature-stable. These advantages make them attractive for some applications. Iron ferrites are ceramic materials with relative permeabilities as high as 5000. Iron oxide is combined in a mix with other materials such as nickel, manganese, zinc, and magnesium. The mix is formed into the desired core shape and then fired in a kiln.

Whatever the method used for making toroid cores, they are more difficult to wind because bobbins cannot be used. Toroidal inductors are usually shuttle-wound. The wire is placed on a shuttle, which is then passed around and through the core as the wire is transferred to the core. The advantage of a toroidal inductor is that its field is almost completely confined to the core. The core flux links every turn for high efficiency. Toroidal inductors exhibit very little flux leakage (leakage was covered in Chap. 10), and they are inherently self-shielding. This means that they do not produce external fields that could interfere with surround-

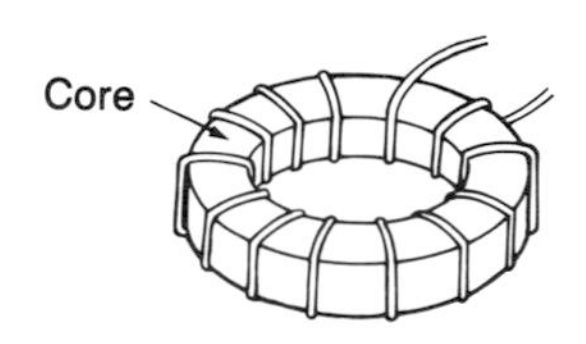


**FIGURE 13-11 Toroidal inductor.**

ing circuits nor are they susceptible to interference from external sources.

Figure 13-12 shows a *pot-core* inductor, which has characteristics similar to the toroid type. It is easier to wind, however, since it uses a bobbin. Pot cores tend to be more expensive than toroid cores. This offsets their winding advantages, and both toroid and pot-core inductors are popular for applications where efficiency and self-shielding are important. Toroidal and pot-core inductors have inductance ratings that typically fall in the microhenry and millihenry ranges.

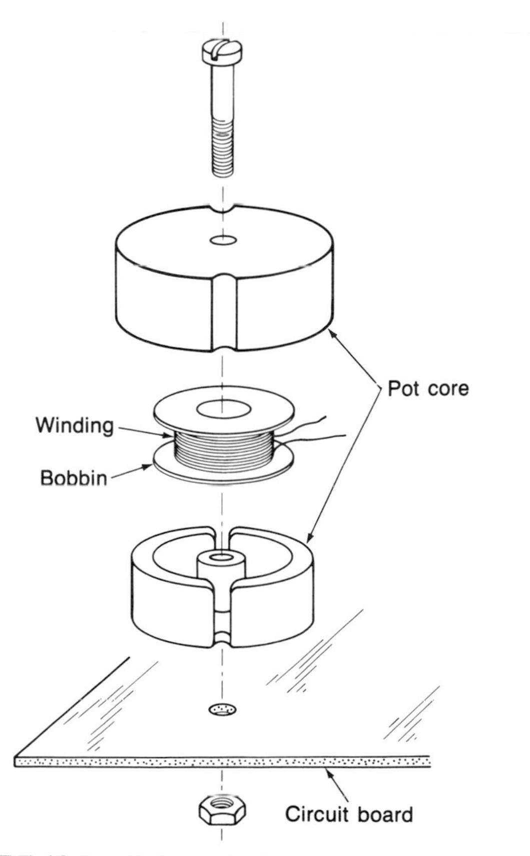


**FIGURE 13-12 Pot-core inductor.**

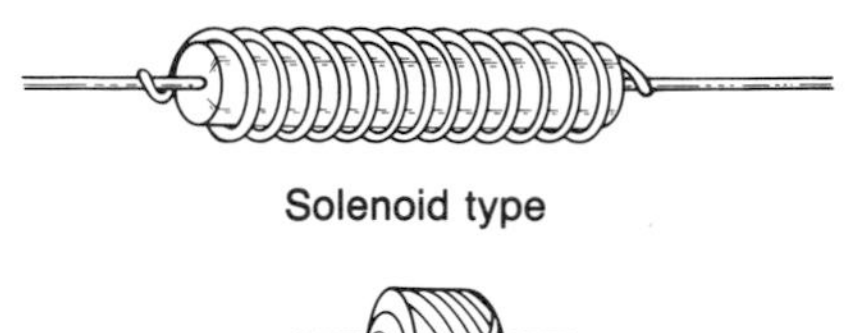


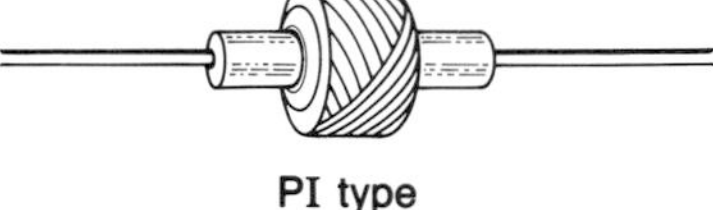


**FIGURE 13-13 RF chokes.**

*Radio frequency* (RF) *chokes* are shown in Fig. 13-13. Radio frequencies are high frequencies with rapid changes in current. These inductors derive their name from their ability to choke off (oppose) high-frequency currents. Radio frequency chokes are wound on fiber or ceramic forms when small values of inductance are needed. Or, they may be wound on powdered iron or iron ferrite forms to achieve greater values of inductance. Solenoid style chokes usually consist of a single layer of wire and typically have a low value of inductance (1 $\mu$H or less). PI style chokes (PI refers to a multilayer coil with a cylindrical shape) have several layers of wire, and some have several PI sections on one form. Note that the angle with which the wire is wound on the PI section alternates from layer to layer. PI style RF chokes are available with inductance ratings in the microhenry and millihenry ranges.

Small inductors are often molded as shown in Fig. 13-14. The axial lead types look quite a bit like carbon composition resistors. In fact, they may use a color code that is similar to the resistor color code. The manufacturer often places one extra wide band on the inductor package so the device will not be mistakenly identified as a resistor. The inductor color code uses the same scheme as the resistor color code with two important exceptions. First, molded inductors are coded in *micro*henry units. Second, the color gold indicates a decimal point when it appears in a position other than the tolerance position.

To read the color code on an axial-lead inductor package, start with the band adjacent to the wide band. Translate each color to a number. The color gold indicates a decimal point or a tolerance of 5 percent if it appears in the last position. In those cases where gold is not used to indicate a decimal point, the third band is used as a multiplier as with resistors.

---

**EXAMPLE 13-6**

An inductor has the following color bands: a wide silver band, followed by gray, gold, red, and silver bands. What is its nominal inductance?

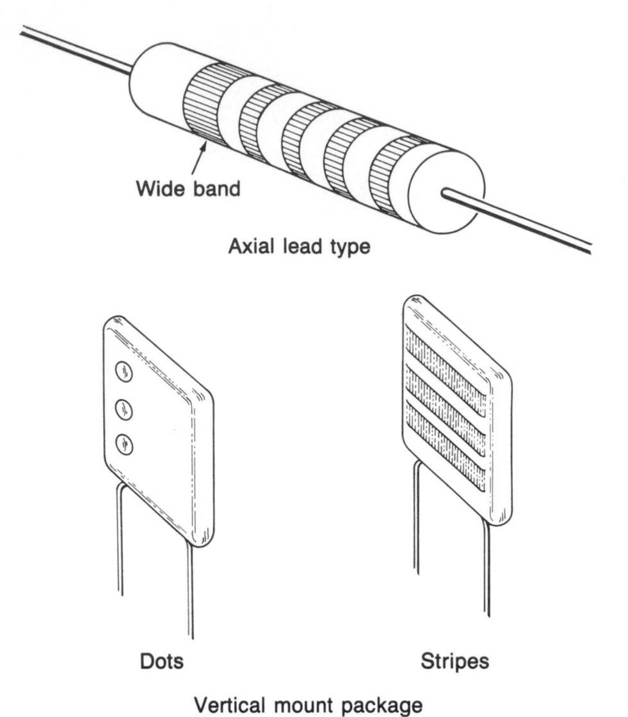


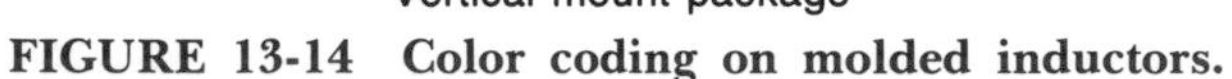


**FIGURE 13-14 Color coding on molded inductors.**

### Solution

The wide silver band is used to indicate that the device is an inductor. It does not represent a number. The other bands translate to

Gray → 8
Gold → .
Red → 2
Silver → 10 percent

The nominal value of the inductor is 8.2 $\mu$H with a tolerance of $\pm 10$ percent.

---

### EXAMPLE 13-7

What is the nominal value of an inductor with a wide brown band followed by red, black, and brown bands?

### Solution

Red → 2
Black → 0
Brown → $\times 10$ (multiplier)

The nominal value of the inductor is 200 $\mu$H. Since there is no tolerance band, the actual value may vary as much as $\pm 20$ percent.

---

Figure 13-14 also shows two molded vertical mount packages. These may be coded with color stripes or dots. The colors are read from top to bottom (the leads are on the bottom). For example, a 15-$\mu$H inductor could have a brown dot on top, a green dot in the middle, and a black dot on the bottom.

Some examples of variable inductors are shown in Fig. 13-15. These use a metal (usually brass) screw to position a ferrite or powdered iron core, or the core itself can be threaded for positioning. Special alignment tools are available for adjusting these inductors. The metal screw type is adjusted with a nonmagnetic rod that has a small metal blade in one end. Since the tool is mostly nonmagnetic material, its interaction with the coil is minimized. The use of ordinary metal screwdrivers is not recommended. Threaded ferrite and powdered iron cores have a hexagonal hole through their centers, and nonmagnetic hex alignment tools are available for adjusting them. The response is the same for both the metal screw and threaded core types. The inductance increases as the core is moved into the winding area and decreases as it is moved out. Maximum inductance is attained when the core is centered within the wire coil. The ferrule shown on the metal screw type is

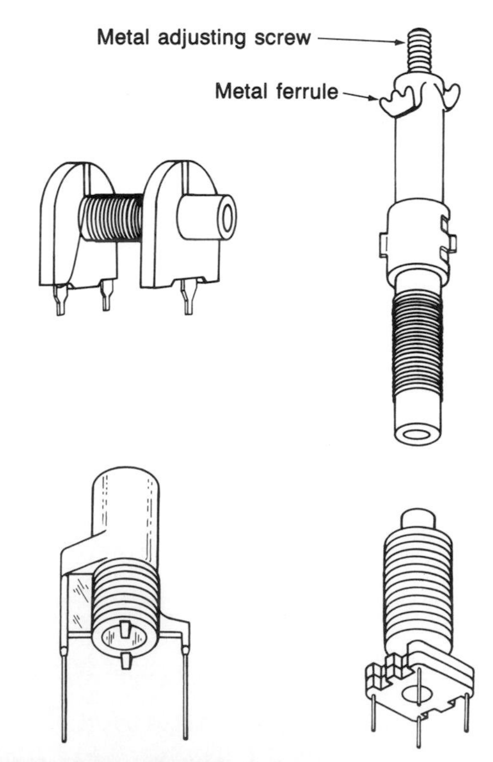


**FIGURE 13-15 Variable inductors.**

intended to physically support the inductor by mounting it into a hole in a chassis or a shield. The other variable inductors shown in Fig. 13-15 are designed to be mounted on printed circuit boards.

Figure 13-16 shows a shielded variable inductor designed for printed circuit board mounting. It is adjusted with a threaded core that surrounds the coil. The inductance increases as the core is brought closer to the coil. A metal shield prevents interaction with other circuits and extraneous fields.

Ferrite beads can be used to increase the inductance of a wire. They are placed on the wire as shown in Fig. 13-17. The permeability of the ferrite is greater than air, and the bead increases the field intensity around a current-carrying conductor. This, of course, increases the inductance of the conductor, but only by a small amount. Ferrite beads are manufactured in a range of sizes and shapes.

Figure 13-18 shows a *printed* inductor. These are formed with copper foil on printed circuit boards. They may also be formed with a conductive film on an insulating substrate. The inductance values of printed inductors are quite small. They are typically used in very high frequency applications.

The last type of inductance to be discussed is not a component or a device. It is an unavoidable and sometimes undesired effect called *stray inductance*. As we have learned, all conductors exhibit some inductance. In many cases, the inductance of a noncoiled conductor can be ignored. However, at very high frequencies stray inductance can produce

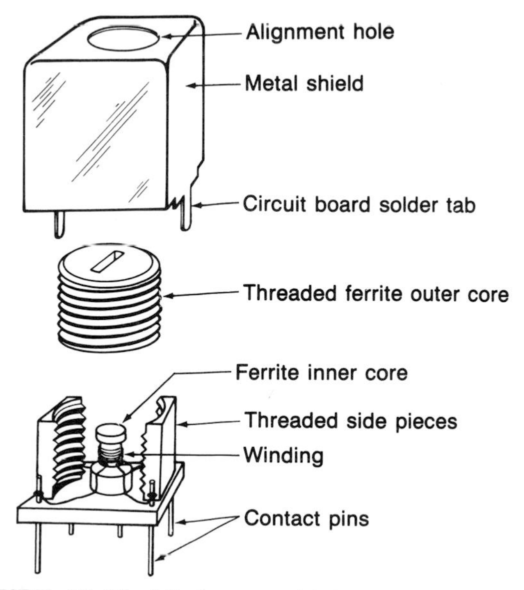


**FIGURE 13-16 Miniature, shielded, adjustable inductor.**

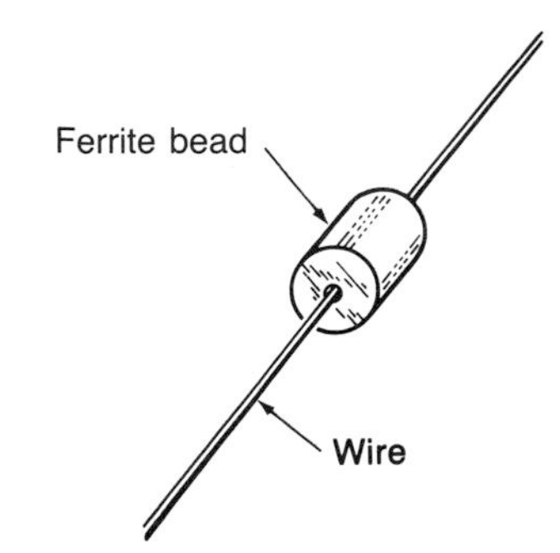


**FIGURE 13-17 Ferrite bead inductor.**

significant effects. The inductance of a straight, round, nonmagnetic wire in free space is given by

$$L = \frac{l\{[\ln(2l/r)] - 0.75\}}{5 \times 10^3}$$

where $L$ = inductance, $\mu$H
$l$ = wire length, mm
ln = natural logarithm
$r$ = wire radius, mm

---

**EXAMPLE 13-8**

What is the inductance of a round copper wire in free space that is 4 mm in diameter and 50 mm long?

**Solution**

The radius of the wire is half the diameter, or 2 mm. Applying the formula,

$$L = \frac{50[(\ln 100/2) - 0.75]}{5 \times 10^3} = 0.0316\ \mu\text{H}$$

---

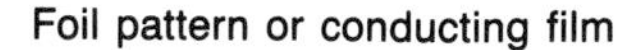


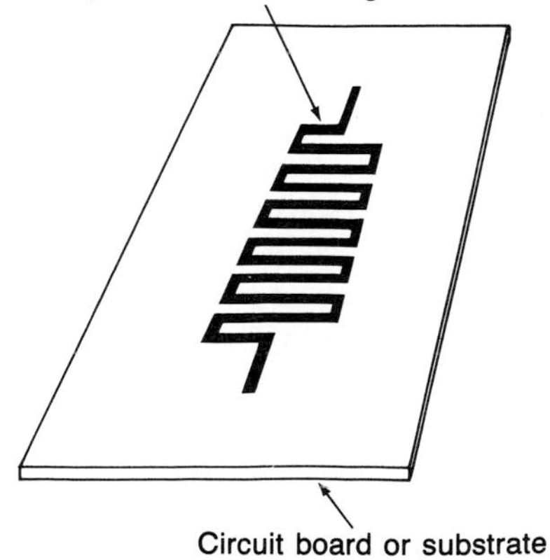

**FIGURE 13-18 Printed inductor.**

An inductance value this small is insignificant in dc and low-frequency ac circuits. However, it can be significant in very high frequency circuits. For example, the wire leads on components must be kept as short as possible in some applications. This is often referred to as *lead dress*. Good practice dictates maintaining the original lead dress when replacing electronic components.

## Self-Test

**11.** True or false? An inductor can be wound on a ceramic or plastic form and still be considered an air-core type.

**12.** True or false? Toroidal inductors are noted for low flux leakage and good self-shielding characteristics.

**13.** True or false? Cup-core inductors are easier to wind than toroidal types.

**14.** True or false? Powdered-iron-core inductors are too easily saturated for many applications.

**15.** What is the nominal inductance of a molded axial-lead inductor with the following color bands: a wide gold band followed by green, gold, blue, and a silver band?

**16.** True or false? Nonmetallic tools are preferred for adjusting variable inductors.

**17.** The inductance of a coil _______ as its core is moved away from its winding.

**18.** True or false? An electronic component with long leads may exhibit enough stray inductance to cause problems in a very high frequency circuit.

## 13-4 SERIES INDUCTORS AND PARALLEL INDUCTORS

When inductors are connected in series, as shown in Fig. 13-19, each of them produces a cemf in response to any change in current. The individual cemf's series-aid to oppose the source voltage. Since the current is constant in a series circuit, the total cemf is proportional to the sum of the individual inductances. Therefore, the total inductance of two or more inductors connected in series is given by

$$L_T = L_1 + L_2 + L_3 + \cdots + L_N$$

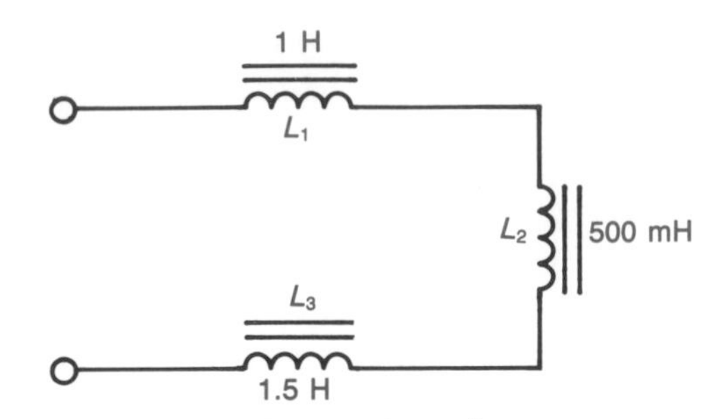


**FIGURE 13-19 Inductors in series.**

The previous equation is based on the assumption that the inductors are *not* magnetically coupled to each other.

### EXAMPLE 13-9

What is the total inductance of the circuit shown in Fig. 13-19?

**Solution**

Adding the individual inductances gives

$$L_T = 1 \text{ H} + 0.5 \text{ H} + 1.5 \text{ H} = 3 \text{ H}$$

When the field of one inductor cuts the turns of another inductor, they are magnetically coupled. The total inductance of two series inductors can be increased or decreased by this coupling:

$$L_T = L_1 + L_2 \pm 2L_M$$

where $L_M$ is the *mutual inductance* of the coupled inductors.

The mutual inductance is positive when the flux from one coil aids the flux of the other coil. It is negative when the flux is in opposition. The mutual inductance of two coils can be determined by

$$L_M = \frac{L_{TA} - L_{TO}}{4}$$

where $L_{TA}$ = total inductance with coils aiding
$L_{TO}$ = total inductance with coils opposing

### EXAMPLE 13-10

A 300-mH inductor is in series with an 800-mH inductor, and they have a mutual inductance of 100 mH. What is their total inductance when they are connected in series-aiding? In series-opposing?

**Solution**

Applying the equation for the series-aiding case,

$$L_T = 300 \text{ mH} + 800 \text{ mH} + 2 \times 100 \text{ mH} = 1.3 \text{ H}$$

And for the series-opposing case,

$$L_T = 300 \text{ mH} + 800 \text{ mH} - 2 \times 100 \text{ mH} = 0.9 \text{ H}$$

---

**EXAMPLE 13-11**

Two series inductors show a total inductance of 2 H when connected in series-aiding and 1 H when connected in series-opposing. What is their mutual inductance?

**Solution**

Applying the equation,

$$L_M = \frac{2 \text{ H} - 1 \text{ H}}{4} = 250 \text{ mH}$$

---

The *coefficient of coupling* is a measure of how tightly two inductors are coupled. The coefficient of coupling is a pure number (no units) that varies from 0 to 1. It may also be expressed as a percentage, in which case it varies from 0 to 100 percent. When the magnetic fields of two coils are completely independent, the coefficient of coupling is 0, as is the mutual inductance of the two coils. When every magnetic line of one coil links every turn of the other coil, the coils are coupled as tightly as possible and the coefficient is equal to 1. The coefficient of coupling (letter symbol $k$) for two inductors can be found with

$$k = \frac{L_M}{\sqrt{L_1 L_2}}$$

---

**EXAMPLE 13-12**

Two 250-mH inductors have a mutual inductance of 250 mH. What is their coefficient of coupling?

**Solution**

Applying the equation,

$$k = \frac{250 \text{ mH}}{\sqrt{250 \text{ mH} \times 250 \text{ mH}}} = 1$$

---

Look at Fig. 13-20. The only way to closely approach $k = 1$ is to wind both inductors on the same high-permeability core. This couples them tightly. Figure 13-20*a* shows that each coil has an inductance of 250 mH. Figure 13-20*b* shows the coils connected in series-aiding. This can be verified with the left-hand rule, which shows that for a given direction of current flow, each coil produces the same flux direction around the core. Note that the total inductance is 1 H for this connection, which is four times the inductance of one of the coils. An equation was presented in Sec. 13-3 which showed that the inductance of a coil varies as the *square* of the number of turns. Connecting the two coils in series doubles the number of turns and the inductance increases by a factor of four. Figure 13-20*c* shows the two coils connected in series opposition. The flux from one coil is equal and opposite to the

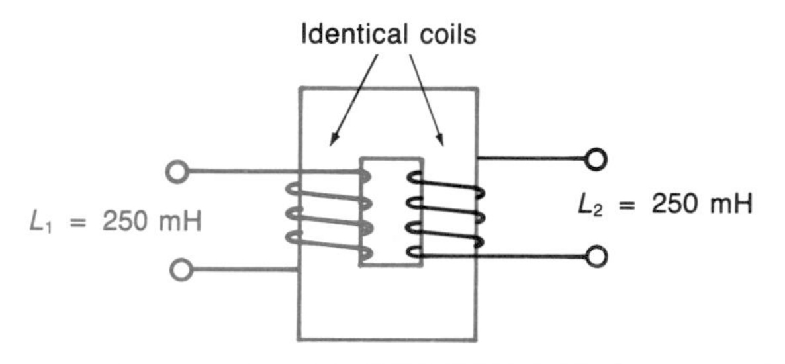


(*a*) Two tightly-coupled inductors

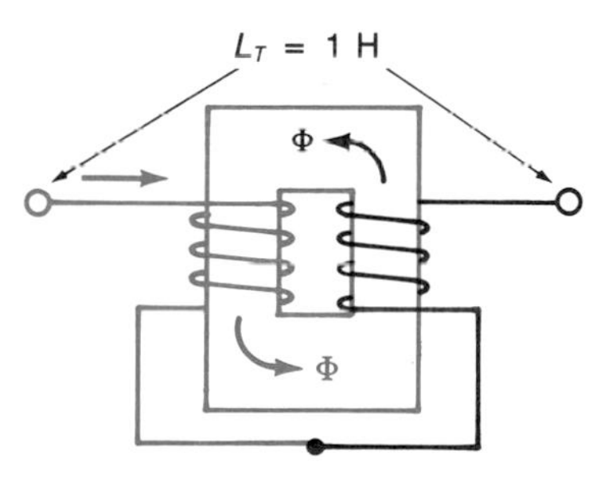


(*b*) Series-aiding

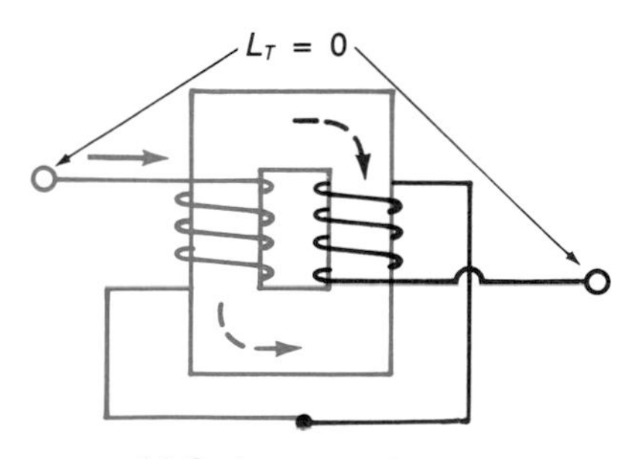


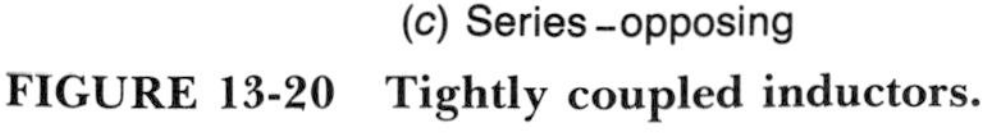

(*c*) Series-opposing

**FIGURE 13-20 Tightly coupled inductors.**

flux produced by the other coil. The cancellation of flux produces a total inductance of 0.

Coupling between inductors is often undesired and $k$ can be made very small by any of three techniques:

**1.** Use separate cores and increase the physical separation between the coils.

**2.** Position the coils so the flux from one coil does not cut the turns on the other. This may be accomplished by orienting the coil axes at right angles to one another.

**3.** Shield the coils or use self-shielding styles such as toroids or cup cores.

Tight coupling is desirable in many types of transformers. Transformers are covered in Chap. 20.

Figure 13-21 shows inductors connected in parallel. In this case the current divides among the inductors. Any given inductor conducts less than the total current, and the rate of current change is less. For example, if three identical inductors are in parallel and the total current is changing at a rate of 9 A/s, the current change in one inductor will be only 3 A/s. The total inductance is less than the smallest inductor in the circuit and is given by

$$L_T = \frac{1}{1/L_1 + 1/L_2 + 1/L_3 + \cdots + 1/L_N}$$

The above equation is valid only when the individual inductors are not coupled to each other ($k = 0$).

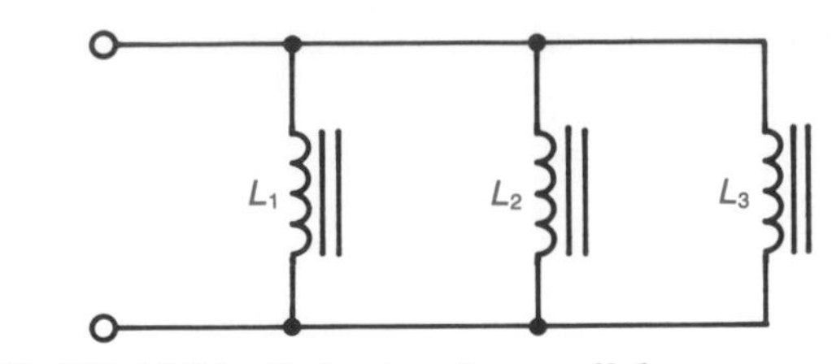


**FIGURE 13-21 Inductors in parallel.**

---

**EXAMPLE 13-13**

What is the total inductance of a 1-H, a 500-mH, and a 1.5-H inductor when they are connected in parallel?

**Solution**

Applying the equation,

$$L_T = \frac{1}{1/1 + 1/0.5 + 1/1.5} = 273 \text{ mH}$$

Note that the total inductance is less than the smallest inductor.

---

When equal-value inductors are connected in parallel, the total inductance is equal to the value of one inductor divided by the number of inductors. When two inductors of different values are connected in parallel, the following equation may be used to find the total inductance:

$$L_T = \frac{L_1 \times L_2}{L_1 + L_2}$$

## Self-Test

**19.** A circuit contains three series inductors with values of 250 mH, 500 mH, and 1 H. Assuming $k = 0$, find the total inductance of the circuit.

**20.** Two series-opposing inductors have values of 1 H and 1.7 H. What is their total inductance if their mutual inductance is 300 mH?

**21.** Two series inductors show a total inductance of 1.4 H when series-aiding and 1 H when series-opposing. What is their mutual inductance?

**22.** What is the coefficient of coupling for Question 21?

**23.** List three methods used to uncouple inductors.

**24.** Two uncoupled 3-H inductors are connected in parallel. What is their total inductance?

---

## SUMMARY

1. Inductance (letter symbol $L$) is that circuit property that opposes any change in current flow.

2. Any current change in a conductor creates a counter-electromotive force (cemf) that opposes the change.

3. Lenz's law predicts that any current resulting from an electromagnetically induced voltage sets up a flux that opposes the inducing flux.

4. Faraday's law predicts that the magnitude of an induced voltage is proportional to the number of conductor turns and the time rate of flux change linking the conductor:

$$V = N \times \frac{\Delta\Phi}{\Delta t}$$

5. The SI unit of inductance is the henry (letter symbol H). A 1-H inductor generates a cemf of 1 V when the current through it changes 1 A in 1 s:

$$L = \frac{-V}{\Delta I/\Delta t}$$

6. An ideal current pulse has zero rise and fall times. Real pulses can only approach the ideal due to circuit inductance.

7. The inductance of a coil is determined by the permeability of its core, its turns, its area, and its length:

$$L = \mu_r \times 4\pi \times 10^{-7} \times \frac{N^2a}{l}$$

*Note:* The dimensions are in meters and the inductance in henrys.

8. The inductance of a single-layer air-core coil is more accurately predicted with an approximation:

$$L \approx \frac{N^2r^2}{25.4r + 22.9l}$$

*Note:* Dimensions are in centimeters and inductance in microhenrys.

9. A coil may be wound on a nonmagnetic form and still be considered an air-core inductor.

10. Many iron-core inductors use bobbin-wound coils.

11. The four most widely applied core materials for inductors are silicon steel, iron ferrites, powdered iron, and air.

12. Toroidal and cup-core inductors are noted for high efficiency and self-shielding characteristics.

13. Small, molded inductors may use a color code, similar to the resistor color code, to indicate their nominal inductance in microhenry units.

14. Small, variable inductors use threaded cores or screws to position a movable core. They should be adjusted with nonmagnetic tools.

15. The inductance of a circuit conductor or of the lead on a component is often referred to as stray inductance.

16. The inductance of a straight, round, nonmagnetic wire in free space is given by

$$L = \frac{l\{[\ln(2l/r)] - 0.75\}}{5 \times 10^3}$$

*Note:* Dimensions are in millimeters and inductance in microhenrys.

17. The term lead dress refers to the length and position of component leads.

**18.** The total inductance of two or more series inductors having no mutual inductance is given by

$$L_T = L_1 + L_2 + L_3 + \cdots + L_N$$

**19.** Magnetically coupled series inductors exhibit mutual inductance which can be either aiding or opposing.

**20.** The total inductance of two series-connected inductors that are coupled is given by

$$L_T = L_1 + L_2 \pm 2L_M$$

**21.** The coefficient of coupling between two inductors is

$$k = \frac{L_M}{\sqrt{L_1 L_2}}$$

*Note:* $k$ is a pure number and varies from 0 to 1.

**22.** Values of $k$ approaching 1 are obtained by winding coils on a single, high-permeability core.

**23.** Values of $k$ approaching 0 are obtained by using separate cores, physically separating coils, controlling coil position, and using shields.

**24.** The total inductance of two or more uncoupled parallel inductors is given by

$$L_T = \frac{1}{1/L_1 + 1/L_2 + 1/L_3 + \cdots + 1/L_N}$$

## CHAPTER REVIEW QUESTIONS

**13-1.** Refer to Fig. 13-1*b*. If a load is connected across the open turn, the left-hand rule can be used to demonstrate that the resulting load current will set up a flux that ______ the inducing flux.

**13-2.** Refer to Fig. 13-1*b*. Suppose the variable resistor is increasing in value. What will happen to the induced voltage across the open turn?

**13-3.** The effects described in Questions 13-1 and 13-2 are predicted by ______ law.

**13-4.** As a current pulse in an inductor approaches the ideal, the inductor cemf approaches ______.

**13-5.** Name a core material that provides moderate permeability and does not saturate easily.

**13-6.** Pot cores, compared to toroids, have the advantage of being ______ wound.

**13-7.** A vertical-mount molded inductor has a red dot at the top, a red dot in the center, and a brown dot at the bottom of its package. What is its nominal inductance rating?

**13-8.** When stray inductance must be minimized, it is best to use conductors that are ______ in diameter.

**13-9.** The coefficient of coupling between two coils wound on a single, high-permeability core may approach ______.

**13-10.** The coefficient of coupling between two pot-core inductors mounted on opposite ends of a circuit board may be expected to approach ______.

## CHAPTER REVIEW PROBLEMS

**13-1.** What magnitude of voltage will be induced in a 50-turn coil that is linked by a flux changing from 0.4 to 0.9 Wb in 50 ms?

**13-2.** What is the inductance of a coil that develops a cemf of 0.7 V when the current through it changes 100 mA in 50 ms?

**13-3.** The current through a 2.5-H inductor changes from 0.2 to 0.65 A in 0.3 s. What is the magnitude of the induced voltage?

**13-4.** Calculate the inductance of a 700-turn coil wound on a core with a relative permeability of 2000 if the core area is $5 \times 10^{-4}$ m$^2$ and the average magnetic path length is 0.3 m.

**13-5.** Use the most accurate approach to approximate the inductance of a 50-turn, single-layer, air-core coil that is 3 cm long and 1 cm in diameter.

**13-6.** What is the free space inductance of a 500-mm-long, round copper wire that is 2 mm in diameter?

**13-7.** What is the total inductance of a 350-mH coil connected in series with a 1.1-H coil, assuming no mutual inductance? When they are parallel-connected?

**13-8.** Two 1-H series inductors have a mutual inductance of 0.5 H. What is their total inductance when series-aiding? When series-opposing?

**13-9.** What is the coefficient of coupling for the inductors in Problem 13-8?

**13-10.** What is the total inductance of three uncoupled, 9-H inductors connected in parallel?

## ANSWERS TO SELF-TESTS

**1.** inductance
**2.** mutual
**3.** counter (or back)
**4.** 11.1 V
**5.** It will increase.
**6.** 1 H
**7.** 86 kV
**8.** inductance
**9.** 0.247 μH, 15.4 percent error
**10.** 85.4 percent error
**11.** true
**12.** true
**13.** true
**14.** false
**15.** 5.6 μH
**16.** true
**17.** decreases
**18.** true
**19.** 1.75 H
**20.** 2.1 H
**21.** 100 mH
**22.** 0.085
**23.** physical separation, physical position, shielding
**24.** 1.5 H

# Chapter 14 Inductors in DC and AC Circuits

Chapter 13 introduced inductance and inductors. In that chapter, you learned that an inductor produces a counterelectromotive force in response to a changing current. The cemf is the induced voltage created by an expanding or collapsing magnetic flux.

In this chapter, we build on these ideas to show how inductors behave in dc and ac circuits. The behavior of inductors appears to be quite different from that of resistors in both ac and dc circuits. However, inductor circuits, like all circuits, must follow universal circuit laws such as Kirchhoff's laws and Ohm's law.

## 14-1 ENERGY STORED IN AN INDUCTOR

When pure inductance (an ideal inductor) takes energy from a current source, it stores that energy in its magnetic field. As the current through an

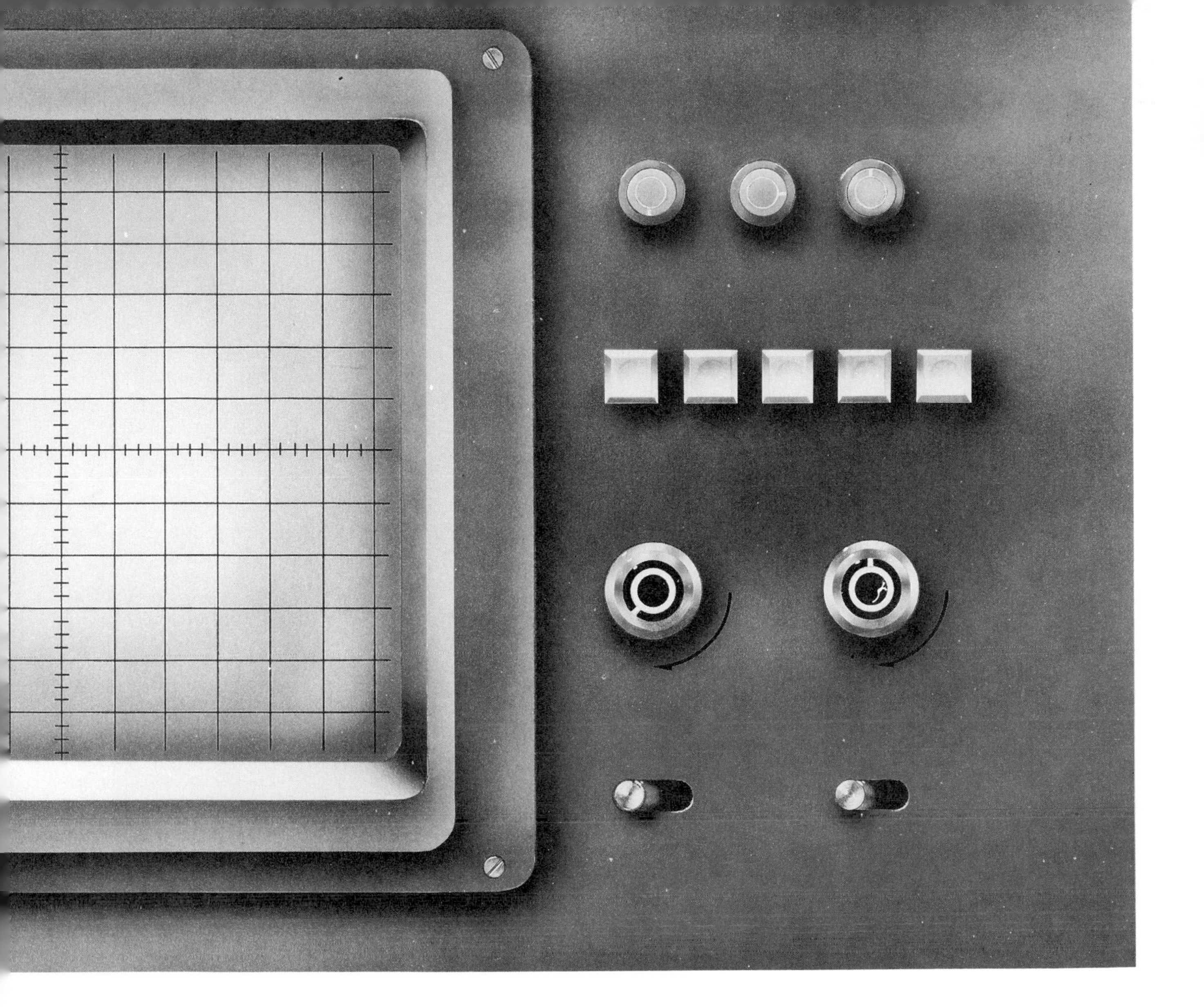

ideal inductor builds from zero to its final (steady-state) value, which is determined by the limits of the current source, the energy taken from the source is used to develop the magnetic flux. As long as the steady-state current continues to flow, the energy taken from the source remains stored in the magnetic field. When the current starts to decay, this stored energy is returned to the current source (or converted to heat in any circuit resistance) as the magnetic field collapses. As long as the inductor core is not saturated, the magnetic field and the flux density are controlled by the inductance and the current through the inductor. Therefore, the energy stored in an inductor is also controlled by the inductance and the current.

The formula for determining the amount of energy an inductor stores can be developed by using the relationships and formulas developed in Chaps. 2 and 3. We know that

$$W = Pt$$
$$P = IV$$

and, therefore,

$$W = IVt$$

where $W$ = work (energy), J
$P$ = power, W
$t$ = time, s
$I$ = current, A
$V$ = voltage, V

Thus, to determine the amount of energy an inductor stores, we need to determine the inductor's current and voltage during the time it is storing energy. Because the inductor stores energy only during the time the current is increasing, we must determine the average current during the time the current is rising. This can be done by referring to Fig. 14-1, which shows the current in an inductor increasing at a constant rate until it reaches its maximum value, which is determined by the current source. Since the current rises linearly, the average current is just $0.5I_{maximum}$ ($0.5I_{max}$) where $I_{max}$ is the final, steady-state current through the inductor. Thus, the appropriate current for the energy formula is $0.5I_{max}$.

The appropriate voltage for this formula is the inductor cemf during the time energy is being stored. In Chap. 13 it was shown that the inductor cemf is determined by $V_L = (\Delta I_L/\Delta t)L$. This expression reduces to $V_L = (I_{max}/t)L$ when the current change is linear, because during the $\Delta t$ that the current is increasing, $I_L$ changes from zero to $I_{max}$. Voltage $V_L$ remains constant as long as the rate of change in $I_L$ remains constant. Therefore, for the time interval when the current is changing in Fig. 14-1, $V_L$ is a constant value.

Now we can substitute into the general energy formula to derive the formula for the energy stored in an inductor. Thus,

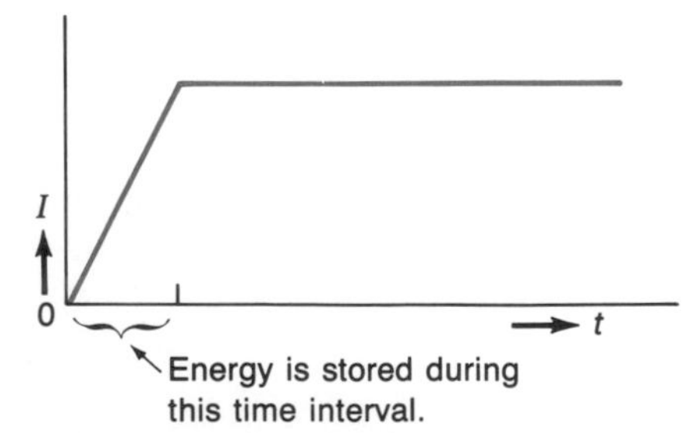

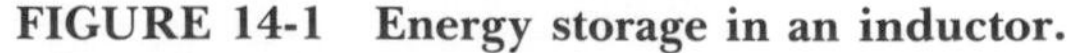

FIGURE 14-1 Energy storage in an inductor.

$W = IVt$ — Substitute $0.5I_{max}$ for $I$

$W = 0.5I_{max}Vt$ — Substitute $V_L = (I_{max}/t)L$ for $V$

$$W = \frac{0.5I_{max}I_{max}Lt}{t}$$

Simplifying yields

$$W = 0.5I_{max}^2 L$$

In this formula for energy stored in an inductor, the subscript max is usually dropped and the formula is written as

$$W = 0.5LI^2 \quad \text{or} \quad W = \frac{LI^2}{2}$$

As shown below, the energy is in its base unit of joule (J) when inductance and current are in their base units of henry (H) and ampere (A):

$W = 0.5LI^2 = 0.5 \text{ H} \cdot \text{A} \cdot \text{A}$ — Substitute the equivalent of H, which is V/(A/s) or (V · s)/A

$= 0.5\dfrac{\text{V} \cdot \text{A} \cdot \text{A}}{\text{A/s}} = 0.5 \text{ V} \cdot \text{A} \cdot \text{s}$ — Substitute the equivalent of A, which is C/s

$= \dfrac{0.5 \text{ V} \cdot \text{C} \cdot \text{s}}{\text{s}} = 0.5 \text{ V} \cdot \text{C}$ — Substitute the equivalent of V, which is J/C

$$W = 0.5\frac{\text{J} \cdot \text{C}}{\text{C}} = 0.5 \text{ J}$$

**EXAMPLE 14-1**

How much energy is stored by a 2-H inductor if the current through the inductor is 500 mA?

**Solution**

$$W = 0.5LI^2 = 0.5 \times 2 \text{ H} \times (0.5 \text{ A})^2 = 0.25 \text{ J}$$

## Self-Test

1. Where is energy stored in an inductor?
2. What happens to the energy stored in an inductor when the current decays to zero?
3. Determine the energy stored in a 400-mH inductor when the inductor current is 300 mA.
4. How much current must a 10-H inductor carry if it is storing 2 J of energy?

## 14-2 CURRENT RISE IN INDUCTIVE CIRCUITS

Because of an inductor's cemf, the current in an inductive circuit cannot instantaneously increase to its maximum value like it does in a circuit containing only resistance. The current rise in an inductive circuit containing either a real inductor or an inductor and a resistor connected to a dc voltage source is illustrated in Fig. 14-2. The resistor in the diagram in this figure can represent either (or both) the internal resistance of the inductor or a physical resistor because a practical inductor behaves like an ideal inductor in series with a small amount of resistance. When a physical resistor ($R$) is in series with the inductor, it can be called an $RL$ circuit. The final, steady-state current level in Fig. 14-2*a* is determined by the total circuit resistance and the amount of dc source voltage. The total circuit resistance consists of (1) the resistance of the inductor's windings, (2) the resistance of the physical resistor (if any), (3) the internal resistance of the source, and (4) the wiring resistance. Unless specifically mentioned, we will assume that these last two resistances are so small compared to the first two that they can be ignored when analyzing inductive circuits. Thus, the final, or maximum, current in Fig. 14-2 is

$$I_{\max} = \frac{V_T}{R} = \frac{10 \text{ V}}{20 \ \Omega} = 0.5 \text{ A}$$

where $V_T$ is total voltage, and $R$ is resistance.

Notice in Fig. 14-2*b* that the slope of the current curve continuously decreases until it is zero. Kirchhoff's voltage law and the cemf formula $[V_L = (\Delta I/\Delta t)L]$ can explain why this slope must decrease as the current level increases. At the instant the switch in the circuit is closed, the current tries to jump to its maximum value, but the first tiny (but rapid) increase in current produces a large cemf across the inductance. Of course, this cemf is series-opposing the source voltage. Thus, the instant the switch is closed, the $\Delta I/\Delta t$ (slope) can be no greater than that needed to produce a cemf equal to the source voltage. For Fig. 14-2, the initial slope would be

$$V_L = V_T = \frac{\Delta I}{\Delta t} L$$

so

$$\frac{\Delta I}{\Delta t} = \frac{V_T}{L} = \frac{10 \text{ V}}{2 \text{ H}} = 5 \text{ A/s}$$

As the current increases from zero, part of $V_T$ drops across $R$. Thus, according to Kirchhoff's voltage law, the cemf must decrease so that the instantaneous values of $V_R$ and the cemf equal $V_T$. By the time the current has reached one-half of its final value (0.25 A for the circuit in Fig. 14-2), the voltage across the resistance is one-half of the source voltage (5 V for the circuit in Fig. 14-2). At this instant (when $I = 0.5I_{\max}$), the slope is

$$V_L = V_T - V_R = \frac{\Delta I}{\Delta t} L$$

so

$$\frac{\Delta I}{\Delta t} = \frac{10 \text{ V} - 5 \text{ V}}{2 \text{ H}} = 2.5 \text{ A/s}$$

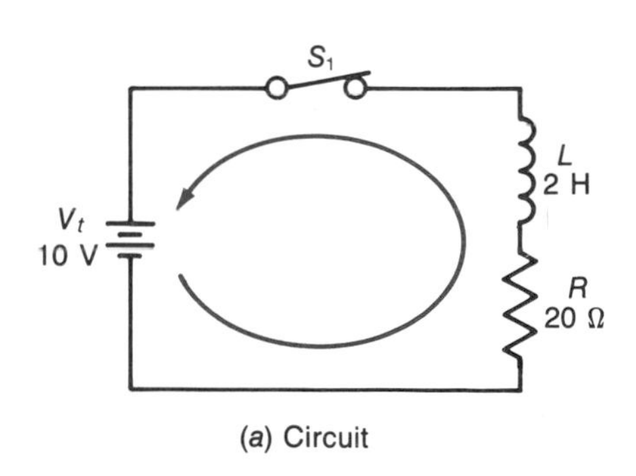


(*a*) Circuit

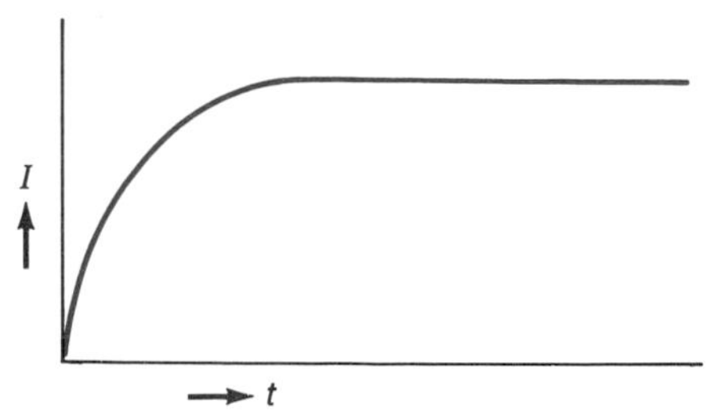


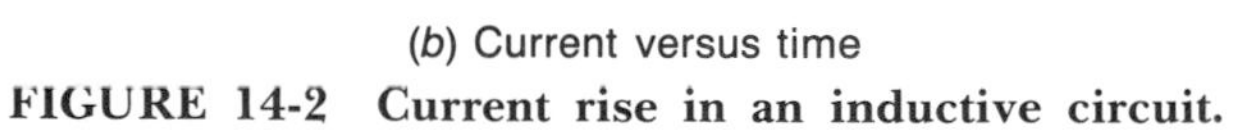

(*b*) Current versus time

**FIGURE 14-2 Current rise in an inductive circuit.**

### EXAMPLE 14-2

What is the initial slope of the current-versus-time curve for the circuit in Fig. 14-3? Also, what is the slope when the current is 90 percent of its maximum value?

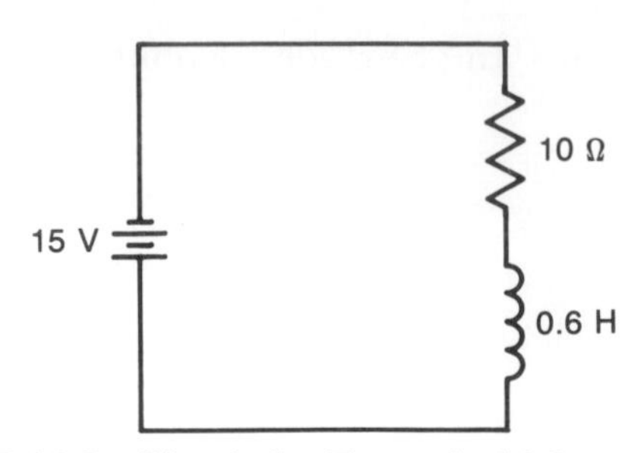


**FIGURE 14-3 Circuit for Example 14-2.**

**Solution**

$$\text{Initial slope} = \frac{V_T}{L} = \frac{15\ \text{V}}{0.6\ \text{H}} = 25\ \text{A/s}$$

$$I_{\text{max}} = \frac{V_T}{R} = \frac{15\ \text{V}}{10\ \Omega} = 1.5\ \text{A}$$

$$90\ \text{percent of}\ I_T = 1.5\ \text{A} \times 0.9 = 1.35\ \text{A}$$

$$V_R\ \text{at 90 percent of}\ I_T = 1.35\ \text{A} \times 10\ \Omega = 13.5\ \text{V}$$

$$V_L\ \text{at 90 percent of}\ I_T = V_T - V_R = 15\ \text{V} - 13.5\ \text{V} = 1.5\ \text{V}$$

$$\text{Slope at 90 percent of}\ I_T = \frac{1.5\ \text{V}}{0.6\ \text{H}} = 2.5\text{A/s}$$

The above example and Fig. 14-2*b* show that the slope of the current curve rapidly approaches zero as the current nears its maximum value. Although theoretically the slope never reaches zero, for practical purposes the slope is essentially zero by the time the current reaches 99 percent of its maximum value. By this time, the current is considered to have reached its steady-state value. Also, at this time the inductor is considered to be storing as much energy as it can for the given circuit because the cemf is essentially zero.

The current-versus-time curve in Fig. 14-2 could also be labeled a resistive voltage–versus–time curve. This must be so because Ohm's law tells us that the voltage across a resistance must be directly proportional to the current through the resistor. Since the resistive voltage is increasing at the same rate as the current, the inductive voltage (cemf) must be decaying at a complementary rate. According to Kirchhoff's voltage law, the instantaneous values of these two voltages ($V_{R_i}$, $V_{L_i}$) must always equal the source voltage. A graph of these three voltages plotted against time is shown in Fig. 14-4. This figure clearly shows that the $V_T = V_{R_i} + V_{L_i}$, where $V_L$ is, as previously discussed, the cemf. It should be remembered that it is not possible to separately measure $V_R$ and $V_L$ when the only resistance in the circuit is the resistance of the inductor's winding. Any measuring instrument connected across the inductor measures the sum of $V_R$ and $V_L$.

## Self-Test

5. What keeps the current in an inductive circuit from instantly reaching its steady-state value?
6. What is the value of the steady-state current when a 3-H inductor with 50 Ω of internal resistance is connected to a 25-V source?
7. What is the initial slope of the current curve in the circuit of Question 6?
8. When the current in an inductive circuit is 80 percent of maximum, what percentage of the source voltage will the inductive (cemf) voltage be?

## 14-3 TIME CONSTANTS FOR RISING CURRENTS

The time required for an inductive circuit to reach its steady-state, or final, current can be specified in *time constants.* The symbol for a time constant is the Greek lowercase letter tau ($\tau$). A time constant $\tau$ can be defined, for a rising current, as the time required for the current to reach its final value *if* the current continues to rise at its initial rate or slope. This definition is illustrated in Fig. 14-5. The advantage of specifying the horizontal axis of a rising current curve in time constants rather than time is that it requires the same number of time constants ($\tau$) for all *RL* circuits to essentially reach their final current. One curve is applicable to all *RL* circuits. Thus, the curve in Fig. 14-5 is often referred to as a *universal time constant* curve.

The formula for calculating $\tau$ can be derived

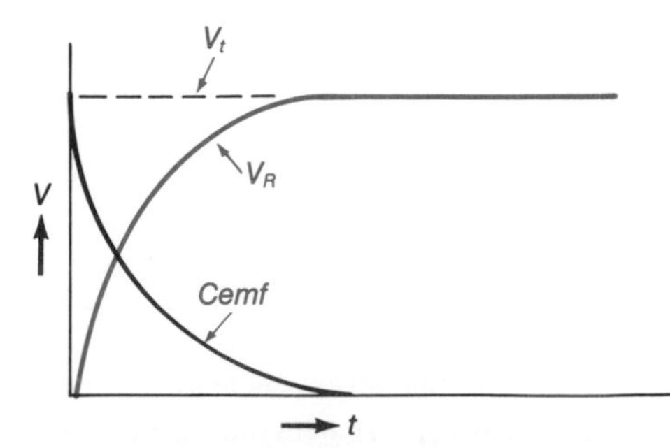


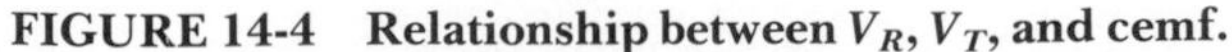

**FIGURE 14-4 Relationship between $V_R$, $V_T$, and cemf.**

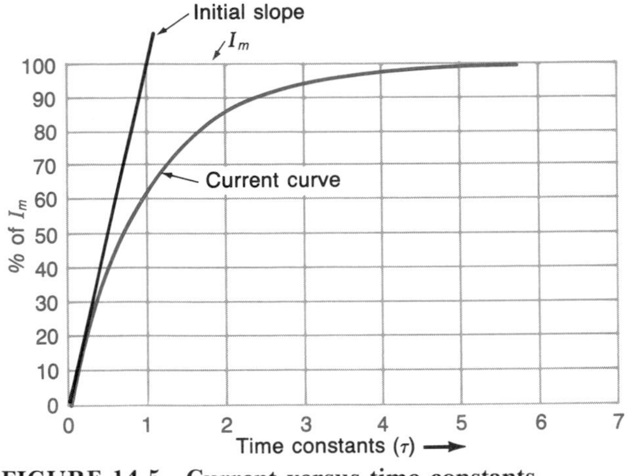


**FIGURE 14-5 Current versus time constants.**

from its definition and the relationships we have been working with. From Fig. 14-5 we can see that the initial slope can be expressed as

$$\text{Initial slope} = \frac{I_{max}}{\tau}$$

We previously established that the initial slope is also equal to $V_T/L$ when the current rise is linear, as it is assumed to be in defining the time constant. Now combining these two expressions for initial slope, we get

$$\frac{V_T}{L} = \frac{I_{max}}{\tau}$$

And rearranging gives

$$\tau = \frac{L}{V_T/I_{max}}$$

And since $R = V_T/I_{max}$,

$$\tau = \frac{L}{R}$$

The appropriate unit for $\tau$ can be established by substituting base units and base unit equivalents into the time constant formula:

$$\tau = \frac{L}{R} = \frac{\text{H}}{\Omega} = \frac{(\text{V} \cdot \text{s})/\text{A}}{\text{V/A}} = \text{s}$$

Thus, when inductance is in henrys and resistance is in ohms, the time constant is in seconds.

The universal time constant curve in Fig. 14-5 is useful for estimating the instantaneous circuit current in an *RL* circuit after a specified number of $\tau$ or a specified time.

---

**EXAMPLE 14-3**

Determine the instantaneous current $I_i$ for the circuit in Fig. 14-3 after it has been connected to the source for 0.1 s.

**Solution**

$$I_{max} = \frac{V_T}{R} = \frac{15\text{ V}}{10\ \Omega} = 1.5\text{ A}$$

$$\tau = \frac{L}{R} = \frac{0.6\text{ H}}{10\ \Omega} = 0.06\text{ s}$$

$$\text{Number of } \tau = \frac{t}{\tau} = \frac{0.1\text{ s}}{0.06\text{ s}} = 1.67\ \tau$$

From the curve,

$$I_i \approx 0.82 I_{max} = 0.82 \times 1.5\text{ A} = 1.23\text{ A}$$

---

The above example shows that the result obtained by interpreting the universal curve may not be very accurate. When greater accuracy is desired, the instantaneous value of the current at any time after the switch in an inductive circuit is closed can be determined with the formula

$$I_i = \frac{V_T}{R}(1 - e^{-t/\tau})$$

where $I_i$ = instantaneous current at time $t$
$V_T$ = dc source voltage
$R$ = total resistance
$e$ = exponential constant ($e = 2.718$)
$\tau$ = time constant ($\tau = L/R$)

The derivation of this formula requires differential calculus. It is derived by solving the voltage equation $V_T = V_{R_i} + V_{L_i} = I_iR + (\Delta L/\Delta t)L$ for $I_i$. Since hand-held calculators have $e^x$ and ln (log base $e$) function keys, solving for $I_i$ is not difficult.

---

**EXAMPLE 14-4**

Rework Example 14-3 using the exponential formula rather than the exponential curve.

**Solution**

$$I_i = \frac{15\text{ V}}{10\ \Omega}(1 - e^{-0.1\text{ s}/0.06\text{ s}}) = 1.5\,(1 - 0.189) = 1.217\text{ A}$$

---

Sometimes $\tau$ is defined as the time required for the current to increase by 63.2 percent of the dif-

| Number of $\tau$ | Percentage of $I_m$ |
|---|---|
| 1 | 63.212 |
| 2 | 86.466 |
| 3 | 95.021 |
| 4 | 98.168 |
| 5 | 99.326 |
| 6 | 99.752 |
| 7 | 99.909 |

**FIGURE 14-6 Percentage of maximum current when the initial current was zero.**

ference between $I_{\max}$ and the current at the beginning of the time constant. The usefulness of this definition can be illustrated by referring to Fig. 14-6, which lists the percentage of $I_{\max}$ at the end of each of the first $7\tau$ when the initial current was zero. These percentages were calculated using the exponential formula. The percentage of $I_{\max}$ at the end of the second $\tau$ could also have been determined by taking 63.2 percent of the remaining percentage of $I_{\max}$ and adding it to the 63.2 percent of $I_{\max}$ that existed at the beginning of the second $\tau$. Thus, at the end of the second $\tau$, the percentage of $I_{\max} = (100 \text{ percent} - 63.2 \text{ percent}) \times (63.2 \text{ percent}) + 63.2 \text{ percent} = 86.46$ percent.

Figure 14-6 shows that the current is over 99 percent of maximum after $5\tau$. This is why the current is usually assumed to be at its maximum value (and the slope at 0) after five time constants.

The exponential equation for $I_i$ can be solved for $t$ so that one can easily determine the time required to reach a specified $I_i$ after an $RL$ circuit is completed. The steps in solving for $t$ are

$$I_i = \frac{V_t}{R}(1 - e^{-t/\tau}) \qquad \text{Multiply by } R/V_T$$

$$\frac{I_iR}{V_T} = 1 - e^{-t/\tau} \qquad \text{Multiply by } -1 \text{ and rearrange}$$

$$e^{-t/\tau} = 1 - \frac{I_iR}{V_T} \qquad \text{Substitute } V_T/V_T \text{ for } 1$$

$$e^{-t/\tau} = \frac{V_T - I_iR}{V_T} \qquad \text{Take the reciprocals}$$

$$e^{t/\tau} = \frac{V_T}{V_T - I_iR} \qquad \text{Take the natural (base } e\text{) log}$$

$$\frac{t}{\tau} = \ln\frac{V_T}{V_T - I_iR} \qquad \text{Finally, multiply by } \tau$$

$$t = \tau \ln\frac{V_T}{V_T - I_iR}$$

[Note: the natural log is designated as ln on many calculators.]

**EXAMPLE 14-5**

For the circuit in Fig. 14-7, determine how long it will take (after the switch is closed) for $I_i$ to reach 75 percent of $I_{\max}$. Also determine $V_{R_i}$ and $V_{L_i}$ when $I_i = 0.75\, I_{\max}$.

**Solution**

$$\tau = \frac{3 \text{ mH}}{10\ \Omega} = 0.3 \text{ ms}$$

$$I_{\max} = \frac{20 \text{ V}}{10\ \Omega} = 2\text{A}$$

$$I_i = 0.75 \times 2 \text{ A} = 1.5 \text{ A}$$

$$t = 0.3 \text{ ms} \times \ln \frac{20 \text{ V}}{20 \text{ V} - (1.5 \text{ A} \times 10\ \Omega)} = 0.416 \text{ ms}$$

$$V_{R_i} = I_iR = 1.5 \text{ A} \times 10\ \Omega = 15 \text{ V}$$

$$V_{L_i} = V_T - V_{R_i} = 20 \text{ V} - 15 \text{ V} = 5 \text{ V}$$

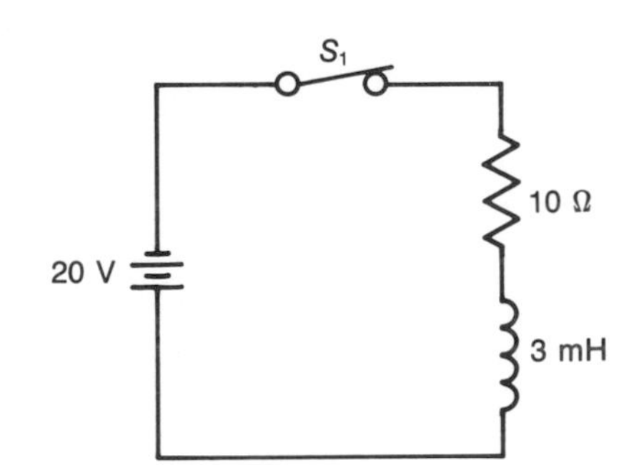


**FIGURE 14-7 Circuit for Example 14-5.**

## Self-Test

**9.** If the current in an $RL$ circuit rose linearly at its initial slope, it would reach $I_{\max}$ in ______.

**10.** Current in an inductive circuit is essentially at its maximum value after ______ $\tau$.

**11.** A 0.7-H inductor has 30 Ω of winding resistance. What is its time constant? How long will it take a dc current to reach its maximum value in this inductor?

**12.** The inductor in Question 11 is connected in series with a 40-Ω resistor and a 35-V source. Determine $I_i$ after the circuit has been connected for 30 ms. Also determine the voltage across the inductor at this time.

**13.** How long after closing the switch will it take the circuit in Fig. 14-7 to reach a current of 1 A?

## 14-4 CURRENT FALL IN INDUCTIVE CIRCUITS

The process of storing energy in an inductor as the current is rising is sometimes referred to as *charging* an inductor. Removing the inductor's energy as the current decays is referred to as *discharging* the inductor.

Because the inductor opposes a decrease in current as well as an increase in current, the current in an inductive circuit cannot instantly fall to zero when an inductive circuit is turned off. When such a circuit is turned off, the current starts to fall, but the falling current allows the inductor's magnetic field to start collapsing. The collapsing magnetic field induces a cemf in the inductor. This cemf is series-aiding to the source because a collapsing field produces a polarity opposite to that produced by an expanding field. (We learned in Sec. 14-3 that an expanding field produces a series-opposing cemf.) This series-aiding cemf attempts to maintain the current at its steady-state value. Of course, the current must continue to decay if the magnetic field is to continue collapsing and producing a series-aiding cemf. As mentioned earlier in this chapter, the inductor is returning its energy as the current decays.

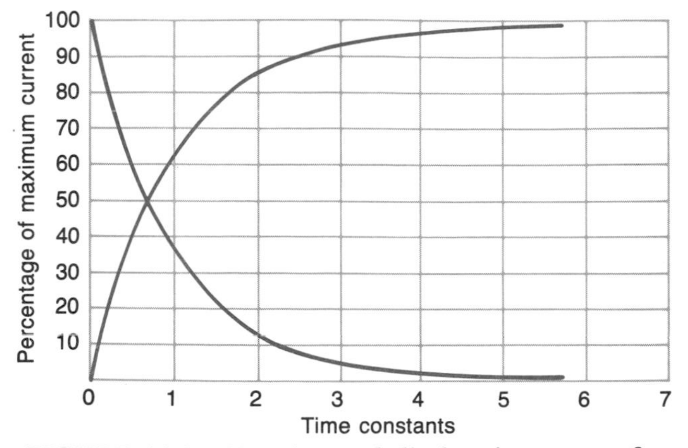


**FIGURE 14-8 Charging and discharging curves for an inductive circuit.**

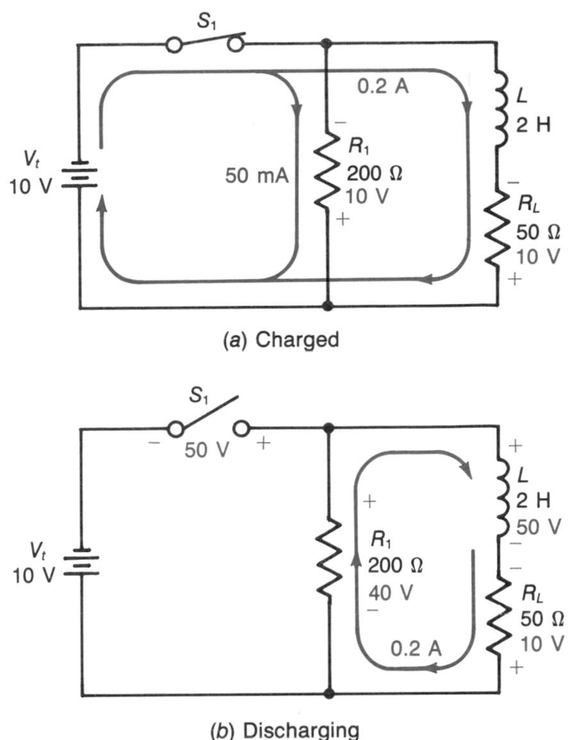


**FIGURE 14-9 Charging and discharging an inductive circuit.**

Figure 14-8 shows the current-versus-time curves for both a rising and a falling current. Notice that the curve for the declining current is exponential, which is complementary to the rising-current curve. Using techniques analogous to those for a rising current, it can be shown that the time constant formula for the discharging inductive circuit is the same as that for the charging circuit ($\tau = L/R$). Notice in Fig. 14-8 that the discharging circuit current decays 63.2 percent during one time constant.

Using differential calculus to solve the voltage equation for a discharging inductive circuit yields the instantaneous current equation:

$$I_i = I_{\text{max}} e^{-t/\tau}$$

In this equation, $I_{\text{max}}$ is the maximum inductor current at the instant the current starts to fall. For a circuit like the one in Fig. 14-9*a* this current would be $I_{\text{max}} = V_T/R_L = 10\text{ V}/50\ \Omega = 0.2\text{ A}$ if switch $S_1$ has been closed for at least $5\tau$. If $S_1$ has been closed for less than $5\tau$, then $I_{\text{max}}$ for the discharge current formula would be the value of $I_i$ determined by the charge current formula.

In Fig. 14-9, $R_L$ represents the winding resistance of a practical inductor and $L$ represents the pure inductance of the practical inductor. Resistor $R_1$ is a physical resistor in parallel with the real inductor. Notice in Fig. 14-9*a* that there is no voltage shown across $L$. This means that the current has reached its steady-state value.

The instant the switch in Fig. 14-9*a* is opened, the voltages, currents, and polarities are as shown in Fig. 14-9*b*. Also, the instant the switch is opened, the inductor becomes the energy and voltage source for the circuit. The inductor's cemf instantly jumps up to the value required to support $I_{max}$ through the total resistance of the discharge path ($R_L + R_1$ for this circuit). Notice in Fig. 14-9*b* that the full cemf of the inductor must appear across the open switch in order to satisfy Kirchhoff's voltage law. Since the peak voltage across the opening switch contacts is only 50 V, the arcing between the contacts as they open is negligible. Thus, the full 0.2 A of peak instantaneous current furnished by the inductor flows through $R_1$ and $R_L$ as shown in Fig. 14-9*b*. The current furnished by the inductor exponentially decays until all of the inductor's stored energy has been used. As shown in Fig. 14-10, both the resistive and the inductive voltages also decay exponentially. Figure 14-9*b* and Fig. 14-10 show that $V_R$ and $V_L$ have opposite polarities. This is to be expected since $V_L$ is the source voltage and $V_R$ is the voltage drop in a discharging circuit.

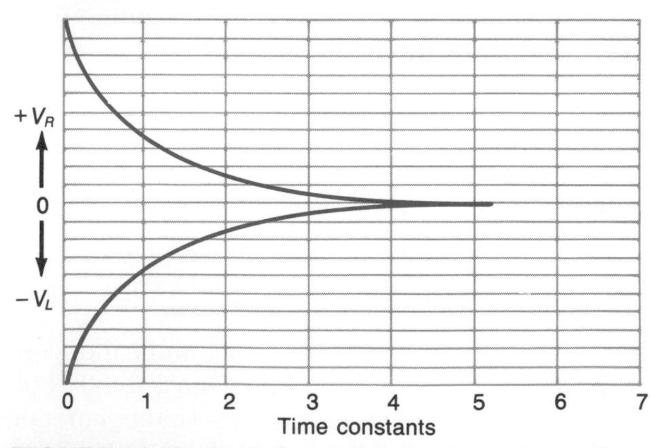


**FIGURE 14-10 Resistive and inductive voltages for a discharging inductive circuit.**

**EXAMPLE 14-6**

For the circuit in Fig. 14-11, determine $I_i$, $V_{R_1}$, and the cemf ($V_L$) after $S_1$ has been closed for $20\tau$ and then opened for 0.002 s.

**Solution**

$$\tau = \frac{L}{R} = \frac{0.8\text{ H}}{200\ \Omega} = 0.004\text{ s}$$

$$I_{max} = \frac{V_T}{R_L} = \frac{15\text{ V}}{50\ \Omega} = 0.3\text{ A}$$

After discharging for 0.002 s,

$$I_i = I_{max}e^{-t/\tau} = 0.3\text{ A } e^{-0.002\text{ s}/0.004\text{ s}} = 182\text{ mA}$$

$$V_{R_1} = I_iR_1 = 182\text{ mA} \times 150\ \Omega = 27.3\text{ V}$$

$$V_{RL} = I_iR_L = 182\text{ mA} \times 50\ \Omega = 9.1\text{ V}$$

$$V_L = V_{R_1} + V_{RL} = 27.3\text{ V} + 9.1\text{ V} = 36.4\text{ V}$$

The formula for finding $I_i$ in a discharging circuit can be solved for $t$ so that the time required for the discharge current to fall to a specified value can also be calculated. The resulting formula is

$$t = \tau \ln \frac{I_{max}}{I_i}$$

**EXAMPLE 14-7**

After opening $S_1$, how many time constants will be required to reduce the current in Fig. 14-11 to 50 mA?

**Solution**

From Example 14-6, $\tau = 0.004$ s and $I_{max} = 0.3$ A, so

$$t = \tau \ln \frac{I_{max}}{I_i} = 0.004\text{ s} \ln \frac{0.3\text{ A}}{0.05\text{ A}} = 0.0072\text{ s}$$

$$\text{Number of } \tau = \frac{t}{\tau}$$

$$\frac{0.0072\text{ s}}{0.004\text{ s}} = 1.8$$

The circuit in Fig. 14-9 shows that the inductor's cemf exceeds the dc source voltage when an inductive circuit is first turned off by opening a switch. In the circuit of Fig. 14-9 the cemf is only five times as great as the source voltage; however, this ratio

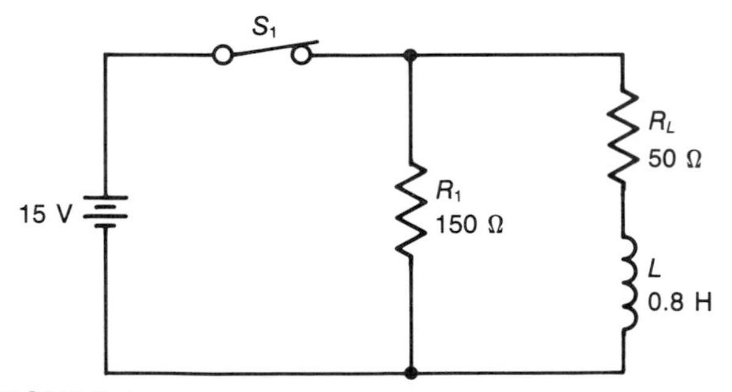


**FIGURE 14-11 Circuit for Example 14-6.**

can be increased by increasing the discharge resistance. For example, if $R_I$ in Fig. 14-9 is increased to 2 kΩ, the voltage across $R_1$ increases to 2 kΩ × 0.2 A = 400 V. The peak cemf, and voltage across $S_1$, will be 400 V + 10 V = 410 V. The cemf-to-source voltage ratio is now 41:1.

When no discharge resistor, such as $R_1$ in Fig. 14-9, is included in a circuit, the cemf can be thousands of volts. When the switch in the circuit in Fig. 14-12 is opened, the inductor's cemf must be large enough to establish an arc (column of ionized air) between the switch contacts as the switch is opened. This arc is maintained until the inductor is essentially discharged. The exact value of cemf, and the voltage across the switch, depends on such factors as the distance between the open contacts and the humidity of the air. Although it takes less voltage to maintain than it does to establish an arc, some idea of the magnitude of the cemf and switch voltage can be gained from the fact that it takes about 30 kV to establish an arc through 1 cm of dry air.

The large cemf produced when an inductive circuit is turned off is often referred to as an *inductive kick* or an *inductive spike voltage*. As we have seen, the value of the inductive kick can be reduced by providing a low-discharge resistance. If these spike voltages are not controlled, they can erode switch contacts and destroy voltage-sensitive devices such as transistors and integrated circuits. A simple way to limit the inductive kick in a circuit is to connect a varistor (voltage-dependent resistor) in parallel with the inductor or inductive device. *Varistors*, which were discussed in detail in Chap. 5, are devices whose resistances drop from a very high value to a very low value at some specified critical voltage. When the varistor's critical voltage is well above the circuit's dc source voltage, the varistor has no effect on circuit operation until the inductor's spike voltage exceeds its critical voltage. Then the varistor conducts and discharges the inductor while limiting the cemf to the varistor's critical voltage.

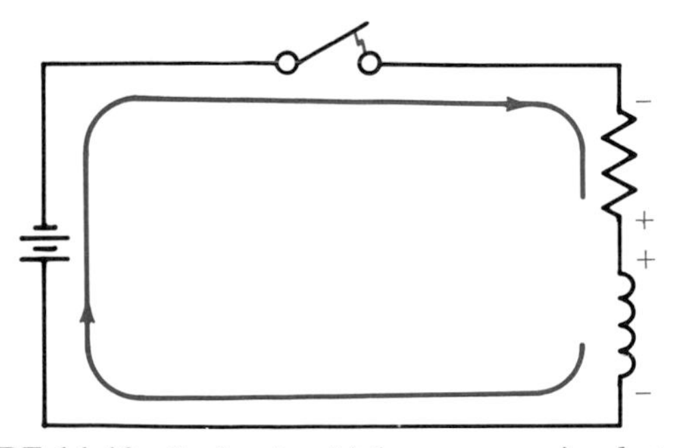


**FIGURE 14-12 Inductive kick causes arcing between switch contacts.**

## Self-Test

**14.** When an inductor is storing energy, it is said to be ________.

**15.** The polarity of the cemf produced by a falling current is ________ the source voltage.

**16.** If $\tau$ is 40 ms, how long does it take an inductive circuit to discharge?

**17.** True or false? When the current is at its steady-state value, the cemf will be maximum.

**18.** True or false? When an inductive circuit is turned off, the cemf may be hundreds of times larger than the dc source voltage.

**19.** When an inductive current changes from charging to discharging, does the direction of the current in the inductor change? Does the polarity of the cemf change?

**20.** What is meant by the term "inductive kick"?

**21.** How can "inductive kick" be limited?

**22.** Change $R_L$ in Fig. 14-11 to 40 Ω, and then determine $V_L$ and $I_i$ after the circuit has been discharging for 0.003 s. Also determine the peak cemf.

**23.** How long will it take for the current in Question 22 to decay to 20 mA?

## 14-5 INDUCTOR ACTION IN AN AC CIRCUIT

Figure 14-13*a* shows an inductor connected to a sinusoidal ac source. In this circuit the inductor is assumed to be an ideal inductor so no series resistance is shown. In many ac circuits, assuming an ideal inductor does not lead to significant error because the inductor's resistance is small compared to the inductor's other type of opposition (reactance) to ac.

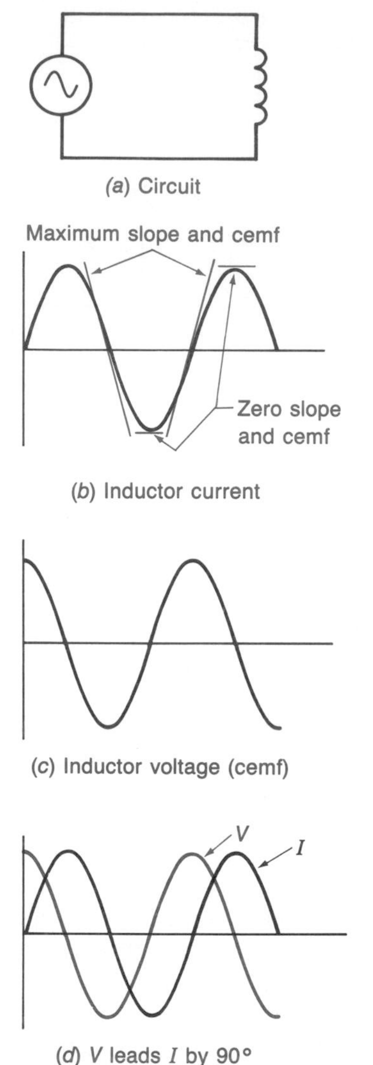


**FIGURE 14-13 Current and voltage in an inductive circuit.**

### Phase Angle

In Chapters 10 and 13 you learned that the amount of flux per unit of time that cuts a conductor determines the magnitude of the induced voltage. The magnitude of the cemf in an inductor is controlled by the same factor. In an ideal inductor, the flux builds and collapses in phase with the rising and falling current. The faster the current is rising or falling, the greater the cemf of the inductor will be at a given instant; i.e., the instantaneous cemf is a function of the instantaneous slope of the current waveform. For a sinusoidal current, the slope is greatest as the waveform crosses through zero, and the slope is zero when the waveform is at either its negative or positive crest (see Fig. 14-13*b*). As shown in Fig. 14-13*c*, this causes the inductor voltage (cemf) to be maximum at the instant the inductor current is zero. In other words, the inductor current and voltage are 90° out of phase. The current and voltage waveforms have been drawn on the same graph in Fig. 14-13*d* to more clearly show this 90° phase relationship. Notice two things in Fig. 14-13*d*. (1) The source voltage and the source current for the simple circuit of Fig. 14-13*a* must be equal to the inductor voltage and inductor current to satisfy Kirchhoff's laws. (2) The *voltage leads* the *current* by 90° in a pure inductance; for example, the voltage reaches its peak positive value 90° before the current reaches its peak positive value. The voltage leads the current because the polarity of the inductor's cemf is always such that it opposes what the current is trying to do. When the current starts its sharp rise from zero toward its positive peak, the inductor voltage (cemf) will already be at its positive peak so that it series-opposes the source voltage and restricts the current rise.

The phase relationship between sinusoidal waveforms is often shown in phasor form as in Fig. 14-14. The angle between the current phasor and voltage phasor is referred to as the *phase angle* or as *angle theta*. The phasor diagram in Fig. 14-14 also shows the voltage leading the current because, by convention, phasors rotate in a counterclockwise direction.

### Power

The power in an ideal inductor circuit is zero. There are several ways to explain why this is so. First, only resistance can convert electric energy into heat energy, and an ideal inductor has no re-

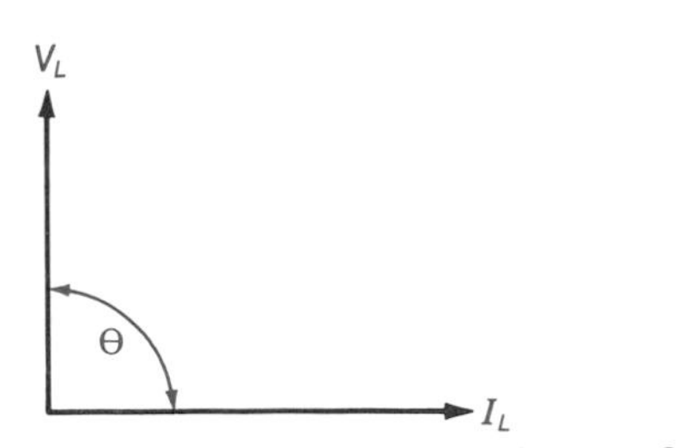


**FIGURE 14-14 Current and voltage phasors for an ideal inductor circuit.**

sistance. Since power equals converted energy divided by time, there can be no power when no energy is being converted. Second, an inductor stores energy while current is rising and returns energy while current is falling. Inspection of the current waveform in Fig. 14-13*b* shows that the time and the average slope of the rising and falling current are identical. Thus, no energy is lost (converted to heat). Third, and last, the power in any circuit can be calculated by the formula $P = IV\cos\theta$. The cosine of 90° (angle $\theta$ for an ideal inductor circuit) is zero. Thus, the power for the inductor circuit must be zero.

Why the formula $P = IV\cos\theta$ must be used when $I$ and $V$ are not in phase is illustrated in Fig. 14-15. In Fig. 14-15*a*, the current and voltage are in phase, and the product of $I$ times $V$ always yields a positive power because $I$ and $V$ are either both positive, negative, or zero at any instant during the complete ac cycle. Thus, the average of all the algebraic sums of the instantaneous powers is simply $I$ times $V$. However, when the phase angle is 45°, as in Fig. 14-15*b*, approximately 25 percent of the time $I$ and $V$ have opposite polarities, which yields a negative power during these times. Now the average of the algebraic sum of the instantaneous powers is considerably less than the product of $I$ and $V$. Specifically, the average of the algebraic sum is $IV\cos\theta$. The cosine of 45° is 0.707, so the power is 70.7 percent of the $IV$ product. This tells us that a combination inductor and resistor circuit in which $I$ and $V$ are 45° out of phase only returns 29.3 percent of the energy it takes from the ac power source each cycle. Figure 14-15*c* shows that when $I$ and $V$ are 90° out of phase, no power is used; the algebraic sum of the instantaneous powers is zero.

Figure 14-16 uses phasor diagrams to show why multiplying the $IV$ product by $\cos\theta$ is necessary to obtain power when $I$ and $V$ are not in phase. You learned in Chap. 12 that adding two right-angle phasors gives a resultant phasor with an angle somewhere between 0 and 90°. The process used to do this is reversible; i.e., a phasor at any angle between 0 and 90° can be split into two smaller phasors: one smaller phasor at 0° and the other smaller phasor at 90°. Thus, the voltage phasor in Fig. 14-16*a* is split into two phasors in Fig. 14-16*b*. The smaller phasor that is in phase with the current phasor represents the part of the total voltage that is resistive, while the smaller phasor that is 90° out of phase with the current phasor represents the inductive part of the total voltage. (The phasors in Fig. 14-16 could represent a practical inductor which has a large amount of resistance for its value of inductance.) Multiplying $V_T$ by $\cos\theta$ gives the magnitude of $V_R$, which when multiplied by the resistive current, yields the power. (In this case, $I_T$ and $I_R$ are the same current.)

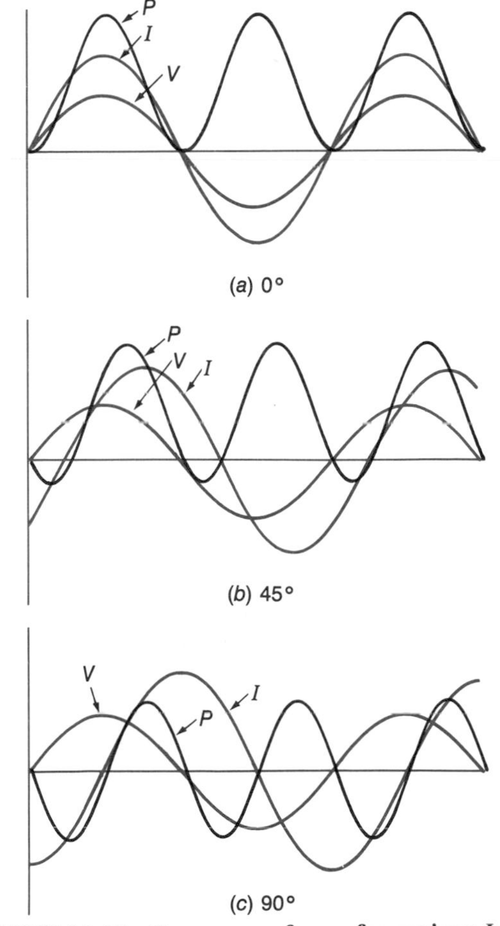


**FIGURE 14-15 Power waveforms for various $I$ and $V$ phase angles.**

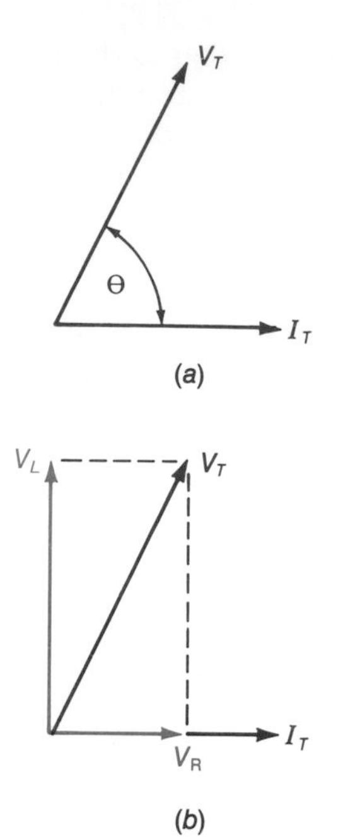


**FIGURE 14-16 Finding the in phase current and voltage.**

## Self-Test

**24.** The ac voltage and current for an ideal inductor circuit are 20 V and 1.8 A. What is the power for this circuit?

**25.** The current in an ideal inductor ______ the voltage by ______ electric degrees.

**26.** How can a circuit have current flow and voltage and yet have no power?

## 14-6 INDUCTIVE REACTANCE

Although an ideal inductor has no resistance, it does have an opposition to ac current. An inductor has a form of opposition to ac which is called *reactance.* Reactance is abbreviated with the letter symbol $X$. The major difference between resistance and reactance is that current flowing through a resistance always causes electric energy to be converted to heat energy, while current flowing through a reactance never causes this conversion. Since reactance is a form of opposition to current, the ohm is also its base unit. The reactance of an inductor is called *inductive reactance* and is symbolized by $X_L$ to distinguish it from the reactance of a capacitor, which is called *capacitive reactance* and abbreviated $X_C$. Since reactance is just another form of opposition, it can be substituted into Ohm's law in place of resistance when calculating currents and voltages in sinusoidal ac circuits. Thus, $X_L = V_L/I_L$ or, in a simple ideal inductor circuit like that in Fig. 14-17*a*, $X_L = V/I$. Of course, both $V$ and $I$ must be in the same type of units (root mean square, average, etc.).

The reactance of an inductor ($X_L$) is the result of the inductor producing a cemf which opposes changes in current. Since a sinusoidal ac voltage source produces a sinusoidal current in a circuit, the inductor's cemf is always opposing the ac current and the ac voltage source that produced it. Thus, reactance limits the circuit current to the amount needed to produce an instantaneous voltage (cemf) which is equal to, and counter to, the instantaneous source voltage. Since, as discussed in Secs. 14-2 and 14-3, the cemf of an inductor is determined by the amount of inductance and the slope of the current change in the inductor, inductive reactance is also determined by these two factors. For a given magnitude of current, the slope of the current change is determined by the frequency. Thus, the amount of inductive reactance is determined by the amount of inductance and the frequency of the current.

The formula for inductive reactance can be derived from the formula and relationships that we have already developed. The ones we will use are

$$T = \frac{1}{f}$$

$$X_L = \frac{V_L}{I_L}$$

$$V_L = L\left(\frac{\Delta I}{\Delta t}\right)$$

$$I_{av} = 0.637 I_p$$

where $T$ = period (time)
$f$ = frequency
$I_{av}$ = average current
$I_p$ = peak current

Figure 14-17 shows that during the first quarter of a cycle the current through the inductor changes from 0 to $I_p$. During the next quarter of a cycle, the current changes from $I_p$ back to 0. The last two quarter cycles are seen to be a repeat of the first two in terms of the magnitude of the $\Delta I$. Thus, during each quarter cycle the $\Delta I$ is equal to $I_p$.

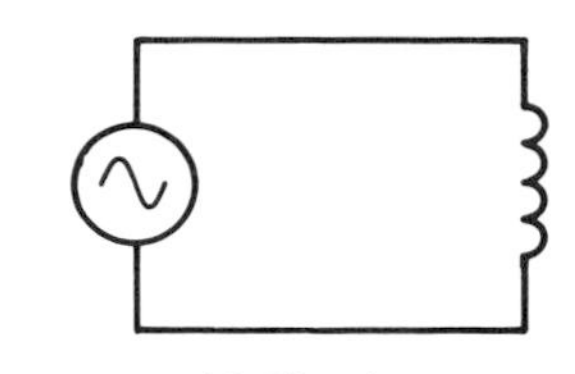

(a) Circuit

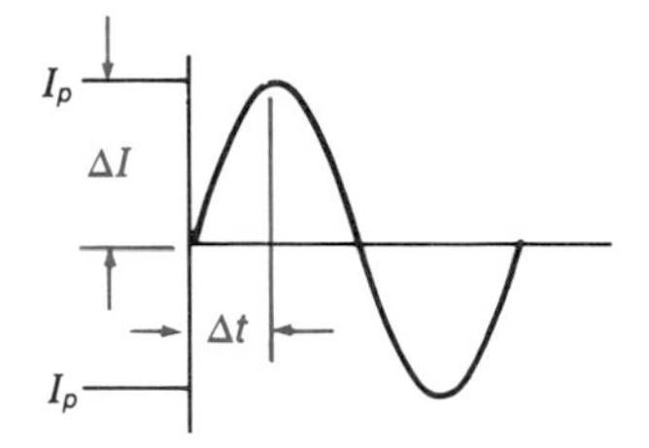


(b) First-quarter cycle

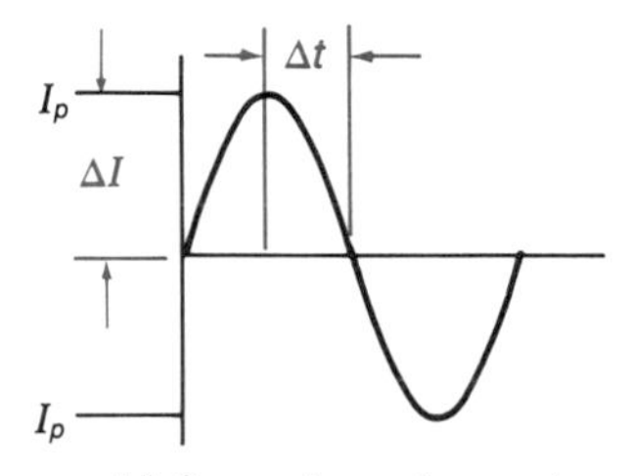


(c) Second-quarter cycle

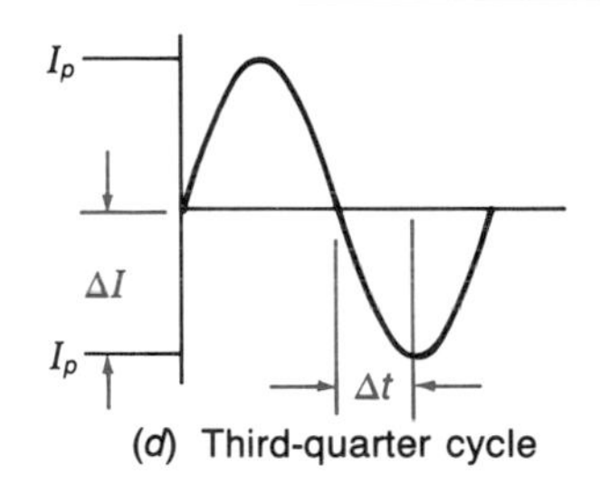


(d) Third-quarter cycle

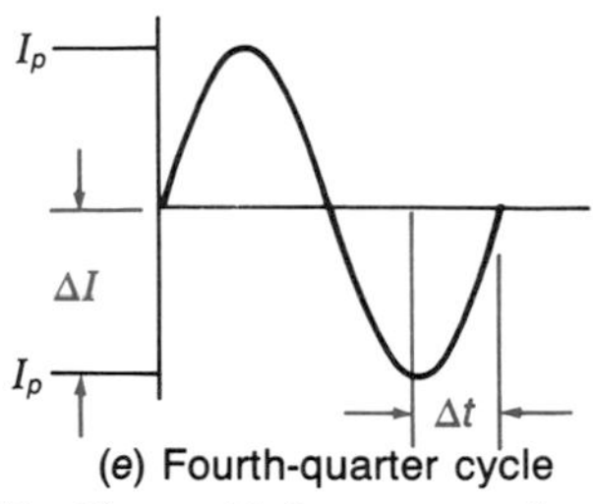


(e) Fourth-quarter cycle

**FIGURE 14-17 Sinusoidal current changes in an inductor circuit.**

Therefore, over a cycle the average $\Delta I = I_p$. Note that this is the average $\Delta I$ because the $\Delta I$ for a sinusoidal current wave is continually changing. Figure 14-17 also shows that the $\Delta t$ needed for the current to go through its change is 0.25 period ($T$). Since $T = 1/f$, the $\Delta t$ can be expressed as $0.25T = 1/4f$. Thus, for a sinusoidal current, we can write

$$\text{Average } \frac{\Delta I}{\Delta t} = \frac{I_p}{1/4f} = 4fI_p$$

Since $V_L = L\ \Delta I/\Delta t$, the average $V_L$ equals $L$ times the average $\Delta I/\Delta t = L4fI_p$. The peak current through the inductor in Fig. 14-17 is $I_p$. Because this is a sinusoidal current, the average inductor current is

$$I_{L,\text{av}} = 0.637I_p$$

Now, by substituting into Ohm's law, we can determine the formula for $X_L$. The steps are

$$X_L = \frac{V_L}{I_L} = \frac{L4fI_p}{0.637I_p} = 6.28fL$$

Notice that the values for both $V_L$ and $I_L$ in the above formula are average values. Also notice that $6.28 \approx 2\pi$, so the formula for $X_L$ is often written as either

$$X_L = 6.28fL \qquad \text{or} \qquad X_L = 2\pi fL$$

In this formula the inductive reactance $X_L$ is in ohms when the frequency $f$ is in hertz and the inductance $L$ is in henrys.

---

**EXAMPLE 14-8**

For the circuit in Figure 14-18*a*, determine $L$ and $X_L$.

**Solution**

$$X_L = \frac{V_L}{I_L} = \frac{10\text{ V}}{0.04\text{ A}} = 250\ \Omega$$

$$X_L = 6.28fL$$

$$L = \frac{X_L}{6.28f} = \frac{250\ \Omega}{6.28 \times 500\text{ Hz}} = 79.6\text{ mH}$$

---

**EXAMPLE 14-9**

For the circuit in Fig. 14-18*b*, determine $X_L$ and $I_L$.

**Solution**

$$X_L = 6.28fL = 6.28 \times 10 \times 10^3\text{ Hz} \times 5 \times 10^{-3}\text{ H} = 314\ \Omega$$

$$I_L = \frac{V_L}{X_L} = \frac{15\text{ V}}{314\ \Omega} = 47.7\text{ mA}$$

---

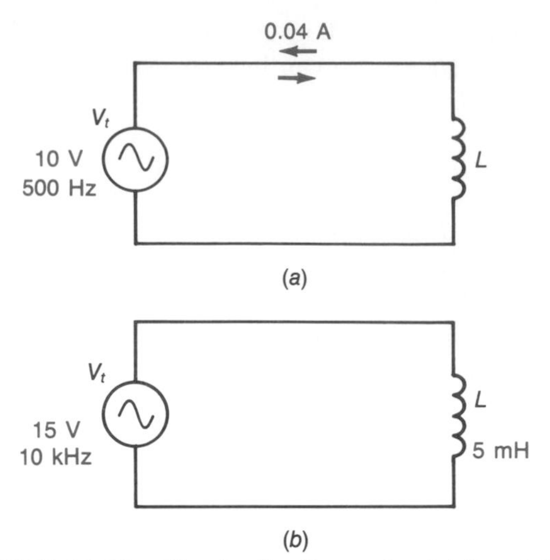


**FIGURE 14-18 Circuits for Examples 14-8 and 14-9.**

## Self-Test

**27.** What is the difference between resistance and reactance?

**28.** Is inductive reactance inversely or directly proportional to frequency? To inductance?

**29.** What is the inductive reactance of a 270-mH inductor at 4 kHz? How much current will flow if this inductor is connected to a source voltage of 25 V at 4 kHz?

**30.** If both the voltage and the frequency of the source were doubled, what would happen to the current in Question 29?

## 14-7 INDUCTIVE REACTANCE IN SERIES AND PARALLEL

Series, parallel, and series-parallel inductive circuits are analyzed using the same laws and rules as are used with similar resistive circuits *if* the inductors are ideal inductors or if the resistances of the inductors are small compared to the reactances. We will assume the latter is the case in this section. By changing $R$ to $X_L$, one can use the series and parallel resistances formula for series and parallel reactances. Thus, for parallel inductor circuits, the total reactance is

$$X_{L_T} = \frac{1}{1/X_{L_1} + 1/X_{L_2} + \cdots + 1/X_{L_n}}$$

or $$X_{L_T} = \frac{X_{L_1} X_{L_2}}{X_{L_1} + X_{L_2}}$$

For series inductors the formula is

$$X_{L_T} = X_{L_1} + X_{L_2} + \cdots + X_{L_n}$$

Of course, the total reactance can also be determined by Ohm's law or the general reactance formula: When $I_T$ and $V_T$ are known, $X_{L_T} = V_T/I_T$, and when $L_T$ is known, $X_{L_T} = 6.28fL_T$. In Chap. 13 you learned that $L_T = L_1 + L_2$ for series inductors and $L_T = (L_1 \times L_2)/(L_1 + L_2)$ for parallel inductors.

Kirchhoff's laws (voltage and current) are as applicable to inductor circuits as they are to resistor circuits. When changing from resistor circuits to inductor circuits, just change the $R$ subscripts to $L$ subscripts. Thus, for the voltage drops in a series inductor circuit, we can write

$$V_T = V_{L_1} + V_{L_2} + \cdots + V_{L_n}$$

and for a parallel inductor circuit, the current formula is

$$I_T = I_{L_1} + I_{L_2} + \cdots + I_{L_n}$$

The following three examples are illustrative of how inductor circuits can be analyzed. Notice in these examples that power is not determined because ideal inductors use no power.

**EXAMPLE 14-9**

For the circuit in Fig. 14-19*a*, determine $L_1$ and $X_{L_T}$.

**Solution**

$$X_{L_2} = 6.28 \times 250 \times 0.8 = 1256\ \Omega$$
$$V_{L_2} = V_T - V_{L_1} = 25\text{ V} - 10\text{ V} = 15\text{ V}$$
$$I_{L_1} = I_T = I_{L_2} = \frac{V_{L_2}}{X_{L_2}} = \frac{15\text{ V}}{1256\ \Omega} = 11.94\text{ mA}$$
$$X_{L_1} = \frac{V_{L_1}}{I_{L_1}} = \frac{10\text{ V}}{11.94\text{ mA}} = 837.5\ \Omega$$
$$X_{L_T} = X_{L_1} + X_{L_2} = 1256\ \Omega + 837.5\ \Omega = 2093.5\ \Omega$$
$$L_1 = \frac{X_{L_1}}{2\pi f} = \frac{837.5}{6.28 \times 250} = 0.53\text{ H}$$

Notice from this example that for series inductor circuits the ratio $L_1/L_2 = V_{L_1}/V_{L_2} = X_{L_1}/X_{L_2}$.

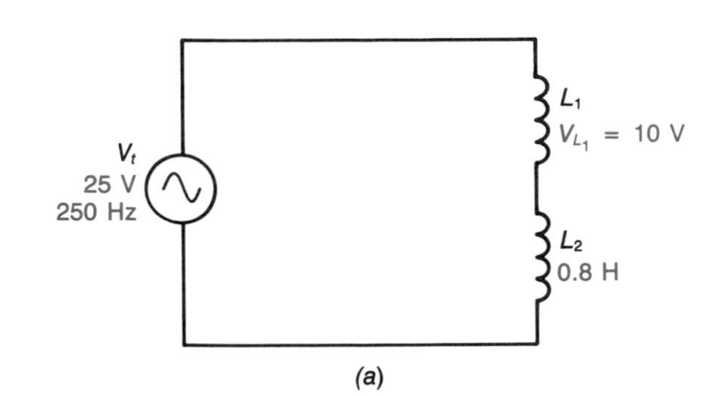


(a)

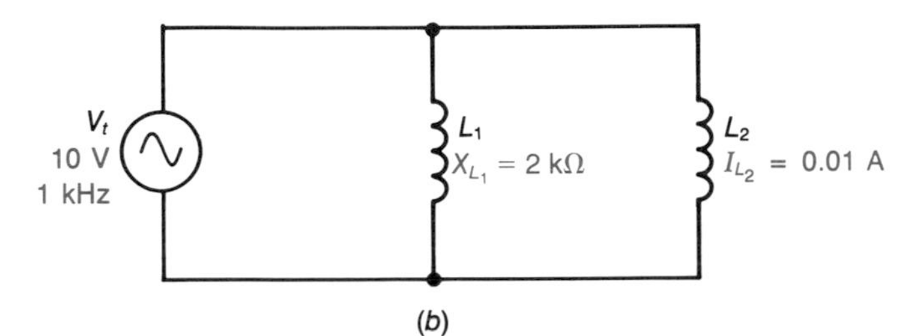


(b)

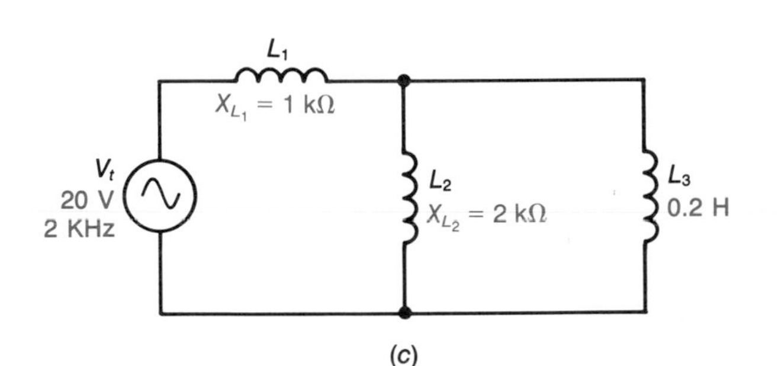


(c)

**FIGURE 14-19 Circuits for Examples 14-9, 14-10, and 14-11.**

**EXAMPLE 14-10**

For the circuit in Fig. 14-19*b*, determine $I_T$, $X_{L_T}$, and $L_T$.

**Solution**

$$I_{L_1} = \frac{10 \text{ V}}{2 \text{ k}\Omega} = 5 \text{ mA}$$

$$I_T = I_{L_1} + I_{L_2} = 5 \text{ mA} + 0.01 \text{ A} = 15 \text{ mA}$$

$$X_{L_T} = \frac{10 \text{ V}}{15 \text{ mA}} = 666.7 \ \Omega$$

$$L_T = \frac{X_{L_T}}{2\pi f} = \frac{666.7}{6.28 \times 1000} = 106 \text{ mH}$$

**EXAMPLE 14-11**

For the circuit in Fig. 14-19*c*, determine $X_{L_T}$, $V_{L_1}$, and $I_{L_3}$.

**Solution**

$$X_{L_3} = 6.28 \times 2000 \times 0.2 = 2512 \ \Omega$$

$$X_{L_2,L_3} = \frac{X_{L_2} X_{L_3}}{X_{L_2} + X_{L_3}} = \frac{2000 \times 2512}{2000 + 2512} = 1113.5 \ \Omega$$

$$X_{L_T} = X_{L_2,L_3} + X_{L_1} = 1113.5 + 1000 = 2113.5 \ \Omega$$

$$I_{L_1} = I_T = \frac{20 \text{ V}}{2113.5 \ \Omega} = 9.46 \text{ mA}$$

$$V_{L_1} = 9.46 \text{ mA} \times 1 \text{ k}\Omega = 9.46 \text{ V}$$

$$V_{L_3} = V_T - V_{L_1} = 20 \text{ V} - 9.46 \text{ V} = 10.54 \text{ V}$$

$$I_{L_3} = \frac{10.54 \text{ V}}{2512 \ \Omega} = 4.20 \text{ mA}$$

## Self-Test

**31.** In Fig. 14-19*a*, what would happen to $V_{L_1}$ if the source frequency were doubled?

**32.** A 0.4-H inductor and a 0.3-H inductor are series-connected to a 45-V, 200-Hz source. Determine $X_{L_T}$ and $I_T$.

**33.** If the same inductors are parallel-connected to the same source, what will be the value of $X_{L_T}$ and $I_T$?

**34.** Change $X_{L_1}$ to 2 kΩ and then determine $I_{L_2}$ in the circuit in Fig. 14-19*c*.

## 14-8 QUALITY FACTOR

Inductors are used in circuits to provide reactance rather than resistance. Unfortunately, it is not possible to make an inductor that does not have some resistance. The *quality factor* (usually just called *quality*) of an inductor refers to the ratio of the inductor's reactance to its resistance. Since $R$ converts energy to heat and $X_L$ stores energy, quality can also be defined as the ratio of the energy stored by the inductor to that of the energy converted by the inductor. Because reactance varies with frequency, the quality of an inductor must also be frequency-dependent. The letter symbol used for the *quality* of an inductor is $Q$. Although this is the same symbol that is used to represent charge, there is little chance of confusing the two uses of $Q$ because charge and quality are used in entirely different contexts. The formula used to calculate quality is

$$Q = \frac{X_L}{R}$$

Notice that quality is a pure number (no units) because the ohm unit of $R$ cancels the ohm unit of $X_L$. Since the desirable property of an inductor is reactance and the undesirable property is resistance, $Q$ can be thought of as a figure of merit for inductors.

The $R$ in the quality formula represents the *effective* or *equivalent resistance* of the inductor. The effective resistance of an inductor is sometimes referred to as the *effective series resistance* or just the series resistance, because it causes a practical inductor to act like an ideal inductor in series with some resistance. The effective resistance of an inductor is not just the ohmic or dc resistance (resistance measured with an ohmmeter) of the inductor's winding. The total equivalent resistance of an inductor may include, in addition to the ohmic resistance of the winding, some resistance due to one or more of the following: (1) skin effect, (2) eddy-current loss, and (3) hysteresis loss.

## Skin Effect

Skin effect, illustrated in Fig. 14-20, refers to the fact that electrons increasingly travel on the surface of a conductor as the frequency increases. Skin effect reduces the effective cross-sectional area of a conductor, which, in turn, increases the effective resistance of the conductor. At very high frequencies, skin effect can cause the effective winding resistance of an inductor to be considerably greater than the ohmic resistance of the winding. Since air-core inductors can be operated at the highest frequencies, skin effect is most important

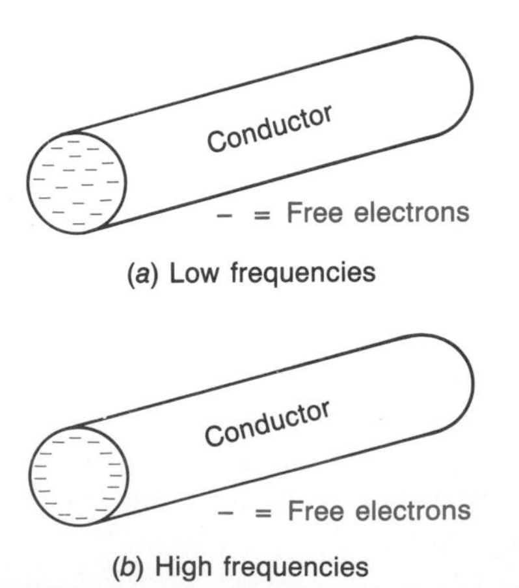


**FIGURE 14-20 Current carrier distribution in a conductor.**

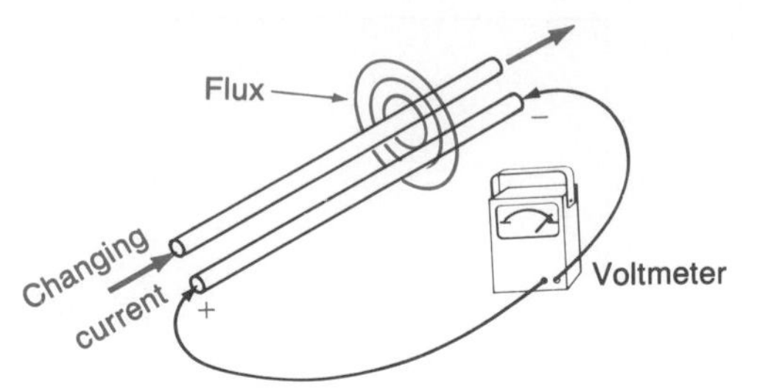


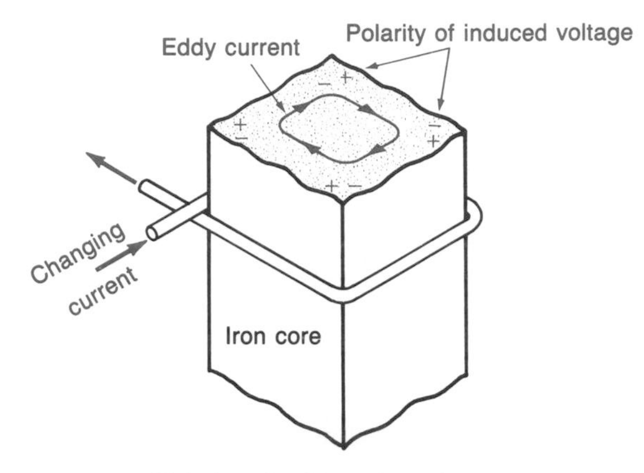


**FIGURE 14-21 Eddy current in an iron core.**

for this type of inductor. At the low frequencies where iron-core inductors are used, skin effect is insignificant.

## Eddy Current

Figure 14-21*a* illustrates a principle that has been used in Chaps. 10 and 13. This figure shows that a current-carrying conductor produces a magnetic flux which induces a voltage into any parallel conductor when the magnitude of the current, and thus flux, changes. (This is the basic phenomenon which accounts for both self- and mutual inductance.) Not only does a conductor with a changing current flow induce a voltage into parallel conductors, but it also induces a voltage into any parallel *conductive material*. Since the iron core of an inductor is a conductive material and each of its four surfaces is parallel to the current-carrying winding, it will have voltages induced into it. As shown in Fig. 14-21*b*, these voltages on the parallel surfaces are series-aiding and cause a current, called the *eddy current*, to flow in the core material. The eddy current in the core produces heat in the core material. This heat is undesirable, so it is referred to as an *eddy-current loss.* Ultimately, this heat energy had to come from the energy and voltage source to which the inductor is connected. Since only resistance can convert electric energy to heat energy, the eddy-current loss is part of the effective resistance of the inductor. It is part of the $R$ in the formula $Q = X_L/R$.

The amount of eddy-current loss is controlled by the amount of induced voltage and the amount of eddy current ($P = IV$). The induced voltage cannot be changed by modifying the core material. It is controlled by the number of turns in the winding and the frequency and magnitude of the winding current. Thus, the way to minimize the eddy-current loss is to reduce the eddy current by making the core resistance as high as practical.

Notice in Fig. 14-21*b* that the eddy current flows through the cross section of the core. Therefore, the core resistance can be greatly increased by constructing the core as shown in Fig. 14-22. Each silicon steel laminate in Fig. 14-22 is oxidized before the core is assembled. The oxide of silicon steel has a high resistance, so very little eddy current flows from laminate to laminate. Also, because each laminate is only a few thousandths of an inch thick, the resistance to current flowing the width of a laminate is also quite large. Thus, laminating a core increases the core resistance, which decreases the eddy current, which reduces the power, which reduces the eddy-current loss, which (finally) reduces the effective resistance of the inductor.

The thinner the laminations are made, the larger the core resistance will be. However, thinner laminations increase the percentage of oxide in the core material, and the oxide of silicon steel has a

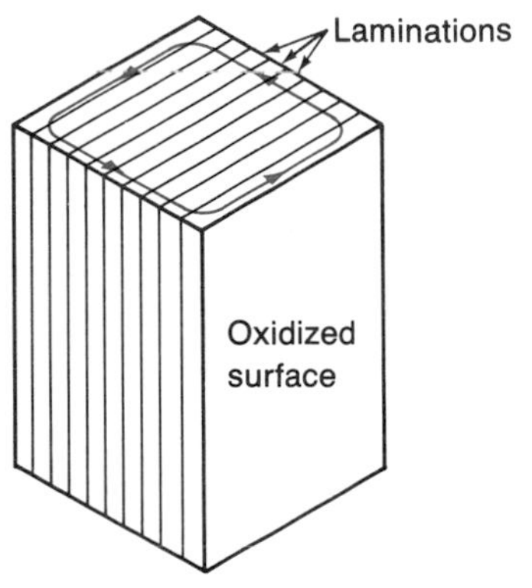


**FIGURE 14-22 Laminated core reduces eddy current.**

low permeability compared to silicon steel. This means that, for a given cross-sectional area, the reluctance of the core increases as the laminations are made thinner. These are some of the factors taken into account when designing an iron-core inductor.

Since the voltage induced in the core of the inductor is partially dependent on the frequency of the current, the eddy-current loss is also frequency-dependent. As the frequency increases, so does the eddy-current loss.

## Hysteresis Loss

*Hysteresis* and *hysteresis loss* were discussed in Chap. 10, "Magnetism." Briefly, hysteresis loss refers to the heat produced in a magnetic material each time the direction of the magnetic flux is reversed.

When an iron-core inductor is connected to an ac source, the hysteresis loss in the core requires energy from the source. Since hysteresis loss occurs twice each cycle, the amount of loss increases as the frequency increases. This is another reason why iron-core inductors are only used at power and audio frequencies.

Hysteresis loss can only be reduced by changing the core material to a material that has a narrower hysteresis loop. Silicon steel is used for the core material in magnetic devices such as inductors, motors, relays, and transformers because it has both a narrow hysteresis loop and high permeability.

*Core loss* is a term used to refer to the energy (heat) loss in the core of magnetic devices. Core loss includes both eddy-current loss and hysteresis loss.

*Copper loss* is a term used to refer to the heat loss in the windings of magnetic devices. Copper loss is minimized by using the largest practical size of conductor. Of course, increasing wire size adds to the cost, bulk, and weight of the magnetic device.

Now we know that both the effective resistance of an inductor and the reactance of an inductor are frequency-dependent and that both increase as the frequency is increased. Thus, as was mentioned previously, $Q$ is frequency-dependent. For each inductor there is a limited range of frequencies where $Q$ is optimum and fairly stable. In this frequency range, the change in $X_L$ is approximately directly proportional to the change in $R$. Below this frequency range, the ohmic resistance dominates

$R$, so $R$ decreases very little as $f$ is further decreased, but $X_L$ continues to decrease in direct proportion to the frequency change. Thus, $Q$ decreases when the frequency is too low. Above the optimum frequency range, the core losses and the skin effect dominate $R$, and $R$ can increase proportionately more than $X_L$. This results in a decrease in $Q$ if the frequency is too high. For these reasons, the $Q$ of an inductor is usually specified for a given frequency or range of frequencies.

**EXAMPLE 14-12**

The effective series resistance of a 5-mH inductor is 15 Ω at 25 kHz. What is the $Q$ of this inductor at 25 kHz?

**Solution**

$$X_L = 2\pi fL = 6.28 \times 25 \times 10^3 \times 5 \times 10^{-3} = 785\ \Omega$$

$$Q = \frac{X_L}{R} = \frac{785}{15} = 52.3$$

## Self-Test

**35.** Quality is a ratio of ______ to ______.

**36.** What is the unit for quality?

**37.** List the factors that determine the effective resistance of an inductor.

**38.** What is eddy current?

**39.** Define copper loss and core loss.

**40.** What is the $Q$ of a 2-H inductor at 120 Hz if its effective resistance is 100 Ω?

## 14-9 MEASURING INDUCTANCE

A number of techniques can be used to determine the inductance of an unknown inductor. Most of them involve measuring a current or a voltage (or the lack of either) in a circuit where other variables, such as frequency, source voltage, or rate of current rise, are carefully controlled. Let us look at several simple ways to measure inductance.

One of the simplest ways to measure inductance is illustrated in Fig. 14-23. If the voltage and frequency of the source are held constant, the current through the ammeter will be inversely proportional to the inductance because $I = V/X_L = V/2\pi fL$. Thus, if the frequency and the voltage of the source are known, the scale on the meter can be calibrated in units of inductance rather than units of current. Since Fig. 14-23 shows an analog meter, this type of circuit could be called an *analog inductance meter.*

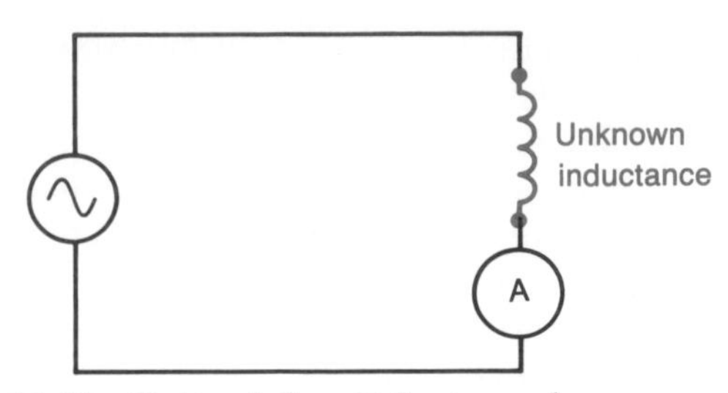


**FIGURE 14-23 Determining inductance by measuring current when $V$ and $f$ are known.**

Figure 14-24 shows a slightly different scheme for measuring inductance. In this scheme a small resistor is used in series with the inductor. If the resistor is very small compared to the reactance of the inductor, the circuit current will still be essentially controlled by the inductance. The voltage across the resistor will, therefore, be inversely proportional to the inductance. In this circuit a sensitive voltmeter, either analog or digital, is connected across the resistor. The voltmeter scale is, again, calibrated in units of inductance. The advantage of this circuit over the previous circuit is that it is easier to measure small values of voltage than small values of current.

By now you should be familiar with the formula $L = V_L/(\Delta I/\Delta t)$. Application of this formula to the circuit in Fig. 14-25 provides another way to mea-

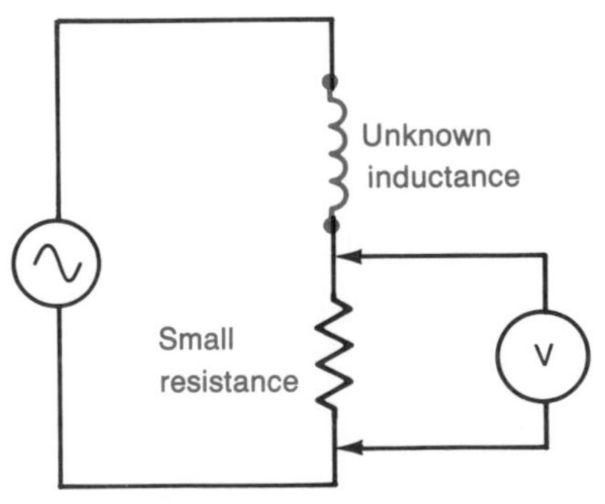


**FIGURE 14-24 The voltage drop across $R$ is directly proportional to $I$ and inversely proportional to $L$.**

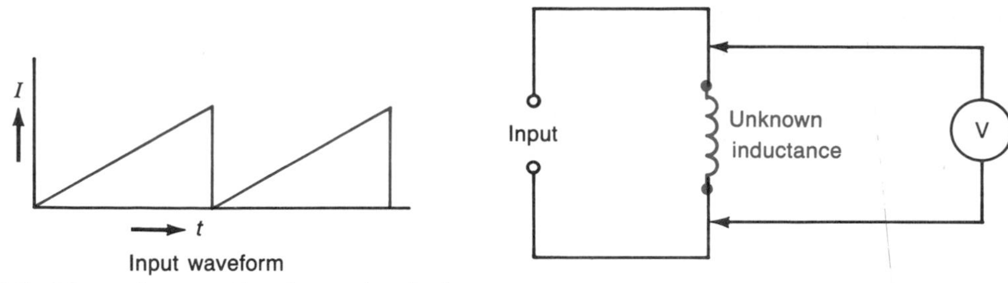


**FIGURE 14-25 Measuring cemf to determine inductance.**

sure inductance. An electronic current source, with a waveform like that in Fig. 14-25, is applied to an unknown inductor and the cemf ($V_L$) produced by the inductor is measured while the current is rising linearly. When the $\Delta I/\Delta t$ of the wave is held constant, $V_L$ is directly proportional to $L$. Thus, the voltmeter, when calibrated in units of inductance, will also have a linear scale. This linear relationship between $V$ and $L$ is one of the main advantages of this measuring technique. This type of arrangement is convenient for a digital readout of the inductance. (The basic ideas of digital readout of electrical quantities are presented in Chap. 16. More detailed information on digital displays is given in Chap. 21.)

Another way of measuring inductance is often referred to as the *comparison technique*. This technique uses a bridge circuit (see Fig. 14-26) to compare the reactance of the unknown inductance ($L_u$) to that of a known, or standard, inductance ($L_s$). (Balanced bridge circuits are presented in Chap. 8.) Inductance $L_s$ is a precision (standard) inductor. When $R_2$ in Fig. 14-26 is adjusted so that the ratio of $R_1/X_{L_s}$ is equal to the ratio $R_2/X_{L_u}$, the bridge is balanced. When the bridge is balanced, the voltage between points A and B is zero. Of course, the current flow between A and B is also zero. For the balanced bridge, we can write the formula

$$\frac{R_1}{X_{L_s}} = \frac{R_2}{X_{L_u}}$$

Solving for $X_{L_u}$ yields

$$X_{L_u} = \frac{R_2\, X_{L_s}}{R_1}$$

Since $X_L = 2\pi fL$ and since $f$ is the same for both $X_{L_u}$ and $X_{L_s}$, the above formula reduces to

$$L_u = \frac{R_2\, L_s}{R_1}$$

A dial, calibrated in units of inductance, is connected to the shaft of $R_2$ in Fig. 14-26. To use this bridge, $R_2$ is adjusted so that the lowest possible reading is obtained on the current meter or voltmeter. Then the inductance is read from the dial attached to $R_2$. One advantage of this circuit is that the source frequency and voltage can vary and not affect the accuracy of the bridge. This is because a change in $f$ has, proportionately, the same effect on both inductors so the two ratios of $R/X$ are not changed.

**EXAMPLE 14-13**

The meter in Fig. 14-26 indicates zero when $R_1 = 200\ \Omega$, $R_2 = 500\ \Omega$, and $L_s$ is a 500-mH inductor. What is the value of $L_u$?

**Solution**

$$L_u = \frac{R_2 L_s}{R_1} = \frac{500 \times 0.5}{200} = 1.25\ \text{H}$$

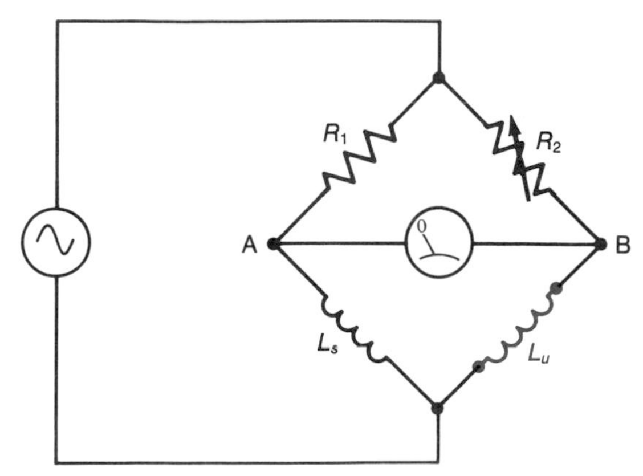


**FIGURE 14-26 Inductance bridge.**

When the $Q$ of the unknown inductor in the bridge of Fig. 14-26 is not equal to the $Q$ of the standard inductor, the bridge will not completely *null* (balance so that the meter indicates zero). Many inductance bridges use very high $Q$ standard inductors connected in series with a small rheostat. This rheostat can be adjusted so that the $Q$'s of the two inductors are equal; then complete balance can be obtained. If the frequency of the source is stable, a dial to indicate the $Q$ of the unknown inductor can be connected to the shaft of the rheostat. Now, when the bridge is nulled, both $Q$ and $L$ can be determined for the unknown inductor.

### Self-Test

**41.** Describe the scale needed to indicate inductance on the meter in Fig. 14-23.

**42.** Would the inductance scale on the meter in Fig. 14-25 be linear or nonlinear? Why?

**43.** What happens to the balance of the bridge in Fig. 14-26 if $f$ increases 20 percent while $V_T$ decreases 30 percent? Why?

## SUMMARY

**1.** An inductor stores energy in its magnetic field while the current is rising and returns energy while the current is falling.

**2.** For a given inductor, the amount of stored energy is determined by the maximum current through the inductor.

**3.** The inductor's cemf stops the current from instantly rising to its maximum value.

**4.** The final value of the current in a dc inductor circuit is controlled by $V_T$ and the inductor's resistance.

**5.** In a dc $RL$ circuit, the sum of the instantaneous values of $V_R$ and $V_L$ equals $V_T$.

**6.** A time constant $\tau$ is the time it would require for the current in a dc $RL$ circuit to increase from zero to $I_{max}$ if the current rise were linear and at the initial slope. It is also the time required for the current to change by 63.2 percent of the largest possible current change.

**8.** An inductor essentially charges or discharges in $5\tau$.

**9.** Time constant $\tau$ is in seconds when $L$ is in henrys and $R$ is in ohms.

**10.** The polarity of the cemf reverses when the current changes from a rising current to a falling current.

**11.** For a falling current, the cemf series-aids the source voltage.

**12.** In a dc inductive circuit, the cemf and the voltage across an opening switch can be hundreds of times greater than the dc source voltage.

**13.** Inductive kick or an inductive voltage spike can erode switch contacts and destroy voltage-sensitive components.

**14.** Inductive kick can be restricted with a varistor or any other device that provides a low-resistance path for the inductor's discharge current.

**15.** In an ideal inductor, $V$ leads $I$ by 90°.

**16.** Reactance does not convert electric energy to heat energy.

**17.** In an ideal inductor, flux changes and current changes are in phase with each other.

**18.** The angle between the current and voltage in a circuit is called the phase angle or angle theta ($\theta$).

**19.** Inductance and reactance use no power.

**20.** Only resistance uses power.

**21.** Inductive reactance is directly proportional to both frequency and inductance.

**22.** Reactance is the result of the inductor's cemf.

**23.** Ohm's law and Kirchhoff's laws can be used for inductor circuits just like they are used for resistor circuits. Just change $R$ to $X_L$ and change the subscript $R$ to the subscript $L$.

**24.** Quality $Q$ is a figure of merit for inductors. It has no units. For a given inductance, a higher $Q$ means a smaller amount of energy will be converted to heat. Quality $Q$ is frequency-dependent.

**25.** The effective resistance of an inductor is determined by ohmic resistance, skin effect, hysteresis loss, and eddy-current loss.

**26.** Eddy current flows in the iron core of an inductor. It is caused by induced voltage. It is minimized by laminating the core material.

**27.** The concentration of current carriers on the outer surface of a conductor at higher frequencies is called the skin effect. Skin effect reduces the effective cross-sectional area of a conductor.

**28.** Hysteresis loss refers to the heat produced in a magnetic material when the magnetic flux reverses.

**29.** Copper loss is the term used to refer to the energy loss in the windings of a magnetic device.

**30.** Core loss is the combined loss due to hysteresis and eddy currents.

**31.** The inductance of an inductor can be determined by (1) measuring its current when $f$ and $V_T$ are known, (2) measuring its cemf when the slope of the current change is known, and (3) comparing its reactance to the reactance of a known amount of inductance.

**32.** The voltage source for an inductance bridge can vary without disturbing the bridge balance.

**33.** The scale for an inductance meter which measures the cemf is linear.

**34.** An inductance bridge, with a stable source $f$, can determine the $Q$ of the unknown inductor.

**35.** Useful formulas are

$$W = 0.5\, I^2 L$$

$$V_L = \left(\frac{\Delta I}{\Delta t}\right) L$$

$$\text{Initial slope} = \frac{V_T}{L}$$

$$I_{\text{max}} = \frac{V_T}{R}$$

$$\tau = \frac{L}{R}$$

$$I_i = \frac{V_T}{R}(1 - e^{-t/\tau}) \qquad \text{for rising currents}$$

$$t = \tau \ln \frac{V_T}{V_T - I_i R} \qquad \text{for rising currents}$$

$$I_i = I_{max} e^{-t/\tau} \qquad \text{for falling currents}$$

$$t = \tau \ln \frac{I_{max}}{I_i} \qquad \text{for falling currents}$$

$$P = IV \cos \theta$$

$$X_L = 2\pi f L = 6.28 f L$$

$$X_{L_T} = \frac{1}{1/X_{L_1} + 1/X_{L_2} + \cdots + 1/X_{L_n}} \qquad \text{for parallel inductors}$$

$$X_{L_T} = X_{L_1} + X_{L_2} + \cdots + X_{L_n} \qquad \text{for series inductors}$$

$$Q = \frac{X_L}{R}$$

$$L_u = \frac{R_2 L_s}{R_1} \qquad \text{for an inductance bridge}$$

## CHAPTER REVIEW QUESTIONS

For the following items, determine whether each statement is true or false.

**14-1.** Energy is stored in an inductor's winding resistance.

**14-2.** After $3\tau$ an inductor is considered to be fully charged.

**14-3.** The cemf reverses polarity when the current in an inductor changes from rising to falling.

**14-4.** An inductor stores energy when the current is rising.

**14-5.** When an inductive circuit is turned off, the cemf will be greatest when the discharge resistance is as small as possible.

**14-6.** Inductive kick occurs when an inductive circuit is turned on.

**14-7.** The magnitude of an inductive voltage spike can be limited by paralleling the inductor with a varistor.

**14-8.** The current leads the voltage by 90° in an ideal inductor.

**14-9.** An ideal inductor uses no power.

**14-10.** Inductive reactance is inversely proportional to frequency.

**14-11.** The base unit for quality is the ohm.

**14-12.** A high-quality inductor has a higher energy-stored-to-energy-converted ratio than does a low-quality inductor.

**14-13.** The iron core of an inductor is laminated to reduce hysteresis loss.

**14-14.** The skin effect contributes to the core loss in a magnetic device.

**14-15.** An inductance meter which measures the inductor current usually has a linear scale.

**14-16.** An inductance bridge can be used to measure the $Q$ of an inductor.

For the following items, fill in the blank with the word or phrase required to correctly complete each statement.

**14-17.** The current rise in an inductive circuit is exponential because of the inductor's ______.

**14-18.** In $1\tau$ the current in an inductor circuit will increase from zero to ______ percent of $I_{max}$.

**14-19.** When $L$ is in millihenrys, $\tau$ will be in ______ when $R$ is in ohms.

**14-20.** A falling current produces a cemf which is ______ the source voltage.

**14-21.** An inductive voltage spike occurs when an inductive circuit is ______.

**14-22.** The angle between the source voltage and the source current is called ______.

**14-23.** The heat in the core of an inductor is produced by ______ and ______.

**14-24.** When the $\Delta I/\Delta t$ is constant, the cemf is directly proportional to ______.

## CHAPTER REVIEW PROBLEMS

**14-1.** What will be the instantaneous $V_R$ when the instantaneous $V_L = 8$ V and $V_T = 20$ V?

**14-2.** A 4-H inductor is conducting 3 A of current. How much energy is it storing?

**14-3.** How much current must a 500-mH inductor conduct to store 0.06 J of energy?

**14-4.** What is the cemf of a 0.7-H inductor when the current is changing at a rate of 0.2 A/s?

**14-5.** A dc $RL$ circuit with $L = 0.36$ H and $R = 40\ \Omega$ is connected through a switch to a 20-V source. Determine the following:

- **a.** The steady-state current
- **b.** The instantaneous current 5 ms after the switch is closed
- **c.** The initial slope of the current rise

**14-6.** A 250-mH inductor is connected to a 10-V, 2-kHz source. Determine $X_L$, $I$, and $P$.

**14-7.** If the inductor in Problem 14-6 has an effective resistance of 35 Ω, what is its $Q$?

**14-8.** For the circuit in Fig. 14-27*a*, determine $X_{L_T}$ and $L_1$.

**14-9.** For the circuit in Fig. 14-27*b*, determine $X_{L_T}$ and $V_{L_3}$.

**14-10.** For the circuit in Fig. 14-27*c*, determine the value of the unknown inductance if the given values balance the bridge.

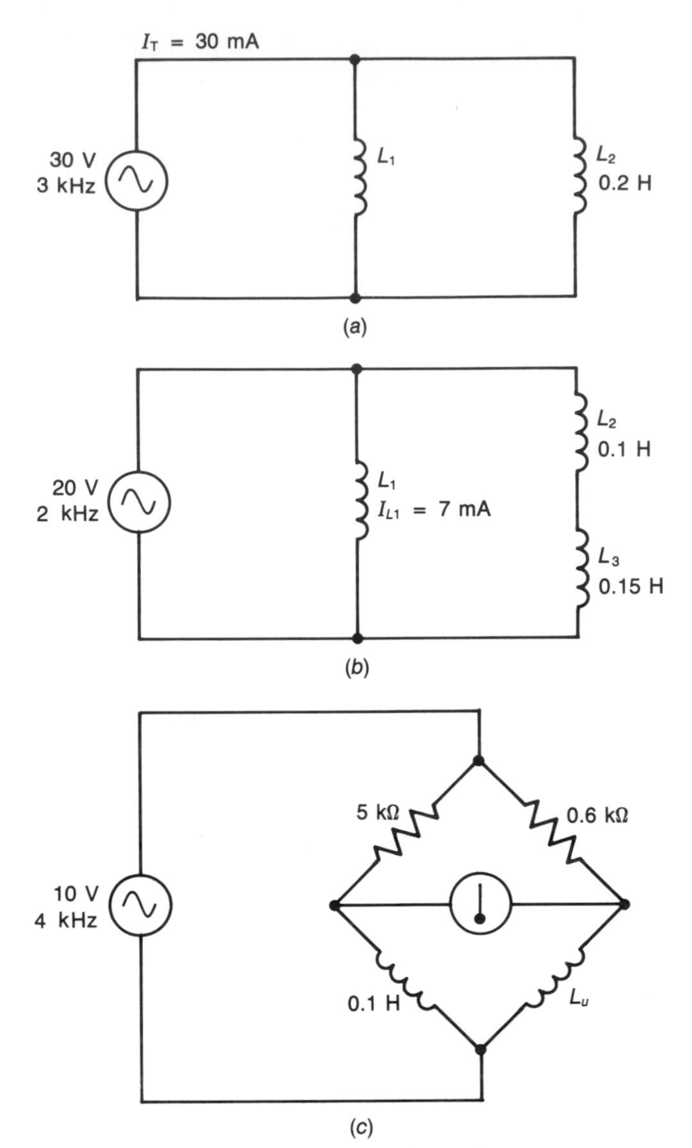


**FIGURE 14-27 Circuits for Review Problems 14-8, 14-9, and 14-10.**

## ANSWERS TO SELF-TESTS

**1.** in the magnetic field of the inductor

**2.** It is returned to the source or dissipated in the resistances of the circuit.

**3.** 18 mJ

**4.** 632 mA

**5.** the inductor's cemf

**6.** 0.5 A

**7.** 8.33 A/s

**8.** 20 percent

**9.** one time constant

**10.** 5

**11.** 23.3 ms, 116.7 ms

**12.** $I_i$ after 30 ms = 475 mA, 16 V

**13.** 208 μs

**14.** charging

**15.** series-aiding

**16.** 0.2 s

**17.** false

**18.** true

**19.** no, yes

**20.** the high cemf produced by a discharging inductor

**21.** by providing a low-resistance discharge path

**22.** $V_L = 34.9$ V and $I_i = 183.6$ mA, peak cemf = 71.25 V

**23.** 12.3 ms

**24.** zero

**25.** lags, 90

**26.** The $I$ and $V$ are 90° out of phase and energy is just being transferred back and forth between the source and the load.

**27.** Resistance converts electric energy to heat energy; reactance does not.

**28.** directly, directly

**29.** $X_L = 6782.4\ \Omega$, 3.69 mA

**30.** no change

**31.** There would be no change in $V_{L_1}$.

**32.** $X_{L_T} = 879.2\ \Omega$ and $I_T = 51.2$ mA

**33.** $X_{L_T} = 215.3\ \Omega$ and $I_T = 209$ mA

**34.** $I_{L_2} = 3.58$ mA

**35.** reactance, resistance

**36.** Quality has no units.

**37.** ohmic resistance, skin effect, eddy-current loss, and hysteresis loss

**38.** the current that flows in the core of an inductor or other magnetic device

**39.** Copper loss is the power dissipated in the windings of a magnetic device, and core loss (eddy-current loss and hysteresis loss) is the power dissipated in the core of a magnetic device.

**40.** 15

**41.** The scale would be reverse-reading (more current represents less inductance) and nonlinear (crowded at the end representing the larger inductances).

**42.** Linear, because, for a linear current rise, $V_L$ is directly proportional to $L$.

**43.** Nothing, because the $R/X$ ratios are unchanged by changes in $f$ or $V_T$?

**44.** 500 mH

# Chapter 15 Capacitance

Capacitance is the electrical property which stores electric energy in the form of an electric field. Capacitance opposes sudden changes in circuit voltages by either releasing part of the stored energy or storing additional energy.

Physical devices which are constructed to provide specified amounts of capacitance are called *capacitors.* At one time capacitors were also called condensers. This term (condenser) is still used in the automotive industry.

To understand and appreciate capacitance and capacitors, one must first deal with electrostatics. *Electrostatics* refers to the characteristics and behavior of the electric fields created by electric charges. The topic of electrostatics was introduced in Chap. 1 where Coulomb's law was used to investigate the force between two charges. The ideas developed there are used in this chapter to help explain the nature and behavior of capacitance.

## 15-1 ELECTRIC FIELDS

An electrostatic field, usually just called an *electric field,* exists around an electric charge and between two electric charges. This electric field, as explained in Chap. 1, causes the forces of attraction and repulsion between unlike and like charges, respectively. As explained in Chap. 1, the formula for the force between charges is

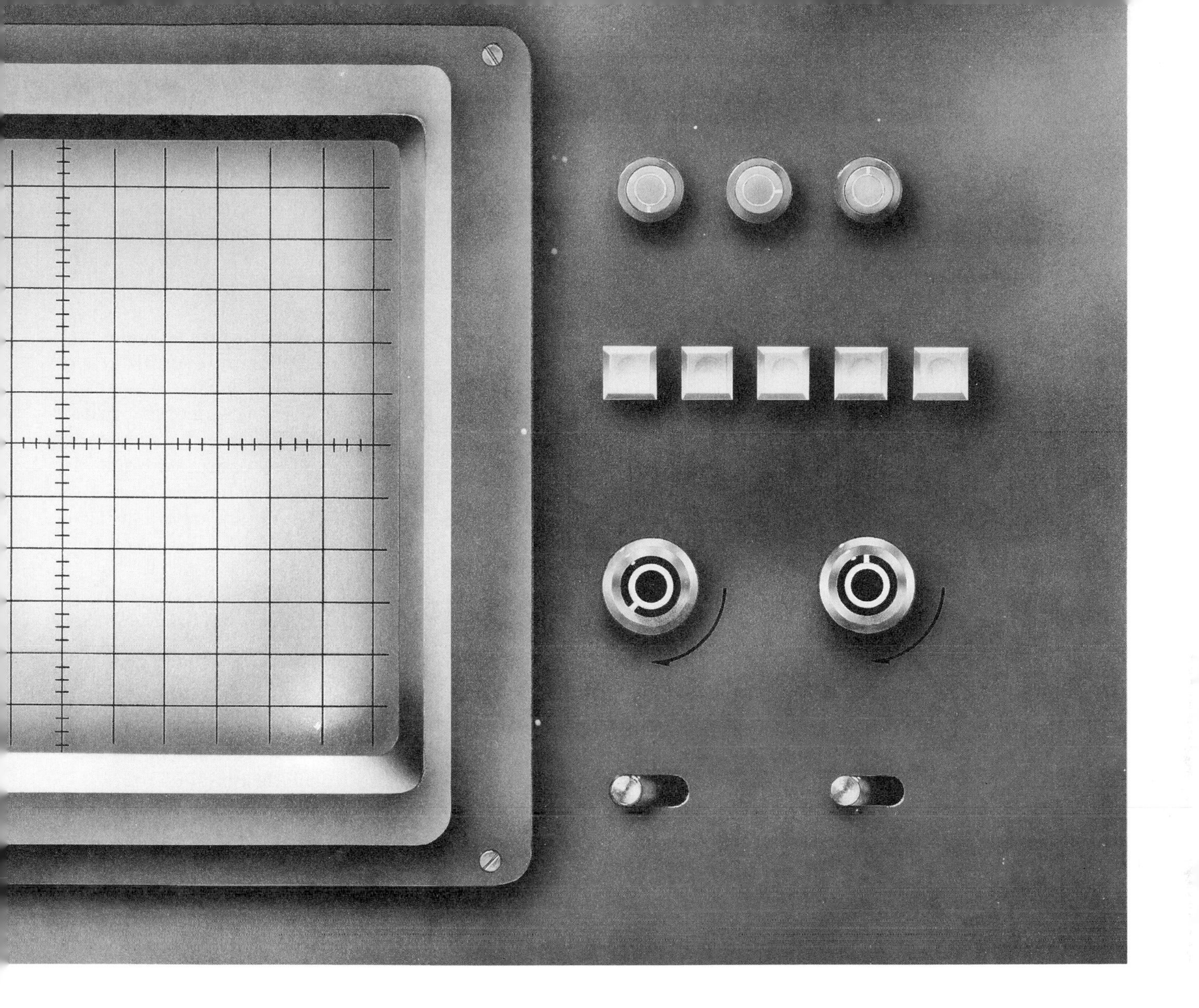

$$F = K\frac{Q_1Q_2}{D^2}$$

where $F$ = force
$D$ = distance between charges
$Q_1$, $Q_2$ = charges
$K$ = constant

The electric field around an isolated electric charge is illustrated in Fig. 15-1. The electric field is represented by the lines which protrude perpendicularly from the surface of the charged sphere. Of course, these lines do not actually exist. They merely represent the invisible force field which we call an electric field. Although these *lines of force* are invisible and/or imaginary, the concept of an electric line of force is extremely useful in quantifying electric fields and the forces these fields exert on other electric charges.

The direction of a line of force is indicated by an arrowhead on the line. The arrowhead indicates in which direction a negative free (mobile) electric charge would move if placed in the electric field. The direction of the lines of force in Fig. 15-1*b* shows that a movable negative charge will be pulled into the electric field associated with a stationary positive charge.

*Electric field intensity* is a term used to define the force of an electric field on an electric charge. Electric field intensity can be expressed in *newtons per coulomb* (N/C) since the newton and coulomb are the base units of force and charge. A little substitu-

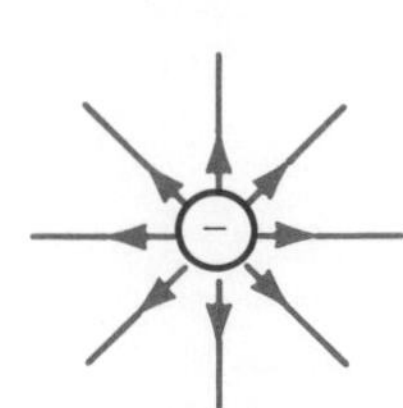

(a) Electric field around an isolated negative charge

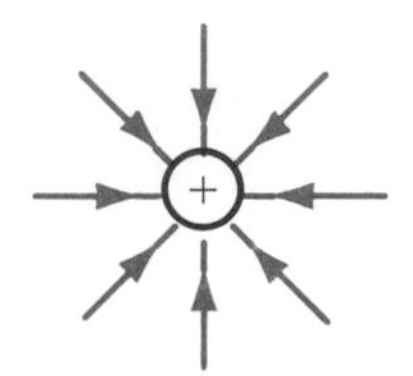

(b) Electric field around an isolated positive charge

**FIGURE 15-1 Arrowheads indicate the direction a negative charge will move when placed in the electric field.**

tion from other base unit relationships will show that newtons per coulomb is equal to *volts per meter* (V/m):

$$\text{Joule (J)} = \text{newton-meter (N} \cdot \text{m)}$$

therefore

$$\text{N} = \frac{\text{J}}{\text{m}}$$

$$\text{V} = \frac{\text{J}}{\text{C}} \qquad \text{(definition of a volt)}$$

So, by substitution,

$$\frac{\text{N}}{\text{C}} = \frac{\text{J/m}}{\text{C}} = \frac{\text{J}}{\text{C} \cdot \text{m}} = \frac{\text{V}}{\text{m}}$$

Thus, electric field intensity can be expressed in volts per meter as well as newtons per coulomb.

In Chap. 1 it was shown that the force between two charges was directly proportional to the size of the charges and inversely proportional to the square of the distance between the charges. Remembering this, you can see that the electric field intensity around an isolated charge, like that in Fig. 15-1*a* or *b*, will decrease as the distance from the charge increases. In fact, it will follow the inverse square law.

The electric lines of force and the electric field intensity created by two unlike charges are quite different than those created by an isolated charge. When the unlike charges are distributed on two parallel conducting plates, like those in Fig. 15-2, the electric lines of force are parallel to each other. They are evenly spaced throughout their entire length. In other words, there exists a uniform electric field. The force on an electric charge placed in their electric field is less dependent on the location of the charge than it is in the fields of Fig. 15-1. This is because the force of attraction to one plate increases as the force of repulsion from the other plate decreases and vice versa. For example, as a positive charge moves toward the negative plate, the force of attraction increases as the square of the distance between the two charges decreases; but, at the same time, the repelling force of the positive plate decreases as the square of the distance between this plate and the positive charge increases.

Since an electric field exists between the charged plates in Fig. 15-3 and since electric field intensity has units of volts per meter, there is a voltage (potential energy difference) between the two charged plates. Assume this voltage is 100 V. This 100 V is uniformly distributed across each electric line of force because the electric field is uniform. As represented in Fig. 15-3 by the dashed line, there is a surface called the *equipotential surface,* which connects all possible points in the electric field that have the same potential. In a uniform electric field like that in Fig. 15-3, the equipotential surfaces are planes which are parallel to the plates. Also, as shown in Fig. 15-3, equipotential surfaces for equally spaced potentials (25 V, 50 V, 75 V, etc.) are equally spaced physically when the electric field is uniform. Thus, all parts of a material inserted between two charged parallel plates are subjected to the same stress.

*Electric flux* is a term used to refer to all of the electric lines of force in an electric field. By convention, it is assumed that each unit of electric

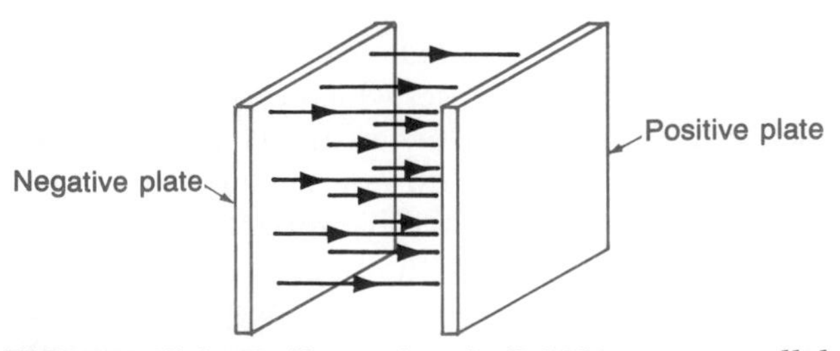


**FIGURE 15-2 Uniform electric field between parallel conductor plates.**

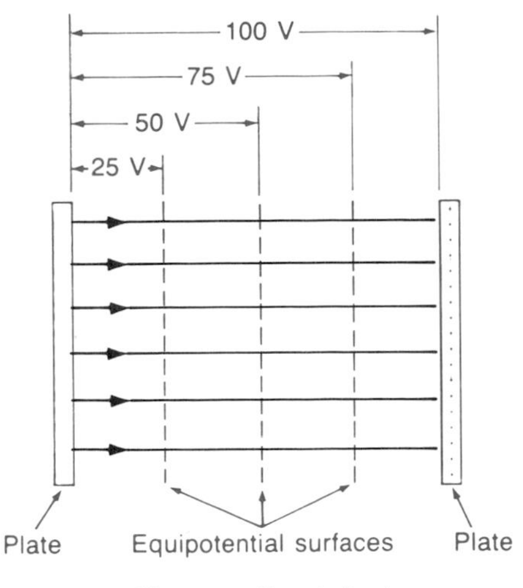


**FIGURE 15-3 Voltage distribution in a uniform electric field.**

charge produces one line of electric force; therefore, the coulomb is used as the unit for electric flux as well as the unit for electric charge.

Electric flux is not a very useful term to describe the effectiveness of an electric field unless the cross-sectional area of the field is also known. A more descriptive term is *electric flux density*. This term specifies the electric flux per unit cross-sectional area. Electric flux density is specified in coulombs per square meter ($C/m^2$). Remember that coulomb, as used here, is the unit of electric flux.

## Self-Test

1. What is an electric field?
2. The direction of a line of force represents the direction a ________ charge moves in the electric field.
3. What is the unit for electric field intensity?
4. What happens to the force between two charges when the distance between them is doubled?
5. Define a uniform electric field.
6. A surface where all points on the surface are at the same potential is called an ________.
7. Collectively, electric lines of force are known as ________.
8. Coulomb per meter squared is the unit for ________.

## 15-2 DIELECTRICS

An insulator placed in an electric force field between two charged plates is called a *dielectric*. The electric field stresses the molecular structure of a dielectric and causes static charges to build up within the dielectric. As seen in Fig. 15-4, the surface of the dielectric nearest the positive plate becomes negatively charged while the surface nearest the negative plate becomes positively charged. For clarity, the electric flux has been omitted from Fig. 15-4; however, it is always present between charged plates. Since a dielectric material is an insulator, no electric charge travels from the dielectric to the plates or vice versa. Thus, the charge on the dielectric material's surfaces is an *induced charge*. It is induced by the charge on the plates forcing a redistribution of the electric charges in the molecules and atoms of the dielectric material. The building of the charge on the dielectric is illustrated in Fig. 15-5. Figure 15-5*a* shows the location of the nucleus, and the average electron path, for an atom in a dielectric which is not in an electric field. Figure 15-5*b* illustrates how the nucleus location and the electron path are modified by the electric field. Notice in Fig. 15-5*b* that positive charges are concentrated on the surface of the dielectric next to the negative plate, while negative charges concentrate next to the opposite plate. Deformation of the molecular structure of the dielectric by the electric field is called *polarization*. The dielectric is *polarized* by the electric field.

The amount of polarization of a dielectric is determined by the type of dielectric material and electric field intensity between the plates. As the electric field intensity is increased, the amount of polarization also increases. However, there is a limit to the amount of polarization, and thus the electric field intensity, that a given dielectric can

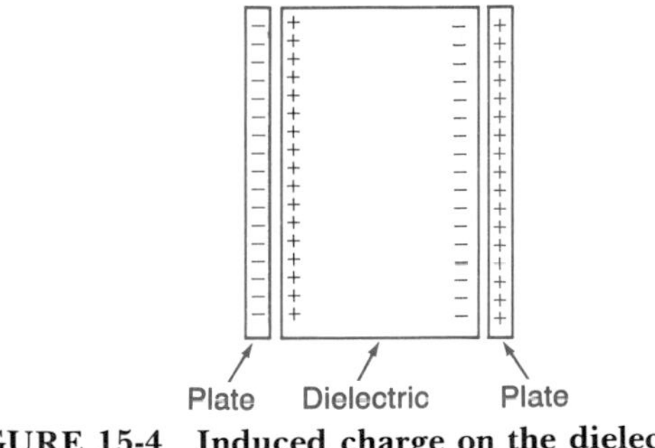
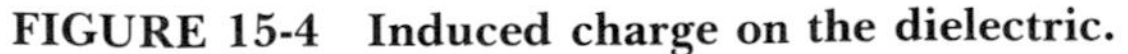


**FIGURE 15-4 Induced charge on the dielectric.**

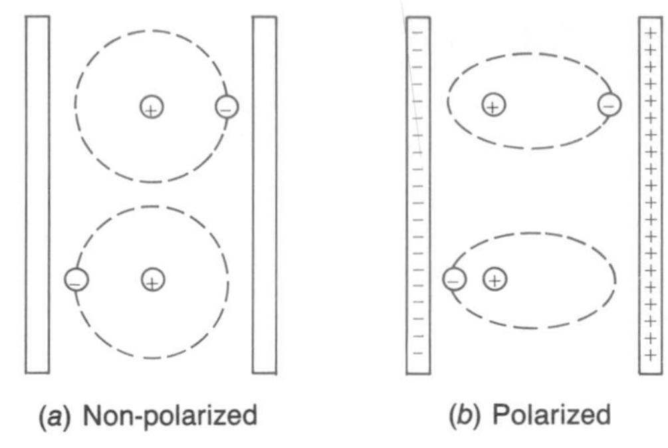


**FIGURE 15-5 Dielectric polarization.**

tolerate. If this limit is exceeded, electrons are pulled free of the parent atoms and a destructive current starts to flow through the dielectric. Because the dielectric resistance is so high, even a small current overheats and destroys the dielectric material.

The *dielectric strength* of an insulating material (dielectric) determines the electric field strength it can withstand before it breaks down and starts conducting. Dielectric strength is specified in volts per millimeter (V/mm). Dielectric strength for materials used in capacitors ranges from about 3 kV/mm (air) to over 50 kV/mm (mica, glass, etc.).

**EXAMPLE 15-1**

Two plates are separated by 0.1 mm of dielectric material which has a dielectric strength of 15 kV/mm. How much voltage can be applied across the plates?

**Solution**

Voltage = dielectric strength × dielectric thickness = 15 kV/mm × 0.1 mm = 1500 V.

When the electric field is removed from a dielectric, the polarization, and thus the induced charges, may not immediately disappear. The phenomenon of a dielectric retaining some of its polarization and induced charge for a period of time after the removal of the electric field is known as *dielectric absorption*. The amount of dielectric absorption depends on the type of dielectric material, the electric field intensity to which the dielectric is subjected, and the length of time the dielectric is polarized. For practical purposes, air has no noticeable dielectric absorption. On the other hand, glass, when subjected to a large electric field intensity for an extended time, has a very noticeable dielectric absorption. The importance of dielectric absorption is illustrated in Sec. 15-4.

*Permittivity* is a term used to describe a dielectric's ability to accommodate electric flux. Thus, permittivity is a function of the medium through which the electric flux must pass. Permittivity is mathematically defined as

$$\text{Permittivity} = \frac{\text{electric flux density}}{\text{electric field intensity}}$$

An appropriate unit for permittivity can be determined by substituting the units for flux density and field intensity into the above formula:

$$\text{Permittivity} = \frac{\text{C/m}^2}{\text{N/C}} = \frac{\text{C}^2}{\text{N} \cdot \text{m}^2}$$

Or, if field intensity is expressed in volts per meter, the unit for permittivity can be

$$\text{Permittivity} = \frac{\text{C/m}^2}{\text{V/m}} = \frac{\text{C}}{\text{V} \cdot \text{m}} = \frac{\text{farad (F)}}{\text{m}}$$

where a *farad* (F) is defined as coulombs per volt.

Permittivity with the units just determined is called *absolute permittivity* and is abbreviated $\epsilon$ (Greek lowercase epsilon). The permittivity of a vacuum or free space has been experimentally determined to be $8.85 \times 10^{-12}$ $\text{C}^2/\text{N} \cdot \text{m}^2$ or F/m. The permittivity of free space is often abbreviated $\epsilon_o$. Rather than express the permittivity of all materials in the above units, materials are often rated for *relative permittivity*. The relative permittivity ($\epsilon_r$) of a material is determined by dividing its absolute permittivity by the absolute permittivity of free space (a vacuum):

$$\epsilon_r = \frac{\epsilon}{\epsilon_o}$$

The units for $\epsilon$ and $\epsilon_o$ cancel, so $\epsilon_r$ is unitless (a pure number).

The relative permittivity of a material is also known as the *dielectric constant* of the material. In this book either term is used so that the reader will be familiar with both. The symbol $k$ is often used as the abbreviation for dielectric constant. Thus $k = \epsilon/\epsilon_o = \epsilon_r$.

Relative permittivity of materials used in capacitors to separate their plates varies from 1.006 for air to greater than 2500 for some ceramic materials. Except for ceramic materials, the common dielectric materials (mica, paper, oil, glass, metal oxides, and plastic materials) have dielectric constants in the 2 to 7 range. The exact value of $k$ varies with temperature and humidity as well as the exact chemical composition of the material.

## Self-Test

**9.** Is a dielectric material an insulator or a conductor?

**10.** What happens to a dielectric material when it is put between two charged plates?

**11.** True or false? Dielectric materials block electric lines of force.

**12.** List the units for each of the following:
  **a.** dielectric strength
  **b.** permittivity
  **c.** relative permittivity

**13.** The relative permittivity of a material is also known as its ______.

**14.** The ability of a dielectric to remain polarized after the electric field is removed is called ______.

**15.** Deformation of the atomic-molecular structure of a dielectric is known as ______.

**16.** How much voltage can be applied between two plates that are separated by 2 mm of a material which has a dielectric strength of 10 kV/mm?

## 15-3 CAPACITANCE AND CAPACITORS

*Capacitance* can be defined in several ways: (1) it is the electrical property which opposes changes in voltage, and (2) it is the electrical property that stores electric energy. The abbreviation for capacitance is $C$, and, as we shall see later, the base unit of capacitance is the *farad* (F), which is a coulomb per volt (C/V).

Physical devices which are constructed to provide a specified amount of capacitance are called *capacitors*. There are three major categories of capacitors: variable, nonpolar fixed (usually just called capacitors), and polar fixed (usually just called electrolytics). The symbols for these three categories are shown in Fig. 15-6. Within these three categories there are many subcategories or types of capacitors, but all of them are represented by one of the three symbols in Fig. 15-6.

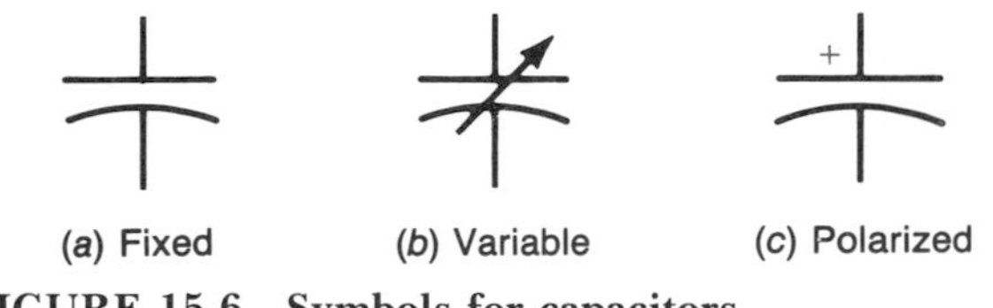

**FIGURE 15-6 Symbols for capacitors.**

A capacitor, as shown in Fig. 15-7, consists of two conductive plates separated by a dielectric. The plates have leads attached to them so that the capacitor can be connected to other circuit components.

Figure 15-8 shows how a capacitor can receive and store energy from another energy source. In Fig. 15-8*a*, the capacitor is discharged (has no charge on its plates) so there is no electric field between its plates and no energy is stored. When the switch is closed as in Fig. 15-8*b*, current flows. The voltage source $V$ forces excess electrons to build up on one plate of the capacitor and removes electrons from the other plate. Thus, one plate receives a negative charge (excess of electrons), while the other plate receives a positive charge (deficiency of electrons). In addition to setting up an electric field between the plates, the charges on the plates induce a charge into the dielectric. During the fraction of a second when the capacitor is charging and $V_C$ is less than $V$, some voltage is dropped across the internal resistances of the source, the circuit conductors, and the capacitor's leads and plates (thus, Kirchhoff's voltage law is satisfied). Current continues to flow until the electric field intensity produces a voltage $V_C$ between the plates which is equal to the source voltage $V$. As shown in Fig. 15-8*c*, the capacitor is fully charged when $V_C = V$. At this time, no current flows. Now the voltage source $V$ can be removed from the cir-

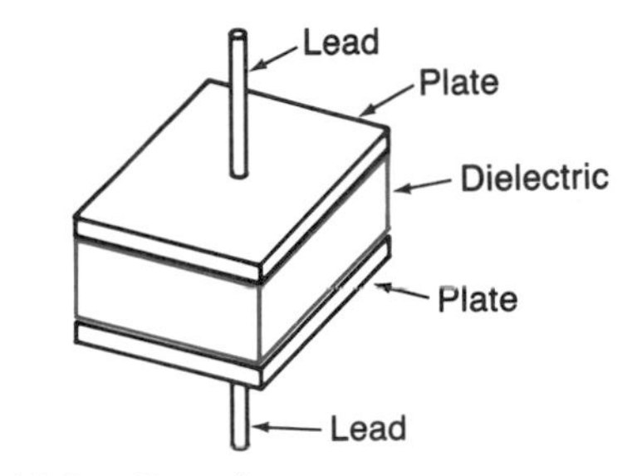

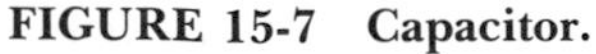
**FIGURE 15-7 Capacitor.**

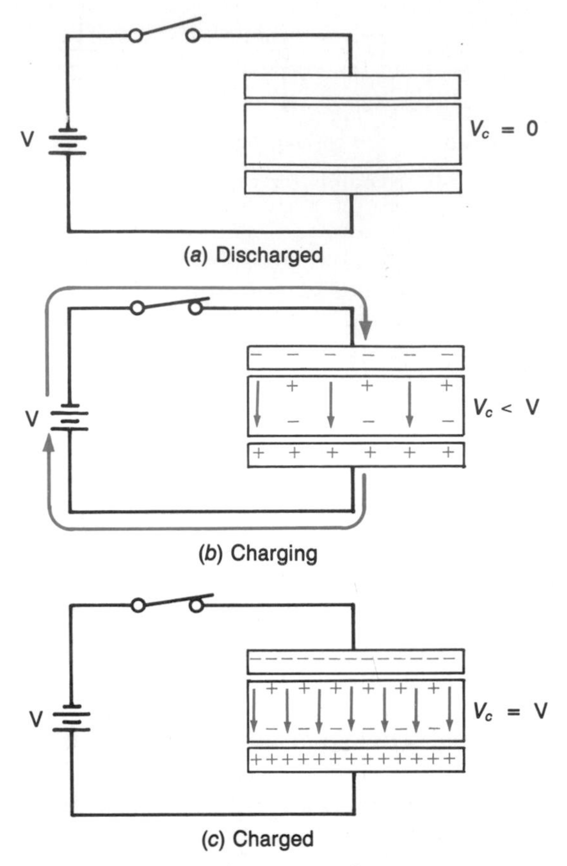


**FIGURE 15-8 Charging a capacitor.**

cuit and the capacitor remains charged. The capacitor is now storing electric energy because there is a voltage between its leads and it has charge stored on its plate ($W = VQ$ as explained in Chap. 2).

The relationship between capacitance $C$, charge $Q$, and voltage $V$ is

$$C = \frac{Q}{V}$$

Why $C = Q/V$ can be seen by looking carefully at several relationships previously developed. First,

$$F = k\frac{Q_1Q_2}{D^2}$$

Second, electric field intensity has units of volts per meter or newtons per coulomb, so we can write

$$\frac{V}{D} = \frac{F}{Q} \quad \text{or} \quad \frac{\text{voltage}}{\text{distance}} = \frac{\text{force}}{\text{charge}}$$

For a capacitor, the distance $D$ between the plates is fixed, so $D$ in the above formulas can be considered a constant for a given value of capacitance $C$.

Now suppose the charge on the plates of a capacitor is doubled by increasing the voltage applied to the capacitor. According to the first formula above, the force $F$ would quadruple because the charge on each plate ($Q_1$ and $Q_2$) doubles while $D^2$ remains unchanged. Now, the second formula above shows that for $F$ to quadruple while $Q$ doubles, $V$ must also double if $D$ remains unchanged. Thus, capacitance is a proportionality constant which relates charge and voltage. That is, $C = Q/V$.

When $Q$ and $V$ are expressed in their base units of coulomb and volt, respectively, $C$ is in its base unit, the farad (F). A farad, of course, is just another name for coulombs per volt. A farad is that amount of capacitance which will store one coulomb of charge (on each plate) when the capacitor is charged to one volt.

---

**EXAMPLE 15-2**

What is the charge on a 450-$\mu$F capacitor which is connected to a 50-V source?

**Solution**

$$C = \frac{Q}{V}$$

Therefore,

$$Q = CV = 450 \times 10^{-6} \text{ F} \times 50 \text{ V} = 22.5 \text{ mC}$$

---

**EXAMPLE 15-3**

How much voltage is required to deposit 0.04 C on the plate of a 1000-$\mu$F capacitor?

**Solution**

$$V = \frac{Q}{C} = \frac{0.04 \text{ C}}{1000 \times 10^{-6} \text{ F}} = 40 \text{ V}$$

---

## Self-Test

**17.** C/V is equal to ________.

**18.** How many capacitor symbols are there?

**19.** What is a capacitor?

**20.** If the charge on a capacitor is doubled, the voltage will ________.

**21.** A 100-$\mu$F capacitor is connected to a 200-V source. Determine the charge on the capacitor.

**22.** How does a capacitor store energy?

## 15-4 FACTORS AFFECTING CAPACITANCE

The capacitance of a capacitor is primarily determined by three factors: (1) the area of the plates, (2) the distance between the plates, and (3) the type of dielectric. Since the dielectric constant of most dielectric materials is temperature-sensitive, the capacitance of a capacitor also is dependent upon the temperature.

When the area of the plates is doubled as in Fig. 15-9*a*, the amount of $Q$ on the plates has to double to maintain the same electric flux density and electric field intensity. (This assumes other factors, dielectric material and distance between plates, remain constant.) Maintaining the same electric field intensity (units of volts per meter) maintains the same voltage across the capacitor. According to the formula $C = Q/V$, $C$ must double when $Q$ doubles if $V$ is to remain constant. Thus, *plate area and capacitance are directly proportional.*

Figure 15-9*b* shows that when the distance between the plates is doubled while the $Q$ is held constant, twice as much voltage is required to maintain the same electric flux density and electric field intensity. Again, the formula $C = Q/V$ shows that $C$ will be half as great if $V$ doubles while $Q$ remains constant. Thus, *$C$ is inversely proportional to the distance between the plates.*

Although the air dielectric in all the capacitors in Fig. 15-9*a* and *b* experiences some polarization, it has not been shown because it would only make the illustrations unnecessarily complex. However, when comparing dielectrics, polarization must be taken into account because the degree of polarization under identical electric field intensity determines the dielectric constant of a material. Figure 15-9*c* illustrates the effects of increasing the dielectric by changing from an air dielectric to a solid dielectric like glass or ceramic. In order to show the induced charges that result from polarization, Fig. 15-9*c* is shown in cross section. With air as a dielectric, some polarization and induced charge result. As shown in Fig. 15-9*c*, the induced charges counteract part of the charges accumulated on the plates. Those plate charges not counteracted by induced charges set up the electric field intensity between the plates. When a solid dielectric with $k = 2$ replaces the air dielectric, the amount of polarization and induced charges increase as seen in Fig. 15-9*c*. To counteract the greater induced

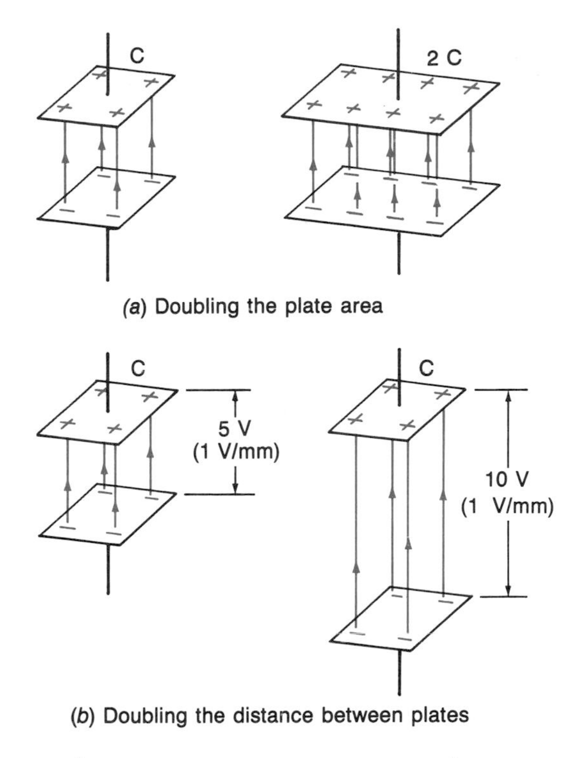


(*a*) Doubling the plate area

(*b*) Doubling the distance between plates

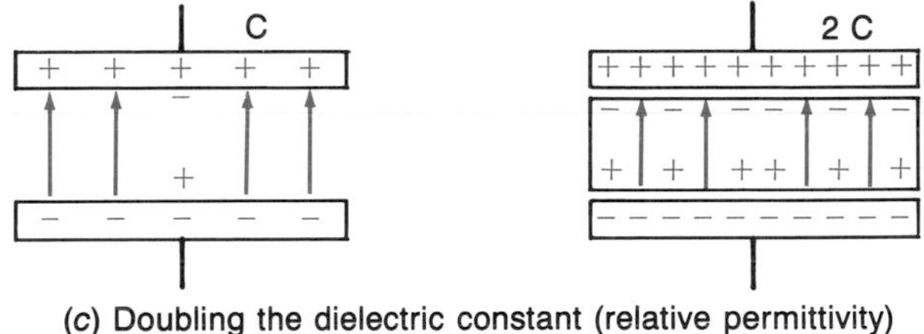


(*c*) Doubling the dielectric constant (relative permittivity)

**FIGURE 15-9 Factors affecting capacitance.**

charge in the higher permittivity dielectric and still maintain the same electric field intensity, the charge on the plates must be doubled. Again, doubling $Q$ while holding $V$ constant requires twice as much capacitance ($C = Q/V$). Thus, *capacitance and permittivity are directly proportional.*

Now that we know the relationship between capacitance, plate area, plate separation, and permittivity, we can write a formula for calculating the capacitance of a parallel plate capacitor. Remember from the discussion of permittivity $\epsilon$ in Sec. 15-2 that $\epsilon_r = \epsilon/\epsilon_o$. Therefore, permittivity of a dielectric can be expressed as $\epsilon = \epsilon_o \epsilon_r$. Thus, the capacitance of a parallel plate capacitor can be determined by the formula

$$C = \frac{\epsilon_o \epsilon_r a}{D}$$

where $\epsilon_o$ = permittivity of free space
$\epsilon_r$ = relative permittivity of dielectric
$a$ = area of one plate, $m^2$
$D$ = distance between plates, m
$C$ = capacitance, F

The farad unit for capacitance (F) can be verified by substituting the specified units for the other quantities into the capacitance formula (remember that $\epsilon$ has units of coulombs per volt-meter).

$$1 \text{ F} = \frac{(\text{C/V} \cdot \text{m}) \times 1 \text{ m} \times 1 \text{ m}}{\text{m}} = 1 \frac{\text{C}}{\text{V}}$$

A farad is equal to one coulomb per volt.

A farad is an extremely large amount of capacitance. Practical values of capacitance are expressed in $\mu$F ($10^{-6}$ F) or pF ($10^{-12}$ F).

**EXAMPLE 15-4**

Two plates, each 10 cm × 10 cm, are separated by 3 mm of mica which has a dielectric constant of 4. Determine the capacitance.

**Solution**

$$a = 10 \times 10^{-2} \text{ m} \times 10 \times 10^{-2} \text{ m} = 100 \times 10^{-4} \text{ m}^2$$
$$D = 3 \times 10^{-3} \text{ m}$$
$$\epsilon_o = 8.85 \times 10^{-12} \text{ F/m} \qquad \text{(from Sec. 15-2)}$$
$$C = \frac{\epsilon_o \epsilon_r a}{D} = \frac{8.85 \times 10^{-12} \times 4 \times 100 \times 10^{-4}}{3 \times 10^{-3}}$$
$$= 1.18 \times 10^{-10} = 118 \text{ pF}$$

The formula for the capacitance of parallel plate capacitors is often given as $C = 8.85ka/10^{12}D$. You can see that this formula is the same as the one used in the above example except that $k$ is substituted for $\epsilon_r$ and $8.85/10^{12}$ is substituted for $\epsilon_o$.

The effect of dielectric absorption on a capacitor is illustrated in Fig. 15-10. A charged capacitor, shown in Fig. 15-10*a*, polarizes the dielectric and sets up an electric field. When the capacitor is discharged (see Fig. 15-10*b*) by momentarily connecting a short across its terminals, electrons flow from the negative plate to the positive plate until the charges in the system are balanced. Because of the absorption effect, some charge on each plate is neutralized by the charges remaining in the dielectric. If the short in Fig. 15-10*b* is removed before the dielectric has had time to completely depolarize, the capacitor develops a voltage between its terminals as the dielectric depolarizes. This is illustrated in Fig. 15-10*c* where part of the absorption charge has disappeared as the dielectric continues to depolarize. The final result, as seen in Fig. 15-10*c*, is that the plates again have a net charge and a weak electric field is reestablished.

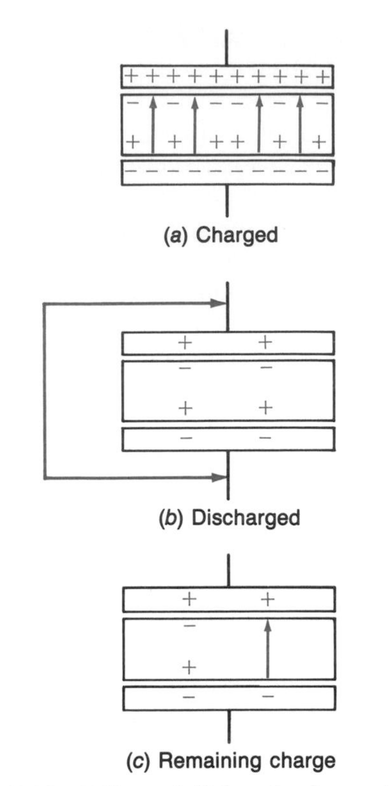

**FIGURE 15-10 Effect of dielectric absorption.**

In a television receiver, the glass of the picture tube (kinescope) is the dielectric for a capacitor whose plates are conductive coatings on the inside and outside of the picture tube. This capacitor is charged to more than 20 kV under normal operation of the receiver, so it has a large amount of dielectric absorption. After this capacitor is discharged by a temporary short, it can build up hundreds of volts between its plates as it slowly depolarizes.

## Self-Test

**23.** List three major factors that determine the capacitance of a capacitor, and specify whether capacitance is directly or inversely proportional to each factor.

**24.** What causes a capacitor to retain some charge even after its leads have been momentarily shorted together?

**25.** Determine the capacitance of two 10 mm × 30 cm plates that are separated by 0.1 mm of dielectric with $k = 4$.

## 15-5 ENERGY STORAGE

Capacitance can be defined as the ability to store energy in an electric field. The amount of energy stored by a capacitor can be calculated with the formula

$$W = 0.5CV^2$$

where $W$ = energy (work), J
$C$ = capacitance, F
$V$ = voltage, V

Before using this formula let us see how it can be derived.

Figure 15-11*a* shows a capacitor being charged by a constant-current source. The instant switch $S_1$ is closed, the voltage across the capacitor is zero and the current in the circuit is equal to the value set by the constant-current source. As time passes the voltage across the capacitor builds up as the capacitor plates take on a charge. Figure 15-11*b* shows that a constant-current source provides a linear buildup of charge $Q$ on the plates. (Since $Q = It$, $Q$ must be a function of time $t$ when current $I$ is constant.) Figure 15-11*c* shows the voltage across the capacitor also builds up linearly. (Since $V = Q/C$, $V$ must follow $Q$ for any given value of $C$.)

After a period of time, the switch in Fig. 15-11*a* is opened. At this time the current, as shown in Fig. 15-11*b* and *c*, drops to zero. However, the accumulated charge and voltage remain on the capacitor because there is no way for the capacitor to discharge.

The instantaneous power from the source in Fig. 15-11*a* can be calculated from the instantaneous current and voltage in Fig. 15-11*c*. Note that the instantaneous power varies, linearly, from zero to $IV$. Thus, the average power over the time the capacitor was charging is $P = 0.5IV$. The energy stored in the capacitor (and taken from the source) is

$$W = Pt = 0.5IVt$$

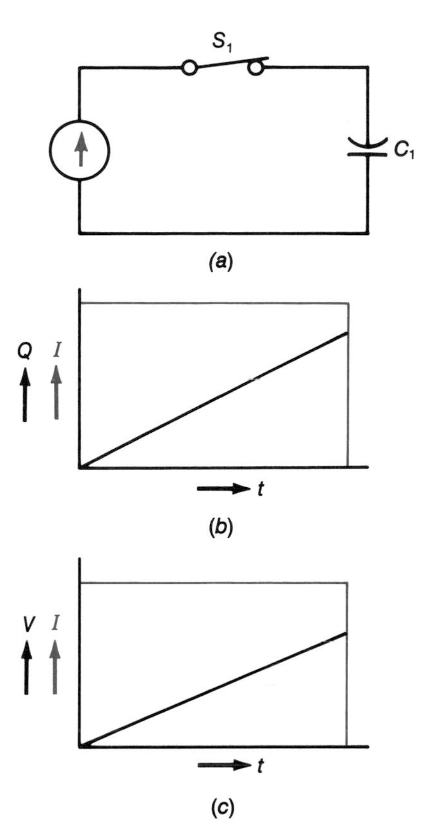


**FIGURE 15-11 Charging a capacitor with a constant current source.**

Substituting $I = Q/t$ yields

$$W = \frac{0.5QVt}{t} = 0.5QV$$

And substituting $Q = CV$ yields

$$W = 0.5CVV = 0.5CV^2$$

### EXAMPLE 15-5

A 40-$\mu$F capacitor is charged to 200 V. Determine the amount of energy stored in the capacitor.

**Solution**

$$W = 0.5CV^2 = 0.5 \times 40 \times 10^{-6} \times 200^2 = 0.8 \text{ J}$$

### EXAMPLE 15-6

A 100-$\mu$F capacitor is charged to 500 V. How much energy is taken from the capacitor when it is discharged to 100 V?

**Solution**

Energy with 500 V is

$$W_1 = 0.5 \times 100 \times 10^{-6} \times 500^2 = 12.5\ \text{J}$$

Energy with 100 V is

$$W_2 = 0.5 \times 100 \times 10^{-6} \times 100^2 = 0.5\ \text{J}$$

Energy removed is

$$W = W_1 - W_2 = 12.5\ \text{J} - 0.5\ \text{J} = 12\ \text{J}$$

Notice in the above example how reducing the voltage by a factor of 5 reduces the energy by a factor of 25. This is to be expected when the formula contains $V^2$.

Small low-voltage capacitors store microjoules of energy. Even large high-voltage capacitors only store a few kilojoules of energy. This is not much energy compared to 1 kWh (3.6 MJ) of energy which only costs between 3 and 15 cents when purchased from an electric power company.

Although the amount of energy stored in a capacitor may be very small, a capacitor can provide a large quantity of power because it can provide a large current for a short period of time. That is, a capacitor, when used as an energy source, has a very low internal resistance. An example may help illustrate the power available from a capacitor. A capacitor is used in an electronic photo flash to fire the xenon flash tube. While the capacitor only stores 20 to 40 J of energy, it is discharged by the xenon bulb in about 1 ms or less. Since $P = W/t$, the power involved in removing 20 J from a capacitor in 1 ms is $P = 20\ \text{J}/(1 \times 10^{-3}\ \text{s}) = 20\ \text{kW}$.

## Self-Test

**26.** Will doubling the capacitance or doubling the voltage have the greater effect on the energy stored in a capacitor? Why?

**27.** A 1000-$\mu$F capacitor is charged to 100 V. Determine the amount of energy stored.

**28.** If the energy stored in the capacitor in Question 27 is removed in 2 ms, what is the power?

## 15-6 CHARACTERISTICS OF CAPACITORS

In addition to capacitance, there are many other characteristics to consider in selecting and using a capacitor. Some of these are (1) tolerance, (2) voltage rating, (3) operating temperature range, (4) temperature coefficient, (5) dissipation factor, and (6) insulation resistance. Most of these characteristics are determined by the dielectric used in the capacitor.

Like resistors, capacitors have tolerances. Except on polarized capacitors, the tolerance is symmetrical, i.e., ±5 percent. Polarized capacitors often have unsymmetrical tolerance with the + tolerance being many times greater than the − tolerance. This is because most polarized capacitors are used in applications where only the minimum capacitance is important.

The voltage rating of a capacitor is usually specified as a *dc working voltage*. It specifies the maximum dc voltage at which a capacitor should be continuously operated (charged) when the temperature is below some specified value. When the temperature exceeds this value, the dc working voltage must be reduced in accordance with the derating curves provided by the manufacturer. Some capacitors also have a *surge voltage rating* which specifies the duration, magnitude, and rate of recurrence of the voltage surges that the capacitor can tolerate.

The materials from which a capacitor is constructed, especially the dielectric material, determine the minimum and maximum temperature at which a capacitor should be operated. Based on many hours of tests, the manufacturer specifies these temperature limits as the *operating temperature range*.

The relative permittivity (dielectric constant) of a dielectric may change as the temperature changes. Of course, any change in the dielectric constant results in a change in capacitance. Therefore, a capacitor has a temperature coefficient rating specified in parts per million per degrees Celsius (ppm/°C). Note that this rating is specified in the same way as it was for resistors. The specifications for a capacitor, or the information printed on a capacitor, may include something like N150, P50, or NP0. These are abbreviations indicating that the temperature coefficient is negative 150 ppm/°C, positive 50 ppm/°C, or 0 ppm/°C, respectively.

The *dissipation factor* of a capacitor is the ratio of the energy converted to heat by the resistances associated with a capacitor over the energy stored in the capacitor. Usually this ratio is converted to, and expressed as, a percentage. Details of dissipation factor are provided in Chap. 16.

The *insulation resistance* of a capacitor refers to the ohmic (dc) resistance between the leads or plates of a capacitor. Of course, this is essentially the resistance of the dielectric material which is classified as an insulator. Insulation resistance is specified in megohms (MΩ). Typical values range from 100 to 10,000 MΩ.

In many applications of capacitors, the values of some of the above characteristics are not critical or even specified. The characteristics most often specified are capacitance, tolerance, and voltage rating.

### Self-Test

**29.** List six characteristics or ratings, in addition to capacitance, that describe a capacitor.

**30.** What determines the voltage rating of a capacitor?

**31.** Is the resistance between the plates of a capacitor usually less than or greater than 100 MΩ?

## 15-7 TYPES OF CAPACITORS

A typical electronics parts catalog lists 20 to 30 types of capacitors. Capacitors can be typed by such things as intended use, type of dielectric, manufacturing technique, physical appearance, etc. This section concentrates on some of the more common types and their characteristics.

### Types of Variable Capacitors

The symbol for the variable capacitor is shown in Fig. 15-6*b*. The capacitance of a variable capacitor can be changed by varying either the effective plate area or the distance between the plates. There are numerous ways of physically varying these two factors. Figure 15-12 illustrates three ways. In Fig. 15-12*a*, the plates occupy opposite halves of a circle so the effective plate area, and thus the capacitance, is minimum. When the rotor is turned 180°, the plates occupy the same half of the circle and provide maximum capacitance. In the ceramic trimmer capacitor, both the stator and the rotor are insulators, and the plate patterns are deposited on them using either tin or silver. The pin and rivet in Fig. 15-12*a* are conductors, and they connect the deposited plates to lead terminals.

Figure 15-12*b* is a compression-type capacitor adjusted for maximum capacitance. The two sets of plates are under tension so when the screw is loosened, the plates spring apart and decrease the capacitance. This type of capacitor has been pretty much replaced by the type shown in Fig. 15-12*a*. These two types are called *trimmer capacitors* because they are used to trim (make fine adjustments) in circuits that must have an exact amount of capacitance to operate properly. Maximum capacitance for trimmer capacitors typically ranges from about 2 to 100 pF.

In the rotary capacitor in Fig. 15-12*c*, the two stator plates are electrically connected together by

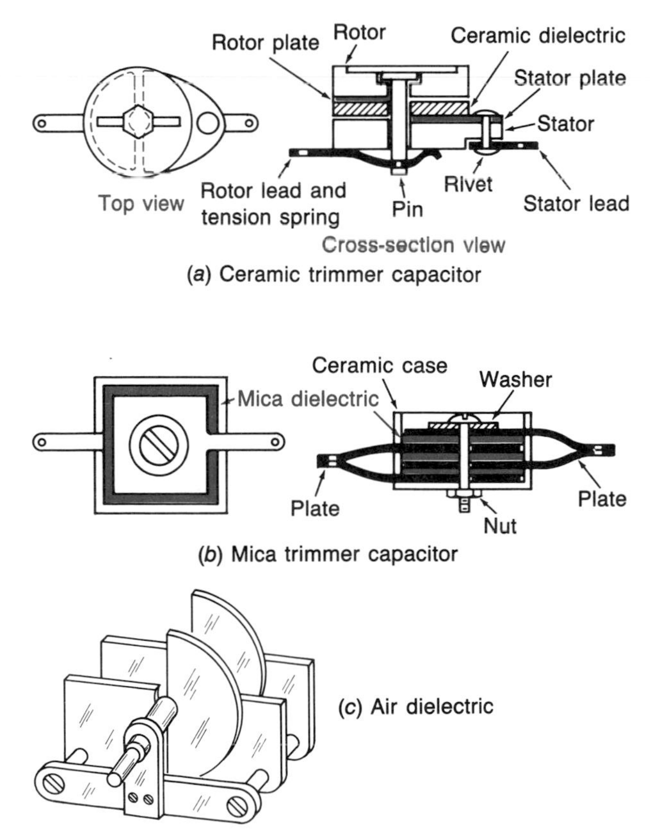

**FIGURE 15-12 Variable capacitors.**

metal spacers. The two rotor plates are electrically connected to the rotatable shaft which is a conductor. The rotor plates and stator plates are electrically isolated by the air dielectric and the ceramic crossbar to which the rotor bracket is attached. The capacitor is shown in the half-capacitance position. Rotating the shaft clockwise increases the effective plate area and the capacitance. Rotary capacitors can have many rotor and stator plates. Smaller versions often use a thin plastic film dielectric between the plates. This thin film minimizes the distance between the plates and increases the dielectric constant. Typical maximum values for rotary capacitors range from a few picofarads to over 500 pF.

The rotary plates of a rotary capacitor are represented by the curved line in the schematic symbol in Fig. 15-6*b*. In an electronic circuit the rotary plates are connected to ground or the lowest possible potential with respect to ground.

## Types of Polarized Capacitors

Polarized capacitors are *electrolytic capacitors.* There are two major types of electrolytic capacitors—aluminum and tantalum.

The essential parts of an *aluminum electrolytic capacitor* are illustrated in the cross-sectional view in Fig. 15-13*a*. From this figure you can see that the plates are aluminum, and the dielectric is *aluminum oxide* ($Al_2O_3$). The electrolyte, which is a conductor, is a chemical compound or mixture in either a dry or a paste form. The separator is made from porous paper or gauze which is easily penetrated and saturated by the electrolyte. The composition of the electrolyte varies, depending on the desired capacitor characteristics and rating, but is usually quite alkaline. Closer inspection of Fig. 15-13*a* shows that the electrolyte is actually part of the negative plate assembly. It is the conductive path from the aluminum oxide (dielectric) surface to the negative aluminum plate.

The electrolyte aids in electrochemically forming and maintaining the aluminum oxide on the positive plate *when the plate is positive with respect to the electrolyte.* If the polarity is reversed, the electrochemical reaction reverses and a large current flows through the capacitor as the oxide is removed. This large current heats the electrolyte and destroys, often with a violent explosion, the capacitor. Thus, the polarity of the electrolyte must be observed. The polarity is marked on the physical capacitor with − and/or + signs by the − and/or + leads. The curved line on the symbol in Fig. 15-6*c* is the negative lead of an electrolytic capacitor.

Figure 15-13*a* also illustrates why electrolytic capacitors provide a much higher (20 to 100 times) capacitance-to-volume ratio than do other types of fixed capacitors. The positive aluminum plate, and sometimes the negative plate also, is highly etched. The effective surface area of an etched surface is many times greater than that of a smooth, flat surface. The oxide, which is electrochemically formed and follows every nook and cranny of the etched surface, is only a few molecules thick. The electrolyte, which is part of the negative plate assembly,

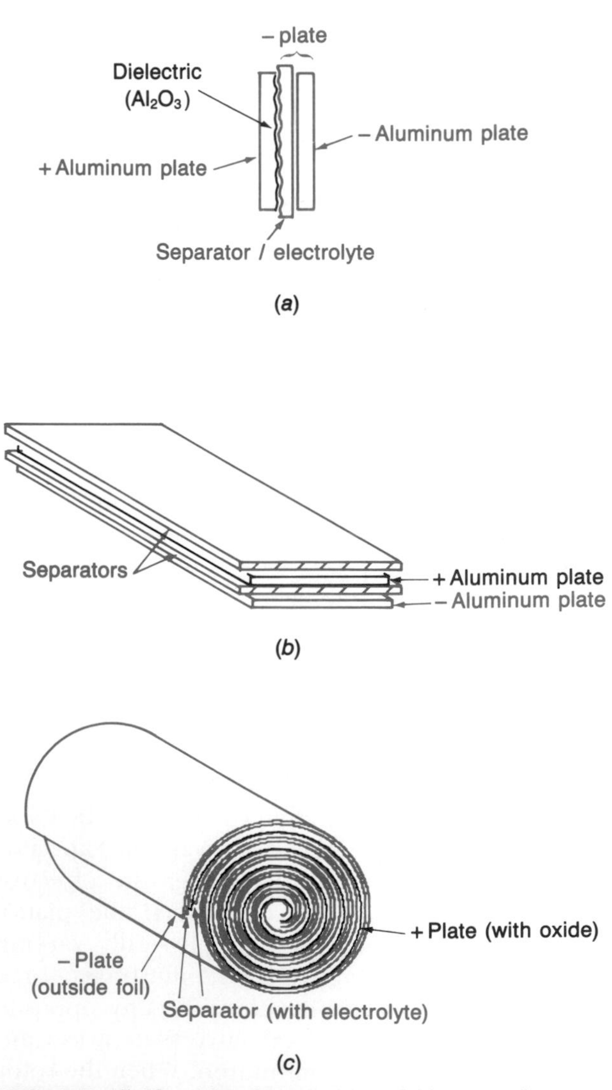


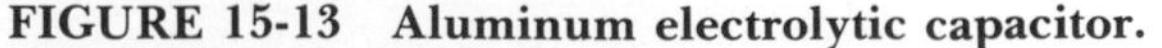

**FIGURE 15-13 Aluminum electrolytic capacitor.**

follows every nook and cranny of the oxide, so the plates are very close together. These two factors, large effective plate area and small distance between plates, give the electrolytic capacitor its high capacitance-to-volume ratio.

The two aluminum plates of the electrolytic capacitor are long narrow strips of aluminum foil which are alternately stacked with two strips of electrolyte-saturated separation material. This stack, shown in Fig. 15-13*b*, is then rolled into a tubular form as in Fig. 15-13*c*. One very important difference should be noted between Fig. 15-13*b* and *a*. The positive plate in Fig. 15-13*b* has an oxide layer on *both* surfaces. If it did not, the electrolyte would short the unoxidized surface of the positive plate to the negative plate.

One characteristic of all rolled (tubular-shaped) capacitor elements like that in Fig. 15-13*c* should be noted. Rolling almost doubles the effective plate area. This is because, except for the last turn of the outside foil, both surfaces of the foil are utilized.

The tubular roll of Fig. 15-13*c* has leads connected to the plates. Then, the completed unit is enclosed in an aluminum can. Quite often the negative plate is electrically and mechanically bonded to the can. The can is usually covered with an insulating sleeve.

A *tantalum electrolytic capacitor* is illustrated in Fig. 15-14. Instead of the rolled foil construction, the tantalum capacitor has a tantalum slug or pellet for its positive plate (or anode as it is often called for a tantalum capacitor). This slug is covered by a very thin layer of *tantalum pentoxide*, which is the dielectric. Because tantalum pentoxide has a higher dielectric constant than does aluminum oxide, the tantalum electrolytic has a higher capacitance-to-volume ratio than does the aluminum electrolytic.

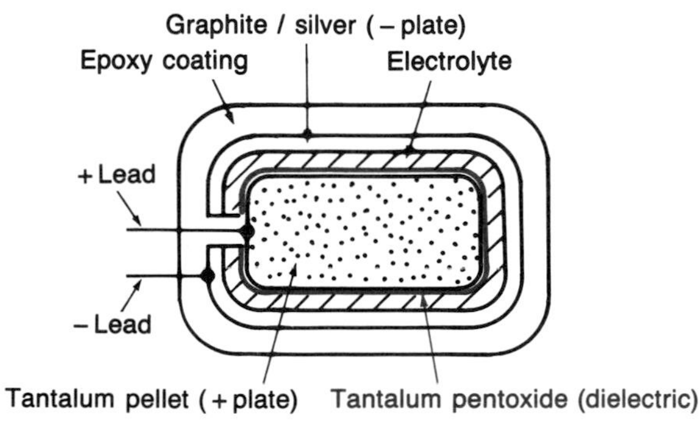


**FIGURE 15-14 Tantalum electrolytic capacitor.**

The electrolyte, in either the solid dry form or the wet gel form, surrounds the pentoxide layer. Finally, the electrolyte is encased by the negative plate (or cathode as it is often called) which is often made of a thin layer of silver lined with a thin layer of graphite. The capacitor element is then encapsulated in an epoxy coating or in a metal case.

Although the tantalum electrolytic capacitor is more expensive than the aluminum electrolytic, it has some advantages over the aluminum electrolytic. In addition to the capacitance-to-volume ratio already mentioned, these advantages include

1. Much longer storage and operating life
2. Much higher insulation resistance
3. Much lower dissipation factor
4. Wider temperature range and more stable capacitance value.

On the other hand, tantalum capacitors are not available in as large a capacitance value as are aluminum capacitors. Also, aluminum capacitors are available in higher working voltages.

Remember, the polarity of polarized capacitors must be observed. Reverse polarity or ac voltage may cause a polarized capacitor to explode and cause serious injury to personnel and damage to equipment.

## Types of Fixed Capacitors (Nonpolar)

When large values of nonpolarized capacitance (larger than a couple microfarads) are needed, *nonpolarized* aluminum electrolytic capacitors are often used. A nonpolarized electrolytic is essentially two electrolytics connected in series by attaching together either the two positive leads or the two negative leads. As seen in Fig. 15-15*a*, one of the two electrolytic capacitors always has correct polarity regardless of the polarity of the applied voltage. The correctly polarized capacitor limits the reverse current through the other capacitor. As illustrated in Fig. 15-15*b*, the nonpolar electrolytic can be made by putting an oxide on both aluminum plates. With this arrangement, either plate can be either polarity, and one of the two oxide-electrolyte interfaces will have correct polarity across it. Because of high dissipation factors, some nonpolar electrolytics are limited to intermittent operation.

Three common physical arrangements of the plates and dielectric material for fixed capacitors

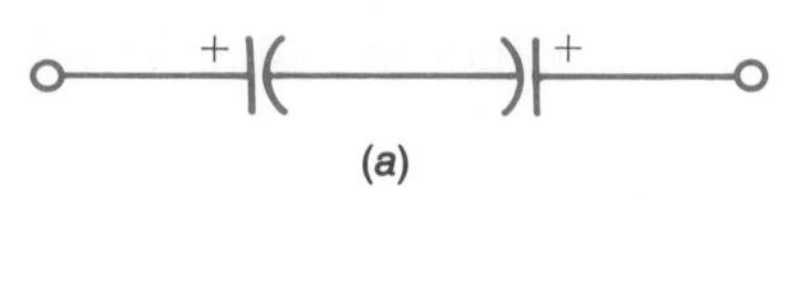


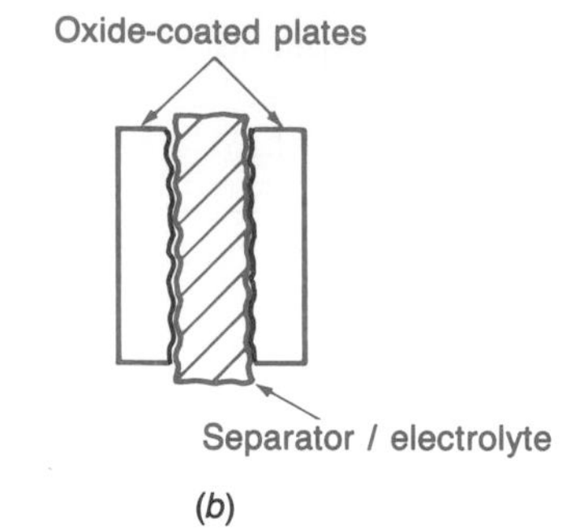


**FIGURE 15-15 Nonpolarized electrolytic capacitor.**

are illustrated in Fig. 15-16. Because of the high dielectric constant of ceramic materials, practical values of capacitance can be obtained with the single-layer arrangement of Fig. 15-16*a*. The thin circular shape of the capacitor shown in Fig. 15-16*a* is called a *disk ceramic* and is a very common type of ceramic capacitor. Common values range from a few picofarads to about 0.5 $\mu$F.

The multilayer structure illustrated in Fig. 15-16*b* is used to increase the plate area without increasing the length and width of the capacitor. It is used when the dielectric material is too hard and brittle to be rolled as in the film capacitor. Mica or ceramic is the commonly used dielectric for this structure. The cross-sectional view in Fig. 15-16*b* shows the leads protruding from the ends of the capacitor (axial leads). For the radial lead configuration, the leads would be rotated 90° before being welded or soldered to the plates. Quite often the plate material for the stacked-plate capacitor is deposited directly on the dielectric material. For example, a silvered mica capacitor has thin silver plates deposited on thin sheets of mica. Stacked-plate ceramic capacitors are referred to as *monolithic* ceramic capacitors because the deposited layers of dielectric and plate materials are bonded together as a single unit by firing in a kiln at a high temperature.

The film capacitor structure is illustrated in Fig. 15-16*c*. The axial lead type is sometimes referred to as a *tubular capacitor*. A wide range of plastic materials is used for the film (dielectric). Paper, or a paper-plastic film sandwich, is also used as dielectric in film capacitors. Common plastic films include polypropylene, polystyrene, polycarbonate, and polyester (polyethylene terephthalate). Some of these films are known by trade names such as Mylar (polyester). Each of these films provides a capacitor with different characteristics, such as maximum operating temperature, insulation resistance, or dissipation loss at high frequencies.

When paper is used as a dielectric, it is sometimes saturated in a special oil which has a high dielectric strength. These capacitors are often called *oil-filled capacitors* and are hermetically sealed in a metal container.

The plates of a film capacitor are made of either aluminum foil or tin foil. The arrangement of the plates and film, before being rolled, is shown in Fig. 15-17. Notice that, when rolled, a different plate is protruding from each end of the tubular roll. This arrangement makes it easy to attach a lead to each of the plates. The lead wire should make contact with as much of the protruding plate as possible in order to minimize the inductance of the capacitor. Referring to the cross-sectional view of Fig. 15-16*c*, you can see that if the lead were only attached to one end of the long foil strip, the charging and discharging current in the plate would have to flow through a multiturn coil to reach the other end of the plate. When the lead contacts all of the protruding plate, the charging current only has to flow across the width of the foil strip. Notice that the plates of the electrolytic capacitor in Fig. 15-13 are not offset because the lead on an electrolytic is usually just connected to the end of the foil strip. Thus, the plate inductance of the electrolytic is not minimized. However, electrolytics are only used at low frequencies, and the small amount of plate inductance has no noticeable effect at low frequencies.

The colored band on one end of an axial-lead film capacitor indicates the lead that is connected to the outside foil of a capacitor. The outside foil is identified in the cross-sectional view of Fig. 15-16*c*. Knowing the outside-foil lead is only important when connecting a capacitor to a low-signal-level electronic amplifier circuit. On the schematic symbol, Fig. 15-6*a*, the outside foil is identified by the curved line.

There are two major types of film capacitors: (1) the *film-and-foil* (usually just called a film capaci-

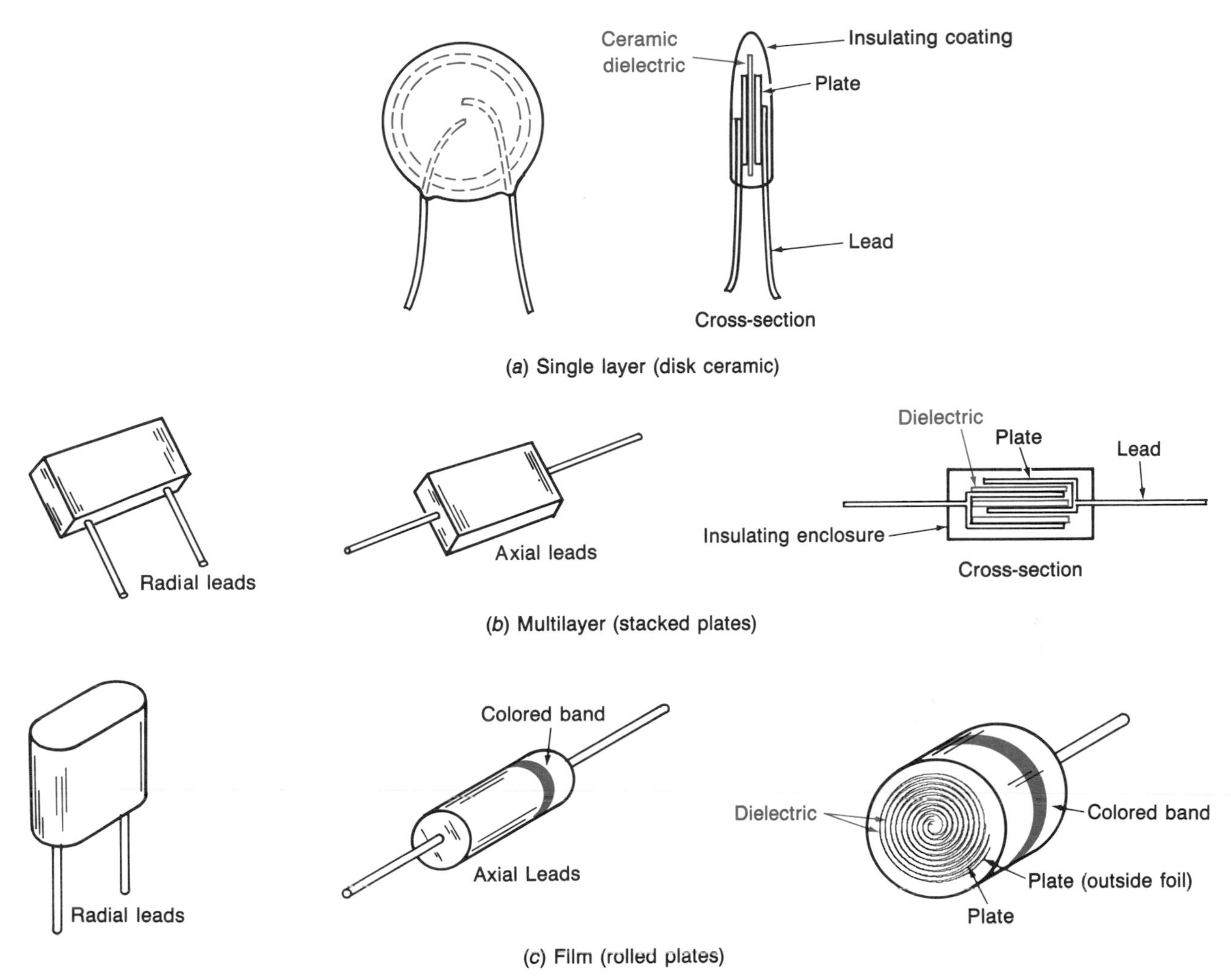


**FIGURE 15-16 Nonpolarized fixed capacitors.**

tor) and (2) the *metalized* film. The film-and-foil type uses two discrete strips of film and two discrete sheets of foil. The metalized film uses two sheets of film on which a thin layer of metal (the plate) has been deposited. Either aluminum or zinc may be deposited (sprayed) on the film. The advantages of the metalized film capacitor are that it is smaller and lighter than the film-and-foil capacitor and is self-healing. It is smaller and lighter because the deposited plate can be made much thinner than a foil plate. It is self-healing because the very thin deposited plate is instantaneously vaporized around any small pinhole punched through the film by a high-voltage spike. Thus, the charred, conducting edges of the pinhole in the film do not make contact with the plates of the capacitor. With a film-and-foil capacitor, the foil is too thick to be vaporized and its charred surface shorts through the charred edge of the film hole to the other plate.

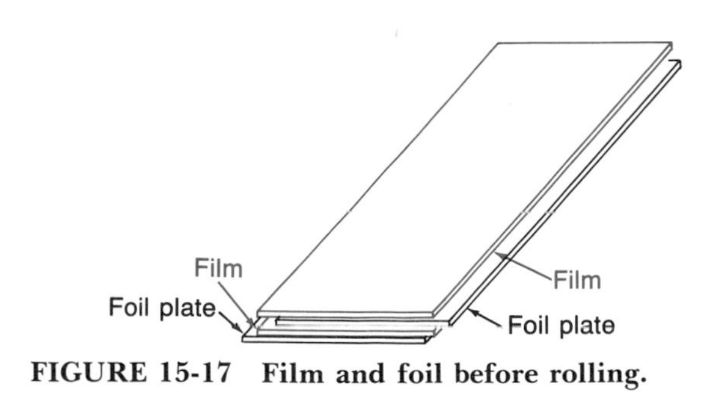


**FIGURE 15-17 Film and foil before rolling.**

## 15-8 STRAY CAPACITANCE

So far this chapter has dealt with the capacitance of capacitors. However, capacitance exists between any two conductors, or conductive parts, that are separated by an insulator, or insulation. This is known as *stray capacitance.* Stray capacitance, like stray inductance, exists everywhere in electric and electronic circuits and components. There is capacitance between the test leads that connect a meter to a circuit, the turns of wire used in an inductor, the contacts of an open switch, the turns of wire in a wire-wound resistor, the elements of a transistor, etc. Fortunately, the magnitude of stray capacitances is usually very small, often less than 1 pF, and they have negligible effect in dc and low-frequency ac circuits. However, in high-frequency circuits, such as those used in radio and television systems, stray capacitance is extremely important. In designing and constructing such circuits every precaution is taken to minimize stray capacitance. When possible, stray capacitance is used as part of the capacitance required by the circuit for proper operation.

The major detrimental effect of stray capacitance and stray inductance is that they limit the maximum operating frequency of a circuit or system. For example, the maximum frequency of ac voltage that can be measured by a volt-ohm-milliammeter (VOM) is limited by stray capacitance and inductance. The highest frequency to which a transistor amplifier can respond is often controlled by the interelement capacitances of the transistor and the capacitances within and between other components and conductors used in the circuit.

### Self-Test

**32.** List three major categories of capacitors.

**33.** List two ways or techniques used to vary the capacitance of a variable capacitor.

**34.** What does the curved line on the capacitor symbol represent for each category of capacitor?

**35.** What is the function of the electrolyte in a polarized capacitor?

**36.** On which plate is the dielectric formed in an electrolytic capacitor?

**37.** What is the dielectric material in a tantalum capacitor?

**38.** What is a trimmer capacitor?

**39.** List these capacitors in order of decreasing capacitance-to-volume ratio: variable capacitor, aluminum electrolytic, tantalum electrolytic, film-and-foil, and metalized foil.

**40.** How is the outside foil on a film capacitor identified?

**41.** List the common dielectric materials used in fixed capacitors.

**42.** Why is a metalized foil capacitor self-healing?

**43.** What is stray capacitance?

**44.** Is stray capacitance most troublesome at high or low frequencies?

## 15-9 INDICATING CAPACITANCE

### Marked Values

There are a number of ways in which the manufacturer of a capacitor may indicate its value. Often the value is printed on the case of the capacitor. The value is indicated in either picofarads or microfarads. Sometimes only the number, without any units, is printed on the capacitor. For example, a small disk ceramic capacitor may be marked 47 with an N150 printed below the 47. This would indicate 47 pF with a 150 ppm negative temperature coefficient. The 47 would have to be picofarads, because disk ceramics are always less than 1 μF. Had the capacitor been marked 0.01, then it would be microfarads because fixed capacitors are always greater than 1 pF. When the value marked on a capacitor is followed by a unit, an M is often used for the μ symbol. This is not particularly confusing when we remember that a capacitor is always less than 1 F.

Some manufacturers always express the capacitance in picofarads by using a three-digit code. In this code, the first two digits are significant and the third is the multiplier. Using this system, 103 would represent 10,000 pF, or 0.01 μF. With this system, 100 would represent 10 pF, not 100 pF. Often these three digits are followed by a letter to indicate some characteristic of the capacitor. This can be confusing when the letter k is used to indicate ±10 percent and M is used to indicate ±20 percent. We usually think of k as meaning kilo and M as meaning mega when they follow a digit.

**EXAMPLE 15-7**
What is the capacitance, in microfarads, and dc working voltage of a capacitor which is labeled 334 k, 100 V?

**Solution**
The code tells us that the capacitance is 330,000 pF ±10 percent, which is equal to 0.33 μF ±10 percent. The dc working voltage is equal to the 100 V as indicated by the 100 V.

### Color Codes

At one time three-dot color codes were very popular for indicating the capacitance of a capacitor. Six-dot color codes were also used to indicate such characteristics as temperature coefficient, voltage rating, and tolerance as well as capacitance. These codes are no longer in common use. They can be found in reference handbooks and older textbooks.

Figure 15-18 shows how some radial-lead tantalum (electrolytic) capacitors are currently being color-coded. The digit values for each color are exactly the same as those used for resistors except for the voltage rating. The voltage value for each color is shown in the figure. With this color code system the capacitance is given in picofarads.

**EXAMPLE 15-8**
A tantalum capacitor has horizontal color bands of silver, red, violet, and green (from top to bottom). One edge of the capacitor has a yellow oval. What are the specifications of the capacitor?

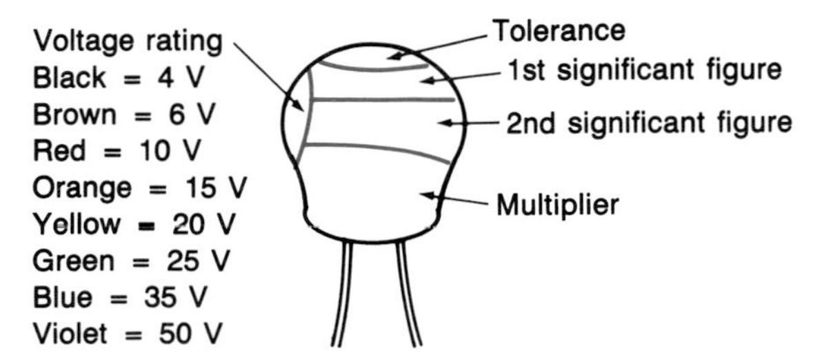


**FIGURE 15-18 Color code for a tantalum capacitor.**

**Solution**
Use the resistor color code in Chap. 5. It indicates the tolerance is ±10 percent and the capacitance is 2,700,000 pF = 2.7 μF. The code in Fig. 15-18 shows that the working voltage is 20 V.

### Self-Test

**45.** Which type of capacitor uses a color code to indicate capacitance?

**46.** A capacitor is coded 473 k. What is its capacitance?

**47.** A disk ceramic capacitor is marked 0.01. What is its capacitance?

## 15-10 SERIES, PARALLEL, AND SERIES-PARALLEL CAPACITORS

The relationship between $C$, $V$, and $Q$ ($C = Q/V$) and Kirchhoff's laws can be used to develop the formulas for the equivalent capacitance of capacitors connected in series or parallel. Once these formulas are known, series-parallel circuits can also be solved.

### Series Capacitors

For a series capacitor circuit like the one in Fig. 15-19, Kirchhoff's voltage law verifies that $V_T = V_{C_1} + V_{C_2} + V_{C_3}$. Furthermore, the current, and thus the charge, in a series circuit is the same throughout the circuit, i.e.,

$$Q_T = Q_{C_1} = Q_{C_2} = Q_{C_3}$$

Since $C = Q/V$, we can rearrange terms and write an expression for the voltage across each capacitor. That is,

$$V_{C_1} = \frac{Q_{C1}}{C_1} = \frac{Q_T}{C_1} \qquad V_{C_2} = \frac{Q_T}{C_2} \qquad V_{C_3} = \frac{Q_T}{C_3}$$

and

$$V_T = \frac{Q_T}{C_T}$$

Now, substituting these expressions into Kirchhoff's voltage laws yields

$$\frac{Q_T}{C_T} = \frac{Q_T}{C_1} + \frac{Q_T}{C_2} + \frac{Q_T}{C_3}$$

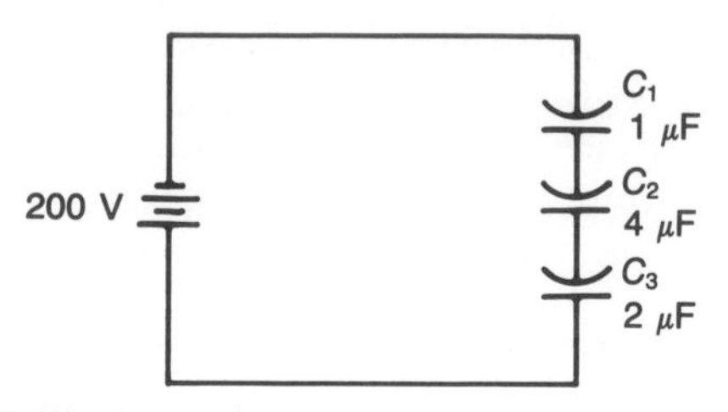


**FIGURE 15-19 Series capacitors.**

Finally, multiplying this equation by $1/Q_T$ provides

$$\frac{1}{C_T} = \frac{1}{C_1} + \frac{1}{C_2} + \frac{1}{C_3}$$

which can be rewritten in the familiar (for parallel resistor networks) "reciprocal of the sum of the reciprocals" form of

$$C_T = \frac{1}{1/C_1 + 1/C_2 + 1/C_3}$$

Of course, the above formula can be expanded for any number of series capacitors. Using the same techniques that were used for parallel resistors provides the shortcut formulas of

$$C_T = \frac{C_1C_2}{C_1 + C_2} \quad \text{and} \quad C_T = \frac{C}{n}$$

**EXAMPLE 15-9**

Determine the total, or equivalent, capacitance for the circuit in Fig. 15-19.

**Solution**

$$C_T = \frac{1}{1/1\ \mu\text{F} + 1/4\ \mu\text{F} + 1/2\ \mu\text{F}} = 0.571\ \mu\text{F}$$

Note from this example that the equivalent capacitance is less than the smallest capacitance in series.

The voltage distribution in a series capacitor circuit can be calculated by the formula

$$V_{C_n} = \frac{C_TV_T}{C_n}$$

where the subscript $n$ identifies any one of the capacitors in the series string. This formula is derived from the same formulas used to derive the equivalent capacitance formula. The derivation is

$$Q_T = C_TV_T \quad \text{and} \quad Q_{C_n} = C_nV_{C_n} = Q_T$$

Thus,

$$C_TV_T = C_nV_{C_n}$$

and rearranging,

$$V_{C_n} = \frac{C_TV_T}{C_n}$$

When only two capacitors are in series, the above formula can be shortened by substituting $C_1C_2/(C_1 + C_2)$ for $C_T$, to

$$V_{C_1} = \frac{C_2V_T}{C_1 + C_2} \quad \text{or} \quad V_{C_2} = \frac{C_1V_T}{C_1 + C_2}$$

**EXAMPLE 15-10**

Fifty volts is applied to a 0.25-μF capacitor connected in series with a 0.5-μF capacitor. Determine the voltage across the 0.5-μF capacitor.

**Solution**

Let $C_1 = 0.25\ \mu\text{F}$ and $C_2 = 0.5\ \mu\text{F}$, then

$$V_{C_2} = \frac{C_1V_T}{C_1 + C_2} = \frac{0.25\ \mu\text{F} \times 50\ \text{V}}{0.25\ \mu\text{F} + 0.5\ \mu\text{F}} = 16.7\ \text{V}$$

Notice that voltage distributes in inverse proportion to the capacitance.

**EXAMPLE 15-11**

Determine the voltage across each capacitor in Fig. 15-19.

**Solution**

From a previous example,

$$C_T = 0.571\ \mu\text{F}$$

$$V_{C_1} = \frac{0.571\ \mu\text{F} \times 200\ \text{V}}{1\ \mu\text{F}} = 114.2\ \text{V}$$

$$V_{C2} = \frac{0.571\ \mu\text{F} \times 200\ \text{V}}{4\ \mu\text{F}} = 28.6\ \text{V}$$

$$V_{C_3} = V_T - V_{C_1} - V_{C_2}$$
$$= 200\ \text{V} - 114.2\ \text{V} - 28.6\ \text{V} = 57.2\ \text{V}$$

As a cross-check,

$$V_{C_3} = \frac{0.571\ \mu\text{F} \times 200\ \text{V}}{2\ \mu\text{F}} = 57.1\ \text{V}$$

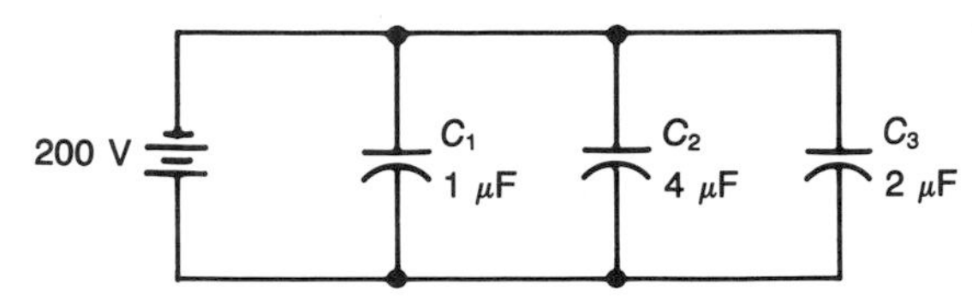


**FIGURE 15-20 Capacitors in parallel.**

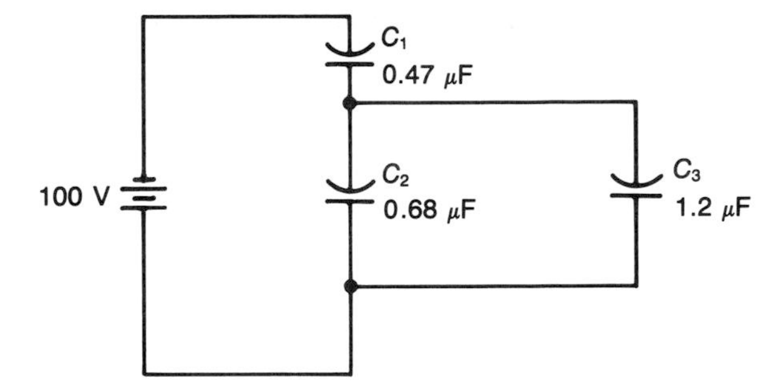


**FIGURE 15-21 Circuit for Example 15-13.**

## Parallel Capacitors

The equivalent capacitance for parallel capacitors like those in Fig. 15-20 is found by

$$C_T = C_1 + C_2 + C_3$$

This formula is derived as follows:

$$V_T = V_{C_1} = V_{C_2} = V_{C_3}$$

and

$$Q_T = Q_{C_1} + Q_{C_2} + Q_{C_3}$$

Substitute $Q = VC$ for $Q$:

$$V_T C_T = V_{C_1} C_1 + V_{C_2} C_2 + V_{C_3} C_3$$

Divide by $V_T$:

$$C_T = C_1 + C_2 + C_3$$

### EXAMPLE 15-12

Determine the total capacitance (equivalent capacitance) for the circuit in Fig. 15-20. Also determine the charge on $C_1$ and $C_2$.

**Solution**

$$C_T = C_1 + C_2 + C_3 = 1\ \mu\text{F} + 4\ \mu\text{F} + 2\ \mu\text{F} = 7\ \mu\text{F}$$
$$Q_{C_1} = V_{C_1} C_1 = 200\ \text{V} \times 1\ \mu\text{F} = 200\ \mu\text{C}$$
$$Q_{C_2} = 200\ \text{V} \times 4\ \mu\text{F} = 800\ \mu\text{C}$$

## Series-Parallel Capacitors

Series-parallel capacitor circuits can be solved by applying the formulas for series and parallel capacitors to the appropriate parts of the circuits. Of course, the general capacitance formula ($C = Q/V$) and Kirchhoff's laws can also be applied.

### EXAMPLE 15-13

Determine the total capacitance, the voltage across $C_1$ and $C_2$, and the charge on $C_3$ in the circuit shown in Fig. 15-21.

**Solution**

$$C_{2,3} = C_2 + C_3 = 0.68\ \mu\text{F} + 1.2\ \mu\text{F} = 1.88\ \mu\text{F}$$
$$C_T = \frac{C_{2,3} C_1}{C_{2,3} + C_1} = \frac{1.88\ \mu\text{F} \times 0.47\ \mu\text{F}}{1.88\ \mu\text{F} + 0.47\ \mu\text{F}} = 0.376\ \mu\text{F}$$
$$V_{C_1} = \frac{C_T V_T}{C_1} = \frac{0.376\ \mu\text{F} \times 100\ \text{V}}{0.47\ \mu\text{F}} = 80\ \text{V}$$
$$V_{C_2} = V_T - V_{C_1} = 100\ \text{V} - 80\ \text{V} = 20\ \text{V}$$
$$Q_{C_3} = V_{C_3} C_3 = 20\ \text{V} \times 1.2\ \mu\text{F} = 24\ \mu\text{C}$$

## Self-Test

**48.** A 10-μF and an 18-μF capacitor are connected in series to a 100-V source. Which capacitor will drop the most voltage? Will the total capacitance be less than 10 μF?

**49.** Calculate the total capacitance, the charge on the 10-μF capacitor, and the voltage on the 18-μF capacitor for the circuit in Question 48.

**50.** Change the value of $C_3$ in Fig. 15-21 to 0.91 μF and determine $C_T$, $Q_{C_3}$ and $V_{C_3}$.

**51.** A 1-μF, a 3-μF, and a 6-μF capacitor are series-connected to a 150-V source. Determine the voltage across the 3-μF capacitor and the total capacitance.

## SUMMARY

1. Electric fields exist around, and between, charged bodies.
2. A positive charge moves in the direction of the electric lines of force in an electric field.
3. Electric field intensity has units of either volts per meter or newtons per coulomb.
4. A uniform electric field has equipotential surfaces that are flat planes. It produces a uniform electric field intensity.
5. Electric flux is the total of the electric lines of force and has units of coulombs.
6. The density of electric flux is specified in coulombs per square meter.
7. A dielectric is an insulator material between two plates which receives an induced charge when the two plates are charged.
8. An electric field polarizes a dielectric by deforming its molecular structure.
9. Dielectric strength is a rating of a material's ability to withstand the stress of an electric field. It has units of volts per millimeter.
10. Dielectric absorption is the property of a dielectric material which causes it to remain partially polarized when the electric field is first removed.
11. Permittivity ($\epsilon$) is a rating of a dielectric's ability to accommodate electric flux. Relative permittivity compares a material's permittivity to that of a vacuum or free space.
12. Dielectric constant ($k$) is another name for relative permittivity.
13. Capacitance opposes changes in voltage; it stores energy; it is a proportionality constant relating $Q$ and $V$.
14. Farad is the base unit of capacitance. One farad is equal to coulombs per volt.
15. A charged capacitor has a voltage between its terminals which is equal to the voltage of the charging source.
16. Capacitance is directly proportional to plate area and dielectric constant and inversely proportional to the thickness of the dielectric (distance between plates).
17. Dielectric absorption can cause a capacitor to retain some charge even after its plates (leads) have been momentarily shorted together.
18. The energy stored in a capacitor is directly proportional to the square of the voltage.
19. Capacitors are rated for capacitance, tolerance, working voltage, temperature coefficient, dissipation factor, operating temperature range, and insulation resistance.
20. Variable capacitors are varied by changing either the effective plate area or the distance between the plates.
21. The curved line on the schematic symbol of a capacitor can represent the negative plate, the outside foil, or the rotor plate of various capacitors.
22. The dielectric in electrolytic capacitors is either aluminum oxide or tantalum pentoxide on the positive plate.

**23.** The electrolyte in an electrolytic capacitor is a conductor, it serves as part of the negative plate, and it maintains the oxide dielectric.

**24.** Nonpolarized electrolytics have an oxide on each plate.

**25.** Fixed, nonpolar capacitors may have single-layer, multilayer, or rolled plates.

**26.** Common dielectrics for fixed capacitors are mica, ceramic, paper, and various plastic film materials.

**27.** Metalized film capacitors are more compact than film-and-foil capacitors. They are also self-healing.

**28.** Stray (undesired) capacitance exists between any two conductive surfaces separated by insulation. It limits the high-frequency response of many devices and circuits.

**29.** Some tantalum (electrolytic) capacitors use a color code to indicate capacitance, tolerance, and working voltage.

**30.** Many fixed capacitors use a three-digit code to indicate capacitance.

**31.** Total series equivalent capacitance is calculated using the reciprocal formula.

**32.** In series capacitor circuits, voltage distributes inversely proportional to the individual capacitances.

**33.** Kirchhoff's laws apply to all types of capacitor circuits.

**34.** Useful formulas are

$$\epsilon_r = \frac{\epsilon}{\epsilon_o}$$

$$\text{Maximum voltage} = \text{dielectric strength} \times \text{dielectric thickness}$$

$$C = \frac{Q}{V}$$

$$C = \frac{\epsilon_o \epsilon_r a}{D} = \frac{8.85ka}{10^{12}D}$$

$$W = 0.5CV^2$$

For series capacitors,

$$C_T = \frac{1}{1/C_1 + 1/C_2 + 1/C_3} \quad \text{or} \quad C_T = \frac{C_1C_2}{C_1 + C_2} \quad \text{or} \quad C_T = \frac{C}{n}$$

$$V_{C_n} = \frac{C_T V_T}{C_n} \quad \text{or} \quad V_{C_1} = \frac{C_2 V_T}{C_1 + C_2} \quad \text{or} \quad V_{C_2} = \frac{C_1 V_T}{C_1 + C_2}$$

For parallel capacitors,

$$C_T = C_1 + C_2 + C_3$$

## CHAPTER REVIEW QUESTIONS

For the following items, determine whether each statement is true or false.

**15-1.** A unit of charge is assumed to produce one electric line of force.

**15-2.** In an electric field, negative charges move in the direction of the electric lines of force.

**15-3.** For a given electric field intensity, electric flux density increases when the dielectric constant increases.

**15-4.** All capacitors are represented by a single schematic symbol.

**15-5.** Doubling the thickness of the dielectric in a capacitor doubles its capacitance.

**15-6.** Changing from an air dielectric to an equal size of glass dielectric increases the capacitance of the capacitor.

**15-7.** In terms of energy stored, doubling $C$ is more effective than doubling $V$.

**15-8.** The voltage rating of a capacitor is determined primarily by the dielectric constant of the dielectric material.

**15-9.** NPO is the abbreviation for "no polarity operation" of a capacitor.

**15-10.** The curved line on a capacitor symbol indicates the positive lead of a capacitor.

**15-11.** Most variable capacitors use the principle that capacitance is directly proportional to the dielectric constant.

**15-12.** The electrolyte in a capacitor is part of the negative plate.

**15-13.** Rolling the plates of a capacitor reduces the effective plate area.

**15-14.** Tantalum capacitors sometimes use a color code to indicate capacitance, voltage rating, and tolerance.

**15-15.** The total equivalent capacitance of series capacitors is always less than the value of the smallest capacitor.

**15-16.** The smallest capacitor in a series capacitor circuit drops the least voltage.

**15-17.** In a parallel capacitor circuit, the largest capacitor receives the most charge from the source.

For the following items, fill in the blank with the word or phrase required to correctly complete each statement.

**15-18.** The unit for electric flux is the ______.

**15-19.** Another name for dielectric constant is ______.

**15-20.** An electric field ______ charges on the surface of a dielectric material.

**15-21.** The device which stores electric energy is the ______.

**15-22.** Coulombs per volt is equal to a ______.

**15-23.** Capacitance opposes any change in ______.

**15-24.** The charge remaining on a capacitor after its leads have been momentarily shorted together is caused by ______.

**15-25.** Doubling the plate area of a capacitor will ______ the capacitance.

**15-26.** The colored band on a capacitor identifies the ______.

**15-27.** ______ capacitors are self-healing.

**15-28.** Stray capacitance is most troublesome in circuits operating at ______ frequencies.

**15-29.** The dielectric of an aluminum electrolytic capacitor is ______ which is formed on the ______ plate.

For the following items, choose the letter that best completes each statement.

**15-30.** The unit for flux density is
  a. Newtons per volt
  b. Coulombs per square meter
  c. Volts per meter squared
  d. Newtons per coulomb

**15-31.** The unit for electric field intensity is
  a. Newtons per volt
  b. Newtons per meter
  c. Volts per meter
  d. Coulombs per meter

**15-32.** The ability of a material to withstand an electric field is known as its
  a. Absorption strength
  b. Polarization strength
  c. Permittivity strength
  d. Dielectric strength

**15-33.** The ability of a material to remain polarized after the removal of the electric field is called
  a. Dielectric absorption
  b. Polarization strength
  c. Dielectric permittivity
  d. Relative permittivity

**15-34.** Capacitor specifications usually do not list
  a. Dielectric constant
  b. Insulation resistance
  c. Dissipation factor
  d. Dc working voltage

**15-35.** Which of the following is not used as a dielectric in capacitors?
  a. Tantalum pentoxide
  b. Mylar
  c. Ferrite
  d. Paper

## CHAPTER REVIEW PROBLEMS

**15-1.** Determine the maximum voltage rating of a capacitor that has 0.02 mm of dielectric thickness if the dielectric strength is 20 kV/mm.

**15-2.** Determine the voltage on a 4-$\mu$F capacitor which is storing 0.01 C of charge.

**15-3.** A capacitor produces 60 V when storing 0.03 C of charge. What is its capacitance?

**15-4.** A capacitor has two plates that are each 15 mm $\times$ 45 cm. The plates are separated by a dielectric which is 0.2 mm thick and has a relative permittivity of 2. Determine its capacitance.

**15-5.** How much energy will a 2000-$\mu$F capacitor store when it is charged to 300 V?

**15-6.** What is the capacitance of a capacitor labeled 333 M?

**15-7.** What is the total equivalent capacitance for a 4-$\mu$F capacitor and an 8-$\mu$F capacitor connected in parallel?

**15-8.** If the capacitors in Problem 15-7 were in series, what would the equivalent capacitance be?

**15-9.** For the circuit in Fig. 15-22 determine $C_T$, $V_{C_2}$, and $Q_T$.

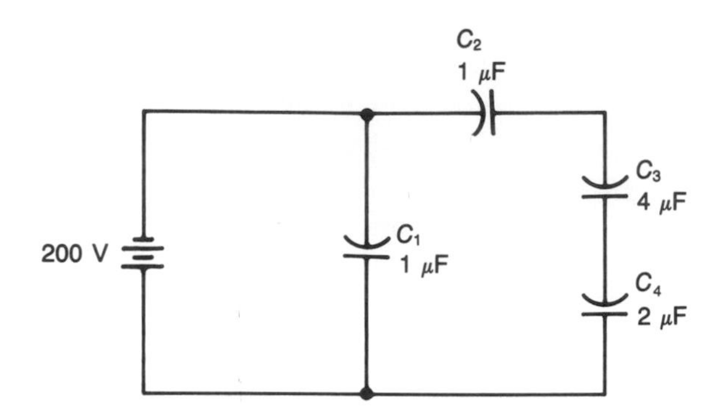


**FIGURE 15-22 Circuit for Review Problem 15-9.**

## ANSWERS TO SELF-TESTS

**1.** an invisible force field that exists around an electric charge

**2.** negative

**3.** newtons per coulomb or volts per meter

**4.** The force becomes one-fourth as great.

**5.** an electric field in which the lines of force are evenly spaced and parallel to each other.

**6.** equipotential surface

**7.** electric flux

**8.** electric flux density

**9.** insulator

**10.** It becomes polarized and receives an induced charge.

**11.** false

**12.** a. volts per millimeter, b. coulombs squared per newton-meter squared or farads per meter, c. unitless

**13.** dielectric constant

**14.** dielectric absorption

**15.** dielectric polarization

**16.** 20 kV

**17.** farad (F)

**18.** three

**19.** A device that has capacitance. It consists of two plates separated by a dielectric.

**20.** double

**21.** 20 mC

**22.** When a source forces charges onto the capacitor's plate, a voltage is established between the plates. Voltage, by definition, is a potential energy difference between two points.

**23.** plate area—directly, distance between plates—inversely, permittivity—directly

**24.** dielectric absorption

**25.** 1062 pF

**26.** voltage, because $W = 0.5CV^2$

**27.** 5 J

**28.** 2.5 kW

**29.** insulation resistance, dissipation factor, temperature coefficient, temperature range, working voltage, and tolerance

**30.** dielectric strength and thickness of the dielectric

**31.** greater than

**32.** variable, fixed polar, and fixed nonpolar

**33.** vary the effective plate area or vary the distance between the plates

**34.** the rotor plate, the negative plate, and the outside foil

**35.** It is part of the negative plate, and it helps form the oxide on the positive plate.

**36.** the positive plate

**37.** tantalum pentoxide

**38.** a small variable capacitor

**39.** tantalum electrolytic, aluminum electrolytic, metalized foil, film-and-foil, and variable

**40.** by a colored band toward the lead end that connects to the outside foil

**41.** mica, ceramic, paper, and plastic films including polyester, polycarbonate, polystyrene, and polypropylene

**42.** because the deposited plate is vaporized around the hole where film failure occurs

**43.** Stray capacitance is the capacitance that exists between any two conductive surfaces which are insulated from each other.

**44.** high frequencies

**45.** tantalum

**46.** 0.047 $\mu$F $\pm$10 percent

**47.** 0.01 $\mu$F

**48.** the 10 $\mu$F, yes

**49.** $C_T = 6.43\ \mu$F, $Q = 643\ \mu$C, $V = 35.7$ V

**50.** $C_T = 0.363\ \mu$F, $Q_{C1} = 20.7\ \mu$C, $V_{C3} = 22.8$ V

**51.** $V = 33.3$ V, $C_T = 0.667\ \mu$F

# Chapter 16 Capacitors in DC and AC Circuits

Chapter 15 introduced you to capacitance. In Sec. 15-10 you were introduced to the behavior of capacitors in dc circuits when the voltage distribution across series capacitors was discussed. This chapter continues on with the behavior of capacitors in dc circuits and then covers the behavior of capacitors in ac circuits. The general principles of measuring capacitance are also covered in this chapter.

## 16-1 CONTROLLED CHARGING AND DISCHARGING

In Chap. 15 it was stated that capacitors require a fraction of a second to charge up to the source voltage because of resistances in the source, capacitor, and circuit leads. It was also stated that during that fraction of a second, Kirchhoff's voltage law must be satisfied. No attempt was made to estimate the instantaneous voltages or currents because the resistance was not known. To further our understanding of capacitance and how it is used in circuits, we need to explore in greater detail how a capacitor charges and discharges.

Figure 16-1*a* shows a capacitor charging through an external resistance. If this external resistance is large enough, we can ignore the internal resistance of the source and the capacitor and use the resistor's value in estimating instantaneous currents and voltages. Suppose the capacitor in Fig. 16-1*a* is discharged and then the switch is closed. At the instant of switch closure, the voltage across

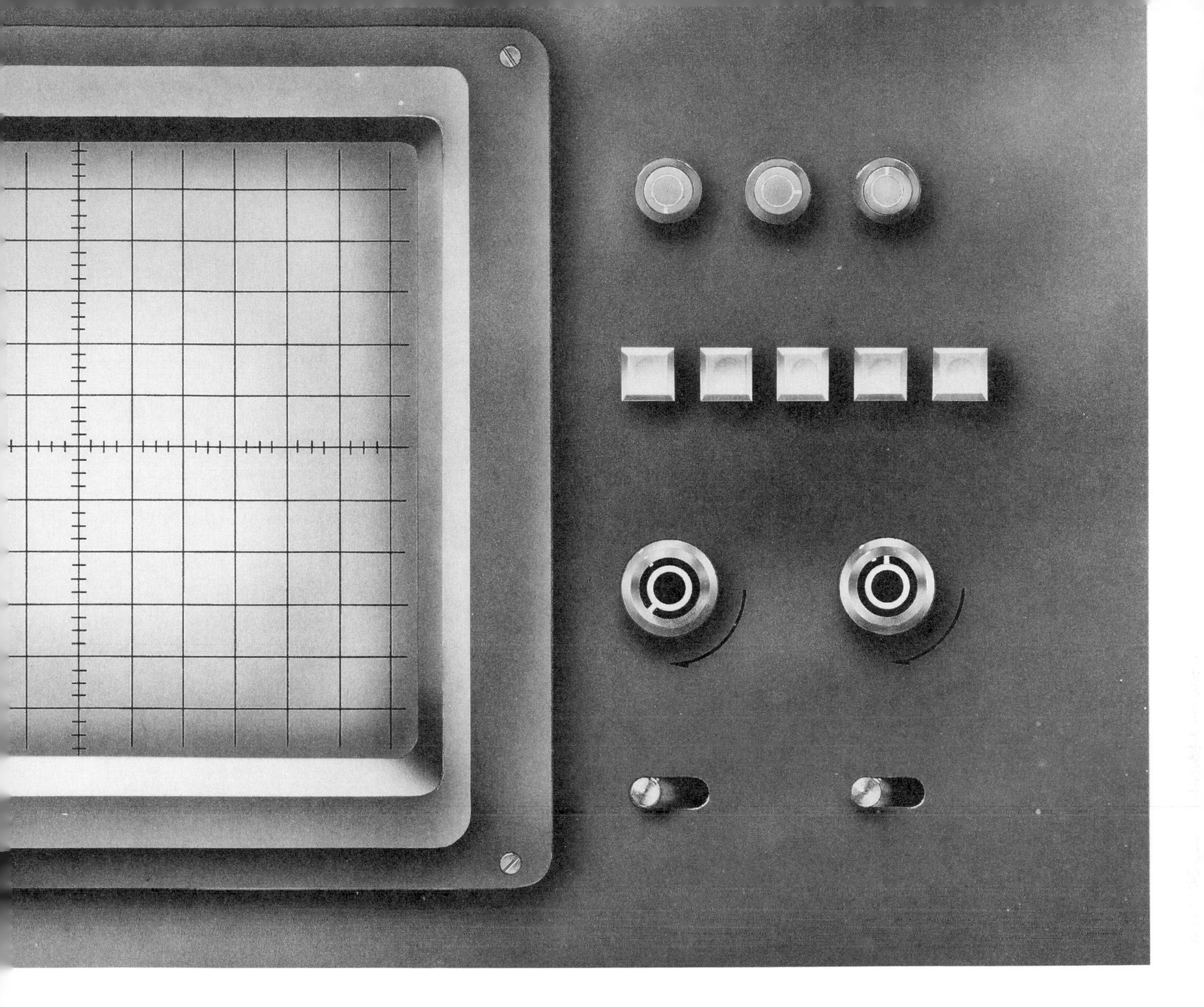

the capacitor is zero. Thus, by Kirchhoff's voltage law, the source voltage $V_T$ appears across the resistor $R$. The instantaneous current $I$ is

$$I = \frac{V_T}{R} = \frac{100\ \text{V}}{1\ \text{M}\Omega} = 100\ \mu\text{A}$$

This is the value of current shown in Fig. 16-1*b* at time zero (the instant the switch is closed). At this current level, charge $Q$ is building up on the capacitor at a rate of $I = 100\ \mu\text{A} = Q/t = 100\ \mu\text{C/s}$. However, as shown in Fig. 16-1*b*, as the charge accumulates on the plates of the capacitor, a voltage builds up across the capacitor. Of course, this voltage ($V_C$) reduces the voltage across the resistor ($V_R = V_T - V_C$), which, in turn, must reduce the circuit current and the rate at which charge is de-

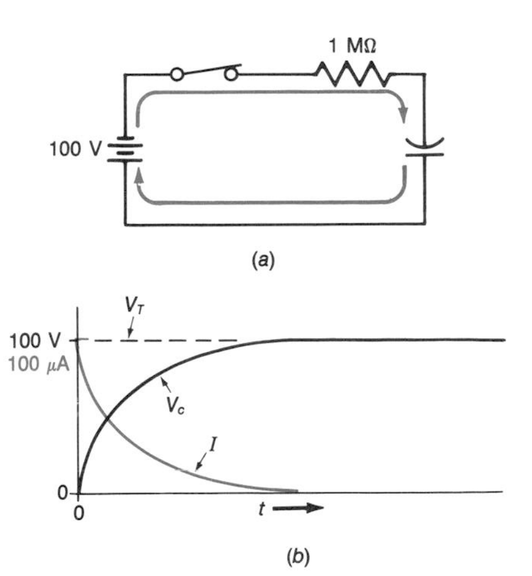


**FIGURE 16-1 Voltage and current waveforms for a charging capacitor.**

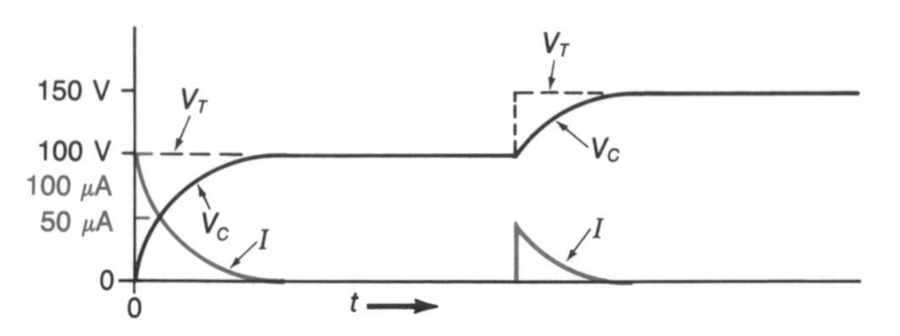


**FIGURE 16-2 Capacitor response to an increase in source voltage.**

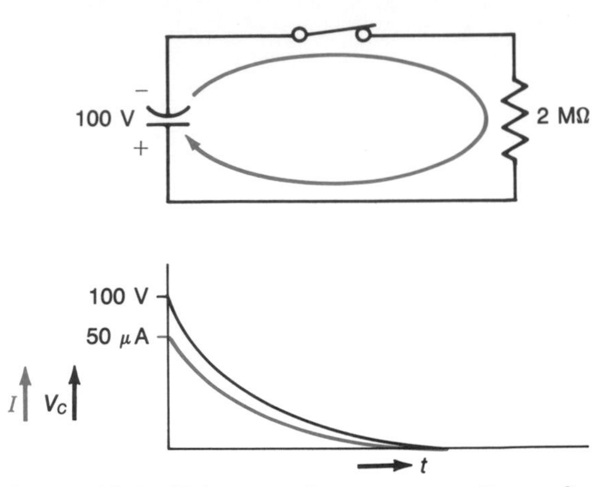


**FIGURE 16-4 Voltage and current waveforms for a discharging capacitor.**

posited on the capacitor plates. If the rate at which charge is accumulating decreases, then the rate at which the capacitor voltage builds up must also decrease. As shown in Fig. 16-1, both the magnitude of the current and the rate at which the capacitor voltage changes continue to decrease until both reach a zero value. At this time, the capacitor is charged to the source voltage and there is no current flowing (assuming the capacitor is ideal and has infinite insulation resistance). As long as the source voltage does not change, nothing further happens in this circuit.

However, if at some later time the source voltage suddenly increases to 150 V, the circuit current again flows until the capacitor is charged up to the source voltage. As seen in Fig. 16-2, the instantaneous current jumps up to 50 μA and then decays back to zero. If, on the other hand, the source voltage suddenly decreases to 50 V, the capacitor discharges until its voltage again equals that of the source. As shown in Fig. 16-3*a*, when the capacitor discharges, it forces current in the reverse direction back through the source. Figure 16-3*b* shows the resulting waveforms.

Notice in Figs. 16-1 through 16-3 that the current only flows during the time it takes for the capacitor to charge or discharge so that its voltage equals the source voltage. This is why it is often stated that a capacitor "blocks dc" or that a capacitor "acts like an open circuit to dc." Also notice from Figs. 16-2 and 16-3 that as long as the source voltage is changing, current is flowing. Although these figures illustrate a sharp step change in the source voltage, the same thing would happen, i.e.,

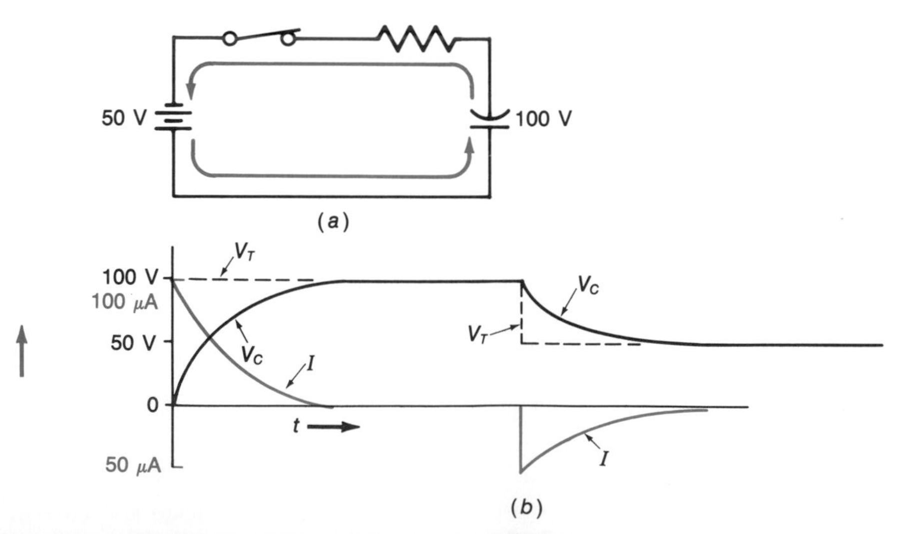


**FIGURE 16-3 Capacitor response to a decrease in source voltage.**

current would flow, if the source voltage were changing in a sinusoidal fashion. The only difference would be in the shape of the waveforms.

Figure 16-4 illustrates how a charged capacitor discharges through a known resistance. The instant the switch is closed, the current jumps to its maximum instantaneous value and decays as the voltage on the capacitor decays. Of course, the current and voltage reach zero at the same instant.

### Self-Test

**1.** A discharged capacitor is in series with a 470-kΩ resistor. What is the peak instantaneous current when this series combination is connected to a 50-V source?

**2.** A capacitor is charged to 100 V and then discharged through a 5-Ω resistor. What is the peak instantaneous current?

**3.** What does it mean to say a capacitor blocks dc?

## 16-2 TIME CONSTANT

To specify the time it takes to charge or discharge a capacitor through a resistor, the *time constant* of the *RC* combination must be determined. For an *RC* circuit, a time constant is the amount of time it would require for the capacitor to charge to the source voltage *if* the current remained constant at its peak instantaneous value. The general formula for determining this time can be obtained by applying the formulas and relationships we already know. We know the following: $V = Q/C$, $Q = It$, $I = V_T/R$, where $I$ is the peak instantaneous current and $V = V_T$ when the capacitor is fully charged. Now we can write

$$V_T = V = \frac{Q}{C} = \frac{It}{C} = \frac{\mathrm{V_T/R \times t}}{\mathrm{C}} = \frac{\mathrm{V_T\,t}}{\mathrm{RC}}$$

and solving for $t$ yields

$$t = RC$$

The symbol for a time constant is $\tau$ (the Greek lowercase letter tau). Since we just determined that the time for a time constant is $R$ multiplied by $C$, the formula relating time constant $\tau$ to $R$ and $C$ is

$$\tau = RC$$

where $\tau$ is in seconds when $R$ is in ohms and $C$ is in farads. While it may not be obvious that multiplying ohms by farads gives seconds, a little substitution of base unit equivalents shows it to be true:

$$\Omega = \frac{\mathrm{V}}{\mathrm{A}} = \frac{\mathrm{V}}{\mathrm{C/s}} = \frac{\mathrm{V \cdot s}}{\mathrm{C}}$$

$$\mathrm{F} = \frac{\mathrm{C}}{\mathrm{V}}$$

Therefore,

$$\Omega \cdot \mathrm{F} = \frac{\mathrm{V \cdot s}}{\mathrm{C}} \times \frac{\mathrm{C}}{\mathrm{V}} = \mathrm{s}$$

As we saw in Sec. 16-1, the current does not remain at its peak instantaneous value. Rather, it decays at an exponential rate while the voltage is building up at an exponential rate. Thus, the capacitor only partially charges in $1\tau$. Through differential calculus it has been determined that the relationship between $\tau$, $t$, $V_T$, and $V_C$ is

$$V_{C,t} - V_T - (V_T - V_{CE})e^{-t/\tau}$$

where $V_{C,t}$ = capacitor voltage at time $t$
$V_T$ = source voltage charging capacitor
$V_{CE}$ = existing voltage on capacitor
$e$ = exponential constant ($e = 2.718$)
$\tau$ = time constant ($\tau = RC$)

When $V_{CE}$ is zero, this formula reduces to

$$V_{C,t} = V_T - V_T e^{-t/\tau}$$

which, when $V_T$ is factored out, becomes

$$V_{C,t} = V_T(1 - e^{-t/\tau})$$

When $t = \tau$, this formula can express $V_{C,\tau}$ (the capacitor voltage after $1\tau$) as a portion of $V_T$. At the end of $1\tau$, when $t = \tau$, the exponent of $e$ will be $-1$ and $V_{C,\tau}$ is

$$V_{C,\tau} = V_T(1 - e^{-1}) = V_T(1 - 0.368) = 0.632\ V_T$$

Thus, in $1\tau$ a discharged capacitor charges to 63.2 percent of the source voltage. When $t = 2\tau$, i.e., after two time constants, $V_{C,2\tau}$ is

$$V_{C,2\tau} = V_T(1 - e^{-2}) = V_T(1 - 0.135) = 0.865\ V_T$$

This shows that $C$ charges to 86.5 percent of the source voltage in $2\tau$. As you might expect from examination of the calculus-derived formula for $V_{C,t}$, the value of $V_{C,2\tau}$ could also be expressed as

$$V_{C,2\tau} = V_{C,\tau} + 0.632(V_T - V_{C,\tau})$$

That is, the capacitor takes on an additional 63.2 percent of the remaining voltage ($V_T - V_{CE}$) each time constant. Using either of the two above formulas (and changing the subscript and exponent numbers) for 3, 4, 5, 6, and 7$\tau$ yields 95, 98.2, 99.3, 99.8, and 99.9 percent, respectively.

Figure 16-5 is a curve which plots capacitor voltage ($V_{C,t}$), in percentage of source voltage, against time expressed in time constants. This curve is very useful for graphically estimating the voltage on a capacitor in an *RC* circuit after a specified time. Just convert time into time constants and read the graph. This graph and the previous calculations show that for practical work, the capacitor can be considered to be charged after 5$\tau$.

The above calculations showed that the capacitor charges another 63.2 percent of the voltage difference between $V_T$ and $V_C$ for each time constant. Because of this fact, a time constant is sometimes defined as "the time it takes a capacitor to charge to 63.2 percent of the available voltage, or to discharge 63.2 percent of its voltage." In this definition, the "available voltage" is the difference between the voltage on the capacitor ($V_{CE}$) and the source voltage ($V_T$).

The calculus formula can also be used to determine the voltage remaining on a discharging capacitor after a period of time. When there is no source voltage (as in the circuit in Fig. 16-4), $V_T$ is zero and the formula reduces to

$$V_{C,t} = V_T - (V_T - V_{CE})e^{-t/\tau} = 0 - (0 - V_{CE})e^{-t/\tau}$$

or

$$V_{C,t} = V_{CE}\, e^{-t/\tau}$$

where $V_{CE}$ is, of course, the voltage to which the capacitor was originally charged. When $t = \tau$, this

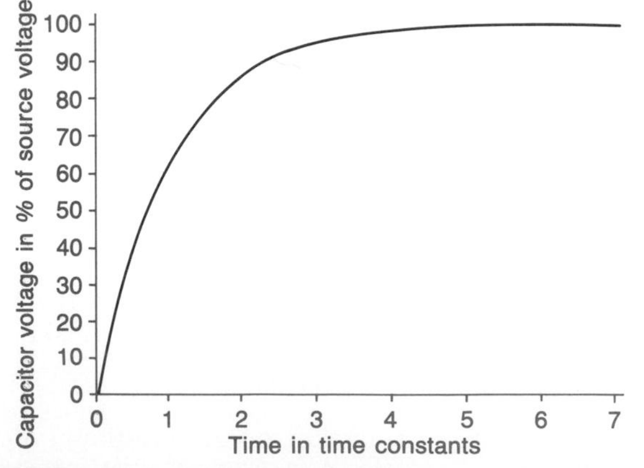


**FIGURE 16-5 Universal charging curve for *RC* circuits.**

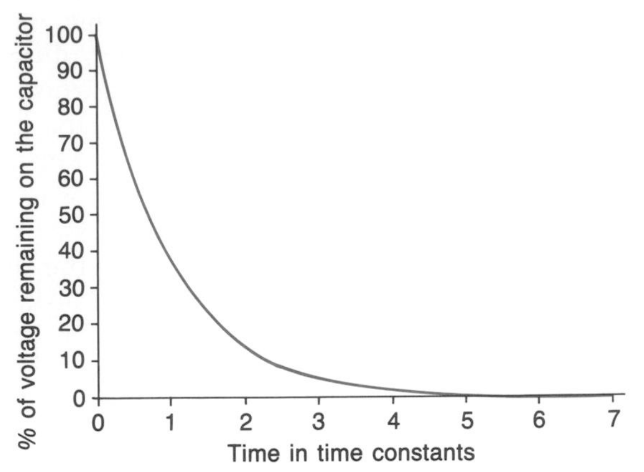


**FIGURE 16-6 Universal discharging curve for *RC* circuits.**

formula shows that $V_{C,t} = V_{CE}\, e^{-1} = 0.368V_{CE}$, which shows that the discharging capacitor loses 63.2 percent of its voltage in 1$\tau$. By the technique used for a charging capacitor, it can be shown that a discharging capacitor loses 63.2 percent of its remaining voltage for each time constant. A universal discharging curve like that in Fig. 16-6 can be used to estimate the voltage remaining on a capacitor.

So far we have learned how to calculate, or estimate from a graph, the instantaneous voltage ($V_{C,t}$) on a capacitor at time *t*. The same formulas and graphs can also be used to determine instantaneous currents in the *RC* circuit. As discussed in Sec. 16-1, current decays exponentially as voltage grows exponentially for a charging capacitor. Thus, Fig. 16-6 is also a curve showing how current decays with time. After 1$\tau$, the current is 36.8 percent of its peak instantaneous value. The formula for the instantaneous voltage of a discharging capacitor can be used for the instantaneous current of a charging capacitor by changing *V* to *I*. Thus, for a charging capacitor

$$I_{C,t} = I_{CE}\, e^{-t/\tau}$$

where $I_{C,t}$ is the instantaneous current charging the capacitor, and $I_{CE}$ is the magnitude of the charging current at the beginning of the time interval. The current created by a discharging capacitor decays in step with the voltage. Therefore, this current formula is also the correct one for the instantaneous current in a discharging *RC* circuit.

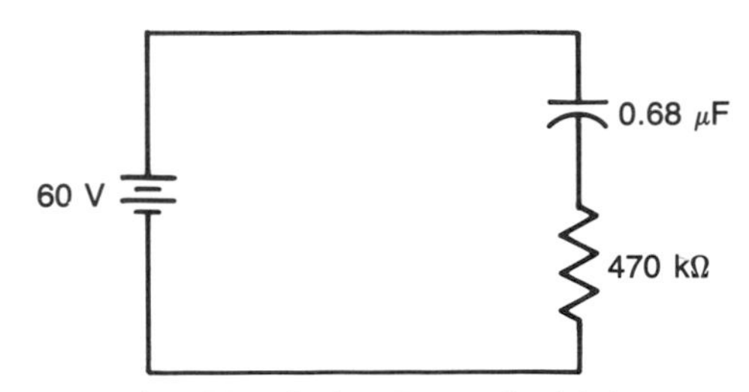


**FIGURE 16-7 Circuit for Example 16-1.**

### EXAMPLE 16-1

For the circuit in Fig. 16-7, determine $\tau$ and $V_{C,t}$ after 0.2 s and $I_{C,t}$ after 0.1 s when the capacitor was originally discharged.

**Solution**

$$\tau = RC = 470 \times 10^3\ \Omega \times 0.68 \times 10^{-6}\ \text{F} = 0.32\ \text{s}$$
$$V_{C,t} = V_T(1 - e^{-t/\tau}) = 60\ \text{V}(1 - e^{-0.2/0.32})$$
$$= 60(1 - 0.535) = 27.9\ \text{V}$$

*Note:* $e^{-0.2/0.32}$ can be solved by entering $-0.625$ $(-0.2/0.32)$ into your calculator and pressing the $e^x$ function key.

$$I_{CE} = \frac{V_T}{R} = \frac{60\ \text{V}}{470\ \text{k}\Omega} = 0.128\ \text{mA}$$
$$I_{C,t} = I_{CE}\, e^{-t/\tau} = 0.128\ \text{mA} \times e^{-0.1/0.32} = 0.0934\ \text{mA}$$

The graphs in Figs. 16-5 and 16-6 can be used to see if this solution contains any gross errors. For $V_{C,t}$ the capacitor charges approximately two-thirds (0.2/0.32) of a time constant. Referring to Fig. 16-5 and projecting up from two-thirds intersects the curve at about 48 percent. Since 48 percent of 60 V is 28.8 V, it appears no gross errors were made in calculating $V_{C,t}$. For the instantaneous current, the capacitor charged about $\frac{1}{3}\tau$. Using the curve in Fig. 16-6, $\frac{1}{3}\tau$ yields about 74 percent of peak current. Again, no gross errors are apparent because 74 percent of 0.128 mA is 0.095 mA, which is close to the calculated value of 0.0934 mA.

### EXAMPLE 16-2

Refer to Fig. 16-4 and assume the capacitance is 1.5 $\mu$F. Determine $V_{C,t}$ and $I_{C,t}$ after 5 s.

**Solution**

$$\tau = 1.5\ \mu\text{F} \times 2\ \text{M}\Omega = 3\ \text{s}$$
$$V_{C,t} = V_{CE}\, e^{-t/\tau} = 100\ \text{V} \times 0.189 = 18.9\ \text{V}$$
$$I_{CE} = \frac{V_{CE}}{R} = \frac{100\ \text{V}}{2\ \text{M}\Omega} = 50\ \mu\text{A}$$
$$I_{C,t} = I_{CE}\, e^{-t/\tau} = 50\ \mu\text{A} \times 0.189 = 9.45\ \mu\text{A}$$

### EXAMPLE 16-3

Refer to Fig. 16-3 and assume $R$ is 1 M$\Omega$ and $C$ is 0.47 $\mu$F. Determine $V_{C,t}$ and $I_{C,t}$, 1 s after the switch is closed. (The existing voltage on the capacitor $V_{CE}$ is 100 V as shown on the diagram.)

**Solution**

$$\tau = 1\ \text{M}\Omega \times 0.47\ \mu\text{F} = 0.47\ \text{s}$$
$$V_{C,t} = V_T - (V_T - V_{CE})e^{-t/\tau}$$
$$= 50\ \text{V} - (50\ \text{V} - 100\ \text{V})e^{-1/0.47}$$
$$V_{C,t} = 50\ \text{V} - (-5.96\ \text{V}) = 55.96\ \text{V}$$
$$I_{CE} = \frac{V_{CE} - V_T}{\text{R}} = \frac{100\ \text{V} - 50\ \text{V}}{1\ \text{M}\Omega} = 50\ \mu\text{A}$$
$$I_{C,t} = I_{CE}\, e^{-t/\tau} = 50\ \mu\text{A} \times 0.119 = 5.96\ \mu\text{A}$$

### EXAMPLE 16-4

The capacitor in Fig. 16-8 is discharged. Determine the final voltage on the capacitor after the switch has been in position 2 for 3 s and then in position 3 for 5 s.

**Solution**

With switch $S_1$ in position 2,

$$\tau = 1\ \text{M}\Omega \times 2\ \mu\text{F} = 2\ \text{s}$$
$$V_{C,t} = V_T(1 - e^{-t/\tau}) = 75\ \text{V} \times 0.777 = 58.3\ \text{V}$$

With $S_1$ in position 3,

$$\tau = 2\ \text{M}\Omega \times 2\ \mu\text{F} = 4\ \text{s}$$
$$V_{C,t} = V_T - (V_T - V_{CE})e^{-t/\tau}$$
$$= 100\ \text{V} - [(100\ \text{V} - 58.3\ \text{V}) \times 0.287] = 88.0\ \text{V}$$

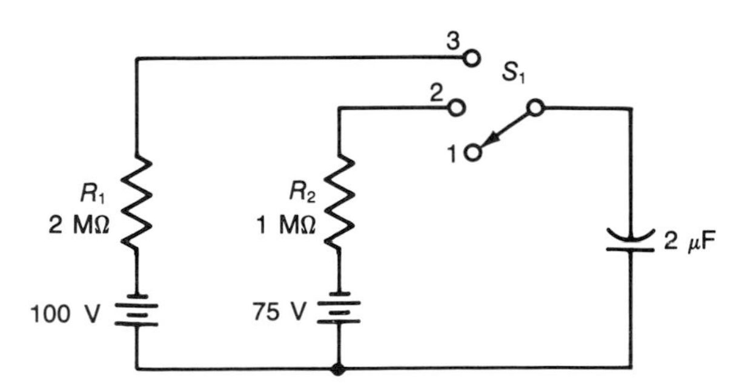


**FIGURE 16-8 Circuit for Example 16-4.**

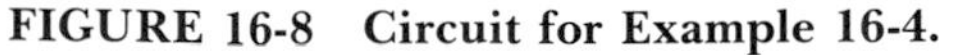

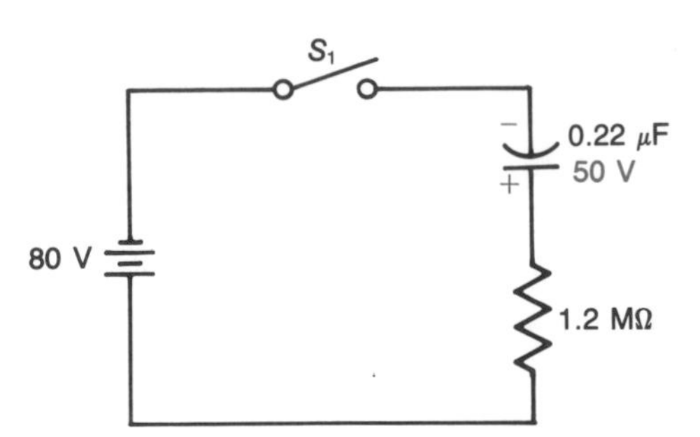


**FIGURE 16-9 Circuit for Self-Test Question 7.**

## Self-Test

**4.** An $RC$ circuit is connected to a 300-V dc source. Determine the voltage across the capacitor after $1\tau$, $1.5\tau$, and $3\tau$.

**5.** Change $R$ in Fig. 16-7 to 330 kΩ and then determine the voltage across the capacitor after the circuit has been connected for 0.1 s. (Assume $C$ was initially discharged.)

**6.** A 3-μF capacitor is charged to 90 V and then discharged through a 560-kΩ resistor for 3 s. Determine the voltage and the current at the end of this discharge time.

**7.** Refer to Fig. 16-9. Determine $V_{C,t}$ after $S_1$ has been closed for 0.2 s and $I_{C,t}$ after $S_1$ has been closed for 2.0 s.

**8.** When the voltage across the capacitor in a dc series $RC$ circuit is increasing, the voltage across $R$ is ______ and the circuit current is ______.

Sometimes it is desirable to determine how long it will take the capacitor in an $RC$ circuit to charge or discharge to a specified voltage. This can be done by solving the calculus-derived formula for $t$. The solution is straightforward because natural logs (ln) are to the base $e$ (2.718). Starting with $V_{C,t} = V_T - (V_T - V_{CE})e^{-t/\tau}$, the steps are

**1.** Multiply by $-1$ and add $V_T$ to both sides. This yields

$$V_T - V_{C,t} = (V_T - V_{CE})e^{-t/\tau}$$

**2.** Divide by $V_T - V_{CE}$ to get

$$\frac{V_T - V_{C,t}}{V_T - V_{CE}} = e^{-t/\tau}$$

**3.** Take the reciprocal of both sides to get

$$\frac{V_T - V_{CE}}{V_T - V_{C,t}} = e^{t/\tau}$$

**4.** Take the natural log and multiply by $\tau$ to get the result in the most usable form:

$$t = \tau \ln \frac{V_T - V_{CE}}{V_T - V_{C,t}}$$

When $C$ is charging from 0 V, the formula is

$$t = \tau \ln \frac{V_T}{V_T - V_{C,t}}$$

And, when $C$ is discharging, the appropriate formula is

$$t = \tau \ln \frac{V_{CE}}{V_{C,t}}$$

Since $\tau = RC$, $RC$ could be substituted into any of the three above formulas. Then, any of the formulas could be solved for $R$ or $C$.

### EXAMPLE 16-5

How long will it take the capacitor in Fig. 16-9 to charge to 66 V after $S_1$ is closed?

**Solution**

$\tau = 1.2\ \text{M}\Omega \times 0.22\ \mu\text{F} = 0.264\ \text{s}$

$$t = 0.264\ \text{s} \ln \frac{80\ \text{V} - 50\ \text{V}}{80\ \text{V} - 66\ \text{V}} = 0.264\ \text{s} \times 0.762 = 0.20\ \text{s}$$

Notice that the answer to this example agrees with Self-Test Question 7, in which it was determined that the voltage on the capacitor would be 65.9 V after 0.2 s.

Figure 16-10 illustrates the importance of the $RC$ time constant when a square wave voltage is applied to a series $RC$ circuit. In Fig. 16-10$a$, the time constant is $\tau = 10 \times 10^3\ \Omega \times 1 \times 10^{-6}\ \text{F} = 0.01\ \text{s} = 10\ \text{ms}$, which is 10 times as long as the period of the waveform. During the 0.5 ms (one-half of the period) that the waveform remains at either +5 V or −5 V, the capacitor builds up very little voltage. Specifically, for the first $\frac{1}{2}$ cycle, its voltage is

$$V_{C,t} = 5\ \text{V}\ (1 - e^{-0.5\ \text{ms}/10\ \text{ms}}) = 0.24\ \text{V}$$

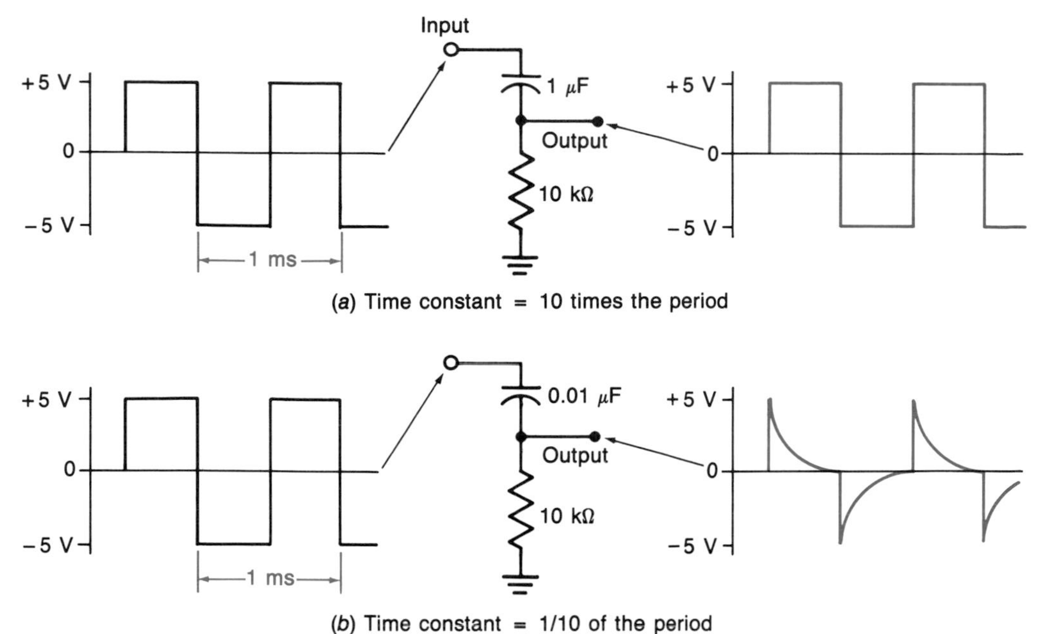


**FIGURE 16-10** ***RC* response to a square wave.**

The output waveform across the resistor has about a $\frac{1}{4}$-V drop from its leading edge to its trailing edge.

In Fig. 16-10*b*, the time constant has been reduced to $\tau = 10 \times 10^3 \times 0.01 \times 10^{-6} = 0.1$ ms. Thus, during one-half of the period, which equals $5\tau$, the capacitor essentially charges to 5 V. The output voltage (voltage across the resistor) is essentially zero by the end of the $\frac{1}{2}$ cycle. As seen in Fig. 16-10*b*, the output from the square wave input is very distorted (due to differentiation) when the time constant is small compared to the period of the waveform.

## Self-Test

**9.** How long will it take a 0.01-μF capacitor, which is charged to 65 V, to discharge to 30 V through a 10-kΩ resistor?

**10.** How long will it take a 10-μF capacitor to charge to 23.5 V through a 2.2-MΩ resistor if the source voltage is 78 V?

**11.** For minimum distortion of a square wave appearing across the resistor, should the time constant of the *RC* circuit be greater than or less than the period of waveform?

## 16-3 CAPACITOR ACTION IN AN AC CIRCUIT

The charging and discharging action of a capacitor in a sinusoidal ac circuit is quite different than in a dc circuit because the source voltage is continuously changing. In a simple circuit like that in Fig. 16-11, the capacitor's voltage must, according to Kirchhoff's voltage law, stay equal to the source voltage. To do so, the capacitor has to charge and store energy whenever the source voltage is increasing in either a positive or a negative direction. When the source voltage is decreasing, the capacitor must discharge and return energy to the source. Thus, as illustrated in Fig. 16-11, the capacitor charges for the first $\frac{1}{4}$ cycle as the source voltage builds from zero to peak positive voltage. To charge the capacitor, current must flow in the direction shown in Fig. 16-11*a*. During this first $\frac{1}{4}$

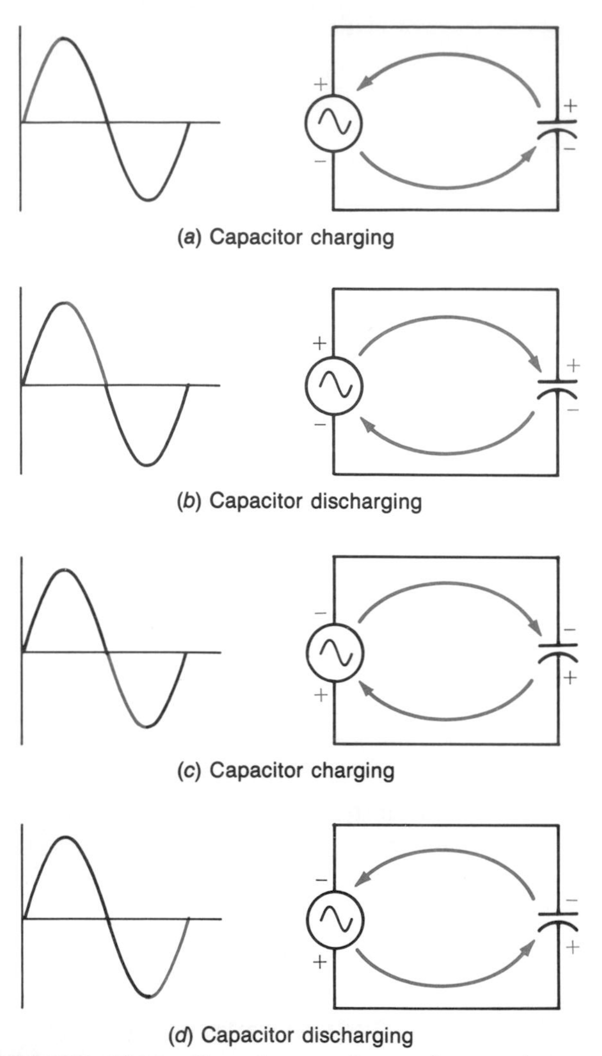


**FIGURE 16-11 Capacitor storing and returning energy.**

cycle, the current starts out at its maximum value and decays, sinusoidally, to zero by the time the voltage has reached its peak value. Why current behaves in this manner can be understood by referring to Fig. 16-12*a*. When the sine wave is going through its zero value, it has its greatest slope, i.e, its voltage is changing most rapidly (per unit of time) at this time. For the capacitor's voltage to follow this rapid increase in source voltage, it must receive a large amount of charge per unit of time. Since current is charge per unit of time ($I = Q/t$), the current is maximum as the source voltage passes through zero. At the other extreme, as illustrated in Fig. 16-12*a*, the change in voltage (slope) is zero when the voltage waveform is at its peak value. Of course, if the voltage is not changing at that instant, the charge on the capacitor is not changing either. Thus, the current is zero when the voltage is at its peak.

Referring back to Fig. 16-11*b* shows what happens during the second $\frac{1}{4}$ cycle. Now the source voltage is decaying so the capacitor has to discharge; therefore, the direction of the current reverses. During this $\frac{1}{4}$ cycle, the capacitor is returning the energy it stored during the first $\frac{1}{4}$ cycle.

Notice in Fig. 16-11 that the direction of the current is the same during the second and third $\frac{1}{4}$ cycles. As the voltage waveform leaves zero at the start of this third $\frac{1}{4}$ cycle, the capacitor is again being charged and is storing energy. Since the polarity across the capacitor reverses at the time the capacitor changes from returning energy to storing energy, the current in the circuit does not change directions.

Finally, during the fourth $\frac{1}{4}$ cycle, the capacitor again discharges as seen in Fig. 16-11*d*. The sequence of events shown in Fig. 16-11 is repeated cycle after cycle.

## 16-4 PHASE ANGLE AND POWER

For the reason discussed in Sec. 16-3, the current in an ideal-capacitor circuit reaches its peak value when the voltage is zero. (An ideal capacitor is one which has a dissipation factor of zero.) As seen in Fig. 16-12*b*, this causes the current to be exactly 90° out of phase with the voltage. Notice that the *current leads the voltage* by 90° because the current reaches its + peak before the voltage reaches its + peak. This current-voltage relationship is shown in phasor form in Fig. 16-12*c*. In this figure, the angle between the current and the voltage is represented by $\theta$ (Greek lowercase theta). This angle is referred to as the *phase angle* or *angle theta*.

The power in an ac circuit may be calculated by

$$P = IV \cos \theta$$

or

$$P = I^2R$$

where $P$ = power
$I$ = current
$V$ = voltage
$R$ = resistance

The last formula is true because only resistance converts electric energy into heat energy. Since an ideal capacitor has no resistance, an ideal-capacitor

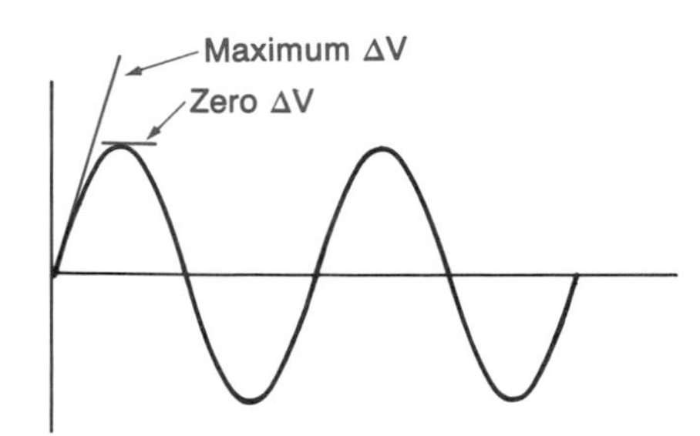


(*a*) Slope of voltage curve determines current

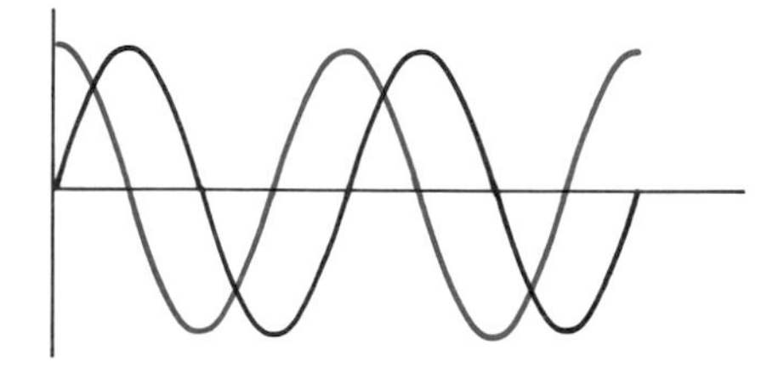

(*b*) Current leads voltage by 90°

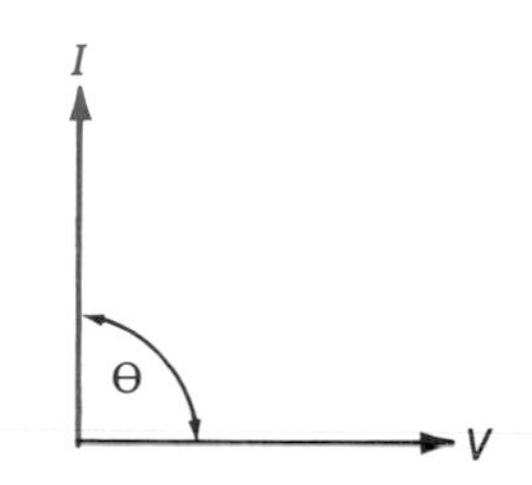


(*c*) Current and voltage phasors

**FIGURE 16-12 Phase relationship between current and voltage.**

circuit uses no power; it just transfers electric energy back and forth between the capacitor and the source.

The first formula, $P = IV \cos \theta$, also shows that a capacitor circuit uses no power when $I_C$ and $V_C$ are 90° out of phase. For a phase angle $\theta$ of 90°, the cosine is zero. When $\cos \theta$ is zero, the power, of course, is zero.

Figure 16-13 graphically illustrates why ideal-capacitor circuits use no power regardless of the current and voltage levels, while resistor circuits use an amount of power based on the level of the currents and voltages. In Fig. 16-13*a*, the product of the current and voltage yields a + power for $\frac{1}{4}$ cycle and a − power for $\frac{1}{4}$ cycle. The algebraic sum of these power pulses is zero. In Fig. 16-13*b*, where $I_R$ and $V_R$ are in phase, every power pulse is positive and the circuit uses maximum power for a given current and voltage. (In Chap. 17 we will see

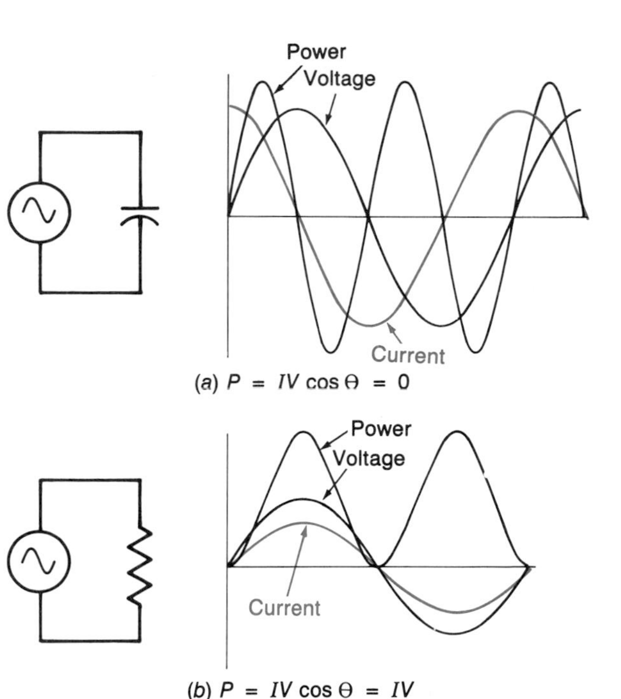


(*a*) $P = IV \cos \Theta = 0$

(*b*) $P = IV \cos \Theta = IV$

**FIGURE 16-13 Power in ac circuits.**

what happens when $R$ and $C$ are combined in an ac circuit.)

## Self-Test

**12.** When a source voltage is increasing in a capacitor circuit, the capacitor is ______ energy.

**13.** When the source voltage in a capacitor circuit is at its peak positive value, the source current will be at its ______ value.

**14.** How many times does a capacitor charge each cycle?

**15.** Does a capacitor cause current to lead or lag voltage?

**16.** What is the phase angle in an ideal-capacitor circuit?

**17.** In a capacitor circuit, the current is 1.75 A and the voltage is 117 V. What is the power?

## 16-5 CAPACITIVE REACTANCE

As we saw in Sec. 16-4, an ideal capacitor uses no power and has no resistance, yet it limits the current flow to the value needed to keep the capacitor's voltage equal to the source voltage. Since the capacitor controls the magnitude of the current, it has some form of opposition to current. The opposition of a capacitor is called *capacitive reactance.* The term *reactance,* as you probably remember, is used instead of resistance because, by definition, resistance converts electric energy to heat energy and a capacitor's opposition does not convert energy. Since the symbol for reactance is $X$, the symbol for capacitive reactance is $X_C$.

When $f$ and $C$ are known, the formula for calculating $X_C$ can be derived from the formulas and relationships that you already know. The formulas we will use are $Q = VC$, $T = 1/f$, $I = Q/t$, $I_p = 1.57 I_{av}$, and Ohm's law, which tells us that $X_C = V_C/I_C$. Figure 16-14 shows that the capacitor must charge to $V_p$ in $\frac{1}{4}$ cycle. Thus, the time $t$ for the capacitor to charge will be $T/4$ where $T$ is the period of the waveform. Now, we can express $t$ in terms of frequency $f$ as follows:

$$t = \frac{T}{4} = \frac{1/f}{4} = \frac{1}{4f}$$

The charge $Q$ on the capacitor after the first $\frac{1}{4}$ cycle shown in Fig. 16-14 can be expressed as

$$Q_C = V_{C,p}C$$

where $V_{C,p}$ is the peak voltage on the capacitor at the end of the first $\frac{1}{4}$ cycle ($V_{C,p} = V_p$). Now that we know $Q$ and $t$ for the first $\frac{1}{4}$ cycle, we can find the average current for the first $\frac{1}{4}$ cycle:

$$I_{C,\text{av}} = \frac{Q_C}{t} = \frac{V_{C,p}C}{1/4f} = V_{C,p}C4f$$

From previous discussions we know the magnitude of the average current is the same for each $\frac{1}{4}$ cycle. Only the direction of the current reverses every $\frac{1}{2}$ cycle. Next, let us convert $I_{C,\text{av}}$ to $I_{C,p}$ so that the capacitor voltage and current are in the same ac units:

$$I_{C,p} = 1.57\, V_{C,p}C4f = 6.28\, V_{C,p}fC$$

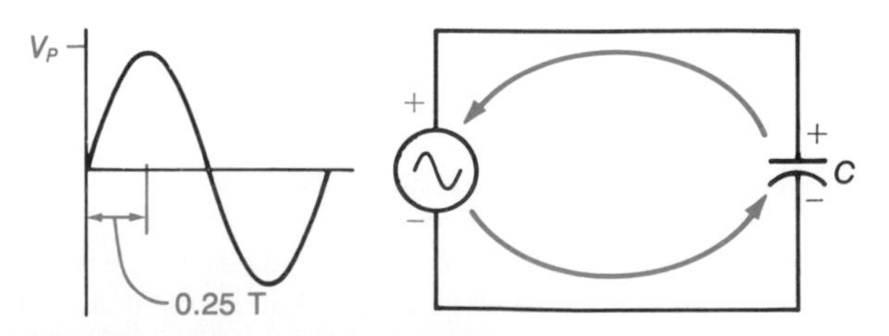


**FIGURE 16-14 The capacitor charges to the peak voltage in one-quarter of the period.**

Now, substituting into Ohm's law,

$$X_C = \frac{V_{C,p}}{I_{C,p}} = \frac{V_{C,p}}{6.28\, V_{C,p}fC} = \frac{1}{6.28\, fC}$$

Since 6.28 is a constant equal to $2\pi$, the capacitive reaction formula can be written as

$$X_C = \frac{1}{2\pi fC}$$

When $f$ and $C$ are in base units of hertz (Hz) and farads (F), respectively, $X_C$ is in ohms, which is the base unit of opposition whether it be resistance or reactance. Proof that the base unit of reactance is the ohm can be shown by substituting other base units for hertz and farad. Since $f = 1/T$ and since $T$ has units of seconds, 1/s can be substituted for hertz. Also we know that F = C/V, V = A · Ω, and A = C/s. Now we can show that 1/Hz · F (from $X_C = 1/2\pi fC$) equals Ω:

$$\frac{1}{\text{Hz} \cdot \text{F}} = \frac{1}{(1/\text{s})\,(\text{C/V})} = \frac{\text{V} \cdot \text{s}}{\text{C}} = \frac{\text{A} \cdot \Omega \cdot \text{s}}{\text{C}}$$
$$= \frac{(\text{C/s})\Omega \cdot \text{s}}{\text{C}} = \frac{\text{C} \cdot \Omega}{\text{C}} = \Omega$$

Notice from the capacitive reactance formula that $X_C$ is inversely proportional to both $f$ and $C$. This relationship makes sense because doubling $f$ allows only half as much time for the capacitor to charge, and, assuming $V$ is not changed, the same $Q$ must move through the circuit. The same $Q$ in half the time means $I$ doubled, and, for $I$ to double, $X_C$ must have been reduced to half of its former value. Doubling $C$, while holding $V$ and $f$ constant, causes $Q$ and $I$ to double, which means that $X_C$ is again half of its former value.

The capacitive reactance can, of course, be determined without the reactance formula if $I_C$ and $V_C$ are known. Just use Ohm's law: $X_C = V_C/I_C$.

**EXAMPLE 16-6**

Determine the current in Fig. 16-14 if the voltage is 20 V, the frequency is 100 Hz, and the capacitance is 0.47 μF.

**Solution**

$$X_C = \frac{1}{2\pi fC} = \frac{1}{6.28 \times 100 \times 0.47 \times 10^{-6}} = 3386\ \Omega$$

$$I_C = \frac{V_C}{X_C} = \frac{20\ \text{V}}{3386\ \Omega} = 5.9\ \text{mA}$$

**EXAMPLE 16-7**

If the frequency in Example 16-6 is changed to 50 Hz, what will $X_C$ be?

**Solution**

Reactance $X_C$ is inversely proportional to $f$, so if $f$ is halved, $X_C$ doubles:

$$X_C = 2 \times 3386\ \Omega = 6772\ \Omega$$

**EXAMPLE 16-8**

How much capacitance is needed to produce 1200 Ω of reactance when the frequency is 2.3 MHz?

**Solution**

$$X_C = \frac{1}{2\pi fC}$$

So, rearranging

$$C = \frac{1}{2\pi fX_C} = \frac{1}{6.28 \times 2.3 \times 10^6 \times 1200} = 57.7\ \text{pF}$$

## Self-Test

**18.** Define capacitive reactance.

**19.** How does $X_C$ differ from $R$?

**20.** Is $X_C$ directly proportional to $C$?

**21.** What is the value of $X_C$ when $f = 455$ kHz and $C = 29$ pF?

**22.** A capacitor is connected to a 15-V, 200-Hz source. If the current is 20 mA, what is the value of $C$?

## 16-6 CAPACITIVE REACTANCES IN SERIES AND PARALLEL

Capacitive reactances in series, parallel, or series-parallel follow the same rules as do resistors in the same configuration. You use the same formulas except you substitute $X_C$ for $R$. Thus, for series reactances the formula is

$$X_{C_T} = X_{C_1} + X_{C_2} + \cdots + X_{C_n}$$

where $C_T$ stands for total capacitance. And, for parallel reactances the appropriate formulas are

$$X_{C_T} = \frac{1}{1/X_{C_1} + 1/X_{C_2} + \cdots + 1/X_{C_n}}$$

or

$$X_{C_T} = \frac{X_{C_1}X_{C_2}}{X_{C_1} + X_{C_2}} \quad \text{or} \quad X_{C_T} = \frac{X_C}{n}$$

Of course, the total capacitive reactance of any combination of reactances can be determined by Ohm's law if $I_T$ and $V_T$ are known. That is, $X_{C_T} = V_T/I_T$. Also, when the total capacitance $C_T$ is known, the capacitive reactance formula can be used. Just put in the appropriate subscripts. That is,

$$X_{C_T} = \frac{1}{2\pi fC_T}$$

**EXAMPLE 16-9**

Determine the total capacitive reactance for the circuit in Fig. 16-15.

**Solution No. 1**

$$X_{C_1} = \frac{1}{6.28 \times 200 \times 1 \times 10^{-6}} = 796\ \Omega$$

$$X_{C_2} = \frac{1}{6.28 \times 200 \times 0.33 \times 10^{-6}} = 2411\ \Omega$$

$$X_{C_T} = \frac{X_{C_1}X_{C_2}}{X_{C_1} + X_{C_2}} = 598\ \Omega$$

**Solution No. 2**

$$C_T = C_1 + C_2 = 1 + 0.33 = 1.33\ \mu\text{F}$$

$$X_{C_T} = \frac{1}{6.28 \times 200 \times 1.33 \times 10^{-6}} = 598\ \Omega$$

Which of the two methods should be used depends on how much additional information is required. If, for example, only the value of $I_T$ were needed,

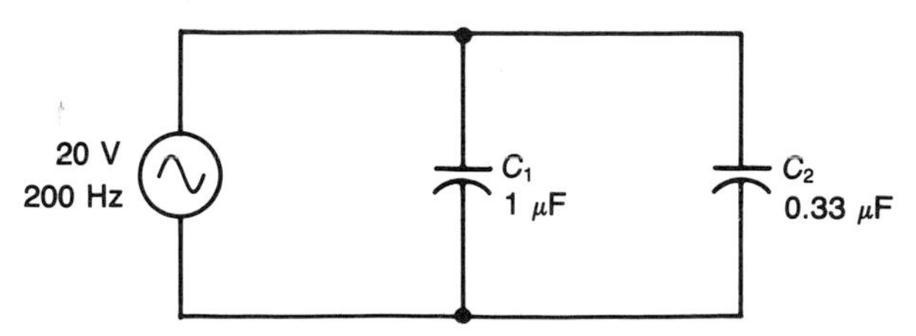

FIGURE 16-15 Circuit for Example 16-9.

the second solution might be best. However, if the values for $I_T$, $I_{C_1}$, and $I_{C_2}$ were all required, the first solution would be advantageous.

**EXAMPLE 16-10**

Determine $X_{C_T}$ for the circuit shown in Fig. 16-16.

**Solution No. 1**

$$C_{3,4} = C_3 + C_4 = 1 + 2 = 3\ \mu\text{F}$$

$$C_{3,4,5} = \frac{C_{3,4}C_5}{C_{3,4} + C_5} = \frac{3 \times 1}{3 + 1} = 0.75\ \mu\text{F}$$

$$C_{1,2} = \frac{C_1C_2}{C_1 + C_2} = \frac{1 \times 2}{1 + 2} = 0.67\ \mu\text{F}$$

$$C_T = C_{3,4,5} + C_{1,2} = 0.75 + 0.67 = 1.42\ \mu\text{F}$$

$$X_{C_T} = \frac{1}{2\pi f C_T} = \frac{1}{6.28 \times 100 \times 1.42 \times 10^{-6}} = 1121\ \Omega$$

**Solution No. 2**

$$X_{C_1} = \frac{1}{2\pi f C_1} = \frac{1}{6.28 \times 100 \times 1 \times 10^{-6}} = 1592\ \Omega$$

$$X_{C_3} = X_{C_5} = X_{C_1}$$

By inverse proportionality,

$$X_{C_2} = 0.5 \times 1592 = 796\ \Omega$$

$$X_{C_4} = X_{C_2}$$

$$X_{C_{3,4}} = \frac{X_{C_3}X_{C_4}}{X_{C_3} + X_{C_4}} = \frac{1592 \times 796}{1592 + 796} = 531\ \Omega$$

$$X_{C_{3,4,5}} = X_{C_{3,4}} + X_{C_5} = 531 + 1592 = 2123\ \Omega$$

$$X_{C_{1,2}} = X_{C_1} + X_{C_2} = 1592 + 796 = 2388\ \Omega$$

$$X_{C_T} = \frac{X_{C_{1,2}}X_{C_{3,4,5}}}{X_{C_{1,2}} + X_{C_{3,4,5}}} = \frac{2123 \times 2388}{2123 + 2388} = 1124\ \Omega$$

Within round-off error, the two solutions agree. Notice that three more calculations are required in the second solution if all capacitors have different values. However, a different value for each capacitor would not require any additional calculations in the first solution.

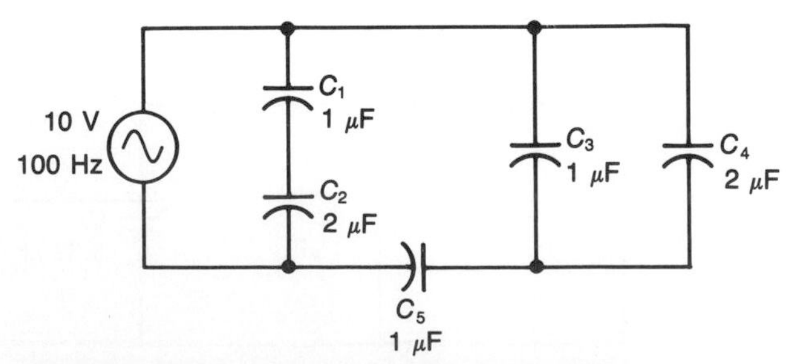


**FIGURE 16-16 Circuit for Example 16-10.**

## Self-Test

**23.** A 0.1-μF capacitor and a 0.068-μF capacitor are series-connected to a 12-V, 1-kHz source. Determine the circuit current and the voltage across the 0.1-μF capacitor.

**24.** Which capacitor in Fig. 16-15 will draw the larger current?

**25.** Will $C_1$ or $C_5$ in Fig. 16-16 develop the larger voltage?

## 16-7 CAPACITOR DISSIPATION FACTOR

Chapter 15 introduced the dissipation factor (DF) of a capacitor. In that chapter dissipation factor was defined as the ratio of the energy converted to heat over the energy stored in the capacitor. Since the energy stored in $\frac{1}{4}$ cycle is controlled by $X_C$ and since the energy converted is controlled by $R$, the dissipation factor is mathematically defined as $\text{DF} = R/X_C$. This ratio is usually expressed as a percentage. The conversion of some of the electric energy into heat energy is caused by the total effective resistance of all of the parts of the capacitor. The total effective resistance is made up of the following resistances:

1. Resistance of the lead wires
2. Resistance of the plates
3. Insulation resistance of the dielectric
4. Resistance due to polarization of the dielectric

This last resistance, resistance due to polarization, is analogous to hysteresis in the core of an inductor. It is often referred to as *dielectric hysteresis* and the energy loss caused by it is called *dielectric hysteresis loss* or just dielectric loss. In an ac circuit, the polarity of the voltage across the capacitor changes every $\frac{1}{2}$ cycle. Thus, the polarization of the dielectric in the capacitor must also change every $\frac{1}{2}$ cycle. In Chap. 15 it was explained that polarization causes the electron orbital paths in the dielectric to be distorted. This distortion slightly increases the energy level of the electrons. Now, as the polarization polarity reverses, it must pass through a neutral (no voltage) point. At this point, the orbital paths, and the energy level of the electrons, have returned to normal. The difference in energy between the energy level of the electrons in the distorted and normal orbital paths is converted to heat. Thus, each time the orbital path of the electron returns from distorted to normal, some electric energy is converted to heat energy. For an ac circuit this happens twice each cycle.

This energy loss in the dielectric material, due to polarization reversals, has the same effect as if it were a small resistance in series with the capacitor. The equivalent circuit for a capacitor, which takes into account the four resistances listed above, is presented in Fig. 16-17. The insulation (or dielectric) resistance is shown in parallel with the ideal capacitor. Since this resistance is extremely high ($> 10^4$ MΩ for most film capacitors), it dissipates very little power even when the capacitor voltage is high. The other resistances, as seen in Fig. 16-17, appear as very small value resistors in series with the leads of the ideal capacitor. Because these resistances are very small, they dissipate very little power even when the capacitor current is quite high. Collectively, the resistances shown in Fig. 16-17 are often called the *equivalent series resistance* (abbreviated ESR in capacitor literature). This is a convenient and accurate term because, in terms of energy loss, all of the resistors shown in Fig. 16-17 could be replaced with a single series resistor.

Two other terms are sometimes used to describe the energy loss in capacitors. They are *quality (Q)* or quality factor and *power factor* (PF). You already encountered quality factor in Chap. 14 on inductance. Quality factor was defined as $Q = X/R$. For a capacitor, it is $Q = X_C/R$. Since DF $= R/X_C$, you can see that $Q$ and $DF$ are reciprocals of each other. If a capacitor has a DF of 1 percent, then the $Q$ would be $1/0.01 = 100$.

Power factor PF is a term used to indicate the portion of the circuit current and the circuit voltage that produces power. Therefore, PF is numerically equal to the cos $\theta$. Its value is controlled by the combined effect of $R$ and $X_C$. A combination of $R$ and $X_C$ is called *impedance (Z)*. (The details of impedance are covered in Chap. 17.) When $R$ and $X_C$ are in series, as they are in a capacitor where $R$ = ESR, the power factor becomes PF $= R/Z$. When the dissipation factor is small, $X_C$ is very large compared to $R$. This makes $Z \approx X_C$. Thus, PF $\approx$ DF for high-quality capacitors.

Since the $X_C$ of a given capacitor is a function of frequency, so is the DF. Not only is $X_C$ a function of frequency, but so is the ESR. In fact, the ESR increases much more rapidly with increased frequency than does $X_C$. Thus, the DF increases as frequency increases. Of course, $Q$ decreases as frequency increases.

A high-quality film capacitor can have a DF of less than 0.1 percent ($Q > 1000$) at 1 kHz. On the other hand, the DF of an electrolytic capacitor can be over 10 percent at 120 Hz ($Q < 10$).

With the small DF of film capacitors, assuming an ideal capacitor for purposes of capacitor circuit analysis leads to very little error. We will continue to make this assumption when solving circuits which use capacitors.

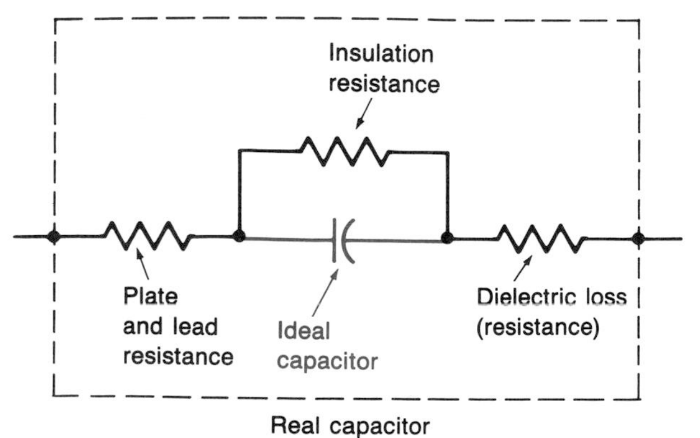


**FIGURE 16-17 Equivalent circuit for a capacitor.**

**EXAMPLE 16-11**

A 0.068-μF capacitor has an ESR of 4 Ω at 1000 Hz. Determine its DF.

**Solution**

$$X_C = \frac{1}{2\pi f C} = \frac{1}{6.28 \times 1000 \times 0.068 \times 10^{-6}} = 2342\ \Omega$$

$$\text{DF} = \frac{R}{X_C} \times 100 = \frac{4\ \Omega}{2342\ \Omega} \times 100 = 0.17 \text{ percent}$$

## Self-Test

**26.** What is the $Q$ of a capacitor that has a DF of 2 percent?

**27.** What happens to the value of $Q$ when $f$ increases?

**28.** What happens to the value of DF when $f$ increases?

**29.** In a given circuit, would a capacitor with a PF of 0.05 or 0.1 get hotter?

**30.** What causes dielectric loss in a capacitor?

## 16-8 MEASURING CAPACITANCE

An ac bridge circuit like the one in Fig. 16-18 can be used to measure the unknown capacitance of the capacitor, labeled $C_u$. The values of $R_2$ and $C_1$ are known, and the value of $R_1$ for any setting of its shaft is accurately indicated by a dial attached to its shaft.

When the bridge is balanced, the voltmeter in Fig. 16-18 indicates zero. The bridge is balanced when $R_1/X_{C_1} = R_2/X_{C_u}$ because, under this condition, the voltage drop across $R_1$ equals the voltage drop across $R_2$, which leaves both leads of the voltmeter at the same potential. The balanced bridge formula can be solved for $C_u$ if we first substitute an equivalent expression for $X_{C_u}$ and $X_{C_1}$. The procedure is

$$\frac{R_1}{X_{C_1}} = \frac{R_1}{1/2\pi f C_1} = R_1 2\pi f C_1$$

and

$$\frac{R_2}{X_{C_u}} = \frac{R_2}{1/2\pi f C_u} = R_2 2\pi f C_u$$

so

$$R_1 2\pi f C_1 = R_2 2\pi f C_u$$

Dividing by $2\pi f$ gives

$$R_1 C_1 = R_2 C_u$$

Rearranging gives

$$C_u = \frac{R_1 C_1}{R_2}$$

When the frequency of the source is a known stable value, the dial on $R_1$ can be calibrated in units of capacitance rather than ohms. Then the operator of the bridge does not need the above formula.

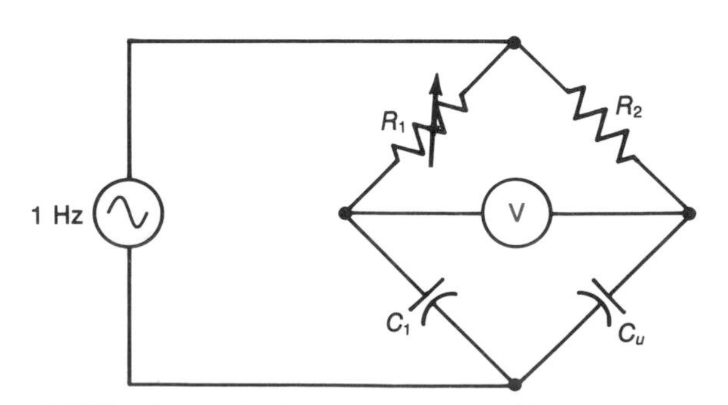


**FIGURE 16-18 Bridge for measuring capacitance.**

Most capacitance-measuring bridges also have a small variable resistor in series with $C_1$. When $C_u$ is of lower quality than $C_1$, which is a very high $Q$ capacitor, some resistance is added in series with $C_1$ so that the $Q$ of the $C_1$ leg of the bridge is the same as the $Q$ of the $C_u$ leg of the bridge. Only when the two $Q$'s are matched will the bridge be in complete balance. The dial connected to the shaft of the resistor in series with $C_1$ can be calibrated to indicate either the $Q$, PF, or DF of the unknown capacitor $C_u$.

**EXAMPLE 16-12**

In Fig. 16-18, $C_1 = 0.01\ \mu\text{F}$, $R_2 = 10\ \text{k}\Omega$, and the bridge balances when $R_1 = 6\ \text{k}\Omega$. Determine the value of $C_u$.

**Solution**

$$C_u = \frac{R_1 C_1}{R_2} = \frac{6 \times 0.01}{10} = 0.006\ \mu\text{F}$$

Digital capacitance meters employ an entirely different approach to measuring capacitance. The basic idea is to measure the *time* it takes for an unknown capacitor to charge to a specified voltage *when* the capacitor is being *charged by a constant-current source*. When charging from a constant current source, the time required for the capacitor to reach a specified voltage is a linear function of capacitance. That is, if current and voltage are fixed, time and capacitance are directly proportional. This relationship is graphically illustrated in Fig. 16-19. Why this relationship exists can be seen by substituting $Q = It$ (from $I = Q/t$) into the basic capacitance formula:

$$C = \frac{Q}{V}$$

Therefore,

$$C = \frac{It}{V}$$

Solving for $t$ yields

$$t = \frac{CV}{I}$$

Thus, for fixed values of $V$ and $I$, $t$ varies directly with $C$.

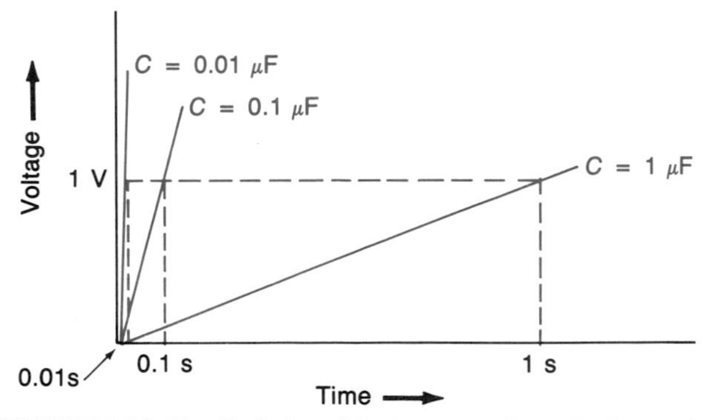


**FIGURE 16-19 Relationship between $V$ and $t$ for various values of $C$ when $I$ is a constant 1 μA.**

As shown in Fig. 16-19, a 1-μF capacitor requires 1 s to charge to a reference (or specified) voltage of 1 V when the constant current is 1 μA. This time is determined by

$$t = \frac{1\ \mu\text{F} \times 1\ \text{V}}{1\ \mu\text{A}} = \frac{(1\ \mu\text{C/V})\ (1\ \text{V})}{1\ \mu\text{C/s}} = 1\ \text{s}$$

The other times in Fig. 16-19 were also calculated using a 1-V reference voltage and a 1-μA current source. Notice from the data given in Fig. 16-19 that the numerical values of $C$ and $t$ are the same—only the units vary. Now, suppose one had a clock that only displayed from 0.000 to 1.999 s and that this clock started at 0.000 s when a 0.01-μF capacitor was connected to a 1-μA constant-current source. Further, suppose the clock was stopped at the exact time the capacitor reached 1 V. The clock would be displaying 0.010 s. This 0.010 could also be labeled microfarad and the device called a capacitance meter rather than a clock.

With the scheme described above, capacitance between 0.001 and 1.999 μF could be measured. The range of measurable capacitances could be altered by changing either the reference voltage or the constant current.

The exact technique used by, and the complete circuit of, a digital capacitance meter are quite complex. The major circuits used in such an instrument are covered in the chapter on digital electronics.

## Self-Test

**31.** What are two methods of measuring capacitance?

**32.** How long does it take to charge a 15-μF capacitor to 20 V with a 30-μA constant-current source?

**33.** The fixed resistor and capacitor in a bridge are 20 kΩ and 0.015 μF. The bridge balances when the variable resistor is at 5 kΩ. What is the unknown capacitance?

## SUMMARY

1. Except for the initial charging (or discharging) current, a capacitor blocks direct current.
2. The peak instantaneous capacitor current in a dc circuit is determined by the voltage and resistance.
3. A time constant ($\tau$) is the time required for a capacitor in an $RC$ circuit to charge or discharge an additional 63.2 percent of the available voltage. $\tau = RC$
4. As the capacitor voltage increases, the capacitor current decreases.
5. The sum of $V_R$ and $V_C$ at any instant must equal $V_T$ in a series $RC$ dc circuit.
6. The time required to charge a capacitor is directly proportional to both resistance and capacitance.
7. For minimum distortion of a square wave, the $RC$ time constant should be much greater than the period of the waveform.
8. A capacitor charges (stores energy) twice and discharges (returns energy) twice for each cycle of alternating current in an ac capacitor circuit.

9. An ideal capacitor uses (converts) no energy even though both $V_C$ and $I_C$ are greater than zero. This is because $I_C$ and $V_C$ are 90° out of phase. Power dissipation is also zero.

10. When source voltage is increasing in absolute value, the capacitor, in a capacitor circuit, is charging. When source voltage is decreasing, it is discharging.

11. In an ac capacitor circuit, the current is zero when the voltage is maximum.

12. In a capacitor circuit current leads voltage by 90°.

13. Capacitive reactance is the opposition of a capacitor to a changing current. It is inversely proportional to both frequency and capacitance. Its base unit is the ohm.

14. Reactance does not convert electric energy to heat energy.

15. Capacitive reactances in series behave the same as do resistances in series, except that they use no power.

16. Capacitive reactances in parallel are like resistances in parallel, except that they use no power.

17. In series capacitor ac circuits, the smallest capacitor drops the most voltage because it has the most reactance.

18. In parallel capacitor ac circuits, the largest capacitor draws the most current because it has the least reactance.

19. Power dissipation in a real capacitor is caused by plate and lead resistance, insulation resistance, and dielectric hysteresis loss (polarization reversal).

20. Collectively all of the effective resistances of a capacitor are called the ESR (equivalent series resistance) of the capacitor.

21. Dissipation factor (DF), quality ($Q$), and power factor (PF) are all used to indicate the energy lost (relative to the energy stored) in the capacitor.

22. DF and PF increase and $Q$ decreases when $f$ increases.

23. Unknown amounts of capacitance can be measured on an ac bridge which has known values of $R$ and $C$.

24. Digital capacitance meters determine capacitance by measuring the time required to charge a capacitor to a specified voltage when the current is a constant value.

25. Some useful formulas are

$$\tau = RC$$

$$V_{C,t} = V_T - (V_T - V_{CE})e^{-t/\tau} \qquad \text{General}$$

$$V_{C,t} = V_{CE}\, e^{-t/\tau} \qquad \text{Discharging through } R \text{ only}$$

$$t = \tau \ln \frac{V_T - V_{CE}}{V_T - V_{C,t}} \qquad \text{General}$$

$$t = \tau \ln \frac{V_T}{V_T - V_{C,t}} \qquad \text{Charging from zero}$$

$$t = \tau \ln \frac{V_{CE}}{V_{C,t}} \qquad \text{Discharging through } R \text{ only}$$

$$P = IV \cos \theta = 0 \qquad \text{for an ideal-capacitor circuit}$$

$$X_C = \frac{1}{2\pi fC}$$

In series,

$$X_{C_T} = X_{C_1} + X_{C_2} + \cdots + X_{C_n}$$
$$V_T = V_{C_1} + V_{C_2} + \cdots + V_{C_n}$$

In parallel,

$$X_{C_T} = \frac{1}{1/X_{C_1} + 1/X_{C_2} + \cdots + 1/X_{C_3}}$$
$$I_T = I_{C_1} + I_{C_2} + \cdots + I_{C_n}$$
$$\text{DF} = \frac{R}{X_C}$$
$$Q = \frac{X_C}{R}$$
$$C_u = \frac{R_1 C_1}{R_2} \quad \text{where } R_1 \text{ is the variable resistor}$$

## CHAPTER REVIEW QUESTIONS

For the following items, determine whether each statement is true or false.

**16-1.** In a dc *RC* circuit, the current is largest the instant the circuit is completed.

**16-2.** Except for short-term charging or discharging currents, a capacitor blocks direct current.

**16-3.** If the current through an ideal capacitor doubles, the power dissipation also doubles.

**16-4.** In a dc series *RC* circuit, $V_R$ plus $V_C$ must equal $V_T$ at all times.

**16-5.** The time required to charge a capacitor is inversely proportional to both *R* and *C*.

**16-6.** For minimum distortion of a square wave, the *RC* time constant should be shorter than the period of the waveform.

**16-7.** The base unit for reactance is the ohm.

**16-8.** PF is the reciprocal of DF.

For the following items, fill in the blank with the word or phrase required to correctly complete each statement.

**16-9.** During ________ a capacitor will charge from 0 to 63.2 percent of the source voltage.

**16-10.** As the voltage across a capacitor increases, the capacitor current ________.

**16-11.** In an ideal capacitor, the current ________ the voltage by ________ electric degrees.

**16-12.** In an ac capacitor circuit, maximum (peak) current occurs when the voltage is ________.

**16-13.** Capacitive reactance is ________ proportional to frequency.

**16-14.** The ________ capacitor in a series circuit drops the largest voltage.

**16-15.** The ________ capacitor in a parallel circuit draws the largest current.

**16-16.** When frequency increases, the quality of a capacitor ________.

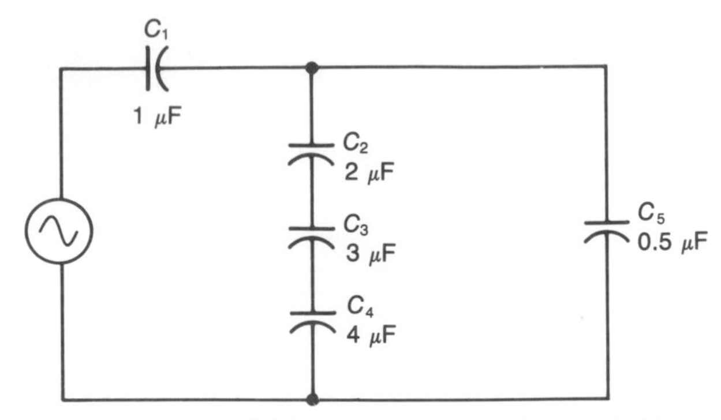

**FIGURE 16-20 Circuit for Review Question 16-20 and Review Problem 16-13.**

**16-17.** A digital capacitance meter measures ______ to determine the unknown capacitance of a capacitor.

**16-18.** Besides quality and dissipation factors, ______ is used to indicate the relative energy loss in a capacitor.

For the following items, provide the information requested.

**16-19.** What is the major difference between reactance and resistance?

**16-20.** Which capacitor in Fig. 16-20 will drop the most voltage?

## CHAPTER REVIEW PROBLEMS

**16-1.** A discharged 2-μF capacitor in series with a 270-kΩ resistor is connected to a 38-V dc source. What is the peak instantaneous current?

**16-2.** Change the capacitance in Problem 16-1 to 4 μF and again determine the peak instantaneous current.

**16-3.** Determine the time constant of a 1000-pF capacitor and a 100-kΩ resistor connected in series to a 50-V dc source.

**16-4.** How long would it take the capacitor in Problem 16-3 to charge from 0 to 30 V?

**16-5.** How long would it take to discharge a 3-μF capacitor from 80 V to 20 V through a 2-MΩ resistor?

**16-6.** A 0.5-μF capacitor has 50 V on it. It is then connected through a 1-MΩ resistor to a 100-V source. How long does it take the capacitor to charge to 85 V?

**16-7.** A 30-μF capacitor is connected through a 100-kΩ resistor to a 60-V source. How much voltage will be on the capacitor after 1.6 s? After 3 s?

**16-8.** A 2-μF capacitor is to be charged to 10 V from a 20-V source in 2 s. Determine the value of resistance needed.

**16-9.** Calculate the reactance of a 0.068-μF capacitor at 250 Hz.

**16-10.** How much capacitance is required in a simple capacitor circuit if the circuit current is to be 10 mA and the source voltage is 20 V, 700 Hz?

**16-11.** A 0.22-$\mu$F capacitor and a 0.33-$\mu$F capacitor are series-connected to a 30-V, 400-Hz source. Determine the total reactance, the total current, and the voltage across the 0.22-$\mu$F capacitor.

**16-12.** If the capacitors in Problem 16-11 were in parallel, what would be the total resistance, the total current, the total resistance, and the current through the 0.22-$\mu$F capacitor?

**16-13.** Determine the total capacitance of the circuit in Fig. 16-20.

**16-14.** Determine the dissipation factor of a 0.1-$\mu$F capacitor at 1 kHz if the ESR is 3 $\Omega$.

**16-15.** Refer to Fig. 16-18. Resistor $R_1 = 3\ \text{k}\Omega$, $R_2 = 5\ \text{k}\Omega$, and $C_1 = 0.033\ \mu\text{F}$. Determine the value of $C_u$.

## ANSWERS TO SELF-TESTS

**1.** 106.4 $\mu$A

**2.** 20 A

**3.** It means no current will flow after the capacitor charges to the value of the source voltage if the source voltage remains constant.

**4.** 189.6 V, 233.1 V, and 285 V

**5.** 21.6 V

**6.** 15.1 V and 26.9 $\mu$A

**7.** 65.9 V and 0 $\mu$A (2 s > 7$\tau$)

**8.** decreasing, decreasing

**9.** 77.3 $\mu$s

**10.** 7.89 s

**11.** greater than

**12.** storing

**13.** zero

**14.** two

**15.** lead

**16.** 90° with $I$ leading $V$

**17.** zero

**18.** the opposition of a capacitor

**19.** Reactance $X_C$ does not convert electric energy to heat energy as $R$ does.

**20.** no

**21.** 12 k$\Omega$

**22.** 1.061 $\mu$F

**23.** 3.05 mA, 4.86 V

**24.** $C_1$

**25.** $C_5$

**26.** 50

**27.** decreases

**28.** increases

**29.** 0.1

**30.** reversal of the polarization polarity (or reversal of the polarity of the capacitor's voltage)

**31.** Use an ac bridge to compare the reactance of the unknown capacitance to that of a known capacitance. Measure the time (which represents capacitance) it takes to charge the unknown capacitance to a reference voltage level with a constant-current source.

**32.** 10 s

**33.** 0.00375 $\mu$F

# Chapter 17 *RCL* Circuits

Up to this point, this book has dealt with the characteristics of resistors, capacitors, and inductors individually in ac circuits. We know that resistance causes no phase shift, that capacitance causes current to lead voltage by 90°, and that inductance causes current to lag voltage by 90°. We also know that reactance does not convert electric energy to heat and that resistance does.

In this chapter we will learn what happens when various combinations and configurations of resistance, capacitance, and inductance are used in an ac circuit. While the major emphasis is on sinusoidal ac, we will look briefly at the response of *RCL* circuits to a pulse-train signal.

## 17-1 SERIES *RC* CIRCUITS

All of the rules and laws that we have used with other types of series circuits also apply to all series *RCL* circuits. However, in applying these rules and laws we have to keep in mind that the current and voltage in these circuits are not in phase. Accounting for this phase shift requires some new techniques for applying old rules and laws.

### Current and Voltage Relationships

The current at any given instant in a series circuit must be the same throughout all parts of the circuit. (It must satisfy Kirchhoff's current law.)

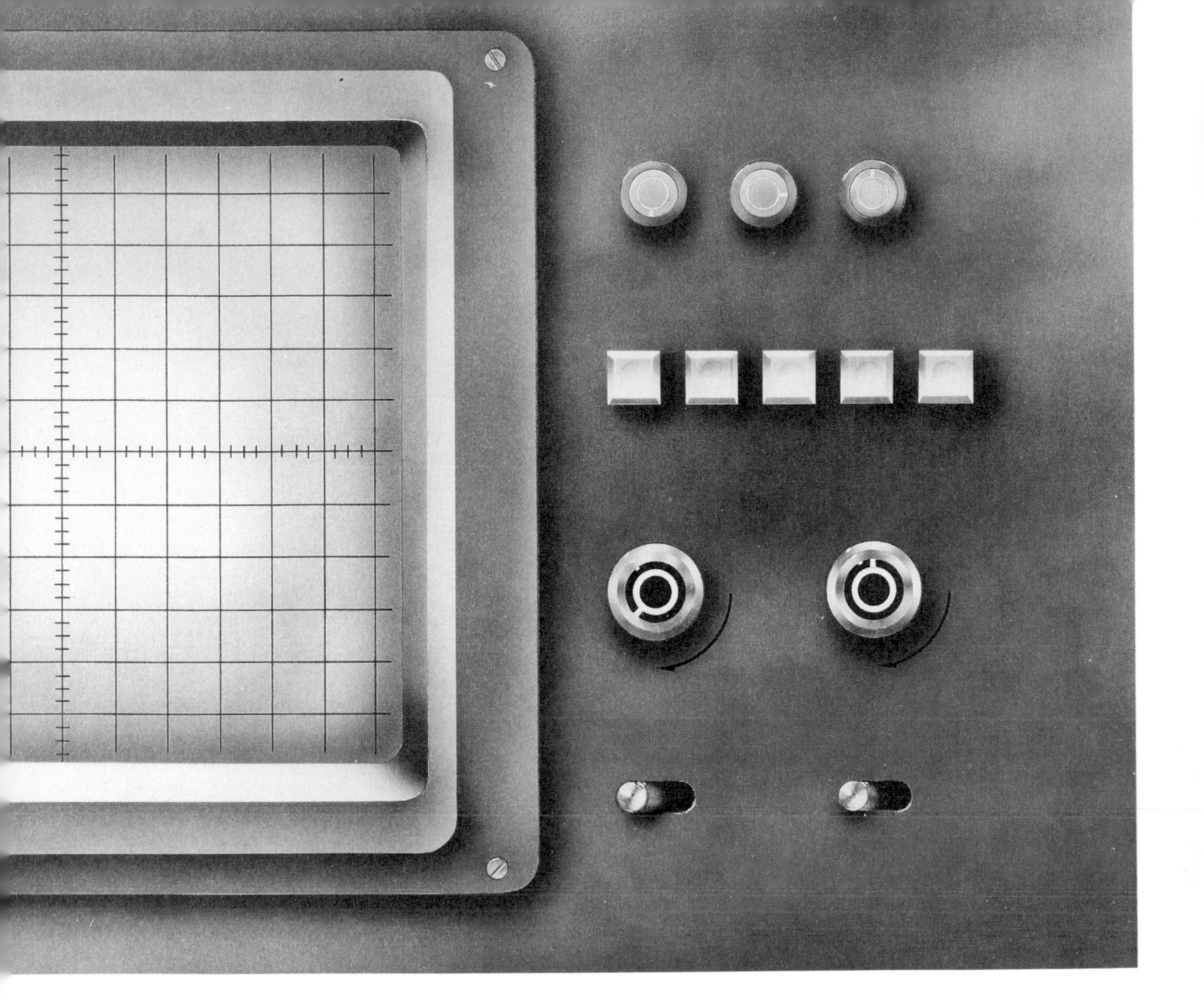

Thus, for the series *RC* circuit in Fig. 17-1*a*, we can write that $I_T = I_C = I_R$. Since current is the only quantity which is common to all parts of a series circuit, we will use it as the reference quantity when showing phase relationships between currents and voltages. Figure 17-1*b* is a sine wave graph for the current and voltages in Fig. 17-1*a*. Notice the following facts from this graph: (1) the resistive voltage is in phase with the current, (2) the capacitive voltage lags the current by 90°, and (3) the total (source) voltage is the instantaneous sum of the resistive and capacitive voltage. This last fact satisfies Kirchhoff's voltage law. Also notice in Fig. 17-1*b* that $V_C$ and $V_R$ have the same magnitude. This means that the values of *C*, *f*, and *R* in Fig. 17-1*a* must be such that $X_C = R$ because Ohm's law tells us that $V_C/X_C = I = V_R/R$.

Drawing sine wave graphs is tedious work, and graphically determining the instantaneous sum of sinusoidal voltages is even more tedious. Therefore, from now on we will use phasor and vector diagrams to represent electrical quantities that are out of phase. Figure 17-1*c* shows in phasor form the same information that Fig. 17-1*b* shows in sine wave form. In Fig. 17-1*c*, the current phasor is used as the reference phasor, so the resistive voltage phasor will be on the same line. (By convention, the reference phasor is drawn on the $+x$ axis.) Since phasors are assumed to rotate counterclockwise, the capacitive voltage phasor must be drawn

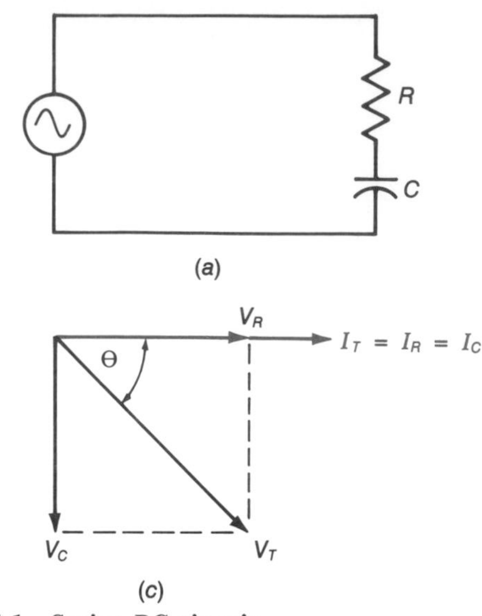


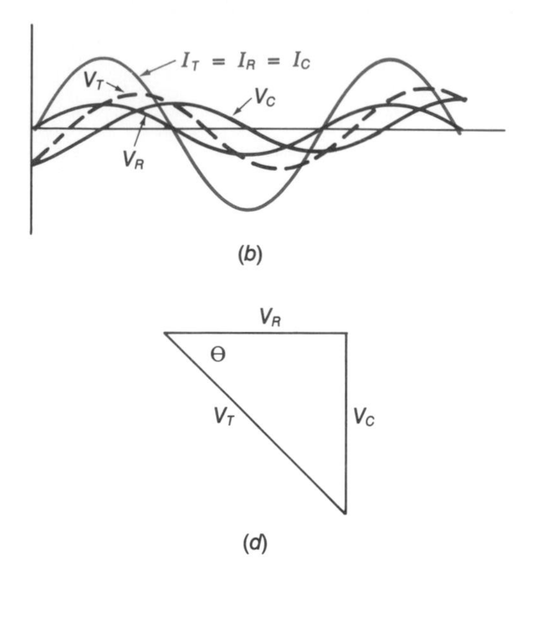


**FIGURE 17-1 Series *RC* circuit.**

down on the $-y$ axis. Of course, since $XR = X_C$, the $V_R$ and $V_C$ phasors must be the same length. Finally, $V_R$ and $V_C$ are graphically added together to obtain $V_T$. Angle theta ($\theta$), which is the phase angle between $V_T$ and $I_T$, is also marked on this phasor diagram. Of course, when $V_R = V_C$, $\theta$ is 45°.

Often it is convenient to draw the three voltage phasors in the form of a triangle as shown in Fig. 17-1*d*. This makes the mathematical analysis, using the characteristics of a right triangle, easier to visualize. (The three voltage phasors always form a right triangle because the reactive voltage is always 90° out of phase with the resistive voltage. Voltage $V_C$ is the reactive voltage when the capacitor is assumed to be ideal.) Now, using the pythagorean theorem, we can find the magnitude of the source voltage by

$$V_T^2 = V_R^2 + V_C^2$$

And solving for $V_T$,

$$V_T = \sqrt{V_R^2 + V_C^2}$$

Using trigonometric functions, angle $\theta$ can be determined by

$$\theta = \arccos\frac{V_R}{V_T} = \arctan\frac{V_C}{V_R} = \arcsin\frac{V_C}{V_T}$$

---

**EXAMPLE 17-1**

Determine $\theta$ and $V_T$ for the circuit in Fig. 17-2*a*. Also draw a phasor diagram for this circuit.

**Solution**

$$\theta = \arctan\frac{20 \text{ V}}{10 \text{ V}} = 63.4°$$

$$V_T = \sqrt{10^2 + 20^2} = 22.4 \text{ V}$$

The phasor diagram is shown in Fig. 17-2*c*.

---

When solving problems that involve adding phasors, it is always worthwhile to make a fairly accurate sketch of given and determined phasors. If one has made a gross error, it will usually be obvious from the phasor diagram sketch. Another worthwhile check for gross errors is to see if the resultant phasor ($V_T$ in this case) is larger than either of the two right-angle phasors, but smaller than the arithmetic sum of them.

The fact that $V_C$ is 90° out of phase and $V_T$ is not in phase with $V_R$ and $I_T$ could be indicated by adding an angle sign (with the degrees inserted) after the value of these quantities. Thus, for Example 17-1, we could write $V_R = 10 \text{ V} \angle 0°$, $V_C = 20 \text{ V} \angle -90°$, $V_T = 22.4 \text{ V} \angle -63.4°$, and $I_T = ?\text{ A} \angle 0°$ (the magnitude of $I_T$ is unknown). However,

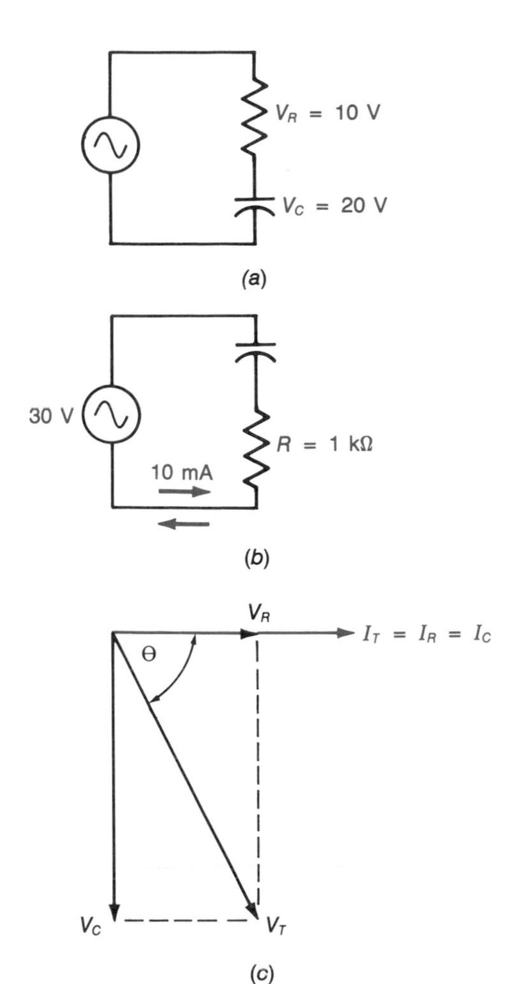


**FIGURE 17-2 Circuits for Examples 17-1 and 17-2.**

for the circuits in this chapter, $V_R$, $V_C$, and $I_T$ will always be at these same angles and $V_T$ will always be at angle $\theta$ so we will not use these angle signs. Also, since $\theta$ is in the fourth quadrant, its value should be negative. However, in the circuits in this chapter the location of $\theta$ will be obvious so we will not use the − sign. The formal $\angle$ (angle) notation will be used in later chapters where the added phasors are not at 90° to each other.

## EXAMPLE 17-2

Determine $\theta$ and $V_C$ for the circuit in Fig. 17-2*b*.

**Solution**

$$V_R = 10 \text{ mA} \times 1 \text{ k}\Omega = 10 \text{ V}$$

$$\theta = \arccos \frac{10 \text{ V}}{30 \text{ V}} = 70.5°$$

$$V_T^2 = V_R^2 + V_C^2$$

Solving for $V_C$,

$$V_C = \sqrt{V_T^2 - V_R^2} = \sqrt{30^2 - 10^2} = 28.3 \text{ V}$$

In this example, $V_C$ could also have been determined by using a trigonometric function:

$$\tan \theta = \frac{V_C}{V_R}$$

Therefore,

$$V_C = \tan 70.5° \times 10 \text{ V} = 28.3 \text{ V}$$

or

$$\sin \theta = \frac{V_C}{V_T}$$

Therefore,

$$V_C = \sin 70.5° \times 30 \text{ V} = 28.3 \text{ V}$$

## Self-Test

**1.** What is the phase relationship between $V_R$ and $V_C$?

**2.** True or false? When $V_C = 2V_R$, the capacitive reactance must be half as large as the resistance in a series *RC* circuit.

**3.** True or false? When angle $\theta$ and either $V_R$, $V_T$, or $V_C$ are known, the remaining voltages can be determined in a series *RC* circuit.

**4.** In a series *RC* circuit, $V_T = 15$ V and $V_R = 10$ V. Determine $V_C$ and $\theta$.

**5.** In a series *RC* circuit, $V_C = 50$ V and $\theta = 30°$. Determine $V_R$ and $V_T$.

## Impedance

The total opposition of a circuit that contains both resistance and reactance is called *impedance*. The symbol for impedance is *Z*. Ohm's law tells us that the total opposition to current, be it resistance, reactance, or impedance, is always equal to $V_T/I_T$. Thus,

$$Z = \frac{V_T}{I_T}$$

When $V_T$ is in volts and $I_T$ is in amperes, $Z$ is in ohms; therefore, the base unit for impedance is the ohm.

**EXAMPLE 17-3**

Determine the impedance for the circuit in Fig. 17-2*b*.

**Solution**

$$Z = \frac{V_T}{I_T} = \frac{30\text{ V}}{10\text{ mA}} = 3\text{ k}\Omega$$

The relationship between impedance, resistance, and reactance for a series *RC* circuit is illustrated in Fig. 17-3. From the values given in Fig. 17-3*a*, we can draw the voltage triangle in Fig. 17-3*b* and also calculate $V_T = \sqrt{30^2 + 40^2} = 50$ V. In a series circuit, individual voltages ($V_T$, $V_R$, and $V_C$) must be directly proportional to individual oppositions ($Z$, $R$, and $X_C$) because $I_T = I_R = I_C$. Thus, a triangle with the same proportions and same angle $\theta$ as the voltage triangle can be drawn, as in Fig. 17-3*c*, and labeled $Z$, $R$, and $X_C$ instead of $V_T$, $V_R$, and $V_C$. This new triangle is referred to as the *impedance triangle*. It is composed of the three vector quantities $Z$, $R$, and $X_C$, which are called vectors rather than phasors because they are not sinusoidal waveforms as are current and voltage. The fact that $Z$, $R$, and $X_C$ have the same phase relationships as $V_T$, $V_R$, and $V_C$ can be illustrated by manipulating the formula which relates these three voltages. Thus,

$$V_T^2 = V_R^2 + V_C^2$$

Dividing by $I_T$ yields

$$\left(\frac{V_T}{I_T}\right)^2 = \left(\frac{V_R}{I_T}\right)^2 + \left(\frac{V_C}{I_T}\right)^2$$

Substituting the Ohm's law equivalent for each term yields $Z^2 = R^2 + X_C^2$ and solving for $Z$ yields $Z = \sqrt{R^2 + X_C^2}$. Of course, the trigonometric functions can be used with the impedance triangle in the same way as they were used with the voltage triangle. For example, $\theta = \arctan\,(X_C/R)$, $X_R = \cos\,\theta \times Z$, etc.

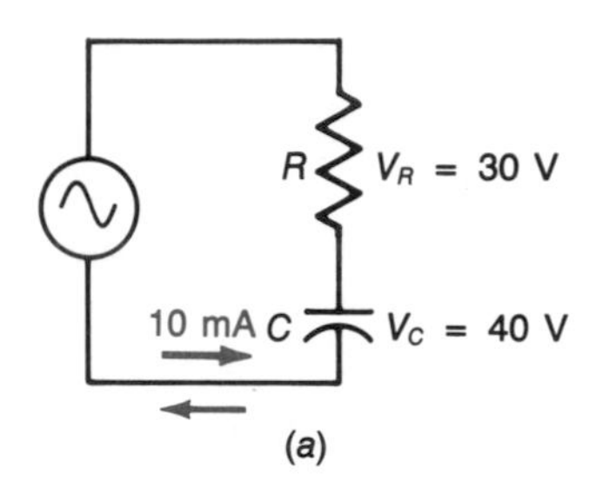


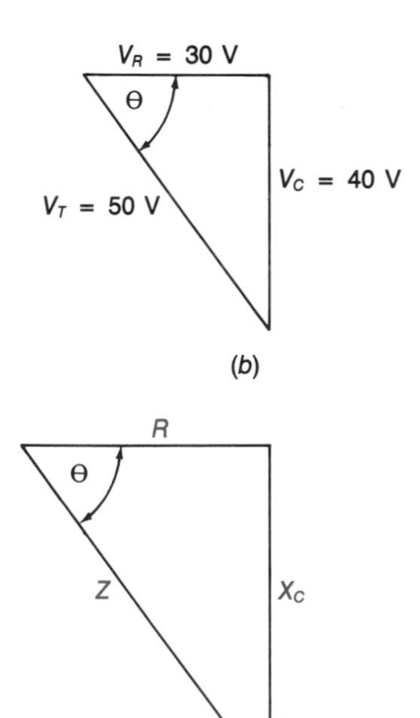


**FIGURE 17-3 Voltage and impedance triangles have the same angles.**

**EXAMPLE 17-4**

Without calculating or using $V_T$, find $Z$ and $\theta$ for the circuit in Fig. 17-3.

**Solution**

$$R = \frac{V_R}{I_T} = \frac{30\text{ V}}{10\text{ mA}} = 3\text{ k}\Omega$$

$$X_C = \frac{V_C}{I_T} = \frac{40\text{ V}}{10\text{ mA}} = 4\text{ k}\Omega$$

$$Z = \sqrt{R^2 + X_C^2} = \sqrt{(3\text{ k}\Omega)^2 + (4\text{ k}\Omega)^2} = \sqrt{25\text{ k}^2\Omega^2}$$
$$= 5\text{ k}\Omega$$

$$\theta = \arccos\frac{R}{Z} = \arccos\frac{3\text{ k}\Omega}{5\text{ k}\Omega} = 53.1°$$

Notice in the calculation of $Z$ that the answer will be in kilohms if both $R$ and $X_C$ are in kilohms also.

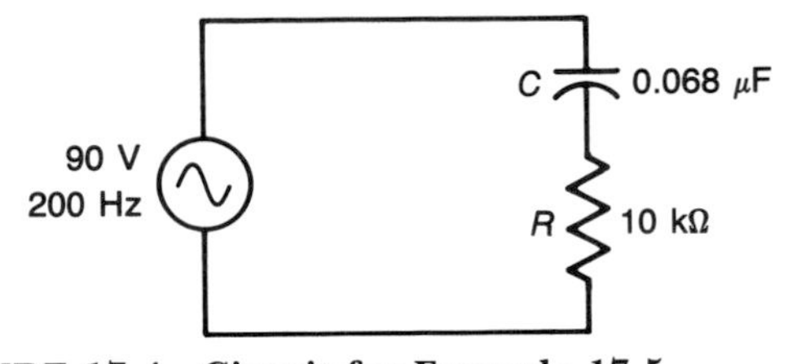


**FIGURE 17-4 Circuit for Example 17-5.**

**EXAMPLE 17-5**

For the circuit in Fig. 17-4 determine, $I_T$, $Z$, $\theta$, and $V_C$.

**Solution**

$$X_C = \frac{1}{2\pi fC} = \frac{1}{6.28 \times 200 \times 0.068 \times 10^{-6}} = 11.7\ \text{k}\Omega$$

$$Z = \sqrt{10^2 + 11.7^2} = 15.4\ \text{k}\Omega$$

$$I_T = \frac{V_T}{Z} = \frac{90\ \text{V}}{15.4\ \text{k}\Omega} = 5.84\ \text{mA}$$

$$\theta = \arctan\frac{X_C}{R} = \arctan\frac{11.7}{10} = 49.5°$$

$$V_C = I_T X_C = 5.84\ \text{mA} \times 11.7\ \text{k}\Omega = 68.3\ \text{V}$$

The last two quantities, $\theta$ and $V_C$, could have been calculated using a number of different combinations of trigonometric functions.

## Self-Test

**6.** What is impedance?

**7.** What is the impedance if $R = 2.2\ \text{k}\Omega$ and $X_C = 4\ \text{k}\Omega$ when $R$ and $C$ are in series?

**8.** Determine $I_T$, $V_R$, $V_C$, and $\theta$ when 100 V at 1 kHz is applied to the $RC$ network in Question 7.

**9.** Determine $X_C$ in a series $RC$ circuit if $\theta = 26°$ and $Z = 30\ \text{k}\Omega$.

**10.** Determine $R$ in a series $RC$ circuit if $Z = 10\ \text{k}\Omega$ and $X_C = 6\ \text{k}\Omega$.

**11.** In a series $RC$ circuit, can $Z$ ever be less than $R$ or $X_C$? As great as $R + X_C$?

**12.** Will increasing the frequency in a series $RC$ circuit cause $\theta$ to increase or decrease? Why?

## Power

The power ($P$) in an $RC$ circuit (or any circuit containing $X$) cannot be determined by $P = I_T V_T$ because the current and the voltage are not in phase. In a series $RC$ circuit only the voltage dropped across the resistor dissipates power. Regardless of the amount of reactive current, the voltage across the reactance dissipates no power. For any circuit containing reactance, the formula for power must include the cos $\theta$, i.e., the complete power formula is $P = I_T V_T \cos\theta$. In a series circuit the cos $\theta$ determines the part of the total voltage which is resistive voltage, because, as we have seen, $\cos\theta = V_R/V_T$ and thus $V_R = \cos\theta\ V_T$.

In a parallel impedance circuit, $V_T = V_R$, but $I_T$ is composed of $I_R$ and $I_X$, where $I_X$ can be either inductive or capacitive current. In the parallel case then, multiplying by cos $\theta$ determines how much of $I_T$ is resistive current. Thus, $P = IV\cos\theta$ always produces the product of the resistive current times resistive voltage.

Since only resistance can dissipate power, the power in any circuit can always be determined by the formulas $P = V_R{}^2/R$ and $P = I_R{}^2R$. If you remember the derivation of these formulas, you can see that either is the same as $P = I_R V_R$. For example, $P = I_R{}^2R = I_R I_R R = (V_R/R)I_R R = V_R I_R$.

**EXAMPLE 17-6**

Determine the power for the circuit in Fig. 17-5.

**Solution (the hard way)**

$$Z = \sqrt{4.7^2 + 3^2} = 5.58\ \text{k}\Omega$$

$$V_T = 15\ \text{mA} \times 5.58\ \text{k}\Omega = 83.7\ \text{V}$$

$$\cos\theta = \frac{4.7\ \text{k}\Omega}{5.58\ \text{k}\Omega} = 0.84$$

$$P = 15\ \text{mA} \times 85.7\ \text{V} \times 0.84 = 1.1\ \text{W}$$

**Solution (the easy way)**

$$P = (15\ \text{mA})^2 \times 4.7\ \text{k}\Omega = 1.1\ \text{W}$$

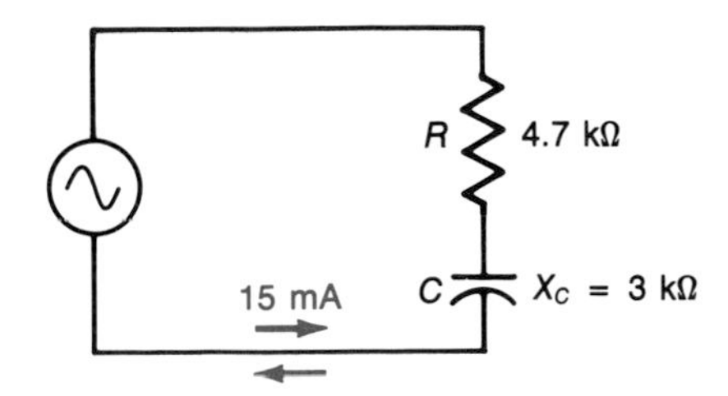


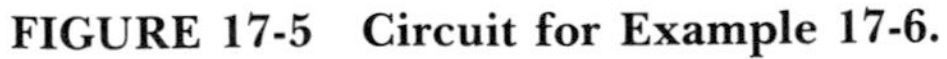

**FIGURE 17-5 Circuit for Example 17-6.**

## Self-Test

**13.** The power dissipated by 1.2 kΩ of reactance when the reactive current is 5 mA is ______ W.

**14.** Determine the power of a series *RC* circuit when $V_T = 70$ V, $R = 22$ kΩ, and $X_C = 30$ kΩ.

All of the major ideas needed to work with any combination of *R*, *C*, and/or *L* in either series or parallel circuits have been developed in this section. With only minor modifications and/or extensions, these ideas are used extensively in the next sections.

## 17-2 SERIES *RL* CIRCUITS

In Chaps. 13 and 14, we learned that inductive reactance and ideal inductors cause the voltage to lead the current by 90°. This tells us that $X_L$ will also lead *R* by 90° because, as always, *R* is used as the reference vector in the impedance triangle. The voltage triangle and the impedance triangle for a typical series *RL* circuit are shown in Fig. 17-6. As shown in Fig. 17-6*c*, the $V_T$ phasor and angle $\theta$ are in the first quadrant, so both have positive angles. As expected, Fig. 17-6*d* shows that the impedance also has a positive angle because the inductive reactance is at +90°.

The formulas for working with series *RL* circuits are derived by the same procedures as were the formulas for series *RC* circuits. They are the same formulas except that $X_L$ is substituted for $X_C$ and $V_L$ is substituted for $V_C$.

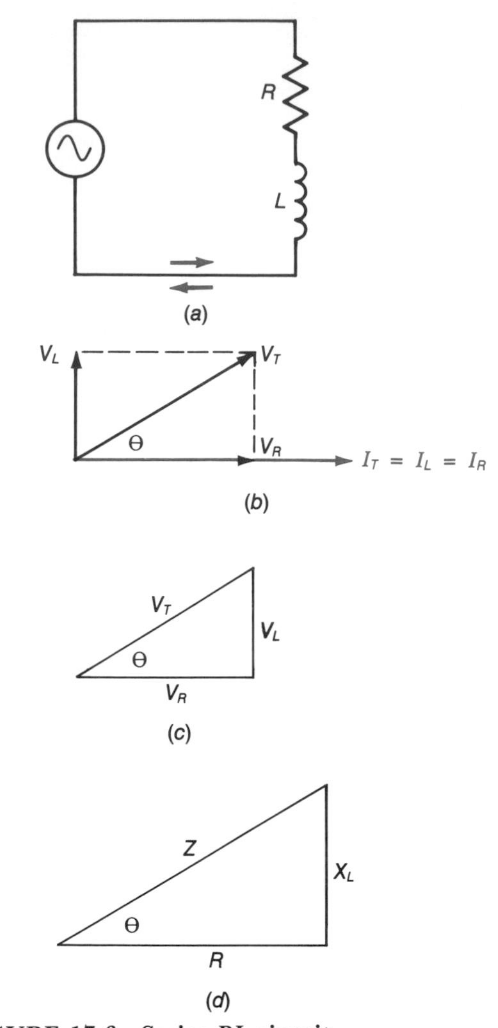


**FIGURE 17-6 Series *RL* circuit.**

### EXAMPLE 17-7

For the circuit in Fig. 17-6*a*, $R = 6.8$ kΩ, $L = 30$ mH, and $V_T = 40$ V at 50 kHz. Determine $I_T$, $V_L$, $V_R$, *Z*, *P*, and $\theta$.

**Solution**

$$X_L = 2\pi fL = 6.28 \times 50 \times 10^3 \times 30 \times 10^{-3} = 9.42\ \text{k}\Omega$$

$$Z = \sqrt{R^2 + X_L^2} = \sqrt{6.8^2 + 9.42^2} = 11.62\ \text{k}\Omega$$

$$I_T = \frac{V_T}{Z} = \frac{40\ \text{V}}{11.62\ \text{k}\Omega} = 3.44\ \text{mA}$$

$$V_R = I_T R = 3.44\ \text{mA} \times 6.8\ \text{k}\Omega = 23.4\ \text{V}$$

$$V_L = I_T X_L = 3.44\ \text{mA} \times 9.42\ \text{k}\Omega = 32.4\ \text{V}$$

$$\theta = \arctan\frac{X_L}{R} = \arctan\frac{9.42\ \text{k}\Omega}{6.8\ \text{k}\Omega} = 54.2°$$

$$P = I_T V_T \cos\theta = 3.44\ \text{mA} \times 40\ \text{V} \times \cos 54.2° = 80.4\ \text{mW}$$

As a quick check for gross errors, let us determine power by $P = V_R^2/R = 23.4^2/6800 = 80.5$ mW. Within round-off error, the two answers for *P* agree.

## Self-Test

**15.** In a series *RL* circuit, the inductive voltage will ______ the source voltage.

**16.** In a series *RL* circuit, angle $\theta$ is in the ______ quadrant.

**17.** Decreasing the frequency in a series *RL* circuit causes angle $\theta$ to ______.

**18.** Increasing the frequency in a series *RL* circuit will cause the power to ______.

**19.** Determine the inductance in a series *RL* circuit if $f = 10$ kHz, $R = 4$ kΩ, and $Z = 5$ kΩ.

**20.** Determine $\theta$, $Z$, $V_R$, and $P$ for the circuit in Fig. 17-6*a* when $X_L = 4$ kΩ, $R = 3$ kΩ, and $V_T = 60$ V.

## 17-3 SERIES *RCL* CIRCUITS

Figure 17-7 shows the various phasor relationships in *RCL* circuits. Because voltage lags current by 90° in a capacitor and leads current by 90° in an inductor, $V_L$ is leading $V_C$ by 180°. When two phasors that are 180° apart are added, the magnitude of the resultant phasor is the arithmetic difference of the magnitudes of the two phasors being added. The angle of the resultant phasor is, of course, that of the largest of the two phasors being added. For the $V_L$ and $V_C$ phasors in Fig. 17-7, the resultant phasor is labeled $V_X$. The resultant reactive voltage ($V_X$) is inductive when $X_L$ is greater than $X_C$, and capacitive when $X_C$ is greater than $X_L$. Since $V_X$ must be directly proportional to $X$ (just as $V_L$ is directly proportional to $X_L$ or $V_C$ is directly proportional to $X_C$), the vector relationships between the vectors $Z$, $R$, and $X$ are the same as the phasor relationships between $V_T$, $V_R$, and $V_X$. Thus, the voltage and impedance triangles in Fig. 17-7*a* or *b* have the same proportions and angles. Since the magnitude of $V_X$ is $V_C - V_L$, the mathematical equivalent of the graphical solution for $V_T$ in Fig. 17-7 is

$$V_T = \sqrt{V_R^2 + V_X^2} = \sqrt{V_R^2 + (V_C - V_L)^2}$$

And, since $X = X_C - X_L$, the formula for the impedance of a series *RCL* circuit is

$$Z = \sqrt{R^2 + (X_C - X_L)^2}$$

Angle $\theta$ can be determined by any of the trigonometric formulas used with series *RC* or *RL* circuits by replacing $X_C$ or $X_L$ with $X$. As shown in Fig. 17-7, $\theta$ is negative (in the fourth quadrant) when the net reactance $X$ is capacitive and positive when $X$ is inductive. In other words, $I$ leads $V$ when $X$ is capacitive, and $V$ leads $I$ when $X$ is inductive.

One very important point to remember about *RCL* circuits is that either $V_C$ or $V_L$ *or both* $V_C$ and $V_L$ can be (and often are) larger than $V_T$. This also

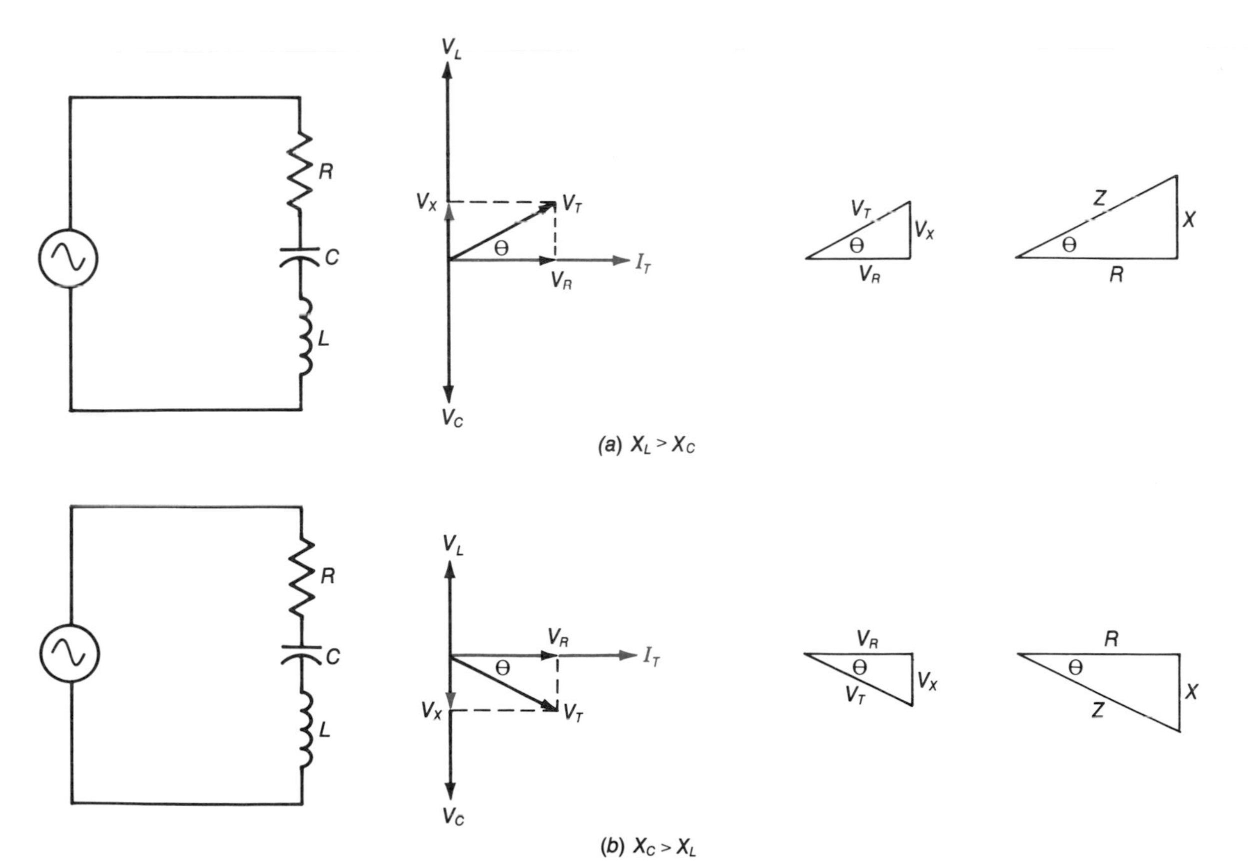


FIGURE 17-7 Series *RCL* circuit.

means, of course, that either or both $X_L$ and $X_C$ can be larger than $Z$. For example, Fig. 17-7*a* shows an instance where both $V_L$ and $V_C$ are larger than $V_T$. In Fig. 17-7*b* $V_C$ is larger than $V_T$, but $V_L$ is smaller than $V_T$. Notice from the triangles in Fig. 17-7 that it is not possible for either $V_R$ or $V_X$ to exceed $V_T$. Likewise, $Z$ will always be greater than either $R$ or $X$.

After looking at the illustrations in Fig. 17-7, you may be wondering what happens when $X_L = X_C$. This is a special condition called *resonance*. It is dealt with in detail in Chap. 18. For now, suffice it to say that some interesting things happen when $X_L = X_C$ and $R$ is reduced to zero.

**EXAMPLE 17-8**

Determine $V_L$, $V_C$, $Z$, $I_T$, $\theta$, and $P$ for the circuit in Fig. 17-7 when $R = 1.2\ k\Omega$, $C = 7800$ pF, $L = 50$ mH, $V_T = 55$ V, and $f = 10$ kHz. Will $I_T$ lead or lag $V_T$?

**Solution**

$$X_C = \frac{1}{2\pi fC} = \frac{1}{6.28 \times 10 \times 10^3 \times 7800 \times 10^{-12}}$$
$$= 2041\ \Omega$$
$$X_L = 2\pi fL = 6.28 \times 10 \times 10^3 \times 50 \times 10^{-3} = 3140\ \Omega$$
$$Z = \sqrt{R^2 + (X_C - X_L)^2} = \sqrt{1.2^2 + 1.101^2} = 1629\ \Omega$$

Note: $(-1.101)^2 = +1.101^2$

$$\theta = \arccos\frac{R}{Z} = \arccos\frac{1200\ \Omega}{1629\ \Omega} = 42.6°$$
$$I_T = \frac{V_T}{Z} = \frac{55\ \text{V}}{1629\ \Omega} = 33.8\ \text{mA}$$
$$P = IV\cos\theta = 33.8\ \text{mA} \times 55\ \text{V} \times 0.74 = 1.38\ \text{W}$$
$$V_L = I_T X_L = 33.8\ \text{mA} \times 3.14\ k\Omega = 106.1\ \text{V}$$
$$V_C = I_T X_C = 33.8\ \text{mA} \times 2.041\ k\Omega = 70.0\ \text{V}$$

Since $X_L$ is greater than $X_C$, $I_T$ will lag $V_T$.

Example 17-8 can be checked for mathematical error with $P = I_R{}^2R = (0.0338\ \text{A})^2 \times 1200\ \Omega = 1.37$ W or by calculating

$$V_R = I_T R \quad \text{and} \quad V_T = \sqrt{V_R{}^2 + (V_L - V_C)^2}$$

**EXAMPLE 17-9**

Determine the value of $\theta$ in Example 17-8 if $f$ is reduced to 5 kHz.

**Solution**

Since $X_L$ is directly proportional to $f$ and since $X_C$ is inversely proportional to $f$, the new values are

$$X_L = \frac{3140\ \Omega}{2} = 1570\ \Omega$$
$$X_C = 2041\ \Omega \times 2 = 4082\ \Omega$$
$$X = X_C - X_L = 4082\ \Omega - 1570\ \Omega = 2512\ \Omega$$

Now use the impedance triangle and a trigonometric function to get

$$\theta = \arctan\frac{X}{R} = \arctan\frac{2512\ \Omega}{1200\ \Omega} = 64.5°$$

Notice from this example that the current now leads the voltage because $X_C > X_L$. Also notice that the angle is actually $-64.5°$ since the vector triangle is in the fourth quadrant.

$$\arctan\frac{-2512\ \Omega}{1200\ \Omega} = -64.5°$$

## Self-Test

**21.** In a series *RCL* circuit, not at resonance, $Z$ will always be greater than ______ or ______.

**22.** The magnitude of $V_X$ is equal to $V_C$ ______ $V_L$.

**23.** In a series *RCL* circuit, can
- **a.** $V_T$ be greater than $V_L$?
- **b.** $V_C$ be greater than $V_T$?
- **c.** $R$ be greater than $Z$?
- **d.** $Z$ be greater than $X$?

**24.** ______ is the term used to describe the circuit condition when $X_L = X_C$.

**25.** For a series *RCL* circuit, $R = 2\ k\Omega$, $X_C = 3\ k\Omega$, and $X_L = 2\ k\Omega$. What is the impedance?

**26.** Determine $Z$, $\theta$, and $I_T$ for a series *RCL* circuit where $R = 2.2\ k\Omega$, $L = 5$ mH, $C = 1000$ pF, $f = 50$ kHz, and $V_T = 20$ V.

**27.** In Question 26, will $I$ lead or lag $V$?

## 17-4 PARALLEL *RC* CIRCUITS

In analyzing parallel resistance-reactance circuits, we must use parallel circuit rules. We know that in a parallel circuit all voltages, instantaneous or otherwise, are equal; i.e., $V_T = V_R = V_C$. Because of this, $V_T$ is used as the reference phasor for parallel circuits.

The $I_C$ and $I_R$ phasors are as shown in Fig. 17-8 because $I_C$ leads $V_C$ by 90° and $I_R$ is in phase with $V_R$ whether $R$ and $C$ are in parallel or series. The total current is, according to Kirchhoff's current law, the sum of $I_R$ and $I_C$. These 90° current phasors can be summed with the formula

$$I_T = \sqrt{I_R^2 + I_C^2}$$

The current triangle in Fig. 17-8*c* shows that $\theta$ can be calculated as follows:

$$\theta = \arctan\frac{I_C}{I_R} = \arccos\frac{I_R}{I_T} = \arcsin\frac{I_C}{I_T}$$

The length of the current phasors in Fig. 17-8 shows that $X_C$ is greater than $R$ in this particular circuit. Notice that in a parallel $RC$ circuit, $\theta < 45°$ when $X_C > R$, while in a series $RC$ circuit, $\theta > 45°$ when $X_C > R$. This is because the parallel circuit is dominated by the smallest parallel opposition, while the series circuit is dominated by the largest opposition.

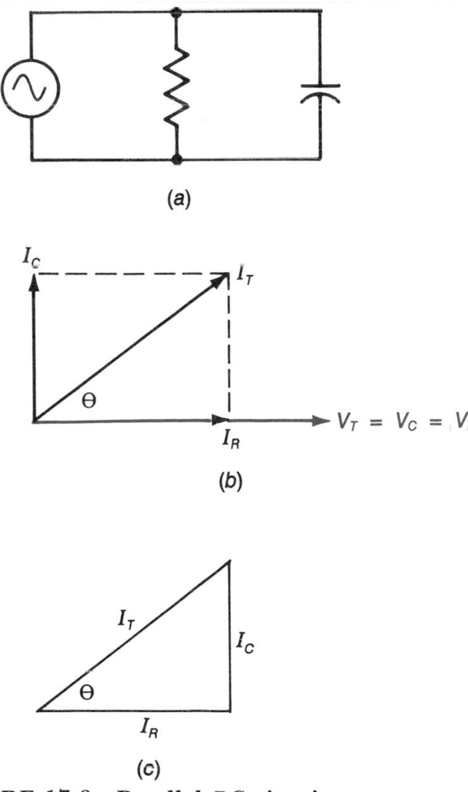


**FIGURE 17-8 Parallel *RC* circuit.**

Notice in Fig. 17-8 that no impedance triangle is drawn. It is not possible to draw an impedance triangle for parallel circuits because parallel oppositions ($X$ and $R$) are inversely proportional to current. That is, when $V$ is constant, doubling $R$ causes $I$ to be halved. The easiest way to determine $Z$ in a parallel circuit is to first determine $I_T$ and then use Ohm's law ($Z = V_T/I_T$) to calculate $Z$. If $V_T$ is not specified, assign any value to $V_T$ and proceed to calculate $I_T$ and $Z$.

**EXAMPLE 17-10**

Determine the impedance, at 400 Hz, of a 10-kΩ resistor in parallel with a 0.03-μF capacitor. Also determine $\theta$.

**Solution**

$$X_C = \frac{1}{6.28 \times 400 \times 0.03 \times 10^{-6}} = 13{,}270\ \Omega$$

Arbitrarily specify $V_T = 10$ V, then

$$I_C = \frac{10\text{ V}}{13.27\text{ k}\Omega} = 0.754\text{ mA}$$

$$I_R = \frac{10\text{ V}}{10\text{ k}\Omega} = 1.0\text{ mA}$$

$$I_T = \sqrt{I_R^2 + I_C^2} = \sqrt{1.0^2 + 0.754^2} = 1.252\text{ mA}$$

$$Z = \frac{V_T}{I_T} = \frac{10\text{ V}}{1.252\text{ mA}} = 7987\ \Omega$$

$$\theta = \arctan\frac{0.754\text{ mA}}{1\text{ mA}} = 37°$$

Notice in this example that $Z$ is less than either $R$ or $X$. This is as it should be for parallel circuits. Also note that $X_C > R$; therefore, $\theta < 45°$.

From the Ohm's law expression of $Z$, other formulas can be derived for calculating the $Z$ of parallel circuits. Two common ones are

$$Z = \frac{1}{\sqrt{(1/R)^2 + (1/X_C)^2}} \quad \text{and} \quad Z = \frac{RX_C}{\sqrt{R^2 + X_C^2}}$$

A derivation of the first formula is

$$I_T = \frac{V_T}{Z}$$

Substitute for $I_T$:

$$\sqrt{I_R^2 + I_C^2} = \frac{V_T}{Z}$$

Substitute for $I_R$ and $I_C$:

$$\sqrt{\left(\frac{V_T}{R}\right)^2 + \left(\frac{V_T}{X_C}\right)^2} = \frac{V_T}{Z}$$

Multiply by $1/V_T$:

$$\sqrt{\left(\frac{1}{R}\right)^2 + \left(\frac{1}{X_C}\right)^2} = \frac{1}{Z}$$

Rearrange and take the reciprocal:

$$Z = \frac{1}{\sqrt{(1/R)^2 + (1/X_C)^2}}$$

This formula just combines into one step what we did in four steps in Example 17-10. The 1 in the numerator and denominator can be thought of as 1 V which represents $V_T$. When viewed this way, the numerator is $V_T$ and the denominator is $I_T$—thus, back to Ohm's law.

Backing up to the next-to-last formula in the above derivation allows us to derive the second parallel impedance formula listed above:

$$\sqrt{\left(\frac{1}{R}\right)^2 + \left(\frac{1}{X_C}\right)^2} = \frac{1}{Z}$$

Square both sides:

$$\left(\frac{1}{R}\right)^2 + \left(\frac{1}{X_C}\right)^2 = \left(\frac{1}{Z}\right)^2$$

Multiply by $R^2X_C^2$:

$$X_C^2 + R^2 = \left(\frac{1}{Z}\right)^2 R^2X_C^2$$

Take the square root:

$$\sqrt{X_C^2 + R^2} = \frac{1}{Z}RX_C$$

Rearrange and take the reciprocal:

$$Z = \frac{RX_C}{\sqrt{R^2 + X_C^2}}$$

Notice that both of these derived formulas add the parallel oppositions vectorally as well as reciprocally.

**EXAMPLE 17-11**

Using the $R$ and $X_C$ values from Example 17-10, determine $Z$ using the formula for $Z$ which was just derived.

**Solution**

$$Z = \frac{RX_C}{\sqrt{R^2 + X_C^2}} = \frac{10 \times 13.27}{\sqrt{10^2 + 13.27^2}} = \frac{132.7}{\sqrt{276}} = 7987\ \Omega$$

This agrees within round-off error with the value determined in Example 17-10.

The impedance for a parallel resistance-reactance circuit can also be found using trigonometric functions. From the current triangle in Fig. 17-8*c*, we can see that cos $\theta$ = adjacent/hypotenuse = $I_R/I_T$. Now, substituting the Ohm's law equivalent, we get

$$\cos\theta = \frac{I_R}{I_T} = \frac{V_T/R}{V_T/Z} = \frac{Z}{R}$$

And solving for impedance we get

$$Z = R\cos\theta$$

Of course, this relationship also shows that $\theta$ can be determined from the values of the parallel opposition. Solving for $\theta$ we get

$$\theta = \arccos\frac{Z}{R}$$

Using the same substitution technique and the other trigonometric functions, these useful formulas can be derived:

$$\theta = \arctan\frac{R}{X_C} = \arcsin\frac{Z}{X_C}$$

$$X_C = \frac{Z}{\sin\theta}$$

$$R = X_C\tan\theta$$

**EXAMPLE 17-12**

For the circuit in Fig. 17-9, determine $V_T$, $I_T$, $I_R$, $C$, $Z$, $\theta$, and $P$.

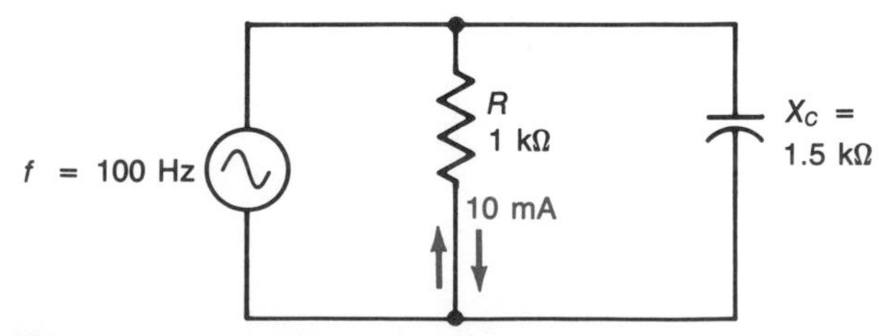


**FIGURE 17-9 Circuit for Example 17-12.**

**Solution**

$$V_T = V_C = V_R = I_R R = 10\text{ mA} \times 1\text{ k}\Omega = 10\text{ V}$$

$$I_C = \frac{10\text{ V}}{1.5\text{ k}\Omega} = 6.67\text{ mA}$$

$$I_T = \sqrt{10^2 + 6.67^2} = 12.02\text{ mA}$$

$$Z = \frac{10\text{ V}}{12.02\text{ mA}} = 832\ \Omega$$

$$\theta = \arctan\frac{6.67}{10} = 33.7°$$

$$C = \frac{1}{2\pi f X_C} = \frac{1}{6.28 \times 100 \times 1500} = 1.06\ \mu\text{F}$$

$$\cos 33.7° = 0.832$$

$$P = 12.02\text{ mA} \times 10\text{ V} \times 0.832 = 0.1\text{ W}$$

Checks:

$$P = (10\text{ mA})^2 \times 1\text{ k}\Omega = 0.1\text{ W}$$

$$Z = R\cos\theta = 1\text{ k}\Omega \times 0.832 = 832\ \Omega$$

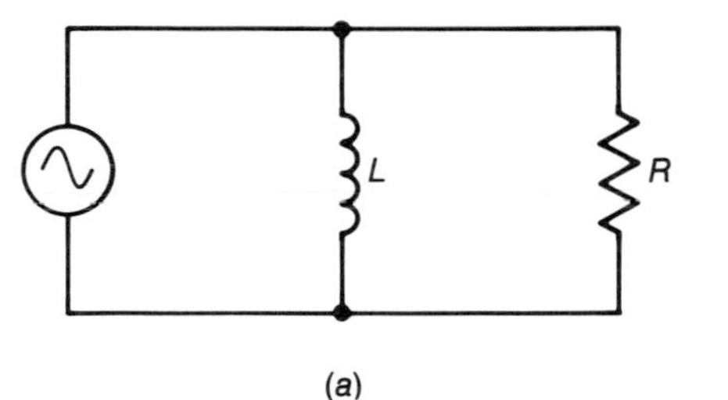


(a)

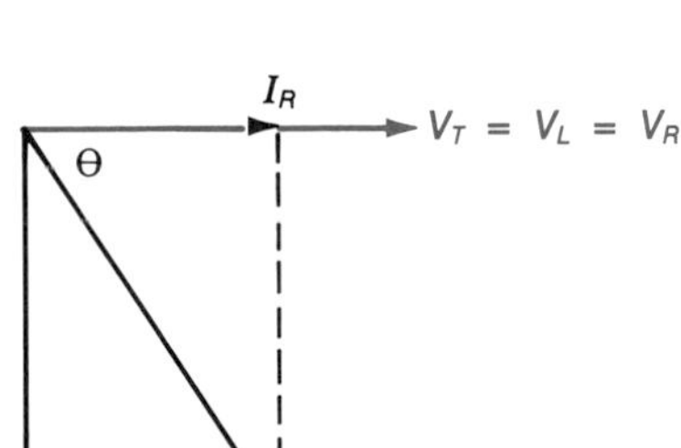


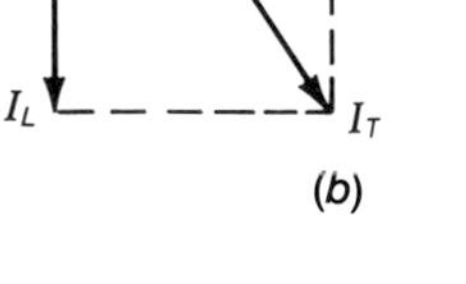


(b)

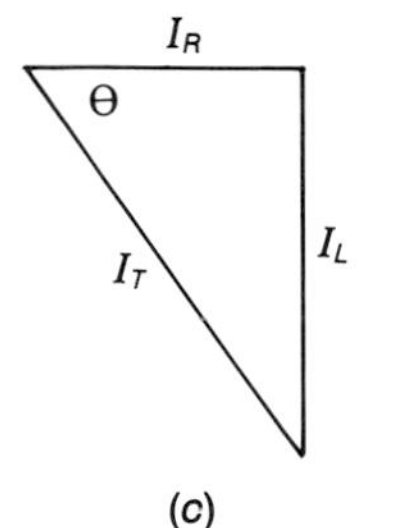


(c)

**FIGURE 17-10 Parallel *RL* circuit.**

## Self-Test

**28.** In a parallel *RC* circuit, $Z$ will be ______ either $R$ or $X_C$.

**29.** In a parallel *RC* circuit with $R > X_C$, $\theta$ will be ______ 45°.

**30.** True or false? $\theta = \arccos(R/Z)$ in a parallel *RC* circuit.

**31.** Determine $\theta$, $P$, $Z$, and $I_T$ for a parallel *RC* circuit when $R = 15\text{ k}\Omega$, $C = 0.1\ \mu\text{F}$, $f = 100$ Hz, and $V_R = 40$ V.

## 17-5 PARALLEL *RL* CIRCUITS

The same formulas and procedures are used to solve parallel *RL* circuits as were used to solve parallel *RC* circuits. Just change $X_C$ to $X_L$ and $I_C$ to $I_L$. Of course, in the *RL* circuit the current lags the voltage as shown in Fig. 17-10*b*. This means that $I_T$ is in the fourth quadrant and that angle $\theta$ is negative.

The same generalizations that held true for the parallel *RC* circuit also are true for the parallel *RL* circuit. That is (1) when $R$ is less than $X_L$, $\theta$ is less than 45°, (2) $Z$ is less than either $X_L$ or $R$, and (3) $I_T$ is greater than either $I_R$ or $I_L$ but less than the arithmetic sum of $I_R + I_L$.

**EXAMPLE 17-13**

For the circuit in Fig. 17-10*a*, $R = 10\text{ k}\Omega$, $I_L = 2$ mA, $L = 20$ mH, and $V_T = 15$ V. Determine $I_T$, $Z$, $\theta$, and $f$.

**Solution**

$$X_L = \frac{15\text{ V}}{2\text{ mA}} = 7.5\text{ k}\Omega$$

$$X_L = 2\pi f L$$

Solving for $f$,

$$f = \frac{X_L}{2\pi L} = \frac{7.5\text{ k}\Omega}{6.28 \times 20 \times 10^{-3}} = 59.7\text{ kHz}$$

$$I_R = \frac{15\text{ V}}{10\text{ k}\Omega} = 1.5\text{ mA}$$

$$I_T = \sqrt{1.5^2 + 2^2} = 2.5\text{ mA}$$

$$Z = \frac{15\text{ V}}{2.5\text{ mA}} = 6\text{ k}\Omega$$

$$\theta = \arctan\frac{2\text{ mA}}{1.5\text{ mA}} = 53.1°$$

The impedance in this example could also be calculated using the formula $Z = RX_L/\sqrt{R^2 + X_L^2}$. This is the formula derived in Sec. 17-4, only $X_L$ has been substituted for $X_C$.

## Self-Test

**32.** Which circuit values could you determine in Example 17-13 if $I_L$ had not been specified?

**33.** Will increasing the $f$ in a parallel *RL* circuit cause $\theta$ to increase or decrease? Why?

**34.** Is it possible for $I_R$ to equal 20 mA in a parallel *RL* circuit when $I_T = 18$ mA? Why?

**35.** Determine $P$, $\theta$, $Z$, and $I_T$ for a parallel *RL* circuit when $V_T = 25$ V at 1 kHz, $R = 6.8$ kΩ, and $L = 0.8$ H.

## 17-6 PARALLEL *RCL* CIRCUITS

Applying the same ideas to the currents in parallel *RCL* circuits as we did to the voltages in series *RCL* circuits allows us to easily work with parallel *RCL* circuits. As illustrated in Fig. 17-11*b*, the current in the inductive branch of the circuit is 180° out of phase with the current in the capacitive branch. Of course, for ideal components, each of these currents will be 90° out of phase with the resistive branch current: $I_L$ will lag by 90° and $I_C$ will lead by 90°. The current formula is

$$I_T = \sqrt{I_R^2 + (I_C - I_L)^2}$$

or

$$I_T = \sqrt{I_R^2 + I_X^2}$$

where $I_X = I_C - I_L$.

Whether $\theta$ is negative or positive depends on the relative values of $I_C$ and $I_L$ and, thus, on the relative values of $X_C$ and $X_L$. When $X_C > X_L$, $I_L$ will dominate and $\theta$ will be negative ($I_T$ will be in the fourth quadrant).

Like series *RCL* circuits, parallel *RCL* circuits can be resonant. That is, one or more of the values of $X$, $L$, and $f$ can be adjusted until $X_L = X_C$. At this point, $I_C = I_L$ and the total current then equals the resistive current. Resonant circuits are dealt with in Chap. 18.

Analogous to the reactive voltages in a series *RCL* circuit, the reactive currents (either or both) can exceed the total current. However, the magnitude of the total current will always exceed that of the net reactive current $(I_C - I_L)$ or the resistive current (except at resonance when $I_T = I_R$).

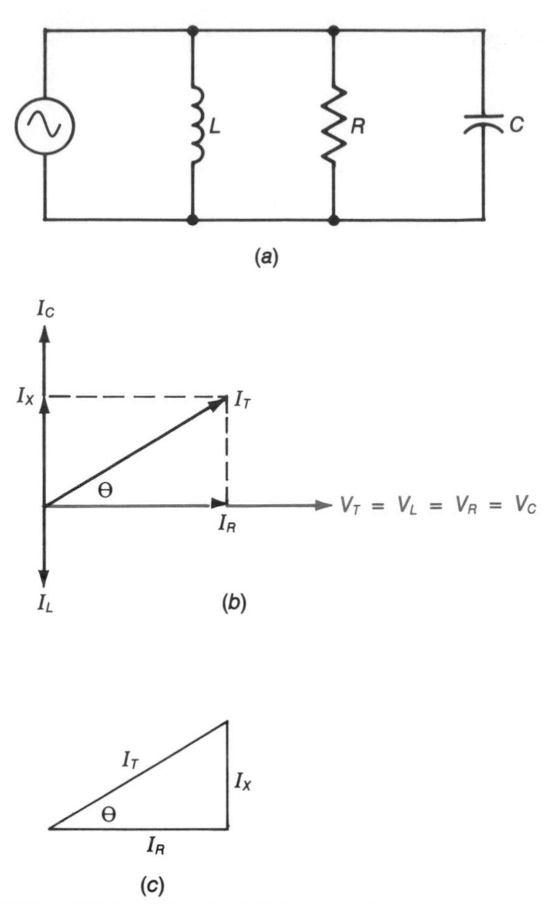


**FIGURE 17-11** **Parallel *RCL* circuit.**

Like any other circuit, the impedance in a parallel *RCL* circuit can be determined by Ohm's law: $Z = V_T/I_T$. With substitution procedures like those used for the parallel *RC* instance, it is possible to derive these formulas for the impedance in a parallel *RCL* circuit:

$$Z = \frac{RX_LX_C}{\sqrt{(RX_L - RX_C)^2 + (X_L^2X_C^2)}}$$

or

$$Z = \frac{1}{\sqrt{(1/R)^2 + (1/X_C - 1/X_L)^2}}$$

As discussed in Sec. 17-14, "Parallel *RC* Circuits," the denominator of this last formula is the total current and the numerator is the source voltage when the source voltage is 1 V. Although these derived formulas are quite manageable with modern hand calculators and personal computers, the Ohm's law method is usually easier to compute and requires fewer total calculations if $I_T$ must also be determined.

**EXAMPLE 17-14**

In a parallel $RCL$ circuit, $R = 5\ \text{k}\Omega$, $L = 10\ \text{mH}$, $C = 320\ \text{pF}$, and $V_T = 30\ \text{V}$ at 100 kHz. Determine $I_C$, $I_X$, $I_T$, $Z$, $P$, and $\theta$. Does $I_T$ lead or lag $V_T$?

**Solution**

$$X_C = \frac{1}{6.28 \times 100 \times 10^3 \times 320 \times 10^{-12}} = 4976\ \Omega$$

$$X_L = 6.28 \times 100 \times 10^3 \times 10 \times 10^{-3} = 6280\ \Omega$$

$$I_R = \frac{30\ \text{V}}{5\ \text{k}\Omega} = 6.0\ \text{mA}$$

$$I_C = \frac{30\ \text{V}}{4976\ \Omega} = 6.03\ \text{mA}$$

$$I_L = \frac{30\ \text{V}}{6280\ \Omega} = 4.78\ \text{mA}$$

$$I_X = 6.03 - 4.78 = 1.25\ \text{mA}$$

$$I_T = \sqrt{6^2 + 1.25^2} = 6.13\ \text{mA (leads)}$$

$$Z = \frac{30\ \text{V}}{6.13\ \text{mA}} = 4894\ \Omega$$

$$\theta = \arctan \frac{1.25}{6.0} = 11.8°$$

$$P = \frac{(30\ \text{V})^2}{5\ \text{k}\Omega} = 180\ \text{mW}$$

## Self-Test

**36.** Is $R$, $X_C$, or $X_L$ larger in Fig. 17-11? Why?

**37.** List three values which could be changed to decrease $\theta$ in Fig. 17-11. Also indicate whether and why the value should be increased or decreased.

**38.** List two ways to decrease the power in Fig. 17-11.

**39.** For the circuit in Fig. 17-11*a*, $V_T = 5.0\ \text{V}$, $f = 10\ \text{kHz}$, $C = 0.2\ \mu\text{F}$, $X_L = 100\ \Omega$, and $I_R = 0.05\ \text{A}$. Determine $P$, $\theta$, $I_T$, $Z$, and $L$.

## 17-7 RESPONSE TO PULSES

The response of an $RC$ circuit to a square wave was mentioned in Chap. 16 where it was shown that the time constant and the frequency controlled the shape of the waveforms across the resistor. Now we will investigate the response of both $RC$ and $RL$ circuits to both symmetrical and asymmetrical pulses.

### *RC* Circuits—Symmetrical Input

The response of an $RC$ circuit to a symmetrical pulse train or square wave varies depending on whether the voltage is viewed across the resistor or the capacitor. When viewed across the capacitor, as in Fig. 17-12*a*, the leading edge of the waveform is rounded off because the capacitor cannot instantly charge and discharge. The longer the $RC$ time constant and the higher the frequency, the greater is the rounding of the waveform.

As seen in Fig. 17-12*a*, very little rounding occurs when the period $T$ of the waveform is very long compared to the time constant $\tau$. In Fig. 17-12*a* the period is $100\tau$, so the input voltage is +5 V for $50\tau$. Since it takes only $5\tau$ to essentially charge a capacitor, the capacitor's voltage, and thus the output, will be at 5 V for nine-tenths of the time the input is at 5 V. Of course, the capacitor also discharges in $5\tau$, so the output will also be zero for 90 percent of the time that the input is zero.

Figure 17-12*b* shows that when $T = 10\tau$, the capacitor just gets charged and discharged during the time the input is +5 V and zero. Thus, the output voltage never reaches a steady state even though its peak-to-peak value is equal to the peak-to-peak value of the input voltage.

When $\tau$ and $T$ are equal, as in Fig. 17-12*c*, the output voltage is less than half the value of the input voltage, and it begins to look like a triangular waveform. The value of the voltage is reduced because during the $0.5\tau$ that the capacitor has to charge, its voltage can only increase to 39.35 percent of the available voltage. Also, during the $0.5\tau$ that it has to discharge, it cannot discharge back to 0 V. Thus, as seen in Fig. 17-12, the changes in the capacitor's voltage will be less than the changes in the source voltage. The output waveform in Fig. 17-12*c* looks triangular because during the first $\frac{1}{2}\tau$ the voltage-time curve for either a charging or discharging capacitor is almost straight. (Refer back to Figs. 16-5 and 16-6 for these curves.)

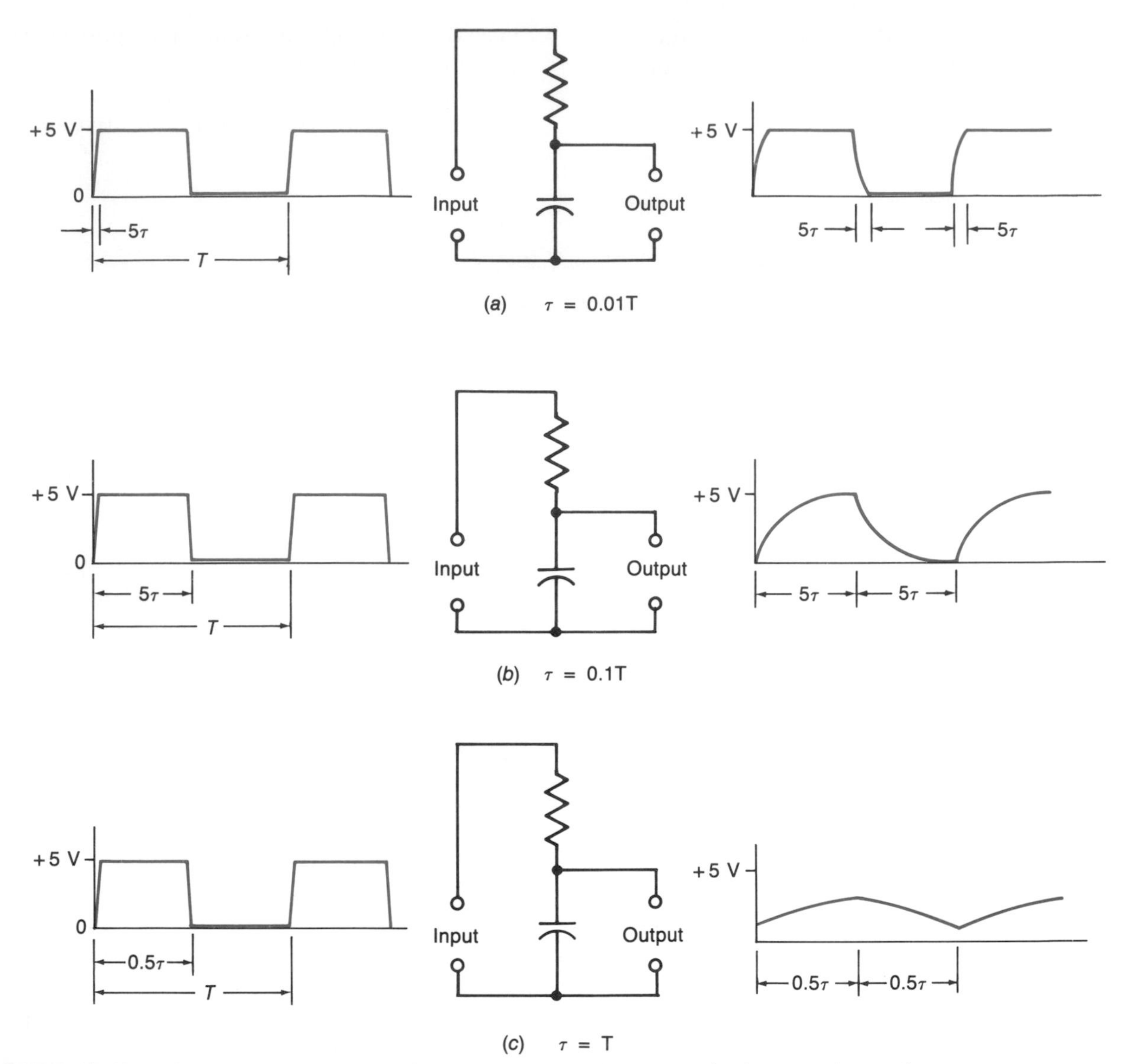


**FIGURE 17-12** ***RC*** **response to symmetrical pulses when the output is the capacitor voltage.**

## EXAMPLE 17-15

For the *RC* circuit in Fig. 17-12, what is the maximum frequency at which a 0.1-$\mu$F capacitor will charge to the value of the input waveform voltage if the resistance is 10 k$\Omega$. Assume the capacitor is fully charged after $5\tau$.

### Solution

$$\tau = RC = 10\ \text{k}\Omega \times 0.1\ \mu\text{F} = 1\ \text{ms}$$

Time to charge the capacitor is

$$5\tau = 5 \times 1\ \text{ms} = 5\ \text{ms}$$

Since the waveform is symmetrical, the capacitor must charge in $0.5T$. Thus,

$$0.5T = 5\tau = 5\ \text{ms}$$

So the period is

$$T = 2 \times 5\ \text{ms} = 10\ \text{ms}$$

Finally,

$$f = \frac{1}{T} = \frac{1}{10\ \text{ms}} = 100\ \text{Hz}$$

The output waveform in Fig. 17-12*c* shows the output after the input has been applied for many cycles. Thus, the detail of how the capacitor's voltage builds up during the first few cycles is lost. In some electronic circuits the input pulse train is only applied to the *RC* circuit for a few cycles before the capacitor is discharged. Then the process of capacitor voltage buildup and discharge is repeated.

The details of the buildup of the capacitor's voltage are illustrated in Fig. 17-13 for a circuit in which the period of the symmetrical waveform is $2\tau$. In this figure the capacitor is completely discharged before the first positive input pulse arrives. The capacitor immediately starts to *integrate* the input voltage. That is, it builds up a voltage which is a function of the frequency (or the period) of the input waveform. *Integrator circuits* refers to *RC* circuits in which the output is taken across the capacitor. (As we will see later in this chapter, integration can be much more pronounced when the input pulses are nonsymmetrical.) In Fig. 17-13*b* and *c* you can see that the values of *f*, *R*, and *C* are such that the capacitor charges and discharges for exactly $1\tau$. Thus, the capacitor's voltage increases another 63.2 percent of the difference between its voltage and the source voltage (10 V) each time the input goes positive. It loses 63.2 percent of its voltage every time the input drops to zero. The table in Fig. 17-13*d* contains the capacitor's voltage at the end of each charge and discharge action for the first six charges and the first six discharges. These voltages were calculated using the formulas given in Chap. 16 for *RC* time constants. Notice from Fig. 17-13*d* that the capacitor's voltage stabilizes at about +7.31 for its peak value and about +2.69 for its minimum value. The values listed in Fig. 17-13*d* are rounded to two decimal places, so the voltages after $6\tau$ of charging and $6\tau$ of discharging are still increasing ever so slightly. This might lead one to believe that eventually, after a few thousand cycles, the peak capacitor voltage should reach the same value as the source voltage. However, this does not happen. Even the best capacitor has less than infinite insulation resistance. This resistance allows enough leakage current to counteract and cancel the minute additional charge the capacitor takes on each cycle after it is essentially charged.

As shown in Fig. 17-14, the output waveforms for an *RC* circuit are entirely different when the output is taken across the resistor. The output is then known as a *differentiated output*. Thus, this type of circuit is often referred to as a *differentiator*. The amount of differentiation is again dependent on the period of the input and the time constant.

We know that a resistor produces a voltage drop only when current flows through it. Therefore, we also know that the circuit in Fig. 17-14 has an output voltage only when the capacitor is charging or discharging. As seen in Fig. 17-14*b* and *c*, the larg-

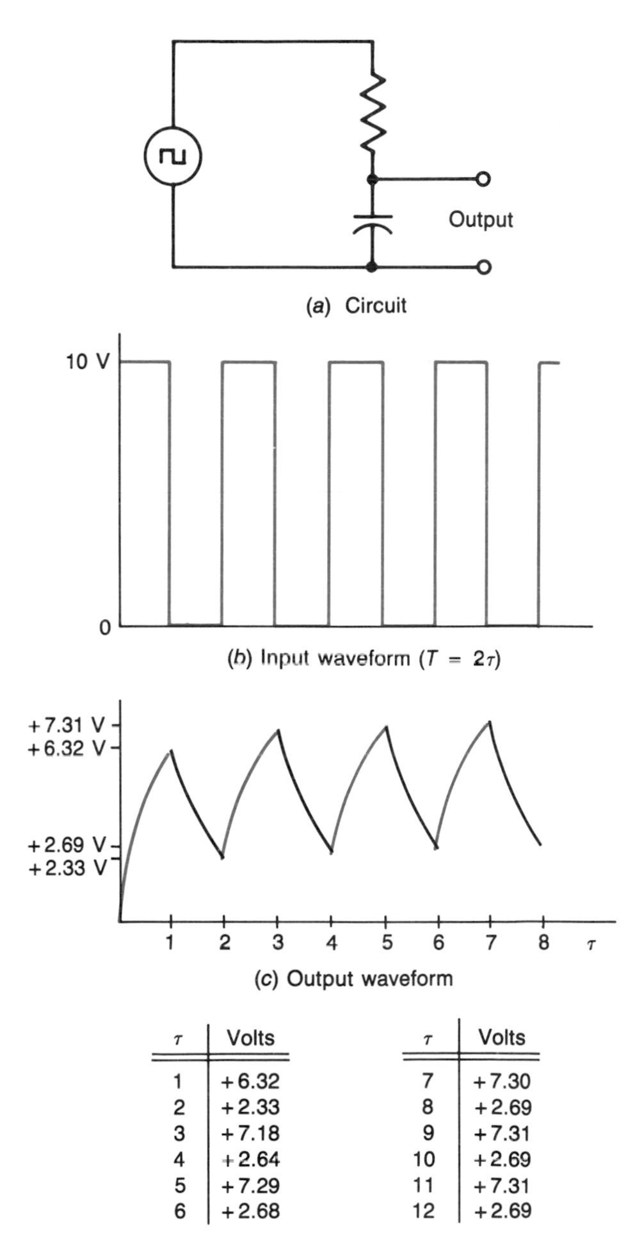


| $\tau$ | Volts | $\tau$ | Volts |
|---|---|---|---|
| 1 | +6.32 | 7 | +7.30 |
| 2 | +2.33 | 8 | +2.69 |
| 3 | +7.18 | 9 | +7.31 |
| 4 | +2.64 | 10 | +2.69 |
| 5 | +7.29 | 11 | +7.31 |
| 6 | +2.68 | 12 | +2.69 |

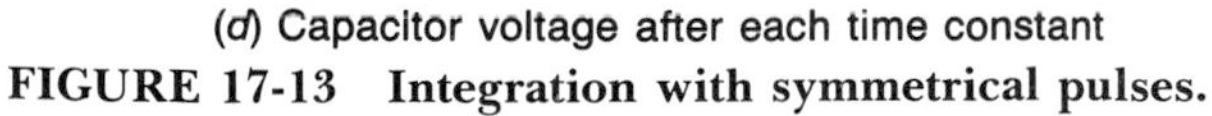

(*d*) Capacitor voltage after each time constant

**FIGURE 17-13 Integration with symmetrical pulses.**

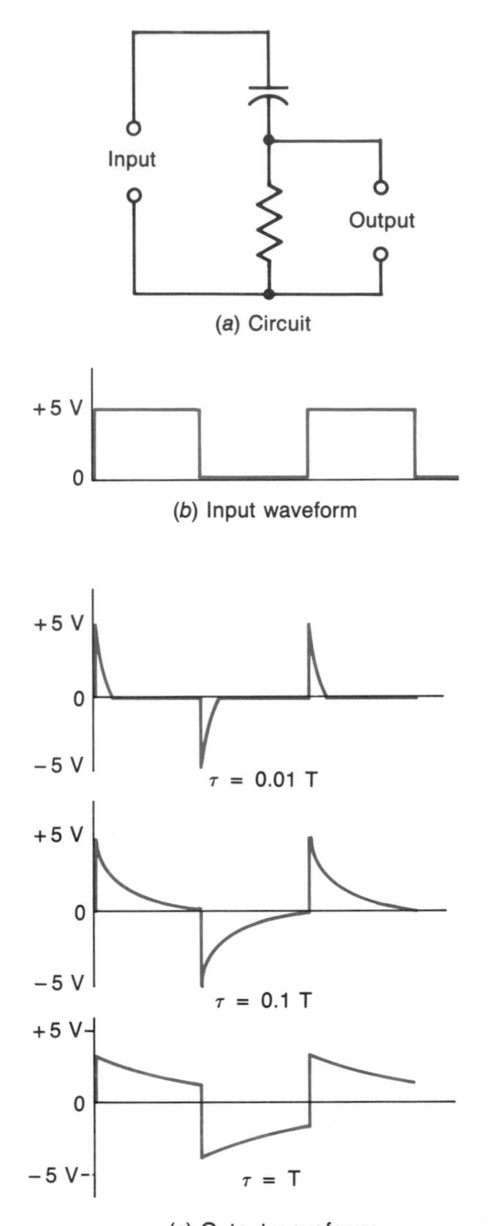


**FIGURE 17-14 Differentiated output taken across the resistor in an *RC* circuit.**

est voltage appears across the resistor the instant the source voltage changes. This is because at this instant the difference between the source voltage and the capacitor voltage is greatest. Thus, the circuit current and the resistor voltage are also greatest at this instant. The top-output waveform in Fig. 17-14*c* shows that the output is a relatively short duration voltage spike when the period is long compared to the time constant. After $5\tau$ the capacitor is through charging, so the output is zero. The middle-output waveform shows that when $\tau = 0.1T$, the capacitor is essentially fully charged or discharged just as the input voltage changes. The bottom waveform shows what happens when the capacitor does not have time to fully charge or fully discharge. The output starts to look more like a square wave, but its amplitude is also reduced. When the time constant is hundreds of times longer than the period, the output is almost a perfect square wave. At this time, the peak-to-peak amplitude equals that of the source.

Notice in Fig. 17-14 that when the capacitor has time to fully charge and discharge ($\tau < 0.1$ T), the peak-to-peak output is twice as large as the peak-to-peak input. Analyzing the circuit action explains why this is so. The resistor voltage must equal the source voltage the instant the source goes to +5 V. (The common input-output lead is the reference for this +5 V.) As the capacitor charges, the voltage across the resistor decays. After $5\tau$, the resistor voltage is zero, where it remains as long as the source is +5 V. Of course, the capacitor is now charged to 5 V with its top terminal being positive with respect to its bottom terminal. Now, when the input suddenly drops to 0 V, the full 5 V contained on the capacitor is applied across the resistor. The top end of the resistor is connected to the − end of the capacitor's voltage, and the bottom end of the resistor is connected, through the internal resistance of the source, to the + end of the capacitor. Thus, the instantaneous resistor voltage is −5 V. Of course, as the capacitor discharges through the resistor and the source resistance, the resistor's voltage decays to zero as does the capacitor's voltage.

Comparing Fig. 17-12 and Fig. 17-14 shows that the capacitor output is most like the input when $T \gg \tau$, while the resistor output is most like the input when $T < \tau$. This should not be too surprising if we remember to apply Kirchhoff's voltage law. Whatever source voltage does not appear across the capacitor must appear across the resistor and vice versa. This can be seen quite clearly in Fig. 17-15 where the three voltages for an *RC* circuit are drawn to the same scale. Notice in this figure that the instantaneous sum of the resistor and capacitor voltage produces a square wave exactly equal to the source voltage.

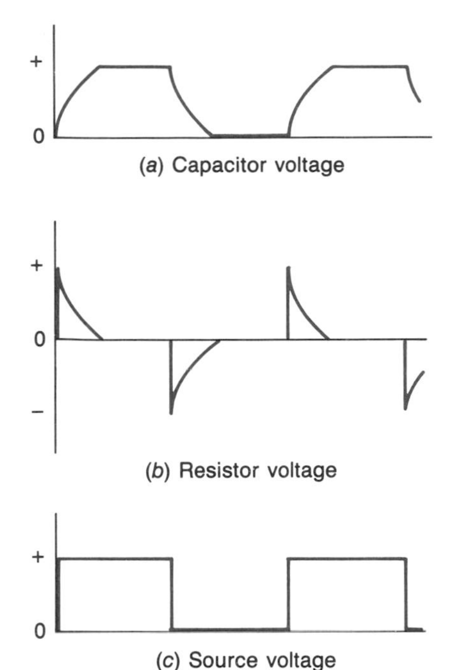


**FIGURE 17-15 The instantaneous sum of the capacitor and resistor voltages equals the source voltage.**

## *RL* Circuit—Symmetrical Input

A series *RL* circuit can also be used to differentiate a waveform. With this circuit (see Fig. 17-16), the differentiated waveform is taken across the inductor rather than the resistor. This is because the inductor produces a counter-electromotive force (cemf) when the current through it is changing. The current change is most dramatic at the instant the source voltage changes from one level to another, so the voltage spikes across the inductor also occur at this time (see Fig. 17-16*c*).

For a given *RL* value, the lower the frequency (longer the period), the more noticeable the differentiation is. When the period is very long compared to the *LR* time constant, the cemf of the inductor decays to zero before the voltage level of the pulse changes again.

Comparison of Figs. 17-14 and 17-16 shows that there is no major difference between the *RL* and *RC* differentiator circuit. Even the phase relationship between the input and output voltage waveforms is the same. With the *RC* circuit, the output goes negative when the input drops to zero. This is because the capacitor sends a discharge current down through the resistor as explained in Sec. 16-1. With the *RL* circuit, the output also goes negative as the input drops to zero. This is because the inductor's cemf is always of such polarity as to oppose any change in current. As the source voltage decays to zero, the circuit current tries to follow the source voltage. In an attempt to maintain the current level, the cemf of the inductor is negative on its upper end and positive on its lower end. This makes the cemf series-aiding the source, which it must be if it is opposing a decaying current.

## Asymmetrical Input

When a nonsymmetrical pulse waveform is applied to a series *RC* circuit, the charge and discharge

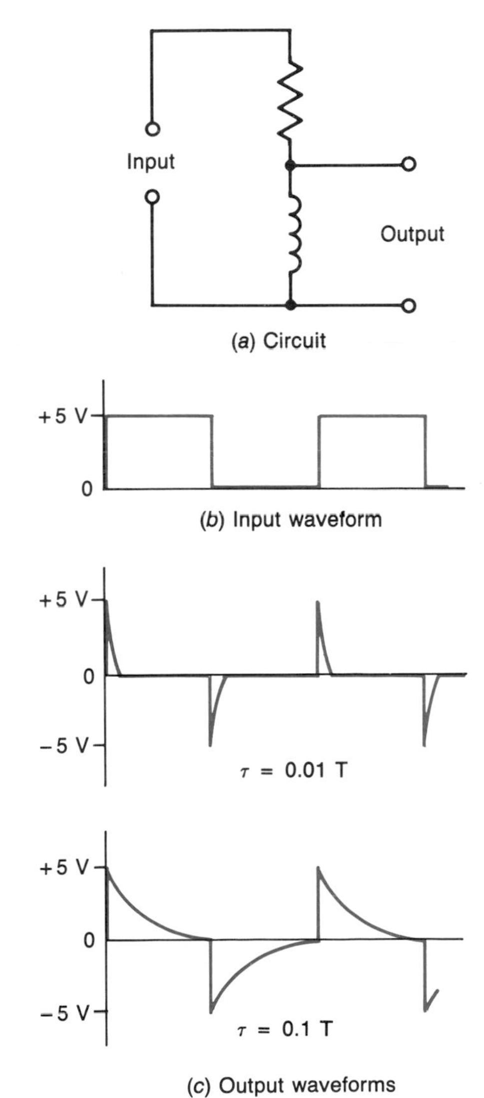


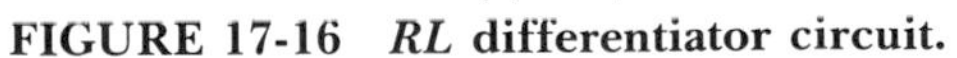
(*c*) Output waveforms

**FIGURE 17-16 *RL* differentiator circuit.**

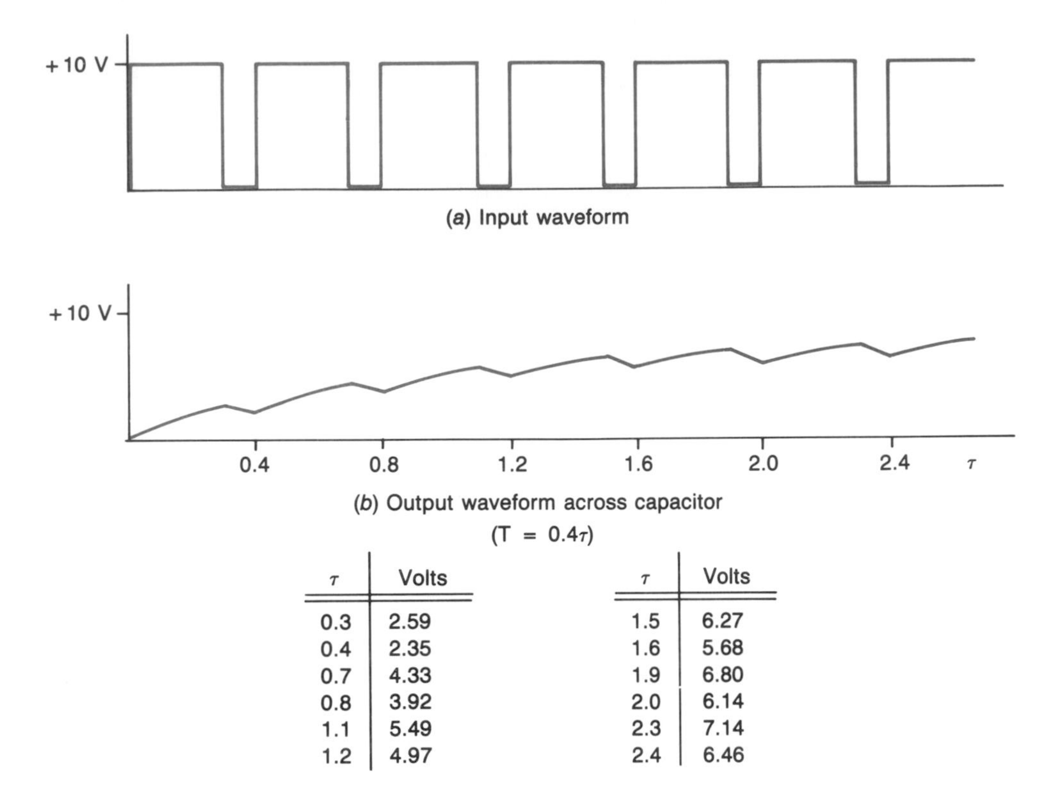


| τ | Volts |
|---|---|
| 0.3 | 2.59 |
| 0.4 | 2.35 |
| 0.7 | 4.33 |
| 0.8 | 3.92 |
| 1.1 | 5.49 |
| 1.2 | 4.97 |

| τ | Volts |
|---|---|
| 1.5 | 6.27 |
| 1.6 | 5.68 |
| 1.9 | 6.80 |
| 2.0 | 6.14 |
| 2.3 | 7.14 |
| 2.4 | 6.46 |

**FIGURE 17-17 Integration of an asymmetrical waveform.**

times for the capacitor will be unequal. If, as in Fig. 17-17, the input is +10 V for a longer time interval than it is zero, the capacitor will integrate the input voltage. At the end of each cycle, the capacitor's voltage will be greater than it was at the beginning of the cycle. This voltage build-up continues until the capacitor reaches its final voltage level or until the capacitor is rapidly discharged by something like a temporarily conducting transistor in parallel with the capacitor. The capacitor voltages listed in Fig. 17-17*c* were easily determined by applying the standard capacitor charge and discharge formulas.

**EXAMPLE 17-16**

Determine the voltage on the capacitor at the end of the second cycle of the waveform in Fig. 17-17 if $T = 0.8\tau$

**Solution**

If $T = 0.8\tau$, then $C$ charges for $0.6\tau$ and discharges for $0.2\tau$. Arbitrarily assign a value to $\tau$. Let us use 10 s. Now the charge time $t_c$ and discharge time $t_d$ are

$$t_c = 10 \text{ s}/\tau \times 0.6\tau = 6 \text{ s}$$
$$t_d = 10 \text{ s}/\tau \times 0.2\tau = 2 \text{ s}$$

At the end of the first charge time,

$$V_C = V_{C,t} = V_T - (V_T - V_{CE})e^{-t/\tau} = 10 \text{ V} - 10 \text{ V } e^{-6 \text{ s}/10 \text{ s}} = 4.512 \text{ V}$$

After discharging for 2 s,

$$V_C = V_{C,t} = V_{CE}\, e^{-t/\tau} = 4.512 \text{ V } e^{-2 \text{ s}/10 \text{ s}} = 3.694 \text{ V}$$

After recharging for 6 s,

$$V_C = 10 \text{ V} - (10 \text{ V} - 3.694 \text{ V})e^{-6 \text{ s}/10 \text{ s}} = 6.539 \text{ V}$$

After another discharge time, which is the end of the second cycle,

$$V_C = 6.539 \text{ V } e^{-2 \text{ s}/10 \text{ s}} = 5.354 \text{ V}$$

## Self-Test

**40.** Which of the waveforms in Fig. 17-18 represents the output for a series *RC* circuit when the input is symmetrical and the output is taken across the

**a.** Capacitor and $\tau = 0.01T$?
**b.** Capacitor and $\tau = T$?
**c.** Resistor and $\tau = 0.1T$?
**d.** Resistor and $\tau = 0.01T$?

**41.** Which of the waveforms in Fig. 17-18 would represent the output for a series *RL* circuit when the output was taken across the

**a.** Inductor and $\tau = 0.1T$?
**b.** Inductor and $\tau = 0.01T$?
**c.** Resistor and $\tau = 0.01T$?
**d.** Resistor and $\tau = T$?

**42.** Which output in Fig. 17-18 is for an asymmetrical input?

**43.** With an integrator circuit, the output is taken across the ________.

**44.** With an *RL* differentiator circuit, the output is taken across the ________.

**45.** With an *RC* differentiator circuit, the output is taken across the ________.

**46.** What value of capacitor is needed to provide an output like that in Fig. 17-18*e* if $R = 5\ \text{k}\Omega$ and $f = 1$ kHz?

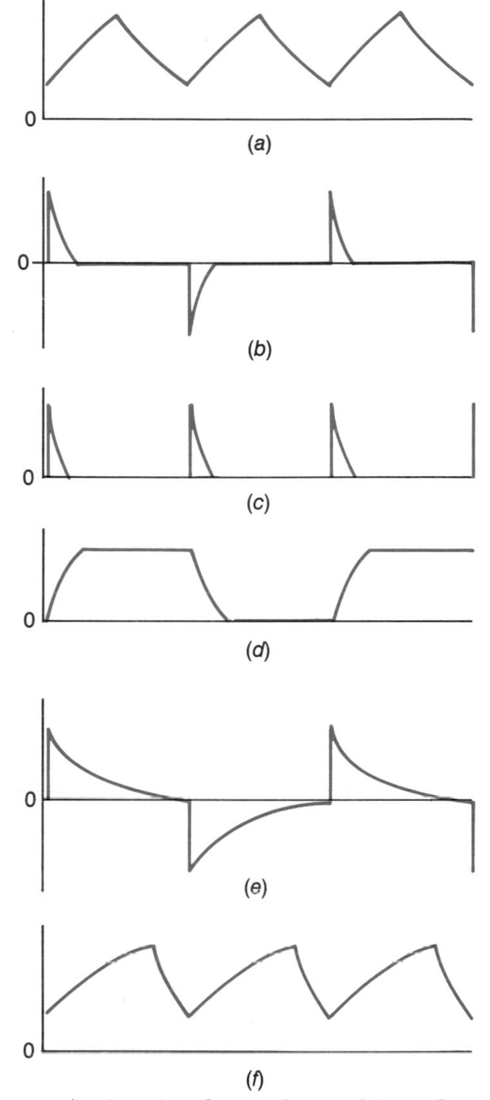

**FIGURE 17-18** **Waveforms for Self-Test Questions.**

## SUMMARY

1. Impedance is a combination of reactance and resistance. It has units of ohms.

2. In a circuit containing reactance, power is not equal to current times voltage.

3. The net reactance $(X_C - X_L)$ in a circuit is symbolized with an $X$.

4. Resonance occurs when $X_L = X_C$. At that time $Z = R$.

5. Resistance and reactance are 90° out of phase, so trigonometric functions can be used with $R$, $X$, and $Z$ ratios.

6. In a series $RC$ circuit,
   a. $V_C$ lags $V_R$ by 90°.
   b. $V_T$ leads $V_C$ and lags $V_R$.
   c. $I_T$ is in phase with $V_R$, leads $V_C$ by 90°, and leads $V_T$.
   d. $Z > R$, $Z > X_C$, and $Z < R + X_C$.
   e. $V_T > V_R$, $V_T > V_C$, and $V_T < V_R + V_C$.
   f. Decreasing $f$ causes $V_C$ to increase, $V_R$ to decrease, and $\theta$ to increase.
7. In a series $RL$ circuit,
   a. $V_L$ leads $V_R$ by 90°.
   b. $V_T$ lags $V_L$ and leads $V_R$.
   c. $I_T$ is in phase with $I_R$, lags $V_L$ by 90°, and lags $V_T$.
   d. $Z > R$, $Z > X_L$, and $Z < R + X_L$.
   e. $V_T > V_R$, $V_T > V_L$, and $V_T < V_R + V_L$.
   f. Increasing $f$ causes $X_L$ to increase and $\theta$ to increase.
8. In a series $RCL$ circuit,
   a. $X_L$ and $X_C$ are 180° out of phase.
   b. $V_L$ and $V_C$ are 180° out of phase
   c. $V_T$ will lead $I_T$ when $X_L > X_C$.
   d. Either, or both, $X_L$ and/or $X_C$ can be greater than $Z$.
   e. $\mathrm{Z} > \mathrm{R}$ except at resonance.
   f. Either, or both, $V_L$ and/or $V_C$ can be greater than $V_T$.
   g. If $V_C > V_L$, increasing $f$ causes $\theta$ to decrease.
9. In a parallel $RC$ circuit,
   a. $Z < R$ and $Z < X_C$.
   b. $I_T > I_R$, $I_T > I_C$, and $I_T < I_R + I_C$.
   c. $I_C$ leads $I_R$ by 90°.
   d. $I_T$ lags $I_C$ but leads $I_R$.
   e. Increasing $f$ causes $I_C$ to increase and $\theta$ to increase.
10. In a parallel $RL$ circuit,
    a. $Z < R$ and $Z < X_L$.
    b. $I_T > I_R$, $I_T > I_L$, and $I_T < I_R + I_L$.
    c. $I_L$ lags $I_R$ by 90°.
    d. $I_T$ leads $I_L$ but lags $I_R$.
    e. Increasing $f$ causes $I_L$ to decrease and $\theta$ to decrease.
11. In a parallel $RCL$ circuit,
    a. $I_L$ and $I_C$ are 180° out of phase.
    b. When $X_L > X_C$, $I_T$ will lead $V_T$.
    c. Either, or both, $I_C$ and/or $I_L$ can be greater than $I_T$.
    d. $Z < R$ except at resonance.
    e. When $X_L > X_C$, increasing $f$ will cause $\theta$ to increase.
12. A series $RC$ or $RL$ circuit can be used to differentiate a square wave or pulse-train input.
13. With an $RC$ differentiator, the output is taken across the resistance.
14. With an $RL$ differentiator, the output is taken across the inductor.
15. A pulse-train input can be integrated by an $RC$ circuit. The output is taken from the capacitor for an integrator circuit.
16. For a differentiator, the amount of differentiation increases when either $\tau$ or $f$ is decreased.

**17.** For an integrator, the amount of integration increases when either $\tau$ or $f$ is increased.

**18.** Some useful formulas are

General for all impedance circuits,

$$Z = V_T/I_T$$
$$P = I_T V_T \cos\theta$$
$$P = I_R V_R$$

For series *RC* circuits,

$$\theta = \arctan\frac{X_C}{R} = \arctan\frac{V_C}{V_R}$$
$$Z = \sqrt{R^2 + X_C{}^2}$$
$$V_T = \sqrt{V_R{}^2 + V_C{}^2}$$

For parallel *RC* circuits,

$$\theta = \arctan\frac{R}{X_C} = \arctan\frac{I_C}{I_R}$$
$$I_T = \sqrt{I_R{}^2 + I_C{}^2}$$
$$Z = \frac{RX_C}{\sqrt{R^2 + X_C{}^2}}$$

For *RL* circuits use the *RC* formulas and change $X_C$ to $X_L$.

For series *RCL* circuits,

$$Z = \sqrt{R^2 + (X_C - X_L)^2}$$
$$V_T = \sqrt{V_R{}^2 + (V_C - V_L)^2}$$
$$\theta = \arctan\frac{X_C - X_L}{R} = \arctan\frac{V_C - V_L}{V_R}$$

For parallel *RCL* circuits,

$$I_T = \sqrt{I_R{}^2 + (I_C - I_L)^2}$$
$$\theta = \arctan\frac{I_C - I_L}{I_R}$$

## CHAPTER REVIEW QUESTIONS

For the following items, determine whether each statement is true or false.

**17-1.** In a series impedance circuit, individual voltages can be arithmetically added to find the total voltage.

**17-2.** Impedance is a combination of resistance and reactance.

**17-3.** A circuit is said to be at resonance when the reactance equals the impedance.

**17-4.** For given values of $L$ and $R$, the amount of differentiation increases when $f$ increases.

**17-5.** In a series *RL* circuit, $I_R$ lags $I_L$ by 90°.

**17-6.** Increasing the frequency in a parallel *RC* circuit will increase $\theta$.

**17-7.** Decreasing $L$ in a series *RL* circuit will cause power to increase.

**17-8.** When $X_L$ is greater than $X_C$ in a parallel *RCL* circuit, $\theta$ will be in the first quadrant.

**17-9.** In a parallel *RC* circuit, $Z$ must be greater than $R$.

**17-10.** In a series *RCL* circuit, either $X_L$ or $X_C$ must be greater than $Z$.

**17-11.** In a series *RCL* circuit in which $X_L$ is greater than $X_C$, $V_T$ will lead $I_T$.

**17-12.** In a parallel *RCL* circuit in which $X_L$ is greater than $X_C$, increasing $f$ causes $\theta$ to decrease.

For the following items, fill in the blanks with the word or phrase required to correctly complete each statement.

**17-13.** In an *RC* differentiator circuit, the output is taken across the ________.

**17-14.** In an *RL* differentiator circuit, the output is taken across the ________.

**17-15.** In an *RC* integrator circuit, the output is taken across the ________.

**17-16.** Resistance and reactance are ________ out of phase.

**17-17.** Capacitive reactance is ________ out of phase with inductive reactance.

## CHAPTER REVIEW PROBLEMS

**17-1.** A 10-V peak-to-peak square wave is connected to an *RC* circuit where $\tau = 0.01T$. What are the peak-to-peak values of $V_R$ and $V_C$?

**17-2.** The frequency of a symmetrical square wave connected to a 0.1-$\mu$F capacitor and a resistor is 1 kHz. What value of $R$ will allow the capacitor to charge and discharge for $2\tau$?

**17-3.** If $V_L = 10$ V and $I_L = 20$ mA, how much power is being dissipated?

**17-4.** In a series *RC* circuit, $I_C = 8$ mA, $V_T = 20$ V, $f = 100$ Hz, and $R = 1$ k$\Omega$. Determine the following:

- **a.** $Z$
- **b.** $P$
- **c.** $\theta$
- **d.** $C$

**17-5.** In a parallel *RCL* circuit, $L = 100$ mH, $f = 10$ kHz, $R = 10$ k$\Omega$, $V_T = 30$ V, and $C = 0.001$ $\mu$F. Determine the following:

- **a.** $\theta$
- **b.** $Z$
- **c.** $I_T$
- **d.** $P$

**17-6.** In a series *RCL* circuit, $I_T = 0.01$ A, $R = 1$ k$\Omega$, $V_C = 10$ V, $V_L = 25$ V, and $L = 2$ H. Determine the following:

- **a.** $V_T$
- **b.** $Z$
- **c.** $\theta$
- **d.** $f$

**17-7.** In a parallel *RC* circuit, $X_C = 5$ k$\Omega$, $R = 6.8$ k$\Omega$, and $V_T = 10$ V at 10 kHz. Determine the following:

- **a.** $Z$
- **b.** $\theta$
- **c.** $I_T$
- **d.** $C$

**17-8.** In a series *RL* circuit, $\theta = 20°$, $R = 5\ k\Omega$, and $I_T = 20$ mA. Determine the following:

**a.** $Z$
**b.** $X_L$
**c.** $V_T$
**d.** $P$

## ANSWERS TO SELF-TESTS

**1.** $V_R$ leads $V_C$ by 90°

**2.** false

**3.** true

**4.** 11.2 V, 48.2°

**5.** 86.6 V, 100 V

**6.** the combined opposition of reactance and resistance

**7.** 4565 Ω

**8.** 21.9 mA, 48.2 V, 87.6 V, and 61.2°

**9.** 13.2 kΩ

**10.** 8 kΩ

**11.** no, no

**12.** Decrease, because $X_C$ will decrease and $\theta = \arctan(X_C/R)$.

**13.** zero

**14.** 77.9 mW

**15.** lead

**16.** first

**17.** decrease

**18.** decrease

**19.** 47.7 mH

**20.** $\theta = 53.1°$, $Z = 5\ k\Omega$, $V_R = 36$ V, and $P = 0.432$ W

**21.** the resistance, the net reactance (or vice versa)

**22.** minus

**23.** a. yes, b. yes, c. no, d. yes

**24.** resonance

**25.** 2.24 kΩ, lead

**26.** 2728 Ω, 36.2°, 7.33 mA

**27.** $I$ leads $V$

**28.** smaller than

**29.** larger than

**30.** false

**31.** 43.3°, 106.7 mW, 10.92 kΩ, 3.66 mA

**32.** only $I_R$

**33.** Decrease, because $X_L$ will increase, which will cause $I_L$ to decrease while $I_R$ remains constant. Thus, more of the total current is resistive, which means $\theta$ is decreasing.

**34.** No, because $I_T$ is the phasor sum of $I_R$ and $I_L$ and these two currents are also 90° out of phase for an ideal inductor.

**35.** 91.9 mW, 53.5°, 4042 Ω, 6.19 mA

**36.** $X_L$ because the $I_L$ is shortest.

**37.** (1) Decrease $f$ because this would decrease $X_L$, increase $I_L$, and decrease $I_X$; (2) decrease $L$ because this would also decrease $X_L$, increase $I_L$, and decrease $I_X$; and (3) decrease $C$ because this would increase $X_C$, decrease $I_C$, and again decrease $I_X$. Of course, if any one of these were decreased too much, $\theta$ would then start to increase in the negative direction.

**38.** (1) reduce $V_T$ and (2) increase $R$

**39.** 0.25 W, 14.4°, 51.6 mA, 96.9 Ω, and 1.6 mH

**40.** **a.** *d* **b.** *a* **c.** *e* **d.** *b*

**41.** **a.** *e* **b.** *b* **c.** *d* **d.** *a*

**42.** *f*

**43.** capacitor

**44.** inductor

**45.** resistor

**46.** 20 μF

# Chapter 18 Resonance and Filters

In Chap. 17, resonance was merely defined as the condition when $X_L = X_C$. In this chapter we will investigate the operation of resonant circuits in some detail. Then we will see how resonant circuits are used as filters. Finally, we will look at the characteristics of some nonresonant filters which utilize the concepts developed in Chap. 17 for *RC* and *RL* circuits.

## 18-1 SERIES-RESONANT CIRCUITS

Resonance can occur in either *RCL* or *LC* (inductor, capacitor) circuits. *Resonance* is technically defined as the condition in an *RCL* or *LC* circuit when $I_T$ is in phase with $V_T$. For all series circuits, and most parallel circuits, the magnitudes of $X_L$ and $X_C$ are equal at resonance. Because $X_C$ and $X_L$ are vector quantities which are 180° out of phase with each other, they counteract and effectively cancel each other. Thus, the impedance of a series-resonant circuit is just the resistance in the circuit. Since we know that $X_L$ is directly proportional to $L$ and $f$ and that $X_C$ is inversely proportional to $C$ and $f$, resonance must also be controlled by $f$, $C$, and $L$. The frequency $f_r$ at which series resonance occurs can be determined from the reactance formulas. At resonance $X_L = X_C$; therefore,

$$2\pi fL = \frac{1}{2\pi fC}$$

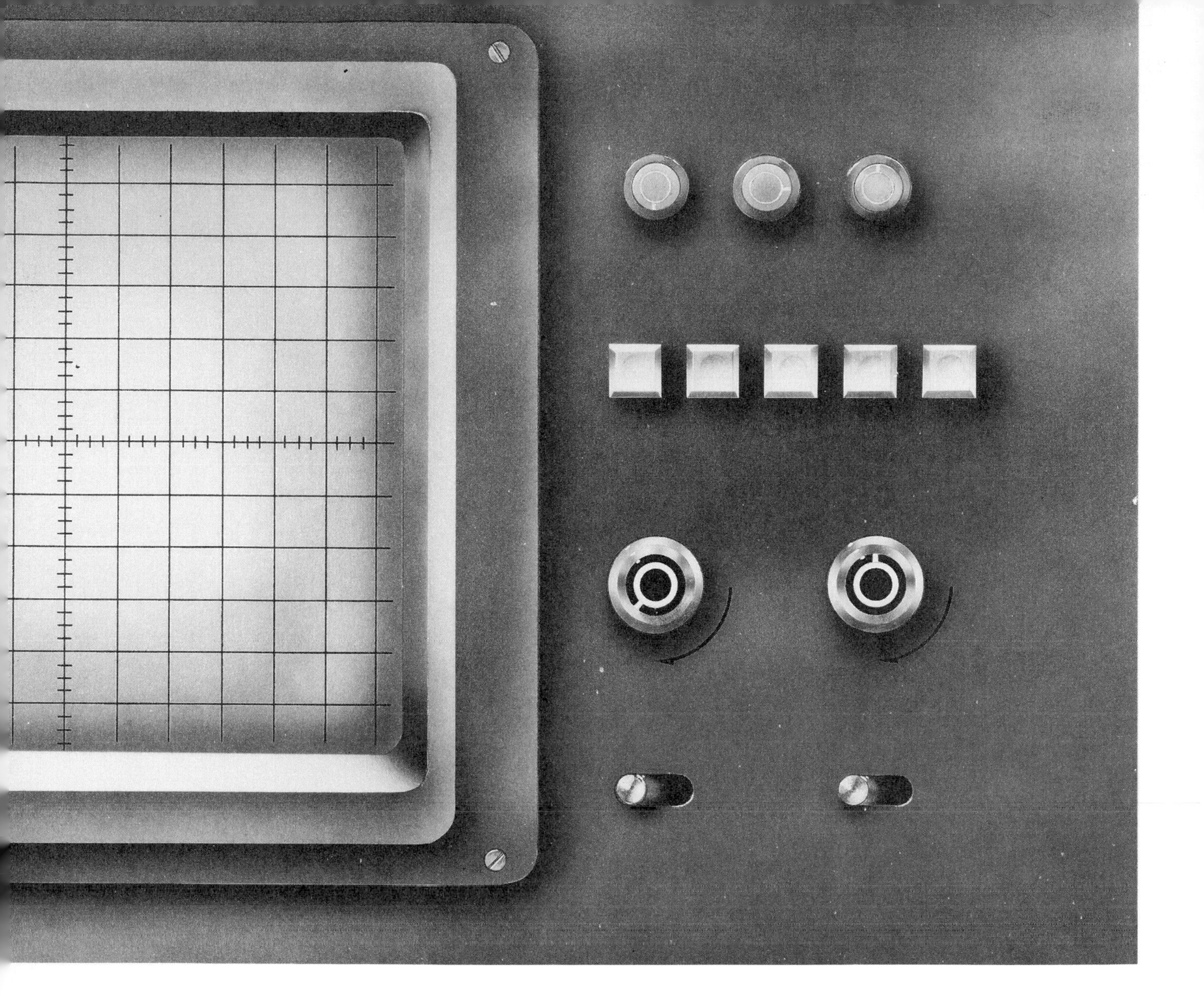

Solving for $f$ (and adding the $r$ subscript) gives

$$f_r = \frac{1}{2\pi\sqrt{LC}}$$

Inspection of this formula shows the following to be true: (1) increasing either the inductance or the capacitance causes the resonance frequency to decrease; (2) for a given value of inductance and capacitance, there is only one resonant frequency; and (3) there are an infinite number of inductor and capacitor combinations for any specified resonant frequency.

The above three points are illustrated in Fig. 18-1, which plots $X_L$ and $X_C$ against $f$ for two values of $L$ and $C$. From this illustration, you can see that

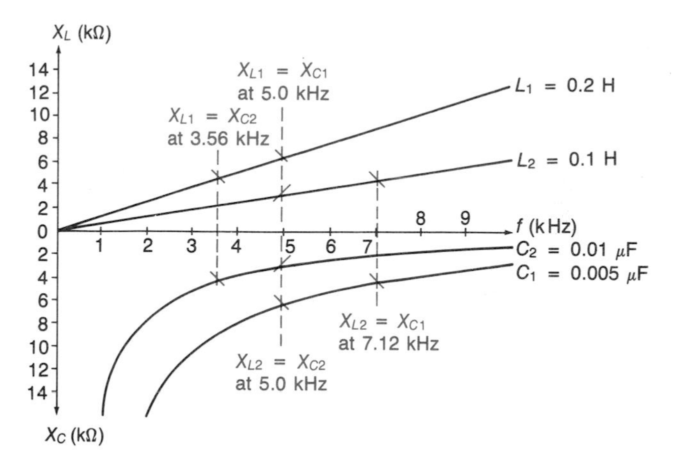


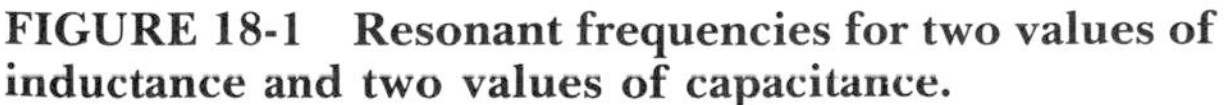
**FIGURE 18-1 Resonant frequencies for two values of inductance and two values of capacitance.**

there is only one frequency at which the reactance of $L_1$ is equal to (and cancels) the reactance of $C_1$. Also notice that the combination of $L_2$ and $C_2$ has the same $f_r$ as does the $L_1$ and $C_1$ combination. Finally, in Fig. 18-1 you can see that increasing the capacitance from $C_1$ to $C_2$ while holding $L$ at $L_1$ causes $f_r$ to decrease.

### EXAMPLE 18-1

What is the resonant frequency of a 50-mH inductor connected in series with a 1000-pF capacitor?

**Solution**

$$f_r = \frac{1}{6.28\sqrt{50 \times 10^{-3} \times 1000 \times 10^{-12}}}$$

$$= \frac{1}{6.28\sqrt{50 \times 10^{-6}}} = 22.5 \text{ kHz}$$

If the capacitance in Example 18-1 were halved to 500 pF, the resonant frequency would increase to $f_r = 1/6.28\sqrt{25 \times 10^{-6}} = 31.8$ kHz. Notice that $f_r$ does not double when $L$ or $C$ is halved. Rather, it increases by a factor of the $\sqrt{2}$.

### EXAMPLE 18-2

What value of capacitance is needed to resonate with a 300-$\mu$H inductor at 400 kHz?

**Solution**

$$f_r = \frac{1}{6.28\sqrt{LC}}$$

Solving for $C$ we get

$$C = \frac{(0.159/f)^2}{L} = \frac{[0.159/(400 \times 10^3)]^2}{300 \times 10^{-6}} = 527 \text{ pF}$$

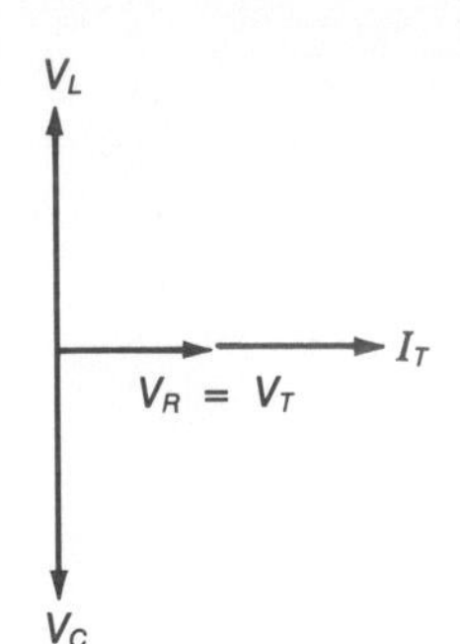

(a) *RCL* with ideal components

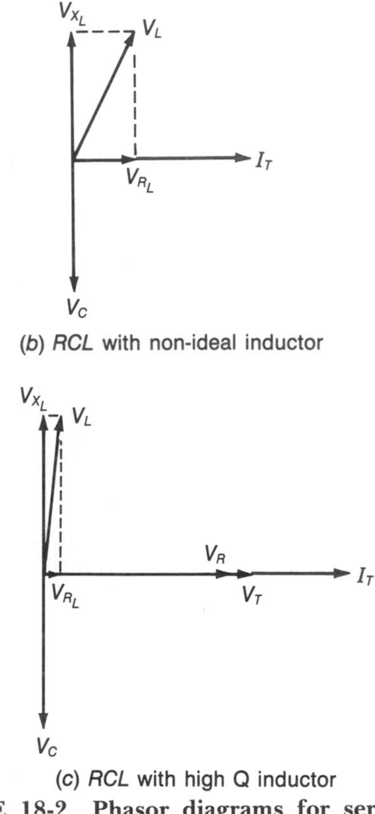


(c) *RCL* with high Q inductor

**FIGURE 18-2 Phasor diagrams for series resonant circuits.**

Why the resistance (either external or series-equivalent) in a series *RCL* does not affect the resonant frequency is illustrated in Fig. 18-2. In order for the source current and voltage to be in phase (which is the meaning of resonance), the reactive voltages must cancel each other. In Fig. 18-2*a* the inductor and capacitor are assumed to be ideal (no series-equivalent resistance), so their voltages are exactly 90° out of phase with the current, and they will cancel each other when $X_C = X_L$. The source voltage then drops across the external resistance. The phasors in Fig. 18-2*b* are for an *LC* circuit with a real inductor with some series-equivalent resistance $R_L$. In this circuit, the voltage which would be measured across the inductor $V_L$ is composed of a reactive voltage and a resistive voltage. The reactive part of the voltage still cancels $V_C$ when $X_L =$

$X_C$ because of the common current. The resistive part $V_{R_L}$ of the inductor voltage $V_L$ equals the source voltage $V_T$ and is in phase with the source current. Figure 18-2*c* shows why in a series *RCL* circuit with a high *Q* (quality) inductor it is permissible to assume an ideal inductor. Notice in this figure that $V_L \approx V_{X_L}$ and $V_T \approx V_R$.

## Self-Test

**1.** What is the relationship between $X_C$ and $X_L$ when a circuit is series resonant?

**2.** True or false? Increasing *C* increases $f_r$.

**3.** True or false? Doubling *L* will make $f_r$ half of its former value.

**4.** True or false? Many values of *L* and *C* can be used to obtain a specified value of $f_r$.

**5.** Determine the resonant frequency of a 2-$\mu$F capacitor and a 1-H inductor.

**6.** Determine the inductance required to provide a resonant frequency of 10 kHz if the capacitance is 0.01 $\mu$F.

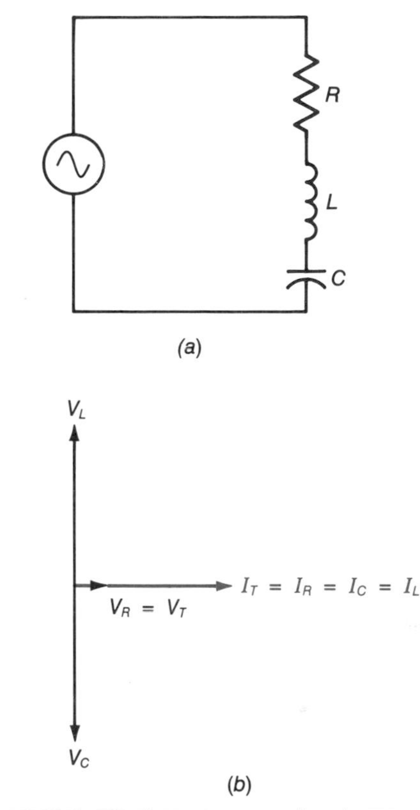


**FIGURE 18-3 Series resonant circuit.**

## Series *LC* Circuits

The resistor in Fig. 18-3*a* represents the effective series resistance of the capacitor and inductor as well as any physical resistor that may be in the circuit. As you know, inductors generally have a much lower *Q* than do capacitors. Therefore, nearly all of the effective series resistance is attributable to the inductor. In our analysis of resonant circuits we will assume all the effective series resistance in an *LC* circuit is due to the inductor.

Separating the effective series resistance from the inductance of the inductor allows us to use a $V_L$ phasor that is 90° from the current. Including the effective series resistance on the *x* axis allows us to account for the fact that the inductor is not an ideal component. If the resistance were not separated, then for real components, the $V_L$ phasor in Fig. 18-3*b* would have to be drawn at an angle less than 180° from $V_C$. While working with phasors that are separated by less than 90° or 180° is not too difficult, it is really unnecessary for understanding series resonance. (The details of analyzing circuits in which the phasors are at any angle to each other are covered in Chap. 19.)

Figure 18-3*b* again shows that at resonance the resistive voltage is equal to the source voltage, and the source voltage and current are in phase (angle theta is 0). Therefore, at resonance, the series *RCL* or *LC* circuit is purely resistive and has minimum opposition to the source current.

For a specified source voltage, the magnitude of the current in a series-resonant circuit is controlled by the resistance in the circuit. When there is no physical resistor in the circuit, the current is controlled by the effective series resistance of the inductor and the internal resistance of the voltage source. For a constant-voltage source (which is the type we will continue to assume), the circuit current is strictly a function of the effective series resistance of the circuit. For practical purposes, then, the current in an *LC* series-resonant circuit is controlled by the resistance of the inductor. With a high *Q* inductor, this resistance can be very small. Thus, the current at resonance can be quite large. Of course, when the current is large, both $V_L$ and $V_C$ are large because the values of $X_L$ and $X_C$ are independent of the current magnitude.

**EXAMPLE 18-3**

Determine $V_L$ and $V_C$ for a series-resonant circuit when $L = 20$ mH, $C = 0.01\ \mu$F, $R = 16\ \Omega$, and $V_T = 4$ V.

**Solution**

$$f_r = \frac{1}{6.28\sqrt{LC}} = \frac{1}{6.28\sqrt{0.02 \times 0.01 \times 10^{-6}}}$$
$$= 11.26 \text{ kHz}$$
$$X_L = 6.28fL = 6.28 \times 11.26 \text{ kHz} \times 0.02 \text{ H} = 1414\ \Omega$$
$$X_C = X_L = 1414\ \Omega \quad \text{Because of resonance}$$
$$V_R = V_T \quad \text{Because of resonance}$$
$$I_T = I_R = \frac{V_R}{R} = \frac{4 \text{ V}}{16\ \Omega} = 0.25 \text{ A}$$
$$V_L = I_L X_L = 0.25 \text{ A} \times 1414\ \Omega = 353.5 \text{ V}$$
$$V_C = V_L = 353.5 \text{ V}$$

## Circuit Quality and Series-Resonant Voltage Rise

Notice in Example 18-3 that the reactive voltages are much larger than the source voltage. This phenomenon of the reactive voltage exceeding the source voltage is known as the *series-resonant voltage rise* (voltage amplification or magnification.) The magnitude of the voltage rise ratio $V_X/V_T$ is a function of the circuit quality $Q$.

The *circuit* $Q$ of a resonant $LC$ circuit is defined as the ratio of the circuit reactance, either inductive or capacitive, to the total series-equivalent resistance $R_S$. Thus, circuit $Q = X/R_S$. Since $X$ stores and releases energy without converting it to heat and since $R$ converts electric energy to heat, $Q$ is a ratio of the energy stored to the energy converted in the circuit. Circuit $Q$ is a pure number just as is the $Q$ of a capacitor or an inductor. Because the inductor usually has so much more resistance than the capacitor in a resonant circuit, the circuit $Q$ is essentially equal to the $Q$ of the inductor. This is why circuit $Q$ is often simply defined as the $Q$ of the inductor. For our work with series-resonant circuits, we will use this definition of circuit $Q$.

By substituting into the $Q$ formula we can see why the series-resonant voltage rise is a function of the circuit $Q$. First, we know that for any circuit $X_L = V_L/I_L$ and $R = V_R/I_R$. Second, we know that for a series-resonant circuit, $I_T = I_R = I_L$ and $V_R = V_T$. Now, substituting into the $Q$ formula, we get

$$Q = \frac{X_L}{R_L} = \frac{V_L/I_L}{V_{R_L}/I_{R_L}} = \frac{V_L/I_T}{V_T/I_T} = \frac{V_L}{V_T}$$

Rearranging gives

$$V_L = QV_T$$

which shows that the voltage rise is numerically equal to $Q$.

The formula $V_L = QV_T$ shows that when a high $Q$ inductor is used in a series $LC$ circuit, the series-resonant voltage rise can be several hundred times. However, when an external resistor is included ($RCL$ circuit), the voltage rise may be less than one. Any time the total series resistance (external resistance plus effective series resistance of $L$ and $C$) exceeds $X_L$ at resonance, the voltage rise will be less than one.

## Self-Test

**7.** For a given source voltage, what determines the current in a series-resonant circuit?

**8.** At resonance, what is the relationship between $V_T$ and $V_{R_L}$ in a series $LC$ circuit?

**9.** What is the value of $\theta$ at resonance?

**10.** What is the value of the net reactive voltage at resonance?

**11.** For a given value of $L$, $C$, and $V_T$, what happens to $V_C$ in a series-resonant circuit if the $Q$ of the inductor is doubled? Why?

**12.** The $Q$ of a series-resonant circuit is 50. What is the resonant voltage rise if $V_T$ is 5 V? If $V_T$ is 10 V?

**13.** Under what conditions will $V_C < V_T$ in a series-resonant $RCL$ circuit?

**14.** In a series-resonant circuit, $L = 10$ mH, $C = 0.05\ \mu$F, $R_L = 20\ \Omega$, and $V_T = 5$ V. Determine $f_r$, $I_T$, and $V_C$.

## Selectivity and Bandwidth

The *selectivity* of a resonant $LC$ circuit refers to the circuit's ability to select one frequency (or a narrow band of frequencies) and reject frequencies higher or lower than the selected frequency. Resonant $LC$ circuits are used, for example, to separate the signal (a narrow band of frequencies) of one radio station from those of all the other radio stations. By making either the inductor or the capacitor variable, the resonant frequency of the $LC$ circuit can be adjusted to match that of the station which is to be received. Figure 18-4 illustrates how the series-resonant voltage rise of a series circuit can be used to emphasize one frequency and discriminate against two others. The three generators in the circuit represent three equal-strength radio station signals which are electromagnetically induced into the inductor and are thus in series with the inductor and capacitor. The 8 Ω of $R_L$ is the effective series resistance of the $LC$. When the capacitor is adjusted to 209 pF, the circuit is resonant at $f_2$ (1100 kHz). At resonance, as shown in Fig. 18-4$b$, the voltage available to the amplifier is 4.3 V. (The voltages and currents shown for resonance assume zero source resistance. In a real circuit the source would have many ohms of resistance, and the resonant voltages and currents would be considerably less than shown.) Figure 18-4$c$ and $d$ shows the voltages at the frequencies above and below resonance, respectively. The phasor diagrams in Fig. 18-4 show that $V_C$ at resonance is 95.6 times greater (4.3/0.045) than $V_C$ at 1600 kHz and 60.6 times greater (4.3/0.071) than $V_C$ at 600 kHz. Since the signals entered the circuit at the same voltage level, the series-resonant circuit has been selective: it has selected the signal at which it is resonant. The voltage across the capacitor at resonance should be $Q$ times as large as the source voltage.

**EXAMPLE 18-4**

From the data given in Fig. 18-4, prove that the resonant voltage rise is equal to $Q$.

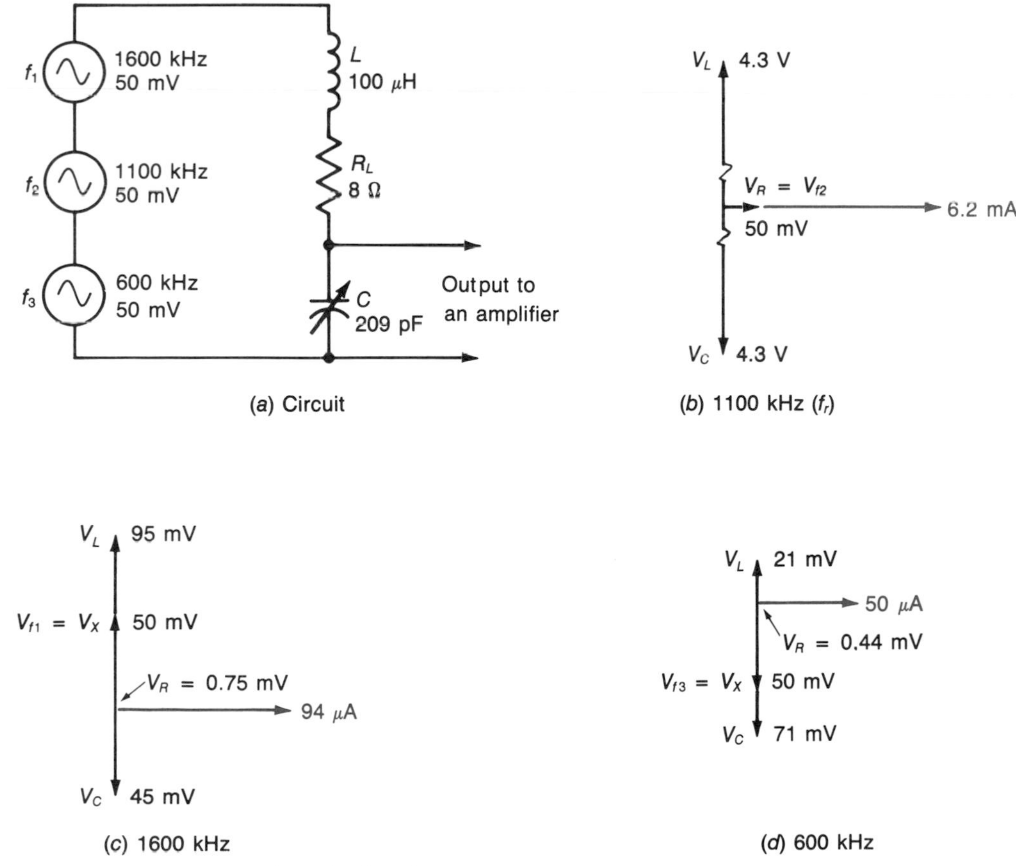

**FIGURE 18-4 Selectivity of a series $LC$ circuit.**

**Solution**

At resonance,

$$X_L = 2\pi fL = 6.28 \times 1100 \times 10^3 \times 100 \times 10^{-6} = 691\ \Omega$$
$$Q = \frac{X_L}{R_L} = \frac{691\ \Omega}{8\ \Omega} = 86.4$$
$$V_C = QV_T = 86.4 \times 50\ \text{mV} = 4.3\ \text{V}$$

Notice in Fig. 18-4 that at frequencies above resonance the $LC$ circuit is inductive and $V_T$ leads $I_T$, while at frequencies below resonance it is capacitive and $V_T$ lags $I_T$. This is expected since $X_C$ is inversely proportional to $f$ while $X_L$ is directly proportional to $f$.

Figures 18-5 and 18-6 show the overall frequency response of an $LC$ circuit. Figure 18-5 shows plots of frequency versus impedance, while Fig. 18-6 shows current versus frequency. The graphs in Fig. 18-5 are for a circuit in which $L$ = 12.67 mH, $C$ = 0.02 $\mu$F, and the effective series resistance is 20 Ω. This $LC$ combination is resonant at 10 kHz. In Fig. 18-5*a*, the frequency axis of the graph is linear at 1 kHz per division. This tends to distort the curve because doubling the resonant frequency requires twice the horizontal distance, as does halving the resonant frequency. Figure 18-5*b* is a response curve for the same series circuit except that its frequency axis is logarithmic, which makes the curve symmetrical around the resonant frequency of 10 kHz. Notice that the impedance of a series $LC$ circuit is minimum (equal to $R$) at the resonant frequency.

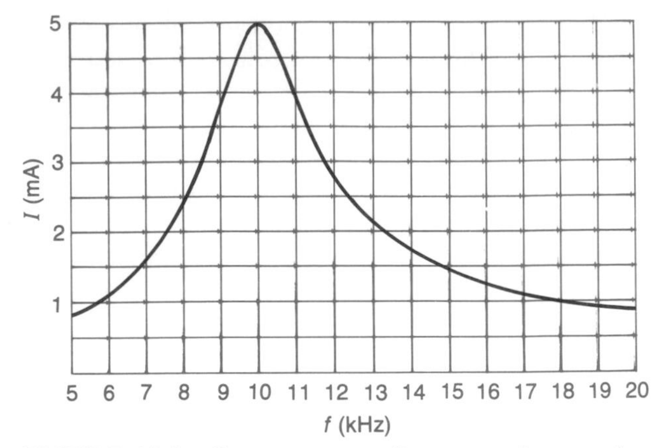


**FIGURE 18-6 Current versus frequency for a series *LC* circuit.**

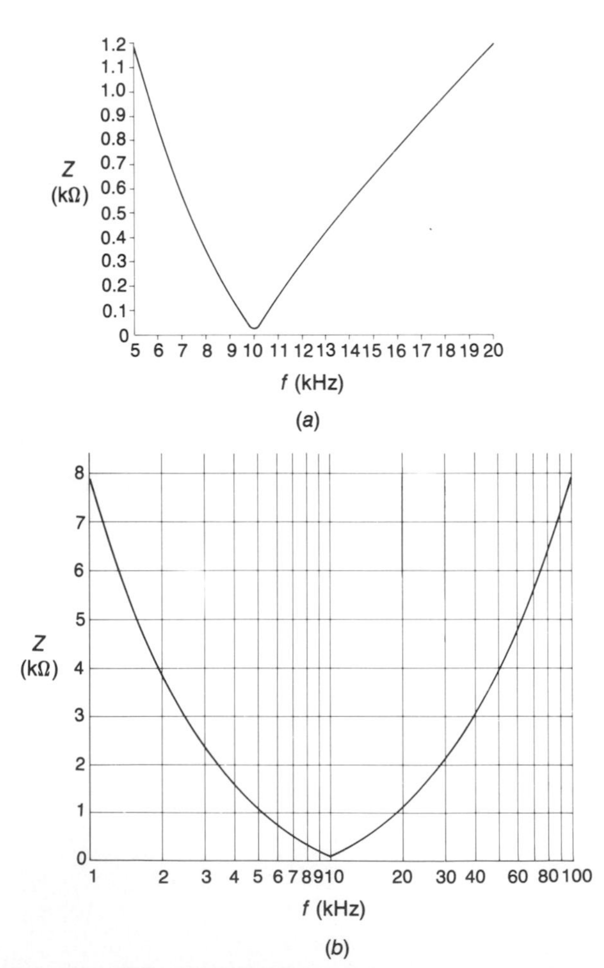


**FIGURE 18-5 Frequency response of a series *LC* circuit.**

The graph in Fig. 18-6 is for the same circuit as the graphs in Fig. 18-5 except that $R$ was increased to 200 Ω. Increasing $R$ from 20 to 200 Ω lowers the $Q$ by a factor of 10. This spreads out the curve so that it is easier to see the details of how the current peaks at the resonant frequency. Since the peak current in Fig. 18-6 is 5 mA and the resistance in the circuit is 200 Ω and $R = Z$ at resonance, the voltage applied to the circuit to obtain the data for this curve must have been $V = IR = 5\ \text{mA} \times 200\ \Omega = 1\ \text{V}$.

The selectivity of an $LC$ circuit is determined by the *bandwidth* (BW) of the circuit. The BW of an $LC$ circuit is defined as the band of frequencies that will provide 50 percent or more of the maximum power the circuit can provide. Maximum power in an $LC$ circuit occurs at resonance because $P = I^2R$ and maximum current flows in an $LC$ circuit at resonance. The upper and lower frequencies ($f_{up}$ and $f_{lo}$) of the bandwidth are those two frequencies at which $I^2$ is half of its maximum value because the $R$ in the circuit does not change. If $I^2$ has to be reduced to $0.5I^2$, then $I$ must be reduced to $\sqrt{0.5I^2} = 0.707I$. Thus, as shown in Fig. 18-7, the bandwidth extends from the lower frequency ($f_{lo}$), at which the current is reduced to 0.707 of its maximum value, to the upper frequency ($f_{up}$), at which the current is again 0.707 of its maximum value. Using these definitions of $f_{up}$ and $f_{lo}$, the BW can be defined mathematically as $\text{BW} = f_{up} - f_{lo}$.

The points on the curve in Fig. 18-7 where the upper and lower BW frequencies intersect the curve (at 70.7 percent of $I$ maximum) are called the *half-power points*. Thus, the BW of a circuit is often referred to as the band of frequencies between the half-power points on the current-versus-frequency curve.

---

**EXAMPLE 18-5**

Determine the BW from the graph in Fig. 18-7.

**Solution**

$f_{lo} \approx 8.8$ kHz
$f_{up} \approx 11.3$ kHz
$\text{BW} = f_{up} - f_{lo} = 11.3 \text{ kHz} - 8.8 \text{ kHz} = 2.5 \text{ kHz}$

---

The BW, and thus the selectivity, of an $LC$ circuit is controlled by the $Q$ of the circuit. The higher the $Q$, the narrower the BW. As illustrated in Figs. 18-8 and 18-9, there are two ways to change the $Q$ of the $LC$ circuit without changing $f_r$. One is to change the effective series resistance while holding $L$ and $C$ constant, and the other is to change the $L$ and $C$ values while holding the resistance constant.

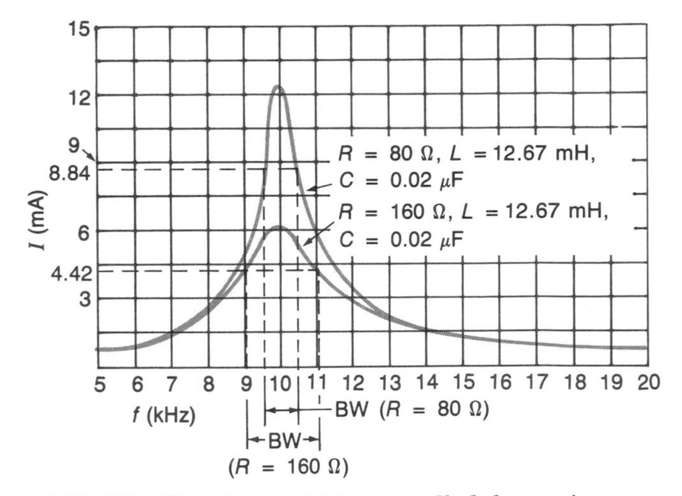


**FIGURE 18-8 Bandwidth controlled by resistance when $L$ and $C$ are constant.**

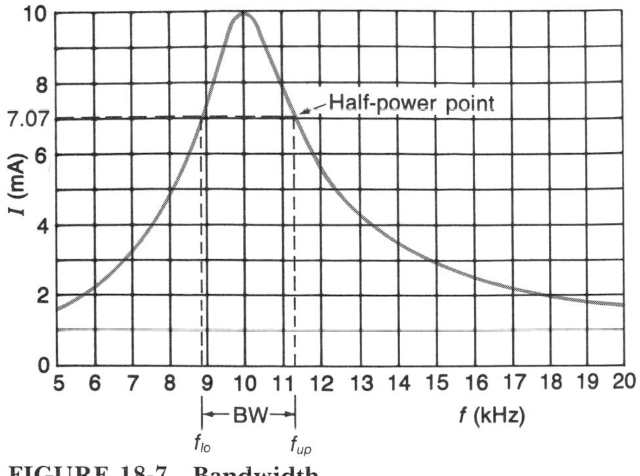

**FIGURE 18-7 Bandwidth.**

Figure 18-8 shows that doubling the resistance while holding $L$ and $C$ constant doubles the BW. Doubling $R$ while not changing $L$ or $f_r$ causes the $Q$ to be half as great because $Q = X_L/R$ and $X_L$ does not change unless $f_r$ or $L$ changes. Thus, BW doubles when $Q$ is halved.

Mathematically, the relationship between $Q$, BW, and $f_r$ can be developed by combining relationships that we already know. We know that at resonance $X_L - X_C = 0$, and $Z = R$. At the half-power point, $R = X_L - X_C$ because then $Z$ will be 1.414 times its resonant value, which will make $I$ equal to 0.707 of its resonant value. Since $X_L$ and $X_C$ are directly proportional and inversely proportional to $f$, respectively, we know that $X_L$ will increase as much as $X_C$ decreases as the frequency is raised above $f_r$. Conversely, $X_L$ will decrease as much as $X_C$ increases as $f$ is decreased from $f_r$. If $X_L - X_C = R$ at the half-power points ($f_{up}$ and $f_{lo}$), then $X_L$ at $f_{up}$, which we will label $X_{L,up}$ must be equal to $X_L$ at resonance ($X_{L,r}$) plus $0.5R$. Therefore, we can write

$$X_{L,up} = X_{L,r} + 0.5R \qquad \text{or} \qquad 0.5R = X_{L,up} - X_{L,r}$$

Using this same reasoning, we know that for the lower half-power point

$$X_{L,lo} = X_{L,r} - 0.5R \qquad \text{or} \qquad 0.5R = X_{L,r} - X_{L,lo}$$

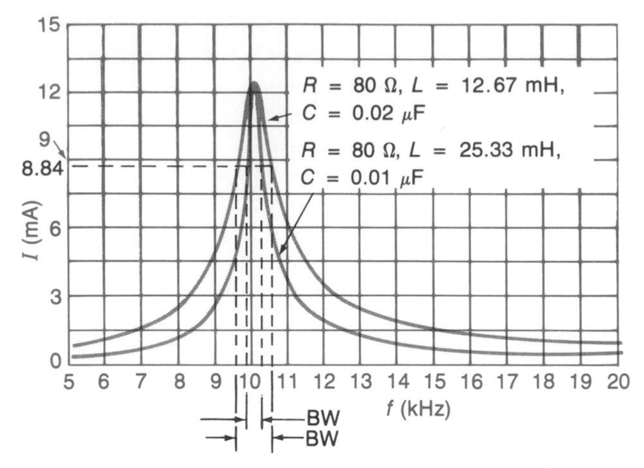


**FIGURE 18-9 Bandwidth controlled by L/C ratio when R is constant.**

Now, adding these two expressions for $0.5R$ yields

$$R = X_{L,\text{up}} - X_{L,\text{lo}}$$

Substituting for $X_L$ gives

$$R = 2\pi f_{\text{up}}L - 2\pi f_{\text{lo}}L = 2\pi L(f_{\text{up}} - f_{\text{lo}})$$

and rearranging, we have

$$f_{\text{up}} - f_{\text{lo}} = \frac{R}{2\pi L} \qquad \text{or} \qquad \text{BW} = \frac{R}{2\pi L}$$

Now substituting $L = X_L/2\pi f_r$ (from $X_L = 2\pi f_r L$) for $L$, we get

$$\text{BW} = \frac{R}{2\pi X_L/2\pi f_r} = \frac{R2\pi f_r}{2\pi X_L} = \frac{Rf_r}{X_L}$$

We know that $Q = X_L/R$, so $1/Q = R/X_L$, which allows $1/Q$ to be substituted for $R/X_L$ to yield the final expression of

$$\text{BW} = \frac{f_r}{Q}$$

Thus, it is clear that BW is inversely proportional to $Q$.

Figure 18-9 illustrates the second way of changing the $Q$ (and thus the BW) of an $LC$ circuit. This technique of changing BW is often referred to as changing the $L/C$ ratio to change the BW. It must be emphasized *that increasing the L/C ratio does not decrease the BW unless the resistance remains constant* or increases proportionately less than the $L/C$ ratio increases. The only time that changing the $L/C$ ratio changes the BW is when it also changes the circuit $Q$. This point is illustrated in Fig. 18-10, which shows the curves for the same $LC$ combinations as Fig. 18-9 except that $Q$ is held constant by doubling $R$ when $L$ was doubled.

**EXAMPLE 18-6**

A series $LC$ circuit is composed of a 15-μH, 10-Ω inductor and a 100-pF capacitor. Determine its BW. Also determine $V_C$ at 3.7 MHz if $V_T = 0.5$ V.

**Solution**

$$f_r = \frac{1}{2\pi\sqrt{LC}} = \frac{1}{6.28\sqrt{100 \times 10^{-12} \times 15 \times 10^{-6}}} = 4.11 \text{ MHz}$$

$$X_L = 2\pi fL = 6.28 \times 4.11 \times 10^6 \times 15 \times 10^{-6} = 387.2\ \Omega$$

$$Q = \frac{X_L}{R} = \frac{387.2\ \Omega}{10\ \Omega} = 38.72$$

$$\text{BW} = \frac{f_r}{Q} = \frac{4110 \text{ kHz}}{38.72} = 106.15 \text{ kHz}$$

At 3.7 MHz,

$$X_L = 6.28 \times 3.7 \times 10^6 \times 15 \times 10^{-6} = 348.5\ \Omega$$

$$X_C = \frac{1}{6.28 \times 3.7 \times 10^6 \times 100 \times 10^{-12}} = 430.4\ \Omega$$

$$Z = \sqrt{R^2 + (X_C - X_L)^2} = \sqrt{10^2 + 81.9^2} = 82.5\ \Omega$$

$$I_T = \frac{V_T}{Z} = \frac{0.5 \text{ V}}{82.5\ \Omega} = 6.06 \text{ mA}$$

$$V_C = I_C X_C = 6.06 \text{ mA} \times 430.4\ \Omega = 2.6 \text{ V}$$

Inspection of the curves in Figs. 18-6 through 18-10 shows that the frequencies of the half-power points are not absolutely equally spaced from the resonant frequency. However, their spacing is

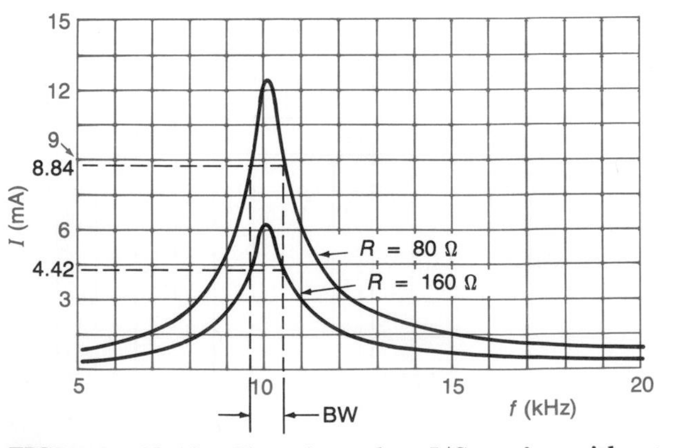


**FIGURE 18-10 Changing the L/C ratio without changing Q.**

even enough that a good approximation of the upper and lower BW frequencies can be obtained by assuming equal spacing.

**EXAMPLE 18-7**
A series $LC$ circuit is resonant at 150 kHz and has a $Q$ of 50. Estimate $f_{up}$ and $f_{lo}$.

**Solution**

$$BW = \frac{f_r}{Q} = \frac{150 \text{ kHz}}{50} = 3 \text{ kHz}$$

$$f_{up} \approx \frac{BW}{2} + f_r = \frac{3}{2} + 150 = 151.5 \text{ kHz}$$

$$f_{lo} \approx f_r - \frac{BW}{2} = 150 - \frac{3}{2} = 148.5 \text{ kHz}$$

## Self-Test

**15.** Define selectivity.

**16.** The selectivity of a circuit is specified by its ________.

**17.** What is the relationship (when other factors remain unchanged) between
  **a.** $R$ and $Q$?
  **b.** $Q$ and BW?
  **c.** $R$ and BW?
  **d.** $Q$ and resonant voltage rise?
  **e.** $L/C$ ratio and BW?

**18.** Define BW.

**19.** Define half-power points.

**20.** What is the BW of a series $LC$ circuit with a $Q$ of 50 and an $f_r$ of 75 kHz?

**21.** A series $LC$ circuit has an inductance of 200 $\mu$H, a capacitance of 300 pF, and a $Q$ of 60. Determine $f_r$, BW, $f_{up}$, and $f_{lo}$.

## 18-2 IDEAL PARALLEL-RESONANT CIRCUITS

Resonance in a parallel $RCL$ circuit which uses an ideal inductor and capacitor is illustrated in Fig. 18-11. The phasor diagram in Fig. 18-11$b$ shows the major characteristics of this ideal circuit: (1) angle $\theta$ is zero, (2) $I_R$ equals $I_T$, and (3) $I_C$ equals $I_L$.

Since $I_C = I_L$ and $V_C = V_L$, $X_C$ must also equal $X_L$. Furthermore, Ohm's law tells us that the impedance at resonance is equal to the parallel resistance ($Z = V_T/I_T = V_R/I_R = R$). As the resistance in this circuit is increased, $Z$ increases while $I_T$ and $I_R$ decrease. However, $I_C$ and $I_L$ remain constant because they are determined by $V_T$, $X_C$, and $X_L$. Now, if parallel resistance $R_P$ is increased to infinity by removing it from the circuit, $I_T$ reduces to zero and $Z$ increases to infinity. Yet, $I_C$ and $I_L$ remain unchanged. The parallel inductance and capacitance are merely transferring energy back and forth between them while current flows in the $L$ and $C$ branches of the circuit. That is, the inductor forces current to flow as its magnetic field collapses. This current charges the capacitor. Then the capacitor takes over and forces the current to flow as it discharges and forces the inductor's magnetic field to rebuild. This process is then repeated again and again. The current flowing in the $L$ and $C$ branches of the circuit is called the *tank current* to distinguish it from the source current ($I_T$). The parallel $LC$ combination is often called a tank circuit. The source current is sometimes referred to as the line current since it is the current in the lines connect-

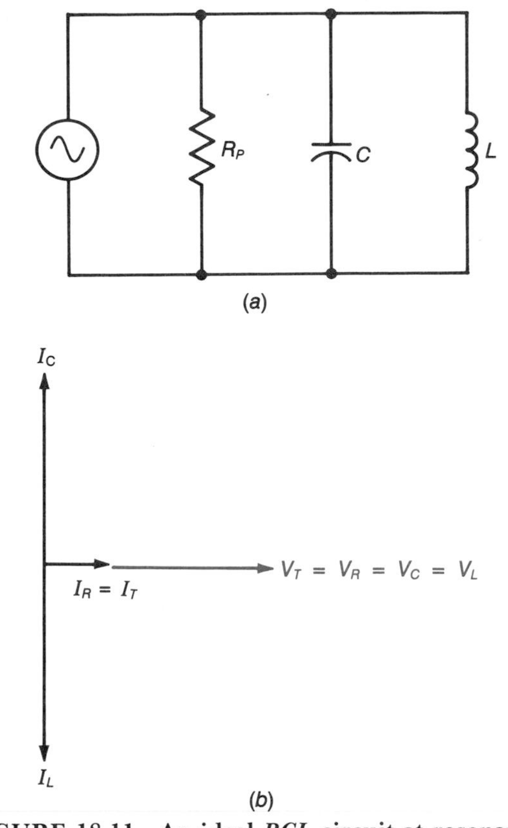

**FIGURE 18-11 An ideal *RCL* circuit at resonance.**

ing the source to the $LC$ tank circuit. In a circuit using real components, the tank circuit converts part of its energy to heat each time energy is transferred between $L$ and $C$. Thus, in a real parallel $LC$ circuit, there is some line current and $Z$ is less than infinite.

Since $X_C$ must equal $X_L$ at resonance for the ideal parallel $RCL$ or $LC$ circuit, the resonant frequency formula is the same as for the series $LC$ circuit. That is,

$$f_r = \frac{1}{2\pi\sqrt{LC}} \quad \text{for ideal components}$$

The formula for the circuit $Q$ of the ideal $RCL$ circuit can be derived by determining the energy stored (which is a function of the reactive power) and the energy converted (which is a function of true power) by the circuit. Thus,

$$Q = \frac{V_T^2/X}{V_T^2/R_P} = \frac{R_P}{X}$$

where $R_P$ is the parallel resistance, and $X$ is the reactance of either the capacitor or the inductor. This is the $Q$ of the circuit, not the inductor. (The $Q$ of an ideal inductor is infinite.) Notice in this formula that $R_P$ and $Q$ are directly proportional.

**EXAMPLE 18-8**

Determine $Q$, $I_L$, and $I_T$ for the circuit in Fig. 18-11 if $R_P = 100\ \text{k}\Omega$, $C = 0.02\ \mu\text{F}$, $L = 200$ mH, and $V_T = 20$ V.

**Solution**

$$f_r = \frac{1}{2\pi\sqrt{LC}} = \frac{1}{6.28\sqrt{0.02 \times 10^{-6} \times 0.2}} = 2517.7\ \text{Hz}$$

$$X_L = 2\pi fL = 6.28 \times 2517.7 \times 0.2 = 3162.2\ \Omega$$

$$Q = \frac{R_P}{X_L} = \frac{100\ \text{k}\Omega}{3162.2\ \Omega} = 31.6$$

$$I_L = \frac{V_L}{X_L} = \frac{20\ \text{V}}{3162.2\ \Omega} = 6.32\ \text{mA}$$

$$I_T = I_R = \frac{V_R}{R} = \frac{20\ \text{V}}{100\ \text{k}\Omega} = 0.2\ \text{mA}$$

Notice from Example 18-8 that the tank current ($I_L$ or $I_C$) is $Q$ times greater than the line current ($I_T$). The phenomenon of $I_{\text{tank}}$ being greater than $I_{\text{line}}$ in a parallel-resonant circuit is known as the *parallel-resonant current rise* or the current amplification (or magnification) of a parallel-resonant circuit. The relationship between $Q$ and resonant current rise is developed from the formula for $Q$ and Ohm's laws. Thus,

$$Q = \frac{R_P}{X_L} = \frac{V_T/I_T}{V_T/I_L} = \frac{I_L}{I_T}$$

and rearranging yields

$$I_{\text{tank}} = QI_{\text{line}}$$

where $I_{\text{line}} = I_T$ and $I_{\text{tank}} = I_L$ or $I_C$

The response of an $RCL$ circuit at a frequency above and below resonance is illustrated in Fig. 18-12. When the frequency is above the resonant frequency as in Fig. 18-12*a*, the circuit is capacitive; $I_T$ leads $V_T$ and $I_C$ is greater than $I_L$. Operating below resonance, as illustrated by the phasors in Fig. 18-12*b*, causes $I_T$ to lag $V_T$ and $I_L$ to be greater than $I_C$. Thus the circuit is inductive.

Complete frequency response curves are provided in Fig. 18-13 for an ideal parallel $RCL$ circuit

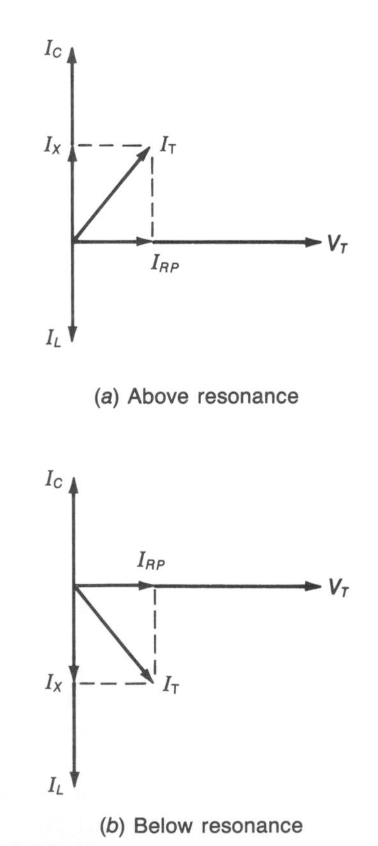


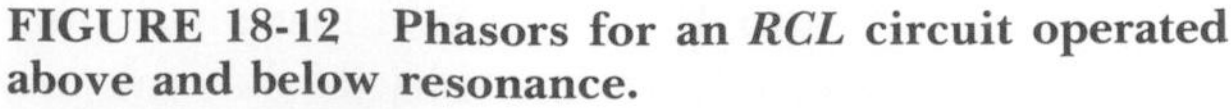

**FIGURE 18-12 Phasors for an $RCL$ circuit operated above and below resonance.**

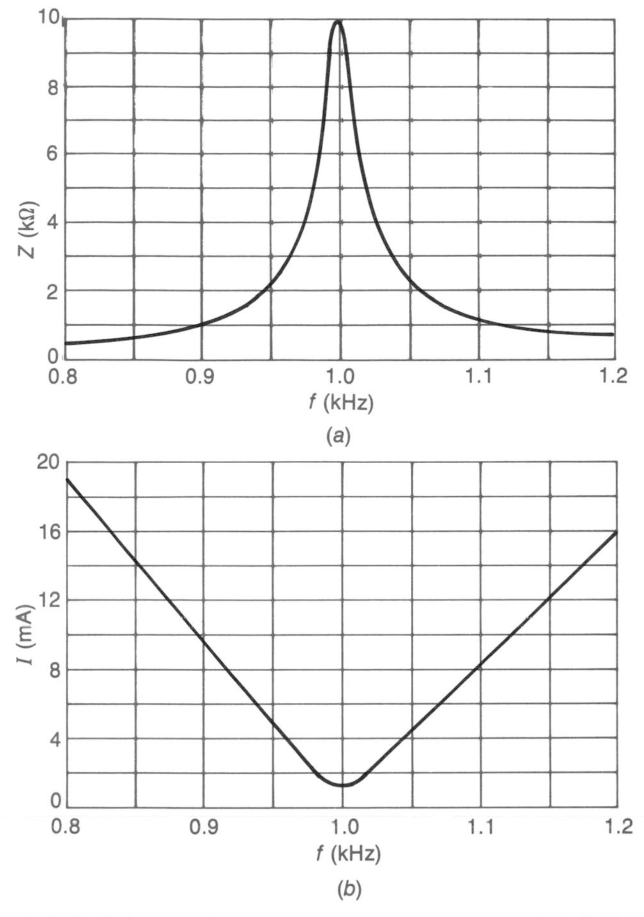


**FIGURE 18-13 Frequency response of a parallel *RCL* circuit.**

where $R_P = 10\ \text{k}\Omega$, $C = 0.68\ \mu\text{F}$, $L = 37.25$ mH, and $V_T = 10$ V. Figure 18-13*a* plots frequency versus impedance while Fig. 18-13*b* plots frequency versus current. As Ohm's law would indicate, these two graphs are the inverse of each other.

The BW of a parallel-resonant circuit can be taken from the frequency-versus-impedance plots as in Fig. 18-14. Using the same procedures as were used with series *LC* circuits, it can be shown that the half-power points ($f_{\text{up}}$ and $f_{\text{lo}}$) occur at 0.707 of *Z* maximum and that the BW is a function of *Q*. Thus, the appropriate BW formula for a parallel-resonant circuit is

$$\text{BW} = \frac{f_r}{Q}$$

Notice in Fig. 18-14 that doubling *Q* halves the BW just as it did with series-resonant circuits.

### EXAMPLE 18-9

For the circuit used in Example 18-8, determine BW, $f_{\text{up}}$, and $f_{\text{lo}}$.

**Solution**

Using data for Example 18-8,

$$\text{BW} = \frac{f_r}{Q} = \frac{2517.7\ \text{Hz}}{31.6} = 79.7\ \text{Hz}$$

$$f_{\text{up}} \approx f_r + 0.5\text{BW} = 2557.6\ \text{Hz}$$

$$f_{\text{lo}} \approx f_r - 0.5\text{BW} = 2477.9\ \text{Hz}$$

Like the selectivity of a series circuit, the selectivity of a parallel circuit is specified by its BW. Since BW is an inverse function of circuit *Q*, the selectivity must be a direct function of *Q*. (The smaller the BW, the better the selectivity.) Doubling the *Q* of a circuit doubles its selectivity.

### EXAMPLE 18-10

Determine *Q* for a resonant parallel *RCL* circuit in which $R_P = 10\ \text{k}\Omega$, $V_T = 50$ V, and $I_C = 50$ mA.

**Solution**

$$I_T = \frac{V_T}{Z} = \frac{V_T}{R_P} = \frac{50\ \text{V}}{10\ \text{k}\Omega} = 5\ \text{mA}$$

*Note:* $R_P = Z$ at resonance.

$$Q = \frac{I_C}{I_T} = \frac{50\ \text{mA}}{5\ \text{mA}} = 10$$

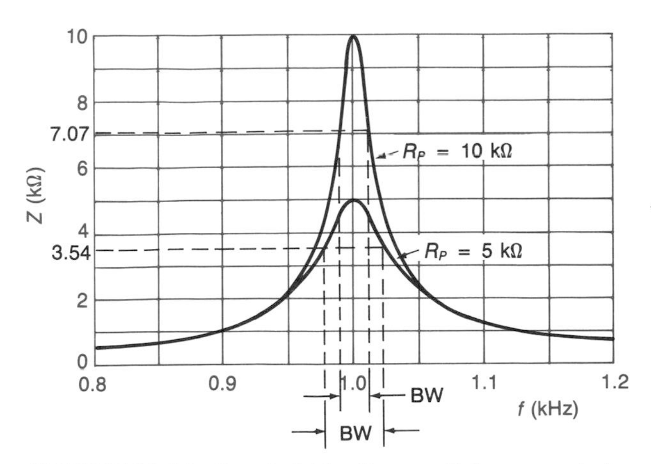


**FIGURE 18-14 Bandwidth of a parallel resonant circuit.**

## Self-Test

**22.** A parallel-resonant $LC$ circuit is known as a ________ circuit.

**23.** At parallel resonance, angle $\theta$ is ________.

**24.** What relationship exists between $Q$ and $R_P$? $R_P$ and BW?

**25.** Define BW for a parallel-resonant circuit.

**26.** A parallel $RCL$ circuit is ________ when operating above its resonant frequency.

**27.** Determine $Q$ for an ideal parallel-resonant $RCL$ circuit when $I_L = 36$ mA and $I_T = 1.2$ mA.

**28.** Determine $I_C$ for a parallel-resonant circuit when $Q = 40$, $R_P = 20\ \text{k}\Omega$, and $V_T = 60$ V.

## 18-3 PRACTICAL PARALLEL *LC* CIRCUITS

Practical $LC$ circuits are those circuits which use real, rather than ideal, components. With real components the equivalent series resistance of the inductor must be accounted for. This resistance, $R_L$ in Fig. 18-15*a*, controls the $Q$ of the circuit when there is no parallel resistor $R_P$. When a parallel resistor is present, then both resistances degrade the $Q$ of the circuit.

The practical $LC$ circuit in Fig. 18-15*a* forms a series-parallel $RCL$ circuit which we have not yet learned to analyze mathematically. (Chapter 19 covers series-parallel $RCL$ circuits of all types.) However, from our knowledge of series $RL$ circuits and phasor diagrams, we can visualize what happens in such a circuit. The phasor diagram in Fig. 18-15*b* shows that the current through the inductor will be less than 90° out of phase with the voltage because of the $R_L$ of the inductor. When $I_L$ is exactly equal in magnitude to $I_C$, the two phasors do not cancel each other and the resultant line current $I_T$ is not in phase with $V_T$. Thus, the circuit is not at resonance. The circuit is slightly capacitive when $I_C = I_L$ because $I_T$ leads $V_T$.

The angle of the $I_L$ phasor in Fig. 18-15*b* is determined by the $Q$ of the inductor. The larger $R_L$ is, the smaller $Q$ will be for a given value of $L$ and $f$. As $Q$ gets smaller, the angle between $I_L$ and $V_T$ also gets smaller. We can easily figure these angles from the impedance triangle for a series $RL$ circuit developed in Chap. 17. This triangle showed that $\tan\theta = X_L/R$. If $Q = X_L/R$, then the $\tan\theta = Q$ and $\theta = \arctan Q$. The $\theta$ in this case is the angle between $I_L$ and $V_T$ because $I_L$ is the total current through the series combination of $L$ and $R_L$ in this branch of the circuit. From this formula we can see that when $Q = 10$, the angle between $I_L$ and $V_T$ is $\arctan 10 = 84°$. When $Q = 5$, the angle is $\arctan 5 = 79°$.

Figure 18-15*c* shows the phasor diagram for the circuit of Fig. 18-15*a* at resonance. In this diagram $I_L$ is greater than $I_C$. Also, $I_L$ is split into its two component parts: $I_{X_L}$ and $I_R$. Splitting $I_L$ into $I_R$ and $I_{X_L}$ just shows how much of the current through the real inductor is caused by its resistance and how much is caused by its reactance. At resonance $I_{X_L}$ is equal to $I_C$ and the resultant current of $I_R$ or $I_T$ is in phase with the source voltage $V_T$. Using the techniques to be developed in Chap. 19 it can be shown that the resonant frequency $f_r$ for a practical $LC$ circuit is

$$f_r = \frac{1}{2\pi\sqrt{LC}}\sqrt{1 - \frac{CR_L{}^2}{L}}$$

The difference between $f_r$ calculated with the above formula and with the $f_r = 1/(2\pi\sqrt{LC})$ formula is very small unless $Q < 10$. Fortunately, most $LC$ circuits use inductors which have a $Q > 10$; therefore, the latter formula is usually used with all parallel $LC$ circuits.

The error introduced by using the simpler $f_r$ formula can be determined by analyzing the diagram in Fig. 18-15*d*, which is for a circuit with a $Q$ of 10. As shown previously, a $Q$ of 10 provides an angle of 84° between $I_L$ and $V_T$. This leaves an angle of 6° between $I_L$ and $I_{X_L}$. Since $I_L$ and $I_{X_L}$ are the hypotenuse and the adjacent side of a right triangle, we can use a trigonometric function to compare the lengths of these phasors. When $Q = 10$, the angle is 6°, so

$$\cos 6° = \frac{I_{XL}}{I_L} \qquad \text{and} \qquad 0.995 I_L = I_{X_L}$$

Thus, we see that assuming $I_L = I_{X_L}$ and using the simpler $f_r$ formula introduces about 0.5 percent error. Even when $Q = 4$, the error in using the simpler formula is only about 3 percent. For this reason, we will use the simpler $f_r = 1/(2\pi\sqrt{LC})$ formula in our analysis of all parallel $LC$ circuits.

The $Q$ for a practical parallel-resonant $LC$ circuit is just $Q = X_L/R$ where $R$ is the resistance of the inductor $R_L$. As expected, this is the same formula

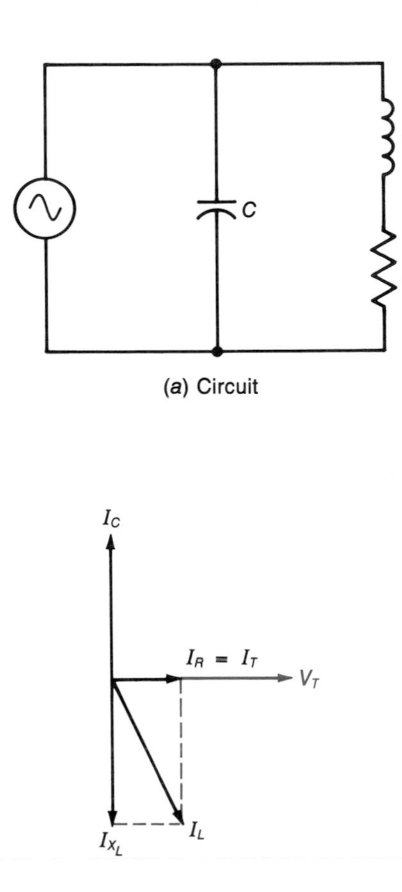


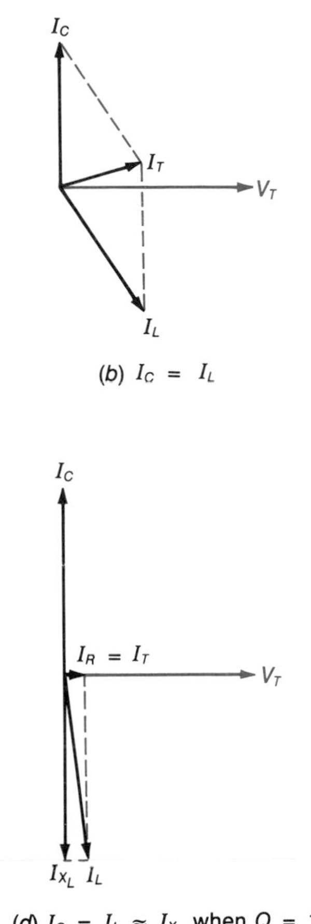


**FIGURE 18-15 Practical parallel *LC* resonant circuit.**

as that used for a practical series-resonant *LC* circuit because in both cases $R$ is in series with $X_L$. When a practical parallel *RCL* circuit like the one in Fig. 18-16 is analyzed, the circuit $Q$ must be determined by accounting for both $R_P$ and $R_L$. This can be done with the following formula:

$$Q = \frac{1}{X_L/R_P + R_L/X_L}$$

In determining the resonant current rise in a practical *LC* or *RCL* circuit, remember that $I_L$ is greater than $I_C$ and that part of $I_L$ is resistive current. Therefore, use the relationship $Q = I_C/I_T$ but not $Q = I_L/I_T$.

Sometimes the BW of a practical parallel *LC* circuit is too narrow to select the desired band of frequencies. In this case, a parallel resistor is added to form a parallel *RCL* circuit like the one in Fig. 18-16. Resistor $R_P$ in this circuit is referred to as a *damping resistor,* and the circuit is said to be damped. Of course, this damping resistor lowers the $Q$ of the circuit so that the BW is increased and a wider band of frequencies is selected by the circuit.

Now that we have examined practical parallel-resonant circuits, we can see that we can treat them like ideal parallel-resonant circuits with the following exceptions: (1) When the $Q$ of the inductor is less than 10, the $f_r$ formula is different, and (2) the $Q$ formula for the *RCL* circuit is different. In summary, the new $f_r$ formula accounts for $I_L$ being less than 90° out of phase with $V_T$, and the new $Q$ formula accounts for the degrading effect of both $R_L$ and $R_P$.

**EXAMPLE 18-11**

For the circuit in Fig. 18-16, $R_P = 15\ \text{k}\Omega$, $C = 0.01\ \mu\text{F}$, $L = 25.3$ mH, and $R_L = 30\ \Omega$. Determine $f_r$ and the circuit $Q$ if the inductor $Q > 10$.

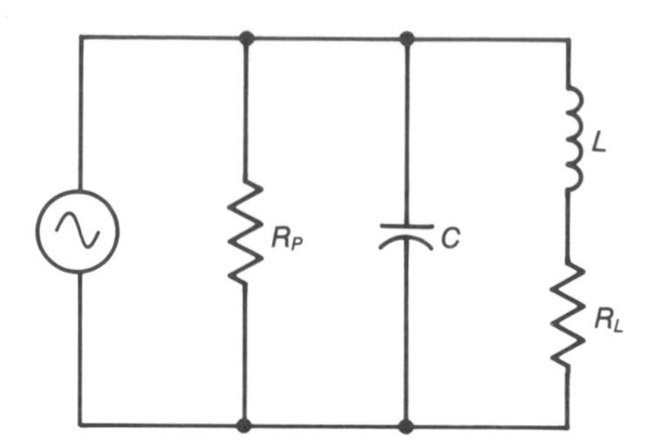


**FIGURE 18-16 Practical parallel *RCL* resonant circuit.**

**Solution**

$$f_r = \frac{1}{2\pi\sqrt{LC}} = \frac{1}{6.28\sqrt{0.0253 \times 0.01 \times 10^{-6}}} = 10 \text{ kHz}$$

$$X_L = 1589\ \Omega$$

$$Q = \frac{1}{X_L/R_P + R_L/X_L} = \frac{1}{1589/15{,}000 + 30/1589} = 8$$

Notice in this example that the $f_r = 1/(2\pi\sqrt{LC})$ is still appropriate even for a circuit $Q$ of less than 10 because the inductor $Q$ is over 10 (approximately 53).

## Self-Test

**29.** In a practical parallel $LC$ circuit, is the circuit capacitive or inductive when $I_L = I_C$?

**30.** In a practical parallel-resonant circuit, is $I_L$ smaller or larger than $I_C$?

**31.** When is it not appropriate to use the formula $f_r = 1/(2\pi\sqrt{LC})$?

**32.** What is a damping resistor?

**33.** What is the circuit $Q$ and the inductor $Q$ of a parallel-resonant $LC$ circuit with $R_L = 50\ \Omega$, $L = 20$ mH, and $f_r = 15$ kHz?

**34.** What is the circuit $Q$ of the circuit in Question 33 if a 20-k$\Omega$ damping resistor is added?

## 18-4 GAIN, ATTENUATION, AND THE DECIBEL

The ratio of the output signal to the input signal of an electric-electronic system is called the *gain* of the system. The output and input signals can be expressed in units of current, voltage, or power. Of course, both signals must be expressed in the same units. Since gain is equal to the signal output divided by the signal input, it is a pure number (i.e., it has no units).

When the gain of a system is less than one (i.e., more signal in than out), the system has *attenuated* the signal. *Attenuation* occurs when more signal is available at the input than the output. Saying that a system produces signal attenuation is another way of saying that the system gain is less than one. Attenuation is also referred to as a signal loss. Thus, we can say that a system has a signal loss when the gain is less than one. Although technically gain can be less than one, in this chapter we will only use the term gain when the gain is greater than one. For gains of less than one, we will use the term attenuation.

Gain, and attenuation, can be expressed as either a linear or a logarithmic function of the output-to-input signal ratio. To distinguish between these two systems, logarithmic gain has been assigned the unit of decibel (dB). Thus, a "gain of 3" is understood to be a linear function of the signal ratio, while a "gain of 3 dB" is understood to be a logarithmic function of the signal ratio.

### Bel and Decibel

The *bel* is the base unit for logarithmic gain (or attenuation if the gain is less than one) in an electric-electronic system. The bel was selected as the name for the base unit of logarithmic gain in honor of Alexander Graham Bell. The bel is too large a unit for most uses, so the *decibel* (one-tenth of a bel) is the common unit used to express logarithmic gain or loss. Mathematically, the decibel is defined as

$$\text{dB} = 10 \log \frac{P_o}{P_i}$$

where log = common (base 10) log
$P_o$ = output power
$P_i$ = input power

From this formula you can see that the decibel will be negative when the gain is less than one, positive when it is greater than one, and zero when it is exactly one. This means that a +dB indicates a gain in signal strength (i.e., the output power is greater than the input power); a −dB represents an attenuation of the signal (less output than input); and 0 dB means the system has neither gain nor attenuation.

### EXAMPLE 18-12

An amplifier receives 0.1 W of input signal and delivers 15 W of signal power. What is its linear and its logarithmic power gain?

**Solution**

$$\text{Power gain} = \frac{P_o}{P_i} = \frac{15\text{ W}}{0.1\text{ W}} = 150$$
$$= 10 \times 2.18 = 21.8\text{ dB}$$

The question in Example 18-12 could have been, "What is the power gain and the decibel power gain?"

In addition to expressing logarithmic gain and loss, the decibel is also used to express the magnitude of the ratio of any two powers. For example, the decibel can be used to express the change in the power gain of an amplifier when only the frequency of the input power is changed. In this case, the power symbols in the formula are changed from $P_o$ and $P_i$ to $P_1$ and $P_2$.

### EXAMPLE 18-13

For the same input power level, an amplifier provides 10-W output at 1000 Hz but only 6-W output at 20 kHz. What is the decibel loss in power gain as the frequency of the input increases from 1 to 20 kHz?

**Solution**

$$\text{Loss of gain (dB)} = 10 \log \frac{6\text{ W}}{10\text{ W}} = -2.22\text{ dB}$$

The power output of the amplifier in Example 18-13 at 20 kHz is −2.22 dB when the 1-kHz power output is used as a reference power. Whenever an output power is specified in decibels, there is always a reference power. In other words, when a reference power is stated (or implied by previous agreement), the output of a device can be specified in decibels.

A logarithmic unit for attenuation or gain was originally conceived of to equate gain or loss of audio signal power to the response of the human ear. The ear responds in a logarithmic fashion to the sound level or the power of an audio signal. For example, most people can just perceive a 1-dB change in the signal power from a speaker. Because the ear responds logarithmically, it does not matter if the power changes from 1 to 1.26 W (a 1-dB change) or from 10 to 12.6 W (also a 1-dB change). Thus, even though the absolute change in power was much greater in the second case (2.6 versus 0.26 W), the effect on the ear was the same—a just-perceivable change in loudness.

### EXAMPLE 18-14

The power output of an audio system is 18 W. For a person to notice an increase in the output (loudness or sound intensity) of the system, what must the output power be increased to?

**Solution**

The power must be increased 1 dB, so

$$1\text{ dB} = 10 \log \frac{P}{18\text{ W}}$$

Dividing by 10 yields

$$0.1\text{ dB} = \log \frac{P}{18\text{ W}}$$

Taking the antilog gives

$$\text{antilog } 0.1\text{ dB} = \frac{P}{18\text{ W}}$$

and solving for $P$ results in

$$P = 18\text{ W antilog } 0.1\text{ dB} = 18\text{ W} \times 1.26 = 22.68\text{ W}$$

Although the decibel system was primarily developed to work with audio systems, it is now extensively used throughout all fields of electricity and electronics. For example, an antenna is often rated in decibels. In this case the output of the antenna is being referenced to the output of an agreed-on standard antenna under specified conditions. Thus, if an antenna has a rated output of 3 dB, its output power will be twice as great as that of the reference antenna under the same signal conditions (a power ratio of 2 yields 3 dB).

Voltage and current ratios and gains can also be expressed in decibels, *providing* that the two voltages or currents occur across or through the same value of resistance. The decibel formula for current or voltage can be derived from the power decibel formula. For voltage, the derivation is

$$\text{dB} = 10 \log \frac{P_o}{P_i}$$

Substituting $V^2/R$ for $P$ gives

$$\text{dB} = 10 \log \frac{V_o^2/R_o}{V_i^2/R_i}$$

And when $R_o = R_i$, we get

$$\text{dB} = 10 \log \frac{V_o^2}{V_i^2} = 10 \log \left(\frac{V_o}{V_i}\right)^2$$

Since $\log X^2 = 2 \log X$, the square can be removed to yield

$$\text{dB} = 20 \log \frac{V_o}{V_i}$$

Using the same process and substituting $I^2R$ for $P$ gives the current ratio formula of

$$\text{dB} = 20 \log \frac{I_o}{I_i}$$

Remember that these two formulas are accurate only when the input and output resistances are equal.

**EXAMPLE 18-15**

The output of a microphone is rated at −52 dB. The reference level is 1 V under specified sound conditions. What is the output voltage of this microphone under the same sound conditions?

**Solution**

$$-52 \text{ dB} = 20 \log \frac{V}{1 \text{ V}}$$

Divide by 20:

$$-2.6 \text{ dB} = \log \frac{V}{1 \text{ V}}$$

Take antilog:

$$\text{antilog}(-2.6 \text{ dB}) = \frac{V}{1 \text{ V}}$$

Solve for $V$:

$$V = 0.0025 \times 1 \text{ V} = 2.5 \text{ mV}$$

## Self-Test

**35.** A circuit which attenuates a signal has a gain of ______.

**36.** The input power to a circuit is 5 W and the output is 4 W. Determine the decibel attenuation.

**37.** A circuit provides −5 dB of attenuation. The input voltage is 10 V. What is the output voltage if $R_i = R_o$?

**38.** What is the decibel power gain of an amplifier that requires 0.2 W input to provide 30 W of output?

**39.** What condition must be satisfied for the formula $\text{dB} = 20 \log (I_o/I_i)$ to be accurate?

## 18-5 FILTERS

*Filters* are circuits which are frequency-selective. By using various combinations of reactance and resistance, filter circuits can be made to either emphasize or deemphasize selected ranges of frequencies.

The general concept of a common type of electric filter is illustrated in Fig. 18-17. Basically, the filter has two series elements across which is connected the input signal. The output signal is taken across just one of the series elements. The impedance of one or both of these elements is frequency-dependent. The input signal consists of a wide

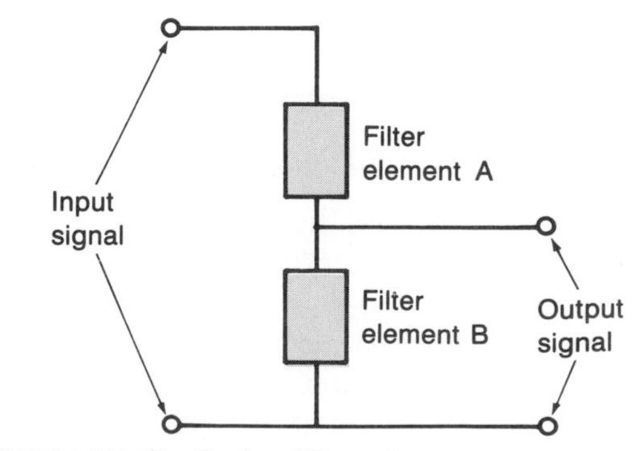

**FIGURE 18-17 Series filter elements.**

range of frequencies from which a narrower range of frequencies is to be selected for the output signal or rejected from the output signal. For example, the input to a filter may be a signal containing all of the frequencies needed for the audio output from an FM radio receiver *plus* many higher frequencies which would only overload the audio system in the receiver. The output of the filter would contain only the audio frequencies to be amplified and sent to the speakers.

## Low-Pass Filters

As shown in Fig. 18-18, either an *RL* or an *RC* circuit can be used as a low-pass filter. Notice that the output is taken across the resistor for the *RL* circuit and across the capacitor for the *RC* circuit. This is because the inductive reactance increases with increasing frequency, while the capacitive reactance decreases with increasing frequency.

In Fig. 18-18*a*, the capacitor is filter element B in Fig. 18-17 and the resistor is filter element A. Notice from these two figures that for the *RL* filter the frequency-dependent component is element A rather than element B. For very low frequencies, $X_L$ is very small compared to $R$, so the output signal $V_o$ is approximately equal to the input signal $V_i$. For very high frequencies, $X_L$ is very large compared to $R$ and $V_o$ is almost zero.

### EXAMPLE 18-16

Determine $V_o$ for the circuit in Fig. 18-18*b* at 1 kHz, 6 kHz, and 50 kHz when $V_i = 10$ V, $L = 72$ mH, and $R - 2700\ \Omega$.

**Solution**

At $f = 1$ kHz,

$$X_L = 2\pi fL = 6.28 \times 1000 \times 0.072 = 452\ \Omega$$
$$Z = \sqrt{X_L{}^2 + R^2} = \sqrt{452^2 + 2700^2} = 2738\ \Omega$$
$$V_o = V_R = I_T R = \frac{V_i}{Z} \times R = \frac{10\ \text{V} \times 2700\ \Omega}{2738\ \Omega} = 9.9\ \text{V}$$

At $f = 6$ kHz,

$$X_L = 6.28 \times 6 \times 10^3 \times 0.072 = 2713\ \Omega$$
$$Z = \sqrt{2713^2 + 2700^2} = 3828\ \Omega$$
$$V_o = \frac{10\ \text{V} \times 2700\ \Omega}{3828\ \Omega} = 7.1\ \text{V}$$

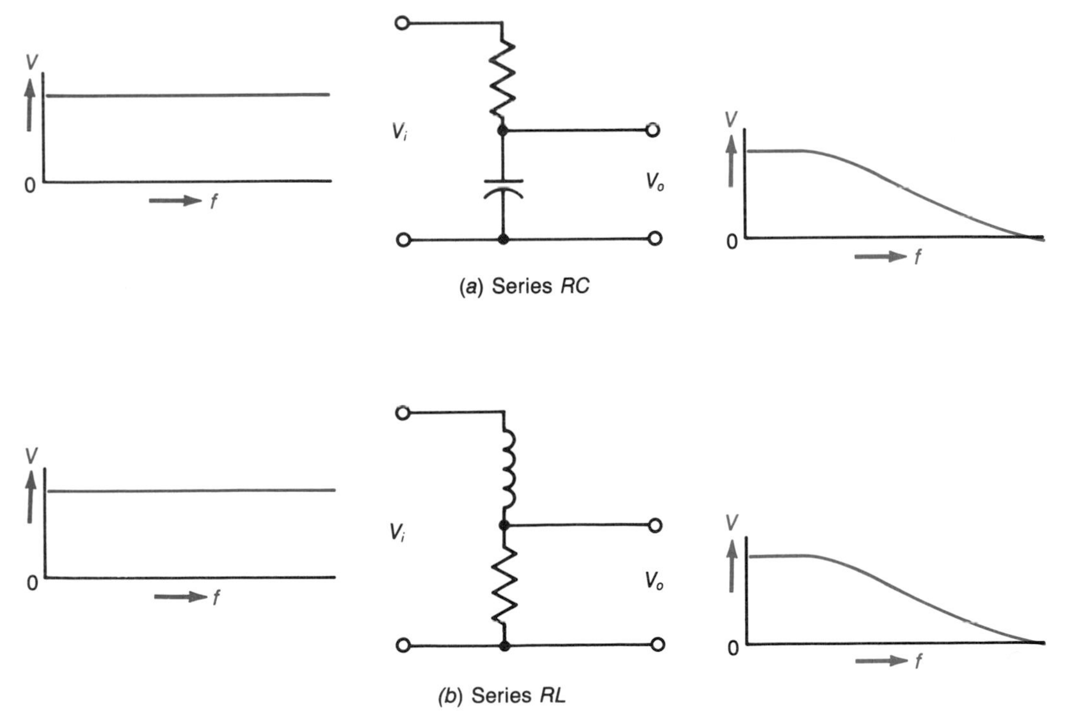


**FIGURE 18-18 Low-pass filters.**

At $f = 50$ kHz,

$$X_L = 6.28 \times 50 \times 10^3 \times 0.072 = 22{,}608\ \Omega$$
$$Z = \sqrt{22{,}608^2 + 2700^2} = 22{,}769\ \Omega$$
$$V_o = \frac{10\ \text{V} \times 2700\ \Omega}{22{,}769\ \Omega} = 1.2\ \text{V}$$

The calculations in Example 18-16 verify that the series *RL* circuit is a low-pass filter when the output is taken across the resistor. In the actual use of a filter like this, the resistor could be the internal or input resistance of any electric-electronic device such as an amplifier. Calculations like those in Example 18-16 show that the series *RC* circuit is also a low-pass filter when the output is taken across the capacitor.

A complete frequency response curve for a low-pass filter is shown in Fig. 18-19. This curve is for the circuit and conditions used in Example 18-16. Notice that the three voltage-frequency points determined in the example agree with the curve in Fig. 18-19. The curve in Fig. 18-19 would also be obtained for the *RC* low-pass filters when $C = 0.01\mu\text{F}$, $R = 2700\ \Omega$, and $V_i = 10$ V. The only required change in Fig. 18-19 would be to change $V_R$ to $V_C$.

The output of a filter can also be expressed in decibels as shown in Fig. 18-20. This plot is for the same *RL* low-pass filter used in Example 18-16. Thus, it is only a different way of showing the information provided by the plot in Fig. 18-19 for the frequency range of 1 to 60 kHz. Since the output in Fig. 18-20 is in decibels, there has to be a reference. The reference for a filter can be the output voltage when the output is its maximum possible value. For the low-pass filter of Fig. 18-18*b* and Example 18-16, $V_o = V_i = 10$ V at 0 Hz (assuming an ideal inductor). Thus, the reference for this *RL* low-pass filter is $V_i$. Now using $V_o$ at various frequencies and the formula dB = 20 log $(V_o/V_i)$, we can determine the decibel attenuation for this filter at these various frequencies.

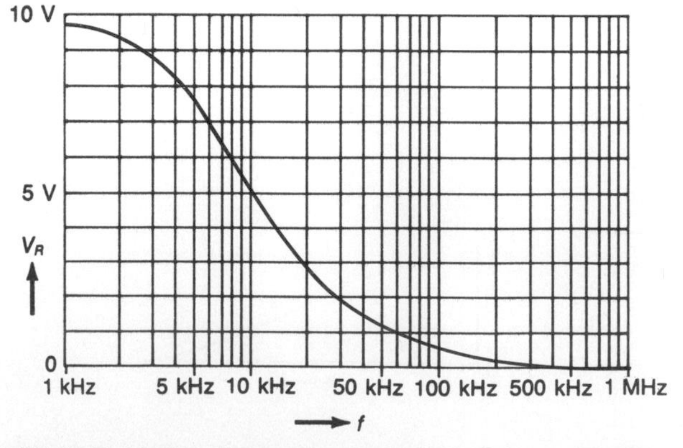


**FIGURE 18-19 Frequency response for an *RL* filter when $R = 2700\ \Omega$ and $L = 0.072$ H.**

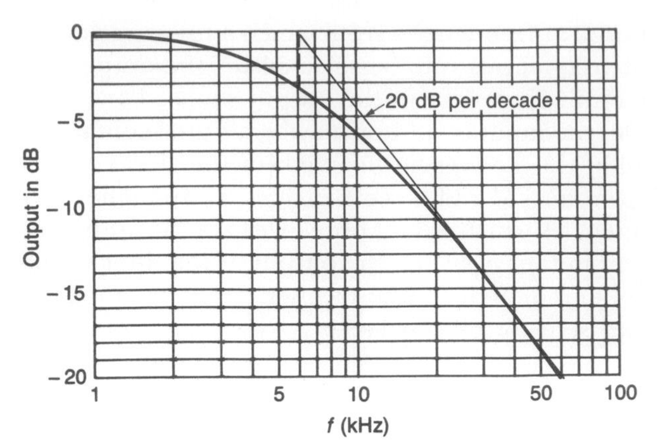


**FIGURE 18-20 Filter attenuation in decibels.**

### EXAMPLE 18-17

Determine the decibel attenuation for the circuit of Example 18-16 at 1 kHz, 6 kHz, and 50 kHz using the data determined in the solution to that example.

**Solution**

At 1 kHz,

$$\text{Attenuation (dB)} = 20 \log \frac{V_o}{V_i} = 20 \log \frac{9.9\ \text{V}}{10\ \text{V}} = -0.09\ \text{dB}$$

At 6 kHz,

$$\text{Attenuation (dB)} = 20 \log \frac{7.1\ \text{V}}{10\ \text{V}} = -2.98\ \text{dB}$$

At 50 kHz,

$$\text{Attenuation (dB)} = 20 \log \frac{1.2\ \text{V}}{10\ \text{V}} = -18.42\ \text{dB}$$

Notice that the values calculated in Example 18-17 agree with those in Fig. 18-20.

An $RC$ or $RL$ filter is said to have a frequency response curve with a *slope* of approximately 6 dB per octave or 20 dB per decade. (Octave and decade as used here mean a 1:2 and 1:10 frequency ratio, respectively.) How close this approximation is can be seen in Fig. 18-20 where a line with a 20 dB per decade slope has been drawn on the frequency response graph from 0 to −20 dB and from 6 to 60 kHz (i.e., 20 dB per decade.) Notice in Fig. 18-20 that a line projected up from the −3-dB point on the frequency response curve intercepts the 20-dB slope line at 0 dB. Thus, the frequency at the −3-dB point of the curve is often called the *break* or *cutoff frequency*. Using this information, an approximate response curve for an $RL$ or $RC$ filter can be sketched on semilog graph paper. The procedure is to calculate the −3-dB frequency and locate it on the graph paper. Then project up to the 0-dB line and mark this point on the 0-dB line. Next, locate on the −20-dB line the frequency which is a decade higher than the −3-dB frequency. Now draw a slope line from this point on the −20-dB line to the previously located point on the 0-dB line. Once the 20-dB per decade line and −3-dB point are located, the approximate curve can be sketched in between the −3-dB point, the slope line, and the 0-dB line.

Calculating the −3-dB point is quite easy because it is always at the frequency where $X$ (either $X_C$ or $X_L$) is equal to $R$. When $X = R$, the voltages $V_X$ and $V_R$ are each $0.707V_T$. Because $P = V^2/R$, the power dissipation of $R$ is 50 percent of the value it was at the frequency when $V_R = V_T$. For this reason, the −3-dB point on the frequency response curve is also known as the *half-power point*.

---

**EXAMPLE 18-18**

Determine the −3-dB frequency and the −20-dB frequency for an $RL$ low-pass filter when $R = 1\ \text{k}\Omega$ and $L = 0.6$ mH.

**Solution**

At −3-dB, $X_L = R$, so rearranging $X_L = 2\pi fL$ and solving for $f$ yields

$$f = \frac{X_L}{2\pi L} = \frac{1000\ \Omega}{6.28 \times 0.6 \times 10^{-3}} = 265.4\ \text{kHz}$$

$$f \text{ for } -20\ \text{dB} = 10 \times 265.4\ \text{kHz} = 2.654\ \text{MHz}$$

---

An example of how a low-pass filter might be used is illustrated by the waveform in Fig. 18-21 which shows a 90-kHz sine wave superimposed on a 3-kHz sine wave. This complex waveform is the result of adding 10 V of a 3-kHz sine wave with 1 V of 90-kHz waveform. When this waveform is applied to the $RL$ low-pass filter we have been using in our example ($R = 2700\ \Omega$ and $L = 72$ mH), the 1 V at 90 kHz is almost missing from the output. We can predict what the output will look like by using the curve in Fig. 18-19. Projecting up from 3 kHz, we see the curve is intercepted at the 9-V level, which represents 90 percent of the 10-V input. Projecting up from 90 kHz, we see the curve is intercepted at about 0.6 V, which represents 6 percent of the 1-V input (0.06 V). Thus, the output would be 9 V at 3 kHz with 0.06 V at 90 kHz superimposed on it.

If we view the 90 kHz as noise and the 3 kHz as the desired signal, then it is meaningful to speak of the *signal-to-noise ratio* (S/N) of the waveform in Fig. 18-21. Before it is filtered, the S/N ratio is 10 V/1 V = 10. After it is filtered, the S/N ratio is increased to 9 V/0.06 V = 150. Thus, while the filter attenuated the signal by about 10 percent, it improved the S/N ratio by a factor of 15.

## High-Pass Filters

Series $RL$ and $RC$ circuits can also be used for high-pass filters as shown in Fig. 18-22. For the $RC$ circuit of Fig. 18-22*a*, the output is taken across the resistor for a high-pass filter. In this circuit, most of the input voltage drops across $C$ at low frequencies (all of it at 0 Hz) because $X_C$ is very large compared to $R$. As the frequency increases, $X_C$ de-

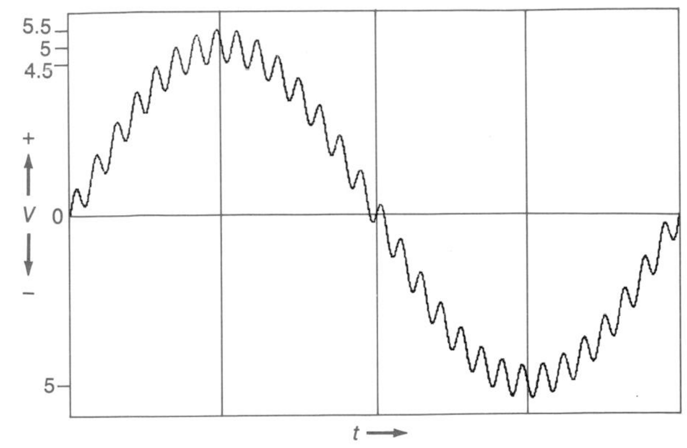

**FIGURE 18-21 Waveform of a noisy signal.**

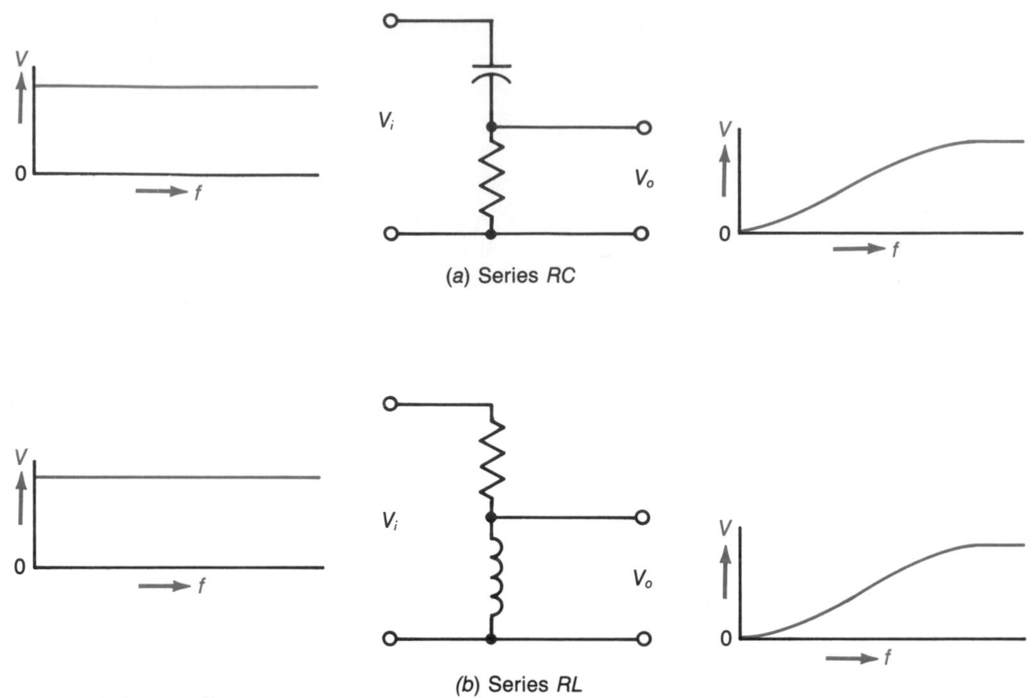


**FIGURE 18-22 High-pass filters.**

creases, so less voltage drops across $C$ and more across $R$. At very high frequencies $X_C$ is so small compared to $R$ that essentially all of the input voltage drops across $R$.

For the *RL* high-pass circuit, see Fig. 18-22*b*, the output must be taken across the inductor. This way, the output is zero at 0 Hz and increases as the frequency increases.

The frequency response of a high-pass filter with $R = 260\ \Omega$, and $C = 1\ \mu F$, and $V_i = 10$ V is plotted on the graphs in Figs. 18-23 and 18-24. These graphs can be used in the same way as were the graphs for low-pass filters. Notice that the $-3$-dB frequency in Fig. 18-24 is 600 Hz and that the output voltage in Fig. 18-23 is 70.7 percent of the input voltage at 600 Hz. These graphs also apply to the high-pass *RL* circuit if the correct combination of $L$ and $R$ are selected.

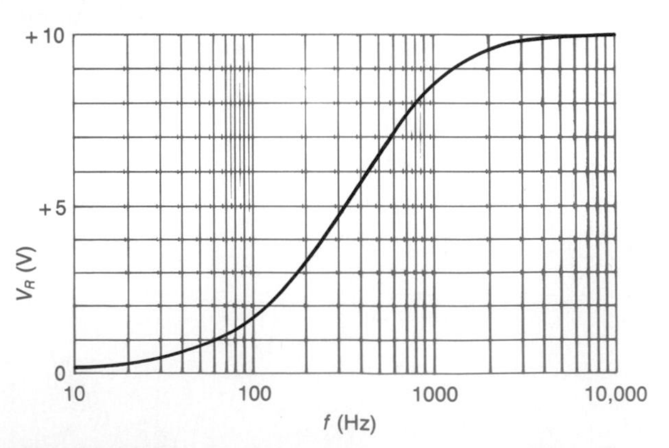


**FIGURE 18-23 Frequency versus output voltage for a high-pass filter.**

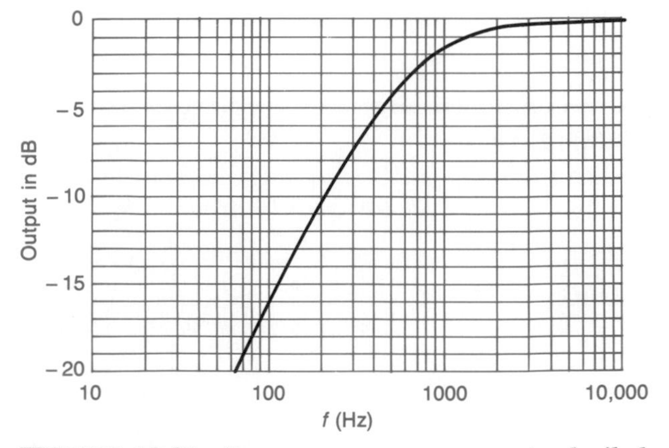


**FIGURE 18-24 Frequency versus output in decibels for a high-pass filter.**

**EXAMPLE 18-19**

Determine the value of $R$ needed to make a filter match the curve in Fig. 18-24 if $L = 100$ mH.

**Solution**

The −3-dB frequency, from Fig. 18-24, is 600 Hz; therefore,

$$X_L = 2\pi fL = 6.28 \times 600 \times 0.1 = 376.8\ \Omega \quad \text{at} \quad -3\text{dB},$$
$$R = X_L, \text{ so } R = 376.8\ \Omega.$$

## Self-Test

**40.** Filters can be defined as circuits which are _______.

**41.** If the output of a filter changes from 5 to 3 V as the frequency increases from 1 to 10 kHz, the filter is a _______ filter.

**42.** For a low-pass $RC$ filter, the output is taken across the _______.

**43.** A low-pass $RL$ filter with $R = 2.2$ kΩ and $L = 50$ mH has an input voltage of 10 V. What is its output voltage at 5 kHz?

**44.** What relationship exists between $X$ and $R$ when the output of a filter is −3 dB?

**45.** Determine the −3-dB and −20-dB frequency for the filter in Question 43. Also, what will the output voltage be at −3 dB?

**46.** What is the approximate slope of an $RL$ or $RC$ filter circuit?

**47.** List three names for the −3-dB frequency of a filter.

**48.** How can a filter improve the S/N ratio?

**49.** For which type of high-pass filter is the output voltage taken across the resistor?

## Resonant Filters

Inserting an $LC$ circuit in series with a resistor produces a resonant filter. As illustrated in Fig. 18-25*a*, the series $LC$ circuit and its series output or load resistor form a series $RCL$ circuit. Therefore, the techniques developed in Sec. 18-1 can be used to quantitatively analyze this *bandpass filter*. When the frequency of the input voltage is the resonant frequency of the $LC$ combination, the output voltage will equal the input voltage (assuming ideal components). Even with real components, the load resistor can be so much greater than the series-equivalent resistance of the inductor and capacitor that $V_o$ is still essentially equal to $V_i$ at resonance. Above and below resonance, the $V_o$ will decrease. When the net reactance is equal to the resistance, the output is at the half-power point.

**EXAMPLE 18-20**

Determine the 0-dB and the upper −3-dB frequency for a bandpass filter when $R = 1$ kΩ, $L = 270$ mH, and $C = 0.022\ \mu$F.

**Solution**

Zero decibels occurs at $f_r$, so

$$f_r = \frac{1}{2\pi\sqrt{LC}} = \frac{1}{6.28\sqrt{0.27 \times 0.022 \times 10^{-6}}} = 2066 \text{ Hz}$$

Find $X_L$ and $Q$ at $f_r$:

$$X_L = 6.28 \times 2066 \times 0.27 = 3503\ \Omega$$
$$Q = \frac{X_L}{R} = \frac{3503}{1000} = 3.5$$
$$\text{BW} = \frac{f_r}{Q} = \frac{2066}{3.5} = 590 \text{ Hz}$$
$$f_{\text{up}} \approx f_r + 0.5\text{BW} = 2066 + (0.5 \times 590) = 2361 \text{ Hz}$$

An exact value for $f_{\text{up}}$ in Example 18-20 can be calculated by making $X$, i.e., $X_L - X_C$, equal to $R$ and solving for $f_{\text{up}}$. This gives

$$R = 2\pi f_{\text{up}}L - \frac{1}{2\pi f_{\text{up}}C}$$

And solving for $f_{\text{up}}$ yields

$$f_{\text{up}} = \frac{RC + \sqrt{R^2C^2 + 4LC}}{4\pi LC}$$

Using this formula and the values given in Example 18-20, $f_{\text{up}}$ is calculated as 2382 Hz.

The *bandstop* or *band-reject* filter in Fig. 18-25*b* uses the high impedance of the parallel $LC$ circuit to reject the band of frequencies at and around its resonant frequency. At the resonant frequency of the parallel $LC$, the resistance of the resistor is so small compared to the impedance of the parallel

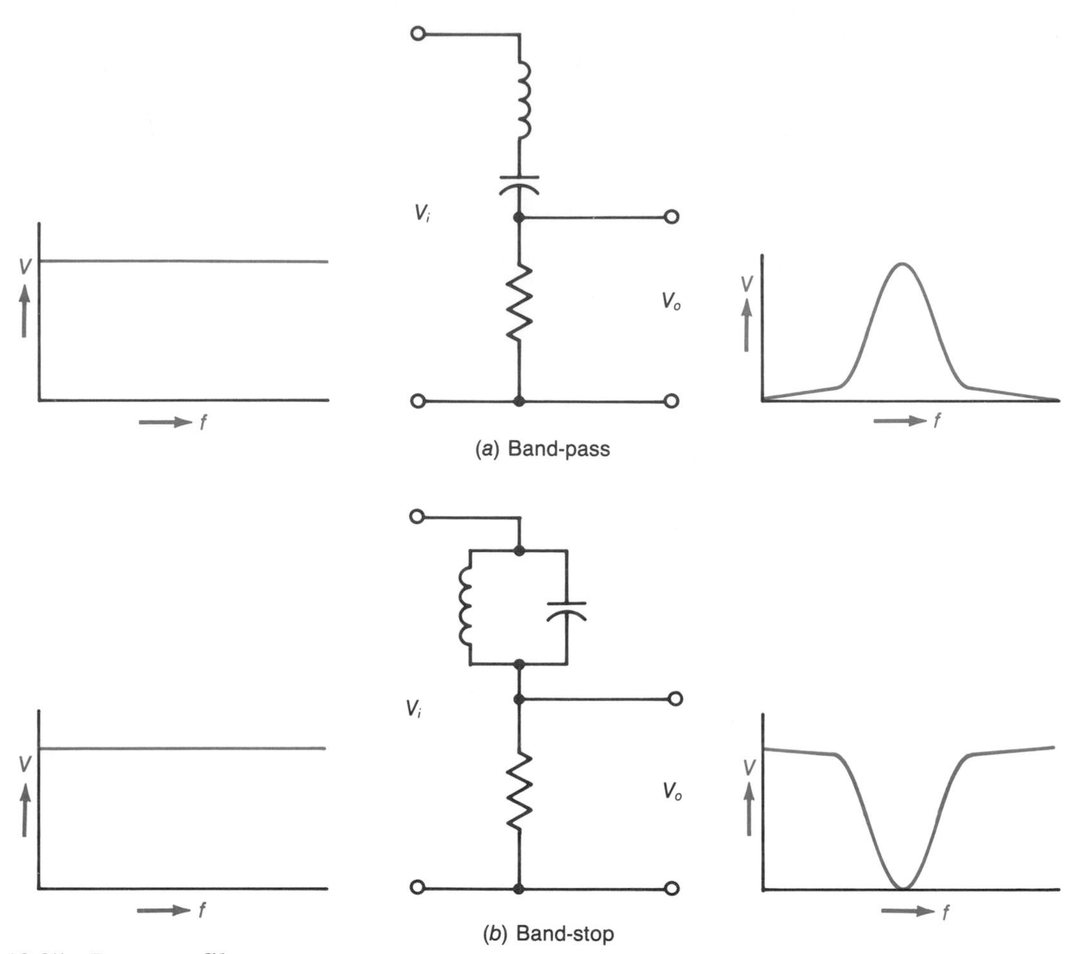


**FIGURE 18-25 Resonant filters.**

$LC$ combination that essentially none of $V_i$ appears across the resistor as $V_o$.

## Multiple Resonant Filters

Figure 18-26 shows two type of *multiple* or *double resonant* filters. These circuits can be used to stop one band of frequencies while passing another band of frequencies. The frequency versus output voltage for the circuit in Fig. 18-26*a* is shown in Fig. 18-27. Selection of component values to obtain the desired reject and pass center frequencies can be done by careful application of the rules and formulas we already know. For example, let us select the inductor values for the circuit in Fig. 18-26*a* when $R = 1\ \text{k}\Omega$, $C = 200\ \text{pF}$, $V_i = 10\ \text{V}$, the reject frequency is 290 kHz, and the pass frequency is 710 kHz. First, we know that a parallel-resonant circuit will provide the reject part of the filter, so $L_1$ and $C$ will determine the reject frequency. Since we know the value of $C$ and $f$, we can solve for $L_1$ using the resonant frequency formula:

$$f = \frac{1}{2\pi\sqrt{L_1C}}$$

Squaring gives

$$f^2 = \frac{1}{4\pi^2L_1C}$$

And rearranging,

$$L_1 = \frac{1}{4\pi^2Cf^2}$$

$$= \frac{1}{4 \times 3.14^2 \times 200 \times 10^{-12} \times (290 \times 10^3)^2} = 1.5\ \text{mH}$$

Next, we need to determine the net reactance (which is equal to $Z$) for the parallel $L_1C$ combination at the pass frequency. Thus,

$$X_L = 6.28 \times 710 \times 10^3 \times 1.5 \times 10^{-3} = 6688\ \Omega$$

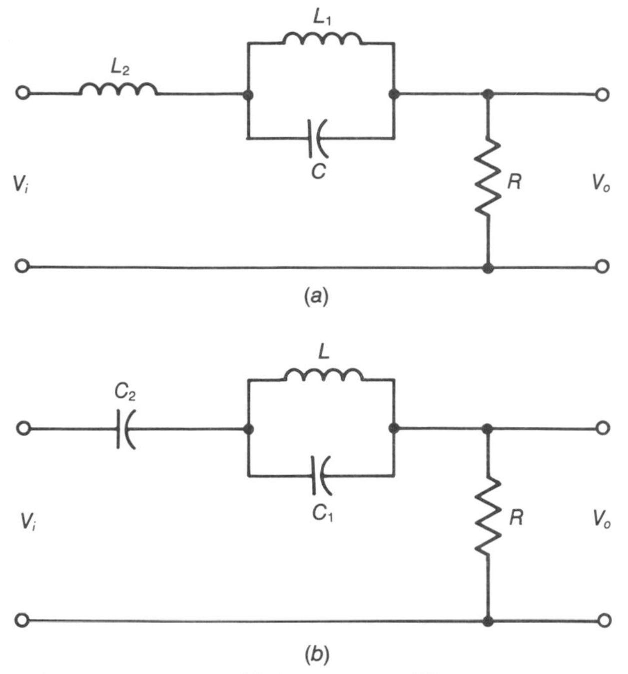

**FIGURE 18-26 Multiple resonant filters.**

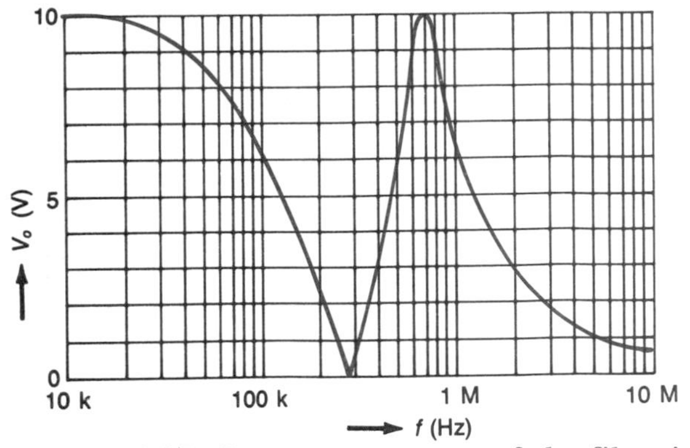


**FIGURE 18-27 Frequency response of the filter in Fig. 18-26a.**

$$X_C = \frac{1}{6.28 \times 710 \times 10^3 \times 200 \times 10^{-12}} = 1121\ \Omega$$

$$Z = \frac{V_T}{I_T} = \frac{V_T}{(V_T/X_C) - (V_T/X_L)}$$

$$= \frac{10}{(10/1121) - (10/6688)} = 1347\ \Omega$$

Since this is a parallel circuit with more $I_C$ than $I_L$, the $Z$ will be capacitive reactance. Now for this reactance ($Z$ of the parallel part of the circuit) to be series-resonant with $L_2$, the reactance of $L_2$ must equal the net capacitive reactance of the parallel $LC$ combination. So at the pass frequency, $X_{L_2} =$ 1347 Ω. Therefore, $L_2$ will be

$$L_2 = \frac{1347\ \Omega}{6.28 \times 710 \times 10^3} = 0.3\ \text{mH}$$

The curve in Fig. 18-27 is for the circuit in Fig. 18-26*a* when it is using the values just calculated. This curve shows that frequencies around 710 kHz are passed by the series-resonant part of the circuit, and frequencies around 290 kHz are rejected by the parallel-resonant part of the circuit. Notice from this curve that frequencies far below the pass and reject frequencies are passed without much attenuation. This is because low frequencies have a low reactance path through $L_2$ and $L_1$. Further, notice from the curve that frequencies far above the series-resonant frequency are largely rejected by the high reactance of $L_2$.

The circuit in Fig. 18-26*b* can be used when the reject frequency is higher than the pass frequency. The procedure for determining component values is the same as that used for the circuit in Fig. 18-26*a*.

## EXAMPLE 18-21

Determine the values of $C_2$ and $C_1$ in Fig. 18-26*b* when $V_i = 10$ V, $R = 1$ kΩ, $L = 1.5$ mH, $f_{\text{reject}} =$ 290 kHz, and $f_{\text{pass}} = 185$ kHz.

### Solution

$$C_1 = \frac{1}{4 \times 3.14^2 \times 1.5 \times 10^{-3} \times (290 \times 10^3)^2} = 201\ \text{pF}$$

Next, determine $Z$ of $L$, $C_1$ at 185 kHz:

$$X_L = 6.28 \times 185 \times 10^3 \times 1.5 \times 10^{-3} = 1743\ \Omega$$

$$X_C = \frac{1}{6.28 \times 185 \times 10^3 \times 201 \times 10^{-12}} = 4282\ \Omega$$

$$Z = \frac{10\ \text{V}}{(10\ \text{V}/1743) - (10\ \text{V}/4282)} = 2940\ \Omega$$

Note that $Z$ is inductive because $I_L > I_C$.

$$X_{C_2} = 2940\ \Omega \qquad \text{for resonance}$$

so

$$C_2 = \frac{1}{6.28 \times 185 \times 10^3 \times 2940} = 293\ \text{pF}$$

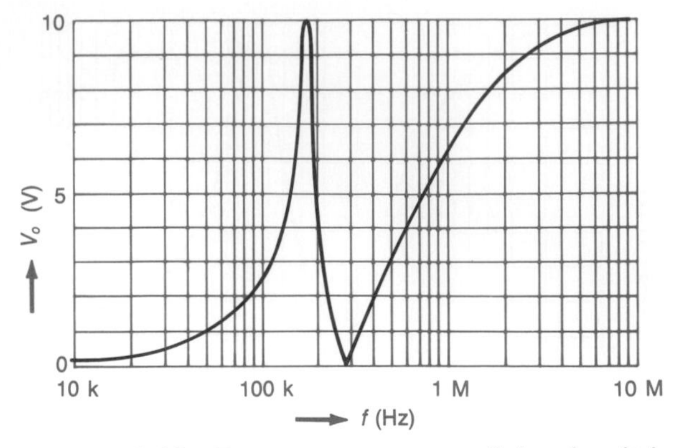


**FIGURE 18-28 Frequency response of the circuit in Fig. 18-26b.**

Figure 18-28 shows the frequency response for the circuit in Fig. 18-26*b* when the values from Example 18-21 are used. Notice that this circuit rejects frequencies far below the resonant frequencies because the series reactance of $C_2$ and $C_1$ is very high at low frequencies compared to the resistance of $R$. Of course, for frequencies far above resonance, the reactance of these two capacitors is very small compared to $R$, so these frequencies, as shown in Fig. 18-28, pass through to the output.

## Self-Test

**50.** A series $RCL$ circuit makes a ______ filter.

**51.** A band-reject filter is a ______ $LC$ combination which stops a band of frequencies.

**52.** Estimate $f_{up}$ and $f_{lo}$ for a bandpass filter when $R = 1\ \text{k}\Omega$, $C = 0.05\ \mu\text{F}$, and $L = 250$ mH.

**53.** A double resonant circuit with a capacitor in series with a parallel $LC$ will ______ frequencies far below the resonant frequencies.

**54.** For the circuit in Fig. 18-26*a*, determine $L_2$ and $f_{reject}$ when $L_1 = 2$ mH, $C = 500$ pF, and $f_{pass} = 190$ kHz.

## SUMMARY

1. At series resonance $X_C$ equals $X_L$.
2. Increasing either $L$ or $C$ decreases $f_r$.
3. Frequency $f_r$ is inversely proportional to the square root of $L$ and $C$.
4. For a given $L$ and $C$, there is but one value of $f_r$.
5. For a given $f_r$, there are an infinite number of $LC$ combinations.
6. The $Q$ (quality) of an $LC$ circuit (either series or parallel) is $Q = X_L/R$ where $R$ is the total resistance in series with the ideal inductor.
7. In a series-resonant circuit, $I_T = V_T/R$, $V_T = V_R$, $V_L = V_C$, and $V_L$ and $V_C$ can be $\gg V_T$.
8. Angle $\theta$ is zero for all resonant circuits.
9. Series-resonant voltage rise is equal to $Q$.
10. Bandwidth (BW) is inversely proportional to $Q$.
11. The smaller the BW, the more selective the circuit.
12. Increasing the $L/C$ ratio decreases the BW if $R$ does not increase proportionately.
13. The half-power points on a frequency response curve mark the limits of the BW. They occur at 0.707 of the maximum height of the curve. They represent the −3-dB frequencies.
14. A parallel $LC$ circuit is called a tank circuit and the reactive current is the tank current.

**15.** A parallel *LC* circuit is capacitive when operated above its resonant frequency.

**16.** A series *LC* circuit is inductive when operated above its resonant frequency.

**17.** In a practical *LC* or *RCL* circuit, $I_L$ is slightly greater than $I_C$ at resonance.

**18.** A damping resistor is used to increase the BW of a parallel *LC* circuit.

**19.** When $I_L = I_C$ in a practical parallel *LC* circuit, the circuit is capacitive.

**20.** At resonance $I_L > I_C$ in a practical parallel *LC* circuit.

**21.** Inductor $Q$ and circuit $Q$ are not the same in either parallel or series *RCL* circuits.

**22.** Circuits with a gain of less than one attenuate a signal.

**23.** The decibel (dB) is equal to 10 times the log of the ratio of two powers or 20 times the log of the ratio of two currents or two voltages.

**24.** A negative decibel indicates attenuation, while a positive decibel indicates a gain > 1.

**25.** Filters can improve the signal-to-noise ratio.

**26.** Filters are frequency-selective.

**27.** Either an *RL* or *RC* circuit can be used as either a high-pass or a low-pass filter, depending on where $V_o$ is taken.

**28.** Reactance $X = R$ when the output of a filter is −3 dB.

**29.** The slope of an *RL* or *RC* filter is 6 dB per octave or 20 dB per decade.

**30.** The −3-dB point on the frequency response curve of a high-pass or low-pass filter is also known as the half-power point, the break frequency, or the cutoff frequency.

**31.** Parallel *LC* filter elements produce band-reject filters, while series *LC* filter elements produce bandpass filters.

**32.** Band-reject and bandpass filters have a parallel *LC* element or a series *LC* element, respectively, in series with a resistor.

**33.** Double resonant filters have both a band-reject and a bandstop characteristic.

**34.** Useful formulas are

$$f_r = \frac{1}{2\pi\sqrt{LC}}$$ Except for parallel *LC* when the inductor $Q < 10$

$$Q = \frac{X_L}{R}$$ All *LC* circuits and series *RCL* circuits

$$Q = \frac{R_P}{X_L}$$ For parallel *RCL* circuits

$$Q = \frac{1}{(X_L/R_P) + (R_L/X_L)}$$ For practical parallel *RCL* circuits

$$V_L = V_C = QV_T$$ For series resonance

$$\text{BW} = \frac{f_r}{Q}$$

$$f_{\text{up}} \approx f_r + 0.5\text{BW}$$

$$f_{\text{lo}} \approx f_r - 0.5\text{BW}$$

$I_C = QI_T$ For parallel circuits

$$dB = 10 \log \frac{P_o}{P_i}$$

$$dB = 20 \log \frac{V_o}{V_i}$$

$$dB = 20 \log \frac{I_o}{I_i}$$

## CHAPTER REVIEW QUESTIONS

For the following items, determine whether each statement is true or false.

**18-1.** In a series-resonant circuit, $X_L = X_C$.

**18-2.** The resonant frequency of a series $LC$ circuit is inversely proportional to both $L$ and $C$.

**18-3.** A specific $f_r$ can be achieved by many different $LC$ combinations.

**18-4.** The quality of an $LC$ circuit is essentially determined by the quality of the inductor.

**18-5.** For a series-resonant circuit, $V_T = V_R$.

**18-6.** At resonance, angle $\theta$ is zero for all combinations of $RCL$ and $LC$ circuits.

**18-7.** $V_L$ can be greater than $V_T$ only when a circuit is series-resonant.

**18-8.** The tank current is greater than the line current in a parallel-resonant circuit.

**18-9.** Angle $\theta$ is zero at resonance only for ideal $LC$ or $RCL$ circuits.

**18-10.** In a practical parallel-resonant circuit, $I_C$ is greater than $I_L$.

**18-11.** Increasing the $L/C$ ratio of a circuit is a sure way of reducing its BW.

**18-12.** When the voltage output of a circuit drops to 70.7 percent of its maximum value, the output is −3 dB.

**18-13.** A series-resonant circuit is called a tank circuit.

**18-14.** A given value of $L$ and $C$ can be resonant at two frequencies.

**18-15.** An ideal parallel $LC$ circuit has more capacitive current than inductive current when operating above its resonant frequency.

**18-16.** In a parallel $RCL$ circuit, inductor $Q$ and circuit $Q$ are equal.

**18-17.** Attenuators have a gain of more than one.

**18-18.** The output of an attenuator circuit can never be a positive decibel output.

**18-19.** The only type of filter that can improve the signal-to-noise ratio is the low-pass filter.

**18-20.** With a bandpass filter, the output is usually taken across the $LC$ combination.

**18-21.** When the output of a low-pass filter is −3 dB, the reactance is twice as great as the resistance.

**18-22.** The break frequency and the cutoff frequency are both equal to the −3-dB frequency.

For the following items, fill in the blank with the word or phrase required to correctly complete each statement.

**18-23.** Decreasing the inductance of an *LC* circuit will cause the resonant frequency to ______.

**18-24.** $I_T = V_T/R$ for a ______ resonant circuit.

**18-25.** The numerical value of the series-resonant voltage rise is equal to ______.

**18-26.** In a practical parallel-resonant *LC* circuit, the formula for quality is ______.

**18-27.** Bandwidth BW is ______ proportional to $Q$.

**18-28.** The selectivity of a circuit is specified by its ______.

**18-29.** If $R_L$ in a series *LC* circuit is doubled, the BW of the circuit will be ______.

**18-30.** The −3-dB points on a frequency response curve coincide with the ______ points.

**18-31.** A series *LC* circuit is ______ when operating below its resonant frequency.

**18-32.** The resistor used to decrease the selectivity of a parallel *LC* circuit is called a ______ resistor.

**18-33.** The decibel is based on the ______ of a power ratio.

**18-34.** With a low-pass *RC* filter, the output is taken across the ______.

**18-35.** Band-reject filters use a ______ *LC* combination.

**18-36.** The slope of a high-pass filter is about ______ per octave or ______ per decade.

**18-37.** At a half-power point, the output of a filter will be ______.

**18-38.** A filter with a capacitor in series with a parallel *LC* element is a ______ filter.

**18-39.** The filter in Question 18-38 will ______ frequencies well below its resonant frequencies.

**18-40.** A band-reject filter is also known as a ______ filter.

## CHAPTER REVIEW PROBLEMS

**18-1.** What is the BW of a circuit in which the half-power points occur at 150 and 180 kHz?

**18-2.** Determine the resonant frequency of a 2-H inductor connected in series with a 1-$\mu$F capacitor.

**18-3.** Determine the inductance required to series-resonate with a 180-pF capacitor at 320 kHz. Also determine the BW and the $Q$ if $R_L = 20\ \Omega$.

**18-4.** A series-resonant circuit has a $Q$ of 50 and source voltage of 5 V. What is the value of $V_C$?

**18-5.** Estimate $f_{\text{up}}$ for a circuit with $Q = 40$ and $f_r = 200$ kHz.

**18-6.** Determine $I_C$ for a parallel-resonant circuit in which $Q = 30$, $R_P = 15\ \text{k}\Omega$, and $V_T = 15$ V.

**18-7.** Determine the $Q$ of a resonant circuit in which $X_L = 10\ \text{k}\Omega$, $R_P = 400\ \text{k}\Omega$, and $R_L = 40\ \Omega$.

**18-8.** What is the decibel attenuation when $P_i = 50$ W and $P_o = 30$ W?

**18-9.** A circuit has a gain of 10 dB. What is $V_o$ if $V_i = 0.1$ V and $R_i = R_o$?

**18-10.** Determine the −3-dB frequency of a low-pass $RL$ filter when $R = 10\ \text{k}\Omega$ and $L = 100$ mH.

**18-11.** Determine the series inductance needed for a bandpass frequency of 600 kHz when the parallel $L$ and $C$ are 300 $\mu$H and 500 pF, respectively.

## ANSWERS TO SELF-TESTS

**1.** They are equal.

**2.** false

**3.** false

**4.** true

**5.** 112.6 Hz

**6.** 25.3 mH

**7.** the total series resistance

**8.** They are equal.

**9.** zero

**10.** zero

**11.** $V_C$ doubles, because doubling $Q$ while holding $L$ and $f$ (and thus $X_L$) constant halves $R$ which doubles $I_C$ and $V_C$

**12.** 50, 50

**13.** when the total series resistance is greater than $X_C$ at resonance

**14.** $f_r = 7.12$ kHz, $I_T = 0.25$ A, and $V_C = 111.8$ V

**15.** the ability of an $LC$ circuit to select a narrow band of frequencies and reject all others

**16.** bandwidth

**17.** **a.** $Q$ is inversely proportional to $R$.
**b.** BW is inversely proportional to $Q$.
**c.** BW is directly proportional to $R$.
**d.** Resonant voltage rise is directly proportional to $Q$.
**e.** BW is inversely proportional to the $L/C$ ratio when $R$ is held constant.

**18.** Bandwidth is the range of frequencies within which the circuit current will be at least 70.7 percent of maximum.

**19.** Half-power points are the points of the $f$ response curve where the power in the circuit is 50 percent of maximum. These points occur at $f_{up}$ and $f_{lo}$.

**20.** 1.5 kHz

**21.** $f_r = 649.75$ kHz, BW = 10.83 kHz, $f_{up} = 655.16$ kHz, and $f_{lo} = 644.34$ kHz

**22.** tank

**23.** zero

**24.** $Q$ is directly proportional to $R_P$; BW is inversely proportional to $R_P$.

**25.** The difference between the two frequencies at which $Z$ is 70.7 percent of $Z$ at resonance.

**26.** capacitive

**27.** 30

**28.** 120 mA

**29.** capacitive

**30.** larger

**31.** when the inductor $Q$ in a parallel $LC$ or $RCL$ circuit is $< 10$

**32.** the $R_P$ added to a parallel-resonant $LC$ circuit to broaden its BW by lowering its $Q$

**33.** $Q_{circuit} = Q_{inductor} = 37.7$

**34.** 8.3

**35.** less than one

**36.** −0.97 dB

**37.** 5.62 V

**38.** 21.76 dB

**39.** $R_i = R_o$

**40.** frequency-selective

**41.** low-pass

**42.** capacitor

**43.** 8.14 V

**44.** They are equal.

**45.** 7 kHz, 70 kHz, 7.07 V

**46.** 20 dB per decade or 6 dB per octave

**47.** break frequency, cutoff frequency, and half-power point

**48.** When the noise frequencies are either much higher or much lower than the desired frequency, a low-pass filter or a high-pass filter, respectively, will attenuate the noise frequency more than the signal frequency.

**49.** the series *RC* filter

**50.** bandpass

**51.** parallel

**52.** $f_{\text{up}} \approx 1743$ Hz and $f_{\text{lo}} \approx 1106$ Hz

**53.** reject or stop

**54.** $L_2 = 4.7$ mH and $f_{\text{reject}} = 159.2$ kHz

# Chapter 19 AC Network Analysis

In Chaps. 12 and 17 you learned how to graphically and mathematically add and subtract right-angle phasors. These techniques are very useful for combining reactance and resistance to obtain the impedance of a series circuit. They are also useful for combining resistive and reactive currents or voltages. However, in series-parallel circuits composed of resistors, capacitors, and/or inductors, the phasors that must be manipulated are no longer 90° out of phase with each other. Also, these phasors must be multiplied and divided as well as added and subtracted.

An area of mathematics called *vector algebra* allows one to add, subtract, multiply, and divide vectors separated by any number of degrees. Since phasors are vectors rotating at a specified angular velocity, this type of algebra can also be applied to phasors. Thus, we can use vector algebra to solve more complex impedance circuits than those dealt with in Chaps. 12 and 17.

You will notice the techniques presented in this chapter are an extension of the techniques used in solving simple series and/or parallel circuits in which the inductors and capacitors were assumed to be ideal components. If these techniques are not fresh in your memory, you should review them in Chap. 17 at this time.

## 19-1 THE *j* OPERATOR

In mathematics a symbol used to represent a mathematical procedure is called an *operator*. The × symbol indicates the multiply operation and the + operator means that addition is to be performed.

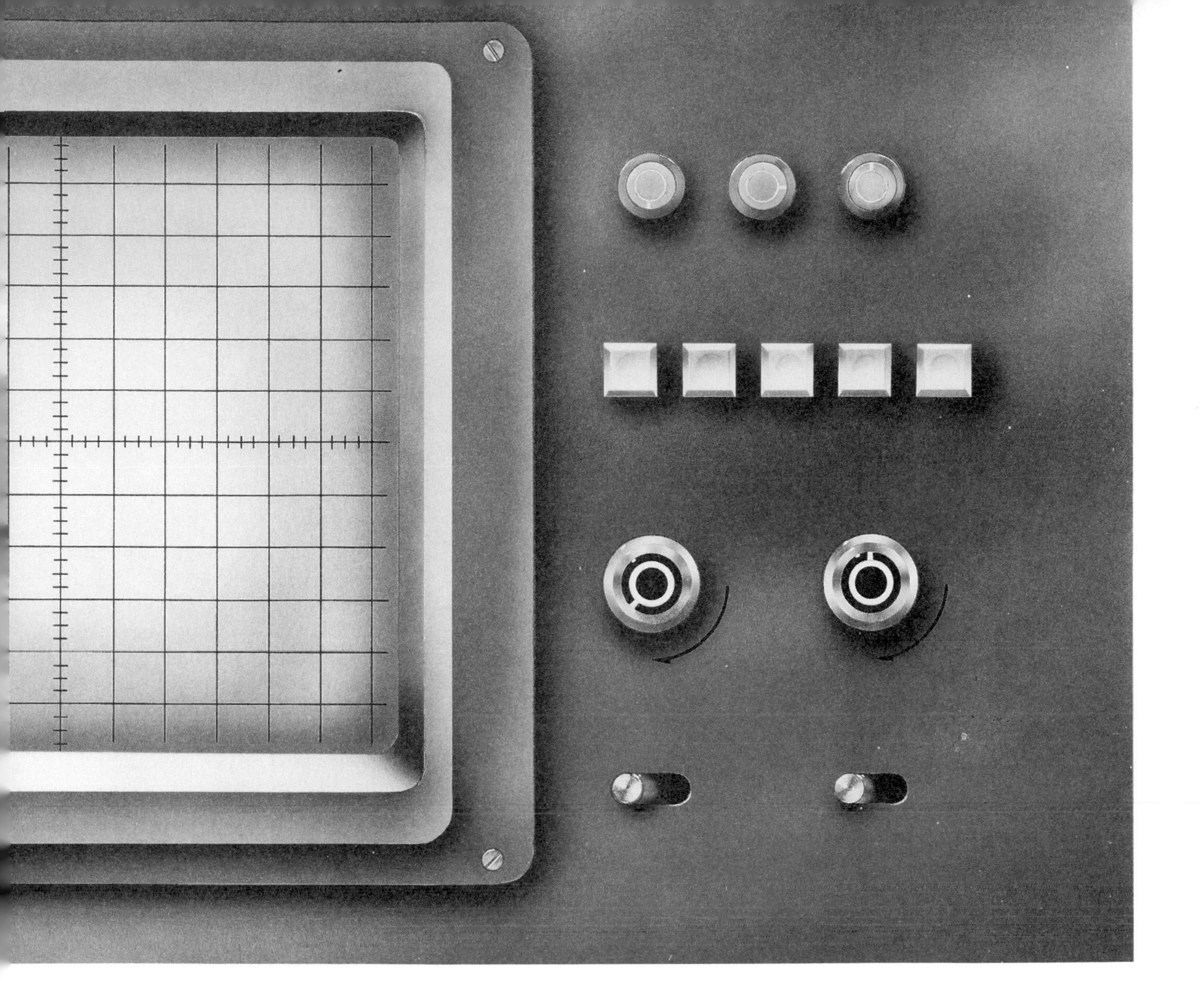

The *j operator* means that a vector or phasor is to be rotated 90° in a counterclockwise direction. The results of a phasor being subjected to the $j$ operator are illustrated in Fig. 19-1. The first $j$ operation rotates the phasor $P$ in Fig. 19-1*a* 90° to the $+y$ axis of the graph. This shifted phasor is labeled $jP$, where $P$ stands for phasor, in Fig. 19-1*b*. The second application of the $j$ operator, shown in Fig. 19-1*c*, yields $jjP = j^2P$. This phasor $j^2P$ is rotated 180° from the $+x$ axis; i.e., it is on the $-x$ axis. Because the $j^2P$ phasor is on the $-x$ axis, it must be equal to $-P$. Using this relationship, we can show that $j$ must be equal to $\sqrt{-1}$:

$$j^2P = -P$$
$$j^2 = -1$$
$$j = \sqrt{-1}$$

Since $j = \sqrt{-1}$, then $j^2 = \sqrt{-1}\sqrt{-1} = -1$. Therefore, the $j^2P$ phasor reduces to $-P$ as shown in Fig. 19-1*c*. The third rotation of the phasor $P$ by the $j$ operator produces $jjjP = j^2jP = -jP$. As shown in Fig. 19-1*c*, the $-j$ operator can be thought of as rotating the phasor 270° counterclockwise or 90° clockwise. The fourth rotation of the phasor by the $j$ operator yields $jjjjP = j^2j^2P = (-1)\ (-1)P = P$.

The $\sqrt{-1}$ is called an *imaginary number* because there is no number which will yield $-1$ when squared. In mathematics, the symbol used to identify an imaginary number is $i$. However, in electricity-electronics, the $j$ symbol is used because $i$ is used for instantaneous current. Thus a $j$ number or a $j$ phasor is called an imaginary number. Notice in Fig. 19-1 that the $+j$ is on the $+y$ axis and the $-j$ is on the $-y$ axis. Thus, when working with pha-

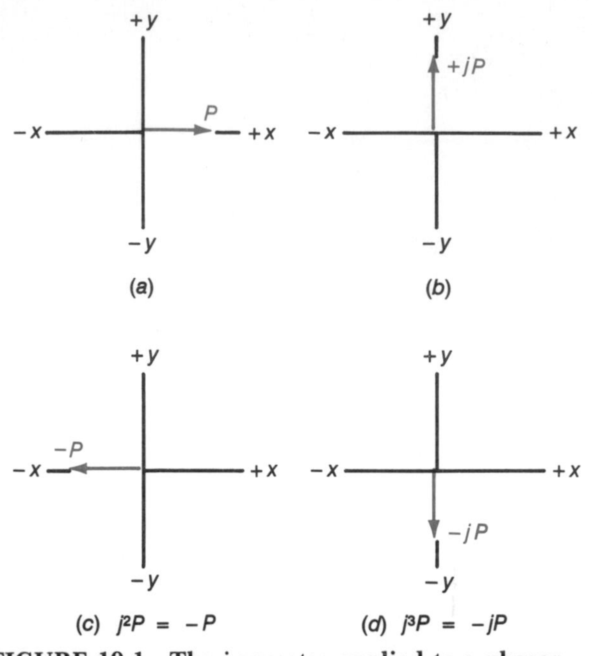


**FIGURE 19-1 The *j* operator applied to a phasor.**

sors, the quantities on the *y* axis are either − or + imaginary numbers. The magnitude of a phasor quantity on this axis is preceded by either a $+j$ or a $-j$. As with other positive numbers and variables, the $+j$ is usually written as $j$ if the + is not needed for clarity. For example, the two phasor quantities shown in Fig. 19-2 would be expressed as $j5.2$ and $-j3.6$.

Phasors on the *x* axis of the graph represent real numbers; i.e., numbers which do not involve the square root (or any even-numbered root) of a negative number. The phasors in Fig. 19-3 are on the real axis. The quantities they represent are +4.2 and −2.6.

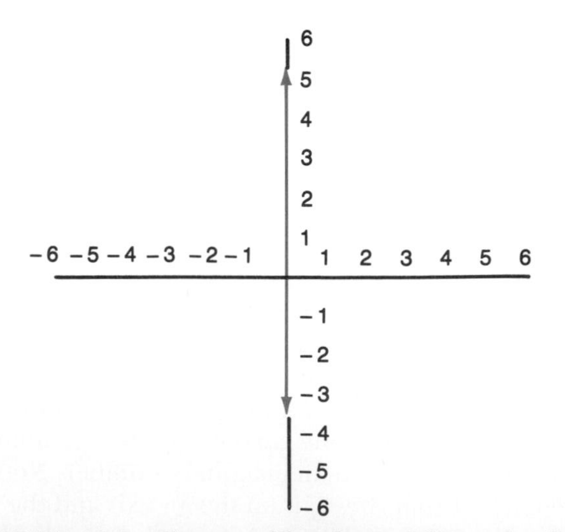


**FIGURE 19-2 Phasors with values of $j5.2$ and $-j3.6$.**

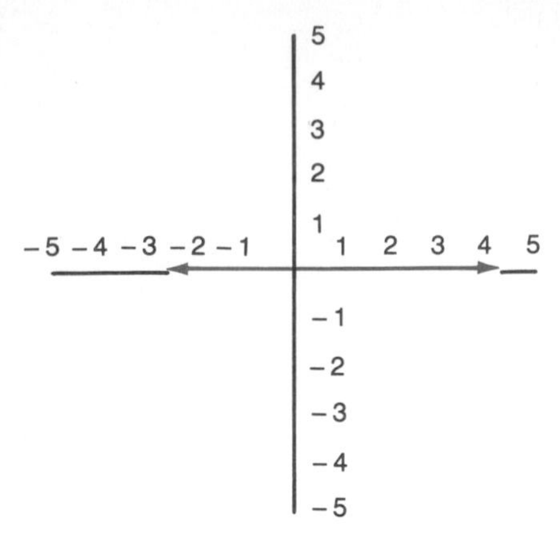


**FIGURE 19-3 Phasors representing real numbers.**

Numbers that contain both real numbers and imaginary numbers are called *complex numbers*. For example, $4 + j3$ is a complex number. The 4 is the real part or component of the complex number and the $j3$ is the imaginary component of the number. This complex number ($4 + j3$) is represented by the phasor in Fig. 19-4*a*. The complex number $-3 - j4$ is diagramed in Fig. 19-4*b*.

## 19-2 RECTANGULAR AND POLAR NOTATION

A phasor can be expressed in either rectangular form or polar form. The *rectangular form* of a phasor specifies the real (*x* axis) component and the imaginary (*y* axis) component of the phasor. In other words, a phasor expressed by its rectangular (*x* and *y*) coordinates is in rectangular form. Of course, the rectangular form of a phasor is a complex number. The phasors in Fig. 19-4 are specified in rectangular form. Notice that neither the magnitude nor the phase angle is known for a phasor in the rectangular form; however, as demonstrated in Sec. 19-3 either can be easily determined using trigonometric functions.

The *polar form* of a phasor is exactly opposite from the rectangular form. That is, in the polar form both the magnitude and the phase angle (with reference to the $+x$ axis) are specified, but neither the *x*-axis component nor the *y*-axis component is known. A phasor in polar form is illustrated in Fig. 19-5. The notation for this phasor is $12\,\underline{/43^\circ}$. With this notation, the 12 specifies the magnitude, the $\underline{/\quad}$ is the symbol for angle, and

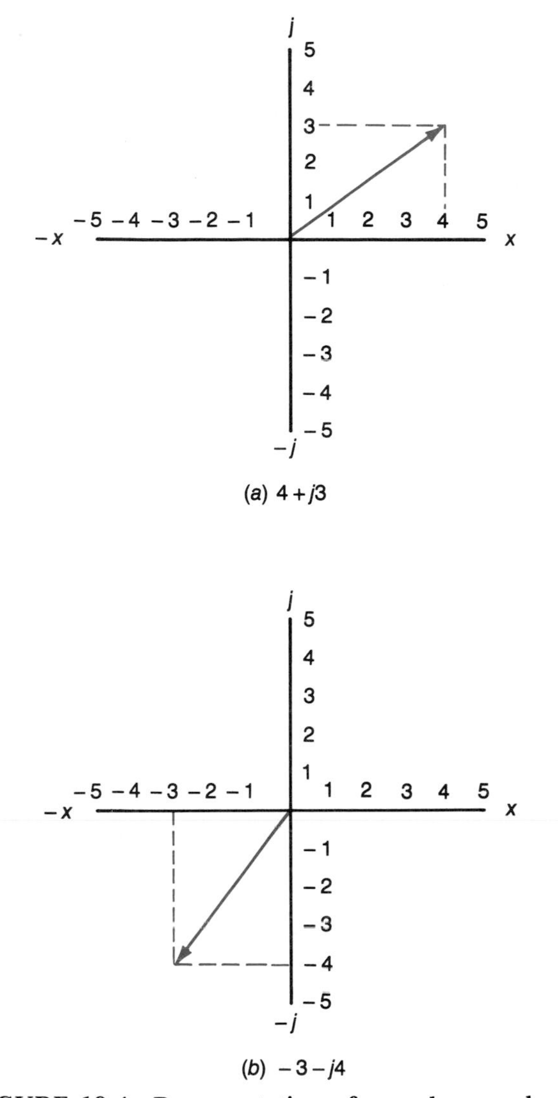


**FIGURE 19-4 Representation of complex numbers.**

the 43° specifies the angle. Since the angle is positive, it means the angle is counterclockwise from the $+x$ axis. Had the 43 been −43, then the phasor in Fig. 19-5 would have been rotated 43° clockwise from the $+x$ axis.

**EXAMPLE 19-1**

Using the appropriate form, specify the phasors shown in Fig. 19-6.

**Solution**

For Fig. 19-6*a*, the phasor is $10 - j15$.

For Fig. 19-6*b*, the phasor is $20\,\underline{/-70^\circ}$ or $\underline{/290^\circ}$.

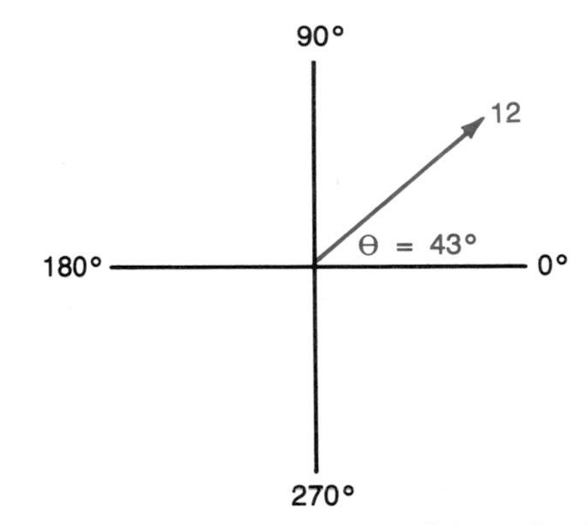


**FIGURE 19-5 A phasor expressed in polar form.**

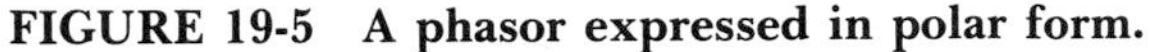

If the phasor in Fig. 19-6*a* were a voltage phasor, it could represent the source voltage for a series *RC* circuit with $V_C = 15$ V and $V_R = 10$ V. If the phasor in Fig. 19-6*b* were a current phasor, it could represent the source current in a parallel circuit with $I_L$ being greater than $I_R$.

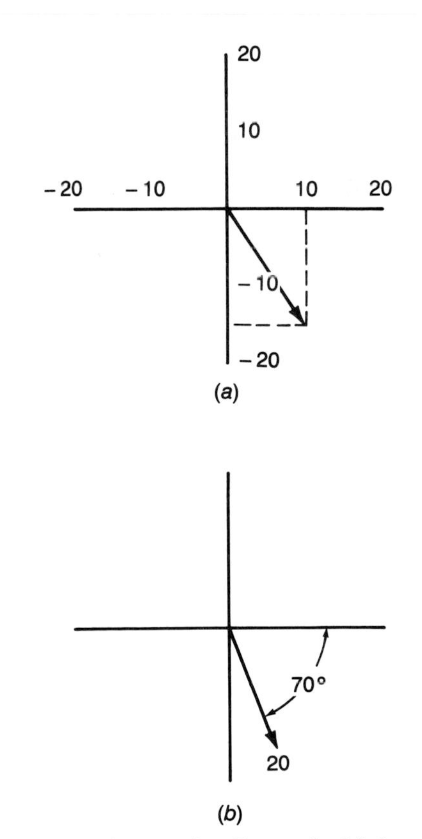

**FIGURE 19-6 Phasors for Example 19-1.**

## 19-3 RECTANGULAR-TO-POLAR CONVERSION

To convert the rectangular form $(10 - j15)$ phasor in Fig. 19-6*a* to its equivalent polar form, one must find the magnitude and phase angle of the phasor. The magnitude can be calculated by using either the pythagorean theorem or trigonometric functions. The phase angle $(\theta)$ is found by trigonometric functions regardless of how the magnitude is found. Assume the phasor in Fig. 19-6*a* is a voltage phasor $V_1$. Its phase angle $\theta$ will be

$$\theta = \arctan\frac{\text{opposite}}{\text{adjacent}} = \arctan\frac{-15\text{ V}}{10\text{ V}} = -56.3°$$

Using the pythagorean theorem, the magnitude is found to be

$$V_1 = \sqrt{10^2 + (-15)^2} = 18\text{ V}$$

Thus, phasor $V_1$ in polar form is $18\text{ V}\,\underline{/-56.3°}$.

Using a trigonometric function, the magnitude is

$$\cos\theta = \frac{\text{adjacent}}{\text{hypotenuse}} = \frac{10\text{ V}}{V_1}$$

Therefore,

$$V_1 = \frac{10\text{ V}}{\cos(-56.3°)} = 18\text{ V}$$

Again, the polar form of $V_1$ is found to be $18\text{ V}\,\underline{/-56.3°}$.

The magnitude of $V_1$ could just as well have been calculated using the sine function. The formula would be

$$V_1 = \frac{-15\text{ V}}{\sin(-56.3°)}$$

---

**EXAMPLE 19-2**

Determine the polar form of the phasor in Fig. 19-4*b*. Name the phasor $I_1$.

**Solution**

Let $\theta_{-x}$ represent the angle between $I_1$ and the $-x$ axis. Then

$$\theta_{-x} = \arctan\frac{-4}{-3} = 53.1°$$
$$\theta = \theta_{-x} + 180° = 233.1°$$
$$I_1 = \frac{4}{\sin 53.1°} = 5$$
$$I_1 = 5\,\underline{/233.1°} \quad \text{or} \quad 5\,\underline{/-126.9°}$$

---

If a phasor diagram of the rectangular phasor being converted is not provided, it is a good idea to draw a free-hand diagram. As you can see from Example 19-2, the phasor diagram made it easy to visualize how to calculate angle $\theta$.

## 19-4 POLAR-TO-RECTANGULAR CONVERSION

Converting a phasor from polar form to a rectangular form involves computing the real ($x$ axis) and imaginary ($y$ axis) components of the polar phasor. If the angle of the polar phasor is $>180°$ but $<360°$, the $j$ (imaginary) component will be negative. The real component will be negative if the angle is $>90°$ but $<270°$. At exactly 180 and 360° there is no $y$ component, and at exactly 90 and 270° there is no $x$ component.

The phasor in Fig. 19-6*b* is in the polar form of $20\,\underline{/-70°}$. If this phasor represents a voltage $V_1$ in volts, then we can express the voltage as $V_1 = 20\text{ V}\,\underline{/-70°}$. To convert this voltage phasor to rectangular form we can calculate the real and imaginary components using the cosine function and the sine function, respectively:

$$V_{\text{real}} = V_1 \cos\theta = 20\text{ V} \times 0.342 = 6.84\text{ V}$$
$$-j = V_1 \sin\theta = 20\text{ V} \times 0.940 = 18.8\text{ V}$$

In rectangular form then, $V_1 = 6.84\text{ V} - j18.8\text{ V}$.

---

**EXAMPLE 19-3**

Voltage $V_2$ is a phasor equal to $15\text{ V}\,\underline{/-40°}$. Convert $V_2$ to its rectangular form.

**Solution**

$$\begin{aligned} V_2 &= V_2\cos\theta - jV_2\sin\theta \\ &= 15\text{ V}\cos 40° - j15\text{ V}\sin 40° \\ &= 11.49\text{ V} - \text{j}9.64\text{ V} \end{aligned}$$

---

The solution shows that $15\text{ V}\,\underline{/-40°} = 11.49\text{ V} - j9.64\text{ V}$.

### Self-Test

1. The ______ indicates that a vector is rotated 90° in a counterclockwise direction.
2. A number preceded by the $j$ symbol represents an ______ number.
3. A complex number has an ______ component and a ______ component.

**4.** Which form of a phasor is expressed as a complex number?

**5.** Convert the phasor in Fig. 19-4*a* to its polar form.

**6.** Convert the phasor in Fig. 19-5 to its rectangular form.

**7.** Convert $-8 + j14$ to its polar form.

**8.** Convert $64 \angle -60°$ to its rectangular form.

## 19-5 PHASOR ALGEBRA

The easiest way to add or subtract phasors is to first convert them to rectangular form if they are in polar form. Once the phasors are in rectangular form, the real $(x)$ components and the imaginary $(y)$ components can be algebraically added or subtracted. The answer (the sum or difference) can be left in rectangular form or it can be converted back to polar form if desired. Drawing a free-hand sketch of the phasors usually aids in this conversion process. Also, the sketch, if it is reasonably accurate, may point out gross errors in calculations and/or procedures.

### Addition of Phasors

The process of adding two phasors is illustrated by Examples 19-4 and 19-5.

**EXAMPLE 19-4**

Voltage $V_1 = -10\ \text{V} + j50\ \text{V}$ and $V_2 = 30\ \text{V} + j20\ \text{V}$. Add these two voltage phasors and express the result in polar form.

**Solution**

First sketch the phasors as in Fig. 19-7. The sum phasor is found by constructing a parallelogram using the two given phasors.

Next, algebraically add $x$ and $y$ components:

$$\begin{aligned} V_1 &= -10\ \text{V} + j50\ \text{V} \\ V_2 &= \phantom{-}30\ \text{V} + j20\ \text{V} \\ \hline V_3 &= \phantom{-}20\ \text{V} + j70\ \text{V} \end{aligned}$$

Finally, convert $V_3$ to polar form:

$$\theta = \arctan \frac{70\ \text{V}}{20\ \text{V}} = 74°$$

$$V_3 = \frac{20\ \text{V}}{\cos 74°} = 72.8\ \text{V} \angle 74°$$

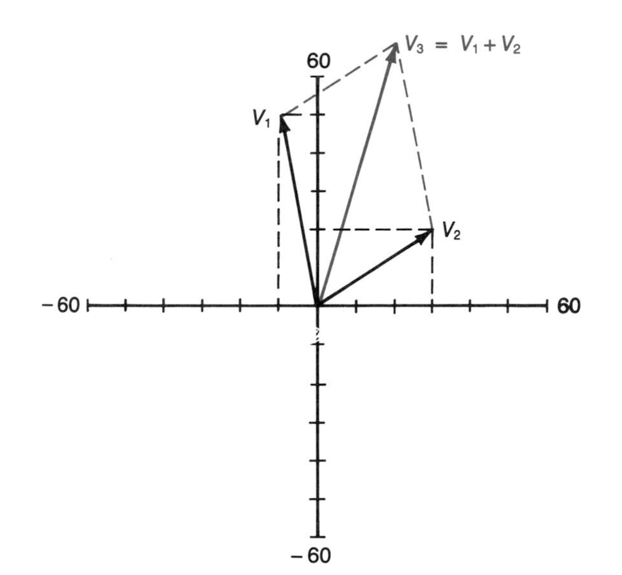


**FIGURE 19-7 Phasors for Example 19-4.**

The calculated value of $V_3$ (the sum of $V_1$ and $V_2$) appears to be in agreement with the sketch in Fig. 19-7.

**EXAMPLE 19-5**

Add $V_1 = 70\ \text{V} \angle 45°$ to $V_2 = 67\ \text{V} \angle -63°$ and record the sum $(V_3)$ in rectangular form.

**Solution**

Sketch the given phasors and the resultant phasor as in Fig. 19-8.

$$\begin{aligned} V_1 &= 70\ \text{V} \cos 45° + j70\ \text{V} \sin 45° = 49.5\ \text{V} + j49.5\ \text{V} \\ V_2 &= 67\ \text{V} \cos 63° - j67\ \text{V} \sin 63° = 30.4\ \text{V} - j59.7\ \text{V} \\ \hline V_3 &= \qquad\qquad\qquad\qquad\qquad\qquad\quad 79.9\ \text{V} - j10.2\ \text{V} \end{aligned}$$

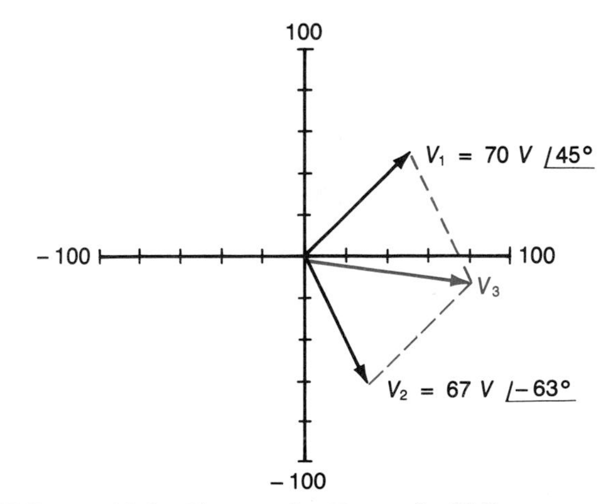

**FIGURE 19-8 Phasors for Example 19-5.**

When more than two phasors must be added, the addition process for adding two phasors is repeated as many times as necessary. That is, phasor 1 is added to phasor 2, the resultant phasor is then added to phasor 3, etc.

### Subtraction of Phasors

Subtracting two phasors is mathematically as easy as adding two phasors. However, sketching a diagram which shows the resultant phasor is a slightly different process. Figure 19-9 shows the graphical solution to subtracting $I_1$ from $I_2$. Note that $I_1$ must be rotated 180° (flipped over) and then *added* to $I_2$. Rotating $I_1$ 180° and adding it to $I_2$ is the graphical equivalent of mathematically subtracting by changing the signs of the subtrahend and then *adding* the subtrahend to the minuend.

**EXAMPLE 19-6**

Subtract $I_1 = 3.6\text{ A}\,\angle{-56.3°}$ from $I_2 = 5.83\text{ A}\,\angle 31°$. Record the difference in polar form.

**Solution**

Convert to rectangular form:

$$\begin{aligned} I_1 &= 3.6\text{ A}\cos -56.3° - j3.6\text{ A}\sin -56.3° \\ &= 2.00\text{ A} - j3.00\text{ A} \\ I_2 &= 5.83\text{ A}\cos 31° + j5.83\text{ A}\sin 31° \\ &= 5.00\text{ A} + j3.00\text{ A} \end{aligned}$$

Sketch the phasors as in Fig. 19-9.

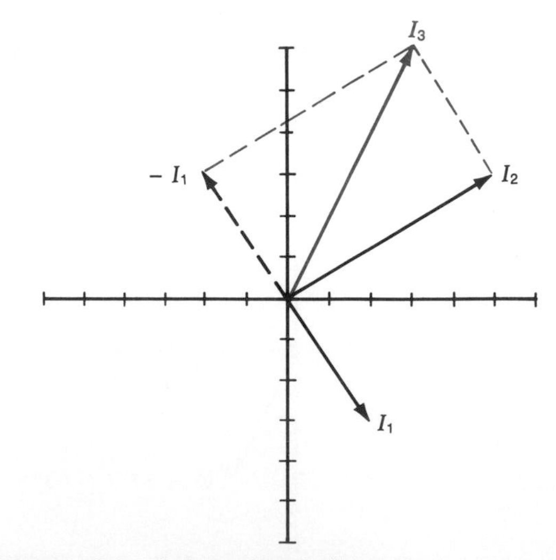


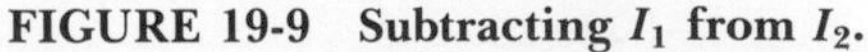
**FIGURE 19-9 Subtracting $I_1$ from $I_2$.**

Subtract $I_1$ from $I_2$ by changing the signs of $I_1$ and adding:

$$\begin{aligned} I_2 &= \phantom{-}5.00\text{ A} + j3.00\text{ A} \\ -I_1 &= -2.00\text{ A} + j3.00\text{ A} \\ \hline I_3 &= \phantom{-}3.00\text{ A} + j6.00\text{ A} \end{aligned}$$

Finally, convert $I_3$ to its polar form:

$$\theta = \arctan\frac{6\text{ A}}{3\text{ A}} = 63.4°$$

$$I_3 = \frac{3\text{ A}}{\cos 63.4°} = 6.7\text{ A}\,\angle 63.4°$$

The calculated value of $6.7\text{ A}\,\angle 63.4°$ for $I_3$ appears to agree with the phasor diagram in Fig. 19-9.

### Multiplication of Phasors

Phasor multiplication is most easily performed when the phasors are in polar form. To multiply polar phasors, just multiply the magnitudes and *algebraically add* the phase angles.

**EXAMPLE 19-7**

Multiply $16\,\angle{-30°}$ by $7 + j2$.

**Solution**

Convert $7 + j2$ to polar form:

$$\theta = \arctan\frac{2}{7} = 15.95°$$

$$7 + j2 = \frac{7}{\cos 15.95°} = 7.28\,\angle 15.95°$$

Multiply the magnitudes of $16\,\angle{-30°}$ and $7.28\,\angle 15.95°$:

$$16 \times 7.28 = 116.48$$

Algebraically add the angles:

$$-30° + 15.95° = -14.05°$$

The product is $116.48\,\angle{-14.05°}$.

**EXAMPLE 19-8**

Determine the voltage across an inductive reactance of $5\text{ k}\Omega\,\angle 90°$ if the current is $4\text{ mA}\,\angle 0°$.

**Solution**

$$V_L = I_L X_L = 4\text{ mA}\,\angle 0° \times 5\text{ k}\Omega\,\angle 90° = 20\text{ V}\,\angle 90°$$

## Division of Phasors

To divide phasors, the magnitudes are divided and the angles are algebraically subtracted. Of course, rectangular phasors must first be converted to polar form.

### EXAMPLE 19-9

Divide $30 - j10$ by $10.54 \underline{/60°}$.

**Solution**

Express $30 - j10$ in polar form:

$$\theta = \arctan\frac{-10}{30} = -18.43°$$

$$30 - j10 = \frac{30}{\cos(-18.43°)} = 31.62 \underline{/-18.43°}$$

Divide the magnitudes:

$$\frac{31.62}{10.54} = 3$$

Algebraically subtract the angles:

$$-18.43 - (+60°) = -78.43°$$

The quotient of $30 - j10$ divided by $10.54 \underline{/60°}$ is

$$3 \underline{/-78.43°}$$

### EXAMPLE 19-10

Find the impedance of a circuit in which $I = 2\text{ A} + j3\text{ A}$ and $V = 10\text{ V} - j6\text{ V}$.

**Solution**

$$\theta = \arctan\frac{3}{2} = 56.3°$$

$$I = \frac{2}{\cos 56.3°} = 3.6\text{ A} \underline{/56.3°}$$

$$\theta = \arctan\frac{-6}{10} = -30.96°$$

$$V = \frac{10}{\cos(-30.96°)} = 11.66\text{ V} \underline{/-30.96°}$$

$$Z = \frac{11.66\text{ V} \underline{/-30.96°}}{3.6\text{ A} \underline{/56.3°}} = 3.24\ \Omega \underline{/-87.26°}$$

## Self-Test

**9.** The rectangular form of phasors is used for ______ and ______ phasors.

**10.** The polar form of phasors is used for ______ and ______ phasors.

**11.** Angles are algebraically ______ when multiplying phasors.

**12.** Angles are algebraically ______ when dividing phasors.

**13.** Add $-4\text{ A} + j5\text{ A}$ to $11\text{ A} \underline{/240°}$.

**14.** Subtract $9\text{ V} \underline{/45°}$ from $8\text{ V} - j7\text{ V}$.

**15.** Multiply $3\text{ A} \underline{/60°}$ by $4\ \Omega \underline{/-30°}$.

**16.** Divide $3\text{ A} + j4\text{ A}$ into $10\text{ V} \underline{/-40°}$.

## 19-6 CONDUCTANCE, SUSCEPTANCE, AND ADMITTANCE

In Chap. 17 you learned how to determine the absolute value of impedance $Z$. That is, the magnitude of $Z$ was calculated, but the angle of $Z$ was not specified. Now that you are familiar with vector algebra, you can specify resistance, reactance, and impedance as factors showing both magnitude and angle. This allows you to solve more complex series-parallel impedance problems. Also, vector algebra allows you to easily solve parallel impedance problems by changing resistance $R$ to conductance $G$, reactance $X$ to *susceptance* $B$, and impedance $Z$ to *admittance* $Y$.

In Chap. 7 you learned that $G$ is the reciprocal of $R$ and has the base unit of siemens S. Analogously, $B$ (susceptance) is the reciprocal of reactance. Susceptance also has siemens as its base unit because reactance has ohms as its base unit. Susceptance, then, is the ability of reactance to accommodate alternating current. Susceptance can be either capacitive susceptance $B_C$ or inductive susceptance $B_L$.

The reciprocal of impedance is called admittance $Y$. It also has siemens as its base unit. Admittance, then, can be used to specify the ability of an $RX$ circuit to support alternating current.

Since $R$, $X$, and $Z$ and their reciprocals $G$, $B$, and $Y$ have no angular velocity (their values are not time-dependent), they are not true phasors. Yet their values are determined by voltage phasors and current phasors. Therefore, for purposes of analyzing ac circuits, they can be treated like a phasor or a vector. Writing $X_C = 70\ \Omega \underline{/-90°}$ merely indicates that this 70 Ω of reactance causes the current to lead the voltage by 90°. Likewise, stating that $Z = 150\ \Omega \underline{/30°}$ indicates that the net $R$ and the net $X$ of the circuit cause the circuit current to lag the applied voltage by 30°.

In working with ac circuits, capacitive reactance is always $\angle -90°$, inductive reactance is always $\angle 90°$, and resistance is always $\angle 0°$. The rationale for the signs of these angles is illustrated in Fig. 19-10, which shows the current and voltage phasors for a series *RCL* circuit. For a series circuit, the current phasor is placed on the reference $+x$ axis. Since we know that $I_C$ leads $V_C$ by 90°, the $V_C$ phasor must be placed on the −90° axis of the graph. In Chaps. 16 and 17 we learned that $X_C = V_C/I_C$ and that for series circuits $I_C = I_T$. Applying these relationships to the phasors in Fig. 19-10 shows that

$$X_C = \frac{V_C\angle -90°}{I_T\angle 0°} = \frac{V_C}{I_T}\angle -90°$$

This same line of reasoning will show that the angle for $X_L$ is +90° and the angle for $R$ is 0°. The angle for $Z$ in Fig. 19-10 should be positive because the circuit is inductive ($V_T$ leads $I_T$). Using Ohm's law and the phasors $I_T$ and $V_T$ we can show that

$$Z = \frac{V_T\angle 26.6°}{I_T\angle 0°} = \frac{V_T}{I_T}\angle 26.6°$$

The angle for $R$, $X_C$, and $X_L$ could just as well have been determined from the phasor diagram for a parallel *RCL* circuit. In this case, the calculation for $X_C$ using phasor currents and voltages would be

$$X_C = \frac{V_C}{I_C} = \frac{V_T\angle 0°}{I_C\angle 90°} = \frac{V_T}{I_C}\angle -90°$$

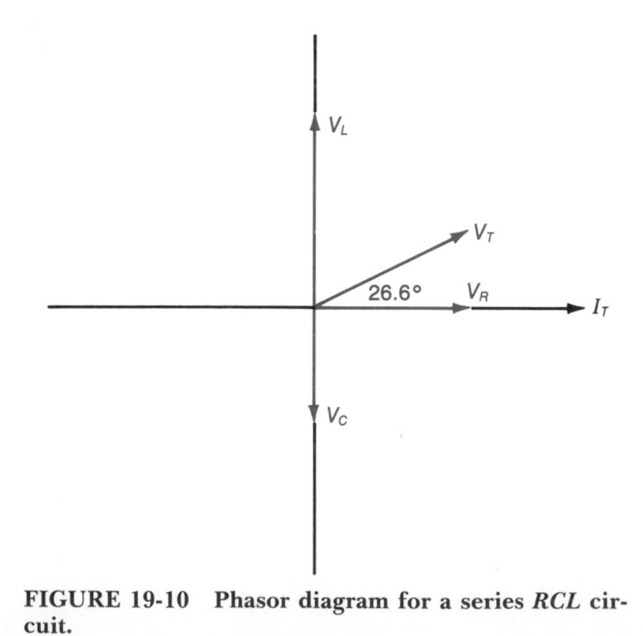


**FIGURE 19-10 Phasor diagram for a series *RCL* circuit.**

Now that we know the angles for $R$, $X_C$, and $X_L$, we can figure the angles for $G$, $B_C$, and $B_L$. We know that $1/R = G$; therefore, $1/R\angle 0° = G\angle 0°$. Likewise, $1/X_C\angle -90° = B_C\angle 90°$ and $1/X_L\angle 90° = B_L\angle -90°$. In the previous paragraph we established that $Z$ for a circuit with a net inductive reactance has a +angle. Thus, for a net inductive behaving circuit $Y$ will have a −angle. Of course, for a capacitive type circuit, the angle for $Z$ will be negative and for $Y$ the angle will be positive.

For a parallel *RCL* circuit, it can be shown that $Y = G + B_C + B_L$. The procedure is as follows: By Kirchhoff's law,

$$I_T\angle \theta = I_R\angle 0° + I_C\angle 90° + I_L\angle -90°$$

By Ohm's law substitutions,

$$\frac{V_T\angle 0°}{Z\angle -\theta} = \frac{V_T\angle 0°}{R\angle 0°} + \frac{V_T\angle 0°}{X_C\angle -90°} + \frac{V_T\angle 0°}{X_L\angle 90°}$$

By multiplying by $1/V_T\angle 0°$,

$$Y\angle \theta = G\angle 0° + B_C\angle 90° + B_L\angle -90°$$

Notice that $B_C$ is always on the $+j$ axis and $B_L$ is always on the $-j$ axis. Therefore, $B_C$ becomes $+jB$ and $B_L$ becomes $-jB$. The above formula can also be written as

$$Y = G + jB - jB$$

**EXAMPLE 19-11**

Determine $Z$ and $I_T$ for the circuit in Fig. 19-11. (Assume $R$, $C$, and $L$ are ideal components.)

**Solution**

$$G = \frac{1}{R} = \frac{1}{2.2\text{ k}\Omega} = 0.455\text{ mS}$$

$$+jB = B_C = \frac{1}{X_C} = 2\pi fC$$
$$= 6.28 \times 5 \times 10^5\text{ Hz} \times 4 \times 10^{-11}\text{ F} = 0.126\text{ mS}$$

$$-jB = B_L = \frac{1}{X_L} = \frac{1}{2\pi fL} = \frac{1}{6.28 \times 5 \times 10^5 \times 1 \times 10^{-3}}$$
$$= 0.318\text{ mS}$$

$$Y = G + jB - jB = 0.455\text{ mS} + j0.126\text{ mS} - j0.318\text{ mS}$$
$$= 0.455\text{ mS} - j0.192\text{ mS}$$

Convert to polar form:

$$Y = 0.455\text{ mS} - j0.192\text{ mS} = 0.494\text{ mS}\angle -22.88°$$

$$Z = \frac{1}{Y} = \frac{1}{0.494\text{ mS}\angle -22.88°} = 2\text{ k}\Omega\angle 22.88°$$

$$I_T = \frac{V_T}{Z} = \frac{10\text{ V}\angle 0°}{2\text{ k}\Omega\angle 22.88°} = 5\text{ mA}\angle -22.88°$$

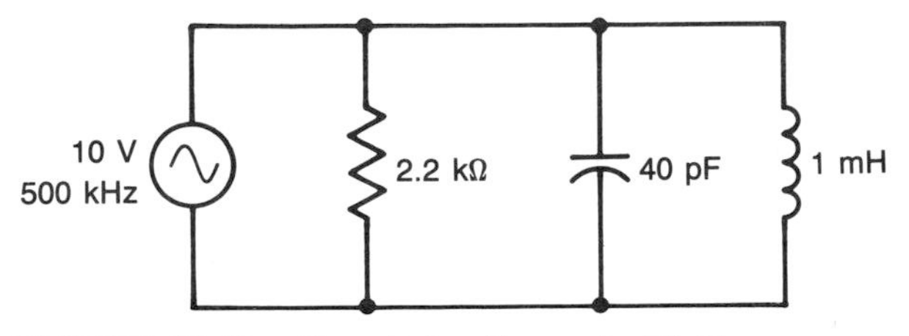


**FIGURE 19-11 Circuit for Example 19-11.**

## Self-Test

**17.** Write the symbol for, and define, each of the following terms:

**a.** Susceptance
**b.** Admittance
**c.** Conductance

**18.** The base unit of susceptance is the ______.

**19.** The symbol for the base unit of admittance is ______.

**20.** Write two formulas for calculating admittance.

**21.** The angle for $X_C$ is ______.

**22.** The angle for $jB$ is ______.

**23.** The term $-jB$ stands for the ______ of ______ reactance.

**24.** Impedance $Z$ for a capacitive circuit will have a ______ angle.

**25.** Change the frequency in Fig. 19-11 to 1 MHz and calculate $Z$, $B_C$, $X_L$, and $I_T$.

**26.** Do the results in Question 25 indicate that the circuit in Fig. 19-11 is inductive or capacitive when operating at 1 MHz?

## 19-7 SERIES AND PARALLEL IMPEDANCES

The oblong boxes in the schematic diagram in Fig. 19-12*a* contain the necessary amounts of $R$ and $X$ needed to produce the angle and magnitude of $Z$ specified. Impedances in series are treated like any other form of opposition in series; i.e., the total opposition equals the sum of the individual oppositions and the source voltage distributes in direct proportion to the individual oppositions. *But, one must use phasor algebra* when working with series (or parallel) impedances. The voltage across individual impedances can be determined by using Ohm's law, Kirchhoff's voltage law, or the voltage division formula for series circuits; just remember to use phasor algebra.

### EXAMPLE 19-12

Determine $Z_T$, $I_T$, $V_{Z_1}$, and $V_{Z_2}$ for the circuit in Fig. 19-12*a*.

**Solution**

$$Z_T = Z_1 + Z_2$$

Convert $Z_1$ and $Z_2$ to rectangular form and add to get $Z_T$:

$$Z_1 = 520\ \Omega \cos 40° + 520\ \Omega \sin 40° = 398.3\ \Omega + j334.2\ \Omega$$
$$Z_2 = 745\ \Omega \cos 10° - 745\ \Omega \sin 10° = 733.4\ \Omega - j129.4\ \Omega$$
$$Z_T = 1131.7\ \Omega + j204.8\ \Omega$$

Convert $Z_T$ to polar form:

$$\theta = \arctan\frac{204.8}{1131.7} = 10.26°$$

$$Z_T = \frac{1131.7\ \Omega}{\cos 10.26°}\angle\theta = 1150.1\ \Omega\angle 10.26°$$

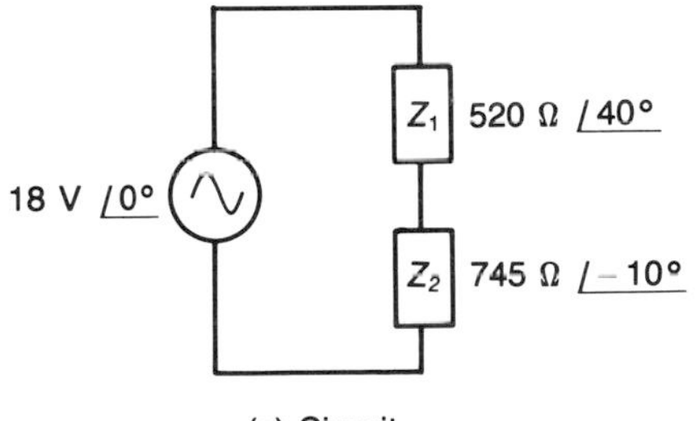


(*a*) Circuit

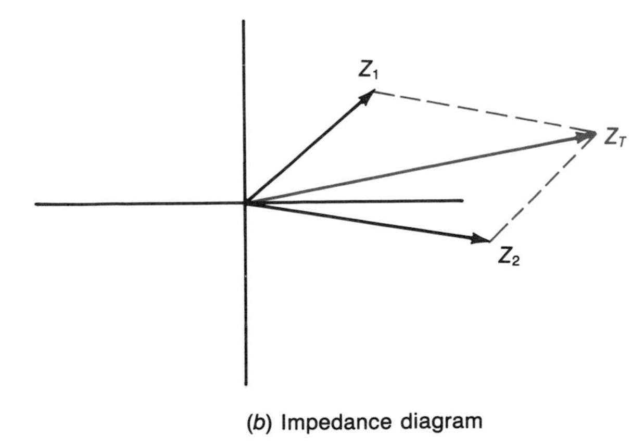


(*b*) Impedance diagram

**FIGURE 19-12 Series impedances.**

Note that $Z_T$ agrees with the sketch in Fig. 19-12*b*. Calculate $I_T$ by Ohm's law:

$$I_T = \frac{V_T}{Z_T} = \frac{18\text{ V}\angle 0°}{1150.1\ \Omega\angle 10.26°} = 15.65\text{ mA}\angle -10.26°$$

Calculate $V_{Z_1}$ by Ohm's law:

$$V_{Z_1} = I_T Z_1 = 15.65\text{ mA}\angle -10.26° \times 520\ \Omega\angle 40° = 8.14\text{ V}\angle 29.74°$$

or by voltage division:

$$V_{Z_1} = \frac{Z_1 V_T}{Z_1 + Z_2} = \frac{520\ \Omega\angle 40° \times 18\text{ V}\angle 0°}{1150.1\ \Omega\angle 10.27°} = 8.14\text{ V}\angle 29.74°$$

Calculate $V_{Z_2}$ by Kirchhoff's voltage law:

$$V_{Z_2} = V_T - V_{Z_1} = 18\text{ V}\angle 0° - 8.14\text{ V}\angle 29.74°$$
$$= 18\text{ V} - 7.07\text{ V} - j4.04\text{ V}$$
$$= 10.93\text{ V} - j4.04\text{ V} = 11.65\text{ V}\angle -20.29°$$

Of course, $V_{Z_2}$ could also have been determined by Ohm's law or the voltage division formula.

Parallel impedances can be handled like other forms of parallel opposition as long as we use phasor algebra. Total impedance can be found by any of three methods. The first method uses Ohm's law to determine the branch currents, Kirchhoff's current law to determine the total current, and, finally, Ohm's law to determine the total impedance. The second method involves converting the individual impedances to admittances, adding these admittances to obtain the total admittance, and, finally, converting back to total impedance. The third method uses the product divided by sum formula of $Z_T = Z_1Z_2/(Z_1 + Z_2)$.

### EXAMPLE 19-13

Using the admittance method, determine the total impedance for the circuit in Fig. 19-13*a*.

**Solution**

$$Y_1 = \frac{1}{Z_1} = \frac{1}{265\ \Omega\angle -60°}$$
$$= 3.77\text{ mS}\angle 60° = 3.77\text{ mS}\cos 60° + 3.77\text{ mS}\sin 60°$$
$$= 1.89\text{ mS} + j3.26\text{ mS}$$

$$Y_2 = \frac{1}{Z_2} = \frac{1}{195\ \Omega\angle 80°} = 5.13\text{ mS}\angle -80°$$
$$= 5.13\text{ mS}\cos 80° - 5.13\text{ mS}\sin 80°$$
$$= 0.89\text{ mS} - j5.05\text{ mS}$$

Sketch the admittance diagram as in Fig. 19-13*b*:

$$Y_T = Y_1 + Y_2$$
$$= (1.89\text{ mS} + j3.26\text{ ms}) + (0.89\text{ mS} - j5.05\text{ mS})$$
$$= 2.78\text{ mS} - j1.79\text{ mS}$$
$$\theta = \arctan\frac{-1.79}{2.78} = -32.8°$$
$$Y_T = \frac{2.78\text{ mS}}{\cos 32.8°}\angle -32.8° = 3.31\text{ mS}\angle -32.8°$$
$$Z_T = \frac{1}{Y_T} = \frac{1}{3.31\text{ mS}\angle -32.8°} = 302\ \Omega\angle 32.8°$$

### EXAMPLE 19-14

Determine the total impedance for Fig. 19-13*a* using the branch current methods.

**Solution**

$$I_1 = \frac{15\text{ V}\angle 0°}{265\ \Omega\angle -60°} = 56.6\text{ mA}\angle 60°$$
$$= 28.3\text{ mA} + j49.0\text{ mA}$$
$$I_2 = \frac{15\text{ V}\angle 0°}{195\ \Omega\angle 80°} = 76.9\text{ mA}\angle -80°$$
$$= 13.4\text{ mA} - j75.7\text{ mA}$$
$$I_T = (28.3\text{ mA} + j49.0\text{ mA}) + (13.4\text{ mA} - j75.7\text{ mA})$$
$$= 41.7\text{ mA} - j26.7\text{ mA} = 49.5\text{ mA}\angle -32.6°$$
$$Z_T = \frac{15\text{ V}\angle 0°}{49.5\text{ mA}\angle -32.6°} = 303\ \Omega\angle 32.6°$$

Within round-off error, these results agree with that obtained in Example 19-13.

### EXAMPLE 19-15

Using the product divided by sum formula, determine the total impedance for Fig. 19-13.

**Solution**

$$Z_T = \frac{265\ \Omega\angle -60° \times 195\ \Omega\angle 80°}{265\ \Omega\angle -60° + 195\ \Omega\angle 80°}$$
$$= \frac{51{,}675\ (\Omega)^2\angle 20°}{(132.5\ \Omega - j229.5\ \Omega) + (33.9\ \Omega + j192.0\ \Omega)}$$
$$= \frac{51{,}675\ (\Omega)^2\angle 20°}{166.4\ \Omega - j37.5} = \frac{51{,}675\ (\Omega)^2\angle 20°}{170.6\ \Omega\angle -12.7°}$$
$$= 303\ \Omega\angle 32.7°$$

Again, the value of $Z_T$ calculated by this method agrees with that found in Examples 19-13 and 19-14.

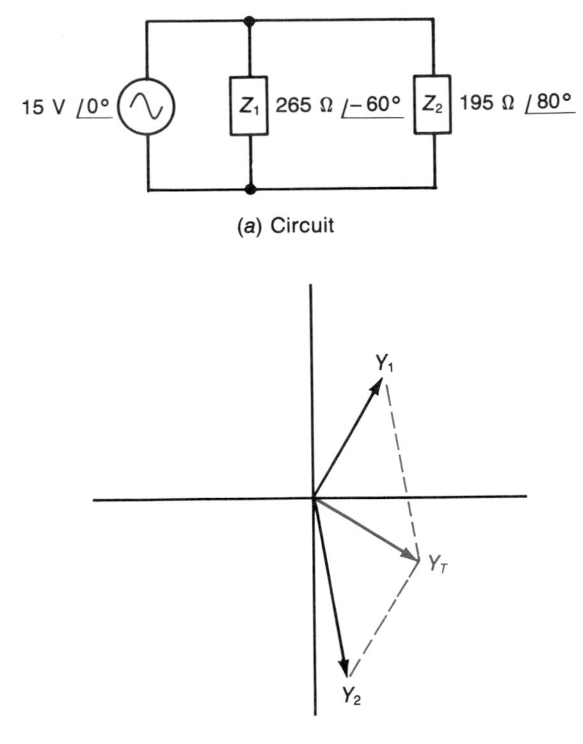


**FIGURE 19-13 Parallel impedance.**

### EXAMPLE 19-16

Find $Z_T$ and $I_{R_1}$ for the circuit shown in Fig. 19-14*a*.

**Solution**

Draw an equivalent circuit as in Fig. 19-14*b*:

$$Y_1 = G_1 - jB = \frac{1}{680\ \Omega} - j\frac{1}{1000\ \Omega}$$
$$= 1.47\text{ mS} - j1.0\text{ mS} = 1.78\text{ mS}\ \underline{/-34.2^\circ}$$
$$Z_1 = \frac{1}{Y_1} = 562\ \Omega\ \underline{/34.2^\circ}$$
$$Z_2 = 1000\ \Omega - j1200\ \Omega = 1562\ \Omega\ \underline{/-50.2^\circ}$$
$$Z_T = 562\ \Omega\ \underline{/34.2^\circ} + 1562\ \Omega\ \underline{/-50.2^\circ}$$
$$= (464.8\ \Omega + j316\ \Omega) + (1000\ \Omega - j1200\ \Omega)$$
$$= 1464.8\ \Omega - j884.1\ \Omega = 1711\ \Omega\ \underline{/-31.1^\circ}$$
$$I_T = \frac{V_T}{Z_T} = \frac{20\text{ V}\ \underline{/0^\circ}}{1711\ \Omega\ \underline{/-31.1^\circ}} = 11.7\text{ mA}\ \underline{/31.1^\circ}$$
$$V_{R_1} = V_{Z_1} = I_T Z_1 = 11.7\text{ mA}\ \underline{/31.1^\circ} \times 562\ \Omega\ \underline{/34.2^\circ}$$
$$= 6.6\text{ V}\ \underline{/65.3^\circ}$$
$$I_{R_1} = \frac{V_{R_1}}{R_1} = \frac{6.6\text{ V}\ \underline{/65.3^\circ}}{680\ \Omega\ \underline{/0^\circ}} = 9.7\text{ mA}\ \underline{/65.3^\circ}$$

The current division formula used for two parallel resistances is also appropriate for two parallel impedances. Using the data from Example 19-13 to 19-15, $I_2$ for Fig. 19-13 could be calculated thus:

$$I_2 = \frac{Z_1 I_T}{Z_1 + Z_2} = \frac{265\ \Omega\ \underline{/-60^\circ} \times 49.5\text{ mA}\ \underline{/-32.6^\circ}}{170.6\ \Omega\ \underline{/-12.7^\circ}}$$
$$= \frac{13{,}117.5\text{ mV}\ \underline{/-92.6^\circ}}{170.6\ \Omega\ \underline{/-12.7^\circ}} = 76.9\text{ mA}\ \underline{/-79.9^\circ}$$

Of course, for the circuit of Fig. 19-13 it is easier, and quicker, to use Ohm's law to calculate $I_2$. However, in a Norton equivalent circuit it is often convenient to use this current division formula.

The current division formula using $Z$ is only appropriate for two impedances in parallel. When more than two impedances are involved, the impedances should be converted to admittances. Then, the current division formula becomes $I_{Y_n} = I_T Y_n / Y_T$.

Now that we know how to handle series and parallel impedances, we can solve series-parallel circuits involving combinations of $R$, $C$, and $L$.

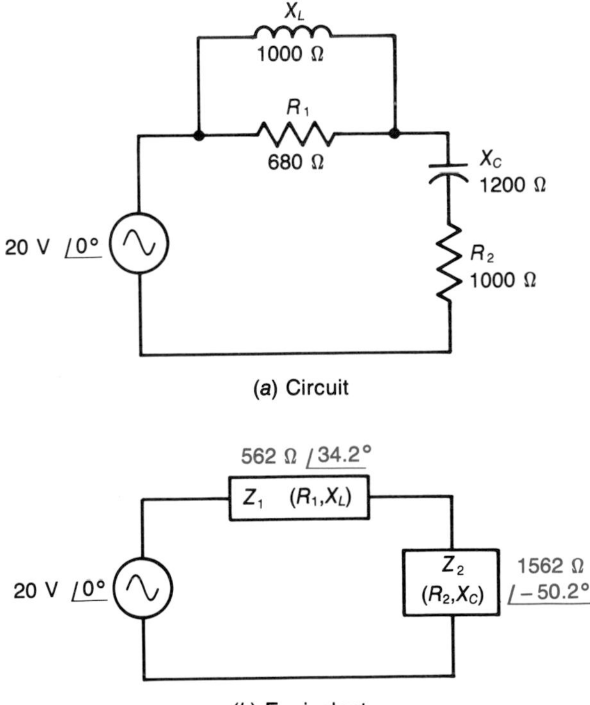


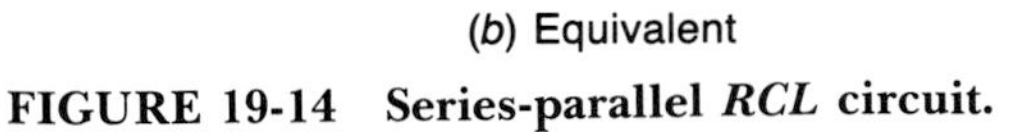

**FIGURE 19-14 Series-parallel *RCL* circuit.**

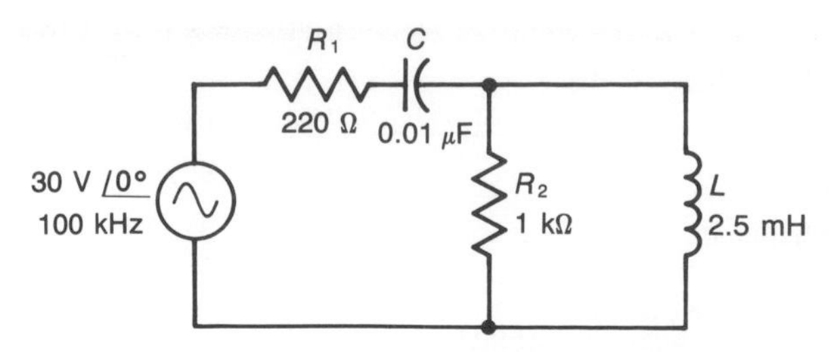


**FIGURE 19-15 Circuit for Self-Test Question 29.**

## Self-Test

**27.** Determine the total impedance of $Z_1 = 3\ \text{k}\Omega\ \underline{/48^\circ}$ in series with $Z_2 = 6\ \text{k}\Omega\ \underline{/80^\circ}$.

**28.** Determine the $Z_T$ for $Z_1 = 200\ \Omega\ \underline{/10^\circ}$ in parallel with $Z_2 = 300\ \Omega\ \underline{/-30^\circ}$.

**29.** Determine $Z_T$, $V_C$, and $I_L$ for Fig. 19-15.

## 19-8 EQUIVALENT SERIES AND PARALLEL IMPEDANCES

Sometimes it is advantageous to convert a parallel impedance into an equivalent series impedance or vice versa. For instance, Fig. 19-15 could be solved as a series *RCL* circuit if the parallel combination of $R_2$ and $L$ were converted to its series-equivalent $R$ and $L$.

Series and parallel impedances will be equivalent when their magnitudes and angles are equal. Figure 19-16 shows that $V_T$ will lag $I_T$ by the same angle in either a series or a parallel circuit if the proper ratios of $R$ to $X$ are used. Of course, the magnitude of $Z$ in either Fig. 19-16*a* or *b* can be adjusted to any value without changing $\theta$ as long as the $R/X$ ratio remains the same.

When $Z_P$ (parallel impedance) and $Z_S$ (series impedance) are equal, the formulas for converting between $R_P$ (parallel resistance) and $R_S$ (series resistance) can be developed by referring to Fig. 19-16. Thus,

$$\cos\theta = \frac{V_{R_S}}{V_T} = \frac{IR_S}{IZ} = \frac{R_S}{Z}$$

Also,

$$\cos\theta = \frac{I_{R_P}}{I_T} = \frac{V/R_P}{V/Z} = \frac{Z}{R_P}$$

Therefore,

$$\frac{R_S}{Z} = \frac{Z}{R_P}$$

And solving for $R_P$,

$$R_P = \frac{Z^2}{R_S} = \frac{(\sqrt{R_S^2 + X_S^2})^2}{R_S} = \frac{R_S^2 + X_S^2}{R_S}$$

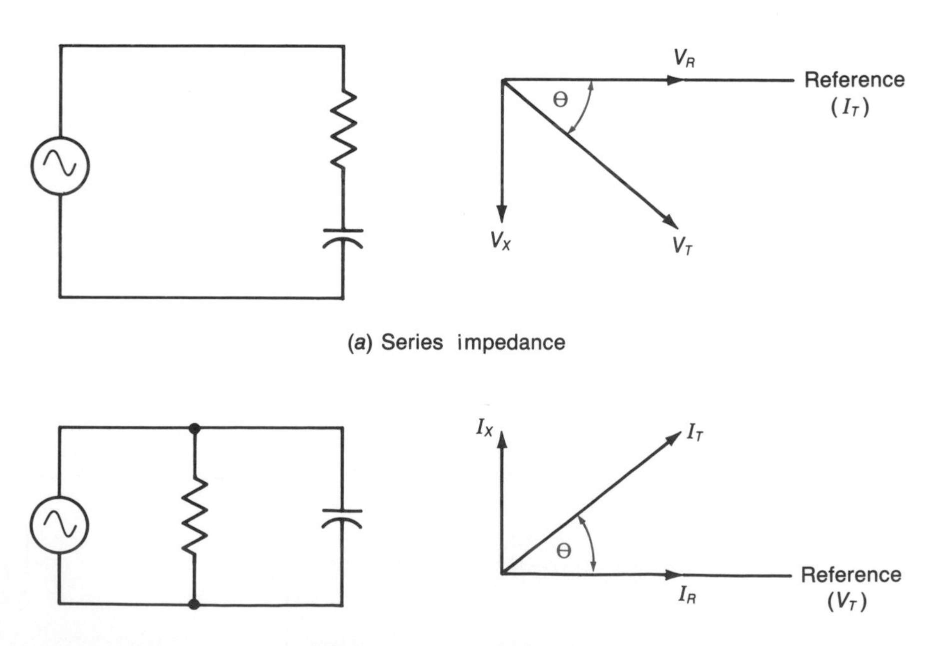


**FIGURE 19-16 Current leads voltage by the same angle in these equivalent *RC* circuits.**

And solving for $R_S$,

$$R_S = \frac{Z^2}{R_P} = \frac{[(R_P/X_P)/\sqrt{R_P^2 + X_P^2}]^2}{R_P}$$
$$= \frac{R_P^2 X_P^2/(R_P^2 + X_P^2)}{R_P}$$
$$= \frac{R_P X_P^2}{R_P^2 + X_P^2}$$

Formulas for converting between $X_P$ and $X_S$ can be derived by the same procedure if one starts with $\sin\ \theta = V_{X_S}/V_T = IX_S/IZ = X_S/Z$ and $\sin\ \theta = I_{X_P}/I_T = (V/X_P)/(V/Z) = Z/X_P$. The derived formulas will be

$$X_P = \frac{X_S^2 + R_S^2}{X_S}$$
$$X_S = \frac{X_P R_P^2}{X_P^2 + R_P^2}$$

The four conversion formulas were developed from *RC* circuits. They could have been developed from *RL* circuits just as readily. They apply to either type of impedance circuit.

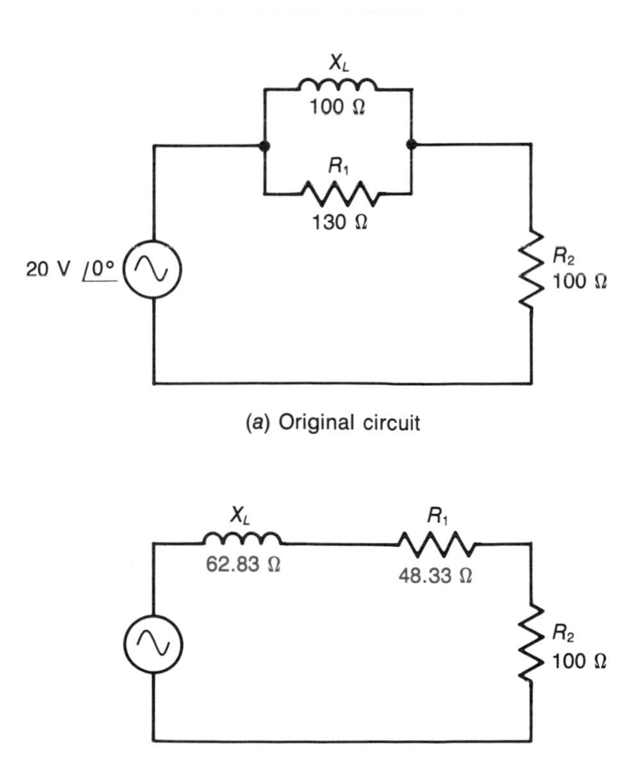


**FIGURE 19-17 Circuit for Example 19-17.**

**EXAMPLE 19-17**

For the circuit in Fig. 19-17*a*, determine $Z_T$ and $V_{R_2}$. Use the parallel-to-series conversion formulas.

**Solution**

$$R_S = \frac{R_P X_P^2}{R_P^2 + X_P^2} = \frac{130\ \Omega \times (100\ \Omega)^2}{(130\ \Omega)^2 + (100\ \Omega)^2}$$
$$= \frac{1.3 \times 10^6\ \Omega^3}{2.69 \times 10^4\ \Omega^2} = 48.33\ \Omega$$
$$X_S = \frac{X_P R_P^2}{X_P^2 + R_P^2} = \frac{100\ \Omega \times (130\ \Omega)^2}{(130\ \Omega)^2 + (100\ \Omega)^2}$$
$$= \frac{1.69 \times 10^6\ \Omega^3}{2.69 \times 10^4\ \Omega^2} = 62.83\ \Omega$$

Draw the series-equivalent components as in Fig. 19-17*b*.

$$Z_T = 100\ \Omega\ \underline{/0^\circ} + 48.33\ \Omega\ \underline{/0^\circ} + 62.83\ \Omega\ \underline{/90^\circ}$$
$$= 148.33\ \Omega + j62.83\ \Omega = 161\ \Omega\ \underline{/23^\circ}$$
$$I_T = \frac{V_T}{Z_T} = \frac{20\ \text{V}\ \underline{/0^\circ}}{161\ \Omega\ \underline{/23^\circ}} = 124\ \text{mA}\ \underline{/-23^\circ}$$
$$V_{R_2} = I_T R_2 = 124\ \text{mA}\ \underline{/-23^\circ} \times 100\ \Omega\ \underline{/0^\circ}$$
$$= 12.4\ \text{V}\ \underline{/-23^\circ}$$

## Self-Test

**30.** A 51-Ω resistor is connected in parallel with a 68-Ω reactance. Determine the equivalent series resistance and reactance.

**31.** Determine the parallel-equivalent resistance and reactance for $Z = 90\ \Omega\ \underline{/30^\circ}$. *Hint:* Use trigonometric functions to determine the series resistance and reactance.

## 19-9 THEVENIN'S AND NORTON'S THEOREMS

The details of Thevenin's and Norton's theorems for dc circuits are covered in Chap. 9. In this chapter, we extend the use of these theorems to ac circuits. The only changes required are that $R_{\text{TH}}$ now becomes $Z_{\text{TH}}$ and both $I_{\text{N}}$ and $V_{\text{TH}}$ are now phasors. Conversion between a Norton and a Thevenin source is still done by the formulas $V_{\text{TH}} = I_{\text{TH}} Z_{\text{TH}}$ and $Z_{\text{TH}} = Z_{\text{N}}$. The procedures for determining $Z_{\text{TH}}$, $V_{\text{TH}}$, and $I_{\text{N}}$ are the same as before. That is, $V_{\text{TH}}$ is the open circuit (load removed) voltage across the output terminals and

$I_{TH}$ is the short-circuited current (load shorted) available at the output terminals. Impedance $Z_{TH}$ is the impedance between the output terminals when the load is removed and all sources, either current or voltage, are replaced with their internal impedance. Remember that for many circuits the internal impedance of the source is so small compared to the load impedances that it is assumed to be zero and the source is replaced by a conductor for purposes of calculating $Z_{TH}$ or $Z_N$.

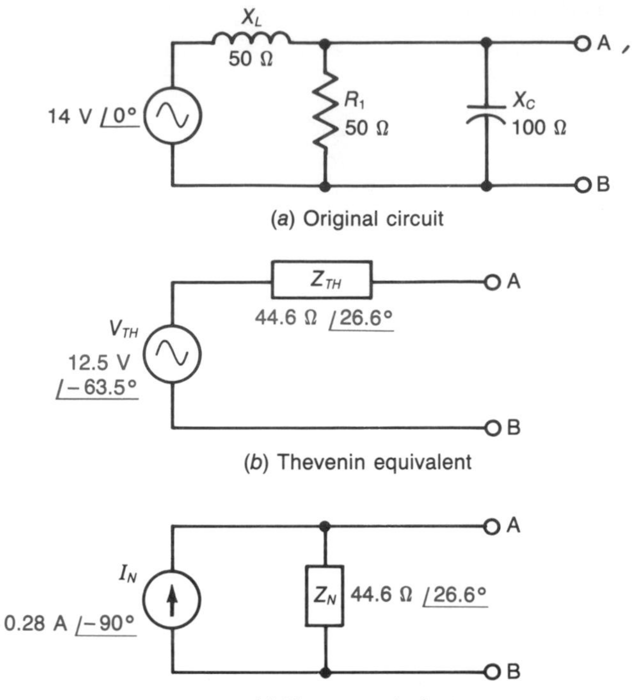


**FIGURE 19-18 Thevenin and Norton equivalent circuits.**

## EXAMPLE 19-18

Determine $Z_{TH}$ and $V_{TH}$ for the circuit shown in Fig. 19-18*a* and draw the Thevenin equivalent circuit.

### Solution

Replacing the source with a conductor puts $X_C$, $R_1$, and $X_L$ in parallel. Thus,

$$Y_{TH} = \frac{1}{50\ \Omega\ \angle 0^\circ} + \frac{1}{50\ \Omega\ \angle 90^\circ} + \frac{1}{100\ \Omega\ \angle -90^\circ}$$
$$= 22.4\ \text{mS}\ \angle -26.6^\circ$$
$$Z_{TH} = \frac{1}{Y_{TH}} = \frac{1}{22.4\ \text{mS}\ \angle -26.6^\circ} = 44.6\ \Omega\ \angle 26.6^\circ$$
$$Z_{R1}, X_C = \frac{50\ \angle 0^\circ \times 100\ \angle -90^\circ}{50\ \angle 0^\circ + 100\ \angle -90^\circ} = 44.7\ \Omega\ \angle -26.6^\circ$$
$$Z_T = Z_{R1}, X_C + X_L = 44.7\ \Omega\ \angle -26.6^\circ + 50\ \Omega\ \angle 90^\circ$$
$$= 50\ \Omega\ \angle 36.9^\circ$$

By the voltage division formula,

$$V_{TH} = \frac{44.7\ \Omega\ \angle -26.6^\circ \times 14\ \text{V}\ \angle 0^\circ}{50\ \Omega\ \angle 36.9^\circ}$$
$$= 12.5\ \text{V}\ \angle -63.5^\circ$$

The Thevenin equivalent circuit is shown in Fig. 19-18*b*.

## EXAMPLE 19-19

Determine the Norton equivalent circuit for Fig. 19-18*a*.

### Solution

Since $Z_N = Z_{TH}$, we can use the value from Example 19-18:

$$Z_N = 44.6\ \Omega\ \angle 22.6^\circ$$

Short-circuiting the A and B terminals puts $X_L$ across the source. Thus,

$$I_N = \frac{14\ \text{V}\ \angle 0^\circ}{50\ \Omega\ \angle 90^\circ} = 0.28\ \text{A}\ \angle -90^\circ$$

The Norton equivalent circuit is shown in Fig. 19-18*c*.

Looking back at Example 19-18 you can see that determining $V_{TH}$ required considerable manipulation of complex numbers. However, finding $I_N$ required only one calculation in Example 19-19. Thus, the easiest way to find the Thevenin equivalent circuit for a circuit like the one in Fig. 19-18*a* is to find the Norton equivalent and then convert to the Thevenin with $V_{TH} = I_N Z_N$.

In dual-source ac circuits the instantaneous polarities of the sources must be indicated to determine whether the sources are basically series-aiding or series-opposing. As shown in Fig. 19-19*a* and *b*, both the magnitude and the phase angle of the net voltage in an ac circuit depend on the instantaneous polarities.

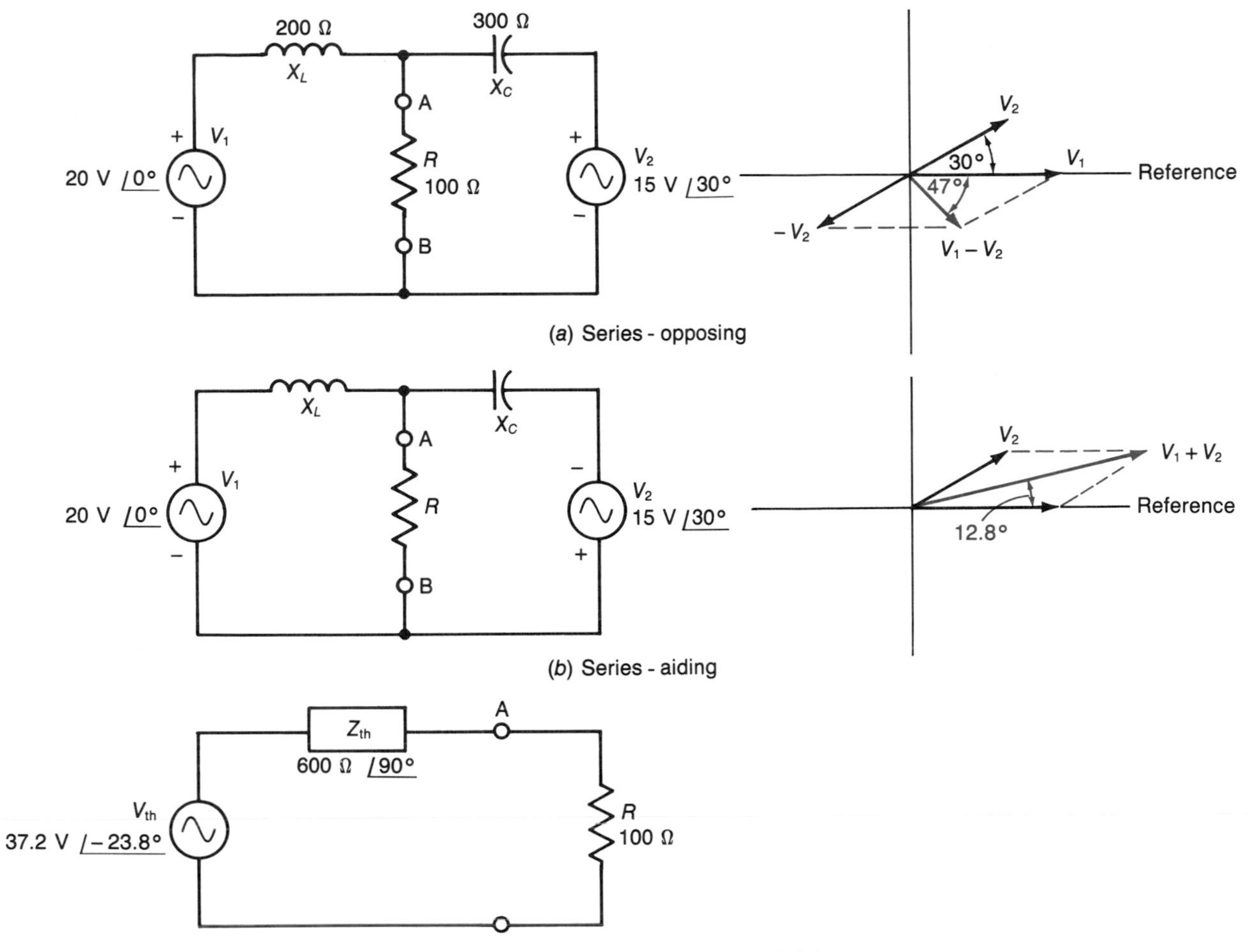


**FIGURE 19-19 Instantaneous polarities on ac sources.**

---

**EXAMPLE 19-20**

Refer to Fig. 19-19$a$. Determine the current through $R$.

**Solution**

Consider $R$ to be the load. Temporarily remove it, then either Nortonize or Theveninize the circuit. Finally, add $R$ as a load to the equivalent circuit and determine $I_R$. Let us Theveninize the circuit:

$$V_1 - V_2 = V_3 = 20\text{ V}\angle 0^\circ - 15\text{ V}\angle 30^\circ = 10.3\text{ V}\angle -46.9^\circ$$

$$Z_T = 200\ \Omega\angle 90^\circ + 300\ \Omega\angle -90^\circ = 100\ \Omega\angle -90^\circ$$

$$V_{X_L} = \frac{X_L V_3}{Z_T} = \frac{200\ \Omega\angle 90^\circ \times 10.3\text{ V}\angle -46.9^\circ}{100\ \Omega\angle -90^\circ} = 20.6\text{ V}\angle 133.1^\circ$$

$$V_{\text{TH}} = V_1 - V_{X_L} = 20\text{ V}\angle 0^\circ - 20.6\text{ V}\angle 133.1^\circ = 37.2\text{ V}\angle -23.8^\circ$$

$$Z_{\text{TH}} = \frac{200\angle 90^\circ \times 300\angle -90^\circ}{200\angle 90^\circ + 300\angle -90^\circ} = 600\ \Omega\angle 90^\circ$$

Draw the Thevenin equivalent as in Fig. 19-19$c$ and determine $I_R$.

$$Z_T = Z_{\text{TH}} + R = 600\ \Omega\angle 90^\circ + 100\ \Omega\angle 0^\circ = 608.3\ \Omega\angle 80.5^\circ$$

$$I_R = \frac{V_{\text{TH}}}{Z_T} = \frac{37.2\text{ V}\angle -23.8^\circ}{608.3\ \Omega\angle 80.5^\circ} = 61\text{ mA}\angle -104.3^\circ$$

---

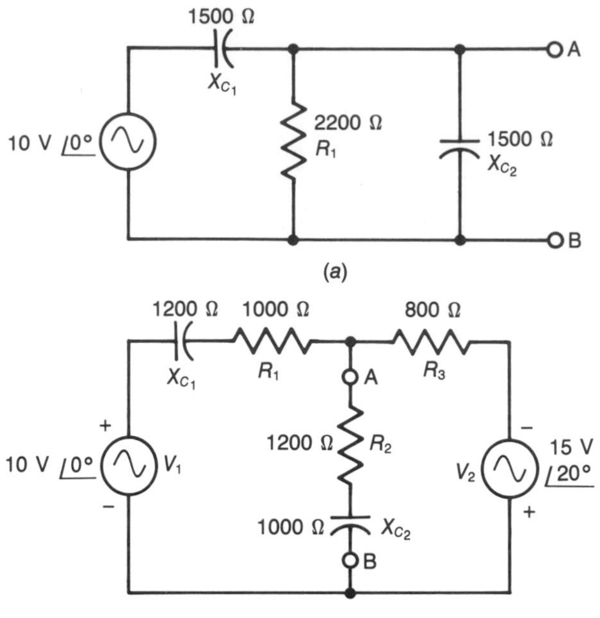


**FIGURE 19-20 Circuits for Self-Test Questions 32 and 33.**

Had we elected to Nortonize the circuit in Example 19-20, the procedure for finding $I_N$ would have been

$$I_1 = \frac{V_1}{X_L} = \frac{20\ \text{V} \angle 0^\circ}{200\ \Omega \angle 90^\circ} = 0.1\ \text{A} \angle -90^\circ$$

$$I_2 = \frac{V_2}{X_C} = \frac{15\ \text{V} \angle 30^\circ}{300\ \Omega \angle -90^\circ} = 0.05\ \text{A} \angle 120^\circ$$

$$I_N = I_1 + I_2 = 0.1\ \text{A} \angle -90^\circ + 0.05\ \text{A} \angle 120^\circ = 0.062\ \text{A} \angle -113.8^\circ$$

You can see that Nortonizing the circuit required fewer calculations for this circuit.

## Self-Test

**32.** For the circuit in Fig. 19-20*a* determine $I_N$, $V_{TH}$, and $R_{TH}$.

**33.** For the circuit in Fig. 19-20*b* determine the voltage between terminals A and B.

## SUMMARY

1. The $j$ operator rotates a phasor 90° counterclockwise.
2. The $-j$ operator rotates a phasor 90° clockwise.
3. A $j$ number is an imaginary number. It is always on the $y$ axis of a graph.
4. Numbers on the $x$ axis of a graph are known as real numbers.
5. A number composed of a real component and an imaginary component is a complex number.
6. A phasor can be expressed in either polar form or rectangular form.
7. Polar form specifies magnitude and phase angle of a phasor, i.e., $20\ \text{V} \angle 30^\circ$.
8. Rectangular form specifies the real ($x$ axis) component and the imaginary ($y$ axis) component of a phasor, i.e., $5\ \text{A} - j3\ \text{A}$.
9. Trigonometric functions can be used to convert from polar to rectangular coordinates and vice versa.
10. Phasor algebra guidelines:

    Addition—Add real $x$ components and add imaginary $y$ components.

    Subtraction—Subtract real components and subtract imaginary components.

Multiplication—Multiply magnitudes and algebraically add angles.

Division—Divide magnitudes and algebraically subtract angles.

**11.** Conductance $G$ is the reciprocal of resistance. Its base unit is the siemen.

**12.** Susceptance $B$ is the reciprocal of reactance. Its unit is siemens, and its angle is $-90°$ for $X_L$ and $+90°$ for $X_C$.

**13.** Admittance $Y$ is the reciprocal of impedance, and it has units of siemens.

**14.** The angle for $X_C$ is $-90°$ and for $X_L$ is $90°$.

**15.** For parallel circuits, $Y = G + jB - jB$.

**16.** For parallel circuits, $Y_T = Y_1 + Y_2 + \cdots + Y_n$.

**17.** Series and parallel impedance problems are solved just like series and parallel resistance problems except that phasor algebra must be used.

**18.** Conversion between parallel and series resistance and reactance can be accomplished by these formulas:

$$R_P = \frac{R_S^2 + X_S^2}{R_S} \quad \text{and} \quad R_S = \frac{R_P X_P^2}{R_P^2 + X_P^2}$$

$$X_P = \frac{X_S^2 + R_S^2}{X_S} \quad \text{and} \quad X_S = \frac{X_P R_P^2}{X_P^2 + R_P^2}$$

**19.** Thevenin's and Norton's theorems are applicable to impedance-type circuits. Variable $V_{TH}$, $Z_{TH}$, and $I_N$ are calculated using phasor algebra.

## CHAPTER REVIEW QUESTIONS

For the following items, determine whether each statement is true or false.

**19-1.** The value of an imaginary number is plotted on the $x$ axis of a graph.

**19-2.** A phasor in polar form is expressed with imaginary and real components.

**19-3.** Phasor addition is usually performed with the phasors in rectangular form.

**19-4.** In dividing phasors, the phase angles are algebraically added together.

**19-5.** The ohm is the base unit of admittance.

**19-6.** The impedance of a series $RC$ circuit will have a negative angle.

**19-7.** For any circuit the Thevenin impedance is equal to the Norton impedance.

**19-8.** With a dual-source ac circuit, the instantaneous polarities of the sources have to be known to Nortonize the circuit.

For the following items, fill in the blank with the word or phrase required to correctly complete each statement.

**19-9.** The operator $-j$ indicates a 90° rotation in a ______ direction.

**19-10.** A number preceded by a $j$ is an ______ number.

**19-11.** The ______ form of a phasor is expressed as a complex number.

**19-12.** When multiplying phasors, the angles are algebraically ______.

**19-13.** Division of phasors is usually done with the phasors in ______ form.

**19-14.** The base unit of susceptance is the ______.

**19-15.** The reciprocal of resistance is ______.

**19-16.** ______ reactance has an angle of 90°.

**19-17.** The susceptance of a ______ has an angle of 90°.

## CHAPTER REVIEW PROBLEMS

**19-1.** Convert 10 A $-\,j5$ A to its other form.

**19-2.** Convert 140 $\Omega \angle -19°$ to its other form.

**19-3.** Add $15 \angle 30°$ to $25 + j40$.

**19-4.** Subtract $15 \angle 30°$ from $20 \angle -40°$.

**19-5.** Multiply $15 \angle 30°$ by $10 \angle -80°$.

**19-6.** Divide $15 \angle 30°$ by $10 \angle -80°$.

**19-7.** For the circuit in Fig. 19-21*a* determine the following: $Y_T$, $Z_T$, $I_T$, and $I_C$.

**19-8.** Determine $Z_T$ for $Z_1 = 700\ \Omega \angle 30°$ in series with $Z_2 = 500\ \Omega \angle -20°$.

**19-9.** Determine $Z_T$ for $Z_1 = 700\ \Omega \angle 30°$ in parallel with $Z_2 = 500\ \Omega \angle -20°$.

**19-10.** Determine the parallel-equivalent resistance and reactance for 500 Ω of resistance in series with 800 Ω of reactance.

**19-11.** Voltage $V_{\mathrm{TH}} = 9\ \mathrm{V} \angle -20°$ and $Z_{\mathrm{TH}} = 50\ \Omega \angle 14°$. Determine $I_{\mathrm{N}}$.

**19-12.** For the circuit in Fig. 19-21*b*, determine $Z_{\mathrm{TH}}$ and $I_{\mathrm{N}}$.

**19-13.** Determine the voltage across a reactance of $200\ \Omega \angle 90°$ connected between terminals A and B in Fig. 19-21*b*.

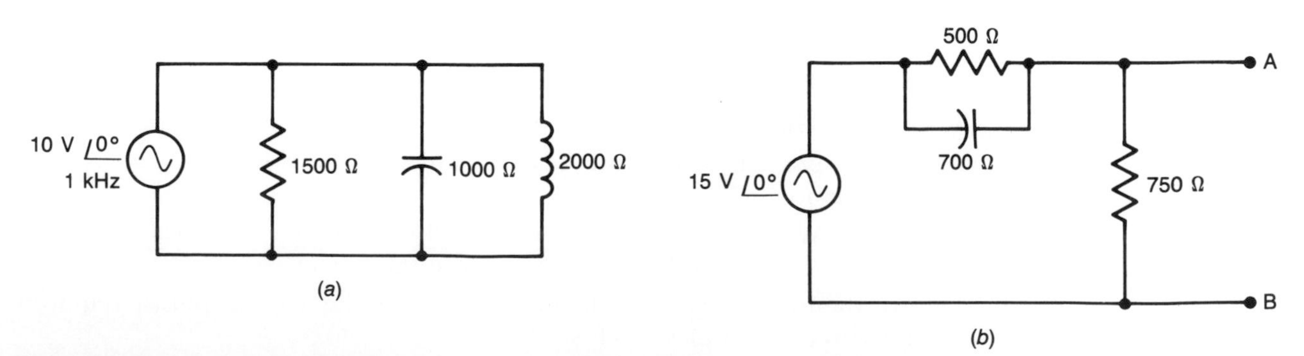


**FIGURE 19-21 Circuits for Chapter Review Problems 19-7, 19-12, and 19-13.**

## ANSWERS TO SELF-TESTS

1. $j$ operator
2. imaginary
3. imaginary, real
4. rectangular
5. $5 \angle 36.87°$
6. $8.78 + j8.18$
7. $16.1 \angle 119.7°$
8. $32 - j55.4$
9. adding, subtracting
10. multiplying, dividing
11. added
12. subtracted
13. $-9.5$ A $- j4.53$ A or $10.5$ A $\angle 205.5°$
14. $1.64$ V $- j13.36$ V or $13.5$ V $\angle -83°$
15. $12$ V $\angle 30°$ or $10.4$ V $+ j6$ V
16. $2\ \Omega \angle -93.1°$ or $-0.11\ \Omega - j2\ \Omega$
17. **a.** $B$, the reciprocal of $X$
    **b.** $Y$, the reciprocal of $Z$
    **c.** $G$, the reciprocal of $R$
18. siemen
19. $Y$
20. $Y = 1/Z$, $Y = G + B_C + B_L$ or $Y = G + jB - jB$
21. $-90°$
22. $+90°$
23. susceptance, inductive
24. negative
25. $Z = 2156\ \Omega \angle -11.44°$, $B_C = 0.251$ S, $X_L = 6283\ \Omega$, $I_T = 4.64$ mA $\angle 11.44°$
26. capacitive
27. $Z_T = 8.5$ k$\Omega \angle 66.9°$
28. $Z_T = 127.4\ \Omega \angle -5.84°$
29. $Z_T = 976.7\ \Omega \angle 17.5°$, $V_C = 4.9$ V $\angle -107.5°$, $I_L = 16.5$ mA $\angle -75°$
30. $R_S = 32.6\ \Omega$, $X_S = 24.5\ \Omega$
31. $R_P = 103.9\ \Omega$, $X_P = 180.0\ \Omega$
32. $I_N = 6.67$ mA $\angle 90°$, $V_{TH} = 4.73$ V $\angle 18.8°$, $Z_{TH} = 710\ \Omega \angle -71.2°$
33. $7.86$ V $\angle -10.2°$

# Chapter 20 Transformers

Chapters 13 and 14 dealt with self-inductance. In these chapters you learned how the self-inductance of an inductor or coil created an opposition (reactance) which controls the amount of current produced by an ac source.

In this chapter you will learn about mutual inductance and transformers. You will learn how the mutual inductance between the coils of a transformer enables a transformer to change (transform) the magnitude of ac voltages and currents. Changing the level (magnitude) of an ac voltage is the major use of transformers. Transformers make it possible for electric energy to be produced at a relatively low voltage, transported long distances at very high voltages (>200 kV), and used in our homes at the safer voltages of 120 and 240 V.

## 20-1 MUTUAL INDUCTANCE AND COUPLING

*Mutual inductance* exists between two coils when the changing magnetic flux in one coil produces a counter-electromotive force (cemf) (voltage) in the second coil. When two coils provide mutual inductance, the coils are *magnetically coupled.* Two, or more, coupled coils are referred to as a *transformer.* The result of mutual inductance is illustrated in Fig. 20-1 where the decreasing current through coil A causes the flux associated with coil A to decay. The collapsing flux from coil A cuts the turns of coil B and induces a voltage into coil B. Although Fig. 20-1 shows a decaying current and a corresponding collapsing flux in coil A, the result would be essentially the same if the current were

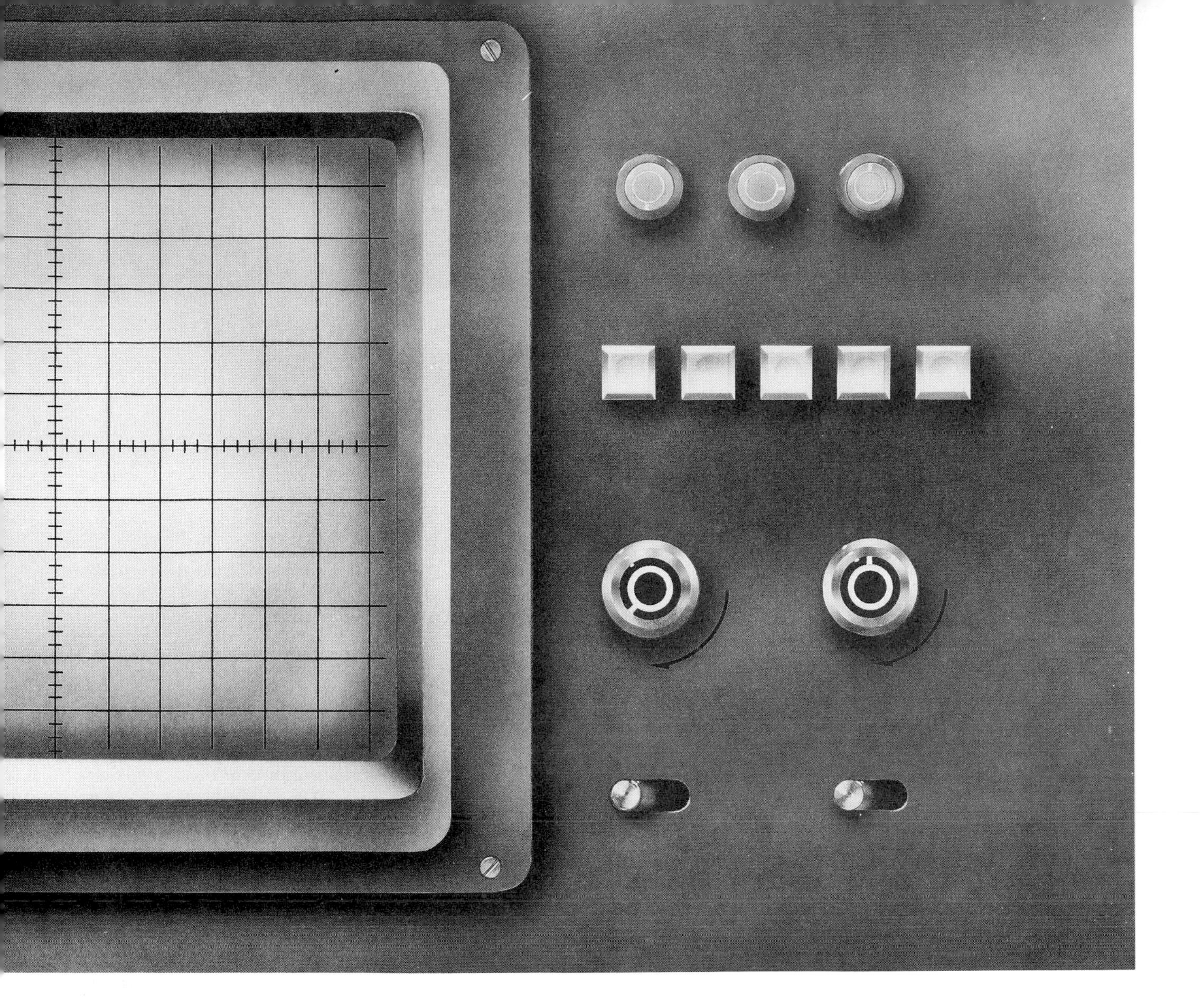

increasing and the flux expanding. A voltage (cemf) would still be induced into coil B. The only difference between the case of the expanding flux and that of the collapsing flux would be the polarity of the induced voltage.

The amount of mutual inductance between two coils is controlled by the inductance of each coil and by the amount of coupling between the two coils. *Coupling* refers to the portion of the flux produced by one coil which cuts, or links, the second coil. The portion of flux which links the coils is specified by the *coefficient of coupling*. The flux which links the coils is called the *mutual flux*. It produces mutual inductance. The flux of a coil that does not link up with the other coil is referred to as *leakage flux*. Leakage flux produces self-inductance but not mutual inductance.

The coefficient of coupling between two coils which do not share a common magnetic core is determined by the distance between the coils and by the axis orientation of the coils. Figure 20-2*a* and *b* shows that as the distance between the coils increases, less of the flux produced by one coil links up with the other coil. This is because the reluctance of the path between the coils is increasing as the distance increases. Figure 20-2*c* shows why axis orientation influences the coupling coefficient. When the axes are perpendicular, the flux from the excited coil (coil with current through it) is parallel to the turns of the other coil. Thus, no voltage is induced into the second coil and the coefficient of coupling is zero.

The coefficient of coupling of two coils wound on a continuous iron core (see Fig. 20-3) is very

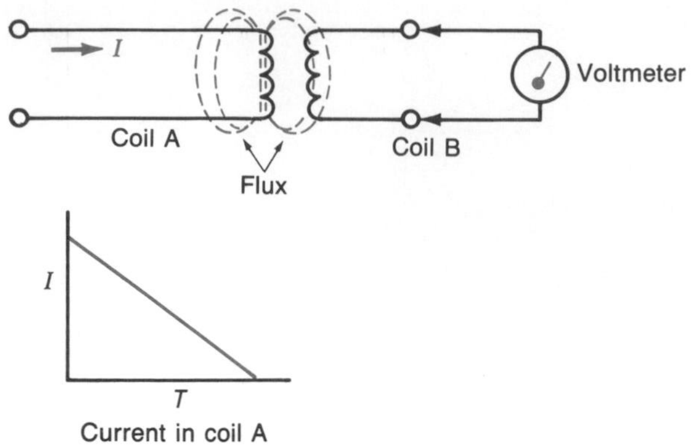

**FIGURE 20-1 Mutual inductance.**

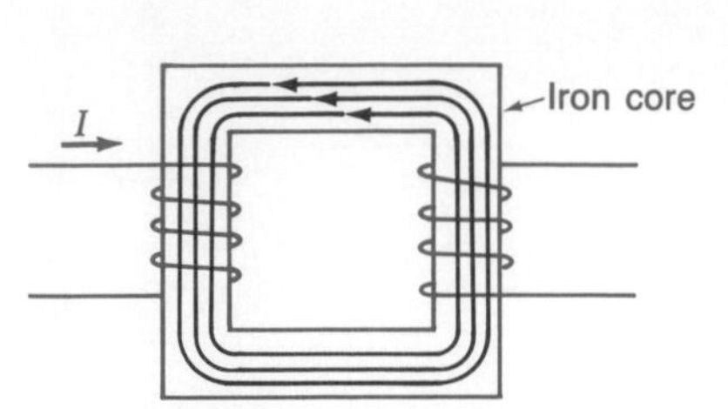

**FIGURE 20-3 Coupling in an iron core.**

close to unity (1). There is very little leakage flux because the reluctance of the iron-core path, which goes through the axis of the other coil, is extremely low compared to the air path that the leakage flux would have to traverse.

The coefficient of coupling is represented by the symbol $k$, which can range from 0 (when no flux links the two coils) to 1 (when all of the flux links the two coils). Of course, with a $k$ of 0 there is no mutual inductance and, therefore, no voltage is induced into the second coil. Sometimes $k$ is expressed as a percentage (0 to 100 percent). Mathematically, $k$ is defined as

$$k = \frac{\phi_m}{\phi_t}$$

where $\phi_t$ is the flux of the coil receiving current, and $\phi_m$ is the flux that links with the other coil. The difference between $\phi_t$ and $\phi_m$ is the leakage flux ($\phi_l$).

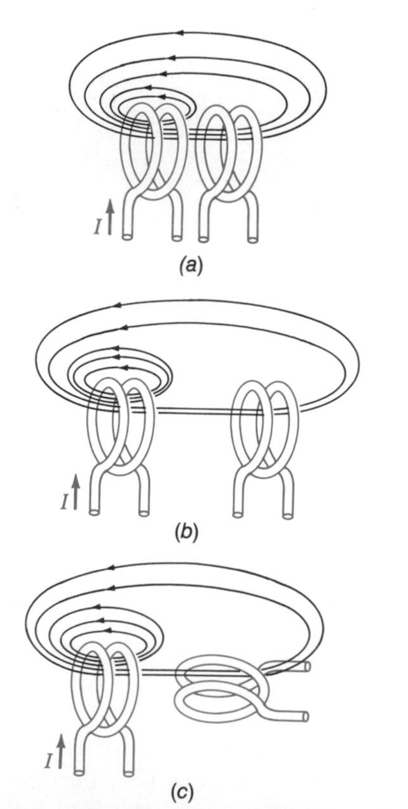

**FIGURE 20-2 Changing the coefficient of coupling.**

**EXAMPLE 20-1**

The coefficient of coupling between two coils is 0.6 or 60 percent. The excited coil produces 0.1 Wb of flux ($\phi_t$). How much flux ($\phi_m$) is coupled to the other coil? What is the value of the leakage flux?

**Solution**

$$\phi_m = k\phi_t = 0.6 \times 0.1 \text{ Wb} = 0.06 \text{ Wb}$$
$$\phi_l = \phi_t - \phi_m = 0.1 \text{ Wb} - 0.06 \text{ Wb} = 0.04 \text{ Wb}$$

Mutual inductance, like self-inductance, is specified in henrys. One henry of mutual inductance means that a current change of 1 A/s in the excited coil will produce 1 V of cemf in the other coil. The formula for determining the mutual inductance $L_m$ of two coupled coils $L_1$ and $L_2$ is

$$L_m = k\sqrt{L_1 L_2}$$

**EXAMPLE 20-2**

The coefficient of coupling between a coil of 2 H and a coil of 0.9 H is 0.7. Determine the mutual inductance.

**Solution**

$$L_m = k\sqrt{L_1 L_2} = 0.7\sqrt{2 \text{ H} \times 0.9 \text{ H}} = 0.94 \text{ H}$$

## Self-Test

1. What is mutual inductance?
2. Define coefficient of coupling.
3. Determine $k$ when $\phi_t = 0.6$ Wb and $\phi_m = 0.5$ Wb.
4. Determine $L_m$ when $k = 0.95$, $L_1 = 240$ mH, and $L_2 = 400$ mH.

## 20-2 BASIC TRANSFORMER ACTION

*Transformers* are the physical devices that operate by mutual inductance. Remembering, and applying, some of the concepts from Chaps. 10, 11, and 13 are essential to understanding how transformers operate. Let us briefly review some of the concepts needed to understand the current and voltage relationships in transformers. As you learned when studying induced voltages in Chaps. 11 and 13, the polarity of the induced voltage changes when either the magnetic flux is reversed or the relative motion between the conductor and the flux is reversed. With mutual inductance, the coils (conductors) are stationary; however, when the flux changes from an expanding to a decaying flux, the direction in which the flux cuts the coil's turns is reversed. Thus, the polarity of the induced voltage reverses whenever the flux changes from expanding to decaying or vice versa. Remember that both the amount and the rate of flux buildup or collapse determines the magnitude of the induced voltage. You also learned in Chaps. 10 and 13 that the flux of a coil is caused by a current flowing in the coil. The magnitude of the flux follows the magnitude of the current, and the direction of the flux reverses when the current reverses if the coil has an air core. Additionally, you learned, in Chaps. 13 and 14 on inductance, that a coil causes the voltage to lead the current by 90 electric degrees.

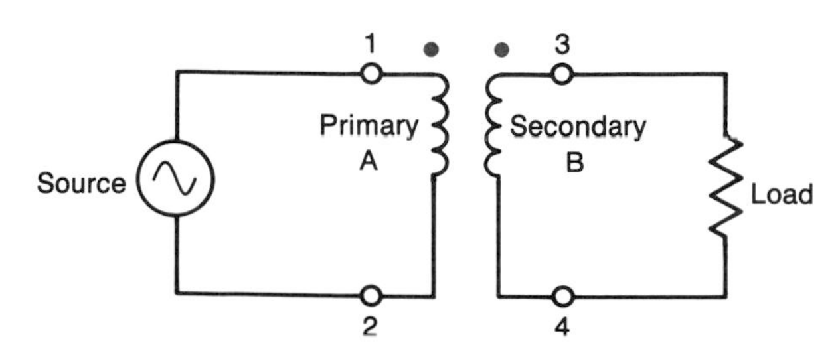


**FIGURE 20-4 Primary and secondary windings.**

### Primary and Secondary

The coupled coils in Fig. 20-4 form an air-core transformer. Coil A is the *primary winding*, and coil B is the *secondary winding*. The primary winding of any transformer is the winding which is connected to the primary power source. The secondary winding is the winding which provides power to any load connected to its terminals. To the load, the secondary winding is the power source; thus, the secondary is a secondary power source.

Transformers are bidirectional devices. Power can be applied to either coil as long as the current and voltage ratings of the coil are not exceeded. In Fig. 20-4, a power source could be applied between terminals 3 and 4 and power taken from terminals 1 and 2. Under this condition, the coil between terminals 3 and 4 would be called the primary.

The windings of a transformer are labeled primary and secondary according to the intended purpose of the transformer. For example, if a transformer is designed to receive 120 V from an outlet and provide a 5-V output, the coil designed to receive 120 V would be labeled primary. However, if a 5-V ac source were available, the other coil (5-V coil) could be used as a primary and the original primary would then be a secondary providing a 120-V output.

### Dot Notation

Figure 20-5*b* shows the voltage and current relationships when an ac voltage is applied to a coil (the primary) which is coupled to another coil (the secondary). The inductance of coil A causes its current and voltage to be 90° out of phase. As shown

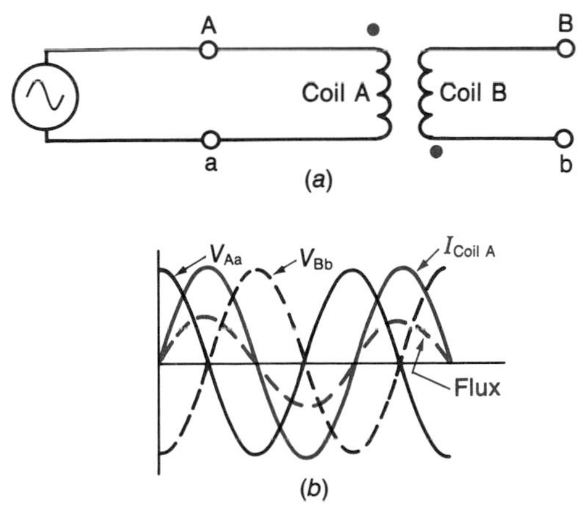


**FIGURE 20-5 Current and voltages in coupled coils.**

in Fig. 20-5*b*, the voltage of the coil $V_{Aa}$ leads the current by 90°. The flux created by the current is, of course, in phase with the current. Since the rate of change in the current, and, therefore, the flux, is greatest as the current goes through zero, the voltage induced into coil B, as well as coil A, is maximum as the current goes through its zero value. The voltage in coil B is 180° out of phase with the voltage in coil A when the two coils are wound in different directions and the same end (start end or finish end) of both coils is used as the reference end for the voltages. When the coils are wound in the same direction, the start end of one coil and the finish end of the other coil must be used as the reference ends for the primary and secondary voltages to be 180° out of phase.

It is customary to place a dot on one end of each transformer coil in a diagram to indicate that all dotted ends have the same instantaneous polarity. Thus, in Fig. 20-5*a* point A and point B have the same instantaneous polarities. It is also customary to place the dot indicators at opposite ends of the primary and secondary coils to show that the transformer is providing 180° of phase shift. Of course, this implies that the bottom ends of the coils in the schematic diagram are the reference points for the voltages. Whether or not a transformer provides 180° of phase shift in an actual electric circuit depends on how the coil lead wires are connected. Reversal of either the primary *or* the secondary leads changes the phase relationship from 0 to 180° or vice versa.

### EXAMPLE 20-3

At a specified instant, terminal 2 in Fig. 20-4 is positive with respect to terminal 1. At the same instant, what polarity is terminal 4 with respect to terminal 3?

#### Solution

By convention, the dots on terminals 1 and 3 must have the same instantaneous polarity. Therefore, the voltages at terminals 2 and 4 are in phase, and terminal 4 is positive with respect to terminal 3. The location of the dots also shows that the primary and secondary voltages are in phase when the bottom ends of both coils are used as the reference points.

### Self-Test

**5.** The winding of a transformer which is connected to the power source is the ______ winding.

**6.** The winding of a transformer which provides power for the load is the ______ winding.

**7.** The start end of a coil may be identified by a ______ on a schematic diagram.

**8.** In Fig. 20-5, the dot notation indicates that the phase angle between the primary voltage and the secondary voltage is ______ .

## 20-3 IDEAL IRON-CORE TRANSFORMER

The schematic symbol for an *iron-core transformer* is shown in Fig. 20-6. This type of transformer is used only at power and audio frequencies.

An *ideal iron-core transformer* would be 100 percent efficient and have a coefficient of coupling equal to 1 and a negligible primary current when the secondary is not loaded. This means the ideal transformer would have (1) no copper loss, i.e., zero coil resistance; (2) no core loss, i.e., no eddy currents or hysteresis; (3) no leakage flux, and (4) an extremely high inductance primary. Large, well-designed, iron-core transformers approach these ideal characteristics. Because real iron-core transformers can be so close to the ideal, it is meaningful to assume an ideal transformer in developing many of the principles that govern the behavior of transformers.

### Voltage Ratio

The ratio of the primary voltage over the secondary voltage is known as the *voltage ratio* of the transformer. The voltage ratio is also referred to as the *transformation ratio,* which is represented by the symbol *a*. The voltage ratio (i.e., transformation ratio) of a transformer is equal to the *turns ratio* of the transformer. This relationship can be expressed as

$$a = \frac{V_p}{V_s} = \frac{N_p}{N_s}$$

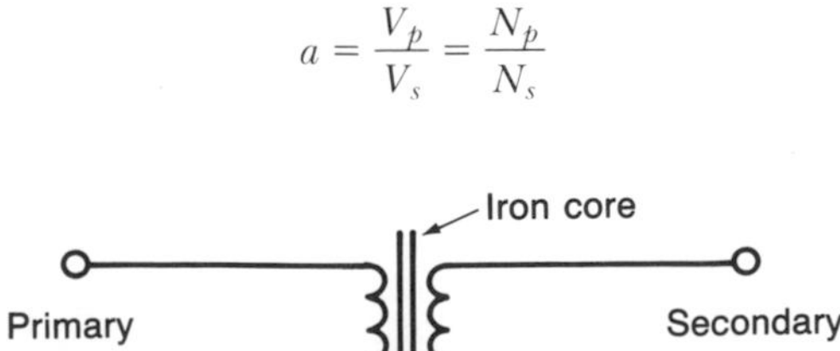


**FIGURE 20-6** Symbol for an iron-core transformer.

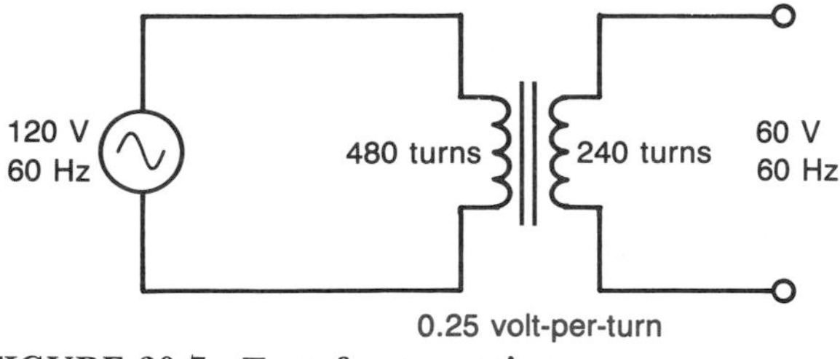

**FIGURE 20-7 Transformer ratios.**

where $N$ is the symbol for turns and the subscripts $p$ and $s$ indicate primary and secondary, respectively.

It is easy to understand why the turns ratio of a transformer equals the voltage ratio if one remembers two things: (1) the flux which produces the cemf in the primary is the same flux that induces a voltage into the secondary, and (2) the cemf of the primary equals the voltage applied to the primary. For the transformer in Fig. 20-7, the 480$N$ in the primary winding must be cut by sufficient flux to produce 120 V of cemf. Thus, each turn of the primary has 0.25 V of cemf induced into it. Now, with 100 percent coupling, the same amount of flux that cuts each turn of the primary also cuts each turn in the secondary. Thus, each turn in the secondary also has 0.25 V induced into it. Therefore, the voltage of the 240-turn secondary in Fig. 20-7 will be one-half of the primary voltage (60 V when the primary voltage is 120 V).

Notice in Fig. 20-7 that the *volts-per-turn ratio* (V/$N$) is 0.25 V/$N$ for both primary and secondary. The volts-per-turn ratio (and its reciprocal the turns-per-volt ratio) will always be the same for all windings of an ideal transformer. You can see that the turns-per-volt ratio and the volts-per-turn ratio are just shorthand ways of expressing the fact that the turns ratio equals the voltage ratio. That is, if

$$\frac{V_p}{V_s} = \frac{N_p}{N_s} \quad \text{then} \quad \frac{V_p}{N_p} = \frac{V_s}{N_s} \quad \text{and} \quad \frac{N_p}{V_p} = \frac{N_s}{V_s}$$

**EXAMPLE 20-4**

A transformer with a transformation ratio of 5 is connected to a 120-V, 60-Hz source. What is the turns ratio and the secondary voltage?

**Solution**

$$\text{Turns ratio} = a = 5$$

$$a = \frac{V_p}{V_s}$$

therefore,

$$V_s = \frac{V_p}{a} = \frac{120 \text{ V}}{5} = 24 \text{ V}$$

**EXAMPLE 20-5**

The turns-per-volt ratio of a transformer is 3$N$/V. The secondary voltage is 360 V and the primary has 720 $N$. Determine the primary voltage and the number of turns in the secondary.

**Solution**

Since the turns-per-volt ratio is the same for all windings,

$$V_p = \frac{720\ N}{3\ N/\text{V}} = 240 \text{ V}$$

$$N_s = 3\ N/\text{V} \times 360 \text{ V} = 1080\ N$$

Transformers are often called *step-up transformers* when the secondary voltage exceeds the primary voltage as in Example 20-5. Conversely, when the secondary voltage is less than the primary voltage, the transformer is referred to as a *step-down transformer*.

## Current Ratio and Power

The ideal transformer is 100 percent efficient; therefore, the primary power $P_p$ must equal the secondary power $P_s$. Furthermore, the primary current and voltage are in phase for the ideal transformer. This is so because due to its infinite primary inductance, the ideal transformer draws no primary current until a load is connected to the secondary. When power is drawn from the secondary, enough in phase primary current is drawn to make the primary power equal to the secondary power. The reason that the primary current increases when the secondary is loaded is because the magnetomotive force (mmf) of the current-carrying secondary coil opposes and cancels part of the mmf of the primary coil. Therefore, the primary current must increase to increase the mmf of the primary coil so that the net mmf remains constant. The net mmf must remain constant to produce the flux needed to create a cemf equal to the source voltage applied to the primary.

Even with a real transformer, the no-load primary current will be quite small compared to the full-load primary current. Because the no-load current is relatively small, the full-load primary current is nearly in phase with the primary voltage even though the no-load current is almost purely inductive (90° out of phase).

The *current ratio* $I_p/I_s$ of an ideal transformer can be determined by remembering that $P_p = P_s$ and then applying Ohm's law and the power formula to the secondary and primary circuits. Of course, it is only meaningful to speak of a transformer's current ratio when a load is applied to the secondary because only then is there any secondary current. Notice that the transformer's voltage ratio is independent of whether or not the transformer is loaded. For this reason, a transformer's voltage ratio is specified more often than is its current ratio.

---

**EXAMPLE 20-6**

Determine the current ratios for the transformer in the circuit shown in Fig. 20-8.

**Solution**

$$\frac{V_p}{V_s} = \frac{N_p}{N_s}$$

Therefore,

$$V_s = \frac{V_p N_s}{N_p} = \frac{100\text{ V} \times 600}{400} = 150\text{ V}$$

$$I_s = \frac{V_s}{R_s} = \frac{150\text{ V}}{50\ \Omega} = 3\text{ A}$$

Since the load is pure resistance, the cos $\theta$ can be dropped from the power formula.

$$P_s = P_p = I_s V_s = 3\text{ A} \times 150\text{ V} = 450\text{ W}$$

$$I_p = \frac{P_p}{V_p} = \frac{450\text{ W}}{100\text{ V}} = 4.5\text{ A}$$

$$\text{The current ratio} = \frac{I_p}{I_s} = \frac{4.5\text{ A}}{3\text{ A}} = 1.5{:}1$$

---

Notice in this example that the current was stepped down by a ratio of 1.5:1 while the voltage was stepped up by a ratio of 1:1.5. The current ratio is always the reciprocal of the voltage ratio, as shown by the following:

$$P_s = P_p$$

And, by substitution,

$$I_s V_s = I_p V_p$$

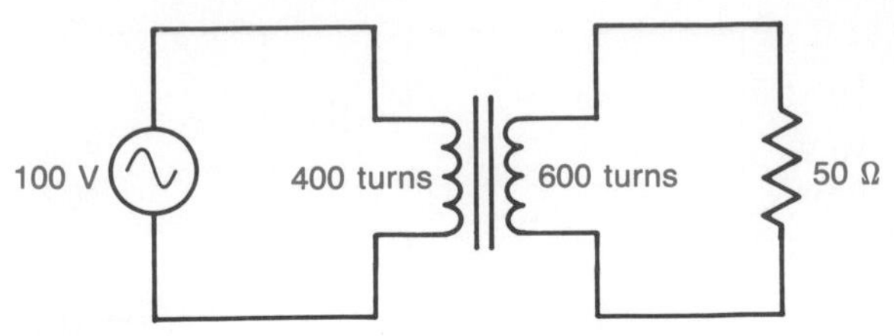

**FIGURE 20-8 Circuit for Example 20-6.**

And, by rearranging,

$$\frac{I_s}{I_p} = \frac{V_p}{V_s}$$

Since the voltage ratio is equal to the turns ratio, the current ratio is also the reciprocal of the turns ratio. That is,

$$\frac{I_s}{I_p} = \frac{N_p}{N_s}$$

---

**EXAMPLE 20-7**

If a 30-Ω resistor is connected to the secondary of the transformer in Fig. 20-7, how much primary current will flow?

**Solution**

$$I_s = \frac{V_s}{R_s} = \frac{60\text{ V}}{30\ \Omega} = 2\text{ A}$$

$$\frac{I_s}{I_p} = \frac{V_p}{V_s}$$

Therefore,

$$I_p = \frac{V_s I_s}{V_p} = \frac{60\text{ V} \times 2\text{ A}}{120\text{ V}} = 1\text{ A}$$

---

## Self-Test

**9.** List the characteristics of an ideal iron-core transformer.

**10.** The voltage ratio of a transformer is also called the ______ ratio.

**11.** Applying a load to the secondary causes the primary current to ______.

**12.** How many turns are required for a 36-V secondary if a 240-V primary has 600 $N$?

**13.** True or false? If the primary winding of a 1:2 step-up transformer has 3 $N$/V, the secondary winding will have 6 $N$/V.

**14.** True or false? The secondary flux of a transformer cancels part of the primary flux.

## Reflected Impedance and Impedance Ratio

In the explanation of current ratio, it was previously shown that a load on the secondary of a transformer causes the transformer's primary to draw more current from the source. Therefore, we can say that the load, whether it be an impedance or a resistance, is *reflected* through the transformer back to the source. The ideal transformer between the source and the load merely adjusts the source voltage to the level needed to operate the load. If the load is pure resistance, the primary current caused by the load will be in phase with the source voltage. If, on the other hand, the load contains some reactance, the primary current resulting from the load will not be in phase with the primary voltage. The phase angle will depend on the relative amounts of $X$ and $R$ in the load impedance.

The turns ratio of a transformer not only determines the voltage and current ratios, it also determines the *impedance ratio*. The fact that a transformer has an impedance ratio tells us that the impedance reflected back to the source is not equal to the secondary load impedance (unless the turns ratio is 1:1). Since the transformer can change both the voltage and the current levels between the source and the load, it is reasonable to expect it to be capable of changing the impedance level also.

The relationship between the impedance ratio and the turns ratio can be established by using Ohm's law and the relationship between the current, voltage, and turns ratios. By Ohm's law, the impedance in the secondary $Z_s$ is $Z_s = V_s/I_s$, and the primary impedance $Z_p$ "seen" by the source is $Z_p = V_p/I_p$. We already know that $a = V_p/V_s = I_s/I_p = N_p/N_s$. By appropriate substitution we can now show that

$$\frac{Z_p}{Z_s} = \frac{V_p/I_p}{V_s/I_s} = \frac{V_pI_s}{V_sI_p} = \left(\frac{V_p}{V_s}\right)\left(\frac{I_s}{I_p}\right) = \left(\frac{N_p}{N_s}\right)\left(\frac{N_p}{N_s}\right) = \left(\frac{N_p}{N_s}\right)^2$$

Thus, the impedance ratio is equal to the turns ratio squared. The impedance reflected back to the source $Z_p$ can be calculated by $Z_p = a^2Z_s$ because the transformation ratio $a$ is equal to the turns ratio.

The reflected impedance formula can also be used when the load on the transformer is pure resistance. Just substitute $R$ for $Z$. The appropriateness of substituting $R$ for $Z$ can be seen by looking at the power relationship in an ideal transformer:

$$P_p = P_s$$

By substituting $P = V^2/R$,

$$\frac{V_p^2}{R_p} = \frac{V_s^2}{R_s}$$

Rearranging and substituting,

$$\frac{R_p}{R_s} = \frac{V_p^2}{V_s^2} = \left(\frac{V_p}{V_s}\right)^2 = \left(\frac{N_p}{N_s}\right)^2 = a^2$$

Thus, $R_p = a^2R_s$. Usually, we use the formula $Z_p = a^2Z_s$ and the term impedance even when the load is pure resistance.

**EXAMPLE 20-8**
Determine the impedance (resistance) reflected back to the source in Fig. 20-8.

**Solution**

$$a = \frac{N_p}{N_s}$$

$$Z_p = a^2Z_s = \left(\frac{400}{600}\right)^2 \times 50\ \Omega = 22.2\ \Omega$$

Notice from Example 20-8 that a step-up transformer reflects an effective impedance back to the source which is less than the actual impedance of the load.

**EXAMPLE 20-9**
Determine the impedance ratio of the transformer in Fig. 20-7.

**Solution**

$$\frac{Z_p}{Z_s} = a^2 = \left(\frac{480}{240}\right)^2 = 4$$

The results of a transformer's ability to transform impedance are illustrated in Fig. 20-9. In Fig. 20-9*a*, a 100-Ω impedance is connected to a gener-

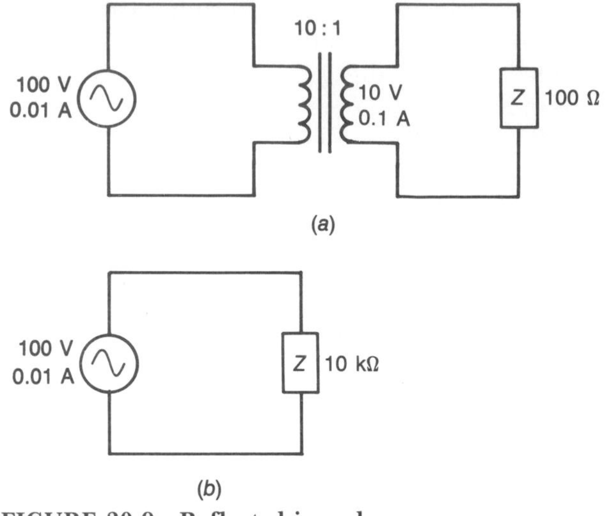


**FIGURE 20-9 Reflected impedance.**

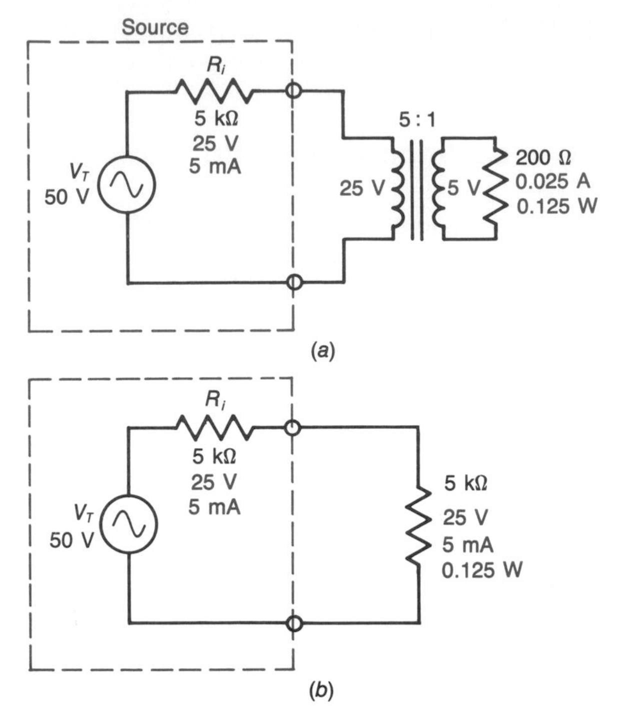


**FIGURE 20-10 Matching source and load resistances.**

ator through a transformer with a 10:1 turns ratio. This provides a 100:1 impedance ratio. The load, through the transformer, draws 0.01 A from the source. Figure 20-9*b* shows that a 10-kΩ impedance connected directly to the generator also draws 0.01 A from the source. The circuits in Fig. 20-9*a* and *b* are equivalent in terms of the load on the generator. This illustrates what is meant by statements such as, "the 100-Ω load is seen by the generator (source) as a 10-kΩ load," or "the 100-Ω load is reflected to the primary as a 10-kΩ load."

In Chap. 9 you studied the maximum power transfer theorem. In that chapter it was shown that transferring maximum power from a source to a load requires that the load resistance (or impedance) has to match (equal) the internal resistance (or impedance) of the source. *Impedance matching,* which is necessary for maximum transfer of power, can be accomplished with a transformer. Because a transformer can be made with any desired $Z$ ratio, it is possible to use a transformer to impedance-match any load to any source. Iron-core impedance-matching transformers are common in audio systems. For example, they can be used to (1) match the internal impedance of a microphone (the source) to the input impedance of an amplifier (the load for the microphone), (2) match the output impedance of one amplifier stage (the source) to the input impedance of the next amplifier stage (the load), or (3) match the output impedance of a power amplifier (the source) to the impedance of a speaker (the load).

As an example of impedance (or resistance) matching, Fig. 20-10*a* shows how maximum power is transferred from a 50-V source with 5 kΩ of internal resistance to a 200-Ω resistance load. The calculations used to obtain the values shown in contrasting color in this figure are as follows:

$$Z_p = a^2 Z_s = \left(\frac{5}{1}\right)^2 \times 200\ \Omega = 5\ \text{k}\Omega$$

Then, by the voltage division formula,

$$V_p = \frac{V_T Z_p}{R_i + Z_p} = \frac{50\ \text{V} \times 5\ \text{k}\Omega}{5\ \text{k}\Omega + 5\ \text{k}\Omega} = 25\ \text{V}$$

$$V_{R_i} = V_T - V_p = 50\ \text{V} - 25\ \text{V} = 25\ \text{V}$$

$$I_{R_i} = \frac{25\ \text{V}}{5\ \text{k}\Omega} = 5\ \text{mA}$$

$$V_s = \frac{V_p}{a} = \frac{25\ \text{V}}{5} = 5\ \text{V}$$

$$I_s = \frac{5\ \text{V}}{200\ \Omega} = 0.025\ \text{A}$$

$$P_s = 5\ \text{V} \times 0.025\ \text{A} = 0.125\ \text{W}$$

The equivalent circuit of Fig. 20-10*b* shows that connecting a 5-kΩ load to the source also transfers 0.125 W to the load.

When the source has actual internal impedance rather than internal resistance, the situation is

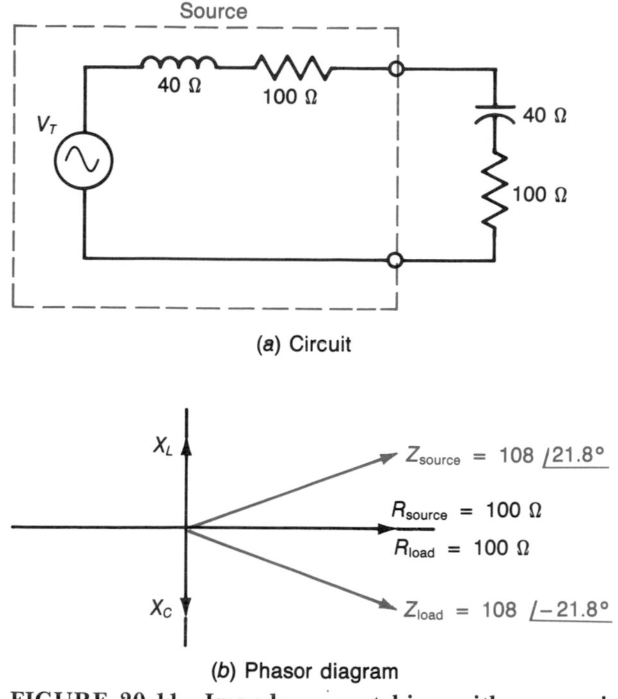


**FIGURE 20-11 Impedance matching with a capacitive load.**

slightly different. For maximum power transfer, the load impedance must be the *conjugate* of the internal impedance. Conjugate means that the phase angle of the load impedance must be equal in magnitude but of the opposite sign to the internal impedance and that the magnitude of the two impedances must be equal. Under these conditions, as seen in Fig. 20-11, the reactance of the load is canceled by the opposite type of reactance of the source. This leaves the resistance of the load equal to the resistance of the source, which is the required condition for maximum power transfer. When a transformer is connected between the source and the load, the magnitude of the impedance is transformed but the phase angle remains unchanged.

## EXAMPLE 20-10

Verify that maximum power is being transferred in Fig. 20-12*a*.

### Solution

$$Z_p = \left(\frac{2}{1}\right)^2 \times 300\ \Omega\ \underline{/-30^\circ} = 1200\ \Omega\ \underline{/-30^\circ}$$

Draw and analyze the equivalent circuit in Fig. 20-12*b*.

$$\begin{aligned} Z_T = Z_i + Z_p &= 1200\ \Omega\ \underline{/30^\circ} + 1200\ \Omega\ \underline{/-30^\circ} \\ &= (1039\ \Omega + j600\ \Omega) + (1039\ \Omega - j600\ \Omega) \\ &= 2078\ \Omega\ \underline{/0^\circ} \end{aligned}$$

The calculation of $Z_T$ shows that the resistive parts of $Z_i$ and $Z_p$ are equal (1039 Ω each) and that the reactive parts of $Z_p$ and $Z_i$ cancel each other. Therefore, the conditions for maximum power transfer have been met.

## EXAMPLE 20-11

Verify that the power of $Z_p$ in Fig. 20-12*b* is equal to the power of $Z_s$ in Fig. 20-12*a*. That is, verify that $P_p = P_s$ when an ideal transformer is loaded by an impedance.

### Solution

Calculate the magnitude of $I_T$ (which is also the primary current $I_p$) using $Z_T$ from Example 20-10.

$$I_T = \frac{V_T}{Z_T} = \frac{100\ \text{V}}{2078\ \Omega} = 48\ \text{mA}$$

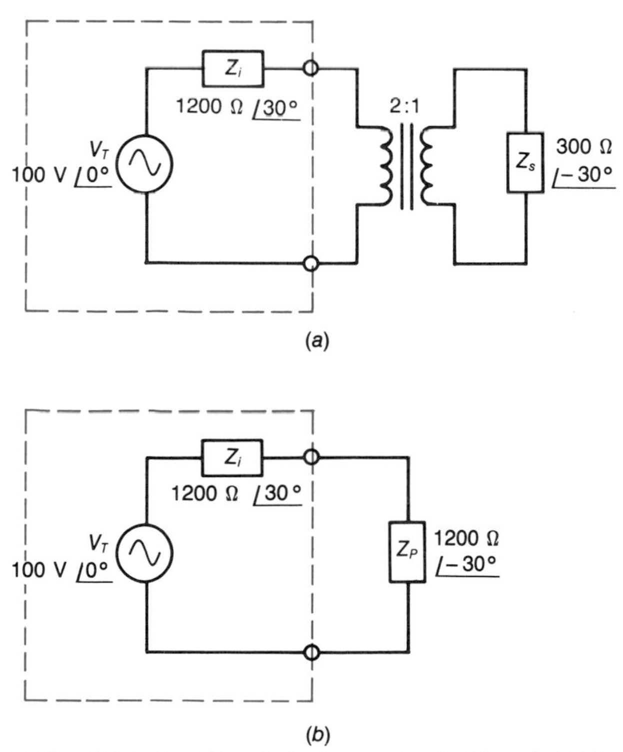


**FIGURE 20-12 Circuit for Examples 20-10 and 20-11.**

Since only the resistance of $Z_p$ uses power, the power of $Z_p$ is

$$P_p = I_p^2 R_p = (48 \text{ mA})^2 \times 1039 \ \Omega = 2.4 \text{ W}$$

The magnitude of the voltage across the primary $V_p$ in Fig. 20-12*a* will be

$$V_p = Z_p I_p = 1200 \ \Omega \times 48 \text{ mA} = 57.6 \text{ V}$$

$$I_s = \frac{V_s}{Z_s} = \frac{28.8 \text{ V}}{300 \ \Omega} = 96 \text{ mA}$$

(Note that $I_s$ could also be determined by the current ratio and $I_T$, which is equal to $I_p$.)

Next, convert $Z_s$ to rectangular form to find its resistive part:

$$Z_s = 300 \ \Omega \angle -30^\circ = 260 \ \Omega - j150 \ \Omega$$

The power of $Z_s$ can now be found by

$$P_s = I_s^2 R_s = (96 \text{ mA})^2 \times 260 \ \Omega = 2.4 \text{ W}$$

Thus, $P_s = P_p$, which is the case for ideal transformers.

## Self-Test

**15.** Define reflected impedance.

**16.** What is the relationship between the impedance ratio and the voltage ratio?

**17.** True or false? The primary impedance $Z_p$ of a step-up transformer will be greater than the secondary load impedance $Z_s$.

**18.** For maximum transfer of power from a source with $50 \ \Omega \angle 40^\circ$ internal impedance, the load should be ______.

**19.** An ideal transformer requires 80 V across the primary to produce a secondary voltage of 10 V. What is the impedance ratio?

**20.** A 10-Ω resistive load is to be impedance-matched by a transformer to a source with 6250 Ω of internal resistance. Determine the turns ratio of the transformer.

**21.** If the open-circuit voltage of the source in Question 20 is 50 V, what will the secondary power be?

## 20-4 REAL IRON-CORE TRANSFORMERS

The formulas and concepts developed for ideal transformers are useful in dealing with real transformers even though real transformers are not 100 percent efficient, do not have unity coupling, and do have significant primary current under no-load conditions. In general, the smaller the power rating of the transformer, the further the transformer's parameters will be from the ideal parameters.

### Magnetizing Current

The *magnetizing current,* also called the *exciting current,* refers to the current that flows in the primary of the transformer when there is no load on the secondary. This is called the magnetizing current because it is the current required to build up the flux in the core of the transformer. Of course, the flux is required to produce the cemf, which must equal the source voltage.

The exciting current is mostly controlled by the inductive reactance of the primary coil. As seen in Fig. 20-13, the current caused by the primary reactance is 90° out of phase with the voltage applied to the primary. However, a small amount of the exciting current is resistive current that is in phase with the voltage. This resistive current is caused by the core losses (hysteresis and eddy current) and the copper loss. Since the exciting current is small (compared to the primary current under load conditions), the copper loss will be negligible. Thus, the resistive current will be essentially the result of core losses. As shown in Fig. 20-13, the resistive current is small compared to the reactive current, so the magnetizing current is nearly 90° out of phase and the cos $\theta$ is very small. Thus, the power used by an unloaded transformer is very small compared to the power the transformer can deliver to a load (less than 1 percent for large transformers).

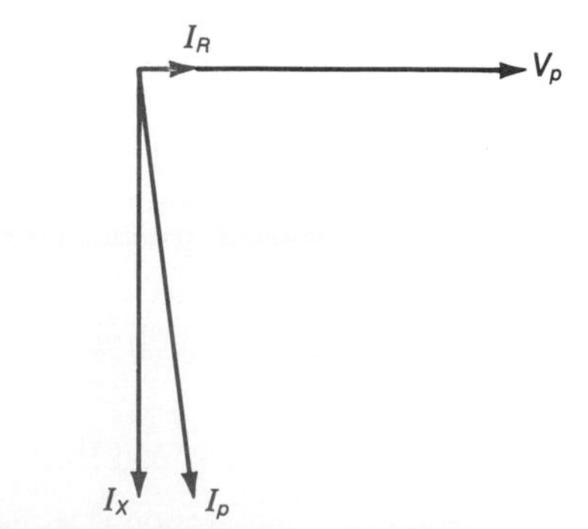


**FIGURE 20-13 Exciting current in the primary of a transformer.**

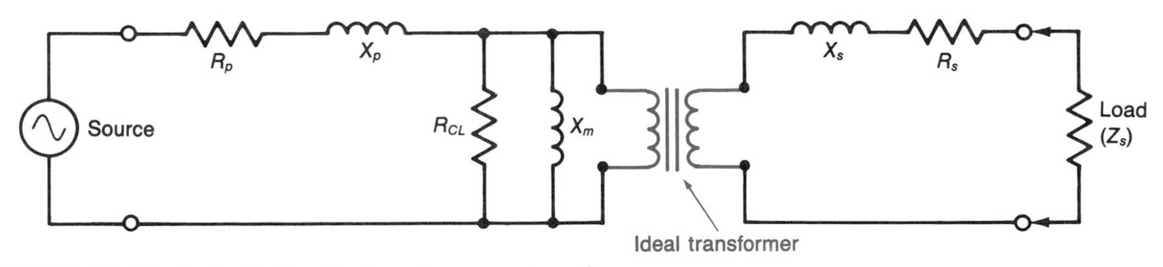


**FIGURE 20-14 Equivalent circuit of an iron-core transformer.**

## The Loaded Transformer

Even though the permeability of an iron core is thousands of times larger than that of air, some of the flux from the coils of the transformer will pass through the air rather than through the core. This is the leakage flux which does not link up with another coil. The leakage flux from the primary coil creates a small reactance which behaves as if it is in series with the primary winding. Likewise, leakage flux from the secondary winding causes a reactance in series with the secondary coil. The reactances caused by the leakage flux are shown in Fig. 20-14 as $X_p$ and $X_s$.

Figure 20-14 shows a simple equivalent circuit for a real iron-core transformer. The resistances of the primary and secondary windings $R_p$ and $R_s$ behave as if they are in series with their respective windings. The core loss (CL) is represented by a resistance $R_{CL}$ in parallel with the mutual reactance $X_m$ which controls the exciting current. All of these real transformer constants behave as if they were connected (as shown in Fig. 20-14) to an ideal transformer.

From the equivalent circuit in Fig. 20-14, one can determine what happens when a purely resistive load is connected to the secondary of a transformer. The secondary current, as mentioned earlier, produces an mmf which opposes the mmf of the primary current. This not only results in an increase in the primary current but also in a change in the phase angle of the current. Since the resistive load current results in an in phase primary current, angle $\theta$ decreases as the load current increases. Figure 20-15 shows that angle $\theta$ is very small when the transformer is fully loaded. This is because the in phase current caused by the load is so much greater than the out-of-phase exciting current. With large transformers, the phase angle can be less than 5° when the load is pure resistance. The cosine of 5° is 0.996; so, dropping the cos $\theta$ from the power formula when working with transformers that have resistive loads may not introduce much error. Of course, if the transformer load is an impedance, the additional primary current caused by the load will not be in phase with the primary voltage. In fact, if the load is pure reactance, the additional primary current will be almost 90° out of phase. If it were not for the increased copper loss caused by the increased currents, the additional primary current would be a full 90° out of phase.

When a load is connected to a transformer, the secondary terminal voltage decreases. Inspection of Fig. 20-14 shows why this happens. The resistance and reactance $R_s$ and $X_s$ in series with the secondary coil of the ideal transformer in the equivalent circuit drop part of the available coil voltage. The secondary voltage of the terminals (voltage across the load) decreases as the load current increases. This is why the loaded secondary voltage

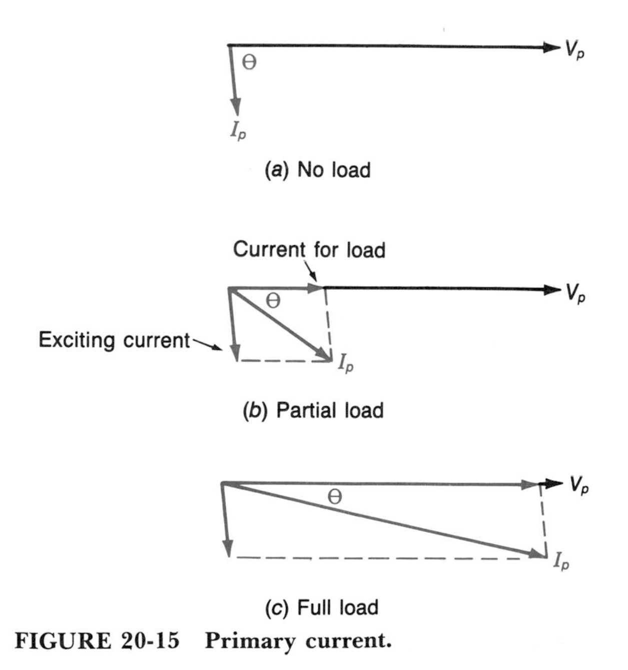


**FIGURE 20-15 Primary current.**

of a transformer is typically 5 to 10 percent less than the open-circuit secondary voltage.

Figure 20-14 also shows why the power loss in a transformer increases as the load current increases. Although the core losses are determined by the exciting current, which is independent of the load current, the copper losses are a function of the load current. Because $P = I^2R$, doubling the load current quadruples the copper loss in the secondary and greatly increases the copper loss in the primary. (Due to the exciting current, the primary current does not double when the load current doubles.)

## Transformer Losses

Core material and construction techniques are chosen to minimize the value of the exciting current, the leakage flux, and the core losses. Transformer cores are constructed from silicon steel laminates. Silicon steel is used because it has a high permeability and a narrow hysteresis loop. The silicon steel is laminated to reduce the eddy currents. (These topics are discussed in Chap. 13 on inductance.) A typical arrangement of core laminates is shown in Fig. 20-16. Notice that the I- and E-shaped laminates are alternately reversed so that there is no continuous air path or gap through the cross section of the core. This construction technique keeps the permeability of the core as high as possible. With this type of construction, the center leg of the E-shaped laminates is fed through the coil form on which the primary and secondary coils have been wound. Except for alternating the I- and E-shaped laminates, the core of a transformer is constructed just like the core of an inductor. Notice that the center leg of the E-shaped laminate is twice as wide as either outer leg. This is necessary because the center leg carries the flux for both outer legs.

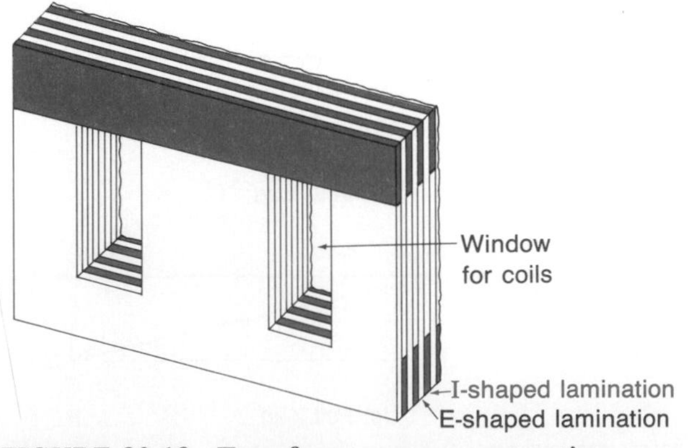


**FIGURE 20-16 Transformer core construction.**

The copper loss, which is caused by current flowing through the resistance of the windings, increases as the transformer is loaded. Under full-load conditions, the copper loss can be at least as significant as the core losses. Copper loss is minimized by using as large a diameter coil wire as practical. The diameter of the wire is limited, however, by the size of the window opening in the core, the number of turns needed in the primary to control the exciting current, and the number of turns needed in the secondary to obtain the desired secondary voltage.

## Efficiency

The efficiency $\eta$ of a transformer is calculated using the general efficiency formula $\eta = P_o/P_i$. The $P_o$ in the formula is the power $P_s$ provided by the secondary to the load. The $P_i$ is the power provided to the primary $P_p$ by the main source. Power $P_i$ is equal to the load power $P_s$ plus the copper loss $P_{\text{copper}}$ plus the core loss $P_{\text{core}}$. Thus, for a transformer, the efficiency formula is

$$\eta = \frac{P_s}{P_p} = \frac{P_s}{P_s + P_{\text{copper}} + P_{\text{core}}}$$

---

**EXAMPLE 20-12**

The secondary of a transformer provides 90 V across an impedance of 15 Ω $\angle 25°$. The core losses are 12 W and the copper losses are 16 W. Determine the percentage of efficiency of the transformer.

**Solution**

$$I_s = \frac{V_s}{Z_s} = \frac{90\text{ V}\angle 0°}{15\ \Omega \angle 25°} = 6\text{ A}\angle -25°$$

$$P_s = I_s V_s \cos\theta = 6\text{ A} \times 90\text{ V} \times \cos 25° = 489\text{ W}$$

$$\eta = \frac{489\text{ W}}{489\text{ W} + 12\text{ W} + 16\text{ W}} = 0.946$$

$$\text{Percentage of } \eta = \eta \times 100 = 94.6\text{ percent}$$

---

## Self-Test

**22.** When there is no load on the secondary, the primary current is called _______ current.

**23.** True or false? Without a load on the transformer, the primary current is in phase with the primary voltage.

**24.** Does each of the following increase, decrease, or remain the same when a load is connected to the secondary of a transformer?

**a.** Primary current
**b.** Angle $\theta$
**c.** Secondary voltage
**d.** Core losses
**e.** Copper losses

**25.** Why is the iron core of a transformer constructed from laminations?

**26.** Determine the efficiency of a transformer that produces 5 A through a $15\text{-}\Omega \angle 20^\circ$ load if the transformer losses (copper + core) are 19 W.

## 20-5 MEASURING CORE AND COPPER LOSSES

A close estimate of the core and copper losses in a transformer can be obtained by measuring the primary power under two conditions: (1) the secondary open-circuited and the primary at rated voltage, and (2) the secondary short-circuited and the primary voltage reduced to the value needed to cause the maximum rated current flow in the windings. In making these tests, either winding can be used as the primary in order to use the available variable voltage source and wattmeter.

The open-circuit test is illustrated in Fig. 20-17. In performing this test, the variable ac source is adjusted so that the voltmeter indicates the voltage rating of the primary. The ammeter indicates the magnetizing (exciting) current. Since there is no load (secondary) power in this test circuit, the wattmeter indicates the power losses of the transformer under no-load conditions. The no-load power loss is almost entirely core loss because no secondary current is flowing and only the exciting current flows in the primary. Thus, the copper loss, under no-load conditions, is usually negligible compared to the core loss. The no-load copper loss can be estimated using the measured exciting current and the measured (or specified) primary ohmic resistance ($P = I^2R$). This calculated copper loss can be subtracted from the wattmeter reading to give a closer estimate of the core loss.

Within the rated power of a transformer, the full-load core loss is essentially equal to the no-load core loss. On first thought, it might seem that the full-load core loss should be larger because of the larger primary current. However, core loss (hysteresis and eddy current) depends on the magnitude and rate of change of the core flux, and these factors remain constant from no-load to full-load conditions. If they did not, the cemf of the primary would not equal the source voltage and the induced secondary voltage would change.

The short-circuit test, illustrated in Fig. 20-18, determines the full-load copper loss. In this circuit, the variable voltage source is increased, from zero output, until the ammeter indicates the rated full-load primary current. Under these conditions, the shorted secondary also has full-load current because of the current ratio of the transformer. Since the secondary is not delivering any power to a load, the wattmeter now indicates the transformer losses under full-load current conditions. Because the secondary is short-circuited, the primary voltage is only a small fraction of its rated value when rated currents are flowing. Therefore, the cemf, the flux needed to produce the cemf, and the exciting current are all a small fraction of their normal, rated-primary-voltage values. With these greatly reduced values, the core loss is very small. For example, when $V_p$ is 10 percent of rated value, the voltage induced into the core and the resulting eddy current are also reduced to one-tenth of their normal values. Since the core resistance has not changed, the eddy-current loss is 1 percent of its normal value ($P = I^2R$). For this reason, essentially all of the indicated power is due to copper loss. Thus, the wattmeter reading for the short-circuit test

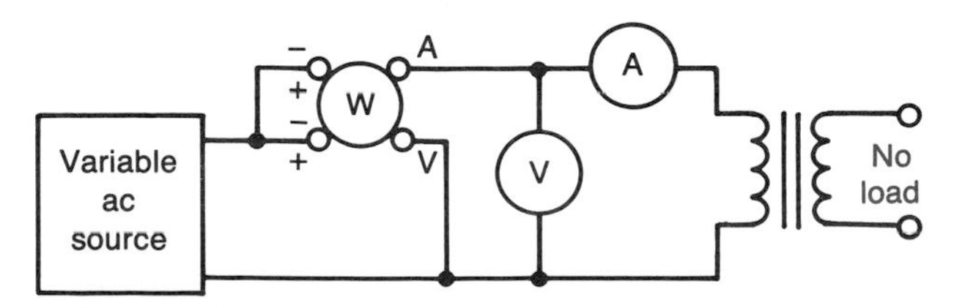


**FIGURE 20-17 Measuring core losses.**

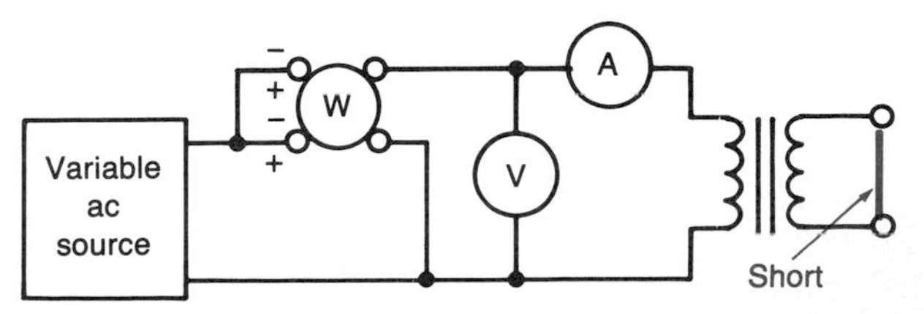


**FIGURE 20-18 Measuring copper losses.**

gives a reasonable estimate of the copper loss under full-load conditions.

The copper loss of a transformer can also be estimated by using the measured winding resistance, the voltage ratings, and the rated secondary current. At power frequencies, the skin effect is negligible. Therefore, the measured resistance and the rated current can be used to determine the copper loss of a winding.

**EXAMPLE 20-13**

A transformer has a 120-V primary and 20-V secondary rated at 4 A. The measured winding resistances are 0.3 Ω and 5 Ω for secondary and primary, respectively. Estimate the copper loss.

**Solution**

$$I_p = \frac{I_s V_s}{V_p} = \frac{4\text{ A} \times 20\text{ V}}{120\text{ V}} = 0.67\text{ A}$$

$$\text{Copper loss, primary} = I_p^2 R_p = (0.67\text{ A})^2 \times 5\ \Omega = 2.2\text{ W}$$

$$\text{Copper loss, secondary} = I_s^2 R_s = (4\text{ A})^2 \times 0.3\ \Omega = 4.8\text{ W}$$

$$\text{Total copper loss} = 2.2\text{ W} + 4.8\text{ W} = 7\text{ W}$$

Notice in this example that the relationships developed for ideal transformers were used. Thus, the result is an estimate, but it is a very useful estimate.

## Self-Test

27. The open-circuit test measures the ______ losses of a transformer.

28. Why are the core losses negligible during the short-circuit test?

29. A transformer has the following ratings: $V_p =$ 120 V, $a = 6$, $R_p = 5\ \Omega$, and $R_s = 0.2\ \Omega$. Determine the copper loss when the load current is 6 A.

## 20-6 EQUIVALENT CIRCUITS

In addition to determining copper and core losses, the data from the short-circuit and open-circuit tests can be used to estimate some of the constants specified in the equivalent circuit of Fig. 20-14 which is redrawn as Fig. 20-19*a*. Because any secondary impedance is reflected back to the primary by the impedance ratio $a^2$ of the ideal transformer, the equivalent circuit of Fig. 20-19*a* can be redrawn as shown in Fig. 20-19*b*. In this figure, $R_s$, $X_s$, and $Z_s$ are referred back to the primary so that the entire transformer can be analyzed by knowing the power, current, and voltage supplied to the primary.

With the open-circuit test, $a^2X_s$ and $a^2R_s$ of Fig. 20-19*b* can be ignored because no current flows through them. Also, as discussed previously, $R_p$ and $X_p$ are so small compared to $R_{CL}$ and $X_m$ that they are insignificant for the open-circuit test. Thus, the open-circuit test reduces to the circuit shown in Fig. 20-19*c*. The current, voltage, and power measurements from the open-circuit test can now be used to analyze this circuit using the standard power and impedance formulas given below:

$$R = \frac{V^2}{P}$$

so

$$R_{CL} = \frac{V_{P_{OC}}^2}{P_{OC}}$$

where $P_{OC}$ is the open-circuit primary power and $V_{P_{OC}}$ is the open-circuit primary voltage.

$$Z_{OC} = \frac{V_{P_{OC}}}{I_{P_{OC}}} = \frac{R_{CL}X_m}{\sqrt{R_{CL}^2 + X_m^2}}$$

Solving for $X_m$ yields

$$X_m = \frac{\sqrt{Z_{OC}^2 R_{CL}^2}}{\sqrt{R_{CL}^2 - Z_{OC}^2}}$$

Note that $X_m$ can also be calculated by using the admittance $Y$, conductance $G$, and susceptance $B$ relationship:

$$B_m = \sqrt{Y^2 - G^2}$$

where

$$B_m = \frac{1}{X_m} \qquad G = \frac{1}{R_{CL}} \qquad \text{and} \qquad Y = \frac{1}{Z} = \frac{I_{P_{OC}}}{V_{P_{OC}}}$$

From the short-circuit text data, we can determine the equivalent winding resistance $R_e$ and the equivalent winding reactance $X_e$ as seen by the source. Referring to the equivalent circuit of Fig. 20-19*d*, we can see that $R_e = R_p + a^2R_s$ and $X_e =$ $X_p + a^2X_s$. As stated previously, $R_{CL}$ and $X_m$ are so large compared to $a^2X_s + a^2R_s$ that they draw an

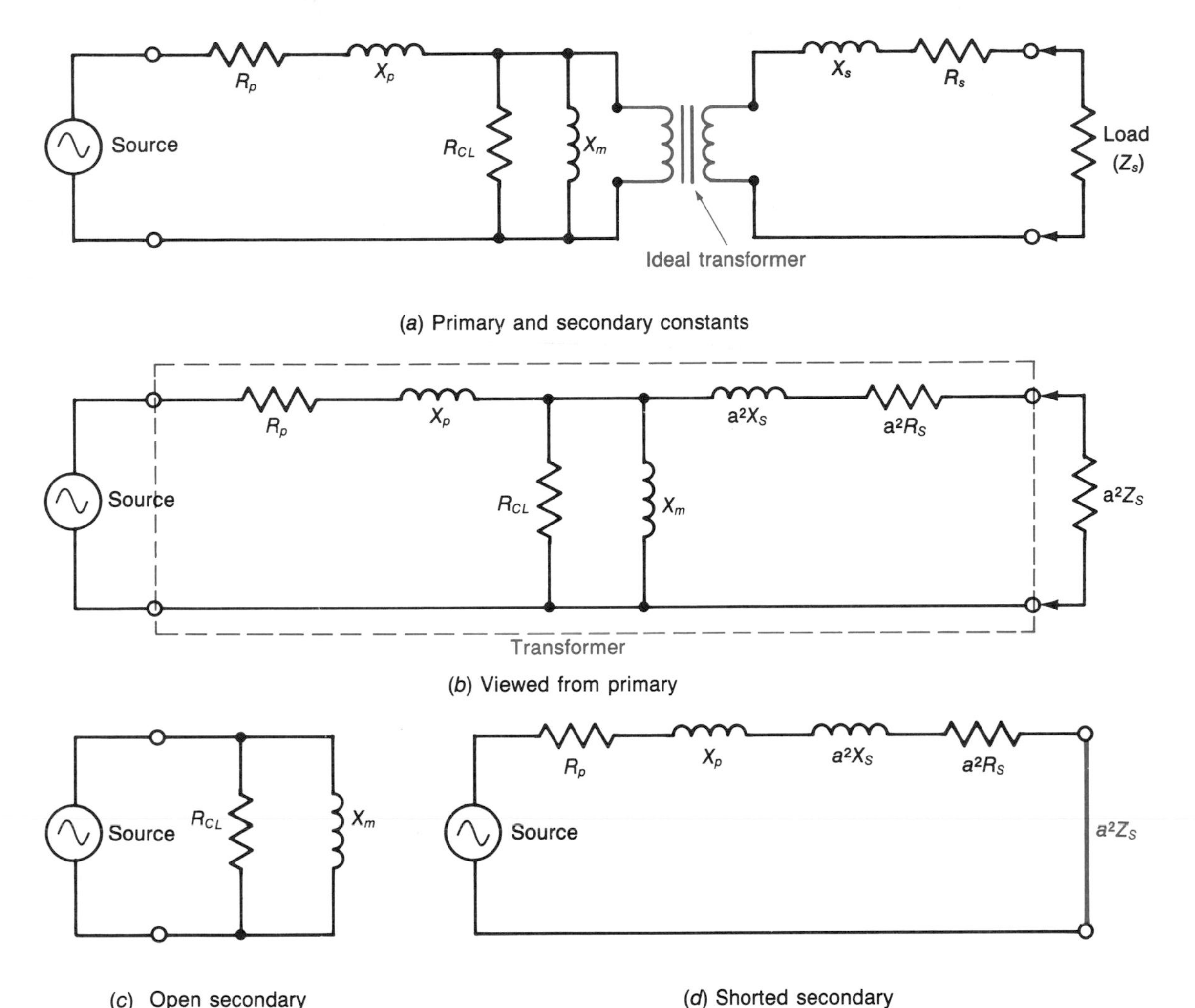


**FIGURE 20-19 Equivalent circuits of a transformer.**

insignificant current under short-circuit test conditions. Thus, they can be omitted from the equivalent circuit of Fig. 20-19*d*. The formulas needed to determine $R_e$ and $X_e$ from the short-circuit (SC) data are

$$R = \frac{P}{I^2}$$

so

$$R_e = \frac{P_{SC}}{I_{P_{SC}}^2}$$

$$Z_{SC} = \frac{V_{P_{SC}}}{I_{P_{SC}}} = \sqrt{R_e^2 + X_e^2}$$

Solving for $X$ yields

$$X_e = \sqrt{Z_{SC}^2 - R_e^2}$$

**EXAMPLE 20-14**

The open-circuit and short-circuit tests on a transformer yielded 3 W, 208 V, 0.05 A and 5 W, 10 V, 1 A, respectively. Determine $R_{CL}$, $X_m$, $R_e$, and $X_e$.

**Solution**

Open circuit:

$$R_{CL} = \frac{(208\ \text{V})^2}{3\ \text{W}} = 14.42\ \text{k}\Omega$$

$$Z_{OC} = \frac{208\ \text{V}}{0.05\ \text{A}} = 4.16\ \text{k}\Omega$$

$$X_m = \frac{\sqrt{4.16^2 \times 14.42^2}}{\sqrt{14.42^2 - 4.16^2}} = 4.34\ \text{k}\Omega$$

Short circuit:

$$R_e = \frac{5\text{ W}}{(1\text{ A})^2} = 5\ \Omega$$
$$Z_{SC} = \frac{10\text{ V}}{1\text{ A}} = 10\ \Omega$$
$$X_e = \sqrt{10^2 - 5^2} = 8.7\ \Omega$$

From the values obtained in Example 20-14 above, you can see that omitting $R_p$ and $X_p$ from the open-secondary equivalent circuit introduces almost no error. Also, omitting $R_{CL}$ and $X_m$ from the shorted-secondary equivalent circuit results in an exceedingly small error.

The open-circuit and short-circuit test data, along with the rated secondary current and voltage, allow one to determine the efficiency of a transformer and angle $\theta$ for the primary voltage and current.

**EXAMPLE 20-15**

The transformer in Example 20-14 provides 20 V at 9.5 A to a pure resistance load. Determine its efficiency. Also determine angle $\theta$ for no-load and full-load conditions.

**Solution**

$$P_o = 20\text{ V} \times 9.5\text{ A} = 190\text{ W}$$

Using data from Example 20-14,

$$\eta = \frac{P_o}{P_i} = \frac{190\text{ W}}{190\text{ W} + 5\text{ W} + 3\text{ W}} = \frac{190\text{ W}}{198\text{ W}} = 0.96$$
$$P = IV\cos\theta$$

Therefore,

$$\cos\theta = \frac{P}{IV}$$

For no-load (open-circuit test) conditions,

$$\cos\theta = \frac{3\text{ W}}{0.05\text{ A} \times 208\text{ V}} = 0.288$$
$$\theta = \arccos\theta = \arccos 0.288 = 73.2^\circ$$

For full-load conditions (use $I_p$ from the short-circuit test),

$$\cos\theta = \frac{198\text{ W}}{1\text{ A} \times 208\text{ V}} = 0.952$$
$$\theta = \arccos 0.952 = 17.8^\circ$$

By definition, $IV$ is the *apparent power,* which is symbolized by an $S$. Therefore, the formula $\cos\theta = P/IV$ used in Example 20-15 is often written $\cos\theta = P/S$.

A simplified, practical equivalent circuit for a fully loaded transformer is shown in Fig. 20-20*b*. The justification for the equivalent circuit of Fig. 20-20*b* can be seen from the values given for the constants in the more complex circuit of Fig. 20-20*a*. These values are the ones determined in Examples 20-14 and 20-15. Notice that the value of $R_{CL} \parallel X_m$ is approximately 20 times as large as $a^2Z_s + a^2R_s + ja^2X_s$. Thus, omitting $R_{CL}$ and $X_m$ in the equivalent circuit of Fig. 20-20*b* introduces about 5 percent error in this case. With this circuit, it is quite easy to determine the $I_p$, $\eta$, and $\theta$ for a loaded transformer even when the load is not pure resistance.

**EXAMPLE 20-16**

A transformer with a 240-V primary and a transformation ratio of 5 is connected to an $8\text{-}\Omega\ \angle 30^\circ$ load. Determine $I_p$, $\eta$, and $\theta$ if $X_e = 10\ \Omega$ and $R_e = 12\ \Omega$.

**Solution**

$$a^2Z_s = 5^2 \times 8\ \Omega\ \angle 30^\circ = 200\ \Omega\ \angle 30^\circ$$
$$Z_t = 10\ \Omega\ \angle 90^\circ + 12\ \Omega\ \angle 0^\circ + 200\ \Omega\ \angle 30^\circ$$
$$= j10\ \Omega + 12\ \Omega + 173.2\ \Omega + j100\ \Omega$$
$$= 185.2\Omega + j110\ \Omega = 215.4\ \Omega\ \angle 30.7^\circ$$
$$I_p = \frac{V_p}{Z_p} = \frac{240\text{ V}\ \angle 0^\circ}{215.4\ \Omega\ \angle 30.7^\circ} = 1.11\text{ A}\ \angle -30.7^\circ$$
$$\theta = 30.7^\circ$$
$$\cos\theta = 0.86$$
$$P_i = I_pV_p\cos\theta = 1.11\text{ A} \times 240\text{ V} \times 0.86$$
$$= 229.1\text{ W}$$

Next, determine the resistive part $R_{s_R}$ of $a^2Z_s$:

$$R_{s_R} = \cos 30^\circ \times 200\ \Omega = 173.2\ \Omega$$
$$P_o = I_p^2R_{s_R} = (1.11\text{ A})^2 \times 173.2\ \Omega = 213.4\text{ W}$$
$$\eta = \frac{213.4\text{ W}}{229.1\text{ W}} = 0.93$$

The $P_o$ in Example 20-16 could have been found by subtracting $I_p^2R_e$ from $P_i$ because only $R_e$ causes inefficiency in the equivalent circuit of Fig. 20-20*b*. Also, the $P_i$ for the example could have been found by $I_p^2(R_e + R_{s_R})$ because only resistance uses

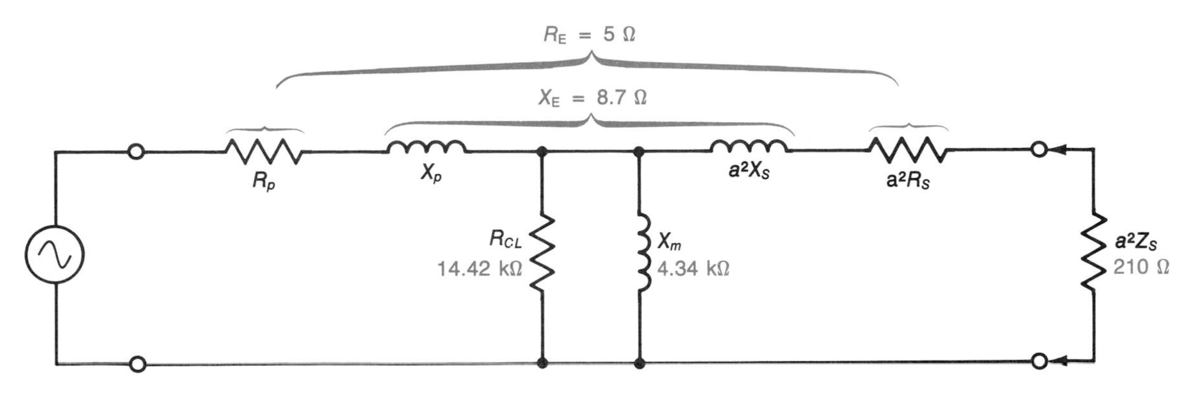


(a) Equivalent circuit including $R_{CL}$ and $X_m$

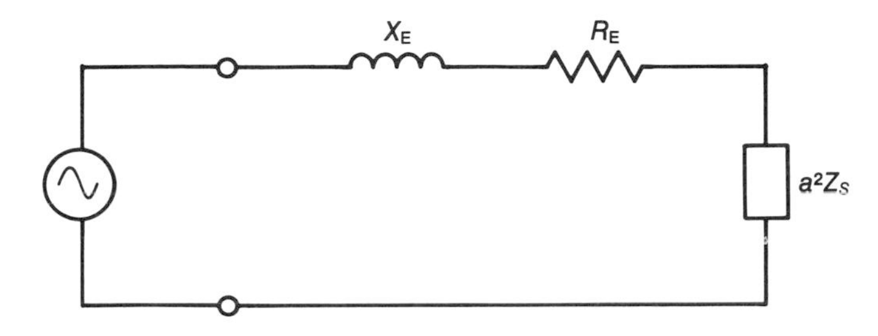


(b) Practical equivalent circuit

**FIGURE 20-20 Equivalent circuit for a fully loaded transformer.**

power. Now we can see that, for the circuit of Fig. 20-20*b*, $\eta$ can be determined by

$$\eta = \frac{P_o}{P_i} = \frac{I_p^2 R_{S_R}}{I_p^2 (R_e + R_{S_R})} = \frac{R_{S_R}}{R_e + R_{S_R}}$$

where $R_{S_R}$ is the resistive part of $a^2 Z_S$. From this expression of efficiency, we can see why the efficiency of a transformer decreases as the load becomes more reactive.

## Self-Test

**30.** Test results on a transformer yielded these values: $I_{P_{SC}} = 0.48$ A, $V_{P_{SC}} = 9$ V, $P_{SC} = 3$ W, $I_{P_{OC}} = 0.04$ A, $V_{P_{OC}} = 120$ V, and $P_{OC} = 2.5$ W. Determine $X_m$, $R_{CL}$, $R_e$, and $X_e$.

**31.** The transformer in Question 30 has a turns ratio of 5 and it is loaded with a 10-Ω $\angle 30°$ impedance. Determine $\theta$, $\eta$, $I_p$, $I_s$, and $V_s$ under loaded conditions.

**32.** Will the efficiency of a transformer be greater when the load is 2 Ω $\angle 0°$ or 2 Ω $\angle 20°$? Why?

**33.** Will angle $\theta$ be larger when the load is 4 Ω $\angle 10°$ or 4 Ω $\angle 60°$?

## 20-7 MULTIPLE AND TAPPED WINDINGS

Transformers can have more than one primary winding and/or secondary winding. Also, either (or both) the primary or the secondary winding can be tapped with one or more taps. Figure 20-21 shows some common arrangements for transformer windings. In Fig. 20-21*a* the primary is tapped so that the transformer will work equally well in locales where the nominal voltage available for the primary is either 110 or 120 V.

The transformer in Fig. 20-21*b* has two primary windings so that it can be operated from either a 120- or a 240-V source. When operated from a 120-V source, the two primaries are connected in parallel. For 240-V operation, the primaries are series-connected. Of course, the two primaries must be properly phased, so that the mmf, and thus the flux, of the two coils aid each other. Also, for the parallel connection, the two windings must have identical voltage and current ratings.

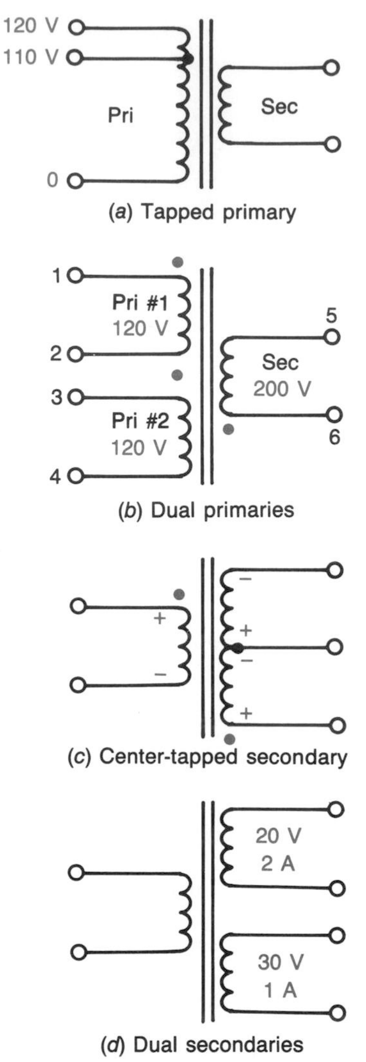


**FIGURE 20-21 Transformer winding configurations.**

When the schematic diagram of the transformer has dot notation, correct phasing is straightforward if we remember that the dotted ends of the coils all have the same instantaneous polarity. Figure 20-22 shows the correct connections for parallel- and series-connected primaries. The polarities shown on the diagrams are instantaneous; of course, they reverse every $\frac{1}{2}$ cycle. The series connection is *series-aiding;* that is, the cemf of one coil aids the cemf of the other coil. Inspection of Fig. 20-22 shows that, with either the series or the parallel connection, accidental reversal of one winding causes the cemf of the two coils to cancel each other. Under this condition, the primary current would be limited only by the coil resistance; the coils would rapidly burn up unless the primaries were properly fused. Figure 20-22 also shows that each primary has the same voltage across it, and thus the same current through it, whether connected in series or parallel. The main difference between the two connections is that the series connection requires half as much current from the source.

With most multiple winding transformers, the manufacturer furnishes a chart (often printed on the coil insulation paper) that shows the correct terminal connection for series and parallel operation of the windings. When such a chart is not available, correct phasing must be experimentally determined.

The technique for determining correct phasing is illustrated in Fig. 20-23. The procedure is as follows:

1. Arbitrarily select one lead from each of the two windings to be phased and connect them together. The two windings are now in series.

2. Observing all safety rules, connect an appropriate ac voltage to one of the remaining windings. This voltage can be any value equal to, or less than, the rated voltage of the selected winding.

3. Again observing safety rules, measure the voltage across the free leads of the series-connected windings as shown in Fig. 20-23*a* and *b*. Remove the applied voltage.

4. If, as shown in Fig. 20-23*a*, the measured voltage is large (the value depends on the voltage ratio of the windings), the windings are *series-aiding* and they are correctly phased for series operation. For parallel operation, the terminal connections for either (but not both) of the series windings must be reversed so that the windings are connected *series-opposing*. In Fig. 20-23*a*, connections to terminals 1 and 2 were reversed. Then the remaining two terminals (terminals 2 and 4 in Fig. 20-23*a*) are connected together to complete the parallel hookup.

5. If the measured voltage of step 3 above is close to zero (exactly zero if the windings are perfectly balanced) as in Fig. 20-23*b*, the windings are connected series-opposing. For parallel operation, the two unconnected terminals (terminals 1 and 4 in Fig. 20-23*b*) are connected together. For series operation, the terminal connections to one of the two windings must be reversed to make the windings series-aiding.

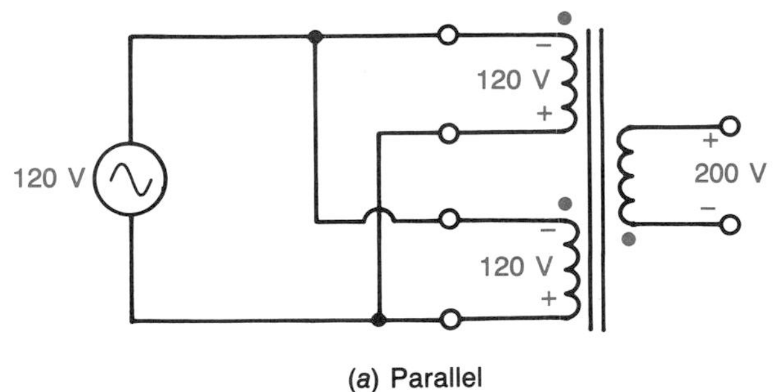


(a) Parallel

(b) Series (aiding)

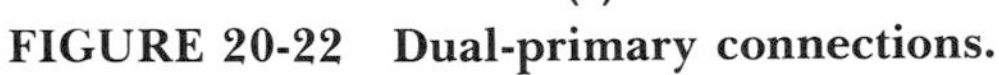

**FIGURE 20-22 Dual-primary connections.**

Some transformers, like the one in Fig. 20-21*d*, have multiple secondaries that are not equal or balanced. Of course, such secondaries cannot be connected in parallel because the differential voltage would force a destructive current flow limited only by the internal resistance of the two windings. However, such windings can be connected either series-aiding or series-opposing. The current drawn from series-connected secondaries must be limited to the smallest current rating of the secondaries. For example, the secondaries of the transformer in Fig. 20-21*d* can provide 50 V at 1 A when connected series-aiding and 10 V at 1 A when connected series-opposing.

Figure 20-21*c* shows a transformer with a center-tapped secondary. The two halves of the secondary winding are series-aiding so that ends of the winding are always opposite polarities with respect to the center tap. This type of transformer is used extensively in circuits where alternating current is converted (rectified) into direct current.

## Self-Test

**34.** To operate a dual-primary transformer from the highest possible source voltage, the primaries are connected ______.

**35.** Do parallel-connected or series-connected primaries require the larger current from a source?

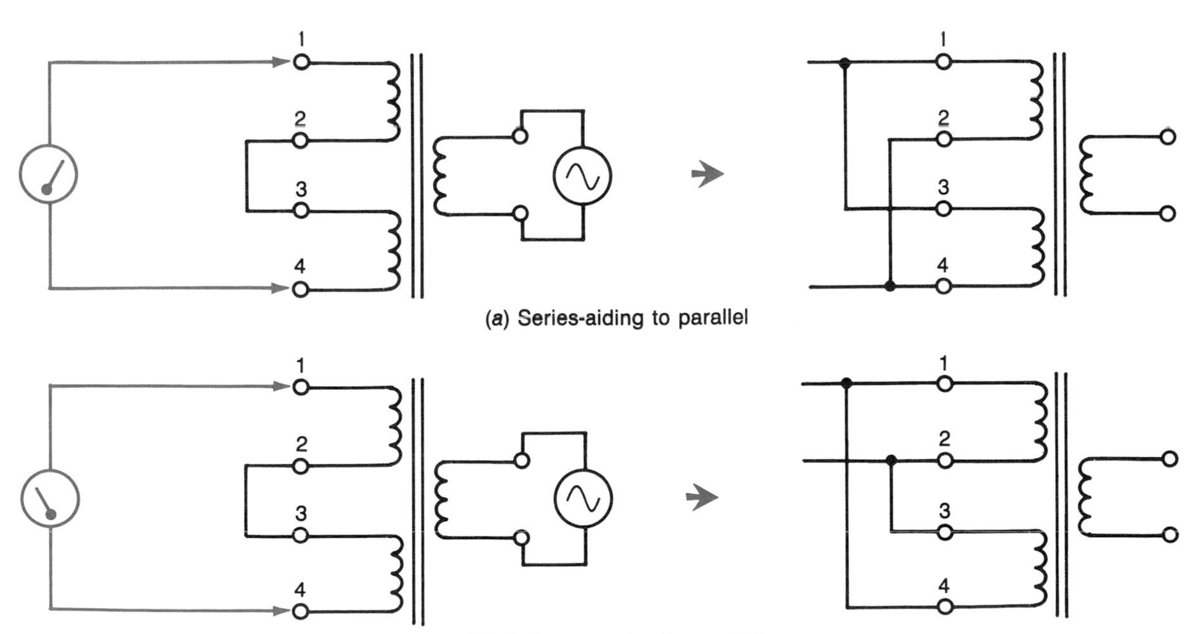


(a) Series-aiding to parallel

(b) Series-opposing to parallel

**FIGURE 20-23 Phasing transformer windings.**

**36.** True or false? For correctly phased parallel windings, the dot ends of the windings should be connected together.

**37.** A 16-V, 4-A winding is connected series-opposing to a 3-V, 3-A winding. How much voltage and current will be available?

## 20-8 TYPES OF TRANSFORMERS

This chapter has emphasized *power transformers* that operate at 60 Hz and change the level of voltage between the primary and the secondary. *Audio transformers* were also mentioned. They are transformers designed to operate at frequencies up to 20 kHz, and they provide impedance matching in many audio systems. Some other common types of transformers are *isolation, auto, constant voltage,* and *bandpass.*

### Isolation Transformers

An *isolation transformer* has a 1:1 turns ratio ($a = 1$). Its sole purpose is to electrically isolate the primary power source from the circuit connected to the secondary of the isolation transformer. All of the other transformers discussed in this chapter also provide isolation, but their main purpose was to efficiently change voltage levels or match impedances. Thus, they are not called isolation transformers.

Isolation transformers are used to isolate the secondary circuit from the ground of the primary power system. This type of isolation is essential for the safety of personnel who service a piece of equipment which does not have a power transformer to provide isolation. Figure 20-24*a* shows that without an isolation transformer there is 120 V between the chassis of the equipment and earth ground. Figure 20-24*b* shows that the isolation transformer removes the shock hazard by eliminating the potential between chassis and ground.

### Autotransformers

The autotransformers shown in Fig. 20-25 do not provide electric isolation. There is always a direct connection from one line of the primary power source to one side of the secondary circuit because the primary and secondary share a common coil.

All of the turns in an autotransformer are wound in the same direction. Thus, the induced voltage in the upper part of the winding in Fig. 20-25*b* is series-aiding the cemf in the lower part of the winding.

The winding of the variable autotransformer in Fig. 20-25*c* is wound on a toroid (donut-shaped) core so that each turn of the winding is exposed on the flat circular surface of the core. The voltage adjustment control on the transformer is mechanically connected to a brush which rides on the exposed turns on the winding. This arrangement allows the secondary voltage to be varied from 0 V to maximum output voltage, which is often about 115 percent of the primary voltage.

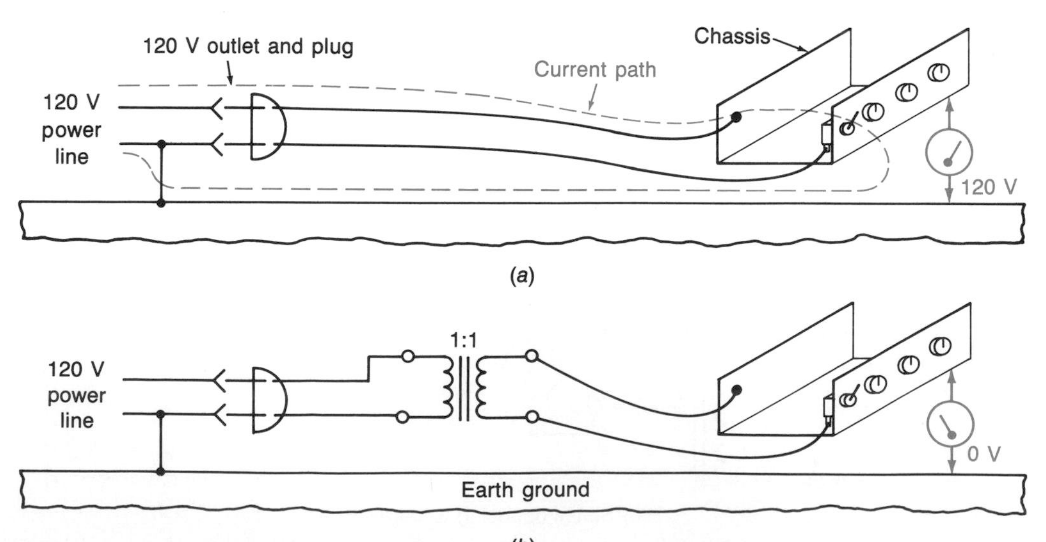


**FIGURE 20-24 Use of an isolation transformer.**

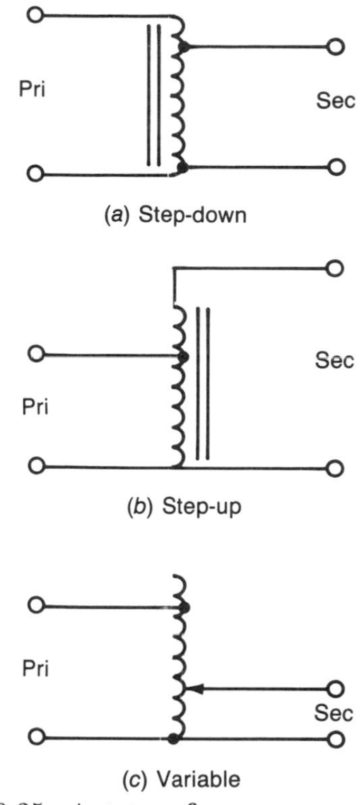


**FIGURE 20-25 Autotransformers.**

When an electrically isolated variable ac power source is needed, the variable autotransformer can be used in conjunction with an isolation transformer. The primary of the autotransformer is connected to the secondary of the isolation transformer.

## Constant-Voltage Transformers

The constant-voltage transformer is designed to provide a constant secondary voltage even when both the primary voltage and the secondary load current change. Of course, there are limits on how much these variables can change without the secondary voltage changing. A typical constant-voltage transformer maintains the secondary voltage within 2 percent of nominal value when the primary voltage varies ±15 percent from its nominal value.

While there are many variations in the construction details of constant-voltage transformers, they all use a capacitor connected to part, or all, of a secondary winding.

One possible arrangement is shown in Fig. 20-26 where the capacitor is connected across part of the tapped secondary winding. The capacitor and

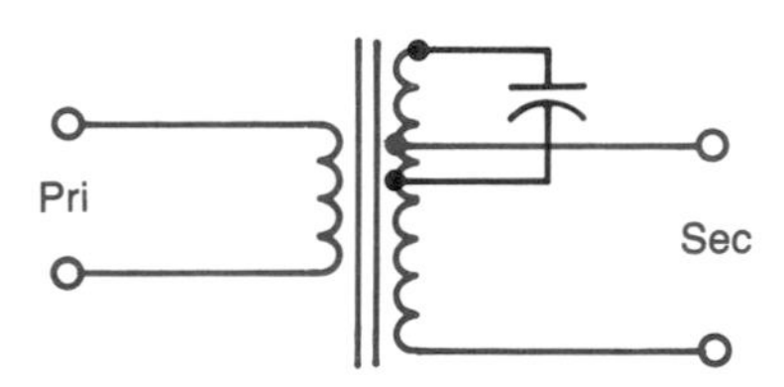


**FIGURE 20-26 Constant-voltage transformer.**

winding inductance form a parallel *LC* circuit which is excited by the 60-Hz voltage induced into the winding. The inductance of the winding is controlled by the core's permeability, which is, in turn, controlled by the amount of flux in the core.

As the primary voltage increases, so does the exciting current and the core flux. The increased core flux decreases the permeability of the core and reduces the inductance of the winding connected to the capacitor. This reduced inductance makes the resonant frequency of the LC circuit closer to 60 Hz so the current in the *LC* circuit increases. This increased *LC* current has very little effect on the primary current because a parallel-resonant *LC* circuit requires only enough power to make up for the losses in the capacitor and inductor. Yet, the current in the *LC* (tank) circuit can be very large. This larger current in the *LC* circuit drives the core into magnetic saturation at the peaks of the sine wave current. Since the saturated core cannot induce additional voltage into the secondary winding, the secondary voltage is held very close to the value it had when the primary voltage was at its lower value.

## Bandpass Transformers

Transformers that operate at frequencies above the audio range are often called *radio frequency* (RF) *transformers*. These transformers have either ferrite or air cores. The ferrite core is located in the center of the coils. It does not provide a complete low-reluctance (high-permeability) path for the flux. Thus, these transformers have a low coefficient of coupling (high levels of leakage flux).

A *bandpass transformer* is an example of an RF transformer. Its schematic symbol is shown in Fig. 20-27*a*. Notice that both the primary and the secondary of the transformer are paralleled with a capacitor. Thus, both the primary and the secondary form resonant circuits. The resonant frequency can be adjusted by changing the location of the ferrite core in each coil of the transformer. The arrowheads on the core symbol indicate that the ferrite core is moveable. Since, as you learned in Chap. 18, resonant circuits are frequency-selective, this transformer is very frequency selective because it has two resonant circuits. It couples a band of frequencies from the primary to the secondary. As shown in Fig. 20-27*b*, frequencies above or below this band are rejected.

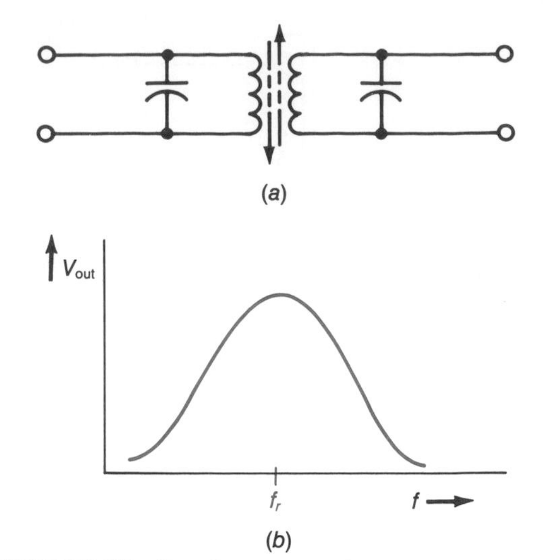


**FIGURE 20-27 Bandpass transformer.**

## Self-Test

**38.** A transformer with $a = 1$ is called an ______ transformer.

**39.** When servicing electric equipment that has no power transformer, the technician may use an ______ transformer.

**40.** An ______ can provide a variable ac output.

**41.** The transformer that does not provide electric isolation is the ______.

**42.** True or false? To operate properly, a constant-voltage transformer must have a constant voltage applied to the primary.

**43.** True or false? A bandpass transformer is a type of audio transformer.

## SUMMARY

1. Two coils linked by flux provide mutual inductance. The flux from one coil produces a cemf in the other coil.
2. When a flux changes from an expanding to a collapsing flux, or when a flux changes direction, the polarity of the cemf is reversed.
3. Coefficient of coupling $k$ is the portion of the total flux that links two coils: $k = \theta_m/\theta_t$.
4. Leakage flux is the flux of a coil that does not link up with another coil: $\theta_l = \theta_t - \theta_m$.
5. The amount of mutual inductance is determined by the coefficient of coupling and the inductance of the two inductors that are coupled together: $L_m = k\sqrt{L_1 L_2}$.
6. A transformer consists of two, or more, magnetically coupled coils.
7. The primary of a transformer receives power from another power source.
8. The secondary of a transformer provides power to a load.
9. The primary voltage may be 180° out of phase with the secondary voltage.
10. On a schematic diagram, the dotted ends of all coils have the same instantaneous polarity.
11. Ideal transformers are 100 percent efficient, have unity coupling, and have very little primary current when the secondary is not loaded: $P_s = P_p$ and $I_s V_s = I_p V_p$.

**12.** The transformation ratio $a$ is equal to the voltage ratio and the turns ratio of an ideal transformer:

$$\frac{V_p}{V_s} = \frac{N_p}{N_s} = a$$

**13.** The turns-per-volt and volts-per-turn ratios are the same for all windings when the coefficient of coupling is one.

**14.** When the current is stepped up by a transformer, the voltage is stepped down and vice versa:

$$\frac{I_s}{I_p} = \frac{V_p}{V_s}$$

**15.** The primary current increases when a load is connected to the secondary.

**16.** Secondary current produces a flux which opposes (or cancels) part of the flux produced by the primary current.

**17.** A transformer reflects its secondary (load) impedance back to its primary and, thus, back to the source connected to the primary.

**18.** The impedance ratio of a transformer is determined by its turns ratio: $Z_p/Z_s = (N_p/N_s)^2 = (V_p/V_s)^2 = a^2$.

**19.** For maximum transfer of power, the load impedance must be the conjugate of the source impedance. A transformer can be used to change the magnitude of $Z$ without changing its phase angle.

**20.** Iron-core transformers are used for impedance matching in audio circuits.

**21.** When a transformer steps up voltage, it steps down the reflected impedance.

**22.** The no-load primary current is called the exciting or magnetizing current. It is controlled by the reactance of the primary, and it is nearly 90° out of phase with the voltage.

**23.** When a load is applied to a transformer, the secondary voltage decreases, the primary current increases, angle $\theta$ decreases, and the copper losses increase. The core losses are independent of the load current.

**24.** Iron cores are constructed from laminated silicon steel to minimize core losses.

**25.** The efficiency of a transformer is $\eta = P_s/(P_s + P_{\text{copper}} + P_{\text{core}})$.

**26.** The copper and core losses of a transformer can be estimated by the short-circuit test and the open-circuit test, respectively. Copper loss can also be estimated from the current ratings and measured resistances of the transformer windings.

**27.** Data from the open-circuit and short-circuit tests can be used to determine the constants needed for an equivalent circuit analysis of a transformer:

$$R_{\text{CL}} = \frac{V_{p\text{OC}}^2}{P_{\text{OC}}}$$

$$X_{\text{m}} = \frac{\sqrt{Z_{\text{OC}}^2 R_{\text{CL}}^2}}{\sqrt{R_{\text{CL}}^2 - Z_{\text{OC}}^2}}$$

$$R_e = \frac{P_{SC}}{I_{pSC}^2}$$

$$X_e = \sqrt{Z_{SC}^2 - R_e^2}$$

**28.** Windings must be matched to be connected in parallel. Dot end is connected to dot end for correct phasing.

**29.** Series-aiding windings provide a voltage equal to the sum of the individual windings. Dot end is connected to nondotted end.

**30.** Isolation transformers have a turns ratio of 1.

**31.** A single coil is used for both the primary and the secondary of an autotransformer.

**32.** A constant-voltage transformer has a parallel *LC* circuit in the secondary.

**33.** A bandpass transformer is an RF transformer used to pass a narrow band of signal frequencies from the primary to the secondary.

## CHAPTER REVIEW QUESTIONS

For the following items, determine whether each statement is true or false.

**20-1.** When the mutual flux changes from an expanding to a collapsing flux, the polarity of the induced voltage reverses.

**20-2.** The primary current is always greater than the secondary current of a transformer.

**20-3.** Leakage flux produces mutual inductance.

**20-4.** The polarity of a cemf only changes when the direction (polarity) of the flux changes.

**20-5.** If a transformer steps down the current, it steps up the voltage.

**20-6.** With a step-up transformer, the reflected impedance will be greater than the load impedance.

**20-7.** For maximum power transfer, the phase angle (both magnitude and polarity) of the load impedance must equal that of the source impedance.

**20-8.** The current in the secondary of a transformer is called the magnetizing current.

**20-9.** Exciting current is in phase with the voltage.

**20-10.** Part of the exciting current in a transformer is caused by hysteresis and eddy-current losses.

**20-11.** The iron-core of a transformer is laminated to reduce hysteresis loss.

**20-12.** A 10-$\Omega \angle 0°$ load will provide greater efficiency for a transformer than will a 10-$\Omega \angle 30°$ load.

**20-13.** Connecting together the dot ends of two windings provides a series-aiding connection.

**20-14.** Parallel-connected primaries require more source current than do series-connected primaries.

**20-15.** A capacitor is used in the primary of a constant-voltage transformer.

**20-16.** Only the isolation transformer provides electric isolation.

**20-17.** Bandpass transformers usually have an iron core.

For the following items, fill in the blank with the word or phrase required to correctly complete each statement.

**20-18.** Power is taken from the ______ winding of a transformer.

**20-19.** Power is applied to the ______ winding of a transformer.

**20-20.** The dot end of one winding will have the ______ instantaneous polarity as the dot end on any other winding of a transformer.

**20-21.** If the turns ratio is 10, the voltage ratio will be ______.

**20-22.** If the turns ratio is 5, the current ratio will be ______.

**20-23.** The load impedance as "seen" by the primary of a transformer is called ______.

**20-24.** During the short-circuit test of a transformer, the ______ losses are considered to be insignificant.

**20-25.** The two halves of a center-tapped winding are series ______.

**20-26.** Matched windings connected series-opposing will provide ______ V.

**20-27.** A ______ can have a variable-voltage, 60-Hz output.

For the following items, choose the letter that best completes each statement.

**20-28.** The flux that links two coils is the
- **a.** Leakage flux
- **b.** Total flux
- **c.** Lost flux
- **d.** Mutual flux

**20-29.** When a load is connected to the secondary of a transformer, the primary current
- **a.** Remains the same
- **b.** Reverses
- **c.** Increases
- **d.** Decreases

**20-30.** The secondary voltage in a transformer is the result of
- **a.** Flux from the primary
- **b.** Flux from the secondary
- **c.** Cemf of the primary
- **d.** Cemf of the secondary

**20-31.** Which of the following cannot be changed from the primary to the secondary of an ideal transformer?
- **a.** Voltage
- **b.** Current
- **c.** Power
- **d.** Impedance

**20-32.** Increasing the load on the transformer causes
- **a.** Copper losses to increase
- **b.** Angle $\theta$ to decrease
- **c.** The secondary voltage to decrease
- **d.** All of the above

For the following items, provide the information requested.

**20-33.** List two factors that control the coefficient of coupling in an air-core transformer.

**20-34.** If the current ratio of a transformer is 4, what is the turns ratio?

**20-35.** Is the transformer in Question 20-34 a step-up or a step-down transformer?

## CHAPTER REVIEW PROBLEMS

**20-1.** Determine the coefficient of coupling if the leakage flux is 0.01 Wb and the mutual flux is 0.1 Wb.

**20-2.** The $k$ of a 10-mH inductance and a 155-mH inductance is 0.8. Determine the mutual inductance.

**20-3.** An ideal transformer has a 600-$N$, 120-V primary and a 20-V secondary. Determine the following:
- **a.** The transformation ratio
- **b.** The number of turns on the secondary
- **c.** The primary current when a 10-$\Omega$ resistor is connected to the secondary
- **d.** The current ratio
- **e.** Power in the primary
- **f.** The volt-per-turn ratio of the secondary

**20-4.** When the transformation ratio of a transformer is 5, what is its impedance ratio?

**20-5.** What turns ratio is required to match a 5-$\Omega$ load to a 2500-$\Omega$ source?

**20-6.** A transformer has a 100-V primary and a transformation ratio of 0.4. How much power will be provided to a 625-$\Omega$ load connected to its secondary?

**20-7.** A transformer produces 95 V across a 20-$\Omega \angle 25°$ load. If the transformer losses are 10 W, what is the percentage of efficiency?

**20-8.** A transformer short-circuit test resulted in $I_p = 1$ A, $V_p = 7$ V, and $P_{SC} = 5$ W. Determine $R_e$ and $X_e$.

**20-9.** The following specifications for a transformer are known: $V_p =$ 240 V, $X_e = 20\ \Omega$, $R_e = 15\ \Omega$, and $a = 4$. Determine $I_p$ when the load is $10\ \Omega \angle 20°$.

**20-10.** How much current and voltage will be provided by a 6-V, 0.5-A winding connected series-aiding to a 10-V, 0.7-A winding?

## ANSWER TO SELF-TESTS

**1.** Mutual inductance is the result of the flux of one coil cutting the turns of another coil.

**2.** Coefficient of coupling is the percentage or portion of the available flux which is involved in mutual inductance.

**3.** 0.83

**4.** 294 mH

**5.** primary

**6.** secondary

**7.** dot

**8.** 180°

**9.** very high primary inductance, 100 percent efficient, 100 percent coupling

**10.** transformation

**11.** increase

**12.** 90

**13.** false

**14.** true

**15.** Reflected impedance is the impedance of the secondary load as seen from the primary of the transformer.

**16.** The impedance ratio equals the square of the voltage ratio.

**17.** false

**18.** $50\ \Omega\ \angle -40°$

**19.** 64

**20.** 25

**21.** 0.1 W

**22.** exciting or magnetizing current

**23.** false

**24.** **a.** increase, **b.** decrease, **c.** decrease, **d.** remain the same, **e.** increase

**25.** to minimize eddy-current loss

**26.** 0.949 or 94.9 percent

**27.** core

**28.** Because the primary voltage is very low.

**29.** 12.2 W

**30.** $X_m = 3.51\ \text{k}\Omega$, $R_{\text{CL}} = 5.76\ \text{k}\Omega$, $R_e = 13\ \Omega$, $X_e = 13.5\ \Omega$

**31.** $\theta = 31.1°$, $I_p = 448\ \text{mA}\ \angle -31.1°$, $\eta = 0.943$, $I_s = 2.24\ \text{A}\ \angle -31.1°$, $V_s = 22.4\ \text{V}\ \angle 1.0°$

**32.** $2\ \Omega\ \angle 0°$, because the core and copper losses will be the same with each load, but the output power for $2\ \Omega\ \angle 0°$ will be greater.

**33.** $4\ \Omega\ \angle 60°$

**34.** series-aiding

**35.** parallel-connected

**36.** true

**37.** 13 V, 3 A

**38.** isolation

**39.** isolation

**40.** autotransformer

**41.** autotransformer

**42.** false

**43.** false

# Chapter 21 Measuring Instruments

This chapter analyzes two of the most commonly used measuring devices—the *multimeter* and the *oscilloscope.* In Chap. 2 you were shown how to measure current, voltage, and resistance. In this chapter you will be shown how a multimeter measures these electrical quantities.

The oscilloscope is an extremely useful and versatile tool for analyzing ac circuits—especially those that contain active devices such as transistors and integrated circuits. The major use of the oscilloscope is to view a voltage waveform, i.e., a plot of voltage versus time. From a displayed waveform, both the frequency and the amplitude of a signal can be determined.

Another important use of the oscilloscope is phase measurement. The phase relationships between ac voltages in an electronic circuit may be very important. The oscilloscope can be used to quickly determine these relationships.

## 21-1 METER SPECIFICATIONS

Many important meter specifications are common to both analog meters and digital meters. These specifications include accuracy, resolution, frequency response, and internal impedance.

The *accuracy* of a measurement refers to how close the measured value is to the true value of the

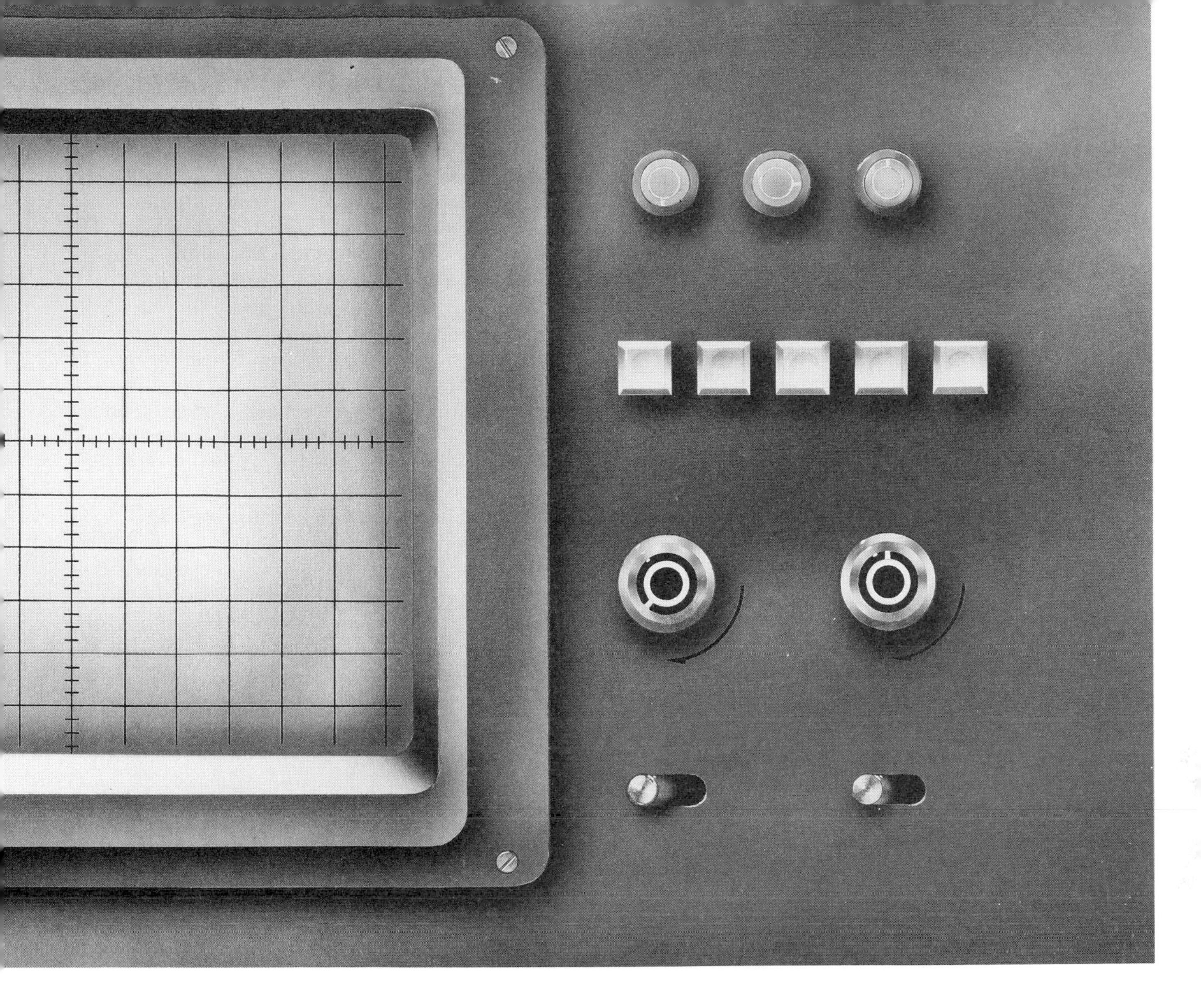

quantity being measured. The accuracy of an analog measuring instrument is usually stated as a percentage of the full-scale reading, i.e., a percentage of the range on which the measurement is made. The accuracy may be different for each function, and, on ac functions, the accuracy may vary with the frequency. With an analog ohmmeter, accuracy is specified as percentage of scale length or degrees of arc. This is because full scale on every ohmmeter range is infinity.

**EXAMPLE 21-1**

An analog dc voltmeter has an accuracy of ±2 percent of full-scale reading. If the meter indicates 28 V on the 50-V range, what are the minimum and maximum values of the voltage being measured?

**Solution**

$$\text{Minimum value} = 28\text{ V} - (50\text{ V} \times 0.02) = 27\text{ V}$$
$$\text{Maximum value} = 28\text{ V} + (50\text{ V} \times 0.02) = 29\text{ V}$$

The accuracy of a digital meter is usually specified as ± (a percentage of the reading + $m$ digits). Digital meters are usually much more accurate than analog meters.

**EXAMPLE 21-2**

A digital meter has an accuracy of ± (0.2 percentage of reading plus 1 digit) on the dc voltage function. The meter indicates 28.00 V dc. What is the maximum value of the measured voltage?

**Solution**

$$\text{Maximum value} = 28\text{ V} + [(28\text{ V} \times 0.002) + 0.01\text{ V}] = 28.07\text{ V}$$

The *resolution* of a measuring instrument refers to the smallest change in the measured quantity that can be detected and accurately specified. On an analog instrument, resolution is determined by the number of divisions on the scale and the range of the instrument. With a digital instrument, the resolution is determined by the range and the number of display digits in the digital readout. Resolution was thoroughly covered in Chap. 2. Review that chapter if you have forgotten the details of this topic.

High resolution can be misleading. It might lead one to assume an indicated voltage is more accurate than it really is. For example, a digital meter with four readout digits will have a resolution of 1 mV on a 2-V ac range. At a frequency of 25 kHz, this meter may only have an accuracy of ± (5 percent of reading + 2 digits). When the meter indicates 1.825 V, what is the possible range of the voltage being measured? The answer to this question can be found by calculating $V_{max}$ and $V_{min}$ (maximum and minimum voltage, respectively):

$$V_{max} = 1.825\text{ V} + [(1.825\text{ V} \times 0.05) + 0.002\text{ V}] = 1.918\text{ V}$$

$$V_{min} = 1.825\text{ V} - [(1.825\text{ V} \times 0.05) + 0.002\text{ V}] = 1.732\text{ V}$$

The true voltage can range from 1.918 to 1.732 V. Thus, the last two digits of resolution are not too meaningful in this case because of the low accuracy at 25 kHz. Had this same reading been obtained on a 2-V dc range where the accuracy might be 0.1 percent, the high resolution would have been much more meaningful.

The frequency of either a current or a voltage must be considered when measuring ac with either a digital or an analog meter. In general, the accuracy of an instrument decreases as the frequency increases. Above some specified frequency a given meter's readings may be so inaccurate that they are useless. Always consult the manufacturer's instruction-operator's manual to determine the frequency response of a given instrument. The limited *frequency response* of any meter is caused by the undesired capacitance and inductance possessed by all electric and electronic components and circuits.

The internal impedance (or internal resistance in the case of direct current) of a meter is another of its important characteristics. For a voltmeter the internal impedance should be as high as possible; for an ammeter it should be as low as possible. Section 21-2 shows that too little internal resistance can cause misleading voltage readings and too much internal resistance can cause misleading current readings.

The dc voltage ranges of a typical digital meter have an internal resistance, usually called *input resistance,* of 10 MΩ. For the ac voltage ranges, the 10 MΩ of input resistance is in parallel with approximately 90 pF. Thus, the input impedance for ac voltage ranges is a function of the frequency of the voltage being measured. Even at 60 Hz the input impedance will be considerably less than 10 MΩ because $X_C$ will be approximately 29.5 MΩ and $Z$ will be about 9.5 MΩ.

The input resistance and the input impedance of the voltage ranges for an analog multimeter vary from range to range. A typical volt-ohm-milliammeter has an input resistance of 100 kΩ on the 5-V dc range and 20 MΩ on the 1000-V dc range. The input resistance of a VOM can be calculated from its *ohms-per-volt* rating. This rating is usually printed on the scale plate of the meter, and it is often referred to as the *sensitivity* of the meter. The input resistance of the meter on any range is found by multiplying the ohms-per-volt rating by the voltage range.

**EXAMPLE 21-3**

Determine the internal (input) resistance of a VOM on the 50-V range if the meter has a sensitivity of 10 kΩ/V.

**Solution**

$$\text{Input resistance} = \text{sensitivity} \times \text{range} = 10\text{ k}\Omega/\text{V} \times 50\text{ V} = 500\text{ k}\Omega$$

The most common sensitivity for a VOM on the dc voltage range is 20 kΩ/V. However, meters with ratings as low as 1 kΩ/V and as high as 100 kΩ/V are also available.

On the ac voltage ranges, the input impedance is typically one-half to one-quarter as great when frequency $f = 60$ Hz. Of course, the impedance decreases as $f$ increases.

The internal resistance and impedance of both the digital ammeter and the analog ammeter vary from range to range. On the 1-A range the internal resistance is typically between 0.2 and 0.5 Ω. On the 100-μA range the internal resistance may be 2 to 3 kΩ.

## Self-Test

1. An ac ammeter has an accuracy of ±3 percent of full scale. On the 10-mA range it is indicating 2.6 mA. What is the minimum value of the measured current?
2. Should one use the 20-V range or the 50-V range of an analog meter to measure and adjust the output of a power supply to 15 V? Why?
3. Why is it important to know the frequency of the ac voltage being measured?
4. Does high resolution ensure high accuracy? Why?
5. Why is the input impedance of a voltmeter dependent on frequency?
6. What is the input resistance of a voltmeter on the 20-V range if its sensitivity is 20 kΩ/V?

## 21-2 METER LOADING

Whenever a meter is connected to a circuit, the internal resistance of the meter changes the circuit resistance and thus the circuit current. Except in parallel circuits, the voltage distribution may also change. These changes caused by connecting the meter to the circuit are called the *loading effect*. Whether or not the loading effect is significant depends on the characteristics of both the meter and the circuit components. If a meter causes a 2 percent change in current in a circuit constructed with 1 percent resistors, the loading is very significant. Had the circuit been constructed with 10 percent resistors, the 2 percent change would not be very significant.

### Voltmeter Loading

The loading effect of a voltmeter is illustrated in Fig. 21-1. Figure 21-1*a* shows the nominal currents and voltages in the circuit before a meter is used to measure $V_{R_1}$. Figure 21-1*b* shows the currents and voltages after an analog meter with 100 kΩ of input resistance is connected across $R_1$ to measure its voltage drop. In Fig. 21-1*b* the circled 100-kΩ resistor labeled $R_V$ represents the voltmeter's internal resistance. Since the specified meter caused the currents and voltages to change 20 percent, it has caused severe loading of the circuit. Had the same voltage been measured with a digital multimeter (DMM) with 10 MΩ of input resistance, the loading effect would have been negligible because 50 kΩ || 10 MΩ ≈ 50 kΩ.

**EXAMPLE 21-4**

The voltage across $R_2$ in Fig. 21-2 is measured with a VOM on the 50-V range. (a) If the meter is rated at 10 kΩ/V, what would the nominal voltage reading be? (b) If the meter has an accuracy of ±3 percent, and the resistors have tolerances of ±10 percent, what would the minimum voltage reading be? (c) What should the theoretical voltage across $R_2$ be?

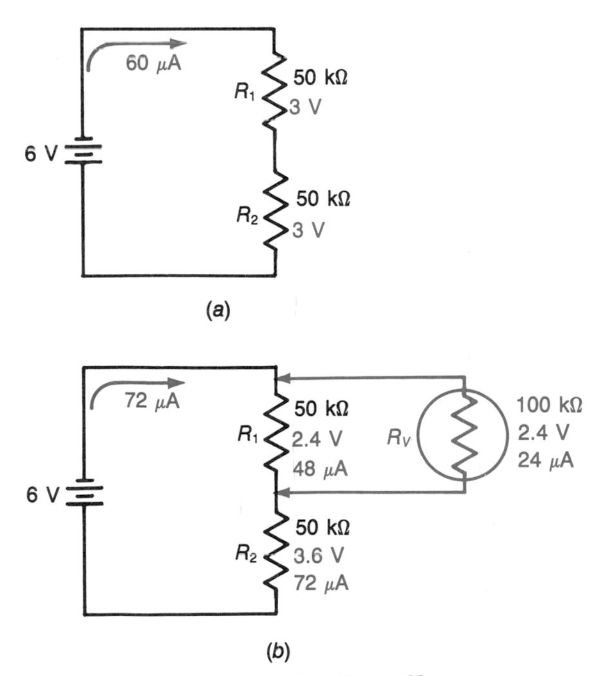

**FIGURE 21-1** Voltmeter loading effect.

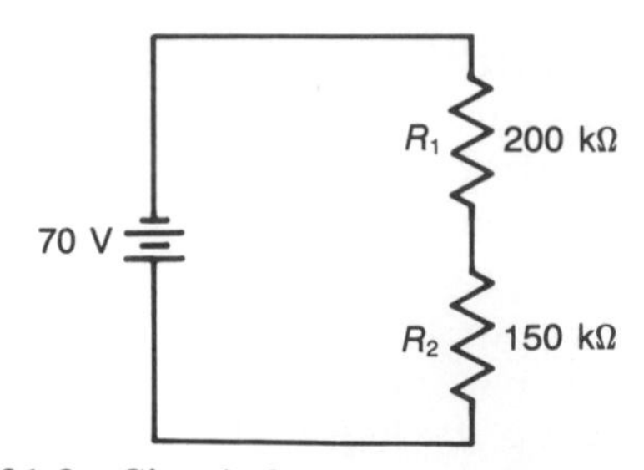


**FIGURE 21-2 Circuit for Example 21-4.**

**Solution**

a. Input resistance $= R_V = 10\ \text{k}\Omega/\text{V} \times 50\ \text{V} = 500\ \text{k}\Omega$

$$R_V \parallel R_2 = \frac{500\ \text{k}\Omega \times 150\ \text{k}\Omega}{500\ \text{k}\Omega + 150\ \text{k}\Omega} = 115.4\ \text{k}\Omega$$

$$V_{R_2} = \frac{115.4\ \text{k}\Omega \times 70\ \text{V}}{115.4\ \text{k}\Omega + 200\ \text{k}\Omega} = 25.6\ \text{V}$$

b. Minimum voltage will drop across $R_2$ when $R_1$ is maximum and $R_2$ is minimum:

$$R_{1,\max} = 200\ \text{k}\Omega + (200\ \text{k}\Omega \times 0.1) = 220\ \text{k}\Omega$$

$$R_{2,\min} = 150\ \text{k}\Omega - (150\ \text{k}\Omega \times 0.1) = 135\ \text{k}\Omega$$

$$R_V \parallel R_{2,\min} = \frac{500\ \text{k}\Omega \times 135\ \text{k}\Omega}{500\ \text{k}\Omega + 135\ \text{k}\Omega} = 106.3\ \text{k}\Omega$$

$$V_{R_{2,\min}} = \frac{106.3\ \text{k}\Omega \times 70\ \text{V}}{106.3\ \text{k}\Omega + 220\ \text{k}\Omega} = 22.8\ \text{V}$$

$$\text{Reading}_{\min} = 22.8\ \text{V} - (50\ \text{V} \times 0.03) = 21.3\ \text{V}$$

c. $$V_{R_2} = \frac{150\ \text{k}\Omega \times 70\ \text{V}}{150\ \text{k}\Omega + 200\ \text{k}\Omega} = 30\ \text{V}$$

---

In Example 21-4 the voltage drops are calculated using the series voltage distribution formula. Notice two things from this example: (1) the effects of meter loading are easy to calculate, and (2) tolerances and meter inaccuracy, under worst-case conditions, can cause measured values to be considerably different from expected values. In this example the measured value ($\text{reading}_{\min} = 21.3$ V) was 17 percent less than the expected value ($V_{R_2} = 25.6$ V).

A practical way to determine if severe meter loading is occurring is to switch the VOM to a range higher than the range required to keep the pointer from pegging. If the reading on the higher range is greater than can be accounted for by meter accuracy and resolution, then significant meter loading was occurring on the lower range.

The loading effect of an ac voltmeter is more difficult to analyze because the input capacitance of the meter, as well as the input resistance, must be taken into account. The results of ac voltmeter loading are shown in Fig. 21-3*b*. In this circuit, $R_V$ and $C_V$ represent the internal resistance and capacitance, respectively, of the ac voltmeter. These results were obtained by solving the circuit impedances using vector algebra. Going through the procedures and calculations used in solving the circuit in Fig. 21-3*b* provides a good review of vector algebra as well as showing how an ac voltmeter loads a circuit and causes phase shifts in the circuit. Two techniques or methods of solving for the voltage drops in Fig. 21-3*b* are discussed. The first method involves solving for the parallel impedance of $R_2$ and the meter and then solving for the total impedance and current. The second method involves converting the parallel resistance and reactance into their series-equivalent resistance and reactance and then solving for the total impedance and current. Using either method, the first steps are to calculate $R_{2,V}$ and $X_C$. Resistor $R_{2,V}$ is the equivalent resistance of $R_2$ and $R_V$ in Fig. 21-3*b*.

$$R_{2,V} = \frac{50\ \text{k}\Omega \times 10{,}000\ \text{k}\Omega}{50\ \text{k}\Omega + 10{,}000\ \text{k}\Omega} \approx 49.8\ \text{k}\Omega$$

$$X_{C_V} = \frac{1}{6.28 \times 20 \times 10^3 \times 90 \times 10^{-12}} \approx 88.5\ \text{k}\Omega$$

The reduced circuit after this calculation is shown in Fig. 21-3*c*.

For the first method, the next step is to find $Z_{R_{2,V}};C_v$, which is the impedance of $R_{2,V}$ in parallel with $X_C$. This can be done by arbitrarily assigning a voltage across $R_{2,V}$ and $C_V$ and then calculating the currents $I_{R_{2,V}}$, $I_{C_V}$, and $I_{R_{2,V};C_V}$. This last current and the arbitrarily assigned voltage can then be used to find $Z_{R_{2,V};C_V}$. For the calculations that follow, the assigned voltage is $10\ \text{V}\ \angle 0°$:

$$I_{R_{2,V}} = \frac{10\ \text{V}\ \angle 0°}{49.8\ \text{k}\Omega\ \angle 0°} = 0.20\ \text{mA}\ \angle 0°$$

$$I_{C_V} = \frac{10\ \text{V}\ \angle 0°}{88.5\ \text{k}\Omega\ \angle -90°} = 0.113\ \text{mA}\ \angle 90°$$

$$\theta = \arctan\frac{I_{C_V}}{I_{R_{2,V}}} = \arctan\frac{0.113\ \text{mA}}{0.20\ \text{mA}} = 29.5°$$

$$I_{R_{2,V}};C_V = 0.20\ \text{mA} + j0.113\ \text{mA} = 0.23\ \text{mA}\ \angle 29.5°$$

$$Z_{R_{2,V}};C_V = \frac{10\ \text{V}\ \angle 0°}{0.23\ \text{mA}\ \angle 29.5°} = 43.5\ \text{k}\Omega\ \angle -29.5°$$

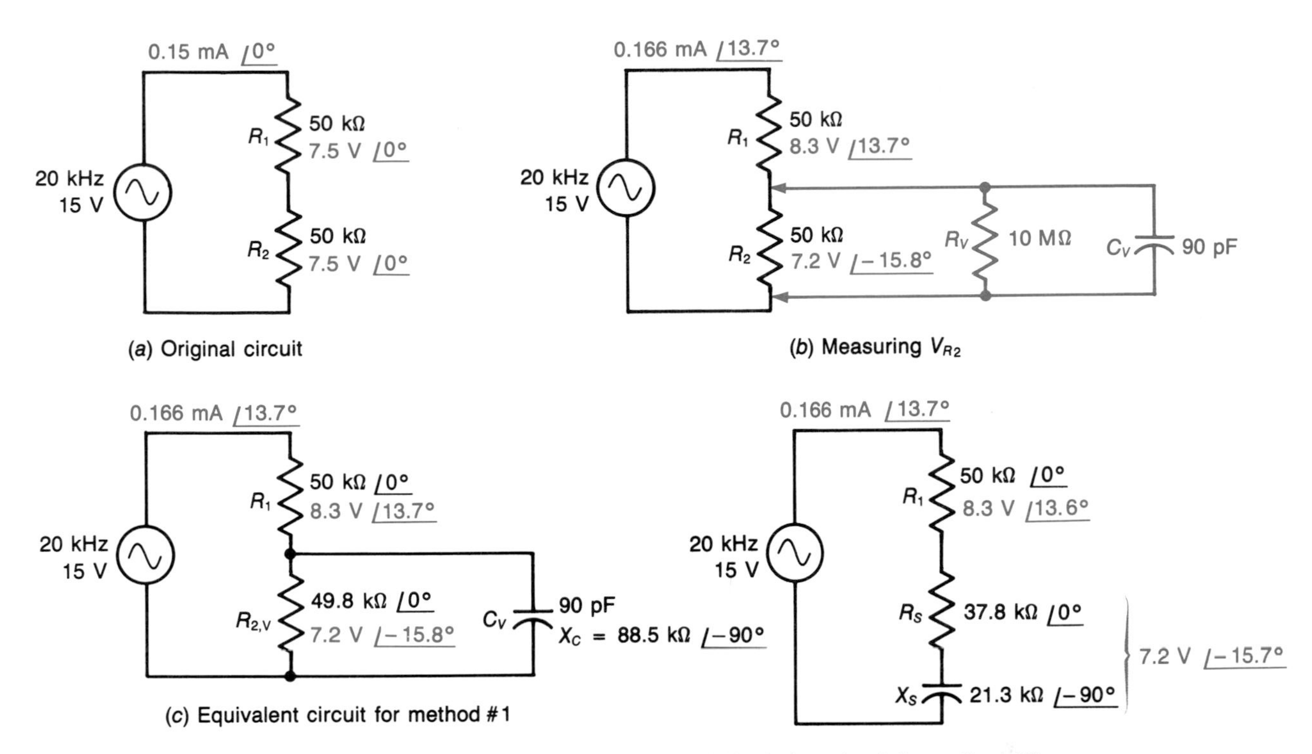


**FIGURE 21-3 Ac voltmeter loading effect.**

Note that the values of the currents calculated above are not the true values in the circuit; they were calculated only to obtain the value of $Z$ and its phase angle. [Impedance $Z$ could also have been determined by converting $R$ to $G$ (conductance) and $X_C$ to $B$ (susceptance).]

The next step in the first method is to find the total impedance $Z_T$ by adding $R_1$ to the rectangular form of $Z_{R_{2,V}};C_V$. The calculations are

$$\begin{aligned} Z_{R_{2,V}};C_V &= 43.5\ \text{k}\Omega \cos 29.5° - j43.5\ \text{k}\Omega \sin 29.5° \\ &= 37.9\ \text{k}\Omega - j21.4\ \text{k}\Omega \\ Z_T &= 37.9\ \text{k}\Omega - j21.4\ \text{k}\Omega + 50\ \text{k}\Omega \\ &= 87.9\ \text{k}\Omega - j21.4\ \text{k}\Omega \\ \theta &= \arctan\frac{-21.4\ \text{k}\Omega}{87.9\ \text{k}\Omega} = -13.68° \approx -13.7° \\ Z_T &= \sqrt{(87.9\ \text{k}\Omega)^2 + (-21.4\ \text{k}\Omega)^2}\ \angle\theta \\ &= 90.5\ \text{k}\Omega\ \angle -13.7° \end{aligned}$$

Now the total current $I_T$ can be determined by Ohm's law:

$$I_T = \frac{15\ \text{V}\ \angle 0°}{90.5\ \text{k}\Omega\ \angle -13.7°} = 0.166\ \text{mA}\ \angle 13.7°$$

Finally, $V_{R_2}$, which is equal to the voltage across $Z_{R_{2,V}};C_V$, can be calculated as follows:

$$\begin{aligned} V_{R_2} &= I_T \times Z_{R_{2,V}};C_V \\ &= 0.166\ \text{mA}\ \angle 13.7° \times 43.5\ \text{k}\Omega\ \angle -29.5° \\ &= 7.2\ \text{V}\ \angle -15.8° \end{aligned}$$

As a check on the calculations used to determine $V_{R_2}$, one could calculate $V_{R_1}$ and then see if the circuit voltages satisfy Kirchhoff's voltage law. This exercise is left for the reader to do.

The second method for finding $V_{R_2}$ involves developing and solving the equivalent circuit shown in Fig. 21-3*d*. First determine $R_s$ and $X_s$ by

$$\begin{aligned} R_s &= \frac{R_p X_p{}^2}{R_p{}^2 + X_p{}^2} = \frac{49.8k \times (88.5k)^2}{(49.8k)^2 + (88.5k)^2} = 37.8\ \text{k}\Omega\ \angle 0° \\ X_s &= \frac{R_p{}^2 + X_p}{R_p{}^2 + X_p{}^2} = \frac{(49.8k)^2 \times 88.5k}{(49.8k)^2 + (88.5k)^2} \\ &= 21.3\ \text{k}\Omega\ \angle -90° \end{aligned}$$

Next, solve for $Z_T$ and $I_T$ as follows:

$$R_{s_1} - R_s + R_1 - 37.8\ \text{k}\Omega + 50\ \text{k}\Omega = 87.8\ \text{k}\Omega\ \angle 0°$$

$$\theta = \arctan\frac{X_s}{R_{s_1}} = \arctan\frac{-21.3\ \text{k}\Omega}{87.8\ \text{k}\Omega}$$
$$= -13.64° \approx -13.6°$$
$$Z_T = \sqrt{(87.8\ \text{k}\Omega)^2 + (21.3\ \text{k}\Omega)^2}\ \underline{/\ \theta} = 90.3\ \text{k}\Omega\ \underline{/\ -13.6°}$$
$$I_T = \frac{15\ \text{V}\ \underline{/\ 0°}}{90.3\ \text{k}\Omega\ \underline{/\ -13.6°}} = 0.166\ \text{mA}\ \underline{/\ 13.6°}$$

Now determine $V_{R_1}$ by Ohm's law and then convert $V_{R_1}$ to rectangular form:

$$V_{R_1} = 0.166\ \text{mA}\ \underline{/\ 13.6°} \times 50\ \text{k}\Omega\ \underline{/\ 0°} = 8.3\ \text{V}\ \underline{/\ 13.6°}$$
$$V_{R_1} = 8.3\ \text{V}\cos 13.6° + j8.3\ \text{V}\sin 13.6°$$
$$= 8.07\ \text{V} + j1.95\ \text{V}$$

Next find $V_{R_2}$ using Kirchhoff's voltage law:

$$V_{R_2} = V_T - V_{R_1} = 15\ \text{V}\ \underline{/\ 0°} - (8.07\ \text{V} - j1.95\ \text{V})$$
$$= 6.93\ \text{V} - j1.95\ \text{V}$$
$$= 7.2\ \text{V}\ \underline{/\ \theta}$$
$$\theta = \arctan\frac{-1.95\ \text{V}}{6.93\ \text{V}} = -15.7°$$

After using either of the two methods to find $V_{R_2}$, we can finally determine the severity of the loading effect caused by using the DMM to measure the voltage across $R_2$ in Fig. 21-3*a*. The percentage of error caused by the loading effect is

$$\text{Percentage of error} = \frac{7.5\ \text{V} - 7.2\ \text{V}}{7.5\ \text{V}} \times 100 = 4\ \text{percent}$$

## Ammeter Loading

Figure 21-4 illustrates the loading effect of an ammeter. The nominal values for the circuit are shown in Fig. 21-4*a*. Figure 21-4*b* shows that measuring the circuit current reduces its value from 0.83 to 0.73 mA if the ammeter has 250 Ω of internal resistance. This 250 Ω of internal resistance is typical for a 1-mA ammeter of either the digital or analog type. The percentage of error caused by the loading effect is:

$$\text{Percentage error} = \frac{0.83\ \text{mA} - 0.73\ \text{mA}}{0.83\ \text{mA}} \times 100$$
$$= 12\ \text{percent}$$

The internal resistance of an ammeter is inversely proportional to the current range for a VOM and for many ranges of a DMM. If the current in the circuit in Fig. 21-4 had been measured on a 5-mA range, $R_A$ would have been only 50 Ω and the measured current would have been 1.5 V/1850 Ω = 0.81 mA. With a VOM, the resolution of the 5-mA range may have limited the reading to 0.8 mA, but this is still a better estimate of the current than the 0.73 mA obtained on the 1-mA range.

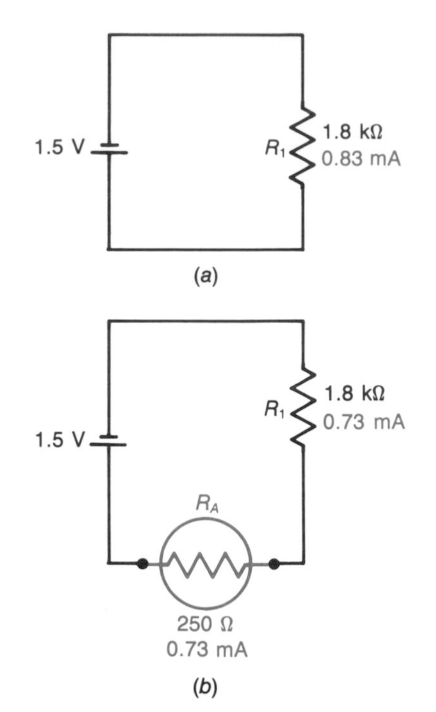


**FIGURE 21-4 Ammeter loading effect.**

## Self-Test

**7.** Which meter would have the greater loading effect on the 1000-V range, a DMM with 10 MΩ input resistance, or a VOM with 20 kΩ/V sensitivity? Why?

**8.** Change $R_1$ in Fig. 21-2 to 20 kΩ and determine $V_{R_1}$ when measured on the 10-V range of a 20-kΩ/V VOM. What percentage of error was caused by the loading effect?

**9.** Change $R_1$ in Fig. 21-4 to 2200 Ω and calculate the percentage of error caused by the loading effect of the 1-mA ammeter.

**10.** A 100-μA meter with ±2 percent accuracy and 2000 Ω of internal resistance is used to measure the current through a 30-kΩ, ±5 percent resistor which is connected to a 3-V source. Calculate the minimum measured current.

## 21-3 DIGITAL METERS

A block diagram for a simplified digital voltmeter (DVM) is illustrated in Fig. 21-5. Although modern DVMs use more sophisticated circuits and techniques than those shown in this diagram, the concepts involved are the same.

Before one can understand how a DVM operates, one must understand the purpose and operation of each block from which it is constructed. A *comparator* is an electronic circuit that compares two input voltages. Depending on the relationship between the two input voltages, the output voltage is one of two possible voltage levels. For the comparator in Fig. 21-5 the output voltage is +3 V unless the input voltage from the ramp generator is equal to or greater than the voltage being measured ($V_i$). Only when $V_i$ is less than the ramp voltage will the output of the comparator be low (essentially at 0 V).

The *ramp voltage generator* in Fig. 21-5 is a circuit in the DVM which, when turned on (enabled) by the timer output, produces an output voltage which linearly increases from 0 to +2 V. When disabled (turned off) by the timer output signal going to 0 V, the output of the ramp generator drops to zero and remains at zero until the generator is again enabled by the timer output going high (+3 V).

The pulse generator in Fig. 21-5 generates 2000 pulses every second. The pulses go from essentially 0 to +3 V. The continuous train of pulses out of the pulse generator is fed to the counter.

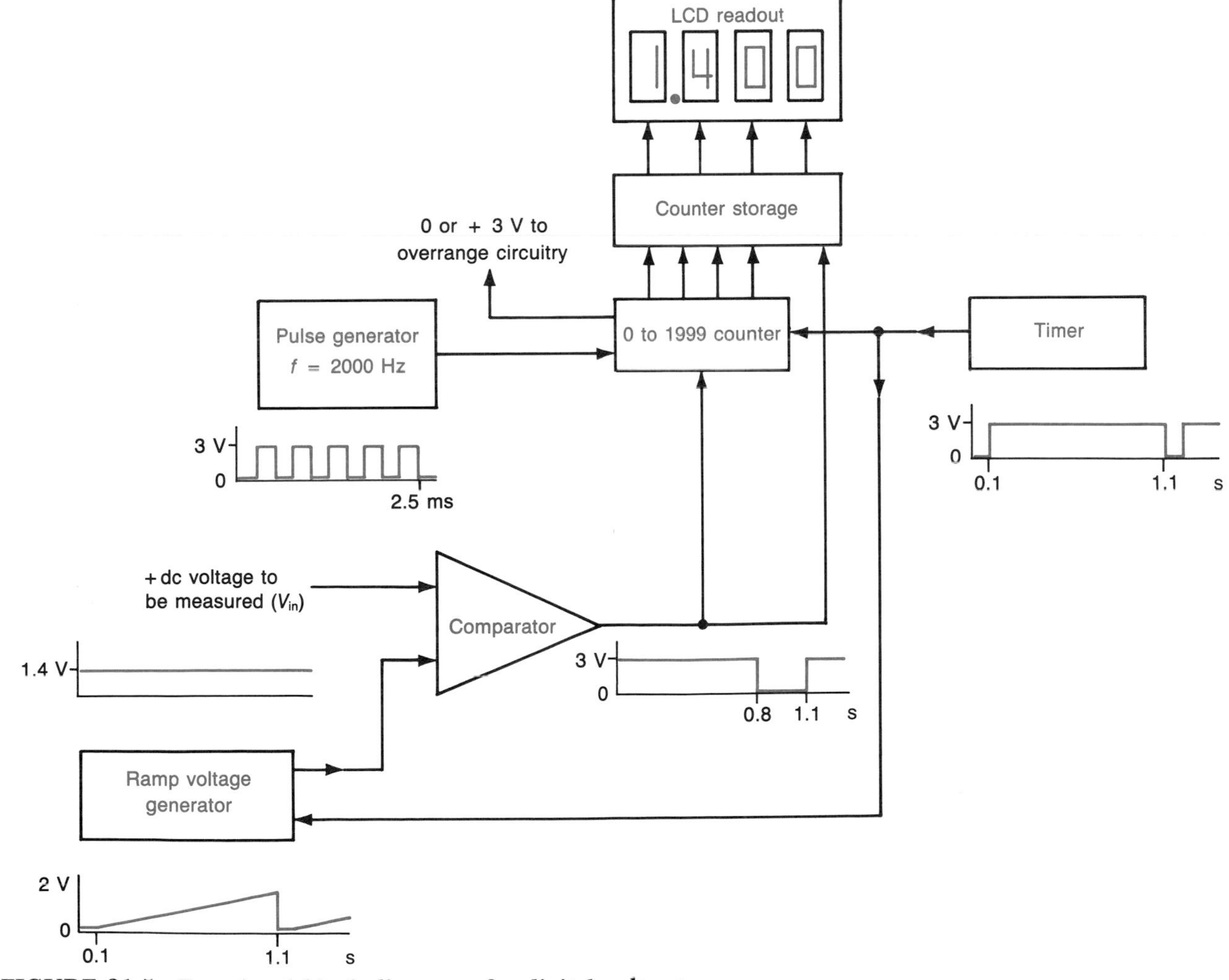


**FIGURE 21-5 Functional block diagram of a digital voltmeter.**

The *counter* counts the pulses it receives from the pulse generator *if* it is enabled (turned on) by *both* the +3-V output of the comparator *and* the +3-V output of the timer. When the output of the comparator goes low, the counter stops counting and retains the count it contained the instant before the comparator output dropped to zero. When the timer output is 0 V, the counter is set to zero so it can start sequentially counting from 0 to 1999 when enabled again. Unless the input voltage to the DVM is greater than the voltage range of the DVM, the counter never receives more than 1999 pulses from the pulse generator while it is enabled. Thus, as shown in Fig. 21-5, one of the outputs from the counter circuitry block is an output line that goes +3 V when the counter receives the two-thousandth pulse while it is enabled. This indicates that the DVM is on too low a range setting. The remaining four outputs from the counter block provide the number (0000 to 1999) contained in the counter at any given instant.

The number contained in the counter is stored in the *counter storage* block only when the storage block is enabled for a fraction of a second while the output of the comparator changes from +3 to 0 V. The number stored in the counter storage block is constantly displayed by the readout block.

The *readout* block display is either seven-segment light-emitting diodes (LEDs) or seven-segment liquid crystal displays (LCDs). The location of the decimal point in the readout changes when the range is changed. The location shown is for the 2-V range. On the 20-V range the decimal point would be located one more digit to the right.

The *timer,* as mentioned in the previous paragraphs, provides the signal which sets and holds the counter at zero count and the ramp voltage at 0 V. When the timer output goes to +3 V, the counter starts to count pulses and the ramp generator's output starts its linear rise from 0 to +2 V. The timer continuously generates the waveshape shown in Fig. 21-5; i.e., the output is high for 1.0 s, low for 0.1 s, high for 1.0 s, etc.

Now we can integrate all of the blocks discussed above to see how the DVM measures an input voltage of 1.400 V. To aid in this integration process, a timing diagram for the waveforms of the various blocks has been produced in Fig. 21-6. At time 0 in this figure, the timer output is just turning on the ramp generator and the counter. Since $V_i$ is greater than the ramp voltage, the comparator output is +3 V and the counter is counting the pulses out of the pulse generator. The counter continues counting until the ramp voltage reaches 1.400 V. Since the ramp voltage increases at a rate of 2 V/s, it takes 0.7 s for the ramp voltage to reach 1.400 V. During this time the counter counts (2000 pulses per second) × 0.7 s = 1400 pulses. At the end of 0.7 s the comparator output goes low; the counter is disabled; the counter storage is temporarily enabled and stores the 1400-count output of the counter; and the readout displays 1.400 V. At the end of 1 s the timer output goes to zero and resets the counter to zero count and the ramp voltage to zero volts. Note that when the ramp voltage is reset to zero, the comparator output returns to +3 V so that the counter can be enabled when the timer output returns to +3 V. Also, note that when the counter is reset to zero, the displayed voltage remains at 1.400 V because the counter storage is disabled except while the comparator's output changes from +3 to 0 V. At the end of 1.1 s the whole process starts over again.

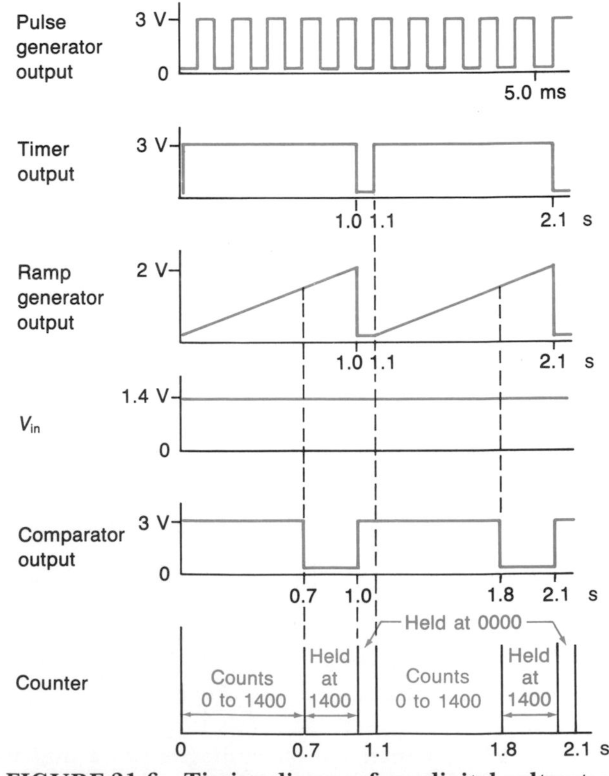

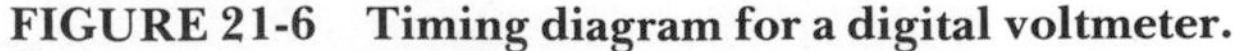


**FIGURE 21-6 Timing diagram for a digital voltmeter.**

Notice from the timing diagram in Fig. 21-6 that the readout is updated every 1.1 s. Thus, if the input voltage changes just after the comparator output goes low, the new input voltage will not be displayed for approximately 1 s.

When the input voltage to the DVM in Fig. 21-5 is 2 V or higher, the counter sends out an over-range signal when it receives the two-thousandth pulse. Thus, even though this DVM is said to have a 2-V range, it only indicates up to 1.999 V before it shows over-range conditions.

The circuit in Fig. 21-7 shows how the input voltage to the DVM can be split up so that any dc voltage up to 2000 V can be measured with the circuit in Fig. 21-5. On the 20-V position of the range switch ($S_1$ in Fig. 21-7) the 1.4 V dropped across the series string of $R_2$, $R_3$, and $R_4$ is sent on to the comparator. Another section of the range switch (not shown) changes the decimal point on the readout so that the 1400 count now represents 14.00 V instead of 1.400 V.

Many DMMs have a basic voltmeter circuit with a 200-mV range. This basic voltmeter is also used to indirectly measure current and resistance. Current measurements are made by measuring the voltage drop across a small-value, high-precision resistor. This resistor is switched across the input leads of the basic voltmeter circuit by the current function switch. When the DMM is inserted in series with the load in which current is being measured, a voltage (which is directly proportional to the current) develops across the precision resistor. For example, if the precision resistor were 100 Ω and 1.4 mA flowed through it, the voltage drop would be 0.14 V. This 0.14 V applied to a 200-mV basic voltmeter would cause a four-digit readout to display 1.400 when the range switch was set to the 2-mA range.

When measuring resistance with a DMM, a low-value, constant-current source is connected to the unknown resistor. Using a constant-current source rather than a constant-voltage source provides a linear relationship between resistance and voltage. Thus, the same basic digital voltmeter system can be used to indirectly measure resistance.

More sophisticated DMMs have autoranging and autopolarity features. With these instruments the operator just has to select the correct function (current, voltage, or resistance) and internal circuitry selects the correct range and indicates the polarity of the applied voltage.

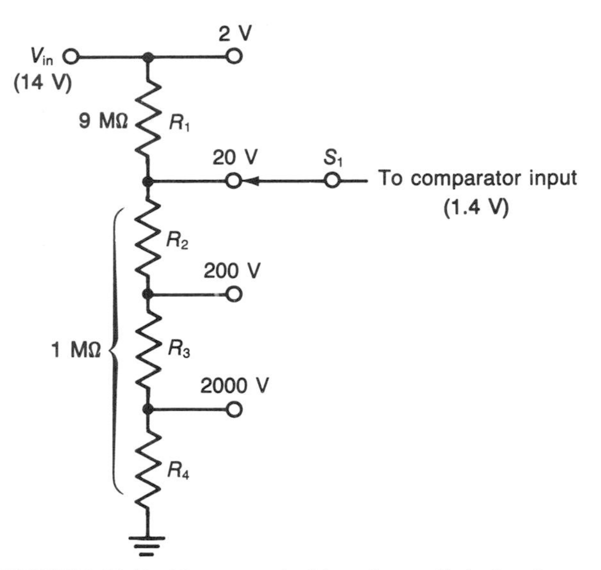


**FIGURE 21-7 Range switching for a digital voltmeter.**

Also, more sophisticated DMMs read out the true rms value of an ac voltage or current regardless of the waveshape. This is in contrast to most other meters which can only read the rms value of a pure sine wave.

## Self-Test

**11.** Refer to Fig. 21-5 or 21-6. If it takes 0.4 s for the comparator output to change after the counter is enabled, what count will be in the counter?

**12.** If the meter in Fig. 21-5 were on the 200-V range, (a) where would the decimal point be located? (b) and what would the resolution be?

**13.** When does the counter storage in Fig. 21-5 update?

**14.** What would be the value of $R_2$ in Fig. 21-7?

## 21-4 ANALOG METERS

The heart of an analog meter is the *meter movement.* A meter movement like the one illustrated in Fig. 21-8 is capable of measuring small values of current. Since it takes a voltage to force current through the coil, the meter movement is also capable of measuring small values of voltage. Of course, if the meter movement can measure current or voltage, it can also indirectly measure resistance as was discussed in Sec. 21-3 on the DMM.

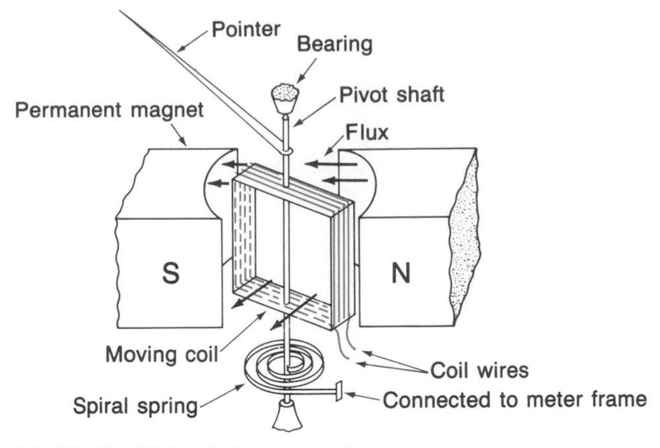


**FIGURE 21-8 Principle of the permanent-magnet moving-coil (PMMC) meter movement.**

The meter movement illustrated in Fig. 21-8 can be called a *d'Arsonval* meter movement or a *permanent-magnet moving-coil* (PMMC) meter movement. Its principle of operation is quite simple. The magnetic flux created by the current-carrying coil is at an angle to the flux of the permanent magnet. Thus, the magnetic poles on the open ends of the coil are not aligned with the poles of the permanent magnet. As you learned in Chap. 10 on magnetism, the opposite poles of these two magnetic fields will be attracted to each other so that the flux of the two magnetic fields can be aligned. This force of attraction creates a clockwise torque on the coil, pivot shaft, and pointer assembly and starts to move the pointer upscale. As the assembly rotates clockwise, the spiral spring winds up and produces a counterclockwise torque on the assembly. The clockwise rotation stops when the counterclockwise torque of the spring balances out the clockwise torque of the magnetic fields. The torque of the magnetic fields is a function of the strength of the moving coil's field, which in turn is a function of the coil's current. Thus, the larger the current flowing through the coil, the further upscale the pointer travels.

All meter movements have three important ratings that make it easy to convert them to multirange ammeters, voltmeters, and ohmmeters: (1) full-scale current $I_{fs}$, (2) full-scale voltage $V_{fs}$, and (3) internal resistance $R_i$. The internal resistance is the resistance of the coil wire. Current $I_{fs}$ is the amount of current required to deflect the pointer to the end of the scale, and $V_{fs}$ is the amount of voltage required to provide $I_{fs}$ through $R_i$. Manufacturers specify at least two of these ratings. Of course, if two of the ratings are known, the third can be calculated by using Ohm's law. Typical values of $V_{fs}$ range from 100 to 500 mV. The full-scale current can be as small as a few microamperes: a VOM often uses a movement with an $I_{fs}$ of 50 $\mu$A. The $R_i$ of a 50-$\mu$A movement is typically 2 to 5 k$\Omega$. Resistance $R_i$ decreases as $I_{fs}$ increases. A typical 1-mA movement may only have 100 to 500 $\Omega$ of $R_i$.

The d'Arsonval meter movement is polarized because reversal of the current in its coil causes a down-scale deflection. Yet, these meter movements are used in VOMs that measure alternating current as well as direct current. The VOM is able to measure alternating current because the VOM contains *rectifiers* (also called *diodes*) which convert alternating current into pulsating direct current. As indicated in Fig. 21-9, a rectifier or diode allows current to flow in only one direction (against the arrowhead in the symbol). Figure 21-9 shows that

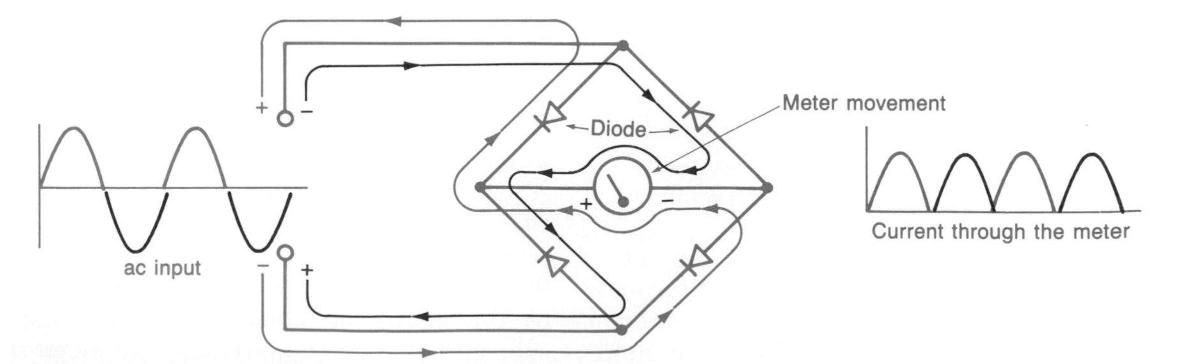


**FIGURE 21-9 Converting alternating current to pulsating direct current so that the PMMC meter movement can be used to measure alternating current.**

while alternating current is applied to the rectifier–meter-movement circuit, only pulses of direct current flow through the meter. These pulses are always in the same direction through the meter movement and always cause the pointer to move up-scale. The scale of the VOM is calibrated to indicate the rms value of sinusoidal alternating current.

## Shunts and Ammeters

A meter movement can be converted to a higher range ammeter by adding a *shunt resistor* in parallel with the meter movement. The shunt resistor carries the excess range current not needed to cause full-scale deflection of the meter movement. Figure 21-10 shows the diagram for a multirange ammeter on the 1-mA range. The value of the shunts can be calculated if two of the three ratings of the meter movement are known. Just treat the meter movement like a resistor through which the current is known and apply Kirchhoff's current law and Ohm's law to the circuit. Applying Kirchhoff's law to Fig. 21-10 shows that the shunt current $I_{R_s}$ is equal to the range current $I_r$ minus the full-scale current $I_{fs}$. Inspection of this figure shows that the voltage across the shunt $V_{R_s}$ is equal to $V_{fs}$ because the shunt and the movement are in parallel. Thus, the formula for calculating the shunt resistance is

$$R_s = \frac{V_{R_s}}{I_{R_s}} = \frac{V_{f_s}}{I_r - I_{fs}}$$

### EXAMPLE 21-5

Determine the shunt resistance needed for the 1-mA range of the ammeter in Fig. 21-10.

**Solution**

$$V_{fs} = I_{fs}R_i = 100\ \mu A \times 3\ k\Omega = 300\ mV$$

$$R_s = \frac{300\ mV}{1\ mA - 100\ \mu A} = 333.3\ \Omega$$

On the ranges where $I_r \gg I_{fs}$, the formula for shunt resistance can be reduced to $R_s = V_{fs}/I_r$. For example, in calculating $R_s$ for the 100-mA range in Fig. 21-10, it makes little difference whether 300 mV is divided by 100 mA or 99.9 mA (100 mA − 100 μA) unless the tolerance of the shunt is to be less than 0.1 percent.

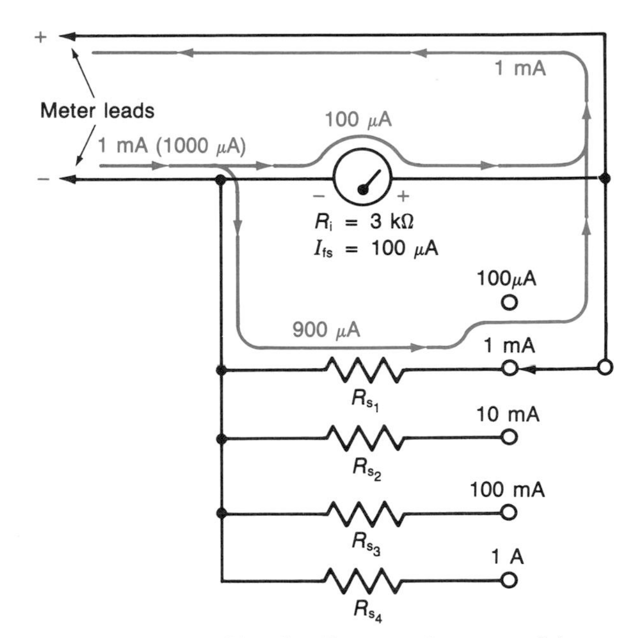


**FIGURE 21-10 Circuit diagram for a multirange ammeter.**

Shunts are made from materials that have small temperature coefficients of resistance. This allows VOMs to be used in environments with widely varying temperatures without dramatic changes in the accuracy of the instruments.

The switch used for changing ranges on an ammeter must be a *make-before-break* switch. This type of switch is also called a *shorting* switch. A possible physical arrangement for a shorting switch is drawn in Fig. 21-11*a*. The switch is shown at the instant it is moving from positive 2 to positive 1. Notice that the movable part (pole) of the switch always makes contact with a new position before breaking contact with the old position. If a nonshorting switch (Fig. 21-11*b*) is used, the meter movement must carry the full-range current for the instant that the pole is not making contact with any position. When switching between high-current ranges, the high current would destroy the coil in the meter movement.

## Multipliers and Voltmeters

The voltage range of a meter movement can be extended (multiplied) beyond the value of $V_{fs}$ by using a *multiplier resistor*. A multiplier resistor is connected in series with the meter movement to limit the current through the meter movement to

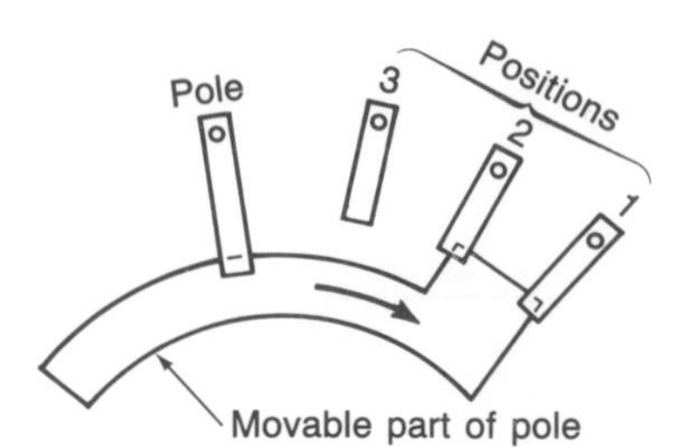


(*a*) Shorting switch

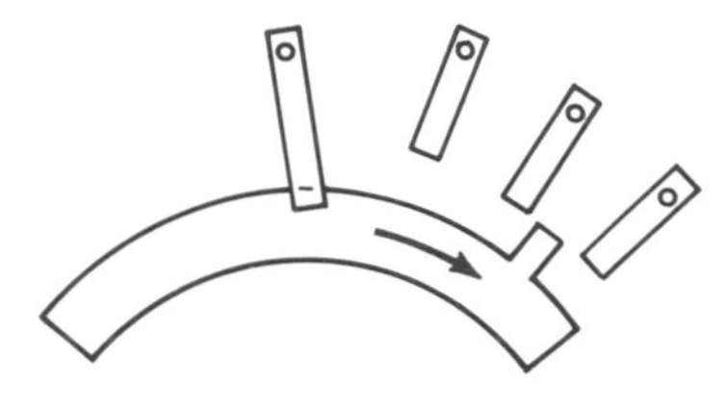

(*b*) Nonshorting switch

**FIGURE 21-11 Rotary switches.**

$I_{fs}$ when the full extended-range voltage is applied. A multirange voltmeter circuit is presented in Fig. 21-12. The voltmeter is shown on the 10-V range. Notice that, internally, a voltmeter is a series circuit. Thus, Kirchhoff's voltage law and Ohm's law can be used to calculate the multiplier resistance $R_m$. Full-range voltage $V_r$ must be applied to the voltmeter leads to cause $I_{fs}$ to flow and force the pointer to full scale. Kirchhoff's voltage law tells us that the voltage drop across the multiplier $V_m$ plus $V_{fs}$ must be equal to $V_r$. Also, since this is a series circuit, we know that the multiplier current $I_m$ must be equal to $I_{fs}$. Now we can use Ohm's law to develop the general formula for calculating the resistance of a multiplier:

$$R_m = \frac{V_m}{I_m} = \frac{V_r - V_{fs}}{I_{fs}}$$

### EXAMPLE 21-6

Calculate the multiplier resistance needed by the voltmeter circuit shown in Fig. 21-12 when it is on the 10-V range.

**Solution**

$$V_{fs} = I_{fs}R_i = 50\ \mu\text{A} \times 5\ \text{k}\Omega = 250\ \text{mV}$$

$$R_m = \frac{10\ \text{V} - 250\ \text{mV}}{50\ \mu\text{A}} = 195\ \text{k}\Omega$$

In the higher voltage ranges $V_r$ will be so much larger than $V_{fs}$ that the formula for $R_m$ can be reduced to $R_m = V_r/I_{fs}$. As a general rule, you can drop $V_{fs}$ from the formula whenever $V_{fs}$ is a smaller percentage of $V_r$ than the percentage tolerance of the multiplier resistor.

The switch for selecting voltmeter ranges should be the nonshorting type. Then, when switching ranges, two multipliers will never be temporarily paralleled, making an effective range smaller than the smallest of the two ranges momentarily shorted together.

The ohms-per-volt rating (sensitivity) of a voltmeter is a function of the $I_{fs}$ of the meter movement. We know that the current for each voltage range must be limited to $I_{fs}$ by the total internal resistance $(R_i + R_m)$ of the voltmeter. Ohm's law tells us that the total internal resistance $R_{i,V}$ *for each volt* must be $R_{i,V} = 1\ \text{V}/I_{fs}$. Therefore, the sensitivity of a voltmeter can be found by the formula

$$\text{Sensitivity} = \frac{1}{I_{fs}}$$

and the unit for sensitivity is ohms per volt.

### EXAMPLE 21-7

Determine the sensitivity of a voltmeter which uses a 100-μA meter movement.

**Solution**

$$\text{Sensitivity} = \frac{1}{I_{fs}} = \frac{1}{100\ \mu\text{A}} = 10\ \text{k}\Omega/\text{V}$$

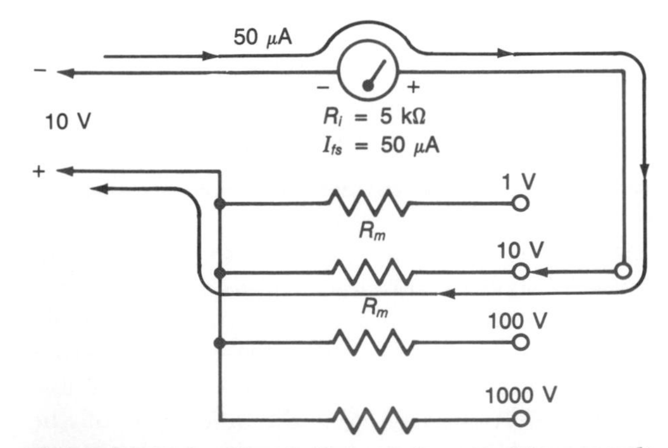


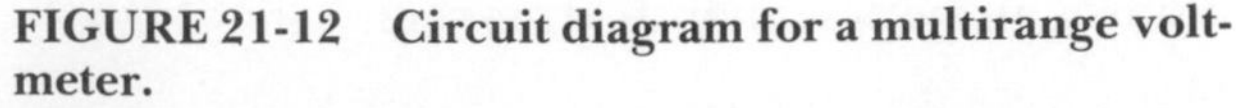

**FIGURE 21-12 Circuit diagram for a multirange voltmeter.**

The internal resistance of a VOM on any dc voltage range can be determined if either $I_{fs}$ or sensitivity is known. When sensitivity is known, just multiply the sensitivity by the range.

---

**EXAMPLE 21-8**

Determine the total internal resistance $R_{i,V}$ of a voltmeter on the 10-V range when the meter movement is rated for 50 $\mu$A of full-scale current.

**Solution**

$$\text{Sensitivity} = \frac{1}{50\ \mu\text{A}} = 20\ \text{k}\Omega/\text{V}$$

$$R_{i,V} = 20\ \text{k}\Omega/\text{V} \times 10\ \text{V} = 200\ \text{k}\Omega$$

---

Notice that this value of $R_{i,V}$ agrees with the value of $R_m$ and $R_i$ from Example 21-6 and Fig. 21-12, respectively; i.e., 200 kΩ = 195 kΩ + 5 kΩ.

## Ohmmeter Circuits

A simple series-type ohmmeter circuit is shown in Fig. 21-13*a*. The rheostat $R_2$ is the ohms-adjust control. It is adjusted so that full-scale current flows when the ohmmeter leads are shorted together, as shown in Fig. 21-13*b*. This represents zero resistance and explains why the resistance scale on the VOM is reverse-reading. The ohms-adjust control is needed to compensate for the decreasing cell voltage as the cell ages. The total internal resistance for this ohmmeter $R_{i,O}$ is the internal resistance of the meter movement $R_i$ plus $R_1$ and $R_2$. Since $R_{i,O}$ must limit the circuit to $I_{fs}$ when zero resistance is being measured, $R_{i,O}$ can be calculated by $R_{i,O} = V_{B_1}/I_{fs}$ where $V_{B_1}$ represents the cell or the battery voltage. For the circuit in Fig. 21-13*a*, this resistance is

$$R_{i,O} = \frac{1.5\ \text{V}}{100\ \mu\text{A}} = 15\ \text{k}\Omega$$

Figure 21-13*c* shows the results of using the ohmmeter of Fig. 21-13*a* to measure an unknown resistance $R_u$ which in this case is 15 kΩ. The total circuit resistance $R_T$ equals $R_{i,O} + R_u$. Thus, $R_T$ = 15 kΩ + 15 kΩ = 30 kΩ. The current through the meter, which is $I_T$, is now reduced to $I_T$ = 1.5 V/30 kΩ = 50 $\mu$A. Thus, half-scale current represents 15 kΩ for the ohmmeter in Fig. 21-13*a*.

Now, let us change $R_u$ in Fig. 21-13*c* to 45 kΩ and again calculate $I_T$. First, $R_T = R_{i,O} + R_u$= 15 kΩ + 45 kΩ = 60 kΩ. Now $I_T$ = 1.5 V/60 kΩ= 25 $\mu$A or quarter-scale current. Thus, quarter-scale current represents 45 kΩ of resistance being measured.

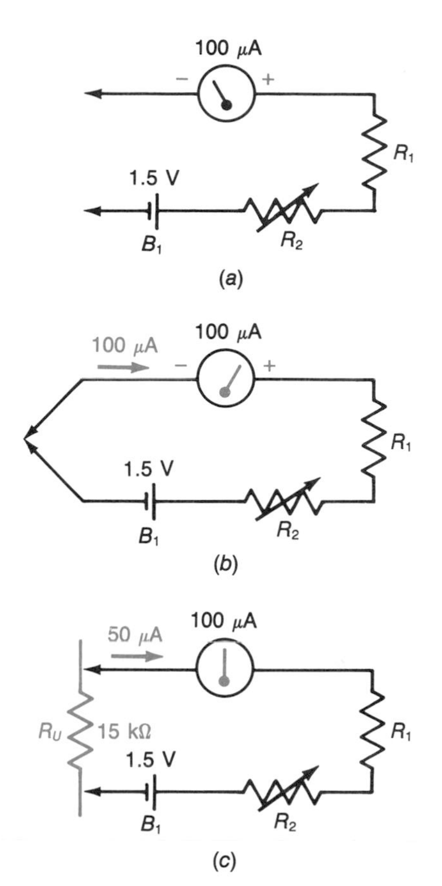


**FIGURE 21-13 Series-type ohmmeter circuit.**

Notice how nonlinear the ohmmeter scale for the circuit in Fig. 21-13*a* will be. The first half of the scale covers from 0 to 15 kΩ; the next quarter of the scale covers from 15 to 45 kΩ; and the last quarter of the scale covers from 45 kΩ to infinity. Now you know why the ohmmeter scale on a VOM is both reverse-reading and nonlinear.

Figure 21-14 shows how to make a lower resistance range by shunting the meter movement. Shunting the movement increases the circuit current required to provide $I_{fs}$ when zero resistance is being measured. This in turn lowers $R_{i,O}$ and the resulting $R_u$ required for half-scale current.

To provide a higher resistance range, the voltage of $B_1$ in Fig. 21-13*a* can be increased. If $B_1$ is increased, then $R_{i,O}$ must be increased to limit the current to $I_{fs}$ when ohms-adjusting for 0 Ω. If $R_{i,O}$ is increased, then $R_u$ required for half-scale current must also be increased. This, of course, represents a higher resistance range.

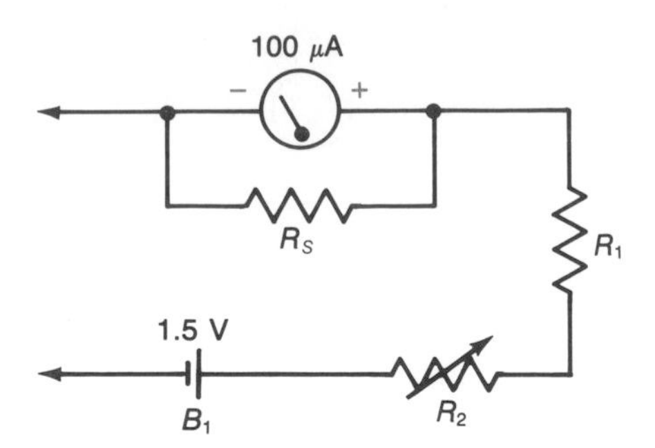


**FIGURE 21-14 A shunt provides a lower resistance range.**

## Self-Test

**15.** What causes the up-scale torque in a PMMC meter movement?

**16.** Why is the d'Arsonval meter movement polarized?

**17.** What is a rectifier?

**18.** Determine $V_{fs}$ for a meter movement which has an $I_{fs}$ of 100 $\mu$A and an $R_i$ of 2 kΩ.

**19.** Determine the shunt resistance needed for the 10-mA and 1-A ranges in Fig. 21-10.

**20.** Change $R_i$ to 4 kΩ and $I_{fs}$ to 50 $\mu$A and then determine the shunt resistance needed for the 1-mA range in Fig. 21-10.

**21.** What is another name for a shorting switch?

**22.** What is a multiplier resistor?

**23.** A 20-$\mu$A, 200-mV meter movement is used to make a 25-V voltmeter. Determine the following: (a) sensitivity, (b) multiplier resistance, and (c) total internal resistance.

**24.** Why is the ohmmeter scale on a VOM a reverse-reading scale?

**25.** A series ohmmeter circuit uses a 3-V battery and a 1-mA meter movement. What is the half-scale resistance for this ohmmeter?

## 21-5 CATHODE-RAY OSCILLOSCOPES

The *cathode-ray oscilloscope,* hereafter referred to as the oscilloscope, is an essential instrument for analyzing electronic circuits. The most common use of the oscilloscope is to produce a graph which represents voltage on the vertical axis and time on the horizontal axis. The oscilloscope is also used to produce graphs that show the phase and/or frequency relationships between two alternating voltages.

### Cathode-Ray Tube

The *cathode-ray tube* (CRT) is the component which displays the output of an oscilloscope. The essential elements of a CRT are illustrated in Fig. 21-15. All the air has been evacuated from the CRT so that the electrons in the electron beam will not hit any gas molecules as they travel from the cathode to the phosphor coating.

When current flows through the *heater* wire (Fig. 21-15), the heater wire heats the cathode to the point at which the *cathode* starts to emit electrons from its end surface. Emitting electrons from a heated surface is called *thermionic emission.*

The *control grid* in Fig. 21-15 controls the intensity of (number of electrons in) the electron beam that strikes the phosphor coating. The number of electrons passing through the control grid opening is controlled by an electrostatic force field created by making the control grid negative with respect to the cathode. The amount of negative voltage on the control grid is controlled by the *intensity* or *brightness* control of the oscilloscope.

The *focus anode, acceleration anode,* and *Aquadag coating* all are positive with respect to the cathode. The acceleration anode is more positive than the focus anode; the Aquadag coating is the most positive of these three elements. The electrons that get through the control grid are accelerated toward the phosphor coating by progressively more positive voltages on these three elements.

The electrostatic fields between the focus anode and the accelerating anode and between the focus anode and the control grid form an *electrostatic lens* that focuses the electron beam to a fine point by the time it strikes the phosphor coating. The focal length of this lens can be varied by varying the voltage on the focus anode. The voltage on the focus anode is varied by the focus control on the oscilloscope. Many oscilloscopes also provide a control for varying the voltage on the acceleration anode. This is called the *astigmatism control.* It is adjusted so that the electron beam remains in focus when it is forced from the center of the tube face (screen). Since the intensity control also varies the electrostatic field between the control grid and the focus anode, on an oscilloscope there is some interaction

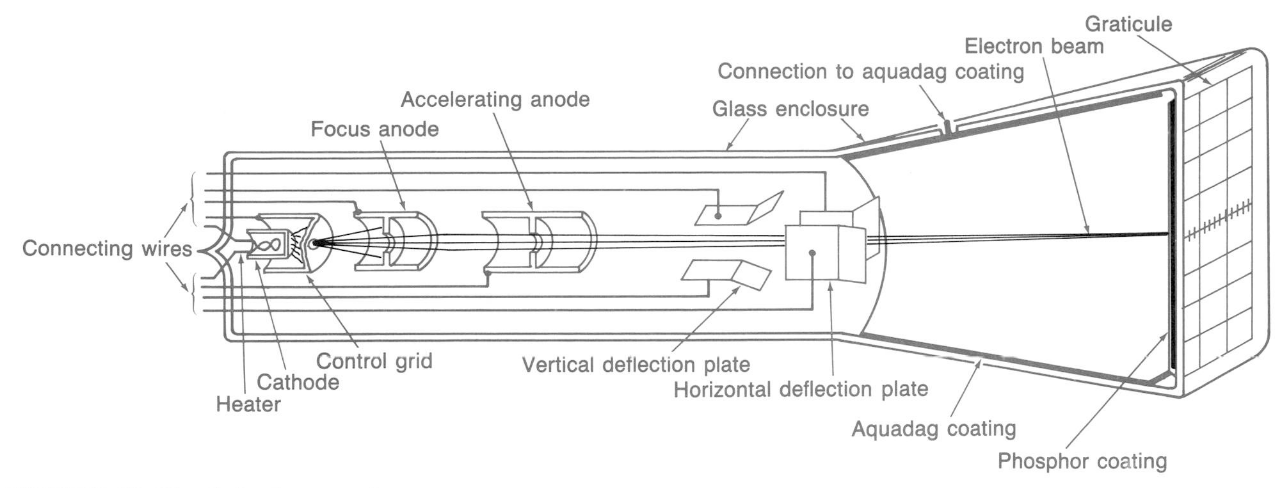


**FIGURE 21-15 Cathode-ray tube.**

between the focus control, the astigmatism control, and the intensity control.

Both sets (horizontal and vertical) of deflection plates have a positive voltage connected to them. However, the voltage to one plate of each set is adjustable so that the location of the electron beam can be controlled. On the oscilloscope, the vertical ($y$-axis) control adjusts the voltage on one of the two vertical plates. If the voltage on the top vertical plate in Fig. 21-15 is made more positive than the voltage on the bottom plate, the electron beam is deflected up. If, on the other hand, the top plate is made less positive, the beam is deflected down. The horizontal ($x$-axis) control on the oscilloscope horizontally positions the electron beam by adjusting the voltage on one of the horizontal plates.

The voltage-deflection function in a CRT is linear. That is, a 10-V change in the voltage on a deflection plate causes the electron beam to deflect the same distance whether the beam is located in the center of the tube or toward any edge of the tube.

The kinetic energy possessed by the electrons that strike the phosphor coating is converted to other forms of energy. Part of it is converted to visible light by the phosphor; part of it is converted to heat; and part of it is imparted to other electrons which are emitted from the surface of the phosphor. Electron emission caused by other electrons striking a material is called *secondary emission*. The electrons freed by secondary emission have a low velocity and are attracted to the Aquadag coating by its positive voltage. If no secondary emission occurred, the phosphor would build up a negative charge and repel the electrons in the electron beam because there is no direct electric connection to the phosphor coating.

The face of the CRT is covered with a *graticule*, which is a group of parallel horizontal and parallel vertical lines on the face of the tube. The space between two lines is called a *major division*. The vertical line and horizontal line that passes through the center of the face of the CRT are marked into five *minor divisions*. Each minor division is, therefore, equal to 0.2 major division. Thus, the deflection of the electron beam can be measured to the nearest two-tenths of a division in either the horizontal or vertical direction. A typical graticule for an oscilloscope has 8 vertical and 10 horizontal divisions.

## Displaying a Waveform

The process of displaying a waveform on the screen (phosphor coating) of a CRT is illustrated in Fig. 21-16. The sine wave is applied to the top vertical plate of the CRT and the bottom plate is held at +150 V. Since the top plate is more positive than the bottom plate at time zero, the vertical plates exert an upward force on the electron beam. The sawtooth type of waveform in Fig. 21-16 is applied to the right-hand horizontal plate and the left-hand plate is held at +150 V. Thus, at time zero, the horizontal plates are forcing the beam to the left. The net result of the horizontal and vertical forces on the beam at time zero is to locate the

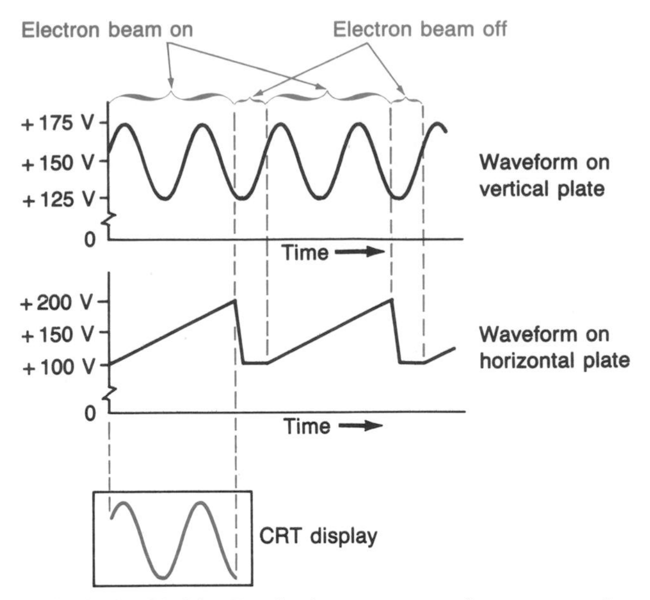


**FIGURE 21-16 Producing a waveform on the cathode-ray tube.**

beam above the center of the CRT screen (face of the tube) and to the extreme left-hand side of the screen. This is the location of the beam at the beginning of the displayed waveform shown in Fig. 21-16.

By the time the horizontal waveform in Fig. 21-16 has increased to +150 V, there is no horizontal force on the beam because both plates are at +150 V. At this same time, however, there is a downward force on the beam because the top vertical plate is less positive (approximately +130 V) than the bottom plate (+150 V). The location of the beam at this time is as shown by the display waveform in Fig. 21-16; i.e., the beam is horizontally centered but well below vertical center.

When the horizontal waveform reaches +200 V, the beam is at the extreme right side of the screen. This ends the trace part of the first cycle of the horizontal waveform. Part of the vertical waveform has been "traced" onto the screen of the CRT. Even though the beam only strikes one spot of the phosphor at any given time, the entire waveform can be seen on the screen because the phosphor continues to glow for a short time after it is struck by the beam. The length of time that the afterglow continues is called the *persistence* of the phosphor coating.

At the time the horizontal waveform reaches +200 V, the electron beam is turned off by applying a large negative voltage to the control grid of the CRT. Then, as shown in Fig. 21-16, the horizontal waveform quickly returns to +100 V and remains at +100 V for a period of time. This part of the cycle when the beam is turned off is called the *retrace* because the electron beam, had it been left on, would be retracing back to the left side of the screen. Turning the beam off during retrace is referred to as *retrace blanking*. Notice from Fig. 21-16 that the part of the vertical waveform occurring during horizontal retrace is never displayed.

The second cycle of the horizontal waveform starts the instant that the vertical waveform has exactly the same value as it had when the first cycle of the horizontal waveform started. This causes the electron beam to trace the exact same pattern during the second horizontal cycle as it did during the first horizontal cycle. If the vertical and horizontal waveforms are not synchronized in this fashion, then multiple, moving displays appear on the screen of the CRT.

The waveform applied to the horizontal plate in Fig. 21-16 is called the *sweep* waveform. The action of the waveform during the trace time is referred to as horizontal sweep. If the vertical waveform is to be displayed without any distortion of its time dimension, then the increasing *(ramp)* part of the sweep waveform must be linear. A linear ramp means that a change in ramp voltage divided by the time required for the change is constant.

Since both the horizontal plate deflection and the ramp voltage are linear, the horizontal distance that the electron beam travels can be calibrated in units of time rather than voltage. Thus, in using an oscilloscope, horizontal deflection is usually specified in microseconds, milliseconds, or seconds per division (μs/DIV, ms/DIV, or s/DIV). The horizontal system that generates the horizontal deflection waveform is often referred to as the horizontal *time base*.

In order to simplify the discussion on displaying a waveform, a waveform was applied to only one vertical and one horizontal plate. In an oscilloscope, waveforms are applied to both vertical and both horizontal plates. However, the waveform applied to the bottom plate is an inverted (180° out of phase) version of the waveform applied to the top plate. Thus, the two waveforms aid each other in vertically deflecting the electron beam. The same thing applies to the waveforms on the horizontal plates.

## Functional Analysis

A functional block diagram for an oscilloscope is shown in Fig. 21-17. Also shown in this figure are the waveforms associated with each block when the oscilloscope is used to view a sinusoidal signal.

The signal to be viewed is applied to the vertical input jack of the oscilloscope where it encounters capacitor $C_1$ paralleled by switch $S_1$. This switch is a coupling selector switch. When $S_1$ is open, the vertical channel of the scope is *ac-coupled.* Only alternating current, or the ac component of a fluctuating dc signal, can pass through the capacitor to the vertical attenuator. When $S_1$ is closed, the oscilloscope is *dc-coupled;* i.e., both alternating current and direct current will pass through the attenuator and amplifier and be applied to the vertical deflection plates of the CRT. As illustrated in Fig. 21-18, the results of ac and dc coupling are the same where the input is a pure ac waveform.

The *vertical attenuator* in Fig. 21-17 is a calibrated step attenuator operated by a rotary switch. This attenuator determines how much of the vertical input signal is applied to the vertical amplifier. The positions on the attenuator switch are marked or labeled in V/DIV (voltage per division). The setting of the attenuator switch specifies how much input voltage is required to deflect the electron beam one major division on the graticule *if* the

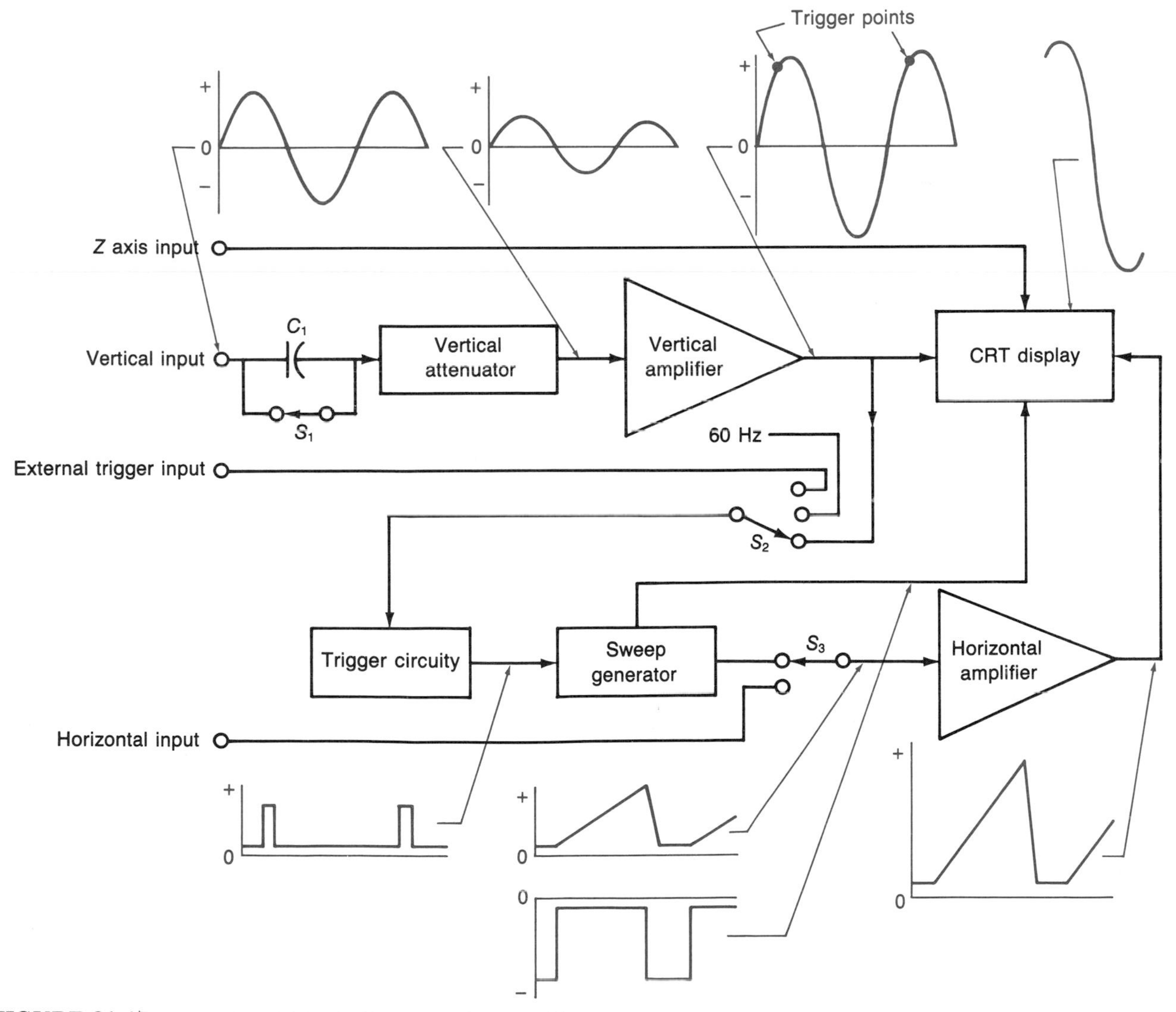


**FIGURE 21-17 Functional block diagram of an oscilloscope.**

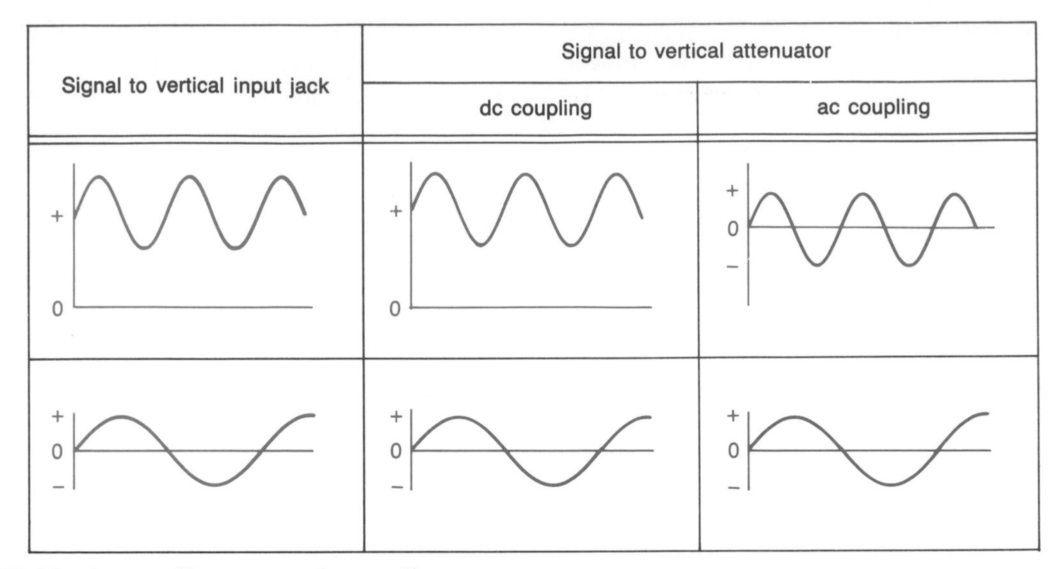


**FIGURE 21-18 Ac coupling versus dc coupling.**

gain control on the vertical amplifier is in its calibrated position. The attenuator switch is sometimes referred to as the *deflection factor* switch.

The vertical amplifier block in Fig. 21-17 takes the vertical signal voltage from the attenuator and increases its amplitude to the level needed to drive the deflection plates in the CRT. A CRT may require more than 100 V of signal on the vertical deflection plates to deflect the electron beam the full height of the graticule. If the input signal is in millivolts, then the amplifier must increase the amplitude of the signal by a large factor to provide full deflection. The factor by which an amplifier increases the signal applied to its input is called the *voltage gain* of the amplifier. An oscilloscope has a control which allows the operator to adjust (vary) the gain of the vertical amplifier. This gain control may be labeled gain, vernier, variable, or calibrate on various scopes.

Whatever it is called, the gain control is used in conjunction with the attenuator (V/DIV or deflection factor) switch to adjust the height of the displayed waveform. *Only when* the gain control is in its calibrate position are the attenuator markings (V/DIV) accurate. For example, if the attenuator is set on 1 V/DIV and the gain control is not in its calibrate position, then it will require more than 1 V of signal at the vertical input to deflect the beam 1 division on the graticule.

The output of the vertical amplifier is connected to the vertical deflection plates of the CRT. As seen in Fig. 21-17, it may also be fed, through $S_2$, to the trigger block.

The function of the trigger circuitry is to produce a waveform (pulse) that determines when the horizontal sweep starts. That is, the trigger pulse determines precisely when the ramp part of the horizontal deflection waveform starts. The trigger circuitry allows the oscilloscope operator to pick any point on a waveform to trigger (start) the horizontal sweep. The waveforms in Fig. 21-17 illustrate the results of selecting a trigger point just before the vertical input signal reaches its positive peak. Because the sweep generator in this figure finishes its ramp buildup in less time than the period of 1 cycle of the vertical input, the generator is triggered by every trigger pulse. Thus, the display is less than 1 cycle.

The trigger point on a waveform is determined by the setting of the *slope* switch and the *level* control. The slope switch can be set for either a +slope or a −slope. As illustrated in Fig. 21-19, the +slope of a waveform is the "positive-going" portion of the waveform. That is, it is the part of the waveform during which the voltage is becoming more positive. Of course, −slope refers to the negative-going part of the waveform. The level control determines which point on the selected slope triggers the sweep generator. Figure 21-19 shows that the level control can select any point between the peak negative and peak positive points of the waveform. Figure 21-20 shows that the displayed waveform

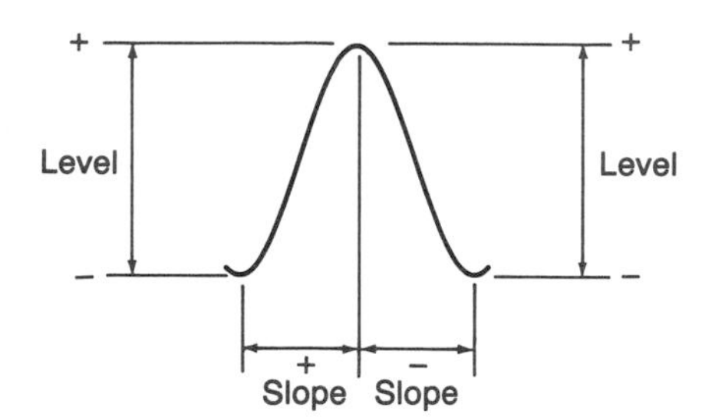


**FIGURE 21-19 The Slope Switch and Level Control select the trigger point.**

starts on the negative-going portion of the negative $\frac{1}{2}$ cycle. This trigger point is selected by having the slope switch set to the −slope position and the level control rotated toward the −level.

As indicated in Fig. 21-17, there are three signal sources that can be used to trigger the sweep. One of the three sources is selected by the *trigger source* switch on the oscilloscope ($S_2$ in Fig. 21-17). The source switch has three positions labeled INT (internal), EXT (external), and LINE (power line). When INT is selected, the trigger signal is the signal that is being displayed. This is the source used most often, and, when using this source, the oscilloscope is said to be internally triggered. With the switch in the external position, the oscilloscope operator can use any signal source to trigger the oscilloscope by connecting the desired signal to the external trigger jack. The external signal source must be either the same frequency as that of the signal being viewed or a submultiple of it. Otherwise, the displayed signal will not be stationary. The LINE position of the switch uses a signal at the power line frequency to trigger the scope. In the United States and many other countries, the line frequency is 60 Hz. Thus, Fig. 21-17 shows 60 Hz connected to one position of the source switch. Line triggering is useful when using an oscilloscope to analyze power control circuits.

The sweep generator in Fig. 21-17 has two outputs. One, called the *blanking pulse,* keeps the electron beam turned off except when the beam is actually tracing (sweeping) across the screen from left to right. As mentioned previously, the blanking pulse is applied to the control grid of the CRT. The other signal, which contains the sweep ramp, is applied to the horizontal deflection plates. Figure 21-20 shows that the sweep generator ignores trigger pulses that occur during the ramp part of its output. If it did not, it would not be possible to have multicycle displays.

The sweep rate (period of the sweep waveform) of the sweep generator is determined by the calibrated sweep-rate and variable sweep-rate controls. The calibrated sweep-rate control operates a rotary switch. Each position of the switch represents a specific (calibrated) sweep rate *if* the variable sweep rate control *and* the horizontal gain control are in their calibrated positions. A typical scope has sweep rates ranging from less than 1 $\mu$s/DIV to over 0.5 s/DIV. The variable sweep-rate control allows the operator to select an infinite number of sweep rates between the discrete settings available with the calibrated sweep-rate control. Of course, when the variable control is not in its calibrate position, the time per division labels on the calibrated control are meaningless.

The horizontal amplifier serves the same purpose as the vertical amplifier; i.e., it increases the amplitude of the signal it receives so that its output signal is sufficient to drive the horizontal plates. The horizontal amplifier receives its input signal from the sweep generator when the horizontal axis

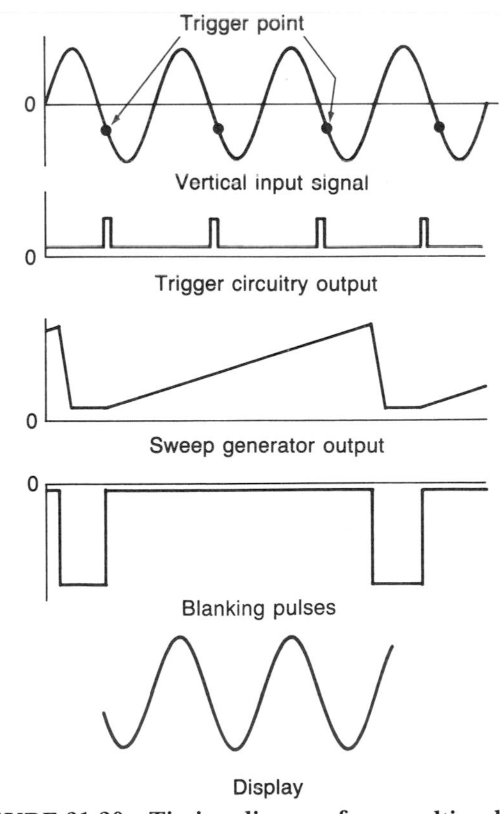


**FIGURE 21-20 Timing diagram for a multicycle display.**

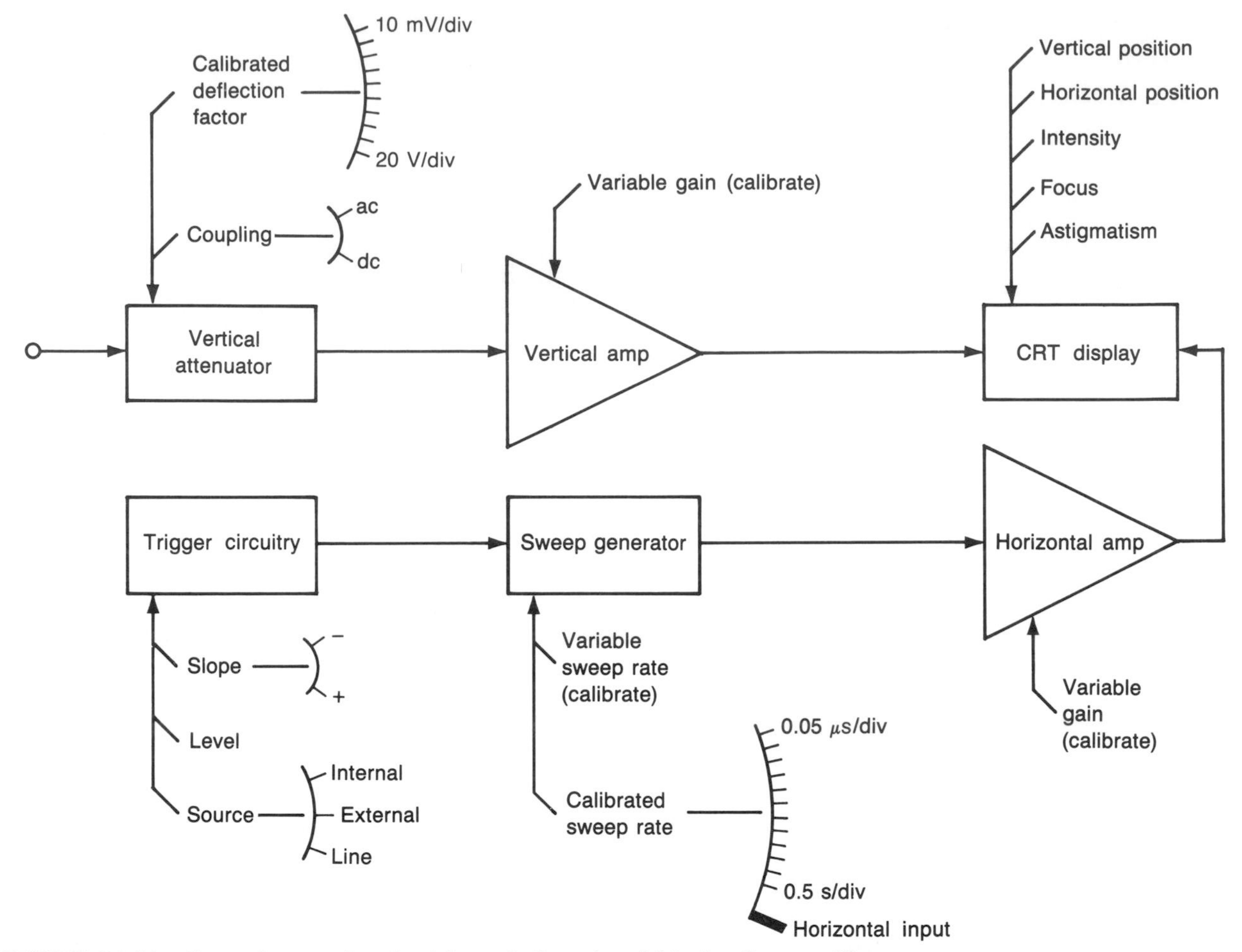


**FIGURE 21-21 Controls associated with each functional block of an oscilloscope.**

of the display is to represent and/or be calibrated in units of time. This is the most common use of the oscilloscope. However, in some measurement applications, the horizontal ($x$) axis represents voltage units just like the vertical ($y$) axis does. For these applications, the horizontal amplifier is connected to a horizontal input jack by a switch ($S_3$ in Fig. 21-17). This switch, $S_3$, is not a separate switch control on the oscilloscope; rather it is a section of the calibrated sweep-rate switch. The last setting of the calibrated sweep-rate switch switches the input to the horizontal amplifier from the sweep generator output to the horizontal input jack.

On some oscilloscopes the horizontal amplifier has a variable gain control just like the vertical amplifier does. It is used to adjust the amount of horizontal deflection when the amplifier is receiving a signal from the horizontal input jack. At all other times, this control should be in the calibrate position; otherwise, the horizontal deflection is not calibrated in the time units shown on the calibrated sweep-rate switch.

The *z-axis* input jack shown in Fig. 21-17 connects to the control grid of the CRT. This jack is sometimes called the *intensity modulation* input. A positive signal on this input causes the displayed waveform to get brighter, and a negative signal, of course, makes the display dimmer. If the signal applied to the $z$-axis input has sufficient amplitude, it can completely turn off the electron beam during its negative $\frac{1}{2}$ cycle. The $z$-axis input is useful for some types of measurements to be explained in Sec. 21-6.

As a summary and review of the major controls on an oscilloscope, Fig. 21-21 shows which controls are associated with each functional block. Switch-type controls are indicated with an expanded leader following its name. All other controls are continuously variable, and they internally operate either a rheostat or a potentiometer.

### Dual-Trace Oscilloscopes

A dual-trace oscilloscope can simultaneously display two waveforms on the screen of the CRT. It has two vertical input jacks, two calibrated step attenuators, and two vertical amplifiers with gain controls. In other words, it has two parallel vertical input setups. These two setups are called channel A and channel B.

The outputs of channels A and B are fed to the two inputs of an electronic switching circuit. The single output of the switching circuit is fed to another vertical amplifier, with a fixed gain, which drives the vertical plates of the CRT. The electronic switch switches back and forth between the inputs it receives from channel A and channel B. Thus, the output of the electronic switch feeds channel A to the vertical amplifier for a period of time; then it switches over and feeds channel B for an equal period of time; and so on. Figure 21-22 illustrates the arrangement of the vertical section of a dual-trace oscilloscope. Notice that each channel has its own vertical position control so that each display can be independently located on the screen.

The electronic switch can be operated in either the *chopped mode* or the *alternate mode*. In the chopped mode, the switch is switching at a frequency many times higher than that of the signal received from either channel A or channel B. In this mode, a small segment of 1 cycle of the channel A signal is displayed and then a small segment of the channel B signal is displayed. The result of this mode of operation is illustrated in Fig. 21-23*a*. In the alternate mode, the electronic switch is synchronized with the horizontal sweep circuit. Thus, the complete channel A waveform is displayed; then the complete channel B waveform is displayed; and so on.

At many frequencies there are no obvious differences in the two modes of operation. However, at very low frequencies the display created by the alternate mode shows excessive flickering because the persistency of the phosphor is not long enough to keep the phosphor glowing until the display is refreshed by another sweep. The chopped mode alleviates the flickering problem. In the chopped mode the segmentation of the signal becomes more obvious at higher frequencies. Thus, use the alternate mode for high frequencies.

The sweep generator of the dual-trace scope can be internally triggered from either channel A or channel B. The sweep is synchronized with whichever channel is selected by the oscilloscope operator. This means that the channel A and the chan-

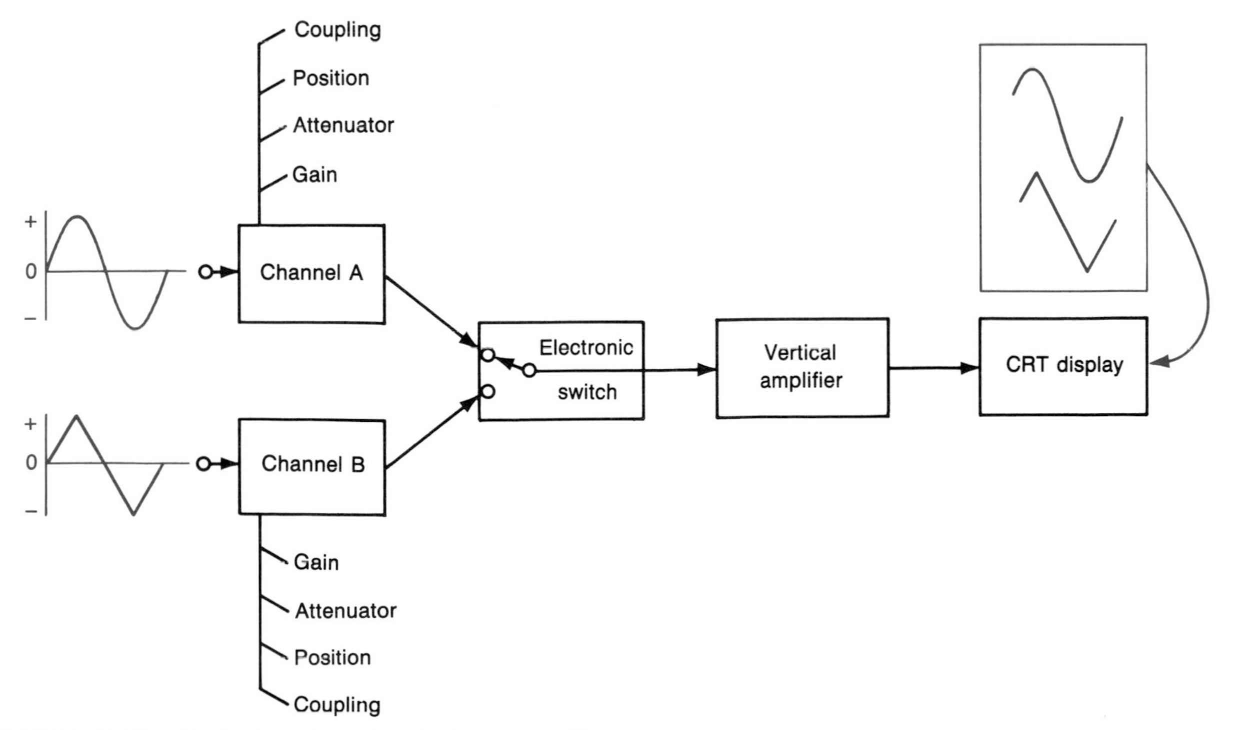


**FIGURE 21-22** **Vertical section of a dual-trace oscilloscope.**

(a) Chopped mode

(b) Alternate mode

**FIGURE 21-23 Dual-trace modes of operation.**

nel B signals must have equal frequencies, or be related by a multiple, for both displayed waveforms to be stationary.

Dual-trace oscilloscopes are very useful for comparing the input and output signals of an amplifier. When the two traces (input and output) are superimposed, it is easy to see if the amplifier has caused a phase shift in the signal or if it has caused any amplitude distortion of the signal.

## Self-Test

**26.** What material is used to coat the inside face of the CRT?

**27.** ______ describes the process of emitting electrons from a heated surface.

**28.** ______ describes the process of emitting electrons from an unheated surface.

**29.** The voltage on the ______ of the CRT controls the intensity of the electron beam.

**30.** The electron beam in the CRT in an oscilloscope can be turned off by the ______ control, the ______ signal, and the ______ signal.

**31.** The size and shape of the spot produced by the electron beam is determined by the ______, ______, and ______ controls.

**32.** What is the function of the Aquadag coating in the CRT?

**33.** What functional block of the oscilloscope keeps the sweep generator synchronized with the vertical input signal?

**34.** For the display shown in Fig. 21-16, what position is the slope switch in if the oscilloscope is on internal trigger? Is the level control turned toward its −limit or its +limit?

**35.** The three positions on the trigger source switch are labeled ______, ______, and ______.

**36.** The deflection factor switch controls the amount of signal going to the ______ amplifier.

**37.** The electron beam is on the ______ axis of the CRT.

**38.** The horizontal input jack is connected to the horizontal amplifier by the ______ switch.

**39.** How can a CRT with a single electron beam display two waveforms?

## 21-6 OSCILLOSCOPE MEASUREMENTS

### Measuring Voltage

One of the most common uses of an oscilloscope is to measure the *amplitude* (*voltage*) of an ac waveform. Alternating current voltage can be measured by following this procedure:

1. Apply the signal to be measured to the vertical input jack. Use ac coupling.
2. Adjust the CRT display controls and the horizontal section controls for a stable, centered display of at least one complete cycle.
3. Turn the vertical gain control to its calibrate position.
4. Adjust the attenuator (deflection factor) switch to obtain the maximum height of display that does not exceed the graticule lines.
5. Using the vertical position control, position the displayed waveform so that the negative peaks of the waveform are on one of the lower horizontal graticule lines.
6. Using the horizontal position control, position the waveform so that one of the positive peaks is on the center vertical graticule line which is marked in minor divisions.
7. Count the number of divisions between the negative and positive peaks of the waveform.

8. Finally, multiply the number of divisions obtained in step 7 by the V/DIV setting of the attenuator switch. The product is the peak-to-peak voltage ($V_{p\text{-}p}$) of the signal applied to the vertical input jack.

**EXAMPLE 21-9**

Determine the rms voltage of the waveform in Fig. 21-24 if the deflection factor switch is on the 2-V/DIV range (setting) and the vertical gain control is set at calibrate.

**Solution**

$$V = 2 \text{ V/DIV} \times 7.4 \text{ DIV} = 14.8V_{p\text{-}p}$$
$$V_{\text{rms}} = 14.8\ V_{p\text{-}p} \times 0.3535 = 5.23V_{\text{rms}}$$

Note that the horizontal section controls do not have to be in the calibrate position when measuring voltage. They can be adjusted to any setting that provides a convenient pattern for measuring the height of the waveform.

## Measuring the Period

Another common use of the oscilloscope is measuring the *period* of the waveform. This measurement can then be used to calculate the frequency of the waveform. To obtain the best resolution of this measurement, do the following:

1. Apply the signal to the vertical input.
2. Turn the horizontal gain control and the variable sweep rate control to their calibrate positions.
3. Adjust the vertical section controls and the CRT controls for a centered display that vertically fills at least three-fourths of the graticule. The vertical gain control does not have to be in the calibrate position.
4. Adjust the calibrated sweep-rate control, and the trigger controls if necessary, to obtain a waveform containing no more cycles than necessary to display one complete cycle.
5. Using the horizontal position control, move the waveform so that the beginning of a complete cycle crosses the center horizontal line at the intersection of one of the left-side vertical lines and the center horizontal line (see Fig. 21-25).
6. Count the number of horizontal divisions occupied by 1 cycle.
7. Finally, multiply the divisions from step 6 by the time per division indicated on the calibrated sweep-rate switch to obtain the period.

Step 5 in the above procedure is not essential, but it does make it easier to read the divisions more accurately in step 6. Also, it is not necessary that the waveform be vertically centered to measure its period. However, a more accurate reading of the divisions can be made when the waveform crosses the center horizontal line at a steep angle.

**EXAMPLE 21-10**

Determine the frequency of the waveform in Fig. 21-25 if the horizontal controls are in the calibrate positions and the calibrated sweep-rate switch is on the 2-ms/DIV range.

**Solution**

$$T = 7.2 \text{ DIV} \times 2 \text{ ms/DIV} = 14.4 \text{ ms}$$
$$f = \frac{1}{T} = \frac{1}{14.4 \text{ ms}} = 69.4 \text{ Hz}$$

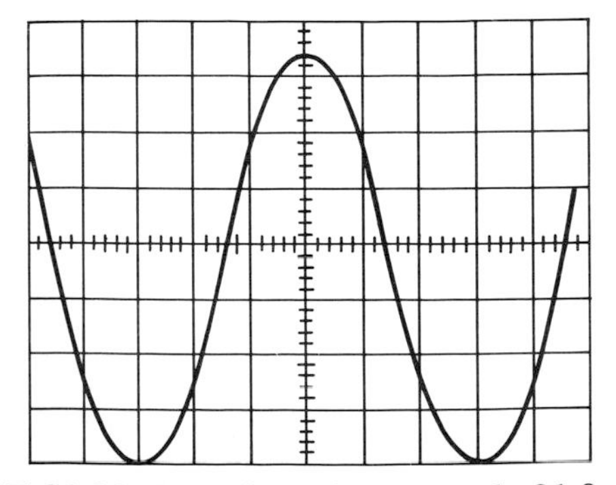

**FIGURE 21-24 Waveform for Example 21-9.**

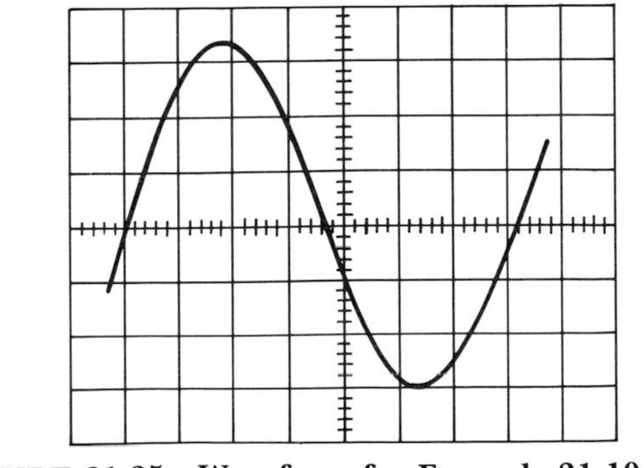

**FIGURE 21-25 Waveform for Example 21-10.**

## Phase Measurement

The phase relationship between two signals can be measured with either a dual-trace or a single-trace oscilloscope. Figure 21-26 illustrates how to measure the phase shift between voltages with a dual-trace oscilloscope. First, vertically center both channels and adjust the vertical gains so that each waveform nearly fills the screen. Second, adjust the sweep-rate controls so that $\frac{1}{2}$ cycle spans 6 divisions; this calibrates the horizontal in electric degrees so that each division represents 180° per 6 DIV = 30°/DIV. Finally, count the number of divisions between equivalent points on the displayed waveforms and multiply by 30°/DIV. This procedure was used to obtain the results in Fig. 21-26. Notice that $\frac{1}{2}$ cycle of the channel A waveform spans 6 divisions; thus, the calibration is 30°/DIV. Further, notice that there are 1.8 divisions between the points where the negative-going waveforms cross the horizontal axis. Therefore, the phase shift is 1.8 DIV × 30°/DIV = 54°. Since the display for channel B crosses the $x$ axis before that for channel A, we can conclude that the signal into channel B leads the signal into channel A by 54°. Three other things should be noted about Fig. 21-26. (1) The oscilloscope is operating in the alternate mode because the displayed waveforms are not chopped. (2) The $x$ axis of the input waveforms is not to the same scale as the $x$ axis of the display waveforms. (3) The scope is internally triggering from channel A.

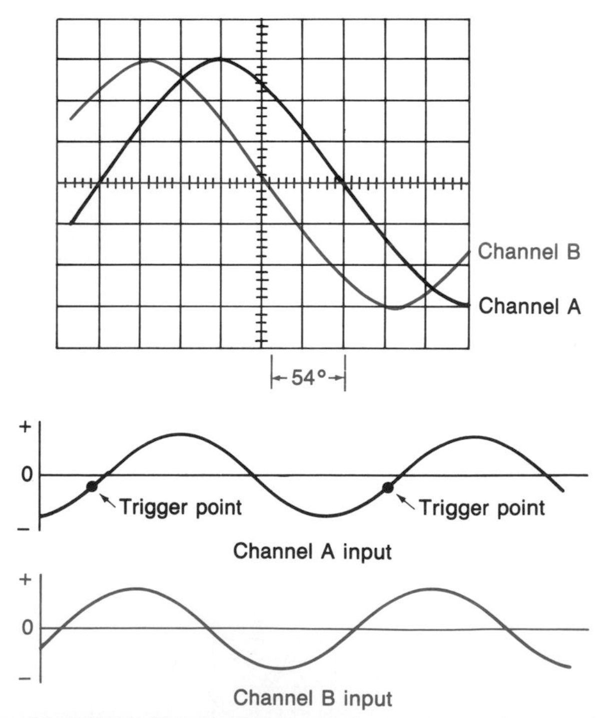


**FIGURE 21-26 Measuring phase shift with a dual-trace oscilloscope.**

Figure 21-27 shows how to connect a single-trace oscilloscope to a circuit to measure the phase difference between the voltages at point A and point B. The single-trace oscilloscope *must be* set for external triggering to measure phase shift. The trigger input can be connected to either point A or point B in a circuit like the one in Fig. 21-27. If connected to point A as in Fig. 21-27, then the waveform displayed by connecting the vertical input to point A can be considered as the reference waveform. The oscilloscope controls would be adjusted to display a waveform like the channel A waveform in Fig. 21-26. When the vertical input is removed from point A, the oscilloscope continues to sweep in step with the voltage at point A even though the waveform display of the voltage at point A disappears. Now, when the vertical input is connected to point B, the voltage waveform at point B is displayed, but the sweep is still synchronized to the voltage at point A. The waveform display for point B is just like the channel B display in Fig. 21-26 except it has less amplitude unless the vertical gain is increased after the vertical input is moved from point A to point B. If one remembers where the point A display crossed the $x$ axis, then it is easy to count the divisions from there to where the point B waveform is crossing the $x$ axis.

You can see that the only differences between measuring phase shift with the dual-trace and the

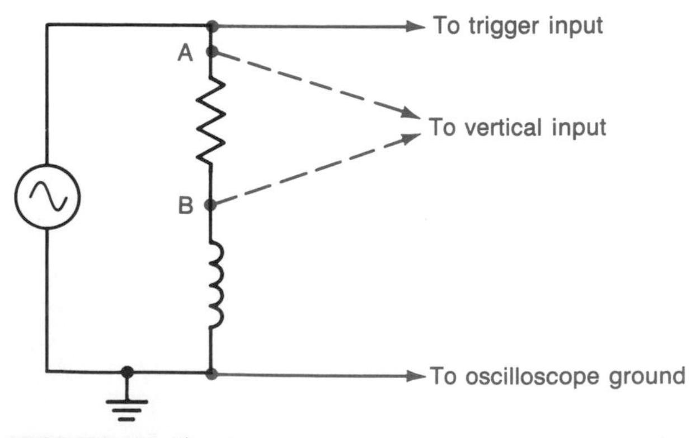


**FIGURE 21-27 Measuring phase shift with a single-trace oscilloscope.**

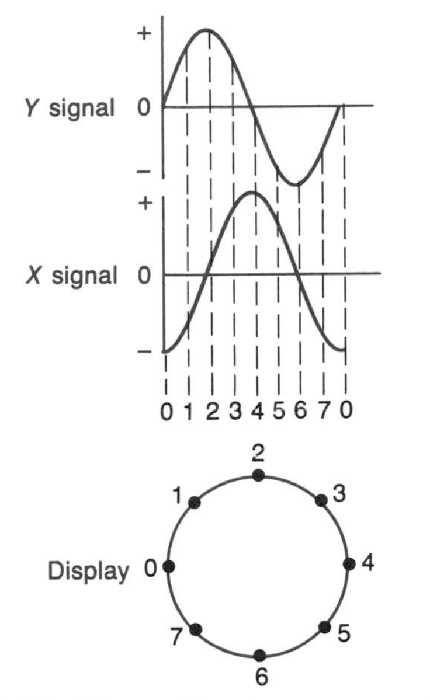


**FIGURE 21-28 How a Lissajous pattern is formed.**

single-trace oscilloscope is that the single-trace requires (1) that the vertical input be manually switched between two points, (2) only one voltage is displayed at a time, and (3) external triggering is required.

## Frequency Measurements

It was previously shown how to determine the frequency of a waveform by measuring the period of the waveform. Now we will see how to measure frequency by using the *x-y* measurement capability of an oscilloscope. The *x-y* measurement refers to a measurement made by not using the internal time base (sweep generator and trigger circuitry) of the oscilloscope. That is, an external signal is fed into the horizontal amplifier as well as the vertical amplifier.

The display that results from feeding both amplifiers with a sinusoidal waveform is called a *Lissajous pattern.* Figure 21-28 shows how two sine waves which are 90° out of phase form a perfect circle. In the figure, points 0 through 7 marked on the display correspond to the eight points on the time axis of the waveforms. At time 0, the maximum negative signal on the horizontal plates of the CRT drives the electron beam to the extreme left of the screen. At this same time, the vertical plates have neither a positive nor a negative signal, so the beam is vertically centered. Following the action of the plate voltages at the other seven times shows that 360 electric degrees of *x-y* signal produce one circular rotation of the beam.

Figure 21-29 shows that a Lissajous pattern can be used to determine either the phase or the fre-

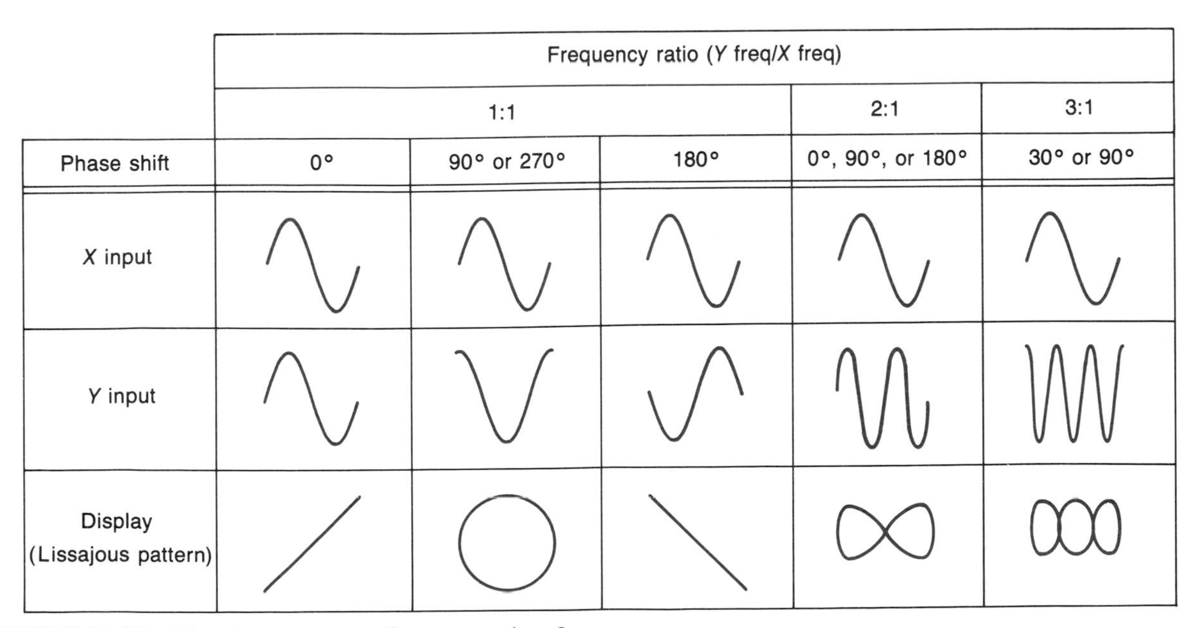


**FIGURE 21-29 Lissajous patterns for measuring frequency.**

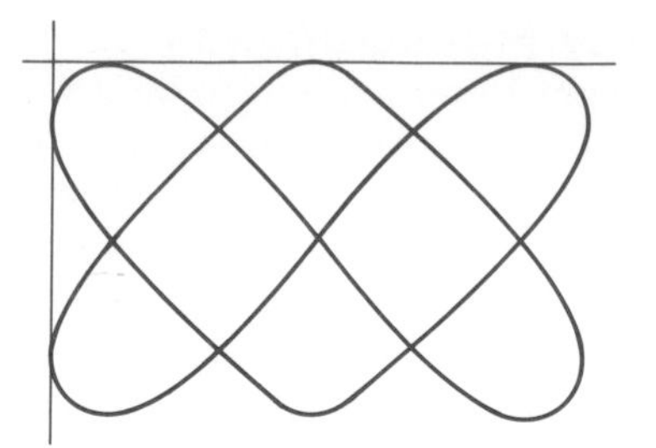

**FIGURE 21-30 Lissajous pattern for a 2:3 frequency ratio.**

quency relationship between two sinusoidal voltages. However, the previous method of determining phase shift is more accurate and easier, so it is usually used. For the displays shown in Fig. 21-29 to remain stable, the frequency and the phase of both the *x* (horizontal) input and the *y* (vertical) input must remain constant. Lissajous patterns are especially useful for checking the dial calibration of a variable-frequency source. For example, the calibration of the low-frequency end of an audio signal generator can be checked against the 60 Hz available from the power distribution system. Just connect the output of a low-voltage power transformer to the input of the horizontal amplifier and the signal generator's output to the vertical amplifier. Adjust the generator's frequency control to get a display like that shown in the lower right-hand corner of Fig. 21-29; the frequency dial on the generator should indicate 180 Hz at this time.

Many other Lissajous patterns can be produced with two sine wave inputs. The frequency ratio of any stable pattern can be determined by the number of places the edges of the pattern would touch a vertical line and a horizontal line. For example, the pattern in Fig. 21-30 touches the vertical line twice and the horizontal line three times. Thus, the vertical-to-horizontal frequency ratio is 2:3.

Figure 21-31 illustrates how to measure frequency by *z*-axis modulating a Lissajous pattern created by a 1:1 frequency ratio. Notice that the Lissajous pattern is a chopped ellipse rather than a chopped circle. It is elliptical because $V_C$ lags $V_T$ by less than 90°. It is chopped into four segments because the *z*-axis frequency is four times as great as

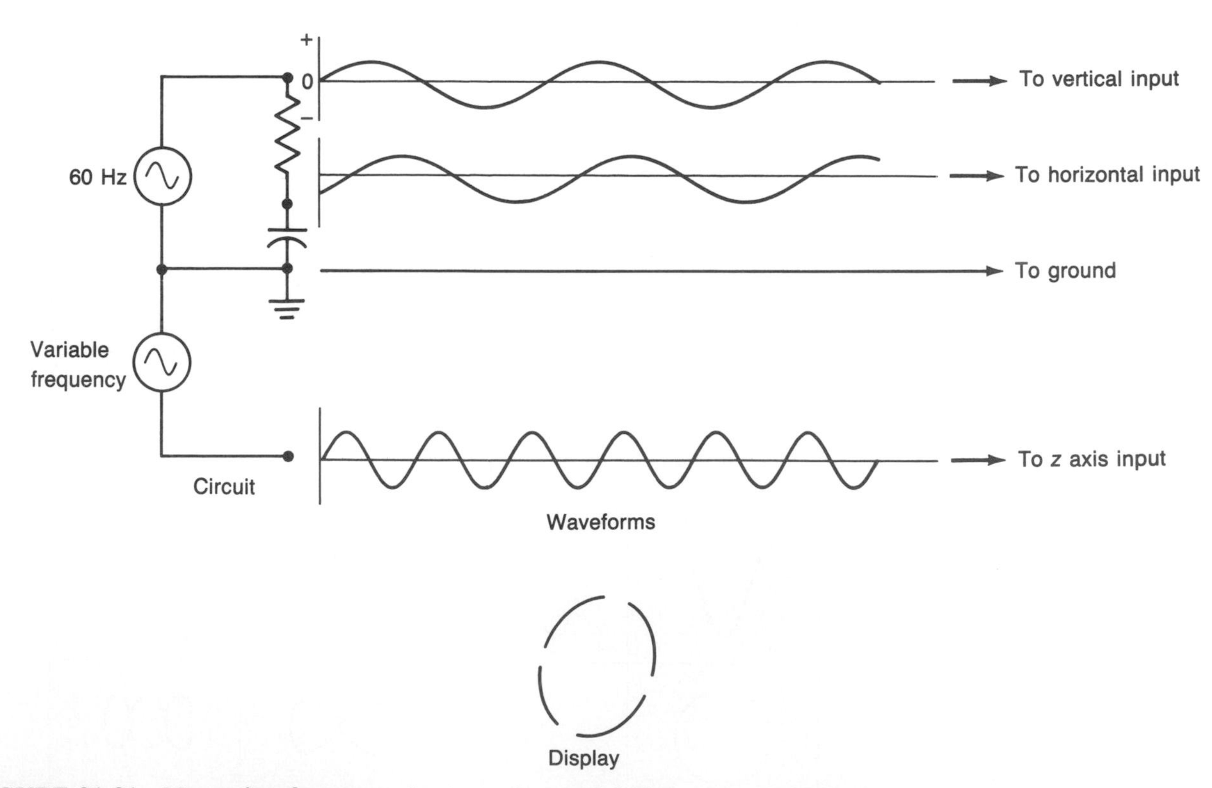


**FIGURE 21-31 Measuring frequency by intensity-modulating a Lissajous pattern.**

the $x$ and $y$ signals. Thus, while the $x$ and $y$ signals cause the electron beam to make one rotation, the $z$ signal turns the beam off four times during its four negative $\frac{1}{2}$ cycles.

When the $z$ and $x$-$y$ signals are not exact multiples of each other, the segmented pattern rotates. Of course, the $z$ signal must be a higher frequency than the $x$-$y$ signals to produce segments; if it were lower, the display would merely blink off and on.

## Self-Test

**40.** Refer to Fig. 21-24. Assume all controls are in the calibrate position, the attenuator is on 5 V/DIV, and the sweep rate is on 50 $\mu$s/DIV. What is the frequency and the peak-to-peak voltage of the waveform?

**41.** Refer to Fig. 21-25. For the oscilloscope settings specified in Question 40, determine the rms voltage and the frequency.

**42.** What type of triggering is used to measure phase shift with a single-trace oscilloscope?

**43.** What is a Lissajous pattern?

**44.** A circular display on an oscilloscope is chopped into 10 parts. If the $x$-$y$ signals are 400 Hz, what is the frequency of the $z$-axis signal?

## 21-7 OSCILLOSCOPE PROBES AND CIRCUIT LOADING

An oscilloscope is connected to a circuit with an oscilloscope *probe*. A *direct* or *X1 probe* is just a flexible, shielded cable with a connector to fit the oscilloscope jack on one end and a probe to connect to the circuit component on the other end. An *X10* (or attenuator) *probe* has a resistor and a variable capacitor built into the body of the probe. The X10 probe has several advantages over the direct or X1 probe.

The inputs to an oscilloscope have an input impedance just like meters do. The vertical input to a typical oscilloscope, when connected to a circuit by an X1 probe, will have about 1 MΩ of input resistance paralleled by about 108 pF of capacitance. As shown in Fig. 21-32*a*, both $R$ and $X_C$ of the oscilloscope's input are in parallel with the component across which a voltage is being measured. Notice in this figure that at less than 1.5 kHz the input capacitance has only 1 MΩ of reactance. Thus, even though $R_2$ in Fig. 21-32 is only 100 kΩ, the scope with the X1 probe is causing significant loading. If you want to calculate the exact amount of loading, you can use the same techniques as were used for meter loading in ac circuits. If the frequency in Fig. 21-32*a* were increased toward the upper end of the audio range, say 15 kHz, $X_C$ would drop to about 100 kΩ and the circuit would be very severely loaded.

Figure 21-32*b* and *c* illustrates the advantages of using an X10 probe when the circuit resistances and/or frequencies are too high for the X1 probe. The 9 MΩ in the probe, in series with the 1 MΩ in the scope, gives an effective input resistance of 10 MΩ; thus, the input resistance has increased by a factor of 10, or X10. Of course, only one-tenth of the voltage across $R_2$ gets into the scope's input. (The other nine-tenths drop across the probe's resistance.) This could be a serious limitation if the voltage being measured were only a few millivolts.

The capacitor in Fig. 21-32*b* is adjusted so that its capacitance (12 pF) is one-ninth as large as the input capacitance of the oscilloscope. The probe capacitor is variable so that the probe can be used with many different oscilloscopes. Even though each oscilloscope may have a slightly different input capacitance, the probe capacitance can be adjusted to provide a 1:9 capacitance ratio for any individual scope. If the capacitance ratio is 1:9, then the reactance ratio is 9:1 as shown in Fig. 21-32*b*.

Since the X10 probe's capacitance of 12 pF is in series with the oscilloscope's 108 pF of capacitance, the effective input capacitance with the X10 probe is $(12\text{ pF} \times 108\text{ pF})/(12\text{ pF} + 108\text{ pF}) = 10.8\text{ pF}$. Thus, the input capacitance is only one-tenth as large with the X10 probe as with the X1 probe. Of course, the capacitive reactance is 10 times as large with the X10 probe. (Compare Fig. 21-32*c* and *a*.)

An X10 probe is *frequency-compensated.* When applied to an X10 probe, this term means that the probe increases the oscilloscope's effective input impedance by a factor of 10 at all operating frequencies. The capacitor in the probe provides frequency compensation because if properly adjusted for an individual oscilloscope, its reactance is nine times greater than the reactance of the scope's capacitance at any operating frequencies. Thus, even

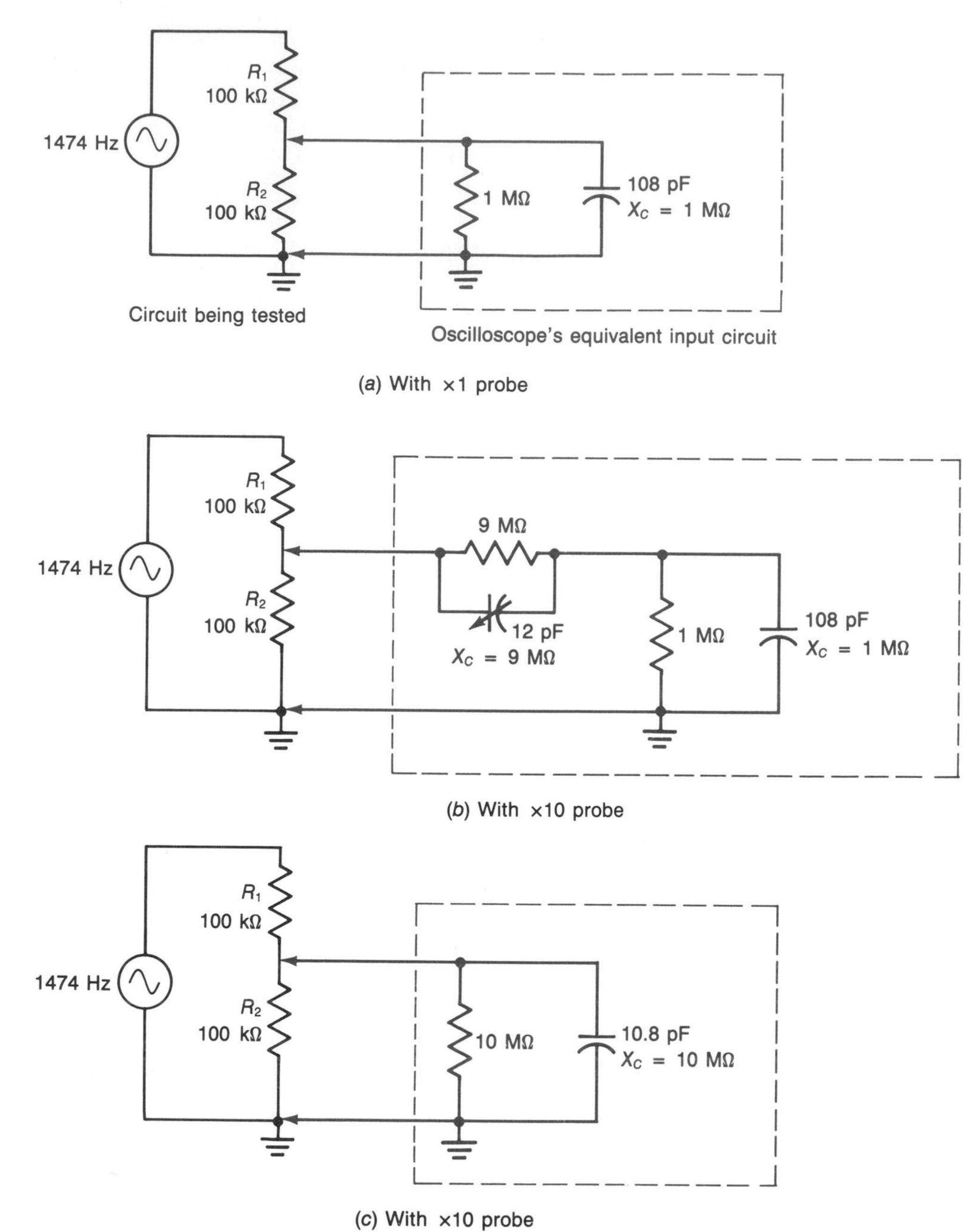


**FIGURE 21-32 How an X10 probe reduces circuit loading.**

though the effective input impedance decreases with frequency when either an X1 or X10 probe is used, the X10 probe provides 10 times as much effective input impedance at any operating frequency. Also, because it is frequency-compensated, the X10 probe effectively divides the measured voltage by 10 at all operating frequencies.

Figure 21-33 shows how an X10 probe would respond to different frequencies if it had no compensating capacitor. Notice that its voltage division factor is very frequency-dependent. When the probe's compensating capacitor is not properly adjusted to match (1:9 ratio) the scope's capacitance, the probe is also frequency-dependent. Of

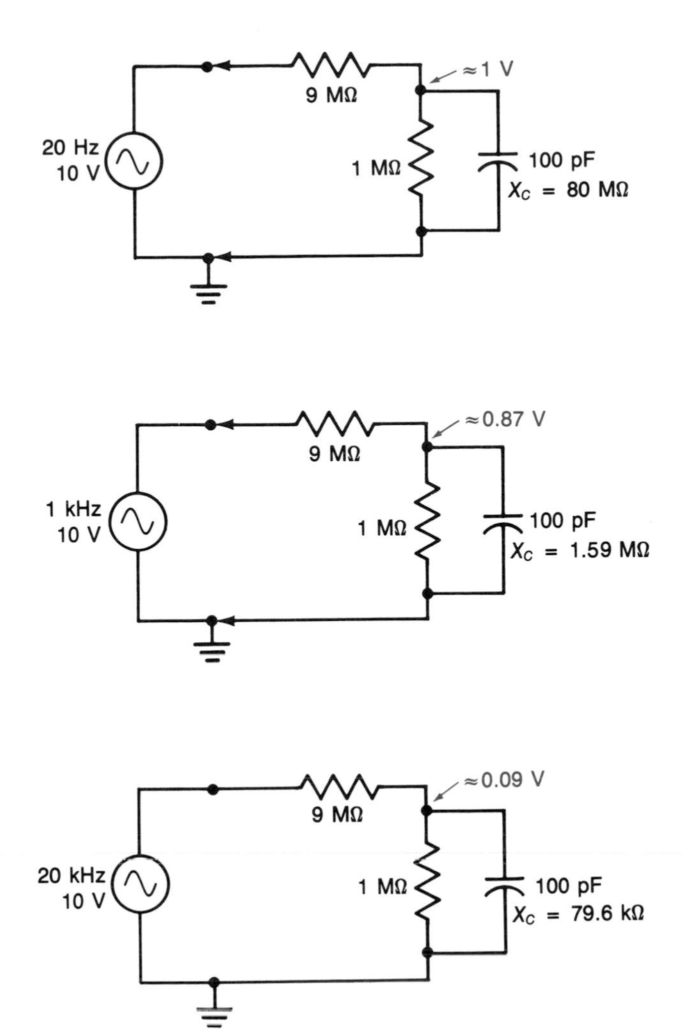


**FIGURE 21-33 Frequency dependency of an uncompensated attenuator.**

course, the frequency dependency would not be as dramatic as that illustrated in Fig. 21-33.

Since a square wave represents a wide range of frequencies, it is used as a test signal to check, and adjust if necessary, the compensation of a probe. Figure 21-34 shows the appearance of the scope display of a square wave when the probe is properly and improperly compensated.

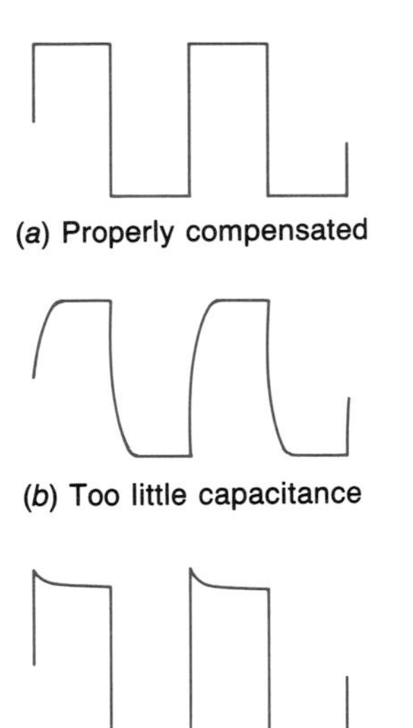

(*a*) Properly compensated

(*b*) Too little capacitance

(*c*) Too much capacitance

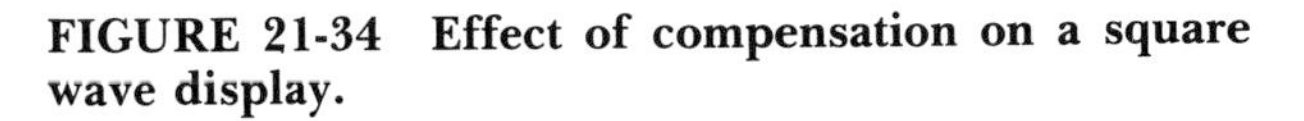
**FIGURE 21-34 Effect of compensation on a square wave display.**

## Self-Test

**45.** When should an X10 probe be used instead of an X1 probe?

**46.** What is a disadvantage of the X10 probe? When might this disadvantage be a limiting factor?

**47.** Why does the X10 probe have a capacitor in the probe? Why is this capacitor variable?

**48.** What type of waveform is used to check the frequency compensation of an attenuator probe?

## SUMMARY

1. Meter specifications include accuracy, resolution, frequency response, and internal impedance.
2. The lower the range, the better the resolution.
3. Voltmeter sensitivity is specified in ohms per volt.

**4.** The lower the input resistance the greater the loading effect of a voltmeter.

**5.** The higher the internal resistance the greater the loading effect of an ammeter.

**6.** Ac meters are frequency-dependent for both accuracy and loading effect.

**7.** The loading effect of the VOM varies with the range setting. In spite of its lower resolution, a higher range may give a more accurate voltage indication.

**8.** The functional blocks of a basic digital voltmeter are the ramp voltage generator, the comparator, the pulse generator, the timer, the counter, the counter storage, and the readout. The voltmeter section of DMM is used to indirectly measure current and resistance.

**9.** On lower ranges, the resolution of a DMM often exceeds its accuracy.

**10.** The PMMC meter movement is also known as the d'Arsonval movement. Its rotational torque is the result of the interaction of two magnetic fields.

**11.** A meter movement can be described by its $I_{fs}$, $V_{fs}$, and $R_i$.

**12.** A rectifier converts alternating current to pulsating direct current.

**13.** Shunts are used to extend the range of an ammeter.

**14.** A shorting (make-before-break) switch is used for range-switching an analog ammeter.

**15.** Multipliers are used to extend the range of a voltmeter.

**16.** A series ohmmeter circuit has a rheostat, a resistor, a battery, and a meter movement connected in series. The rheostat is the ohms adjust. The ohmmeter scale is reverse reading and nonlinear.

**17.** Major parts of a CRT are the heater, cathode, control grid, focus anode, accelerating anode, vertical deflection plates, horizontal deflection plates, phosphor screen, and Aquadag coating.

**18.** A CRT uses both thermionic and secondary emission.

**19.** A time-base signal is applied to the horizontal plates while the signal to be displayed is applied to the vertical plates.

**20.** The functional blocks of an oscilloscope are the trigger circuitry, sweep generator, horizontal amplifier, vertical attenuator, vertical amplifier, and CRT display.

**21.** Input to the $z$ axis is used to intensity-modulate the electron beam.

**22.** Ac coupling separates the ac component of a signal from the dc component of the signal.

**23.** A dual-trace oscilloscope provides for simultaneously displaying two signals. The signals can be displayed in either the chop mode or alternate mode.

**24.** Oscilloscopes are commonly used to measure voltage, period, frequency, and phase.

**25.** To measure voltage, the vertical gain control must be in its calibrate position.

**26.** To measure time (period) the horizontal gain and the variable sweep controls must be in their calibrate positions.

**27.** Lissajous patterns can be used to measure either frequency or phase.

**28.** An X10 probe reduces circuit loading by increasing the effective input resistance and decreasing the effective input capacitance of the oscilloscope.

**29.** An X10 probe is frequency-compensated.

**30.** Some formulas for analog meters are

$$\text{Sensitivity} = \frac{1}{I_{\text{fs}}}$$

$$\text{Input resistance} = \text{sensitivity} \times \text{range}$$

$$R_s = \frac{V_{\text{fs}}}{I_r - I_{\text{fs}}}$$

$$R_m = \frac{V_r - V_{\text{fs}}}{I_{\text{fs}}}$$

## CHAPTER REVIEW QUESTIONS

For the following items, determine whether each statement is true or false.

**21-1.** The input impedance of a VOM on ac voltage ranges is usually greater than the input resistance on comparable dc voltage ranges.

**21-2.** The internal resistance of a VOM is the same on all current ranges.

**21-3.** The internal resistance of a VOM is different on each voltage range.

**21-4.** High resolution ensures high accuracy.

**21-5.** The input impedance of a DMM changes with each voltage range.

**21-6.** If the top vertical plate in a CRT is negative with respect to the bottom vertical plate, the electron beam is deflected downward.

**21-7.** Use dc coupling on the oscilloscope to view both the ac and dc components of a signal.

**21-8.** An X10 probe multiplies the input to the probe by a factor of two.

**21-9.** Adjusting a compensated probe is done by adjusting the variable resistor in the probe.

**21-10.** An X10 probe should be used when measuring signals in a low-frequency, low-resistance circuit.

For the following items, fill in the blank with the word or phrase required to correctly complete each statement.

**21-11.** The d'Arsonval movement is also known as the ______ movement.

**21-12.** Alternating current can be changed to pulsating direct current by a ______.

**21-13.** A make-before-break switch is also called a ______ switch.

**21-14.** The cathode in a CRT produces free electrons by a process called ______.

**21-15.** The part of the CRT on which a waveform is displayed is the ______.

**21-16.** The ______ switch connects the horizontal amplifier to the *x* axis (horizontal) input jack.

**21-17.** The *z* axis input is used to intensity ______ the electron beam.

**21-18.** ______ triggering is used to measure phase shift with a single-trace oscilloscope.

**21-19.** ______ patterns can be used to determine the frequency ratio between the *x*-axis and *y*-axis signals.

**21-20.** An X10 probe is ______-compensated.

For the following items, choose the letter that best completes each statement.

**21-21.** When an ammeter is inserted in a circuit, the circuit current will
- **a.** Increase
- **b.** Decrease
- **c.** Remain the same

**21-22.** The comparator in a DVM compares the input voltage with the voltage from the
- **a.** Ramp generator
- **b.** Counter storage
- **c.** Pulse generator
- **d.** Timer

**21-23.** The readout on a DVM changes when counter storage receives a signal from the
- **a.** Timer
- **b.** Comparator
- **c.** Counter
- **d.** Pulse generator

**21-24.** Which of these controls does not change the shape of the electron beam in the CRT?
- **a.** Focus
- **b.** Astigmatism
- **c.** Gain
- **d.** Intensity

**21-25.** In the CRT, secondarily emitted electrons are collected by the
- **a.** Vertical plates
- **b.** Horizontal plates
- **c.** Astigmatism element
- **d.** Aquadag coating

**21-26.** The blanking signal in an oscilloscope comes from the
- **a.** Input signal
- **b.** Trigger circuit
- **c.** Vertical amplifier
- **d.** Sweep generator

For the following items, provide the information requested.

**21-27.** List two ways to reduce the center-scale resistance of an ohmmeter.

**21-28.** For the display in Fig. 21-35*a*, which position is the slope switch on?

**21-29.** List the three signal sources that can be used to trigger the sweep generator.

## CHAPTER REVIEW PROBLEMS

**21-1.** Determine the minimum and maximum values of a voltage which causes a VOM to indicate 2.07 V on the 2.5-V range if the meter accuracy is ±3 percent.

**21-2.** A voltmeter uses a 20-$\mu$A meter movement. What is its input resistance on the 50-V dc range?

**21-3.** What is the internal resistance of an ammeter on the 10-mA range if the meter movement is rated $I_{fs} = 200\ \mu A$ and $V_{fs} = 200$ mV? What is the shunt resistance?

**21-4.** A series ohmmeter uses a 100-$\mu$A, 1-k$\Omega$ meter movement, a 3-V battery, and the necessary resistors. What is the center-scale resistance?

**21-5.** Determine the multiplier resistance needed to convert a 40-$\mu$A, 2.5-k$\Omega$ meter movement into a 10-V voltmeter.

**21-6.** The display in Fig. 21-35*a* was obtained with a properly calibrated oscilloscope with its switches set to 10 V/DIV and 0.1 ms/DIV. Determine the rms voltage and the frequency.

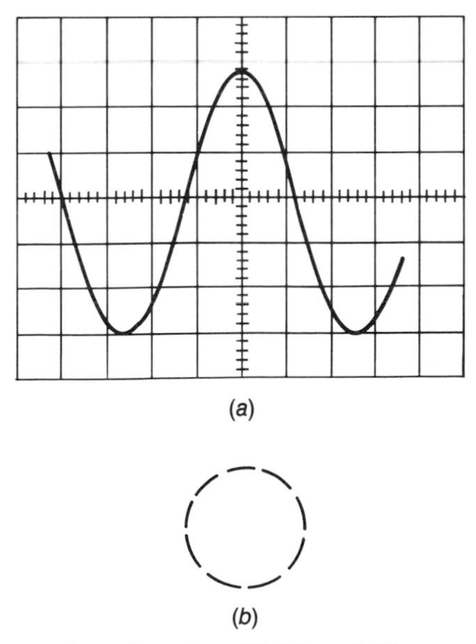


**FIGURE 21-35 Displays for Chapter Review Question 21-28 and Chapter Review Problems 21-6 and 21-7.**

**21-7.** What frequency is represented by the display in Fig. 21-35*b* if the *x*-*y* signals are 60 Hz?

**21-8.** Two 910-kΩ resistors are series-connected to a 50-V dc power source. How much voltage would be measured across either resistor with a DMM with 10 MΩ of input resistance?

## ANSWER TO SELF-TESTS

**1.** 2.3 mA

**2.** 20 V, because the maximum possible error will be less ($X$ percent of 20 is less than $X$ percent of 50).

**3.** because the accuracy of an ac voltmeter decreases with an increase in frequency

**4.** No, because high resolution may provide more significant figures than the accuracy will support.

**5.** because the internal capacitance in the meter is in parallel with the internal resistance and $X_C$ decreases as $f$ increases.

**6.** 400 kΩ

**7.** The VOM, because its input resistance would be 20 MΩ on the 1000-V range, whereas the DMM has 10 MΩ on all ranges.

**8.** 7.57 V, 8.13 percent [(8.24 V − 7.57 V)/8.24 V] × 100

**9.** 10.3 percent [(0.68 mA − 0.61 mA)/0.68 mA] × 100

**10.** 87.6 $\mu$A

**11.** 800 ($2000/1 = X/0.4$)

**12.** to the left of the last digit on the left, 0.1 V

**13.** when the comparator output drops from +3 V to 0 V

**14.** 900 kΩ (140 V/10 MΩ = 1.4 V/$R_{3,4}$ and $R_2$ = 10 MΩ − 9 MΩ − $R_{3,4}$)

**15.** the interaction of a permanent-magnetic field and an electromagnetic field

**16.** because reversal of the current in the coil causes the magnetic field around the coil to reverse polarity and create a counterclockwise torque.

**17.** a device which allows current to flow in one direction only

**18.** 200 mV

**19.** 30.3 Ω, 0.30 Ω

**20.** 210.53 Ω

**21.** make-before-break

**22.** a resistor connected in series with a meter movement to extend the range of a voltmeter

**23.** sensitivity = 50 kΩ/V, $R_m$ = 1240 kΩ, $R_{i,V}$ = 1250 kΩ

**24.** because $I_{fs}$ represents 0 Ω

**25.** 3 kΩ

**26.** phosphor

**27.** thermionic emission

**28.** secondary emission

**29.** control grid

**30.** intensity (or brightness), blanking, *z* axis (or intensity modulation)

**31.** intensity, focus, astigmatism

**32.** It helps accelerate the electrons in the electron beam, and it collects the electrons that are emitted from the phosphor screen.

**33.** the trigger circuitry

**34.** +slope, +limit

**35.** internal, external, line

**36.** vertical

**37.** *z*

**38.** calibrated sweep rate

**39.** by alternating switching (electronically) the beam from one waveform to the other waveform.

**40.** $V_{p\text{-}p}$ = 37 V, $f$ = 3333 Hz

**41.** $V_{rms}$ = 11.3 V, $f$ = 2778 Hz

**42.** external

**43.** a display created by applying sinusoidal signals to both plates of an oscilloscope.

**44.** 4 kHz

**45.** anytime the X1 probe would cause significant circuit loading

**46.** It divides the measured voltage by a factor of 10; when the measured voltage is very small.

**47.** so that its voltage division ratio can be frequency-independent; so that the probe can be used with oscilloscopes with differing input capacitance

**48.** square wave

# Chapter 22 Three-Phase Circuits

Polyphase electricity is usually used whenever large amounts of ac power are to be produced and transported. Although electricity can be produced with many different numbers of phases, that which is produced for commercial distribution is three-phase. Once the three-phase electricity is transported from the main power plant to the locale where it is to be used, part of it is distributed to industrial-commercial users in the three-phase form. The remainder of it is split into three single-phase systems for distribution to residential areas.

This chapter introduces the basic concepts of three-phase systems. It deals with the generation and measurement of three-phase electricity. Also, typical circuit connections for three-phase systems are presented.

## 22-1 REFERENCE POINT AND SUBSCRIPTS

Chapter 6 discussed the use of *reference points* and *double subscripts* in specifying voltages. Correct use of double subscripts makes it easier to understand and analyze three-phase circuits which are essentially circuits with three sources. These three sources may or may not use a common reference point. Before we start to work with three-phase circuits, let us see how to use double subscripts and

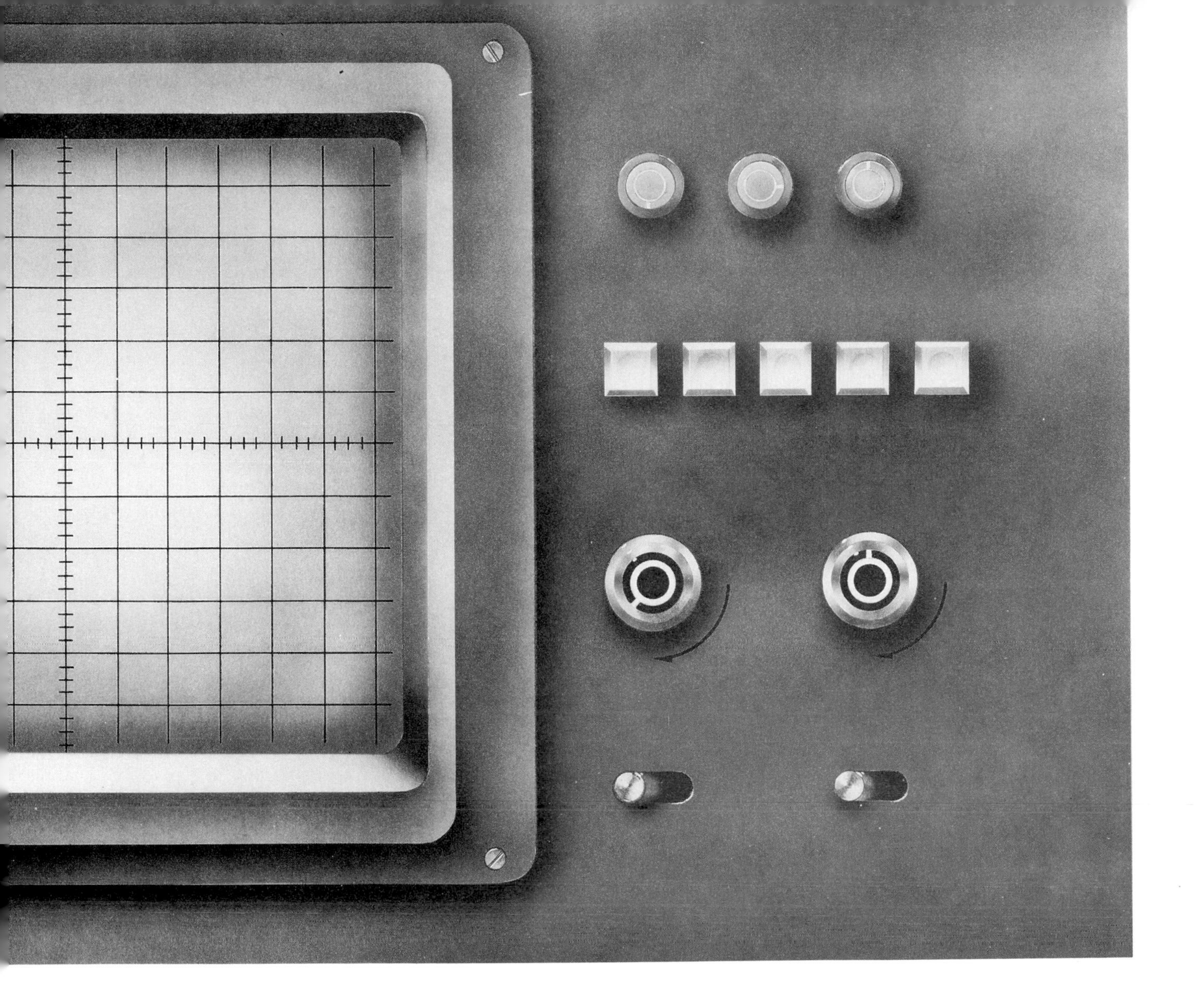

phasors to represent single-phase sources in which the reference point can be either of two points.

The secondary voltage of the transformer in Fig. 22-1*a* is available between points A and a. If point a is used as the reference, then the secondary voltage can be abbreviated as $V_{Aa}$. (*Note:* Some textbooks specify the reference point with the first subscript, but we will stay with the standards stated in Chap. 6.) If $V_{Aa}$ is located on the reference axis of a phasor diagram as in Fig. 22-1*b*, we can then write $V_{Aa} = 10 \text{ V} \angle 0°$. Now if we consider point A as the reference, the secondary voltage is $V_{aA}$. Obviously, the two voltages $V_{Aa}$ and $V_{aA}$ have the same magnitude but *not the same phase angle*. Point a is maximum positive with respect to point A 180° after

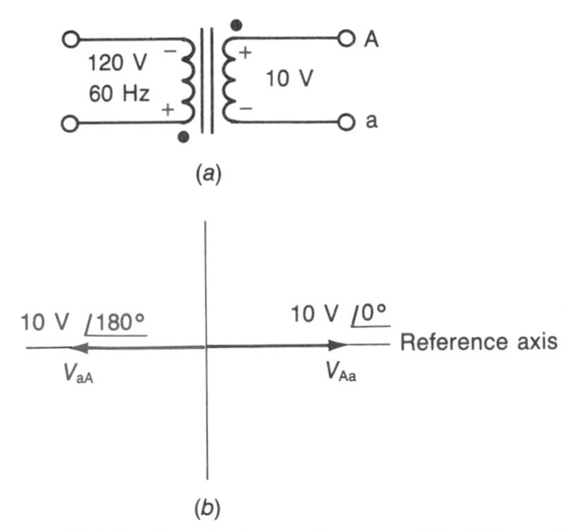


**FIGURE 22-1 Specifying voltages with different reference points.**

point A is maximum positive with respect to point a. Thus, if $V_{aA}$ is put on the reference axis of the phasor diagram, then $V_{Aa}$ will be on the $-x$ axis. In mathematical terms, $V_{aA} = -V_{Aa}$. That is, for the instantaneous polarity shown in Fig. 22-1*a*, $V_{Aa} = +10$ V, $V_{aA} = -10$ V, and $+10 \text{ V} = -(-10 \text{ V})$.

Figure 22-2*a* shows the two secondaries of a transformer connected series-aiding. If point C is the reference, then the two secondary voltages would be specified as $V_{Aa} = 6$ V and $V_{Bb} = 10$ V. These two voltages are shown as phasors in Fig. 22-2*b*. Putting $V_{Bb}$ on the reference axis was an arbitrary decision. However, once that decision was made, $V_{Aa}$ had to be plotted on the $-x$ axis because point A is not positive (with respect to point C) until 180° after point B is positive. Thus, $V_{Aa} = 6 \text{ V} \angle 180°$ and $V_{Bb} = 10 \text{ V} \angle 0°$. Now let us change the reference point in Fig. 22-2*a* to point A and determine the voltage between points B and A. This can be done by adding the two secondary voltages. Thus, $V_{BA} = V_{Bb} + V_{aA}$. Notice the reversal of the subscripts in the last term of this equation. This reversal occurred because the reference point was changed from point C to point A. Since we previously established that $V_{aA} = -V_{Aa}$, we can substitute $-V_{Aa}$ for $V_{aA}$ in the above formula. Thus, $V_{BA}$ for Fig. 22-2*a* is

$$V_{BA} = V_{Bb} - V_{Aa} = 10 \text{ V} \angle 0° - 6 \text{ V} \angle 180° = 16 \text{ V} \angle 0°$$

The $V_{BA}$ phasor is shown in Fig. 22-2*c* along with $V_{Aa}$ and $V_{Bb}$. (In this chapter the intermediate steps involved in phasor algebra are not shown. If you have forgotten how to do phasor algebra, review Chap. 19.)

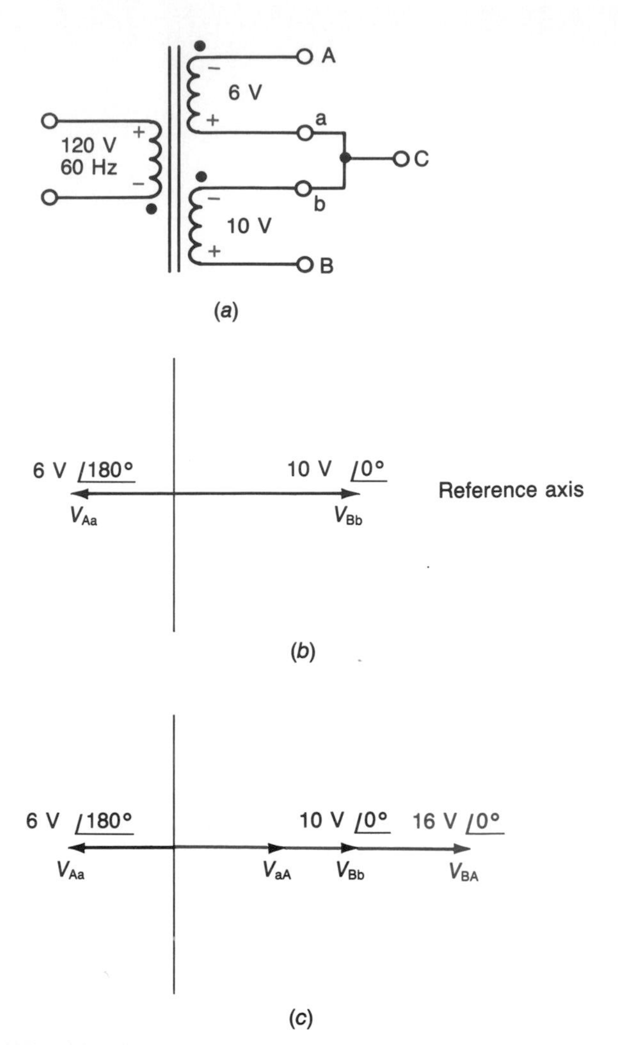


**FIGURE 22-2 Combining phasors that have different reference points.**

### EXAMPLE 22-1

Using the phasors of Fig. 22-2*b* and the schematic diagram of Fig. 22-2*a*, determine the value of $V_{AB}$.

**Solution**

$$V_{AB} = V_{Aa} + V_{bB} = V_{Aa} - V_{Bb}$$
$$= 6 \text{ V} \angle 180° - 10 \text{ V} \angle 0° = 16 \text{ V} \angle 180°$$

Notice that the term whose subscripts are reversed contains the same letter subscripts as the reference subscript of the new voltage. This is always the case when the reference is changed on series-connected voltages.

## 22-2 THREE-PHASE GENERATION

Generating a three-phase (abbreviated 3-$\phi$) alternating current is illustrated in Fig. 22-3*a*, which is a cross-sectional view of a simple 3-$\phi$ generator. Each of the phase coils Aa, Bb, and Cc is electrically insulated from the other phase coils and has a voltage induced into it by the rotating magnetic field. At the instant shown in Fig. 22-3*a*, $V_{Aa}$ is at its peak positive value. This voltage $V_{Aa}$ is called the phase A voltage in the waveform graph in Fig. 22-3*b*. As the magnetic field in Fig. 22-3*a* is rotated 120°, the phase B voltage $V_{Bb}$ changes from $-85$ V to 0 to $+170$ V and the phase C voltage $V_{Cc}$ changes from $-85$ to $-170$ V and back to $-85$ V. The next 120° of rotation, which produces another

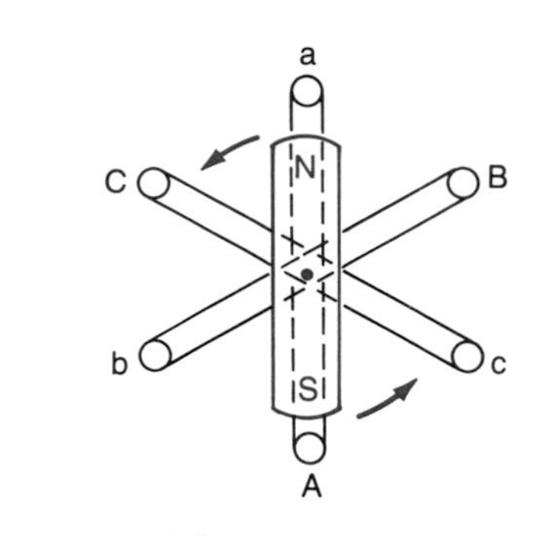


(*a*) Generator principle

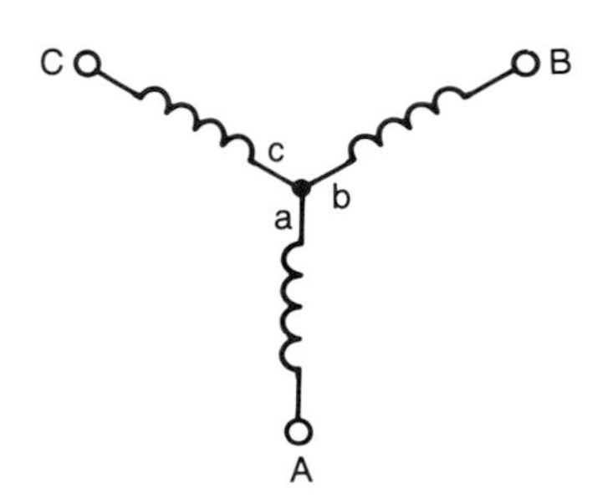


(*d*) Schematic symbol (wye-connected)

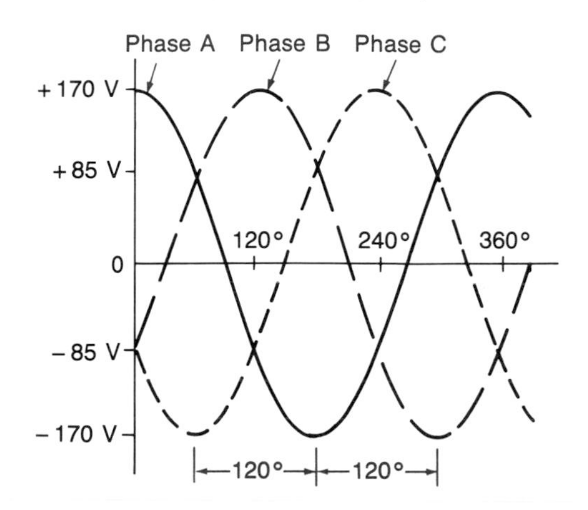


(*b*) Waveform graph

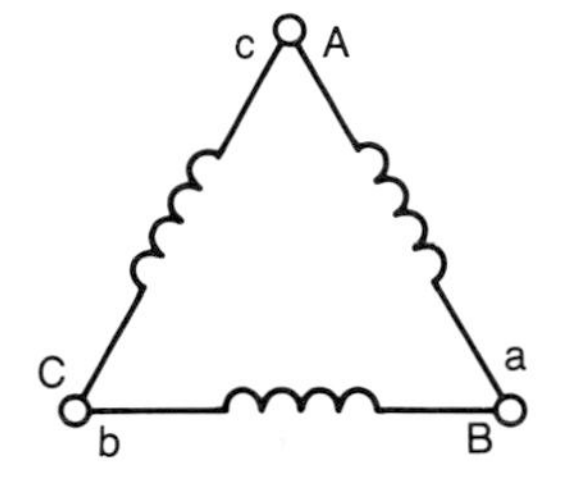


(*e*) Schematic symbol (delta-connected)

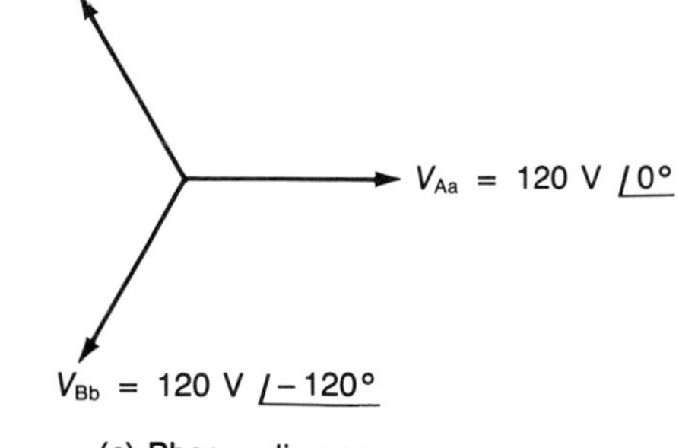


(*c*) Phasor diagram

**FIGURE 22-3 Three-phase alternating current.**

120 electric degrees for each phase, causes phase C to increase to its peak positive value. Constant-speed rotation of the magnetic field in Fig. 22-3*a* produces, as seen in Fig. 22-3*b*, three sinusoidal waveforms which are 120° out of phase with each other.

Careful inspection of Fig. 22-3*b* shows two important points about three-phase voltage (or current). First, the algebraic sum of the instantaneous values of the three waveforms is always zero. Second, at no time are the three phases all negative or all positive. This second condition must be true if the first statement is true.

The phasor diagram for the three voltages produced by the three-phase generator of Fig. 22-3*a* is shown in Fig. 22-3*c*. In this diagram $V_{Cc}$ is drawn at 120° because $V_{Cc}$ leads $V_{Aa}$ by 120° (or lags by −240°) and $V_{Aa}$ was selected as the reference phasor.

The three coil windings in a three-phase source are always connected together so that the three-phase voltage can be carried on three conductors or wires. As shown in Fig. 22-3*d* and *e* the coils can be connected in either a *wye* or a *delta* configuration. The schematic symbol for a three-phase source consists of the three phase coils connected in one of these two configurations.

## Self-Test

1. Voltage $V_{CD} = 48\text{ V}\ \underline{/30°}$. What does $V_{DC}$ equal?
2. Voltage $V_{Ee} = 15\text{ V}\ \underline{/0°}$ and $V_{Ff} = 12\text{ V}\ \underline{/180°}$. These voltages are connected in series with e connected to f. Determine $V_{FE}$.
3. How many electric degrees separate the phases in a three-phase system?
4. The algebraic sum of the instantaneous phase voltages in a three-phase circuit is ______.
5. If the instantaneous values of phases A and B are +60 V and −40 V, respectively, then phase C must be ______ V.

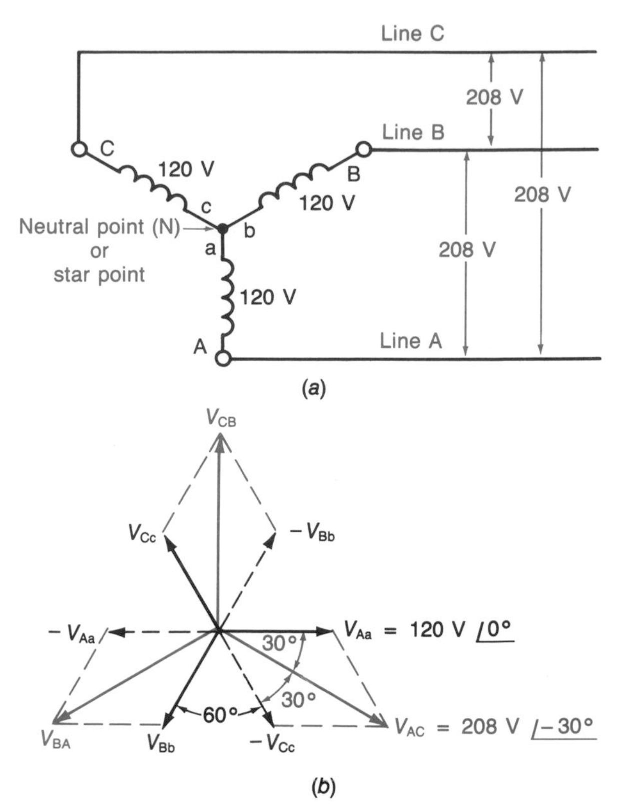


**FIGURE 22-4 Wye connection.**

## 22-3 THE WYE CONNECTION

The *wye* (Y) *connection* of a three-phase winding is shown in Fig. 22-4*a*. The junction of the three phase coils is often referred to as the *neutral point* N because this point is often connected to earth ground. Thus, it is neutral with respect to earth ground. The neutral point can also be referred to as the *star point*. For example, the junction point of the three windings in a wye-connected, three-phase motor is often called the star point instead of the neutral point. A neutral wire (connected to the neutral or star point) is included in many wye-connected power distribution systems.

The line voltages of a wye-connected three-phase system are larger than the phase voltages of the system. As shown in Fig. 22-4*b*, the line voltages are the phasor sum of two series-connected phase voltages. However, when adding two phase voltages to obtain a line voltage, one must remember that the phase voltages are referenced to the neutral point, but a line voltage is referenced to another line. Thus, to graphically determine $V_{AC}$, as in Fig. 22-4*b*, the phasor $V_{Cc}$ must be "flipped over" before being added to $V_{Aa}$. Using the ideas developed in Sec. 22-1, the three line voltages can be calculated as follows:

$$V_{AC} = V_{Aa} - V_{Cc} = 120\text{ V}\ \underline{/0°} - 120\text{ V}\ \underline{/120°} = 208\text{ V}\ \underline{/-30°}$$
$$V_{BA} = V_{Ba} - V_{Aa} = 120\text{ V}\ \underline{/-120°} - 120\text{ V}\ \underline{/0°} = 208\text{ V}\ \underline{/-150°}$$
$$V_{CB} = V_{Cc} - V_{Bb} = 120\text{ V}\ \underline{/120°} - 120\text{ V}\ \underline{/-120°} = 208\text{ V}\ \underline{/90°}$$

Notice from the above calculations that the line voltages are lagging the phase voltages by 30°. However, if we reverse the reference lines and calculate $V_{CA}$, $V_{AB}$, and $V_{BC}$, the line voltages lead the phase voltages by 30°. For instance,

$$V_{CA} = V_{Cc} - V_{Aa} = 120\text{ V}\ \underline{/120°} - 120\text{ V}\ \underline{/0°} = 208\text{ V}\ \underline{/150°}$$

Thus, $V_{CA}$ leads $V_{Cc}$ by 30°.

The results of the above calculations can be used to determine the relationship between the line voltage $V_l$ and the phase voltage $V_p$ in a wye-connected three-phase system. The relationship is $V_l/V_p$ = 208 V/120 V = 1.73. Regardless of the phase voltage, the line voltage is always 1.73 times larger than the phase voltage. The 1.73 conversion factor can be derived by analyzing the phasor diagram in Fig. 22-5. This figure shows the phasors used to determine a line voltage $V_l$ from two phase voltages. Since a parallelogram was constructed to find $V_l$, an isosceles triangle was formed between points N, $V_p$, and $V_l$. Projecting a line perpendicular from the base (phasor $V_l$) to the end of the $V_p$ phasor divides this isosceles triangle into two identical right triangles. This projected line also divides $V_l$ exactly in half. Now we can use trigonometric functions to find the relationship between $V_l$ and $V_p$:

$$\text{Adjacent side} = \cos \theta \times \text{hypotenuse}$$

So,

$$0.5V_l = \cos 30° \times V_p = 0.866 \times V_p$$

And (multiplying by 2),

$$V_l = 1.73V_p$$

Thus the line voltage of a wye-connected system is always 1.73 times larger than the phase voltage.

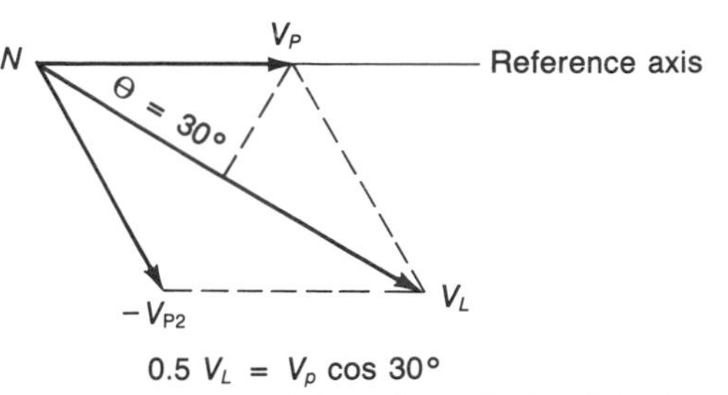


**FIGURE 22-5 Determining the relationship between phase and line voltage.**

**EXAMPLE 22-2**

A wye-connected three-phase system has a line voltage of 480 V. What is the phase voltage?

**Solution**

$$V_l = 1.73\ V_p$$

Therefore,

$$V_p = \frac{V_l}{1.73} = \frac{480\text{ V}}{1.73} = 277\text{ V}$$

Inspection of Fig. 22-4 shows that the line current in a wye system must be equal to the phase current. If a load which draws 10 A is connected between lines C and B, then 10 A flows in line C, line B, phase Cc, and phase Bb.

A four-wire wye system is illustrated in Fig. 22-6. The four-wire system has three line wires and a neutral wire. A 120/208-V four-wire wye system can provide the following voltages: (1) 120-V, single-phase between neutral and any one of the lines, (2) 208-V, single-phase between any two of the lines, and (3) 208-V, three-phase between the three lines.

Notice that the single-phase loads, lamps and motors in Fig. 22-6, are distributed so that the total current for each phase of the source is as close to equal as practical. As in any other type of circuit, the currents in a three-phase system must satisfy Kirchhoff's current law. In three-phase circuits the reference ends of the phase coils are considered negative with respect to the other ends of the coils for the purpose of applying Kirchhoff's laws. Therefore, for a wye circuit like that in Fig. 22-6, the $I_N$ neutral wire current is the current entering the junction which joins the neutral wire and the three loads $Z_A$, $Z_B$, and $Z_C$. The currents leaving this junction are $I_{Z_A}$, $I_{Z_B}$, and $I_{Z_C}$. Therefore, by Kirchhoff's current law, we know that $I_N = I_{Z_A} + I_{Z_B} + I_{Z_C}$.

When the single-phase, 120-V loads are equal, as in Fig. 22-6, the neutral wire carries no current. This is because the phasor sum of the currents for any two of the 120-V loads (if all loads are equal) is equal to the current for the third load. Or, saying it another way, the phasor sum of the three currents is zero. For example, the magnitude of the current for each lamp in Fig. 22-6 is $I = P/V$ = 500 W/120 V = 4.17 A. Since the voltage for each lamp is at a different angle, the current through each lamp will also be at a different angle. Thus, $I_{Z_A} = 4.17\text{ A} \angle 0°$, $I_{Z_B} = 4.17\text{ A} \angle 120°$, and $I_{Z_C} = 4.17\text{ A} \angle -120°$. Adding these three currents to determine $I_N$ yields

$$\begin{aligned} I_N &= 4.17\text{ A} \angle 0° + 4.17\text{ A} \angle 120° + 4.17\text{ A} \angle -120° \\ &= (4.17\text{ A} + j0\text{ A}) + (-2.085\text{ A} + j3.611\text{ A}) \\ &\quad + (-2.085\text{ A} - j3.611\text{ A}) = 0 \end{aligned}$$

Thus, for equal 120-V loads (balanced systems),

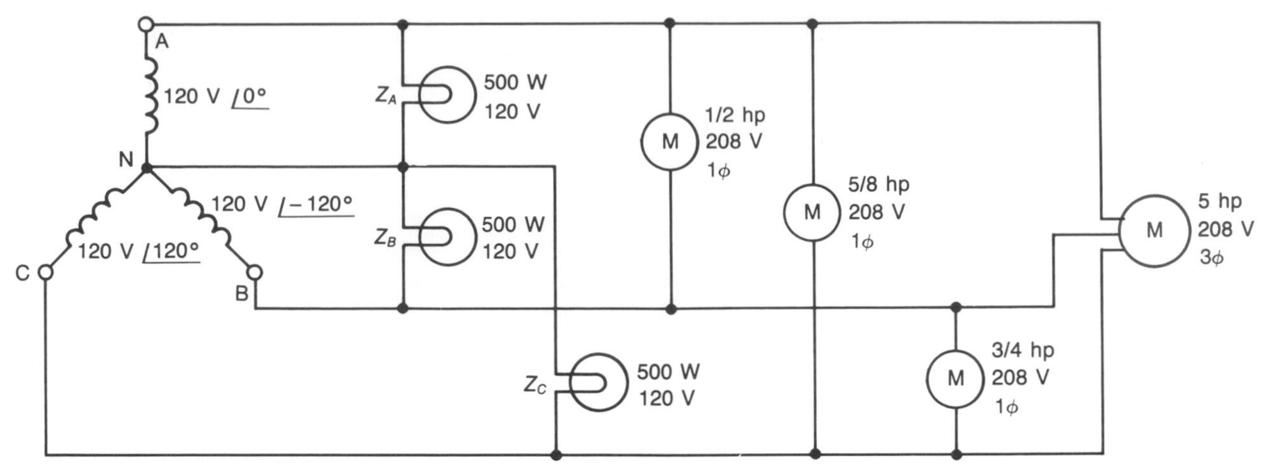


**FIGURE 22-6 Four-wire wye system.**

Kirchhoff's current law shows that there is no current flow in the neutral wire.

When the 120-V loads in Fig. 22-6 are unequal, the neutral wire carries some current. The exact amount of current for unequal loads can be determined by applying Kirchhoff's current law.

### EXAMPLE 22-3

Determine the neutral-wire current in Fig. 22-6 when $Z_A = 700$ W, $Z_B = 900$ W, and $Z_C = 400$ W.

**Solution**

$$I_{Z_A} = \frac{700\text{ W}}{120\text{ V}} \underline{/0^\circ} = 5.83\text{ A} \underline{/0^\circ}$$

$$I_{Z_B} = \frac{900\text{ W}}{120\text{ V}} \underline{/-120^\circ} = 7.50\text{ A} \underline{/120^\circ}$$

$$I_{Z_C} = \frac{400\text{ W}}{120\text{ V}} \underline{/120^\circ} = 3.33\text{ A} \underline{/-120^\circ}$$

$$I_N = 5.83\text{ A} \underline{/0^\circ} + 7.50\text{ A} \underline{/120^\circ} + 3.33\text{ A} \underline{/-120^\circ}$$
$$= 3.64\text{ A} \underline{/83.4^\circ}$$

## Self-Test

**6.** The junction of the three-phase coils in a wye-connected source is called the ________ or ________ point.

**7.** Which part of a wye-connected source is often connected to earth ground?

**8.** Voltage $V_{Bb} = 100\text{ V} \underline{/-120^\circ}$ and $V_{Cc} = 100\text{ V} \underline{/120^\circ}$. Determine $V_{BC}$.

**9.** Does $V_{BC}$ lead or lag $V_{Bb}$ in Question 8? By how many degrees?

**10.** List the voltages available from a four-wire wye system that has a phase voltage of 277 V.

**11.** Can the current in the neutral wire of a four-wire wye system be zero when the loads are not balanced? Explain.

**12.** The currents for the three single-phase loads connected to the neutral wire in a wye system are $12\text{ A} \underline{/20^\circ}$, $3\text{ A} \underline{/160^\circ}$, and $8\text{ A} \underline{/-70^\circ}$. Calculate the neutral-wire current.

## 22-4 THE DELTA CONNECTION

Figure 22-7 shows the *delta* (Δ) *connection* for both a three-phase source and a three-phase load. Notice that the phase coils of the source are connected in a closed loop. This configuration of the source is possible because, as shown in Fig. 22-3*b* and *c*, the instantaneous and phasor sums of the three-phase voltages are zero. Thus, no current flows in the phase coils unless a load is connected to the source as in Fig. 22-7*a*.

When the three-phase coils of the source are connected in a delta, they must be "properly phased." That is, the reference end of one coil (end with the lowercase letter) must be connected to the nonreference end of another coil; otherwise, the phasor sum of the three coil voltages will be twice as great as the voltage of any one coil. Of course, if the delta were connected under this con-

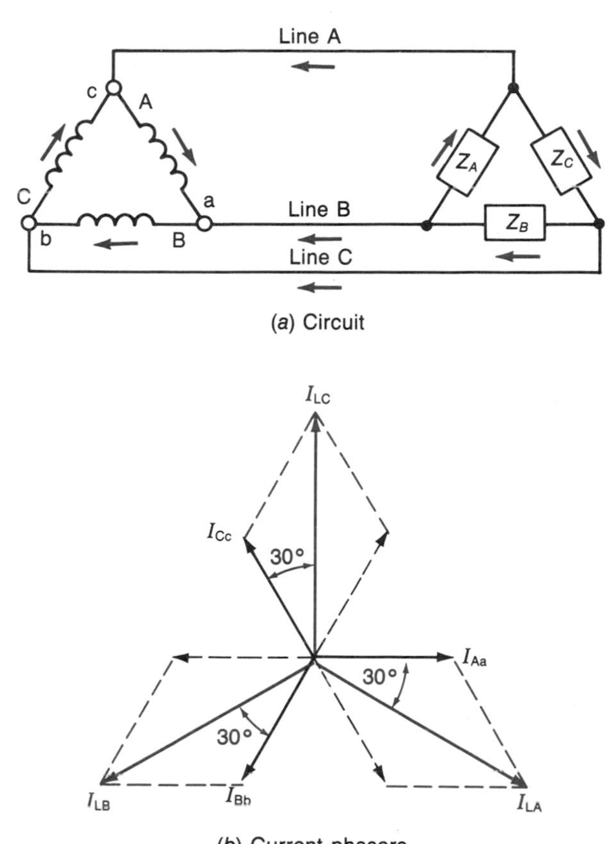


**FIGURE 22-7 Balanced delta (Δ) system.**

dition, an extremely high short-circuit current would flow in the phase coils and, unless the system were protected by appropriate fuses, the coils would overheat and be ruined. Proper phasing of the source coils can be checked, as shown in Fig. 22-8, before the last connection is made to form the delta ring. If the coils are properly phased, the voltmeter reads 0 V when the phase coils (and thus the phase voltages) are exactly equal. If the phase coils are not perfectly balanced, the voltmeter may indicate a very small voltage. However, if one of the coils is improperly phased (lowercase letter connected to lowercase letter), the voltmeter indicates two times the phase voltage.

Inspection of Fig. 22-7 shows that the line voltage is equal to the phase voltage in the delta system. For example, $V_{AB} = V_{Aa}$. Of course, the load voltage is also equal to the phase voltage and the line voltage.

Further inspection of Fig. 22-7 shows that the line current in the delta system is greater than the phase current because one line carries current from two phase coils. For example, current in line B ($I_{lB}$) is composed of phase-coil-A current ($I_{Aa}$) and phase-coil-B current ($I_{Bb}$). When the delta system is balanced, i.e., $Z_A = Z_B = Z_C$, the line current is 1.73 times larger than the phase current.

A phasor diagram for the line and phase currents in a balanced delta system is provided in Fig. 22-7*b*. Notice that determining a line current like $I_{lB}$ involves adding the phasors $I_{Bb}$ and $-I_{Aa}$. This reversal of phasor $I_{Aa}$ is necessary because, as shown by the arrows in Fig. 22-7*a*, the current $I_{Aa}$ is entering junction Ba, while $I_{Bb}$ is leaving junction Ba. The same geometric-trigonometric analysis used to show that $V_l = 1.73V_p$ in the wye system can be used to show that $I_l = 1.73I_p$ in the delta system.

Applying Kirchhoff's current law to the junction of a line and two phase coils or to the junction of a line and two loads yields the following formulas for determining line currents in Fig. 22-7:

$$I_{la} = I_{Aa} - I_{Cc} = I_{Z_A} - I_{Z_C}$$
$$I_{lB} = I_{Bb} - I_{Aa} = I_{Z_B} - I_{Z_A}$$
$$I_{lC} = I_{Cc} - I_{Bb} = I_{Z_C} - I_{Z_B}$$

These formulas are appropriate for either balanced or unbalanced loads.

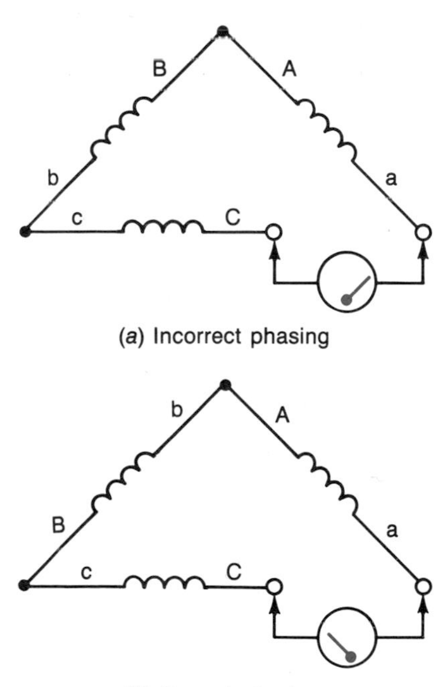


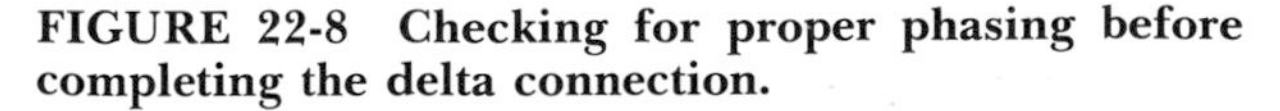
**FIGURE 22-8 Checking for proper phasing before completing the delta connection.**

**EXAMPLE 22-4**

Refer to Fig. 22-7. $V_{Aa} = 440\text{ V}\,\underline{/0^\circ}$, $V_{Bb} = 440\text{ V}\,\underline{/-120^\circ}$, $V_{Cc} = 440\text{ V}\,\underline{/120^\circ}$, and $Z_A = Z_B = Z_C = 44\ \Omega\,\underline{/0^\circ}$. Determine $I_{lA}$.

**Solution**

$$I_{Aa} = \frac{440\text{ V}\,\underline{/0^\circ}}{44\ \Omega\,\underline{/0^\circ}} = 10\text{ A}\,\underline{/0^\circ}$$

$$I_{Cc} = \frac{440\text{ V}\,\underline{/120^\circ}}{44\ \Omega\,\underline{/0^\circ}} = 10\text{ A}\,\underline{/120^\circ}$$

$$I_{lA} = 10\text{ A}\,\underline{/0^\circ} - 10\text{ A}\,\underline{/120^\circ} = 17.3\text{ A}\,\underline{/-30^\circ}$$

Notice two things from this example: (1) the line current is 1.73 times larger than the phase current, and (2) the $-30^\circ$ angle agrees with the phasor diagram shown in Fig. 22-7*b*.

## Self-Test

**13.** If each phase voltage in Fig. 22-8*a* is exactly 120 V, what magnitude of voltage will the meter indicate?

**14.** Which coil in Fig. 22-8*a* should be reversed to provide correct phasing?

**15.** What relationship exists between a line current and a phase current in a balanced delta system?

**16.** What relationship exists between a phase voltage and a line voltage in a balanced delta system?

**17.** The line current in a balanced delta system is 30 A. What is the magnitude of the load current?

**18.** What is the phase relationship between the line currents in a delta system?

## 22-5 WYE-DELTA AND DELTA-WYE SYSTEMS

The previous examples of three-phase systems have used the same configuration (wye or delta) for both the source and the load. A matched source-load configuration is not necessary. A wye source can feed a delta load or vice versa as shown in Fig. 22-9. When a wye load is connected to a delta source, the load must be balanced.

A source of either configuration can, and usually does, feed a combination of load configurations.

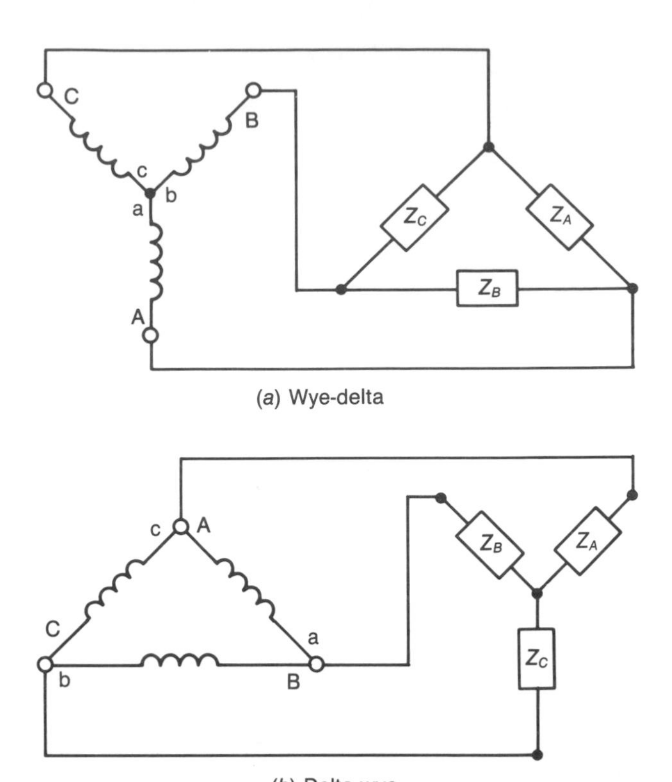


**FIGURE 22-9 Wye-delta and delta-wye systems.**

For example, the four-wire wye source in Fig. 22-10 is loaded by two motors. One motor has its three windings connected in a wye connection and the other motor's winding are delta-connected. Each winding in the wye-connected motor is designed for 120 V, which is the phase voltage of the 120/208-V wye source. The neutral wire need not be connected to this wye-connected motor because it is a balanced load; i.e., each winding is identical to the other two windings. The delta-connected motor must have individual phase windings capable of handling 208 V because, as seen in Fig. 22-10, each winding is connected between two lines. Thus, each winding receives the 208-V line voltage from the 120/208-V source.

As long as the mixed systems are balanced, the same relationships between phase and line voltages and currents exist in mixed-configuration as in single-configuration systems. For example, if the load is wye-connected and the source is delta-connected, the line voltage (phase voltage of the source) is the phasor sum of two phase voltages in the load. Similarly, with a delta load, the line current (phase current of a wye source) is the phasor sum of two of the phase currents in the load.

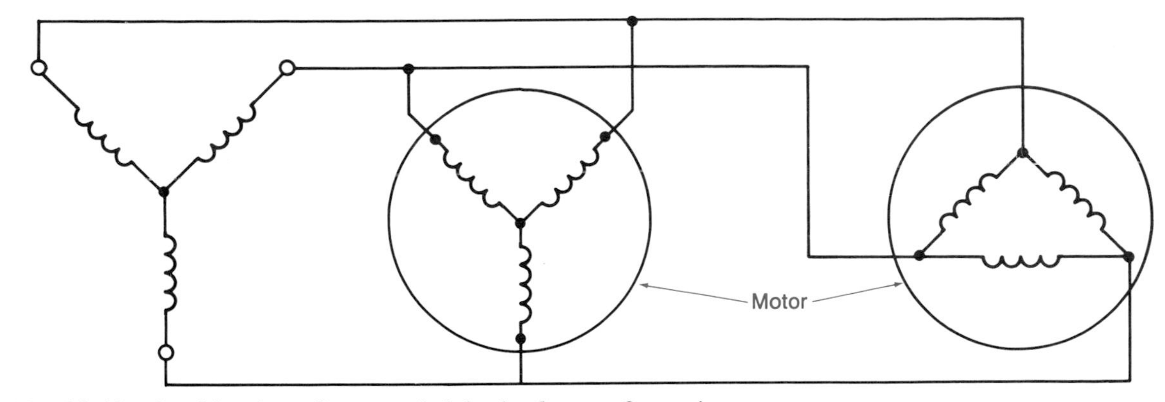


**FIGURE 22-10 Combination of wye and delta loads on a four-wire wye source.**

### EXAMPLE 22-5

Refer to Fig. 22-9*a*. The line voltages $V_l$ are 480 V and $Z_A = Z_B = Z_C = 24\ \Omega\angle 30°$. Determine the currents for line A ($I_A$) and load $Z_A$ ($I_{Z_A}$). Also determine the magnitude of the phase voltage $V_p$ for the source.

**Solution**

Arbitrarily select one line voltage, say $V_{AC}$, to be on the reference axis and have an angle of 0°. Then,

$$I_{Z_A} = \frac{V_{AC}}{Z_A} = \frac{480\ \text{V}\angle 0°}{24\ \Omega\angle 30°} = 20\ \text{A}\angle -30°$$

$$I_{Z_B} = \frac{V_{BA}}{Z_B} = \frac{480\ \text{V}\angle -120°}{24\ \Omega\angle 30°} = 20\ \text{A}\angle -150°$$

$$I_A = I_{Z_A} - I_{Z_B} = 20\ \text{A}\angle -30° - 20\ \text{A}\angle -150° = 34.67\ \text{A}\angle 0°$$

$$V_p = \frac{V_l}{1.73} = \frac{480\ \text{V}}{1.73} = 277\ \text{V}$$

## 22-6 THREE-PHASE TRANSFORMER CONNECTIONS

Three-phase voltages can be transformed by using either a three-phase transformer or three identical single-phase transformers. In either case there is a separate primary winding and a separate secondary winding for each phase voltage.

The structure of a three-phase transformer is illustrated in Fig. 22-11. The primary winding and the secondary winding for each phase are wound on separate legs of the core. The flux in one leg of the core is 120° out of phase with the flux in the other two legs. (This out-of-phase condition results from the phase currents being separated by 120°.) Because of this 120° phase shift, the flux in one leg of the core is always supported by the flux in one of the other legs of the core.

Notice in Fig. 22-11, that the start end of each winding is identified by a dot. Identification of the start end facilitates connecting the primaries and secondaries in either a wye or a delta configuration. When the start ends are not identified, connecting the coils to form either configuration is a trial-and-error process. If correct phasing is in doubt for the delta connection, check it by the method previously discussed and illustrated in Fig. 22-8. Correct phasing of a wye connection can be determined by measuring the line voltages. Of course, all three line voltages are equal if the windings are properly phased.

The primary windings and the secondary windings of a three-phase transformer (or three single-phase transformers) can be either wye- or delta-connected. Also, the primary and secondary do not have to be connected in the same configuration. Thus, there are four possible combinations for

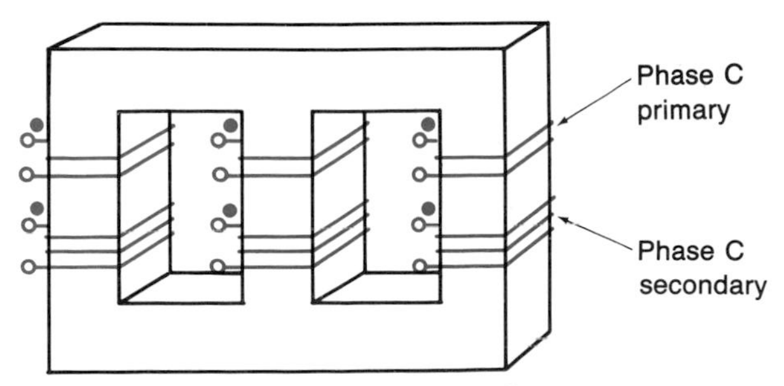


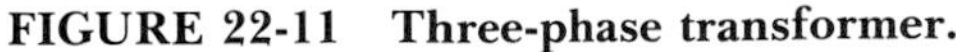
**FIGURE 22-11 Three-phase transformer.**

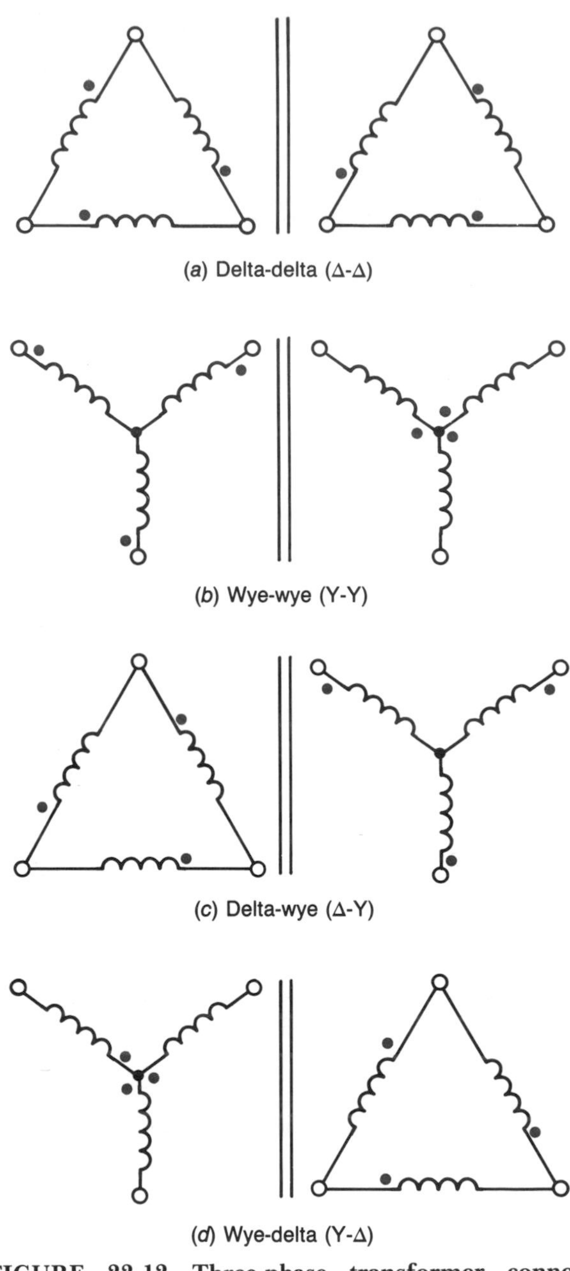


**FIGURE 22-12 Three-phase transformer connections.**

connecting a three-phase transformer: delta-delta, wye-wye, delta-wye, and wye-delta. Each of these combinations is shown in schematic form in Fig. 22-12. This figure shows the correct phasing for each configuration. For the wye connection, either the dot end of all coils (Fig. 22-12*d*) or the nondot end of all coils (Fig. 22-12*c*) forms the star point. For the delta connection, neither the dot ends nor the undotted ends are ever connected together; i.e., a dot end is always connected to an undotted end.

## Self-Test

**19.** True or false? The source and load of a three-phase system must be connected in the same configuration (wye or delta).

**20.** True or false? Both the primary and the secondary windings of a three-phase transformer must be connected in the same configuration.

**21.** True or false? A delta-connected motor should not be connected to a three-phase line that already has a wye-connected motor connected to it.

**22.** For a delta connection, the undotted end of the coil should be connected to the ______ end of another coil.

**23.** What is the result of improper phasing of a wye connection?

## 22-7 CALCULATING POWER IN THREE-PHASE SYSTEMS

Power in a three-phase system, like power in a single-phase system, is a function of current, voltage, and angle $\theta$. Although the total power for any three-phase load is the sum of the powers used by each phase, the procedure used in calculating total power is different for balanced and unbalanced loads.

### Balanced Load

Since each phase load in a balanced load uses the same amount of power, the total power for a balanced load can be determined by calculating the power of one phase and multiplying the result by three. The formula for determining the power $P$ of any one of the phases is $P_p = I_p V_p \cos \theta_p$, where the subscript $p$ means phase. Angle $\theta_p$ is the angle between the phase current and the phase voltage, *not* the angle between the line current and the line voltage.

**EXAMPLE 22-6**
Determine the power for the three-phase load in Fig. 22-13*a* when $V_p = 208$ V and $Z_A = Z_B = Z_C = 30\ \Omega \angle 40°$.

**Solution**

$$V_p \text{ for load} = \frac{V_p \text{ for source}}{1.73} = 120\text{ V} = V_{Z_A} = V_{Z_B} = V_{Z_C}$$

Arbitrarily select $V_{Z_A}$ as the reference phasor; then,

$$I_{Z_A} = I_p = \frac{V_{Z_A}}{Z_A} = \frac{120\text{ V} \angle 0°}{30\ \Omega \angle 40°} = 4\text{ A} \angle -40°$$

$$\theta_p = 0° - 40° = -40°$$

$$P_p = I_p V_p \cos\theta = 4\text{ A} \times 120\text{ V} \times \cos(-40°) = 367.7\text{ W}$$

$$P = 3P_p = 3 \times 367.7\text{ W} = 1103.1\text{ W}$$

**EXAMPLE 22-7**
Refer to Fig. 22-13*b*. Assume $V_l = 208$ V and $Z_A = Z_B = Z_C = 15\ \Omega \angle -20°$. Determine the load power.

**Solution**

$$V_l = V_p = 208\text{ V} \angle 0°$$

$$I_p = \frac{208\text{ V} \angle 0°}{15\ \Omega \angle -20°} = 13.87\text{ A} \angle 20°$$

$$\theta_p = 20° - 0° = 20°$$

$$P_p = 208\text{ V} \times 13.87\text{ A} \times \cos 20° = 2710.98\text{ W}$$

$$P = 3P_p = 8132.94\text{ W}$$

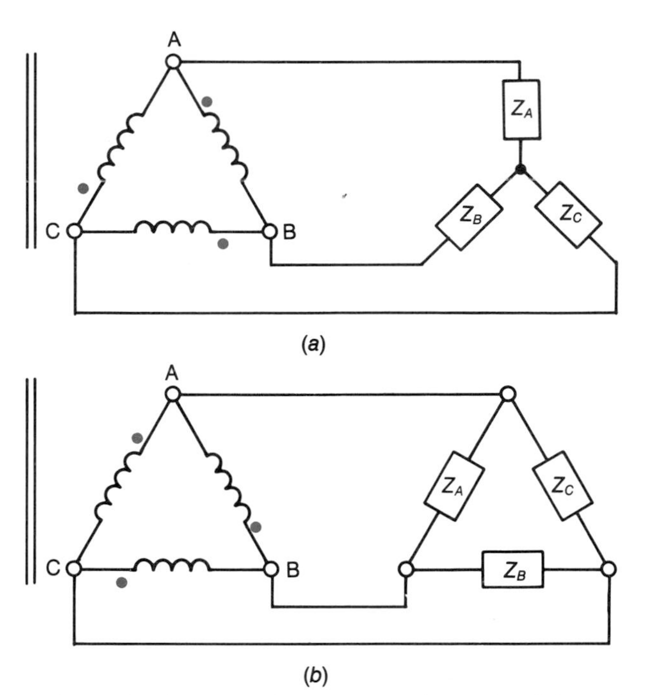


**FIGURE 22-13** Circuits for Examples 22-6, 22-7, and 22-8.

Notice that $\theta_p$ in both Examples 22-6 and 22-7 was equal to the angle of the phase load. This is true in all three-phase circuits. The reason $\theta_p = \angle Z_p$ is because $\angle I_p = \angle V_p - \angle Z_p$ and $\theta = \angle V_p - \angle I_p$. Rearranging $\angle I_p = \angle V_p - \angle Z_p$ yields $\angle Z_p = \angle V_p - \angle I_p$. Thus $\theta_p = \angle Z_p$.

## Unbalanced Load

With an unbalanced load, the power for each phase must be determined in order to find the total power. Thus, for either a delta-connected load or a wye-connected load, the three-phase currents must be calculated. For the wye-connected load, the phase voltage must also be calculated if only the line voltage is specified.

**EXAMPLE 22-8**
Determine the power for Fig. 22-13*b* under the following conditions: $Z_A = 20\ \Omega \angle 0°$, $Z_B = 30\ \Omega \angle 80°$, $Z_C = 40\ \Omega \angle -20°$, $V_{Aa} = 120\text{ V} \angle 0°$, and $V_{Cc} = 120\text{ V} \angle 120°$.

**Solution**

$$I_{Z_A} = \frac{120\text{ V} \angle 0°}{20\ \Omega \angle 0°} = 6\text{ A} \angle 0°$$

$$I_{Z_B} = \frac{120\text{ V} \angle -120°}{30\ \Omega \angle 80°} = 4\text{ A} \angle -200° \text{ or } 4\text{ A} \angle 160°$$

$$I_{Z_C} = \frac{120\text{ V} \angle 120°}{40\ \Omega \angle -20°} = 3\text{ A} \angle 140°$$

$$P_{pA} = 6\text{ A} \times 120\text{ V} \times \cos 0° = 720\text{ W}$$

$$P_{pB} = 4\text{ A} \times 120\text{ V} \times \cos 80° = 83.35\text{ W}$$

$$P_{pC} = 3\text{ A} \times 120\text{ V} \times \cos 20° = 338.29\text{ W}$$

$$P = 720\text{ W} + 83.35\text{ W} + 338.29\text{ W} = 1141.64\text{ W}$$

**EXAMPLE 22-9**
In Fig. 22-14 the line voltage is 480 V. When $V_{Aa}$ is 0°, $V_{Cc}$ is 120°, $Z_A = 50\ \Omega \angle 10°$, $Z_B = 40\ \Omega \angle 20°$, and $Z_C = 60\ \Omega \angle 40°$. Determine the power.

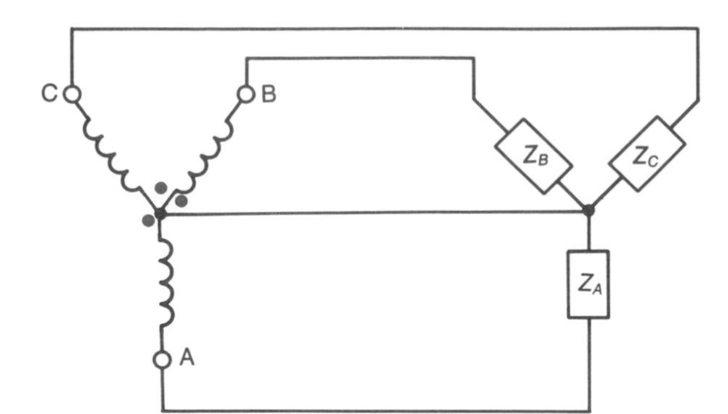


FIGURE 22-14 Circuit for Example 22-9.

**Solution**

$$V_p = \frac{V_l}{1.73} = 277 \text{ V}$$
$$I_{ZA} = \frac{277 \text{ V} \angle 0^\circ}{50\ \Omega \angle 10^\circ} = 5.54 \text{ A} \angle -10^\circ$$
$$P_{pA} = 5.54 \text{ A} \times 277 \text{ V} \times \cos 10^\circ = 1511.3 \text{ W}$$
$$I_{ZB} = \frac{277 \text{ V} \angle -120^\circ}{40\ \Omega \angle 20^\circ} = 6.93 \text{ A} \angle -140^\circ$$
$$P_{pB} = 6.93 \text{ A} \times 277 \text{ V} \times \cos 20^\circ = 1803.8 \text{ W}$$
$$I_{Z_C} = \frac{277 \text{ V} \angle 120^\circ}{60\ \Omega \angle 40^\circ} = 4.62 \text{ A} \angle 80^\circ$$
$$P_{pC} = 4.62 \text{ A} \times 277 \text{ V} \times \cos 40^\circ = 980.3 \text{ W}$$
$$P = 1511.3 \text{ W} + 1803.8 \text{ W} + 980.3 \text{ W}$$
$$= 4295.4 \text{ W}$$

## 22-8 TYPES OF POWER AND POWER FACTOR

In this chapter (and preceding chapters) we have defined power as $P = IV \cos \theta$. This formula determines the *true* or *real power* in any circuit—direct current, single-phase alternating current, or three-phase alternating current. True power specifies the rate at which electric energy is being converted to heat energy. Since only resistance can convert electric energy into heat energy, the true power is always the product of the resistive (real) current and the resistive (real) voltage. As illustrated in Fig. 22-15, multiplying by the cos $\theta$ merely determines what portion of the current or voltage is resistive. Unless otherwise specified, the term power (and the symbol $P$) always refers to true power.

Two other types of power are used in electricity and electronics: *apparent power* and *reactive power*.

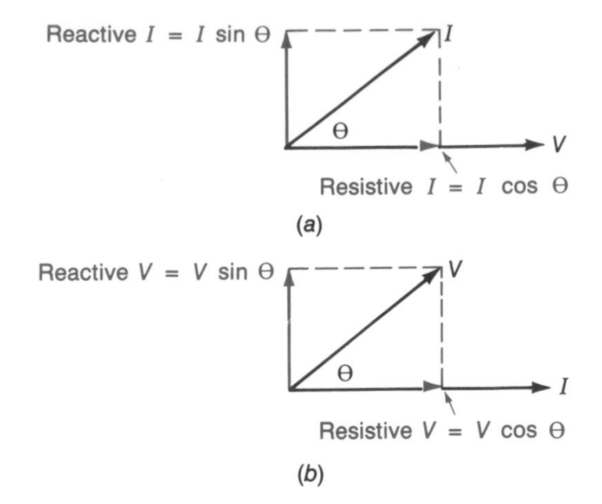


FIGURE 22-15 Reactive and resistive components of phasors.

Reactive power, which is identified by the symbol $Q$, never converts any electric energy into heat energy. Therefore, the base unit for reactive power $Q$ cannot be the watt because the watt represents a conversion rate of a joule per second. The base unit for reactive power is the *reactive voltampere*, which is abbreviated *var*. The reactive power is the product of reactive current and reactive voltage: $Q = I_X V_X$, where $X$ stands for reactance. For pure capacitance, $Q = I_C V_C$, and for pure inductance, $Q = I_L V_L$. When a circuit has impedance (a combination of resistance and reactance), the reactive power is found by the formula $Q = IV \sin \theta$. Again, Fig. 22-15 shows that multiplying by the sin $\theta$ determines the reactive current when voltage is on the real axis and the reactive voltage when current is on the real axis. (Figure 22-15*a* could represent $I$ and $V$ in a parallel $RC$ circuit and Fig. 22-15*b* could represent $I$ and $V$ in a series $RL$ circuit.)

**EXAMPLE 22-10**

The phase voltage in a three-phase circuit is $90 \text{ V} \angle -50^\circ$ and the corresponding phase current is $2.5 \text{ A} \angle -10^\circ$. Determine $P$ and $Q$.

**Solution**

$$\theta = -50^\circ - (-10^\circ) = -40^\circ$$
$$P = IV \cos \theta = 2.5 \text{ A} \times 90 \text{ V} \times \cos 40^\circ = 172.4 \text{ W}$$
$$Q = IV \sin \theta = 2.5 \text{ A} \times 90 \text{ V} \times \sin 40^\circ = 144.6 \text{ var}$$

Since the current is leading the voltage in Example 22-10, the phase load must have a net capacitive reactance. Therefore, the reactive power is capacitive and could be symbolized as $Q_C$

*Apparent power* is a combination of true power and reactive power. It is represented by the symbol $S$ and has a unit of voltampere, which is abbreviated V · A. The formula for apparent power is $S = IV$. Notice that apparent power ignores the phase relationship between current and voltage. When angle $\theta$ is 0°, the apparent power equals the real power. When angle $\theta$ is 90°, the apparent power equals the reactive power.

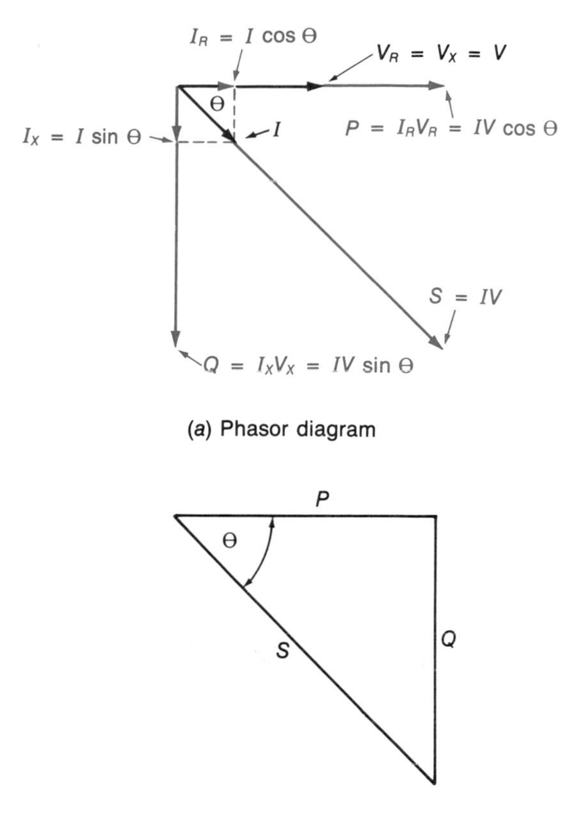

(*b*) Power triangle

**FIGURE 22-16 Power triangle shows the relationship of real, reactive, and apparent power.**

**EXAMPLE 22-11**

Determine the apparent power for the circuit in Example 22-10.

**Solution**

$$S = IV = 2.5 \text{ A} \times 90 \text{ V} = 225 \text{ V} \cdot \text{A}$$

The relationship between the three types of power (real, reactive, and apparent) is illustrated in Fig. 22-16. The phasor diagram in Fig. 22-16*a* is for a circuit with a net reactance which is inductive. Since this diagram uses the circuit voltage as the reference phasor, the circuit current is partitioned into its reactive and resistive components. These three currents $I_X$, $I$, and $I_R$, when multiplied by the voltage, yield $Q$, $S$, and $P$, respectively. Note that the angle of each power will be the same as the angle of the current used in calculating each power. This is so because the common voltage is at 0° and multiplying phasors requires that angles be added.

The power triangle in Fig. 22-16*b* is made up of the three power phasors from the phasor diagram in Fig. 22-16*a*. This triangle clearly shows that as $Q$(reactive power) decreases, $S$ (apparent power) approaches $P$ (real power) in both magnitude and phase. Because $Q$, $S$, and $P$ always form a right triangle, the trigonometric functions and the pythagorean theorem can be used in manipulating these three types of power.

**EXAMPLE 22-12**

The measured phase power in a circuit is 500 W. The phase current is 5 A and the phase voltage is 120 V. Determine the reactive power and angle $\theta$.

**Solution**

$$S = IV = 5 \text{ A} \times 120 \text{ V} = 600 \text{ V} \cdot \text{A}$$
$$S^2 = P^2 + Q^2 \qquad \text{Pythagorean theorem}$$

Therefore,

$$Q = \sqrt{S^2 - P^2} = \sqrt{600^2 - 500^2} = 331.7 \text{ var}$$
$$\theta = \arccos \frac{P}{S} = \arccos \frac{500}{600} = 33.6°$$

Solving $P = IV \cos \theta$ for $IV$ yields $IV = P/\cos \theta$. Since $S = IV$, we can combine these two equations to get $S = P/\cos \theta$ and $P = S \cos \theta$. The $\cos \theta$ determines how much of the apparent power $S$ is real power $P$; therefore, the $\cos\theta$ is often called the *power factor.* Power factor (PF) = $\cos\theta$. Since $\cos\theta$ ranges from 0 for a pure reactive load to 1 for a pure resistive load, PF also ranges from 0 to 1. Sometimes PF is expressed in percent (0 to 100 percent). The PF can be either a *leading* PF or a *lagging* PF. If the net reactance in a circuit is capacitive, the PF is leading; if the net reactance is inductive, it is lagging. Thus, leading and lagging refers to the angle of the current with reference to the voltage.

The PF of an electric system is very important to power companies and large industries because PF determines how efficiently the power distribution equipment (transformers, power lines, etc.) is used. For example, a transformer can only handle a specified amount of load current (because of the wire size used in its winding) regardless of the PF of the load. Suppose a transformer secondary is rated at 480 V and 30 A (14.4 kV · A) and is connected to a 30-A load with a PF of 0.95. The power delivered to the load is 14.4 kV · A × 0.95 = 13.68 kW. If an identical transformer is connected to a 30-A load with a PF of 0.7, the power is 14.4 kV · A × 0.7 = 10.08 kW. Since the transformers are providing the same amount of current and voltage (i.e., apparent power) to the two loads, the power loss is the same in both transformers. Yet, the load with the higher PF is providing more power in the desired form than is the lower PF load. Thus, the higher PF system is more efficient.

Improving the PF (making it closer to 1) of a system is called *PF correction.* The PF can be corrected by adding an appropriate type of reactive load to the system. Since most large loads are inductive, the PF is usually lagging. Therefore, a capacitive reactance load is usually needed to correct the PF. Adding a capacitor in parallel with an inductive load corrects the PF.

**EXAMPLE 22-13**

A 480-V, 60-Hz motor draws 10 A and operates at a PF of 0.85. How much parallel capacitance is required to change the PF to unity (1.0)?

**Solution**

$$S = IV = 10\text{ A} \times 480\text{ V} = 4.8\text{ kV}\cdot\text{A}$$
$$\theta = \arccos \text{PF} = \arccos 0.85 = 31.79^\circ$$

Refer to the power triangle in Fig. 22-16 and note that

$$Q = S\sin\theta = 4.8\text{ kV}\cdot\text{A} \times \sin 31.79^\circ = 2.53\text{ kvar}$$
$$I_X = \frac{Q}{V_X} = \frac{2.53\text{ kvar}}{480\text{ V}} = 5.27\text{ A}$$
$$X_L = \frac{V_X}{I_X} = \frac{480\text{ V}}{5.27\text{ A}} = 91.1\ \Omega$$

For a PF of 1, $X_L = X_C$; so

$$C = \frac{1}{2\pi f X_C} = \frac{1}{6.28 \times 60\text{ Hz} \times 91.1\ \Omega} = 29.1\ \mu\text{F}$$

Notice in this example that the reactive power of the motor, which is inductive and causes a lagging PF, is canceled by the reactive power of the capacitor, which causes a leading PF. Figure 22-17 shows the results of complete PF correction as in Example 22-13. In this figure $Q_L$ is the reactive power of the motor and $Q_C$ is the reactive power of the capacitor. As seen in Fig. 22-17*b*, complete PF correction causes both $Q$ and $\theta$ to become zero and $S$ to be equal to $P$.

Example 22-13 is for a single-phase load. With a three-phase load, each phase would need to be corrected by a separate capacitor.

One final observation about PF correction. Notice that fully correcting an inductive load by adding a parallel capacitor creates a low-quality parallel-resonant circuit. Remember that in a parallel-resonant circuit the inductive current cancels the capacitive current and $\theta$ goes to 0°.

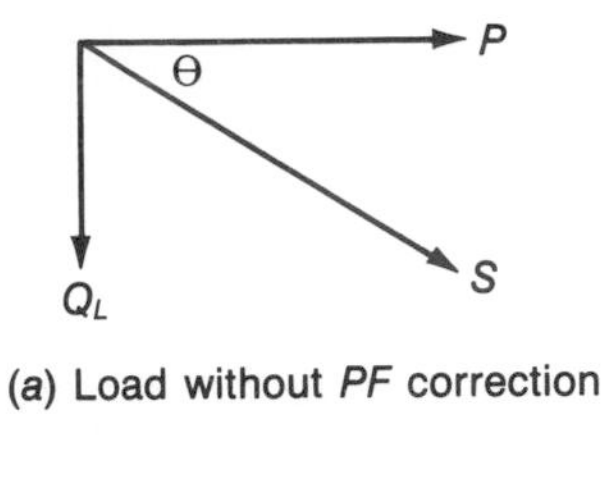


(*a*) Load without *PF* correction

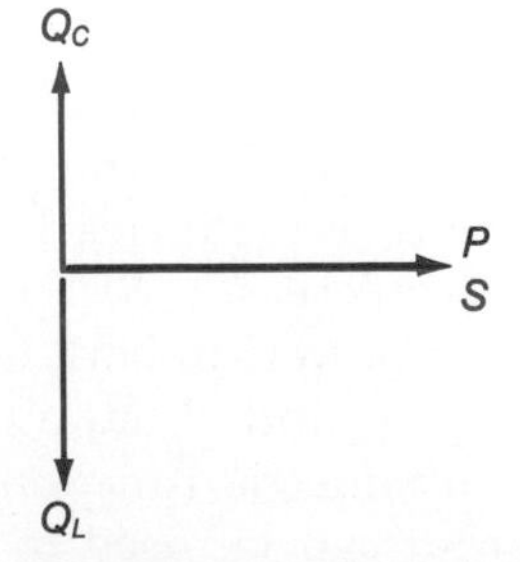


(*b*) Load with *PF* correction

**FIGURE 22-17 Power factor correction.**

## Self-Test

**24.** True or false? In determining the phase power for a balanced load the cosine of the angle between the line current and the line voltage is used in the formula $P_p = IV \cos \theta$.

**25.** Assume the following values for the circuit in Fig. 22-14: $V_{Aa} = 120\ \text{V} \angle 0°$, $V_{Bb} = 120\ \text{V} \angle -120°$, $V_{Cc} = 120\ \text{V} \angle 120°$, $Z_A = 50\ \Omega \angle 20°$, $Z_B = 40\ \Omega \angle -40°$, and $Z_C = 60\ \Omega \angle 80°$. Determine $P$ and $I_N$.

**26.** Why would not the circuit in Fig. 22-13*a* be used with an unbalanced load?

**27.** List three types of power. Also write the symbol and base unit for each type.

**28.** A 700-W load draws 8 A from a 120-V source. Determine $Q$, $S$, $\theta$, and PF.

**29.** When the PF = 1, what is the relationship between $S$ and $P$?

**30.** How can the PF of an inductive load be improved?

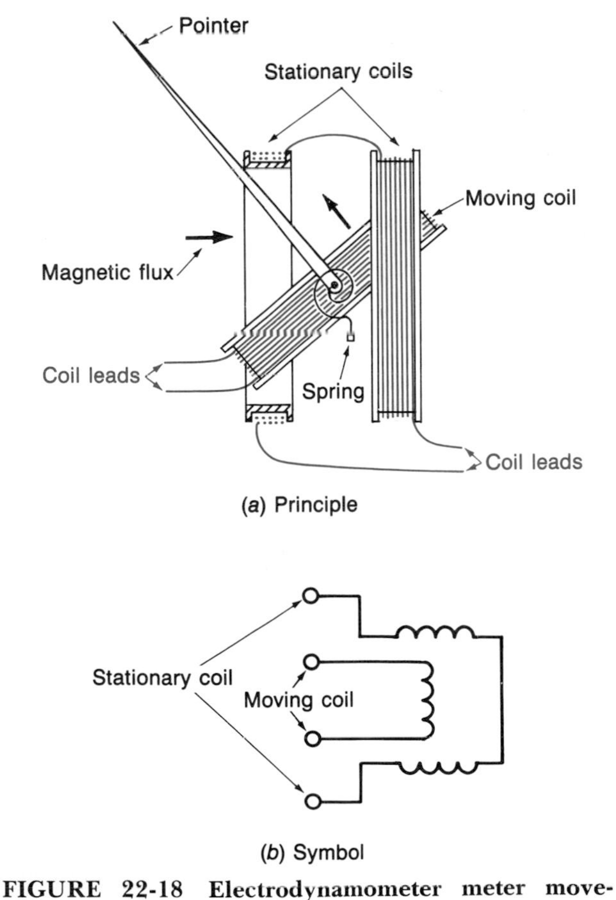


**FIGURE 22-18 Electrodynamometer meter movement.**

## 22-9 MEASURING AC POWER

An *electrodynamometer meter movement* is used for measuring power because it automatically compensates for the value of angle $\theta$; i.e., it indicates true power. The structure of an electrodynamometer meter movement is illustrated in Fig. 22-18*a*. Torque in this meter movement is caused by the interaction of two magnetic fields. One of these fields is produced by a mechanically stationary split coil. The other field is produced by a moving coil; i.e., a coil attached to a shaft which is mounted in jeweled bearings. Also attached to this shaft is a pointer and a countertorque spiral spring (see Fig. 22-18*a*).

When both coils in Fig. 22-18*a* have current flowing in the same direction, their magnetic fields repel each other, causing a clockwise rotational force on the moving coil. Clockwise rotation moves the pointer up-scale and winds the spiral spring so that it produces a counterclockwise torque. When the two torques are equal, the pointer comes to rest and indicates the true power. If the current in either coil (but not both coils) reverses, one of the magnetic fields reverses. Then the fields will attract, and the pointer will be forced down-scale. When both currents (and thus both fields) reverse, the fields still repel each other and the torque is still clockwise.

When the electrodynamometer is used to measure power, it is called a *wattmeter*. Using a wattmeter to measure power in a single-phase circuit is illustrated in Fig. 22-19. As a wattmeter, the stationary coil of the electrodynamometer carries the current for the load for which power is being measured. The moving coil and a series resistor are connected in parallel across the source voltage. Thus, the current in the moving coil is a function of the load voltage, while the current in the stationary coil is a function of the load current. Therefore, the magnitude of the instantaneous torque is a function of the magnitude of the instantaneous load voltage and the instantaneous load current. The instantaneous direction of the torque is a function of the instantaneous polarity of the load voltage and the instantaneous direction of the load current. One end of the voltage coil and one end of the current coil is identified with a ± symbol. When the ± ends of the coils are connected as in Fig. 22-19, the pointer moves up-scale as long as the current and voltage are less than 90° out of phase. This is correct phasing for a wattmeter's coil.

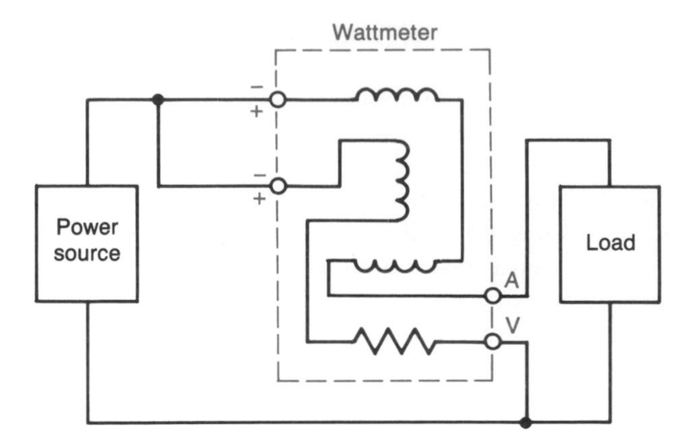


(*a*) Wattmeter connections

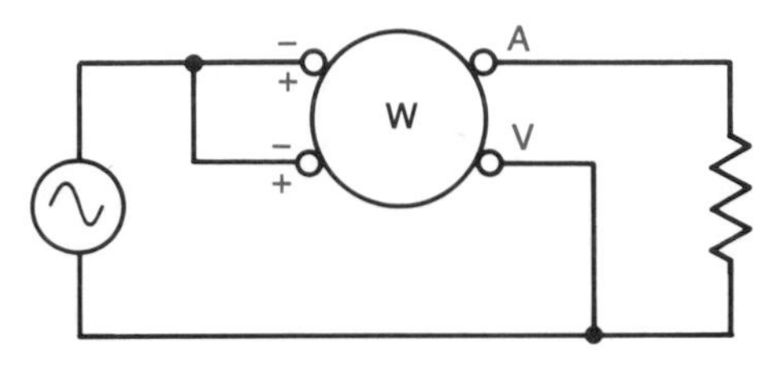


(*b*) Symbol

**FIGURE 22-19 The wattmeter.**

Figure 22-20*a* shows that when *I* and *V* are in phase, the polarity of the voltage and the direction of the current change at the same instant. Therefore, the torque is always in the same direction, and, if the coils are properly phased, the torque is always clockwise (up-scale). When the current and voltage are out of phase, they do not change polarity and direction at the same time. This causes every other torque pulse to be counterclockwise (down-scale). When *I* and *V* are 30° out of phase, as in Fig. 22-20*b*, the down-scale torque is much smaller than the up-scale torque. The net torque is up-scale, which will indicate some true power, but not as much as when the current and voltage are in phase. Figure 22-20*c* shows that when current and voltage are 90° out of phase, the net torque is zero and the indicated power is zero. Thus, the electrodynamometer (wattmeter) automatically compensates for PF.

A single wattmeter can be used to measure the power of any three-phase load (wye or delta and balanced or unbalanced). Figure 22-21 shows how to connect a wattmeter to measure the phase power of either a wye or delta load when the star point is accessible for the wye load and the junctions of the lines and phase loads are accessible for the delta load.

### EXAMPLE 22-14

The wattmeter in Fig. 22-21*a* indicates 1420 W. Determine the power of the three-phase load.

**Solution**

Since there is no neutral wire, the wye load must be balanced; therefore, the power is

$$P_T = 3P_p = 3 \times 1420 \text{ W} = 4260 \text{ W}$$

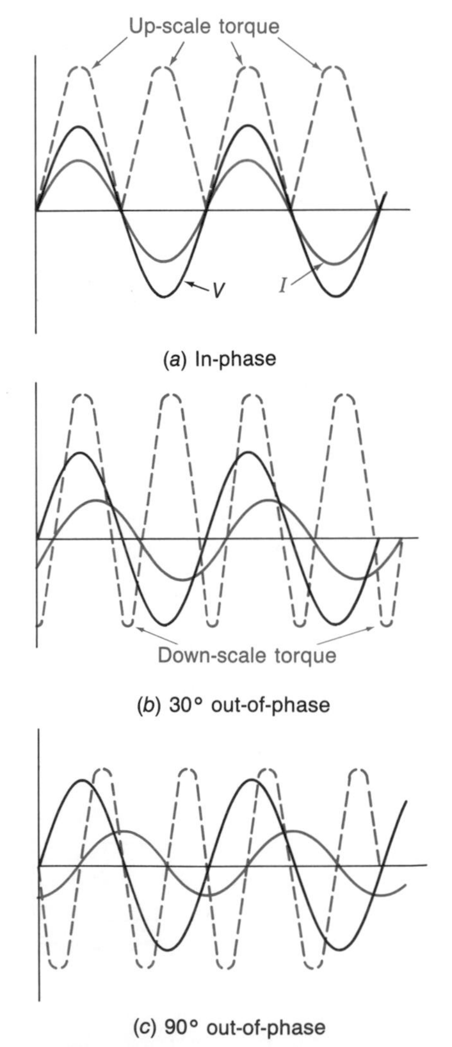

**FIGURE 22-20 A wattmeter indicates true power.**

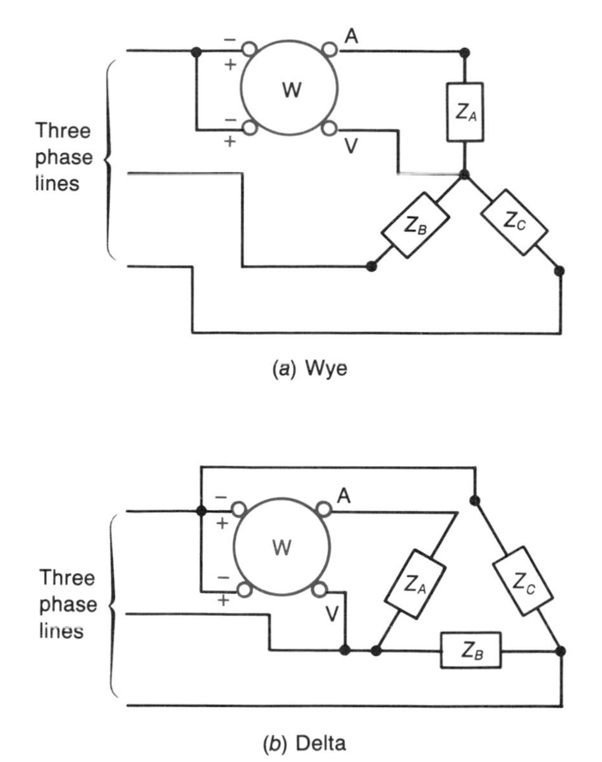


(*a*) Wye

(*b*) Delta

**FIGURE 22-21 Measuring phase power.**

If a wattmeter is normally phased (i.e., the ± ends of the current and voltage coils connected together) and it indicates a reverse reading, the current and voltage are more than 90° out of phase. In such cases, reverse the terminals of either the voltage coil or the current coil and the reading will be forward and of the correct value. Figure 22-22 illustrates why more than 90° of phase shift causes a reverse reading. In Fig. 22-22*a* you can see that the reverse (down-scale) torque pulses exceed the forward torque pulses. Figure 22-22*b* shows that reversal of the current coil leads corrects the situation and causes up-scale deflection of the pointer.

When the three-phase load is unbalanced, or when the star point of a balanced wye load or the phase-line junction of a balanced delta load is not accessible, two wattmeter readings are necessary to determine the total power of the load. The connections for obtaining the two readings are shown in Fig. 22-23. This method of determining the power of a three-phase load is often referred to as the *two-wattmeter method* because two wattmeters are often used so that simultaneous readings can be obtained. This is advantageous if the power of the load is variable. However, when the load power is stable, a single wattmeter can be connected first as $W_1$ and then as $W_2$. The time between the reading of $W_1$ and $W_2$ makes no difference when the load is stable.

When using the two-wattmeter method, it does not matter which two lines are interrupted to insert the current coils of the wattmeters. However, the voltage coils of both wattmeters must return to the uninterrupted line. When *normal phasing* (Fig. 22-23*a*) causes both meters to indicate forward readings, the two readings are *added* together to obtain the total load power. When normal phasing causes one wattmeter to indicate in the reverse direction, either the voltage coil or the current coil must be reversed as in Fig. 22-23*b*, where the *voltage coil* is *reversed*. Now, the smallest power reading is *subtracted* from the largest power reading to obtain the total power for the three-phase load. Remember, the two-wattmeter method is appropriate for either balanced or unbalanced loads. It does not matter whether the loads are wye- or delta-connected.

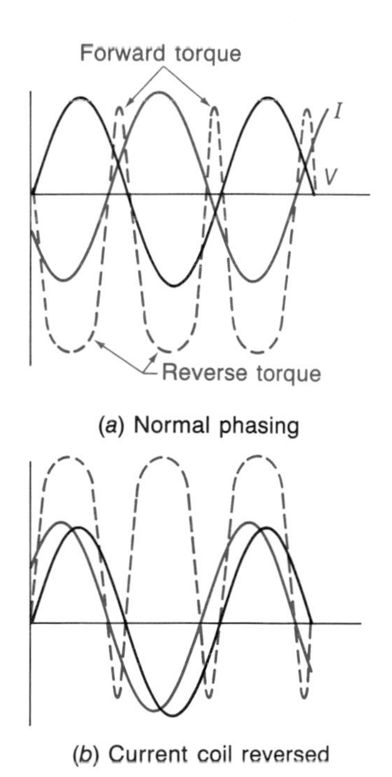


(*a*) Normal phasing

(*b*) Current coil reversed

**FIGURE 22-22 Correcting a reverse reading on a wattmeter.**

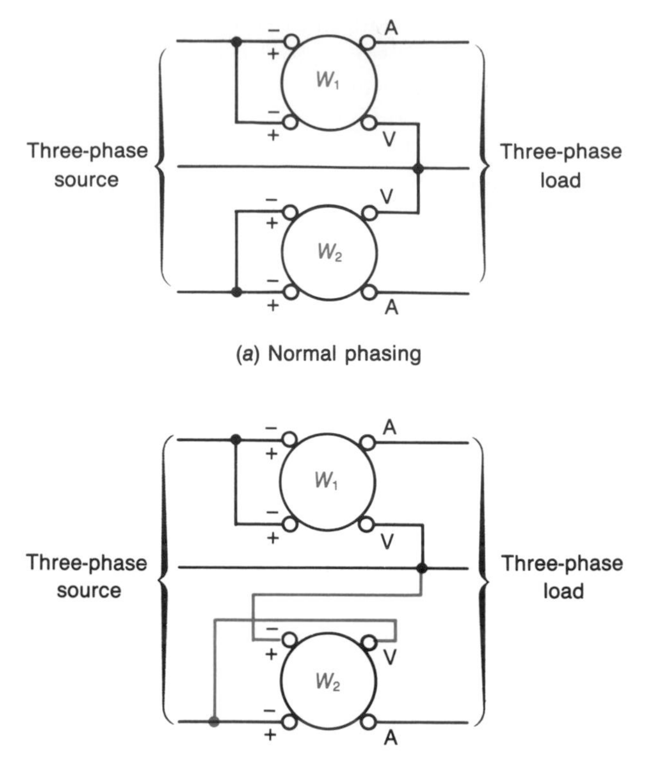


**FIGURE 22-23 Two-wattmeter method of measuring power of a three-phase load.**

**EXAMPLE 22-15**

Refer to Fig. 22-23*a*. Wattmeter $W_1 = 1220$ W, and $W_2 = 860$ W. What is the power of the load?

**Solution**

Since both meters are normally phased,

$$P_T = W_1 + W_2 = 1220 \text{ W} + 860 \text{ W} = 2080 \text{ W}$$

**EXAMPLE 22-16**

In Fig. 22-23*b*, $W_1$ indicates 600 W and $W_2$ indicates 1040 W. Determine the load power.

**Solution**

Wattmeter $W_2$ is reversed-phased, so

$$P_T = W_2 - W_1 = 1040 \text{ W} - 600 \text{ W} = 440 \text{ W}$$

## Self-Test

**31.** An ______ meter movement is used in a wattmeter.

**32.** True or false? Wattmeters indicate true power only if $I$ and $V$ are less than 90° out of phase.

**33.** True or false? Two wattmeters are needed to measure the load power in an unbalanced three-phase system.

**34.** Under what condition should the meter readings be subtracted when using the two-wattmeter method of measuring power of a three-phase load?

## 22-10 ADVANTAGES OF THREE-PHASE SYSTEMS

Three-phase systems have a number of advantages over single-phase systems. Some of these advantages are (1) more efficient use of copper wire for transporting power; (2) more constant power from, and torque on, motors and generators; and (3) less ripple in the dc output when alternating current is rectified (converted to direct current).

The first advantage listed above can be illustrated by comparing equal loads (with equal voltages) in a three-phase and a single-phase system. To operate three 100-W, 120-V lamps from a 120-V single-phase system requires a parallel hookup as shown in Fig. 22-24*a*. The current carried by each conductor connecting the load to the source is

$$I = \frac{P}{V} = \frac{300 \text{ W}}{120 \text{ V}} = 2.5 \text{ A}$$

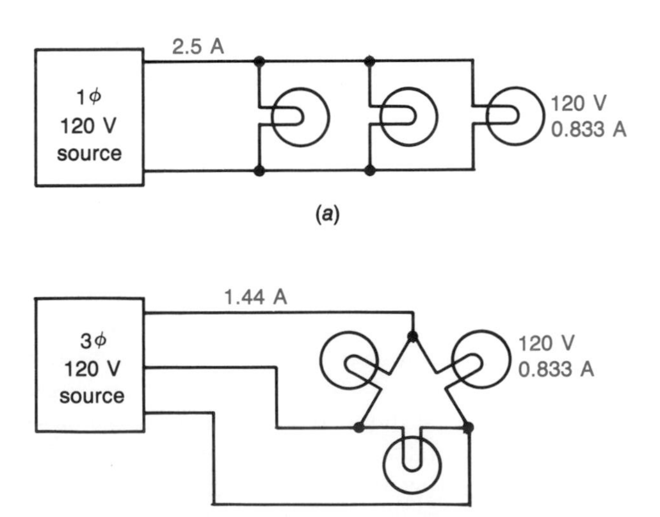


**FIGURE 22-24 Single-phase versus three-phase.**

At 500 circular mils per ampere (500 cmils/A), each conductor must have a cross-sectional area of 1250 cmils. If the load and source are 10 ft apart, the total amount of copper required would be 25,000 cmils · ft.

With the three-phase system in Fig. 22-24*b*, each line connecting the source to the load must carry 1.73 times the phase current in this balanced load. The phase current is 100 W/120 V = 0.833 A. Therefore, the line current is 0.833 A × 1.73 = 1.44 A. At 500 cmils/A, each conductor is 720 cmils in cross-sectional area. The total copper for the three lines is 30 ft × 720 cmils = 21,600 cmils · ft. Compare this to the 25,000 cmils · ft required for the single-phase system. At the same voltage, the single-phase system requires 25,000/21,600 = 1.16 times as much copper for a given load. This is a very significant savings when long distances and large powers are involved.

The second advantage (uniform power and torque) of the three-phase system is illustrated in Fig. 22-25. With a unity-PF load, the power of a single-phase system pulses from zero to maximum and back to zero each $\frac{1}{2}$ cycle. Thus, the torque (and therefore the stress) on the shaft of a generator or a motor pulses twice each cycle. As shown in

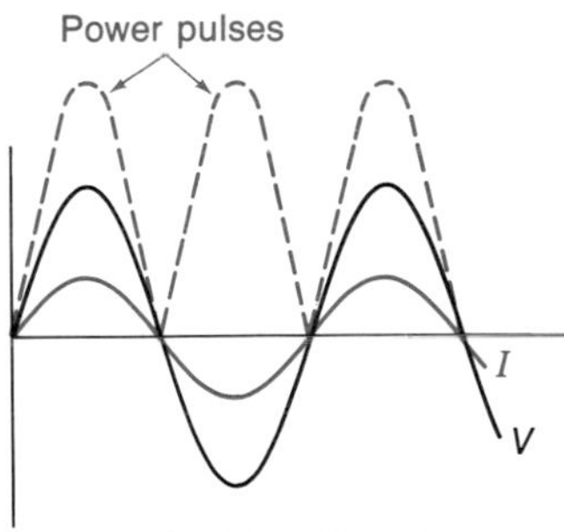

(*a*) Single-phase power

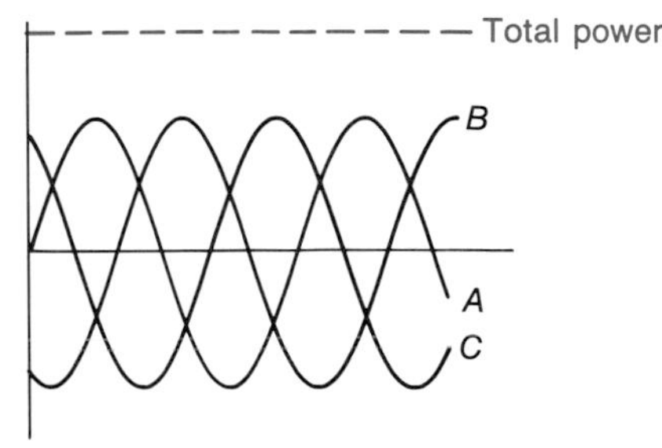

(*b*) Three-phase power

**FIGURE 22-25 Single-phase versus three-phase power.**

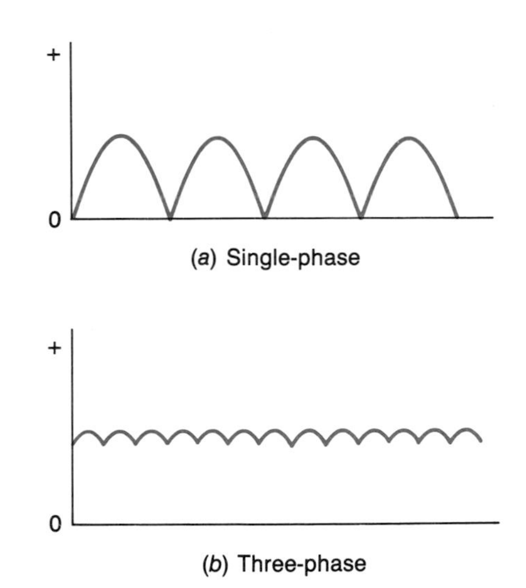


**FIGURE 22-26 Comparison of rectified single-phase and three-phase alternating current.**

Fig. 22-25*b*, the total power (i.e., the sum of the phase powers) is constant in a balanced three-phase system. Thus, a motor or generator shaft is under uniform torque rather than pulsating torque. The reason for the total power being constant is that the sum of the sin² of any three angles separated by 120° is always 1.5. The total power at any instant ($P_{T,\text{inst}}$) is

$$P_{T,\text{inst}} = \frac{(\sin\theta_1 V)^2}{R} + \frac{(\sin\theta_2 V)^2}{R} + \frac{(\sin\theta_3 V)^2}{R}$$

where $\theta_1$ angle of phase A
$\theta_2 = \theta_1 + 120°$
$\theta_3 = \theta_1 + 240°$
$V$ = peak phase voltage
$R$ = resistance of phase load

Factoring out $V/R$ in the above equation yields

$$P_{T,\text{inst}} = \frac{[(\sin\theta_1)^2 + (\sin\theta_2)^2 + (\sin\theta_3)^2]V}{R}$$

Since the sum of the sin² of the angles in the brackets in the last equation is always 1.5, the total instantaneous power remains constant.

The third advantage (purer direct current after rectification) of three-phase over single-phase is obvious from inspection of Fig. 22-26. When single-phase voltage is full-wave-rectified, it creates two pulses of direct current for each cycle. Note in Fig. 22-26*a* that the dc pulse drops to zero twice each cycle. It can require a large capacitor to change this pulsating direct current to an essentially pure direct current. By contrast, the full-wave-rectified three-phase voltage has very little ripple in its dc output because the next phase takes over and provides the output voltage shortly after the previous phase has reached its peak value. Under the same load conditions, much less capacitance is needed to provide an essentially pure dc voltage when three-phase voltage is rectified than when single-phase voltage is rectified.

## Self-Test

**35.** List three advantages of a three-phase system over a single-phase system.

**36.** The instantaneous power of a balanced three-phase load is 2000 W when phase A is at its peak voltage. What will the instantaneous power be 30° later?

## SUMMARY

1. $V_{Aa}$ is 180° out of phase with $V_{aA}$.
2. Three-phase is abbreviated 3-$\phi$.
3. In a three-phase system, the phases are separated by 120°.
4. The algebraic sum of the instantaneous voltages of a three-phase system equals zero. This allows the three phases to be delta-connected.
5. Three-phase sources and loads can use either the delta or the wye connection.
6. With a balanced wye connection, the line voltage is 1.73 times as great as the phase voltage. These two voltages are 30° out of phase. The line current equals the phase current.
7. The star point, or neutral point, is the junction of the three phases in a wye connection. All reference ends or all nonreference ends of the source coils must be connected to the star point for correct phasing.
8. A neutral wire is often connected to the star point. This wire keeps the phase voltage on the load equal to the phase voltage of the source when the load is unbalanced.
9. The neutral wire carries current only when the load is unbalanced.
10. With a balanced delta connection, the line current is 1.73 times as great as the phase current. The line and load voltages are equal with either balanced or unbalanced loads.
11. Correct phasing for the delta connection requires that two reference ends of the source coils never be connected together.
12. The source and the loads of a three-phase system need not use the same (wye or delta) connections.
13. An unbalanced wye-connected load requires a neutral wire from the wye-connected source.
14. Three-phase voltage can be transformed using a single-core three-phase transformer or three single-phase matched transformers.
15. Transformer primaries and secondaries can be either delta-connected or wye-connected.
16. Phasing of transformer connections follows the same rules as phasing of a generator.
17. The power of a three-phase load is the sum of the three phase powers.
18. Three types of power are true or real power $P$, reactive power $Q$, and apparent power $S$.

19. Power types $P$, $Q$, and $S$ form a right triangle.

20. Units for $P$, $Q$, and $S$ are W, var, and V · A, respectively.

21. The cos $\theta$ is called the power factor (PF).

22. Power factor determines how efficiently the power distribution system is used.

23. A lagging PF can be corrected by adding capacitance in parallel with the load.

24. The electrodynamometer meter movement is used in a wattmeter. This movement has a split stationary coil and a moving coil.

25. A wattmeter indicates true power.

26. A single wattmeter can be used to measure three-phase power. Only one reading is necessary when the load is balanced and both the phase voltage and phase current are accessible.

27. A wattmeter gives a reverse reading when $I$ and $V$ are more than 90° out of phase and the meter is normally phased.

28. When using the two-wattmeter method, the readings are added unless one of the meters requires reverse phasing of its coils.

29. Compared to single-phase systems, three-phase systems make more efficient use of copper, provide smoother direct current when rectified, and provide smooth power and torque for rotating machinery.

30. Useful formulas are

$$V_{Ca} = V_{Cc} - V_{Aa}$$
$$V_l = 1.73V_p$$
$$I_l = 1.73I_p$$
$$P_p = I_pV_p \cos \theta_p$$
$$\theta_p = \angle Z_p$$
$$P_T = 3P_p \quad \text{for balanced loads}$$
$$P = IV \cos \theta = I_RV_R$$
$$Q = IV \sin \theta = I_XV_X$$
$$S = IV$$
$$\text{PF} = \cos \theta$$

## CHAPTER REVIEW QUESTIONS

For the following items, determine whether each statement is true or false.

22-1. The voltage of phase A must be 120° out of phase with the voltage of phase B in a three-phase system.

22-2. The junction of any two phase coils in a delta connection is called a star point.

22-3. The line current in a balanced delta system is 17.3 A when the phase current is 10 A.

22-4. When a system is balanced, the neutral wire carries one-third of the current for each phase.

22-5. The dot ends of two phase windings should never be connected together in a delta-connected transformer.

22-6. If a wye-connected source is incorrectly phased, one of the line voltages will be 0 V.

**22-7.** A single-core transformer with a leg for each phase must be used to transform the voltage level of a three-phase voltage.

**22-8.** If the primary of a three-phase transformer is delta-connected, the secondary must also be delta-connected.

**22-9.** Connecting a capacitor in parallel with a load can improve a leading PF.

**22-10.** The base unit for reactive power is the var.

**22-11.** For most efficient use of power distribution equipment, the cos $\theta$ should be unity.

**22-12.** Wattmeters indicate apparent power.

**22-13.** A wattmeter with normal phasing reads in a reverse direction if the current and voltage are more than 45° out of phase.

**22-14.** Two wattmeters are required to measure the total power in an unbalanced delta load.

For the following items, fill in the blank with the word or phrase required to correctly complete each statement.

**22-15.** The abbreviation for three-phase is ______.

**22-16.** At 80° phase A is −138 V and phase B is +154 V. The voltage of phase C is ______.

**22-17.** Another name for the neutral point is the ______.

**22-18.** When a delta-connected source supplies a wye-connected load, the load should be ______.

**22-19.** The ______ is usually grounded in a four-wire wye system.

**22-20.** Connecting together the coil ends identified by uppercase letters provides correct phasing for a ______ connection.

**22-21.** An ac voltmeter connected in series with the coils of a correctly phased delta-connected source should indicate ______ V.

**22-22.** The true power can never exceed the ______ power.

**22-23.** A wattmeter uses an ______ meter movement.

**22-24.** Three advantages of a three-phase system over a single-phase system are ______, ______, and ______.

## CHAPTER REVIEW PROBLEMS

**22-1.** If $V_{Bb} = 120 \text{ V} \angle 60°$, what will $V_{bB}$ equal?

**22-2.** The line voltage of a four-wire wye system is 400 V. What is the phase voltage?

**22-3.** Calculate the neutral-wire current in a four-wire wye system in which the line currents are $9 \text{ A} \angle -60°$, $15 \text{ A} \angle 40°$, and $12 \text{ A} \angle 130°$.

**22-4.** Determine $V_{BC}$ when $V_{Cc} = 20 \text{ V} \angle 30°$ and $V_{Bb} = 20 \text{ V} \angle 150°$.

**22-5.** In Problem 22-4 is $V_{BC}$ leading or lagging $V_{Bb}$?

**22-6.** In a balanced three-phase system, the phase voltage is $200 \text{ V} \angle 30°$ and the phase current is $8 \text{ A} \angle -20°$. Determine the following:

**a.** The PF

**b.** The total load power
**c.** The reactive power
**d.** The apparent power

**22-7.** In Problem 22-6 is the PF leading or lagging?

**22-8.** A wye load (with neutral wire) has the following voltages and impedances: $V_{Aa} = 208\text{ V}\,\underline{/0°}$, $V_{Bb} = 208\text{ V}\,\underline{/120°}$, $V_{Cc} = 208\text{ V}\,\underline{/240°}$, $Z_A = 52\ \Omega\,\underline{/30°}$, $Z_B = 104\ \Omega\,\underline{/-60°}$, and $Z_C = 26\ \Omega\,\underline{/10°}$. Determine the load power.

**22-9.** In a single-phase circuit, a wattmeter indicates 200 W. The circuit current is 3 A and the circuit voltage is 115 V. Determine $Q$, $S$, and PF.

**22-10.** The two-wattmeter method is used to measure the power in a three-phase system. When both meters are normally phased, the meter readings are 2400 W and 1260 W. What is the load power?

**22-11.** If the wattmeter which indicates 2400 W in Problem 22-10 had required reverse phasing to obtain a forward reading, what would the load power have been?

## ANSWERS TO SELF-TESTS

**1.** $V_{DC} = 48\text{ V}\,\underline{/-150°}$ or $48\text{ V}\,\underline{/210°}$

**2.** $V_{FE} = 27\text{ V}\,\underline{/180°}$

**3.** 120°

**4.** zero

**5.** −20 V

**6.** neutral, star

**7.** the neutral or star point

**8.** $V_{BC} = 173\text{ V}\,\underline{/-90°}$

**9.** lead, 30°

**10.** single-phase 277 V, single-phase 480 V, and three-phase 480 V.

**11.** yes, any time the phasor sum of the three load currents equals zero.

**12.** $11.45\text{ A}\,\underline{/-12°}$

**13.** 240 V

**14.** coil Bb

**15.** The line current is 1.73 times larger than the phase current, and the line current is displaced 30° from the phase current.

**16.** They are equal.

**17.** 17.34 A

**18.** There is 120° between each line current.

**19.** false

**20.** false

**21.** false

**22.** dotted

**23.** The line voltages are not equal.

**24.** false

**25.** $P = 588\text{ W}$, $I_N = 4.98\text{ A}\,\underline{/-30°}$

**26.** Because the load phase voltage would also be unbalanced. Unbalanced wye loads are not connected to delta sources.

**27.** power $P$, W; apparent power $S$, V · A; reactive power $Q$, var.

**28.** $Q = 657$ var, $S = 960$ V · A, $\theta = 43.2°$, and PF = 0.73

**29.** They are equal.

**30.** by adding the correct amount of capacitance in parallel with the load

**31.** electrodynamometer

**32.** false

**33.** false

**34.** when one of the meters requires reverse phasing to obtain an up-scale reading

**35.** more efficient use of conductors, constant power and torque for balanced loads, and closer to pure dc output from a rectifier circuit

**36.** 2000 W

# Chapter 23 Introduction to Discrete Electronics

D*iscrete electronics* refers to electronic circuits and products that use individual parts which usually contain only a single device such as a transistor, a capacitor, or a resistor. This is in contrast to integrated circuits in which individual parts may contain thousands of transistors.

When you have finished this chapter you will know how semiconductor materials can form a PN junction, which is the heart of many solid-state devices and circuits. You will understand the operation and use of one of the simpler semiconductor discrete devices—the diode. You will know how to connect a diode to other components to produce a power supply. Finally, you will be able to explain how devices such as capacitors and zener diodes are used to convert pulsating direct current into almost pure direct current.

## 23-1 SEMICONDUCTOR MATERIALS

In a pure semiconductor crystal, sometimes called an *intrinsic crystal,* the four valence electrons of an atom are covalently bonded to the valence electrons of four adjacent atoms. This is illustrated in Fig. 23-1, which shows only the nucleus and the valence electrons of the atoms of a silicon crystal. An intrinsic crystal (semiconductor) is not a conductor because it requires a relatively large in-

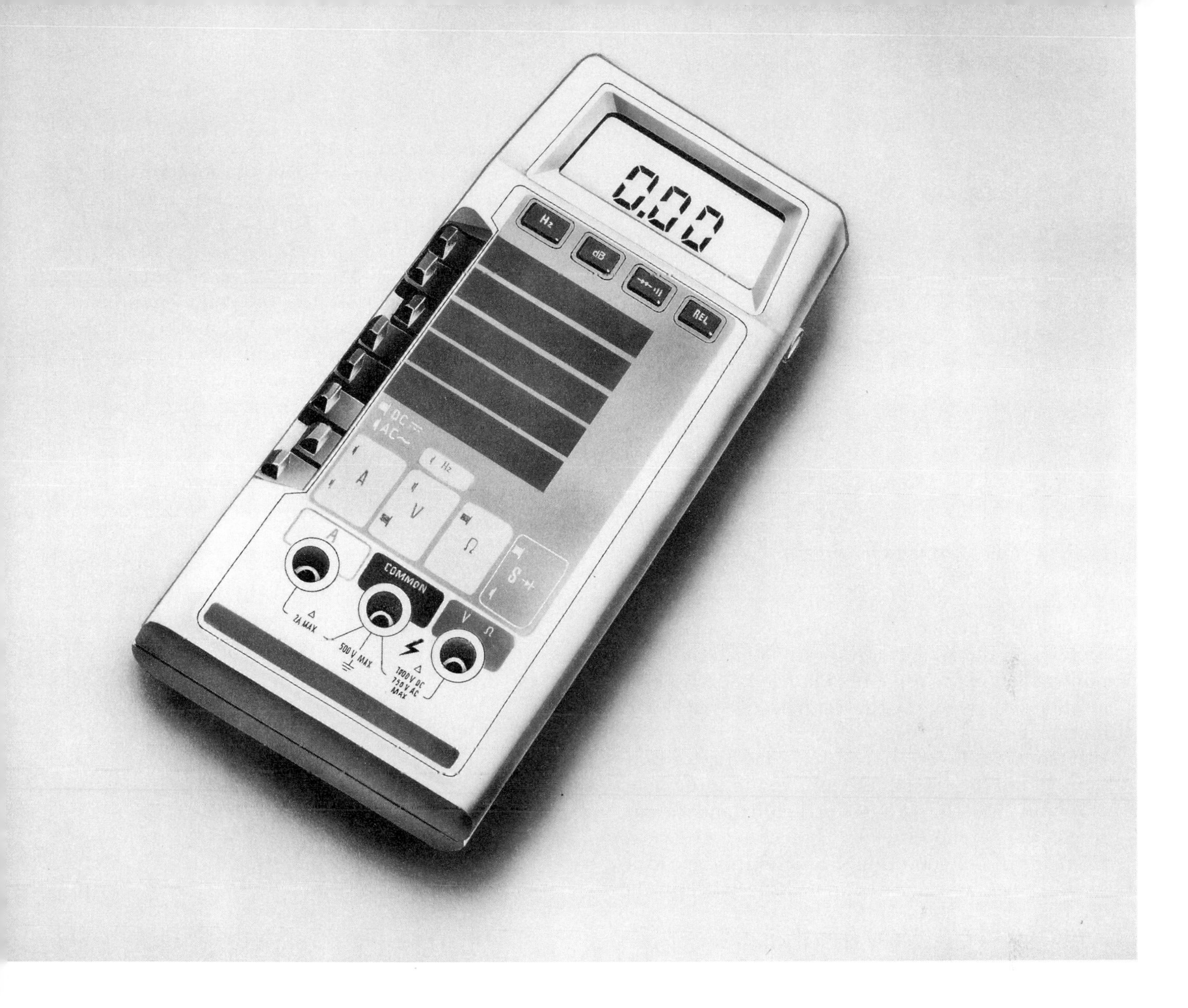


crease in energy level to cause an electron to break a covalent bond and move to the conduction band. On the other hand, an intrinsic crystal is not an insulator because, at any temperature above absolute zero, there are a significant number of electrons that break their covalent bonds as the crystal receives thermal energy.

### Electron-Hole Pair

The energy level of an electron involved in covalent bonding has to be increased in order for the electron to be raised to the conduction level. This gain in energy can be accomplished by subjecting the crystal to an external source of energy such as heat, light, or radiation energy. When the electron gains sufficient energy to break the covalent bond, an *electron-hole pair* is created. As shown in Fig. 23-2, the electron-hole pair provides two current carriers in the crystal: a free electron with a negative charge and a hole with an effective positive charge. The hole exists in the region of the crystal where the covalent bond is broken; it is positive because the atom from which the free electron came is now a positive ion.

In addition to having different electric charges, holes and free electrons are different in the way in which they move through the crystal as current carriers. The free electron can exist anywhere throughout the crystal; it is free to roam about in

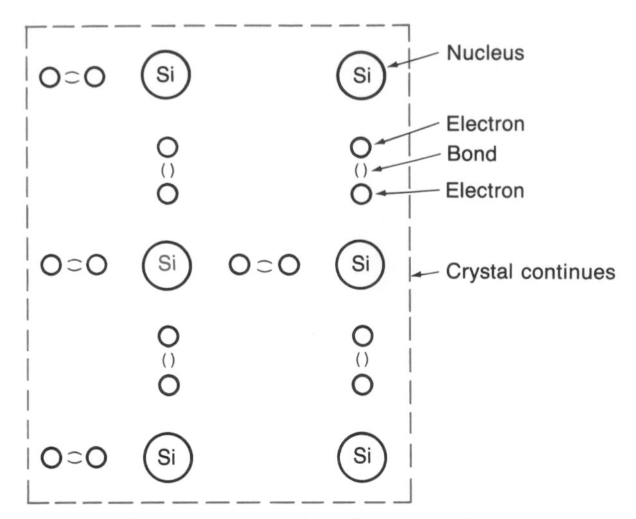


**FIGURE 23-1 Covalent bonding in a silicon crystal.**

the crystal structure. When subjected to an electric force field, the free electron can move toward the positive end of the force field independently of the locations of other charge carriers (electrons and holes).

In contrast, holes are restricted to discrete locations within the crystal (the ion, and thus the hole, is bound into the crystal's structure and cannot freely wander about). The hole can only move from one discrete location to another discrete location as holes are created and destroyed within the crystal.

How a hole travels in a crystal which is subjected to an electric force field (voltage) is illustrated in Fig. 23-3. In Fig. 23-3*a*, electron B gains sufficient energy to break its covalent bond and slide into hole A. As seen in Fig. 23-3*b*, the recombining of electron B and hole A eliminates hole A as the covalent bond is reestablished. However, at this same time, hole B has appeared in the location where electron B broke its original covalent bond. The end result, illustrated in Fig. 23-3*c*, is that a hole has traveled from location A to location B. Notice in this example that the energy level of the electron did not have to be raised to the conduction band before it could move into the adjacent hole. Rather, the electron stayed within the energy levels of the valance band as it moved between adjacent atoms. The only energy increase needed was that required to break the covalent bond. This is a good reason to refer to this type of charge movement as *hole current* so that it is easily distinguished from current carried by free electrons in the conduction band.

With an intrinsic crystal, there are as many holes as there are free electrons. Therefore, the hole current and the electron current are equal, but

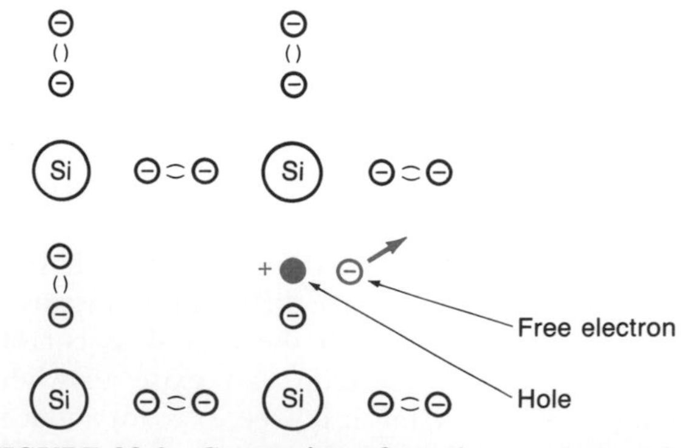


**FIGURE 23-2 Generation of an electron-hole pair.**

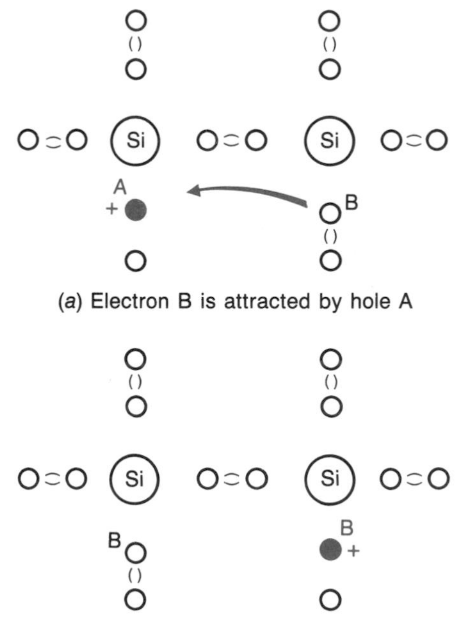


(*a*) Electron B is attracted by hole A

(*b*) Electron B breaks its covalent bond, generates hole B, and eliminates hole A.

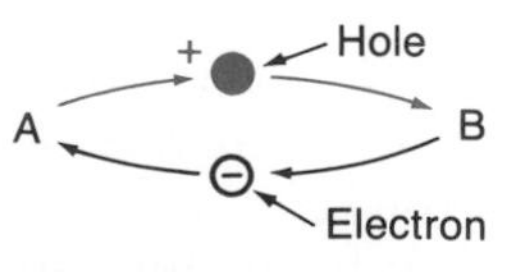


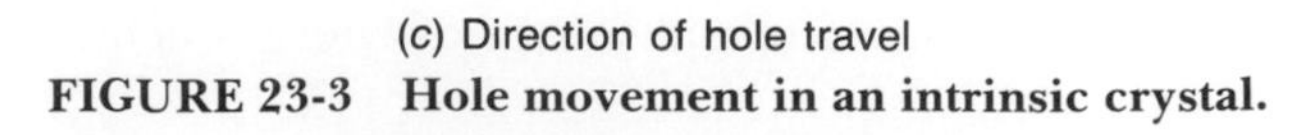

(*c*) Direction of hole travel

**FIGURE 23-3 Hole movement in an intrinsic crystal.**

they flow in opposite directions. It is as correct to speak of hole carriers and hole current as it is to speak of electron carriers and electron current. This is one justification for using conventional current flow, especially in textbooks on solid-state electronics, instead of electron flow.

## Temperature Characteristics of Semiconductors

Semiconductor materials have negative temperature coefficients because a warm crystal generates more electron-hole pairs than does a cool crystal. Of course, the greater the number of electron-hole pairs, the lower the resistance of the crystal.

Although two semiconductor materials may both have negative temperature coefficients, their temperature characteristics may be quite different. For example, at room temperature, germanium has about 1000 times as many current carriers (electron-hole pairs) as does silicon; therefore, at this temperature, silicon is more of an insulator than is germanium. However, germanium has a smaller temperature coefficient than does silicon. To double the conductivity of silicon requires approximately a 6°C increase in temperature, while doubling germanium's conductivity requires approximately a 10°C increase. Even though silicon has a higher temperature coefficient, it has almost entirely replaced germanium in semiconductor devices because of its much higher resistivity at room temperatures. Even at the elevated temperatures at which semiconductors are sometimes operated, the resistivity of silicon is still many times greater than that of germanium.

## Doping Intrinsic Crystals

To be useful in electronic devices such as diodes and transistors, semiconductor materials must be *doped* to increase their conductivity. Doping is the process of adding a few *impurity atoms* (atoms not possessing four valence electrons) to a pure semiconductor element such as silicon. Impurity elements used for doping semiconductors have either *trivalent* (three valence electrons) or *pentavalent* (five valence electrons) atoms. The level of doping in a semiconductor crystal varies with the intended use of the crystal. Typically it ranges from one impurity atom for every million to 100 million semiconductor atoms. Although this may seem like very light doping, it has a very significant effect on the resistivity of the crystal. For example, doping a pure silicon crystal at a ratio of 1:10 million decreases the crystal's resistivity from approximately 60 kΩ · cm to approximately 2 Ω · cm. This dramatic decrease in resistivity means that the doped crystal has also received a dramatic increase in current carriers. The additional carriers created by doping the crystal are much much less temperature-dependent than are the electron-hole pairs. Since these new carriers are now the dominant carriers, the doped crystal is much less temperature-sensitive than was the intrinsic crystal from which it was created.

## N-Type Crystal

Figure 23-4 illustrates the results of doping silicon (a semiconductor element) with arsenic (a pentavalent impurity element). In this figure only the nucleus and valence electrons of the atoms are shown. Pentavalent atoms, in this case, arsenic atoms, are called *donor atoms* because the fifth valence electron of the donor atom is not involved in covalent bonding. This electron then is much easier to free from the parent atom than is a covalently bonded electron. The pentavalent atom has donated a free electron to the crystal; therefore, the doped crystal has a lower resistivity than does the pure crystal.

Crystals which are doped with donor atoms are called *N-type crystals* because the great majority of the available current carriers are negative, i.e., electrons. The number of electron-hole pairs is very small compared to the number of free electrons donated by the pentavalent atoms. Therefore, in the N-type crystal, electrons are called *majority carriers* and holes are known as *minority carriers*.

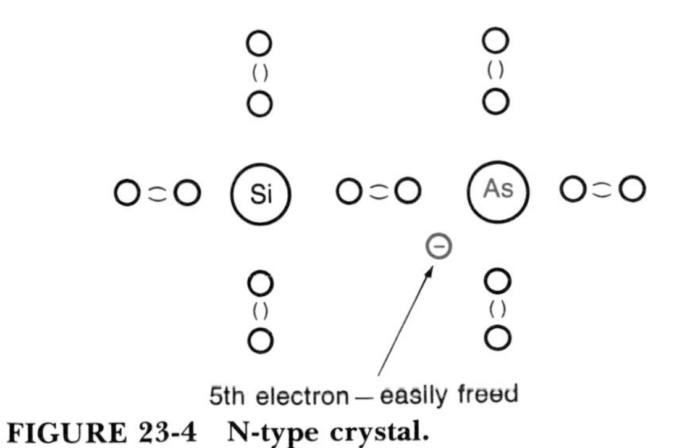


**FIGURE 23-4 N-type crystal.**

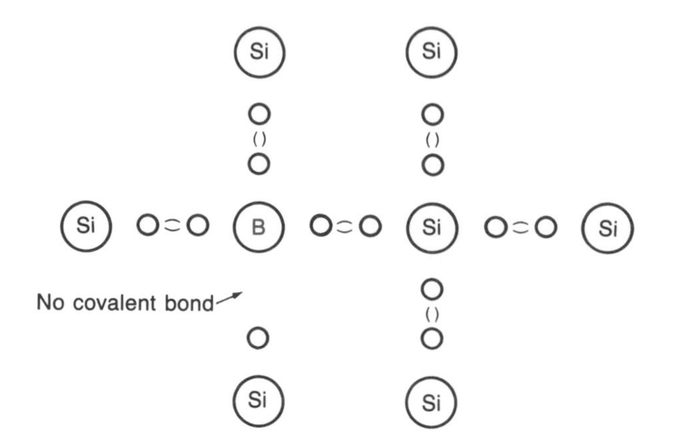


(*a*) Silicon doped with boron leaves one electron unbonded.

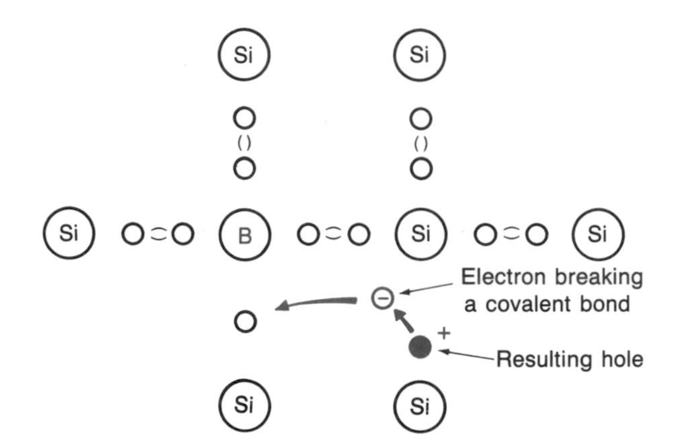


(*b*) An electron is attracted into the unbonded region of the crystal.

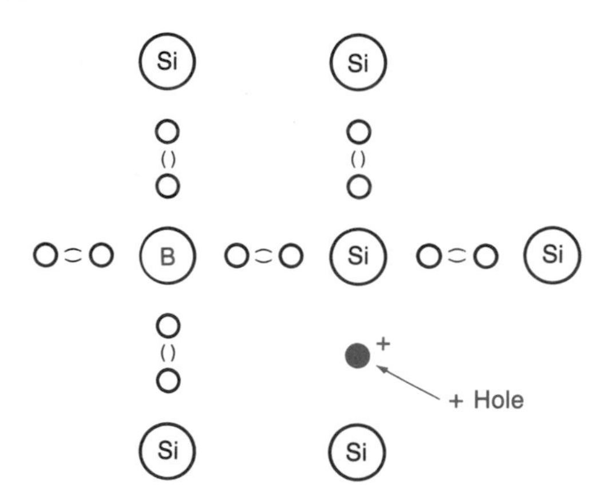


(*c*) A positive hole which is unpaired by a free electron.

**FIGURE 23-5 Adding a trivalent element to a semiconductor crystal.**

Arsenic, antimony, and phosphorus are some of the pentavalent elements used as impurities in doping N-type crystals.

## P-Type Crystal

The *P-type crystal* is formed by doping the intrinsic crystal with a trivalent element such as boron, indium, gallium, or aluminum. Referring to Fig. 23-5*a* will show that the area of the crystal containing a trivalent atom is short one electron for the number of electrons required to tie up all of the electrons of adjacent atoms in covalent bonds. Thus, one of the silicon atoms has an electron which is not covalently bonded. This area of the unbonded electron has a strong affinity (attraction) for another electron so that the covalent bonding pattern of the crystal can be completed and each semiconductor atom (silicon in this case) can have eight shared valence electrons, which is a stable state for atoms. Note that this attraction for another electron is due to incomplete bonding rather than a positive charge.

Now suppose that an electron from a silicon atom, located next to the boron atom, breaks its covalent bond and is attracted to the unbonded electron next to the boron atom (see Fig. 23-5*b*). The two electrons will form a new covalent bond, and all the electrons will again be locked up in covalent bonds. It should be noted that pulling an electron from the silicon atom to complete the bonding with the boron atom requires only a small increase in energy level because the energy level of the valence band of the boron atom is only slightly greater than that of the silicon atom. As seen in Fig. 23-5*c*, the end result is that a + hole now exists in the crystal to serve as a current carrier. Furthermore, note that this hole is not matched by a free electron. This creation of + holes occurs throughout the crystal wherever a trivalent atom is located. Thus, the crystal will contain many positive holes for which there are no free electrons. Of course, there are negative boron ions locked in the crystal structure which electrically balance out the positive holes.

The trivalent atoms used for doping P-type crystals are called *acceptor* atoms because they accept an extra electron to complete the covalent bonding.

In addition to the positive holes created by doping the crystal, the P-type crystal will, of course, have additional current carriers caused by electron-hole pairs that are thermally generated. As in the N-type crystal, the number of these electron-hole pairs is small compared to the number of carriers (holes in the case of the P-type crystal) created by the doping process. Thus, for the P-type crystal, holes are the majority carriers, and electrons are the minority carriers.

## Self-Test

**1.** Select the word or phrase from the right-hand column that is described by the statement in the left-hand column.

| | |
|---|---|
| **a.** Thermally generated current carriers. | Pentavalent element |
| **b.** Elements with four valence electrons | Intrinsic crystal |
| **c.** Contains electrons at the highest energy level | Electron-hole pairs |
| **d.** A pure germanium crystal | Doping |
| **e.** Produces holes in a crystal | Semiconductors |
| **f.** Produces free electrons in a crystal | Conduction band |
| **g.** Adding impurity elements | Trivalent element |

**2.** Is a boron atom a donor or an acceptor atom?

**3.** Which element, silicon or germanium, has the largest (a) temperature coefficient? (b) resistivity?

**4.** Does a pentavalent element produce a P-type material or an N-type material?

## 23-2 PN JUNCTION

Combining P-type material and N-type material into a single crystal produces a *PN junction* at the interface of the two types of material. The interface cannot be simply mechanical-physical; rather, it must be a region in a continuous crystal where the doping atoms change from trivalent to pentavalent.

The structure of a PN junction is illustrated in Fig. 23-6, which shows *only* the holes in the P-type material and the free electrons in the N-type material. Notice that in the junction region of Fig. 23-6 the holes and electrons are replaced by + and − signs. These + and − signs represent electric charges in the junction areas. Further, notice that the + signs are in the N-type material and the −

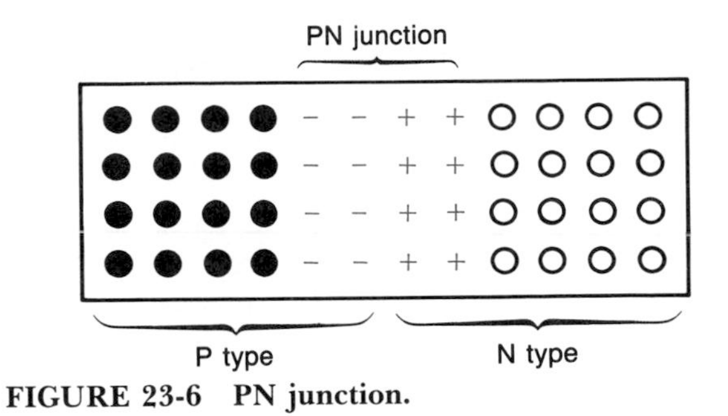


**FIGURE 23-6 PN junction.**

signs are in the P-type material. Although there are negatively charged and positively charged areas in the PN junction, the whole crystal is neutral because the positive and negative charges are equal.

How the negative and positive charges are built up in the PN junction is illustrated in Fig. 23-7. As the PN junction is being formed, electrons from the N-type material are attracted to the holes in the P-type material. When the electron from the arsenic atom in Fig. 23-7 breaks free and crosses into the P-type material, it leaves a + ion behind in the N-type crystal. When the electron from the arsenic atom is recombined with a hole in the P-type material, it eliminates the hole (and the + ion associated with it). The net effect is that the P-type crystal now has an excess negative charge locked into the crystal structure at the junction. At the same time the N-type crystal, at the junction surface, also has an excess positive charge locked into its structure.

The remaining free electrons in the N-type material near the junction surface have two forces acting on them. One force is that of the remaining holes in the P-type material trying to pull them across the junction. The other force is that of the negative charge (built up on the P side of the junction) repelling the free electrons. Free electrons from the N side will continue to cross over to the P side and cancel holes until the two forces are equal. Once the forces of attraction and repulsion on the free electrons have equalized, the junction is formed in its final state.

Notice in Fig. 23-7*b* that there are no current carriers (free electrons or holes) left in the junction region of a PN junction. Because the junction is depleted of carriers, the junction region is often referred to as the *depletion region.* Except for the occasional electron-hole pair that is generated by thermal energy, the depletion region can com-

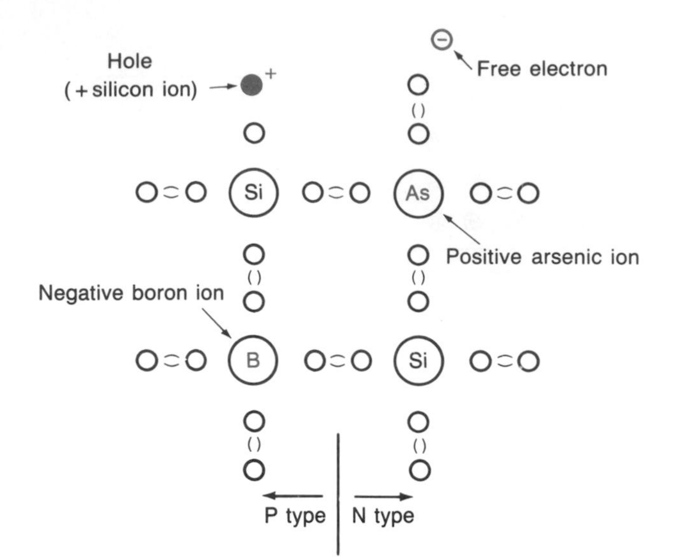


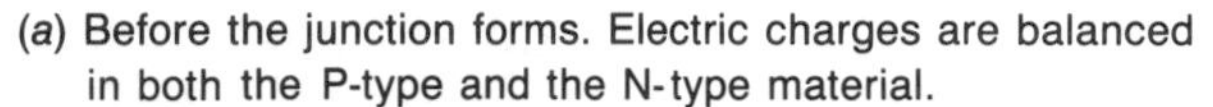
(*a*) Before the junction forms. Electric charges are balanced in both the P-type and the N-type material.

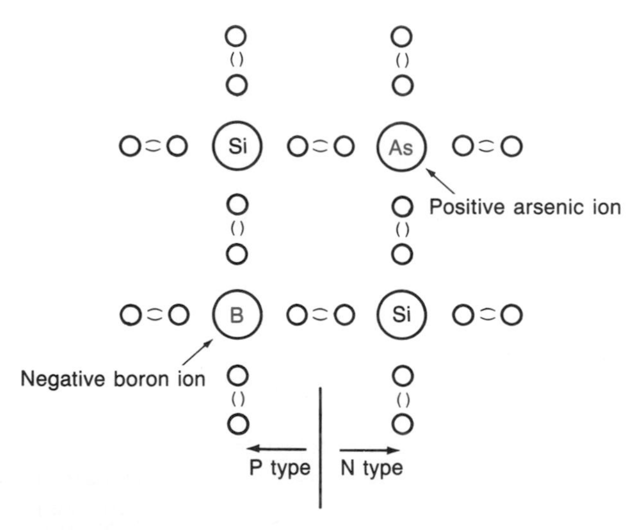


(*b*) After the junction forms. The P side has a net negative charge and the N side has a net positive charge.

**FIGURE 23-7 Electric charges forming in the PN junction region.**

pletely stop current in a circuit which contains a series PN junction.

The accumulation of opposite electric charges in the junction region causes a potential (voltage) to develop across the junction. This junction voltage is called a *potential barrier* because the junction potential is a barrier to current flowing through the junction. The potential is about 0.7 V for a silicon PN junction and about 0.2 V for a germanium PN junction.

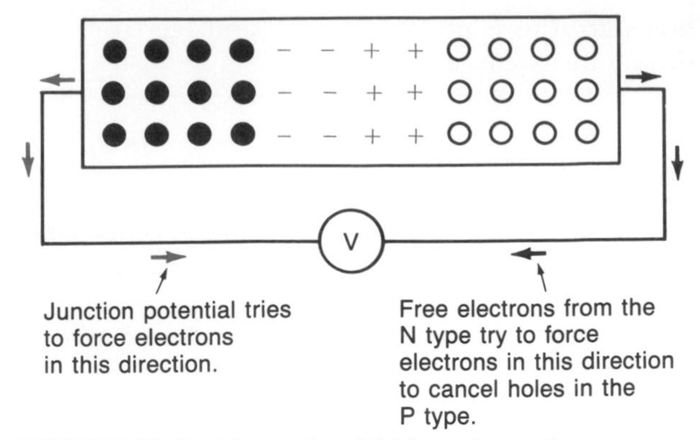


**FIGURE 23-8 Measuring PN-junction voltage.**

Although the PN junction produces a voltage (potential barrier), the value of this voltage cannot be measured with a voltmeter. Why this voltage cannot be measured is illustrated in Fig. 23-8. When the N-type material with its free electrons is externally connected, through the voltmeter, to the P-type material with its positive holes, there is a force (i.e., a potential) trying to push electrons in a clockwise direction. In other words, there is a potential between the external ends of the crystal that is equal in magnitude, but opposite in polarity, to the potential barrier of the junction.

## Reverse Bias

When a voltage is applied across a PN junction with the polarity shown in Fig. 23-9*b*, the junction is *reverse-biased*. Note that reverse bias requires that the P-type material be connected to the negative terminal of the source voltage, and the N-type to the positive terminal.

Reverse-biasing a PN junction causes the depletion region of the junction to widen. Widening of the depletion region, as illustrated in Fig. 23-9*b*, occurs because the external voltage forces additional electrons into the P side of the crystal and pulls additional electrons out of the N side of the crystal. As the depletion region widens, the junction potential increases. The widening of the depletion region, and the movement of electrons into and out of the crystal, stops when the junction potential equals the potential of the external voltage source. Once these two potentials are equal, there is no current flow in the circuit except for a very minute current supported by a few electron-hole pairs which are thermally generated in the deple-

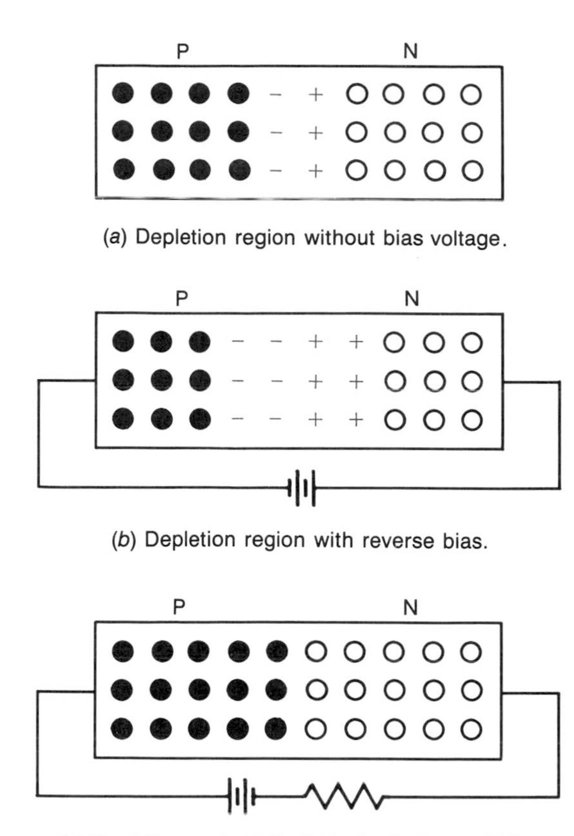


(*a*) Depletion region without bias voltage.

(*b*) Depletion region with reverse bias.

(*c*) Depletion region eliminated with forward bias

**FIGURE 23-9 Biasing a PN junction.**

tion region. For a silicon PN junction at room temperature, this current is typically less than a nanoampere. The minute current that flows under reverse-biased conditions is called *leakage current* or *reverse current*. Because the leakage current is carried by thermally generated electron-hole pairs, its magnitude is essentially controlled by the temperature of the crystal. Until the applied voltage exceeds a certain critical level, the amount of reverse current is nearly independent of the applied voltage. If all of the available electron-hole pairs are already being used as current carriers, increasing the voltage does not increase the current.

As the value of the reverse bias is increased, a critical value is reached when the reverse current dramatically increases. This condition is referred to as *reverse breakdown* of the junction. Unless sufficient external series resistance is included in the circuit, the reverse current increases to the point where it overheats and destroys the crystal. The dramatic increase in current at the critical value of reverse voltage is caused by the production of current carriers in addition to the thermally generated electron-hole pairs. Two phenomena can free additional carriers in the depletion region. One is the *avalanche effect* and the other is the *zener effect*. Avalanche is started by a thermally generated free electron, which is accelerated by the strong voltage gradient in the depletion region, colliding with covalently bonded electrons. The collision provides the covalently bonded electrons with enough energy so that they move to the conduction band. These new free electrons are then accelerated and collide with other covalently bonded electrons, and so on. The zener effect, which usually occurs at lower voltages than does avalanche, is caused by the electric force field in the depletion region being strong enough to tear electrons free of their covalent bonds and raise them to the conduction band. Which phenomenon, zener or avalanche, causes junction breakdown and the resulting high current is determined by such factors as doping concentration and junction geometry. PN junctions which are designed to be continuously operated in the reverse breakdown condition are known as *zener diodes* regardless of whether the breakdown is caused by the avalanche or zener effect.

## Junction Capacitance

A reverse-biased junction possesses capacitance. In fact, specially designed PN junctions (called *varactor diodes*) are used in some electronic circuits solely for their capacitance. If you remember that a capacitor is nothing but two conductive plates separated by an insulator, you can understand why a PN junction provides capacitance. Refer to Fig. 23-10 and you can see that the junction is two poor conductors (P-type material and N-type material) separated by a fair insulator (the depletion region). Also, notice in Fig. 23-9*b* that reverse bias causes

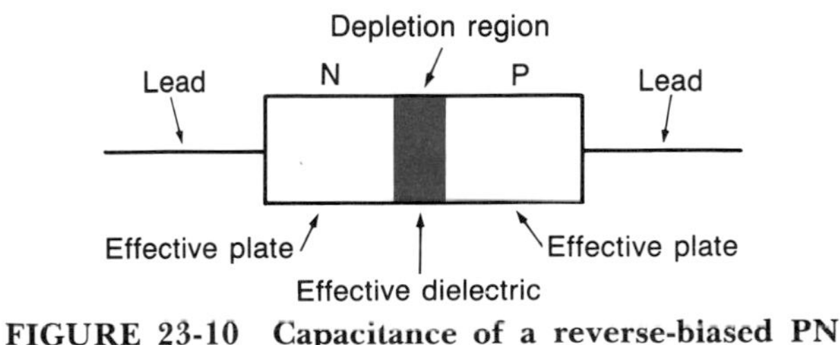


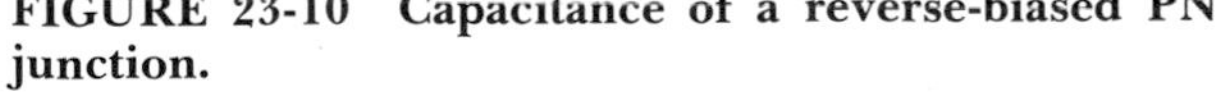

**FIGURE 23-10 Capacitance of a reverse-biased PN junction.**

the depletion region to become wider. Widening the depletion region has the same effect on junction capacitance as increasing the thickness of the dielectric (insulator) of a capacitor has on its capacitance, i.e., either one reduces the capacitance. Now you can understand why a reverse-biased junction can be used like a variable capacitor. Increasing the reverse bias decreases the capacitance and vice versa.

## Forward Bias

*Forward bias* (see Fig. 23-9*c*), eliminates the depletion region and allows the PN junction to conduct current. The depletion region is eliminated because the forward bias forces electrons in the N-type material and holes in the P-type material into the depletion region. Forcing holes into the depletion region cancels the negative charge that was built up at the junction edge of the P-type material. Likewise, the electrons cancel the positive charge at the edge of the N-type material.

Once the negative and positive charges that formed the depletion region have been eliminated, the holes from the P-type material and the electrons from the N-type material continue to meet at the junction of the two materials. As they meet, they combine and are eliminated. Thus, no new charge areas build up, and the electrons and holes continue to meet and combine at the junction as long as forward bias is applied. The current carriers in the forward-biased junction are majority carriers which are plentiful. Therefore, the current can become destructively large if not controlled. The resistor in Figure 23-9*c* is needed to limit the forward current to a safe value.

When the forward bias is removed, the depletion region is reestablished. The areas of negative and positive charges build up the same way as they did when the PN junction was first formed.

## Self-Test

**5.** Are the following statements true or false:

**a.** A PN junction is formed by mechanically holding a P-type crystal against an N-type crystal.

**b.** A positive charge builds up on the P side of a PN junction.

**c.** The potential barrier of a silicon PN junction is greater than that of the germanium PN junction.

**d.** The depletion region of a PN junction has a very high resistance much like that of an insulator material.

**e.** Increasing the reverse bias on a PN junction narrows the depletion region.

**f.** Leakage current in a junction is caused by thermally generated electron-hole pairs.

**g.** To forward-bias a PN junction, the P-type material is connected to the positive terminal of the external voltage source and the N-type material is connected to the negative terminal.

**6.** Why is a resistor placed in series with a PN junction when forward bias is applied to the junction?

**7.** What is the function of a varactor diode?

**8.** What happens to the depletion region of a PN junction when the junction is forward-biased?

## 23-3 DIODES

A *diode* is a component that allows current to freely flow through it in one direction but essentially stops any current from flowing in the reverse direction. From this description of a diode, you can see that a PN junction has all the characteristics of a diode. Thus, all of the previous discussion of the PN junction and its characteristics applies to the diode. The legend for Fig. 23-10 could just as well be "Capacitance of a reverse-biased diode."

Nearly all diodes used in electronics today are silicon PN junctions. Silicon is preferred because of its lower leakage current. There are also a few diodes made from germanium PN junctions. They are primarily used when the lower potential barrier voltage of the germanium junction is advantageous.

Figure 23-11 shows a cutaway view of a typical diode that is capable of carrying 2 A of current. One of the diode leads connects to the P-type material and the other to the N-type material. The silicon PN junction is very small: about the diameter of the lead wire and a few thousandths of an inch thick. The body of this plastic-encased diode is about $\frac{1}{8}$ in. in diameter and about $\frac{5}{16}$ in. long. The band on the left end of the body identifies the *cath-*

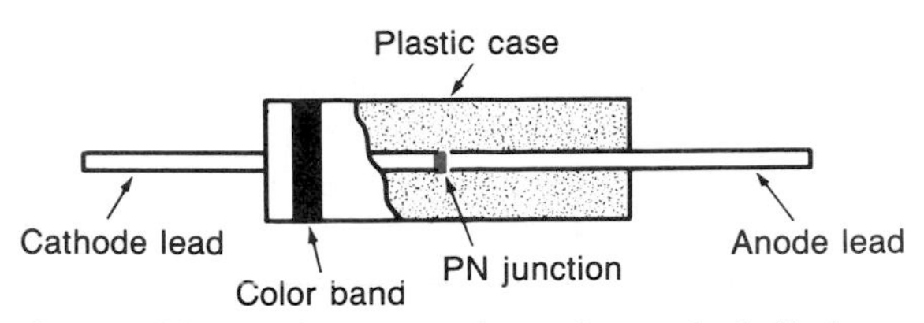


**FIGURE 23-11 Cutaway view of a typical diode.**

*ode* end of the diode. The cathode end of a diode is N-type material, and the anode end is always the P-type material. Thus, to forward-bias a diode, the anode must be positive with respect to the cathode.

The schematic symbol for a diode is shown in Fig. 23-12*a*. The cathode is represented by the vertical straight line and the anode by the triangle. Figure 23-12*b* and *c* illustrate forward-biasing and reverse-biasing, respectively.

Diodes are available in a variety of physical shapes and sizes, and the junction may be enclosed in plastic (as in Fig. 23-11), glass, ceramic, or metal. Some typical shapes are illustrated in Fig. 23-13. The first four diodes in Fig. 23-13 are drawn with the cathode on the left. The last two [(E) and (F)]

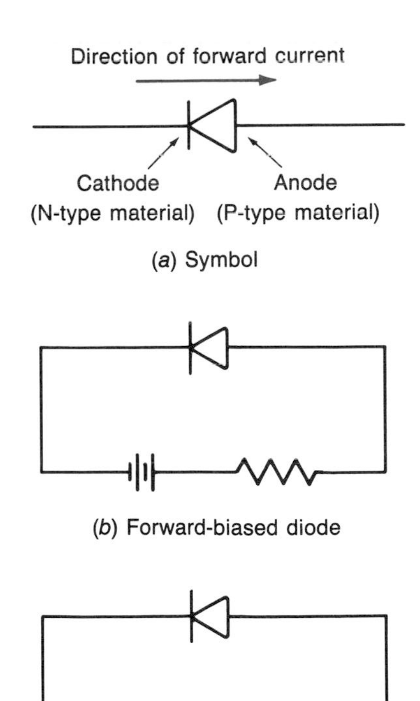


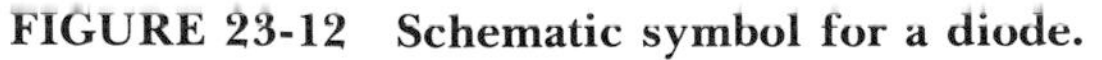

**FIGURE 23-12 Schematic symbol for a diode.**

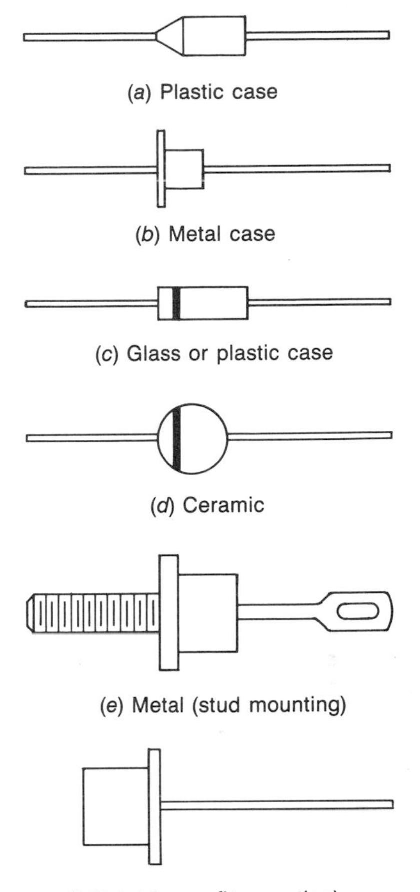


**FIGURE 23-13 Common diode case styles.**

can be obtained with either the cathode or the anode connected to the case. Often the diode symbol is printed on the case to indicate which lead is the cathode and which is the anode.

## Testing Diodes with an Ohmmeter

The cathode and anode leads of a diode can always be determined by testing the diode with an *ohmmeter*. First, however, one has to know the polarity of the ohmmeter leads. For most ohmmeters, the common (black) lead is negative. If in doubt, check the ohmmeter polarity by measuring the voltage between the ohmmeter leads with a voltmeter as shown in Fig. 23-14. If the pointer moves up-scale, the negative leads of the two meters are connected together, as are the positive leads. Also, the ohmmeter polarity can be determined by measuring the forward resistance and the reverse resistance

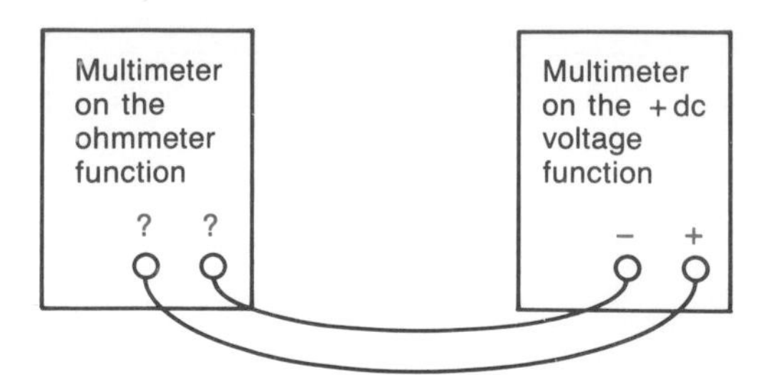


**FIGURE 23-14 Determining the polarity of an ohmmeter's leads by measuring the lead voltage.**

of a diode for which the cathode and anode leads are positively identified. The ohmmeter in Fig. 23-15*a* is indicating less than 10 Ω; therefore, the diode must be forward-biased (or shorted). To be forward-biased, the negative lead of the ohmmeter has to be connected to the cathode and the positive lead connected to the anode of the diode. If the resistance reading is very high (>100 MΩ) as indicated in Fig. 23-15*b*, then the diode must be reverse-biased (or open). Therefore, the negative lead of the ohmmeter must be connected to the anode. Obviously, if the ohmmeter polarity is known, the above reasoning can be used to identify the cathode and anode leads of an unmarked diode.

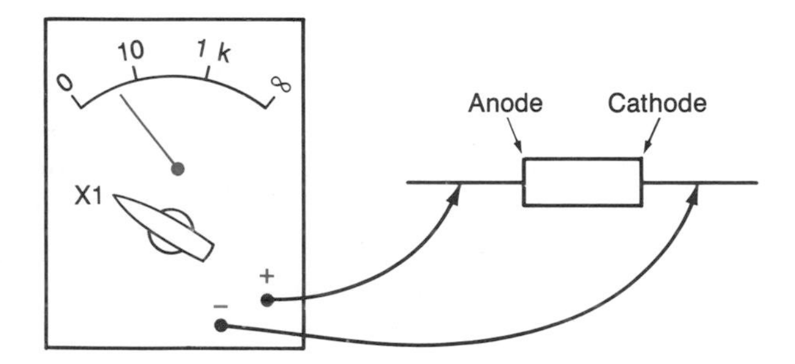


(*a*) Diode is forward-biased by the ohmmeter.

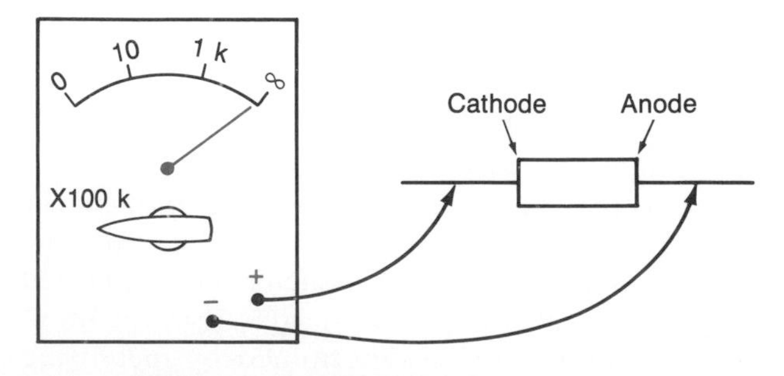


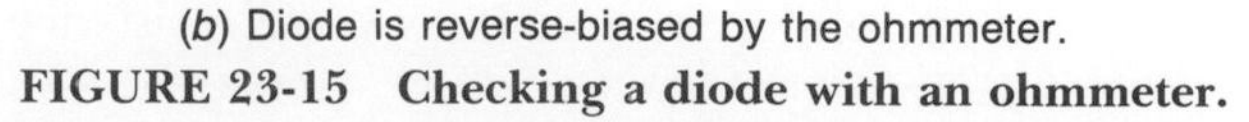

(*b*) Diode is reverse-biased by the ohmmeter.

**FIGURE 23-15 Checking a diode with an ohmmeter.**

A faulty diode can be quickly detected with the ohmmeter. If the diode's resistance remains very large when the ohmmeter leads are reversed, the diode is open. Conversely, if the diode's resistance is very small regardless of the lead polarity, the diode is shorted.

## Types of Diodes

A number of specific types of diodes are manufactured for specific applications in electricity and electronics. Some of the more common types are rectifier diodes, zener diodes, varactor diodes, switching diodes and signal diodes. All of these diodes except the zener diode are represented by the symbol shown in Fig. 23-12. The zener diode symbol, shown in Fig. 23-16, uses a different line configuration to represent the cathode.

*Rectifier diodes* are used in circuits which convert alternating current into pulsating direct current. These diodes are generally used in low-frequency, high-current circuits. Thus, the junction capacitance is of little concern with these diodes.

*Zener diodes* are used to regulate dc voltages. That is, a zener diode can maintain a nearly constant dc voltage under widely varying load conditions. Zener diodes operate in the reverse breakover mode.

*Varactor diodes* are designed to have a relatively large value of junction capacitance. They are operated with reverse bias, but the bias is kept well below the breakover value. Varactors are often used in parallel with *LC* circuits to tune (change the frequency of) the circuit to a specified frequency.

*Switching diodes* are often used in digital electronic circuits. They are designed to operate at high frequencies. That is, they must be able to turn on (conduct in the forward direction) and turn off very rapidly. Thus, junction capacitance is reduced to a minimum in this type of diode because the diode cannot stop conducting when reverse bias is

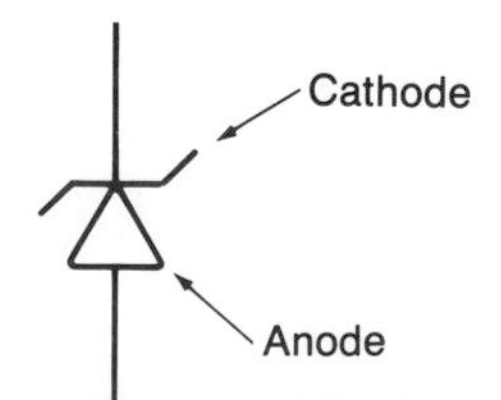


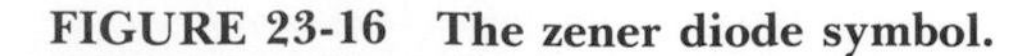

**FIGURE 23-16 The zener diode symbol.**

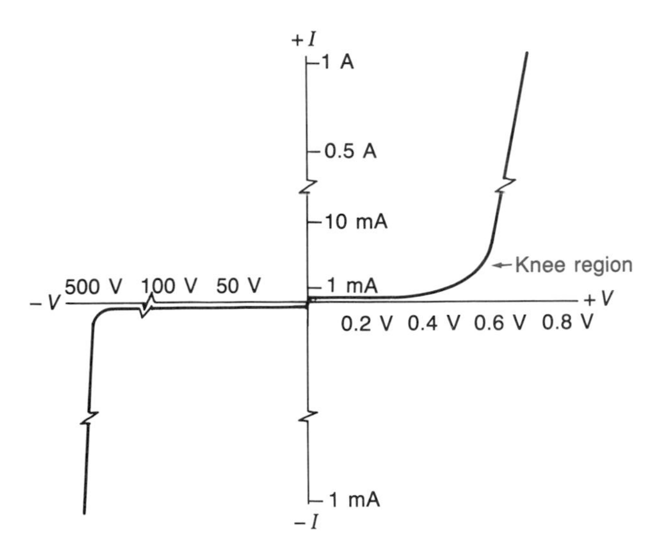


**FIGURE 23-17 Characteristic curve for a silicon diode.**

applied until the junction capacitance has been charged.

*Signal diodes* are low-current, low-forward-voltage, low-capacitance diodes. They are used in high-frequency circuits such as detectors for am and fm radio receivers.

## Diode Curves

The current-versus-voltage curve for a typical silicon diode is shown in Fig. 23-17. In this figure, reverse-bias current and voltage are represented by the $-I$ and $-V$ symbols, while forward-bias conditions are indicated by $+I$ and $+V$. Notice that the diode conducts very little current in the forward direction until the voltage is increased to about 0.6 V; then, the curve bends quite sharply, and the current increases dramatically as the voltage is increased to 0.7 V. The area of the curve where the slope is changing very rapidly is referred to as the *knee* region. It is the region where the applied, external voltage is reaching the value of the potential barrier of the junction. In this voltage range the majority carriers are increasingly combining in the junction.

Also notice in Fig. 23-17 that the value of the reverse current is small compared to the forward current. Before the diode breaks over and conducts in the reverse direction, the reverse leakage current is in the picoampere range. Once the diode breaks over, the current must be kept small enough to ensure that the dissipated power stays within the diode's power rating. Most diodes are not designed to be operated in the reverse breakover mode.

## Static and Dynamic Opposition

A graph which plots voltage versus current, such as the diode curve of Fig. 23-17, describes the resistance of the device, or, if reactance is also involved, the impedance of the device. Since the concept of dynamic and static opposition is the same for either resistance or impedance, let us use resistance to keep the discussion as simple as possible.

*Static resistance* refers to the opposition of a device when a steady-state (pure dc) voltage is applied. *Dynamic resistance* refers to the opposition of a device when a changing voltage (either fluctuating dc or ac voltage) is applied. With a nonlinear device, such as a diode, the two are quite different.

So far in your study of electricity you have been concerned with calculating static resistance with the formula $R = V/I$. Dynamic resistance, sometimes called *ac resistance*, is calculated with the formula $r - \Delta V/\Delta I$, where $r$ is the symbol for dynamic resistance, and $\Delta$ is the symbol which means "a change in." Thus, $\Delta V$ is read as "the change in voltage." Referring to Fig. 23-18*a* to obtain $I$ and $V$ values for these formulas shows that the static and dynamic resistances of a resistor are the same. Also, it will be seen that the static resistance is the same at points A, B, and C. The static resistance at points A, B, and C is, respectively,

$$R = \frac{50\text{ V}}{1\text{ mA}} = 50\text{ k}\Omega \qquad R = \frac{100\text{ V}}{2\text{ mA}} = 50\text{ k}\Omega$$

and
$$R = \frac{150\text{ V}}{3\text{ mA}} = 50\text{ k}\Omega$$

The dynamic resistance as the voltage changes between points A and B, B and C, and A and C is, respectively,

$$r = \frac{100\text{ V} - 50\text{ V}}{2\text{ mA} - 1\text{ mA}} = 50\text{ k}\Omega$$

$$r = \frac{150\text{ V} - 100\text{ V}}{3\text{ mA} - 2\text{ mA}} = 50\text{ k}\Omega$$

and
$$r = \frac{150\text{ V} - 50\text{ V}}{3\text{ mA} - 1\text{ mA}} = 50\text{ k}\Omega$$

Using data from Fig. 23-18*b* shows that for a zener diode (a nonlinear device), the dynamic and static resistances are quite different. Also, for the

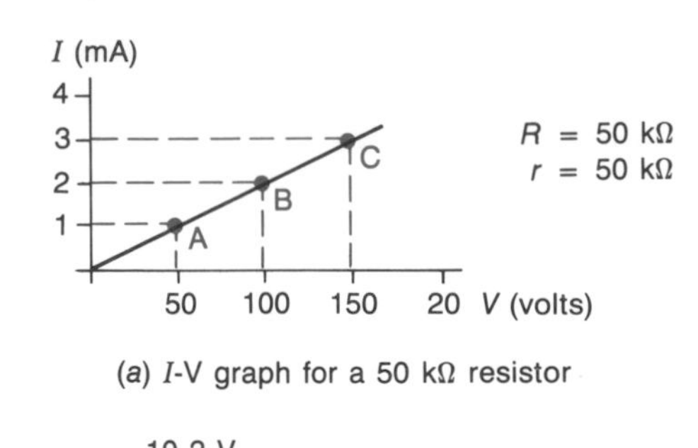

(a) I-V graph for a 50 kΩ resistor

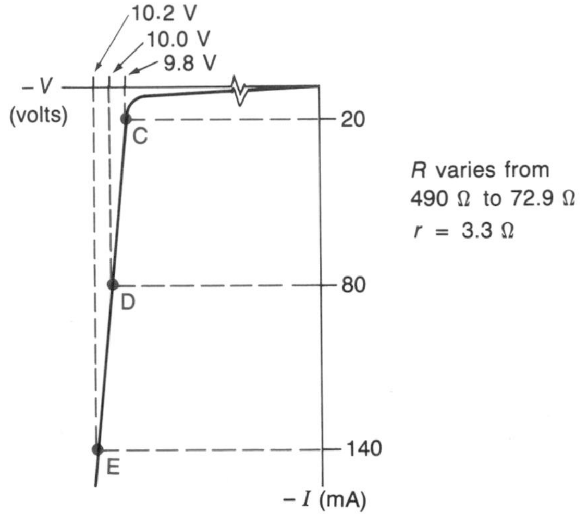

(b) I-V graph for a reverse-biased 10-V zener diode.

**FIGURE 23-18 Static and dynamic resistance of linear and nonlinear devices.**

nonlinear device, the static resistance changes when the current or voltage change. Determining the static resistance at points C, D, and E in Fig. 23-18*b* yields these values:

At point C, $$R = \frac{9.8\ \text{V}}{20\ \text{mA}} = 490\ \Omega$$

At point D, $$R = \frac{10\ \text{V}}{80\ \text{mA}} = 125\ \Omega$$

At point E, $$R = \frac{10.2\ \text{V}}{140\ \text{mA}} = 72.9\ \Omega$$

The dynamic resistance calculated from the data in Fig. 23-18*b* is

$$r = \frac{10\ \text{V} - 9.8\ \text{V}}{80\ \text{mA} - 20\ \text{mA}} = 3.3\ \Omega$$

$$r = \frac{10.2\ \text{V} - 10\ \text{V}}{140\ \text{mA} - 80\ \text{mA}} = 3.3\ \Omega$$

and $$r = \frac{10.2\ \text{V} - 9.8\ \text{V}}{140\ \text{mA} - 20\ \text{mA}} = 3.3\ \Omega$$

The above calculations show that as long as the slope of the line in a graph does not change, the dynamic resistance is constant regardless of the location and the magnitude of the change in voltage or current. However, when the slope of the line is changing, as in the knee region of a forward-biased diode, the values of both the static and dynamic resistances are dependent upon the segment of curve used in calculating them. For example, calculating static and dynamic resistances for the expanded knee region of the forward-biased-junction curve in Fig. 23-19 yields the following values:

At and around point A,

$$R = \frac{0.55\ \text{V}}{11\ \text{mA}} = 50\ \Omega$$

and $$r = \frac{0.6\ \text{V} - 0.5\ \text{V}}{18\ \text{mA} - 7\ \text{mA}} = 9.1\ \Omega$$

At and around point B,

$$R = \frac{0.65\ \text{V}}{28\ \text{mA}} = 23.2\ \Omega$$

and $$r = \frac{0.7\ \text{V} - 0.6\ \text{V}}{50\ \text{mA} - 18\ \text{mA}} = 3.1\ \Omega$$

Notice that both resistances decrease as the current increases. This fact is important when one is involved in such things as designing transistor amplifiers.

## Diode Specifications and Ratings

There are numerous ratings and specifications for diodes. Some of them apply to many types of diodes, while others are specific to a particular type

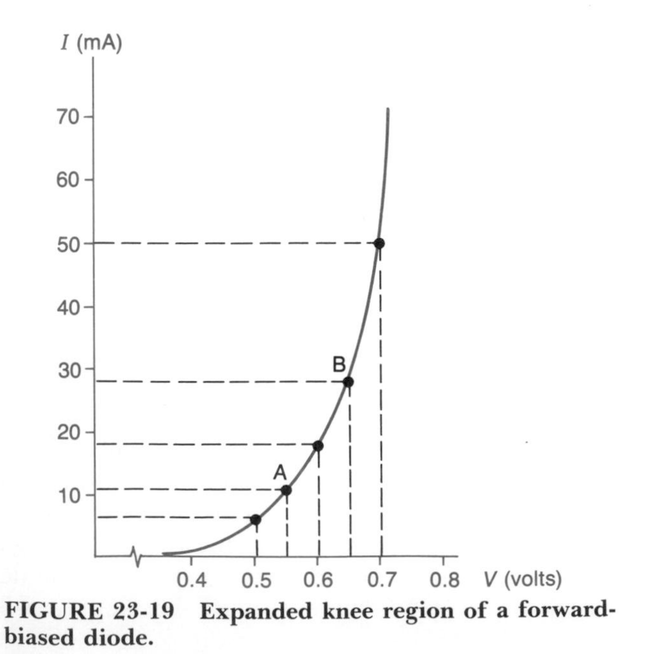

**FIGURE 23-19 Expanded knee region of a forward-biased diode.**

of diode. As a sample of the many ratings, we will look at the more important ratings needed to work with rectifier and zener diodes.

The rectifier diode, sometimes referred to as a *power diode,* has four important ratings: $I_f$, PIV or PRV, $I_{fm}$ or $I_s$, and $I_{fpr}$. Current $I_f$ is the symbol for the average current the diode can continuously carry in the forward direction. If $I_f$ is exceeded for extended periods of time, the diode overheats and the semiconductor crystal is destroyed. If the current overload is severe, the diode, both the crystal and the plastic or glass case, can break into two pieces. The abbreviation PIV stands for "peak inverse voltage," and PRV is the abbreviation for "peak reverse voltage." Both PIV and PRV specify the maximum amount of reverse-bias voltage the diode can withstand with no danger of avalanche (breakover) occurring. It is important not to exceed this rating because rectifier diodes are not made to operate in the reverse breakover mode. Current $I_{fm}$ or $I_s$ is the maximum instantaneous forward current (the surge current) that the diode should be subjected to. Current $I_{fm}$ is typically 10 to 30 times as large as is $I_f$. Current $I_{fm}$ is further restricted by a specified duration, usually the period of 60 Hz. Current $I_{fpr}$ is the peak repetitive current in the forward direction. The repetition rate and duration are usually specified for rectifying 60 Hz. Current $I_{fpr}$ is typically 5 to 10 times as large as $I_f$. In many common applications, the diode's $I_{fm}$ rating will not be exceeded if its $I_f$ rating is observed; therefore, it is quite common to only specify the $I_f$ and PIV rating of a diode.

The major ratings for zener diodes are $V_z$, $P_z$, and $r_z$. Voltage $V_z$ (the reverse breakdown voltage of the zener) is specified with a tolerance which is typically in the 1 to 10 percent range. Power $P_z$ is the power rating of the zener diode. It is numerically equal to the product of $V_z$ and $I_z$ (the current through the reverse-biased zener.) Therefore, if $V_z$ and $P_z$ are specified, one can calculate the maximum current the diode can safely handle. A zener can dissipate its rated power $P_z$ only if specified ambient and/or case temperatures are not exceeded. If these temperatures are exceeded, then $P_z$ must be derated as specified in the manufacturer's literature. The $r_z$ rating specifies the dynamic or ac resistance of the zener. The smaller this value is, the smaller will be the voltage change across the zener for a given current change through the zener. Although zener diodes behave like a rectifier diode when forward-biased, they are not rated for forward characteristics because they only operate in the reverse breakover mode.

**EXAMPLE 23-1**

The ratings for a zener diode are $V_z = 20\ \text{V} \pm 5$ percent, $P_z = 1$ W at 50°C ambient, and $r_z = 15\ \Omega$. Determine (a) the maximum reverse current at 40°C ambient, and (b) the change in voltage across the zener when the zener current changes 20 mA.

**Solution**

(a) $P_z = V_z I_z$ therefore $I_z = \dfrac{P_z}{V_z} = \dfrac{1\ \text{W}}{20\ \text{V}} = 50\ \text{mA}$

(b) $r_z = \dfrac{\Delta V_z}{\Delta I_z}$ therefore $\Delta V_z = \Delta I_z r_z$

$= 20\ \text{mA} \times 15\ \Omega$

$= 300\ \text{mV}$

## Self-Test

**9.** Why are silicon diodes more prevalent than germanium diodes?

**10.** To reverse-bias a diode, should the anode be negative or positive with respect to the cathode?

**11.** An ohmmeter connected to a diode indicates 6 Ω. Is the diode being reverse-biased or forward-biased by the ohmmeter?

**12.** In Question 11, is the negative lead of the ohmmeter connected to the cathode or the anode of the diode?

**13.** Which of the diodes discussed in Sec. 23-3 operates in the reverse breakdown mode?

**14.** Determine $r$ for the $I$-$V$ graph in Fig. 23-20.

**15.** Will the value of $R$ be the same as the value of $r$ for Fig. 23-20?

**16.** Will the value of $R$ be the same at any point on the graphed line in Fig. 23-20?

**17.** List four common ratings for a rectifier diode.

**18.** How much reverse current can a zener safely conduct at 40°C ambient temperature if its ratings are $P = 2$ W at 50°C ambient and $V_z = 5\ \text{V} \pm 1$ percent?

**19.** Should $r_z$ be as large as possible or as small as possible to provide the most stable voltage across a zener through which $I_z$ fluctuates?

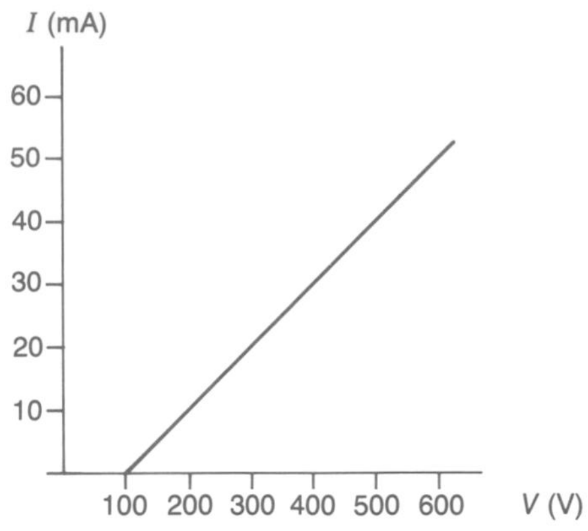


**FIGURE 23-20 Current-voltage graph for Self-Test Questions 14, 15, and 16.**

## 23-4 RECTIFICATION

The process of converting alternating current (or alternating voltage) into pulsating direct current (or pulsating direct voltage) is known as *rectification.* Rectification is accomplished with diodes. Circuits which provide rectification are called *rectifier circuits.* Rectifier circuits can provide either *half-wave rectification* or *full-wave rectification.*

### Half-Wave Rectification

The result of half-wave rectification is illustrated in Fig. 23-21*a*, and the circuit which performs the rectification is drawn in Fig. 23-21*b*. The ground symbol in Fig. 23-21*b* is the reference point for voltages referred to in the discussion which follows.

The terminal of the generator which connects to the anode of $D_1$ alternately swings from +10 to −10 V. When this terminal exceeds +0.7 V, the diode is forward-biased and current flows through $R_1$, dropping a voltage (with the indicated polarity) across $R_1$. Referring back to the diode characteristic curves of Fig. 23-17 shows that a forward-conducting diode drops about 0.6 to 0.8 V with the larger value occurring with larger currents. For typical low-current rectifiers, the diode drops about 0.7 V, which is the value we will use in our analysis of rectifier circuits. The voltage across the forward-biased diode will be represented by the symbol $V_D$. Thus, if the forward-biased diode drops 0.7 V, Kirchhoff's voltage law shows that 9.3 V must drop across the resistor when the generator swings to +10 V. Since the diode and resistor are in series, the resistance of the forward-conducting diode must be small compared to the resistance of the load $R_1$.

When the generator terminal drops to less than +0.7 V, the diode essentially stops conducting. The diode remains in the nonconducting state through the negative $\frac{1}{2}$ cycle because it is reverse-biased. The only current flowing is the leakage, or reverse, current which is so small it can be ignored. For example, if the leakage current is 8 pA and the load resistor $R_1$ is 100 Ω, then the negative voltage drop across $R_1$ would be

$$V_{R_1} = 8 \times 10^{-12}\ \text{A} \times 1 \times 10^{2}\ \Omega = 8 \times 10^{-10}\ \text{V}$$

Compared to the positive voltage of 9.3 V across $R_1$, this $8 \times 10^{-10}$ V can safely be ignored.

For half-wave rectifier circuits, the PIV rating of the diode must be at least equal to the peak voltage of the ac source being rectified. The diode $D_1$ in Fig. 23-21*b* must be able to block the voltage of the generator without going into reverse breakover. Therefore, the PIV rating of $D_1$ must be at least 10 V, which is the peak output of the generator.

**EXAMPLE 23-2**

A 25-V ac source is to be rectified. What is the minimum PIV rating for the diode?

**Solution**

$$V_p = 1.414\ V_{\text{rms}} = 1.414 \times 25\ \text{V} = 35.4\ \text{V}$$
$$\text{Minimum PIV} = V_p = 35.4\ \text{V}$$

where $V_p$ is peak voltage and $V_{\text{rms}}$ is root mean square voltage.

The calculated value of PIV in Example 23-2 is the minimum theoretical value. In actual practice, a diode with a 50-V PIV rating would be used. This higher rating would allow for fluctuations in the ac source voltage and for noise pulses which might be induced into the conductors of the ac source.

The forward current rating $I_f$ of the diode must also be considered. It must equal or exceed the average value of the dc current in the circuit. In Fig. 23-21, the peak current $I_p$ would be

$$I_p = \frac{V_p - V_D}{R} = \frac{10\ \text{V} - 0.7\ \text{V}}{100\ \Omega} = 93\ \text{mA}$$

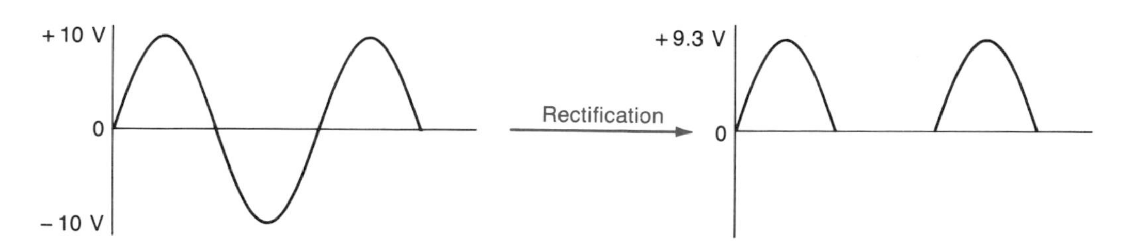


(*a*) Rectifying a 20 $V_{P\text{-}P}$ sinusoidal waveform yields a +9.3 $V_P$ pulsating dc waveform.

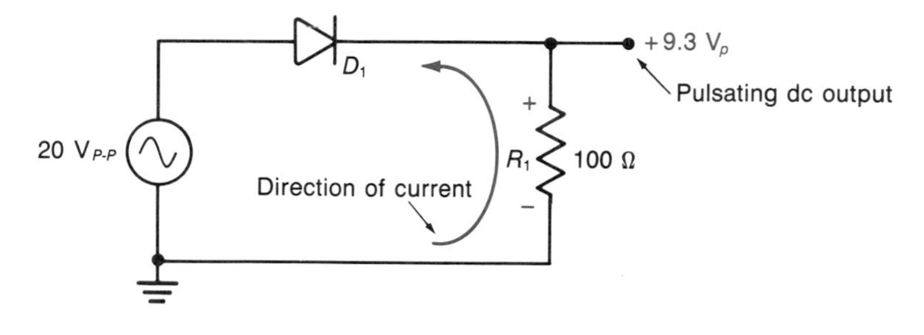


(*b*) Half-wave rectifier circuit

**FIGURE 23-21 Half-wave rectification.**

The pulse of dc current is sinusoidal; therefore, the average value of the pulse would be

$$I_{\text{av of pulse}} = I_p \times 0.637 = 93 \text{ mA} \times 0.637 = 59.2 \text{ mA}$$

However, the current flows for only one-half of the cycle, so the average pulse current must be divided by 2 to obtain the average over the total time between pulses. Thus, the average current $I_{dc}$ of the circuit, and the minimum $I_f$ of the diode, would be

$$I_{dc} = \frac{0.637\ I_P}{2} = \frac{0.637 \times 93 \text{ mA}}{2} = 29.6 \text{ mA}$$

Again, this is the minimum value for a theoretical circuit. To allow for the tolerances in a real circuit, the $I_f$ rating should be not less than 40 mA. The dc current function of a VOM reads average current. Thus, if the dc current in the circuit of Fig. 23-21*b* were measured with a VOM, it should indicate approximately 29.6 mA.

The average voltage (dc voltage) out of the circuit in Fig. 23-21*b* can be calculated in the same way as was the average current. That is,

$$\begin{aligned} V_{\text{av of pulse}} &= (V_p - V_D)\ 0.637 \\ &= (10 \text{ V} - 0.7 \text{ V}) \times 0.637 = 5.92 \text{ V} \end{aligned}$$

This voltage pulse must be averaged over the complete time between pulses, so the above value must be divided by 2. Thus, the dc output voltage for the circuit of Fig. 23-2*b* will be

$$\begin{aligned} V_{dc} &= \frac{(V_P - V_D)\ 0.637}{2} \\ &= \frac{(10 \text{ V} - 0.7 \text{ V})\ 0.637}{2} \\ &= 2.96 \text{ V} \end{aligned}$$

The dc output voltage can be measured with the VOM on the dc voltage function.

Noticc that it is unnecessary to calculate both $V_{av}$ and $I_{av}$ by the above formulas because we can use Ohm's law and the value of $R_1$ to calculate $I_{av}$ if $V_{av}$ is known or vice versa. For example, if we have calculated that $V_{av}$ for Fig. 23-21*b* is 5.92 V, then by Ohm's law we find that

$$I_{av} = \frac{V_{av}}{R_1} = \frac{5.92 V_{av}}{100\ \Omega} = 0.0592 \text{ A} = 59.2 \text{ mA}$$

In summary, the formulas for figuring half-wave-rectifier-circuit values are

$$V_{dc} = \frac{0.637\ (V_P - V_D)}{2}$$

where $V_{dc}$ = dc output voltage
$V_p$ = peak value of ac voltage being rectified
$V_D$ = voltage drop across the conducting diode

$$I_{dc} = \frac{0.637 I_P}{2} = \frac{0.637(V_P - V_D)}{2R_L}$$

where $I_{dc}$ = average dc current
$I_p$ = peak current of pulse
$R_L$ = load resistance ($R_1$ in Fig. 23-21*b*)

$$\text{PIV} \geq V_p$$

where PIV is the peak inverse voltage rating of the diode, and $V_p$ is the peak value of the ac voltage being rectified.

$$I_f \geq I_{dc}$$

where $I_f$ is the average forward current rating of the diode.

The generator shown in Fig. 23-21*b* could represent an electronic generator, an electromagnetic generator, or a 120-V outlet. Also, the generator symbol could be replaced with the secondary winding of a transformer. In fact, in electronic systems the ac source to be rectified is most often the output of a transformer such as $T_1$ in Fig. 23-22.

**EXAMPLE 23-3**

For the circuit in Fig. 23-22, calculate the following: PRV for $D_1$, $V_{dc}$, and $I_{dc}$.

**Solution**

$$\text{PRV} \geq V_p = 1.414\ V_{rms} = 1.414 \times 15\text{ V} = 21.2\text{ V}$$

$$V_{dc} = \frac{0.637(V_P - V_D)}{2} = \frac{0.637(21.2\text{ V} - 0.7\text{ V})}{2}$$

$$= 6.53\text{ V}$$

$$I_{dc} = \frac{V_{dc}}{R_L} = \frac{6.53\text{ V}}{82\ \Omega} = 79.6\text{ mA}$$

## Full-Wave Rectification

Full-wave rectification can be provided with two diodes and a center-tapped transformer as shown in Fig. 23-23, or it can be accomplished with four diodes and a nontapped transformer (see Fig. 23-24).

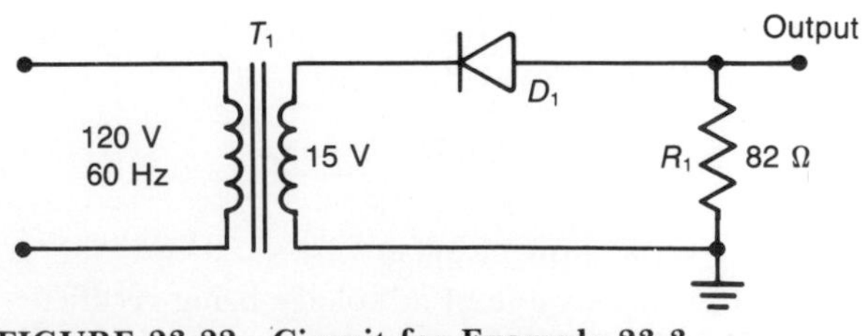


**FIGURE 23-22 Circuit for Example 23-3.**

Figure 23-23*a* shows the direction and path of current flow for the $\frac{1}{2}$ cycle when the polarity of the transformer is as marked. Notice that only $D_1$ is conducting and that only the top half of the transformer is providing power. This is because $D_2$ is reverse-biased.

During the second $\frac{1}{2}$ cycle (see Fig. 23-23*b*), the polarities of the transformer windings are reversed. Therefore, $D_1$ is now reverse-biased and $D_2$ allows the current to flow in the indicated direction and path. Notice that the current through $R_1$ is in the same direction for each $\frac{1}{2}$ cycle.

Careful examination of either Fig. 23-23*a* or *b* shows that the voltage across the reverse-biased diode is equal to the peak voltage of the full secondary winding minus the voltage drop of the conducting diode. The appropriate formula is PIV $\geq V_{P,\text{sec}}$, where $V_{P,\text{sec}}$ is the peak voltage of the full secondary. The reader can verify this formula by applying Kirchhoff's voltage law to the loop composed of the secondary of $T_1$, $D_1$, and $D_2$. For the circuit of Fig. 23-23, the minimum PIV for each diode would be

$$\text{PIV} \geq V_{P,\text{sec}} = 1.414 \times 80\text{ V} = 113\text{ V}$$

Again, in actual practice the diodes would have a PIV rating of at least 150 V, which is the first commonly available value greater than 113 V.

Notice in Fig. 23-23 that the peak value of the output voltage is determined by the peak voltage between the center tap and one end of the secondary. This peak voltage, which is one-half of the full secondary peak voltage, will be identified as $V_{P,\text{ct}}$. Therefore, for this circuit the peak of the output will be

$$V_{P,\text{out}} = V_{P,\text{ct}} - V_D = (1.414 \times 40\text{ V}) - 0.7\text{ V} = 55.9\text{ V}$$

The average output voltage (dc output) of the full-wave rectifier is twice as great as that of the half-wave rectifier. Figure 23-23 shows that there is an output voltage pulse for each $\frac{1}{2}$ cycle of the ac input. Half-wave rectification (Fig. 23-21) produces only one output pulse for each cycle of the ac input; this is why the output of the full wave is twice as great as the output of the half-wave rectifier. For the circuit shown in Fig. 23-23, the dc output is

$$V_{dc} = 0.637(V_{P,\text{ct}} - V_D) = 0.637 \times 55.9\text{ V} = 35.6\text{ V}$$

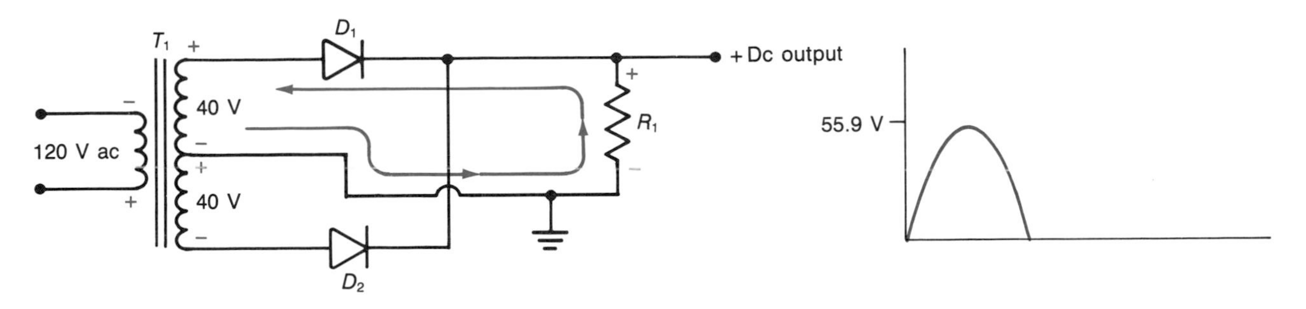


(*a*) During one half-cycle, $D_1$ conducts and $D_2$ is cutoff (reverse-biased).

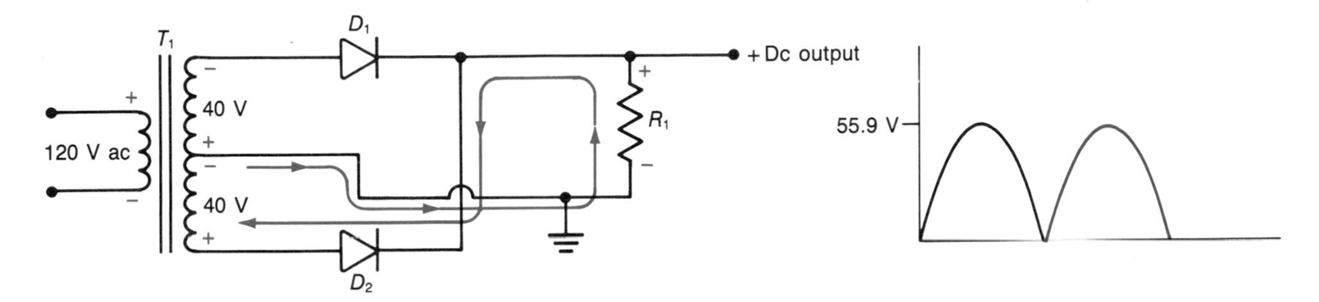


(*b*) During the other half-cycle, $D_2$ conducts and $D_1$ is cutoff.

**FIGURE 23-23 Full-wave rectifier with center-tapped transformer.**

Once $V_p$ and $V_{dc}$ have been calculated, Ohm's law and the value of the load resistance ($R_1$ in Fig. 23-23) can be used to determine $I_{dc}$ and $I_p$.

Notice from Fig. 23-23 that each diode only conducts every other current pulse. Since each diode carries one-half of the load current, the average forward current rating of the diode can be half as great in full-wave circuits as in half-wave circuits. The formula for $I_f$ in full-wave circuits is

$$I_f \geq \frac{I_{dc}}{2} = \frac{V_{dc}}{2R_L}$$

Figure 23-24 shows a full-wave, *bridge rectifier circuit.* Notice that this circuit provides twice as much dc voltage as does the previous full-wave circuit when both circuits use the same transformer. The bridge rectifier circuit does not use the center tap of the transformer and it requires four diodes.

During one $\frac{1}{2}$ cycle, two of the diodes in Fig. 23-24 conduct and allow the full secondary voltage to force current through the load resistor $R_1$. The remaining two diodes are reverse-biased and thus prevent the diode bridge from short-circuiting the transformer secondary. Since two diodes are always conducting, the peak output voltage is

$$\begin{aligned} V_{P,\text{out}} &= V_{P,\text{sec}} - 2V_D = (1.414 \times 80 \text{ V}) - (2 \times 0.7 \text{ V}) \\ &= 111.7 \text{ V} \end{aligned}$$

and the average output voltage $V_{dc}$ is

$$V_{dc} = 0.637(V_{P,\text{sec}} - 2V_D) = 0.637 \times 111.7 \text{ V} = 71.2 \text{ V}$$

Even though two diodes are always reverse-biased in Fig. 23-24, each diode must be able to block the peak voltage of the full secondary. Why this is so can be seen by applying Kirchhoff's voltage law to either the loop composed of $D_1$, $D_4$, and the secondary of $T_1$, or the loop composed of $D_2$, $D_3$, and the secondary of $T_1$. In either case, the forward voltage drop of the conducting diode plus the reverse voltage drop of the blocking diode must equal the full secondary voltage. Thus, the formula for calculating the PIV rating of the diodes is the same for both types of full-wave rectifier circuits.

Although a given transformer can supply twice as much dc voltage with bridge rectification as with center-tapped rectification, the maximum power taken from the transformer must be the same in both circuits. This is because the transformer can only transfer its rated power from the primary to the secondary without saturating the iron core.

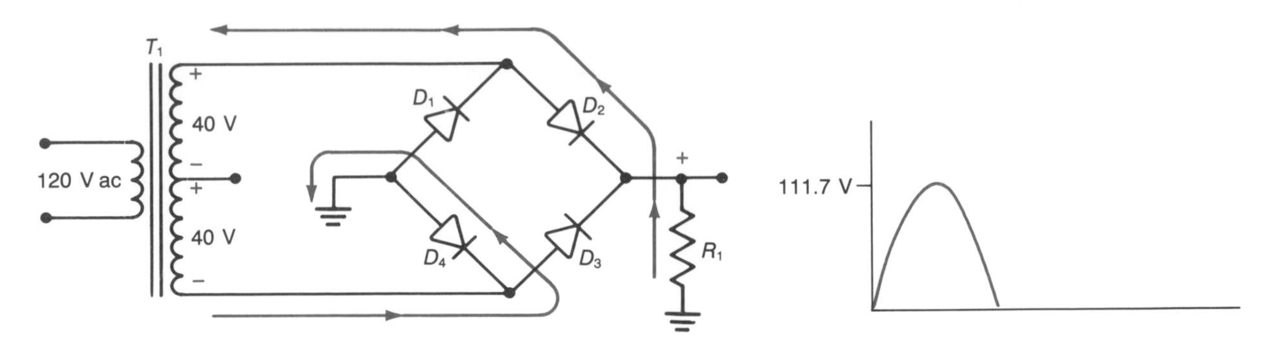


(*a*) Both $D_2$ and $D_4$ conduct. The full secondary voltage (80 V) is being rectified.

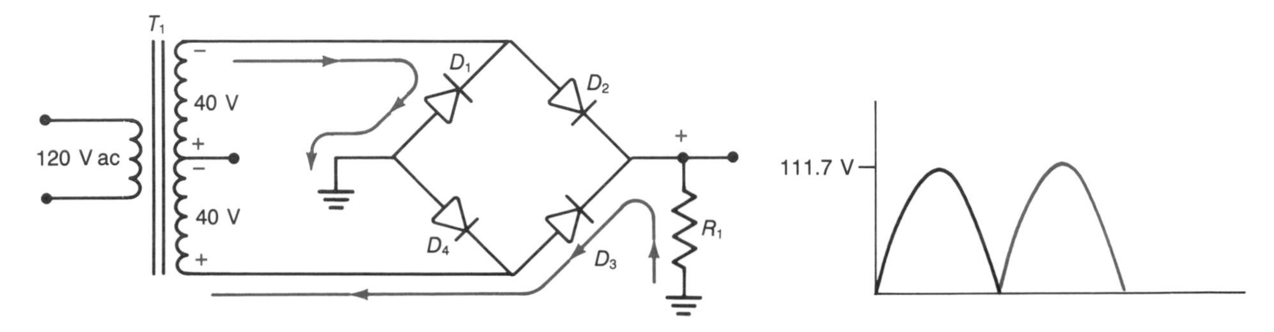


(*b*) During alternate half-cycles, $D_1$ and $D_3$ conduct.

**FIGURE 23-24 Full-wave, bridge rectifier.**

Therefore, for a given transformer, the maximum load current $I_{dc}$ is half as large for a bridge rectifier circuit as it is for a rectifier circuit which uses the center tap of the secondary.

---

**EXAMPLE 23-4**

Change the secondary voltage in Fig. 23-24 to 10 V each side of the center tap, and assign $R_1$ a value of 100 Ω. Now determine $V_{dc}$, $I_f$, and PIV.

**Solution**

$$\text{PIV} \geq V_{p,\text{sec}} \geq 1.414 \times 20 \text{ V} = 28.3 \text{ V}$$

$$V_{dc} = 0.637(V_{p,\text{sec}} - 2V_D) = 0.637(28.3 \text{ V} - 1.4 \text{ V}) = 17.1 \text{ V}$$

$$I_f = \frac{I_{dc}}{2} = \frac{V_{dc}}{2R_L} = \frac{17.1 \text{ V}}{200 \ \Omega} = 85.5 \text{ mA}$$

---

If one is willing to ignore the voltage drop of the diodes in a rectifier circuit, $V_{dc}$ for full-wave rectifiers can be estimated by the formula $V_{dc} = 0.9 \text{ V}_{rms}$, where $\text{V}_{rms}$ is the effective, or rms, value of the ac voltage being rectified. The 0.9 is obtained by multiplying 1.414 by 0.637. These numbers are, of course, the constants used to convert rms to peak and peak to average, respectively. When working with half-wave rectifiers, the formula would be $V_{dc} = 0.45 \text{ V}_{rms}$.

All of the formulas used in this section assume an ideal voltage source (transformer) in which the terminal voltage is independent of the load current. However, from your study of transformers, you know that the internal resistance of the transformer causes its terminal voltage to decrease as the load current increases. Further, you should remember that transformer voltages are usually specified at full-rated current. Thus, with small-load currents, $V_{dc}$ will be slightly higher than the value calculated with the formulas in this section on rectifier circuits.

Rectifier circuits and their associated control and protection devices (switches and fuses) are sometimes referred to as *dc power supplies*. They provide pulsating direct current to the loads connected to them.

## Self-Test

**20.** A full-wave bridge rectifier and a half-wave rectifier are used to rectify the secondary voltage

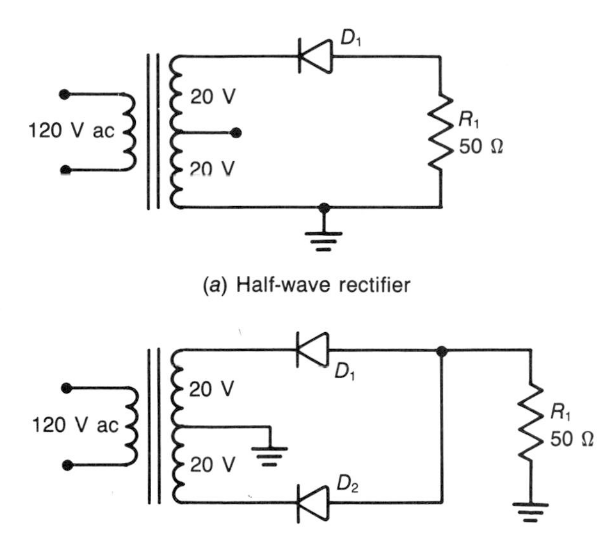


(*b*) Full-wave rectifier

**FIGURE 23-25 Circuit for Self-Test Questions 21, 22, and 23.**

of identical transformers. Compare them in terms of $V_{dc}$ and the minimum PIV for each diode.

**21.** For the circuit in Fig. 23-25*a*, calculate $V_{dc}$ and the minimum PIV and $I_f$ for $D_1$.

**22.** When compared to Fig. 23-25*b*, will Fig. 23-25*a* have a larger, a smaller, or the same value of

**a.** $I_f$
**b.** PIV
**c.** $I_{fm}$
**d.** $V_{dc}$

**23.** Will the output voltages of the circuits in Fig. 23-25 be negative or positive?

**24.** Will the output voltage of the circuit in Fig. 23-22 be negative or positive?

**25.** What type of voltage is provided by rectifier circuits?

**26.** Why should the PIV rating of a diode used in a rectifier circuit always be greater than the calculated PIV?

## 23-5 FILTERS

The pulsating direct current provided by rectifier circuits is adequate for uses such as charging batteries and running some dc motors. For most electronic applications, however, the direct current provided by a power supply must be as close to pure direct current as practical. Changing the pulsating direct current provided by a rectifier circuit into a purer direct current is accomplished by adding a *filter* to the output of the rectifier.

The simplest form of filter for a power supply is a capacitor. The result of adding a capacitor filter to a half-wave rectifier is illustrated in Fig. 23-26. Notice in Fig. 23-26*b* that the capacitor maintains a voltage across the load during the time that the rectifier is not conducting. The capacitor converts the pulsating direct current into fluctuating direct current. The fluctuating direct current of Fig. 23-26*b* can be thought of as being composed of two parts. One part, referred to as the *dc component*, has an average value equal to the average height of the waveform. The other part, called the *ac component*, has a peak-to-peak value equal to the maximum height of the waveform minus the minimum height of the waveform.

The idea of a fluctuating direct current containing an ac and a dc component may be clarified by referring to Fig. 23-27. This figure shows that the voltage at the right-hand end of the resistor, which fluctuates from +6 to +4 V, can be obtained by adding a 2-V peak-to-peak ac voltage to a 5-V dc voltage. The bracket around the battery symbol in Fig. 23-27 indicates that the waveform shown for the battery does not use the ground symbol as a reference point. Rather, the negative end of the battery is the reference point for the dc waveform shown in this figure.

The ac component of the fluctuating dc output of a filtered power supply is called the *ripple voltage*. For a half-wave power supply, the frequency of the ripple voltage is equal to the frequency of the alternating current being rectified. As shown in Fig. 23-28, 1 cycle of ac input produces 1 cycle of ripple in the dc output.

The filter capacitor in Fig. 23-29*a* is able to maintain a voltage across the load because the capacitor charges very rapidly when the diode is conducting and discharges very slowly when the diode is cut off (reverse-biased).

Careful examination of the waveforms in Fig. 23-29*a* shows that for the first few degrees of the positive $\frac{1}{2}$ cycle of the ac input, the capacitor's voltage exceeds the ac input voltage and the diode is reverse-biased. However, later in the positive $\frac{1}{2}$ cycle, the ac input voltage exceeds the capacitor's voltage, the diode is forward-biased, and current flows in the directions shown on the schematic in

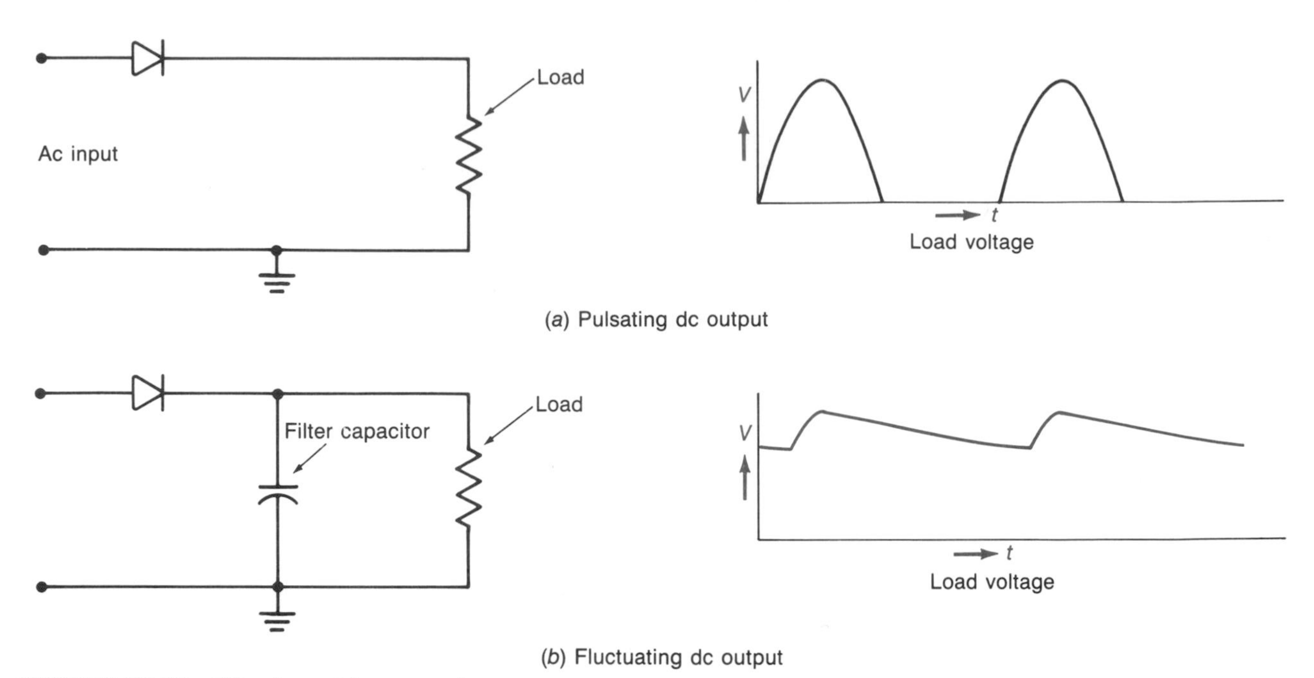


**FIGURE 23-26 Filtering with a capacitor.**

Fig. 23-29*a*. The ac source is providing charging current for the capacitor as well as current for the load $R_1$.

These currents continue to flow until the input wave just passes 90° and the diode becomes reverse-biased. During the time the diode is conducting, the capacitor's voltage remains within about 0.7 V of the instantaneous value of the ac voltage. Notice in Fig. 23-29*a* that the only resistance through which the capacitor must charge is the internal resistance of the conducting diode. This resistance is usually less than 1 Ω. It is composed primarily of the dynamic resistance of the doped crystal and the resistance of the interface between the crystal and the diode leads. The internal resistance of the diode can be estimated by

$$r_D = \frac{\Delta V_D}{\Delta I_D} \approx \frac{V_f - 0.7\ \text{V}}{I_f}$$

where $V_f$ is the voltage drop across the diode when it is conducting its rated, average, forward current $I_f$. In this formula the 0.7 V is the forward voltage

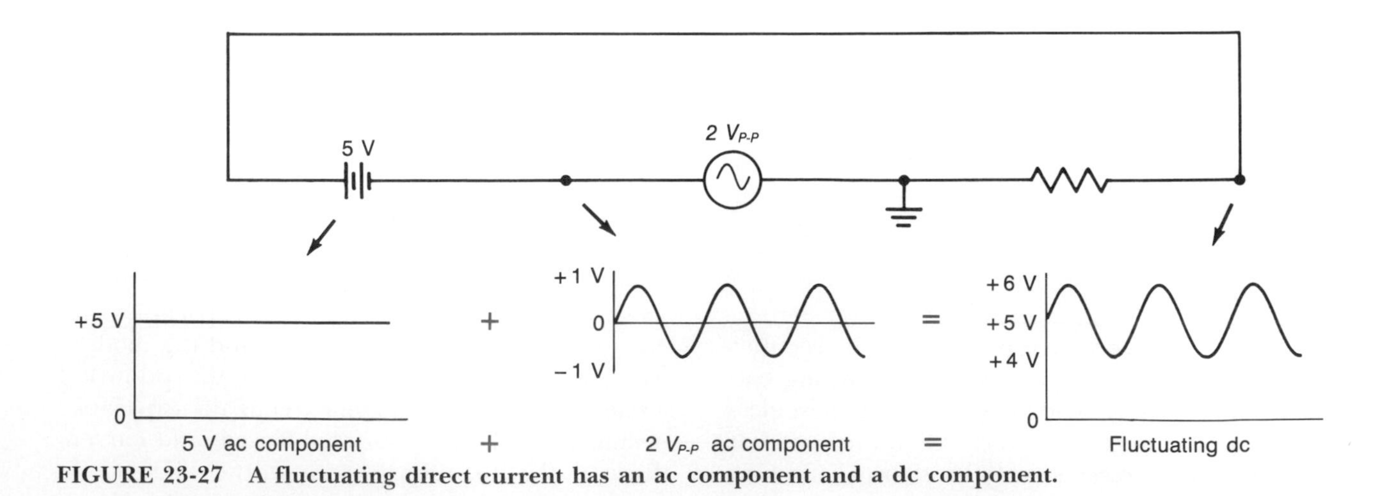


**FIGURE 23-27 A fluctuating direct current has an ac component and a dc component.**

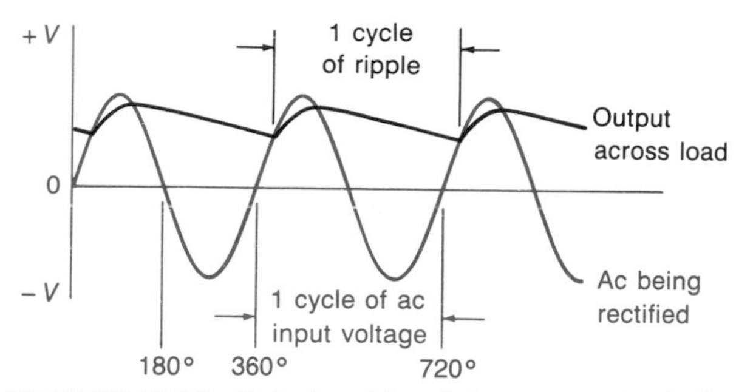


**FIGURE 23-28 Relationship of the output ripple frequency to the ac input frequency.**

at which the diode starts significant conduction. Thus, $V_f - 0.7$ V is equal to $\Delta V$ and $I_f$ is equal to $\Delta I$. The value of $V_f$ and $I_f$ can be found on the manufacturer's data sheet for the specified diode. Current $I_f$ and $V_f$ for a typical diode used in a circuit like Fig. 23-29*a* would be 1 A and 0.9 V, respectively. The estimated value of $r_D$ would be

$$r_D = \frac{0.9 \text{ V} - 0.7 \text{ V}}{1 \text{ A}} = 0.2 \ \Omega$$

Thus, the time constant $\tau$ for charging capacitor $C_1$ in Fig. 23-29*a* would be

$$\tau = RC = 0.2 \ \Omega \times 0.001 \text{ F} = 0.0002 \text{ s} = 0.2 \text{ ms}$$

Compared to the 16.7-ms period of the 60-Hz sine wave being rectified, this 0.2 ms is a relatively short time.

By referring to Fig. 23-29*b*, you can see that the capacitor discharges through the load for a much longer time than it charged through the diode. Even though $C_1$ discharges for a much longer time than it charges, it loses only a small part of its voltage because of the long discharge time constant. For the circuit of Fig. 23-29*b*, the discharge time constant is

$$\tau = RC = 100 \ \Omega \times 0.001 \text{ F} = 0.1 \text{ s} = 100 \text{ ms}$$

Remember that a time constant is the time required for the capacitor to lose 63.2 percent of its voltage. Thus, $C_1$ will not lose much of its voltage when it has only 16.7 ms to both charge and discharge. Notice in Fig. 23-29*b* that there are no polarity signs on the transformer secondary. This is because the diode is reverse-biased, and the capacitor is discharging anytime the top end of the secondary is less positive than the top end of the filter capacitor. Note from the waveform drawing that

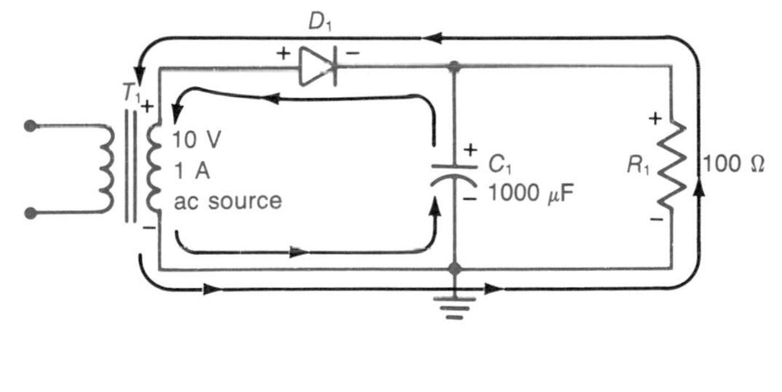


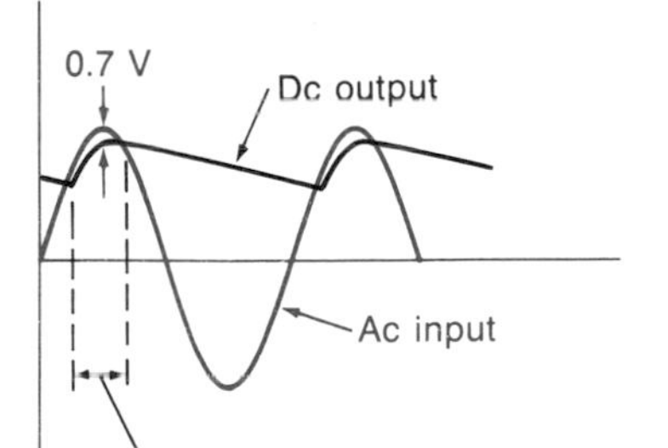

(*a*) Capacitor charges

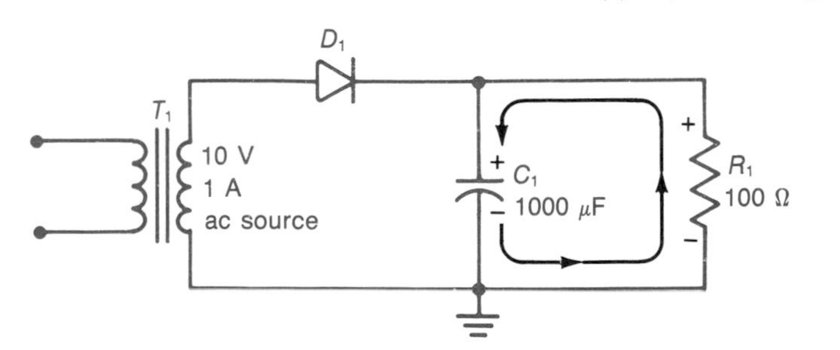


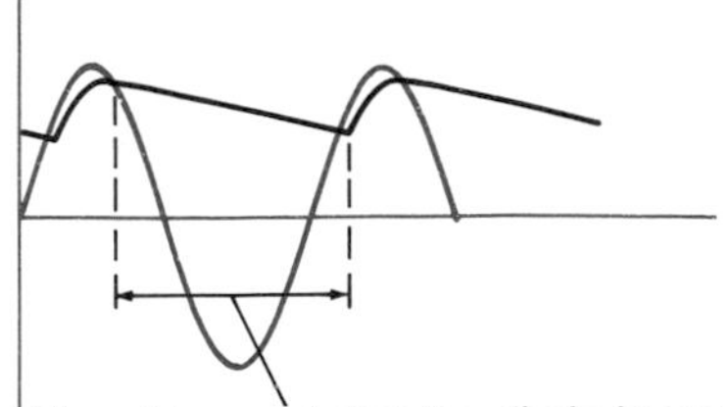


(*b*) Capacitor discharges

**FIGURE 23-29 Charging and discharging of a filter capacitor.**

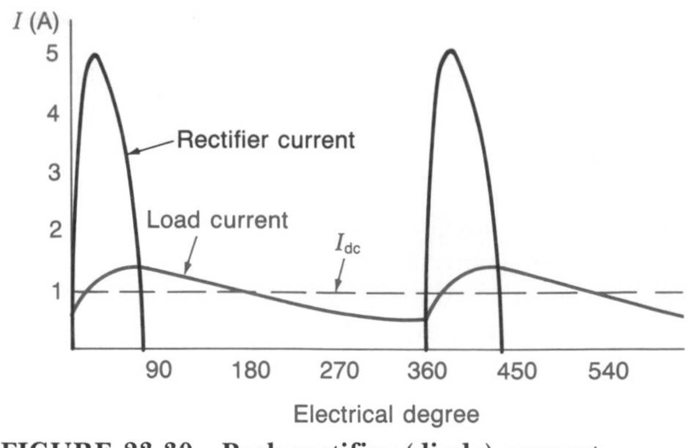


**FIGURE 23-30 Peak rectifier (diode) current.**

the diode is reverse-biased during a large portion of the positive $\frac{1}{2}$ cycle as well as all of the negative $\frac{1}{2}$ cycle of the ac source.

## Peak Rectifier Current

Once a filter capacitor is added to a rectifier circuit, the selection of an appropriate diode for the circuit becomes more complex. One factor which makes it more complex is the peak rectifier (diode) current illustrated in Fig. 23-30. The *peak rectifier current* of a filtered power supply is always greater than the peak load current or the average load current; how many times greater depends upon the ratio of capacitor discharge time to capacitor charge time.

During the time the rectifier is conducting, the capacitor is charging. During this time the capacitor must take on enough charge so that it can provide load current (by discharging through the load) during the time the diode is cut off. In Fig. 23-30, it appears that the capacitor charges for about one-fifth of the time (72°) and discharges for about four-fifths of the time (288°). For 60-Hz ripple, this means the capacitor charges for 3.3 ms and discharges for 13.3 ms. Further, from Fig. 23-30 it appears that $I_{dc}$ is about 1 A. Since the relationship between current, charge, and time is $I = Q/t$ or $Q = It$, the discharging capacitor loses about the following charge:

$$Q = It = 1 \text{ A} \times 13.3 \text{ ms} = 13.3 \text{ mC}$$

Therefore, in 3.3 ms of charging time, the capacitor must increase its charge by 13.3 mC. This represents a current of

$$I = \frac{Q}{t} = \frac{13.3 \text{ mC}}{3.3 \text{ ms}} = 4 \text{ A}$$

The diode must conduct this 4-A charging current plus the slightly greater than 1-A load current flowing at this time. Thus, the peak current through the diode will be greater than 5 A while the load current is only 1 A. Note that this is a repetitive peak current which occurs at the beginning of each cycle of ripple.

Figure 23-31 shows that the less ripple a power supply has, the higher will be the ratio of capacitor discharge to capacitor charge time. This means that, for a given value of load current, the peak rectifier current increases as the amount of ripple decreases. By the time the rms value of the ripple is reduced to 1 percent of the value of the dc output voltage, the peak diode current is approximately 24 times larger than the load current. Many diodes only handle 5 to 10 times as much peak repetitive current as average current. This is one reason why a power supply rated at $I_{dc} = 0.8$ A may use a diode rated at $I_f = 2$ A. The higher rating is

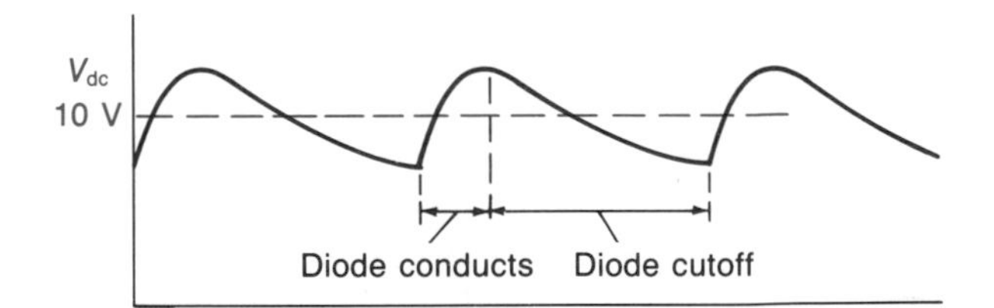

(*a*) Output waveform of a power supply

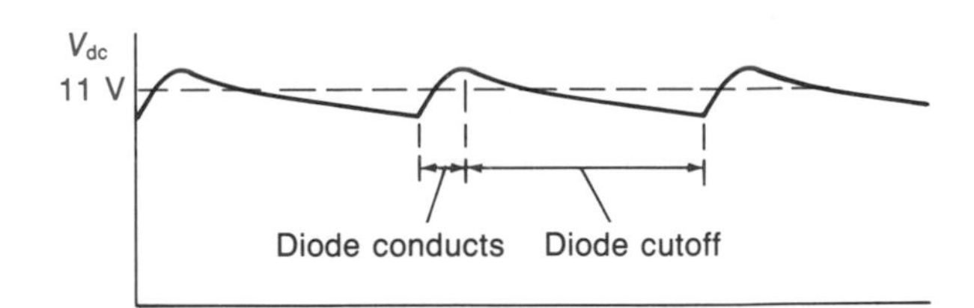


(*b*) Output waveform when either the filter capacitance or load resistance is increased

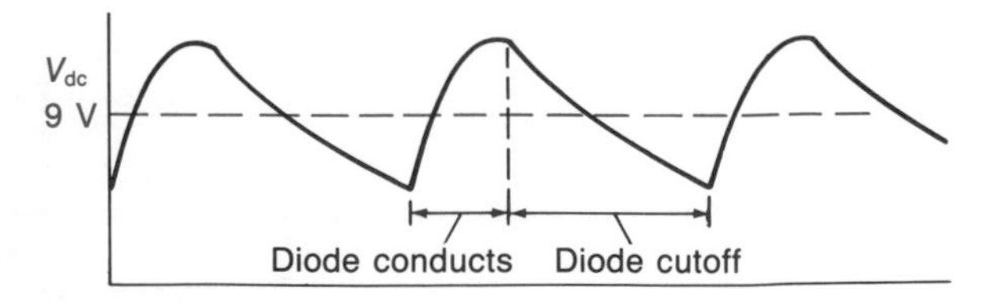


(*c*) Output waveform when either the filter capacitance or load resistance is decreased

**FIGURE 23-31 Effects of changing the filter capacitance and/or load resistance.**

needed to accommodate the large peak repetitive current.

## Effects of Resistance and Capacitance

The waveform in Fig. 23-31*a* represents the output of a filtered power supply with some value of load resistance $R$ attached to its output terminals. The filter capacitance $C$ and the load resistance $R$ determine the slope of the discharge part of the curve. The slope of this part of the curve affects the amplitude of the ripple voltage, the value of the dc output voltage, and the number of degrees during which the capacitor charges and discharges. Figure 23-31*b* shows that decreasing the load current (increasing $R$) on a power supply has the following effects:

1. Decreases the ripple voltage
2. Increases the dc output voltage
3. Increases the capacitor discharge time

The same effects would result from the filter capacitance increasing while the resistance of the load remained unchanged. Figure 23-31*c* shows that decreasing the load resistance, or decreasing the capacitance, causes more ripple, less $V_{dc}$, and more time for the capacitor to charge.

## Ripple Factor

One of the important ratings of a power supply is its *ripple factor.* Ripple factor is defined as

$$\text{rf} = \frac{V_{rip}}{V_{dc}}$$

where $V_{rip}$ is the rms value of the ripple voltage, and $V_{dc}$ is the average dc output voltage. Since the purpose of a filtered power supply is to provide as pure a dc voltage as practical, a low rf is desirable. For a given load current, rf is a function of filter capacitance. Power supplies which provide load currents of a few amperes may require filter capacitors with capacitances of 5000 or 10,000 μF to keep the rf at an acceptable level. Power supplies providing even a few milliamperes use filter capacitors greater than 1 μF. Therefore, filter capacitors are usually electrolytic capacitors which are polarized. These capacitors can explode if reverse polarity is applied to them.

Quite often, the ripple characteristic of a power supply is expressed as a *percentage of ripple* rather than as a ripple factor. Of course, percentage of ripple = rf × 100. Therefore, a power supply with an rf of 0.05 would have 5 percent ripple in the output.

---

**EXAMPLE 23-5**

A power supply has a rated output of 12 V at 1 A with an rf of 0.08. What is the rms value of the ripple voltage?

**Solution**

$$\text{rf} = \frac{V_{rip}}{V_{dc}}$$

Therefore,

$$V_{rip} = \text{rf} \times V_{dc} = 0.08 \times 12 \text{ V} = 0.96 \text{ V}$$

---

If ripple voltage is measured with an oscilloscope, the measured value will be greater than 2.8 times the rms value because the ripple voltage is not sinusoidal. Multiplying the rms value of the ripple by 3.3 instead of 2.8 gives a fair approximation of the peak-to-peak value.

## Voltage Regulation

Another specification for a power supply is its *voltage regulation.* Voltage regulation VR refers to the change in output voltage caused by a change in load current. The formula for calculating VR is

$$\text{VR} = \frac{V_{NL} - V_{FL}}{V_{FL}}$$

where $V_{NL}$ is the dc output voltage when no load (or the minimum load) is connected to the supply, and $V_{FL}$ is the dc output voltage when the rated (full load) current is being drawn from the supply.

The change in output voltage under different load conditions is also known as the "percentage of voltage regulation." Percentage of VR is equal to VR times 100. From the VR formula, you can see that the smaller the percentage of VR, the less the change in output voltage. For most applications, VR is kept as small as practical.

One factor which causes poor VR is inadequate filtering. As shown in Fig. 23-31, as the ripple increases, the dc output voltage drops. Another factor, often a major factor, in determining VR is the internal resistance of the ac source. For instance, the secondary voltage of a transformer secondary (a common ac source) often drops 5 to 10 percent when the load is increased from zero to rated current. This drop is caused by internal resistance. Another factor is the $V_f$ of the diode. Voltage $V_f$ can vary from about 0.7 V when conducting a few milliamperes to well over 1 V when conducting several amperes. With a 5-V power supply, this fraction of a volt can significantly degrade VR.

### EXAMPLE 23-6

A power supply produces 23 V when it is providing its rated current of 300 mA. When the 300-mA load is removed, the output increases to 27 V. What is the percentage of VR?

**Solution**

$$\text{Percentage of VR} = \frac{V_{\text{NL}} - V_{\text{FL}}}{V_{\text{FL}}} \times 100$$

$$= \frac{27\text{ V} - 23\text{ V}}{23\text{ V}} \times 100 = 17.4 \text{ percent}$$

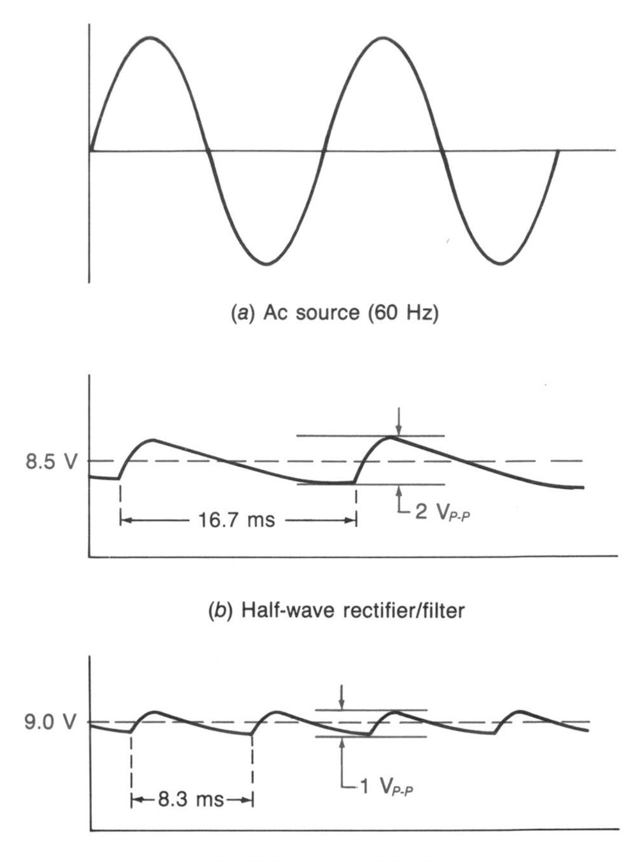

**FIGURE 23-32 Comparison of half-wave and full-wave filtered power supplies with equal values of load current and capacitance.**

## Full-Wave Filter

So far all of the illustrations and examples of filtering have been concerned with filtering half-wave rectifiers. All of the general ideas developed for half-wave filters apply equally well to full-wave filters; only a few of the details must be changed.

Figure 23-32 shows that the ripple frequency for a full-wave filter is twice the frequency of the alternating current being rectified. Thus, the filter capacitor in the full-wave circuit has to discharge only half as long as it does in the half-wave circuit. Or, looking at the comparison from another point of view, the same amount of capacitance leaves half as much ripple in a full-wave as in a half-wave power supply.

Also notice in Fig. 23-32 that the full-wave rectifier provides a slightly higher dc output voltage. With full-wave, the capacitor only has time to discharge from 9.5 to 8.5 V, while with the half-wave, which provides about twice as much discharge time, the capacitor can discharge from 9.5 to about 7.5 V. This smaller amount of ripple with the full-wave provides a higher $V_{\text{dc}}$ output.

For the same amount of ripple voltage, the rectifiers in the full-wave circuit have only about half as high a peak-current-to-average-current ratio. This is because the capacitor is half as large and has to provide load current for only half as much time in the full-wave as in the half-wave circuit.

## PIV and DCWV Ratings for Filter Circuits

Figure 23-33 shows the voltages and polarities needed to figure the minimum PIV rating for the diodes and minimum direct current working voltages DCWV ratings for the capacitors in full-wave and half-wave circuits. In all three circuits, the filter capacitor charges up to the peak value of the ac source. The capacitor voltage must remain at this peak value when there is no load through which the capacitor can discharge. Thus, the DCWV of the capacitor must be at least as large as the peak voltage of the alternating current being rectified. To allow for tolerance of the transformer and ac

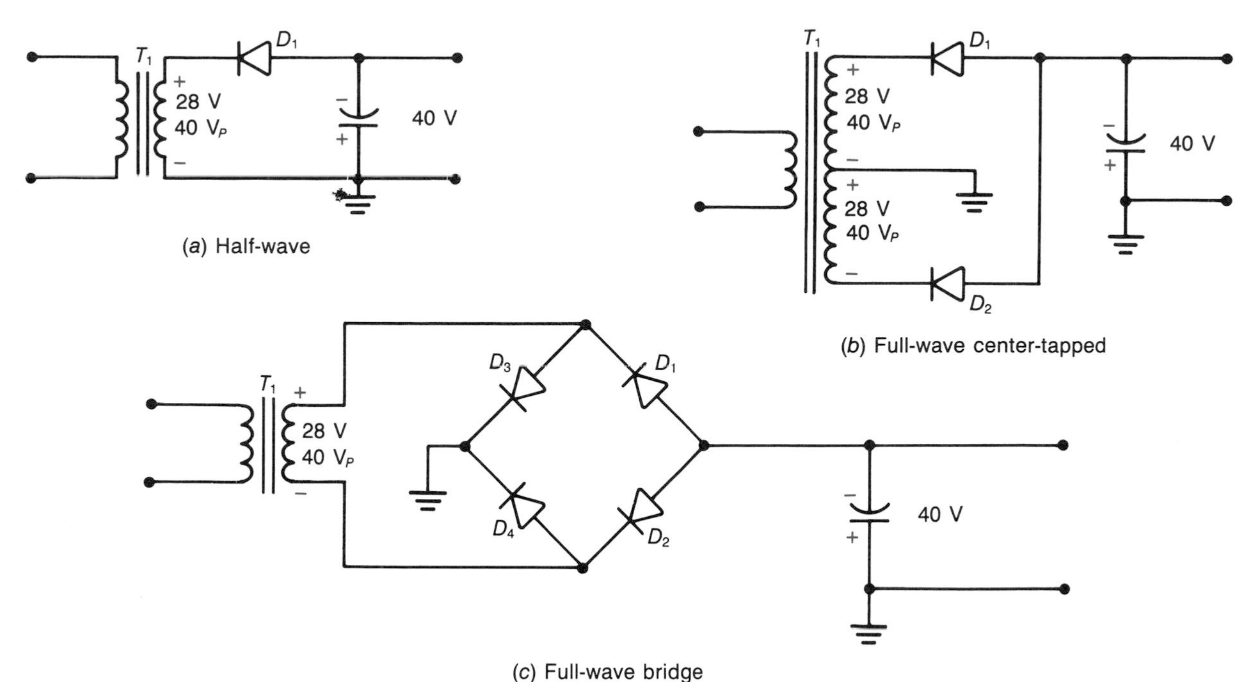


**FIGURE 23-33 Comparison of DCWV and PIV for three common configurations.**

line voltage, it is common practice to add 15 to 20 percent to the calculated DCWV.

The PIV for a diode in a power supply circuit can be determined by applying Kirchhoff's voltage law to each circuit. In Fig. 23-33*a*, it can be seen that the instantaneous voltage of the transformer secondary is series-aiding the voltage on the capacitor. The sum of these two voltages is expressed across $D_1$. Therefore, $D_1$ must be able to block two times the peak voltage of the ac source. The minimum PIV rating for $D_1$ in Fig. 23-33*a* would have to be 80 V. Notice this is roughly three times the rms voltage of the ac source.

Applying Kirchhoff's voltage law to the loop which consists of the top half of the secondary of $T_1$, the capacitor, and $D_1$ shows that the reverse voltage across $D_1$ is also 80 V for the circuit in Fig. 23-33*b*. Thus, the diode in a full-wave, center-tapped supply also needs a PIV rating equal to two times the peak voltage of *the voltage which the diode is* rectifying. Note that in this circuit, each diode rectifies only one-half of the total secondary voltage.

Diodes in the bridge-type power supply (Fig. 23-33*c*) need a PIV rating equal to the peak voltage of the ac source. Using the loop consisting of $D_3$, $D_4$, and the secondary of $T_1$ shows that approximately 40 V appears across $D_4$ because $D_3$ is forward-biased. Or, using a loop composed of the secondary, $D_4$, the capacitor, and $D_1$ shows that the combined voltage (80 V) of the capacitor and the secondary is shared by two reverse-biased diodes.

### EXAMPLE 23-7

Determine the PIV rating for $D_1$ and the DCWV rating for $C_1$ in the circuit shown in Fig. 23-29*a*.

**Solution**

$$V_p = 1.414 \times V_{\text{rms}} = 1.414 \times 10 = 14.14 \text{ V}$$
$$\text{PIV} = 2 \times V_p = 2 \times 14.14 \text{ V} = 28.28 \text{ V}$$
$$\text{DCWV} = V_p = 14.14 \text{ V}$$

## Bleeder Resistor

The power supplies shown in Fig. 23-33 do not include any path through which the filter capacitor can discharge. When the ac power to the supply is turned off, the capacitor remains charged. This can be a real hazard, especially on a high-voltage power supply. When a load is permanently connected to the power supply, this hazard is eliminated as long as the load remains operative.

Some power supplies, especially those intended for use with intermittent and/or temporary loads, contain a *bleeder resistor*. As shown in Fig. 23-34, the

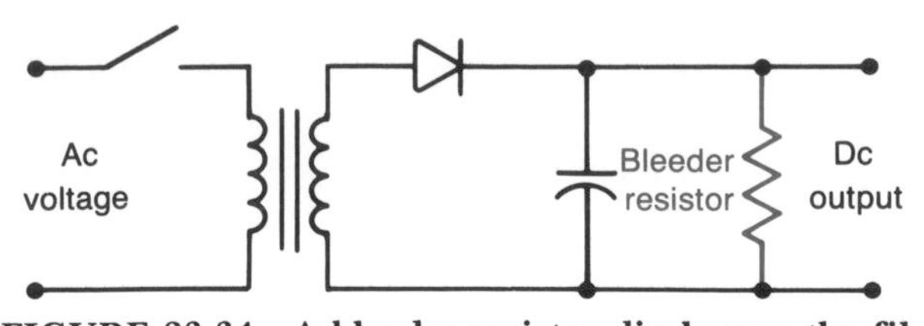


**FIGURE 23-34 A bleeder resistor discharges the filter capacitor when the power supply is turned off.**

bleeder resistor is in parallel with the filter capacitor. The bleeder resistor is included in the circuit to discharge the filter capacitor when the power supply is turned off, even if no load is connected to the power supply. The value of the bleeder resistor is usually large enough so that it draws less than 5 percent of the output current capability of the power supply. The bleeder resistor must be small enough so that the capacitor is discharged in a reasonable amount of time. While the bleeder resistor may improve voltage regulation by keeping the output voltage below the peak of the ac input voltage, it is generally included primarily as a safety factor to discharge the filter capacitor.

## Surge Current

When a power supply is first turned on, it is possible for a very large current, called a *surge current,* to flow through the diode. Figure 23-35 shows the conditions under which maximum surge current

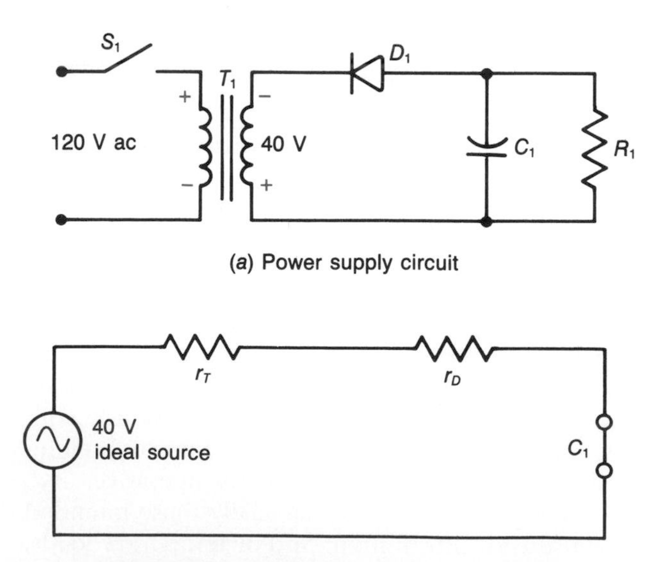


**FIGURE 23-35 Surge current in a power supply.**

($I_{fm}$ or $I_s$) flows. Before power is applied to the circuit, the capacitor is completely discharged by $R_1$. Now suppose the switch is closed at the instant the ac input is at its peak value. At that instant, the capacitor acts like a short circuit because it has no charge. The only thing limiting the instantaneous current is the internal resistance of the forward-biased diode $r_D$ and the internal resistance of the transformer $r_T$. The internal resistance of the diode was discussed in Sec. 23-5, and it was shown to be less than 1 Ω. The internal resistance of a transformer varies considerably, and it depends primarily on the transformer's current and voltage rating. A 10-V, 1-A transformer may have an internal resistance of 0.5 Ω, while a 60-V, 0.04-A transformer's internal resistance may be 70 Ω. A transformer's internal resistance can be estimated by the formula

$$r_T \approx \frac{V_{NL} - V_{FL}}{I_{FL}}$$

where $V_{NL}$ = secondary voltage when no load is connected to transformer
$V_{FL}$ = secondary voltage when rated current is being drawn from secondary
$I_{FL}$ = rated secondary current

For a typical transformer, $V_{NL}$ is 5 to 10 percent higher than $V_{FL}$.

### EXAMPLE 23-8

Refer to Fig. 23-35. Assume $D_1$ has an internal resistance $r_D$ of 0.4 Ω and that $T_1$ has an internal resistance $r_T$ of 4 Ω. Determine the surge current $I_s$.

**Solution**

$$V_{p,\text{sec}} = V_{\text{sec}} \times 1.414 = 40\text{ V} \times 1.414 = 56.6\text{ V}$$

$$I_s = \frac{V_{p,\text{sec}}}{r_D + r_T} = \frac{56.6\text{ V}}{0.4\ \Omega + 4\ \Omega} = 12.9\text{ A}$$

In Example 23-8, the transformer with 4-Ω internal resistance would be rated for no more than 1 A. Thus, the power supply would provide slightly less than 1 A of direct current. The surge current would be more than 13 times as large as the dc current. Most diodes capable of 1 A of dc can handle 13 A of surge current.

Also notice in Example 23-8 that the surge current was largely controlled by the internal resistance of the transformer secondary. This is usually

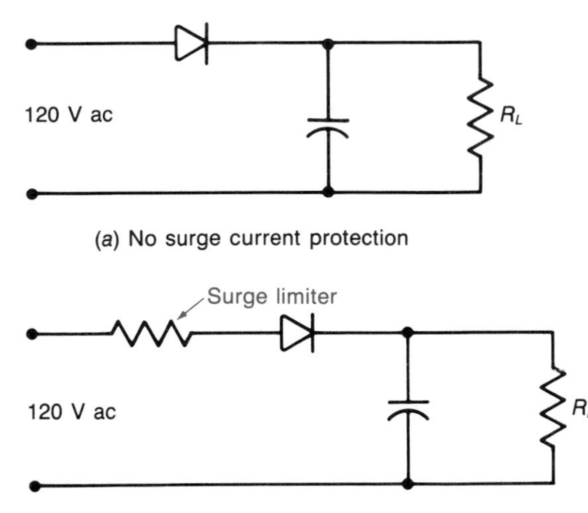


(*a*) No surge current protection

(*b*) Surge current limited by resistor

**FIGURE 23-36 Surge limiter resistor protects the diode from excessive surge current when power is applied to the circuit.**

the case when the transformer's current rating is closely matched to the output current rating of the power supply. However, sometimes the current capability of the ac source is very much larger than the output current rating of the power supply. In such cases, the internal resistance of the ac source is so low that the surge current can destroy the diode. For example, the circuit in Fig. 23-36*a* is using a 120-V outlet as its ac source. The internal resistance of this source is less than 1 Ω, so the surge current could be over 80 A. As shown in Fig. 23-36*b*, a *surge limiter* resistor can be added to the circuit to limit the surge current to a safe level.

## Self-Test

**27.** A filter capacitor converts pulsating direct current into ________ direct current.

**28.** The ac component of the output of a filtered power supply is called the ________ voltage.

**29.** In a half-wave, filtered power supply, does the diode conduct before or after the ac input reaches 90°?

**30.** Does a filter capacitor charge or discharge for the major part of an ac cycle?

**31.** Is the peak load current larger than the peak rectifier current?

**32.** Why would a diode with an $I_f = 5$ A rating be used in a power supply that is rated at 1.5 A?

**33.** What will happen to each of the following if the filter capacitance in a power supply is increased and the load remains the same?

**a.** Dc output
**b.** Ripple
**c.** Peak rectifier current

**34.** What is the percentage of ripple of a power supply that delivers $15V_{dc}$ with an ac component of 0.5 V?

**35.** The output of a power supply drops from 40 to 37 V when a full load is added. What is the percentage of voltage regulation?

**36.** What is the ripple frequency of a full-wave power supply that is powered by a 60-Hz ac source?

**37.** A 30-V ac source is to be rectified with a bridge rectifier and then filtered with a capacitor.

**a.** Determine the minimum PIV rating of the diode.
**b.** Determine the minimum DCWV for the capacitor.

**38.** For the same load current and ripple factor, will the half-wave or the full-wave rectifier have the larger peak rectifier current?

**39.** Why is a bleeder resistor sometimes included in a power supply?

**40.** What is a "surge limiter resistor" and when is it used?

## 23-6 ZENER REGULATOR

The voltage regulation of a filtered power supply is too poor for many electronic circuits. Voltage regulation can be greatly improved by adding a zener regulator to the output of a filtered power supply.

A *zener regulator circuit,* see Fig. 23-37*a*, consists of a zener diode in series with a resistor. The voltage across the zener diode remains very close to 5 V even when the voltage applied to the circuit (the output from the filtered power supply) varies from 7 to 9 V. As shown in Fig. 23-37*c* and *d*, 1.9 V of the 2-V variation in the input voltage to the regulator circuit appears as a voltage variation across the series resistor. Figure 23-37*b* shows that the other 0.1 V of variation appears across the zener as the current through the zener changes from 18.8 to 81.2 mA.

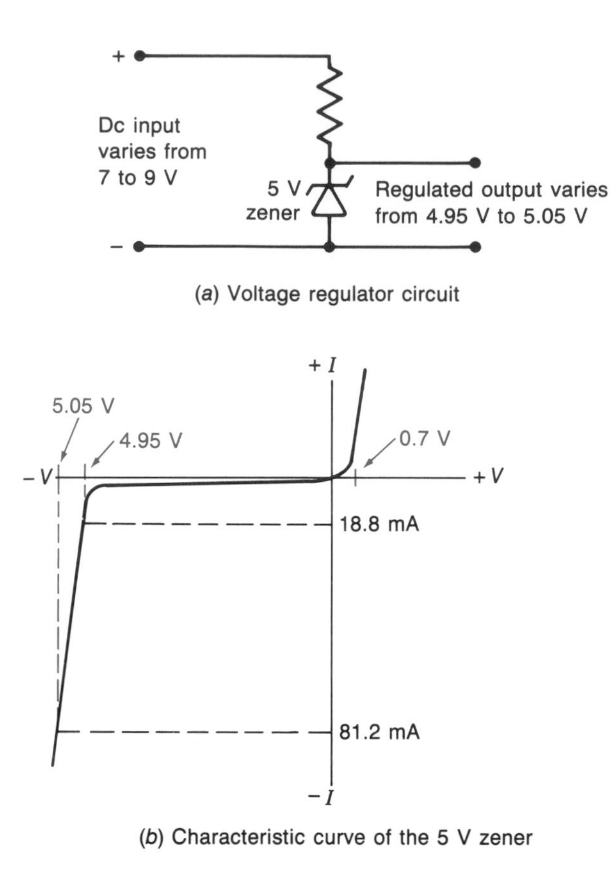


(a) Voltage regulator circuit

(b) Characteristic curve of the 5 V zener

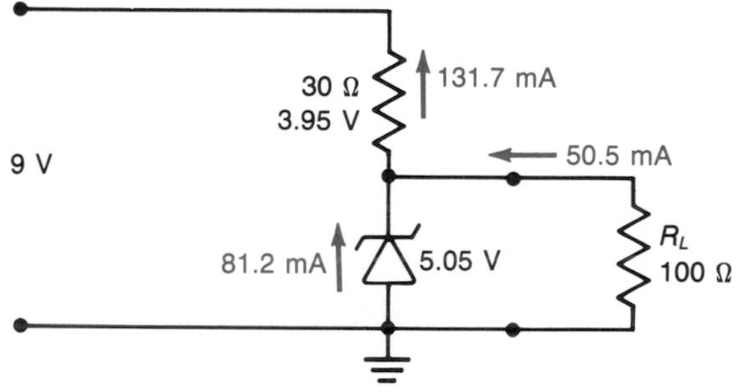


(c) Circuit conditions when the dc input is 9 volts

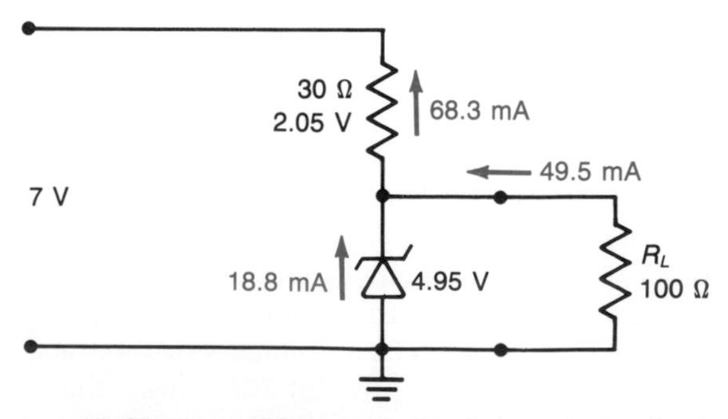


(d) Circuit conditions when the dc input is 7 volts

**FIGURE 23-37 Zener regulator.**

The percentage of voltage regulation VR of the input voltage to zener regulator is

$$\text{Percentage of VR} = \frac{V_{NL} - V_{FL}}{V_{FL}} \times 100$$

$$= \frac{9\text{ V} - 7\text{ V}}{7\text{ V}} \times 100 = 28.6 \text{ percent}$$

The percentage of regulation for the output from the zener regulator is

$$\text{Percentage of VR} = \frac{5.05\text{ V} - 4.95\text{ V}}{4.95\text{ V}} \times 100$$

$$= 2 \text{ percent}$$

This represents a very significant improvement in voltage regulation.

The illustration of Fig. 23-37 shows how the changing zener current compensates for changes in the input voltage. A zener regulator also compensates for changes in load resistance attached to the regulator output. For example, if $R_L$ changed from 100 to 110 Ω, the zener current would increase by the amount that the load current decreased. Thus, the voltage across the load and the voltage across the 30-Ω resistor would remain the same.

A zener regulator can also reduce the amount of ripple in the output of a power supply. The varying input voltage to the zener regulator of Fig. 23-37*a* could be a pure dc voltage that occasionally

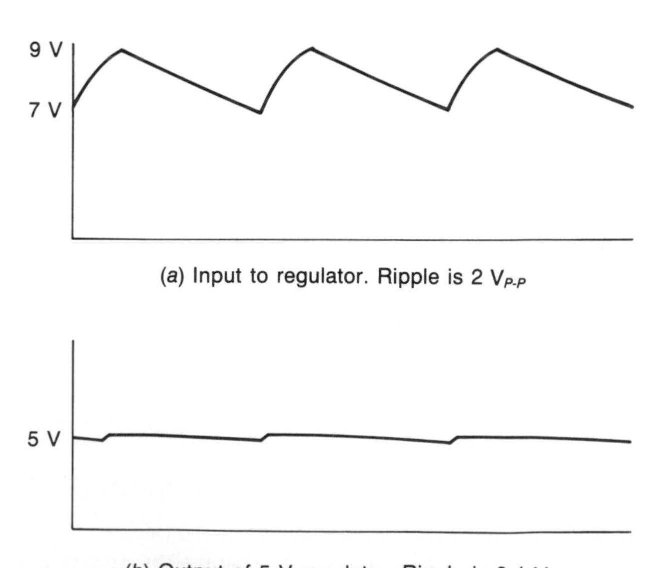


(a) Input to regulator. Ripple is 2 $V_{P\text{-}P}$

(b) Output of 5 V regulator. Ripple is 0.1 $V_{P\text{-}P}$

**FIGURE 23-38 A zener regulator can greatly reduce the ripple in a power supply.**

changes from 9 to 7 V. In this case there would be no ripple into or out of the regulator circuit. However, the input to the regulator could also be poorly filtered direct current as shown in Fig. 23-38*a*. In this case the zener regulator would reduce the ripple from $2V_{p\text{-}p}$ (9 V − 7 V) to $0.1V_{p\text{-}p}$ (5.05 V − 4.95 V).

## Self-Test

**41.** What type of diode is used for voltage regulation?

**42.** Is the load connected in series or in parallel with the diode used in a voltage regulator?

**43.** Can a voltage regulator circuit reduce the ripple factor of a power supply?

## SUMMARY

1. Intrinsic crystals are covalently bonded.
2. When an electron breaks a covalent bond, an electron-hole pair of current carriers is created.
3. Semiconductors have negative temperature coefficients.
4. Pentavalent impurity atoms (donor atoms) produce N-type crystals in which electrons are the majority carriers.
5. Trivalent impurity atoms (acceptor atoms) produce P-type crystals in which holes are the majority carriers.
6. A PN junction is a region in a crystal where the doping atoms change from trivalent to pentavalent.
7. Formation of a PN junction produces a depletion region which is void of current carriers.
8. Reverse bias widens the depletion region and reduces the junction capacitance.
9. Forward bias eliminates the depletion region of a PN junction and allows current to freely flow through the junction.
10. The leakage current in a reverse-biased PN junction is caused by electron-hole pairs which are thermally generated.
11. Compared to a germanium diode, a silicon diode has less leakage current and a greater potential barrier voltage.
12. The cathode of a diode is N-type material and the anode is P-type material.
13. Five common types of diodes are rectifier, zener, varactor, switching, and signal.
14. A diode has both a static resistance and a dynamic resistance.
15. Important diode ratings include PIV or PRV, $V_f$, $I_f$, $I_s$ or $I_{\text{fm}}$, $I_{\text{fpr}}$, $P_z$, $r_z$, and $V_z$.
16. A zener diode operates in the reverse breakdown region.
17. Rectification converts alternating current into pulsating direct current.
18. Filtering changes pulsating direct current into fluctuating direct current.
19. Full-wave rectification is easier to filter than is half-wave rectification.

20. A bridge rectifier produces full-wave rectification.

21. The ripple frequency of full wave is twice the frequency of the alternating current being rectified.

22. For all nonfiltered rectifier circuits, the minimum PIV rating of the diode is equal to peak voltage of the full secondary of the transformer.

23. Filter capacitors are polarized (electrolytic).

24. The ripple voltage of a power supply is the ac component of the output voltage.

25. In filtered power supplies, the diodes conduct for only a few degrees of the input ac cycle.

26. All other factors being equal, peak rectifier current increases as the value of the filter capacitor is increased.

27. Ripple factor decreases as filter capacitance increases.

28. The output voltage of a power supply decreases as the load current increases.

29. Other factors being equal, full-wave, filtered circuits have less ripple, twice the ripple frequency, slightly higher dc output, and smaller peak rectifier currents than do half-wave filtered circuits.

30. A bleeder resistor is included in a power supply to discharge the filter capacitor when no load is connected to the power supply.

31. Surge currents are largest when a power supply is turned on and when the ac source has very little internal resistance.

32. Surge currents can be controlled with surge limiter resistors.

33. A zener diode in series with a resistor forms a voltage regulator circuit.

34. A zener regulator can improve both voltage regulation and ripple factor.

35. Useful formulas include the following:

**a.** For a nonfiltered, half-wave rectifier circuit,

$$I_p = \frac{V_p - V_D}{R_L}$$

$$I_{dc} = \frac{0.637 I_p}{2} = \frac{0.637(V_p - V_D)}{2R_L}$$

$$I_f \geq I_{dc}$$

$$V_{dc} = \frac{0.637(V_p - V_D)}{2}$$

$$\text{PIV} \geq V_p = 1.414\ \text{V}_{rms}$$

**b.** For a nonfiltered, center-tapped rectifier circuit,

$$I_p = \frac{V_{p,ct} - V_D}{R_L}$$

$$I_{dc} = 0.637 I_p = \frac{0.637(V_{p,ct} - V_D)}{R_L}$$

$$I_f \geq \frac{I_{dc}}{2} = \frac{V_{dc}}{2R_L}$$

$$V_{dc} = 0.637(V_{p,ct} - V_D)$$

$$\text{PIV} \geq V_{p,sec}$$

**c.** For a nonfiltered, bridge rectifier circuit,

$$I_p = \frac{V_{p,\text{sec}} - 2V_p}{R_L}$$

$$I_{\text{dc}} = 0.637 I_p = \frac{0.637(V_{p,\text{sec}} - 2V_D)}{R_L}$$

$$I_f \geq \frac{I_{\text{dc}}}{2} = \frac{V_{\text{dc}}}{2R_L}$$

$$V_{\text{dc}} = 0.637(V_{p,\text{sec}} - 2V_D)$$

$$\text{PIV} \geq V_{p,\text{sec}}$$

**d.** For all filtered circuits,

$$\text{VR} = \frac{V_{\text{NL}} - V_{\text{FL}}}{V_{\text{FL}}}$$

$$\text{rf} = \frac{V_{\text{rip}}}{V_{\text{dc}}}$$

$$r_D \approx \frac{V_f - 0.7\ \text{V}}{I_f}$$

$$r_T \approx \frac{V_{\text{NL}} - V_{\text{FL}}}{I_{\text{FL}}}$$

$$I_s = \frac{V_p}{r_D + r_T}$$

**e.** For a half-wave filtered circuit,

$$\text{DCWV} \geq V_{p,\text{sec}}$$

$$\text{PIV} \geq 2V_{p,\text{sec}}$$

**f.** For a full-wave, center-tapped filtered circuit,

$$\text{DCWV} \geq V_{p,\text{ct}}$$

$$\text{PIV} \geq 2V_{p,\text{ct}}$$

**g.** For a full-wave, bridge-filtered circuit,

$$\text{DCWV} \geq V_{p,\text{sec}}$$

$$\text{PIV} \geq V_{p,\text{sec}}$$

## CHAPTER REVIEW QUESTIONS

For the following items, determine whether each statement is true or false.

**23-1.** A silicon crystal which has had boron atoms added to it is known as an intrinsic crystal.

**23-2.** An electron-hole pair is generated when a covalent bond is broken in a semiconductor material.

**23-3.** Germanium has a larger temperature coefficient than does silicon.

**23-4.** In a PN junction, the N side of the junction has a positive charge.

**23-5.** The potential barrier of a PN junction can be measured with a digital voltmeter.

**23-6.** The potential barrier of a germanium diode is greater than the potential barrier of a silicon diode.

**23-7.** Junction capacitance increases as reverse bias increases.

**23-8.** Forward bias reduces the depletion region in a PN junction.

**23-9.** Leakage current is caused by the avalanche effect.

**23.10.** A resistor has more static resistance than it has dynamic resistance.

**23-11.** $I_s$ is about twice as large as $I_f$ for most rectifier diodes.

**23-12.** In a filtered, bridge-rectifier power supply, the minimum PIV rating of each diode is equal to the peak voltage of the ac source.

**23-13.** In a filtered power supply that is providing current to a load, the peak rectifier current always exceeds the peak load current.

**23-14.** In a power supply, more time is spent charging than discharging the filter capacitor.

For the following items, fill in the blank with the word or phrase required to correctly complete each statement.

**23-15.** The majority carriers in a P-type crystal are _______.

**23-16.** The sharing of valence electrons by adjacent atoms is known as _______.

**23-17.** The process of adding impurity atoms to a semiconductor crystal is known as _______.

**23-18.** The current that flows when a PN junction is reverse-biased is called _______ current.

**23-19.** The most common semiconductor material used for diodes is _______.

**23-20.** An unfiltered rectifier circuit provides _______ direct current.

**23-21.** The ripple in the output of a power supply is sometimes called the _______ component of the output.

**23-22.** A zener diode can reduce both the _______ and _______ of the power supply.

For the following items, choose the letter that best completes each statement.

**23-23.** Current carriers can be produced in a semiconductor material when the crystal is subjected to

**a.** Heat
**b.** Light
**c.** Radiation
**d.** Any of the above

**23-24.** Atoms used in producing N-type crystals are

**a.** Pentavalent atoms
**b.** Tetravalent atoms
**c.** Trivalent atoms
**d.** Covalent atoms

**23-25.** The diode which normally operates in the reverse breakdown mode is the

**a.** Signal diode
**b.** Rectifier diode
**c.** Zener diode
**d.** Switching diode

**23-26.** An ohmmeter indicates 3 Ω when its − lead is connected to the banded end of a diode and its + lead is connected to the other end of the diode. This indicates the diode is

- **a.** Open
- **b.** Shorted
- **c.** Forward-biased
- **d.** Reverse-biased

**23-27.** A resistor used to control large, nonrepetitive, rectifier currents is called a

- **a.** Bleeder resistor
- **b.** Current resistor
- **c.** Load resistor
- **d.** Surge resistor

**23-28.** A resistor used to discharge the filter capacitor when the power supply is turned off is called a

- **a.** Bleeder resistor
- **b.** Current resistor
- **c.** Load resistor
- **d.** Surge resistor

## CHAPTER REVIEW PROBLEMS

**23-1.** Calculate the dynamic resistance for the $I$-$V$ graph in Fig. 23-39.

**23-2.** A half-wave rectifier circuit provides 5 A of direct current. Determine the minimum $I_f$ and $I_{fpr}$ ratings for the diode.

**23-3.** What is the maximum reverse current for a 5-W, 10-V zener diode?

**23-4.** For the circuit of Fig. 23-40, determine the following:

- **a.** Minimum PRV rating of $D_1$
- **b.** Minimum $I_f$ rating of $D_2$
- **c.** $V_{dc}$ output
- **d.** Polarity of the output

**23-5.** Add a 1000-$\mu$F capacitor in parallel with $R_L$ in Fig. 23-40. Now determine the following:

- **a.** The minimum PRV rating of $D_2$
- **b.** Ripple frequency
- **c.** DCWV

**23-6.** Will adding a filter capacitor to Fig. 23-40 cause each of the following to increase or decrease?

- **a.** $I_f$
- **b.** $V_{dc}$
- **c.** $I_{fpr}$

**23-7.** The output of a power supply changes from 63 to 56 V when a load is applied. Determine the percentage of voltage regulation.

**23-8.** The ripple voltage of the output of a 30-V power supply is 1.8 V. What is the ripple factor?

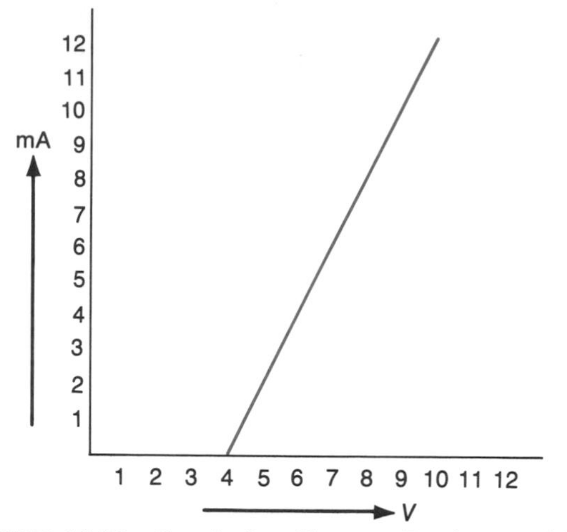


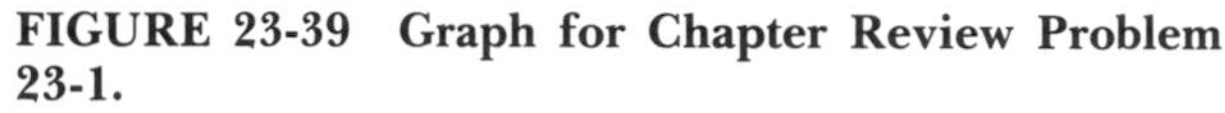


**FIGURE 23-39** **Graph for Chapter Review Problem 23-1.**

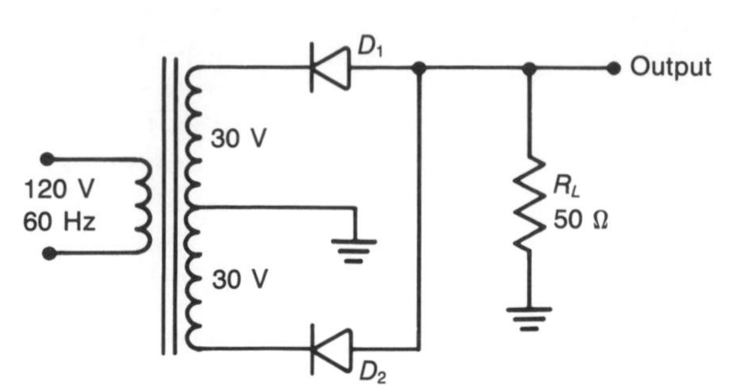


**FIGURE 23-40** **Circuit for Chapter Review Problems 23-4 through 23-6.**

## ANSWERS TO SELF-TESTS

**1.** **a.** electron-hole pairs
**b.** semiconductors
**c.** conduction band
**d.** intrinsic crystal
**e.** trivalent element
**f.** pentavalent element
**g.** doping

**2.** acceptor atom

**3.** **a.** silicon
**b.** silicon

**4.** N-type material

**5.** **a.** false
**b.** false
**c.** true
**d.** true
**e.** false
**f.** true
**g.** true

**6.** to protect the junction from excess current when the barrier voltage is exceeded in the forward direction or the breakover voltage is exceeded in the reverse direction

**7.** to provide variable capacitance which is controlled by a dc voltage (reverse-bias voltage)

**8.** The depletion region is eliminated.

**9.** Silicon diodes are more prevalent because they have less leakage current.

**10.** negative

**11.** forward-biased

**12.** cathode

**13.** zener diode

**14.** 10 kΩ

**15.** no

**16.** no

**17.** PRV (peak reverse voltage), $I_f$ (average forward current), $I_{fm}$ (maximum peak forward current), $I_{fpr}$ (peak repetitive forward current)

**18.** 400 mA

**19.** small as possible

**20.** The bridge rectifier would have approximately twice as much output voltage $V_{dc}$ as the half-wave rectifier. The diodes would have the same PIV rating for both rectifier circuits.

**21.** $V_{dc} = 17.8$ V, PIV = 56.6 V, and $I_f = 356$ mA

**22.** **a.** larger (approximately twice as large)
**b.** same
**c.** larger (approximately twice as large)
**d.** same

**23.** negative

**24.** negative

**25.** pulsating direct current

**26.** to allow for tolerances in line voltage and transformer voltage, and to allow for noise pulses carried on the alternating current to be rectified

**27.** fluctuating

**28.** ripplc

**29.** before

**30.** discharge

**31.** no

**32.** so that the $I_{fpr}$ rating of the diode will be large enough to handle the peak current when the capacitor charges

**33.** **a.** increase, **b.** decrease, **c.** increase

**34.** 3.3 percent

**35.** 8.1 percent

**36.** 120 Hz

**37.** **a.** 42.42 V, **b.** 42.42 V

**38.** half-wave

**39.** to discharge the filter capacitor

**40.** It is a resistor, in series with the rectifier diode, which limits surge current when power is applied to a power supply. It is used when the internal resistance of the ac source is insufficient to limit the surge current to a value that the diode can tolerate.

**41.** zener diode

**42.** parallel

**43.** yes

# Chapter 24 Electronic Amplification

Electronic amplification can be performed by transistors—either bipolar transistors or field-effect transistors. After developing your understanding of these devices, this chapter will introduce you to the more common amplifier circuits in which transistors are used. Finally, you will develop an understanding of the more important specifications used in describing amplifiers.

## 24-1 BIPOLAR TRANSISTOR

*Bipolar junction transistors* (BJT), often called *junction transistors* or *bipolar transistors,* are composed of two PN junctions in a monolithic crystal. As shown in Fig. 24-1 either the P-type material or the N-type material is common to both junctions. Therefore, there are two types of bipolar transistors: NPN and PNP. The material which is common to both junctions is the *base* B of the transistor. A PNP transistor has an N-type base, while the NPN has a P-type base. As seen in Fig. 24-1, the other two parts of the transistor are called the *emitter* E and the *collector* C. Although the emitter and collector are made of the same type of material, they are not interchangeable because of differences in their shape and size. The three parts of the transistor serve the following functions: (1) the emitter emits (provides a source of) current carriers, (2) the base controls the current carriers by determining the number that get to the collector,

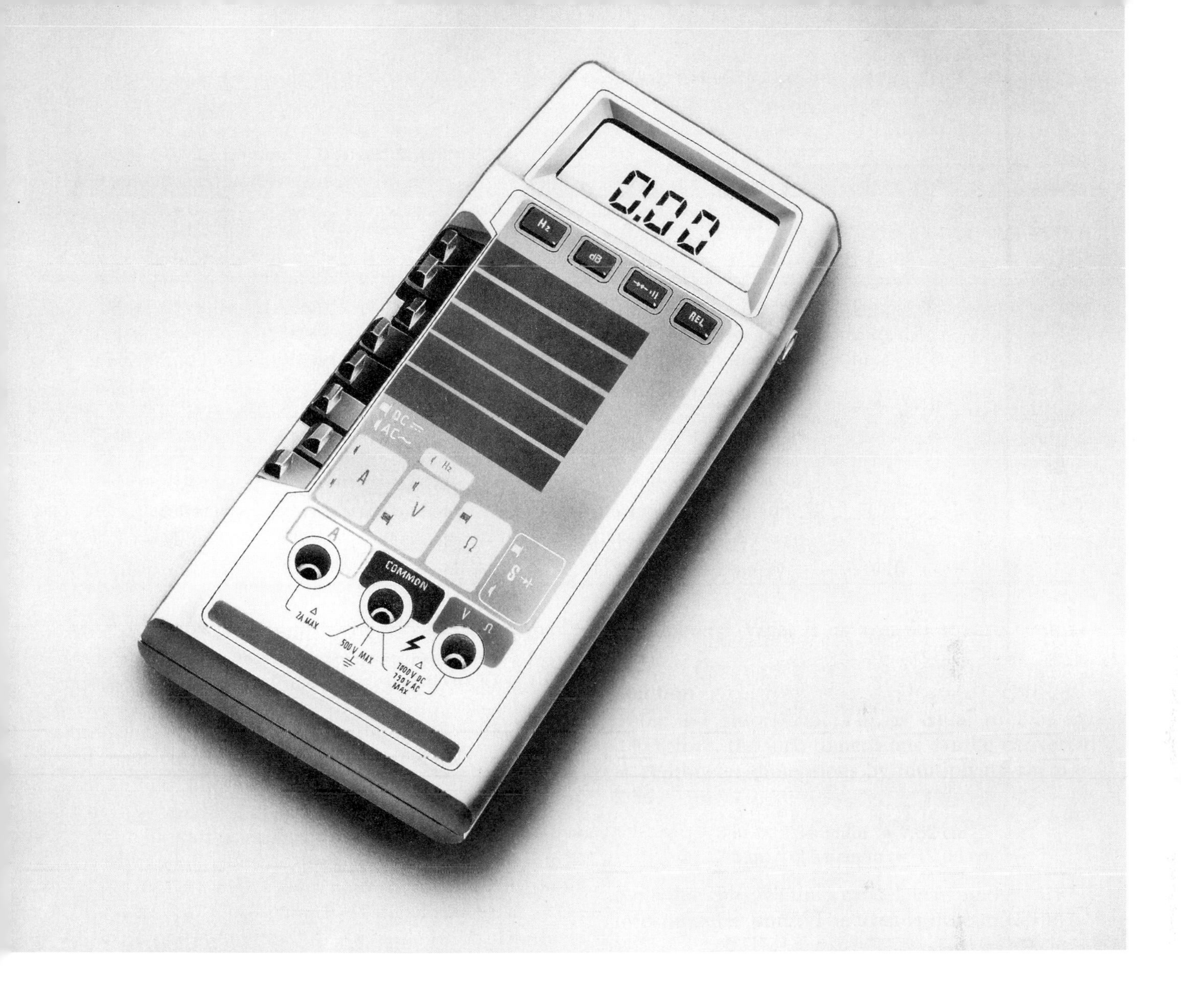


and (3) the collector collects (receives) the current carriers emitted by the emitter and controlled by the base. How these functions are carried out are detailed later on in this chapter.

The two-junction transistors in Fig. 24-1 are called *bipolar* transistors because two polarities of current carriers are involved in their operation. For example, in the NPN transistor, current flowing from the emitter to the collector is carried by electrons (negatively charged), while base current is carried by holes (positively charged).

## Transistor Biasing

To be useful as amplifiers, transistors must have dc voltages applied to them to bias their junctions. Figure 24-2 shows that the base-emitter junction is forward-biased, and the base-collector junction is reverse-biased. This is the correct biasing for bipolar transistors. In Fig. 24-2*a* the forward-biased junction conducts current like any other forward-biased diode does. That is, the depletion region is eliminated and current is supported by majority carriers combining in the junction region. The magnitude of the current is controlled by the applied voltage. For a small silicon transistor, which is by far the most common type, the voltage is generally 0.6 to 0.7 V and the current is in the microampere to milliampere range.

The reverse bias applied to the base-collector junction in Fig. 24-2*b* establishes a junction-depletion region which limits current to the thermally

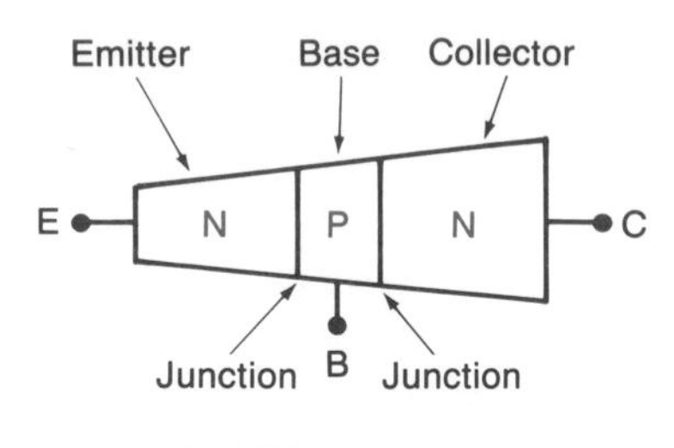


(*a*) NPN transistor

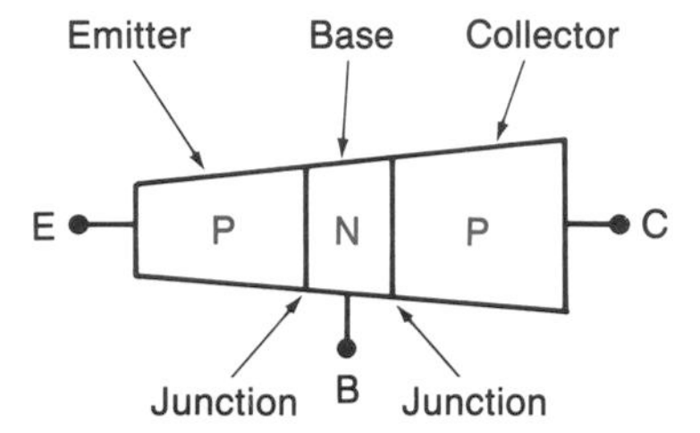


(*b*) PNP transistor

**FIGURE 24-1 Two types of bipolar (junction) transistors.**

generated leakage current. This current is identified as $I_{CBO}$, which means the current from collector C to base B when the emitter is open O-circuited. Current $I_{CBO}$ is typically less than 1 $\mu$A for small silicon transistors. The reverse voltage applied to the base-collector junction typically ranges from a few volts up to over 100 V.

Figure 24-2*c* shows the currents that flow when both junctions are biased at the same time, which is the way that transistors are operated. First, notice in this figure that the leakage current is now routed through the forward-biased emitter-base junction. It must be because electrons cannot travel in two directions at the same time in the base-lead wire. At lower temperatures this leakage current is so small compared to the other currents in the transistor that it can be ignored. Second, notice in Fig. 24-2*c* that there is a majority current flowing through the reverse-biased junction. This current is many times larger than the majority current that flows out of the base of the transistor. How this majority current (collector current) manages to travel through a depletion region requires some explanation.

## Transistor Structure

The first requirement for understanding collector current is to know some details about the structure of a transistor. As illustrated in Fig. 24-3, the base region is very thin. It is typically less than 0.001 in. between the junctions. The base is very lightly doped compared to either the collector or the emitter. The emitter is even more heavily doped than is the collector. Also notice in Fig. 24-3 that the collector is larger than the emitter. It is larger so that it can collect the current carriers that

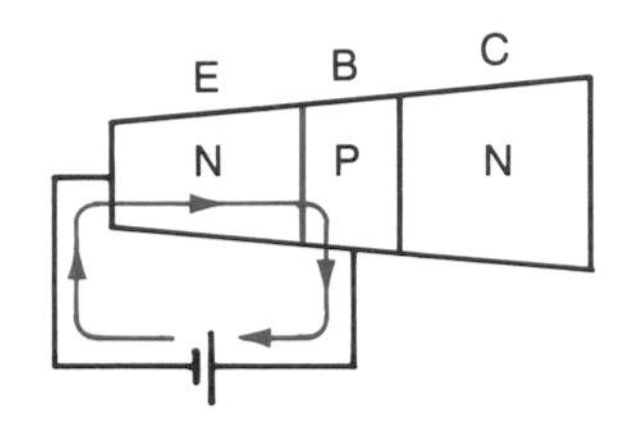


(*a*) Base-emitter is forward-biased. Substantial current is supported by majority carriers.

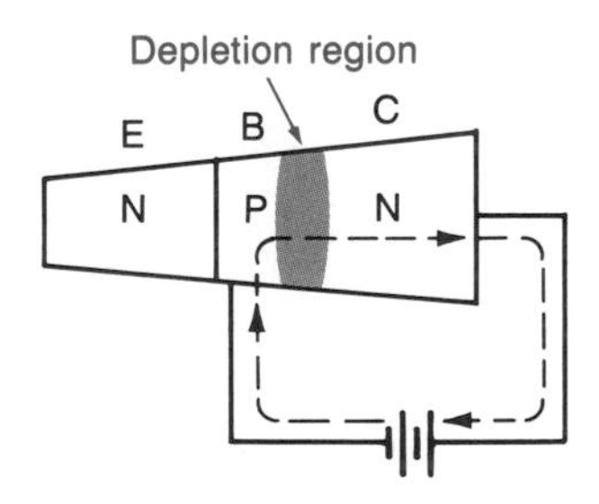


(*b*) Base-collector is reverse-biased. Minute leakage current is supported by minority carriers.

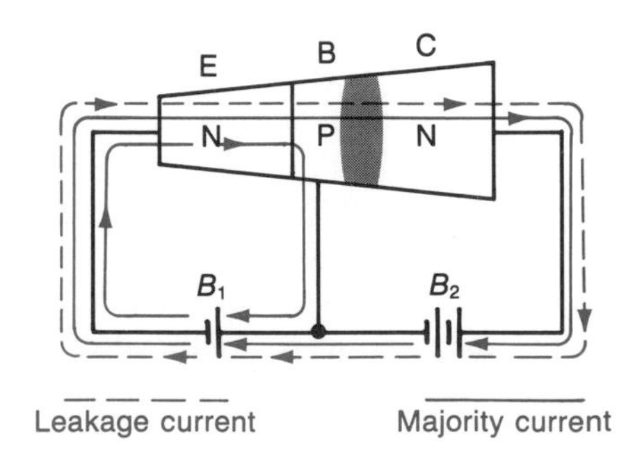


(*c*) Simultaneous biasing of both junctions

**FIGURE 24-2 Biasing a bipolar transistor.**

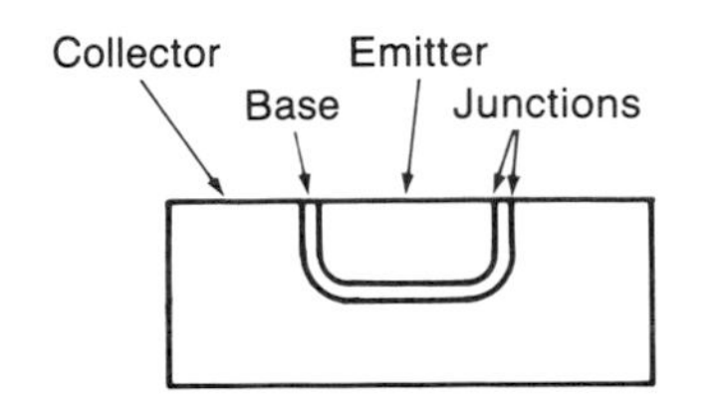


**FIGURE 24-3 Cross-sectional view of an NPN transistor crystal.**

spread out as they are injected into and diffuse through the base. Also, the collector is larger because it must dissipate the heat generated by the carriers traveling from the emitter to the collector.

## Transistor Action

Refer back to Fig. 24-2. Notice that the voltage source $B_1$ not only forward-biases the base-emitter junction, but it also series-aids $B_2$ in applying an electric force field from the emitter to the collector. Since the base-emitter barrier has been removed by forward bias, this electric force field forces (emits) many electrons into the base region of the transistor. But remember that the base is very thin and very lightly doped; therefore, there are very few holes in the base region compared to the number of electrons emitted by the highly doped emitter. Thus, only a few of the electrons combine with holes and become base current. The rest of the electrons emitted into the base region are forced through the reverse-biased base-collector junction by the strong electric force field set up by $B_1$ and $B_2$. These electrons are collected by the collector and become the majority collector current shown in Fig. 24-2*c*.

If the voltage of $B_1$ in Fig. 24-2*c* is slightly reduced, then the base-emitter barrier is only partially removed and fewer electrons are emitted into the base. Therefore, the collector current is reduced. Of course, reducing the amount of base-emitter voltage also reduces the amount of base current, just like reducing the forward bias on a diode reduces diode current. Therefore, we can conclude that *the base current controls the collector current in a bipolar transistor.* So far, the NPN structure has been used exclusively in explaining transistors. The PNP type could have been used just as well if the voltage polarities were reversed and holes substituted for electrons.

## Current Relationships

Again refer back to Fig. 24-2*c*. This time apply Kirchhoff's current law to the junction connecting $B_1$, $B_2$, and the base. Notice that the base current $I_B$ and the collector current $I_C$ enter the junction and only the emitter current $I_E$ leaves the junction. Therefore, we can write that $I_E = I_B + I_C$. This relationship is extremely useful in analyzing transistor circuits.

It was previously pointed out that most of the emitter current goes through the base and becomes collector current, but a small fraction splits off and becomes base current. This is illustrated in Fig. 24-4. The ratio of collector current to base current is called *beta* ($\beta$). Mathematically, $\beta$ is defined as

$$\beta = \frac{I_C}{I_B}$$

Beta, of course, is a pure number because the current units cancel out. Beta, then, describes how well the base current controls the collector current. It is sometimes referred to as "the current gain" of the transistor because a small current in the base controls a large current in the collector. Beta ranges from about 10 for high-power transistors to over 200 for small low-power transistors. The $\beta$ defined above and used throughout this text is known as dc $\beta$. Another $\beta$, ac $\beta$, is sometimes used. It is defined as $\Delta I_C/\Delta I_B$. Alternating current $\beta$ and dc $\beta$ are usually about the same value for a given transistor at frequencies in the audio range. Both $\beta$'s have very large tolerances, so dc $\beta$ suffices for low-frequency circuits.

---

**EXAMPLE 24-1**

A transistor has 5.00 mA of emitter current and 40.0 $\mu$A of base current. Determine the collector current and the $\beta$ of the transistor.

**Solution**

$$I_E = I_C + I_B$$

therefore,

$$I_C = I_E - I_B = 5.00 \text{ mA} - 0.04 \text{ mA} = 4.96 \text{ mA}$$

The $\beta$ of the transistor is

$$\beta = \frac{I_C}{I_B} = \frac{4.96 \text{ mA}}{0.04 \text{ mA}} = 124$$

---

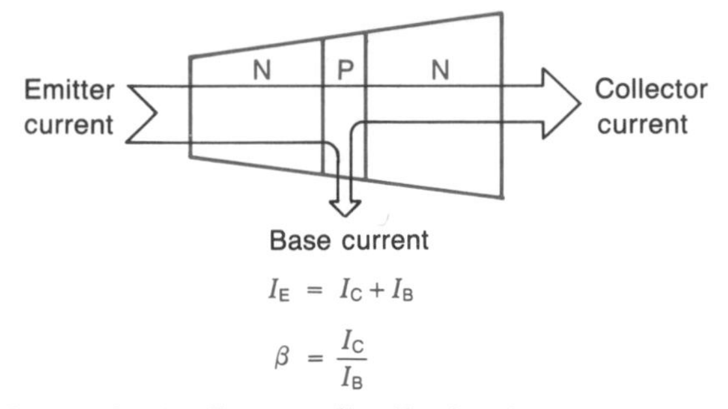


**FIGURE 24-4 Current distribution in a transistor.**

From Example 24-1, you can see why $I_C$ and $I_E$ are sometimes used interchangeably when using approximate methods to analyze transistor circuits. Many current meters do not have sufficient accuracy and precision to detect the difference between 5.00 and 4.96 mA.

**EXAMPLE 24-2**

A transistor has a $\beta$ of 80, and it is desired to have approximately 40 mA of emitter current. How much should the base current be?

**Solution**

There are methods to find a precise answer to this question, but we have not discussed them at this time. So, let us approximate the answer with $I_C \approx I_E$:

$$I_C \approx I_E = 40 \text{ mA}$$

$$\beta = \frac{I_C}{I_B}$$

Therefore,

$$I_B = \frac{I_C}{\beta} = \frac{40 \text{ mA}}{80} = 0.5 \text{ mA}$$

Another current ratio which is sometimes specified for a transistor is alpha ($\alpha$). Mathematically, $\alpha$ is defined as

$$\alpha = \frac{I_C}{I_E}$$

Since $I_C$ is always less than $I_E$, $\alpha$ is always less than one.

## Self-Test

**1.** Why are two-junction transistors called bipolar transistors?

**2.** What type of material is used for the base of a PNP transistor?

**3.** The part that is common to both junctions of a transistor is the ______ of the transistor.

**4.** Which junction of a transistor is forward-biased?

**5.** Which junction of a transistor is reverse-biased?

**6.** Why is it incorrect to say that "the emitter-collector junction is not biased"?

**7.** Is the emitter or the collector of a transistor physically larger?

**8.** Which is larger, the base-emitter voltage or the emitter-collector voltage?

**9.** What does $I_{CBO}$ stand for?

**10.** What are the characteristics of the base that allow most of the emitter current to become collector current?

**11.** In a transistor, the ______ current is always greater than the collector current.

**12.** Determine the collector current of a transistor in which the base current is 1.2 mA and the beta is 70.

**13.** Determine the emitter current for the transistor in Question 12.

**14.** Eighty microamperes of base current causes 12 mA of collector current in a transistor. Determine the $\beta$ for the transistor.

## Schematic Symbols

The pictorial drawing of the doped crystal that has been used so far to represent a transistor is fine for explaining how a transistor operates. However, it is too complex and bulky to be used in schematic diagrams. The *schematic symbols* for bipolar (junction) transistors are drawn in Fig. 24-5. Quite often the circle in the symbol is omitted—especially in complex diagrams. Notice that the only difference between NPN and PNP is the direction of the arrowhead on the emitter lead. Fortunately, the same current-flow conventions are used for transistors as for diodes; i.e., (1) electron flow is against the

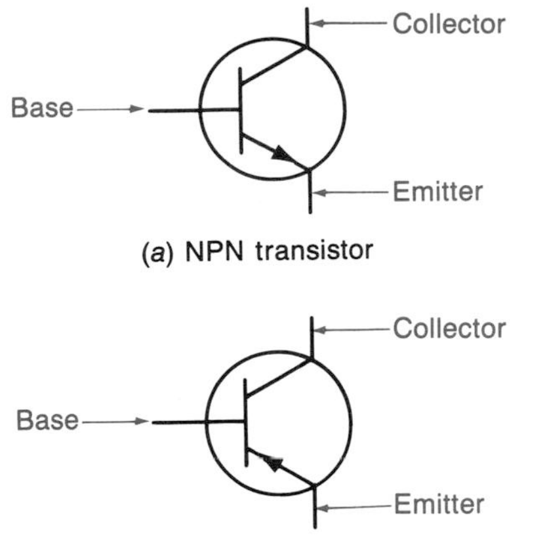


**FIGURE 24-5 Schematic symbols for bipolar transistors.**

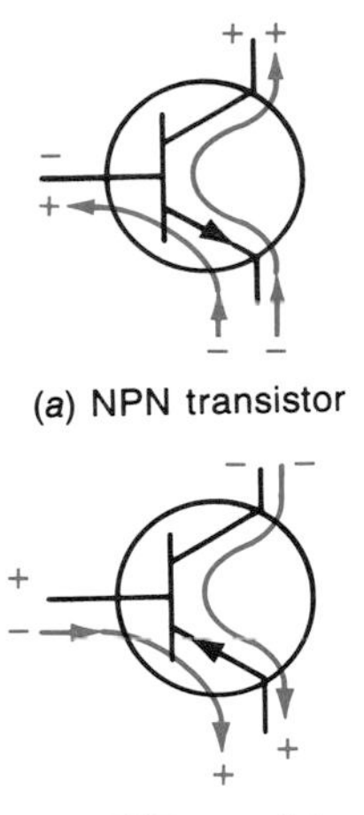


**FIGURE 24-6 Voltage polarities and direction of current (electron) flow in NPN and PNP transistors.**

arrowhead, and (2) forward bias requires that the P-type material be positive with respect to the N-type material. Correct polarities, and resulting directions of current, for both the PNP and the NPN transistors are shown in Fig. 24-6. Notice that the bases are labeled both − and + in this figure. This means that, for the NPN, the base is + with respect to the emitter, but negative with respect to the collector. If the emitter is used as the reference point, then, for the PNP, both the base and the collector are negative; however, the collector is more negative than the base.

## Transistor *I-V* Curves

A family of *collector-characteristic curves* for a low-power (often referred to as a *small-signal*) transistor are illustrated in Fig. 24-7. Notice that there is a curve for each value of base current—thus, the title "family of. . . ." Figure 24-7 shows the *average curves* for all transistors with a specified type number. The curves for an individual transistor with this specified type number may vary a considerable amount from these average curves. Also notice that the graph in Fig. 24-7 is labeled $T_c = 25°C$, where $T_c$ stands for case temperature. This tells us that the curves are correct when the case temperature of the transistor is 25°C. If the temperature increases, the curves are shifted up and a given base current provides more collector current. This increase in collector current is caused by an increase in leakage current $I_{CBO}$. Of course, $I_{CBO}$ is temperature-sensitive. Careful examination of these curves helps in further developing your understanding of transistors and transistor circuits.

From characteristic curves, one can see that the parameters and ratings of a transistor depend on what portion of the curve the transistor is operating. For example, suppose $\beta$ is calculated at the intersection of $V_{CE} = 4$ V and $I_B = 100\ \mu A$ (see Fig. 24-8). Projecting left to the $I_C$ scale shows $I_C = 10$ mA. Therefore, at this intersection,

$$\beta = \frac{I_C}{I_B} = \frac{10\text{ mA}}{0.1\text{ mA}} = 100$$

At the intersection of $V_{CE} = 16$ V and $I_B = 100\ \mu A$, the collector current is 12.5 mA and

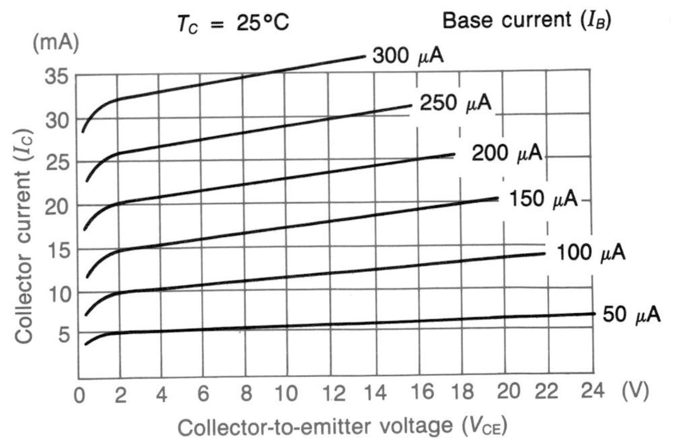


**FIGURE 24-7 Family of collector-to-emitter curves for an NPN transistor.**

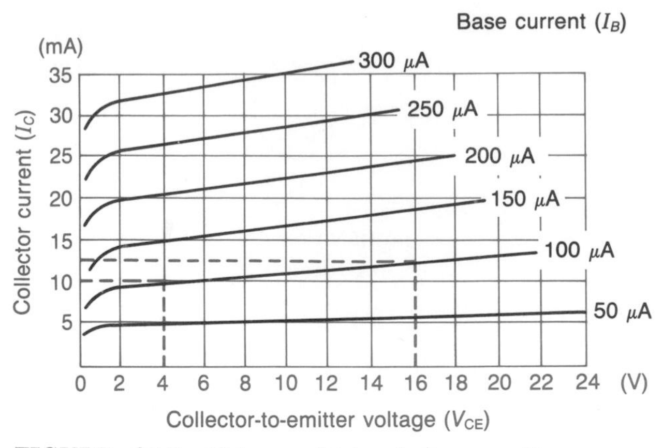


**FIGURE 24-8 Values obtained from collector-to-emitter curves can be used to determine $\beta$ and $r_{ce}$.**

$$\beta = \frac{I_C}{I_B} = \frac{12.5 \text{ mA}}{0.1 \text{ mA}} = 125$$

Because $\beta$ varies as $I_C$ varies, as shown above, it is often specified at some value of $I_C$. Also, because of the variation between transistors with the same type number, $\beta$ is usually specified as either a minimum value or a range. For example, $\beta$ may be specified as 20 to 100 at $I_C = 200$ mA, or it may be specified as 25 minimum at 1 A. In all cases the temperature is assumed to be 25°C unless otherwise specified.

Another useful parameter that can be obtained from the curves in Fig. 24-8 is the *collector-to-emitter dynamic resistance* $r_{ce}$ of the transistor. This dynamic resistance is defined as

$$r_{ce} = \frac{\Delta V_{CE}}{\Delta I_C} \mid I_B$$

where $\mid I_B$ means that $I_B$ is held constant.

**EXAMPLE 24-3**

From the curves in Fig. 24-8, determine $r_{ce}$ when $I_B = 100$ μA and $V_{CE}$ changes from 4 to 16 V.

**Solution**

The construction (dashed) lines in Fig. 24-8 show that $I_C$ varies from 10 to 12.5 mA as $V_{CE}$ changes from 4 to 16 V if $I_B$ is held constant at 100 μA. Therefore,

$$r_{ce} = \frac{16 \text{ V} - 4 \text{ V}}{12.5 \text{ mA} - 10 \text{ mA}} = \frac{12 \text{ V}}{2.5 \text{ mA}} = 4800\ \Omega$$

Inspection of the curves in Fig. 24-8 shows that $r_{ce}$ will be less if $I_B$ is increased because the slope of the curve is steeper as $I_B$ increases. The dynamic resistance of the collector-emitter of a transistor is important in determining the output characteristics of a transistor amplifier under various load conditions. Many transistors have values of $r_{ce}$ that are much larger than the transistor represented by the curves in Fig. 24-8.

Another transistor curve, which is useful in understanding transistor circuit behavior, is shown in Fig. 24-9. This curve is for the same transistor as are the curves of Figs. 24-7 and 24-8. As expected, the curve in Fig. 24-9 looks like a forward-biased diode curve because the base and emitter of a transistor form a PN junction just like the PN junction of a diode. The major difference between the two junctions is that the base is lightly doped, which tends to make the dynamic resistance of the BE junction a little higher than that of a rectifier diode.

**EXAMPLE 24-4**

Determine the dynamic resistance $r_{be}$ of the BE junction graphed in Fig. 24-9 as the base current varies between 50 and 250 μA

**Solution**

$$r_{be} = \frac{\Delta V_{BE}}{\Delta I_B} = \frac{0.70 \text{ V} - 0.65 \text{ V}}{250\ \mu\text{A} - 50\ \mu\text{A}} = 250\ \Omega$$

Like the CE curves of Fig. 24-7, this BE curve is influenced by temperature. At a higher temperature it requires less $V_{BE}$ to produce a given value of $I_B$.

## Transistor Ratings

To successfully use transistors, one has to be familiar with some of their more important ratings: $\beta$, $I_{CBO}$, and $\alpha$ have already been discussed. Some additional ratings are discussed below.

**1.** Power dissipation $P_D$, $P_T$, or $P_C$. This is the maximum power the transistor can dissipate. It is calculated with the formula $P_D = I_C \times V_{CE}$. Power transistors have a metal case or metal tab which must be connected to a heat sink. Large power

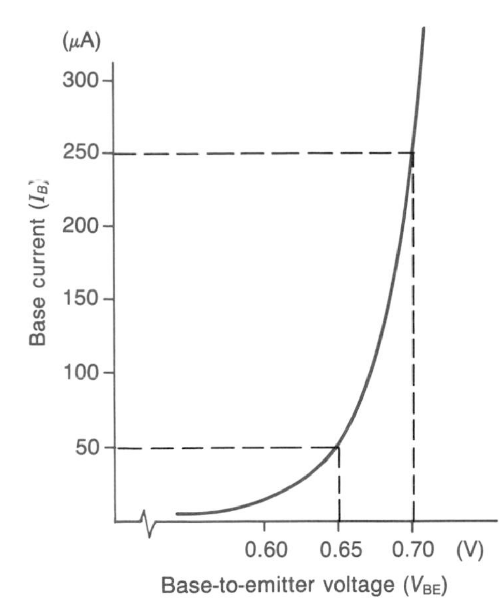


**FIGURE 24-9 Transistor base-to-emitter curve.**

transistors can dissipate their rated power only under ideal conditions—low ambient temperature, unobstructed or forced airflow, and a nearly infinite heat sink. Power $P_D$ must be derated when conditions are less than ideal. Manufacturers provide derating curves for their transistors.

**2.** Collector-base breakover voltage $V_{CBO}$ or $BV_{CBO}$. This is the maximum reverse-bias voltage that should be applied to the collector-base junction. A larger voltage may cause avalanche to occur which usually destroys the transistor.

**3.** Collector-emitter breakover voltage $V_{CEO}$ or $BV_{CEO}$. Exceeding this voltage between the collector and emitter can also avalanche the BC junction and destroy the transistor.

**4.** Emitter-base breakover voltage $V_{EBO}$ or $BV_{EBO}$. In most transistor circuits, the BE junction is always forward-biased and this rating is not pertinent. It specifies the maximum reverse bias that should be applied to the BE junction.

**5.** Maximum collector current $I_{C,max}$. This is the maximum collector current the transistor can accommodate. A transistor should not operate at $I_{C,max}$ unless $V_{CE}$ is very low; otherwise, $P_D$ is exceeded and the transistor is destroyed.

**6.** Maximum operating frequency $f_T$. This is the maximum frequency at which the transistor is useful. This frequency limitation is caused by such things as junction capacitance and transit time (time required by current carriers to move out of, or into, the depletion region of a junction).

### EXAMPLE 24-5

Using the curves of Fig. 24-7, calculate the collector power $P_C$ of the transistor if $I_B = 200\ \mu A$ and $V_{CE} = 10$ V.

**Solution**

From the graph of Fig. 24-7, $I_C = 22.5$ mA when $I_B = 200\ \mu A$ and $V_{CE} = 10$ V. Thus, $P_C = I_C \times V_{CE} = 22.5\text{ mA} \times 10\text{ V} = 225\text{ mW}$.

### EXAMPLE 24-6

A transistor has a $P_D$ rating of 600 mW. Because of adverse operating conditions, it is to be derated 50 percent. Determine the maximum $V_{CE}$ when $I_C$ is 40 mA.

**Solution**

$$P_C = P_D \times 0.5 = 600\text{ mW} \times 0.5 = 300\text{ mW}$$
$$P_C = I_C \times V_{CE}$$

Therefore,

$$V_{CE} = \frac{P_C}{I_C} = \frac{300\text{ mW}}{40\text{ mA}} = 7.5\text{ V}$$

## Self-Test

**15.** On the symbol for a PNP transistor, does the arrow point toward or away from the base?

**16.** With an NPN transistor, should the base be negative or positive with respect to the collector?

**17.** What causes a transistor to be temperature-sensitive?

**18.** Does $\beta$ change as $I_B$ changes?

**19.** Does $\beta$ change as the temperature changes?

**20.** Refer to Fig. 24-7 and determine $\beta$ when $I_B = 250\ \mu A$ and $V_{CE} = 7$ V.

**21.** Refer to Fig. 24-7 and determine $r_{ce}$ when $I_B = 50\ \mu A$ and $V_{CE}$ changes from 4 to 24 V.

**22.** Compare the dynamic resistance of the BE junction of a small-signal transistor to that of a rectifier diode.

**23.** Why do transistors have maximum voltage ratings?

**24.** List three voltage ratings of a transistor.

**25.** What factors limit the high-frequency response of a transistor?

**26.** What happens to a transistor if it is operated at $BV_{CEO}$ and $I_{C,max}$ simultaneously?

## 24-2 TRANSISTOR AMPLIFIERS

The purpose of an *amplifier* is to convert power from a dc source, such as a battery or a dc power supply, into signal power. For example, a microphone produces a weak signal (microwatts of power). As illustrated in Fig. 24-10, the microphone signal can be applied to the input of an amplifier and amplified. The signal coming out of the amplifier will contain the same frequencies and waveform variations as the input signal, but it will be much stronger (watts of power). The increase in the power level of the signal comes from the power supply that powers the amplifier. The amplifier symbol in Fig. 24-10 may represent a single transistor amplifier (single stage) or multiple transistor amplifiers (multiple stages) connected together. Amplifiers can increase the signal power by amplifying the signal voltage, the signal current, or both the current and the voltage. An amplifier's ability to amplify is expressed as the *gain* of the amplifier. The *voltage gain* $A_V$ of an amplifier is the ratio of the output signal voltage $V_o$ over the input signal voltage $V_i$. Thus, if the input signal is 20 mV and the output signal is 1.5 V, the voltage gain would be

$$A_V = \frac{V_o}{V_i} = \frac{1.5\text{ V}}{0.02\text{ V}} = 75$$

Notice that gain is a pure number because the units always cancel out. The formulas for figuring *power gain* and *current gain* are

$$A_P = \frac{P_o}{P_i} \quad \text{and} \quad A_I = \frac{I_o}{I_i}$$

The power gain of an amplifier is also equal to the product of the current gain and the voltage gain. Thus $A_P = A_V \times A_I$. From this relationship it is apparent that either $A_V$ or $A_I$ can be less than one and still have $A_P$ greater than one, which is a requirement for an amplifier.

A single-stage amplifier can be either linear or nonlinear. A linear amplifier produces an output signal which contains an amplified reproduction of all of the input signal. Both the negative and positive $\frac{1}{2}$ cycles are amplified equally. The amplifier in Fig. 24-10 is a linear amplifier. The output of a nonlinear amplifier does not amplify all parts of the input signal equally. A nonlinear amplifier may clip (cut off) part of (or all of) either the negative or positive $\frac{1}{2}$ cycle of the waveform.

Nonlinearity is a form of *distortion.* More specifically, it is a form of *amplitude distortion.* Other forms of distortion are *frequency distortion* and *phase distor-*

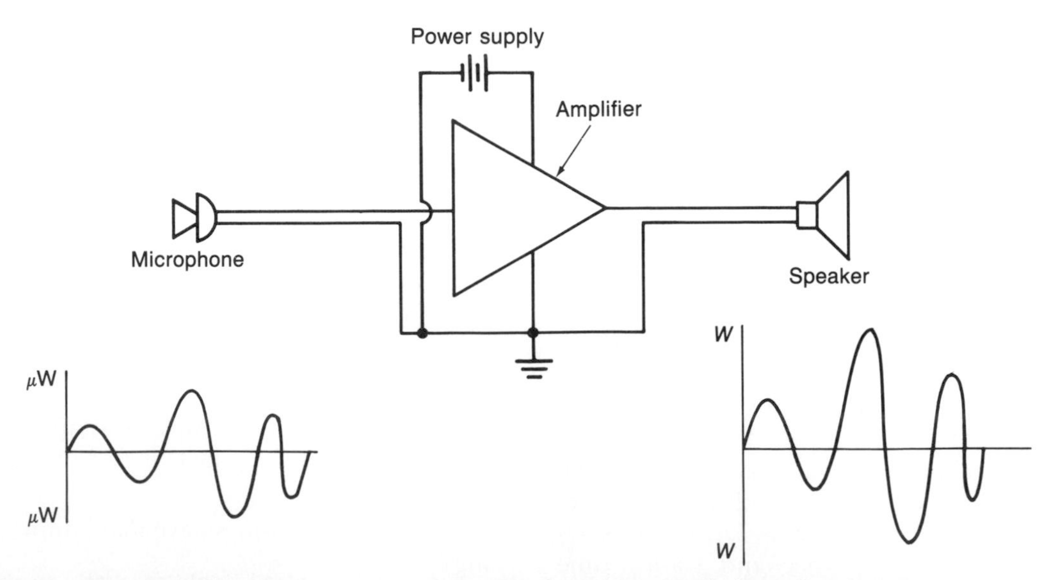


**FIGURE 24-10 The amplifier converts power from the dc power supply into signal power.**

*tion.* Frequency distortion results when one frequency, or band of frequencies, is amplified more than another. Phase distortion is caused by one frequency, or band of frequencies, being phase-shifted more than another as a complex signal is being amplified.

Transistor amplifiers usually contain resistors as well as transistors. The resistors are needed to control current and to convert current changes into voltage changes. In Fig. 24-11*a*, two voltage sources are used to operate the transistor; the base current can be changed only by changing the voltage of voltage source $B_1$. This is not a useful circuit for an amplifier. In Fig. 24-11*b*, one voltage source has been eliminated, but the transistor still operates normally because the base is held positive with respect to the emitter by the voltage applied to the base through the resistor. The base current in Fig. 24-11*b* can now be controlled by changing the value of the resistor. The voltage across the BE junction will still be about 0.7 V and the remainder of the $B_2$ voltage will drop across the resistor.

The circuit in Fig. 24-11*b* still has a major drawback: changing $R_B$ changes $I_B$, and thus $I_C$, but there is no way to convert the change in $I_C$ into a change in $V_{CE}$ ($V_{CE}$ must equal $B_2$). This drawback is eliminated in Fig. 24-11*c* by inserting a resistor in series with the collector lead. Now if $I_C$ changes, $V_{CE}$ has to change also. For example, if $I_C$ increases, Ohm's law tells us that $V_{R_C}$ must also increase ($V_{R_C} = I_C \times R_C$). If $V_{R_C}$ increases, then Kirchhoff's voltage law tells us the $V_{CE}$ must decrease ($V_{CE} = V_{B_2} - V_{R_C}$).

The circuit in Fig. 24-11*c* will work as an amplifier, but it is not a schematic diagram for an amplifier because it does not use the correct symbol for a transistor. Therefore, the circuit is redrawn in Fig. 24-12*a* using the appropriate symbol. This diagram is correct, but it can be further simplified, as is usually done, by eliminating the power supply (battery) symbol as in Fig. 24-12*b*. In Fig. 24-12*b*, the power supply is assumed to be connected be-

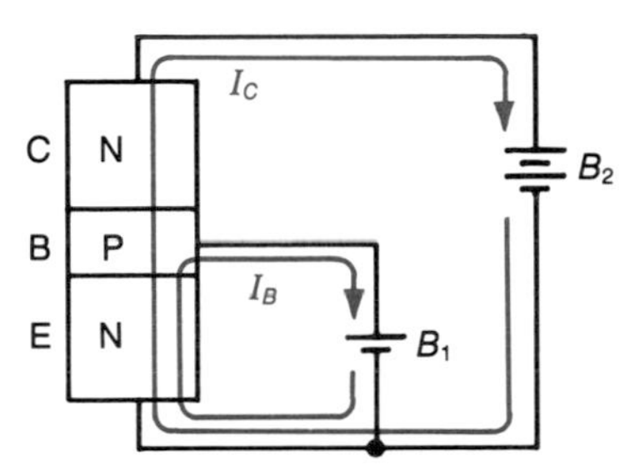


(*a*) NPN transistor with two voltage supplies

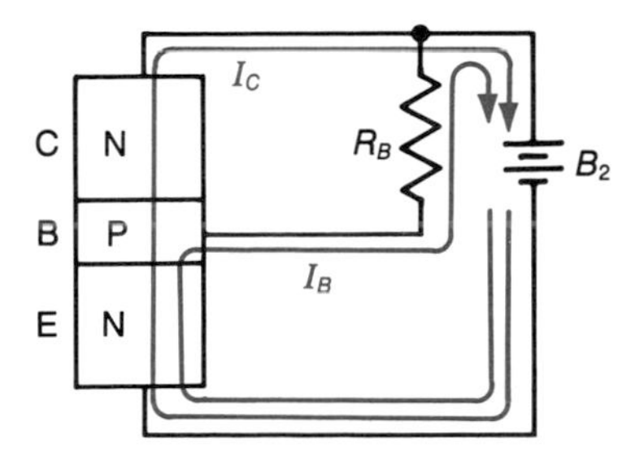


(*b*) One voltage supply eliminated

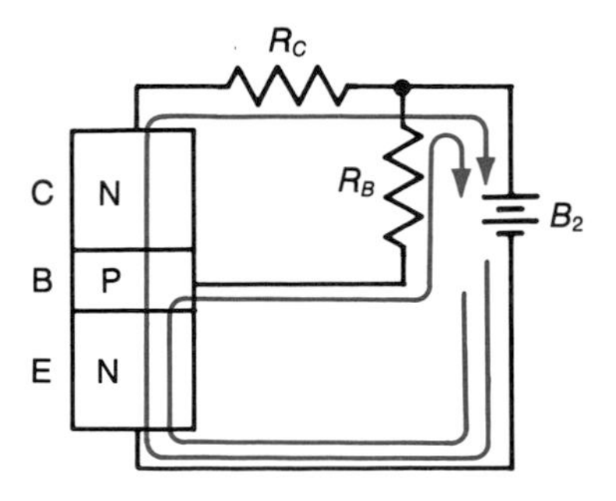


(*c*) $R_C$ allows $V_{CE}$ to change as $I_C$ changes

**FIGURE 24-11 Adding resistors to convert a transistor circuit into a transistor amplifier circuit.**

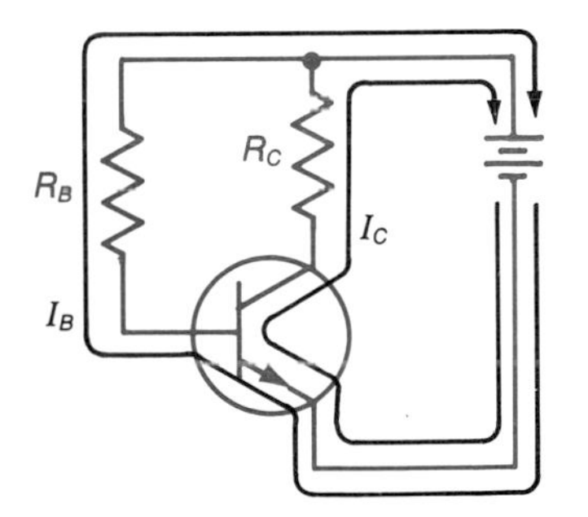


(*a*) Transistor amplifier with power supply shown

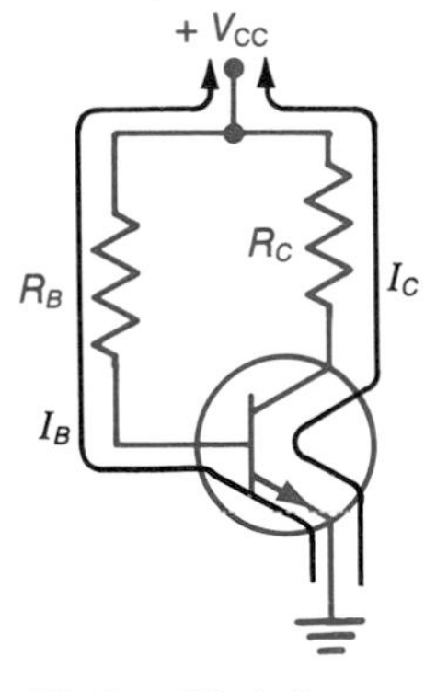


(*b*) Simplified diagram

**FIGURE 24-12 Schematic diagrams of a transistor amplifier.**

tween $V_{CC}$ and the ground (common) symbol. If Fig. 24-12 used a PNP transistor, the battery symbol would be reversed and $V_{CC}$ would be marked negative.

## 24-3 COMMON-EMITTER AMPLIFIER NO. 1

The simplest form of a *common-emitter* amplifier is shown in Fig. 24-13*a*. This simple amplifier is seldom used because of its extreme dependence on both the temperature and the $\beta$ of the transistor. Changes in either one can change this amplifier from a linear to a nonlinear amplifier. It is only included as an example because its operation is easy to understand and analyze.

With a common-emitter amplifier the input is applied to the base (between base and ground or common) and the output is taken from the collector (between collector and ground or common). Since the emitter is connected to ground or common, the emitter is common to both the input and the output—thus, the name "common-emitter."

The capacitors connected to the input and output in Fig. 24-13*a* are included to isolate the dc voltages at the base and the collector from the input and output devices which will be connected to the amplifier. These capacitors are often referred to as *coupling capacitors* because they couple the amplifier to input and output devices.

To fully understand how an amplifier operates, one must first determine the voltages and currents in the circuit before the signal to be amplified is connected to the amplifier. These voltages and current are called the *quiescent values* or *dc values*. The quiescent values can be approximated from the component values given in Fig. 24-13*a* if we remember the major characteristics of a transistor. Figure 24-13*b* shows the current paths, and resulting polarity of the voltage drops, in the circuit. Applying Kirchhoff's voltage law to the base side of the circuit yields the formula $V_{CC} = V_{R_B} + V_{BE}$. Rearranging the formula yields $V_{R_B} = V_{CC} - V_{BE}$. We know that a forward-biased junction drops about 0.7 V; therefore, for the circuit of Fig. 24-13,

$$V_{R_B} \approx 24 \text{ V} - 0.7 \text{ V} \approx 23.3 \text{ V}$$

Now we can use Ohm's law to find $I_{R_B}$, which is the base current $I_B$ for the transistor.

$$I_B = I_{R_B} = \frac{V_{RB}}{R_B} = \frac{23.3 \text{ V}}{154 \text{ k}\Omega} \approx 0.15 \text{ mA}$$

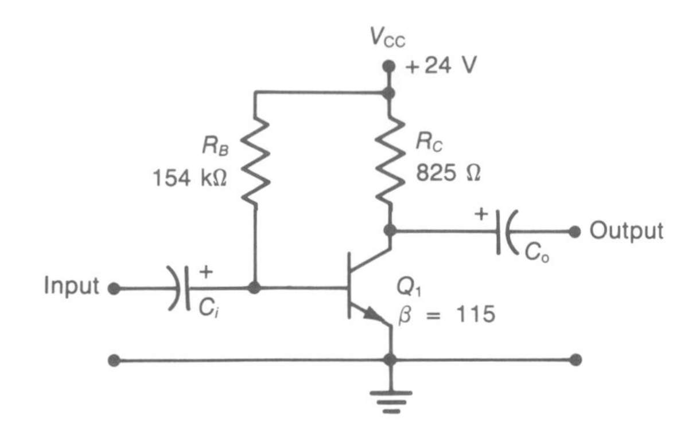


(*a*) Common-emitter amplifier with coupling capacitors

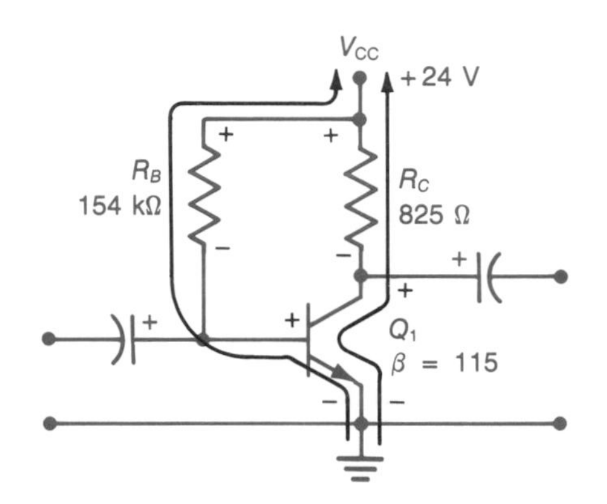


(*b*) Quiescent currents and polarties of voltage drops

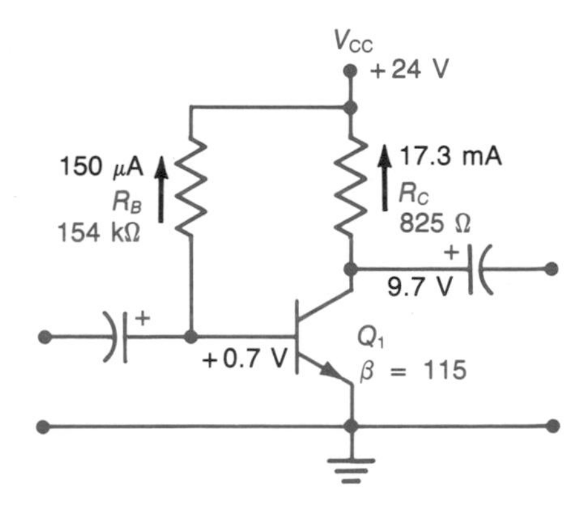


(*c*) Quiescent voltages referenced to common

**FIGURE 24-13 Common-emitter amplifier under quiescent conditions.**

Next, we can use this value of $I_B$ and the value of $\beta$ given in Fig. 24-13 and calculate the value of $I_C$:

$$\beta = \frac{I_C}{I_B}$$

Therefore,

$$I_C = \beta \times I_B = 115 \times 0.15 \text{ mA} = 17.3 \text{ mA}$$

Using this value of $I_C$ and the given value of $R_C$, let us determine $V_{R_C}$ (the voltage across the collector resistor):

$$V_{R_C} = I_C \times R_C = 17.3 \text{ mA} \times 825 \ \Omega = 14.3 \text{ V}$$

Finally, applying Kirchhoff's voltage law to the collector side of the circuits allows us to determine the collector-to-emitter voltage $V_{CE}$:

$$V_{CE} = V_{CC} - V_{R_C} = 24 \text{ V} - 14.3 \text{ V} = 9.7 \text{ V}$$

We now know all of the quiescent voltages and currents for the circuit of Fig. 24-13*b*. These values are shown on the circuit diagram in Fig. 24-13*c*. Notice on this diagram that the voltages are referenced to the common (ground) symbol. (Any voltage with a + or a − sign must have a reference point.) For example, the +9.7 V printed by the collector tells us that the collector is 9.7 V positive with respect to common. In symbolic form we would write $V_C = +9.7$ V. The base voltage for this circuit would be written as $V_B = +0.7$ V.

In Fig. 24-14*a* an ac signal source $G_1$ has been connected to the input of the common-emitter amplifier of Fig. 24-13. Assume the switch $S$ in the signal source is open. Then $C_i$ charges up to 0.7 V after a period of time. This is the quiescent voltage of the capacitor. Once the capacitor has charged, there is no voltage drop across $R_S$ because no current is flowing through it.

For the discussions which follow, it is necessary to realize that $C_i$ and $C_o$ in Fig. 24-14*a* are both large electrolytic capacitors. They are large enough so that the *R-C* time constants of their charge and discharge paths are very long compared to the period of the ac signal applied to the input of the transistor. Under this condition, the voltage across these capacitors remains essentially constant even though charging and discharging currents flow each half cycle of the input signal. (If a capacitor only discharges for one-tenth of a time constant, it does not lose much of its voltage.)

Now let us look at the changes in voltages and currents that occur in Fig. 24-14*a* when the signal source is turned on. In other words, let us look at the ac or signal conditions for the circuit. Suppose the signal source swings to its peak positive value of 0.01 V. In order to satisfy Kirchhoff's voltage law at this instant, either $C_i$ must have discharged down to 0.69 V or $V_{BE}$ must have increased to 0.71 V. Although $C_i$ has been discharging as shown in Fig. 24-14*b*, its voltage has not decreased to 0.69 V because of the long time constant men-

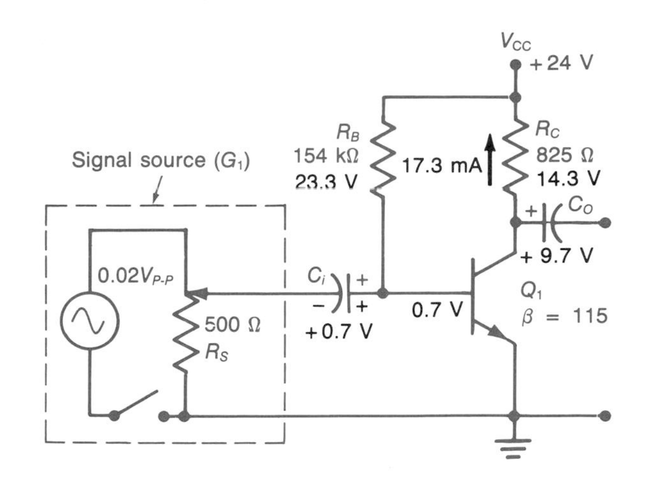


(*a*) Quiescent voltage on the input coupling capacitor.

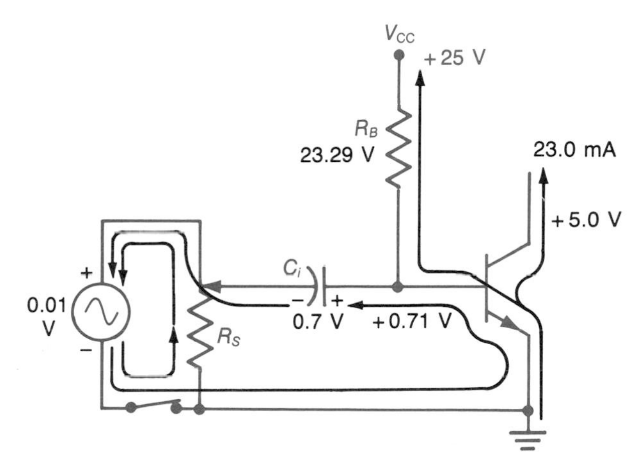


(*b*) $C_i$ aids the signal source and increases $I_B$

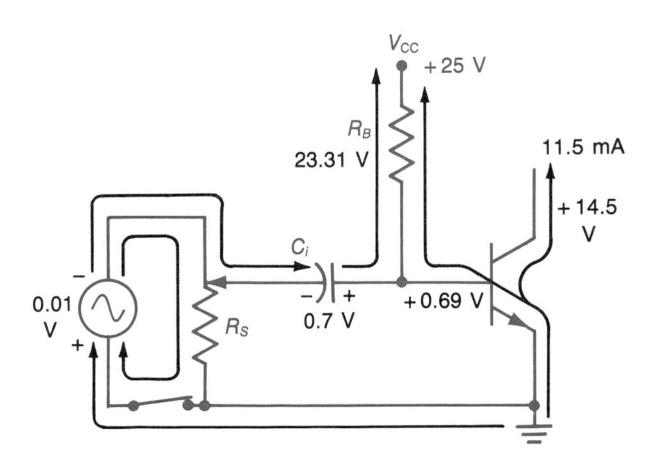


(*c*) $C_i$ charges through $R_B$ increasing $V_{RB}$ and decreasing $V_{EB}$

**FIGURE 24-14 Applying an input signal to a CE amplifier.**

tioned above. Therefore, $V_{BE}$ has increased to 0.71 V because of the additional current $I_B$ through the BE junction. Of course, $V_{R_B}$ has decreased to 23.29 V. Referring to the curve in Fig. 24-15 shows that increasing $V_{BE}$ of transistor $Q_1$ to 0.71 V causes $I_B$ to increase to 200 $\mu$A. If $I_B$ increases to 200 $\mu$A, then $I_C$ increases to $I_C = I_B \times \beta = 200\ \mu\text{A} \times 115 = 23.0$ mA. This means that $V_{R_C}$ must increase to $V_{R_C} = 23.0\ \text{mA} \times 825\ \Omega = 19.0$ V, and $V_{CE}$ must decrease to $V_{CE} = 24\ \text{V} - 19.0\ \text{V} = 5.0$ V. Let us summarize what has happened so far:

1. The input signal increased from 0 to +0.01 V.
2. Voltage $V_B$ increased from +0.70 to +0.71 V.
3. Current $I_B$ increased from 150 to 200 $\mu$A.
4. Current $I_C$ increased from 17.3 to 23.0 mA.
5. Voltage $V_C$ decreased from +9.7 to +5.0 V.

Notice that while $V_B$ swings in a positive direction (+0.70 to +0.71 V), $V_C$ swings in a negative direction (+9.7 to +5.0 V). For a CE amplifier, the input and output are 180° out of phase. As the input signal returns to zero, all voltages and currents return to their quiescent values.

On the negative $\frac{1}{2}$ cycle of the input signal, the source goes to −0.01 V. Now the signal source voltage and the capacitor voltage $V_{C_i}$ are series-opposing. Thus, $V_{EB}$ must decrease to 0.69 V to comply with Kirchhoff's voltage law. As shown in Fig. 24-14*c*, $C_i$ is charging through $R_B$ during the time that the input signal is on its negative swing. Again, referring to Fig. 24-15 shows that at +0.69 V on the base, $Q_1$ has 100 $\mu$A of $I_B$. Using the same procedures as before, we find

$$I_C = 115 \times 100\ \mu\text{A} = 11.5\ \text{mA}$$
$$V_{R_C} = 11.5\ \text{mA} \times 825\ \Omega = 9.5\ \text{V}$$
$$V_{CE} = 24\ \text{V} - 9.5\ \text{V} = 14.5\ \text{V}$$

Summarizing for the negative $\frac{1}{2}$ cycle of the input, we see that a negative-going signal caused $V_B$ to decrease, $I_B$ to decrease, $I_C$ to decrease, and $V_C$ to increase.

Putting the results of the negative and positive $\frac{1}{2}$ cycles together gives a more complete picture of how this amplifier behaves. As illustrated in Fig. 24-16, the collector voltage swings from +14.5 to +5 V while the base swings from +0.69 to +0.71 V. These changes in dc voltages represent the ac component of the waveform and can there-

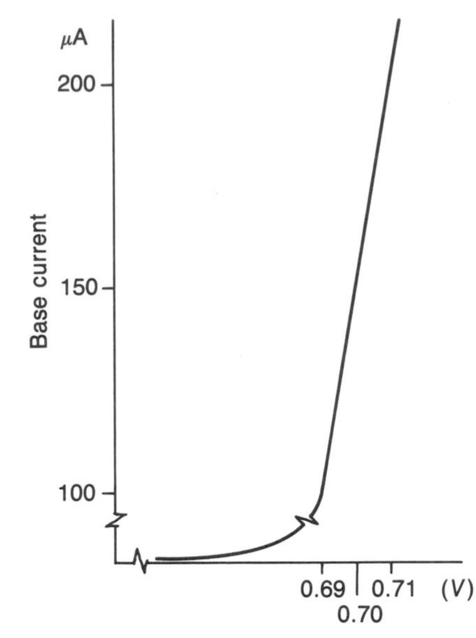


**FIGURE 24-15 Base-to-emitter curve for $Q_1$ in Fig. 24-14.**

fore be used in the gain formula to calculate the gain of this amplifier stage. Thus,

$$A_V = \frac{V_o}{V_i} = \frac{14.5\ \text{V} - 5\ \text{V}}{0.71\ \text{V} - 0.69\ \text{V}} = 475$$

This calculation of voltage gain is not completely correct because we used the same value of $\beta$ (115) in all the calculations of $I_C$ and $V_C$. Yet, we know from previous examination of collector characteristic curves that $\beta$ is not the same for all values of $I_C$.

In the above calculation of $A_V$ we used the change in $V_C$ for $V_o$. We could also have used the change in $V_{R_C}$ because it changed from 9.5 to 19 V while $V_C$ changed from +14.5 to +5 V. From this evidence, you can see why it is proper to speak of the output signal developing across $R_C$. The change in $V_{R_C}$ is caused by the change in $I_C$, and the change in $I_C$ is the output signal current (ac component) of the collector current. Therefore, we can use the following expression for the output signal voltage: $V_o = \Delta I_C \times R_C$.

The input signal voltage $V_i$ is equal to the change in $V_{EB}$. Notice in Fig. 24-14 that the $\Delta I_C$ and the $\Delta I_B$ both flow through the emitter-base junction. Thus, the input signal voltage can be expressed as $V_i = (\Delta I_C + \Delta I_B) \times r_e$, where $r_e$ is the dynamic re-

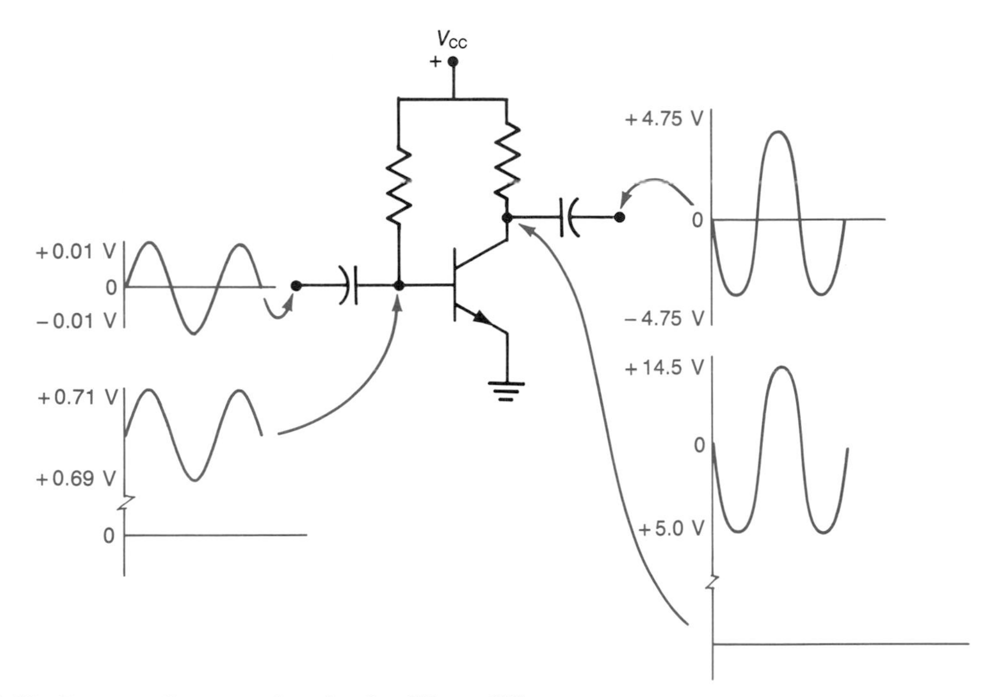


**FIGURE 24-16 Input and output signals of a CE amplifier.**

sistance of the EB junction and is defined as $r_e = \Delta V_{EB}/\Delta I_E$. The $\Delta I_C$ is $\beta$ times as great (115 times in this case) as the $\Delta I_B$. Thus, using the approximation of $\Delta I_C + \Delta I_B \approx \Delta I_C$ leads to only a small error. Using this approximation, we can write $V_i \approx \Delta I_C \times r_e$. Now, by substituting the $V_i$ and $V_o$ relationships we just developed into the voltage gain formula, we get

$$A_V = \frac{V_o}{V_i} \approx \frac{\Delta I_C \times R_C}{\Delta I_C \times r_e} = \frac{R_C}{r_e}$$

The derived formula of $A_V \approx R_C/r_e$ gives us an easy way to estimate the voltage gain if we can find an easy way to estimate $r_e$. An easy way to estimate $r_e$ is to use the formula $r_e = 0.026\ \text{V}/I_{CQ}$, where $I_{CQ}$ is the quiescent collector current. The 0.026 V is a constant derived from a theoretical analysis of a PN junction. This formula takes into account the fact that the slope of the BE junction curve increases as the emitter current increases. Unfortunately, while the $r_e$ formula is easy to use, it is also very inaccurate. It generally predicts too small a value for $r_e$. This results in predicting too great a gain for an amplifier—sometimes the predicted gain using this estimate of $r_e$ is as much as double the actual gain of the amplifier. The overestimate of gain increases as the value of $I_{CQ}$ increases. Using the data from Fig. 24-14, the estimate of $r_e$ would be

$$r_e = \frac{0.026\ \text{V}}{0.0173\ \text{A}} = 1.5\ \Omega$$

and the estimated gain would be

$$A_V = \frac{825\ \Omega}{1.5\ \Omega} = 550$$

The advantage of using the formula $A_V = R_C/r_e$ is that it does not require the use of characteristic curves and it requires relatively few calculations.

Another way to view the operation of an amplifier is to construct a *load line* on the collector characteristic curves. This has been done in Fig. 24-17 for the circuit of Fig. 24-13. The slope of the load line is determined by the value of $R_C$, the collector load resistor. The location of a load line is found by answering two questions:

**1.** What would be the value of $V_{CE}$ if $I_C$ were reduced to zero? When $I_C \approx 0$, the transistor is said to be in *cutoff* or at cutoff. So, the question is: What is the value of $V_{CE}$ when the transistor is cut off?

**2.** What would be the value of $I_C$ if $V_{CE}$ were reduced to zero? The condition when $V_{CE} \approx 0$ is called *saturation*. Again, the question could be re-

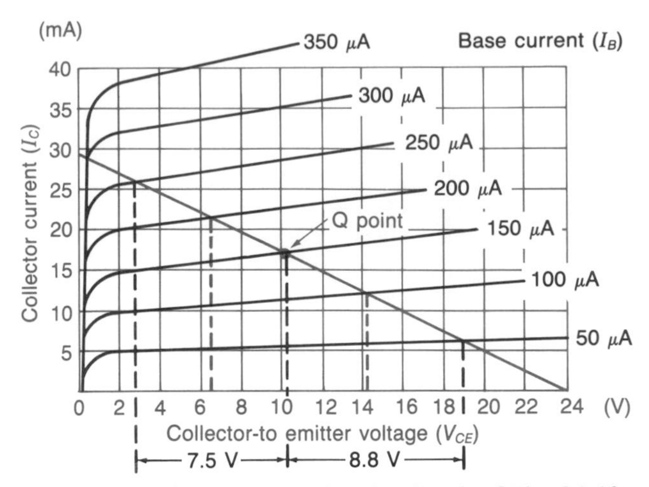


**FIGURE 24-17 Load line for the circuit of Fig. 24-13.**

worded to: What is the value of $I_C$ when the transistor is saturated?

The answer to question 1 is found by inspection of Fig. 24-13*a*. If $I_C = 0$, then $V_{R_C} = 0$, therefore, at cutoff, $V_{CE} = V_{CC} = 24$ V. Thus, one end of the load line is at $I_C = 0$ and $V_{CE} = 24$ V. Likewise, question 2 can be answered by referring to Fig. 24-13*a* and determining $I_C$ when $Q_1$ is saturated. If $V_{CE} = 0$, then $V_{R_C} = V_{CC} = 24$ V and $I_C = I_{R_C} = V_{R_C}/R_C = 24\ \text{V}/825\ \Omega = 29$ mA. The other end of the load line is at $V_{CE} = 0$ and $I_C = 29$ mA. Figure 24-13*c* shows the dc (quiescent) voltages and currents for the circuit. Using these values, the quiescent point, sometimes called the operating point, of the circuit is marked on the load line at the intersection of $I_{CQ}$ (17.3 mA), $V_{CEQ}$ (9.7 V), and $I_{BQ}$ (150 μA). Remember that $V_{CEQ}$ and $I_{CQ}$ were calculated using a $\beta$ of 115. If $\beta$ is not exactly 115 at the Q point, then the three values (17.3 mA, 9.7 V, and 150 μA) may not exactly intersect on the load line. The transistor is operating in the linear region as long as the operating point is never forced to either cutoff or saturation.

Inspection of Fig. 24-17 shows how the transistor operates. If the base current is forced (by the input signal) to increase to 200 μA, then the operating point moves up the load line to the intersection with the 200-μA curve. Projecting down from this intersection shows the $V_{CE}$ decreases to about 6.5 V. When the negative-going input signal forces $I_B$ to 100 μA, the operating point slides down the load line and $V_{CE}$ increases to 14 V.

The load line also illustrates how amplitude distortion can occur in an amplifier. From Fig. 24-17 you can see that a 100-μA increase in $I_B$ (250 μA − 150 μA = 100 μA) causes a 7.5-V decrease in $V_{CE}$, but a 100- μA decrease causes an 8.8-V increase in $V_{CE}$. Thus, equal positive and negative $\frac{1}{2}$ cycles of $I_B$ caused unequal $\frac{1}{2}$ cycles of $V_C$. The actual distortion may not be quite as large as this example indicates because it usually requires a little less $\Delta V_{BE}$ to cause a 100-μA increase in $I_B$ than to cause a 100-μA decrease in $I_B$. From the load line in Fig. 24-17, you can see that an input signal voltage which causes a $\Delta I_B$ of 300 μA drives the operating point from one end of the load line to the other end of the load line. The collector voltage swings from approximately zero to $V_{CC}$ (the transistor is driven from saturation to cutoff). The output signal is as large as it can get. Suppose the input voltage is increased so that $I_B$ swings to 350 μA on the positive $\frac{1}{2}$ cycle. The collector voltage cannot respond to the increase in $I_B$ from 300 to 350 μA. Thus, the negative $\frac{1}{2}$ cycle of the output is clipped as shown in Fig. 24-18. On the negative half of the input waveform the base current is not only driven to zero, but the signal continues to drive the base in a negative direction; it may even reverse-bias the BE junction. The result, as seen in Fig. 24-18, is that the positive $\frac{1}{2}$ cycle of the output signal is also clipped.

Both halves of the signal in Fig. 24-18 were clipped because the transistor was driven into both cutoff and saturation. If the Q point in Fig. 24-17 were moved to the intersection with the 200-μA curve, then the transistor would reach saturation before it reached cutoff and the negative $\frac{1}{2}$ cycle would be clipped before the positive $\frac{1}{2}$ cycle.

You should, by now, have a feel for how a transistor amplifier operates. However, as mentioned before, the simple amplifier of Fig. 24-13 is not too practical because its Q point is dependent on the $\beta$ of the transistor. For example, if $Q_1$ in Fig. 24-13 were replaced with a transistor with a $\beta$ of 200, the transistor would be saturated and the circuit would be useless as a linear amplifier.

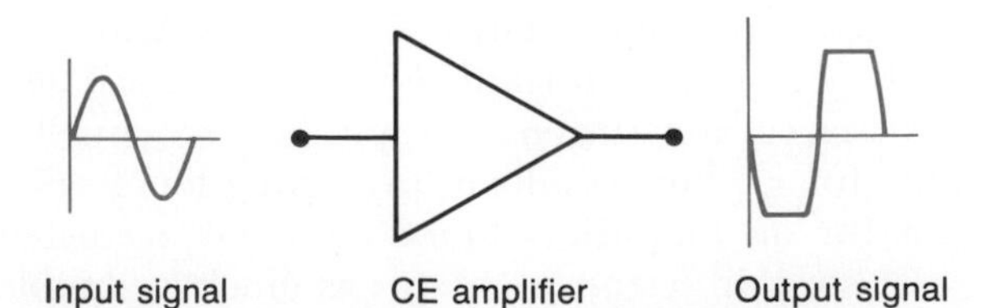


**FIGURE 24-18 An overdriven linear amplifier causes both halves of the output to be clipped when the Q point is in the center of the load line.**

### EXAMPLE 24-7

Prove that $Q_1$ in Fig. 24-13 would be saturated if its $\beta$ were 200.

**Solution**

$$V_{R_B} = V_{CC} - V_{EB} = 24\text{ V} - 0.7\text{ V} = 23.3\text{ V}$$

$$I_B = I_{R_B} = \frac{V_{RB}}{R_B} = \frac{23.3\text{ V}}{154\text{ k}\Omega} = 0.15\text{ mA}$$

$$I_C = \beta \times I_B = 200 \times 0.15\text{ mA} = 30\text{ mA}$$

$$V_{R_C} = I_C R_C = 30\text{ mA} \times 825\ \Omega = 24.8\text{ V}$$

Since $V_{R_C}$ cannot exceed $V_{CC}$, the transistor is saturated and $I_C$ is less than 30 mA.

Load-line analyses are not very practical either, except to see how an amplifier operates. When $\beta$ changes so do the characteristic curves. Of course, as pointed out before, $\beta$ changes dramatically from one transistor to another with the same type number, and $\beta$ changes as temperature changes.

## Self-Test

27. What is an amplifier?

28. Can an amplifier have a power gain > 1 if its current gain is < 1?

29. An amplifier has a 300-mV input signal and a 225-mV output signal. What is its voltage gain?

30. An amplifier has a voltage gain of 50 and a current gain of 30. What is its power gain?

31. The input to a CE amplifier is a sine wave. The output is clipped on the positive ½ cycle.
  a. What type of distortion is occurring?
  b. Is the distortion caused by saturation or cutoff?

32. What does $V_{CC}$ stand for?

33. As long as the transistor stays in the linear region (Q point between cutoff and saturation), does changing the value of $R_B$ or $R_C$ have the most influence on the value of $I_C$?

34. Assuming the transistor stays in the linear region of operation, what would be the effects of increasing $R_C$?

35. What is the phase relationship between the input and output signals of a CE amplifier?

36. What happens to an amplifier when it is overdriven (too much input signal)?

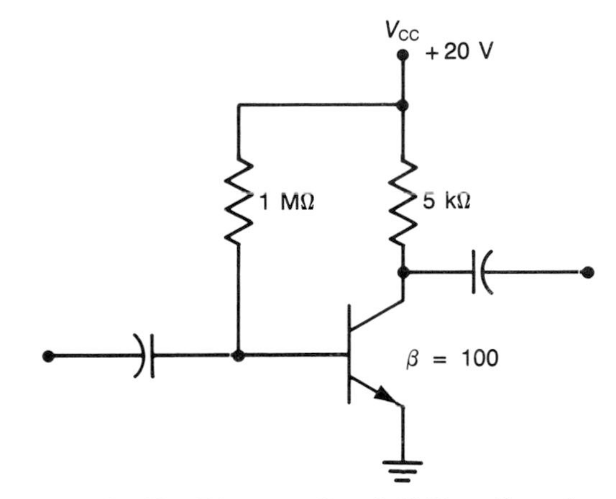


**FIGURE 24-19 Diagram for Self-Test Question 38.**

37. What is $r_e$ the abbreviation for?

38. For Fig. 24-19, determine the following:
  a. $V_C$
  b. $V_B$
  c. $A_V$

## 24-4 COMMON-EMITTER AMPLIFIER NO. 2

We are now ready to work with a more practical amplifier: one that is much less dependent on $\beta$, and one that is connected to a load. Adding a load to an amplifier changes the way the collector circuit responds to a signal, but not the way in which the base circuit responds. Therefore, the techniques developed for the previous circuit are applicable, with only minor modification, to the circuit in Fig. 24-20*a* which has a load resistor $R_L$ connected to its output.

The circuit in Fig. 24-20*a* is often referred to as a *β-independent circuit* because many of its characteristics are nearly independent of the $\beta$ of the transistor. Changing $\beta$ by a factor of 2 has a negligible influence on the quiescent conditions, voltage gain, or maximum unclipped output signal of the amplifier.

Circuits like the one in Fig. 24-20*a* are designed so that the current through $R_{B2}$ and $R_{B1}$ is so large compared to $I_B$ that $I_B$ can be ignored when figuring the voltage distribution across $R_{B2}$ and $R_{B1}$. Using this approximation and the $I_C \approx I_E$ approximation, we can calculate the quiescent voltages and currents for the circuit:

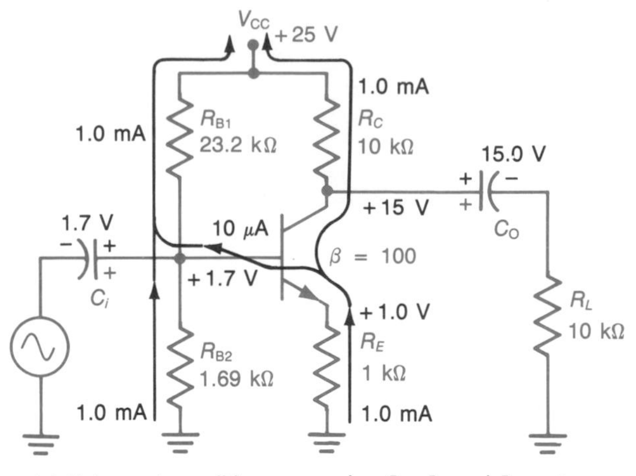


(*a*) Quiescent conditions assuming $I_E \approx I_C$ and $I_B \ll I_{R_{B2}}$

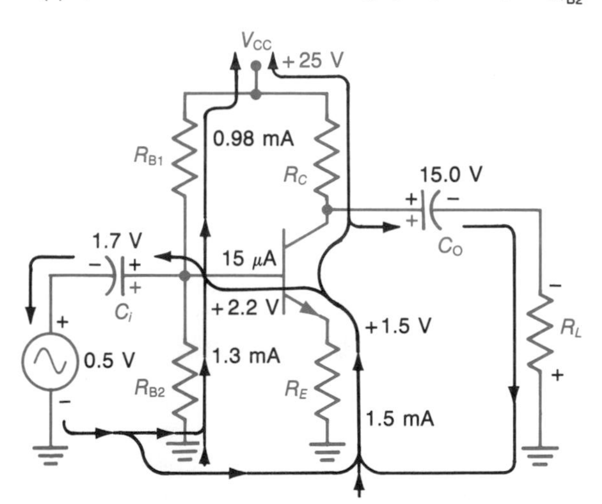


(*b*) Currents and voltages when the input signal swings to +0.5 V

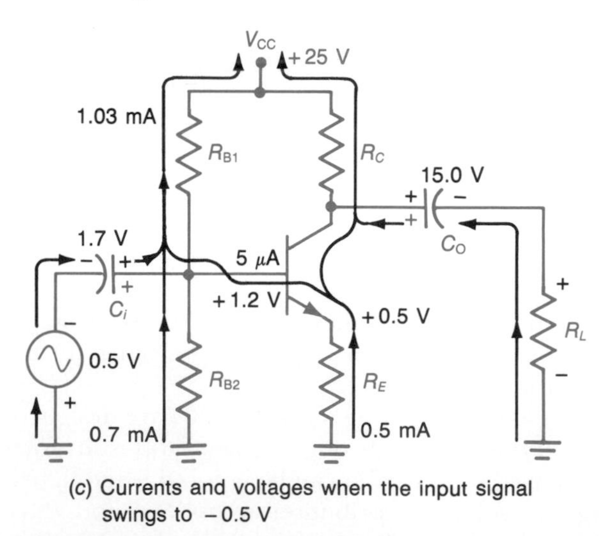


(*c*) Currents and voltages when the input signal swings to −0.5 V

**FIGURE 24-20 Beta-independent CE amplifier with a load connected to its output.**

$$V_B = V_{R_{B2}} = \frac{R_{B2} \times V_{CC}}{R_{B2} + R_{B1}} = \frac{1.69\ \text{k}\Omega \times 25\ \text{V}}{1.69\ \text{k}\Omega + 23.2\ \text{k}\Omega}$$
$$= +1.7\ \text{V}$$
$$I_{R_{B2}} = \frac{V_{R_{B2}}}{R_{B2}} = \frac{1.7\ \text{V}}{1.69\ \text{k}\Omega} = 1.0\ \text{mA}$$
$$V_E = V_B - V_{BE} = +1.7\ \text{V} - 0.7\ \text{V} = +1.0\ \text{V}$$
$$I_C \approx I_E = \frac{V_E}{R_E} = \frac{1.0\ \text{V}}{1\ \text{k}\Omega} = 1.0\ \text{mA}$$
$$V_{R_C} = I_C \times R_C = 1.0\ \text{mA} \times 10\ \text{k}\Omega = 10.0\ \text{V}$$
$$V_C = V_{CC} - V_{R_C} = 25\ \text{V} - 10.0\ \text{V} = +15.0\ \text{V}$$

As discussed for the previous circuit, the coupling capacitors do, under quiescent conditions, charge up to the value of $V_B$ and $V_C$ through the internal resistance of the signal source and the load. Let us calculate $I_B$ to be sure it is small compared to $I_{R_{B2}}$:

$$I_B = \frac{I_C}{\beta} = \frac{1.0\ \text{mA}}{100} = 0.01\ \text{mA}$$

Since $I_{R_{B2}}$ is 1.0 mA, you can see that ignoring $I_B$ introduced approximately 1 percent error into our calculation of $V_B$.

Now let us determine what happens when the signal source changes to +0.5 V as shown in Fig. 24-20*b*. Remember that the coupling capacitor is so large that it cannot lose an appreciable amount of voltage in the time the signal takes to complete the positive ½ cycle. This means that $V_B$ must increase to +2.2 V. When $V_B$ increases to +2.2 V, the current through $R_{B_2}$ increases to $I_{R_{B2}}$ = 2.2 V/1.69 kΩ = 1.3 mA. This increase from 1.0 to 1.3 mA is provided by the signal source. The other voltage and current changes shown in Fig. 24-20*b* were calculated as follows:

$$I_{R_{B1}} = \frac{V_{R_{B1}}}{R_{B_1}} = \frac{V_{CC} - V_{R_{B2}}}{R_{B_1}} = \frac{22.8\ \text{V}}{23.2\ \text{k}\Omega} = 0.98\ \text{mA}$$
$$V_E = V_B - V_{BE} = 2.2\ \text{V} - 0.7\ \text{V} = +1.5\ \text{V}$$
$$I_C \approx I_E = \frac{V_E}{R_E} = \frac{1.5\ \text{V}}{1\ \text{k}\Omega} = 1.5\ \text{mA}$$
$$I_B = \frac{I_C}{\beta} = \frac{1.5\ \text{mA}}{100} = 15\ \mu\text{A}$$

Calculation of $V_C$ and $V_{R_L}$ for Fig. 24-20*b* is complex because $I_C$ splits up to become $I_{R_C}$ and $I_{R_L}$. Solving for these values can be accomplished by thevenizing the collector side of the circuit as shown in Fig. 24-21. In Fig. 24-21*a*, the collector side of the circuit is redrawn showing $V_{CC}$ and the voltage of $C_o$ as batteries. (Remember, a charged capacitor is an electric energy source.) Also in this figure, the collector-to-emitter section of the transistor is replaced by $R_{CE}$. Resistance $R_{CE}$ represents

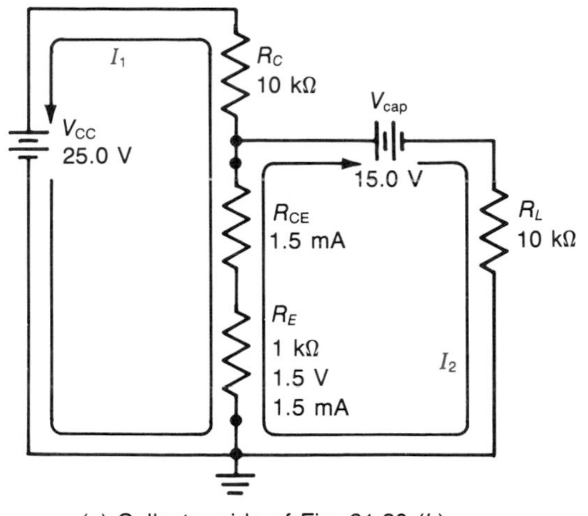


(a) Collector side of Fig. 31-20 (b)

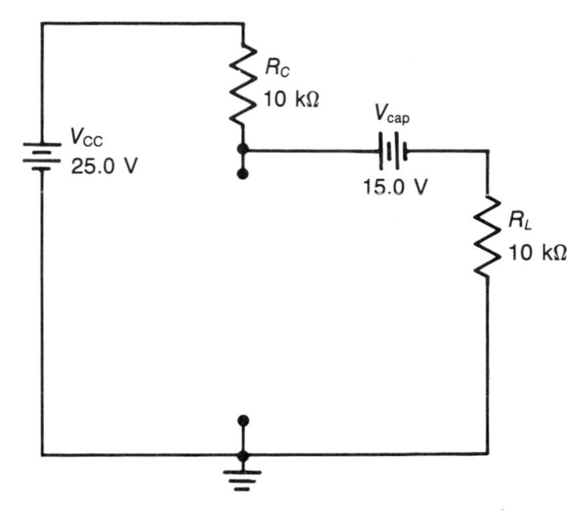


(b) Consider $R_{CE} + R_E$ to be the load and remove them

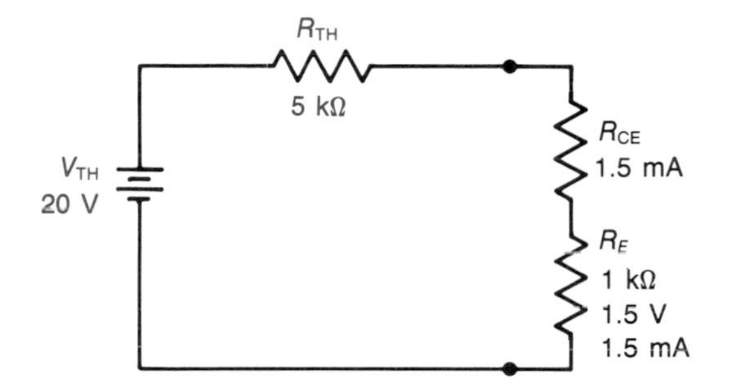


(c) Thevenize circuit (b) and connect the load

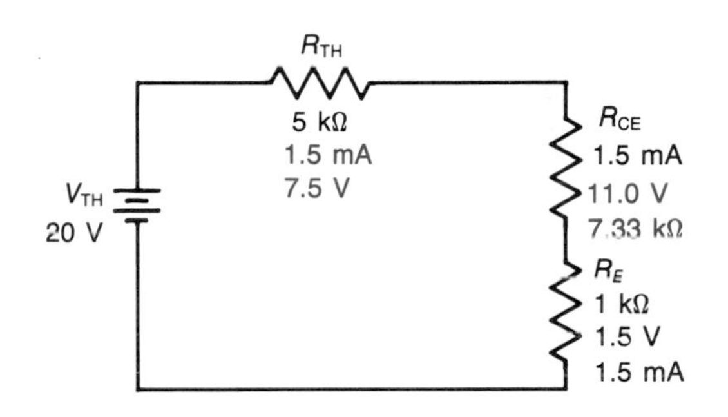


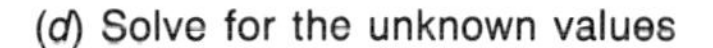
(d) Solve for the unknown values

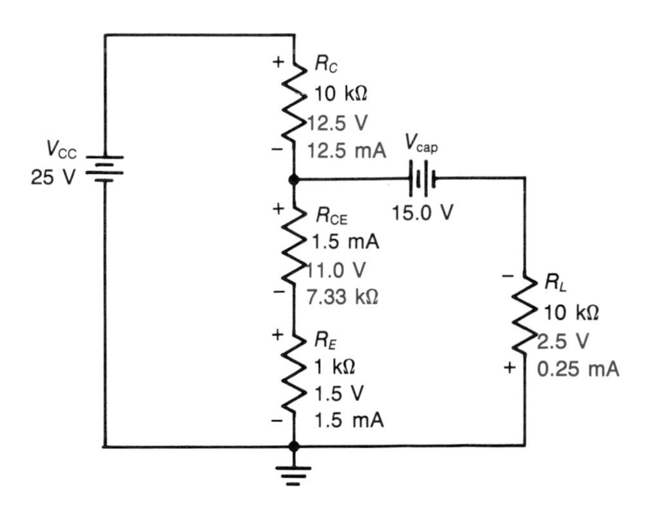


(e) Final results

**FIGURE 24-21 Analysis of the collector side of the circuit in Fig. 24-20 when the input signal voltage is positive and the capacitor is discharging.**

the dc resistance from collector to emitter at the instant the collector current is 1.5 mA. In solving the circuit of Fig. 24-21*a* we will consider the series combination of $R_{CE}$ and $R_E$ to be the load connected to a complex circuit. Removal of this load yields the circuit in Fig. 24-21*b* which can be thevenized as follows:

$$R_{TH} = R_C \,||\, R_L = 10\ \text{k}\Omega \,||\, 10\ \text{k}\Omega = 5\ \text{k}\Omega$$

where || is the symbol for in parallel with, and $R_{TH}$ is thevenized resistance.

$$V_{TH} = V_{CC} - V_{R_C} = V_{CC} - \left(\frac{V_{CC} - V_{cap}}{2}\right)$$
$$= 25\ \text{V} - 5\ \text{V} = 20\ \text{V}$$

where $V_{cap}$ is the voltage on the output coupling capacitor $C_o$.

Next, the load is connected to the Thevenin circuit (Fig. 24-21*c*) and the thevenized circuit is solved to obtain $R_{CE}$ and $V_{R_{CE}}$ as in Fig. 24-21*d*. The key to solving Fig. 24-21*d* is to remember that $I_{R_E}$ and $V_{R_E}$ are known from Fig. 24-20*b*. Then the calculations are

$$V_{R_{TH}} = I_{R_{TH}} \times R_{TH} = I_{R_E} \times R_{TH} = 1.5\ \text{mA} \times 5\ \text{k}\Omega$$
$$= 7.5\ \text{V}$$
$$V_{R_{CE}} = V_{TH} - V_{R_{TH}} - V_{R_E} = 20\ \text{V} - 7.5\ \text{V} - 1.5\ \text{V}$$
$$= 11\ \text{V}$$
$$R_{CE} = \frac{V_{R_{CE}}}{I_{R_{CE}}} = \frac{11\ \text{V}}{1.5\ \text{mA}} = 7.33\ \text{k}\Omega$$

The values just calculated ($V_{R_{CE}}$ and $R_{CE}$) can now be used to finish solving the unknown values in the original complex circuit of Fig. 24-21*a*. This is done in Fig. 24-21*e*. The necessary calculations are

$$V_{R_L} = V_{cap} - V_{R_E} - V_{R_{CE}} = 15.0\ \text{V} - 1.5\ \text{V} - 11.0\ \text{V}$$
$$= 2.5\ \text{V}$$
$$I_{R_L} = \frac{V_{R_L}}{R_L} = \frac{2.5\ \text{V}}{10\ \text{k}\Omega} = 0.25\ \text{mA}$$
$$V_{R_C} = V_{CC} - V_{R_E} - V_{R_{CE}} = 25\ \text{V} - 1.5\ \text{V} - 11.0\ \text{V}$$
$$= 12.5\ \text{V}$$
$$I_{R_C} = \frac{V_{R_C}}{R_C} = \frac{12.5\ \text{V}}{10\ \text{k}\Omega} = 1.25\ \text{mA}$$

With the information obtained from Fig. 24-21, we can now calculate the collector voltage for Fig. 24-20*b*:

$$V_C = V_{CE} + V_{R_E} = V_{R_{CE}} + V_{R_E}$$
$$= 11\ \text{V} + 1.5\ \text{V} = +12.5\ \text{V}$$

The results of the input signal swinging to −0.5 V are shown in Figs. 24-20*c* and 24-22. From Fig. 24-20*c* you can see that (1) both coupling ca-

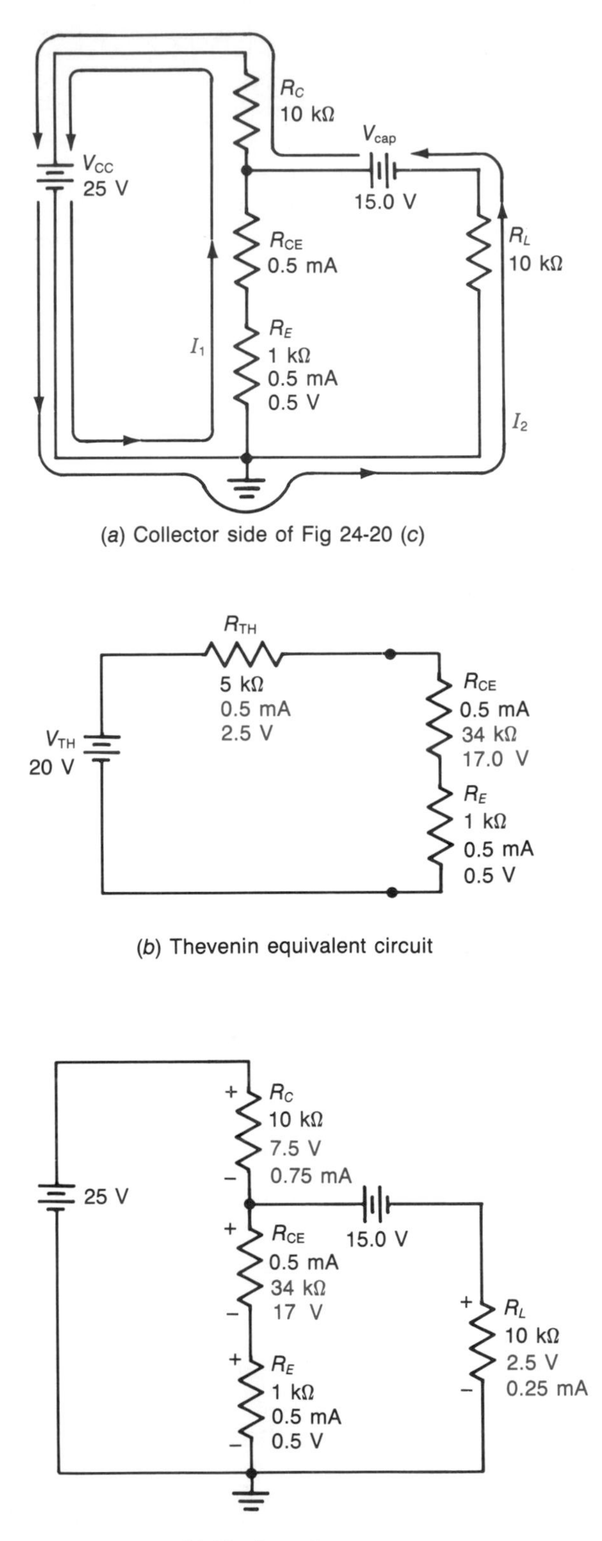


**FIGURE 24-22 Analysis of the collector side of the circuit in Fig. 24-20 when the input signal voltage is negative and the capacitor is charging.**

pacitors are now charging, (2) the signal source is forcing current through $R_{B1}$ rather than $R_{B2}$, and (3) the load current is equal to $I_{R_C} - I_C$ rather than $I_C - I_{R_C}$. Calculation of the collector voltage, load current, and load voltage is again done by thevenizing the circuit as shown in Fig. 24-22. The procedures used to calculate the values in Figs. 24-20*c* and 24-22 are the same as those used for Figs. 24-20*b* and 24-21. Verification of the values is left to the reader. Notice that $V_C$ for Fig. 24-20*c* is +17.5 V.

Now that we have calculated all the currents and voltages for Fig. 24-20, let us figure out what they "tell us." First, the 1-$V_{p\text{-}p}$ input signal resulted in a 5-$V_{p\text{-}p}$ output signal ($V_{R_L}$ ranged from +2.5 V to −2.5 V). Therefore, the voltage gain is $A_V = V_o/V_i = 5\text{ V}/1\text{ V} = 5$. Second, the signal current flows through $R_{B1}$, $R_{B2}$, and the series combination of $R_E$ and $r_e$. The signal current through $R_E$ and $r_e$ shows up as the change in base current. Thus, the current from the signal source $I_i$, which is the input signal current for the amplifier stage, is

$$
\begin{aligned}
I_i &= \Delta I_B + \Delta I_{R_{B1}} + \Delta I_{R_{B2}} \\
&= (15\ \mu\text{A} - 5\ \mu\text{A}) + \\
&\quad (1.03\text{ mA} - 0.98\text{ mA}) + (1.3\text{ mA} - 0.7\text{ mA}) = \\
&0.66\text{ mA}
\end{aligned}
$$

Notice that most of $I_i$ flows through $R_{B_2}$ and very little flows through the BE junction. The output current $I_o$ is the 0.5 $\text{mA}_{p\text{-}p}$ which flows through $R_L$. Therefore, the current gain for this circuit is

$$A_i = \frac{I_o}{I_i} = \frac{0.5\text{ mA}_{p\text{-}p}}{0.66\text{ mA}_{p\text{-}p}} = 0.76$$

Notice the $I_o$ can also be determined by

$$
\begin{aligned}
I_o = \Delta I_{R_L} = \Delta I_C - \Delta I_{R_C} &= (1.5\text{ mA} - 0.5\text{ mA}) \\
&- (1.25\text{ mA} - 0.75\text{ mA}) = 0.5\text{ mA}_{p\text{-}p}
\end{aligned}
$$

This shows that the signal current available from the transistor $\Delta I_C$ splits up and flows through both $R_L$ and $R_C$. Third, the power gain can be figured from the input and output currents and voltages as follows:

$$A_P = \frac{P_o}{P_i} = \frac{V_o \times I_o}{V_i \times I_i} = \frac{5\text{ V}_{p\text{-}p} \times 0.5\text{ mA}_{p\text{-}p}}{1\text{ V}_{p\text{-}p} \times 0.66\text{ mA}_{p\text{-}p}} = 3.8$$

As stated in the introduction to amplifiers, the power gain is also equal to the product of the voltage and current gains. Thus, we can check our calculations by

$$A_P = A_I \times A_V = 0.76 \times 5 = 3.8$$

We have now figured $A_V$ and $A_I$ the hard way; but, in the process we developed an understanding of how ac (signal) currents and voltages are mixed with and separated from dc voltages and currents. This allows us to understand (*and appreciate*) the easier ways of dealing with transistor amplifiers. Since the signal current from the transistor $\Delta I_C$ divides between $\Delta I_{R_C}$ and $\Delta I_{R_L}$, we know that $R_L$ and $R_C$ are in parallel as far as ac currents and voltages are concerned. Thus, the ac load $R_{L,ac}$ that the transistor's collector must drive is $R_L \parallel R_C$. Therefore, the signal output voltage from the amplifier is

$$V_o = \Delta I_C\ (R_L \parallel R_C)$$

Now let us look at another way of specifying the input signal voltage. Inspection of Fig. 24-20 shows that the series combination of $R_E$ and $r_e$ is in parallel with $R_{B2}$. Since the input signal voltage appears across $R_{B2}$ ($\Delta V_{R_{B2}} = 1\text{ V}_{p\text{-}p}$), it also appears across this combination. We also know that the signal voltage across this combination is caused by $\Delta I_E$, which is approximately equal to $\Delta I_C$. Thus, we can express the input signal voltage as $V_i = \Delta I_E(R_E + r_e) \approx \Delta I_C(R_E + r_e)$. Often, $R_E$ is so much larger than $r_e$ that essentially all of the input signal appears across $R_E$. If $r_e \ll R_E$, then $V_i \approx \Delta I_C \times R_E$. The expressions just developed for $V_i$ and $V_o$ can be used to express the voltage gain as

$$A_V = \frac{V_o}{V_i} = \frac{\Delta I_C\ (R_L \parallel R_C)}{\Delta I_C\ (R_E + r_e)} = \frac{R_L \parallel R_C}{R_E + r_e}$$

If $R_E \gg r_e$, then

$$A_V = \frac{R_L \parallel R_C}{R_E}$$

For the circuit in Fig. 24-20, $R_E = 1000\ \Omega$ and $r_e \approx 0.026\text{ V}/1\text{ mA} = 26\ \Omega$. Therefore, the circuit gain would be

$$A_V = \frac{R_L \parallel R_C}{R_E} = \frac{10\text{ k}\Omega \parallel 10\text{ k}\Omega}{1\text{ k}\Omega} = 5$$

Next let us look at a different way to figure the input signal current. The signal current will be

$$I_i = \frac{V_i}{Z_i}$$

where $Z_i$ is the impedance of all of the paths through which input signal current flows. When the circuit is only amplifying low frequencies (like audio frequencies), $Z_i$ is the same as $R_i$ because the reactance of the junction capacitance is so large compared to the resistances that it can be ignored. Then, $Z_i$ or $R_i$ is the ac load that the signal source must drive. Notice in Fig. 24-20 that the input signal current took three paths: through $R_{B1}$, $R_{B2}$, and $R_E$ in series with $r_e$. The sum of the change in currents through these three paths equals the input signal current. Thus, these three paths are in parallel as shown in Fig. 24-23. Notice in Fig. 24-23 that $r_e$ and $R_E$ are multiplied by B + 1. This is because the input signal drives the base and causes changes in base current. The $\Delta I_B$ is amplified B + 1 times by the transistor and appears as $\Delta I_E$ in $R_E$ and $r_e$. Thus, to a source connected to the base lead, the opposition in the emitter leg of the circuit appears to be B + 1 times its true value. When $r_e \ll R_E$, as it is in our case, it can be omitted and just $R_E$ used in the third parallel branch of Fig. 24-23. For the circuit of Fig. 24-20,

$$R_i = R_{B1} \parallel R_{B2} \parallel [(R_E + r_e)(\beta + 1)]$$
$$\approx 23.2\ \text{k}\Omega \parallel 1.69\ \text{k}\Omega \parallel (1\ \text{k}\Omega \times 101) = 1.56\ \text{k}\Omega$$

Assuming the circuit is operating at low frequencies, $R_i \approx Z_i$ and the input current is

$$I_i = \frac{V_i}{Z_1} = \frac{1\ \text{V}_{p\text{-}p}}{1.56\ \text{k}\Omega} = 0.64\ \text{mA}_{p\text{-}p}$$

This value of $I_i$ is within round-off error of the value we obtained by $I_i = \Delta I_{R_{B1}} + \Delta I_{R_{B2}} + \Delta I_B$.

The output signal current $I_o$ can be found using the values of $R_L$ (given), $V_i$ (given), and $A_V$ already calculated. Thus,

$$V_o = A_V \times V_i = 5 \times 1\ \text{V}_{p\text{-}p} = 5\ \text{V}_{p\text{-}p}$$

$$I_o = \frac{V_o}{R_L} = \frac{5\ \text{V}_{p\text{-}p}}{10\ \text{k}\Omega} = 0.5\ \text{mA}_{p\text{-}p}$$

Now, the current gain is found to be

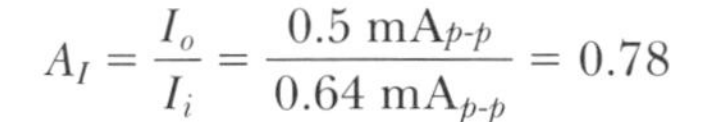

$$A_I = \frac{I_o}{I_i} = \frac{0.5\ \text{mA}_{p\text{-}p}}{0.64\ \text{mA}_{p\text{-}p}} = 0.78$$

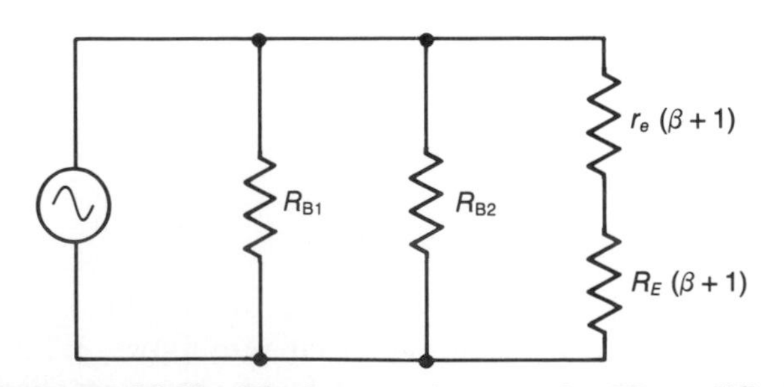


**FIGURE 24-23 The input resistance of a CE amplifier stage.**

which is again within round-off error of the former value.

## 24-5 NEGATIVE FEEDBACK

Putting a resistor $R_E$ in the emitter lead, as in Fig. 24-20, reduces the voltage gain of the transistor amplifier because $R_E$ provides negative feedback. *Negative feedback* is the term used to indicate that part of the output signal is canceling part of the input signal. As seen in Fig. 24-20*b*, the voltage on the emitter goes more positive when the input signal, and thus the base voltage, goes positive. In other words, the signal voltage on the emitter is in phase with the signal voltage of the base. The signal voltage on the emitter will be within a few millivolts in amplitude of the signal on the base. These few millivolts of signal across the BE junction are all that is needed to cause the collector current to fluctuate and create an output signal current. Because of the negative feedback, $V_B$ can swing 1 $\text{V}_{p\text{-}p}$ without driving the transistor into either cutoff or saturation.

Negative feedback also reduces amplitude and frequency distortion in an amplifier. So, even though negative feedback reduces gain, it is often used because of the beneficial reduction of distortion.

All of the gains in Fig. 24-20 can be increased by bypassing part (or all) of $R_E$ with a large capacitor as shown in Fig. 24-24. Adding the large bypass capacitor provides a low-reactance path for the ac (signal) current. Thus, the signal current bypasses part of $R_E$ and only the unbypassed part of $R_E$ develops signal voltage. Bypass capacitors do not change the quiescent conditions because they allow current to pass only when the current is changing. Also, bypassing 600 Ω of $R_E$ does not change the input signal current significantly because the effective resistance to the signal is still 101 × 400 Ω = 40.4 kΩ. The voltage gain for Fig. 24-24 increases to

$$A_V = \frac{10\ \text{k}\Omega \parallel 10\ \text{k}\Omega}{400\ \Omega} = 12.5$$

and the output voltage is

$$V_o = A_V \times V_i = 12.5 \times 1\ \text{V}_{p\text{-}p} = 12.5\ \text{V}_{p\text{-}p}$$

This gives an output current of

$$I_o = \frac{V_o}{R_L} = \frac{12.5\ \text{V}_{p\text{-}p}}{10\ \text{k}\Omega} = 1.25\ \text{mA}_{p\text{-}p}$$

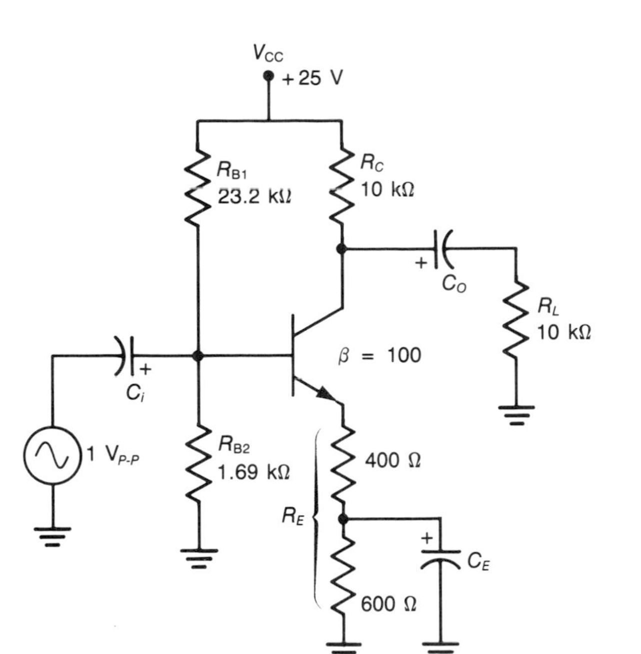


**FIGURE 24-24 CE amplifier with part of $R_E$ bypassed by $C_E$.**

and a current gain of

$$A_I = \frac{I_o}{I_i} = \frac{1.25 \text{ mA}_{p\text{-}p}}{0.64 \text{ mA}_{p\text{-}p}} = 1.95$$

Finally, the new power gain is

$$A_P = A_I \times A_V = 1.95 \times 12.5 = 24.4$$

The power gain without the bypass capacitor was only 3.8. Thus, $A_P$ was increased by more than a factor of 6 by bypassing part of $R_E$.

**EXAMPLE 24-8**

For the circuit in Fig. 24-25, determine $A_V$, $A_I$, and $A_P$.

**Solution**

Since all of $R_E$ except 200 Ω is bypassed, $r_e$ may be significant in this circuit; so, first determine $r_e$:

$$V_B = \frac{10 \text{ k}\Omega \times 20 \text{ V}}{10 \text{ k}\Omega + 56 \text{ k}\Omega} = 3.0 \text{ V}$$

$$V_E = 3.0 \text{ V} - 0.7 \text{ V} = 2.3 \text{ V}$$

$$I_C \approx \frac{2.3 \text{ V}}{5.2 \text{ k}\Omega} = 0.44 \text{ mA}$$

$$r_e \approx \frac{0.026 \text{ V}}{0.44 \text{ mA}} = 59 \text{ }\Omega$$

Since 59 Ω is very significant compared to 200 Ω, include $r_e$ in the $A_V$ and $Z_i$ formulas:

$$Z_i = 10 \text{ k}\Omega \parallel 56 \text{ k}\Omega \parallel [151(59 \text{ }\Omega + 200 \text{ }\Omega)] \approx 7 \text{ k}\Omega$$

$$I_i = \frac{V_i}{Z_i} = \frac{0.05 \text{ V}}{7 \text{ k}\Omega} \approx 7.2 \text{ }\mu\text{A}$$

$$A_V = \frac{12 \text{ k}\Omega \parallel 22 \text{ k}\Omega}{59 \text{ }\Omega + 200 \text{ }\Omega} \approx 30$$

$$V_o = A_V \times V_i = 30 \times 0.05 \text{ V} = 1.5 \text{ V}$$

$$I_o = \frac{1.5 \text{ V}}{22 \text{ k}\Omega} = 68.2 \text{ }\mu\text{A}$$

$$A_I = \frac{68.2 \text{ }\mu\text{A}}{7.2 \text{ }\mu\text{A}} = 9.5$$

$$A_P = 9.5 \times 30 = 285$$

## 24-6 VOLTAGE AMPLIFIERS

In many applications, only the voltage gain of the small-signal amplifier is of interest. Therefore, small-signal amplifiers are sometimes referred to as *voltage amplifiers*. Voltage amplifiers are used to increase the few millivolts available from a signal source to the few volts needed to drive a power amplifier. As shown in Fig. 24-26, voltage amplifiers can be *cascaded* together to achieve the overall voltage gain needed. Notice that the overall gain is the product of the individual gains. In a cascaded (multistage) amplifier, the $Z_i$ of the second stage is the $R_L$ for the first stage. In Fig. 24-26, the −20 in

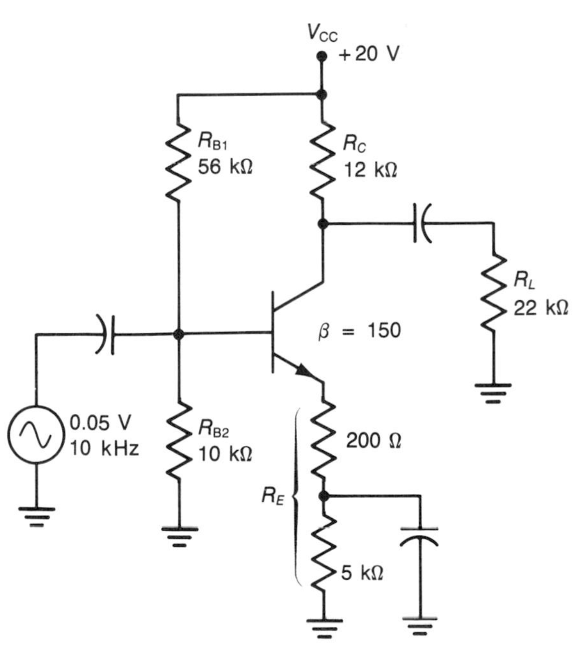


**FIGURE 24-25 Circuit for Example 24-8.**

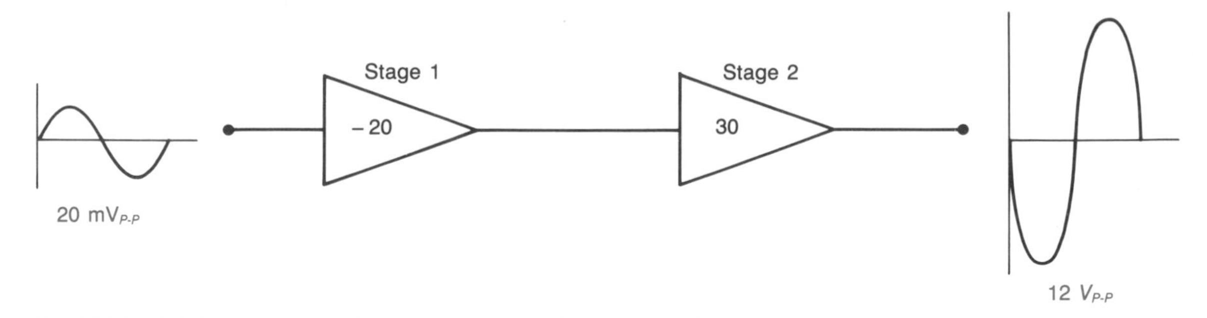


**FIGURE 24-26 Multistage amplifier with an overall voltage gain of 600.**

the symbol for the first stage tells us that the gain is 20 and that the output is 180° out of phase with the input.

## Self-Test

**39.** Is $I_C$, $A_V$, or $Z_i$ more dependent on $\beta$ in the $\beta$-independent circuit? Why?

**40.** Which current can be ignored when figuring $V_B$ for a $\beta$-independent circuit?

**41.** Does the voltage across an output coupling capacitor change when a signal is being amplified?

**42.** True or false? All of the collector signal current flows through the load resistor $R_L$.

**43.** True or false? All of the input signal current flows through the BE junction of the transistor.

**44.** True or false? Thevenin's theorem was used to find the quiescent collector voltage in the $\beta$-independent circuit.

**45.** What happens to each of the following when part of $R_E$ is bypassed?

- **a.** Voltage gain
- **b.** Quiescent $V_E$
- **c.** $Z_i$
- **d.** Distortion

**46.** True or false? The ac load for Fig. 24-25 is 22 kΩ.

**47.** In Fig. 24-25 change $R_{B1}$ to 47 kΩ and the 200-Ω part of $R_E$ to 100 Ω. Now determine the following:

- **a.** $I_C$ (quiescent)
- **b.** $V_{CE}$ (quiescent)
- **c.** $A_V$

## 24-7 DUAL-SUPPLY AMPLIFIER CIRCUITS

Sometimes an amplifier is operated by two dc power supplies. This arrangement is quite common in electronic devices that contain operational amplifiers (presented in Chap. 25) and/or large power amplifiers. In such cases, a small-signal amplifier like the one in Fig. 24-27 may also utilize both power supplies.

In Fig. 24-27*a* both batteries are included in the diagram as are the paths and directions of quiescent current.

Figure 24-27*b* shows the CE dual-supply circuit with the battery symbols replaced by $V_{CC}$ and $V_{EE}$. Voltage $V_{EE}$ is the abbreviation for the emitter supply voltage. Notice that this is a common-emitter configuration because the input is to the base and the output is from the collector. As shown, the emitter is not at common for either dc or ac voltages; however, connecting a large bypass capacitor from the emitter to common, as shown in Fig. 24-27*c*, would put the emitter at common for ac voltages and currents.

In both Fig. 24-27*b* and *c*, the dashed lines represent the ac (signal) current paths in the circuit. Remember that the fluctuating part (ac component) of the dc current is the signal current in the transistor. The only place that a pure alternating current (current that reverses direction) flows is in $R_L$ and the coupling capacitors. (This statement is true for single-supply circuits as well as dual-supply circuits.) Notice in Fig. 24-27*c* that the signal (ac) current bypasses the emitter resistor $R_E$ because the capacitor's reactance at the signal frequency is very small compared to the value of $R_E$.

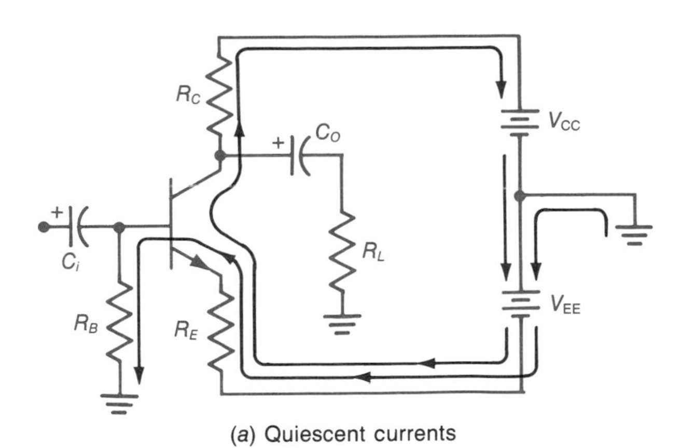


(a) Quiescent currents

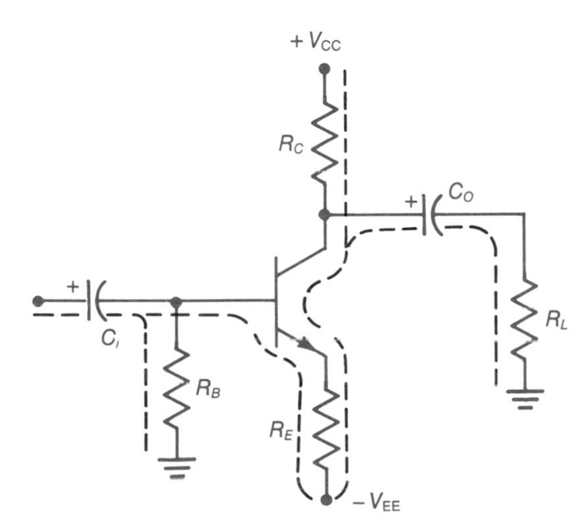


(b) Signal (ac) currents

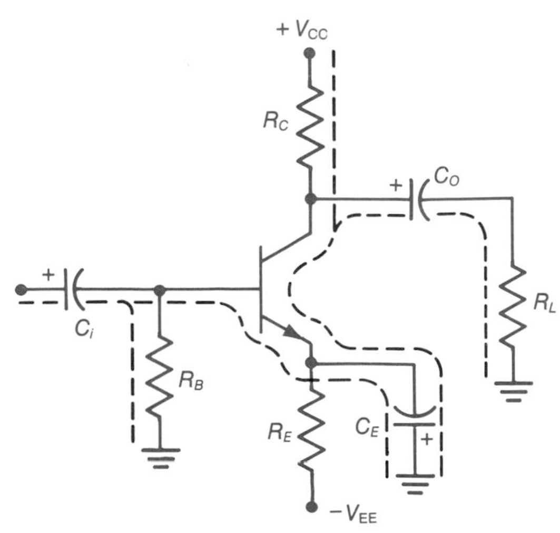


(c) Signal currents with emitter resistor bypassed

**FIGURE 24-27 CE amplifier with dual power supplies.**

## 24-8 OTHER CIRCUIT CONFIGURATIONS

A bipolar transistor can be connected in either a *common-base* (CB) or a *common-collector* (CC) configuration as well as a CE connection. The CC configuration (also known as the *emitter follower*) is shown in Fig. 24-28*a*. The most important characteristic of the CC amplifier is its impedance-matching capability. Compared to the CE and CB configuration, the CC has a high input impedance $Z_i$ and a low output impedance $Z_o$. Output impedance refers to the internal ac (dynamic) impedance of the amplifier between its output terminal and common. Since the transistor amplifier is the signal source for the load it drives, the output impedance of an amplifier is equivalent to the internal resistance of a battery or a generator. Because of its impedance features, the CC can be used to transfer power from a high-impedance source to a low-impedance load. In other words, the common-collector circuit is used for impedance matching to get maximum transfer of power from a high-

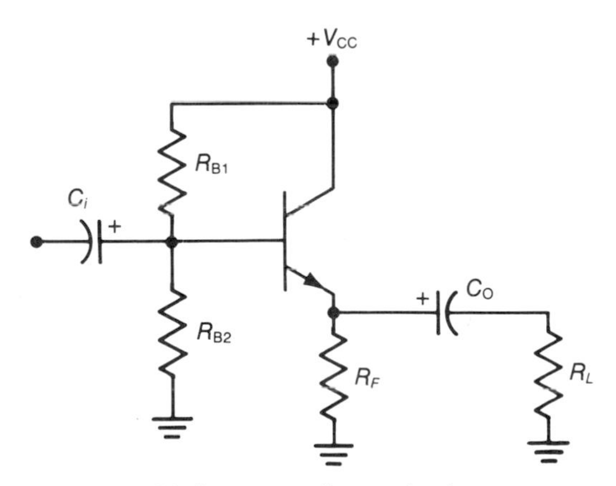


(a) Common-collector circuit

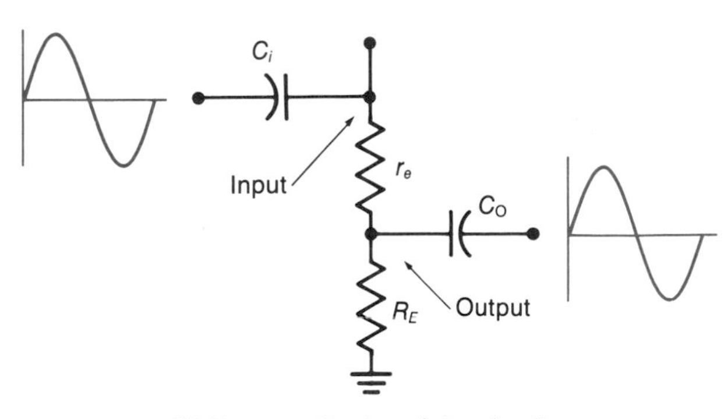


(b) Base-emitter leg of the circuit

**FIGURE 24-28 Common-emitter circuit.**

impedance source to a low-impedance load. In this sense, it acts like a transformer; in fact, it is sometimes referred to as an *electronic transformer*. A properly designed CC amplifier can have an impedance ratio greater than 500:1.

Notice in Fig. 24-28*a* that the collector is connected to $V_{CC}$, which is many dc volts positive with respect to common. However, for the signal, $V_{CC}$ is at common because a well-regulated power supply has very little internal dynamic resistance. Another way to remember that $V_{CC}$ is at common for signal voltages is to recall that the low reactance of the filter capacitor in the power supply bypasses signal (ac) currents.

Two other characteristics of the CC amplifier are (1) its voltage gain is always less than 1, and (2) its output is in phase with its input. Why the CC amplifier has these characteristics can be seen in Fig. 24-28*b*, which is a partial diagram in which the BE junction has been replaced with its dynamic resistance $r_e$. In this figure it is obvious that the input signal voltage is equal to the output signal voltage *plus* the signal voltage that develops across $r_e$. When $R_E$ is large compared to $r_e$, the voltage gain will be almost 1, but still slightly less than 1. Also note in Fig. 24-28*b* that a positive-going input signal increases both $I_B$ and $I_C$, which results in a positive-going voltage at the output. Thus, no phase reversal occurs in the CC amplifier.

A common-base (CB) amplifier is shown in Fig. 24-29. This configuration is not very popular because it has a very low input impedance. It is primarily used for high-frequency amplifiers and oscillators (devices that convert direct current into alternating current). The major characteristics of the CB amplifier are (1) low input impedance, (2) good high-frequency response, (3) no phase shift of the signal, and (4) a current gain of slightly less than 1.

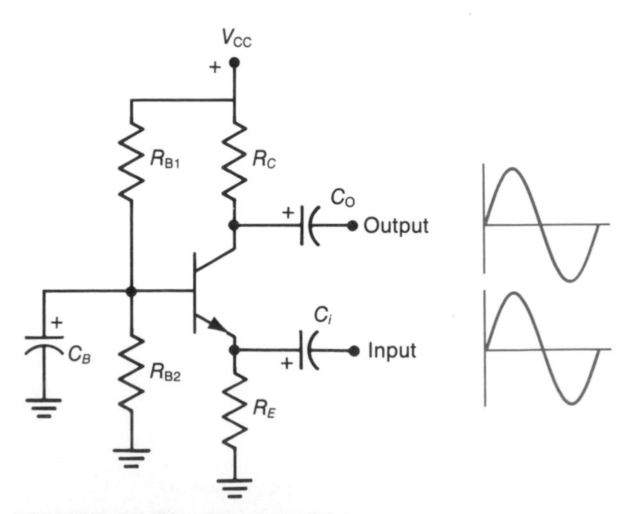

**FIGURE 24-29 Common-base circuit.**

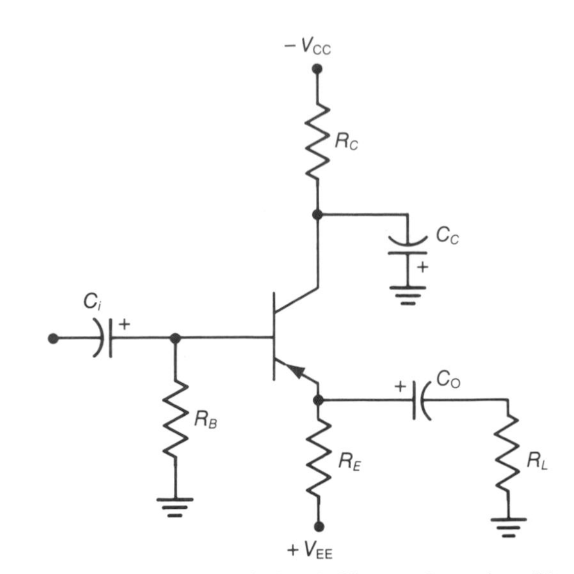


**FIGURE 24-30 Circuit for Self-Test Question 50.**

Notice in Fig. 24-29 that the base is bypassed to common for the signal by capacitor $C_B$. Thus, the name common base. Also notice in this figure that the input signal current can flow through $r_e$ and the low reactance of $C_B$. This is what makes $Z_i$ so very low.

Either the common base or the common collector can, of course, use PNP transistors. Also, either one can be operated with a dual power supply. While the circuit diagram may look quite different, the basic operation and characteristics of the circuits remain unchanged.

## Self-Test

**48.** In which resistor in Fig. 24-28 does the current periodically reverse?

**49.** Which circuit configuration

**a.** Has an $A_V < 1$?

**b.** Has an $A_I < 1$?

**c.** Provides a 180° phase shift of the signal?

**d.** Is used for impedance matching?

**50.** Which circuit configuration is shown in Fig. 24-30? Why?

## 24-9 FIELD-EFFECT TRANSISTORS

There are two major types of field-effect transistors: the *JFET* (junction field-effect transistor) and the *MOSFET* (metal-oxide semiconductor field-effect transistor). The MOSFET is sometimes called an *IGFET* (insulated-gate field-effect transistor.) Also, the JFET is sometimes referred to as a JGFET (junction-gate field-effect transistor). MOSFETs are further divided into two general classes: the depletion type and the enhancement type. The depletion type is used primarily in analog circuits, and the enhancement type is used mostly for digital applications.

The schematic diagrams for various FETs (field-effect transistors of either type) are shown in Fig. 24-31. The gate G, source S, and drain D elements indicated in this figure are equivalent to the base, emitter, and collector, respectively, of the bipolar junction transistor. The B symbol indicates the active bulk substrate which is usually connected to the source either internally or externally. Sometimes the S symbol is used to indicate the bulk substrate. The JFET and the MOSFET are available with either an *N channel* or a *P channel.*

The *channel* of the FET is that part of the transistor between the source and the drain. The transistor current flows through the channel. Since the channel is either N-type or P-type material, all of the transistor current is carried by either electrons (N-channel) or holes (P-channel). Thus the FET is unipolar, whereas the junction transistor is bipolar.

The major difference between the junction transistor and the field-effect transistor is that the latter does not draw appreciable gate current like the former draws base current. Both the static and dynamic resistance from gate to source is very large (many megohms), especially for the MOSFET. The drain current in the FET is controlled by the gate-to-source voltage. Thus, the FET is known as a *voltage-controlled device,* while the junction transistor (BJT) is called a *current-controlled device.*

### JFET

One possible structure for an N-channel JFET is illustrated in Fig. 24-32. Notice that the gate lead connects to P-type material and that this P-type material forms a PN junction with the N-type

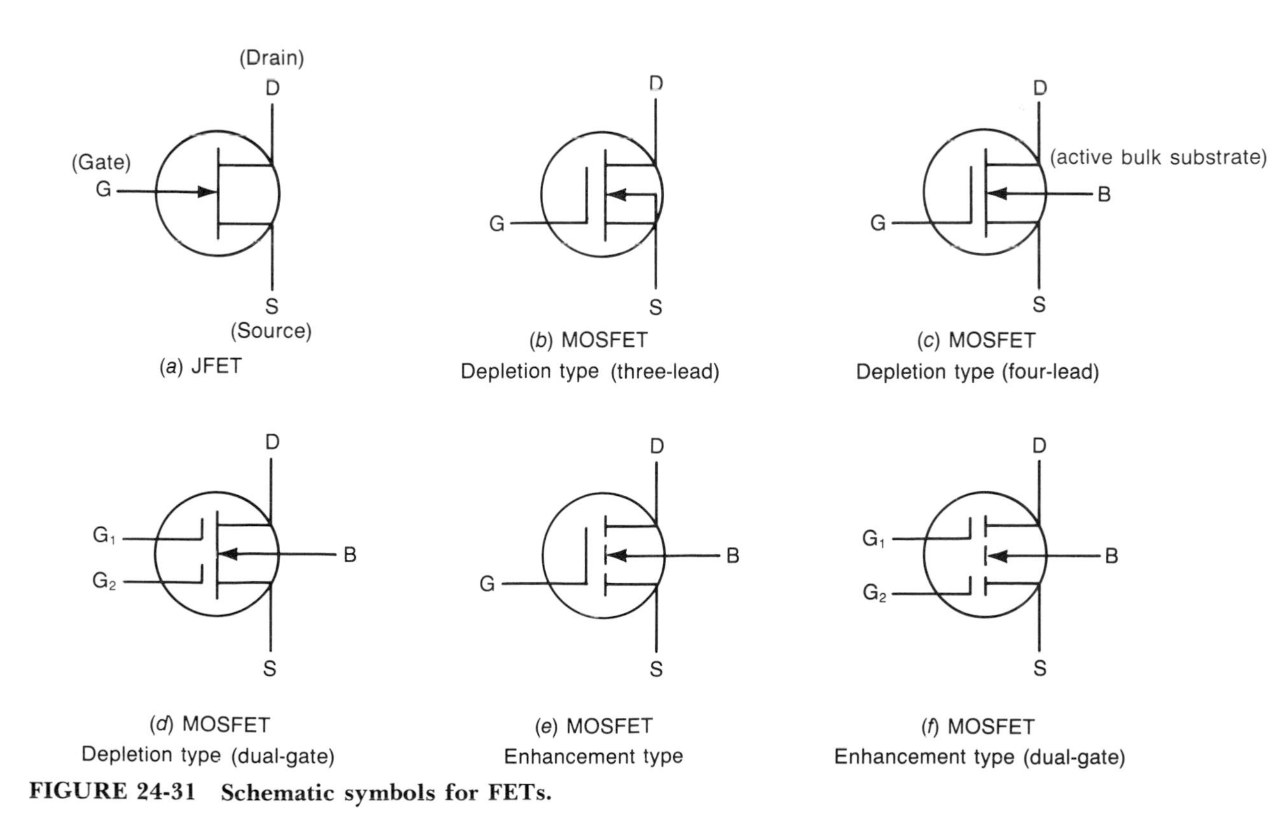


**FIGURE 24-31 Schematic symbols for FETs.**

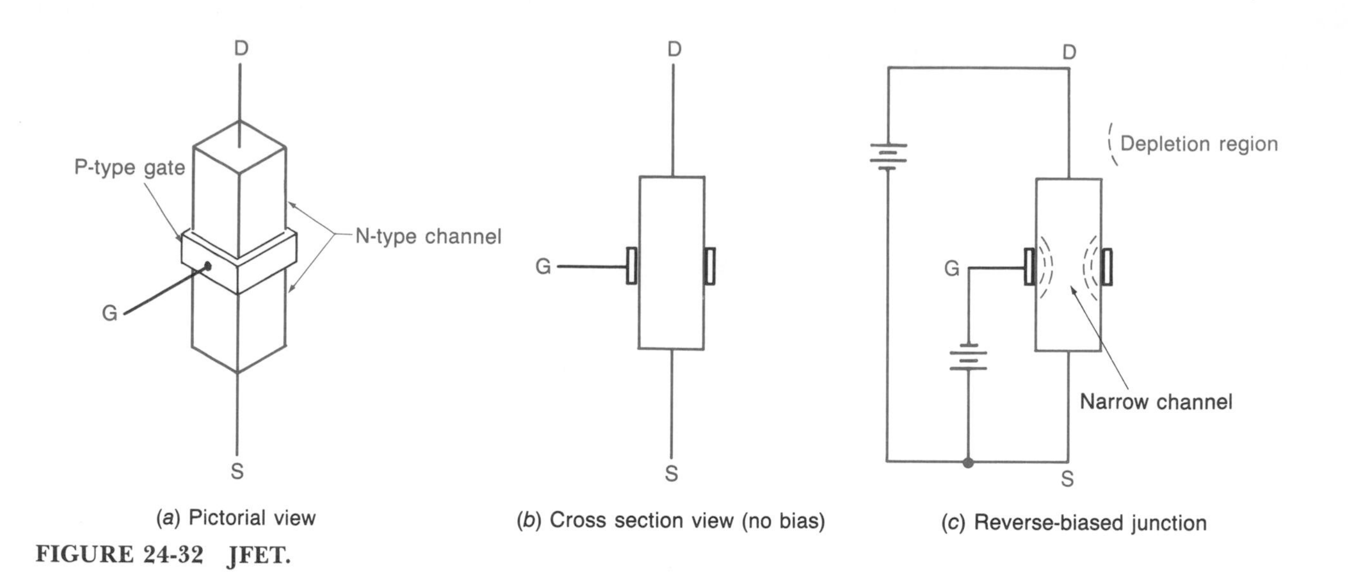


**FIGURE 24-32 JFET.**

channel material. When no bias is applied (see Fig. 24-32*b*), the full cross-sectional area of the channel is available to conduct current when the drain is made positive with respect to the source. As seen in Fig. 24-32*c*, reverse-biasing the gate-to-source junction forms a depletion region which narrows the channel cross-sectional area. If sufficient reverse-bias is applied, the depletion region spreads completely across the channel and cuts off the channel. That is, the channel is devoid of major carriers (electrons for the N channel) and channel current is essentially zero (only a minute leakage current flows).

Correct voltage polarity for the operation of a JFET is shown in Fig. 24-32*c*. Either voltage source may be reduced to zero, but the polarity of neither should be reversed. For the JFET to operate as a linear amplifier, the drain-to-source voltage can never drop to zero.

A family of drain characteristic curves for the N-channel JFET is shown in Fig. 24-33*a*. These curves could be obtained from a simple circuit like that in Fig. 24-32*c* by inserting an ammeter in the drain lead and varying the source-drain voltage while holding the source-gate voltage constant. Of course, no ammeter is needed in the gate lead because the gate is reverse-biased and draws only picoamperes of leakage current. The curves in Fig. 24-33*a* show that the JFET allows maximum drain current when the gate voltage is zero. As the gate voltage becomes negative (with respect to the source), the channel is depleted of some carriers and drain current $I_D$ decreases. Finally, at some negative gate voltage, cutoff is reached and $I_D$ is zero.

Notice in Fig. 24-33*a* that the spacing between the curves is not equal. A 1-V change in $V_{GS}$ from 0 to $-1$ V gives a much greater change in $I_D$ than does a 1-V change from $-3$ to $-4$ V. Since the $\Delta I_D$ is converted to the output voltage of a JFET amplifier, this unequal spacing of the curves leads to amplitude distortion.

Another way to graphically display the characteristics of a FET is with a transconductance curve. This curve (see Fig. 24-33*b*) can be obtained from the drain curves of Fig. 24-33*a* by holding $V_{DS}$ constant and obtaining the values of $I_D$ and $V_{GS}$ for each curve. The plotting points in Fig. 24-33*b* were obtained by holding $V_{DS}$ constant at 8 V. This is called a *transconductance curve* because it graphs the ac conductance ($\Delta I_D/\Delta V_{GS}$) for a signal being transferred from the input ($\Delta V_{GS}$) to the output ($\Delta I_D$) of the FET. Figure 24-33*b* clearly shows that the transconductance of a FET is not a constant and the transconductance increases as $I_D$ increases. The transconductance of a FET is a useful parameter for calculating the gain of a FET amplifier.

Figure 24-33 shows average curves for a specific type number. The curves for individual JFETs with the same type number vary tremendously. For example, the drain current for $V_{GS} = 0$ (abbreviated $I_{DSS}$) can be anywhere from 4 to 12 mA for

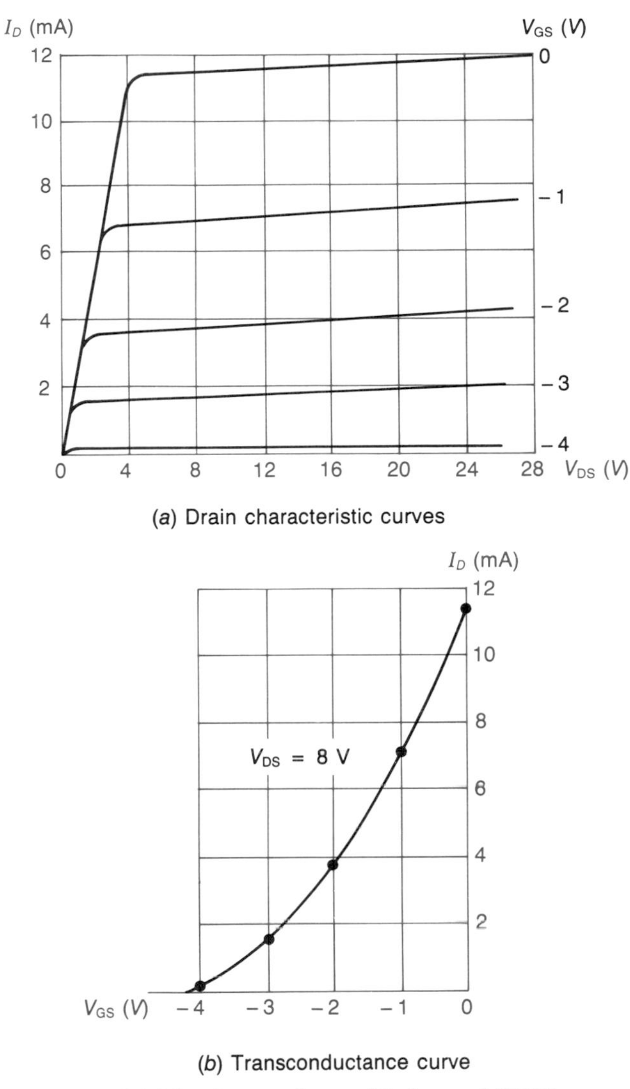


**FIGURE 24-33 Curves for an N-channel JFET.**

JFETs with the same type number. Such large variation in $I_{DSS}$ makes it a little difficult to design a JFET amplifier that operates equally well with all transistors that have the same type number.

Field-effect transistor amplifiers tend to have small voltage gains compared to bipolar transistor amplifiers. A comparison of Figs. 24-33*b* and 24-9 shows why this is so. Notice in Fig. 24-9 that a 50-mV change in $V_{EB}$ causes a change in $I_B$ of 200 $\mu$A, which, with a $\beta$ of 110, causes a 22-mA change in collector current. Figure 24-33*b* shows that a 4-V change in $V_{GS}$ causes only an 11-mA change in drain current.

A JFET operates only in the depletion mode. The *depletion mode of operation* means that the channel of the FET can only be depleted of carriers by changing the gate-to-source voltage. The *enhancement mode of operation* means that the channel carriers can be increased by changing $V_{GS}$. Do not confuse *depletion type* and *depletion mode.* Depletion type is a category of MOSFET. When MOSFETs are discussed later in this section, we will see that a depletion-type MOSFET can operate in either the depletion mode or the enhancement mode. The enhancement-type MOSFET can operate only in the enhancement mode.

Like the bipolar transistor, the FET can also be connected in three different amplifier configurations: the common source (CS), the common drain (CD), and the common gate (CG). In terms of their major characteristics and uses, these three configurations are comparable to the CE, CC, and CB, respectively. Because the CS amplifier is so much more prevalent than the other two FET configurations, we will use it to illustrate how FET amplifiers operate.

Figure 24-34*a* shows a simple CS amplifier. In this figure $V_{DD}$ is the drain supply voltage (comparable to $V_{CC}$), and only the quiescent (dc) current is shown. Note that $R_G$, the gate resistor, has no current through it because the gate source junction is reverse-biased by the voltage dropped across $R_S$ (the source resistor). How $V_{R_S}$ reverse-biases the junction is easy to understand if you remember that no quiescent current flows in $R_G$ and thus there is no dc voltage across $R_G$. Quiescently, the gate is at ground (common), and therefore the gate is negative with respect to the source by the magnitude of $V_{R_S}$.

**EXAMPLE 24-9**

Refer to Fig. 24-34*a*. If $V_{GS}$ = 2 V and $R_S$ = 510 Ω, what is the value of $I_D$?

**Solution**

$$V_{R_S} = V_{GS} = 2\text{ V}$$

$$I_D = I_S = \frac{V_{R_S}}{R_S} = \frac{2\text{ V}}{510\ \Omega} = 3.9\text{ mA}$$

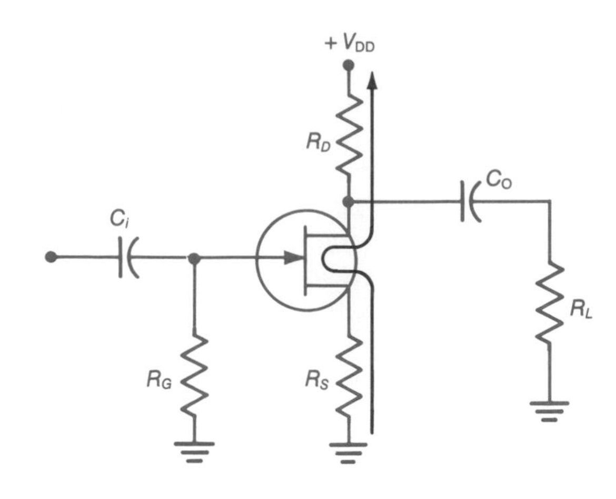


(a) Quiescent current

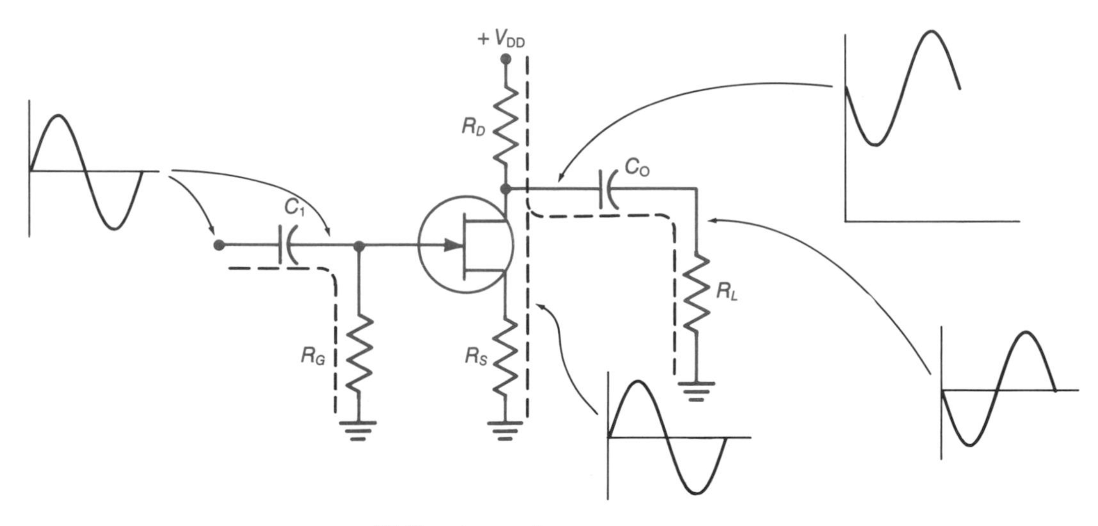


(b) Signal currents

**FIGURE 24-34 Currents in a common-source amplifier.**

Figure 24-34*b* shows the signal current paths for the CS amplifier when it is operating in the audio range of frequencies. Notice that the input signal flows through $R_G$ only. Therefore, the $Z_i$ of this circuit is equal to $R_G$, which can be more than a megohm. The signal current through $R_G$ is true alternating current. When the gate end of $R_G$ is positive, $V_{R_G}$ series-opposes $V_{R_S}$, causing $V_{SG}$ to decrease, $I_D$ to increase, and $V_D$ to decrease. When the input voltage reverses, $V_{R_G}$ aids $V_{R_S}$, causing a reverse effect on $V_{SG}$, $I_D$, and $V_D$. As with the bipolar transistor circuit, the output signal current in the transistor is fluctuating direct current, and the current in $R_L$ is true alternating current. The gain of the CS amplifier in Fig. 24-34 is reduced by the negative feedback (sometimes called *degenerative feedback*) provided by signal voltage across $R_S$. Bypassing $R_S$ with a capacitor increases the gain in the same way as did bypassing $R_E$ in Fig. 24-24.

## Self-Test

**51.** In which mode does a JFET operate?

**52.** What type of bias is applied to the PN junction in a JFET?

**53.** Is the FET or the bipolar transistor capable of the larger voltage gain?

**54.** List two major types of FETs.

**55.** A FET uses holes as current carriers. Is it a P-channel or an N-channel FET?

**56.** Define "cutoff."

**57.** Name the three elements (leads) of a JFET.

**58.** Refer to Fig. 24-33:

**a.** Would the $A_V$ of this FET be greater if biased at $V_{GS} = -1$ V or $V_{GS} = -3$ V?

**b.** If $V_{GS} = -2$ V and $V_{DS} = 16$ V, what is the value of $I_D$?

**59.** Which component in Fig. 24-34 causes degenerative feedback?

## MOSFET

Figure 24-35 illustrates the structure of the two types of MOSFETs. As indicated by the symbols in Fig. 24-31, the gate does not form a junction with the channel in either type of MOSFET. The gate for the MOSFET is a metal film deposited on a layer of silicon dioxide. The silicon dioxide insulates the gate from the channel. Because silicon dioxide is a good insulator and the surface area of the gate is very small, the input resistance of a MOSFET is thousands of megohms. When the gate is biased, i.e., a voltage applied between the gate and the source, the electric field from the gate changes the carrier concentration in the channel area.

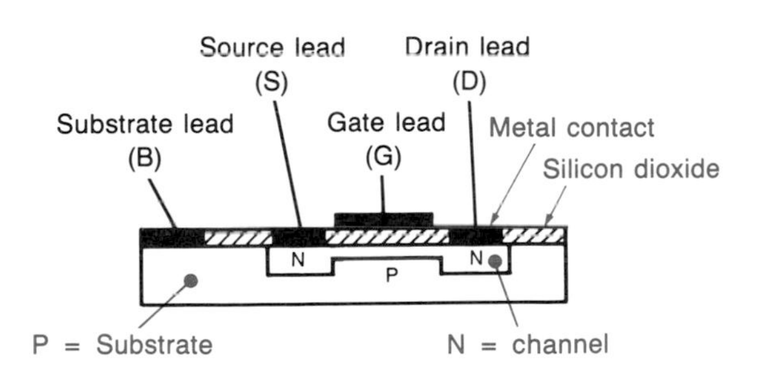


(*a*) Cross-sectional view of an N-channel, depletion-type MOSFET

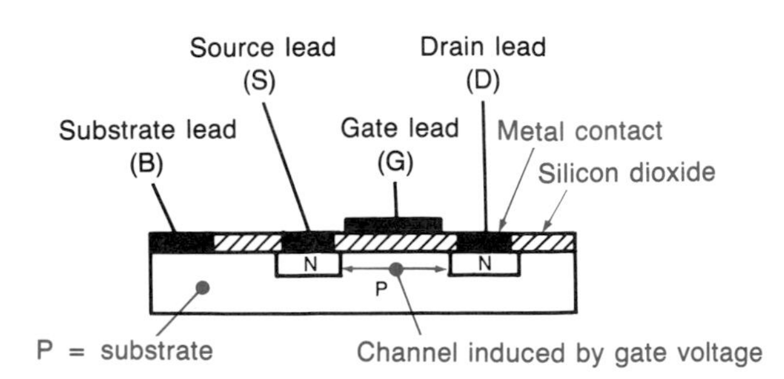


(*b*) Cross-sectional view of an N-channel, enhancement-type MOSFET

**FIGURE 24-35 Illustrative drawings of the structure of metal-oxide semiconductor field-effect transistors.**

As shown by the drain curves in Fig. 24-36, the depletion-type MOSFET can operate in the linear region with either a positive or a negative gate voltage. With an N channel, a positive gate voltage pulls more carriers (electrons) into the channel, increases the drain current, and puts a FET in the enhancement mode of operation. A negative gate voltage pushes carriers out of the N channel and forces the FET into the depletion mode of operation.

Refer back to Fig. 24-31 and notice that the symbol for the enhancement-type MOSFET uses a dashed line to represent the channel. The symbol portrays the incomplete channel in the physical device illustrated in Fig. 24-35*b*. Since the channel is incomplete, no drain current can flow until the channel is completed (enhanced) by application of the correct polarity of gate voltage. The drain curves in Fig. 24-36*b* show that the gate voltage must be positive (with respect to the source) to turn on (enhance) the N-channel, enhancement-type MOSFET. Since this type of MOSFET does not have a complete channel to begin with, it is impossible for it to operate in a depletion mode. The enhancement-type MOSFET is usually used in circuits that have no quiescent gate voltage or drain current. The input signal is then used to turn on the MOSFET.

A common-source (CS) amplifier using a depletion-type MOSFET is shown in Fig. 24-37. Since this type of FET can operate in either mode, no quiescent gate voltage is required for a linear amplifier. Thus, the circuit does not require a source resistor or a bypass capacitor to eliminate the negative feedback created by a source resistor.

All of the MOSFETs discussed so far are limited to low-power (less than 1 W) applications. Also, all of them produce significant amplitude distortion because of their nonlinear transconductance curves or unequal spacing of drain curves. The *VMOS* (vertical MOSFET) is a type of enhancement MOSFET that overcomes these shortcomings. It can control high-power circuits (drain current in amperes rather than milliamperes) and has much more linear characteristic curves. VMOS field-effect transistors are increasingly being used in power amplifiers and power switching circuits which formerly used bipolar transistors.

The silicon dioxide insulator between the gate and channel in a MOSFET is very thin and can be damaged by excess gate voltage. Even static voltages which build up on objects, the human body, or the FET itself can damage this gate insulation. Therefore, MOSFETs must be handled with care. Static charges should be discharged to ground before handling MOSFETs or installing them in a circuit. Many MOSFET devices have zener diodes internally connected between the source and gate so that excess voltage causes a zener to conduct and thus limit the voltage between the gate and source. These protected devices are not so easily damaged by high gate voltages.

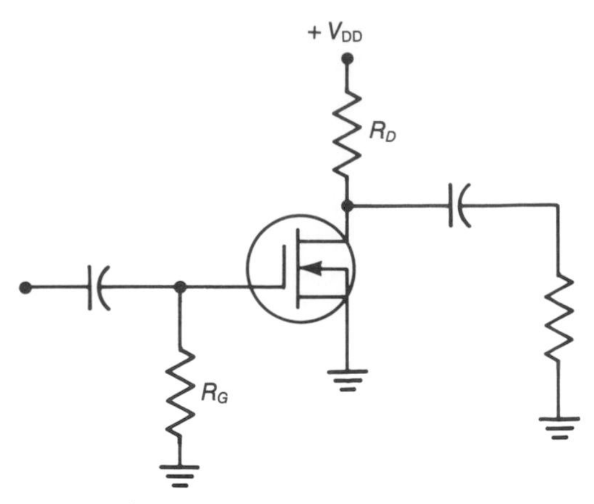


**FIGURE 24-37 A CS amplifier using a depletion-type MOSFET.**

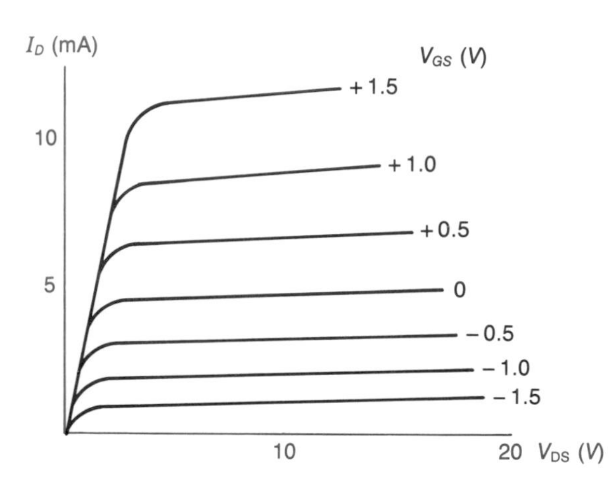

(a) Depletion-type MOSFET

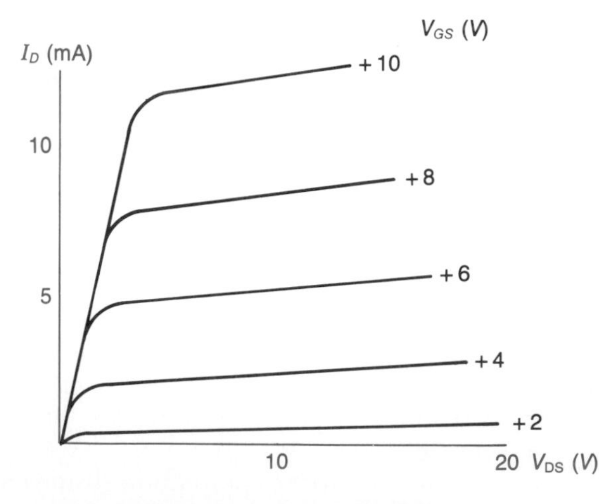


(b) Enhancement-type MOSFET

**FIGURE 24-36 Drain characteristic curves for MOSFETs.**

## Self-Test

**60.** Does the gate of a MOSFET form a PN junction with the channel?

**61.** Which type of MOSFET can operate in either the depletion mode or the enhancement mode?

**62.** Does the JFET or the MOSFET have the higher input resistance? Why?

**63.** Which type of MOSFET symbol uses a broken (dashed) line for the channel?

**64.** Which type of MOSFET can be used as a linear amplifier when $V_{GS} = 0$?

**65.** What is a VMOS transistor?

## SUMMARY

**1.** Bipolar junction transistors (BJTs) have two PN junctions and use both holes and electrons as current carriers.

**2.** The base of a BJT controls the current carriers emitted by the emitter and collected by the collector.

**3.** Silicon transistors are more common than germanium transistors.

**4.** The emitter is the most heavily doped element of the transistor, and the base is the most lightly doped element.

**5.** Beta has no units.

**6.** Current $I_{CBO}$ is the leakage current in the reverse-biased base-collector junction.

**7.** Collector characteristic curves graph the relationships between $I_B$, $I_C$, and $V_{CE}$.

**8.** Important ratings for junction transistors are $\beta$, $I_{CBO}$, $P_D$, $BV_{CBO}$, $BV_{CEO}$, $BV_{EBO}$, $I_{C,max}$, and $f_T$.

**9.** Voltage gain is more often specified than is power gain or current gain.

**10.** Common forms of distortion are amplitude, frequency, and phase.

**11.** Resistance $R_C$ converts a $\Delta I_C$ into a $\Delta V_{CE}$.

**12.** The common element in an amplifier configuration is the one which is not used for either input or output.

**13.** Coupling capacitors keep input and output devices from upsetting the amplifier quiescent voltages.

**14.** Coupling and bypass capacitors should have sufficient capacitance so that their dc voltages do not change appreciably under signal conditions.

**15.** A transistor is in cutoff when $I_C$ or $I_D$ is zero.

**16.** A transistor is in saturation when $V_{CE} \approx 0$ or $V_{DS} \approx 0$.

**17.** The ends of a load line are at $I_{C,max}$ when the transistor is saturated and the value of $V_{CC}$ when the transistor is cutoff.

**18.** Overdriving an amplifier causes its output to be clipped.

**19.** Voltage $V_{C,min}$ and $V_{C,max}$ under signal conditions can be determined by thevenizing the output circuits.

**20.** Unbypassed $R_E$ causes degenerative feedback which reduces gain and distortion.

**21.** Common-emitter amplifiers phase-shift 180° and can have $A_V$ and $A_I > 1$.

**22.** CC amplifiers cause no phase shift, have an $A_V < 1$, and are used for impedance matching.

**23.** Small-signal amplifiers are often referred to as voltage amplifiers.

**24.** The overall voltage gain of cascaded amplifiers is equal to the product of the gains of the individual amplifiers.

**25.** Some amplifiers use two dc suppies—$V_{CC}$ and $V_{EE}$.

**26.** A bypass capacitor keeps a point in a circuit at signal ground even though the point is not at dc ground (or common).

**27.** A FET has higher input resistance than does a BJT, but it provides smaller voltage gain.

**28.** A FET has either a junction gate (JFET) or an insulated gate (MOSFET). It can have either a P channel or an N channel.

**29.** A JFET operates in the depletion mode, an enhancement MOSFET operates in the enhancement mode, and a depletion MOSFET operates in both modes.

**30.** Cutoff occurs when $V_{GS}$ reduces $I_D$ to zero.

**31.** A JFET is never operated with the gate junction forward-biased.

**32.** Transconductance for a FET is equal to $\Delta I_D/\Delta V_{GS}$. Its value increases at higher levels of $I_D$ and varies from transistor to transistor with the same type number.

**33.** Silicon dioxide insulates the gate from the channel in a MOSFET.

**34.** VMOS field-effect transistors are low-distortion, high-power, enhancement-type MOSFETs.

**35.** Useful formulas include the following:

$$I_E = I_B + I_C$$
$$\beta = \frac{I_C}{I_B}$$
$$I_C \approx I_E$$
$$r_{ce} = \frac{\Delta V_{CE}}{\Delta I_C} \mid I_B$$
$$r_{be} = \frac{\Delta V_{BE}}{\Delta I_B}$$
$$P_D = I_C \times V_{CE}$$
$$A_V = \frac{V_o}{V_i}$$
$$A_P = \frac{P_o}{P_i}$$
$$A_I = \frac{I_o}{I_i}$$
$$A_P = A_V \times A_I$$
$$V_{BE} \approx 0.7 \text{ V}$$
$$R_{L,ac} = R_C \,||\, R_L$$
$$r_e \approx \frac{0.026 \text{ V}}{I_C}$$
$$A_V \approx \frac{R_{L,ac}}{R_E + r_e}$$
$$V_B = V_E + V_{BE}$$
$$V_C = V_{CC} - V_{R_C}$$
$$V_E = V_{R_E}$$
$$V_B = \frac{R_{B2} \times V_{CC}}{R_{B1} + R_{B2}}$$
$$R_i \approx Z_i \quad \text{at low frequencies}$$
$$R_i \approx R_{B1} \,||\, R_{B2} \,||\, [(R_E + R_e)(B + 1)]$$
$$I_D = I_S$$
$$R_i = R_G$$

## CHAPTER REVIEW QUESTIONS

For the following items, determine whether each statement is true or false.

**24-1.** When used to describe a transistor, "bipolar" means "two junctions."

**24-2.** $I_{CBO}$ is the abbreviation for "common-base output current."

**24-3.** Emitter current is larger than collector current.

**24-4.** The power gain of an amplifier is less than one if the current gain is less than one.

**24-5.** If the negative $\frac{1}{2}$ cycle of the output of a CE amplifier is clipped, it indicates that the amplifier is driven into saturation.

**24-6.** With a CE amplifier, increasing $R_C$ by 20 percent causes a 20 percent decrease in $I_C$.

**24-7.** The amount of $V_{EB}$ required for $I_B = 200\ \mu A$ decreases as the junction temperature increases.

**24-8.** The input and output signals of a CE amplifier are 180° out of phase.

**24-9.** The dynamic emitter resistance $r_e$ increases as $I_C$ increases.

**24-10.** Negative feedback decreases the $A_V$ of an amplifier.

**24-11.** Negative feedback increases the distortion in an amplifier.

**24-12.** The dc voltage across a coupling capacitor changes 1 V when the signal voltage is 1 V.

**24-13.** In a loaded amplifier, not all of the collector signal current flows through the load resistance $R_L$.

**24-14.** A JFET amplifier is capable of a larger $A_V$ than is a BJT amplifier.

**24-15.** The transconductance of a FET is larger at $V_{GS} = 0.5$ V than at $V_{GS} = -3.0$ V.

**24-16.** A JFET has a higher input resistance than does a MOSFET.

For the following items, fill in the blank with the word or phrase required to correctly complete each statement.

**24-17.** The base-emitter junction of a transistor is operated with ______ bias.

**24-18.** The base-collector junction of a transistor is operated with ______ bias.

**24-19.** The ______ of a transistor has less doping than the collector of a transistor.

**24-20.** For an NPN transistor symbol the arrow points ______ the base.

**24-21.** For a PNP transistor the base is ______ with respect to the collector.

**24-22.** When $V_{CE} = V_{CC}$, the transistor is ______.

**24-23.** The voltage gain of a BJT amplifier can be increased by bypassing the ______ resistor.

**24-24.** A JFET operates in the ______ mode.

**24-25.** The ______ type of MOSFET operates in both the enhancement and the depletion mode.

**24-26.** The FET which is capable of controlling many watts is the ______.

**24-27.** The $V_{GS}$ voltage which reduces $I_D$ to zero is called the ______ voltage.

**24-28.** Holes are the carriers in a ______ channel FET.

**24-29.** FET is the abbreviation for ______.

**24-30.** The MOSFET which has a broken line in its symbol is the ______ type.

For the following items, choose the letter that best completes each statement.

**24-31.** The material used for the base of an NPN transistor is

**a.** Depletion-type material
**b.** N-type material
**c.** P-type material
**d.** Enhancement-type material

**24-32.** The maximum collector-to-emitter voltage for a transistor is indicated by

**a.** $V_{CBE}$
**b.** $BV_{CEO}$
**c.** $V_{CBO}$
**d.** $BV_{CBE}$

**24-33.** In a $\beta$-independent circuit, which of the following is most affected by $\beta$?

**a.** Input resistance
**b.** Base voltage
**c.** Collector current
**d.** $V_{CE}$

**24-34.** Which circuit configuration has a voltage gain of less than one?

**a.** CB
**b.** CE
**c.** CS
**d.** CC

**24-35.** Which circuit configuration has no phase shift?

**a.** CD
**b.** CE
**c.** CS
**d.** None of the above

**24-36.** The transistor that can operate in the linear region with zero bias is the

**a.** JFET
**b.** BJT
**c.** Enhancement-type MOSFET
**d.** Depletion-type MOSFET

## CHAPTER REVIEW PROBLEMS

**24-1.** A transistor has a $\beta$ of 75 and a collector current of 15 mA. Determine its base current and its emitter current.

**24-2.** Determine $P_C$ for a transistor with $V_{CE} = 10$ V and $I_C = 80$ mA.

**24-3.** Using the curves in Fig. 24-38, determine the following:

**a.** $r_{ce}$ when $I_B = 200\ \mu A$

**b.** $\beta$ when $V_{CE} = 12$ V and $I_B = 250\ \mu A$

**24-4.** For the circuit in Fig. 24-39, determine the following:

**a.** The peak-to-peak output voltage when $I_B$ swings between 50 and 150 $\mu A$

**b.** Quiescent collector current

**c.** Voltage gain

**d.** Input resistance

**24-5.** An amplifier has an $A_V$ of 40 and $A_I$ of 15. What is its $A_P$?

**24-6.** For the circuit in Fig. 24-40, determine the following:

**a.** Ac load resistance

**b.** $V_{CE}$

**c.** $V_C$

**d.** Input resistance

**e.** Voltage gain

**f.** Power gain

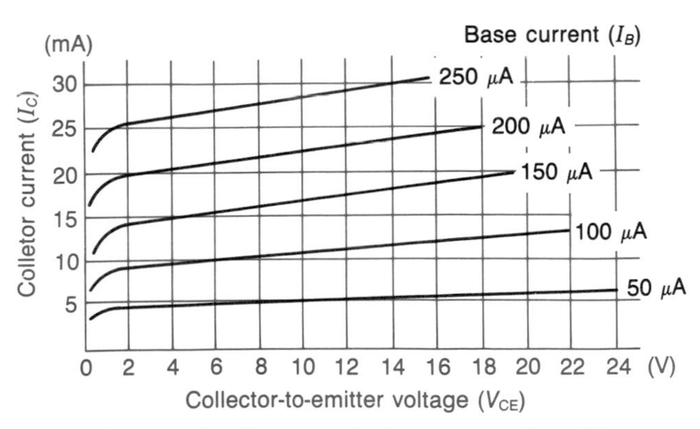


**FIGURE 24-38 Characteristic curves for Chapter Review Problem 24-3.**

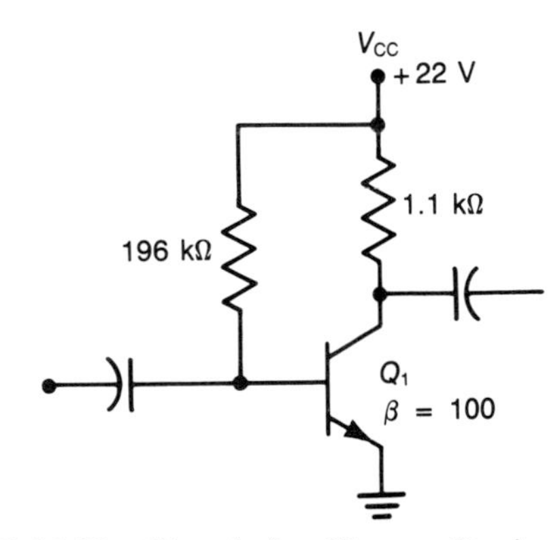


**FIGURE 24-39 Circuit for Chapter Review Problem 24-4.**

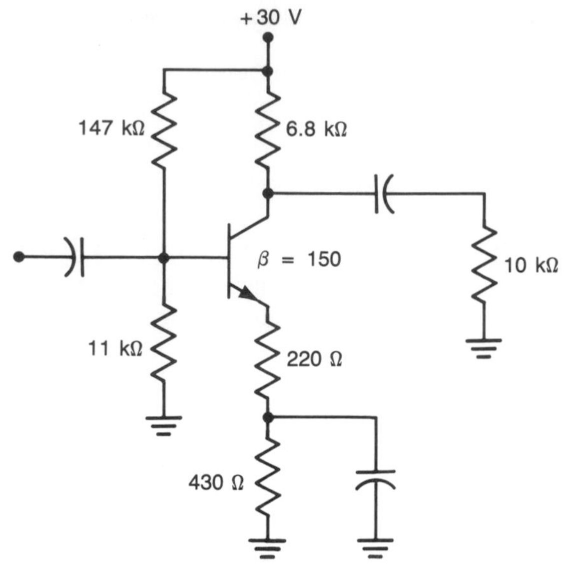


**FIGURE 24-40 Circuit for Chapter Review Problem 24-6.**

## ANSWERS TO SELF-TESTS

**1.** because both holes and electrons are used as majority current carriers

**2.** N-type material

**3.** base

**4.** base-emitter

**5.** base-collector

**6.** because there is no emitter-collector junction

**7.** collector

**8.** the emitter-collector voltage

**9.** the collector-to-base current when the emitter is disconnected

**10.** The base is lightly doped and very thin.

**11.** emitter

**12.** 84 mA

**13.** 85.2 mA

**14.** 150

**15.** toward

**16.** negative

**17.** leakage current in the BC junction ($I_{CBO}$).

**18.** yes

**19.** yes

**20.** 110

**21.** 13.3 kΩ (20 V/1.5 mA)

**22.** The BE junction has a much larger dynamic resistance.

**23.** to protect the junctions from breaking down when they are reverse-biased

**24.** $BV_{CEO}$ or $V_{CEO}$, $BV_{CBO}$ or $V_{CBO}$, and $BV_{EBO}$ or $V_{EBO}$

**25.** junction capacitance and transit time

**26.** It becomes too hot and fractures the crystal.

**27.** a device which increases the power of a signal

**28.** yes

**29.** 0.75

**30.** 1500

**31.** **a.** amplitude
**b.** cutoff

**32.** The collector supply voltage

**33.** $R_B$

**34.** **a.** The slope of the load line would decrease.
**b.** $V_{CE}$ would decrease and the Q point would move toward saturation.
**c.** $I_C$ would decrease a very slight amount because of the slight slope of the collector characteristic curve.
**d.** The voltage gain would increase.

**35.** They are 180° out of phase.

**36.** It goes into nonlinear operation. One-half or both halves of the signal will be clipped.

**37.** the dynamic resistance of the emitter-to-base junction when $I_E$ is changing value

**38. a.** 10.35 V
**b.** 0.7 V
**c.** 370

**39.** $Z_i$, because the resistance of one branch of the input network is multiplied by $\beta + 1$

**40.** $I_B$

**41.** no

**42.** false

**43.** false

**44.** false

**45. a.** increase
**b.** no change
**c.** decreases (how much depends on the value of $R_{B_1}$, which often dominates $Z_i$)
**d.** increases

**46.** false

**47. a.** 0.55 mA
**b.** 10.6 V
**c.** 53

**48.** $R_L$

**49. a.** common collector
**b.** common base
**c.** common emitter
**d.** common collector

**50.** common collector, because (1) $R_C$ is bypassed, (2) the output is from the emitter, and (3) the input is to the base

**51.** depletion mode

**52.** reverse bias

**53.** bipolar transistor

**54.** JFET and MOSFET

**55.** P channel

**56.** the condition when $V_{GS}$ depletes the channel of carriers

**57.** source, gate, drain

**58. a.** $V_{GS} = -1$ V
**b.** 4 mA

**59.** $R_S$

**60.** no

**61.** depletion-type

**62.** MOSFET, because there is no reverse-biased junction through which current can leak

**63.** enhancement-type

**64.** depletion-type

**65.** It is an enhancement-type MOSFET which is designed to handle higher power and provide more linear operation than other MOSFETs.

# Chapter 25 Integrated Circuits and Operational Amplifiers

Integrated circuits are used extensively in both linear and digital circuits and systems. To be knowledgeable in electronics today you must have a basic understanding of ICs: how they are manufactured and how they are used. The first part of this chapter will help you develop this basic understanding of integrated circuits.

The *operational amplifier* is one of the most common linear IC devices. It is used to perform an almost endless variety of electronic functions. Therefore, the last part of this chapter is devoted to learning about operational amplifiers and some of their important applications.

## 25-1 INTEGRATED CIRCUITS

Devices which have many transistors, diodes, resistors, and/or capacitors fabricated on a single piece of *substrate* (supporting material) are known as *integrated circuits* (ICs). Two broad classifications of ICs are *monolithic ICs* and *hybrid ICs*. Monolithic ICs are so much more common than hybrid ICs that the term IC or integrated circuit by itself is understood to refer to monolithic ICs.

Monolithic, meaning "one stone," identifies those ICs in which all the resistors, diodes, transistors, and capacitors are contained in one monolithic silicon crystal. Within this single crystal, the resistors, transistors, and so on, are electrically insulated from each other by reverse-biased PN junctions.

Monolithic ICs are often classified as *small-scale integration* (SSI), *medium-scale integration* (MSI), *large-scale integration* (LSI), and *very large scale integration* (VLSI). The dividing lines between these four categories are not hard and fast. Sometimes

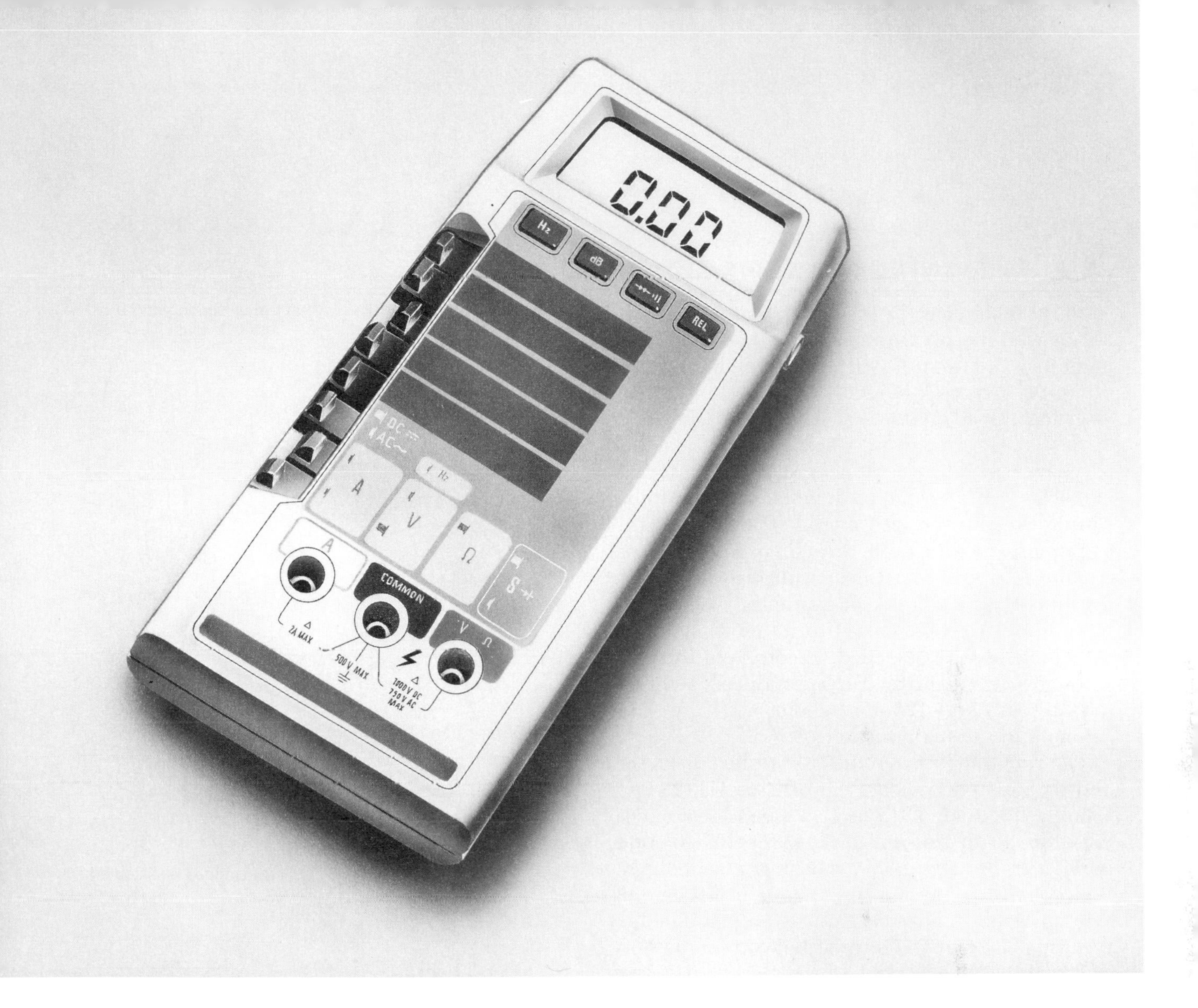


the categories are based on the number of circuit functions, such as logic gates, contained on a single chip. On this basis, the division is often as follows: SSI has fewer than 12 functions; MSI between 12 and 100; LSI between 101 and 1000; and VLSI more than 1000 gates or functions. Other times, these terms are used to indicate the number of circuit elements or components (i.e., resistors, transistors, etc.) contained in the single crystal (called a *chip*). Using this classification, typical numbers are less than 50, 51 to 500, 501 to 10,000, and greater than 10,000 for SSI, MSI, LSI, and VLSI, respectively.

Monolithic ICs are also classified by (1) the type of transistors embedded in the crystal, (2) the way in which the transistors are connected together, and (3) the techniques used in manufacturing them. These classifications are most meaningful when used with ICs designed for digital logic circuits, which are discussed in Chap. 26. A couple of examples of this type of classification are *transistor-transistor logic* (TTL) and *metal-oxide semiconductor* (MOS) logic. Transistor-transistor logic ICs use bipolar-junction transistors and are most often in the MSI and LSI categories. Metal-oxide semiconductor ICs are constructed with MOSFETs and are most common in the LSI and VLSI categories.

Hybrid ICs are constructed on a substrate which is an insulator. The heart of a hybrid IC is a monolithic IC chip which is fastened to the hybrid IC substrate. The monolithic chip is then connected to other components which have been fabricated directly on the substrate. Depending on how the other components (resistors, capacitors, etc.) are fabricated, hybrid ICs are classified as either *thin-film* or *thick-film*. These classifications are detailed in Sec. 25-3 after we have seen how monolithic ICs are constructed.

## 25-2 FABRICATING MONOLITHIC ICS

As illustrated in Fig. 25-1*a*, the fabrication of an IC begins with the preparation of a *silicon ingot*. The ingot may be either N-type or P-type silicon; most often it is P-type. The ingot is grown (formed) by inserting a small crystal (*seed* crystal) into a crucible of molten, doped silicon. Then the seed crystal is very slowly withdrawn from the silicon melt. Capillary action causes some of the molten silicon to be withdrawn with the seed crystal. The slowly withdrawn melt solidifies and, through capillary action, continues to pull more melt with it as it is withdrawn. By controlling such variables as temperature and rate of ingot withdrawn, the diameter of the ingot can be controlled. Ingots over 150 mm (about 6 in.) in diameter can be produced; 100 mm (about 4 in.) and 125 mm (about 5 in.) are very common in the manufacture of ICs.

The rough ingot is ground to finished diameter and the tapered ends are cut off (see Fig. 25-1*b*). Using a fine diamond saw, the trued ingot is cut into thin, about 200-$\mu$m-thick, wafers like the one in Fig. 25-1*c*. The *wafer*, which is a monolithic doped crystal, is then ground flat, polished on one side, and thoroughly cleaned in preparation for receiving the many ICs to be built up on its polished surface. The wafer becomes the substrate for the ICs fabricated on it.

The squares drawn on the surface of the wafer in Fig. 25-1 indicate the area needed for an individual IC. A typical IC in the MSI or LSI category requires an area about 2.5 mm by 2.5 mm (about 0.1 in. by 0.1 in.). Thus, a large wafer produces hundreds of individual, identical ICs. The ICs are formed on the wafer by the *photolithographic process*, which is explained in some detail very shortly.

After the ICs are processed on the wafer, lines are scribed, with a diamond-tipped scribe, on the wafer in order to separate the individual ICs. The wafer is then broken into small pieces called *dice* or *chips*. The individual chip (or die), shown in Fig. 25-1*e*, contains a complete IC circuit. The circuit is then electrically tested before the chip is attached with epoxy cement to the body and pin section of the IC package shown in Fig. 25-1*f*. Next, the very fine wires seen in Fig. 25-1*f* are bonded (welded) between the metal pins of the IC package and the metal contacts deposited on the IC chip during the photolithographic process. Finally, the IC package is completed by attaching a cap or lid over the body on which the chip is mounted. The cap or lid is hermetically sealed to the body of the IC package. This seal can be formed with epoxy cement. The finished IC package shown in Fig. 25-1*g* is known as a *dual in-line package* (DIP), which is a popular way of packaging ICs. Such packages often have 14, 16, 24, or more pins.

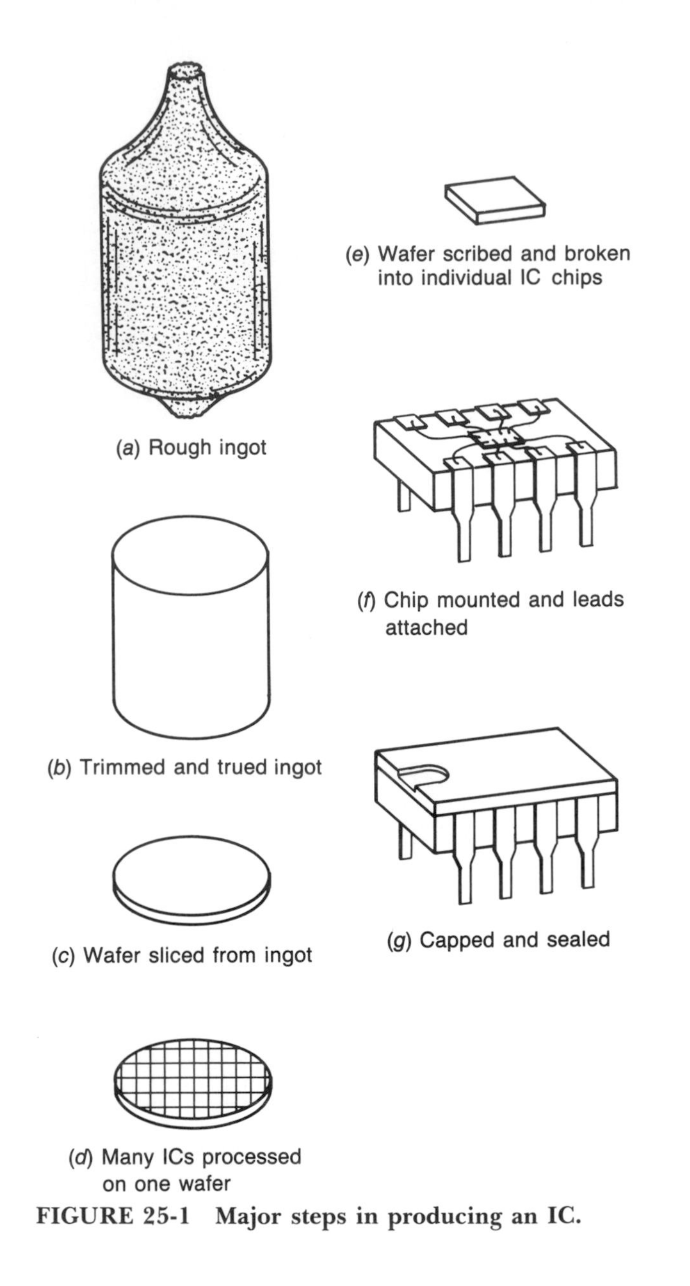


**FIGURE 25-1 Major steps in producing an IC.**

Now that we have an overview of the IC fabrication process, we can go back and look at some details of how circuit elements are put on the wafer. After the P-type wafer has been polished and cleaned, a thin (about 2- to 10-$\mu$m) N-type layer is deposited on the polished surface of the wafer as shown in Fig. 25-2*b*. This layer, called the *epitaxial* layer, is added by subjecting the wafer, at an elevated temperature, to vaporized N-type materials. With this process, the N-type layer essentially grows on top of the substrate (P-type wafer) and becomes a continuation of the monolithic crystal. Since the crystal is monolithic, the joining of the P-type substrate and the N-type epitaxial layer forms a PN junction. The actual circuit elements (transistors, resistors, etc.) will be contained in the N-type epitaxial layer. By keeping this PN junction reverse-biased when the finished IC is in operation, the circuit elements are electrically insulated from the substrate. The hashed lines in Fig. 25-2*b* and other figures which follow have no significance except to identify different materials.

The *silicon dioxide* ($SiO_2$) layer shown in Fig. 25-2*c* is added by exposing the wafer of Fig. 25-2*b* to oxygen gas or water vapor while the temperature is in the range of about 800 to 1400°C. The combination of heat and oxygen oxidizes the N-type material to a depth of about 0.1 to 1 $\mu$m. The depth is controlled by temperature and time.

The next step, shown in Fig. 25-2*d*, is to add a layer (approximately 1 $\mu$m in thickness) of *photosensitive emulsion* over the silicon dioxide. The emulsion is often sprayed on. The type of emulsion used in the examples which follow is the type in which the areas that have been exposed to ultraviolet light will not be removed by photographic developing chemicals.

Oxidizing the wafer and coating it with *photoresist* (emulsion) are the first steps in what is often referred to as the photolithographic process. The continuation of the process is highlighted in Fig. 25-3, which shows only a very small section of the wafer. The next step is to cover the wafer with a photomask and expose the covered wafer to ultraviolet light. As indicated in Fig. 25-3*a*, the areas under the transparent part of the photomask are exposed to the light, while the opaque part of the

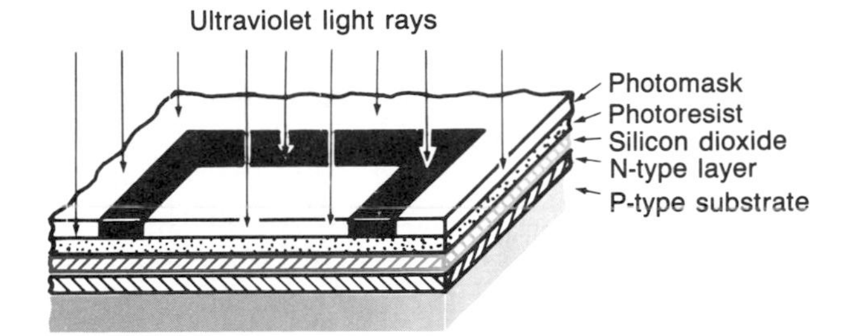


(*a*) Pattern transferred from mask to photoresist

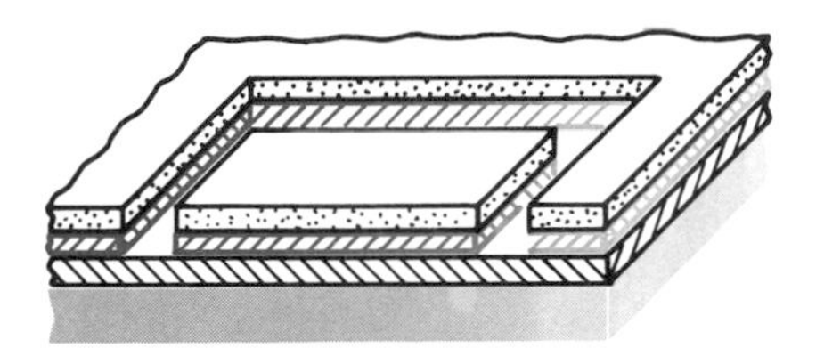

(*b*) Photoresist developed and silicon dioxide etched

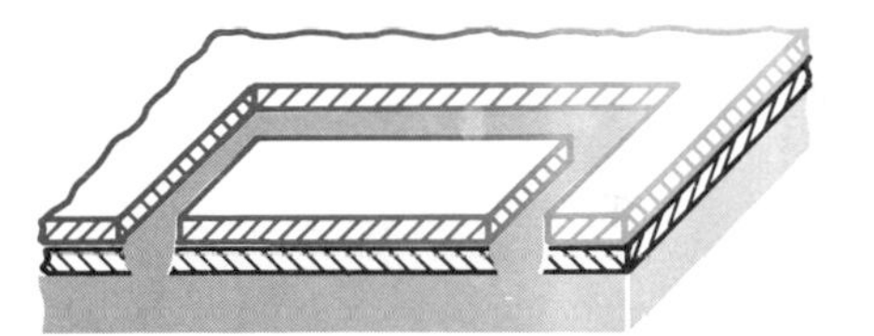

(*c*) Photoresist removed and P-type impurity diffused into the N-type layer

**FIGURE 25-3 Photolithographic process.**

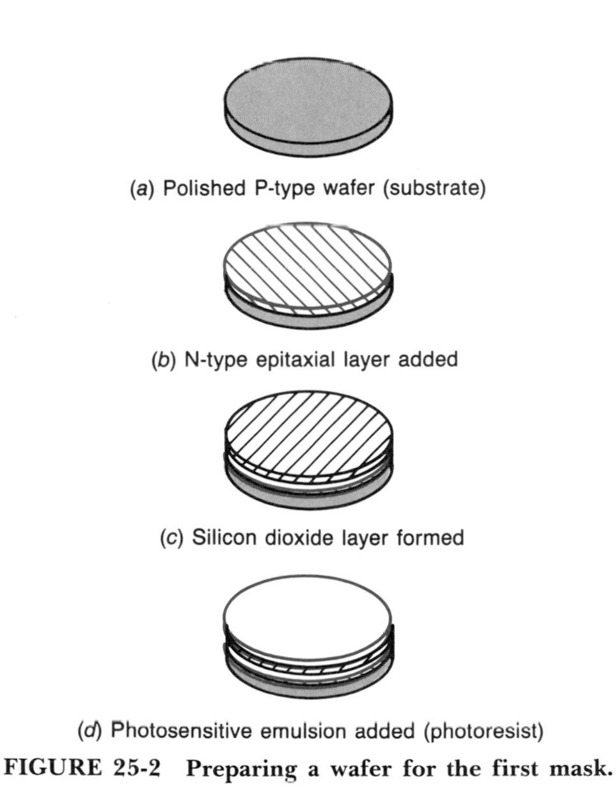

(*a*) Polished P-type wafer (substrate)

(*b*) N-type epitaxial layer added

(*c*) Silicon dioxide layer formed

(*d*) Photosensitive emulsion added (photoresist)

**FIGURE 25-2 Preparing a wafer for the first mask.**

mask blocks the light and keeps some areas from exposure. When the mask is removed and the photoresist developed, the unexposed areas of the photoresist are washed away. Then, when the wafer is subjected to an acid bath, or acid vapors, the silicon dioxide not covered by photoresist is etched away so that an area of the N-type epitaxial layer is exposed (see Fig. 25-3*b*). Next, the rest of the photoresist is removed by a solvent.

Now, when the wafer is exposed at a high temperature to vaporized boron (or any other trivalent dopant), the boron atoms diffuse into the N-type layer. This *diffusion process* is continued until there are more trivalent atoms in the exposed part of the epitaxial layer than there are pentavalent atoms. Figure 25-3*c* shows the end result: the exposed part of the epitaxial layer is changed from N-type to P-type material. The P-type material now joins the P-type substrate and creates an isolated pocket or island of N-type material. Repeating the complete photolithographic process (oxidize, cover with resist, expose through a mask, develop, etch, remove resist, and diffuse) a number of times with different masks allows circuit elements to be diffused into the isolated N-type pockets created by the first mask.

Figure 25-4*b* shows a top view of a small part of an IC which contains the circuit elements (components) shown in Fig. 25-4*a*. A cross-sectional view of this same part of the IC is shown in Fig. 25-4*c*. The cutting line for the cross-sectional view is indicated in Fig. 25-4*b*. It should be noted that the ar-

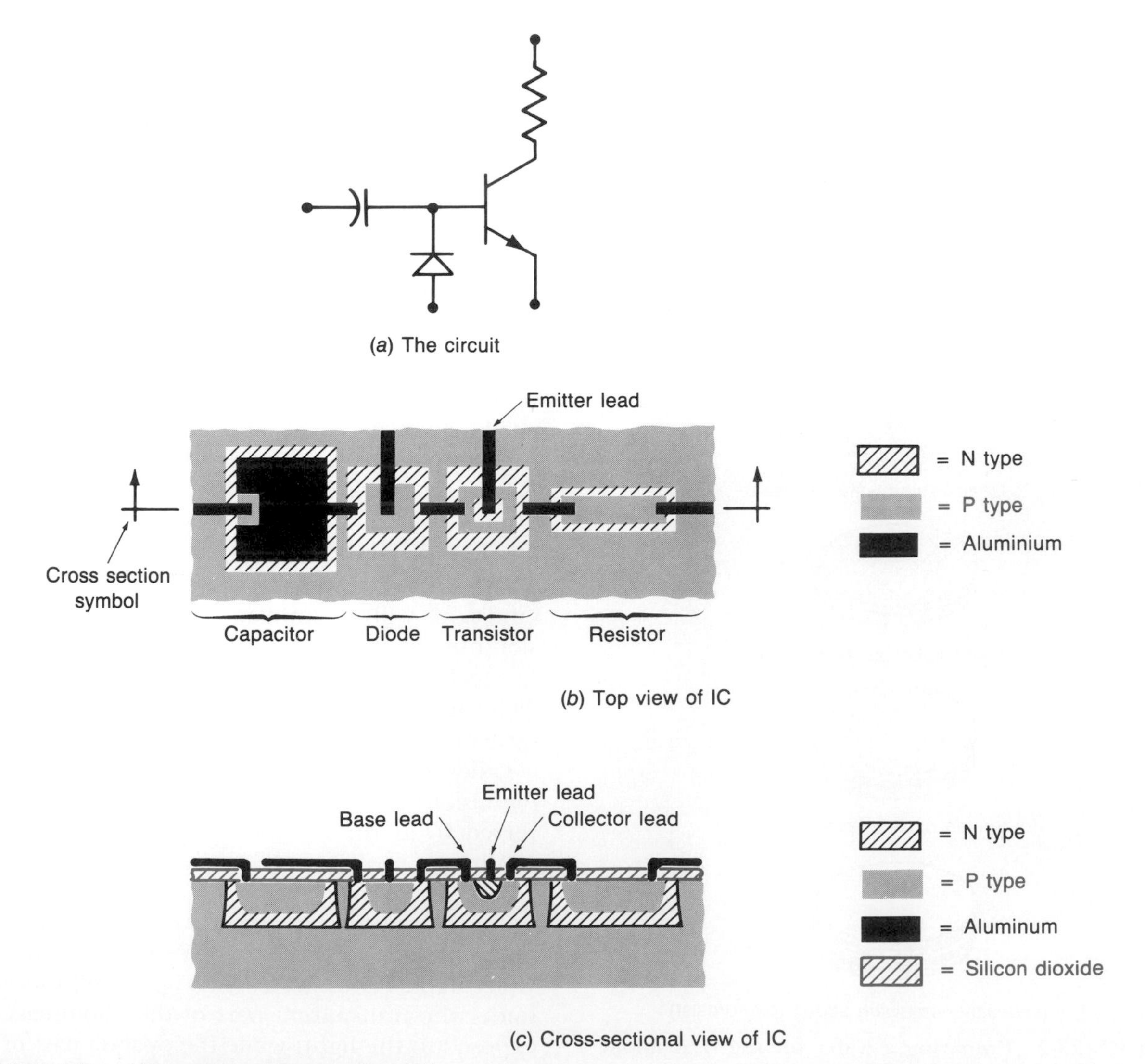


**FIGURE 25-4 Illustration of part of an IC.**

rangement of the circuit elements and relative sizes of the elements are for illustrative purposes only.

In Fig. 25-4 you can see that one plate of the capacitor is P-type material and the other plate is aluminum foil deposited on top of silicon dioxide. Silicon dioxide, which is a good insulator, serves as the dielectric. The N-type material below the P-type plate isolates the capacitor from all other elements and from the substrate. It does this by forming PN junctions with both the P-type plate and the P-type substrate. Regardless of the polarity of any voltage applied between the capacitor plate and the substrate, one of these two junctions will be reverse-biased.

The diode in Fig. 25-4 is also isolated from the substrate by a reverse-biased PN junction. Notice that the same N-type pocket that forms this junction is also the cathode of the diode. The cathode of the diode is connected to the base of the transistor by the thin aluminum foil conductor shown in Fig. 25-4*b* and *c*.

Inspection of Fig. 25-4*c* shows that the transistor is an NPN bipolar device. Its collector is connected to the resistive element which consists of an isolated P-type material. The resistance is determined by the physical dimensions of the P-type island and by the level of doping.

The layouts of the part of the photomasks needed to produce the section of the IC detailed in Fig. 25-4 are shown in Fig. 25-5. Below each mask in this figure is a cross-sectional view which shows the results of the photolithographic process using that particular mask. In these views the photoresist material is not shown. As you follow through the steps in Fig. 25-5, remember that the wafer surface is reoxidized and rephotosensitized before the next mask is used. The first three masks in this figure outline areas that are to be doped by the diffusion process. The fourth mask outlines the small openings to be made in the silicon dioxide so the metal contacts can be made with various P-type and N-type regions. The contacts are made when the heated wafer is exposed to an atmosphere of a vaporized metal such as aluminum. As the vaporized metal is deposited on the wafer, it penetrates through the small openings etched in the dioxide and makes contact with the active parts of the IC. In the process, the entire surface of the wafer is covered with a thin film of metal. The fifth mask is then used to outline the areas of the metal to be left on the wafer to serve as interconnecting conductors, capacitor plates, and contact pads on which lead wires are eventually welded. The unnecessary metal film is then etched off the wafer.

Although not shown in Fig. 25-5, many ICs are sealed with a final coat of silicon dioxide which is vapor-deposited over the entire surface including the metal leads. This final layer of dioxide is called the *passivating layer*. Since this passivating layer would cover the contact pads, a final mask would be needed to expose these pads so that wire leads could be attached.

Many techniques besides the ones described in this section are used in fabricating ICs. For example, the doping impurity element may be added by a process called *ion implantation*. This process uses electric fields to accelerate ions of the impurity element and to direct these accelerated ions into selected areas of the crystal. The process provides good control over the depth and concentration of the implanted ions. Another example is the use of lasers to form the openings in the silicon dioxide layer. One final example is the use of very heavily doped areas which are indicated on cross-sectional views by $N^+$ or $P^+$ symbols. These heavily doped areas or pockets are used for such things as lowering the resistivity of a collector, improving the crystal-to-metal contacts, and controlling the voltage gradient within a junction.

## Self-Test

**1.** List two types of integrated circuits.

**2.** What is LSI the abbreviation for?

**3.** True or false? TTL integrated circuits use MOSFETs.

**4.** True or false? Transistors and diodes are the only circuit elements used in ICs.

**5.** True or false? The ingot for making ICs is grown from pure undoped silicon.

**6.** The thin slice of crystal cut from an ingot is called a ______.

**7.** Define substrate.

**8.** An individual MSI circuit covers about ______ square inch of surface area.

**9.** The ______ process is used to fabricate ICs on a wafer.

**10.** After a wafer is scribed, it is broken into pieces called ______.

**11.** What does DIP stand for?

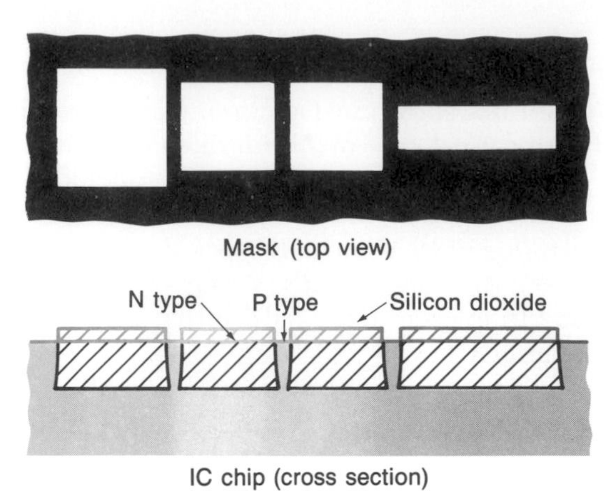


(*a*) Component isolation and formation of collector and other N-pockets

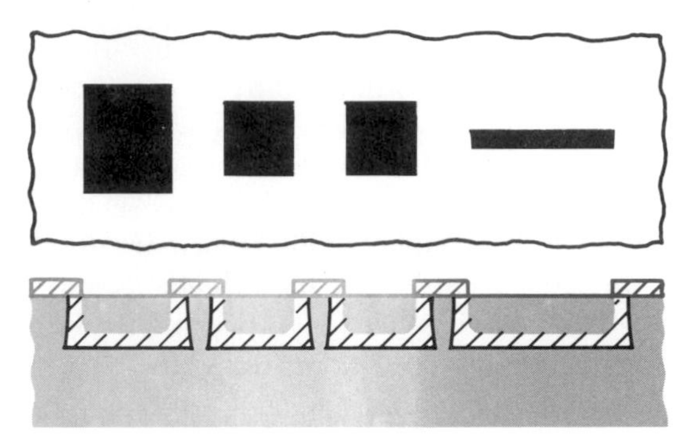

(*b*) Formation of base and other P-pockets

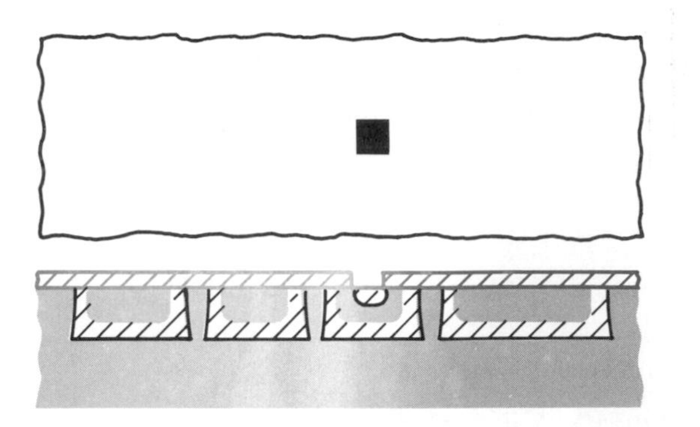

(*c*) Formation of emitter

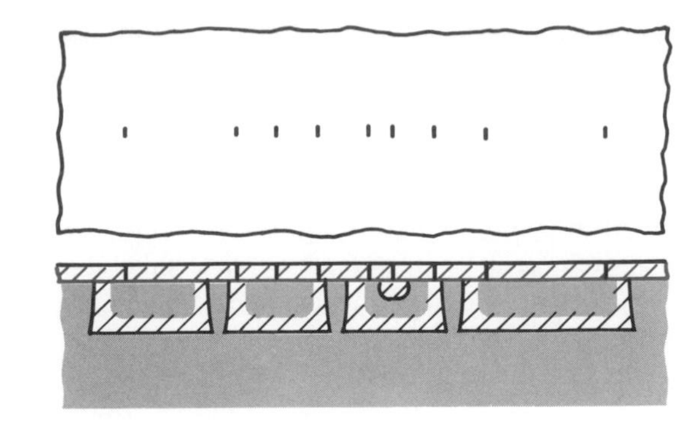

(*d*) Formation of metal contacts

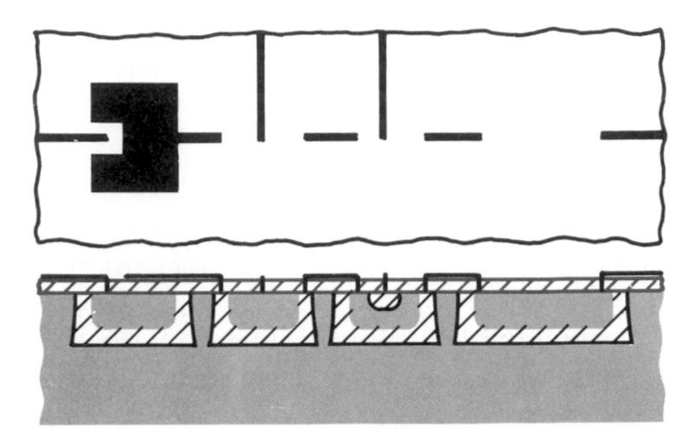

(*e*) Formation of metal leads and capacitor plate

**FIGURE 25-5 Steps in producing the IC shown in Fig. 25-4.**

**12.** The circuit elements of an IC are fabricated in the ______ layer of the wafer.

**13.** How are circuit elements electrically isolated in an IC?

**14.** None of the layers deposited on the substrate of a monolithic IC are more than 10 ______ thick.

**15.** The ______ process is often used to place impurity atoms in selected regions of a wafer.

**16.** The dielectric material for an IC capacitor is ______.

**17.** List three uses of the aluminum foil pattern left on the surface of an IC.

**18.** The ______ layer is the final layer of silicon dioxide deposited on an IC.

**19.** In addition to the diffusion process, crystals can be doped by ______.

**20.** What does $P^+$ stand for?

## 25-3 FABRICATING HYBRID ICS

The substrate for hybrid ICs is some form of insulating material (such as glass, alumina, or quartz) which is able to withstand the high temperatures (up to 1000°C) involved in the fabrication process. The substrate is usually between 250 and 750 μm (0.01 to 0.03 in.) thick. Circuit elements, conductors, resistors, and capacitors are then built up on the surface of the substrate. These circuit elements can be built up on both surfaces of the substrate if necessary. The substrate can also be used as a capacitor's dielectric if the plates of the capacitor are deposited on opposite sides of the substrate. The components are built up using either thick-film or thin-film techniques. Once these elements are built up, other circuit components, including monolithic IC chips, or even a packaged monolithic IC, can be attached to the substrate and electrically connected to the built-up elements by wire bonding (welding) or soldering. The completed circuit is then packaged to form a hybrid integrated circuit.

Figure 25-6 illustrates how a small section of a thick-film hybrid IC might appear. However, the illustration is not drawn to scale. Notice in this illustration that one interconnecting conductor can cross over another conductor. The two conductors are insulated from each other by the same material (often ceramic) used for the dielectric in the capacitor. The capacitance between the crossing conductors is very small because the conductors can be as narrow as 125 μm (0.005 in.).

Although conductors can be made 0.005 in. wide, they are usually much wider than that. The conductors, as well as the resistors and dielectric materials, are typically 12 to 50 μm thick.

The resistor in Fig. 25-6 is laid out in a zigzag pattern to minimize the substrate surface area it occupies. With this configuration, resistances in the megohm range are practical.

As seen in Fig. 25-6, the dielectric of the capacitor extends slightly beyond the top plate to ensure adequate electric separation of the bottom and top plates of the capacitor. The dielectric material being quite thin (12 to 50 μm) allows for usable capacitance values with small plate areas.

The IC chip in Fig. 25-6 can be adhered to the substrate with an epoxy. Then it can be connected to the thick-film conductors by the wire-bonding technique used in fabricating monolithic ICs.

### Thick-Film Process

In the thick-film process the circuit elements are built up on the substrate by repeated silk screening operations. Silk screening is a printing process in which a thin paste material (ink, paint, photoresist, etc.) is forced through the unobstructed openings

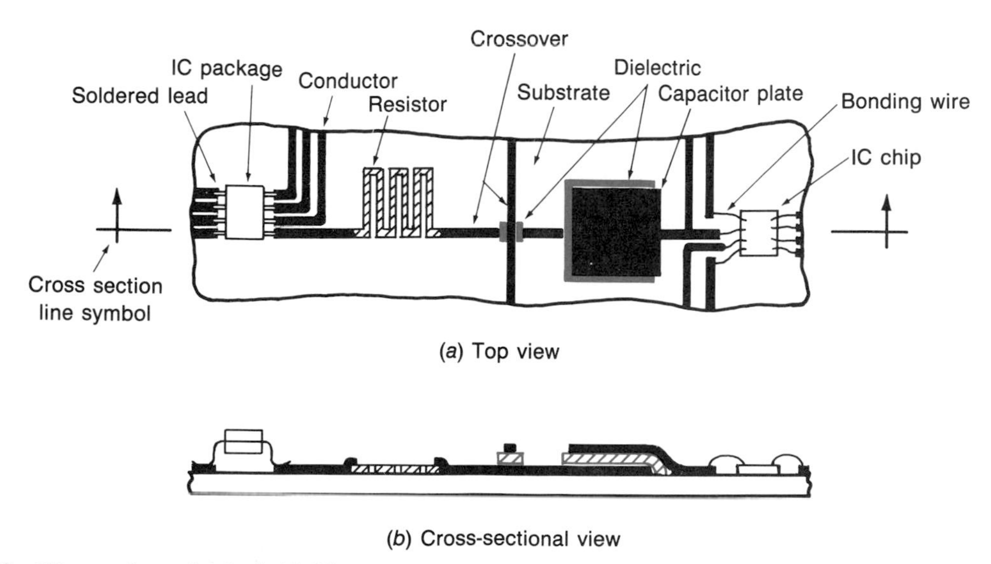

**FIGURE 25-6 Illustration of a hybrid IC.**

in a fine-mesh screen. The paste which goes through the screen pores adheres to the surface of the receiving material which is placed in contact with the screen. The original screens for this process were made from silk; thus, the name "silk screening." Present screens are made from a variety of materials, including stainless steel.

The composition of the paste which is screened onto the substrate determines which circuit element, resistor, conductor, or dielectric, is added. All three types of paste contain (1) a solvent which evaporates to dry the paste and turn it into a solid material; (2) a binding material (such as glass) which melts and fuses together the active particles in the paste when the paste is heated (fired); and (3) the active material which makes the paste a conductor, resistor, or dielectric element. The active material for the dielectric paste is a ceramic compound. For the conductor paste, the low-resistivity metals (gold, silver, copper, platinum, etc.) are the active material. The resistor paste includes various compositions of metals and metal oxides. All of the active materials are finely ground (powdered) before they are added to the paste binder and solvent.

The pattern to be silk-screened onto the substrate is put on the screen by the photoresist process. The screen is filled with photoresist and then exposed through a photomask to ultraviolet light. When developed, the unexposed photoresist washes out, leaving open pores for the paste to pass through.

After a pattern is screened onto the substrate, the paste is dried and then fired at temperatures up to about 1000°C. Then another pattern can be screened on, dried, and fired.

When precise resistance values are needed in a hybrid circuit, the cross-sectional area of the screened resistor can be decreased by trimming the resistor pattern with a laser beam while the resistance is being measured. Of course, when this technique is used, the deposited resistance must be less than the final desired resistance because reducing the cross-sectional area increases the resistance.

### Thin-Film Process

Less than 20 percent of the hybrid ICs are made by the thin-film process. However, thin-film techniques are used extensively for hybrid circuits operating in the microwave frequency range.

Producing the circuit elements for a thin-film IC involves repeated use of the technique that was used to put the metal conductor pattern on top of the monolithic IC. In fact, this is referred to as a thin-film technique because the metal film is very thin and it is built up on top of the monolithic crystal rather than being a continuation of the crystal as all of the preceding layers were. Producing a thin-film hybrid circuit like the one in Fig. 25-6 would require depositing a thin film (usually less than 1 $\mu$m thick) of a resistive material over one entire surface of the substrate. Then this resistive film would be covered with a layer of photoresist. Next, the photoresist would be exposed to ultraviolet light through a photomask containing the resistor pattern to be left on the substrate. After developing and curing the photoresist, all of the resistive film except for that protected by the cured photoresist would be chemically etched away; the thin-film resistor would now be on the substrate. Repeating this photolithographic process with layers of conductive and dielectric materials would produce the other circuit elements shown in Fig. 25-6. Connecting other components to the thin-film circuit elements is done in the same way as it was with thick-film hybrid circuits.

## 25-4 INTEGRATED COMPONENT CHARACTERISTICS

The characteristics of integrated components must be taken into account by the designer of ICs. For example, a diffused resistor like the one shown in Fig. 25-4 has some limiting characteristics. It has both a large tolerance and a large temperature coefficient. Thus, a diffused resistor would not be acceptable in circuits requiring stable, precision resistors. On the other hand, thin-film and thick-film resistors would be a good choice for such circuits. These resistors can be made from materials which have very small temperature coefficients, and they can be laser-trimmed to an exact value after they are on the substrate.

Although not suited for circuits requiring precision resistors, diffused resistors work very well in circuits requiring matched resistors. In some types of circuits, it is not too important whether each of a pair of resistors is 1000 or 1200 Ω. Also, it does not matter too much if the resistance changes from 1100 to 1200 Ω as long as both resistors change to the same value. Diffused resistors work well in this type of circuit because both resistors can be diffused at the same time during the fabrication process. Therefore, the resistors will have matched characteristics.

Another difference between diffused and film resistors is the practical range of resistance obtainable. The diffused resistor has a rather restricted range because its resistance, for a given surface area, must be controlled by doping levels. Film resistors can be produced in a wider range of resistances by controlling the resistivity of the film material.

Because of the amount of chip or substrate surface area which can be devoted to a single capacitor, integrated capacitors are limited to the picofarad range. Capacitors in hybrid circuits can be larger than those in the monolithic circuits. This is so for two reasons: (1) the substrate area of the hybrid is larger than that of the monolithic, and (2) hybrid circuits can use materials with higher dielectric constants than can monolithic circuits.

The high-frequency response of the monolithic IC is limited by the reverse-biased PN junctions used to isolate the IC's components. Remember that a reverse-biased junction has a small amount of capacitance. At high frequencies, the reactances of these junction capacitances are small enough to affect the operation of the circuit. For example, the junction capacitance between the diffused resistor and the substrate in Fig. 25-4 allows part of the collector signal current to flow to the substrate which is usually connected to ground. This lost signal current results in reduced gain for the amplifier stage.

Since hybrid ICs do not rely on reverse-biased junctions for circuit isolation, their high-frequency response can be excellent. Through careful design of mask patterns, the stray capacitance in hybrid ICs can be held to a very low value.

One final characteristic of ICs needs to be mentioned—reliability. Systems constructed with ICs have far smaller failure rates than do systems which use discrete components. Integrated circuit systems require fewer individual components, interconnecting leads, and solder joints per function than do discrete component systems. The fewer of these items in a system, the smaller is the chance of system failure.

## Self-Test

**21.** The substrate for a hybrid IC is an ______ material.

**22.** True or false? The circuit elements of a hybrid IC are diffused into the substrate.

**23.** True or false? Hybrid ICs often use monolithic ICs in their circuits.

**24.** True or false? Transistors are often built up on a substrate using thick-film techniques.

**25.** True or false? Crossing conductors in a hybrid circuit are insulated from each other with a thin layer of dielectric material.

**26.** Why are thick-film resistors laid out in a zigzag pattern?

**27.** List and define the three types of material contained in a thick-film paste.

**28.** How are precision resistors obtained in hybrid ICs?

**29.** Which hybrid circuit, the thick-film or the thin-film, uses the photolithographic process?

**30.** Does the monolithic or hybrid circuit have the better frequency response? Why?

## 25-5 THE DIFFERENTIAL AMPLIFIER

The *differential amplifier*, shown in Fig. 25-7*a*, is a very versatile circuit. It is a commonly used circuit configuration in multistage, linear ICs. It is used for the input (first) stage, and usually several following stages, of the operational amplifiers (op amps) discussed in Sec. 25-6.

No base-to-ground resistors are shown in Fig. 25-7 because IC differential amplifiers do not include them. A dc path from ground to base must be provided by the signal sources, or other elements, connected to the inputs; otherwise, there can be no base current to bias the transistors in the linear region.

The differential amplifier (DA) has two inputs; so, it can simultaneously receive two input signals. It amplifies the instantaneous difference between the two input signals. When signals are applied to both inputs of the DA, Fig. 25-7*b*, it is called a *dual-ended* or *double-ended* input. As shown in Fig. 25-7*c*, the DA operates equally well in a single-ended con-

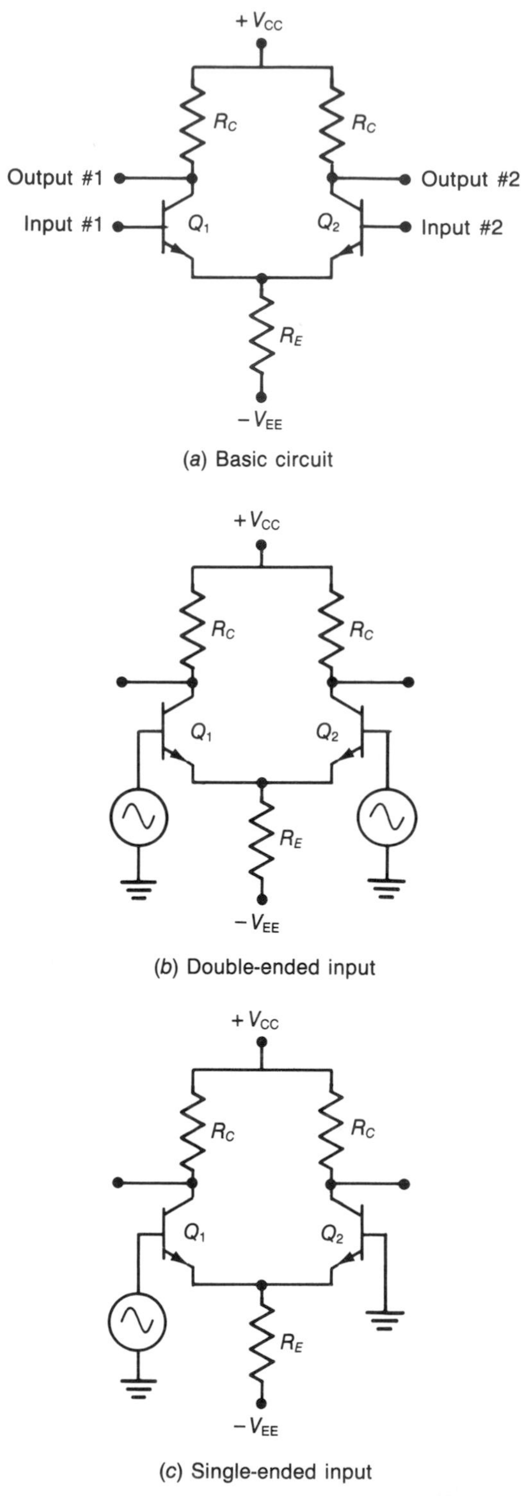


**FIGURE 25-7 Differential amplifier circuits.**

figuration. For single-ended operation, the single signal is applied to one input and the other input is grounded. The signal sources in Fig. 25-7 could be the outputs of other amplifiers or the outputs of any type of input *transducer*. Input transducers are devices that convert changes in heat, pressure, light, etc., into electric signals. Again, the sources must provide dc paths to ground to bias the transistors for linear operation. When the sources do not provide dc paths, base resistors $R_B$ must be added.

The DA in Fig. 25-7 can provide an output from either output terminal to ground. When the output is taken between an output terminal and ground, it is often called a *single-ended output*. The output can also be taken between the two output terminals. When the output is taken between the two output terminals, it is called the *differentiated output* or dual-ended output. This output has no reference to ground, so it is referred to as a *floating output*. The differential output is twice as large as either single-ended output. Figure 25-8 shows that the single-ended outputs are identical in amplitude, but 180° out of phase. This is why the differential output is twice as large as the single-ended output. When an integrated circuit DA is designed for single-ended operation, it has only one output terminal and often one of the collector resistors is also eliminated.

Figure 25-8 summarizes the general operation of an ideal DA by showing various input signal combinations and the resulting output signals. In Fig. 25-8*a*, the input signals are in phase and exactly equal in amplitude. Such signals are known as *common-mode signals*. They are rejected (not amplified) by the differential amplifier because the difference ($V_1 - V_2$) between the two input signals is zero. (The difference between the two input signals is often referred to, as it is in Fig. 25-8, as the *differential input*.) Figure 25-8*a* shows that the single-ended outputs are zero for a common-mode input. These are the idealized single-ended outputs. Even the most sophisticated DA has small single-ended outputs. However, even for an average DA, the single-ended outputs are usually less than the common-mode input signal, i.e., the common-mode voltage gain $A_{CM}$ is less than one. The differential output, the output taken between the collectors, is essentially zero for the common-mode signal as shown in Fig. 25-8*a*.

Figure 25-8*b* through *d* shows that the *differential-mode signals* produce a *differential signal* which is amplified by the DA. Dual-ended input signals do not have to be 180° out of phase, as shown in Fig. 25-8*b*, to produce a differential input signal. As

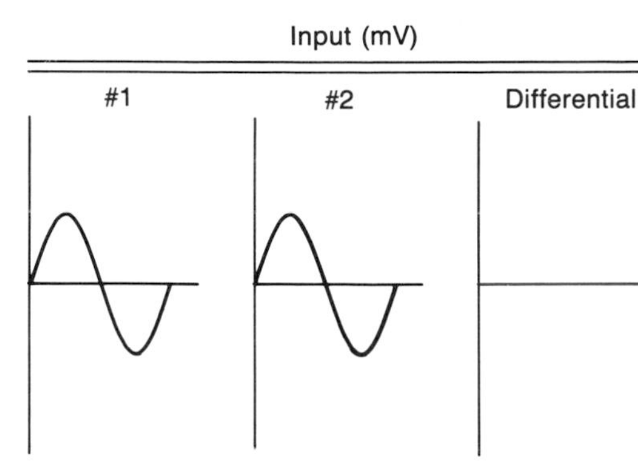


(*a*) Dual-ended input (common-mode signals)

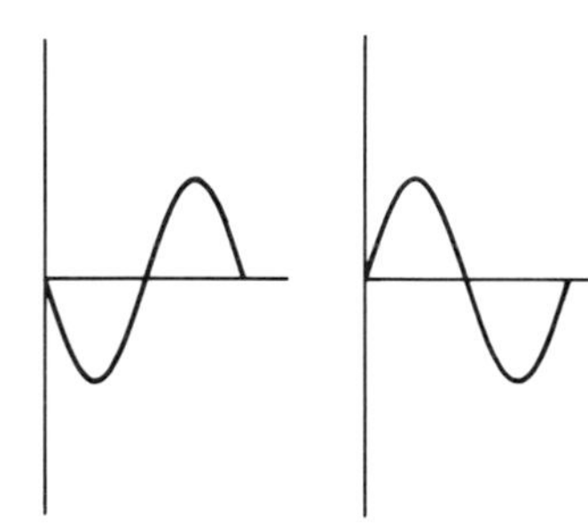
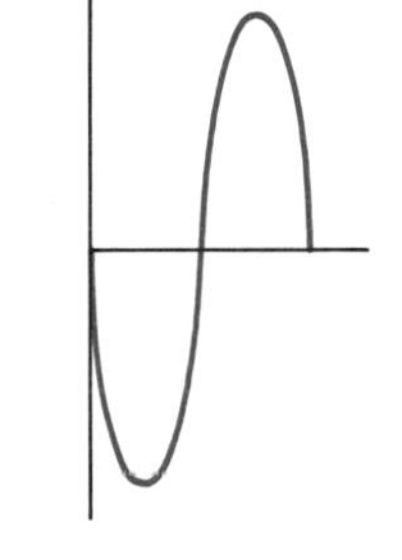
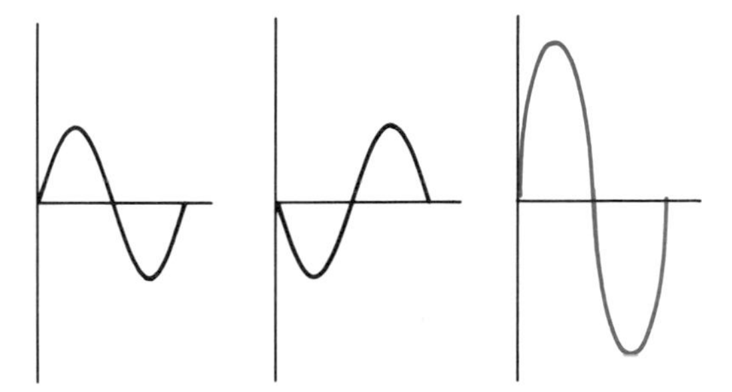

(*b*) Dual-ended input (differential-mode signals)

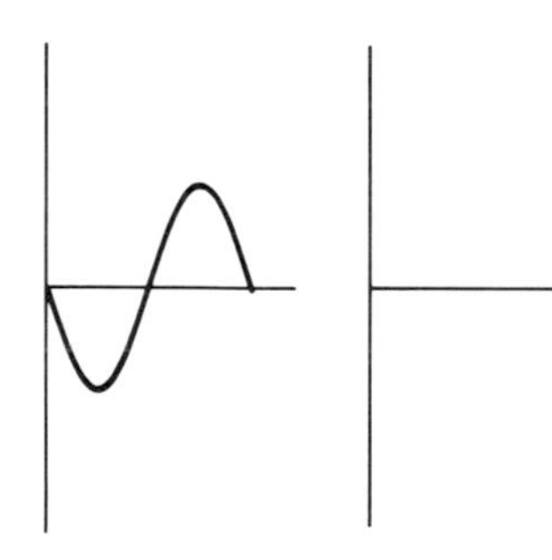
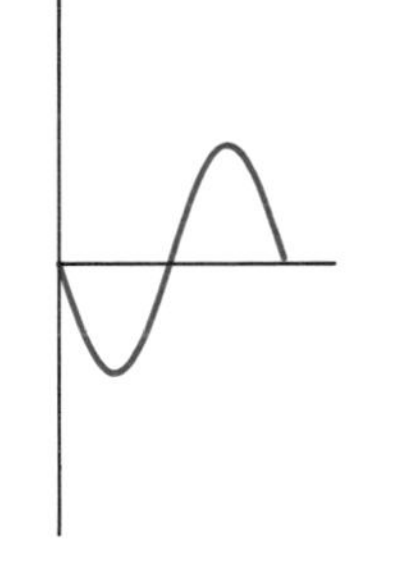
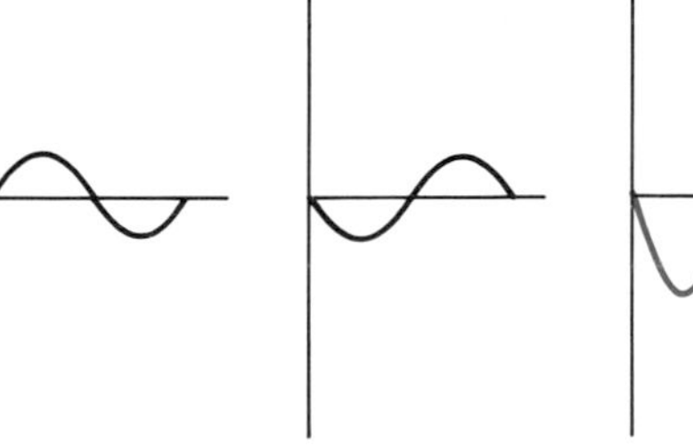

(*c*) Single-ended input (differential-mode signal)

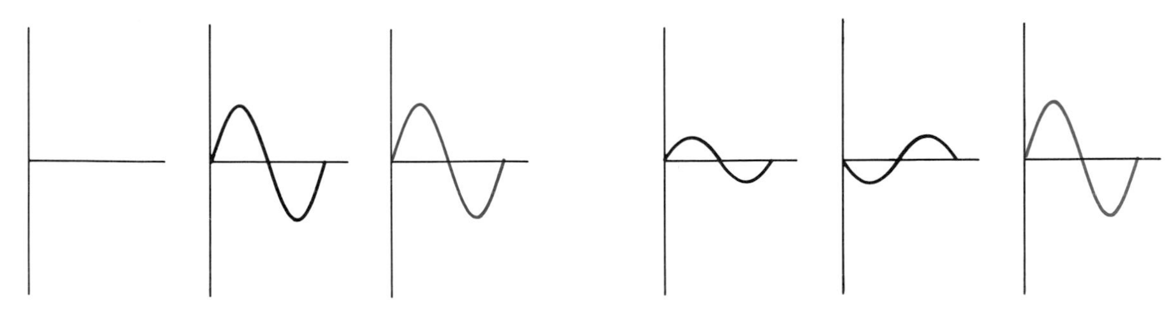

(*d*) Single-ended input (differential-mode signal)

**FIGURE 25-8 Differential amplifier signals for various operating conditions.**

long as the signals have different amplitudes or any phase difference, some differential input signal exists. Notice that when one input is grounded, the signal to the other input is the differential signal. The differential voltage gain $A_D$ is very large. For a single-stage DA, the gain for a single-ended output can be over 200. Thus, the gain for the differential output can be over 400. Since the output from a DA is usually taken as a single-ended output, we will assume the output, and the gain $A_D$, is for a single-ended output unless stated otherwise. The phase relationships of the differential outputs shown in Fig. 25-8 are arbitrary. Whether the first $\frac{1}{2}$ cycle of the signal is negative or positive depends on whether output no. 1 or output no. 2 is used as the reference point.

Figure 25-8 illustrates a very important characteristic of the DA (and also of the op amp which uses a DA as its first stage): the phase relationship between the signal out of one of the output terminals and the single-ended input signal depends on which input terminal receives the input signal. For instance, if the output is taken from output no. 2, then Fig. 25-8*c* shows that it is 180° out of phase with the input when input no. 1 receives the signal. However, as seen in Fig. 25-8*d*, output no. 1 is in phase with the input when input no. 2 receives the input signal. Therefore, when only one output terminal is available, as in Fig. 25-9, it is meaningful to label one input an *inverting input* and the other a *noninverting input.* The inverting input, which provides a 180° phase shift, is often identified with a − sign. The noninverting input is then represented by the + sign. Why these inputs result in different phase outputs will become clear when the detailed operation of the DA is covered later on in this section.

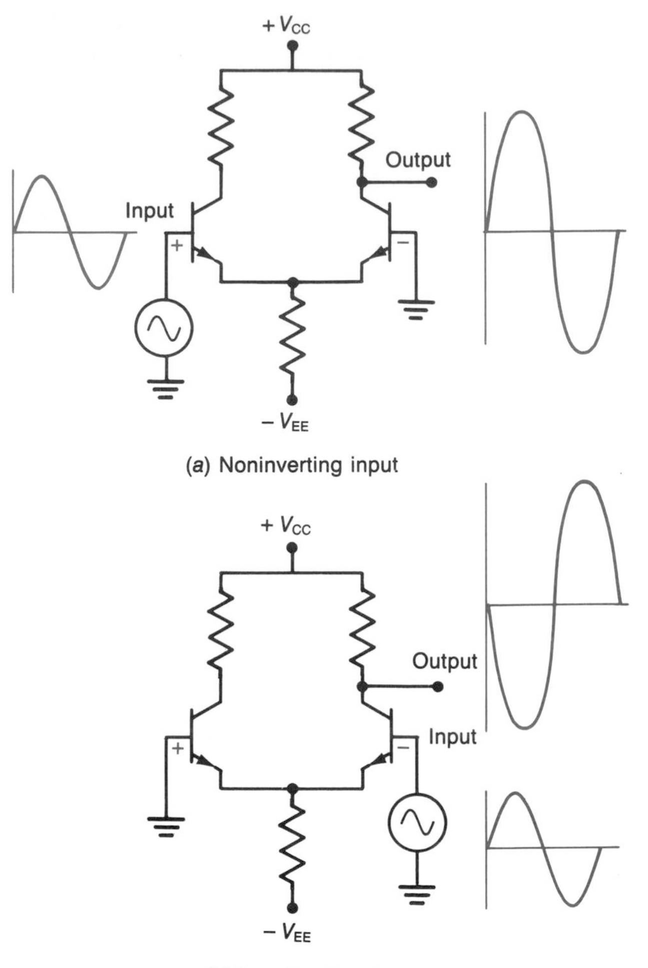


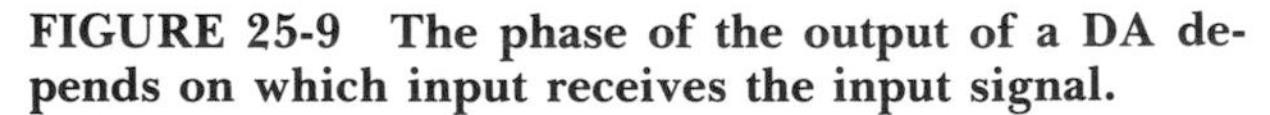
(*b*) Inverting input

**FIGURE 25-9 The phase of the output of a DA depends on which input receives the input signal.**

## Common-Mode Rejection Ratio

An important parameter of a DA, and also of an op amp, is its *common-mode rejection ratio* (CMRR). The CMRR can be calculated with the formula

$$\text{CMRR} = \frac{A_D}{A_{CM}}$$

where $A_D$ and $A_{CM}$ are the differential and common-mode gains, respectively. A single-stage DA can have a CMRR of over 400. Very often, the CMRR is expressed in decibels (dB). The appropriate formula in this case is

$$\text{CMRR}_{dB} = 20 \log_{10} \frac{A_D}{A_{CM}} = 20 \log_{10} \text{CMRR}$$

The importance of a high CMRR is illustrated in Fig. 25-10. In this figure the signal from a transducer is sent through a lead wire to the no. 1 input of the DA. However, a low-frequency *hum* voltage is also induced into the lead wire by a building and collapsing magnetic field which is produced by an adjacent conductor carrying 60-Hz current. The resultant waveform is shown in Fig. 25-10*a*. If this waveform were amplified by a conventional amplifier, the 60-Hz hum in the output would be stronger than the desired signal. In the DA, how-

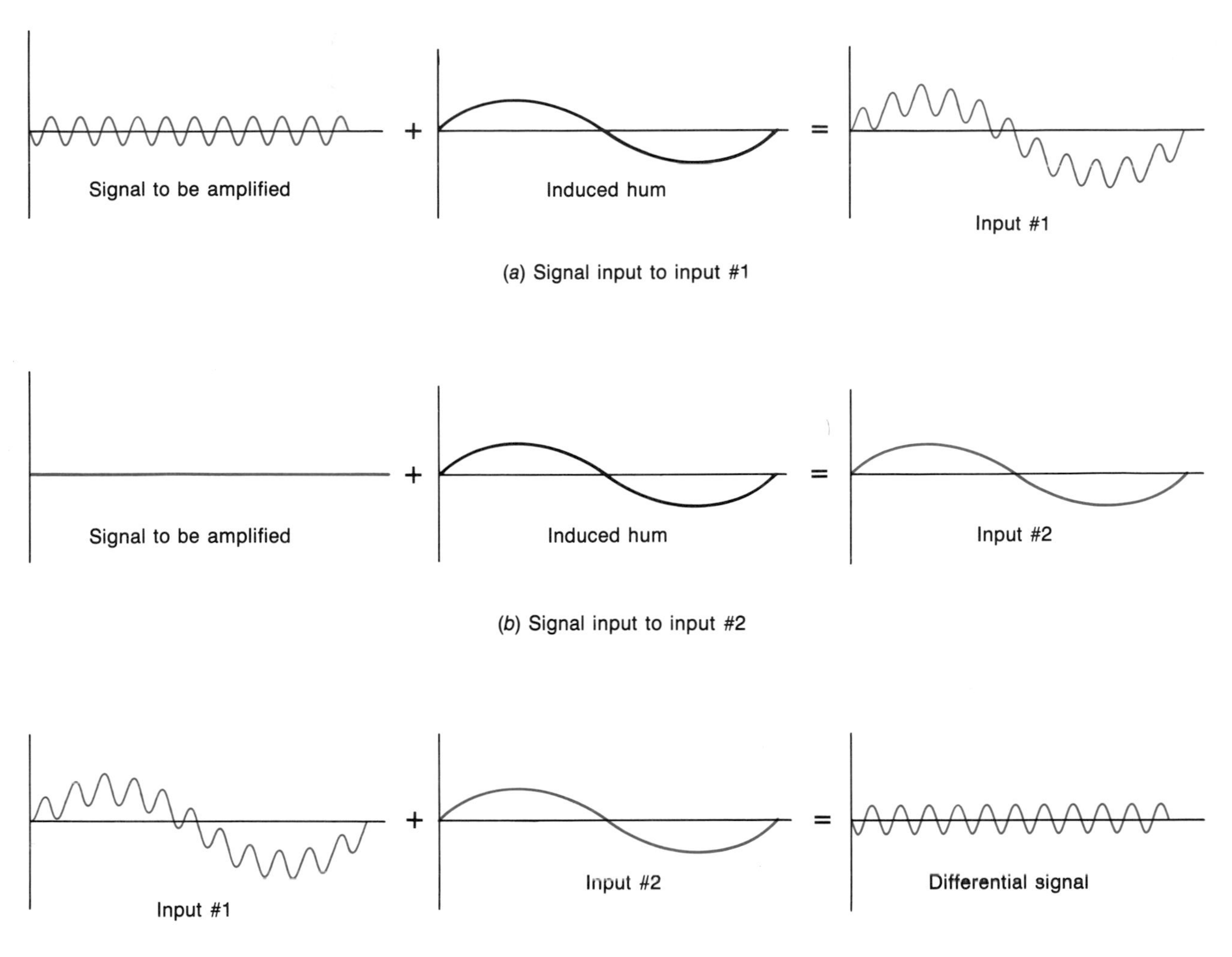


**FIGURE 25-10 A DA rejects hum and static voltages induced into its leads.**

ever, a second lead wire connected to input no. 2, and physically parallel to the input no. 1 lead wire, has the same phase 60-Hz hum induced into it. This is the only voltage applied to input no. 2 (see Fig. 25-10*b*). As shown in Fig. 25-10*c*, the hum components of the two inputs form a common-mode signal which is largely rejected by the DA. The common-mode signal is amplified by an $A_{CM}$ of less than 1, and the differential signal is amplified by an $A_D$ which may be greater than 200. If the two signals were equal at the input terminals, the desired signal would be over 200 times larger than the hum signal at the output terminal. If either $A_{CM}$ is decreased or $A_D$ is increased, then CMRR will increase, and the relative difference in the strength of the common-mode output and differential output will be greater. Thus, the larger the CMRR, the greater is the separation of the desired and undesired signals.

**EXAMPLE 25-1**

What is the CMRR and the $CMRR_{dB}$ of a DA which has a $A_{CM}$ of 0.5 and a $A_D$ of 300?

**Solution**

$$\text{CMRR} = \frac{A_D}{A_{CM}} = \frac{300}{0.5} = 600$$

$$\text{CMRR}_{dB} = 20 \log_{10} \text{CMRR} = 20 \log_{10} 600 = 20 \times 2.78 = 55.6 \text{ dB}$$

**EXAMPLE 25-2**

The DA in Example 25-1 has a 10-mV signal connected to its inverting input and a 2-mV common-mode signal connected to both inputs.

**a.** What will be the phase of the desired output signal?

**b.** What will be the voltage of the desired output signal?

**c.** What will be the voltage of the common-mode output signal?

**Solution**

**a.** 180° out of phase with the input voltage

**b.**
$$A_D = \frac{V_{o,D}}{V_{i,D}}$$

therefore

$$V_{o,D} = A_D V_{i,D} = 300 \times 0.01 = 3\ \text{V}$$

**c.** $V_{o,CM} = A_{CM} V_{i,CM} = 0.5 \times 0.002\ \text{V} = 0.001\ \text{V}$

## Quiescent Analysis of the DA

Figure 25-11*a* shows the quiescent currents and voltages for a DA with the specified component values. In this figure, $R_{B1}$ and $R_{B2}$ can be either physical resistors or the internal resistances of the signal sources. A common approximation used in analyzing this type of circuit is that base voltages $V_{B1}$ and $V_{B2}$ are zero with respect to ground. This is a very reasonable approximation because the base current is very small and the internal dc resistance of the signal source (which replaces $R_{B1}$ and/or $R_{B2}$) is usually very small. For example, in Fig. 25-11*a* the base current for either transistor would be $I_B = I_C/\beta = 0.05\ \text{mA}/200 = 0.25\ \mu\text{A}$. Even with the 10-kΩ base resistors in Fig. 25-11*a*, the base voltage would only be $V_B = I_B \times R_B = 0.25\ \mu\text{A} \times 10\ \text{k}\Omega = -2.5\ \text{mV}$. Compared to the −15 V of $V_{EE}$, this is a negligible amount of voltage.

Assuming that $V_B = 0$, and that $V_{BE} = 0.7$ V (which is the forward-biased junction voltage), we can use Kirchhoff's voltage law to find the voltage across $R_E$. Using the loop consisting of $V_{EE}$, the base-emitter junction of $Q_1$ and $R_{B1}$, we can write

$$\begin{aligned} V_{RE} &= V_{EE} - V_{EB} - V_{R_{B1}} \\ &= 15\ \text{V} - 0.7\ \text{V} - 0\ \text{V} = 14.3\ \text{V} \end{aligned}$$

In this formula the absolute value of $V_{EE}$ is used because we are determining the voltage dropped across $R_E$. The next step in determining the quiescent value in Fig. 25-11*a* is to find $I_{R_E}$, which is $I_{R_E} = V_{R_E}/R_E = 14.3\ \text{V}/143\ \text{k}\Omega = 0.1$ mA. This current will divide between $Q_1$ and $Q_2$. If $Q_1$ and $Q_2$ are identical transistors, which they should be for a DA, then the current splits equally. Assuming identical transistors, the emitters' currents are

$$I_{E_1} = I_{E_2} = \frac{I_{R_E}}{2} = 50\ \mu\text{A}$$

In Chap. 24, we established that $I_C \approx I_E$; thus, for the circuit in Fig. 25-11*a*, we can calculate $V_{R_{C1}}$ and $V_{R_{C2}}$ by $V_{R_{C1}} = V_{R_{C2}} = I_{C_1} \times R_{C_1} = 50\ \mu\text{A} \times 150\ \text{k}\Omega = 7.5$ V. Now that we have found the voltage drops across $R_C$ and $R_E$, we can calculate $V_C$ and $V_E$ using the following formulas:

$$\begin{aligned} V_{C_1} &= V_{C_2} = V_{CC} - V_{R_{C1}} = +15\ \text{V} - 7.5\ \text{V} = +7.5\ \text{V} \\ V_E &= V_{EE} - V_{R_E} = -15\ \text{V} - (-14.3\ \text{V}) = -0.7\ \text{V} \end{aligned}$$

Finally, $V_{CE}$ must be equal to the difference between $V_C$ and $V_E$, which is 8.2 V. Since $V_{CE}$ and $V_{R_C}$ are nearly equal, the transistors are biased almost in the middle of their active (linear) regions.

Notice in Fig. 25-11*a* that there is no potential difference between the two collectors. Therefore, the differential dc output for a balanced DA is zero. In some applications of DAs the dc differential output is the output signal. In these applications the input signal would also be a dc voltage, and the amplifier could then be referred to as a *dc amplifier*. This is one reason why the transistors and collector resistors should be as closely matched as possible. Since it is fairly easy to produce matched pairs in an IC, DAs are very popular circuits for linear ICs.

## Response to Common-Mode Signals

The results of applying a 1-$V_P$ common-mode signal to the DA are shown in Fig. 25-11*b*. The procedures for determining the currents and voltages are the same as those for quiescent conditions except that $V_B$ is no longer zero. The calculations are

$$\begin{aligned} V_{R_E} &= V_{EE} - V_{EB} - V_B = 15\ \text{V} - 0.7\ \text{V} - 1\ \text{V} = 13.3\ \text{V} \\ I_{R_E} &= \frac{13.3\ \text{V}}{143\ \text{k}\Omega} = 0.093\ \text{mA} \\ I_{C_1} &= I_{C_2} = 0.093\ \text{mA} \times 0.5 = 46.5\ \mu\text{A} \\ V_{R_{C1}} &= V_{R_{C2}} = 46.5\ \mu\text{A} \times 150\ \text{k}\Omega \approx 7\ \text{V} \\ V_{C_1} &= V_{C_2} = 15\ \text{V} - 7\ \text{V} = +8\ \text{V} \end{aligned}$$

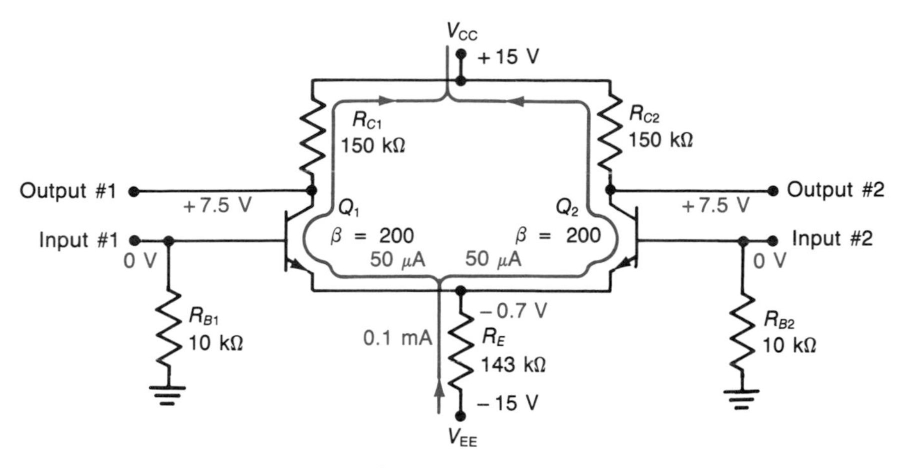


(*a*) Quiescent conditions

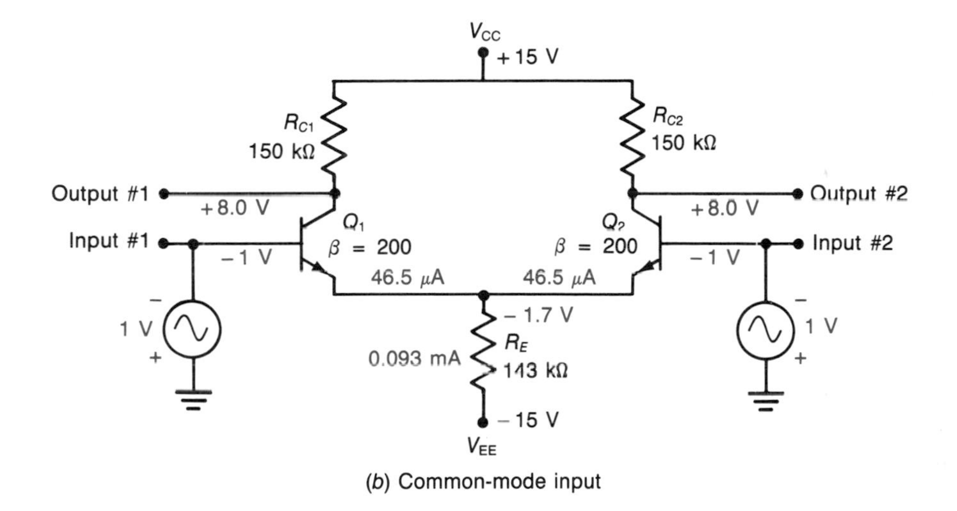


(*b*) Common-mode input

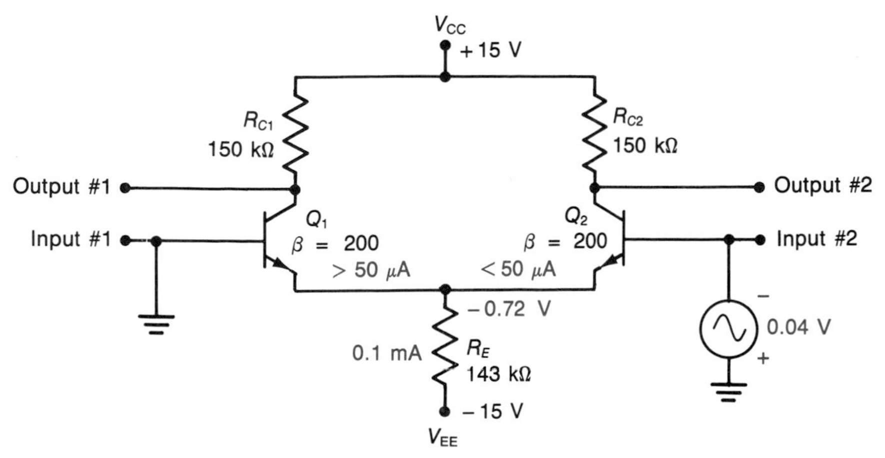


(*c*) Single-ended differential input

**FIGURE 25-11 Voltages and currents in a DA under various conditions.**

Notice in Fig. 25-11*b* that the differential output is still zero. Even though both $V_{C_1}$ and $V_{C_2}$ increased by 0.5 V, the potential difference between the collectors is zero. Thus, the *voltage gain for a differential output* is zero if the transistors and collector resistors are perfectly matched. The $A_{CM}$ for the single-ended output from either output no. 1 or output no. 2 to ground is easy to calculate from the values given in Fig. 25-11*a* and *b*:

$$A_{CM} = \frac{\Delta V_{C_1}}{\Delta V_{B_1}} = \frac{8\text{ V} - 7.5\text{ V}}{1\text{ V} - 0\text{ V}} = 0.5$$

Although the calculated currents and voltages in Fig. 25-11*b* clearly show how the DA responds to common-mode signals, they are not needed to calculate $A_{CM}$. Inspection of Fig. 25-11*b* shows that both transistors are connected in the CE (common-emitter) configuration with an unbypassed emitter resistor which is shared by the two transistors. Since signal currents for both transistors flow through $R_E$, $R_E$ appears to be twice its actual value to either of the transistors. Therefore, the gain formula for an unbypassed CE circuit, which was explained in Chap. 24, will use $2R_E$ in place of $R_E$. Thus, the formula for $A_{CM}$ is

$$A_{CM} \approx \frac{R_{C_1}}{2R_E + r_e}$$

Since $r_e$ is, in most cases, very small compared to $R_E$, it is usually dropped from the formula. This allows us to estimate $A_{CM}$ using only this formula and the component values. For Fig. 25-11*b* we get

$$A_{CM} = \frac{R_{C_1}}{2R_E} = \frac{R_{C_2}}{2R_E} = \frac{150\text{ k}\Omega}{2 \times 143\text{ k}\Omega} = 0.52$$

which is within 4 percent of the value obtained above.

## Response to Differential Signals

Figure 25-11*c* shows a DA with a signal-ended input. Since the base of $Q_1$ is tied to ground, the base-to-emitter voltage of $Q_1$ is determined by $V_E$ of the transistors. Because of the large $A_D$ of this circuit, a 26-mV change in $V_{EB}$ causes a 7.5-V change in $V_C$. With a quiescent $V_C$ of 7.5 V, a change of $V_C$ of 7.5 V drives the transistor to the limits of its linear range. Thus, we can conclude that $V_E$ of the transistors stays within about $\pm 26$ mV of $-0.7$ V. This also means that $V_{R_E}$ only varies $\pm 26$ mV from its quiescent value of 14.3 V. If the $\Delta V_{R_E}$ is no more than 52 mV, then the $\Delta I_{R_E}$ can be no more than $\Delta I_{R_E} = \Delta V_{R_E}/R_E =$ 52 mV/143 kΩ = 0.36 μA. This is a very small change compared to the quiescent value of $I_{R_E}$ (100 μA). Now you can see why it is often stated that "a large value of $R_E$ provides a constant-current source for the transistors in a differential amplifier." Of course, as illustrated in Fig. 25-11*b*, this statement does not hold true if large common-mode signals are applied to the DA.

If the current supplied to the emitters of the DA remains constant when a differential signal is applied, how can the DA amplify the signal? The answer to this question is that the input signal determines how much of the constant current goes to $Q_1$ and how much is routed to $Q_2$. How the signal accomplishes this can be determined by analysis of Fig. 25-11*c*. Note in this figure that the input signal voltage is developed across the dynamic resistances of the base-emitter junctions of $Q_1$ and $Q_2$ connected in series. This means that the instantaneous 0.04 V of input signal splits equally across the two junctions if $Q_1$ and $Q_2$ are identical transistors. For the instantaneous signal polarity shown in Fig. 25-11*c*, the 0.02 V of signal impressed on the junction of $Q_2$ reduces $V_{EB_2}$ to 0.68 V. Applying Kirchhoff's voltage law to the $Q_2$ side of the circuit shows that $V_{R_E}$ must decrease to 14.28 V and $V_E$ must increase to $-0.72$ V when $V_{EB_2}$ decreases to 0.68 V. The calculations used to show this are

$$\begin{aligned} V_{R_E} &= V_{EE} - V_{EB} - V_{signal} \\ &= 15\text{ V} - 0.68\text{ V} - 0.04\text{ V} = 14.28\text{ V} \\ V_E &= V_{EE} - V_{RE} = -15\text{ V} - (-14.28\text{ V}) = -0.72\text{ V} \end{aligned}$$

Of course, a decrease in $V_{EB_2}$ causes a decrease in $I_{B_2}$ and $I_{C_2}$. The decrease in $I_{C_2}$ causes $V_{C_2}$ to become more positive. When the emitter voltage is $-0.72$ V, then $V_{EB_1}$ is also 0.72 V. This increase in $V_{EB_1}$ from 0.70 to 0.72 V causes $I_{B_1}$ and, therefore, $I_{C_1}$ to increase. An increase in $I_{C_1}$ causes $V_{C_1}$ to become less positive; i.e., $V_{C_1}$ is going in a negative direction.

Notice that the above analysis again shows that a differential input causes the collector (output) signals to be 180° out of phase; when $V_{C_2}$ is going in a positive direction, $V_{C_1}$ is going in a negative direction. Looking at Fig. 25-11*c* from another point of view shows why the outputs are out of phase. Transistor $Q_2$ in this figure is acting as a CE amplifier as far as output no. 2 is concerned; the input is on the base and the output is from the collector. We know from Chap. 24 that a CE amplifier causes 180° of phase shift. However, further inspection of Fig. 25-11*c* shows that the small change in $V_E$ caused by current changes in $Q_2$ provides the input signal to the emitter of $Q_1$. In terms of providing a signal to $Q_1$, $Q_2$ appears to be a CC amplifier because the signal is taken from the emitter. The CC amplifier causes no phase shift, so the signal applied to the emitter of $Q_1$ is in phase with the input signal. Transistor $Q_1$ is connected as a CB amplifier; i.e., the input signal is applied to the emitter and the output is taken from the collector. Since the CB amplifier causes no phase shift, the output signal on the collector of $Q_1$ is still in phase with the input signal. With equal collector resistors, the voltage gains of the CB and CE are essentially equal. Since the input signal to the CB and CE are equal, the output signals at terminals nos. 1 and 2 are also equal.

Estimating the voltage gain $A_D$ for Fig. 25-11*c* is quite easy and straightforward when two points are understood: (1) The resistance of $R_E$ is so large compared to the dynamic resistance $r_e$ of the two base-emitter junctions that essentially none of the signal current flows through $R_E$; therefore, $Q_2$ can be treated like a CE amplifier with a fully bypassed emitter resistor. (2) The signal current flows through the $r_e$ of both $Q_1$ and $Q_2$; therefore, $r_e$ must be replaced by $2r_e$ in the gain formula for $Q_2$. Thus, the gain formula for a single-ended output is $A_D = R_C/2r_e$, where $R_C$ is either $R_{C_1}$ or $R_{C_2}$ and $r_e$ is either $r_{e_1}$ or $r_{e_2}$. For the circuit of Fig. 25-11*c* the voltage gain can be calculated after $r_e$ is determined. The calculations are

$$r_e = \frac{26\text{ mV}}{I_C} = \frac{26\text{ mV}}{0.05\text{ mA}} = 520\ \Omega$$

$$A_D = \frac{R_C}{2r_e} = \frac{150\text{ k}\Omega}{1040\ \Omega} = 144$$

Knowing the gain of Fig. 25-11*c*, we can now determine how much input signal the DA can handle before clipping occurs. We know the emitters must remain about −0.7 V with respect to ground and that the collector-to-emitter voltage is less than 1 V when a transistor is saturated. Therefore, the collector voltage can swing from its quiescent value of 7.5 V to about 0 V at saturation. When the transistor is cut off, no voltage drops across $R_C$ so the collector swings to 15 V. Therefore, if all components are perfectly matched, the collector voltage can swing 15 $V_{p\text{-}p}$. This is the maximum output voltage. We can now use the general gain formula to find the maximum peak-to-peak input voltage:

$$V_{i,p\text{-}p} = \frac{V_{o,p\text{-}p}}{A_D} = \frac{15\ V_{p\text{-}p}}{144} = 104\text{ mV}_{p\text{-}p}$$

This 104-m$V_{p\text{-}p}$ input signal is shared by two junctions, so the peak signal across one junction is 26 mV. This is why it was stated in the opening of this discussion on differential input signals that a 26-mV change in $V_{EB}$ drives the collector voltage to its limit.

The input impedance of the DA of Fig. 25-11*c* is determined by the values of $r_e$ and $\beta$ of the transistors. Since the input signal flows through the junctions of both transistors, the formula is $Z_i = \beta \times 2r_e$. Using the value of $r_e$ previously determined, the input impedance for Fig. 25-11 is $Z_i = 200 \times 2 \times 520\ \Omega = 208\text{ k}\Omega$. This is a high value for a circuit which uses bipolar transistors and has a high voltage gain. However, a large value of $Z_i$ is common for DAs. Some of the bipolar DAs have over 1 MΩ of $Z_i$. Of course, DAs constructed from FETs have hundreds of megohms of input impedance.

## Active Collector and Emitter Resistances

The discussion of, and formulas for, $A_D$, $A_{CM}$, and CMRR show the advantages of having large values of $R_E$ and $R_C$ and small values of $r_e$. However, we know that a large value of $R_E$ produces a small value of $I_C$, which, in turn, produces a large value of $r_e$. Also, we know that producing large values of $R$ in a monolithic IC requires too much chip area.

A solution to this problem is to replace the resistors $R_C$ and $R_E$ with transistors that are biased in the center of their linear region. Then the collector-to-emitter resistances of these transistors serve as the emitter resistors and the collector resistors. An example of how a transistor can replace $R_E$ is shown in Fig. 25-12. In this figure, $R_1$ and $D_1$ bias $Q_3$ in the center of its linear region. No signal is applied to either the emitter or the base of $Q_3$; therefore, $Q_3$ (which is an active device) is used as an active resistance.

The advantage of an active resistance is that it has two types of resistance: a dynamic or ac resistance $r$, and a static or dc resistance $R$. For $Q_3$ in Fig. 25-12, the static resistance is $R_{CE} = V_{CE}/I_C$ and the dynamic resistance is $r_{ce} = \Delta V_{CE}/\Delta I_C$. The values needed to make these calculations can be obtained from the collector-emitter characteristic curves as

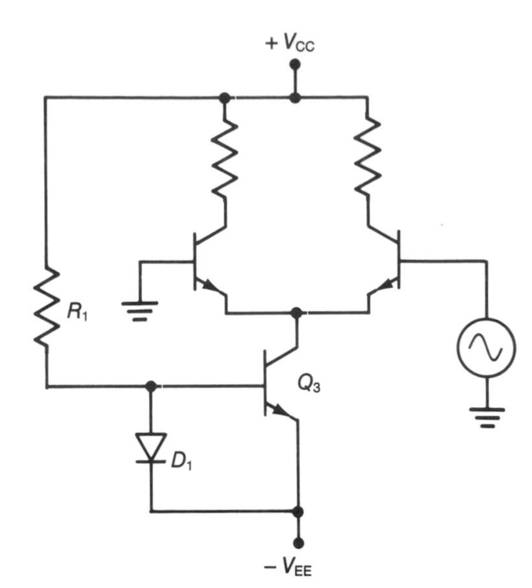


**FIGURE 25-12 Replacing $R_E$ with a transistor provides an active emitter resistance which improves the performance of the DA.**

discussed in Chap. 24. The important point to remember is that $r_{ce}$ is always much larger than $R_{CE}$.

In Fig. 25-12, the value of $R_{CE}$ (in conjunction with the value of $V_{EE}$) determines the quiescent collector currents in the circuit. Quiescent $I_C$ determines the value of $r_e$, which in turn determines $A_D$ ($A_D = R_C/2r_e$). However, the value of $r_{ce}$ is the appropriate resistance to use for determining $A_{CM}$ because it is the ac resistance, and ac resistances are what determine gain in any circuit. Thus, for the active-emitter-resistance circuit of Fig. 25-12, the common-mode gain formula is $A_{CM} = R_C/2r_{ce}$.

The net effect of the active emitter resistance, then, is to decrease $A_{CM}$ by providing a large ac resistance while at the same time increasing $A_D$ by providing a small dc resistance. Of course, this increases the CMRR because CMRR = $A_D/A_{CM}$.

Using active collector resistors produces similar results. The small value of $R_{CE}$ keeps the quiescent collector voltage in the center of collector voltage limits, while the much larger value of $r_{ce}$ produces a large $A_D$.

### EXAMPLE 25-3

For the circuit in Fig. 25-13, determine $Z_i$, $A_D$, $A_{CM}$, and CMRR.

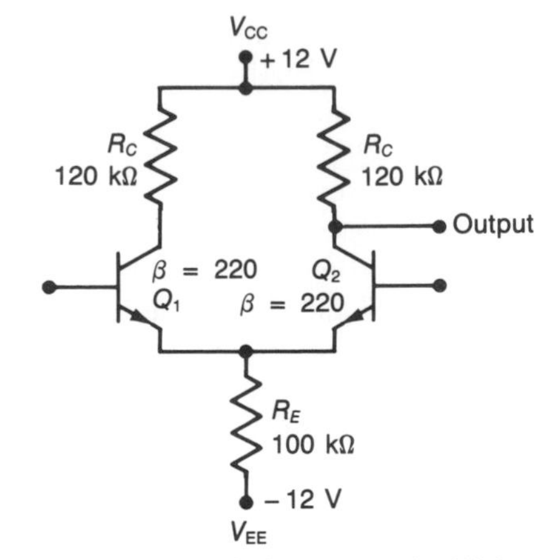


**FIGURE 25-13 Circuit for Example 25-3.**

**Solution**

$$V_{R_E} = V_{EE} - V_{EB} = 12\text{ V} - 0.7\text{ V} = 11.3\text{ V}$$
$$I_{R_E} = \frac{V_{R_E}}{R_E} = \frac{11.3\text{ V}}{100\text{ k}\Omega} = 113\ \mu\text{A}$$
$$I_C = \frac{I_{R_E}}{2} = \frac{113\ \mu\text{A}}{2} = 56.5\ \mu\text{A}$$
$$r_e = \frac{26\text{ mV}}{56.5\ \mu\text{A}} = 460\ \Omega$$
$$Z_i = \beta \times 2r_e = 220 \times 2 \times 460\ \Omega = 202\text{ k}\Omega$$
$$A_D = \frac{R_C}{2r_e} = \frac{120\text{ k}\Omega}{920\ \Omega} = 130$$
$$A_{CM} = \frac{R_C}{2R_E} = \frac{120\text{ k}\Omega}{200\text{ k}\Omega} = 0.6$$
$$\text{CMRR} = \frac{130}{0.6} = 217$$

## Self-Test

**31.** Define differential output.

**32.** Why is $A_{CM}$ so much smaller than $A_D$?

**33.** Why is $I_{R_E}$ more constant when the input signal is differential than when it is common-mode?

**34.** Why is it desirable to have a small $A_{CM}$ and a large $A_D$?

**35.** Compare the amplitude of the differential output signal to the single-ended output signal.

**36.** A DA is receiving a common-mode signal. If $I_{C_1}$ is increasing, what is happening to $I_{C_2}$ and $V_{C_2}$?

**37.** If a DA is receiving a differential signal and $I_{C_2}$ is decreasing, what is happening to $I_{C_1}$?

**38.** What is the advantage of using an active emitter resistance in a DA?

**39.** Refer to Fig. 25-13. Will the output be in phase or out of phase with an input signal applied to the base of $Q_1$?

**40.** Refer to Fig. 25-13. Which input, the base of $Q_1$ or the base of $Q_2$, is the inverting input?

**41.** Change $R_E$ in Fig. 25-13 to 150 kΩ and then determine $Z_i$, $A_D$, and CMRR.

**42.** If CMRR = 412, what is the value of $CMRR_{dB}$?

## 25-6 THE OPERATIONAL AMPLIFIER

The schematic symbol for an operational amplifier (op amp) is shown in Fig. 25-14. Notice in this symbol that the op amp has an inverting input, a noninverting input, and a single-ended output. For the sake of simplicity, $V_{EE}$ and $V_{CC}$ terminals are often omitted from the symbol. Since two or more op amps are often contained in a single IC package, eliminating these terminals on the symbol eliminates unnecessary duplication.

An op amp is a multiple-stage amplifier. Each stage is cascaded to the next stage, but no coupling capacitors are used. Instead the stages are *direct-coupled;* i.e., a conductor connects the output of one stage to the input of the next stage. This means that the quiescent voltage of the output of one stage must be the same as the quiescent voltage of the input of the next. Direct coupling allows the op amp to amplify either ac or dc signals.

The first stage of the op amp is a DA which provides the inverting and noninverting inputs. This stage is followed by one or more amplifiers which

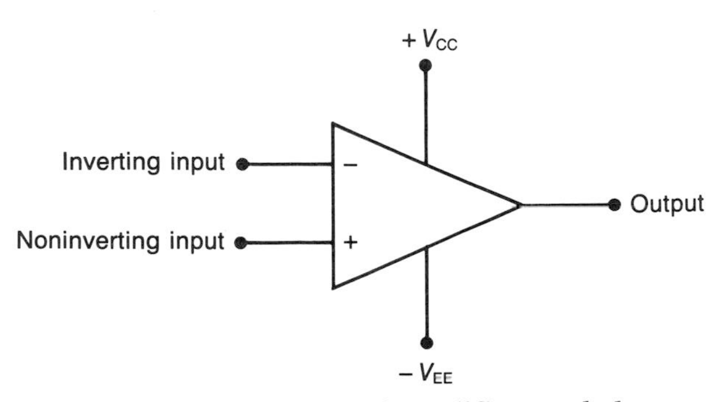

**FIGURE 25-14 Operational amplifier symbol.**

may also be DAs. The last stage of the op amp is an amplifier like the one shown in Fig. 25-15*a*. The quiescent voltages on the bases of $Q_1$ and $Q_2$ in Fig. 25-15*a* are supplied by the previous stage which is direct-coupled to the output stage. When $Q_1$ and $Q_2$ have identically quiescent values of $R_{CE}$, the quiescent dc output voltage is 0, or very close to 0, V. (Some op amps have special terminals that allow an external potentiometer to adjust the quiescent output to exactly zero.) Since both the output and the input terminals of an op amp are at 0 V, two op amps can be direct-coupled.

The circuit of Fig. 25-15*a* allows an input signal to drive the output in either a positive or negative direction. With most op amps the output can swing to within a couple of volts of $+V_{CC}$ and $-V_{EE}$. How the output voltage changes is illustrated in Fig. 25-15*b* and *c*. In Fig. 25-15*b*, a positive-going signal turns $Q_1$ (an NPN transistor) on and drives it toward saturation. The same positive-going signal turns $Q_2$ (a PNP transistor) off and drives it toward cutoff. This causes current to flow up through the load, providing a positive output. Figure 25-15*c* shows that a negative-going signal turns $Q_2$ on and $Q_1$ off. This causes a current through the load in the opposite direction and provides a negative-going output. Circuits like the one in Fig. 25-15, where one series transistor is shutting off while the other is turning on, are known as *push-pull amplifiers.* They are a very popular arrangement for the output stage of many linear systems.

Note that both $Q_1$ and $Q_2$ in Fig. 25-15 are operating as common-collector amplifiers. Thus, the output stage of an op amp has a low-output impedance; it can drive low values of load resistances. Low-output impedance and zero dc output voltage are two very important characteristics of an op amp.

Because the op amp is a multistage amplifier, it has a very large *open-loop voltage gain* at low frequencies. Open-loop voltage gain, abbreviated $A_{V,OL}$, refers to the gain when no negative feedback is provided by the external circuit components connected to the inputs and output of the op amp. Gain $A_{V,OL}$ in an op amp is equivalent to $A_D$ in a DA except that it is much larger. The $A_{V,OL}$ of a typical general-purpose op amp, like the type 741, is about 200,000. A high-performance op amp may have an $A_{V,OL}$ of over 1 million. Often the $A_{V,OL}$ of an op amp is expressed in decibels. The conversion formula is $A_{V,OLdB} = 20 \log_{10} A_{V,OL}$.

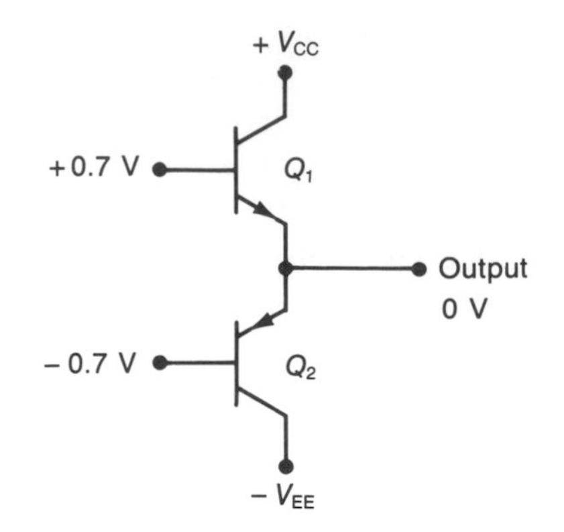


(*a*) Quiescent conditions

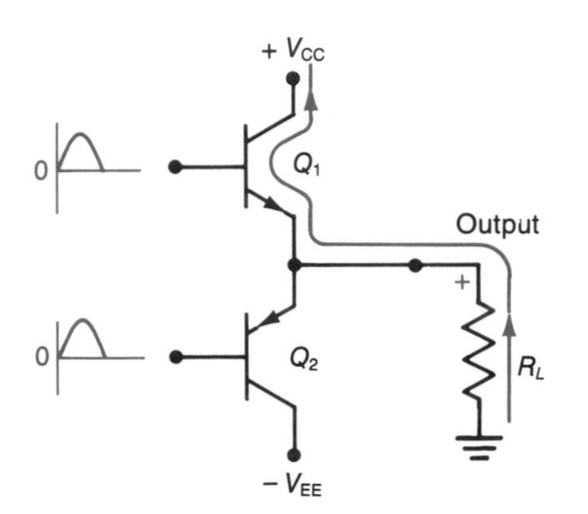


(*b*) Positive-going output when bases are driven in a positive direction

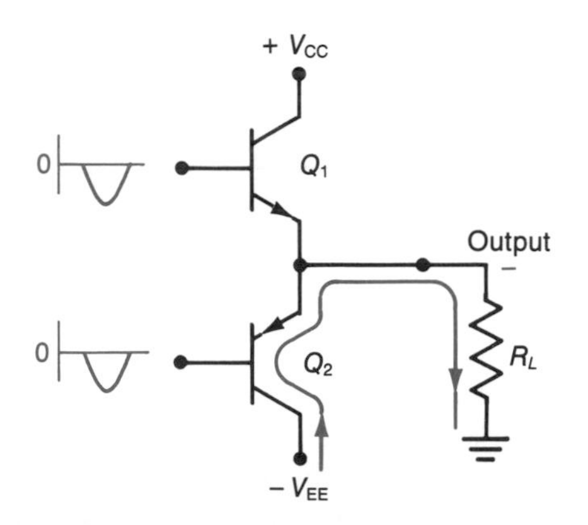


(*c*) Negative-going output when bases are driven in a negative direction

**FIGURE 25-15 Output stage of an op amp.**

Because of the very high open-loop gain, a few microvolts of differential input signal is all that is needed to drive the output of the op amp into saturation.

**EXAMPLE 25-4**

An op amp has an open-loop gain of 100 dB, and its linear output voltage can swing 12 $V_{p\text{-}p}$. What is the maximum peak-to-peak differential input signal that can be applied without clipping the output signal?

**Solution**

$$A_{V,\mathrm{OL,dB}} = 20 \log_{10} A_{V,\mathrm{OL}}$$

Therefore

$$\frac{A_{V,\mathrm{OL,dB}}}{20} = \log_{10} A_{V,\mathrm{OL}}$$

and

$$A_{V,\mathrm{OL}} = \mathrm{antilog}_{10} \frac{A_{V,\mathrm{OL,dB}}}{20} = \mathrm{antilog} \frac{100}{20}$$

$$= 100{,}000 = 1 \times 10^5$$

$$A_{V,\mathrm{OL}} = \frac{V_o}{V_i}$$

Therefore

$$V_i = \frac{V_o}{A_{V,\mathrm{OL}}} = \frac{12\ \mathrm{V}_{p\text{-}p}}{1 \times 10^5} = 120\ \mu\mathrm{V}_{p\text{-}p}$$

This example tells us that for purposes of analyzing an op amp circuit we can consider the voltage difference between the two inputs to be zero. Thus, if one input is grounded, the second input must be essentially at ground (within a few microvolts) or the op amp will be saturated. The second input is said to be at *virtual ground* because of the first input being grounded. If one input is at +1 V, the other input must be at 1 V, etc. This is an important concept to remember if one wants to understand how an op amp operates.

A complete op amp circuit uses negative feedback so that the circuit gain is much less than $A_{V,\mathrm{OL}}$. The feedback is applied by resistors connected to the input and output terminals. These resistors form a circuit loop (path) called a *feedback loop*. The gain of the op amp stage with feedback is therefore called the *closed-loop gain* and is abbreviated $A_{V,\mathrm{CL}}$, or just $A_V$. We will use $A_V$ to represent the gain with feedback.

Since the input stage of the op amp is a DA, we know that the input impedance of the op amp is quite large. For the bipolar DA, $Z_i = \beta \times 2_{r_e}$. When FETs are used in the DA, the input impedance is exceedingly large (hundreds of megohms).

The CMRR of an op amp is very large because the overall gain of two cascaded amplifiers is the product of the two gains. When DA amplifiers are cascaded (as they are in an op amp), the overall common-mode gain gets smaller while the overall differential gain gets larger. For example, when two DAs with individual $A_{CM} = 0.3$ and $A_D = 100$ are cascaded together, the overall $A_{CM} = 0.3 \times 0.3 = 0.09$ and the overall $A_D = 100 \times 100 = 10{,}000$. If you figure out the individual CMRR and the overall CMRR, you will see that the overall CMRR is equal to the product of the individual CMRRs.

In summary, the major characteristics of the op amp, which we will use in explaining how op amp circuits operate, are

1. Inverting and noninverting inputs that must remain at virtually the same voltage
2. Very large input impedance ($\geq$1 M$\Omega$)
3. Very small output impedance ($\leq$200 $\Omega$)
4. Very large open-loop gain ($\geq$100,000)
5. Very large CMRR ($\geq$90 dB)

Before we look at some op amp circuits and applications, we will examine two limitations associated with the op amp. These two limiting factors are frequency response and slew rate.

The *frequency response* curve for a typical general-purpose op amp is shown in Fig. 25-16. Notice that the $A_{V,OL}$ starts to roll off at about 10 Hz and that the gain is down to 1 (0 dB) at a frequency of

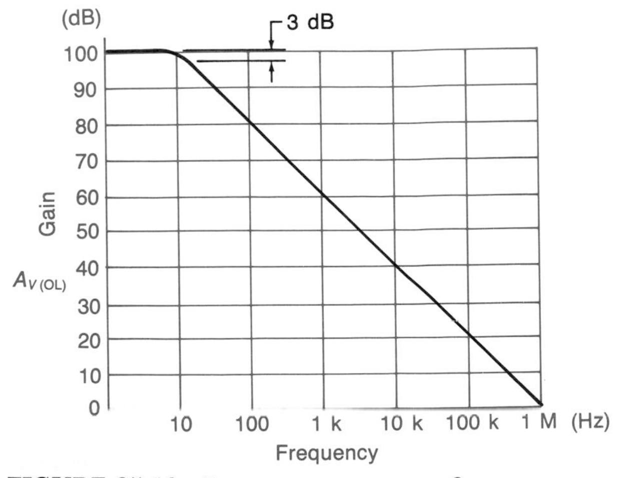


**FIGURE 25-16 Frequency response of an op amp.**

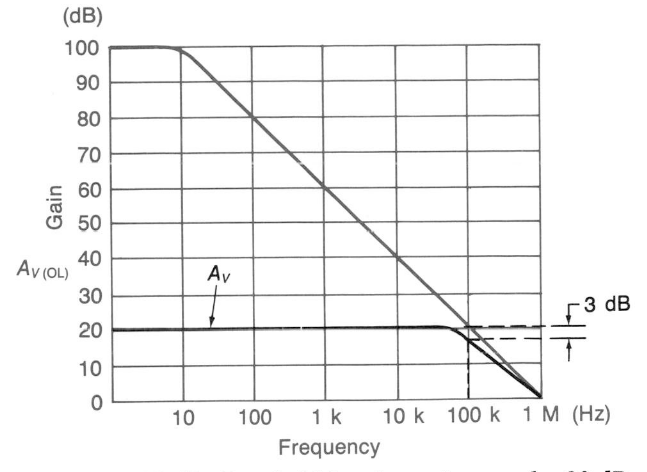


**FIGURE 25-17 Bandwidth when $A_V$ equals 20 dB (gain of 10).**

1 MHz. It should be noted that high-performance, high-frequency op amps are available for which roll-off in gain does not start until 100 Hz and the unity gain extends to 15 MHz.

The frequency range over which the gain remains essentially flat (within 3 dB) is known as the *bandwidth* of an amplifier. Thus, the bandwidth for the op amp in Fig. 25-16 is only 10 Hz (0 to 10 Hz). If the gain of the op amp is reduced by connecting a negative feedback loop, the bandwidth of an op amp circuit can be greatly increased. As shown in Fig. 25-17, if the gain $A_V$ is set at 20 dB, the bandwidth increases to 100 kHz (0 to 100 kHz).

## EXAMPLE 25-5

Determine the bandwidth of the op amp with a frequency response like that in Fig. 25-16 if the $A_V$ is adjusted to 100 by negative feedback.

### Solution

First convert the gain of 100 to gain in decibels:

$$A_{V,\text{dB}} = 20 \log_{10} A_V = 20 \times 2 = 40 \text{ dB}$$

Next, project a line horizontally to the right from 40 dB on the vertical axis until the line intersects the $A_{V,OL}$ plot. At this intersection, drop a line vertically until it intersects the horizontal axis. The frequency at which this last line intersects the horizontal axis is the frequency of the bandwidth when the gain is 100 (40 dB). Thus, the bandwidth is 10 kHz.

The frequency response of the op amp is limited by its internal capacitance. Some of this internal capacitance is the result of the many reverse-biased junctions used to isolate the components in an IC as discussed earlier. Some is caused by the junctions in the transistors themselves. The rest is due to a capacitor built into the op amp to stabilize it at high frequencies. To "stabilize an amplifier" means to keep the amplifier from converting to an *oscillator.* (An oscillator is a circuit which generates its own frequency.) Some op amps do not have an internal capacitor. Instead, terminals are provided on the op amp IC so that the circuit designer can add external capacitors and resistors to achieve the desired frequency response.

Notice in Fig. 25-16 that the $A_{V,\mathrm{OL}}$ has a slope (roll-off) of 20 dB per decade of frequency; i.e., changing the frequency by a factor of 10 causes a 20-dB change in gain. Most op amps that have internal compensating (stabilizing) capacitors have a roll-off of 20 dB per decade. A gain of 20 dB is the same as a gain of 10, so we can also say that changing the gain by 10 causes a ten-fold change in frequency. This relationship allows us to use the following formula for determining bandwidth when the roll-off is 20 dB per decade:

$$\text{Bandwidth} = \frac{f_{\text{unity}}}{A_V}$$

where $f_{\text{unity}}$ is the frequency at which the open-loop gain is one (1 MHz in Fig. 25-16) and $A_V$ is the closed-loop gain. Had we used this formula in solving Example 25-5, we would have obtained the same answer; i.e.,

$$\text{Bandwidth} = \frac{1\ \text{MHz}}{100} = 10\ \text{kHz}$$

The technique used in Example 25-5 is appropriate for a frequency response curve with any slope, whereas the formula is useful only for a slope of 20 dB per decade.

The internal capacitance of an op amp also causes the op amp to have a limited *slew rate.* Slew rate refers to how rapidly the output voltage of the op amp can increase or decrease. Thus, slew rate has units of volts per microsecond (V/μs). The typical slew rate for a general-purpose op amp is about 0.5 V/μs; but op amps with slew rates ranging from about 0.25 to about 100 V/μs are available.

How capacitance and resistance determine slew rates is illustrated in Fig. 25-18. In this figure, $R_1$ could represent the collector resistor of a common-

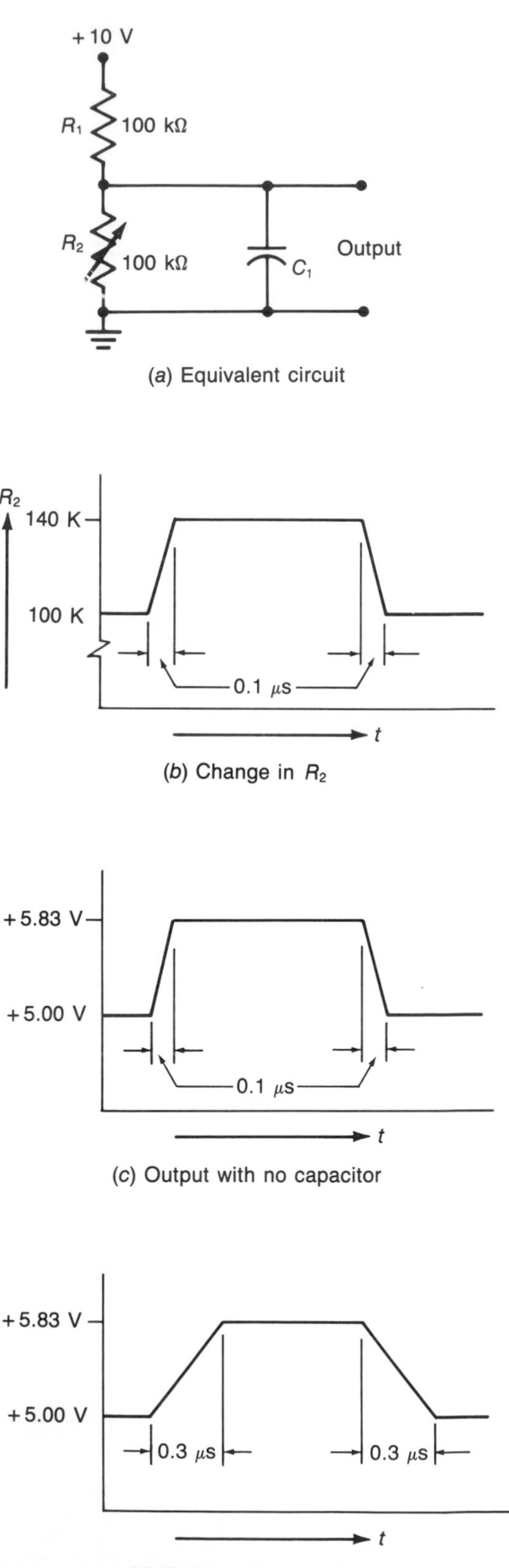

(d) Output with capacitor

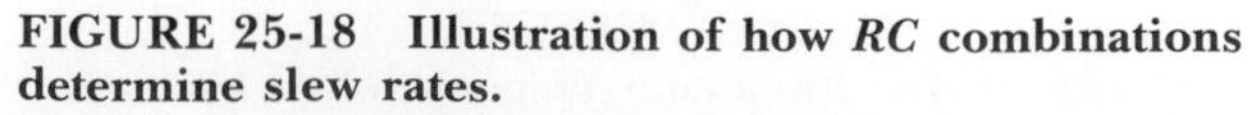
**FIGURE 25-18 Illustration of how *RC* combinations determine slew rates.**

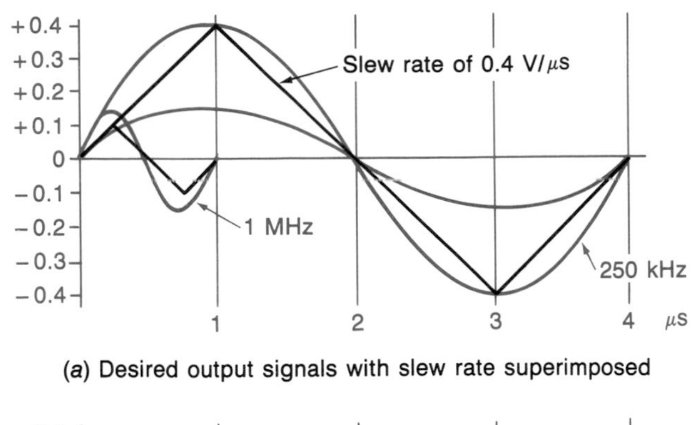


(*a*) Desired output signals with slew rate superimposed

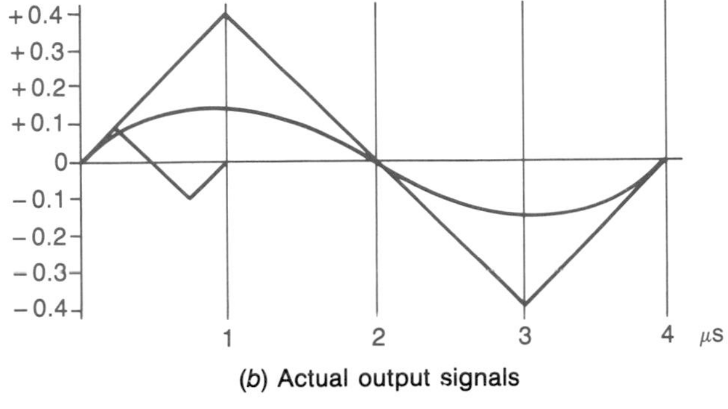


(*b*) Actual output signals

**FIGURE 25-19 Slew-rate distortion.**

emitter amplifier; $R_2$ could represent the static resistance ($R_{CE}$) of the collector-emitter of the transistor; and $C_1$ could represent either (or both) the capacitance of the reverse-biased isolation junction or a built-in stabilization capacitor. When a signal is applied to a transistor's input, its $R_{CE}$ ($R_2$ in Fig. 25-18*a*) will change. This change in $R_{CE}$ ($R_2$) for a square-wave input signal is illustrated in Fig. 25-18*b*. If no capacitor were present in Fig. 25-18*a*, the output voltage, as seen in Fig. 25-18*c*, would rise and fall exactly in time with the changes in the resistance of $R_2$. However, when the capacitor is in the circuit, the output voltage can change only as rapidly as the capacitor's voltage changes; the rate of change in voltage across the capacitor is controlled by the *RC* time constants in the circuit. When the time constants are long compared to the time it takes the signal to change, the output signal change does not occur as rapidly as the input signal change. The results can be seen by comparing Fig. 25-18*b* and *d*. While the input signal caused $R_2$ to change 40 kΩ in 0.1 μs, it took 0.3 μs for the output to adjust to the new resistance ratio in the circuit.

The effects of the slew rate of an op amp are determined by a combination of both the frequency and the amplitude of the output signal. Figure 25-19 shows the interaction between three different sine-wave signals and the slew rate. The 1-MHz signal, even though small in amplitude, is severely distorted. As seen in Fig. 25-19*b*, the 0.3-$V_{p\text{-}p}$, 1-MHz sine wave has been changed to a 0.2-$V_{p\text{-}p}$ triangular waveform. Also notice in Fig. 25-19 that the 0.3-$V_{p\text{-}p}$, 250-kHz sine wave is faithfully reproduced, while the 0.8-$V_{p\text{-}p}$, 250-kHz sine wave is converted to a triangular waveform.

In summary, Fig. 25-19 shows that any time the slope of the signal exceeds the slope of the slew rate, the signal will be distorted. This is referred to as *slew-rate distortion.*

## Self-Test

**43.** Why can an op amp amplify either an ac signal or a dc signal?

**44.** How can the quiescent dc output of an op amp be zero?

**45.** Can two op amps be direct-coupled? Why?

**46.** An op amp has an $A_V$ of 30 dB and an output of 10 $V_{p\text{-}p}$. What is the peak-to-peak value of the input voltage?

**47.** Should each of the following characteristics of an ideal op amp be as low as practical or as high as practical?

**a.** Slew rate
**b.** Input impedance
**c.** Output impedance
**d.** CMRR
**e.** Bandwidth
**f.** $A_{V,\text{OL}}$

**48.** Is a 1-V, 5-kHz or a 1-V, 30-kHz signal more likely to be distorted in an op amp? Why?

**49.** An op amp has an $f_{\text{unity}}$ of 5 MHz and a roll-off of 20 dB per decade. What is its frequency bandwidth when $A_V$ = 60 dB?

## 25-7 OPERATIONAL AMPLIFIER CIRCUITS

### Inverting Amplifier

An *inverting amplifier,* which uses an op amp as its active device, is shown in Fig. 25-20. The values of $R_1$ and $R_F$ determine the voltage gain $A_V$ of the circuit, and the value of $R_1$ determines the input impedance $Z_i$. For the inverting amplifier, $Z_i = R_1$ and $A_V = R_F/R_1$ where $R_F$ is the feedback resistor.

Verification of these relationships can be obtained by studying Fig. 25-20.

In Fig. 25-20*a*, the inverting amplifier is shown inside the dotted line; it consists of the op amp, two resistors, a − voltage supply, and a + voltage supply. When the signal is reduced to zero, both input terminals and the output terminal of the op amp are at 0 V. There is no current through or voltage across any of the resistors in Fig. 25-20*a*.

Figure 25-20*b* shows the instantaneous voltages resulting from an instantaneous signal of −1 V when $R_1 = 1\ \text{k}\Omega$ and $R_F = 10\ \text{k}\Omega$. With these resistances, the gain of the circuit is

$$A_V = -\frac{R_F}{R_1} = \frac{10\ \text{k}\Omega}{1\ \text{k}\Omega} = -10$$

Therefore, the instantaneous output $V_o$ for a −1 V instantaneous input $V_i$ should be

$$V_o = A_V V_i = -10 \times -1\ \text{V} = 10\ \text{V}$$

Furthermore, as shown in Fig. 25-20*b*, the inverting (−) input of the op amp should be at essentially 0 V because of the virtual ground created by grounding the + input. How close the − input actually is to signal ground can be easily figured. Referring back to Fig. 25-16 shows that the $A_{V,\text{OL}}$ of the op amp at 1 kHz is 60 dB (1000). Therefore, the instantaneous differential input (at the negative terminal of the op amp) needed to provide the instantaneous 10-V output must be 10 V/1000 = 10 mV. Ten millivolts is only 1 percent of the 1-V input to the inverting amplifier circuit, so assuming a virtual ground introduces about 1 percent error in analyzing the circuit. Remember that a virtual ground does not imply a low impedance between the − and + terminals; instead, it implies a very minute signal current (shown by a broken line in Fig. 25-20*b*) flowing between them. The impedance between the − and + terminals is still $\beta \times 2r_e$ (usually ≥1 MΩ) when the input DA is bipolar. Because of this very high impedance between the terminals, essentially all of the current flowing through $R_1$ is the feedback signal current $I_{FB}$ which also flows through $R_F$ and the signal source. As seen in Fig. 25-20*b* (and Fig. 25-20*c*), the instantaneous voltage drops across $R_1$ and $R_F$ are directly proportional to their resistances because the same amount of signal current flows through each of them. Applying Kirchhoff's voltage law to the loop (in Fig. 25-20*b*) that contains the signal source, $R_1$, $R_F$, and $R_L$ shows that the indicated voltage drops are correct. Furthermore, it can be seen that when $A_V = 10$, the ratio of $R_F$ over $R_1$ must equal 10 for the voltage drops to satisfy Kirchhoff's law. Now, we know why $A_V = R_F/R_1$ for the inverting amplifier circuit.

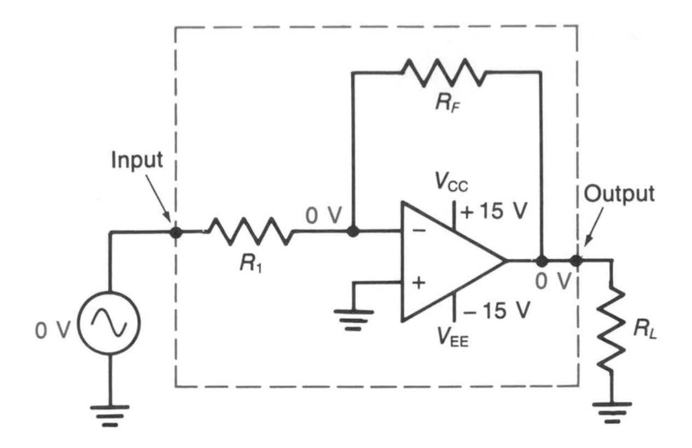

(*a*) Voltages when input signal is zero volts

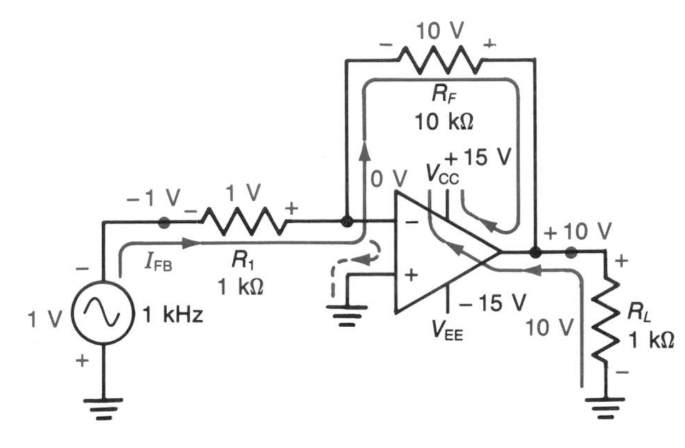

(*b*) Voltages when $A_V$ = 10 and input is −1 V

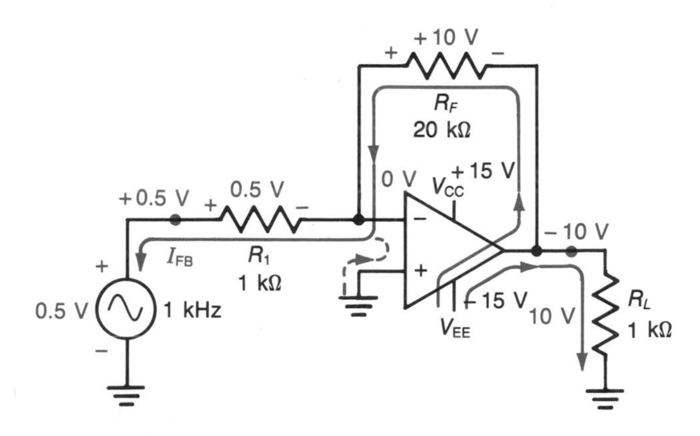

(*c*) Voltages when $A_V$ = 20 and input is +0.5 V

**FIGURE 25-20 Operation of an inverting op amp circuit.**

Since the inverting input in Fig. 25-20*b* is virtually at ground, the input impedance for the inverting amplifier must be equal to $R_1$. Thus, even though the input impedance of the op amp itself is very high, the input impedance to the inverting amplifier is quite low. This is a major limitation of the inverting amplifier.

Figure 25-20*c* shows the instantaneous voltages and the direction of signal current at an instant when the input is +0.5 V. In this figure, the feedback resistor $R_F$ has been changed to 20 kΩ. Notice that changing the resistor ratio to 20:1 requires that the $A_V$ increase to 20 if the inverting terminal is to be at virtual ground and Kirchhoff's voltage law is to be satisfied.

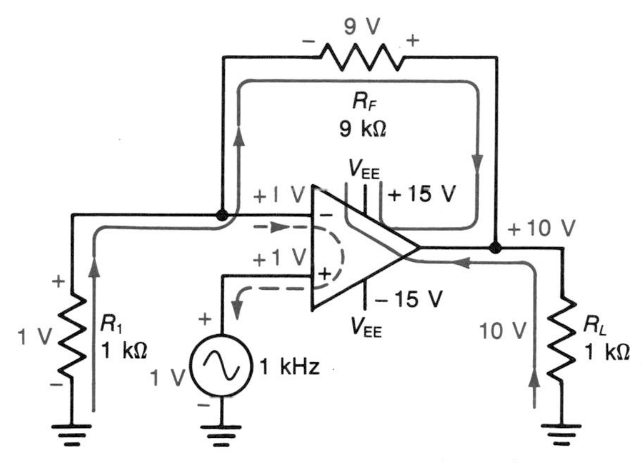


**FIGURE 25-21 Operation of a noninverting op amp circuit.**

### EXAMPLE 25-6

Determine the value of resistors needed to construct an inverting op amp circuit with an input impedance of 10 kΩ and a gain of 26 dB.

**Solution**

$$Z_i = R_1 = 10\ \text{k}\Omega$$

$$A_V = \text{antilog}\,\frac{A_{V,\text{dB}}}{20} = \text{antilog } 1.3 = 20$$

$$R_F = A_V R_1 = 20 \times 10\ \text{k}\Omega = 200\ \text{k}\Omega$$

## Noninverting Amplifier

The *noninverting amplifier,* shown in Fig. 25-21, has a voltage gain which is also determined by the values of $R_1$ and $R_F$. The formula is

$$A_V = \frac{R_F}{R_1} + 1$$

Studying Fig. 25-21 shows why this is the appropriate gain formula for the noninverting amplifier. Because of the large $A_{V,OL}$, the two inputs of the op amp must be within a few microvolts of each other. Thus, when the signal drives the + terminal to +1 V, the − terminal must also go to approximately +1 V. If the amplifier has a gain of 10, then the output goes to +10 V as indicated in Fig. 25-21. Thus, if $V_{R_1}$ must be 1 V, then $V_{R_F}$ must be 9 V to satisfy Kirchhoff's voltage law for the loop composed of $R_1$, $R_F$, and $R_L$. For the reasons described for the inverting amplifier, the instantaneous voltages across $R_F$ and $R_1$ must be directly proportional to the resistance of $R_F$ and $R_1$. Since the ratio of $V_{R_F}$ to $V_{R_1}$ is 9:1, then the ratio of $R_F$ to $R_1$ must also be 9 to 1. The ratio of $R_F$ to $R_1$ will always be 1 less than is the ratio of the input signal voltage to the output signal voltage. This is why the +1 is included in the gain formula for the noninverting amplifier.

The major advantage of the noninverting amplifier over the inverting amplifier is that it has a much larger input impedance $Z_i$. The formula for determining $Z_i$ for the noninverting amplifier is

$$Z_i = \frac{A_{V,\text{OL,OF}}\ Z_{i,\text{op amp}}}{A_V}$$

where $A_{V,\text{OL,OF}}$ is the open-loop gain at the operating frequency.

### EXAMPLE 25-7

Determine $Z_i$ for the circuit of Fig. 25-21 when the op amp has the frequency response shown in Fig. 25-16 and an input impedance of 1 MΩ.

**Solution**

$$A_V = \frac{R_F}{R_1} + 1 = \frac{9\ \text{k}\Omega}{1\ \text{k}\Omega} + 1 = 10$$

From Fig. 25-16,

$$A_{V,\text{OL}} \text{ at 1 kHz} = 60\ \text{dB} = 1000$$

$$Z_i = \frac{1000 \times 1\ \text{M}\Omega}{10} = 100\ \text{M}\Omega$$

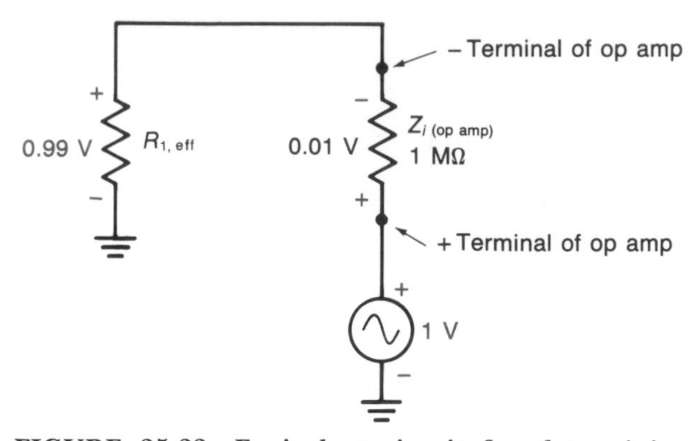


**FIGURE 25-22 Equivalent circuit for determining the effective value of $R_1$ in Fig. 25-21.**

The very large $Z_i$ of the noninverting amplifier is created by the feedback current through $R_1$, making $R_1$ appear to the signal source to be many times larger than its real value. The apparent or effective value of $R_1$ ($R_{1,\text{eff}}$) can be calculated from the equivalent circuit values shown in Fig. 25-22. In the figure, the 0.01 V shown across $Z_{i,\text{op amp}}$ was calculated using $A_{V,\text{OL}}$ and the output signal (10 V/1000 = 0.01 V). Kirchhoff's voltage law was then used to obtain 0.99 V across $R_{1,\text{eff}}$. (Notice again that assuming the − and + terminals of the op amp are at the same potential only introduced a 1 percent error.) Since Fig. 25-22 is a series circuit, we can solve for $R_{1,\text{eff}}$ as follows:

$$R_{1,\text{eff}} = \frac{V_{R_{1,\text{eff}}} Z_{i,\text{op amp}}}{V_{Z_{i,\text{op amp}}}} = \frac{0.99\text{ V} \times 1\text{ M}\Omega}{0.01\text{ V}} = 99\text{ M}\Omega$$

This $R_{1,\text{eff}}$ plus the $Z_{i,\text{op amp}}$ make up the $Z_i$ for the circuit in Fig. 25-21. Now we know why the input impedance for Example 25-7 turned out to be 100 MΩ.

## Summing Amplifier

The op amp circuit shown in Fig. 25-23 is called a *summing amplifier,* or a mixer amplifier. When all of the resistors in the circuit are equal, the instantaneous output voltage is the algebraic sum of the instantaneous input voltages. How the summing circuit operates is shown in Fig. 25-24. Remembering that the inverting input is at virtual ground is the key to understanding this circuit. For this input to be at virtual ground, $R_1$ must drop 4 V. Therefore, the current through $R_1$ must be 4 V/1 kΩ = 4 mA in the direction shown in Fig. 25-24. The same reasoning shows the current through $R_2$ to be 1 mA in the opposite direction. Now, applying Kirchhoff's current law to the junction of $R_1$, $R_2$, and $R_\text{F}$ yields

$$I_{R_\text{F}} = I_{R_1} - I_{R_2} = 4\text{ mA} - 1\text{ mA} = 3\text{ mA}$$

Knowing the current through $R_\text{F}$ allows us to determine that $V_{R_\text{F}}$ is 3 V. Of course, because of the virtual ground, $V_{R_\text{F}}$ and the output voltage are the same. Notice that the −3 V output is the inverted algebraic sum of 4 V + (−1 V). A formula that incorporates all the steps we used in finding the output voltage is

$$V_o = -\left(\frac{V_1}{R_1} + \frac{V_2}{R_2}\right) R_\text{F}$$

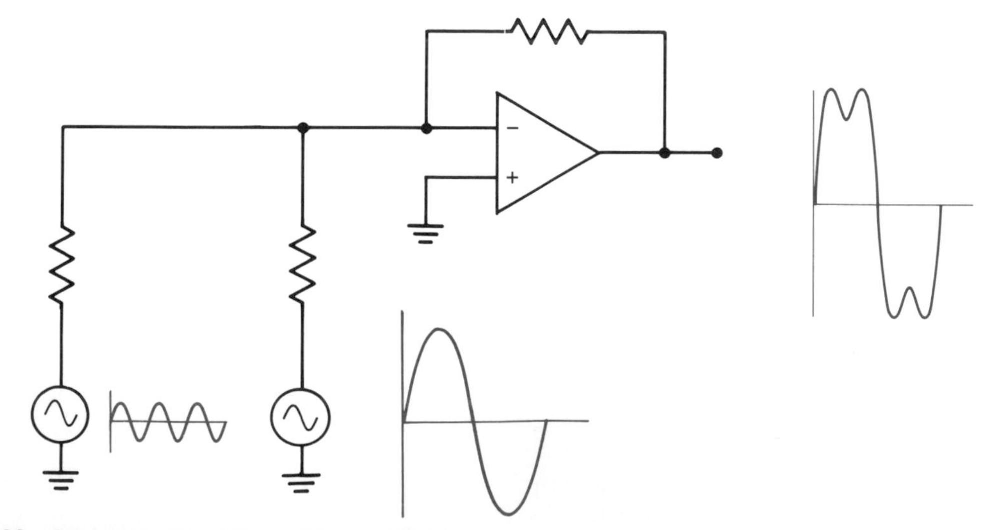

**FIGURE 25-23 Summing amplifier with two inputs.**

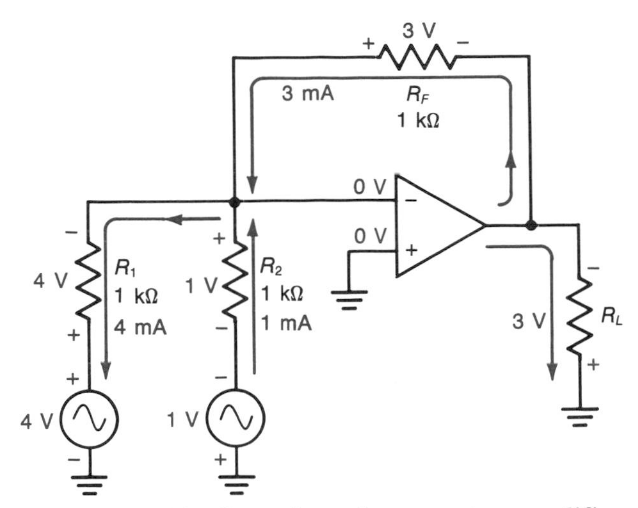


FIGURE 25-24 **Operation of a summing amplifier.**

In using this formula, remember to keep track of the polarities so that the currents at the junction are properly summed (see Example 25-8).

The number of inputs to the summing amplifier is not limited to two. Also, the resistor values do not have to be equal. By using unequal resistor values, each input can be weighted or scaled. If potentiometers are used in place of resistors, the scaling factor can be changed at will by the user of the circuit. Such circuits are very useful for mixing and balancing audio signals in a recording studio.

**EXAMPLE 25-8**

In Fig. 25-24, change $R_1$ to 2 kΩ and $R_F$ to 4 kΩ. Now determine the output voltage.

**Solution**

$$V_o = -\left(\frac{V_1}{R_1} + \frac{V_2}{R_2}\right) R_F = -\left(\frac{4\text{ V}}{2\text{ k}\Omega} + \frac{-1\text{ V}}{1\text{ k}\Omega}\right) 4\text{ k}\Omega$$
$$= -[2\text{ mA} + (-1\text{ mA})]\ 4\text{ k}\Omega = -4\text{ V}$$

## Comparator Amplifier

Another useful op amp circuit is the *comparator amplifier* diagramed in Fig. 25-25*a*. This amplifier is different from the previous op amp circuits in two ways: (1) it uses no feedback, and (2) it is operated in a nonlinear mode. Because the gain of this circuit is the open-loop gain, the slightest voltage (microvolt) difference between the − and + inputs drives the op amp into saturation. When the + input voltage exceeds the − input voltage, the output goes positive to within a couple of volts of $+V_{CC}$. When the − input exceeds the + input, the output goes to within a couple of volts of $-V_{EE}$. The voltages being compared can be alternating current, direct current, or a combination of the two.

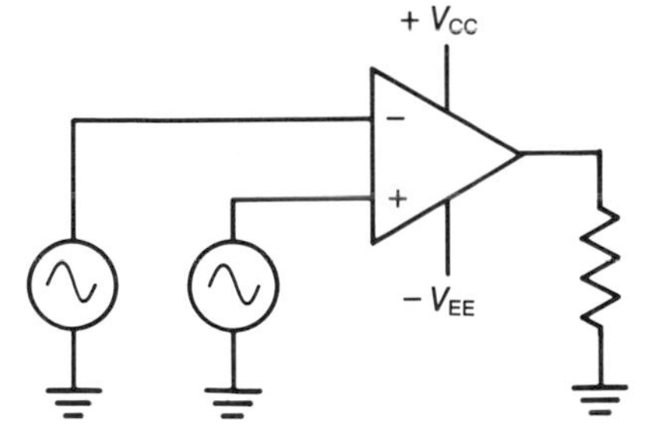


(*a*) Comparator amplifier

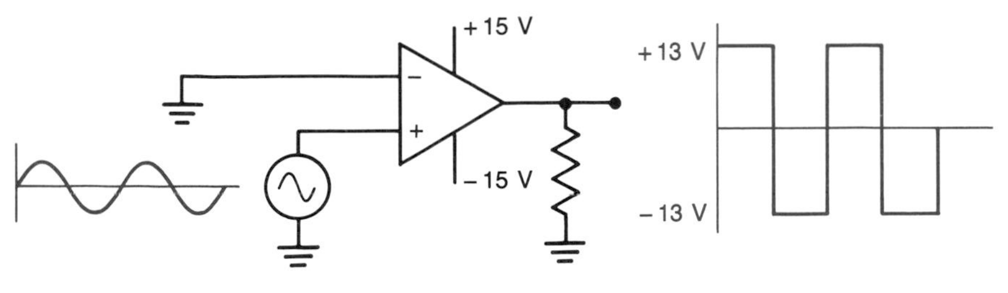


(*b*) Squaring circuit

FIGURE 25-25 **Op amp used in a comparator circuit.**

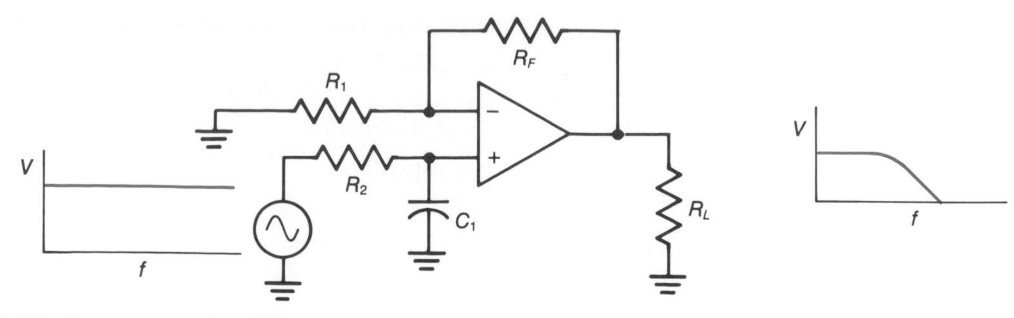

**FIGURE 25-26 Low-pass active filter.**

A comparator can be used to produce a symmetrical square wave from a sine wave as illustrated in Fig. 25-25*b*. Because of the high value of $A_{V,\mathrm{OL}}$, the output rapidly changes from $-13$ to $+13$ V and vice versa. Also, this change in output occurs essentially at the time the ac input goes through 0 V. Therefore, this circuit is also known as a zero voltage detection circuit. It is used as part of a system which turns on and turns off high current loads, especially inductive loads, only when the voltage or current is zero.

## Active Filter

Filters which use active devices such as transistors or op amps are called *active filters.* A low-pass filter is shown in Fig. 25-26. In this circuit, $R_1$ and $R_F$, as usual, determine the gain at low frequencies. At low frequencies, the reactance of $C_1$ is very large, so nearly all of the input signal appears across it. Thus, all the input signal is amplified. As the frequency increases, $X_C$ gets smaller (remember that $X_C = 1/2\pi fC$); less signal appears across $C_1$; more signal appears across $R_2$. At the frequency at which $X_C = R$, the output will be down 3 dB from its low frequency level. From then on the output decreases at a rate of 6 dB per octave (20 dB per decade). At some higher frequency, $X_C$ will be so small that the + terminal of the op amp is essentially grounded and there is no input signal to be amplified by the op amp.

## Single-Supply Amplifier

If an op amp circuit does not need to amplify a dc signal, it can be operated from a single power supply. A noninverting amplifier using a single supply is shown in Fig. 25-27. In this circuit, $R_1$ must be dc-isolated from ground so that the inverting terminal's quiescent voltage can match the voltage on the noninverting terminal regardless of the $R_F/R_1$ ratio needed to achieve the desired voltage gain. The quiescent voltage on the + terminal is determined by the $R_2$ and $R_3$ voltage divider. These re-

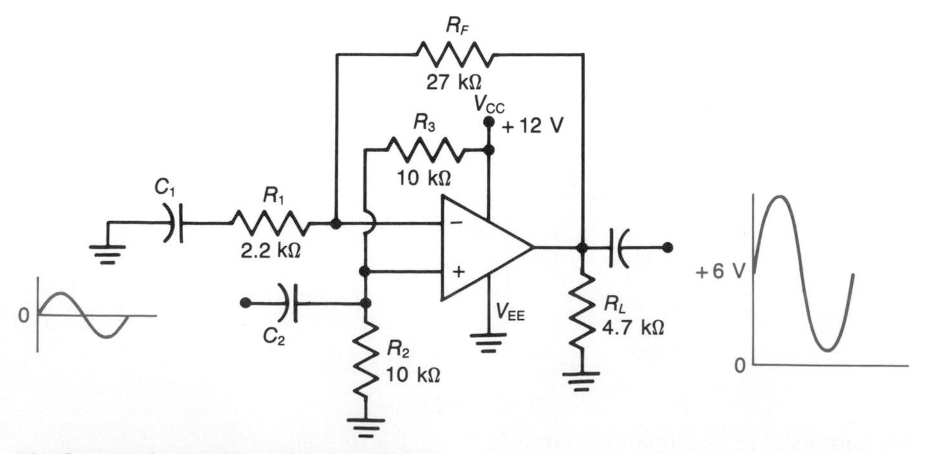

**FIGURE 25-27 Single-supply noninverting amplifier.**

sistors are usually equal values so that the + terminal is held at 0.5 $V_{CC}$. This also causes the quiescent dc output voltage to be 0.5 $V_{CC}$ so that its 0.5 $V_{CC}$ will be applied through $R_F$ to the − terminal of the op amp. Of course, there is no quiescent current through either $R_F$ or $R_1$.

When a signal is applied to the + terminal (through coupling capacitor $C_2$), the output voltage swings above and below 0.5 $V_{CC}$. The changing output voltage causes the feedback current (through $R_F$, $R_1$, and $C_1$) needed to make the − terminal track (follow) the voltage changes occurring at the + terminal.

A major limitation of this single-supply noninverting amplifier is that the input impedance is equal to $R_3$ in parallel to $R_2$. Thus, the circuit cannot have an input impedance in the multimegohm range. Another drawback to single-supply circuits is that the capacitors must be quite large if low frequencies are to be amplified. The reactances of the capacitors, at the lowest frequency to be amplified, must be at least as small as the resistances they are isolating from ground.

## Self-Test

**50.** What is the phase relationship between the signals on the − terminal and the output terminal of an op amp circuit?

**51.** Does the inverting or noninverting amplifier have the larger input impedance? Why?

**52.** True or false? The load resistor $R_L$ determines the gain of an inverting amplifier.

**53.** True or false? If the + terminal is +4 V and the − terminal is −4 V, the op amp will be saturated.

**54.** Compare the quiescent output voltage of the dual-supply and single-supply op amp circuits.

**55.** Which op amp circuit does not use feedback?

**56.** A summing amplifier can sum more than two signals.

**57.** True or false? The signals applied to a summing amplifier must be in phase.

**58.** List two uses of a comparator amplifier.

**59.** For the circuit in Fig. 25-27, determine the following:

**a.** $A_V$

**b.** $Z_i$

**c.** The quiescent voltage on the inverting terminal

**60.** Change $R_2$ in Fig. 25-24 to 2 kΩ and then determine the output voltage.

**61.** Change $R_1$ in Fig. 25-21 to 3 kΩ and then determine $A_V$.

## SUMMARY

1. Integrated circuits can be either monolithic or hybrid.
2. Reverse-biased junctions isolate circuit elements in a monolithic IC.
3. Monolithic ICs may be classified as SSI, MSI, LSI, or VLSI.
4. Hybrid ICs may be classified as thick-film or thin-film.
5. Hundreds of IC chips are processed on a single wafer.
6. The photolithographic process is used in fabricating both monolithic and thin-film hybrid ICs.
7. Silicon dioxide is an insulating material used for the dielectric in IC capacitors.
8. An epitaxial layer is deposited on the substrate of an IC wafer.
9. N- and P-type dopants may be diffused into the epitaxial layer of an IC wafer.
10. Ion implantation is a process used to dope crystals.
11. A passivating layer of silicon dioxide is often used to seal off the surface of an IC.

**12.** The circuit elements in a hybrid IC are built up on the surface of the substrate.

**13.** Silk screening is a process used to place conductive, resistive, and dielectric paste on the substrate of a thick-film IC.

**14.** Precise values of resistance can be obtained by laser trimming a deposited or screened resistor.

**15.** Thin-film hybrid ICs have good high-frequency response.

**16.** Both the input and output of a DA can be dual-ended.

**17.** A DA has a differential voltage gain and a common-mode voltage gain.

**18.** Common-mode rejection ratio (CMRR) is an important rating of both the DA and the op amp.

**19.** The quiescent base voltages for a DA are 0 V.

**20.** The quiescent collector voltages for a DA are equal.

**21.** With a differential input, $I_{R_E}$ remains constant in a DA.

**22.** With a single-ended input, the DA looks like a CC amplifier driving a CB amplifier.

**23.** Differential amplifiers have a large value of $Z_i$.

**24.** The performance of a DA can be improved by using active resistances.

**25.** The operational amplifier has an inverting (−) and noninverting (+) input.

**26.** The input stage of an op amp is a DA and the output stage is a push-pull amplifier.

**27.** Most op amp circuits use dual power supplies.

**28.** With dual supplies, the quiescent voltages at the input and output terminals of an op amp are 0 V.

**29.** Unless the op amp is saturated, its input terminals will be at the same voltage level.

**30.** The internal stages of an op amp IC are direct-coupled.

**31.** Important characteristics of an op amp are as follows: $Z_i \geq 1$ MΩ, $Z_o \leq$ 200 Ω, $A_{V,\mathrm{OL}} \geq$ 100 dB, CMRR $\geq$ 90 dB, slew rate $\geq$ 0.25 V/μs, and a frequency roll-off of 20 dB per decade.

**32.** Slew-rate distortion is a function of both the frequency and the amplitude of the output signal.

**33.** Inverting amplifiers have much lower input impedance than do noninverting amplifiers.

**34.** The virtual ground on the − terminal of the inverting amplifier keeps its $Z_i$ equal to $R_1$.

**35.** Feedback current in the noninverting amplifier makes $R_1$ appear to the signal source to be much larger than its true value.

**36.** Op amp circuits can be operated from a single power supply.

**37.** Summing amplifiers can add or mix two or more signals.

**38.** Comparator amplifiers use no feedback and are operated in saturation.

**39.** Op amps can be used in active filters.

**40.** Some useful formulas are

**a.** For all amplifiers,

$$r_e = \frac{26\text{ mV}}{I_C}$$
$$A_{V,\text{dB}} = 20 \log_{10} A_V$$

**b.** For differential amplifiers,

$$\text{CMRR} = \frac{A_D}{A_{CM}}$$
$$A_{CM} = \frac{R_C}{2R_E + r_e}$$
$$A_D = \frac{R_C}{2r_e}$$
$$Z_i = \beta \times 2r_e$$

**c.** For operational amplifiers,

$$\text{Bandwidth} = \frac{f_{\text{unity}}}{A_V} \quad \text{if roll-off is 20 dB per decade}$$

**d.** For inverting amplifiers,

$$Z_i = R_1$$
$$A_V = -\frac{R_F}{R_1}$$

**e.** For noninverting amplifiers,

$$Z_i = \frac{A_{V,\text{OL,OF}}\, Z_{i,\text{op amp}}}{A_V}$$
$$A_V = \frac{R_F}{R_1} + 1$$

**f.** For inverting summing amplifiers,

$$V_o = -\left(\frac{V_1}{R_1} + \frac{V_2}{R_2} + \cdots + \frac{V_n}{R_n}\right) R_F$$

## CHAPTER REVIEW QUESTIONS

For the following items, determine whether each statement is true or false.

**25-1.** Monolithic ICs are fabricated on an insulating substrate.

**25-2.** Circuit elements in hybrid ICs are insulated from each other by silicon dioxide.

**25-3.** For a thick-film hybrid IC, the components are silk-screened onto the substrate.

**25-4.** Diffused resistors can be laser-trimmed to provide exact values of resistance.

**25-5.** For a monolithic IC, the collector of a transistor is diffused into the epitaxial layer.

**25-6.** The epitaxial layer is diffused onto the substrate.

**25-7.** Ion implantation is a process used to connect lead wires to an IC chip.

**25-8.** The thin-film technique is used to deposit aluminum foil on a monolithic IC.

**25-9.** A wafer is usually less than 1 in. in diameter.

**25-10.** Dual-ended inputs to a DA are always common-mode inputs.

**25-11.** The common-mode gain of a DA should be much greater than the differential gain.

**25-12.** The dc base voltages of a DA with no signals applied will be approximately 0 V.

**25-13.** The input impedance of a DA is usually less than 10 kΩ.

**25-14.** The differential output of a DA is twice as large as the single-ended output.

**25-15.** The voltage gain of an op amp without feedback is called the open-loop voltage gain.

**25-16.** The input impedance of most op amps is greater than 100 MΩ.

**25-17.** The CMRR of a DA is usually larger than the CMRR of an op amp.

**25-18.** The differential signal between the two inputs of an op amp is usually between 1 and 2 V.

**25-19.** The frequency bandwidth of an op amp circuit increases as the amount of degenerative feedback increases.

**25-20.** An inverting amplifier has a larger $Z_i$ than does a noninverting amplifier.

For the following items, fill in the blank with the word or phrase required to correctly complete each statement.

**25-21.** The material used to protect some areas of silicon dioxide from being etched away is ______.

**25-22.** The final protective layer of silicon dioxide on an IC is called the ______ layer.

**25-23.** On a cross-sectional view of an IC, heavily doped P-type regions are indicated with a ______ symbol.

**25-24.** Large values of resistance are easier to obtain with ______ techniques than with diffusion techniques.

**25-25.** The conductive particles in a thick-film paste which has been fired are held together by the ______ material.

**25-26.** Unless a common-mode signal is received, the current through the ______ resistor in a DA remains constant.

**25-27.** The performance of a DA can be improved by using ______ resistances in the emitter and collector legs of the transistors.

**25-28.** If both collector currents in a DA are increasing, the DA must be receiving a ______ signal.

**25-29.** The first stage in an op amp is a ______.

**25-30.** The last stage in an op amp is a ______.

**25-31.** The dc output voltage of an op amp is determined by the dc ______ voltages.

**25-32.** When the noninverting input of an op amp is grounded, the inverting input is at ______.

**25-33.** The stages in an op amp are ______.

**25-34.** A typical op amp has a frequency roll-off of ______ dB per decade.

**25-35.** Most op amp circuits use ______ feedback.

**25-36.** ______ distortion causes an op amp with a sine-wave input to produce a triangular output.

**25-37.** The ______ amplifier circuit uses an op amp without feedback.

For the following items, choose the letter that best completes each statement.

**25-38.** Which of the following abbreviations refers to the density of circuit elements in an integrated circuit?

**a.** MOS
**b.** DIP
**c.** TTL
**d.** SSI

**25-39.** Which of these circuit elements are not found in ICs?

**a.** Resistor
**b.** Active device
**c.** Inductor
**d.** Capacitor

**25-40.** Which of these ICs has the best high-frequency response?

**a.** Monolithic with BJTs
**b.** Monolithic with MOSFETs
**c.** Thin-film hybrid
**d.** Thick-film hybrid

**25-41.** The open-loop gain of most general-purpose op amps is between

**a.** 120 and 90 dB
**b.** 89 and 60 dB
**c.** 59 and 30 dB
**d.** 29 and 10 dB

**25-42.** For equal amplitudes of input signals, the output of an op amp is more likely to be distorted at

**a.** Low frequencies
**b.** Midfrequencies
**c.** High frequencies
**d.** Both **a** and **c**

## CHAPTER REVIEW PROBLEMS

**25-1.** An internally compensated op amp has an $f_{unity}$ of 10 MHz and is operated with an $A_V$ of 80 dB. What will be the upper limit of its bandwidth?

**25-2.** A differential amplifier has a $CMRR_{dB}$ of 60 dB and an $A_D$ of 500. What is its common-mode gain?

**25-3.** For the circuit in Fig. 25-28*a*, determine the following:

**a.** $A_{CM}$
**b.** $A_D$
**c.** CMRR
**d.** $Z_i$
**e.** $V_{C,quiescent}$

**25-4.** Is the signal in Fig. 25-28*a* connected to the inverting or noninverting input?

**25-5.** For the circuit in Fig. 25-28*b*, determine the following:

**a.** $A_V$
**b.** The quiescent voltage on the − terminal
**c.** The quiescent voltage across $R_F$
**d.** $Z_i$
**e.** The instantaneous voltage across $R_F$ when the signal voltage is +1 V
**f.** The instantaneous voltage on the − terminal when the signal voltage is +1 V

**25-6.** For the circuit in Fig. 25-28*c*, determine the output voltage for the input conditions shown.

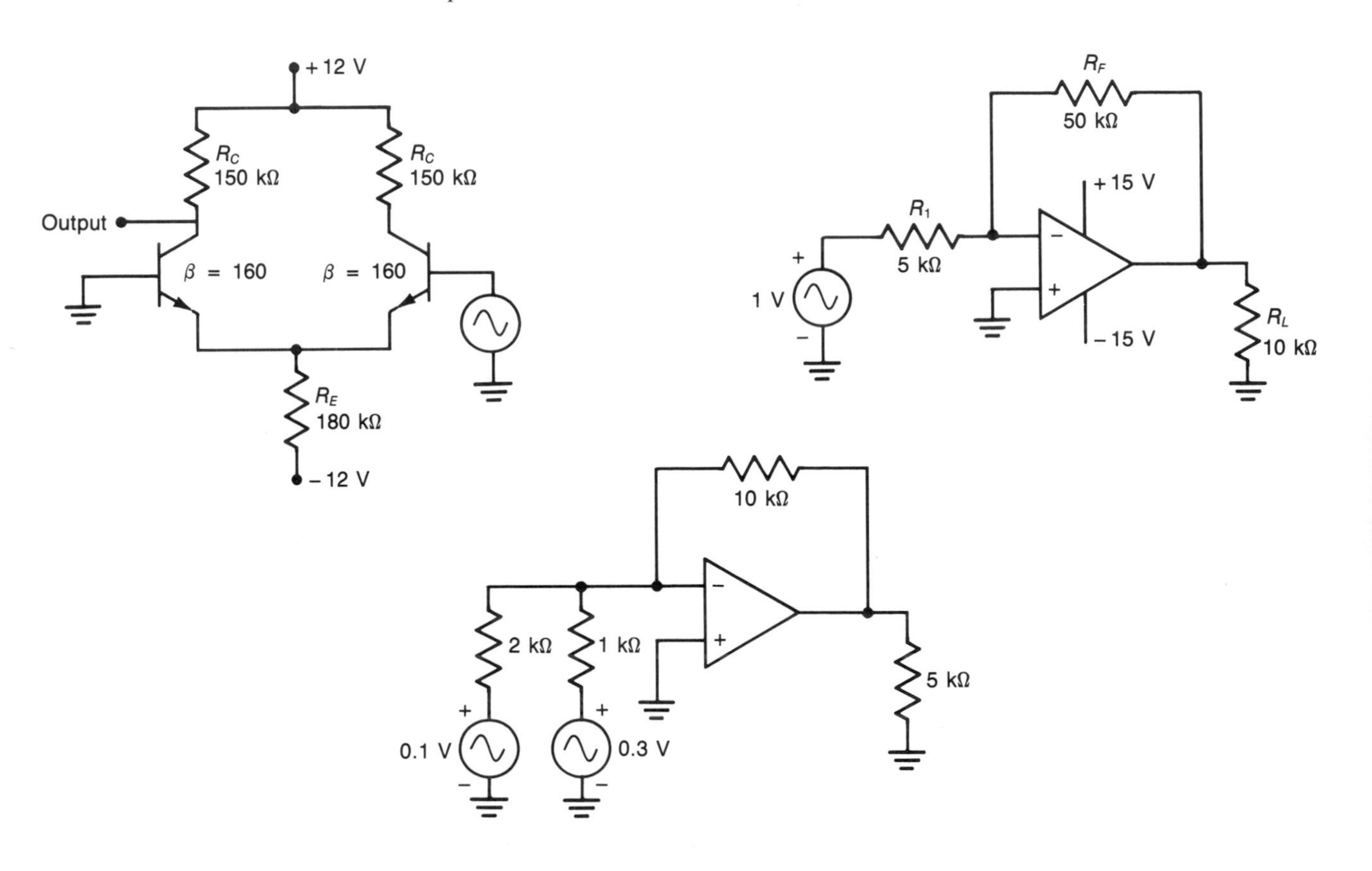


**FIGURE 25-28 Circuits for Chapter Review Problems.**

## ANSWERS TO SELF-TESTS

1. monolithic and hybrid
2. large-scale integration
3. false
4. false
5. false
6. wafer
7. Substrate is the material on which an IC is fabricated.
8. 0.01
9. photolithographic

**10.** chips (or dice)

**11.** dual in-line package

**12.** epitaxial

**13.** by reverse-biased PN junctions

**14.** micrometers

**15.** diffusion

**16.** silicon dioxide

**17.** (1) capacitor plate, (2) interconnecting conductors, and (3) lead-wire pads

**18.** passivating

**19.** ion implantation

**20.** a P-type region that is heavily doped

**21.** insulating

**22.** false

**23.** true

**24.** false

**25.** true

**26.** to minimize the surface area they cover

**27.** solvent—aids in drying the paste
binder—fuses and holds the particles of active ingredients together when the paste is fired
active material—the finely ground particles which make the paste a conductor, resistor, or dielectric

**28.** by laser trimming

**29.** thin film

**30.** hybrid, because its components are not isolated by reverse-biased junctions

**31.** an output taken between the collectors of a DA

**32.** because in the common-mode, signal current flows through $R_E$ and produces negative feedback

**33.** because a differential signal only drives the base(s) of the transistor(s) a few millivolts − or + with respect to ground

**34.** because this yields a large CMRR which makes it easier to separate the undesired common-mode signal from the desired differential signal

**35.** The differential output signal is twice as large.

**36.** $I_{C_2}$ is increasing and $V_{C_2}$ is decreasing.

**37.** $I_{C_1}$ is increasing.

**38.** It improves the CMRR by providing a relatively low $R$ and relatively high $r$.

**39.** in phase

**40.** base of $Q_2$

**41.** $Z_i = 304$ kΩ, $A_D = 87$, CMRR = 217

**42.** $\text{CMRR}_{\text{dB}} = 52.3$ dB

**43.** because it does not use coupling capacitors between stages

**44.** by using a push-pull output stage with equal values of $V_{CC}$ and $V_{EE}$

**45.** yes, because both inputs and outputs are at dc ground

**46.** 316 $\text{mV}_{p\text{-}p}$

**47.** **a.** high as possible
**b.** high as possible
**c.** low as possible
**d.** high as possible
**e.** high as possible
**f.** high as possible

**48.** 1 V, 30 kHz; because the higher the frequency the greater is the possibility of slew-rate distortion (because of too low of a slew rate). Also, 30 kHz may be outside the bandwidth, in which case frequency distortion would occur.

**49.** 5 kHz

**50.** 180° out of phase

**51.** the noninverting, because the − terminal is not grounded in the noninverting amplifier so the + terminal is not at virtual ground

**52.** false

**53.** true

**54.** For a dual supply, the output is zero and for a single supply it is 0.5 $V_{CC}$.

**55.** comparator circuit

**56.** true

**57.** false

**58.** zero voltage detection and squaring

**59.** **a.** 13.3
**b.** 5 kΩ
**c.** +6 V

**60.** −3.5 V

**61.** 4

# Chapter 26 Introduction to Digital Circuits

You are now ready for an introduction to the world of digital electronics. This introduction deals primarily with the major building blocks from which complex digital systems are constructed. This chapter is not concerned with the internal operation of the building blocks; rather, it concentrates on what the blocks can do.

Circuits which have discrete levels of input and output voltages are known as *digital* circuits. These discrete levels represent the digits in a number system. Therefore, digital systems are systems which process information that is represented by discrete levels. The number system we use in our daily lives is the *decimal (base 10) system.* However, digital circuits with 10 discrete levels of input and output would be quite complex and difficult to maintain. Therefore, digital systems process information which is expressed in the *binary (base 2) number system.* By using the binary system, the digital circuit only needs two discrete signal levels—one to represent the digit or figure 0 and one to represent the digit or figure 1. Any two signal voltage levels could be used to represent the 0 and 1 digits, and many different levels have been used as digital circuits have progressed from using vacuum tubes to using metal-oxide semiconductors (MOS) integrated circuits (ICs).

Digital circuits and systems are constructed with ICs. Each IC contains one or more digital logic

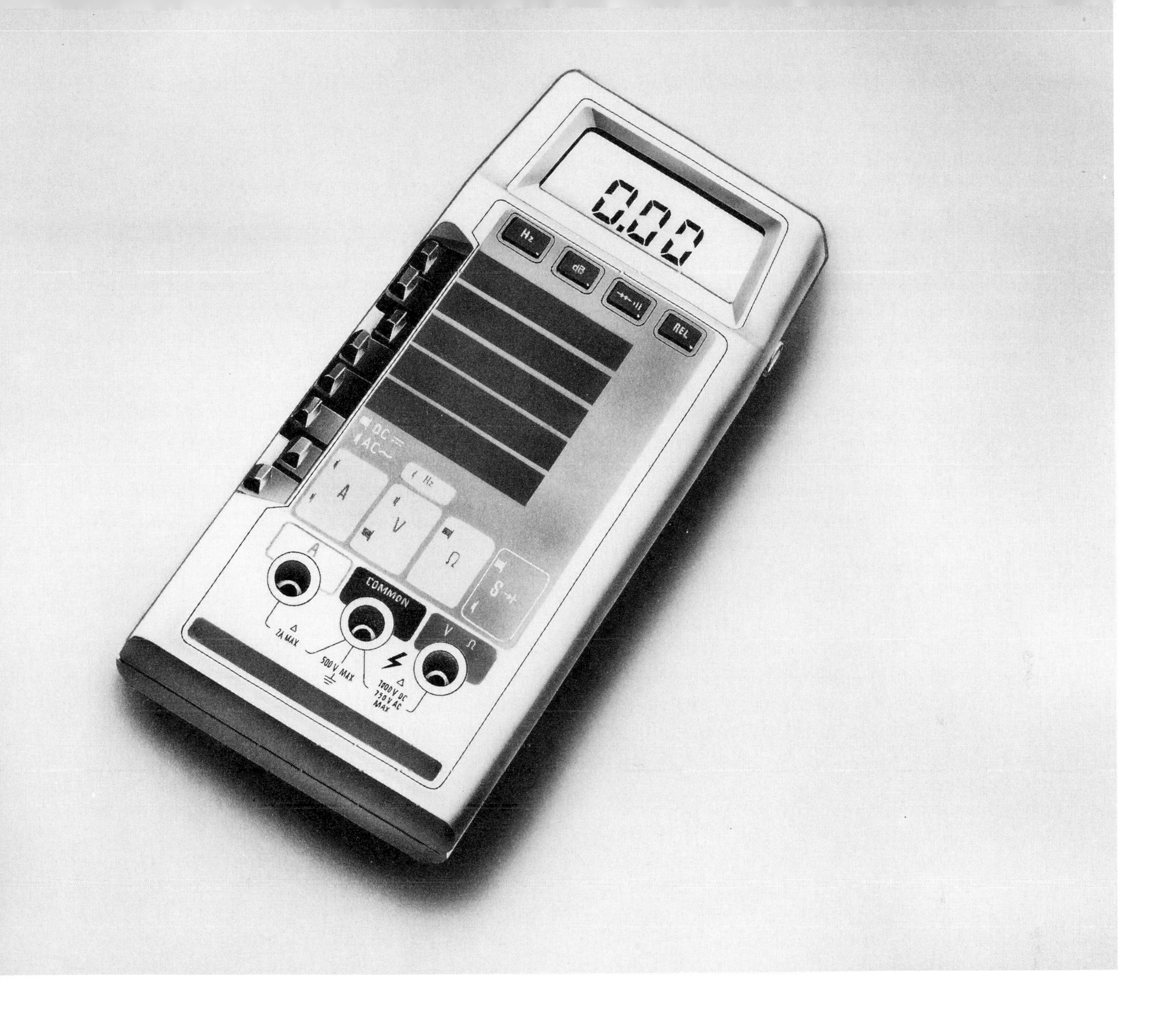


functions (gates). Most digital ICs are housed in a DIP (dual in-line package) with 14, 16, 24, or 40 pins.

## 26-1 A LITTLE TERMINOLOGY

The field of digital electronics has developed a lot of terminology that is not clearly defined and which can be very confusing to beginners. *Logic* refers to a rational decision-making process. In digital electronics, logic means that an exact, definable relationship exists between the input(s) and the output(s) of a digital circuit. This relationship allows a digital logic circuit to make a clearly defined decision and express that decision as an output voltage level. The decision is based on the information (voltage levels) provided to the logic circuit's inputs.

A specific input-output relationship is called a *logic function.* Each logic function is given a specific name. For example, the AND logic function requires that all of the inputs to a logic circuit be at a high voltage level in order for the output to be at a high voltage level. The digital circuit which performs the AND logic function is called an *AND gate.* Thus, *gate* is just another term for circuit. A *logic gate* is a digital circuit which performs a *logic function.*

The term *logic level* refers to the voltage level of an input or the output of a digital circuit. If we agree that a high positive voltage level represents a binary one, then stating that the output logic level is one is the same as stating that the output voltage level is high. But, how high is high?

The voltage levels which represent the binary digits 1 and 0 depend on the *logic family* of ICs used in the digital system. Logic family refers to a group of ICs, all manufactured using the same fabrication technique, which provides all of the common logic functions. One of the most popular logic families is the transistor-transistor logic (TTL) family which uses bipolar junction transistors for the active devices in the IC. For this family, the high output is a positive voltage between +2.4 and +5.0 V. The low output is between 0 and +0.4 V. For the inputs, high and low are +2.0 to +5.0 V and 0 to +0.8 V, respectively. The 0.4-V difference between input and output low levels and the input and output high levels is the *noise margin* voltage. Thus, if the low output of one logic gate is connected to the input of a second logic gate, the second logic gate interprets the input as low even if 0.4 V of noise signal is induced into the conductor connecting the two gates. All the circuits discussed in this chapter use TTL logic.

Two types of digital logic systems are possible—*positive* logic and *negative* logic. The major difference between the two is whether the digit 1 is represented by a high voltage or a low voltage. The most commonly used system is positive logic. In this chapter we use positive logic at all times. With positive logic, the digit 1 is represented by a high voltage (>+2.0 V) and the digit 0 by a low voltage (<+0.8 V).

A digital signal is a signal which rapidly changes from one logic level to the opposite logic level. As seen in Fig. 26-1, the changes may be symmetrical (i.e., the same amount of time at each logic level) or nonsymmetrical. Digital signals used for timing and synchronizing a complex digital system like a computer are called *clock signals*.

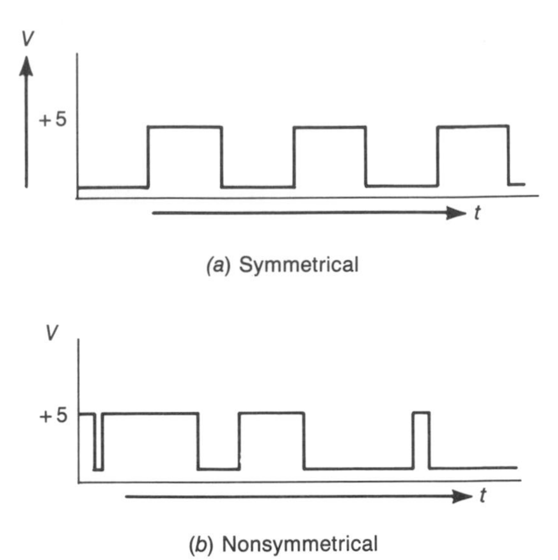


**FIGURE 26-1 Digital signals.**

## 26-2 BINARY NUMBERS

The binary number system uses only two digits (0 and 1) to express all the numbers that can be expressed in the decimal or any other number system. While the binary system may seem strange to those who are only used to the decimal system, it follows the same rules as the decimal system.

All number systems center around a *radix point* as shown in Fig. 26-2. Digits to the left of the radix point represent a whole number, while digits to the right of the radix point represent a fraction. In the base 10 system, the radix point is called the *decimal point*. For the binary system, it is called the *binary point*.

As shown in Fig. 26-2, the first place (column or position) to the left of the radix point is the *units* or *ones* column in any number system. A digit in this place or column is not weighted. It is taken at face value; i.e., it is multiplied by the base raised to the zero power or exponent. Thus, in the decimal system, an 8 in this place is $8 \times 10^0 = 8$; in the binary system a 1 is $1 \times 2^0 = 1$.

In all other places to the left of the radix point, the digit occupying a place is weighted. In base 10 (decimal), the second place to the left is weighted by multiplying the digit by $10^1$. A one in this column represents $1 \times 10^1 = 10$ and a two represents $2 \times 10^1 = 2 \times 10 = 20$ (or two 10s). Thus, in base 10 the second column is often called the *tens column*. In base 2, the second column is called the *twos column* because $1 \times 2^1 = 2_{10}$; i.e., a 1 in the twos column of a binary number is *equal to 2 in the base 10 system*. This is why the subscript 10 is used after the 2. Subscripts are used with numbers if there is any chance of confusing which number system a string of digits represents. In the above case, the 10 subscript is not really necessary because 2 is not a digit in the base 2 system, and we have been dis-

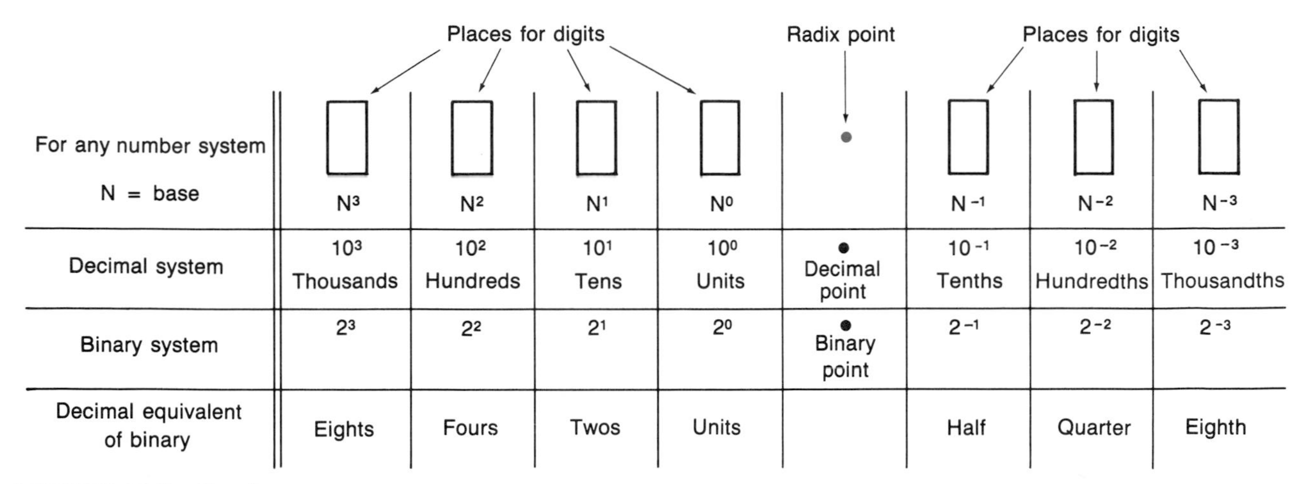


**FIGURE 26-2 Number systems.**

cussing base 8 and base 10 and the number 146 appeared, there could be confusion because 146 is a number in either system. In that case, in order to indicate the number system to which this number belongs, we should write $146_{10}$ or $146_8$.

In terms of its base 10 equivalent, the $2^2$ place is called the fours column. As shown in Fig. 26-2, the $2^3$ column is the eights column, etc.

The digits to the right of the radix point represent tenths ($10^{-1}$), hundredths ($10^{-2}$), etc., for the decimal system and $2^{-1}$, $2^{-2}$, etc., for the binary system. When $2^{-1}$ is converted to its decimal equivalent, it is equal to one-half; $2^{-2}$ is equal to one-fourth, etc. Thus, the places to the right of the binary point are often referred to as halves, quarters, eighths, etc.

In the binary system a digit is often called a *bit.* Bit, then, is just shorthand for "binary digit." The last bit on the left of a binary number is called the *most-significant bit* (MSB) because it has the largest weighting factor. Likewise, the last digit on the right is known as the *least-significant bit* (LSB).

Once you understand the structure of number systems, conversion from binary to decimal is quite simple.

**EXAMPLE 26-1**

Convert the number $1010.11_2$ to base 10.

**Solution**

As shown in Fig. 26-3, start at the right-hand end of the string of binary digits and list in a vertical column the base 10 equivalent of each binary digit. Then add the vertical column.

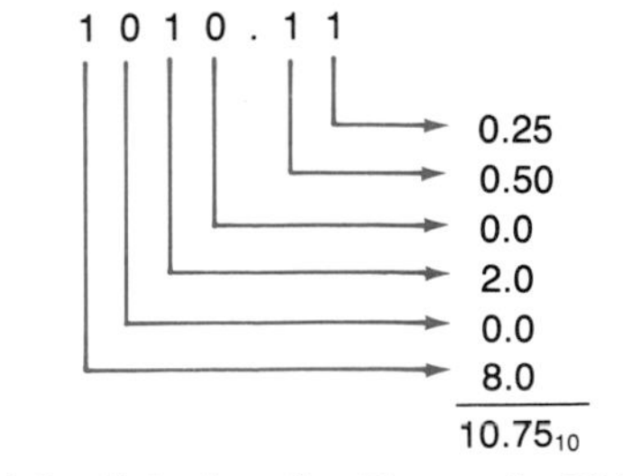

**FIGURE 26-3 Solution for Example 26-1.**

Converting from decimal to binary can be accomplished by repeated subtraction of the base 10 equivalent of the powers of two. Start with the largest power of two which does not yield a negative result when subtracted from the base 10 number. When succeeding powers of two cannot be subtracted without resulting in a negative number, enter a 0 in the binary answer and continue to the next smaller power of 2. An example illustrates the process.

**EXAMPLE 26-2**

Convert the number $35.5_{10}$ to binary.

**Solution**

Figure 26-4 provides the binary-decimal equivalents needed for this problem. It also shows the

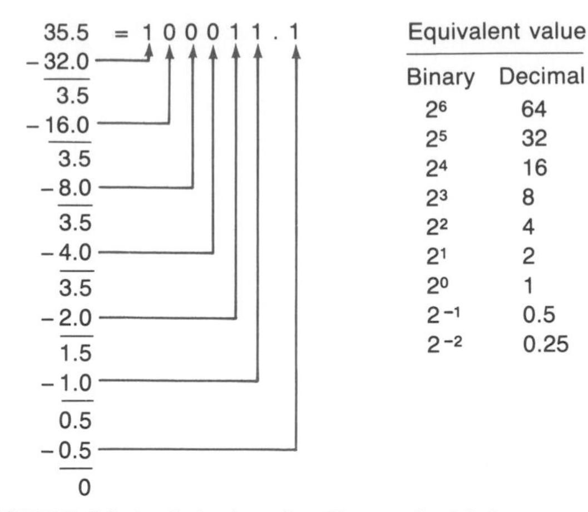


**FIGURE 26-4 Solution for Example 26-2.**

step-by-step solution. Remember to add a zero to the answer whenever a power of two cannot be subtracted.

---

Addition in the binary number system is illustrated in Fig. 26-5. In Fig. 26-5*a* the addition of single-digit binary numbers is illustrated. Notice that the addition of two 1s produces a one to carry into the twos column and leaves a zero in the units column. Adding three 1s produces a one to carry to the twos column and leaves a one in the units column. Addition of multidigit numbers is shown in Fig. 26-5*b*. The same carrying rules apply to multidigit numbers as apply to single-digit numbers. That is, two 1s in any column result in a carryover to the next column and a zero remaining; three 1s in a column produce a carryover and leave a one remaining. Figure 26-5*c* shows how to check your binary addition by converting to equivalent decimal numbers.

A simple process for binary subtraction is illustrated in Fig. 26-6. Notice in the last example in Fig. 26-6*a* that the answer is negative because there is no twos column from which to borrow two 1s. The borrowing procedure is illustrated in the last example in Fig. 26-6*b*. Borrowing from the fours column leaves a zero in that column and provides two 2s for the twos column. In the answer, this yields a one in the twos column. Next, borrowing from the eights column provides two 4s and leaves a one in the fours column of the answer. Figure 26-6*c* illustrates the procedure to follow when it is

```
                     1 ← Carry        1 ← Carry
0       0            1                1
0       1            1                1
—       —            —                —
0       1           10               11
```

(*a*) Adding single-digit numbers

```
1 ← Carry         1 1 1 1 ← Carries
1 0 0 1             1 0 1 1
0 0 0 1             1 1 0 1
———————           —————————
1 0 1 0           1 1 0 0 0
```

(*b*) Adding multidigit numbers

```
1   1            Check
  1 0 1 1 0      = 22₁₀
  1 1 1 0 1      = 29₁₀
———————————      ——————
1 1 0 0 1 1      = 51₁₀
```

(*c*) Checking by equivalent decimal numbers

**FIGURE 26-5 Addition of binary numbers.**

necessary to borrow but the next column contains a zero. In the example, the borrow from the eights column provides a one for the fours column, a one for the twos column, and two 1s for the units column.

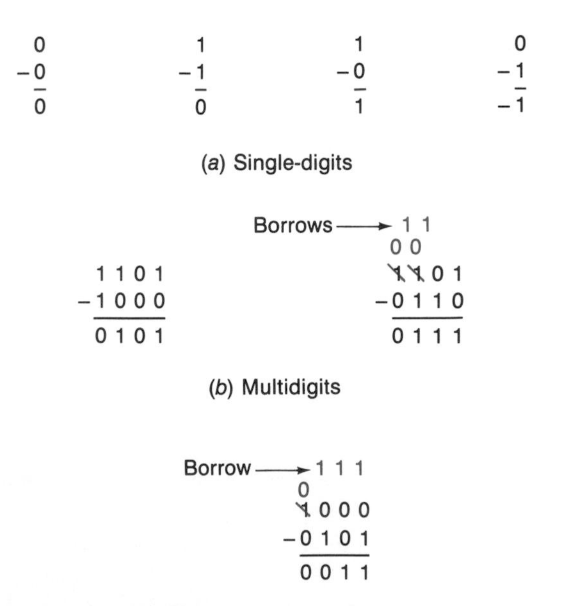


**FIGURE 26-6 Subtraction of binary numbers.**

Figure 26-7 shows how to multiply binary numbers. Notice in Fig. 26-7*b* that there are only two possible partial products—zero and the multiplicand itself. Of course, the multiplicand is shifted left when appropriate.

The digital logic circuits which perform binary multiplication and binary subtraction use different techniques than those illustrated in Figs. 26-6 and 26-7. However, the end results are always the same regardless of which technique is used. If you continue your study of electronics in the area of digital electronics and computers, you will learn about other techniques of doing binary arithmetic.

```
   0        0        1        1
 × 0      × 1      × 0      × 1
 ---      ---      ---      ---
   0        0        0        1
```

(*a*) Single digits

```
    1010
     101
    ----
    1010
   0000
  1010
  ------
  110010
```

(*b*) Multidigits

**FIGURE 26-7 Multiplication of binary numbers.**

## Self-Test

**1.** What is a digital circuit?

**2.** Define the term logic function.

**3.** Define the term logic gate.

**4.** What does TTL stand for?

**5.** Does a high voltage represent a 0 or a 1 in a positive logic system?

**6.** What is a clock signal?

**7.** Convert the following binary numbers to decimal:
**a.** 1011 **b.** 01010.101 **c.** 10001.1

**8.** Convert the following decimal numbers to binary:
**a.** 15 **b.** 18.5 **c.** 38.375

**9.** Perform the following binary arithmetic:
**a.** 101 + 1110 **b.** 1010.101 + 101.1
**c.** 1011 − 1001 **d.** 1101 − 101.1
**e.** 101 × 11 **f.** 1111 × 101

## 26-3 DECISION-MAKING GATES

The logic gates, or circuits, which perform logic functions are often referred to as *decision-making gates.* Each type of logic gate is represented by a logic (schematic) symbol. The relationship between the input (or inputs) and the output of each type of gate is given by a *truth table.*

There are three basic logic gates and logic functions. These three are the INVERT gate, the OR gate, and the AND gate. The INVERT gate is often referred to as an *inverter* and sometimes it is called a NOT gate. Various combinations of the three basic logic gates are so commonly used that they have also been given specific names. Examples are NAND, NOR and exclusive-OR.

### INVERT Gate and Function

Figure 26-8 provides the logic symbol, truth table, and boolean expression for the INVERT gate. The inverter always *complements* the input; i.e., a binary 1 (>+2.0 V) input always provides a binary 0 (<+0.4 V) output. The inputs and outputs of gates are labeled with a letter, and the letter is referred to as a variable. The choice of *A* and *X* for the variable labels in Fig. 26-8 was arbitrary. Any letters could have been used to represent the input variable and the output variable. The truth table shows that if the input *A* is 0, then the output *X*

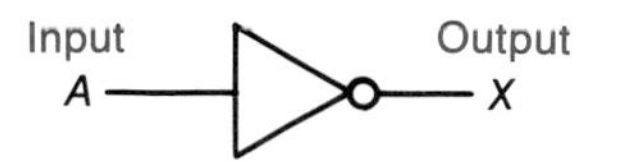


(*a*) Symbol

| A | X |
|---|---|
| 0 | 1 |
| 1 | 0 |

(*b*) Truth table

$X = \overline{A}$

(*c*) Boolean expression

**FIGURE 26-8 INVERT gate.**

will be 1; i.e., the input is *complemented*. A complemented variable is also referred to as an INVERTed variable or a NOTed variable or a *negated* variable.

*Boolean algebra* is a form of algebra used to express logical relationships. A boolean expression can be written to describe each logic gate. For the INVERT (NOT) gate in Fig. 26-8, the expression is $X = \overline{A}$. This expression is verbalized as "$X$ = $A$-bar, $X$ equals $A$-not, or $X$ equals not $A$." The expression $X = \overline{A}$ tells us that the output $X$ is the complement of the input $A$.

The symbol for the INVERT gate is really a combination of two symbols. The triangle indicates that the INVERT gate is an amplifier. Thus the gate provides more output current and power than it requires at its input. The circle on the point of the triangle indicates that the output is an inverted (complemented) version of the input.

Figure 26-9 shows how an inverter (NOT gate) responds to a digital signal. The input signal $A$ and the output signal $X$ are placed one above the other for ease in viewing input/output signal changes. Notice that the output signal is 180° out of phase with the input.

## AND Gate and Function

The logic symbol, truth table, and boolean expression for several AND gates are shown in Fig. 26-10. Notice that the output of the AND gate will be high (1) only when all of the inputs are high (1). The boolean expression for the two-input AND gate in Fig. 26-10*a* is $X = A \cdot B$. The dot between the $A$ variable and the $B$ variable is the symbol for "and."

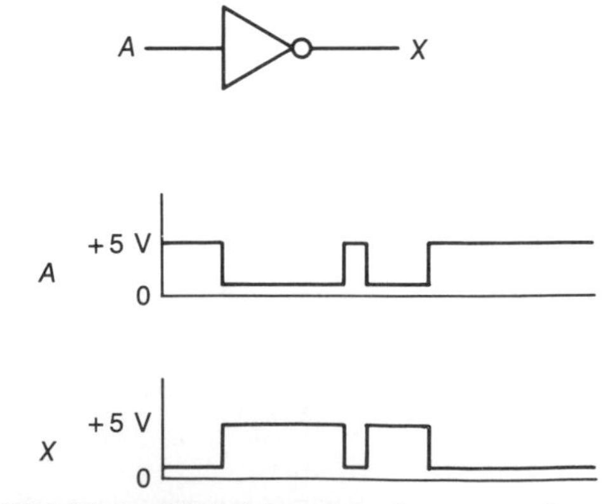


**FIGURE 26-9 Digital input and output of an INVERT gate.**

This expression is read as "$X$ equals $A$ and $B$." Quite often the dot is omitted, and the expression is written as $X$=$AB$. However, the expression "$X$ = $AB$" is still read as "$X$ equals $A$ and $B$." The expression for the three-input AND gate of Fig. 26-10*b* could be written as $X = A \cdot B \cdot C$ or $X = ABC$. In this chapter we use the latter form of the boolean expression.

Notice from Fig. 26-10 that the number of entries in the truth table is a function of the number of variables. The number of possible input combinations (rows) in a truth table is equal to $2^n$ where the exponent $n$ is the number of input variables. Thus, truth tables for gates with more than four inputs get rather large. For example, an eight-input gate would require $2^8 = 256$ rows for its truth table.

Truth tables are easy to set up if one uses a systematic approach. Once the number of rows has been determined, just start the first row with all zeros and then make the next row one binary number larger than the preceding row. For example, the rows in Fig. 26-10*b* are the binary equivalents of 0 to $7_{10}$. Careful examination of Fig. 26-10*b* shows that a vertical list of consecutive binary numbers develops some unique patterns for each column. The LSB (least-significant bit) column, the column under variable $C$, has alternate zeros and ones. The $2^1$ column (the column under variable $B$) has groups of two 0s and two 1s which alternate. In the $A$ column the zeros and ones are in groups of four. In a four-column table, the zeros and ones would be in groups of eight for the MSB column.

The response of an AND gate to digital input signals is illustrated in Fig. 26-11. In the figure, the voltage levels have been omitted from the signal waveforms. The waveforms are assumed to change from $<+0.4$ V to $>+2.4$ V; i.e., from logic 0 to logic 1.

## OR Gate and Function

The OR gate produces a 1 (high) output when one or more of its inputs is 1 (high). Figure 26-12 shows that the OR function has a high output except when all of its inputs are low. The boolean expression for the OR function of Fig. 26-12*a* is $X = A + B$. This expression is read "$X$ equals $A$ or $B$" not "$X$ equals $A$ plus $B$."

Like the AND gate, the OR gate can have more than two inputs. The four-input OR gate of Fig.

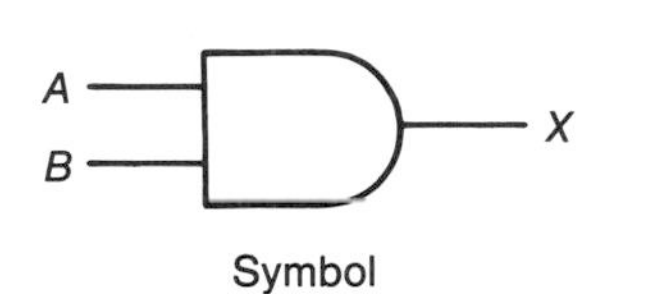


Truth table

| A | B | X |
|---|---|---|
| 0 | 0 | 0 |
| 0 | 1 | 0 |
| 1 | 0 | 0 |
| 1 | 1 | 1 |

$X = A \bullet B$ or $X = AB$

Boolean expression

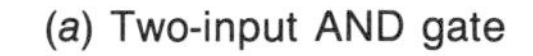

(*a*) Two-input AND gate

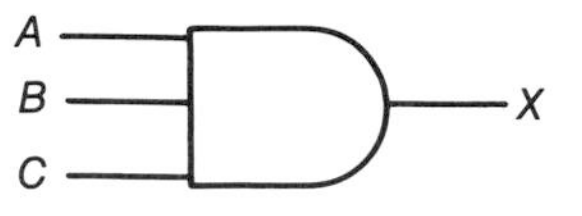


Truth table

| A | B | C | X |
|---|---|---|---|
| 0 | 0 | 0 | 0 |
| 0 | 0 | 1 | 0 |
| 0 | 1 | 0 | 0 |
| 0 | 1 | 1 | 0 |
| 1 | 0 | 0 | 0 |
| 1 | 0 | 1 | 0 |
| 1 | 1 | 0 | 0 |
| 1 | 1 | 1 | 1 |

$X = A \bullet B \bullet C$ or $X = ABC$

Boolean expression

(*b*) Three-input AND gate

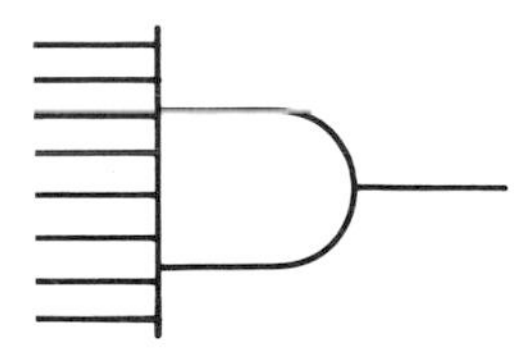

(*c*) Eight-input AND gate

**FIGURE 26-10 AND gate symbols and truth tables.**

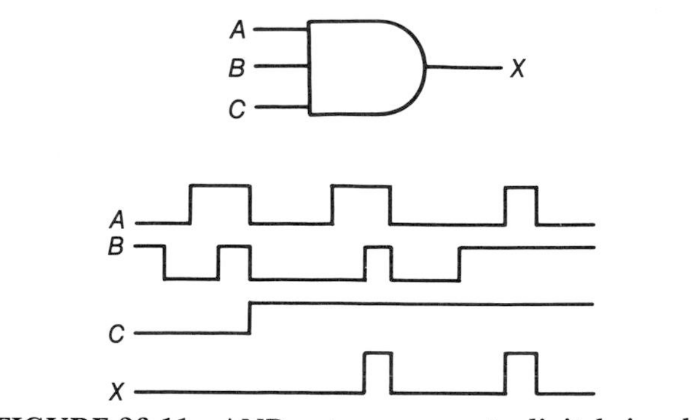


**FIGURE 26-11 AND gate response to digital signals.**

26-12*b* has 16 input combinations. All but one of these combinations produces a high output.

The response of the OR gate to a set of digital signals is shown in Fig. 26-13. Since at least one of the inputs is at logic 1 at all times, the output remains high (logic 1) for the time period shown.

## Exclusive-OR Gate

Many digital systems require a circuit which provides a high (1) output only when one input is high (1) and the other input is low (0). Because this circuit was used so often, it was given the name *exclusive-OR*. Exclusive-OR is often abbreviated XOR.

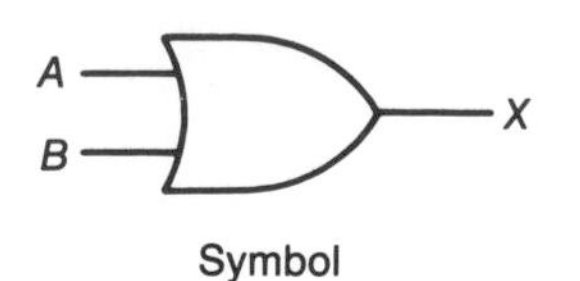


Truth table

| A | B | X |
|---|---|---|
| 0 | 0 | 0 |
| 0 | 1 | 1 |
| 1 | 0 | 1 |
| 1 | 1 | 1 |

$X = A + B$

Boolean expression

(*a*) Two-input OR gate

A
B
C
D
X
Symbol

Truth table

| A | B | C | D | X |
|---|---|---|---|---|
| 0 | 0 | 0 | 0 | 0 |
| 0 | 0 | 0 | 1 | 1 |
| 0 | 0 | 1 | 0 | 1 |
| 0 | 0 | 1 | 1 | 1 |
| 0 | 1 | 0 | 0 | 1 |
| 0 | 1 | 0 | 1 | 1 |
| 0 | 1 | 1 | 0 | 1 |
| 0 | 1 | 1 | 1 | 1 |
| 1 | 0 | 0 | 0 | 1 |
| 1 | 0 | 0 | 1 | 1 |
| 1 | 0 | 1 | 0 | 1 |
| 1 | 0 | 1 | 1 | 1 |
| 1 | 1 | 0 | 0 | 1 |
| 1 | 1 | 0 | 1 | 1 |
| 1 | 1 | 1 | 0 | 1 |
| 1 | 1 | 1 | 1 | 1 |

$X = A + B + C + D$

Boolean expression

(*b*) Four-input OR gate

**FIGURE 26-12 OR gate symbols and truth tables.**

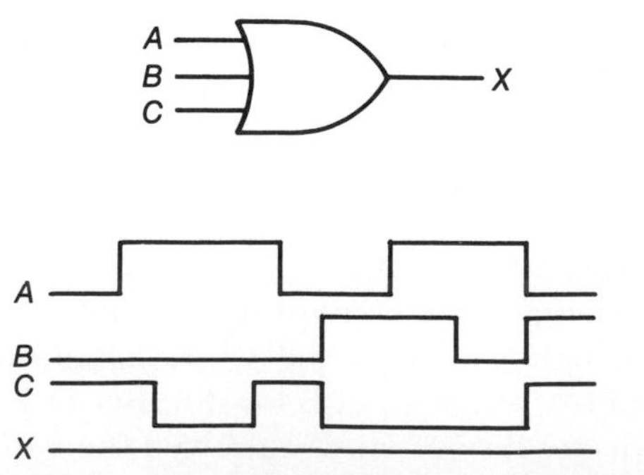


**FIGURE 26-13 OR gate response to digital signals.**

Figure 26-14 summarizes the operation of the exclusive-OR. The abbreviated boolean expression for the exclusive-OR is $X = A \oplus B$; however, the full expression, $X = \overline{A}B + A\overline{B}$, is more descriptive of the circuit. $X = \overline{A}B + A\overline{B}$ describes how the function is accomplished with NOT, AND, and OR gates as shown in Fig. 26-15. In this figure, you can see that a one on both $A$ and $B$ will result in a zero input on each AND gate. These zero inputs on the AND gates result in a zero output for the circuit.

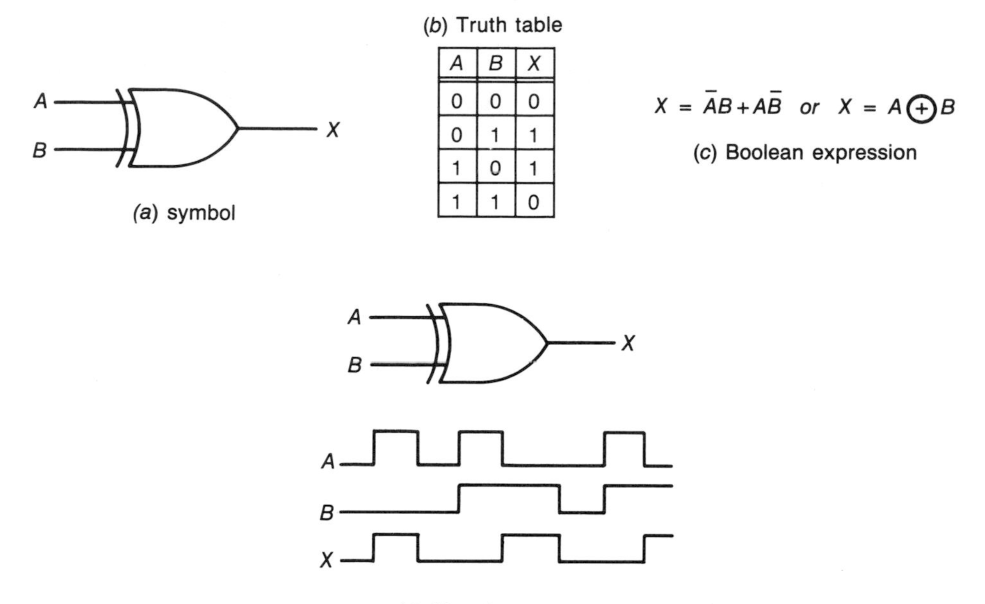


| A | B | X |
|---|---|---|
| 0 | 0 | 0 |
| 0 | 1 | 1 |
| 1 | 0 | 1 |
| 1 | 1 | 0 |

**FIGURE 26-14 Exclusive-OR (XOR) gate.**

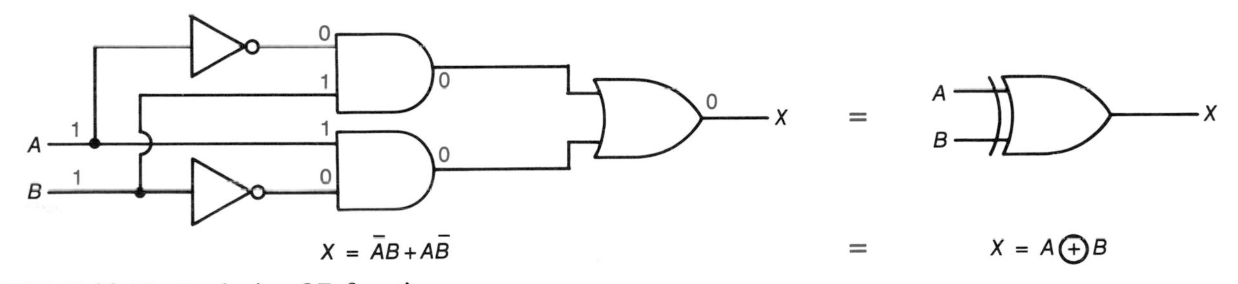


**FIGURE 26-15 Exclusive-OR function.**

## NAND, NOR and Exclusive-NOR Gates

The NAND, NOR, and exclusive-NOR (XNOR) functions are performed by adding an INVERT gate to an AND, OR, and exclusive-OR, respectively. Figure 26-16 shows that each of these combined gates has its own symbol, truth table, and boolean expression. Although the NAND gate is a "not and" gate, it is always called a "nand" gate. The same pronunciation scheme is used for the "nor" and "exclusive-nor" gates. Notice in Fig. 26-16 that the truth tables for the NOR, NAND, and XNOR gates are like the truth tables for the OR, AND, and XOR except that the zeros and ones are reversed in the output columns. The complemented output column is, of course, caused by the inverter. Also, notice in Fig. 26-16 that the complemented (inverted) output of these gates is indicated in the boolean expression by the bar over the complete output including the + or ⊕ symbol.

Figure 26-17 shows the responses of the NAND, NOR, and exclusive-NOR gates to waveform inputs. Careful study of this figure will aid one in understanding and remembering the operation of these circuits.

## Buffer-Driver Gates

A regular TTL gate, such as AND, NOR, etc., can provide a maximum of 16 mA of output current. It can drive a maximum of 10 inputs of other TTL gates. When this limitation of the regular TTL gate is too restrictive, a *buffer-driver* gate can be

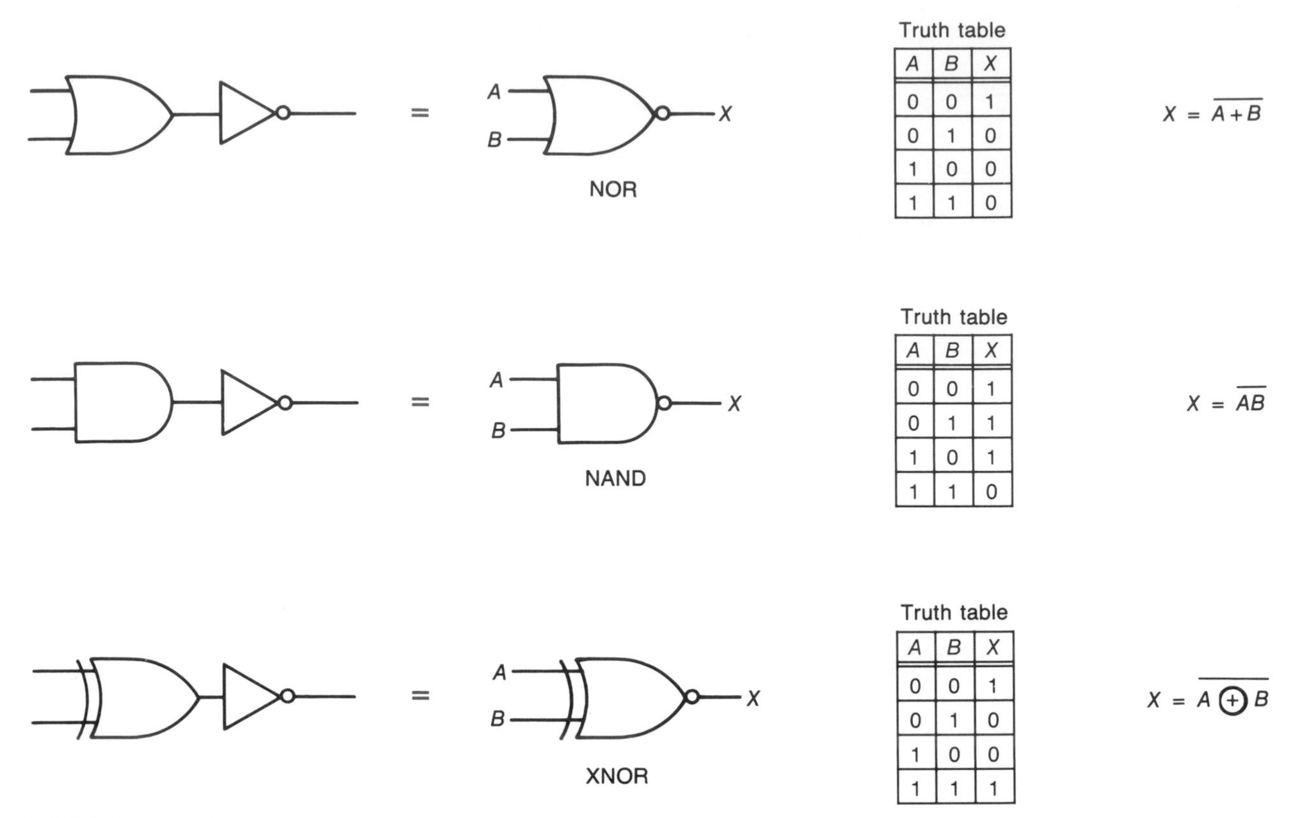


**FIGURE 26-16 NOR, NAND, and XNOR gates.**

used. Buffer-driver gates can drive up to 25 inputs and can provide output currents of up to 40 mA. Often, these gates are just called buffers instead of buffer-drivers.

Many digital circuits use noninverting buffers. The output logic level of such buffers is always the same as the input logic level. Figure 26-18 shows the symbol and logic expression for a noninverting buffer. Note that the symbol is the same as for the INVERT gate except that the circle (often called the *bubble*) is omitted.

## Tristate Gates

In complete digital systems, such as computers, a conductor may carry the digital signal for more than one circuit. Such a conductor is called a *bus*. Usually, multisignal conductors are grouped together to form a four-conductor bus, eight-conductor bus, etc. Of course, a conductor can only carry one signal at a time, so *tristate gates* have been developed to effectively connect and disconnect various subsystems to a bus.

Tristate gates are often called three-state gates. The symbols for some tristate gates are shown in Fig. 26-19. The tristate gate has a *control* or *enable* input as well as its regular input. When the enable input is high (logic 1), the gate in Fig. 26-19*a* behaves as a regular INVERT gate. However, when the enable (control) input is low (logic 0), the output of this gate is floating; it is neither high nor low. The output of the nonenabled three-state gate presents a very high impedance to the bus connected to its output; effectively, the nonenabled tristate gate is disconnected from the bus.

Notice the circle (or bubble) on the enable input of the gate in Fig. 26-19*b*. The bubble tells us that the *active input* for this gate is a logic 0 (low input). For this gate to act as an inverter, the enable input must be low.

Since the control line on the gate in Fig. 26-19*a* lacks a bubble, this gate has an *active high* control input. This is why that control had to be high for this gate to act as an inverter. Figure 26-20 shows how three tristate inverters can be used to allow three different signals to use the same bus. In this

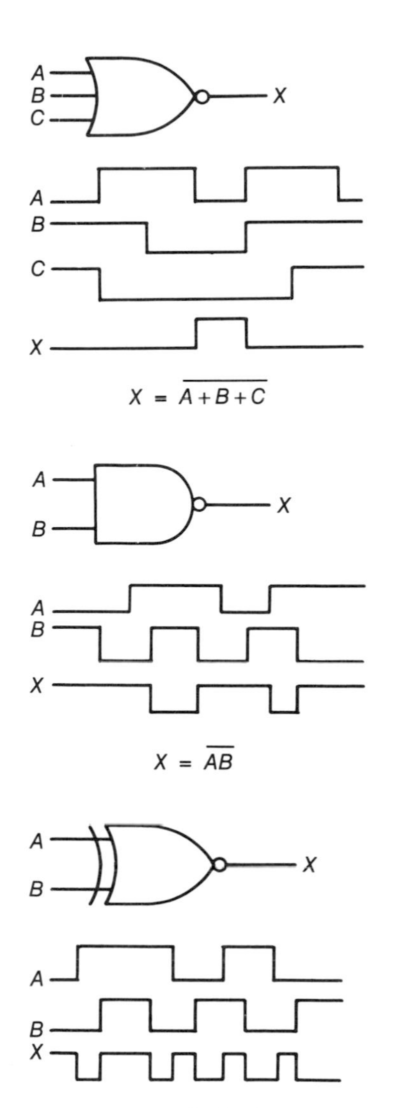

**FIGURE 26-17 Waveform responses of NOR, NAND, and XNOR gates.**

figure, you can see that signal no. 2 is presently connected to the bus. These inverters have active high enable or control inputs; thus, only the second inverter is activated.

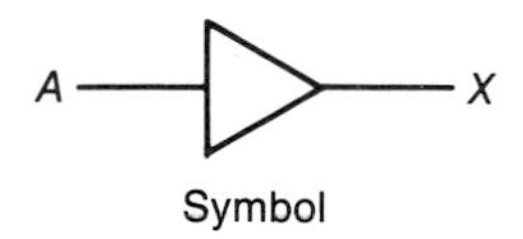


Symbol

$X = A$

Boolean expression

**FIGURE 26-18 Noninverting buffer.**

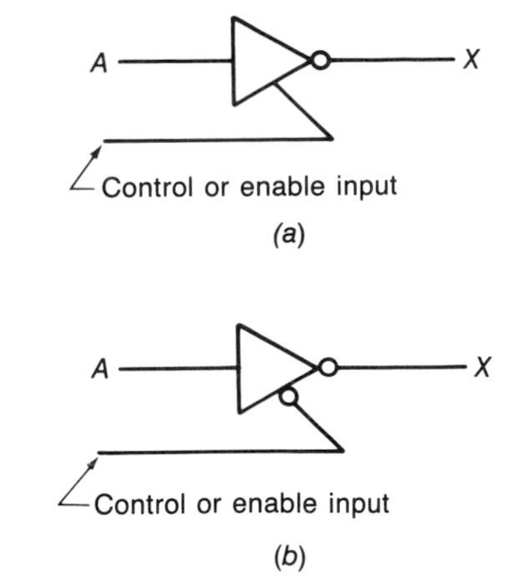


**FIGURE 26-19 Tristate inverters.**

## Structure of Logic Gates

Transistor-transistor logic gates use bipolar transistors, diodes, and resistors to perform their logic functions. Many TTL gates use a multiemitter transistor, like the one shown in Fig. 26-21, for the input stage of the gate circuit. This multiemitter transistor is not used as an amplifier. Instead, as seen in Fig. 26-21*c*, it is used as three diodes with their anodes connected together. If you remember that a multiemitter transistor is used as a group of diodes, then understanding the operation of a TTL gate is not difficult because all other transistors are operated in either cutoff or saturation. The collector-emitter (CE) voltage of a saturated transistor is typically less than 0.4 V. For a cutoff transistor, the CE voltage approaches $V_{CC}$ if no load is connected to the output of the transistor.

Figure 26-22*a* and the equivalent circuit in Fig. 26-22*b* show how the input of a TTL gate operates when a logic 1 is applied to both inputs. The junctions of the transistors in this figure only require about 0.6 V to forward-bias them because these transistors are designed to handle only a few milliamperes of current. Thus, to forward-bias the three series-connected junctions (EB of $Q_3$, EB of $Q_2$, and CB of $Q_1$) in Fig. 26-22*a* requires only +1.8 V. Since a logic 1 holds the two emitters of $Q_1$ above +2.0 V, the EB junctions of $Q_1$ are reverse-biased and no current flows in these junctions except a minute leakage current.

A complete circuit diagram of a NAND gate is shown in Fig. 26-22*c*. In this circuit, $D_1$ and $D_2$ are input protection diodes. Under normal operation they are reverse-biased when the inputs are either

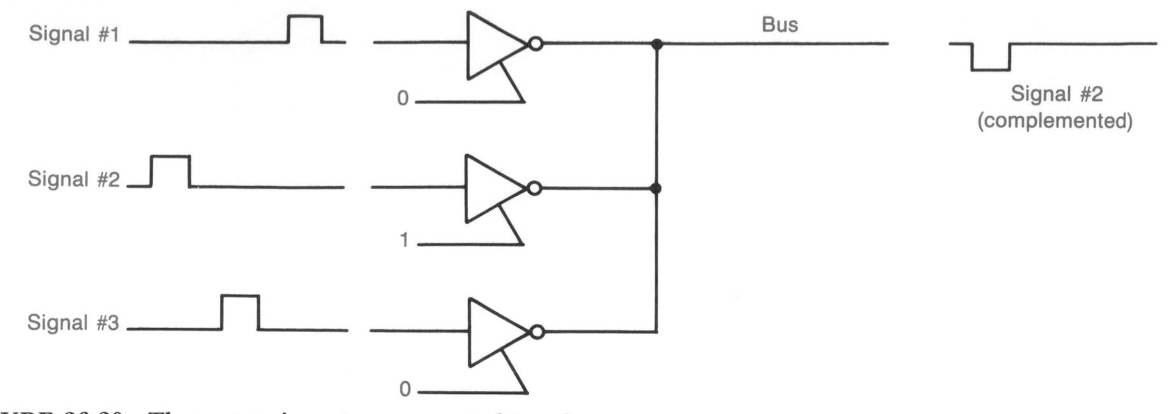


**FIGURE 26-20 Three-state inverters connected to a bus.**

logic 1 or logic 0. Thus, the only current they draw is a few picoamperes of leakage current. If the input signal ever tries to swing more than about 0.5 V negative, then $D_1$ and $D_2$ conduct and protect the EB junctions of $Q_1$ from excess reverse bias.

As shown by the voltages and current paths in Fig. 26-22*c*, logic 1s on the two inputs cause $Q_2$ and $Q_3$ to be driven into saturation. Notice in Fig. 26-22*c* that $Q_4$ is in cutoff because the difference in potential between the collectors of $Q_2$ and $Q_3$ is not sufficient to forward-bias both $D_3$ and the EB junction of $Q_4$. The collector voltage of the saturated output transistor $Q_3$ produces a logic 0 output. The output current from the collector of $Q_3$ is the input current required by the input of the gate (or gates) connected to the output of this NAND.

Figure 26-22*d* shows the current paths and key voltages produced when either (or both) input is held at a logic 0. Notice that when either (or both) EB junction is forward-biased by a logic 0 on an input lead, the base of $Q_1$ can be no more than +1.2 V. This is insufficient voltage to forward-bias three junctions; so, $Q_2$ and $Q_3$ are cut off. This allows $Q_4$ to be driven into saturation and forces the collector of $Q_3$ to be more than +2.4 V; i.e., a logic 1. Note that the load current drawn by $Q_4$ is only the leakage currents from the reverse-biased junctions of the input of the gate (or gates) connected to this NAND.

## Self-Test

**10.** Write the boolean expression for each of the following gates:
**a.** NOT **b.** AND **c.** OR **d.** NAND **e.** NOR **f.** XOR **g.** XNOR **h.** Noninverting buffer.

**11.** Identify the logic gate for each truth table shown in Fig. 26-23.

**12.** How many input rows would there be in the truth table for a five-input OR gate?

**13.** What shape is used on a logic symbol to indicate a variable is complemented or inverted?

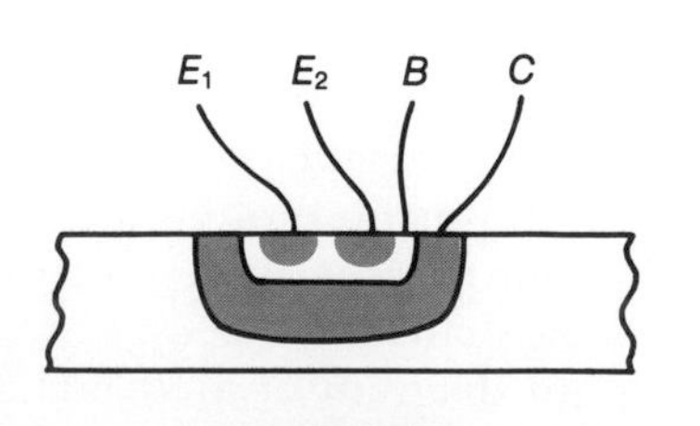


(*a*) Structure

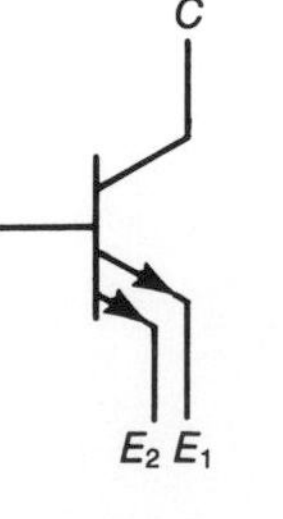


(*b*) Symbol

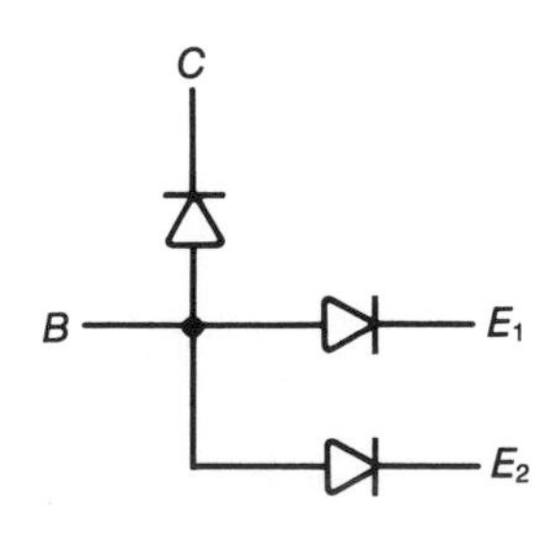


(*c*) Diode equivalent

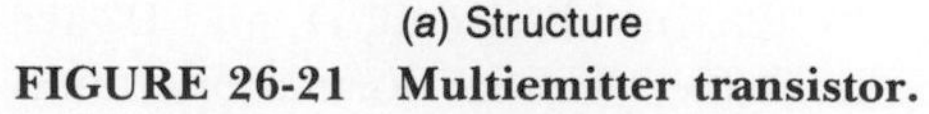

**FIGURE 26-21 Multiemitter transistor.**

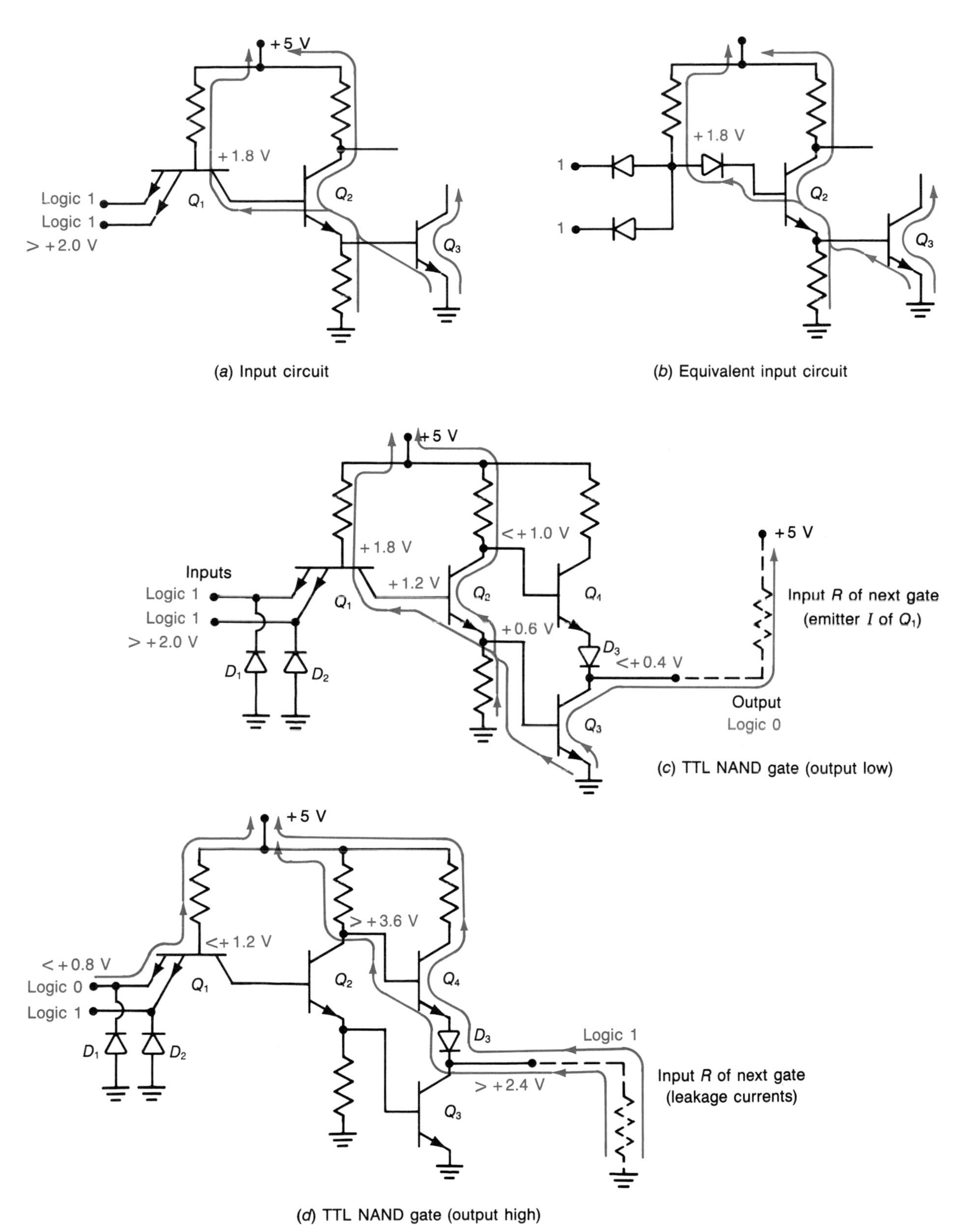


**FIGURE 26-22 Circuit diagram of a typical TTL NAND gate.**

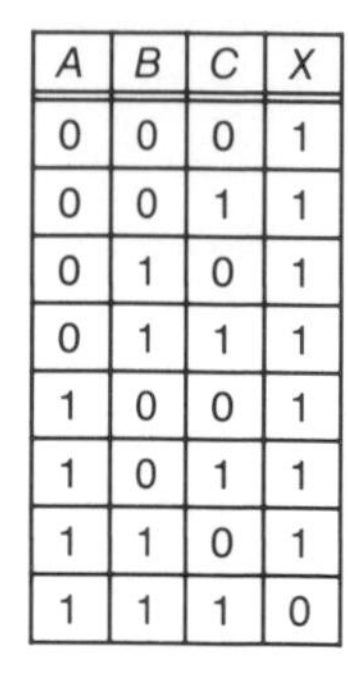

| A | B | C | X |
|---|---|---|---|
| 0 | 0 | 0 | 1 |
| 0 | 0 | 1 | 1 |
| 0 | 1 | 0 | 1 |
| 0 | 1 | 1 | 1 |
| 1 | 0 | 0 | 1 |
| 1 | 0 | 1 | 1 |
| 1 | 1 | 0 | 1 |
| 1 | 1 | 1 | 0 |

(*a*)

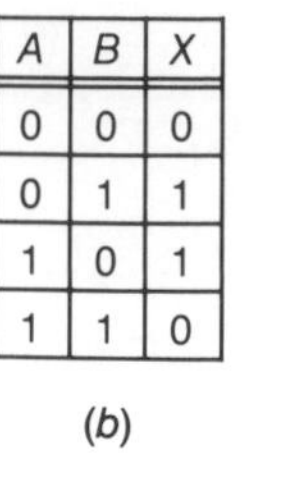

| A | B | X |
|---|---|---|
| 0 | 0 | 0 |
| 0 | 1 | 1 |
| 1 | 0 | 1 |
| 1 | 1 | 0 |

(*b*)

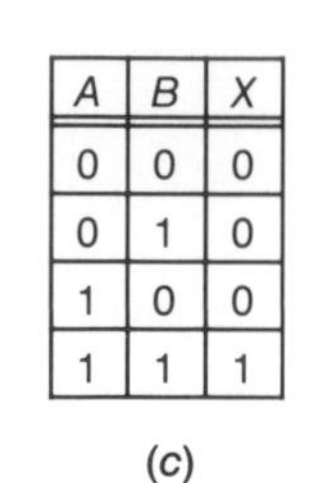

| A | B | X |
|---|---|---|
| 0 | 0 | 0 |
| 0 | 1 | 0 |
| 1 | 0 | 0 |
| 1 | 1 | 1 |

(*c*)

**FIGURE 26-23 Truth tables for Self-Test Question 11.**

**14.** What is another name for the INVERT gate?

**15.** Using words, state the meaning of $X = A + B$.

**16.** What is the sequence of the zeros and ones in the $2^2$ column of a truth table?

**17.** What is a buffer?

**18.** What type of gates can be used to connect signals to a bus?

**19.** How is an active low indicated?

**20.** A multiemitter transistor is used as a group of ______ in a TTL gate.

**21.** Transistors in TTL gates are operated in either ______ or ______.

## 26-4 COMBINATIONAL LOGIC

Some digital logic problems can be solved by connecting together two or more of the gates from Sec. 26-3. When the resulting circuit has no feedback and no memory, it is often referred to as a *combinational logic* circuit. A digital circuit has no memory if its output logic level is solely dependent on its present input logic level(s).

Figure 26-24*a* shows a combinational logic diagram for a circuit which implements the boolean expression of $X = A\overline{B}C + \overline{C}$. On this diagram the boolean expression for the output of each gate is written on the output line. Of course, the output of the first gate becomes the input of the next gate, etc. Although it may not be obvious at first glance, the circuit in Fig. 26-24*a* can be simplified to the circuit in Fig. 26-24*b*. Notice that the circuit in Fig. 26-24*b* requires only a two-input AND gate rather than a three-input AND gate. Closer examination

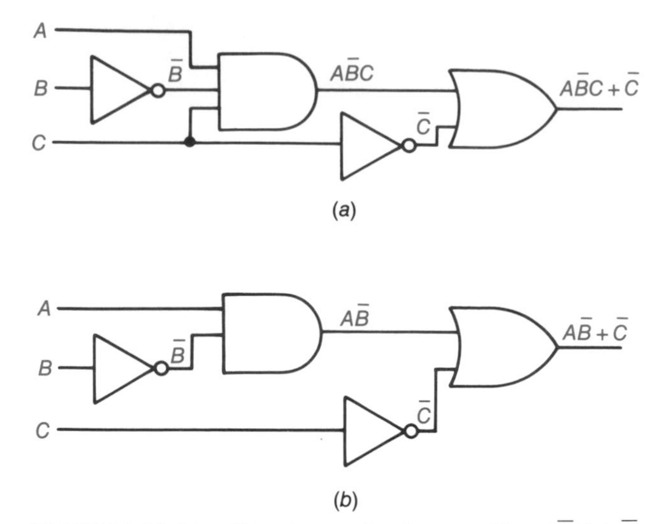


**FIGURE 26-24 Circuits to implement $X = A\overline{B}C + \overline{C}$ and $X = A\overline{B} + \overline{C}$.**

of the expression $X = A\overline{B}C + \overline{C}$ shows why the circuit in Fig. 26-24*a* can be simplified. If either the term $A\overline{B}C$ or the term $\overline{C}$ is a logic 1, then $X$ is a logic 1 also. If $\overline{C}$ is not a logic 1, than $C$ must be a logic 1. Therefore, if $A$ is a logic 1 and $\overline{B}$ is a logic 1, then it does not matter whether $C$ is a logic 0 or a logic 1. If $C$ is a logic 1, then the $A\overline{B}C$ term will produce a logic 1 output. If $C$ is a logic 0, then $\overline{C}$ will be a logic 1 and the $\overline{C}$ term will produce a logic 1 output. Thus, the boolean expression $X = A\overline{B}C + \overline{C}$ reduces to $X = A\overline{B} + \overline{C}$.

The expanded truth table in Fig. 26-25 provides proof that $X = A\overline{B}C + \overline{C}$ is equal to $X = A\overline{B} + \overline{C}$. The first three columns of this truth table list all possible combinations of the three variables $A$, $B$, and $C$. The next four columns show all possible outputs from the two INVERT gates and the two AND gates in Fig. 26-24. Finally, the last two columns are the possible outputs for the two boolean expressions. Notice that the two expressions always provide identical outputs for any given combination of the $A$, $B$, and $C$ variables.

Further explanation of the truth table in Fig. 26-25 may help you in developing expanded truth tables for boolean expressions. The fourth and fifth columns are just complements of the second and third columns. The sixth column is the result of ANDing the first, third, and fourth column. You can see that only in the sixth row do all three of these columns have a one entry. Thus, only in the sixth row does the $A\overline{B}C$ column have a one entry. The seventh column is the result of ANDing

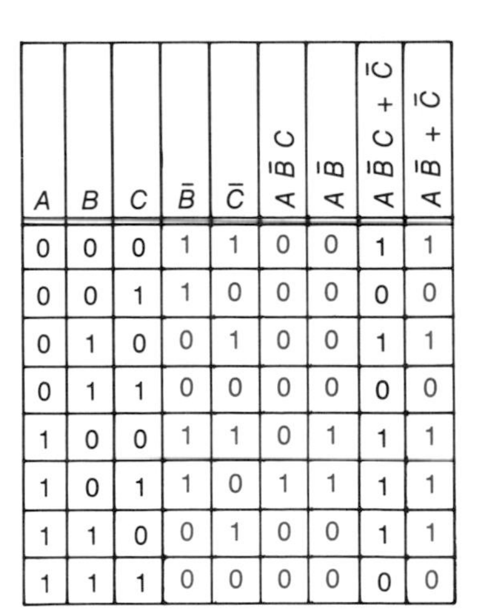

| $A$ | $B$ | $C$ | $\bar{B}$ | $\bar{C}$ | $A\bar{B}C$ | $A\bar{B}$ | $A\bar{B}C + \bar{C}$ | $A\bar{B} + \bar{C}$ |
|---|---|---|---|---|---|---|---|---|
| 0 | 0 | 0 | 1 | 1 | 0 | 0 | 1 | 1 |
| 0 | 0 | 1 | 1 | 0 | 0 | 0 | 0 | 0 |
| 0 | 1 | 0 | 0 | 1 | 0 | 0 | 1 | 1 |
| 0 | 1 | 1 | 0 | 0 | 0 | 0 | 0 | 0 |
| 1 | 0 | 0 | 1 | 1 | 0 | 1 | 1 | 1 |
| 1 | 0 | 1 | 1 | 0 | 1 | 1 | 1 | 1 |
| 1 | 1 | 0 | 0 | 1 | 0 | 0 | 1 | 1 |
| 1 | 1 | 1 | 0 | 0 | 0 | 0 | 0 | 0 |

**FIGURE 26-25 Expanded truth table which shows that $X = A\bar{B}C + \bar{C}$ is equivalent to $X = A\bar{B} + \bar{C}$.**

the first and fourth column. The eighth column ($X = A\bar{B}C + \bar{C}$) is produced by ORing columns five and six. Finally, the last column is the result of ORing columns five and seven.

The reduction, or simplification, of a boolean expression is usually not as simple as the one above. To handle more difficult cases, a more structured procedure is needed. To be proficient at reducing digital circuits and boolean expressions, one has to know the rules of boolean algebra as well as Karnaugh mapping techniques. These topics are covered in textbooks which are devoted to digital electronics. They are not included in this introductory chapter.

To become more familiar with the relationship between logic circuits and boolean expressions, let us write the expression for the circuit shown in Fig. 26-26*a*. In this figure, the gates have been labeled G1, G2, etc., just for easy identification. Inspection of the circuit shows that the outputs of G1 and G2 are ANDed together by G3. The output of G1, as shown in Fig. 26-26*b*, is $AB$, and the output of G2 is $\overline{C + D}$. The final output $X$ is, therefore, $X = (AB)(\overline{C + D})$. Since the bar across the $C + D$ ties these two variables together as one term of the expression, the parentheses around this term are not required. However, they can be left in the expression if it helps to visualize the circuit. The parentheses around the $AB$ term can also be removed without changing the expression. However, if these parentheses are removed, then the digital circuit could use a three-input AND gate as shown

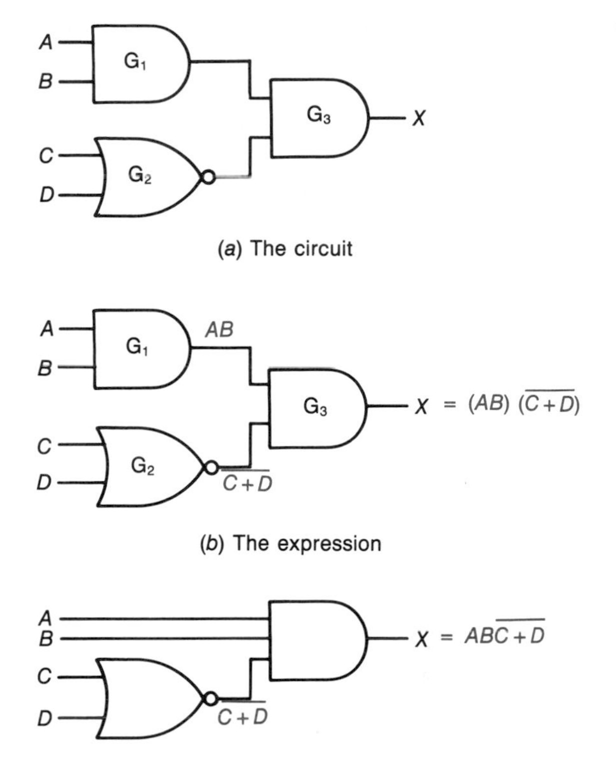


**FIGURE 26-26 Developing boolean expressions.**

in Fig. 26-26*c*. If one does not care how the expression is implemented, then the parentheses around the $AB$ can be removed.

## EXAMPLE 26-3

Write the boolean expression for the digital circuit shown in Fig. 26-27.

### Solution

The output of G2, which is $A + \bar{B}$, is NANDed with $C$. This yields $X = \overline{(A + \bar{B})C}$. Note that the parentheses around $A + \bar{B}$ is required; otherwise, the expression would indicate that $A$ is NORed with $\bar{B}C$.

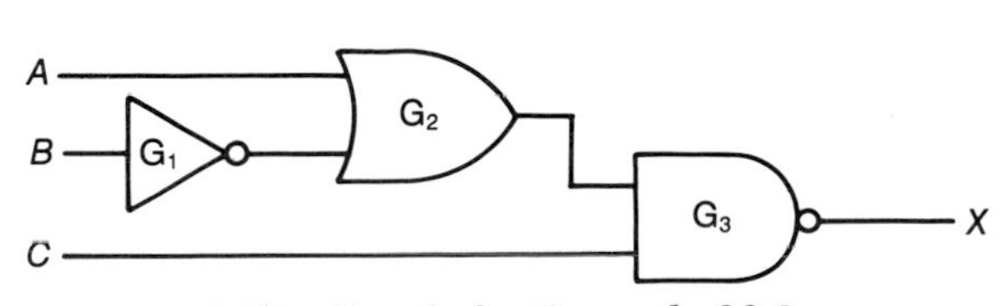


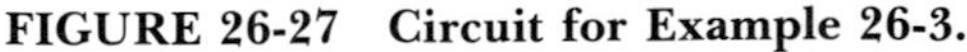

**FIGURE 26-27 Circuit for Example 26-3.**

Now let us develop a logic circuit which will implement the boolean expression $X = \overline{(A \oplus B)C + D}$. The expression tells us that $A$ is XORed with $B$; then, the output of the XOR is ANDed with $C$; and finally, the output of the AND is NORed with $D$. The resulting circuit is shown in Fig. 26-28.

## BCD-to-Decimal Decoder

Many digital circuits, including some combinational logic circuits, are designed to handle *binary-coded decimal* (BCD) data. Binary-coded decimal is a code which uses a 4-bit binary number to represent each digit in the decimal system. This is a convenient code for taking decimal information from a device like a calculator keyboard and converting it to binary information for processing by digital circuits. It is equally useful for converting the binary output from digital circuits to decimal information for displaying on an output device such as a 7-segment light-emitting diode (LED) or liquid crystal display (LCD).

The BCD code is shown in Fig. 26-29*a*. Each of the decimal digits (0 through 9) is represented by its binary equivalent. Since it takes 4 binary digits to represent decimal digits 8 and 9, each decimal digit is assigned a 4-bit binary number even though the binary equivalent may require fewer than four binary places. This way, circuits which use BCD always handle the string of binary bits in four-place groups. Groups of binary bits are often referred to as *words*. In the case of BCD, the *word length* is 4 bits. In computer circuits, 8-, 16-, and 32-bit words are used.

Figure 26-29*b* shows the string of binary digits (three 4-bit words) which represent the decimal number 529. It should be noted that BCD is not a true binary; it is only a code. If the string of binary digits in Fig. 26-29*b* were treated as a binary number, it would be equal to 1321, base 10. To avoid confusion between BCD and true binary, a BCD string is often separated into groups of 4 binary bits, or a subscript of BCD is sometimes attached to the string (see Fig. 26-29*c*). Finally, it should be noted that the four-place binary digits shown in Fig. 26-29*d* are illegal (invalid) in the BCD code.

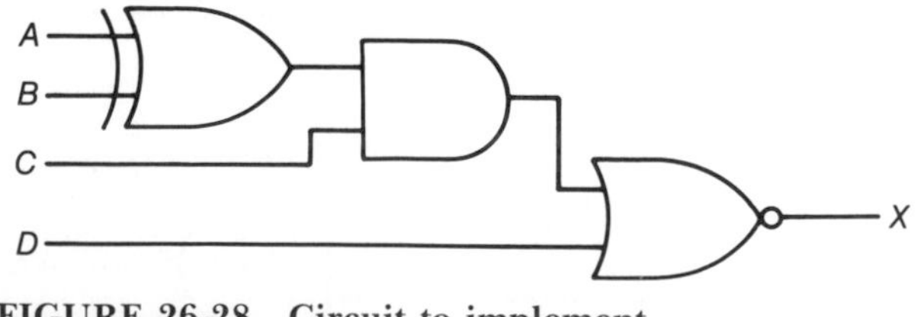


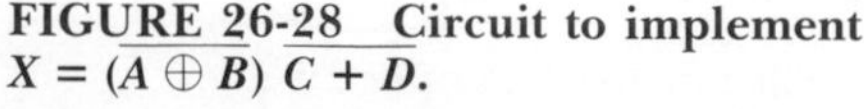

**FIGURE 26-28 Circuit to implement $X = \overline{(A \oplus B)\,C + D}$.**

| BCD code | Decimal digit |
|---|---|
| 0000 | 0 |
| 0001 | 1 |
| 0010 | 2 |
| 0011 | 3 |
| 0100 | 4 |
| 0101 | 5 |
| 0110 | 6 |
| 0111 | 7 |
| 1000 | 8 |
| 1001 | 9 |

(*a*) The code

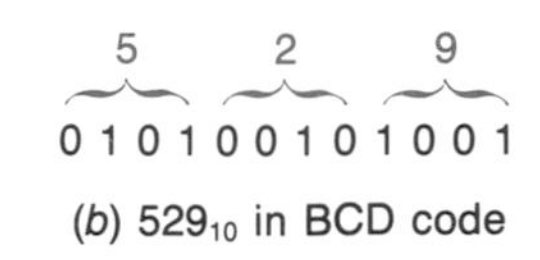


(*b*) $529_{10}$ in BCD code

0 1 0 0 0 0 0 0 0 0 1 0 or $010000000010_{BCD}$

(*c*) Indicating a binary string is BCD

1 0 1 0 1 0 1 1 1 1 0 0 1 1 0 1 1 1 1 0 1 1 1 1

(*d*) Groups of bits which are not allowed in the BCD code

**FIGURE 26-29 BCD (binary-coded decimal) code.**

### EXAMPLE 26-4

What decimal number is represented by the BCD string in Fig. 26-29*c*?

**Solution**

Finding the decimal equivalent of each group of 4 binary bits yields $402_{10}$.

Sometimes the leading zero (or zeros) in the MSB (most significant bit) of a BCD code is omitted. Thus, in decoding BCD it is wise to start with the LSB and decode in groups of 4 binary bits. For example, 39 base 10 in BCD might be written $111001_{BCD}$. If one started decoding with the MSB, then the first 4 bits would result in an invalid code and only 2 bits would be left for the last word. Obviously, the two leading zeros have been omitted from this BCD string.

Figure 26-30*a* shows how a four-input NAND gate can be used to decode the BCD word representing decimal 8. Note that the symbol *D* is used to indicate the MSB of the code word. Only when the BCD code word 1000 is inputted will the output of this circuit go low. Thus, the output of the circuit is an active low. Figure 26-30*b* shows how this simple decoder could be used to activate a LED when the BCD code word for $5_{10}$ is received on the inputs. (A LED emits light only when its forward bias is about 1.5 V.) The resistor in this circuit protects the LED and the NAND gate from drawing too much current.

The logic diagram for a complete BCD-to-decimal decoder is given in Fig. 26-31*a*. If you study this diagram, you will see that it is composed of 10 circuits like the one in Fig. 26-30 except that four inverters are shared by all of the NAND gates. The output of the inverter receiving the MSB of the code word (D line) drives one input line for 8 of the 10 NAND gates.

Notice that only one of the 10 output lines in Fig. 26-31*a* can be active (low) at any one time. Therefore, this decoder circuit is sometimes referred to as a *1-of-10 decoder*. This circuit is also referred to as a *4-line-to-10-line decoder* because it has 4 input lines and 10 output lines. The zeros and ones shown in Fig. 26-31*a* are the logic levels when the circuit is receiving the BCD code for $6_{10}$.

The logic symbol for a BCD-to-decimal decoder is drawn in Fig. 26-31*b*. The bubbles on the output lines indicate that an output line is low only when the code that corresponds to that particular decimal number is received. The logic symbol, rather than the complete logic diagram, is used in the circuit diagrams of digital systems.

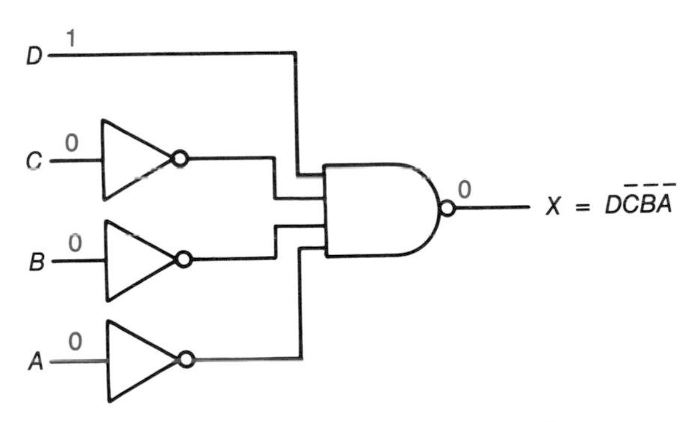


(*a*) Decoding the BCD code word for $8_{10}$

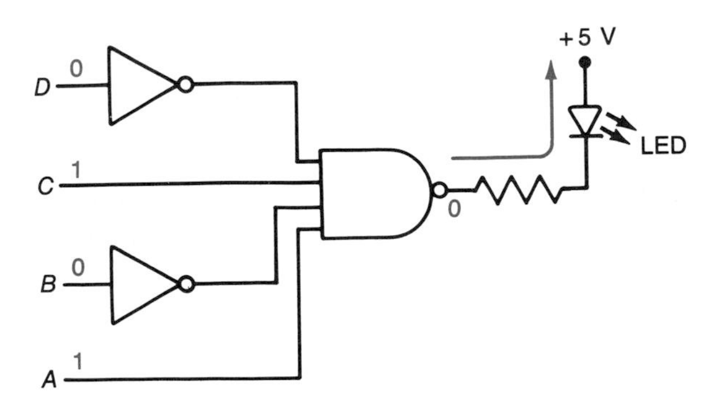


(*b*) LED used to indicate when BCD code word for $5_{10}$ is received

**FIGURE 26-30 Decoding a single BCD code word.**

## BCD-to-Seven-Segment Decoder

A seven-segment LED is shown in Fig. 26-32*a*. Each segment (a through g) is a LED that can be turned off or on independently of the other segments. As shown in the logic diagram of Fig. 26-32*b*, the anodes of all seven segments are tied together so that only one +5-V line is needed to operate the seven segments. This arrangement is called a *common anode* and is used when the LED is to be driven by an active low decoder. A common cathode LED is used when the decoder output is an active high.

By selectively turning on various segments of the seven-segment LED, any of the base 10 digits (0 through 9) can be displayed. The decoder in Fig. 26-32*b* is receiving the BCD code word for $7_{10}$. (*Note: D* is the MSB of the code word.) The input forces the a, b, and c output lines low and turns on the a, b, and c LED segments. As shown in Fig. 26-32*c*, the result of turning on these segments is that the LED displays a $7_{10}$. The resistors in this decoder circuit are included to limit the decoder and LED segment currents to a safe value.

## Multiplexers

A *multiplexer* is another type of combinational digital circuit that is readily available in an IC package. Multiplexing allows a single conductor (bus line) to alternately carry signals from a variety of signal sources. Multiplexers are also called *data selectors*. They operate like an electrically controlled rotary

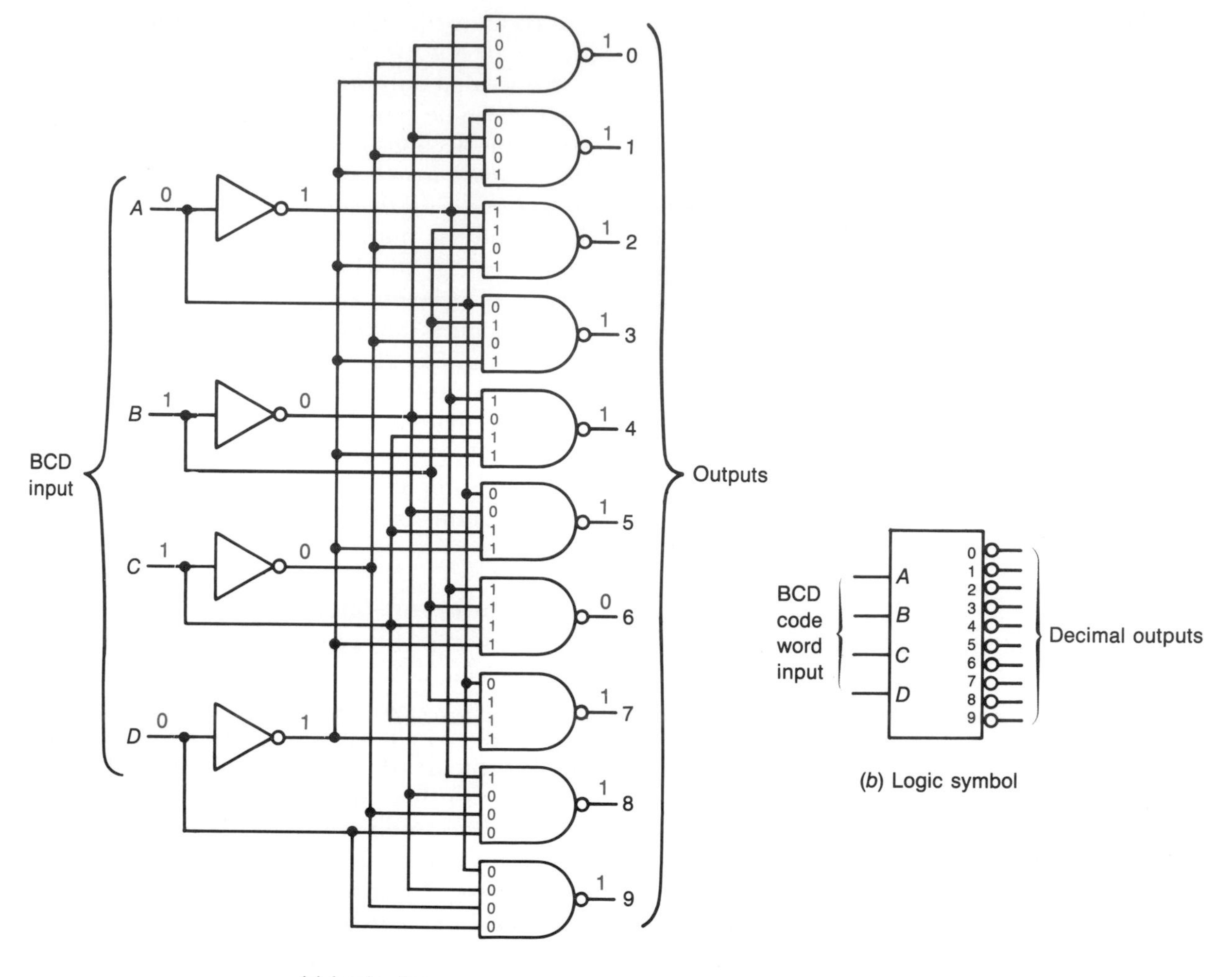


**FIGURE 26-31 BCD-to-decimal decoder.**

switch. The output line is like a pole of a rotary switch, and the input lines are like the positions of a rotary switch. Data selectors are available with as many as 16 input lines.

A complete logic diagram and the logic symbol for a four-line-to-one-line multiplexer is illustrated in Fig. 26-33*a*. Which input line is selected is determined by the 2-bit binary number (or word) applied to the two select lines. Remember that there are $2^2 = 4$ possible combinations available with 2 bits.

The *enable*, or *strobe*, line in Fig. 26-33*a* is a type of on-off control for the entire circuit. Unless this line has a logic 0 on it, no input data from any input line can get through to the output. You can see that if the enable inverter puts a zero on one input of each AND gate, then the output of each AND gate will be zero regardless of the other inputs. Look at the inverter symbol on the enable line and notice that the bubble is drawn on the input side of the inverter. This is an alternate symbol for an inverter. It is used when one wants to indicate an active-low input. Notice that the logic symbol in Fig. 26-33*b* also has a bubble on the enable E input to indicate that the circuit is enabled by a logic 0.

The logic 0s and logic 1s on the diagram in Fig. 26-33*a* show that a binary 01 on the select lines allows the signal applied to $I_2$ to pass through $G_2$ and $G_5$ and become the output signal. All other input signals are blocked by one or more logic zero(s) on the inputs of the other three AND gates.

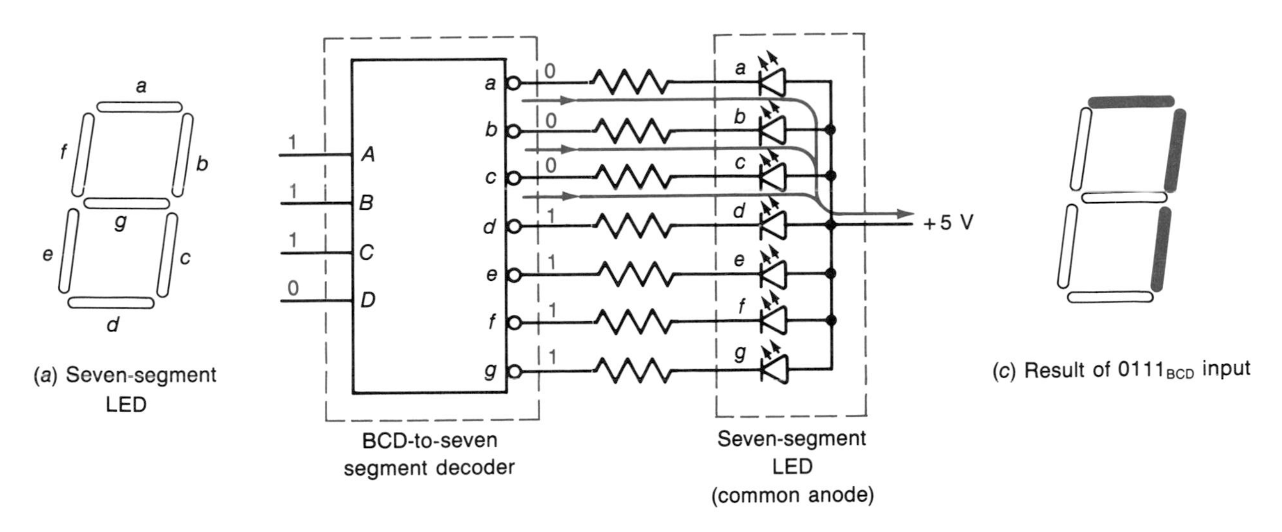


FIGURE 26-32 BCD-to-seven-segment decoder.

## Demultiplexers

A *demultiplexer* is a data-routing circuit. A single input signal can be routed to any one of the output lines. Demultiplexers can have up to 16 output lines.

Notice the logic symbol $G_5$ connected to the data and enable lines in Fig. 26-34*a*. This is an alternate symbol for a NOR gate. It is often referred to as the *dual* of a NOR gate. Sometimes it is called a negative-AND. Regardless of what it is called, it

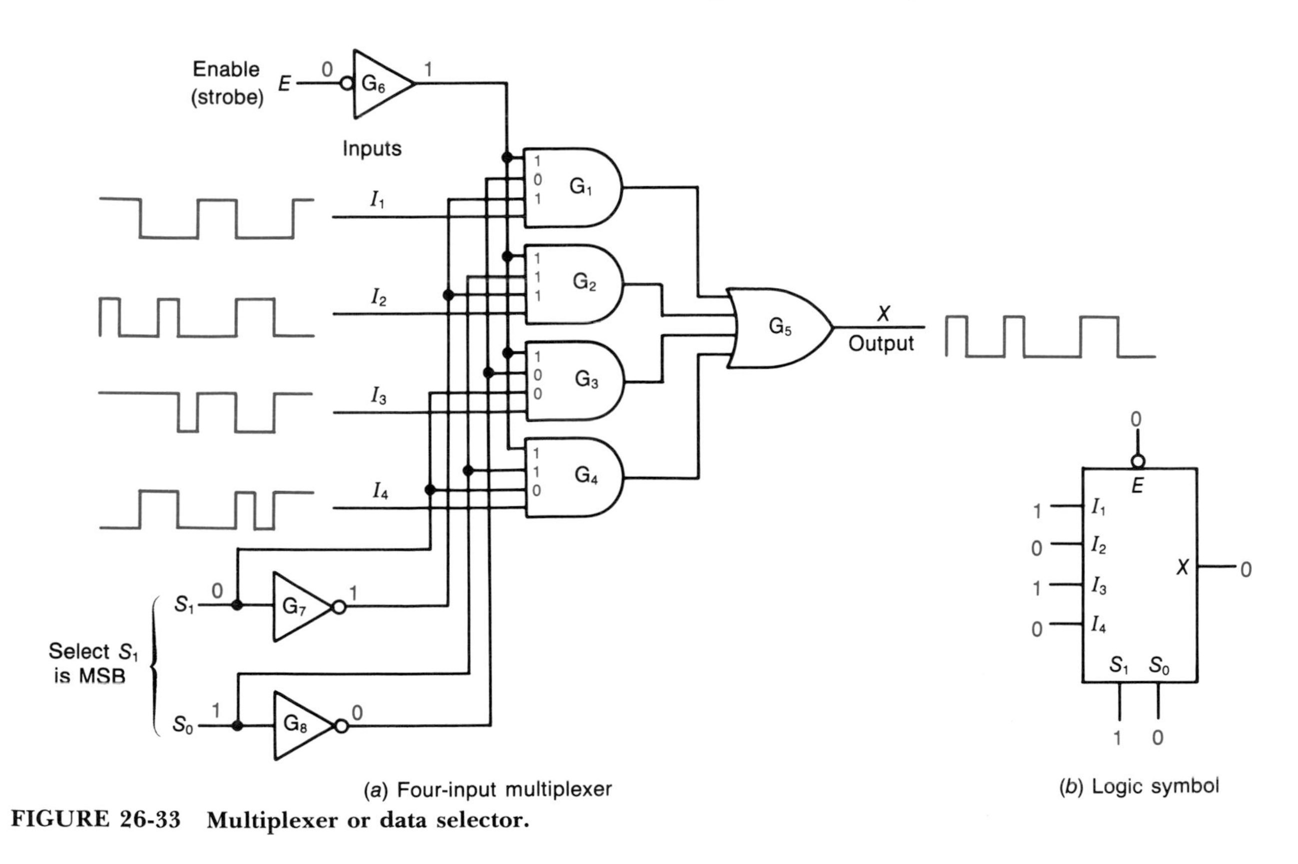


FIGURE 26-33 Multiplexer or data selector.

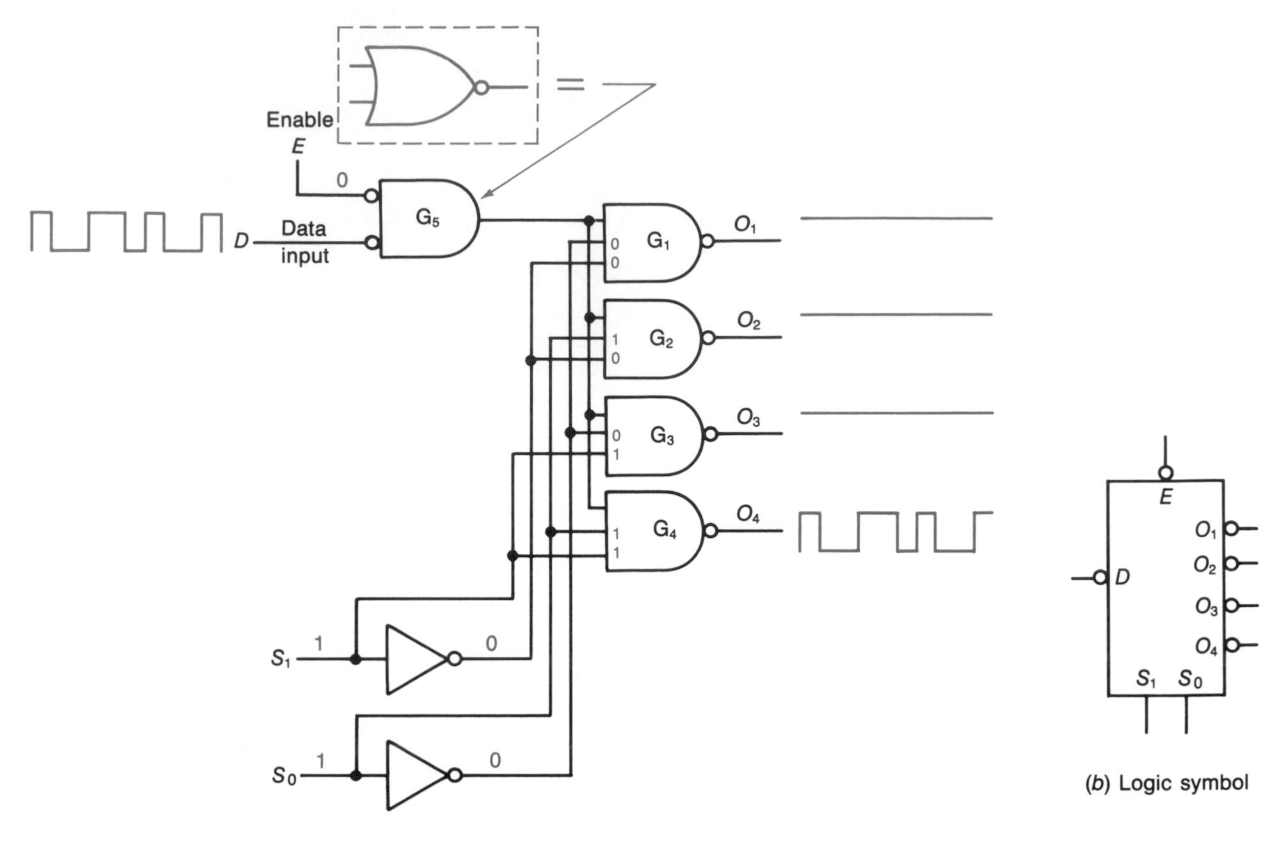


**FIGURE 26-34 Demultiplexer.**

performs exactly the same function as a NOR gate. It is used to indicate that the inputs are active low and that the output is active high. This symbol tells the observer that the output will be high only when both inputs are low. The NOR symbol also tells the observer the same thing *if* the observer remembers the truth table for the NOR function.

The dual, or alternate, symbol for an OR, NOR, AND, or NAND can be found by following these three rules:

1. Change the OR symbol to an AND symbol or vice versa.
2. If an input or output line has a bubble, remove it.
3. If an input or output line has no bubble, add a bubble.

Apply these three rules to the negative-AND in Fig. 26-34*a* and you will find its dual is the standard NOR gate.

Now we can see how the circuit in Fig. 26-34*a* operates. The $11_2$ on the select lines $S_1$ and $S_0$ puts a zero on one, or more, inputs of all but one of the NAND gates $G_4$. With a logic 0 on the enable line, the output of $G_5$ (the negative-AND) goes high whenever the input signal goes low. This high out of $G_5$ causes the output of $G_4$ to be low. Thus, when the input signal (data) is logic 0, the output of $G_4$ is logic 0. Whenever the input signal is logic 1, the output of $G_5$ is logic 0 and $G_4$ produces a high output. Note that the output is in phase with the input because both $G_5$ and $G_4$ invert the input signal. Also notice in Fig. 26-34*a* that the outputs of all of the deselected gates remain at logic 1.

A demultiplexer can also be used as a decoder. For example, if both inputs to $G_5$ in Fig. 26-34*a* were continuously held at logic 0, then the circuit would decode a 2-bit word applied to the select lines into its base 10 equivalent.

If a decoder circuit has an enable line, then it can be used as a demultiplexer. For example, if each of the 10 NAND gates in Fig. 26-31 had a fifth input line connected to a common enable line, then it could be used as a 1-line-to-10-line demultiplexer. The four BCD input lines would become the select lines and the enable line would become the data input line.

## Number or Word Comparators

In digital systems, especially computers, it is often necessary to compare two binary numbers to determine whether they are equal. A 4-bit *comparator* is shown in Fig. 26-35. This circuit could be expanded to handle longer words (numbers) by adding more XNORs and more inputs to the AND gate. With two equal binary numbers applied to the inputs, as in Fig. 26-35*a*, the output is logic 1 indicating that the *A* number is equal to the *B* number. Visualize changing a digit in either the *A* number or the *B* number and the output of one of the XNOR gates will go to a logic 0. Of course, this would cause the output of the AND to also go to a logic 0, which would indicate that *A* is not equal to *B*.

The logic symbol for a more sophisticated comparator is shown in Fig. 26-36. This comparator indicates whether *A* is larger or smaller than *B* when the two numbers are not equal. Such circuits, and the information they provide, are part of the circuitry that allows a computer to perform many complex decision-making tasks.

## Concluding Remarks

This section on combinational logic has introduced you to some of the basic ideas of this type of digital circuitry. A number of codes besides BCD are used in digital circuitry. Many combinational circuits are available for coding base 10 data into these codes and for decoding these codes back to base ten.

A number of other specific circuits which use combinational logic, such as parity checkers and priority encoders, have not been discussed. Readers interested in these circuits can refer to books devoted only to digital circuits. An important class of combinational logic circuits, called *arithmetic logic circuits*, is presented in Sec. 26-6 of this chapter.

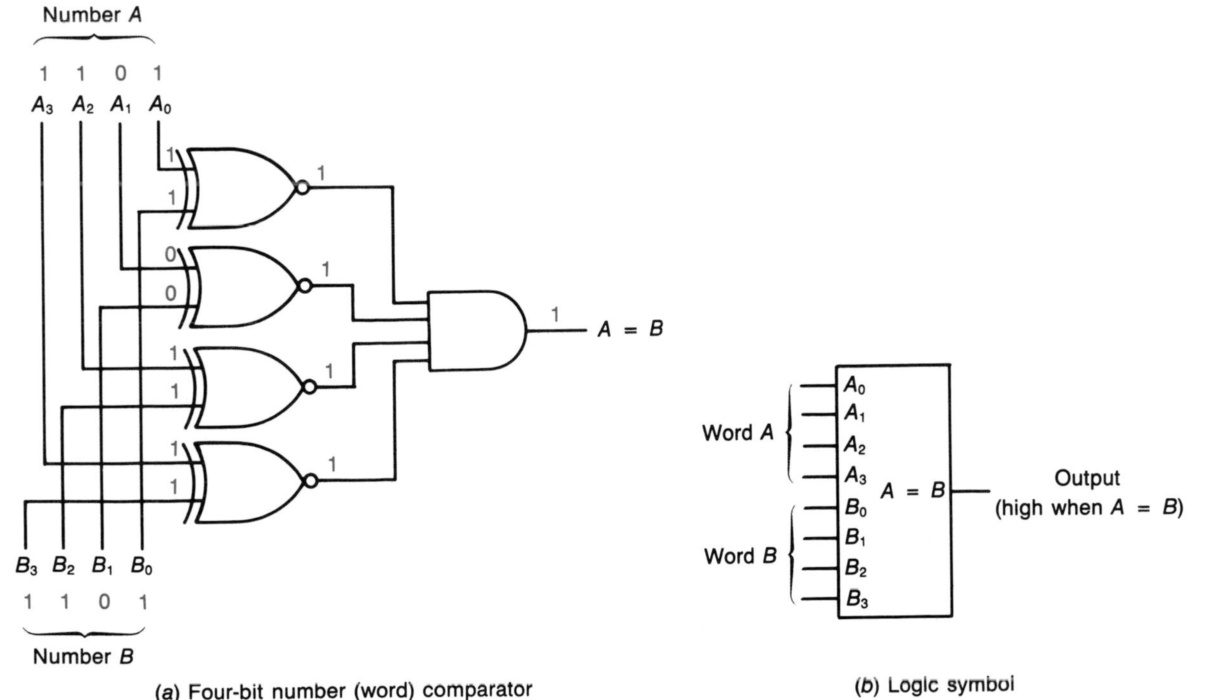


(*a*) Four-bit number (word) comparator

(*b*) Logic symbol

**FIGURE 26-35 Four-bit comparator with one output.**

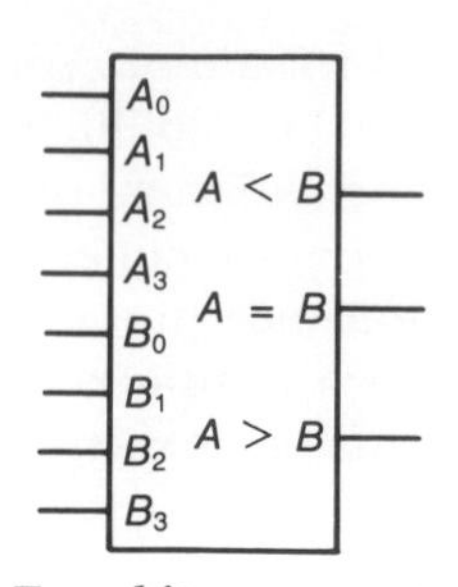


**FIGURE 26-36 Four-bit comparator with three outputs.**

## Self-Test

**22.** True or false? Combinational logic has both memory and feedback.

**23.** Draw logic diagrams which will implement each of the following boolean expressions (do not simplify):

**a.** $X = ABC + \overline{D}$ **b.** $Y = (\overline{E}G + B)\ H$

**c.** $W = \overline{\overline{AB} + C}$ **d.** $Z = \overline{A \oplus BC}$

**24.** Write the boolean expressions for the logic diagrams shown in Fig. 26-37. Do not reduce or simplify the expression.

**25.** What is BCD the abbreviation for?

**26.** Write the BCD code for 829.

**27.** What is the decimal number for 10000111000 BCD?

**28.** The code word 0011 is applied to a BCD-to-seven-segment decoder. Which segments of the LED will be emitting light?

**29.** The input to the select lines of Fig. 26-33*a* are changed to $S_1 = 0$ and $S_0 = 0$. Which signal will appear at the output?

**30.** Refer to Fig. 26-34*a*. The following logic levels are applied: $S_1 = 0$, $S_0 = 0$, D = 1, E = 1. What will be the logic level for $0_1$, $0_2$, $0_3$, and $0_4$?

**31.** Refer to Fig. 26-36. If word *A* is 1011 and word *B* is 1101, what logic level will each output have?

## 26-5 SEQUENTIAL LOGIC

*Sequential logic* circuits have both feedback and memory. The output of a sequential logic circuit depends on both past and present input/output conditions. Some of the major types of sequential logic circuits that we will deal with in this section are flip-flops, registers, and counters-dividers. More complex memory circuits and systems are dealt with in Sec. 26-7 of this chapter.

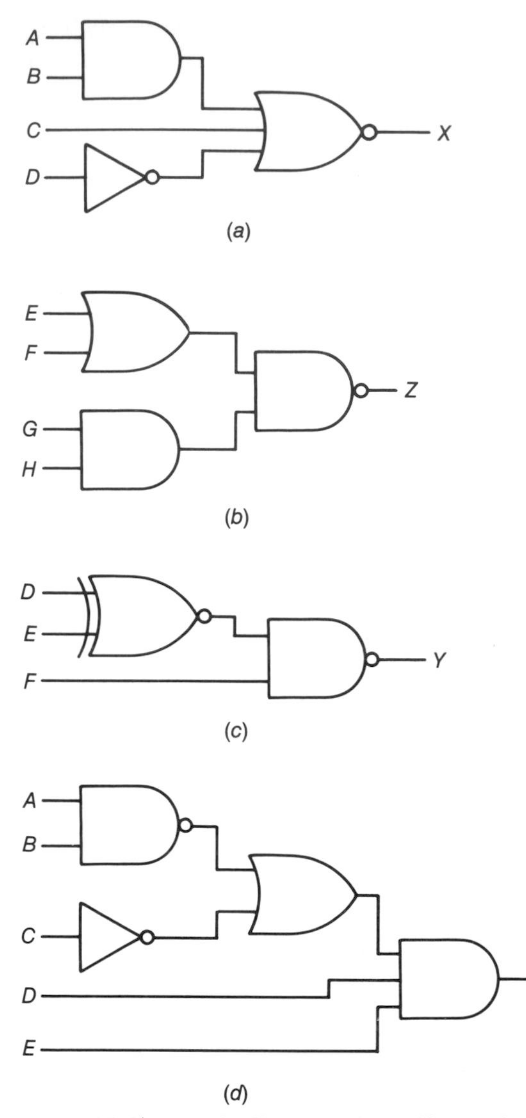


**FIGURE 26-37 Logic diagrams for Self-Test Question 24.**

### Flip-Flops

A *flip-flop* is a sequential logic circuit which can store a single bit of information. Two or more FFs can be connected together to store a multibit word such as a BCD code word discussed in Sec. 26-4. Flip-flops are also used to make counters and dividers.

An FF has two or more inputs and two outputs. The two outputs are always complements of each other.

*SR Flip-Flop* The logic diagram, logic symbol, and truth table for a *set-reset* (SR) *flip-flop* (FF) is shown in Fig. 26-38. This is the simplest form of an FF. It has feedback from the two outputs (Q and $\overline{Q}$) to the two inputs (S and R). This feedback provides the FF with memory.

As shown in Fig. 26-38, an FF has two outputs, usually labeled Q and $\overline{Q}$, which are complements of each other. When multiple FFs are contained in an IC, sometimes only one output is connected to a pin on the IC package. By custom, the "output of an FF" is considered to be the output level of the Q output unless the output of $\overline{Q}$ is specifically specified. Thus, if one states that "the output of the FF in Fig. 26-38 is high," it is understood that Q is logic 1 and $\overline{Q}$ is logic 0.

The terms *set* and *reset* have more general meanings than just the input lines of an SR FF. When an FF (any FF) is set, the Q output is logic 1. When an FF is reset, the Q output is a logic 0. Of course the $\overline{Q}$ output will be the opposite logic state in both cases. Unfortunately, the term preset is used interchangeably with the term set, especially with more advanced types of FFs. The term reset is often replaced with the term *clear*. Remember, setting or presetting an FF causes the output Q to be a logic 1, while resetting or clearing causes the output to be a logic 0.

Referring to the logic levels in Fig. 26-38 will help in understanding how the SR FF operates. When a logic 1 is applied to input S, the output of $G_2$ is a 0 regardless of the logic level on the other input line of $G_2$. The 0 output of $G_2$ is fed back to one input of $G_1$. Now, if the R input to $G_1$ is also a logic 0, the output of $G_1$ is a logic 1. Thus, when the set input is high, the Q output is also high. The logic symbol in Fig. 26-38*b* shows this same relationship because neither the S line nor the Q line has a bubble to indicate either an active-low input or output.

Notice in the truth table of Fig. 26-38*c* that the S = 1, R = 1 row drives both outputs to a logic 0. This is an invalid output for an FF. Thus, a 1,1 input is not allowed for this FF. An input which results in an invalid output is an illegal, prohibited, disallowed, or invalid input. The resulting outputs can also be called invalid, illegal, or prohibited.

Refer back to Fig. 26-38*a* and visualize what will happen to the other logic levels when the set input is changed to logic 0. Since one input to $G_2$ is still a 1, the output of $G_2$ will remain a 0. Thus, both inputs to $G_1$ remain 0, and the Q output remains 1. This is why the truth table in Fig. 26-38*b* lists "no change" for the outputs when R and S are both logic 0. Notice from the truth table that when the input changes from either 0, 1, or 1,0 to 0,0, the output does not change. The circuit "remembers" what the output had been with the previous inputs and retains that output. Thus, an FF has memory.

The NOR gates in Fig. 26-38 can be replaced with NAND gates. This results in a $\overline{SR}$ FF. The logic symbol for such an FF has bubbles on the $\overline{S}$ and $\overline{R}$ input lines to indicate that a logic 0 on the $\overline{S}$ line will cause the Q output to be a logic 1. With $\overline{SR}$ FF, the disallowed input combination is $\overline{S}$ = 0, $\overline{R}$ = 0, and the no-change input is $\overline{S}$ = 1 and $\overline{R}$ = 1. You can understand why the disallowed input is $\overline{S}$ = 0, $\overline{R}$ = 0 if you remember that a 0 on any input of a NAND gate results in a 1 output. Thus, when both $\overline{S}$ and $\overline{R}$ are 0, Q and $\overline{Q}$ are both a logic 1. This, of course, is an invalid condition because there is no way to determine whether the

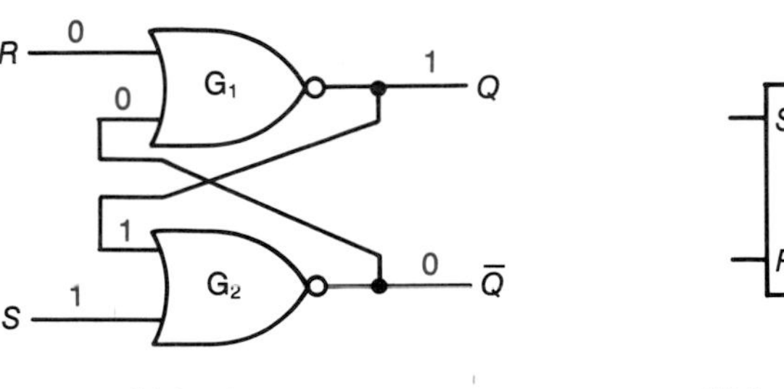


(*a*) Logic diagram

S Q
R Q̄

(*b*) Logic symbol

| S | R | Q | $\overline{Q}$ | Remarks |
|---|---|---|---|---|
| 0 | 0 | NC | NC | No change from previous state |
| 1 | 0 | 1 | 0 | Set |
| 0 | 1 | 0 | 1 | Reset |
| 1 | 1 | 0 | 0 | Next state unpredictable |

(*c*) Truth table (with remarks)

**FIGURE 26-38 SR (set-reset) FF (flip-flop).**

outputs will go to 0,1 or 1,0 when the inputs go from 0,0 to 1,1.

*Clocked FFs* With the simple SR FF, the outputs changed at the same time as the R and S inputs changed. With a clocked FF, the outputs do not change until a clock pulse is received on the CK (clock) input.

There are two broad categories of clocked FFs—*edge-triggered* and *master-slave*. Edge-triggered FFs can be either positive-edge-triggered or negative-edge-triggered. The master-slave FFs and the negative-edge-triggered flip-flops have identical truth tables and, except in a limited number of applications, respond identically to input signals. Therefore, we will limit our study of clocked FFs to the edge-triggered types.

Figure 26-39 shows how the clock signal is shaped by the pulse-shaping circuit in an edge-triggered FF. The output of the shaping circuit is a very narrow pulse that occurs as the clock pulse is making its positive transition for a positive-edge-triggered FF or its negative transition for a negative-edge-triggered FF. Wave shaping the clock input ensures that the outputs of the FF will only change when the clock pulse is in transition rather than anytime the clock pulse is high.

The logic diagram, logic symbol, and waveform analysis of a triggered FF are presented in Fig. 26-40. The > symbol on the clock input line tells us

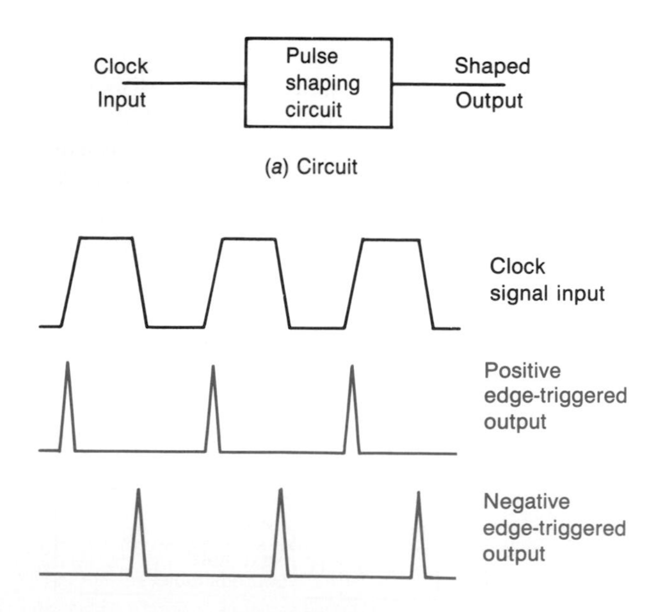


**FIGURE 26-39 Pulse-shaping circuit.**

that this is an edge-triggered FF. The bubble on the clock line tells us it is a negative-edge-triggered FF.

Two sets of zeros and ones (indicated by different colors) are shown on the logic diagram in Fig. 26-40*b*. The set with a one on the clock input line indicates the logic states while the clock is high. The other set shows the logic states as the clock is going through its high-to-low (negative) transition. As the clock finishes its negative-going transition, the following changes occur: (1) the clock line is at logic 0, (2) the output of the wave shaper goes back to zero, and (3) the output of $G_1$ changes back to a logic 1. None of these changes affects the output of the FF.

Notice that the addition of the clock circuitry (the pulse shaper, $G_1$, and $G_2$) has changed logic conditions so that NAND gates now provide an SR rather than an $\overline{SR}$ FF. The invalid SR conditions are still S = 1 and R = 1.

A waveform analysis for the negative-edge-triggered SR FF is illustrated in Fig. 26-40*c*. For waveform analysis, the FF is assumed to start in the reset (Q = 0) condition. Notice four things in this illustration: (1) S and R are never both high *when the FF is being triggered;* (2) the output does not change when S and R are both low; (3) the output changes only when the clock pulse is in transition from a high to a low state; and (4) the Q output is always the same as the S input *after* the clock has made its high-to-low transition. A little reflection will show that three of the above four statements are essentially a reiteration of the truth table given in Fig. 26-38.

When the logic state of the inputs (S and R in Fig. 26-40) is effective only when the clock signal is in transition, the inputs are said to be *synchronous inputs*. Clocked FFs can also have *asynchronous* (often called direct) inputs. An asynchronous input takes priority over all synchronous signals. The output responds immediately to an asynchronous input regardless of what the clock signal is doing. Figure 26-41 provides the details of an SR FF with two direct inputs. The input labeled PS (preset or direct set) is a direct input which, when activated by a logic 0, immediately forces the FF to the set (Q = 1) condition. The CR (clear or direct reset) input is the direct input for resetting the FF. It is also an active low direct input.

Referring to the logic diagram in Fig. 26-41 shows how a zero on the CR line resets (Q = 0) the

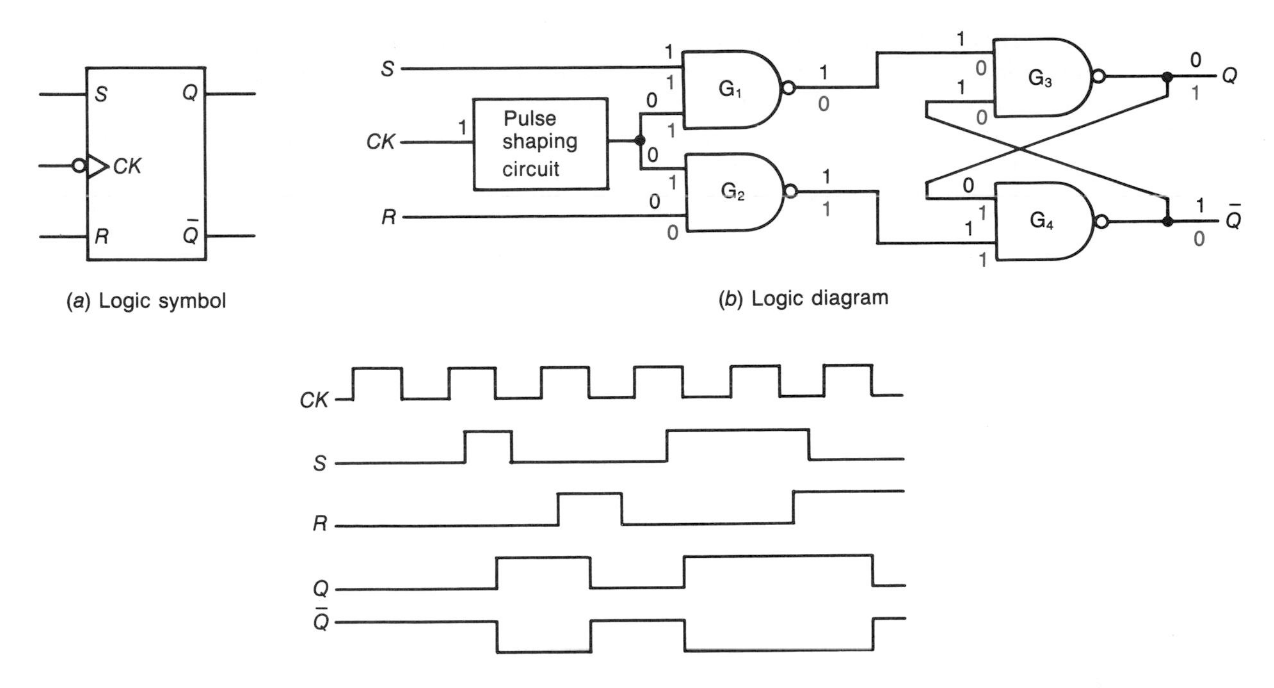


(a) Logic symbol

(b) Logic diagram

(c) Waveform analysis of a negative-edge-triggered S – R F – F

**FIGURE 26-40 Negative-edge-triggered SR flip-flop.**

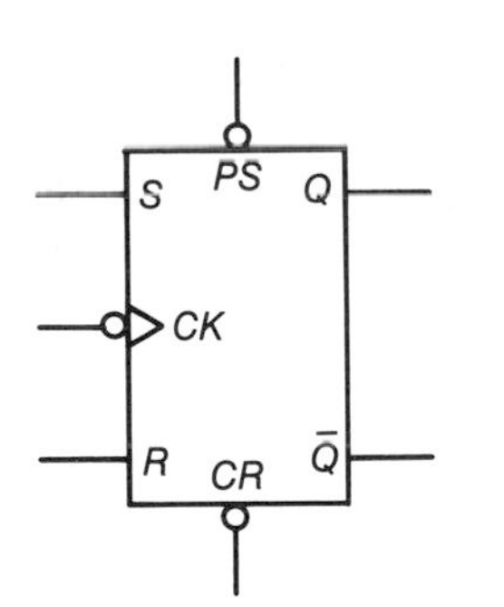


(a) Logic symbol

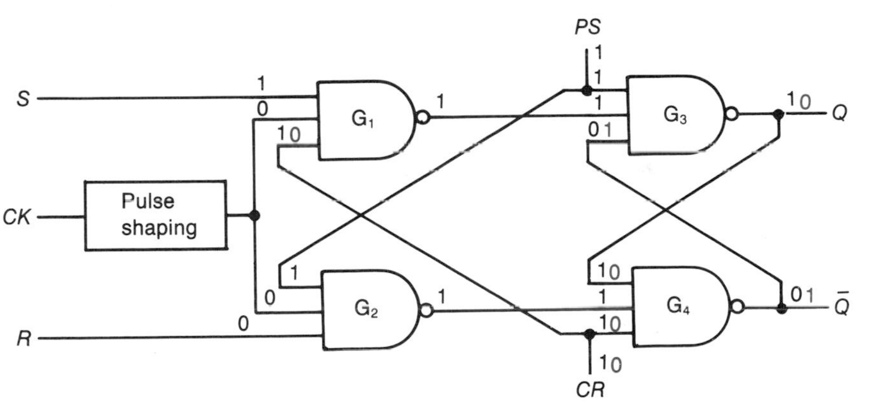


(b) Logic diagram

| S | R | CK | PS | CR | Q | $\bar{Q}$ | Remarks |
|---|---|---|---|---|---|---|---|
| X | X | X | 1 | 0 | 0 | 1 | Direct reset |
| X | X | X | 0 | 1 | 1 | 0 | Direct set |
| X | X | X | 0 | 0 | 1 | 1 | Invalid |
| 1 | 0 | ↓ | 1 | 1 | 1 | 0 | Set |
| 0 | 1 | ↓ | 1 | 1 | 0 | 1 | Reset |
| 0 | 0 | ↓ | 1 | 1 | 0 | 1 | No change |
| 1 | 1 | ↓ | 1 | 1 | 1 | 1 | Invalid |

(c) Truth table

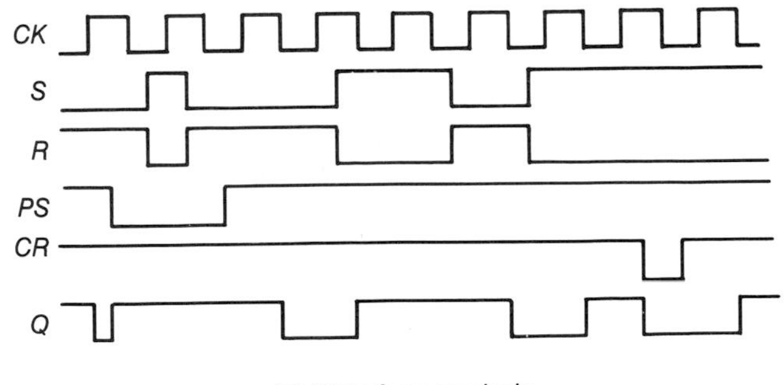


(d) Waveform analysis

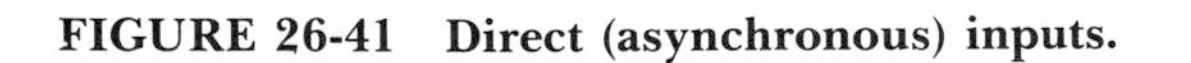

**FIGURE 26-41 Direct (asynchronous) inputs.**

FF. The ones and zeros which are the color of those on the S and R lines indicate the logic states before the CR line was changed to a logic 0. The other colored ones and zeros show the changes which occur when the CR (direct reset) is activated by a logic 0. Notice that a zero on the CR line forces NAND gates $G_1$ and $G_4$ to a one output, regardless of the logic levels of the other inputs to these gates. Also, notice that all three inputs to $G_3$ remain at logic 1 regardless of what logic level the clock and the S and R inputs are.

The truth table in Fig. 26-41 shows that the asynchronous inputs (S and R) are still not allowed to be logic 1s at the same time. Also, notice that the asynchronous inputs (CR and PS) are not allowed to be zero at the same time because an FF cannot be set and reset at the same time. The *X* in this truth table indicates that it does not matter whether the logic level is low or high. The ↓ indicates a negative-going transition of the clock pulse.

The preeminence of the direct inputs (CR and PS) is further illustrated by the waveform analysis in Fig. 26-41*d*. Again, notice that the outputs change without waiting for a clock pulse when a direct input goes low.

We have looked at the SR FF in some detail because it is a relatively simple circuit which illustrates the major characteristics of all types of FFs. In actual practice, the D-type and the JK-type FFs are much more prevalent than are SR FFs.

*D Flip-Flops* As shown in Fig. 26-42, the D FF can be constructed by connecting an inverter gate between the S and R lines of an SR FF. What was the S input line is now the D (data) input line. The operation of the D FF is very simple; whatever logic level is on the D input will be transferred to the Q output when the clock goes through its active transition.

Data FFs, like all other types of clocked FFs, can have direct inputs. The clock can be either negative-edge- or positive-edge-triggered.

*JK Flip-Flops* The advantage of the JK FF over the SR FF is that there is no disallowed or invalid combination of its synchronous inputs (J and K). Elimination of the disallowed input combination is achieved by adding feedback from the output gates to the input gates. The logic symbol, truth table, and waveform analysis for a JK FF are given in Fig. 26-43. Notice that the J and K inputs of the JK FF are like the S and R inputs of the SR FF. The only difference between the two sets of inputs is that J and K can both be logic high during the clock transition.

Notice that when J and K are both high, the FF output changes on *each* negative clock transition. When the FF output changes on each clock cycle, the FF is operating in the toggle T mode. Sometimes an FF is permanently connected to operate in the T mode. Such an FF is sometimes referred to as a *T FF* or a *toggle FF*.

The JK FF is very versatile. It can be made into a D FF by connecting an inverter between its J and K inputs; it can replace the SR FF without modification; and it is a toggle FF when the J and K inputs are tied to a logic 1 voltage. Also, JK FFs are available with direct inputs (CR and PS) and with either positive-edge triggering or negative-edge triggering.

*Latch Flip-Flop* A latch FF (sometimes referred to as a latch) is an FF used for temporary data storage. For example, four latches can store the BCD word for a BCD-to-seven-segment decoder which controls a LED display. The code word stored in the latches is displayed while the rest of the digital circuitry performs its functions to generate the next code word to be decoded and displayed.

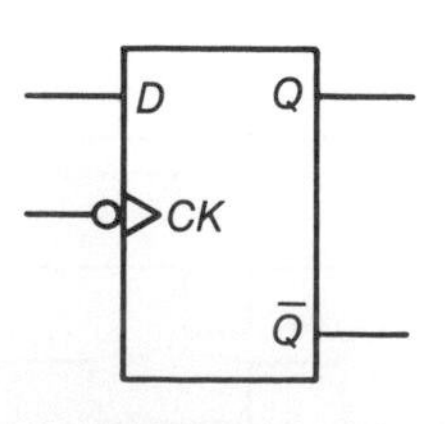


(*a*) Logic symbol

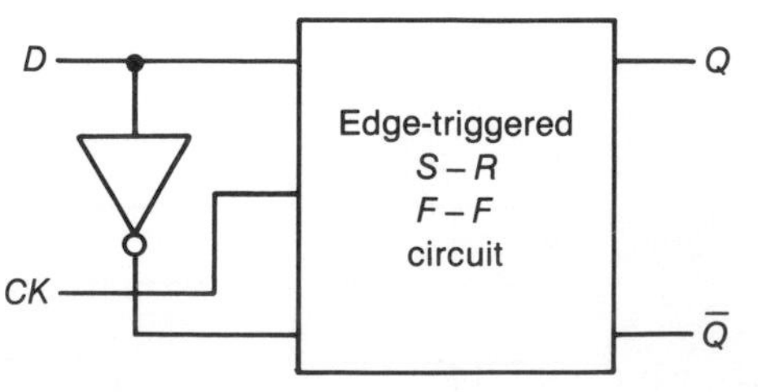


(*b*) Simplified logic diagram

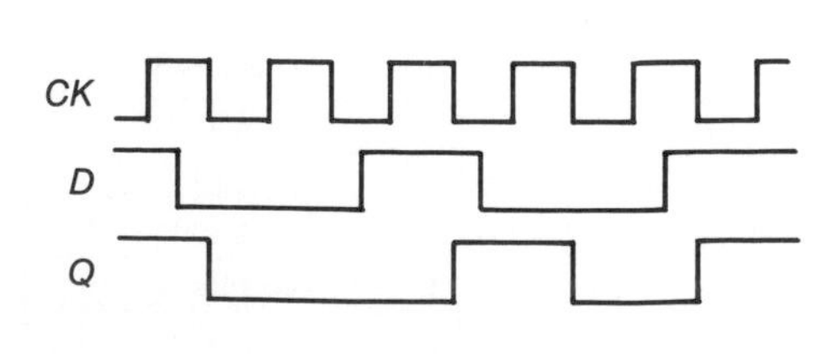


(*c*) Waveform analysis

**FIGURE 26-42 D flip-flop.**

| J | K | CK | Q | $\bar{Q}$ | Remarks |
|---|---|---|---|---|---|
| 0 | 1 | ↓ | 0 | 1 | Reset |
| 1 | 0 | ↓ | 1 | 0 | Set |
| 0 | 0 | ↓ | 1 | 0 | No change |
| 1 | 1 | ↓ | 0 | 1 | Toggles |

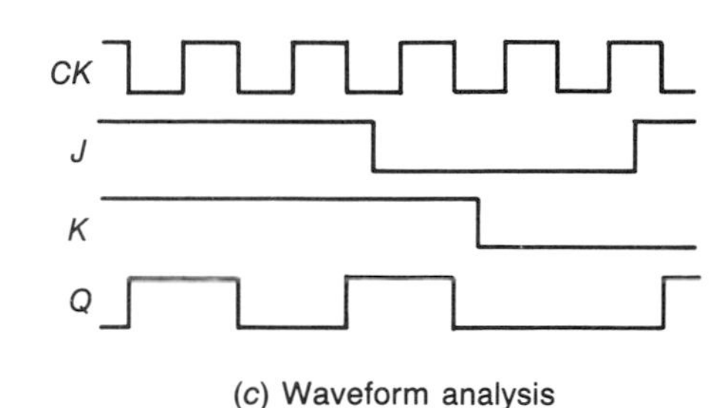


(*a*) Logic symbol (*b*) Truth table (*c*) Waveform analysis

**FIGURE 26-43 JK flip-flop.**

When the latches are strobed (a strobe pulse is like a clock pulse), the latches will "latch onto" the next code word and store it for decoding and display.

The most common latch is the D latch. As shown in Fig. 26-44, the D latch is like a D FF except that the pulse-shaping circuit is removed and the control line is now labeled gate G instead of clock. The control line is called a gate line rather than a clock line because the output can respond to the D input *anytime the gate is high*. This is why a latch FF is sometimes referred to as a gated FF. The gate input of a latch is also called the strobe input or the enable input.

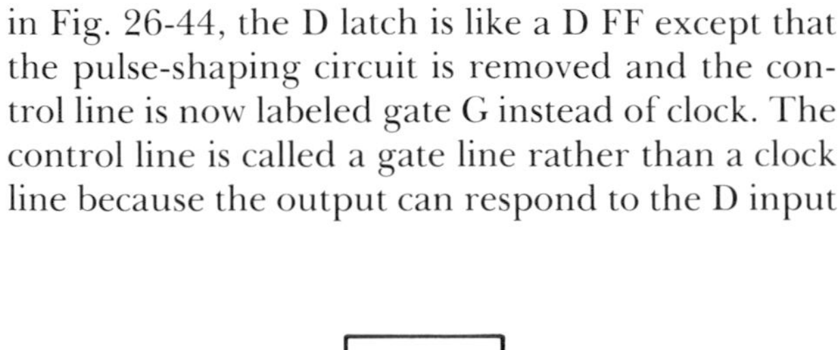

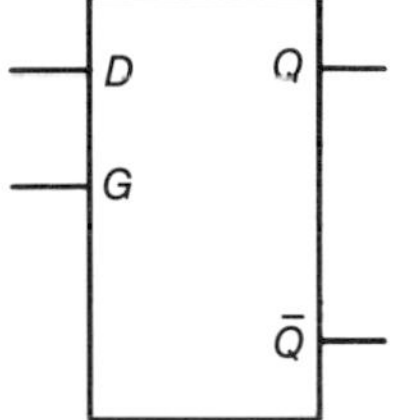


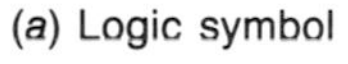

(*a*) Logic symbol

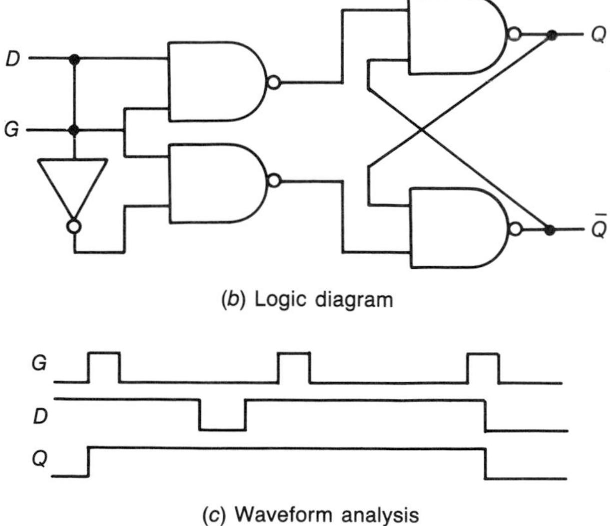


(*b*) Logic diagram

(*c*) Waveform analysis

**FIGURE 26-44 D-latch flip-flop.**

The action of the D latch is illustrated in the waveform diagram in Fig. 26-44. Notice that the gate (strobe or enable) signal is not a symmetrical square wave. Data is transferred to the output only while the strobe is high.

The logic symbol for the D latch shows that the gate is active high because the bubble is not present. The absence of the > symbol shows that the gate is not edge-triggered.

## Self-Test

**32.** Describe each FF shown in Fig. 26-45.

**33.** Draw the Q output for each set of waveforms in Fig. 26-46. (The bar over the input symbol indicates an active low.)

**34.** Which FF has only one input in addition to the clock input?

**35.** Which FF has no disallowed input combinations?

**36.** What is the major difference between a clocked RS FF and a simple RS FF?

**37.** What are FFs used for?

**38.** When can the output of an active high D latch change if the latch has no direct inputs?

## Counters and Frequency Dividers

Flip-flops make ideal base 2 (binary) counters. A single FF counts from 0 to 1. A two-stage binary counter (two FFs cascaded together) can count from 0 to $11_2$ or 0 to $3_{10}$.

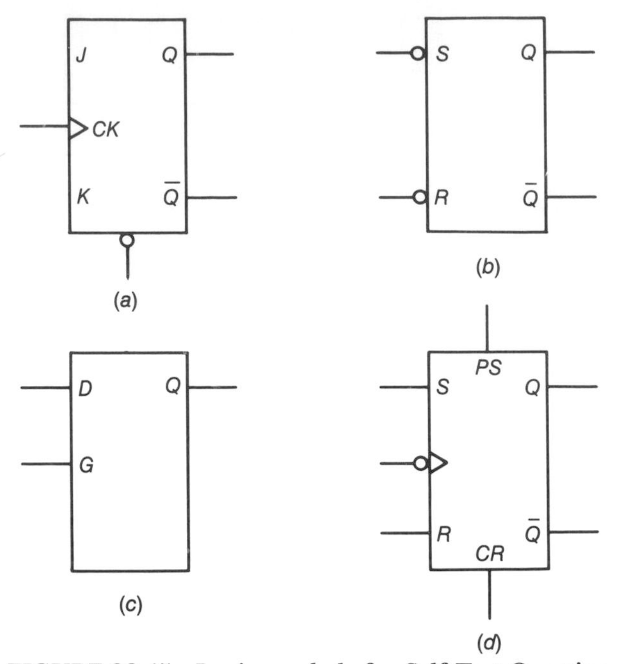


**FIGURE 26-45 Logic symbols for Self-Test Question 32.**

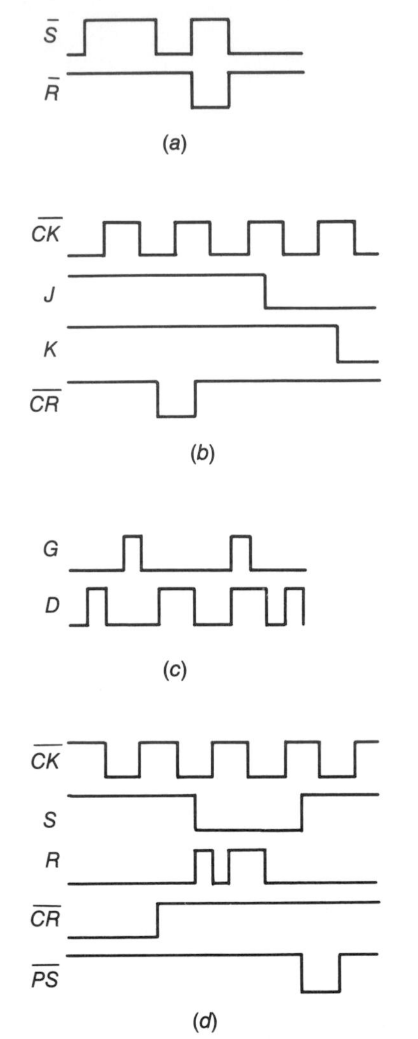


**FIGURE 26-46 Waveforms for Self-Test Question 33.**

The count of a binary counter is equal to $2^n$ where $n$ equals the number of FFs cascaded together. Thus, three FFs provide a count of 8, four FFs a count of 16, etc. The count of a counter is called its *modulus*. A counter with three cascaded FFs is a modulus 8 counter. The counter in Fig. 26-47 is a modulus 4 counter.

The counter in Fig. 26-47 counts the number of clock pulses received on the CK input line. The ones on the J and K inputs indicate that these inputs are permanently connected to a logic 1. Thus, these FFs operate as toggle FFs. Flip-flop $FF_A$ toggles every time the clock goes through a negative transition. Flip-flop $FF_B$ toggles every time that $Q_A$ goes from a logic 1 to a logic 0. The state table in Fig. 26-47 shows that the modulus 4 counter has four distinct output combinations. Notice that $Q_A$, the output of the FF which receives the clock signal input, represents the LSB of the counter's output.

Binary counters are also named by the number of binary bits their count represents. For example, the counter in Fig. 26-47 is a 2-bit counter because its outputs represent all possible combinations of a 2-bit binary word.

The waveforms in Fig. 26-47*d* show that the counter counts four clock pulses and then starts over again. Notice that the frequency of the $Q_A$ output signal is half as great as the clock frequency being counted. Thus, an FF divides the clock frequency by 2. The output of $FF_A$ is the clock input for $FF_B$. The second FF again divides by 2, so the frequency of the $Q_B$ output signal is one-fourth as great as the original clock frequency. Therefore, counters are also named for their frequency dividing ability. A three-stage binary counter can be called a *divide-by-8 counter*.

In summary, then, a counter like the one in Fig. 26-47 can be referred to by four different names. It can be called a two-stage binary counter, a 2-bit binary counter, a modulus 4 counter, or a divide-by-4 counter.

A *decade counter* (modulus 10 counter) uses four FFs. Four FFs would, if used as a binary counter,

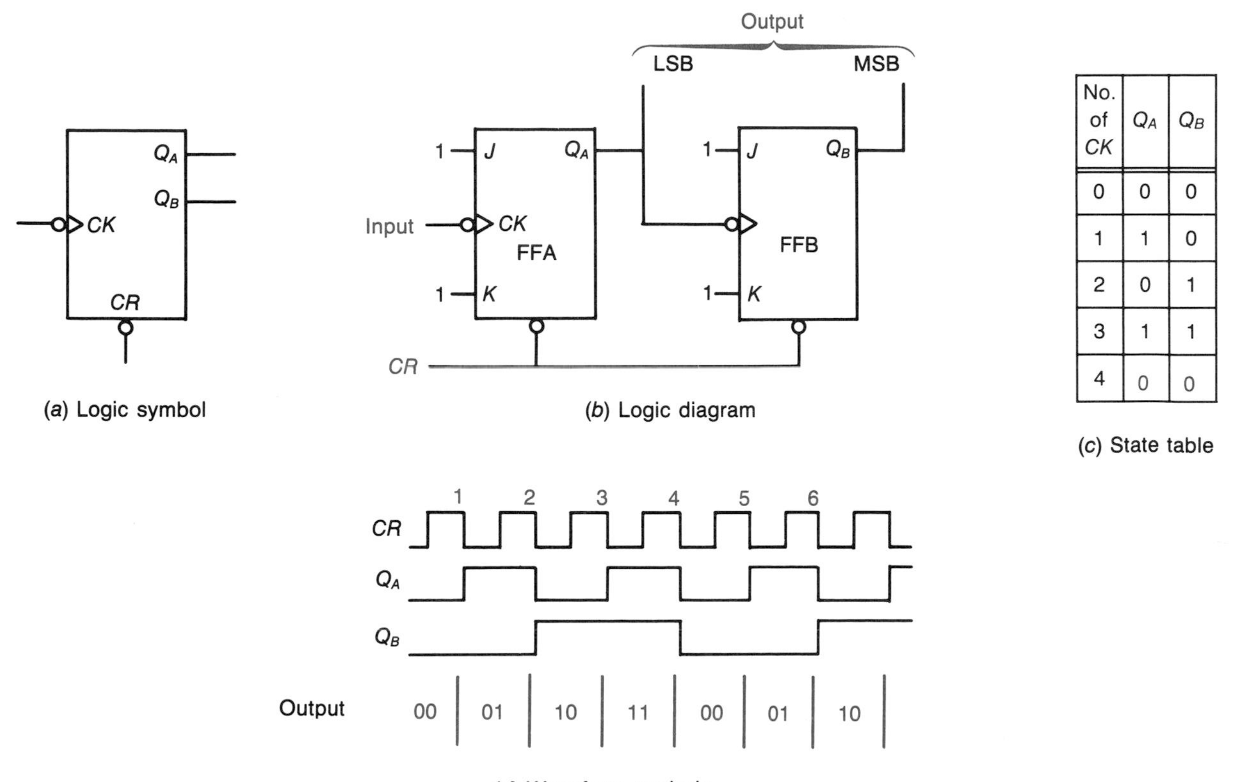


**FIGURE 26-47 Modulus 4 counter.**

have 16 possible output combinations. Therefore, six output counts must be eliminated or skipped over to produce a decade counter. A technique for eliminating the last six states is shown in Fig. 26-48*b*. The two-input NAND gate decodes the count of $10_{10}$ and immediately resets the counter (all FFs are reset). Thus, the counter has 10 stable states: 0 through 9. This is a rather crude way to make a modulus 10 counter. The circuits used in ICs are much more sophisticated, but they are still based on using four FFs and skipping 6 of the 16 available states.

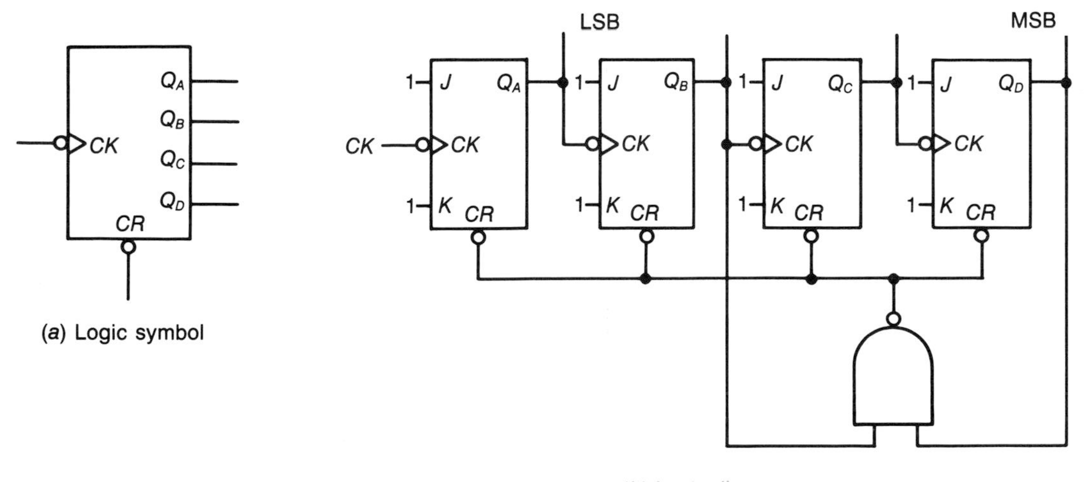


**FIGURE 26-48 Decade (modulus 10) counter.**

The decade counter in Fig. 26-48 provides a BCD output. Therefore, its count (output) could be connected to a BCD-to-seven-segment decoder and then displayed on a seven-segment LED. If the output of the counter is to be periodically sampled, which is common practice, the four output lines of the counter would go to four D latches. Then, the D latches would be periodically strobed. The four latches could be in a single IC. Figure 26-49 shows the logic diagram of such a circuit. The ones and zeros on this diagram indicate the logic levels at the instant the strobe returns to a logic 0.

Decade counters are also called *divide-by-10 counters*. They are commonly used as frequency dividers. For example, a one-pulse-per-second strobe signal can be obtained from a standard 60-Hz signal by putting the 60-Hz signal through a decade counter which is cascaded to a modulus 6 counter. The decoded output of the decade (divide-by-10) counter would be 6 Hz, which would be the input to the modulus 6 (divide-by-6) counter. Finally, the decoded output of the modulus 6 counter would be 1 Hz.

All of the counters we have looked at are classified as asynchronous, or *ripple*, counters. The output changes of these counters are not synchronized with the input signal being counted. The output of $FF_B$ cannot change until its CK input receives a signal from $FF_A$; i.e., the signal ripples through the cascaded FFs. Thus, the name ripple counters. Ripple counters work well for counting and dividing lower-frequency signals, but for high-frequency signals *synchronous* counters are used. Synchronous counters are sometimes classified as *parallel* counters because the CK lines of all FFs are tied in parallel and simultaneously receive the signal to be counted. Thus, all Q output lines of the counter are synchronized with the input signal.

Numerous other types of counters are available in IC packages. For example, some counters can be preset to any desired number within their count range. Counters are also available that can either count up or count down (count forward or count backward) depending on which logic level is applied to a count-control line.

## Self-Test

**39.** Refer to Fig. 26-47. What will be the logic level of $Q_A$ and $Q_B$ after seven input clock pulses? (Assume the counter was cleared before the first pulse arrived.)

**40.** Refer to Fig. 26-48. Assuming the counter starts in a cleared state, what will the logic levels of the Q outputs be after five clock pulses?

**41.** A binary counter uses five FFs. (a) What is its modulus? (b) List four names for this counter.

**42.** A 500-kHz signal is applied to a three-FF counter. The decoded output of this counter is then applied to a decade counter. What is the output frequency of the decade counter?

**43.** What is the major advantage of a synchronous counter over a ripple counter?

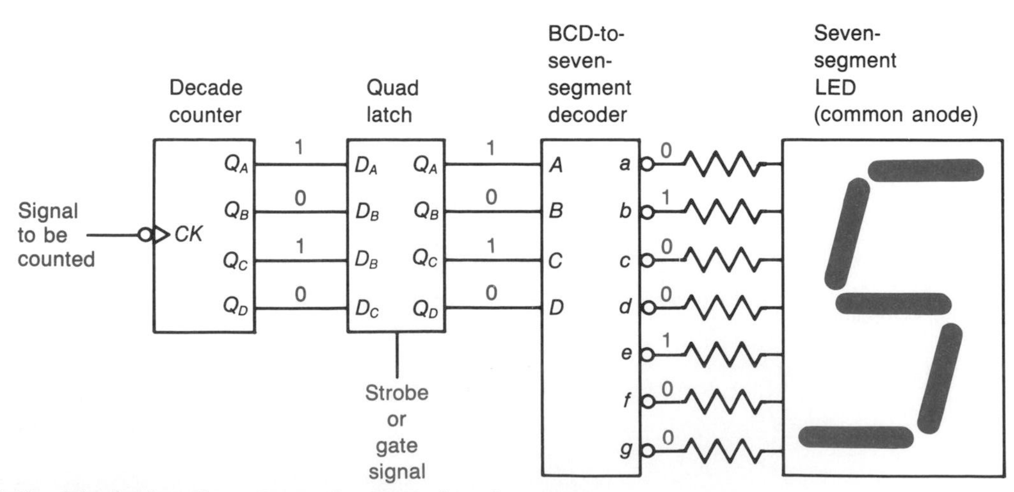


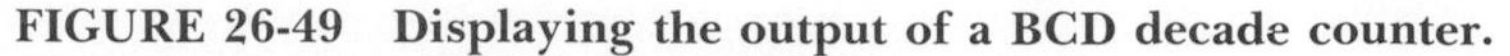

**FIGURE 26-49 Displaying the output of a BCD decade counter.**

## Registers and Shift Registers

A *register* is a digital circuit which stores a multibit binary word. It consists of a group of properly connected FFs. Registers commonly store words with a length of either 4, 8, 16, or 32 bits. The examples in this section use 4-bit words. Longer words use the same type of circuits; they just require an additional FF for each additional bit of word length. Registers are used for temporary storage of data that is being processed by a digital system.

There are two broad categories of registers: storage registers and shift registers. *Storage registers,* often referred to just as registers, are only capable of storing binary words. As you can see in Fig. 26-50, a 4-bit register consists of four FFs with a common clock line and a common clear (reset) line. The output of one FF does not connect to the input of any other FF.

Storage registers always have parallel data inputs (parallel loading) and parallel data outputs. *Parallel in* (parallel input) means that all bits of the data word are loaded into the register during one clock pulse. *Parallel out* means that all bits of the stored word are simultaneously available at the output of the register.

Inspection of Fig. 26-50 shows that the only difference between a 4-bit D latch and a 4-bit D register is that the register is controlled by an edge-triggered clock input and the latch is controlled by a gate input. As shown in an earlier section, a gated input control allows data to enter anytime its control signal is active.

Most IC storage registers have additional control logic on the data input lines so that new data is not loaded during each clock pulse. Also, many of these registers have tristate buffered outputs so that the output can be connected to a bus in a digital system. It should be mentioned that some IC manufacturers classify their storage registers as a quad D FF or an octal D FF.

*Shift registers* are more versatile than are storage registers. A shift register can shift the binary data (words) within a register as well as store the data. This is because the output of one FF is connected to the input of the next FF as shown in Fig. 26-51*a*.

Shift registers are often classified by the way data is put into and taken out of the register. This leads to four classes: serial in–serial out, serial in–parallel out, parallel in–serial out, and parallel in–parallel out. Serial in and serial out means that the data word is put into and taken out of the register 1 bit at a time, i.e., 1 bit for each clock pulse. Obviously, serial transfer of a data word is slower than is parallel transfer, but it is necessary when data words are transferred through a two-conductor cable such as a telephone line. The serial in–parallel out shift register is sometimes referred to as a serial-to-parallel converter because it can receive serial data, like that provided by a telephone line, and convert it to parallel data. The parallel

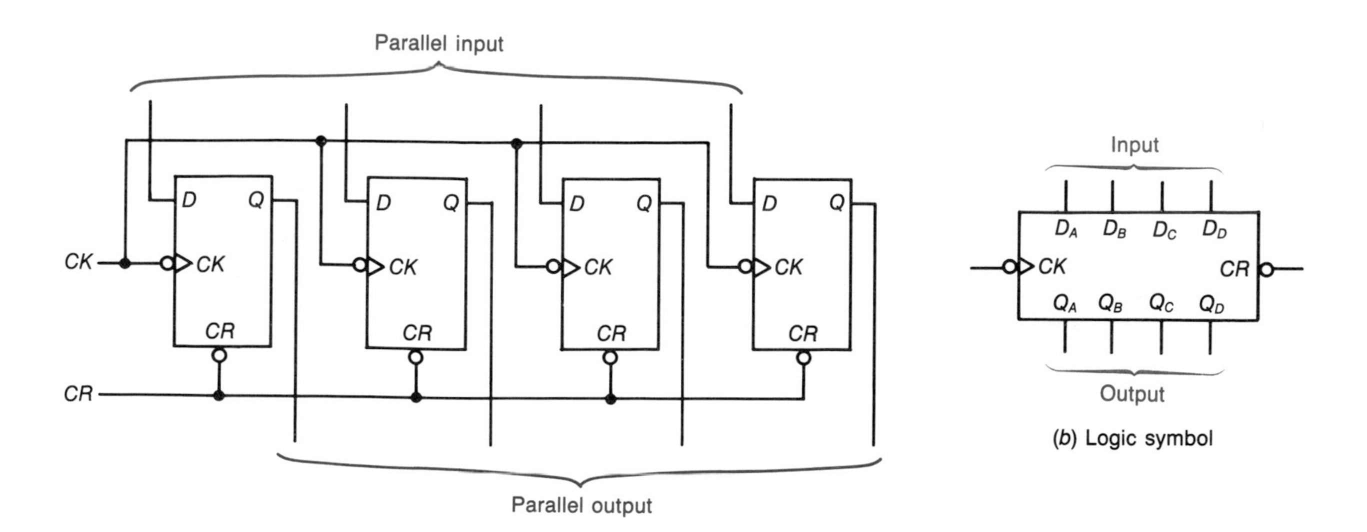


(*a*) Logic diagram

(*b*) Logic symbol

**FIGURE 26-50 Four-bit storage register.**

data can then be processed by a digital system such as a microcomputer.

A parallel in–serial out shift register serves as a parallel-to-serial data converter. It can be used to interface the parallel output of a digital system to a two-conductor bus or cable.

A shift register like the one in Fig. 26-51 loses the data word when it operates in the serial out mode. When it is necessary to have a serial output but not lose the data word, a *recirculating shift register* is used. A recirculating register has feedback paths from the Q and $\overline{Q}$ outputs of the last FF to the S and R inputs of the first FF. Of course, such a register has additional control logic to ensure that the register is not receiving a new data word while the present data word is being recirculated.

In analyzing the operation of Fig. 26-51, it is important to know, and remember, that the output of $FF_A$ does not ripple through all the other FFs when a clock pulse arrives. Instead, the output of $FF_A$ is transferred to the output of $FF_B$; the previous output of $FF_B$ is transferred to the output of $FF_C$; etc. On the same clock pulse, the logic level of the serial input is transferred to the Q output of $FF_A$. The next clock pulse causes the data to transfer (shift) to the right one more position. Thus, it takes four clock pulses to shift in (serial in) a new 4-bit data word. Of course, these same four clock pulses caused the old 4-bit data word to appear, 1 bit at a time, at the serial out $Q_D$ terminal.

Notice in Fig. 26-51 that the data word is available at the parallel output terminals the instant the word has been loaded into the register by either serial loading or parallel loading. Parallel loading of this register requires three steps. First, the data word is put on lines A, B, C, and D. Second, the register is cleared (all Q outputs changed to logic 0) by a logic low strobe on the CR line. Third, the PE line is strobed with a logic 1 pulse. This high pulse enables all the NAND gates so that those gates with a logic 1 on their other input will have a logic 0 output. A logic 0 output from one of these NAND gates presets (Q = 1) the FF to which the NAND gate is connected.

The shift register in Fig. 26-51 can only shift the data one place to the right when it is clocked. Some shift registers, available in an IC package, have additional logic control gates which allow the data to be shifted either to the right or to the left.

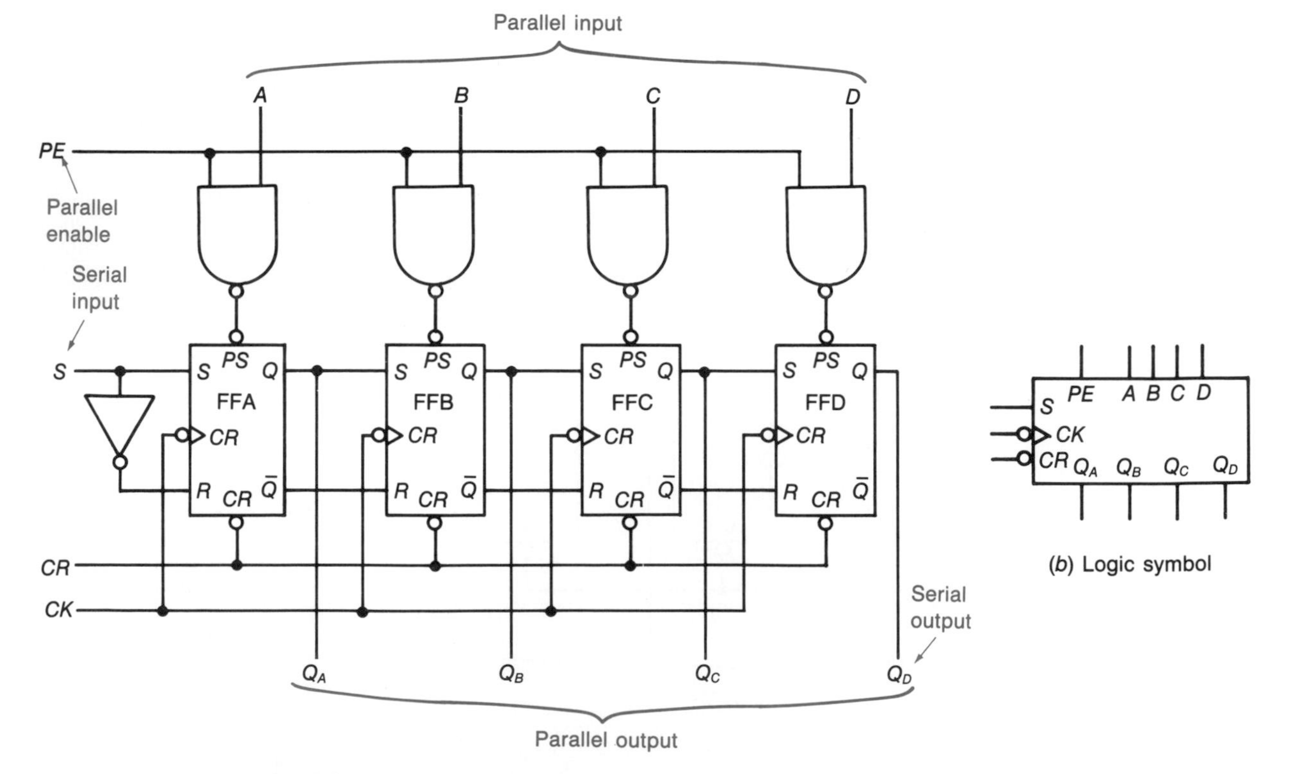


(a) Logic diagram

(b) Logic symbol

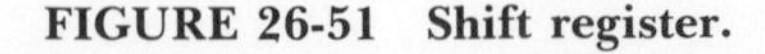
**FIGURE 26-51 Shift register.**

## Self-Test

**44.** What is a register?

**45.** List, and differentiate between, two types of registers.

**46.** What type of output must a storage register have?

**47.** What type of register is used to convert parallel data to serial data?

**48.** Refer to Fig. 26-51. The binary word 1011 is parallel-loaded into the register (D is the MSB). The register is then clocked twice. Will $Q_D$ be a logic 0 or a logic 1? Why?

## 26-6 ARITHMETIC CIRCUITS

*Arithmetic circuits* use combinational logic to perform binary arithmetic operations such as addition, subtraction, and multiplication. Individual ICs are available that will perform a single arithmetic operation. Also available are single LSI and VLSI packages that will do all of the arithmetic operations (in addition to other functions) for a microcomputer. Such ICs are called *microprocessors.*

As an example of arithmetic circuits, we will analyze some adder circuits. Figure 26-52*a* shows a combinational logic circuit capable of adding two single-digit binary numbers. This circuit is called a *half adder* (*HA*) because it does not have an input for a carry in from a previous column of a multidigit number. However, it does have a carry output, so it is useful as an adder for the $2^0$ column of two multidigit numbers. The logic levels indicated on the logic diagram indicate that when the circuit adds 1 and 1, it provides a sum of 0 and a 1 carry.

Figure 26-52*b* shows a full adder (FA) circuit. This circuit has three inputs. Two inputs are for the two digits from any column of two multidigit binary numbers and one input is for carry in from a previous column. Thus, it can be used to add digits in the $2^1$ or higher column of two binary numbers. The zeros and ones on this circuit show the sum and carry outputs when two 1s and a one

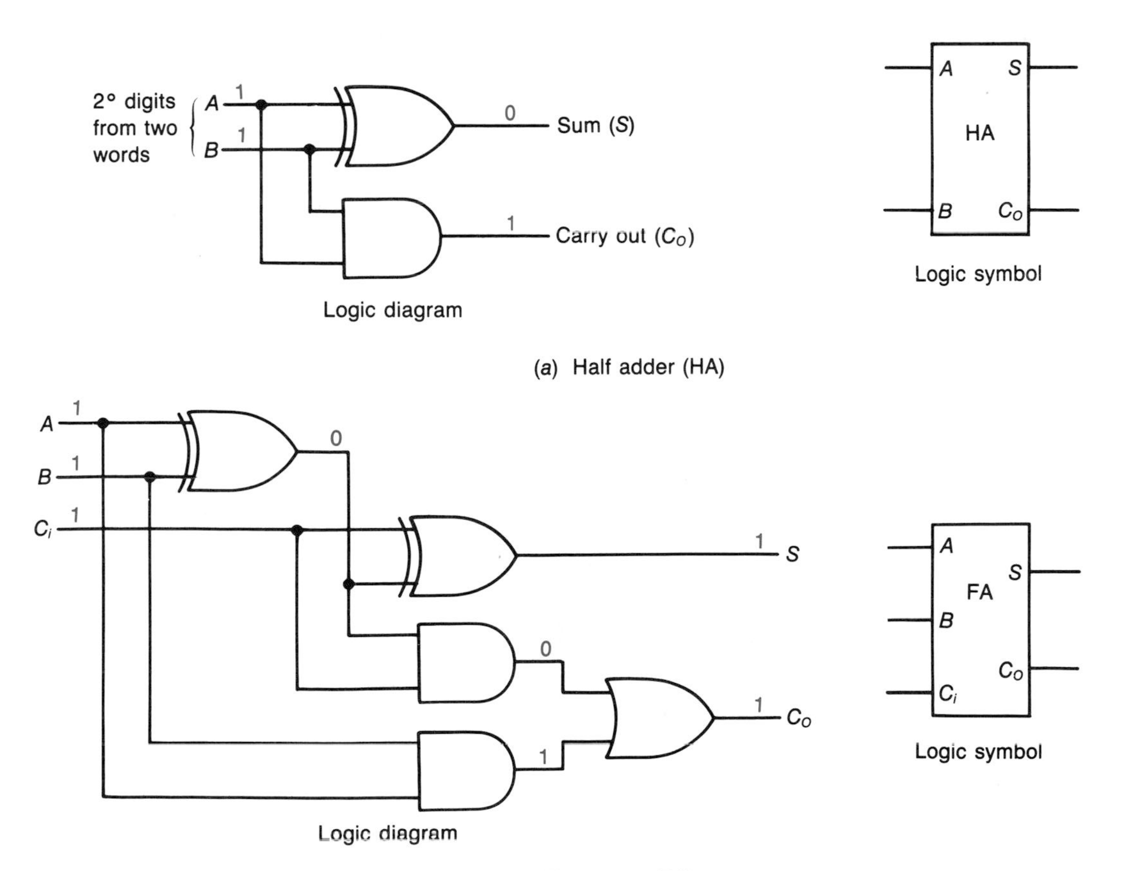


**FIGURE 26-52 Adder circuits.**

carry in are added. Refer back to Fig. 26-5 if you have forgotten the details of binary arithmetic but would like to try adding other combinations of binary digits with this circuit.

The circuit in Fig. 26-53 shows how a half adder and three full adders can be used to add two 4-bit binary words ($A_3$-$A_0$ and $B_3$-$B_0$) The two 4-bit binary words could be obtained from the parallel outputs of two 4-bit registers. The sum and the carry from the adder could be parallel-loaded into a 5-bit register for temporary storage.

Multiplication can be done by adding the first two partial products and then adding this sum (with carry) to the next partial product. This process is repeated until all partial products have been added.

Referring back to Fig. 26-7*b* you can see that the partial products are always either a string of zeros or the multiplicand shifted left an appropriate number of places. Thus, if one of the registers providing input numbers to the adder is a shift register, the partial products can be created by loading and shifting the multiplicand the correct number of places. After the addition of two partial products, the sum and carry outputs can be transferred back to an input register and added to another partial product to produce a new sum and carry.

## Self-Test

**49.** What is the difference between an HA and an FA?

**50.** Refer to Fig. 26-53. If word A is 1011 and word B is 1101, what would be the logic level of the $C_0$ of $FA_1$, $FA_2$, and $FA_3$?

## 26-7 MEMORY CIRCUITS

Both an FF and a register have memory. However, individual FFs and registers are not usually classified as memory because their storage capacity is small and they are only used for temporary storage while data is being manipulated by arithmetic circuits or transferred from one part of a digital system to another part of the system.

There are two broad categories of memory for digital systems: semiconductor and magnetic. In modern digital systems, *magnetic memory* is used for long-term storage of large amounts of data. Magnetic memory devices include magnetic tapes, floppy (flexible) disks, and hard disks. Binary ones and zeros are represented in these devices by the polarity of the magnetic flux in minute isolated areas of the magnetic material contained on the surface of the disks or tapes. Magnetic memory is nonvolatile; i.e., stored data is not lost when the system is turned off or electric power is temporarily lost.

Compared to *semiconductor memory*, magnetic memory is slow; i.e., it takes a relatively long time to store and/or retrieve data. For this reason, the working memory (memory used to temporarily hold data, instructions, and programs) in computers is semiconductor memory.

Semiconductor memory can be classified as either *random-access memory* (RAM) or *read-only memory* (ROM).

*Random-access memory* refers to a large group (often called an *array*) of memory cells (like FFs) arranged so that any one cell can be selected at any time. Further, datum (a zero or a one) can be either written into or read out of the selected cell;

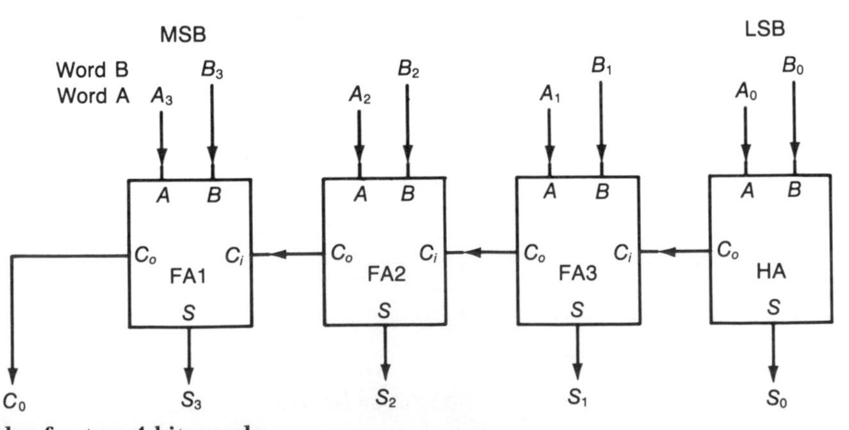


**FIGURE 26-53 Adder for two 4-bit words.**

i.e., datum can be stored in the cell or datum can be retrieved from the cell.

The process of selecting an individual cell is called *addressing the cell.* Each cell has its own unique binary address. A 4-bit binary number is able to address $2^4$, or 16, different memory cells. A 10-bit binary number can address 1024 memory cells. In digital electronics, an array of 1024 memory cells is referred to as 1K of memory. One K of memory is capable of storing 1024 single-bit binary words. Since single-bit binary words are seldom used, memory arrays are connected together in groups of 4, 8, 16, or 32. They are connected so that the same cell location in each array is selected by a given address number. Thus, a RAM with eight of these 1024 cell arrays could be referred to as a 1024 × 8 RAM or a 1K × 8 RAM. Such a RAM could store 1024 (1K) 8-bit words. In a digital system which processes binary numbers in 8-bit words, this amount of memory would just be referred to as 1K of RAM. Remember that K used to describe memory size is an abbreviation for 1024 words. Do not confuse it with k which is the abbreviation for the prefix kilo.

An illustration of a RAM array is shown in Fig. 26-54. While this circuit is not the exact type used in large arrays, it illustrates all the principles involved in larger arrays. It shows how individual memory cells can be addressed and how data can be written into (stored) or read out of (retrieved) a selected cell. The diagram in Fig. 26-54 is capable of storing 16 single-bit words. Increasing the word length to 4 bits would require three more arrays, each containing 16 D latches with decoding gates for the G inputs and the Q outputs. All the X, Y, W, and R lines of all four arrays would be connected in parallel. The same address decoder would be used for all four arrays; i.e., $Y_0$ would go to latch 0, 4, 8, and 12 of each of the four 16 × 1 arrays.

The logic states indicated in Fig. 26-54 are those required to write a 1 into D latch number 5. The binary equivalent of 5 is 0101 which is the binary number on the four address lines. The two LSBs of the address are decoded to select the desired column *Y* of the array of D latches. The two MSBs are decoded to select the desired row *X* of the array. Notice that only the D latch at the intersection of the selected row *X* and column *Y* will have logic 1s on two of the three inputs of the NAND gates which control its G input and its Q output. Since the write W line is also high, the logic 1 on the data line will be stored in memory cell number 5 and be available at the Q output.

Notice in Fig. 26-54 that while writing into the memory (W line high), the outputs of all cells, including the addressed cell, are effectively disconnected from the data line by the three-state buffers connected to the Q outputs of the latches. Thus, there is no data on the data-out line while data is being stored. After the write line returns to a logic 0, the read R line can be changed to a logic 1. The data output of the selected cell will then be put on the data-out line by the enabled three-state buffer.

A wide range of RAMs are available in IC packages. Most of the larger RAMs incorporate ICs that utilize metal-oxide semiconductor (MOS) technology. Two hundred fifty-six K × 1-bit memory arrays are now available in a single IC package (chip). Eight of these IC chips provides 256K of RAM for a microcomputer that handles data in 8-bit binary words.

Read-only memory (ROM) can be addressed in the same way as RAM; i.e., the individual cells can be addressed in any order (randomly). The major difference between RAM and ROM is that the operator of a digital system (such as a microcomputer) *cannot* store (write data into) a ROM. The operator can only read (retrieve) data from the ROM.

Read-only memory is nonvolatile. It is used to provide binary data which will be used over and over again but never needs to be changed by the operator. The instructions (binary words) which allow a computer to receive information from an input device, such as a keyboard, are contained in ROM. Also, a programming language like BASIC can be put in ROM and be available to the computer operator whenever the computer is on.

Each cell in a ROM IC has a logic 0 or a logic 1 permanently stored in it by the manufacturer of the IC. The purchaser tells the manufacturer the data words to be stored at the various addresses.

There are many modified types of ROMs available. For example, a PROM (programmable read-only memory) is a type of ROM which allows the purchaser to program (store data words in) the ROM one time. Once the data is put in the ROM, it can never be changed. An EPROM (eraseable PROM) is a type of ROM which, if the right equipment is available, can be programmed, erased, and reprogrammed by the purchaser. Thus, the EPROM provides semipermanent or quasipermanent programming. EPROMs which can be

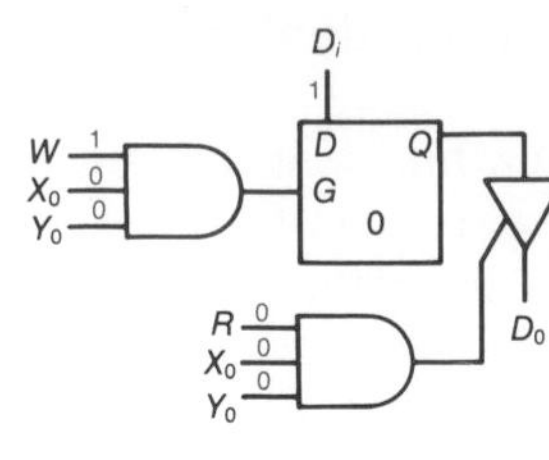


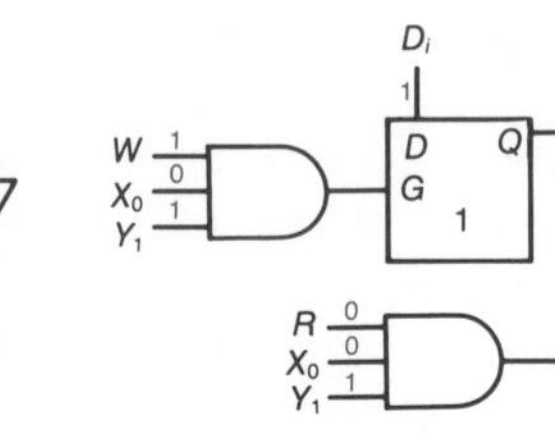


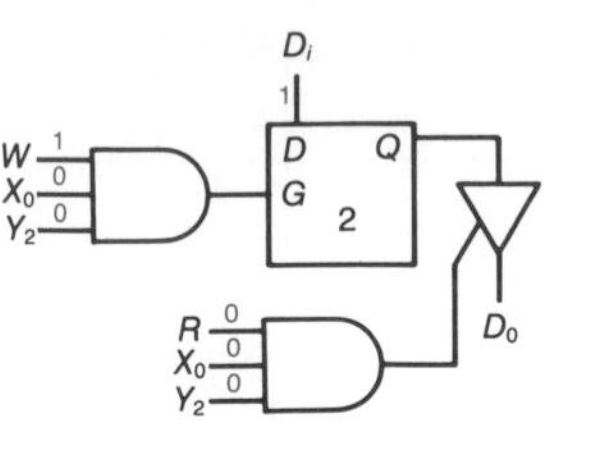


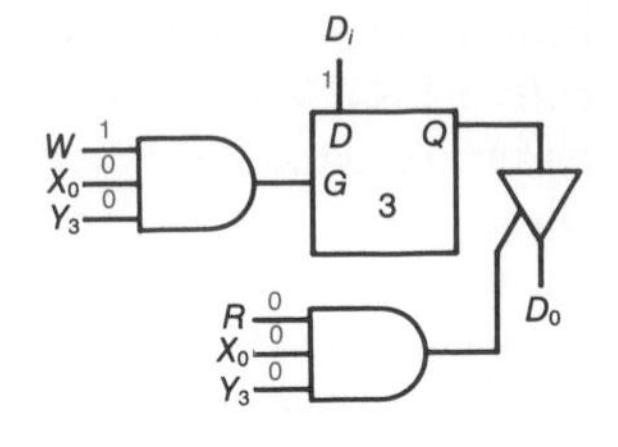


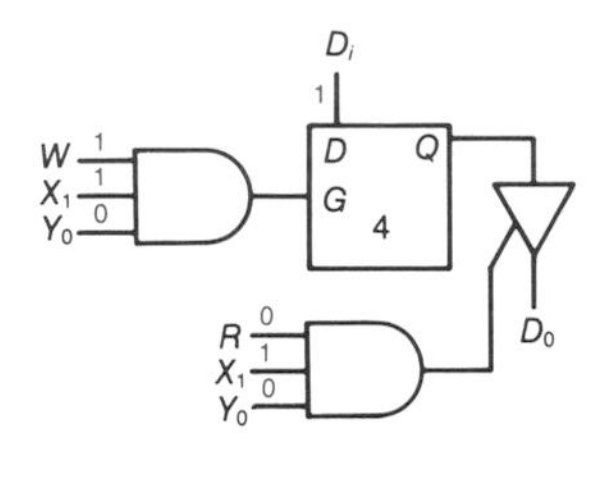


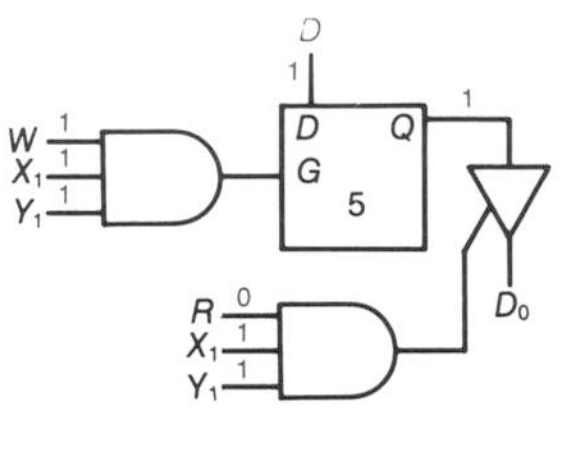


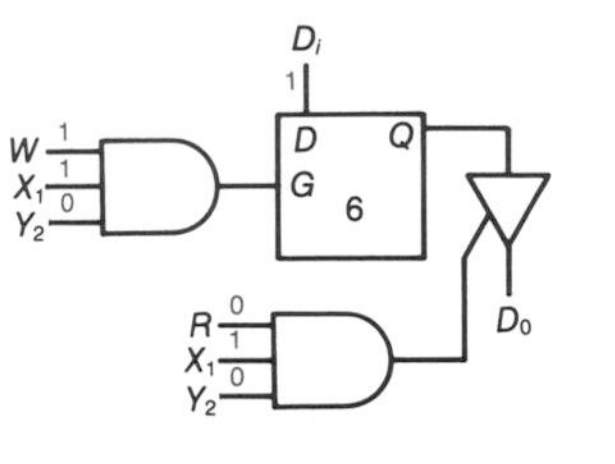


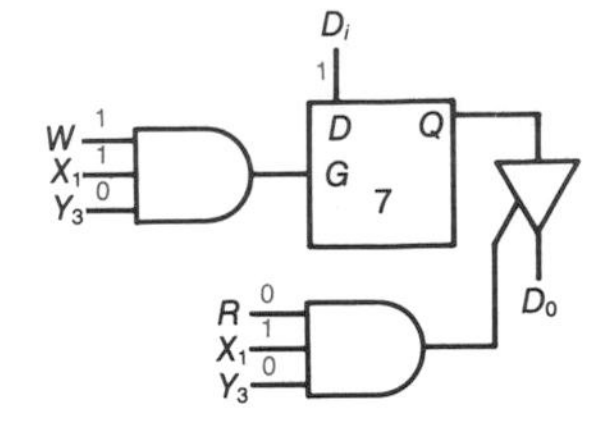


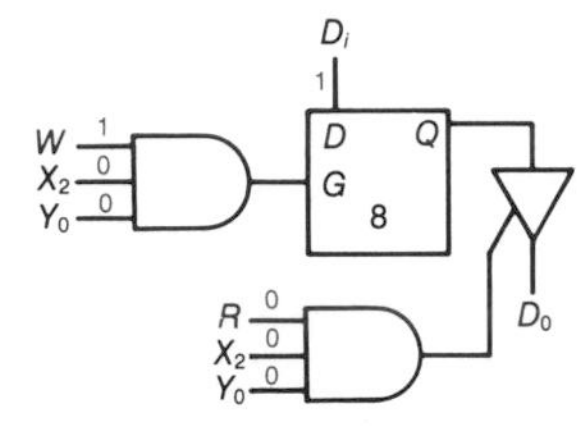


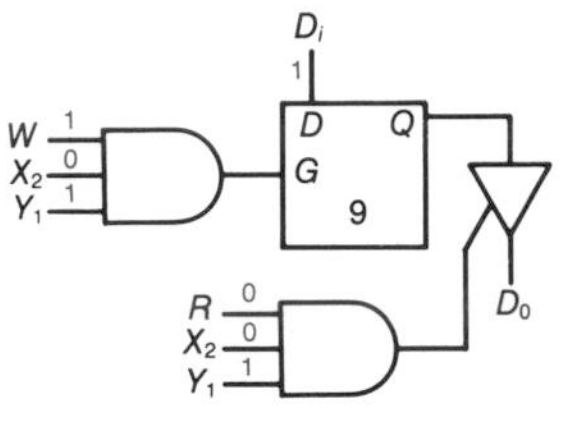


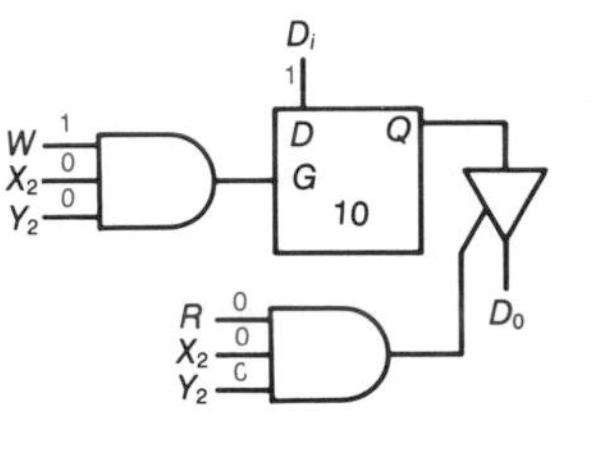


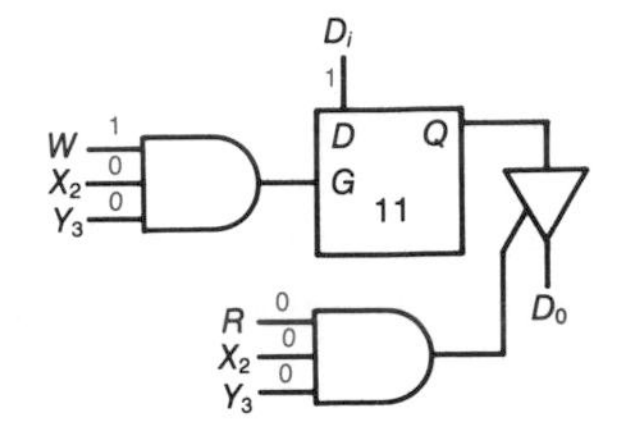


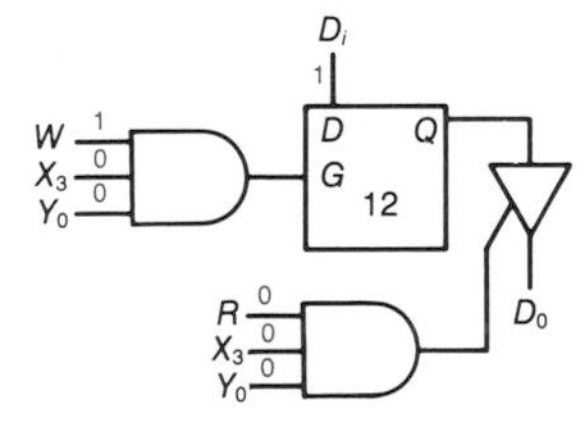


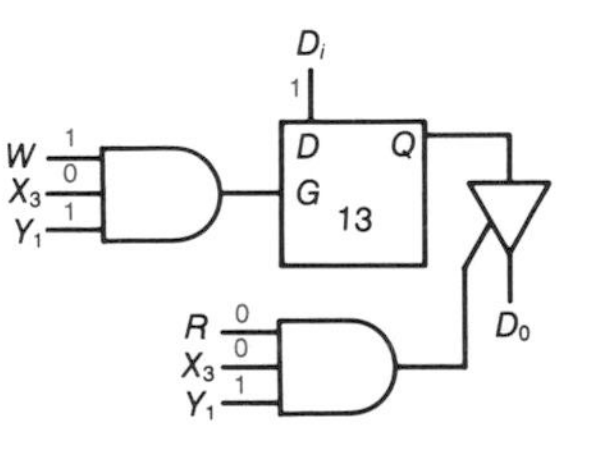


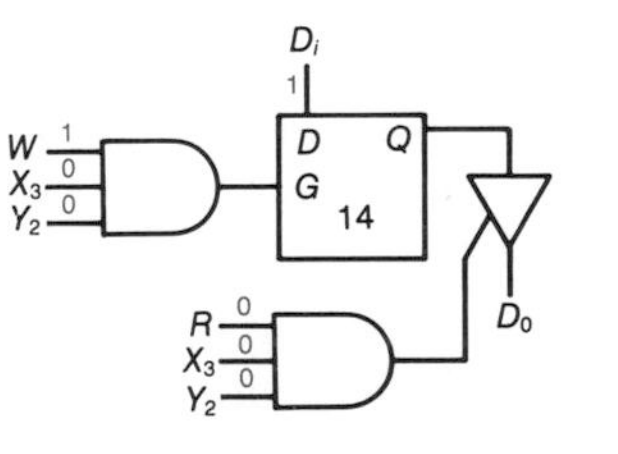


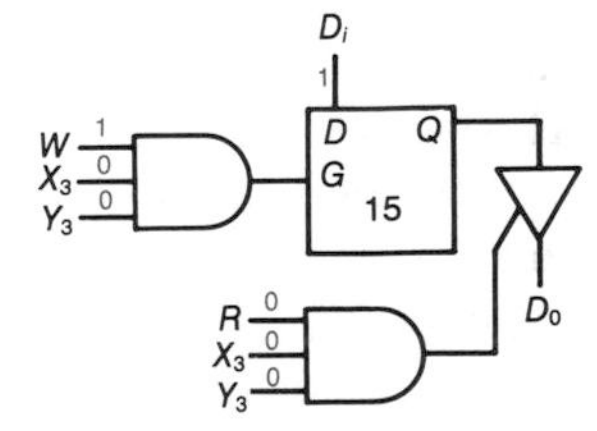


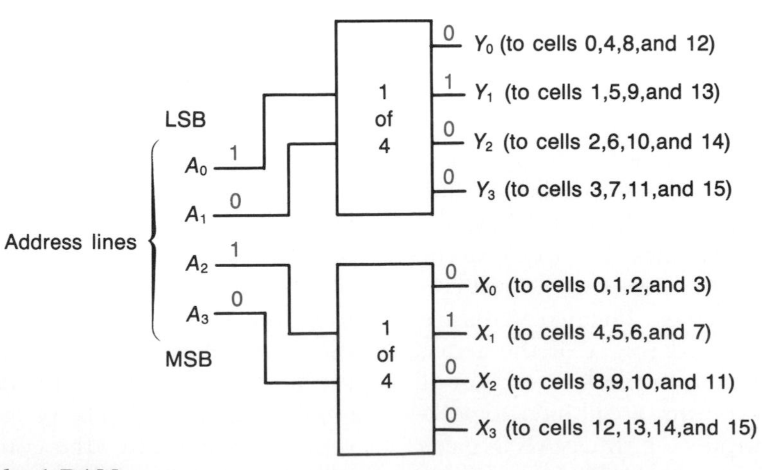


**FIGURE 26-54 A 16-by-1 RAM.**

erased electrically are called EEPROMs (electrically EPROM). Those that are erased by ultraviolet light are called UV EPROMs.

### Self-Test

**51.** Differentiate between RAM and ROM.

**52.** Is it more difficult to address ROM than it is to address RAM?

**53.** Describe how to read the data stored in cell number 14 in Fig. 26-54.

**54.** How many words of memory can be addressed by 16 address lines? How many K of memory would this be?

## 26-8 MICROCOMPUTER SYSTEMS

A simplified block diagram of a microcomputer system is shown in Fig. 26-55. The heart of a microcomputer is the microprocessor. This unit, which is a single IC, contains a number of rather distinct subsystems. Some of the major subsystems are indicated in Fig. 26-55 and discussed below.

The arithmetic logic unit (ALU) contains many of the circuits discussed in earlier sections. It performs all the arithmetic operations required of the computer as well as logic functions such as ANDing, ORing, etc.

The microprocessor uses a number of registers. Some examples are (1) an 8-bit register used to temporarily store intermediate results produced by arithmetic operations, (2) an 8-bit register used to hold data going to or coming from the eight-line data bus, and (3) a 16-bit register used to hold the address of the data word being sent to or retrieved from memory.

The program counter keeps track of which instruction, out of a long sequence of numbered instructions, the computer will execute next. A sequence of numbered instructions is called a program. Thus, the term *program counter*.

The control unit of the microprocessor in Fig. 26-55 orchestrates the overall computer system. It sends control signals (logic 1s and 0s) to the other subsystems of the microprocessor. For instance, it tells the program counter when to increment, it tells the registers when to load a word or when to output a word, etc. The control unit also sends control signals to memory. It enables the ROM when data is needed from ROM. It enables RAM when data is to be either read from or written into RAM.

Notice that the signal direction arrows on the buses in Fig. 26-55 indicate that input/output devices can send signals to the control unit as well as receive control signals from the control unit. For example, an output device, such as a printer, might send back to the control unit to indicate that it is ready to receive more data.

The control unit is also responsible for ensuring that only one device is using the data bus at any given time. It does this by controlling (enabling or disabling) the three-state buffers which connect and disconnect the eight output lines of a device to and from the eight lines of the data bus.

### Self-Test

**55.** Is the address bus in a digital system unidirectional or bidirectional?

**56.** How does an output device indicate to the microprocessor that it is ready for more data?

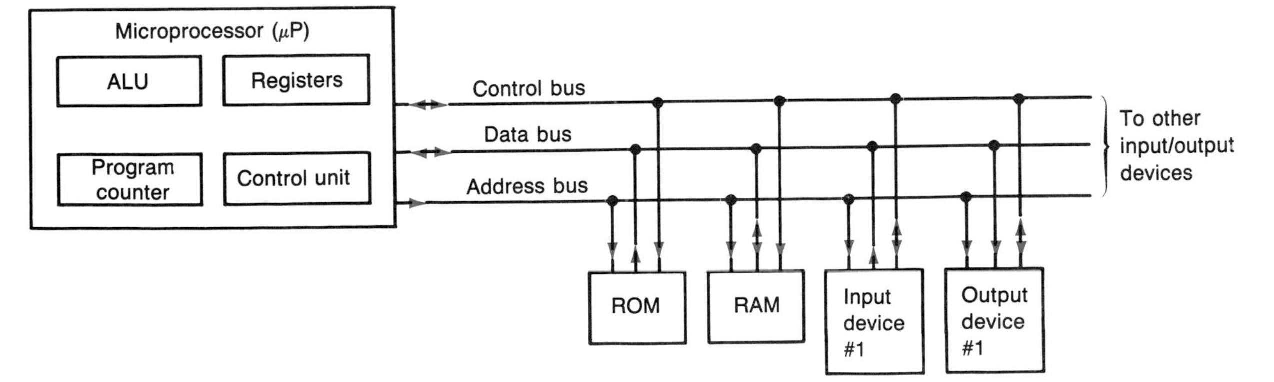


**FIGURE 26-55 Simplified block diagram of a computer system.**

## SUMMARY

1. Digital circuits have discrete levels of input and output voltages. They are usually ICs.
2. Logic gates perform logic functions.
3. The TTL logic family uses bipolar transistors. The input transistors often have multiple emitters.
4. A positive logic system uses the higher voltage level to represent a logic 1.
5. The procedures for binary arithmetic operations are the same as for decimal arithmetic.
6. A truth table shows the logic relationships between inputs and outputs of digital circuits.
7. The following gates are common decision-making gates: AND, OR, NOT, NAND, NOR, XOR, and XNOR. Each of these gates will implement a specific boolean expression.
8. NOT and INVERT are the same function.
9. The output of a disabled three-state (tristate) gate is neither a logic 1 nor a logic 0.
10. Active low inputs and outputs are indicated by bubbles (circles) on a logic symbol.
11. Combinational circuits use two or more decision-making gates. They are used to implement boolean expressions, decode, encode, multiplex, demultiplex, compare words, and perform arithmetical operations.
12. Multiplexers are also called data selectors.
13. Sequential circuits have memory which is the result of feedback. Representative circuits in this category are flip-flops, registers, and counters-dividers.
14. The SR and $\overline{SR}$ FFs are nonclocked. The output immediately follows the input charges.
15. Clocked FFs include the D, the clocked SR, and the JK. The JK and the D have no disallowed states.
16. FFs that change output states on every clock pulse are called toggle (T) FFs. JK FFs can be operated in the toggle mode.
17. A D latch has a gate or enable input rather than a clock input like a D FF has.
18. Direct inputs override all other inputs.
19. A group of cascaded FFs form a binary counter with a $2^n$ modulus. A counter also divides by $2^n$.
20. The "natural" or binary count of a counter can be modified to form a decade counter.
21. Counters can be either asynchronous (ripple) or synchronous.
22. Registers are made from FFs. Registers can be either storage registers or shift registers.
23. Shift registers can convert serial data to parallel data or vice versa.

**24.** Half adders can only add digits from the $2^0$ column of two binary numbers.

**25.** Memory for digital circuits can be either magnetic or semiconductor. Two major types of semiconductor memory are RAM and ROM.

**26.** An 8-bit word computer with 64K of RAM can store 65,536 eight-bit words.

**27.** Some types of ROM (EPROM) can be erased and reprogrammed.

**28.** The major IC in a microcomputer is the microprocessor chip. It controls (and utilizes) the ROM, RAM, input devices, and output devices.

## CHAPTER REVIEW QUESTIONS

For the following items, determine whether each statement is true or false.

**26-1.** Digital circuits and systems are usually constructed with discrete components.

**26-2.** When $1011_2$ is added to $1110_2$ there will be a carry from the $2^2$ to the $2^3$ column.

**26-3.** Transistors in a logic circuit are usually operated in the linear mode.

**26-4.** In Fig. 26-56*a*, the output will be a logic 1.

**26-5.** In Fig. 26-56*b*, $S_1$ must be a logic 0.

**26-6.** In Fig. 26-56*c*, word B cannot be $1101_2$.

**26-7.** In Fig. 26-56*d*, the Q output will be a zero.

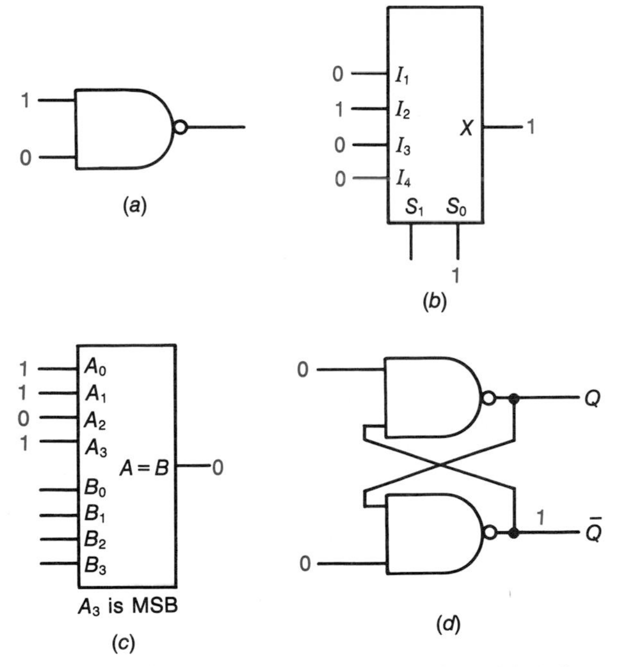


**FIGURE 26-56** Logic symbols and diagrams for Chapter Review Questions 26-4 through 26-7.

**26-8.** A decade counter uses four FFs.

**26-9.** The JK FF is in the toggle mode when J = 0 and K = 0.

**26-10.** Synchronous counters are slower than are asynchronous counters.

**26-11.** ROM cannot be randomly addressed.

**26-12.** Data can be either stored in or retrieved from RAM.

For the following items, fill in the blank with the word or phrase required to correctly complete each statement.

**26-13.** A digital circuit that performs a logic function is called a ______.

**26-14.** ______ is the abbreviation for the logic family which uses bipolar transistors.

**26-15.** ______ logic uses a high-level voltage to represent a logic 1.

**26-16.** A constant-frequency square wave used to trigger certain digital circuits is called a ______.

**26-17.** A ______ gate can provide more output current than an AND or OR gate can.

**26-18.** A ______ on the input of a logic symbol indicates an active low input.

**26-19.** ______ gates are often used to connect subsystems to a common bus.

**26-20.** A ______ can be used as a data selector.

**26-21.** The ______ clocked FF has a disallowed state.

**26-22.** An FF that can change output levels anytime the gate is high is called a ______.

**26-23.** ______ allows a logic circuit to have memory.

**26-24.** A ______ counter is an asynchronous counter.

**26-25.** A parallel in–serial out shift register converts ______ data to ______ data.

**26-26.** A storage register must have a ______ output.

**26-27.** A ______ can add a carry in from a previous column.

**26-28.** Eight address lines (an 8-bit word) can address ______ memory cells.

**26-29.** A 1024-cell memory array is also known as a ______ memory array.

**26-30.** The IC which directs and controls a computer system is the ______.

**26-31.** The three buses connected to RAM and ROM in a computer system are the ______, ______, and ______ buses.

## CHAPTER REVIEW PROBLEMS

**26-1.** Write the boolean expression for each of the following logic gates:
**a.** NOT **b.** NOR **c.** XNOR **d.** NAND

**26-2.** Convert the following binary numbers to decimal numbers.
**a.** 101101 **b.** 101.101

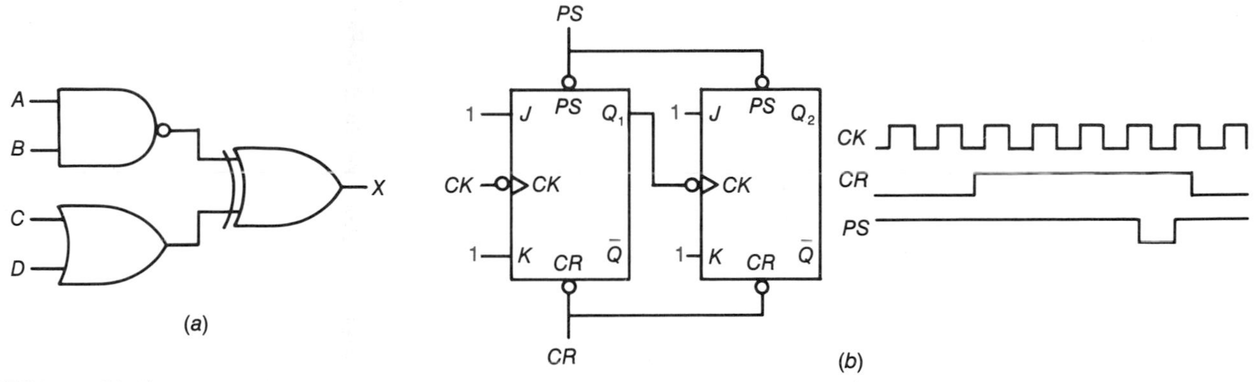


**FIGURE 26-57 Logic diagrams for Chapter Review Problems 26-5 and 26-7.**

**26-3.** Convert the following base 10 numbers to base 2:
**a.** 53 **b.** 36.75

**26-4.** Construct a truth table for a three-input NAND gate.

**26-5.** For the diagram in Fig. 26-57*a*, write the boolean expression.

**26-6.** Draw a logic diagram to implement the expression $y = \overline{AB} + \overline{C + D}$.

**26-7.** Draw the $Q_1$ and $Q_2$ output waveforms for the logic diagram in Fig. 26-57*b*. Include the CK waveform (as a timing reference) in your answer.

**26-8.** What is the modulus of a binary counter that uses three FFs?

**26-9.** What is the output frequency when a 600-kHz signal is applied to a decade counter which is cascaded to a modulus 4 counter?

## ANSWERS TO SELF-TESTS

**1.** an electronic circuit with discrete levels of input and output voltages

**2.** a digital circuit with specified input/output relationships

**3.** a digital circuit which performs a logic function.

**4.** transistor-transistor logic

**5.** one

**6.** a digital signal used to time and synchronize a digital system

**7.** **a.** 11 **b.** 10.625 **c.** 17.5

**8.** **a.** 1111 **b.** 10010.1 **c.** 100110.011

**9.** **a.** 10011 **b.** 10000.001 **c.** 10 **d.** 111.1 **e.** 1111 **f.** 1001011

**10.** **a.** $X = A$ **b.** $X = AB$ or $X = A \cdot B$ **c.** $X = A + B$ **d.** $X = \overline{AB}$ **e.** $X = \overline{A + B}$ **f.** $X = A \oplus B$ **g.** $X = \overline{A \oplus B}$ **h.** $X = A$

**11.** **a.** three-input NAND **b.** XOR **c.** AND

**12.** 32

**13.** circle or bubble

**14.** NOT gate

**15.** $X$ equals $A$ or $B$. Or, the output is low unless one of the two inputs is high.

**16.** four 0s followed by four 1s

**17.** a gate with greater output current capacity than a regular TTL gate

**18.** tristate or three-state gates

**19.** by a bubble or circle

**20.** diodes

**21.** cutoff or saturation

**22.** false

**23.** see Fig. 26-58

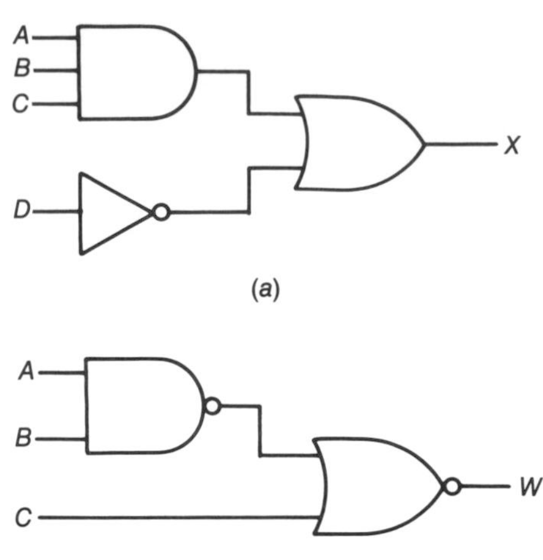

FIGURE 26-58 Answers for Self-Test Question 23.

**24. a.** $X = \overline{AB + C + \overline{D}}$ **b.** $Z = \overline{(E + F)(GH)}$ **c.** $\overline{\overline{D} \oplus EF}$ **d.** $X = (AB + \overline{C})DE$

**25.** binary-coded decimal

**26.** 1000 0010 1001

**27.** 438

**28.** a, b, c, d, and g

**29.** $I_1$

**30.** All outputs will be 1 because $G_5$ disables all of the NAND gates.

**31.** $A < B = 1$, $A = B$ will be 0, and $A > B = 0$.

**32. a.** a positive-edge-triggered JK FF with an active low clear input
**b.** a $\overline{SR}$ FF
**c.** a D latch with an active high gate and no $\overline{Q}$ output
**d.** a negative-edge-triggered SR FF with active high direct-set and direct-reset inputs

**33.** see Fig. 26-59. (One input waveform is shown as a timing reference.)

**34.** the D FF

**35.** the JK FF

**36.** The clocked FF cannot respond to the R and S inputs until the correct (negative or positive) edge of the clock signal arrives.

**37.** Flip-flops are used to (1) store data, (2) count, and (3) divide.

**38.** only when the gate (strobe or enable) is high

**39.** Both $Q_A$ and $Q_B$ will be logic 1.

**40.** $Q_A = 1$, $Q_B = 0$, $Q_C = 1$, and $Q_D = 0$

**41. a.** 32
**b.** five-stage counter, 5-bit counter, modulus 32 counter, and divide-by-32 counter.

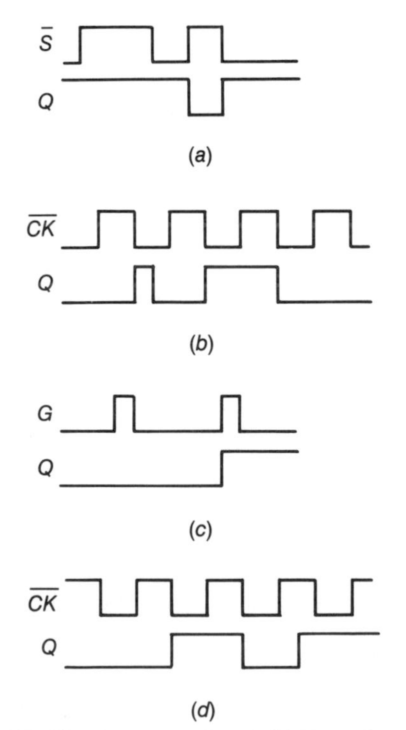


FIGURE 26-59 Answers for Self-Test Question 33.

**42.** 6.25 kHz

**43.** It can count or divide higher clock frequencies.

**44.** a group of FFs connected together to store a multidigit binary word

**45.** Storage and shift. The shift register can shift a word as well as store it. The storage register can only store a word.

**46.** parallel output

**47.** a parallel in–serial out shift register

**48.** one, because the one loaded into $FF_B$ will be shifted into $FF_D$ by the two clock pulses

**49.** The FA can accommodate a $C_i$ and an HA cannot.

**50.** All carry outs would be at a logic level 1.

**51.** The computer operator can only retrieve data from ROM. RAM allows the operator to both retrieve and store data.

**52.** no, both use random access

**53.** Proceed as follows: (1) Return the W line to logic 0, (2) change the address word to 1110, and (3) change the R line to logic 1.

**54.** 65,536; 64K

**55.** unidirectional

**56.** by sending a control signal to the microprocessor control unit via the control bus

# APPENDIX

## Metal Film Resistors

±5% and ±2% Devices

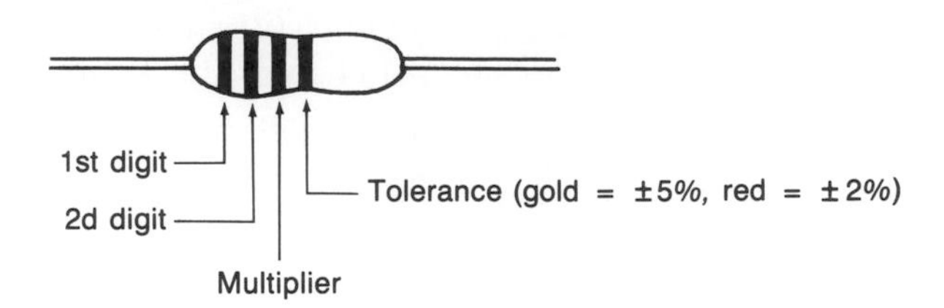


Standard Resistance Values, ±5%, ±2%

| | | | | | |
|---|---|---|---|---|---|
| 10 | 15 | 22 | 33 | 47 | 68 |
| 11 | 16 | 24 | 36 | 51 | 75 |
| 12 | 18 | 27 | 39 | 56 | 82 |
| 13 | 20 | 30 | 43 | 62 | 91 |

±1% Devices

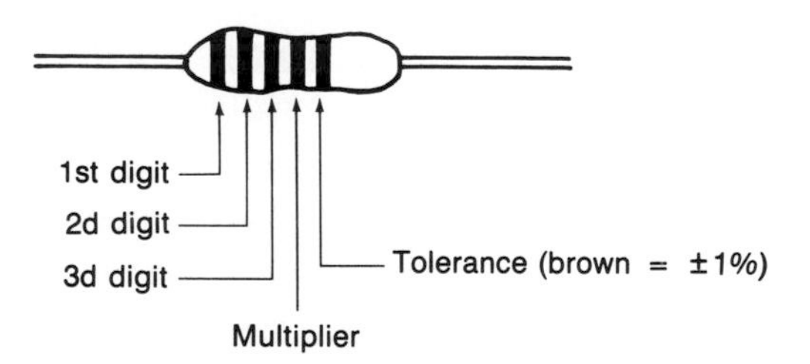


Standard Resistance Values, 1% (E96)

| | | | | | |
|---|---|---|---|---|---|
| 10.0 | 14.7 | 21.5 | 31.6 | 46.4 | 68.1 |
| 10.2 | 15.0 | 22.1 | 32.4 | 47.5 | 69.8 |
| 10.5 | 15.4 | 22.6 | 33.2 | 48.7 | 71.5 |
| 10.7 | 15.8 | 23.2 | 34.0 | 49.9 | 73.2 |
| 11.0 | 16.2 | 23.7 | 34.8 | 51.1 | 75.0 |
| 11.3 | 16.5 | 24.3 | 35.7 | 52.3 | 76.8 |
| 11.5 | 16.9 | 24.9 | 36.5 | 53.6 | 78.7 |
| 11.8 | 17.4 | 25.5 | 37.4 | 54.9 | 80.6 |
| 12.1 | 17.8 | 26.1 | 38.3 | 56.2 | 82.5 |
| 12.4 | 18.2 | 26.7 | 39.2 | 57.6 | 84.5 |
| 12.7 | 18.7 | 27.4 | 40.2 | 59.0 | 86.6 |
| 13.0 | 19.1 | 28.0 | 41.2 | 60.4 | 88.7 |
| 13.3 | 19.6 | 28.7 | 42.2 | 61.9 | 90.9 |
| 13.7 | 20.0 | 29.4 | 43.2 | 63.4 | 93.1 |
| 14.0 | 20.5 | 30.1 | 44.2 | 64.9 | 95.3 |
| 14.3 | 21.0 | 30.9 | 45.3 | 66.5 | 97.6 |

**Greek Alphabet and Common Designations**

| Name | Uppercase | Lowercase | May Be Used to Designate |
|---|---|---|---|
| Alpha | A | $\alpha$ | Area, angles, coefficients |
| Beta | B | $\beta$ | Flux density, coefficients |
| Gamma | $\Gamma$ | $\gamma$ | Conductivity, specific gravity |
| Delta | $\Delta$ | $\delta$ | Density, variation |
| Epsilon | E | $\epsilon$ | Base of natural logarithm |
| Zeta | Z | $\zeta$ | Impedance, coordinates, coefficients |
| Eta | H | $\eta$ | Hysteresis coefficient, efficiency |
| Theta | $\Theta$ | $\theta$ | Temperature, phase angle |
| Iota | I | $\iota$ | |
| Kappa | K | $\kappa$ | Dielectric constant, susceptibility |
| Lambda | $\Lambda$ | $\lambda$ | Wavelength |
| Mu | M | $\mu$ | Amplification factor, permeability, micro |
| Nu | N | $\nu$ | Reluctivity |
| Xi | $\Xi$ | $\xi$ | |
| Omicron | O | $o$ | |
| Pi | $\Pi$ | $\pi$ | Ratio of circumference to diameter |
| Rho | P | $\rho$ | Resistivity |
| Sigma | $\Sigma$ | $\sigma$ | Summation, coefficients |
| Tau | T | $\tau$ | Time constant, time displacement |
| Upsilon | $\Upsilon$ | $\upsilon$ | |
| Phi | $\Phi$ | $\phi$ | Magnetic flux, angles |
| Chi | X | $\chi$ | |
| Psi | $\Psi$ | $\psi$ | Dielectric flux, phase difference |
| Omega | $\Omega$ | $\omega$ | Ohms, angular velocity |

**Selected Symbols and Constants**

| Symbol | Meaning |
|---|---|
| $=$ | $A$ equals $B$ |
| $\neq$ | $A$ does not equal $B$ |
| $\cong$ | $A$ approximately equals $B$ |
| $>$ | $A$ is greater than $B$ |
| $<$ | $A$ is less than $B$ |
| $\geq$ | $A$ is equal to or greater than $B$ |
| $\leq$ | $A$ is equal to or less than $B$ |
| $\lvert\ \rvert$ | Absolute value of |
| $\equiv$ | $A$ equals $B$ by definition |
| $\Sigma$ | The sum of terms |
| $\therefore$ | Therefore |
| $\pi$ | 3.14159 (ratio of circumference to diameter) |
| $e$ | 2.71828 (base of natural logarithm) |
| $k_e$ | $8.9876 \times 10^9$ $(\mathrm{N} \cdot \mathrm{m}^2)/\mathrm{C}^2$ (electric constant) |
| $k_m$ | $1 \times 10^{-7}$ $\mathrm{N}/(\mathrm{A} \cdot \mathrm{m})$ (magnetic constant) |
| $e$ | $1.602 \times 10^{-19}$ C (electric charge) |
| F | $9.648 \times 10^4$ C/mol (faraday constant) |
| $c$ | $2.9979 \times 10^8$ m/s (speed of light) |
| $\mu_o$ | $4\pi \times 10^{-7}$ H/m (magnetic permeability of space) |
| $\epsilon_o$ | $8.85 \times 10^{-12}$ F/m (dielectric permittivity of space) |

**Powers of 10 with Prefixes and Symbols**

| Power | Prefix | Symbol |
|---|---|---|
| $10^{18}$ | Exa | E |
| $10^{15}$ | Peta | P |
| $10^{12}$ | Tera | T |
| $10^{9}$ | Giga | G |
| $10^{6}$ | Mega | M |
| $10^{3}$ | Kilo | k |
| $10^{2}$ | Hecto | h |
| $10^{1}$ | Deka | da |
| $10^{-1}$ | Deci | d |
| $10^{-2}$ | Centi | c |
| $10^{-3}$ | Milli | m |
| $10^{-6}$ | Micro | $\mu$ |
| $10^{-9}$ | Nano | n |
| $10^{-12}$ | Pico | p |
| $10^{-15}$ | Femto | f |
| $10^{-18}$ | Atto | a |

**Selected Relationships Between Base Units**

| Unit | Abbreviation | | Equivalent | |
|---|---|---|---|---|
| Ampere | A | = | coulomb/second | C/s |
| Volt | V | = | joule/coulomb | J/C |
| Ohm | Ω | = | volt/ampere | V/A |
| Watt | W | = | joule/second | J/s |
| Joule | J | = | newton × meter | N × m |
| Farad | F | = | coulomb/volt | C/V |
| Henry | H | = | volt/ampere/second | V/A/s |
| Siemen | S | = | ampere/volt | A/V |

# GLOSSARY

**Ac resistance** The resistance a component or circuit provides to an alternating current; often called dynamic resistance

**Acceptor atom** A trivalent impurity atom

**Accuracy** Specification of the maximum disagreement between true and measured values

**Active device** A component that amplifies or rectifies such as a transistor or a diode

**Admittance *Y*** The reciprocal of impedance $1/Z$

**Airgap** That portion of a magnetic circuit where the flux must pass through air

**Alternating current (ac)** Current that periodically reverses in direction due to polarity reversals at the source

**Alternation** One-half cycle of alternating current or voltage

**Alternator** An alternating current generator

**American wire gage (AWG)** A system of sizing electric conductors based on circular mil area

**Ammeter** An instrument used to measure current flow

**Ampere (A)** The SI unit of current flow (one coulomb per second)

**Ampere-hour (A · h)** The unit for the capacity of a cell or battery

**Ampere-turn (A · t)** The unit for magnetomotive force

**Amplitude** The peak or maximum value of an alternating current or voltage

**Analog** Quantities that can take on a continuous range of values (nondigital)

**AND gate** Output is 1 only when all inputs are 1

**Apparent power *S*** A combination of reactive power and true power, $S = IV$

**Arc** An electric discharge due to the breakdown or ionization of air or a gas

**Armature** The part of a generator where the voltage is induced (also the moving part of a motor or relay)

**Atom** The smallest part of an element capable of existing alone or in combination

**Average** The sum of the instantaneous values of an electrical quantity divided by the number of values

**Bandpass filter** A circuit which passes a band of frequencies while rejecting frequencies above or below the selected band

**Band-reject filter** A filter circuit which rejects (or stops) a band of frequencies in an electric system

**Bandwidth (BW)** The range of frequencies passed (or rejected) by a resonant circuit; those frequencies between the half-power points on a response curve

**Base** An element of a BJT

**Battery** A combination of two or more cells

**Beta (β)** Current gain $I_C/I_B$ of a BJT

***BH* curve** A graph that relates flux density $B$ to magnetizing force $H$

**Bilateral** A component or circuit where the resistance is not related to the applied polarity (resistors are bilateral but diodes are not)

**Bimetallic** A two-metal structure that bends when heated or cooled due to different coefficients of expansion

**BJT** Bipolar junction transistor: uses two polarities (holes and electrons) of current carriers

**Black box** A conceptual approach to network or device analysis based on what can be measured at external terminals

**Bleeder resistor** A resistor used in a power supply to discharge the filter capacitor and improve voltage regulation

**Boolean algebra** Algebra used to express logical relationships

**Branch** A path for current that is in parallel with one or more other paths

**Break frequency** The frequency at which the circuit response curve is −3 dB

**Bridge** A series-parallel arrangement of components most often used for measurements

**Bridge rectifier** A rectifier circuit which provides full-wave rectification without using a center-tapped transformer

**Brush** A sliding contact in a motor or adjustable transformer that is often made from graphite

**Cable** Either a stranded conductor (single-conductor cable) or a combination of conductors insulated from each other (multiple-conductor cable or coaxial cable)

**Capacitance *C*** Electric property that stores electric energy and opposes changes in voltage

**Capacitive reactance** A capacitor's opposition to ac

**Capacity** The rated load for a machine or device or the amount of storage for data systems

**CB amplifier** A BJT amplifier with an emitter input and a collector output

**CC amplifier** A BJT amplifier with a base input and an emitter output

**CE amplifier** A BJT amplifier with a base input and a collector output

**Cell** A single-unit source of electric energy such as an electrochemical cell or a photovoltaic cell

**Celsius (C)** The mksa temperature scale (formerly called centigrade)

**cemf** Counter-electromotive force: the voltage induced into a conductor by a changing current

**Channel** An element of a FET

**Charge** The energy measured in electrostatic units assigned to a particle such as an electron (may also refer to an excess or deficiency of electrons)

**Chassis** A metal frame or box that houses a circuit

**Choke** An alternate name for an inductor

**Circuit** A combination of electric conductors and devices

**Circuit breaker** An overcurrent device that can be reset (restored) after it acts

**Circular mil (cmil)** The cross-sectional area of a round wire that is 1 mil (one-thousandth of an inch) in diameter

**Closed circuit** A complete path for the flow of current (as opposed to an open circuit where current cannot flow)

**CMRR** Common-mode rejection ratio: specifies how well a DA rejects common-mode signals

**Coaxial cable** A two-conductor cable where the inner conductor is insulated from and completely surrounded by the outer conductor

**Coefficient of coupling *k*** A measurement of how much of the magnetic flux of one circuit links another circuit

**Coercive force** The force required to reduce the residual magnetism of a material to zero

**Coil** An alternate name for an inductor

**Collector** An element of a BJT

**Color code** A system that uses colors to indicate the value of devices or to identify leads and connections

**Combinational logic** A multiple-gate logic circuit which does not have memory

**Common** The reference or neutral point in a circuit or system

**Commutator** The segmented rotating contact in a generator or motor

**Comparator** An electronic circuit whose output changes when its two inputs are equal

**Complex number** A number composed of real and imaginary components

**Conductance *G*** The ability of a circuit or device to support current flow (the reciprocal of resistance)

**Conductivity ($\sigma$)** A measure of the conductance of a specified volume of a material (the reciprocal of resistivity)

**Conductor** A material that readily supports the flow of current (may also be used as an alternate word for wire)

**Constant current** A source of electric energy that supplies a fixed value of current flow regardless of the load resistance

**Constant voltage** A source of electric energy that supplies a fixed value of voltage regardless of its load resistance

**Continuity** A complete path for current to flow

**Conventional current** A conceptual flow of current opposite in direction to electron current

**Copper loss** The $I^2R$ loss in the windings of an electromagnetic device

**Core** The material that supports magnetic flux

**Core loss** The energy loss in a magnetic core material which is caused by hysteresis and eddy current

**Corona** Ionization of the air or gas around a point of high potential

**Cosine (cos)** The trigonometric function that relates the length of an adjacent side of a right triangle to the hypotenuse

**Cosine wave** A periodic waveform that follows the cosine function

**Coulomb (C)** The SI unit of electric charge

**Counter-electromotive force (cemf)** An emf produced by self-induction that opposes the applied emf

**Covalent bond** Valence electrons shared by adjacent atoms

**CRT** Cathode-ray tube

**Current *I*** The rate of charge movement in a circuit

**Current divider** An arrangement of parallel branches that divide the flow of current

**Cutoff** Condition of a transistor when $I_C$ or $I_D = 0$

**Cutoff frequency** See "Break frequency"

**Cycle** A complete set of values for a periodic waveform (two alternations)

**DA** Differential amplifier: amplifies the difference between two input signals

**Damping resistor** A resistor added to an *LC* circuit to increase the circuit bandwidth

**D'Arsonval movement** A type of meter movement which uses a permanent magnet and a moving coil

**Dc resistance** The resistance a component or circuit provides to a direct current; often called ohmic resistance or static resistance

**Decade** A ratio of 10:1

**Decibel (dB)** The logarithmic expression of a power ratio. Can also be expressed in terms of voltage or current

**Degauss** Demagnetize

**Degree** The angular measure equal to $\frac{1}{360}$ of a circle

**Delta ($\Delta$)** A series loop of three components (may be called a pi network); also the symbol for "a change in"

**Delta connection** A three-phase system where the line and phase voltages are equal

**Depletion region** A region in the PN junction which is void of current carriers

**Determinant** The numeric value of a square arrangement of numbers called a matrix

**Diamagnetic** A classification for materials having a permeability less than one

**Dielectric** An electric insulator

**Dielectric absorption** Retention by a dielectric material of some of its polarization after removal of the electric field
**Dielectric constant** $K$ See "Relative permittivity"
**Dielectric polarization** Distortion of the molecular structure of a dielectric material by an electric field
**Dielectric strength** Rating of a material's ability to withstand an electric field
**Differentiator** An *RL* or *RC* circuit which converts a square-wave pulse into a spiked pulse
**Digital** Quantities that can take on only discrete values
**Digital circuit** A circuit with discrete levels of input and output voltages
**DIP** Abbreviation for dual-in-line package
**Direct current (dc)** Curent that flows in only one direction
**Discrete component** An individually packaged device
**Dissipation factor (DF)** Ratio of the energy lost to the energy stored in a capacitor; equal to $1/Q$
**DMM** Abbreviation for digital multimeter
**Domain** A group of adjacent atoms that respond as a cell to a magnetic field
**Donor atom** A pentavalent impurity atom
**Doping** Adding impurity atoms to a pure crystal
**Drain** An element of a FET
**Dynamic resistance** See "Ac resistance"

**Eddy current** Current flow (induced) in the core of an electromagnetic device
**Effective resistance** That amount of resistance that would produce the same heat loss as that produced by all of the heat-producing phenomena (hysteresis, dielectric absorption, etc.) associated with an electric component or circuit
**Effective value** The heating or root-mean-square value of a voltage or current
**Efficiency ($\eta$)** The ratio of output power to input power (also the ratio of output work to input work)
**Electric field intensity** Force (per unit charge) of an electric field
**Electric flux** The lines of force in an electric field
**Electric flux density** Electric flux per unit cross-sectional area
**Electricity** Energy due to the motion of charges that can be transformed into heat, light, motion, etc. (also static electricity which deals with the accumulation of charges)
**Electrode** A contact or terminal that supports the flow of current
**Electrodynamometer** A meter movement used to measure power
**Electrolyte** A liquid or a paste with free electrons that supports current flow
**Electromagnet** A magnet that produces flux from current flow
**Electromotive force (emf)** The potential difference produced by a source of electric energy
**Electron-hole pair** Result of freeing an electron from a covalent bond
**Electronics** Pertaining to circuits based on active devices
**Electrons** The negatively charged particles of an atom
**Electrostatic** Dealing with the accumulation of charges and the forces between charged bodies
**Element** A substance containing only one type of atom
**Emitter** An element of a BJT
**Energy** $W$ The ability to do work
**Equivalent circuit** A simplified version of a circuit that aids in its analysis
**ESR** Equivalent series resistance: the effective resistance of a capacitor
**Exciting current** See "Magnetizing current"
**Exponent** Alternate name for power (the number of times a number is to be multiplied times itself)

**Fahrenheit (F)** The U.S. Customary temperature scale
**Farad (F)** Base unit of capacitance
**Faraday's law** For magnetic induction, the induced emf is proportional to the time rate of magnetic flux linked with the circuit
**Feedback** Process of returning part of the output of an amplifier to the input of an amplifier
**Ferrite** A magnetic material based on iron oxide in a ceramic binder
**Ferromagnetic** A classification for magnetic materials with a permeability many times greater than one
**FET** Field-effect transistor: uses the force of an electric field to control channel current; uses only one type of current carrier
**Field** A region containing electrostatic or magnetic lines of force
**Field winding** A coil in a motor or generator used to establish magnetic lines of force
**Filter** A frequency-selective circuit
**Flip-flop (FF)** A logic circuit which can store a logic 1 or a logic 0
**Flux ($\Phi$)** Magnetic or electric lines of force
**Flux density** $B$ The amount of flux per unit area
**Flux leakage** Lines of force that stray from the intended path
**Forward bias** A voltage applied to a PN junction so as to eliminate the depletion region
**Frequency** $f$ The number of cycles completed per second
**Fuel cell** An electrochemical source of energy that is fueled by external gases
**Full-scale current** Current required by a meter movement to cause the meter pointer to move to the end of the scale

**Full-scale voltage** Voltage required by a meter movement to cause the meter pointer to move to the end of the scale
**Function generator** An electronic source for several periodic waveforms such as sine, triangular, and square
**Fuse** An overcurrent device that melts when it acts

**Gate** An element of a FET; also a logic gate
**Gauss (G)** The cgs unit of magnetic flux density
**Generator** A source of electric energy (often refers specifically to a rotating machine)
**Giga (G)** A prefix that designates 10 to the ninth power
**Ground** Common or neutral point in a circuit (may also refer to a connection to the earth)

**Half-power point** The frequencies in a response curve that occur at 0.707 of the maximum curve height; the −3 dB points
**Hall effect** The phenomenon where a voltage is generated across the faces of a current-carrying conductor in a transverse magnetic field
**Harmonic** An integer multiple of a fundamental frequency
**Henry (H)** The SI unit of inductance
**Hertz (Hz)** The SI unit of frequency (one cycle per second)
**High-pass filter** A circuit which discriminates against frequencies below the cutoff frequency
**Hole** The current carrier in a P-type crystal
**Horsepower (hp)** The American customary unit of power (1 hp = 746 W)
**Hybrid IC** An IC, fabricated on an insulator substrate, which incorporates a monolithic IC and other components
**Hydrometer** An instrument for measuring specific gravity
**Hypotenuse** The side of a right triangle opposite to the 90° angle
**Hysteresis** In magnetic core materials, the lagging behind of flux from the force that produces it
**Hysteresis loss** Heat produced in a magnetic material when the magnetic flux reverses

**$I_{CBO}$** Leakage current in the base-collector junction
**Ideal inductor** An inductor which has no resistance
**Imaginary number** A *j* number; a number on the *y* axis of a graph
**Impedance *Z*** A combination of reactance and resistance
**Impedance ratio** The turns ratio squared or the voltage ratio squared
**Inductance *L*** That circuit property which opposes any change in current
**Induction** The generation of voltage or current by flux change
**Inductive kick** Large cemf produced when an inductive circuit is opened
**Inductive reactance $X_L$** An inductor's opposition to alternating current
**Inductor** A component (often a coil) that exhibits inductance
**Infinite (∞)** Having no limits
**Instantaneous value** The value of an electrical quantity at a given instant in time
**Insulator** A material with very high resistance to current flow
**Integrated circuit (IC)** A device with many transistors fabricated on a single piece of substrate
**Integrator** An *RC* circuit in which the capacitor's voltage builds up in response to a square-wave input
**Intensity modulation** See "Z-axis modulation"
**Internal resistance** The resistance inside an electric source
**INVERT gate** Same as NOT gate
**Ion** An atom with a net charge (ionized gases and solutions support the flow of current)
**Isolation transformer** A transformer with a turns ratio of one

***J* operator** The mathematical procedure that rotates a phasor or vector 90°
**JFET** Junction field-effect transistor: has one junction and one type of current carrier
**Joule (J)** The SI unit of work (or energy)
**Junction** A connection point for two or more devices or branches

**Kilo (k)** A prefix that designates 10 to the third power
**Kilowatthour (kWh)** The unit of work sold by electric utility companies
**Kirchhoff's laws** The laws that govern how voltages and currents are distributed in electric circuits

**Ladder network** A circuit whose schematic shows components arranged like rungs on a ladder
**Lag** When one waveform is delayed when compared to another waveform
**Laminated** Built up from thin sheets
**Lead** When one waveform is advanced when compared to another waveform (also may be used to refer to a wire or cable)
**Leakage current** Undesired current through a reverse-biased junction
**Leakage flux** Flux that does not link two coils
**Lenz's law** The law which states that an induced emf will oppose the polarity of the force that produced it
**Line of force** A representation of a force that shows its directions

**Linear** A straight-line relationship (one quantity or effect is in direct proportion to another)
**Lissajous patterns** CRT displays used to measure frequency and/or phase
**Load** A device that converts electric energy to another form such as heat
**Loading effect** Changes in a circuit quantity caused by measuring the quantity
**Logarithm** The power to which some base number is raised
**Logic gate** A circuit which performs a logic function
**Loop** Any closed path in a circuit
**Low-pass filter** A circuit which discriminates against frequencies above the cutoff frequency

**Magnet** A device that attracts magnetic materials
**Magnetic field strength** An alternate term for magnetizing force
**Magnetic flux (Φ)** All of the lines in a magnetic circuit
**Magnetic pole** An area of concentration for magnetic flux
**Magnetize** To activate a magnetic material by applying a magnetizing force
**Magnetizing current** The no-load primary current
**Magnetizing force (H)** The magnetomotive force per unit of length
**Magnetomotive force ($\mathscr{F}$)** The magnetic force produced by current flow
**Magnitude** The value of a quantity without regard to phase angle
**Make before break** See "Shorting switch"
**Maxwell (Mx)** The cgs unit of magnetic flux
**Mega (M)** A prefix that designates 10 to the sixth power
**Megger®** An instrument to test the integrity of insulation materials (a high potential ohmmeter)
**Mesh current** An assumed or assigned loop current used in one approach to circuit analysis
**Meter** The SI unit of length (also a measuring instrument)
**Micro (μ)** A prefix designating 10 to the negative sixth power
**Microprocessor (μP)** An IC capable of performing most, if not all, of the functions of a microcomputer
**Milli (m)** A prefix designating 10 to the negative third power
**Monolithic IC** An IC fabricated on a single, continuous silicon crystal
**MOS** Metal-oxide semiconductor
**Motor** A machine that converts electric energy to rotational mechanical energy
**Multimeter** An instrument for measuring two or more electrical quantities
**Multiplexer** Digital circuit which selects one source of data from two or more sources
**Multiplier** A resistor that extends the range of a voltmeter (also a color band or dot that positions the decimal point in a component's value)
**Mutual inductance $L_m$** The inductance shared by two or more magnetically coupled coils

**N-type crystal** Crystal created by adding pentavalent atoms
**NAND gate** Output is 0 only when all inputs are 1
**Nano (n)** A prefix designating 10 to the negative ninth power
**Negative (−)** Having an excess of electrons
**Net reactance** Absolute value of $X_L - X_C$
**Network** An alternate name for a circuit
**Neutral** The common or reference point in a circuit (may be grounded)
**Neutral point** See "Star point"
**Neutron** A particle with no charge located in the nucleus (center) of an atom
**Node** A point in a circuit where two or more devices or branches connect (also called a junction)
**Nominal** The ideal value of a component or quantity
**Nonlinear** A relationship other than directly proportional (not a straight-line relationship)
**NOR gate** Output is 1 only when all inputs are 0
**Normally closed (NC)** A device that normally (when not activated) provides a complete circuit
**Normally open (NO)** A device that normally (when not activated) provides an open circuit
**NOT gate** A gate in which the output is the complement of the input
**Null** Zero (may also refer to a balanced condition such as in a bridge circuit)

**Octave** A frequency ratio of 2:1
**Ohm (Ω)** The SI unit of resistance
**Ohm's law** The law that relates current, voltage, and resistance in an electric circuit
**Ohmic resistance** The resistance of a circuit or component as indicated by an ohmmeter
**Ohmmeter** An instrument for measuring resistance
**Ohms per volt** Unit for specifying voltmeter sensitivity
**Op amp** Operational amplifier: a very high gain amplifier with an inverting and a noninverting input
**Open circuit** One with no complete path for current flow
**OR gate** Output is 1 when any input is 1
**Oscilloscope** An instrument that displays a graph of voltage versus time
**Output** The current, voltage, or power developed by a circuit or device
**Overload** A load that demands more than the rated current or power

**P-type crystal** Crystal created by adding trivalent atoms
**Parallel** A connection where components or devices see the same voltage rise or drop (also called shunt)
**Paramagnetic** A classification for materials having a permeability slightly greater than one
**Parameter** An electrical quantity or characteristic
**Passive device** One that does not amplify or rectify such as a resistor, inductor, or capacitor
**Peak value** Another name for the maximum value of an electrical quantity
**Period *T*** The time required to complete 1 cycle
**Periodic waveform** A time-varying voltage or current that repeats after a defined time interval (period)
**Permanent magnet** One that retains its magnetic force for long periods of time
**Permeability ($\mu$)** A measure of the ability of a material to support magnetic flux
**Permittivity ($\epsilon$)** Rating of the ability of a dielectric material to accommodate flux
**Phase angle ($\theta$)** The angle of lead or lag between a phasor and the reference position
**Phasing** Determining the correct connections for the coils in a transformer, motor, or generator
**Phasor** A line representing the magnitude and phase angle of an electrical quantity
**Photocell** A light-sensitive transducer
**Photoconductive cell** A transducer which shows a decrease in resistance with an increase in light input
**Photolithographic** The process used in fabricating an IC
**Photon** A particle of light
**Photovoltaic cell** A device that converts light energy to electric energy (also called a solar cell)
**Pico** A prefix designating 10 to the negative twelfth power
**Piezoelectric** A material that produces electric energy when mechanically stressed
**PIV** Peak inverse voltage: rating of a diode
**PMMC** Permanent-magnet moving-coil meter movement
**PN junction** Region in a crystal where doping changes from acceptor to donor atoms
**Polar form** A phasor specified by its magnitude and angle
**Polarity** That property of electric charge determined by an excess (negative) or a deficiency (positive) of electrons (also that magnetic property pertaining to the behavior of poles)
**Polarization** The undesired build up of hydrogen gas on one of the electrodes of an electrochemical cell
**Positive ( + )** A deficiency of electrons
**Potential** The ability of an electric charge to do work
**Potential difference *V*** A measure of the work required to move a given quantity of charge between two points (commonly called voltage)
**Potentiometer** A three-terminal variable resistor
**Power *P*** The rate of expending energy (or of doing work)
**Power factor (PF)** True power divided by apparent power; equal to the cos $\theta$
**Power supply** A source of electric energy
**Primary cell** An electrochemical cell that cannot be restored by electric charging
**Primary coil** The coil of a transformer which is connected to a power source
**Proton** A positively charged particle located in the nucleus (center) or an atom
**PRV** Peak reverse voltage: rating of a diode

**Quality factor *Q*** A figure of merit for inductors and capacitors ($Q = X/R$); a ratio of energy stored to energy lost

**Radian (rad)** An angle of 57.3° (there are $2\pi$ rad in a circle)
**RAM** Random-access memory: memory designed to be "written into" as well as "read out of"
**Range** The maximum value that an instrument can measure
**Reactance *X*** Opposition of capacitors and inductors which does not convert electric energy to heat energy
**Reactive power *Q*** The product of reactive current multiplied by reactive voltage
**Real number** A number on the $x$ axis of a graph
**Rectangular form** A phasor specified by its real ($x$ axis) and imaginary ($y$ axis) components
**Rectification** The process of converting alternating current to pulsating direct current
**Rectifier** A device which converts alternating current to pulsating direct current
**Register** A group of FFs that can store a binary number (word)
**Regulation** The ability of a source to maintain a constant output (often expressed as a percentage)
**Relative permeability ($\mu_r$)** The ratio of the permeability of a material to the permeability of a vacuum
**Relative permittivity** The permittivity of a material compared to the permittivity of free space which is $\approx 1$
**Relay** An electromagnetic switching device that uses a current flow in a coil to open or close contacts
**Reluctance $\mathfrak{R}$** A material's opposition to magnetic flux
**Residual magnetism** The magnetism left in a core or material after the magnetizing force reaches zero
**Resistance *R*** The opposition to the flow of current
**Resistivity ($\rho$)** The resistance of a specified volume of a given material
**Resistor** A component designed to oppose the flow of current
**Resolution** The smallest change in a quantity that a given meter can measure
**Resonance** The circuit condition when $X_L = X_C$

**Resonant circuit** A circuit in which $X_L = X_C$
**Resonant frequency $f_r$** The frequency at which $X_C = X_L$
**Retentivity** The ability of a material to retain magnetism
**Reverse bias** A voltage applied to a PN junction so as to widen the depletion region
**Rheostat** A two-terminal variable resistor
**Ripple voltage** The ac component of the output voltage of a dc power supply
**ROM** Read-only memory: memory designed to be "read out of" but not "written into"
**Root mean square (rms)** The effective or heating value of an alternating voltage or current

**Saturation** A condition where the output from a device or a circuit no longer continues increasing with increasing input
**Secondary cell** An electrochemical cell that can be restored by electric charging
**Secondary coil** The coil of a transformer which provides power to a load
**Secondary emission** The releasing of electrons from an unheated surface by bombardment with other electrons
**Self-inductance *L*** That circuit property which opposes any change in current
**Sensor** A device that converts some physical property into an electric signal
**Sequential logic** A multiple-gate logic circuit which has memory
**Series** A type of circuit where the components are connected end to end in a chainlike fashion
**Shield** A partition, container, or covering to prevent external interference
**Short** A low-resistance path (often refers to a fault condition)
**Shorting switch** A switch in which contact with a new position is made before contact with the old position is broken
**Shunt** A resistor used to extend the current range of a meter movement
**SI** The international system of measurement based mainly on the metric mksa system
**Siemen (S)** The SI unit for conductance, admittance, and susceptence (the reciprocal of an ohm)
**Signal** A current or voltage, often of low amplitude and changing with time
**Silicon dioxide** An insulating material used in fabricating MOSFETs and IC capacitors
**Silk screening** A process used in fabricating hybrid ICs
**Sine (sin)** The trigonometric function that relates the opposite side of a right triangle to the hypotenuse
**Sine wave** A periodic waveform that follows the sine function (may be called a sinusoidal wave)
**Skin effect** Phenomenon of current carriers concentrating on the outer surface of a conductor
**Slew rate** Specifies how rapidly the output of an op amp can change
**S/N ratio** Signal-to-noise ratio. A number equal to $V_{signal}/V_{noise}$
**Solar cell** A device that converts light energy to electric energy
**Solder** An alloy (usually tin and lead) with a low melting point used to make permanent electric connections
**Solenoid** An electromagnet with a moving core used to activate some mechanism
**Source** A device that converts energy into electric energy; also an element of a FET
**Specific gravity** The ratio of the weight of a given volume of some material to the weight of the same volume of water
**SRVR** Series-resonant voltage rise: the ratio of $V_L$ or $V_C$ to $V_T$ in a series-resonant circuit
**Star point** The junction of the three phases in a wye connection
**Static** Without motion (static electricity deals with the accumulation of charges rather than charge motion)
**Static resistance** See "Dc resistance"
**Stator** The stationary part of a motor or generator
**Surge current** Large current required to charge a filter capacitor
**Susceptance *B*** The reciprocal of reactance $1/X$
**Switch** A device or component that provides on-off control in a circuit

**Tangent (tan)** The trigonometric function that relates an opposite side of a right triangle to an adjacent side
**Tank circuit** A parallel *LC* circuit
**Taper** In a variable resistor, the relationship between resistance and angle of rotation
**Temperature coefficient ($\alpha$)** The relationship between temperature and some electric measure
**Tesla (T)** The SI unit of flux density
**Thermionic emission** The releasing of electrons from a heated surface such as the cathode of a CRT
**Thermistor** A device manufactured to have a temperature-dependent resistance
**Thermocouple** A junction of dissimilar metals that converts heat energy to electric energy
**Thermopile** A combination of thermocouples for increased output
**Theta ($\theta$)** A Greek letter used to designate an angle
**Three-phase ($3\phi$)** Three sinusoidal voltages that are 120° out of phase with each other
**Three-state logic** Logic circuits in which the output can effectively be disconnected from a bus line
**Time constant ($\tau$)** Time required for *I* and *V* in an *RL* or *RC* circuit to change by 63 percent of the available *I* and *V*

**Tolerance** The maximum error of a value, often expressed as a percentage

**Toroid** A doughnut-shaped coil

**Transducer** A device that converts energy from one form to another

**Transformation ratio *a*** The voltage ratio or the turns ratio of a transformer

**Transformer** A device that couples electric energy from one circuit to another via a moving electromagnetic field

**Transient** A momentary event or change

**Trimmer** An adjustable component for trimming or adjusting some circuit value

**Tristate logic** See "Three-state logic"

**Troubleshooting** The process of locating faults in circuits or equipment

**True power *P*** Power which converts electric energy into another form of energy ($P = IV \cos \theta$)

**Truth table** A table which shows the output for all possible input combinations to a logic circuit

**TTL** Transistor-transistor logic: a type of logic family

**Turns ratio** The primary-coil turns divided by secondary-coil turns

**Turns-per-volt ratio** The coil turns divided by the coil voltage

**Unbalanced load** Condition in a three-phase system when all three phase loads are not equal

**Valence electron** An electron normally positioned in the outermost orbit of an atom

**var** Reactive volt-ampere: the base unit of reactive power

**Vector** A line representing the magnitude and the direction of a quantity

**VMOS** Vertical metal-oxide semiconductor: a high-power MOSFET

**Volt (V)** The SI unit of electromotive force

**Volt-ampere (V · A)** Base unit for apparent power

**Voltage *V*** The amount of potential difference between two points

**Voltage divider** A series arrangement of components to develop one or more voltages that are less than a source voltage

**Voltage drop *V*** A drop in potential difference caused by an opposition to the flow of current

**Voltage ratio** The primary voltage divided by the secondary voltage

**Voltage rise *V*** An increase in potential difference caused by a source of electric energy

**Voltage spike** See "Inductive kick"

**Voltmeter** An instrument for measuring potential difference

**Voltmeter sensitivity** The ohms-per-volt rating of a voltmeter

**VOM** Volt-ohm-milliampere: a type of analog multimeter

**Watt (W)** The SI unit of power

**Watthour (W · h)** A unit of work or energy

**Wattmeter** A meter which measures true power

**Waveform** The time graph of an electrical quantity such as voltage or current

**Wavelength (λ)** The distance a wave travels during one period

**Weber (Wb)** The SI unit of magnetic flux

**Wheatstone bridge** A series-parallel arrangement of components most often used for measurements

**Winding** Turns of wire used in some electromagnetic device such as a transformer or motor

**Wiper** A moving or sliding contact

**Work** Force × distance (or power × time)

**Wye (Y)** A connection of components or windings (may also be called a T connection)

**Wye connection** A three-phase system where the line and phase currents are equal

**XNOR gate** Output is 0 only when the two inputs are at different logic levels

**XOR gate** Output is 1 only when the two inputs are at different logic levels

**Z axis** The axis parallel to the length of the CRT

**Z-axis modulation** Chopping the CRT display into segments by alternately turning the electron beam on and off

**Zener diode** A diode used for voltage regulation

# ANSWERS TO ODD-NUMBERED REVIEW QUESTIONS AND PROBLEMS

## Chapter 1 Review Questions

**1-1.** SI
**1-3.** protons
**1-5.** negative ion
**1-7.** free
**1-9.** semiconductor
**1-11.** induction
**1-13.** Coulomb

## Chapter 1 Review Problems

**1-1.** 3.96
**1-3.** 240
**1-5.** $1.14 \times 10^6$
**1-7.** $8 \times 10^1$
**1-9.** $3.88 \times 10^{-6}$ ampere
**1-11.** 150 volts
**1-13.** 13.5 kilovolts
**1-15.** 7
**1-17.** 4
**1-19.** 11.1 kilohms

## Chapter 2 Review Questions

**2-1.** (A) resistor (B) cell (C) battery (D) lamp (E) ground (F) capacitor (G) light-emitting diode (H) diode (I) SPST switch (J) transformer (K) variable inductor (L) variable capacitor
**2-3.** common
**2-5.** at the top
**2-7.** electron
**2-9.** Ampere
**2-11.** 300 mA
**2-13.** it is not the range for best resolution
**2-15.** polarity
**2-17.** Joule
**2-19.** rise
**2-21.** yes
**2-23.** from point A to ground
**2-25.** the circuit does not have to be opened
**2-27.** no
**2-29.** 50 ohms
**2-31.** energized

## Chapter 2 Review Problems

**2-1.** 10 A
**2-3.** 22.5 C
**2-5.** 6.8 kV
**2-7.** 1.2 kilohm
**2-9.** 11 A
**2-11.** 5 A

## Chapter 3 Review Questions

**3-1.** 60
**3-3.** trip
**3-5.** manual

## Chapter 3 Review Problems

**3-1.** 150 W
**3-3.** 31.4 A
**3-5.** 7 A
**3-7.** 1.492 kW
**3-9.** $1.80
**3-11.** 66%
**3-13.** 53 ohms

## Chapter 4 Review Questions

**4-1.** leakage
**4-3.** more
**4-5.** National Electrical Code
**4-7.** lowest
**4.9.** oxide

## Chapter 4 Review Problems

**4-1.** 3 kV
**4-3.** 0.28 ohm
**4-5.** 3249
**4-7.** 10.49 ohms
**4-9.** 0.32 ohm

## Chapter 5 Review Questions

**5-1.** size
**5-3.** carbon composition
**5-5.** logarithmic
**5-7.** false
**5-9.** 1 kilohm, 500 ohm
**5-11.** to allow either increasing or decreasing resistance with CW rotation
**5-13.** true
**5-15.** a horizontal line
**5-17.** metallic oxide varistor
**5-19.** thermistor
**5-21.** false

## Chapter 5 Review Problems

**5-1.** 21 mA
**5-3.** 99.6 kilohms
**5-5.** brown-green-gold-gold
b. brown-brown-black-gold
c. red-red-brown-silver
d. brown-gray-yellow-silver
e. brown-black-blue-gold

## Chapter 6 Review Questions

**6-1.** current
**6-3.** 48
**6-5.** CW
**6-7.** CCW
**6-9.** positive
**6-11.** sum
**6-13.** zero
**6-15.** greater
**6-17.** yes

## Chapter 6 Review Problems

**6-1.** 3.490 ohms
**6-3.** 1.38 V
**6-5.** 5.33 V
**6-7.** $-50 - IR_1 - IR_2 - IR_3 = 0$
**6-9.** the direction of flow is wrong
**6-11.** 0 to 8 V
**6-13.** 8 V
**6-15.** −8 V
**6-17.** each resistor dissipates 13.6 mW, $P_T$ = 40.9 mW
**6-19.** 268 mW
**6-21.** 6 V

## Chapter 7 Review Questions

**7-1.** current
**7-3.** most
**7-5.** $R_1$, $R_2$, and $R_3$ are in parallel
**7-7.** they will operate normally
**7-9.** $R_1$

## Chapter 7 Review Problems

**7-1.** $I_1$ = 100 mA, $I_2$ = 200 mA, $I_3$ = 50 mA
**7-3.** 35 mA
**7-5.** 1 S
**7-7.** 40 mS
**7-9.** 2.64 mS
**7-11.** 167 mS
**7-13.** $I_1$ = 0.75 A, $I_2$ = 0.25 A
**7-15.** 5.56 ohms
**7-17.** 388 ohms
**7-19.** 21.3 A

## Chapter 8 Review Questions

**8-1.** regulation
**8-3.** Wheatstone
**8-5.** resistance
**8-7.** R-2R ladder network
**8-9.** $R_1$

## Chapter 8 Review Problems

**8-1.** 25 ohms
**8-3.** 25 W
**8-5.** 0.5 A
**8-7.** $R_B$ = 200 ohms, $P_B = \frac{1}{2}$ W, $R_2$ = 9.09 ohms, $P_2$ = 2 W
**8-9.** $V_1$ = 50 V, $V_2$ = 100 V
**8-11.** both will be zero

## Chapter 9 Review Questions

**9-1.** infinity
**9-3.** current
**9-5.** 3d
**9-7.** true
**9-9.** true
**9-11.** simultaneous equations are not required
**9-13.** false

## Chapter 9 Review Problems

**9-1.** 120 mA, 150 ohms
**9-3.** 222
**9-5.** $70I_1 - 20I_2 - 10I_3 = -15$
$-20I_1 + 70I_2 = 15$
$-10I_1 + 40I_3 = -10$
**9-7.** 8.56 V, positive on the right end
**9-9.** $0.41V_A - 0.4V_B = -2$
$-0.4V_A + 0.57V_B - 0.15V_C = 0$
$-0.15V_B + 0.35V_C = 1$
**9-11.** 1.56 A, from left to right
**9-13.** 1.43 A and 76.3 mA
**9-15.** 5.4 V
**9-17.** 4.22 V
**9-19.** 4 kilohms
**9-21.** 1.67 mA
**9-23.** 0.833 mA
**9-25.** 7.94 V, 588 ohms
**9-27.** 588 ohms
**9-29.** 0.5 W
**9-31.** a 150-ohm resistor in each leg of the pi network

## Chapter 10 Review Questions

**10-1.** repulsion
**10-3.** false
**10-5.** clockwise
**10-7.** right
**10-9.** it will increase
**10-11.** magnetization
**10-13.** 4
**10-15.** average
**10-17.** 2 and 5
**10-19.** narrower
**10-21.** permeability
**10-23.** false
**10-25.** false

## Chapter 10 Review Problems

**10-1.** 1.28 kg
**10-3.** 10 webers/s
**10-5.** $5.31 \times 10^{-4}$ weber
**10-7.** $1 \times 10^3$ ampere-turns/meter
**10-9.** 1,000
**10-11.** $6.75 \times 10^8$ ampere-turns/weber
**10-13.** 3.35 milliteslas, 1.68 milliteslas
**10-15.** 0.97 A

## Chapter 11 Review Questions

**11-1.** true
**11-3.** increase
**11-5.** sine
**11-7.** false
**11-9.** alternating
**11-11.** pulsating dc
**11-13.** it will drop
**11-15.** electrolyte
**11-17.** secondary
**11-19.** polarization
**11-21.** greater
**11-23.** true
**11-25.** energy
**11-27.** Peltier

## Chapter 11 Review Problems

**11-1.** 53.0 V
**11-3.** 173 Wh
**11-5.** 14

## Chapter 12 Review Questions

**12-1.** alternations
**12-3.** rms
**12-5.** 90 degrees, B
**12-7.** 225 degrees (or −135 degrees)
**12-9.** pulse
**12-11.** sine

## Chapter 12 Review Problems

**12-1.** 7.66 V
**12-3.** zero, 38.2 V
**12-5.** 113 V
**12-7.** 0.2 microsecond
**12-9.** 0.436
**12-11.** 3,142 radians per second
**12-13.** 44.7 V
**12-15.** $1.02\ \text{A}\angle 78.7°$
**12-17.** 16.7%
**12-19.** 180 Hz

## Chapter 13 Review Questions

**13-1.** opposes
**13-3.** Lenz's
**13-5.** powdered iron
**13-7.** 220 $\mu$H
**13-9.** 1

## Chapter 13 Review Problems

**13-1.** 500 V
**13-3.** 3.75 V
**13-5.** 7.68 μH
**13-7.** 1.45 H, 266 mH
**13-9.** 0.5

## Chapter 14 Review Questions

**14-1.** F
**14-3.** T
**14-5.** F
**14-7.** T
**14-9.** T
**14-11.** F
**14-13.** F
**14-15.** F
**14-17.** cemf
**14-19.** ms
**14-21.** Opened
**14-23.** Hysteresis loss, eddy-current loss

## Chapter 14 Review Problems

**14-1.** 12 V
**14-3.** 0.49 A
**14-5.** a. 0.5A
b. 213 mA
c. 55.6 A/s
**14-7.** 89.7
**14-9.** $X_{LT} = 1496\ \Omega$, $V_{L_3} = 8$ V

## Chapter 15 Review Questions

**15-1.** T
**15-3.** T
**15-5.** F
**15-7.** F
**15-9.** F
**15-11.** F
**15-13.** F
**15-15.** T
**15-17.** T
**15-19.** Relative permittivity
**15-21.** Capacitor
**15-23.** Voltage
**15-25.** Double
**15-27.** Metalized foil
**15-29.** Aluminum oxide, positive
**15-31.** C
**15-33.** A
**15-35.** C

## Chapter 15 Review Problems

**15-1.** 400 V
**15-3.** 500 μF
**15-5.** 90 J
**15-7.** 12 μF
**15-9.** 1.571 μF, 114.2 V, 314.2 C

## Chapter 16 Review Questions

**16-1** T
**16-3** F
**16-5** F
**16-7** T
**16-9** One time constant
**16-11** Leads; 90
**16-13** Inversely
**16-15** Largest
**16-17** Time
**16-19** *R* converts energy, *X* does not

## Chapter 16 Review Problems

**16-1** 141 μA
**16-3** 0.1 ms
**16-5** 8.3 s
**16-7** 24.8 V; 37.9 V
**16-9** 9367 Ω
**16-11** 3016 Ω; 9.95 mA; 18 V
**16-13** 0.587 μF
**16-15** 0.0198 μF

## Chapter 17 Review Questions

**17-1** F
**17-3** F
**17-5** F
**17-7** T
**17-9** F
**17-11** T
**17-13** Resistor
**17-15** Capacitor
**17-17** 180°

## Chapter 17 Review Problems

**17-1** $V_R = 20\ V_{p\text{-}p}$; $V_C = 10\ V_{p\text{-}p}$
**17-3** Zero
**17-5** a. 43.9° b. 7.2 kΩ c. 4.2 mA d. 90 mW
**17-7** a. 4028 Ω b. 53.7° c. 2.48 mA d. 0.0032 μF

## Chapter 18 Review Questions

**18-1** T
**18-3** T
**18-5** T
**18-7** F
**18-9** F
**18-11** F
**18-13** F
**18-15** T
**18-17** F
**18-19** F
**18-21** F
**18-23** Increase
**18-25** Q
**18-27** Inversely
**18-29** Doubled
**18-31** Capacitive
**18-33** Log
**18-35** Parallel
**18-37** −3 dB
**18-39** Reject

## Chapter 18 Review Problems

**18-1** 30 kHz
**18-3** L = 1.373 mH; BW = 2.32 kHz; Q = 138
**18-5** 202.5 kHz
**18-7** 34.5
**18-9** 0.32 V
**18-11** 265 $\mu$H

## Chapter 19 Review Questions

**19-1** F
**19-3** T
**19-5** F
**19-7** T
**19-9** Clockwise
**19-11** Rectangular
**19-13** Polar
**19-15** Conductance
**19-17** Capacitor

## Chapter 19 Review Problems

**19-1** 11.2 A $\angle -26.6°$
**19-3** 38 + $j$47.5 or 60.8 $\angle 51.3°$
**19-5** 150 $\angle -50°$ or 96.4 − j115
**19-7** $Y_T$ = 0.833 mS $\angle 36.87°$ or 0.666 mS + $j$0.50 mS
$Z_T$ = 1200 Ω $\angle -36.9°$ or 959.6 Ω − $j$720.5 Ω
$I_T$ = 8.33 mA $\angle 36.9°$ or 6.66 mA + $j$5.00 mA
$I_C$ = 10.0 mA $\angle 90°$ or 0 + $j$10 mA
**19-9** 321 Ω $\angle 0.6°$ or 321 Ω + $j$3.36 Ω
**19-11** 180 mA $\angle -34°$ or 149.2 mA − $j$100.6 mA
**19-13** 7.55 V $\angle 82.6°$

## Chapter 20 Review Questions

**20-1** T
**20-3** F
**20-5** T
**20-7** F
**20-9** F
**20-11** F
**20-13** F
**20-15** F
**20-17** F
**20-19** Primary
**20-21** Ten
**20-23** Reflected impedance
**20-25** Aiding
**20-27** Autotransformer
**20-29** C
**20-31** C
**20-33** Axis orientation and distance between coils
**20-35** Step-up

## Chapter 20 Review Problems

**20-1** 0.91
**20-3** a. 10, b. 100, c. 0.2 A, d. 0.1, e. 40 W, f. 0.2 V/N
**20-5** 22.4:1
**20-7** 97.6%
**20-9** 1.32 A $\angle -24.3°$

## Chapter 21 Review Questions

**21-1** F
**21-3** T
**21-5** F
**21-7** T
**21-9** F
**21-11** PMMC
**21-13** Non-shorting
**21-15** Phosphor coating or screen
**21-17** Modulate
**21-19** Lissajous
**21-21** B
**21-23** B
**21-25** D
**21-27** Shunt the meter movement; reduce the battery voltage
**21-29** Internal signal, external signal and line frequency

## Chapter 21 Review Problems

**21-1** $V_{min} = 1.995$ V; $V_{max} = 2.145$ V
**21-3** $R_i = 20\ \Omega$; $R_s = 20.4\ \Omega$
**21-5** 247.5 kΩ
**21-7** 480 Hz

## Chapter 22 Review Questions

**22-1** T
**22-3** T
**22-5** T
**22-7** F
**22-9** F
**22-11** T
**22-13** F
**22-15** 3-$\phi$
**22-17** Star point
**22-19** Neutral or star point
**22-21** Zero
**22-23** Electrodynamometer

## Chapter 22 Review Problems

**22-1** 120 V $\angle 240°$ or $\angle -120°$
**22-3** 13.8 A $\angle 53.1°$
**22-5** Leading
**22-7** Lagging
**22-9** Q = 281.1 var
S = 345 V · A
PF = 0.58
**22-11** 1140 W

## Chapter 23 Review Questions

**23-1** F
**23-3** F
**23-5** F
**23-7** F
**23-9** F
**23-11** F
**23-13** T
**23-15** Holes
**23-17** Doping
**23-19** Silicon
**23-21** ac
**23-23** D
**23-25** C
**23-27** D

## Chapter 23 Review Problems

**23-1** 500 Ω
**23-3** 0.5 A
**23-5** a. 84.8 V
b. 120 Hz
c. 42.4 V
**23-7** 12.5%

## Chapter 24 Review Questions

**24-1** F
**24-3** T
**24-5** T
**24-7** T
**24-9** F
**24-11** F
**24-13** T
**24-15** T
**24-17** Forward
**24-19** Base
**24-21** Positive
**24-23** Emitter
**24-25** Depletion
**24-27** Cutoff
**24-29** Field-effect transistor
**24-31** C
**24-33** A
**24-35** A

## Chapter 24 Review Problems

**24-1** $I_B = 200\ \mu$A; $I_C = 15.2$ mA
**24-3** a. 3.1 kΩ, b.116
**24-5** 600

## Chapter 25 Review Questions

**25-1** F
**25-3** T
**25-5** T
**25-7** F
**25-9** F
**25-11** F
**25-13** F
**25-15** T
**25-17** F
**25-19** T
**25-21** Photoresist
**25-23** P+
**25-25** Binder
**25-27** Active
**25-29** DA
**25-31** Input
**25-33** Direct-coupled
**25-35** Negative
**25-37** Comparator
**25-39** C
**25-41** A

## Chapter 25 Review Problems

**25-1** 1 kHz

**25-3** a. 0.42
b. 90.6
c. 216
d. 274 kΩ
e. +7.275 V

**25-5** a. 10
b. 0 V
c. 0 V
d. 5 kΩ
e. 10 V
f. 0 V

## Chapter 26 Review Questions

**26-1** F
**26-3** F
**26-5** T
**26-7** F
**26-9** F
**26-11** F
**26-13** Logic gate
**26-15** Positive
**26-17** Buffer
**26-19** Tri-state or three-state
**26-21** S-R
**26-23** Feedback
**26-25** Parallel; series
**26-27** Full-adder
**26-29** 1K × 1
**26-31** Control; data; address (in any order)

## Chapter 26 Review Problems

**26-1** a. $X = \overline{A}$
b. $X = \overline{A + B}$
c. $X = \overline{A \oplus B}$
d. $X = \overline{AB}$

**26-3** a. 110101
b. 100100.11

**26-5** $X = \overline{AB} \oplus (C + D)$

**26-7** See Fig. 26-1*c*

**26-9** 15 kHz

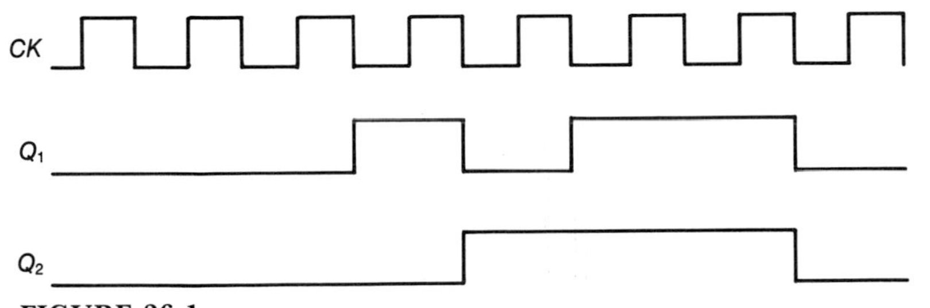


**FIGURE 26-1*c***

# Index